FROM RIVER TO ROCK RECORD

The preservation of fluvial sediments and their subsequent interpretation

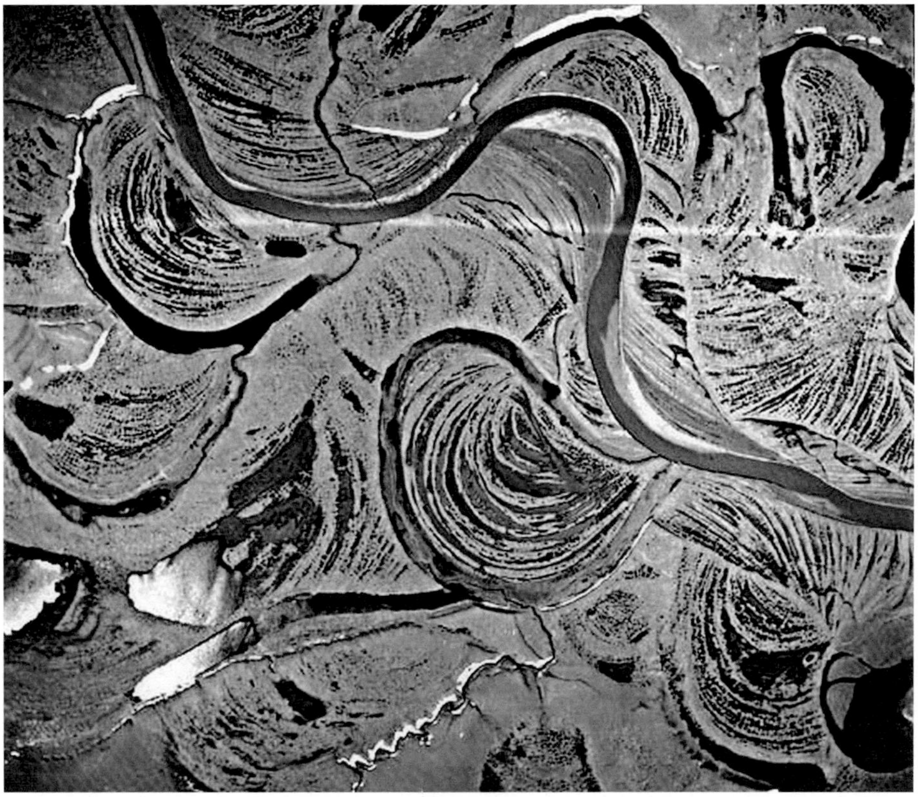

Edited by:

STEPHANIE K. DAVIDSON, SOPHIE LELEU, AND COLIN P. NORTH

Department of Geology and Petroleum Geology, University of Aberdeen, Aberdeen AB24 3UE, U.K.

SEPM Special Publication 97

Tulsa, Oklahoma, U.S.A. *May, 2011*

This volume is the result of presentations made at the *From River to Rock Record* conference and core workshop held on 12 – 14 January 2009 at the University of Aberdeen Aberdeen, Scotland, U.K.

SEPM and the authors are grateful to the following for their generous contributions to the cost of publishing

From River To Rock Record: The Preservation Of Fluvial Sediments And Their Subsequent Interpretation

Areva SA
BP Exploration plc
British Society for Geomorphology
Canadian Natural Resources Ltd
Geological Survey of Norway
Shell Exploration and Production Ltd
Talisman Energy UK Ltd
Total E&P UK Ltd
Tullow Oil plc
University of Nebraska – Lincoln
University of Padova
U.S. Geological Survey

Contributions were applied to the cost of production, which reduced the purchase price, making the volume available to a wider audience

This book includes a ⟨CD/DVD⟩ which contains a complete digital copy of this volume, with many color figures substituted for printed black and white versions, and some supplemental material associated with specific articles in the volume.

SEPM (Society for Sedimentary Geology) is an international not-for-profit Society based in Tulsa, Oklahoma. Through its network of international members, the Society is dedicated to the dissemination of scientific information on sedimentology, stratigraphy, paleontology, environmental sciences, marine geology, hydrogeology, and many additional related specialties.

The Society supports members in their professional objectives by publication of two major scientific journals, the Journal of Sedimentary Research (JSR) and PALAIOS, in addition to producing technical conferences, short courses, and Special Publications. Through SEPM's Continuing Education, Publications, Meetings, and other programs, members can both gain and exchange information pertinent to their geologic specialties.

For more information about SEPM, please visit **www.sepm.org**.

ISBN 978-1-56576-305-0
CD/DVD version 978-1-56576-307-4

SEPM (Society for Sedimentary Geology)
4111 S Darlington, Suite 100
Tulsa, Oklahoma 74135-6373, U.S.A.

Printed in the United States of America

From River To Rock Record: The Preservation Of Fluvial Sediments And Their Subsequent Interpretation

Stephanie K. Davidson, Sophie Leleu, and Colin P. North, Editors

CONTENTS

Introduction

Lessons from the Modern

Short Term: Autogenically Dominated?

Medium Term: Complex Control Interactions

Long Term: Allogenically Dominated?

Introduction

FROM RIVER TO ROCK RECORD: THE PRESERVATION OF FLUVIAL SEDIMENTS AND THEIR SUBSEQUENT INTERPRETATION

STEPHANIE K. DAVIDSON, SOPHIE LELEU, AND COLIN P. NORTH

Department of Geology and Petroleum Geology, School of Geosciences, University of Aberdeen, Kings College, Aberdeen AB24 3UE, Scotland, U.K.

INTRODUCTION

It is our impression that, over the last couple of decades, fluvial geomorphology and fluvial sedimentary geology have been developing in parallel, rather than in conjunction as might be expected. Though we tell our students that these are interdependent disciplines, perusal of the literature suggests there has been a failure of each domain to inform and recognize the needs of the other. Many facts have been collected, from both modern and ancient examples, yet our understanding of the sedimentary record of rivers seems not to have advanced significantly. It is still a struggle to map the architecture observed in the preserved rock record to the genetic components recorded from active modern rivers. Determination of the lateral extent and variation of lithofacies remains exceedingly problematic, especially in the subsurface.

We contend there are two fundamental causes of this dichotomy and the problems it has brought about. Firstly, whilst there is now a good understanding of the instantaneous processes of sediment transport and bedform development, there remains a poor appreciation of the spatial distribution and longer term temporal evolution of bedforms, and the way these bedforms interact with the channel banks and floodplain. Secondly, current sedimentary models fail to take into account the many erosional and depositional modifications that occur as the sedimentary architecture develops, and imply that the observed planform of modern rivers is preserved largely intact. In reality, the rock record is a mosaic of remnants from the original rivers, some small, some large, but few of them with the mutual arrangement they had at the time of deposition.

Of chief concern is the need to remember that interpretation of the rock record is based on what is known about modern settings—typically, we see only what we expect to see and we overlook features with which we are not familiar. In the light of present-day geomorphological knowledge, published fluvial sedimentary facies models, particularly the ones in textbooks, appear over-simplistic but they are still in regular use. For example, there is little recognition in these models of the role of soil and vegetation in modifying river behavior, and few of them take into account the complex interaction between autogenic and allogenic controls of river behavior and sedimentology.

About this volume

This volume is the result of our attempt to bridge the gap between the geomorphology and sedimentary geology domains in order to understand better how sediments in modern rivers become preserved in the rock record, and to improve interpretation from that record of the history of past environmental conditions.

The catalyst for the volume was a conference with the same title as this publication that was hosted at the University of Aberdeen School of Geosciences, in Aberdeen, Scotland, on 12–14 January 2009. By means of a one-day core workshop and two days of presentations and discussion in a single forum, we brought together a broad spectrum of geomorphology and sedimentology researchers, from academia and industry and from five different continents. This interdisciplinary mix of experience considered and discussed ideas and examples ranging through timescales from the annual movement of individual river bars up to sequence stratigraphic analysis of major sedimentary basins spanning millions of years. Though it will be no surprise that we didn't resolve all the outstanding problems, all participants undoubtedly left with a wider appreciation of the issues and new ideas for future research and collaboration.

The articles in this volume are a mixture of novel concepts, new evaluations of the received wisdom about rivers and their sediments, and improved understanding derived from recent experience with interpreting the rock record. In most cases, the main theme of each article is the same as that delivered at the conference, but the analysis and presentation has since been reworked and improved significantly to reflect the views shared at that meeting and the significant input from reviewers and editors.

The collection commences with a group of articles that highlight how, even after decades of research, critical evaluation of modern rivers is still important for informing us how to read the rock record. Ethridge reviews several myths and misconceptions about the interpretation of ancient fluvial channel deposits, and shows how the high degree of variability in plan-view morphology of Holocene rivers is not reflected in the published accounts of ancient deposits. He recommends that emphasis should shift from interpreting river planform to analysis of the preserved bar forms and the temporal evolution of river systems. Long compares the Precambrian and Early Paleozoic record of fluvial systems, the time before rooted vascular plants were common, with accepted knowledge of the wide variety of modern systems. He concludes that the stereotype of all pre-vegetation rivers being braided is a gross oversimplification, and shows that these early examples are much more diverse in character than traditionally described. Archer et al. examine why river terraces, a feature geomorphologists take for granted along modern valleys, are rarely included in sedimentological interpretations. This is partly explained by the difficulty of detecting terraces with seismic surveys, which they demonstrate with forward seismic modeling. They go on to show the importance of terraces in understanding the true complexity of sequence boundaries flooring incised valleys, and how the failure to include them in subsurface reconstructions may explain problems with predicting fluid flow.

The next group of articles focuses primarily on the relatively short-term evolution of river successions during which autogenic processes dominate. Fielding et al. examine the sedimentary record of tropical and subtropical rivers and show that the strongly seasonal character of the discharge produces outcomes that would not be predicted from the standard fluvial facies models in the literature. Leclair presents new data and analyses which overturn conventional assumptions about paleohydrological interpretation of ancient rivers from their dune deposits. She shows how peak flow deposits are not preferentially preserved, and dune height does not scale systematically with flow depth. Ghosh and Sarkar explore the criteria for recognition of

From River To Rock Record: The Preservation of Fluvial Sediments and their Subsequent Interpretation
SEPM Special Publication No. 97, Copyright © 2011
SEPM (Society for Sedimentary Geology), ISBN 978-1-56576-305-0, p. 3–5.

overbank sub-environments, and demonstrate the value of paleosols and pedogenic profiles. Smith et al. describe in detail fine-grained counter-point bars and eddy accretion deposits in modern meander belts. These are fluvial genetic elements that have been poorly studied in the past and are absent from standard river models, yet which can make up substantial portions of what previously would have been identified as sand-rich point bars. Smalley examines the variation in transport direction that can be deduced from image logs in a horizontal borehole, and shows how this can be used to constrain environmental interpretation. Cain and Mountney describe the downstream changes in fluvial system character and preserved deposits for a large internally-draining dryland ancient river, changes which, again, are not represented in commonly used facies models.

The third group of articles investigates examples formed over longer time periods and in situations with significant interaction between autogenic and allogenic processes. McKie describes the complexity present in a sand-rich subsurface succession that at first sight would seem not to be problematic. There are, however, significant variations in the vertical proportion of fine-grained deposits that impact detrimentally on hydrocarbon recovery. Prediction of these variations is important but made difficult because it is unclear if they are due to large scale shifting of the fluvial system or climatic control. Like Ethridge, McKie emphasizes the need for more information at the barform scale instead of the channel-belt scale as traditionally collected. Stouthamer et al. elegantly summarize the impact of river avulsion on sediment architecture for the Rhine-Meuse delta region and show how through the Holocene the inherent autogenic avulsion pattern has been modified by natural base level change and anthropogenic increase in sediment discharge. Flaig et al. document in detail the character of a tidally influenced river system that developed at high latitude during a period of greenhouse climatic conditions. Eilertsen and Corner describe an example of autogenic deep scouring, and subsequent filling, that could easily be mistaken for allogenic incision driven by base level fall. Hornung and Hinderer explore the differences between lacustrine and marine controls on delta sedimentology with their high-resolution sequence stratigraphic analysis of an exceptionally exposed Triassic succession in China. Hillier et al. show that calcretes previously considered as of pedogenic origin, reflecting past soil conditions, may instead be non-pedogenic and of groundwater origin, which has profound implications for environmental and climatic interpretations. Weissman et al. present a review of satellite imagery that suggests that many more ancient fluvial successions may have formed as part of fan-shaped distributary fluvial systems than previously recognized, and that, repeating a recurrent theme in this volume, current fluvial facies models do not adequately account for the character of rivers which are part of such systems where sediment is distributed radially rather than in linear valleys.

The final group of articles considers situations which are dominated by allogenic controls over longer time periods, and attempts to unravel the impact of tectonic, base level, and climatic controls on the fluvial sedimentary architecture. Monegato and Stefani document the impact of active tectonics on the sedimentary record of a single river system and illustrate the complexity of incision and deposition that can result. At a slightly larger scale, Nádor and Sztanó detail the variation in channel density and stacking induced by fluctuations in the rate of basin subsidence, and show in their example that, unlike the predictions of previously published conceptual models, rivers did not always flow towards the subsidence maximum. Dubiel and Hasiotis present an elegant example that demonstrates clearly the impact of climate on the fluvial sedimentary record, and the evolution of a single system during long term climate change from monsoonal humid to more arid conditions. Their study demonstrates the importance of including detailed analysis of both soil type and trace fossils. Gibling et al. round out the volume with a re-evaluation of the problem of distinguishing natural alluvial valleys, created by autogenic erosion, from incised valleys formed during periods of base level fall or climate change. By considering examples that developed large distances inland, they show that climatic effects as the cause of deep valley incision are commonly under-estimated in the rock record. Conversely, they show with other examples that base-level fall need not necessarily produce deeply incised valleys in coastal settings.

Progress made?

The final article by Gibling et al. neatly emphasizes the over-arching messages of the entire volume, that current fluvial lithofacies and sequence stratigraphic models do not yet fully capture the true character of rivers both modern and ancient, and that much more work still needs to be done. One notable example is the failure to make full use of the information represented by overbank processes and sediments.

Another general message from these articles is the benefit that can be gained from substantive integration of geomorphological analysis with study of the preserved sediments. It is crucial to base interpretation of the rock record on what is known about modern river character, rather than relying on speculation based on out-of-date conceptual models. A prime example of where sedimentary geology is out of step with geomorphological understanding is the preoccupation of the former community with classifying ancient rivers as either braided or meandering. Geomorphologists have long appreciated that such pigeon-holing is pointless and that rivers are much more variable, both in space and time, than interpreters of the sedimentary record generally acknowledge. Geomorphologists have at their fingertips a great deal of quantitative information that is potentially of use to the sedimentary community. Unfortunately, a common problem in the past has been that geomorphologists do not consider the temporal evolution and preservation potential of features they are describing, so producing a mass of detail the relevance of which to the rock record is difficult to determine.

What is abundantly clear is that the processes and process-interactions by which a river produces a sedimentary succession are still not fully understood or predictable. We believe this volume usefully illustrates the current state of knowledge and will provide a stimulus for further research, particularly work that integrates geomorphological and sedimentological approaches and emphasizes cross-disciplinary communication.

ACKNOWLEDGMENTS

We are extremely grateful to the following organizations for financial sponsorship of the "From River to Rock Record" conference and core workshop:

Areva SA

BP Exploration plc

British Society for Geomorphology

Canadian Natural Resources Ltd

SEPM Society for Sedimentary Geology

Shell Exploration and Production Ltd

Talisman Energy UK Ltd

Total E&P UK Ltd

Tullow Oil plc

The editors of this SEPM Special Publication would like to thank the following reviewers for their invaluable contribution to this volume. The high standard of the science and presentation is in large part due to their objective and independent assessment:

Jan Alexander
Kathryn Amos
Bill Arnott
Andres Aslan
Philip Ashworth
Whitney Autin
Brian Bluck
Michael Blum
Sylvie Bourquin
Amy Brock
Richard Collier
Rick Donselaar
Russell Dubiel
Frank Ethridge
Christopher Fielding
Martin Gibling
Elizabeth Gierlowski-Kordesch
Gary Hampson
Trevor Hoey
John Holbrook
Brian Jones
Stuart Jones
Richard Langford
Darrel Long
Susan Marriott
Peter McCabe
Paul McCarthy
Andrew Miall
Nigel Mountney
Greg Nadon
Tobi Payenberg
Marta Pérez-Arlucea
Guy Plint
Matthew Pranter
Esther Stouthamer
John Swenson
Robert Tye
Brian Willis
Brian Zaitlin

The editors would also like to thank John B. Southard for his skillful and diligent technical editing of all the articles, Michele Tomlinson for her hard work steering the volume through production, and all the other SEPM staff and officers involved in the gestation of this volume.

Lessons from the Modern

INTERPRETATION OF ANCIENT FLUVIAL CHANNEL DEPOSITS: REVIEW AND RECOMMENDATIONS

FRANK G. ETHRIDGE
Department of Geosciences, Colorado State University, Fort Collins, CO 80523-1482, U.S.A.
Email: fredpet@warnercnr.colostate.edu

ABSTRACT: Holocene rivers have a high degree of morphological variability, and many researchers see a continuum of channel forms that are transitional between end-member planforms such as braided, meandering, or straight. Individual rivers also show a high degree of longitudinal and vertical (through time) variability, as a result of changes in base level, climate, tectonics, tributary contribution, and/or valley slope. This high degree of variability in plan-view morphology of Holocene rivers is generally not reflected in interpretations of ancient fluvial deposits, which continue to be interpreted as meandering or braided. Many of these interpretations are suspect because of myths about modern rivers and because the characteristics described bear little relationship to plan-view morphology. Modern fluvial facies models are built around the nature and type of channel bars. Knowledge of the architecture of bar forms and their occurrence in Holocene river channels has been greatly enhanced in the past two decades with data gained from ground-penetrating radar (GPR) studies. In spite of our expanded data base for modern rivers and their deposits we still need additional data on width-to-thickness ratios and architecture of deposits of aggrading river systems.

A review of the literature suggests that data on channel bar deposits is lacking in most studies of ancient fluvial deposits. The few research efforts that describe paleochannel bars in 3D and 2D exposures of ancient fluvial deposits suggest that there was considerable variability in paleochannel form within a given formation. Our focus in the study of ancient fluvial deposits should shift from interpreting planforms to description and interpretation of preserved bar forms and how river systems evolved. In this regard many ancient fluvial sheet sandstones need to be re-examined.

KEY WORDS: river, fluvial deposits, fluvial channel, morphology, paleochannel

INTRODUCTION

In spite of a number of previous attempts to point out the difficulties involved in interpreting paleochannel patterns of ancient alluvial deposits (Bridge 1985, 1993, 2006; Brierley and Hickin 1991; Bristow 1996; Miall 1995) numerous studies continue to attempt these interpretations with inadequate data. This paper reviews and updates the subject of paleochannel interpretation and emphasizes the importance of cross-disciplinary research between sedimentology and geomorphology in light of the mismatch between the variety of modern fluvial planforms and the paucity of sedimentological facies models. This is accomplished by reviewing some classifications of river channels and by examining morphologic variability in modern river channels and channel belts. This section is followed by a review of fluvial facies models and myths regarding rivers. The bulk of the paper is an examination of what works and what does not work in our attempts to interpret ancient river channel and channel-belt deposits. This effort involves an examination of the use of limited data in outcrops and subsurface, 2D and 3D outcrop exposures, and 3D quantitative seismic geomorphology (QSG) in evaluating ancient fluvial deposits. The question of whether the identification of plan-view morphology of paleochannels is an important or even necessary goal of fluvial research is also raised. This question is logical, given that modern rivers have considerable variation in plan form in space and time and thresholds between various planforms are poorly defined. Furthermore some criteria used to interpret fluvial deposits, such as grain size, sandstone-body dimensions, and proportions of channel and overbank deposits, are not strongly dependent on channel plan form.

CLASSIFICATION AND VARIABILITY IN MODERN RIVER CHANNELS AND CHANNEL BELTS

Classification

Most sedimentologists and geomorphologists think of fluvial channel classification in terms of plan-view morphology, and from this viewpoint there are three basis channel types: straight, meandering, and braided (Leopold and Wolman 1957; Lane 1957). To this classification Brice (1982) added the distinction between stable, equiwidth, sinuous channels and unstable, wide-bend point-bar, meandering channels, and between braided, point-bar channels and braided channels without point bars (Fig. 1). Along with Miall (1977), Brice included anabranching (anastomosing) channel systems in his classification scheme and suggested that each single channel type had a corresponding multiple channel type (Ethridge and Schumm 2007). Wandering is a term added to the mix by Neill (1973) and Church (1983), who defined it as a river that exhibits an irregularly sinuous channel, sometimes splitting around channel islands and in some places braided, with seasonal or perennial side channels. Bridge (2003, 2006) suggested that the term anabranching (or anastomosing) refers to channel belts and not specifically to channels, and, as such, it does not belong in the same classification with channels. Therefore it is best placed in a classification scheme along with tributary and distributary channel belts. Anabranching rivers are discussed by Knighton and Nanson (1993), Nanson and Knighton (1996), Makaske (2001), and Nanson and Gibling (2003) and will not be discussed as a separate channel-belt type here. A few modern rivers even show channel belts that contain braided, meandering, and possibly anastomosing segments within the same valley reach, as shown in the Sagavanirktok River, northern Alaska (Bridge 2003, 2006), and the Waimakariri River, South Island, New Zealand (Fig. 2). The term wandering might also be applied to these rivers.

An attempt to determine if the morphologic pattern of alluvial rivers is related to sediment type resulted in the classification scheme of Schumm (1977), which is based on sediment load (bed load, mixed load, and suspended load) and channel pattern. This classification was developed from a database limited to sand-bed steams from the North American Great Plains and the Murrumbidgee River, Australia, and does not include anastomosing channels. The basic patterns represent end members in a broad spectrum of channel types that include five basic bed-load and mixed-load channel patterns and three types of suspended-load patterns (Fig. 3). These patterns reflect an increase

From River To Rock Record: The Preservation Of Fluvial Sediments And Their Subsequent Interpretation
SEPM Special Publication No. 97, Copyright © 2011
SEPM (Society for Sedimentary Geology), ISBN 978-1-56576-305-0, p. 9–35.

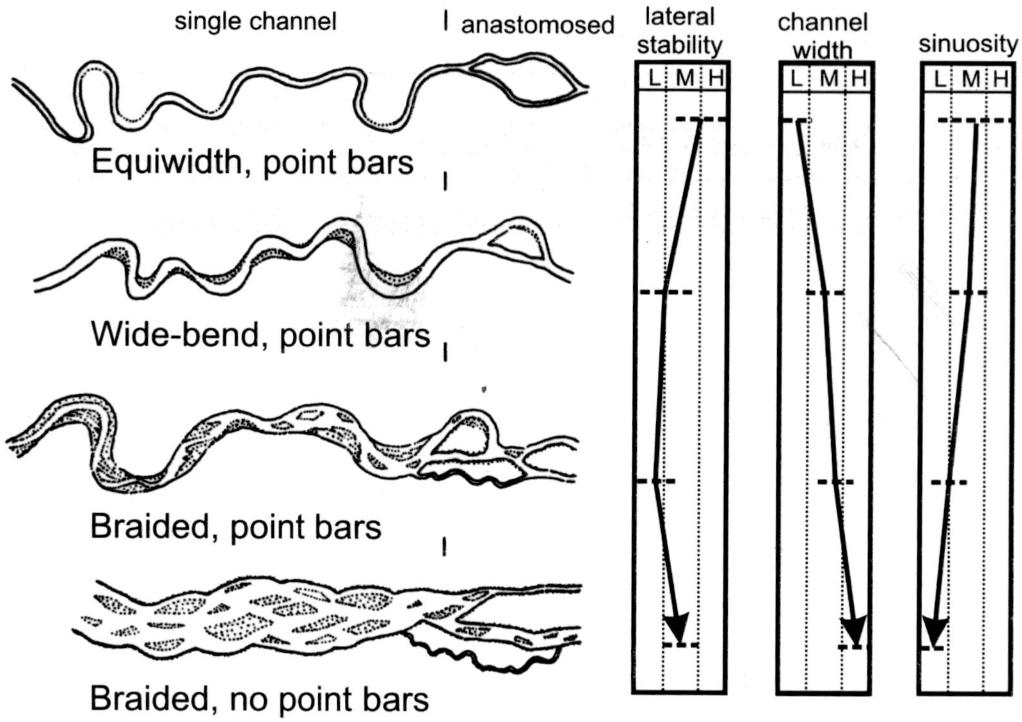

FIGURE 1.—Summary of alluvial river types based on plan-form morphology (modified from Brice 1982). Letters L, M, and H stand for low, medium, and high in terms of lateral stability, channel width, and sinuosity. Horizontal lines indicate variability in lateral stability, channel width and, sinuosity for each river type.

followed by a decrease in sinuosity as valley slope and bedload increases (Fig. 3). Bed-load channels have the highest width–depth ratio. Suspended-load channels are the narrowest and deepest and may have the highest sinuosity, and are very stable. Mixed-load patterns show a wide range of sinuosity and bar types, ranging from truly meandering channels with point bars (pattern 8, Fig. 3) to wandering sinuous-braided channels (pattern 9, Fig. 3) that is similar to Brice's braided point-bar channel pattern (Fig. 1). The database for these patterns is analogue, experimental, flume studies, and field observations (Schumm 1977). Church (2006) used Schumm's classification as a starting point to examine the relations between sediment transport and river morphology on the basis of the Schields number, a dimensionless measure of the capacity of the channel to move sediment of a given grain size.

Fluvial geomorphologists and engineers working on problems of stream restoration have devised methods of classification and characterizing channels that range from form-based to process-based systems. Among the form-based classifications is the classification of natural rivers by Rosgen (1994). This system consists of an organization of data on channels features into discrete combinations and levels. Level I delineation results in nine channel types based on channel relief, pattern, shape, and dimension. A more detailed morphological description of channel features at level II results in further subdivision of channels into some forty-plus types. The complexity of this system alone would make its application to the rock record questionable at best. Simon et al. (2005) suggested that Rosgen's scheme may provide a consistent and reliable means for describing channel morphology, but that its use for predicting river behavior is unjustified. Woolfe and Balzary (1996) proposed a planform-independent, process-based classification of fluvial channels based on relative rates of channel and inter channel sedimentation. They argue that while combinations of channel and inter channel sedimentation that are primarily aggradational dominate rock-record deposits, all combinations can be preserved. Goodwin (1999) and Bledsoe et al. (2008) reviewed systems for stream-channel classification. Goodwin (1999) also included ten recommendations for improving future classifications.

Many geomorphologists, sedimentologists, and engineers acknowledge that alluvial channels form a continuum rather than discrete planform types (Knighton and Nanson 1993; Bridge 2003; Simon et al. 2005; Bledsoe et al. 2008), with many natural channels displaying transitional patterns between planforms. On the other hand, some researchers also see a sudden change in planform type with an imposed increase in flow strength suggesting the existence of geomorphic thresholds (Schumm 1977; Bull 1979; Carson 1984). Thresholds and subsequent change in channel morphology are an integral part of the analysis of a river's response to an external or internal stimulus.

A number of empirical and theoretical approaches have been developed to investigate transitions between braided and meandering. Bridge (1993) provides a review of most of these approaches. Van den Berg (1995) examined 225 data sets from 192 modern rivers to investigate the possible prediction of alluvial channel patterns of perennial rivers. His data suggest that braided and meandering channel patterns can be differentiated using plots of specific stream power versus median grain size of river bed material. His data also show that both channel types occur in rivers with a very wide range of median grain sizes, ranging from fine sand to cobbles (Fig. 4). Lewin and Brewer (2001) conclude that the use of a single-stage stream-power measure and bed-material size is not likely to achieve discrimination. Differences of opinion are argued by Van den Berg and Bledsoe (2003) and by Lewin and Brewer (2003). Tooth and Nanson (2004) and Tooth and McCathy (2004) suggest that traditional discrimination diagrams

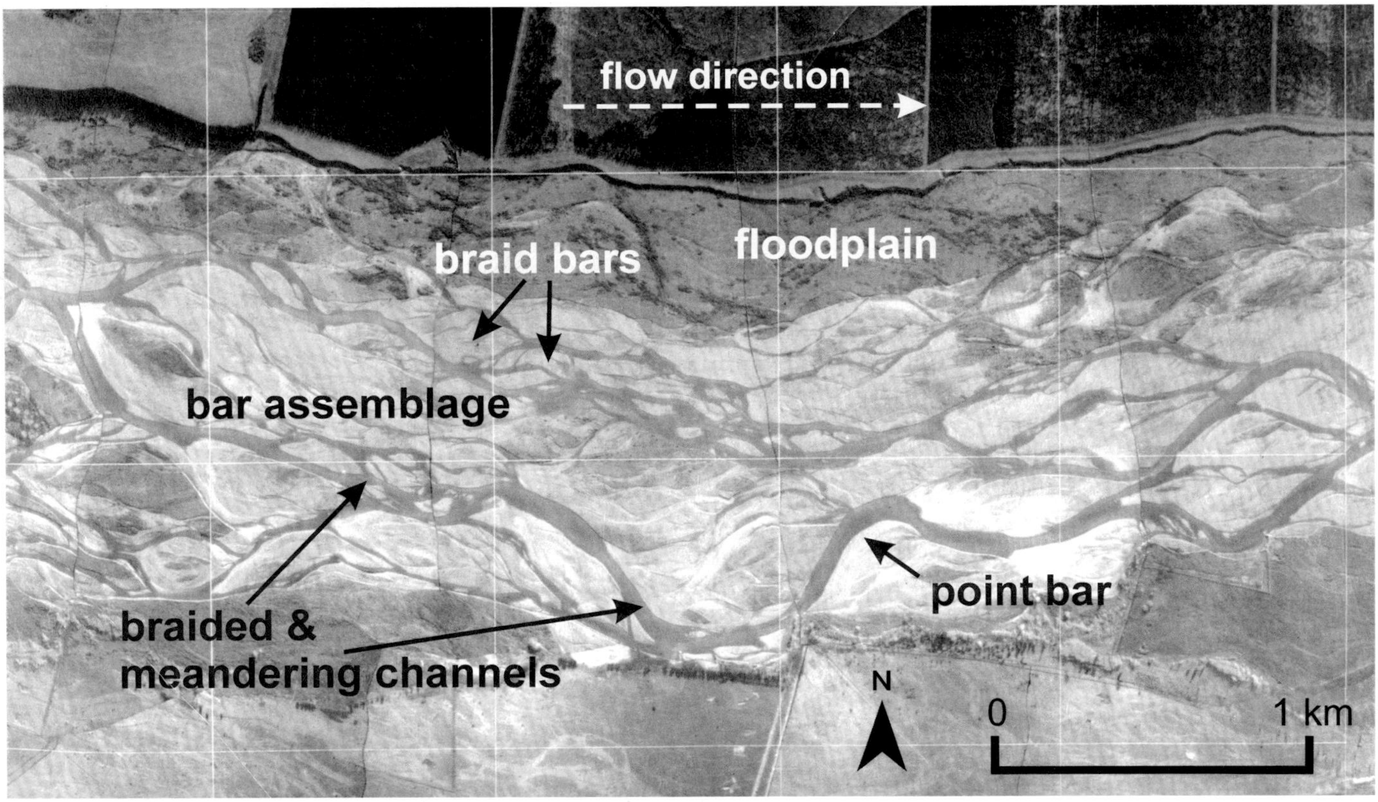

FIGURE 2.—Complex system of braided, meandering, and anastomosing (?) channels in a reach of the Waimakariri River, south island, New Zealand (photo courtesy of Stan Schumm).

for the braided-to-meandering transition do not always work for other transitions in arid ephemeral streams of Australia or in the Okavango wetlands of Botswana. Bledsoe et al. (2008) provide an overview and a discussion of thresholds.

Variability

Modern alluvial rivers (mobile-channel belts of Gibling 2006) demonstrate a great deal of variability and complexity along their courses, within a given reach and through time, as reviewed by Schumm et al. (2000) and Schumm (2005). River variability and complexity include upstream (tectonics, lithology, and climate), downstream (base level and length), and local (tributaries, active tectonics, valley morphology, floods, and vegetation) controls (Schumm, 2005). A few examples are reviewed here to demonstrate how upstream, downstream or local controls can change the plan-view morphology of Holocene rivers.

The gravel-bed Jefferson River in southwestern Montana flows north along the strike of a tectonic valley that has been deformed by syntectonic uplift and tilting (Fig. 5). The area is characterized by active extensional and strike-slip faulting. Changes in channel pattern, dimension, and shape have produced differences in sediment transport necessary to balance the effects of the uplift and tilting. Jorgensen (1990) divided the study reach that included the uplift area into six reaches (Fig. 5). Within these reaches depth varies little, width and width-depth ratio are lower across the uplift (reaches 4 and 5), cross-section area and bankfull discharge increase through the uplift area, and sediment storage is lowest in this area. Within the uplift area the modern channel shows a decrease in sinuosity. Above and below this area the Jefferson has adopted a sinuous planform. Jorgenson (1990) examined three additional rivers, where syntectonic uplift and possibly tilting were important factors: a suspended-load stream, a mixed-load stream, and another bedload stream like the Jefferson. The two finer-grained rivers showed an increase in sinuosity in reaches of steepened valley slope, whereas two gravel-bed rivers are straighter in the axis of uplift and more sinuous in depositional and control reaches.

Leeder (1993) discussed tectonic controls on drainage-basin development, river-channel migration and alluvial architecture. He stresses the importance of tectonic tilting on changing gradients and resulting avulsions, migrations, and incision of fluvial channels. Schumm et al. (2000) concluded that syntectonic river response can be recognized by changes in channel form, but that the response is extremely variable depending on sediment load and on river type and size.

The high variability in form and process related to tectonic activity in modest- to small-size rivers discussed by Schumm et al. (2000) is also observed in the world's largest rivers. Major changes in morphology along the courses of the Amazon and Ganges rivers are very different and may be related to active tectonics. Along the course of the Amazon in Brazil the river traverses the axes of several structural highs (arches). As the river crosses these highs the surface water gradient increases. The river incises, the alluvial valley narrows, and the river course straightens. Downstream of these highs the gradient flattens, the valley widens, and large-scale river movement or avulsion results in scroll bars or anabranches. This is especially true downstream of the Jutaí arch, where valley-floor tilting has also been proposed as the primary mechanism of river channel change (Mertes et al. 1996; Mertes and Dunne 2007). In contrast the Ganges River in India is predominantly braided over long reaches within its alluvial valley;

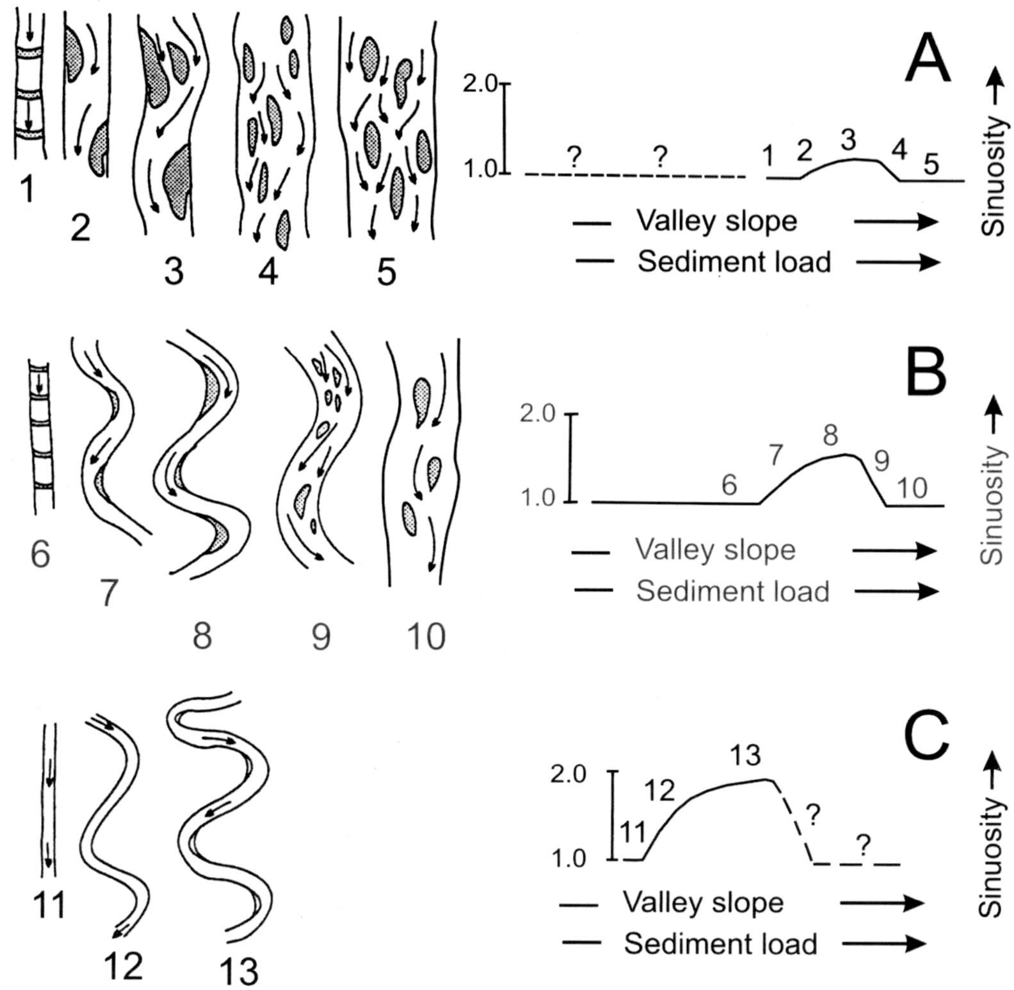

FIGURE 3.—A range of alluvial channel patterns for A) bed-load channels, B) mixed-load channels, and C) suspended-load channels. Patterns within each group are related to changes of valley slope, stream power, and sediment load. Patterns are from analog, experimental, flume studies conducted at Colorado State University (redrawn from Schumm 1977).

however, where the valley narrows, the river meanders, and it has built large point bars several kilometers square in area (Singh 2007).

The North Platte, South Platte, and Arkansas Rivers originate in the Rocky Mountains of Colorado. The South and North Platte Rivers join near North Platte, Nebraska to form the Platte River, which flows across Nebraska to join the Missouri River near Omaha, Nebraska. All of these rivers have undergone significant changes in form and process in the last one hundred plus years, as a result of human activities (Williams 1978; Nadler and Schumm 1981; Crowley 1983; Schumm 2005; Fotherby 2009). Most of these studies emphasize the reduction in width and changes in planform through time and conclude that these changes are a result of reduction in mean annual discharge, reduced peak discharges, and development of perennial flow that is the equivalent of a significant climate change (Schumm 2005). Fotherby's (2009) research focused more on spatial changes among eleven reaches of the Platte River in central Nebraska between Lexington and Chapman. In the two upstream-most reaches the Platte has a plan view morphology that is either wandering or meandering. In each of the other nine reaches the Platte is braided, anastomosing with some braiding, or braided with some anastomosing (Fig. 6). For these nine reaches the main factor controlling fully braided reaches appears to be valley confinement (width between confining features). No other possible factors (including stream power or sediment-transport factors) show any correlation among stream reaches and planform morphology. Discharge in the main channel, slope, and/or sediment transport are controlling factors in reaches one and two, which display wandering and meandering patterns respectively.

The Arkansas River in southern Colorado reveals a type of metamorphosis similar to that of the South Platte and Platte River except in the vicinity of Old Bent's Fort where the river has undergone a metamorphosis from a wide, sandy, braided river to a much narrower, highly sinuous river with abundant exposed bar and swale topography from active lateral accretion (Fig. 7). This change from braided to meandering is due in part to the introduction of abundant finer-grained sediments from an upstream tributary (Nadler and Schumm 1981).

Whatever the principal factor controlling metamorphosis on the South Platte, Platte, and Arkansas Rivers, the fact remains that these rivers demonstrate a high degree of variability in planform morphology both spatially and temporally along their courses, mostly as a result of hydrologic changes. The South Platte and Platte have coarse-grained meandering and low-sinuosity, single channels with alternate bar plan-view patterns in Colorado (Nadler and Schumm 1981; Crowley 1983)

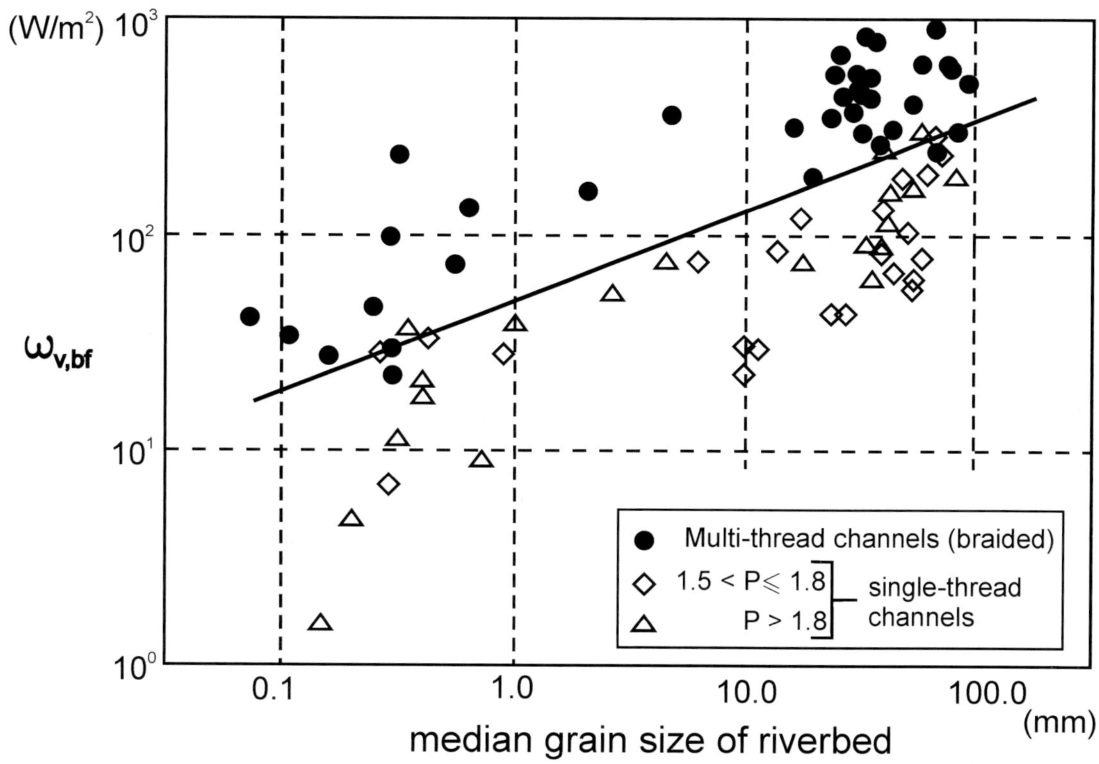

FIGURE 4.—Channel pattern in relation to medium grain size of bed material and specific stream power using bankfull discharge. Data from smaller than 10 m^3/s and low sinuosity single-thread channels ($1.3 < P \leq 1.5$) are excluded (redrawn and modified from van den Berg 1995).

and sandy braided and anastomosing patterns in Nebraska (Crowley 1983; Schumm 2005; Fotherby 2009). Exceptions include areas where there is a reduction in discharge and slope or an increase in suspended load, relative to other river reaches, and a meandering pattern develops (Nadler and Schumm 1981; Fotherby 2009).

Temporal change in plan form in the Platte, South Platte, and Arkansas rivers took place over a period of less than 100 years. More specifically the change from braided to meandering at Old Bent's Fort on the Arkansas was probably accomplished in half that time. Lunt et al. (2004) document temporal changes in the Sagavanirktok River, northern Alaska, utilizing aerial photographs. In their study area the main channel evolved from meandering to braided between 1949 and 2001. Given the rapidity of these transformations in a geologic sense it is reasonable to assume that many ancient channel-belt deposits contain deposits formed under several different planforms.

In summary, classifications of fluvial channels are based principally on plan-view morphology with anywhere from 3 to 40-plus end-member forms. Some geomorphologists, sedimentologists, and engineers recognize a continuum of channel forms with many rivers in transition between one form and another. But the concept of geomorphic thresholds is central to our understanding of modern rivers and their response to disturbance. Examples reviewed here illustrate the high degree of variability in planform and in erosion and deposition that modern rivers experience within and between reaches along their courses and through time. Furthermore small and large rivers appear to be equally variable and similar factors do not necessarily produce the same type of planform change in different river systems. Is this variability recognized and documented in deposits of ancient river systems? To answer this question I review fluvial facies models, discuss myths regarding river channels, and present a brief review of some examples of studies of ancient river deposits.

FLUVIAL FACIES MODELS

Much has been written about fluvial facies models and the interpretation of paleochannel patterns for ancient fluvial deposits. Some important reviews include those of Bridge (1985, 1993, 2006), Brierley (1996), Bristow (1996), Miall (1977, 1988, 1992, 1995, 1999), North (1996), and, of course, the textbooks by Miall (1996) and Bridge (2003). For fluvial systems the concept of facies models began with the development of the concept of lateral accretion and the fining-upward vertical cycle that were developed simultaneously by Bernard et al. (1962) and Allen (1963) for meandering rivers (Miall 1996). Facies models for braided streams began with the work of Doeglas (1962) who studied two modern Holocene braided rivers in France and provided criteria for differentiating braided-river and turbidite deposits. Ore (1964) reviewed available criteria for distinguishing braided and meandering river deposits in general, and Moody-Stuart (1966) listed characteristics of high-sinuosity and low-sinuosity river deposits in terms of presence or absence.

Bluck (1971) and Jackson (1976), working on the modern Endrick and Wabash rivers respectively, documented variations in the type of vertical lithofacies sequence preserved depending on the nature of bend migration and the location of the sequence within a single bend. They questioned the validity of a single vertical profile for the recognition of ancient meandering-river point-bar deposits. Even though the Endrick is a small fraction of the size of the Wabash, sedimentary facies in both were similar. Jackson (1978) in a review of lithofacies models for meandering alluvial streams concluded that many criteria in common use at that time to identify ancient meandering-stream deposits, including epsilon cross stratification, do not occur in all meandering streams, and can occur in non-meandering streams. Collinson (1978) also concludes that no one vertical sequence

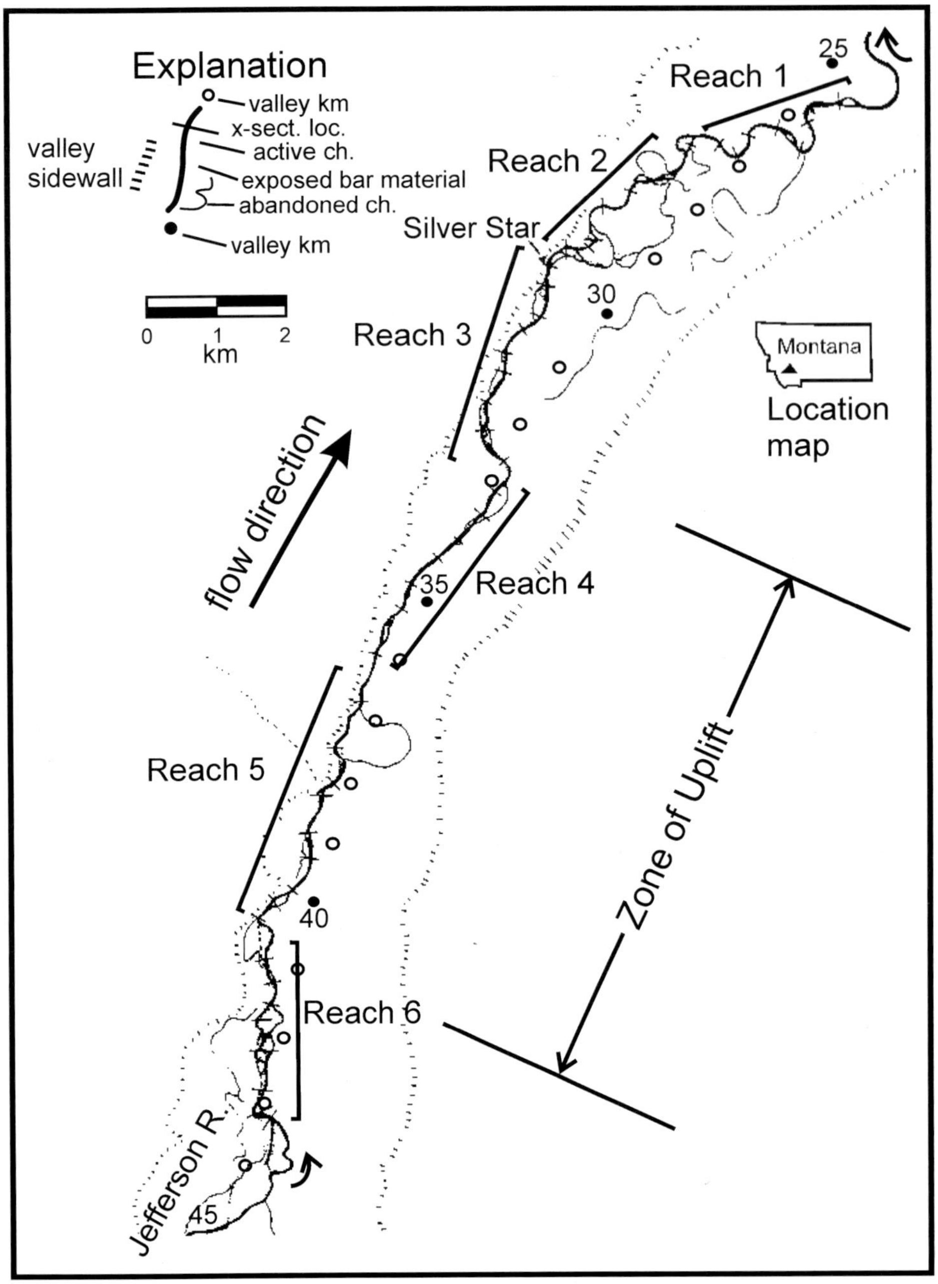

FIGURE 5.—Planform variability in the gravel-bed Jefferson River, southwestern Montana, across an uplifted area. Insert shows location of study reach within the state of Montana (modified from Jorgensen 1990 and Schumm et al. 2000).

typifies a particular type of channel. Because of the ambiguity involved in suggesting a planform for ancient river deposits Friend (1983) recommended a descriptive classification of outcrops based on the shape and interrelationships of sediment bodies.

Allen (1983) recognized and described bars, bar-complexes, and sandstone sheets in the braided stream deposits of the Lower Devonian Brownstones in the Welsh Borders. While this research effort cannot strictly be described as the development of facies models from a study of modern braided streams, this early and important paper, along with the observations made and the conclusions reached, is very important to later development of facies models. Allen described eight depositional features (architectural elements) that characterize the internal architecture of the sandstone sheets comprising the Brownstones, and he illustrates them with extensive use of lateral profiles. He recognized the importance of lateral accretion as a local component of braided streams, the presence of gravel cores that initiated sand deposition, and the formation of channel bars. Allen concluded that the mere presence of lateral-accretion bedding is an unreliable criterion for interpretation of channel pattern, and suggested that the relative horizontal extent of this structure may be of greater importance.

This suggestion that lateral accretion is an important component of bar building in braided channels was demonstrated for a small braid bar

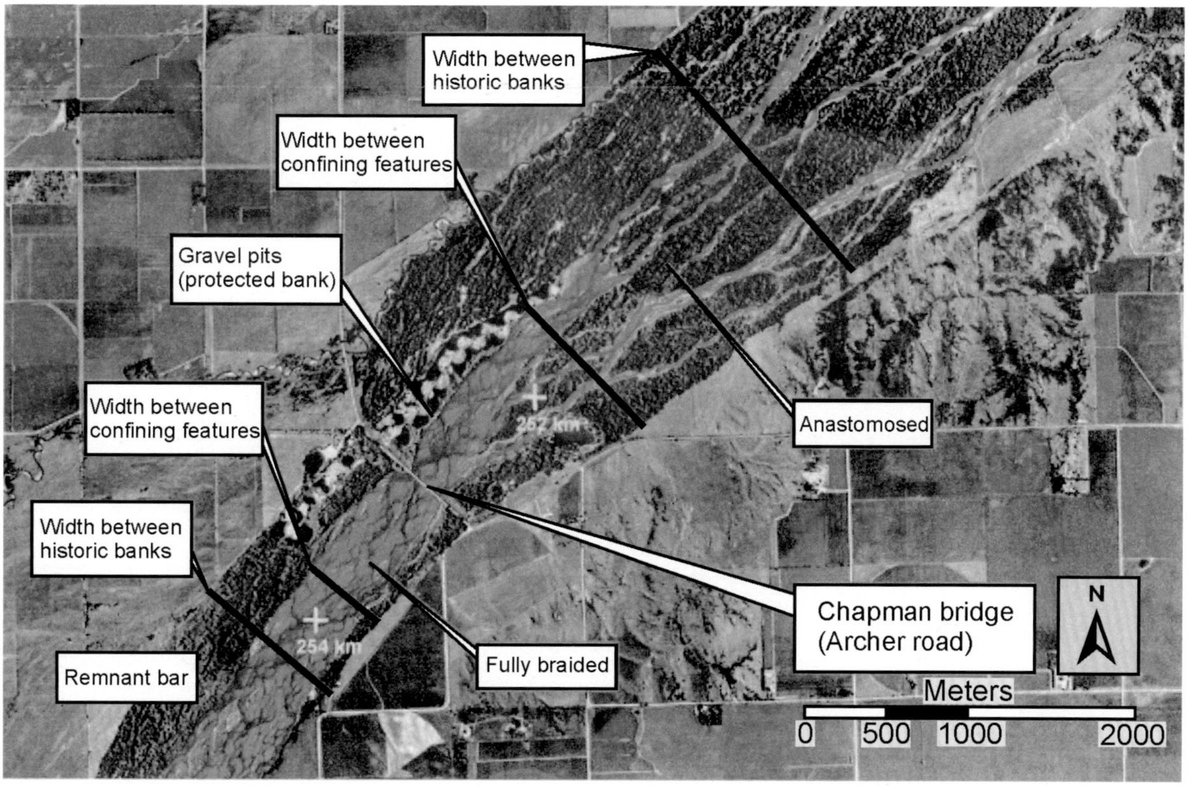

FIGURE 6.—Annotated grayscale aerial photo that illustrates transition from braided to anastomosing planform at river mile 251 on the Platte River, downstream from Grand Island, Nebraska. Note the decrease in valley confinement (increase in width between historical banks and an increase in width between confining features) downstream from the Chapman Bridge (modified from Fotherby 2009; see electronic supplement for color version of Figure S6).

(approximately 15 m long, 5 m wide, and with a relief of 1 m) in the South Platte River, at Goodrich, Colorado by Crews (1983, unpublished report) and Harvey et al. (1985). The bar, exposed by a temporary diversion of flow to another channel was dissected and a fence diagram was constructed (Fig. 8A). These profiles revealed a gravel ridge with crude horizontal stratification that formed the core of the bar, and large, planar and compound foresets that dip away from the axis, reflecting an asymmetric growth outward of the bar by slipface progradation and lateral accretion respectively (Fig. 8A, B). The upper portion of the bar contained a variety of ripple stratification.

Bristow (1987) also documented the presence of lateral-accretion deposits in much larger bars in the Brahmaputra River upstream of its confluence with the Ganges. The Brahmaputra in the study area is a multichannel river with planforms ranging from braided to meandering to anastomosing. Bar types included lateral (point), diagonal, medial, and tributary. A summary of data collected from these bars revealed that the primary types of deposition were lateral accretion (57%), downstream accretion (29%), and upstream accretion (14%).

Allen's use of architectural elements has been adopted by Miall, who standardized and codified the technique that involves the recognition and organization of scales of bounding surfaces, lithofacies, and elements (Miall 1985, 1988, 1993, 1996). Its use has been the subject of some debate (Bridge 1993; as discussed by Bristow 1996). In spite of these criticisms the technique has been widely adopted to the description and interpretation of fluvial deposits in outcrop. Like Miall, Bridge also suggests the importance of examining large-scale outcrops, preferably in three dimensions, in order to interpret paleochannel patterns. Bridge suggested use of the proportion of channel fill in relation to lateral accretion; mean grain size of channel fill relative to lateral accretion; paleocurrent variance; and reconstruction of paleochannel width/depth from channel bar and fill deposits (Bridge 1985). With the recognition of the importance of lateral accretion in the formation of braid bars some of these criteria were later modified (Bridge 2003). Bristow (1996) suggested that the paleochannel patterns might well depend on the relative abundance of channel-scale depositional features rather than their mere presence or absence.

One of the primary problems in developing facies models from modern rivers has been the examination and documentation of deposits below the water level. Most early facies models of rivers relied on the description of bar-top deposits. To overcome this problem ground-penetrating radar (GPR) coupled with continuous cores has been used as a principal technique for studying the sedimentology of small and large Holocene rivers since the 1990s (Gawthorpe et al. 1993; Bridge et al. 1995; Leclerc and Hickin 1997; Bridge et al. 1998; Fielding et al. 1999; Bristow et al. 1999; Vandenberghe and van Overmeeren 1999; Skelly et al. 2003; Best et al. 2003; Lunt and Bridge 2004; Lunt et al. 2004; Sambrook Smith et al. 2005, 2006).

Four of these studies were in meandering rivers, the Madison River in Montana, the South Esk in Scotland (Bridge et al. 1995), the North Thompson in British Columbia (Leclerc and Hickin 1997), and the Burdekin in north Queensland, Australia (Fielding et al. 1999). Radar profiles in all four systems show large-scale inclined strata that result from episodic point-bar accretion and abrupt lateral changes that resulted from periods of erosion and deposition (Bridge et al. 1995). However, the inclined strata in the coarse-grained, meandering Burdekin River, characterized by extreme discharge fluctuations, are

FIGURE 7.—Aerial photograph of the Arkansas River near Bent's Old Fort, southeastern Colorado (1977), showing bar and swale topography from active lateral accretion (photo from Nadler and Schumm 1981; See electronic supplement for color version of Figure S7).

so irregular and at such a low angle that they would be difficult to recognize in the rock record, and deposits of such rivers in the rock record might easily be mistaken for braided rivers (Fielding et al. 1999). Research on the Burdekin River has led to the development of a facies model for fluvial systems in the seasonal tropics and subtropics (Fielding et al. 2009; Fielding et al. this volume).

Most of the remaining papers listed above cover GPR studies of braided or transitional braided-meandering rivers. Rivers studied range from relatively moderate-sized sandy rivers (Calamus and Niobrara rivers in Nebraska and the Saskatchewan in Canada) to giant sandy rivers (Jamuna (Brahmaputra) in Bangladesh) and gravelly rivers (Sagavanirktok in Alaska). Deposits of these various braided channels reveal similar internal architecture in spite of the variability of scales and grain sizes, suggesting a scale invariance in fluvial processes (Sambrook Smith et al. 2005).

Although terminology is not the same among the various research papers, the following meso-scale and macro-scale sedimentary structures are common. Medium- to small-scale trough, cross stratification is the dominant sedimentary structure in most sand-bed and sand-gravel-bed rivers. They are recognized in GPR profiles as hummocky or wavy or discontinuous undular or trough-shaped reflectors. High-angle clinoforms or steeply dipping planar or subplanar cross stratification are produced by bar migration (Best et al. 2003; Sambrook Smith et al. 2006), 2-D dune migration (Skelly et al. 2003) or as unit bars (Fig. 9) (Lunt et al. 2004; Bridge and Lunt 2006). Curved, convex-up, reflectors are interpreted as the erosional base of main or cross-bar channels and confluence scours (Lunt et al. 2004). Large-scale, low-angle inclined strata are produced by growth and migration (accretion) of compound bars. Downstream, upstream, lateral, oblique, and vertical accretions are documented to varying degrees in the rivers studied (Fig. 10).

Of these rivers the only one documented to have undergone significant channel aggradation is the Niobrara that has aggraded between 2.3 and 2.9 meters, related to a base-level rise in the past 60 years. As a result of this aggradation more than eight crevasse splays and several avulsions occurred between 1993 and 1997 (Ethridge et al. 1999; Bristow et al. 1999). A vibracore and GPR study of two of these crevasse splays provides a three-dimensional view of their internal architecture. Both splays are sand rich, one is dominated by wide, shallow channels with cross-stratification, while the other contains sigmoidal sediment packages formed during progradation into standing water locally cut by smaller channels. GPR profiles show a continuous, subhorizontal reflector from a buried clay- and organic-rich buried soil horizon from the previous floodplain surface at the base of these splay deposits.

With the exception of the Brahmaputra River (Coleman 1969) crevasse-splay deposits like those in the braided Niobrara River have not been described in detail. A base-level rise and subsequent channel aggradation control both the style of floodplain sedimentation and the preservation potential of floodplain deposits such as these crevasse splays (Bristow et al. 1999). They form a significant component of preserved overbank deposits in inferred ancient braided river systems such as the Jurassic Morrison Formation in Utah (Kjemperud et al. 2008).

One feature that is especially helpful in the identification of braid-bar deposits in the rock record is the recognition of large-scale

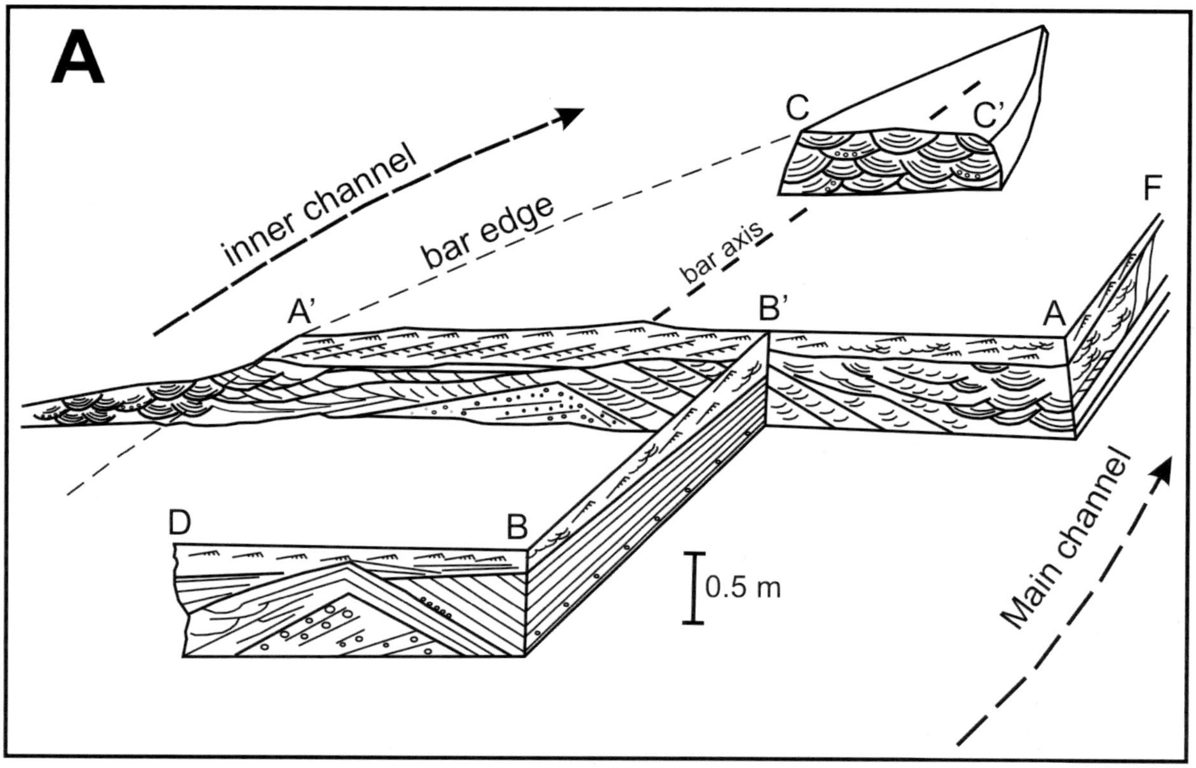

FIGURE 8.—A) Fence diagram of small compound, sandy, braid bar in the South Platte River at Goodrich, Colorado, showing a gravel core with crude horizontal stratification, planar foresets (section D-B) dipping toward the main channel, and the compound foresets with trough cross bed sets within the foreset laminae (section A–A'). Crews interpreted the latter two facies as representing slipface progradation and lateral accretion, respectively (redrawn from Crews 1983, unpublished report). Also note a wide variety of ripple stratification in the upper bar area. B) Close-up of gravel core with steeply dipping planar foresets (right) and compound foresets (left); river flow is into the photo (from Harvey, et al. 1985).

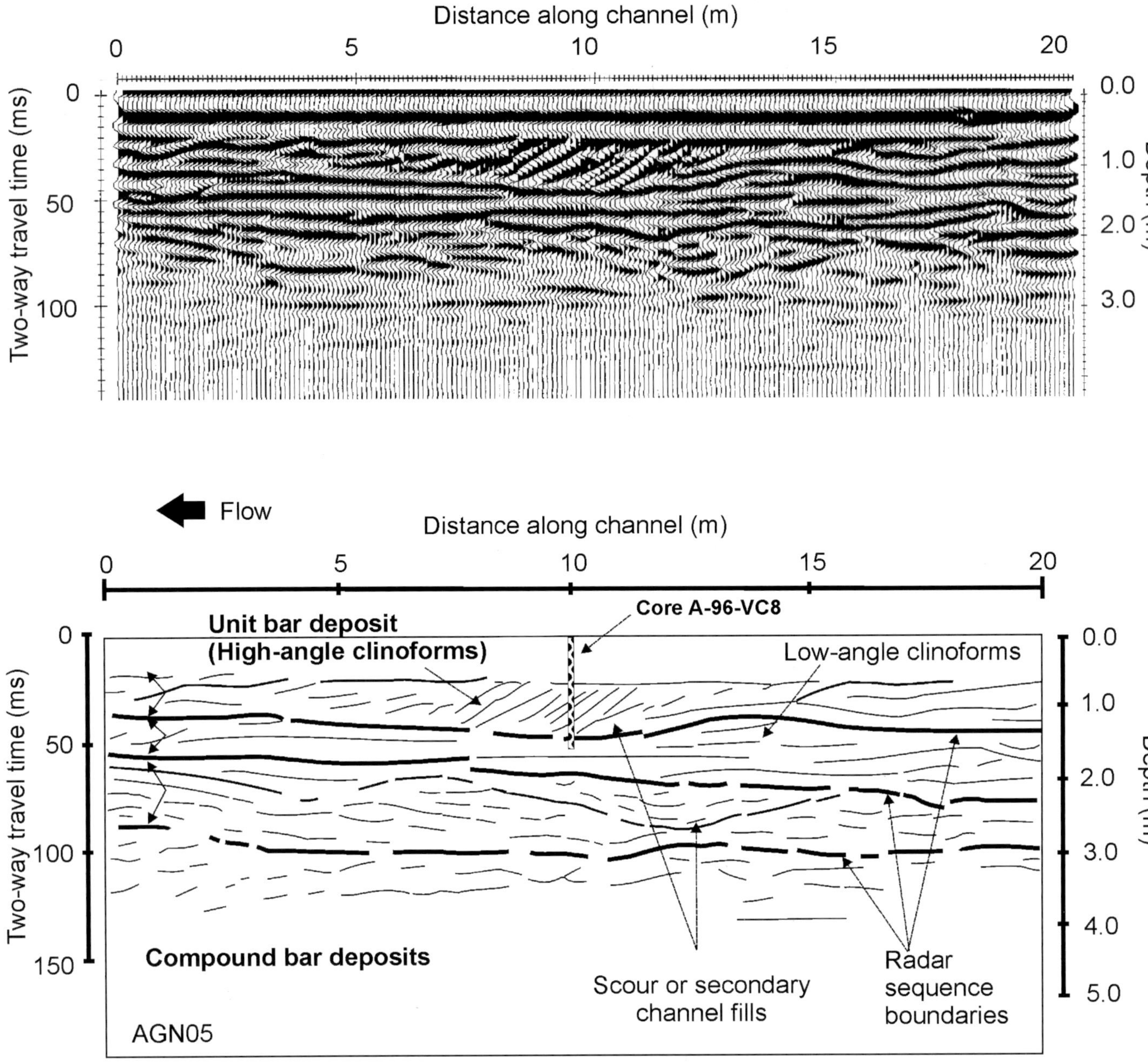

FIGURE 9.—Two hundred MHz profile and interpretive sketch from the middle of a detailed GPR grid roughly parallel to flow in the sandy, braided, aggrading lower Niobrara River, northeast Nebraska. Radar sequence boundaries, defined on the basis of reflection truncations, separate large-scale braid-bar or macroform elements. Downlap on the upper radar sequence boundary at a distance of 8–12 m probably defines a unit bar (Bridge 2006) and onlap on the middle radar sequence boundary defines upstream accretion (modified from Skelly et al. 2003).

stratification dipping in opposite directions away from the middle of these bars in outcrops that are oriented perpendicular to general paleoflow directions (Lunt et al. 2004). Heller et al. (2007) suggested that the distinction between ancient forced bars (point bars and alternate bars) and free bars (mid-channel braid bars) might be achieved by using an index of the distance of bar migration scaled to clinoform width. While both bars clearly coexist in meandering and braided rivers, the dominance of free bars suggests a braided stream pattern and forced bars a meandering pattern. To date no data have been published for this suggested criterion.

Recent fluvial facies models presented by Bridge and Lunt (2006) and Bridge (2006) consist of maps, cross sections, and vertical facies logs. For differentiating braided and meandering paleochannel deposits the delineation and characterization of large-scale channel bars (macroforms) in two-dimensional outcrop exposures (Fig. 11), coupled with detailed vertical profiles are necessary. With only the thickness of medium- to large-scale cross stratification from vertical profiles in outcrop and continuous cores, one can estimate paleochannel depth from the well known relationships between cross-bed thickness and flow depth (Bridge and Tye 2000; Leclair and Bridge 2001). This information, however, tells us little about plan-view morphology of the ancient fluvial channels. For estimating channel-belt width in

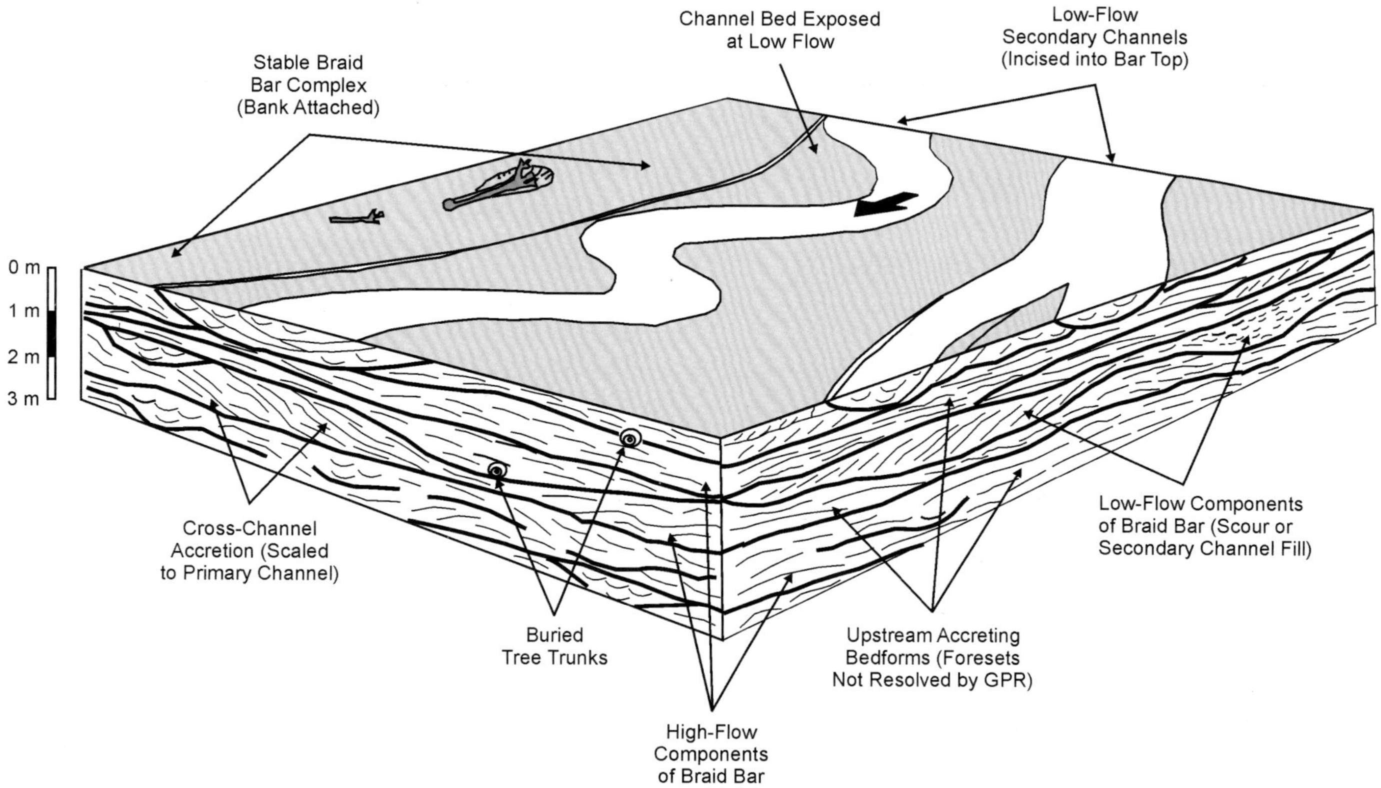

FIGURE 10.—Three-dimensional architectural model for channel-belt deposits of the lower Niobrara River, northeast Nebraska, based on GPR profiles and continuous cores. Although downstream, upstream, lateral, and vertical accretions are noted in GPR profiles, braid-bar deposits are dominated by lateral and upstream accretion. Distinct secondary channels in the deposits represent low-flow modification of the braid bar complex (from Shelly et al. 2003).

subsurface deposits Bridge and Tye (2000) made use of empirical equations (Bridge and Mackey 1993). These equations, however, have considerable standard errors, and there is little justification that they provide any better an estimate of channel-belt width than other techniques or that Bridge and Tye's correlations of sandstone bodies in the Travis Peak Formation at North Appleby field, East Texas, are any better than those suggested by Miall (2006) in his review on reconstructing the architecture of preserved fluvial record.

In summary, fluvial facies models have evolved from lithofacies diagrams that emphasized vertical succession of textures and sedimentary structures to panel diagrams that stress the importance of large-scale macroforms reflecting the remnants of channel bars that are scaled to bankfull depth. Our knowledge of channel bars has been greatly enhanced by numerous studies of modern rivers using GPR technology, which permits description of large-scale macroforms below the ground-water table. This technology is generally coupled with continuous cores to ground truth our interpretations of GPR reflectors. Additional GPR studies are needed in rivers that can be demonstrated to have undergone significant aggradation.

MYTHS REGARDING RIVER CHANNELS

Bridge (2006) provided a list of misconceptions about river channels. From this list he concluded that: (a) most classifications of river channels are flawed (see discussion above of anastomosing or anabranching with straight, meandering, and braided channels); (b) discharge variability is not necessarily greater for braided rivers than for single-channel rivers; (c) rivers that transport large amounts of bed load relative to suspended load do not necessarily have relatively low sinuosity and a high degree of braiding; (d) width/thickness (W/T) ratio of braided channel belts is not much greater than for meandering channel belts; and (e) not all rivers prior to development of vegetation on land were necessarily braided because of a high supply of bed load and unstable banks. Bridge provides anecdotes, but little data to justify his conclusions. In this review I will discuss his conclusions, provide supporting data, and expand the discussion.

Data do exist to support Bridge's conclusion that discharge variability is not necessarily greater in bed-load streams. Van den Berg's (1995) plot of annual flood discharge divided by mean discharge (a measure of discharge variability) versus valley slope shows no discrimination between braided and meandering channels (Fig. 12), and he concludes that high discharge variability does not promote braiding. Since bed-load streams with variable discharge have often been equated with braiding it is important to note this exception. Parts of the coarse-grained Burdekin River, Queensland, Australia, are undergoing meander migration. Yet this river is subject to extreme discharge fluctuations (Fielding et al. 1999). Based on a detailed study of the sediments, bed forms, bars, and GPR data these authors suggested that sand and gravel deposits of this river could easily be mistaken for deposits of a braided stream without the benefit of large outcrop exposures (> 200 m at right angles to flow directions; Alexander and Fielding 1997; Fielding et al. 1999).

The very limited data available on the W/T ratio of Holocene channel-belt deposits support the conclusion that braided channel belts do not have a higher W/T ratio than do meander belts. Bridge (2006) gives a W/T ratio of 375 for the meandering Mississippi River and 250

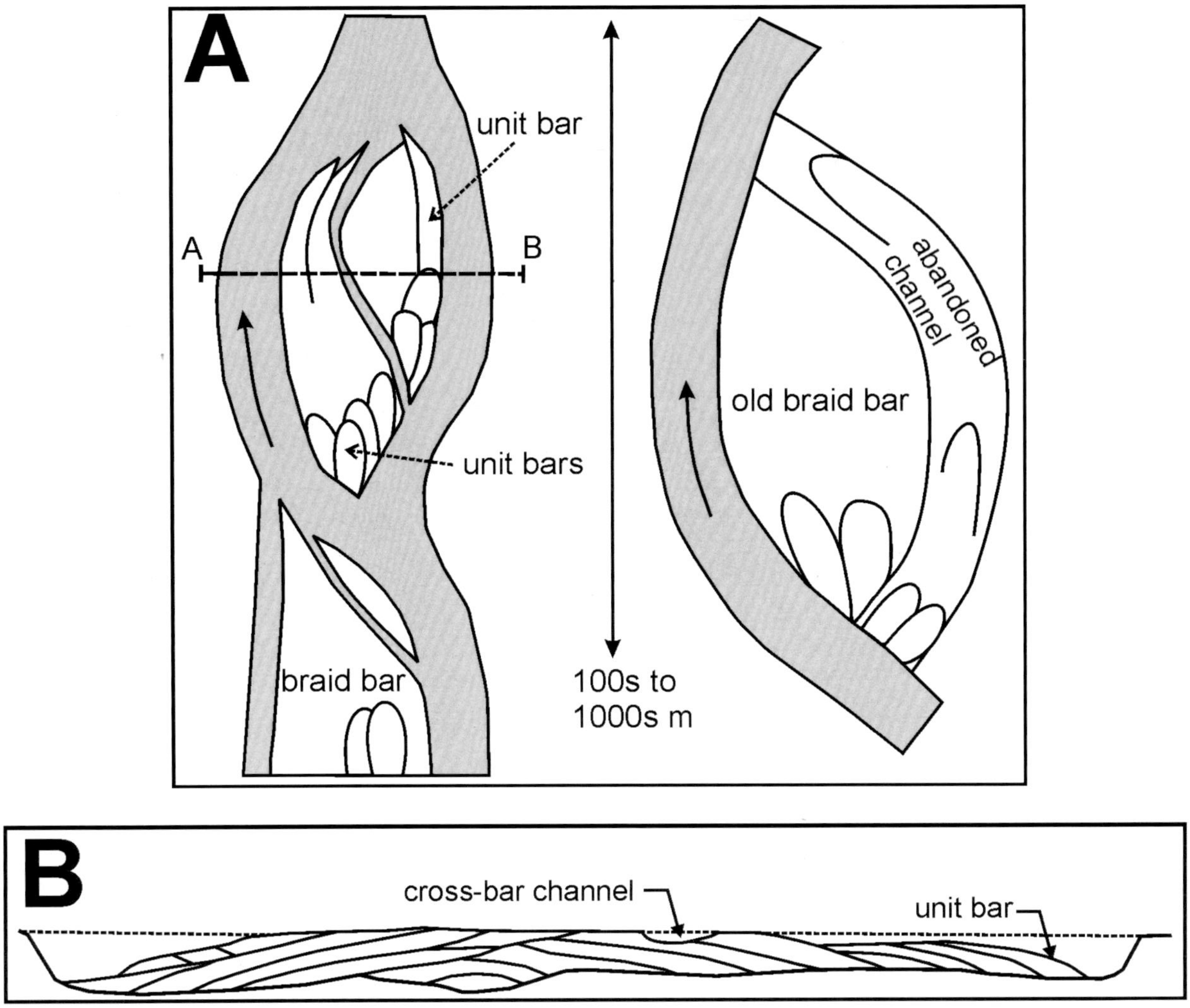

FIGURE 11.—Modern fluvial facies models consist of maps, lateral profiles, and detail vertical lithologic logs. A) Example of map portion of fluvial facies model showing active and abandoned channels, and compound and unit bars in a braided river. B) Single simplified profile (cross section A–B) showing large-scale inclined strata associated with migration of compound bars and unit bars. For simplicity, details of internal sedimentary structures, vertical lithologic logs, and additional maps and cross sections shown in the original models are not included (both figures redrawn and simplified from Bridge 2006).

for the braided Brahmaputra to support his conclusion, but he does not give a reference for either figure. Table 1 summarizes the limited data that are available from three large Holocene rivers. This table also summarizes data on channel-deposit proportion (CDP; similar to net-to-gross data used in the subsurface by petroleum geologists) for two large rivers that have been cored and for which cross sections distinguishing channel and floodplain deposits can be constructed. These data show considerable overlap in W/T ratios for all three river deposits and overlap in CDP values for the Mississippi and the Rhine-Meuse (Bristow 1987; Ashworth et al. 2000; Miall 2006; Gouw 2008; Gouw and Autin 2008). It is clear that data on the W/T ratio and the CDP of many more Holocene rivers systems are needed before attempts are made to justify interpretations of ancient channel-belt deposits on the basis of either of these criteria.

The general acceptance of a braided fluvial style for most pre-vegetation Precambrian rivers systems is clear from even a cursory review of the literature (Sønderholm and Tirsgaard 1998; Long 2004; Eriksson et al. 2006). One of the few exceptions appears to be an often quoted paper by Sweet (1988), who suggested a meandering origin, in part, for a Proterozoic stream deposits in central Australia, based on field evidence for lateral-accretion deposits. Flashy runoff, lower bank stability, and broad, shallow channels with abundant bedload and high rates of channel migration are taken as a given for pre-land-vegetation rivers, and these assumptions certainly cloud interpretations. With a few exceptions (Rainbird 1992; Long 2002) most studies of Precambrian fluvial systems lack any description or interpretation of channel bar structures, which are an important component of modern fluvial facies models. Thus most of these interpretations of ancient planforms are suspect.

The general acceptance that point-bar deposits have a higher variance of paleocurrent directions than do braid-bar deposits may not be valid. Shakla et al. (1999) in a study of braid-bar and point-bar deposits in the Holocene Ganga River, India, showed that, while paleocurrent patterns, variance, and azimuthal dispersion vary in different parts of both braid bars and point bars, the variance values are slightly higher and the azimuthal dispersion values are much higher in

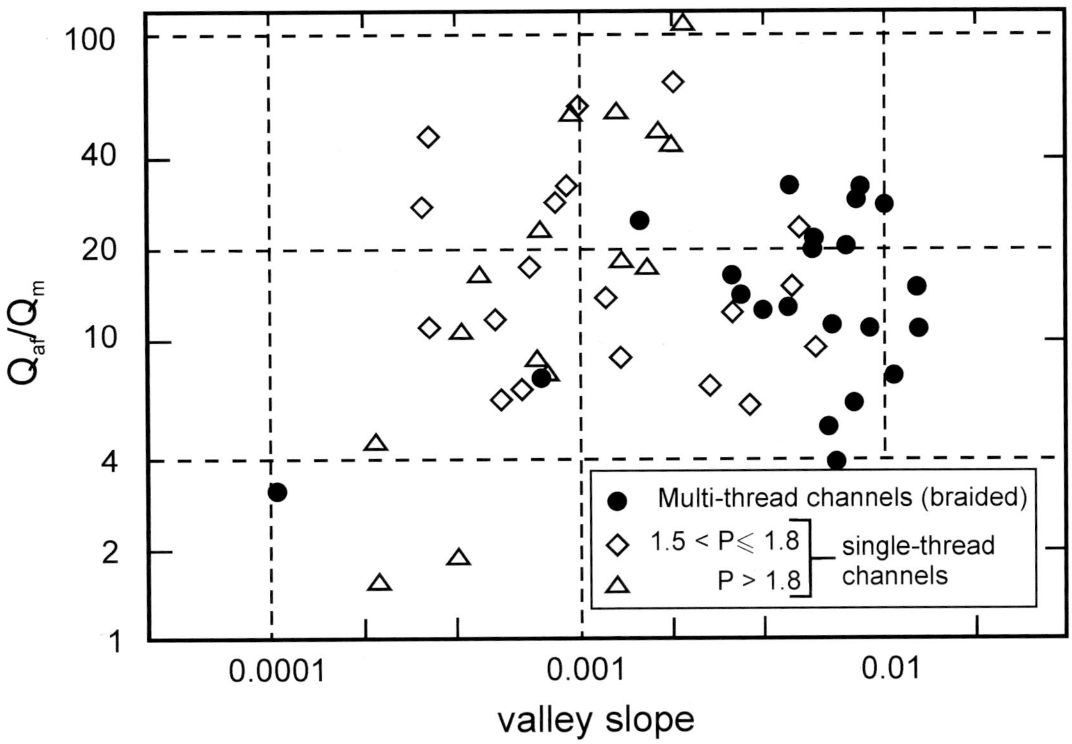

FIGURE 12.—Channel parameters (braided versus single-thread channels) related to a measure of discharge variability (annual flood discharge [Q_{af}] /mean discharge [Q_m]) and valley slope (redrawn and modified from van den Berg 1995).

the braid bar (240° versus 180°; Fig. 13). These data suggest that these parameters might not be useful in distinguishing braid bars and point bars in the rock record (Shukla et al. 1999). Even if distinction of point bars and braid bars is not possible there are still important uses for paleocurrent analysis as outlined by Miall (1996). This is especially true in determining the orientation of 2-D outcrop exposures relative to regional paleoflow direction, and in assessing changes in paleoflow direction through a stratigraphic succession, and reconstructing tributary or distributary patterns.

In summary a number of myths regarding modern river systems and deposits continue to cloud our judgment in attempts to interpret ancient fluvial deposits. These myths include, but are probably not limited to, the idea that highly variable discharge, high width-to-depth ratios, and low paleocurrent variances are equated with braided rivers and their deposits. The general acceptance that most pre-vegetation rivers were braided has generally not been verified with data from channel bar deposits.

ANCIENT FLUVIAL CHANNEL DEPOSITS

General

This section includes a brief review and discussion of research on the interpretation of paleochannel morphology. Examples are first discussed where data on the nature of channel bars is missing and the data presented are deemed insufficient to infer channel morphology. This is followed by a review of some research that includes detailed discussion of bar morphologies from large 2D outcrops and which provides a better possibility of inferring paleochannel morphology. The final part of this section includes a brief review of what we have learned from 3D seismic geomorphology of subsurface fluvial systems and some limitations.

Paleochannels Inferred from Limited Outcrop and Subsurface Data

Campbell (1976) in a study of the sheet sandstones of the Westwater Canyon Member of the Jurassic Morrison Formation concluded that, even though sheet sandstones could form by the coalescing of point-bar or braided-stream deposits, the latter was more likely. He recognized that fining-up cycles present in the Westwater Canyon were present in both meandering-stream and braided-stream deposits, but did not describe large-scale architectural elements that form the basis for our study of bar forms in paleo channel deposits. His often-cited interpretation of these deposits may have fostered a common belief of many geologists to this day that equates the sheet-like character of some ancient fluvial deposits to braided streams. However, the alluvial architecture of fluvial deposits is not directly related to planform morphology of the river, but it is related to avulsion frequency, sedimentation rate, and width of the channel belt relative to the width of the basin (Mackey and Bridge 1995; Heller and Paola 1996; Gibling 2006).

The thickness and width of ancient fluvial deposits are reviewed by Fielding and Crane (1987), Cowan (1991), Robinson and McCabe (1997), Reynolds (1999), and Gibling (2006). An examination of the papers cited by these authors suggests that many ancient fluvial deposits, interpreted as braided, do have high width-to-thickness ratios, but in fact there is insufficient evidence regarding bar morphology and architecture to support their interpretation as braided channels. Three examples with supposed sheet-like character will suffice to make this point. Two inferred braided and low-sinuosity rivers in the log-log plot of Gibling (2006, his Fig. 6) with the greatest widths are the Neogene Ogallala Group (thickness up to 475 m and width up to 1300 m) and

TABLE 1.—Width to thickness ratio and CDP for a few Holocene river deposits.

Modern River	Channel-belt width (m)	Channel-belt thickness (m)	W/T	Channel-belt width (m)	Channel-belt thickness (m)	W/T	Max floodplain width (m)	CDP*	Comments
Mississippi River	15000	40	375						Data from Bridge (2006)
L. Miss. River (V)**				3700	32	116	87000	0.44–0.60	Meander belt (data from Gouw and Autin, 2008)
				7300	32	228			
				2200	8–22	110–275			
				22000–26900	12–20	733-2242			
L. Miss River (N)**				14900	29	514	41400	0.54-.060	Meander belt (data from Gouw and Autin, 2008)
				1200	5–20	60–240			
				2200	5–20	110–440			
				2400	4–20	120–600			
				2800	34	80			
L. Miss River (B)**				4800	30	160	81900	0.25–0.28	Meander belt (data from Gouw and Autin, 2008)
				7900	26–30	263–304			
				5700	18	317			
				6800	15	453			
				5000	16–30	167–313			
				1300	34	38			
Range for L. Miss. River				1200–26900	4–34	28–2242	41400-8700	0.25–0.60	Meander belt (data from Grow and Autin, 2008)
Brahmaputra	10000	40	250						First-order channel belt (data from Bridge, 2006)
				10000–20000	50	200–400	no data	no data	Braided, meandering, and anastomosing patterns (data from Bristow, 1987)
Rhine-Meuse (delta)				40–3200	5–9	10–440	1500–5900	0.29–0.72	Anastomosing and Meandering (data from Gouw, 2008)

*CDP = channel deposit proportion (range based on min. and max. channel-belt width and thickness)
**L. Miss River (V) = Cross section Vicksburg; (N) =cross section Natchez; (B) = cross section Baton Rouge

the Lower Silurian Tuscarora Formation (thickness up to 200 m and width up to 1100 km).

Remnants of the Ogallala Group exist across the High Plains (Fig. 14) of the western USA, but the rivers that produced some of these deposits never covered an expanse this large. Fluvial sands and gravels of the Ogallala Group filled or partially filled paleovalleys (Fig. 15A; Diffendal 1990; Gustavson 1996; Bailey 2001). In the southern and central High Plains of the New Mexico and Texas the Ogallala fluvial deposits occasionally reached a thickness of 180 m, but mostly filled only the lower 10 to 40 m of paleovalleys (Figure 15B) and did not overtop divides (Gustavson 1996). These fluvial sands and gravels are capped by eolian coarse silt and sand, which also covers the upland areas between paleovalleys (Gustavson 1996). In the southern portion of the High Plains, fluvial sands and gravels are interpreted as braided-stream deposits, but to the north in Nebraska the paleovalley deposits are inferred to be products of meandering rivers with occasional braided reaches based on somewhat dated vertical lithofacies models. In northwestern Nebraska the Whitclay Gravel Beds, which constitute a small portion of the Ogallala Group, have recently been interpreted as probably braided, although limited exposures prevented a full assessment of channel and bar relationships and internal architecture (Fielding et al. 2007). Channels, estimated to be up to four meters deep and tens of meters wide, filled a small structurally controlled drainage resulting from Miocene uplift of the Black Hills (Fielding et al. 2007). Presumably the majority of the fluvial systems that produced the bulk of the Ogallala fluvial sands and gravels were related to late Cenozoic

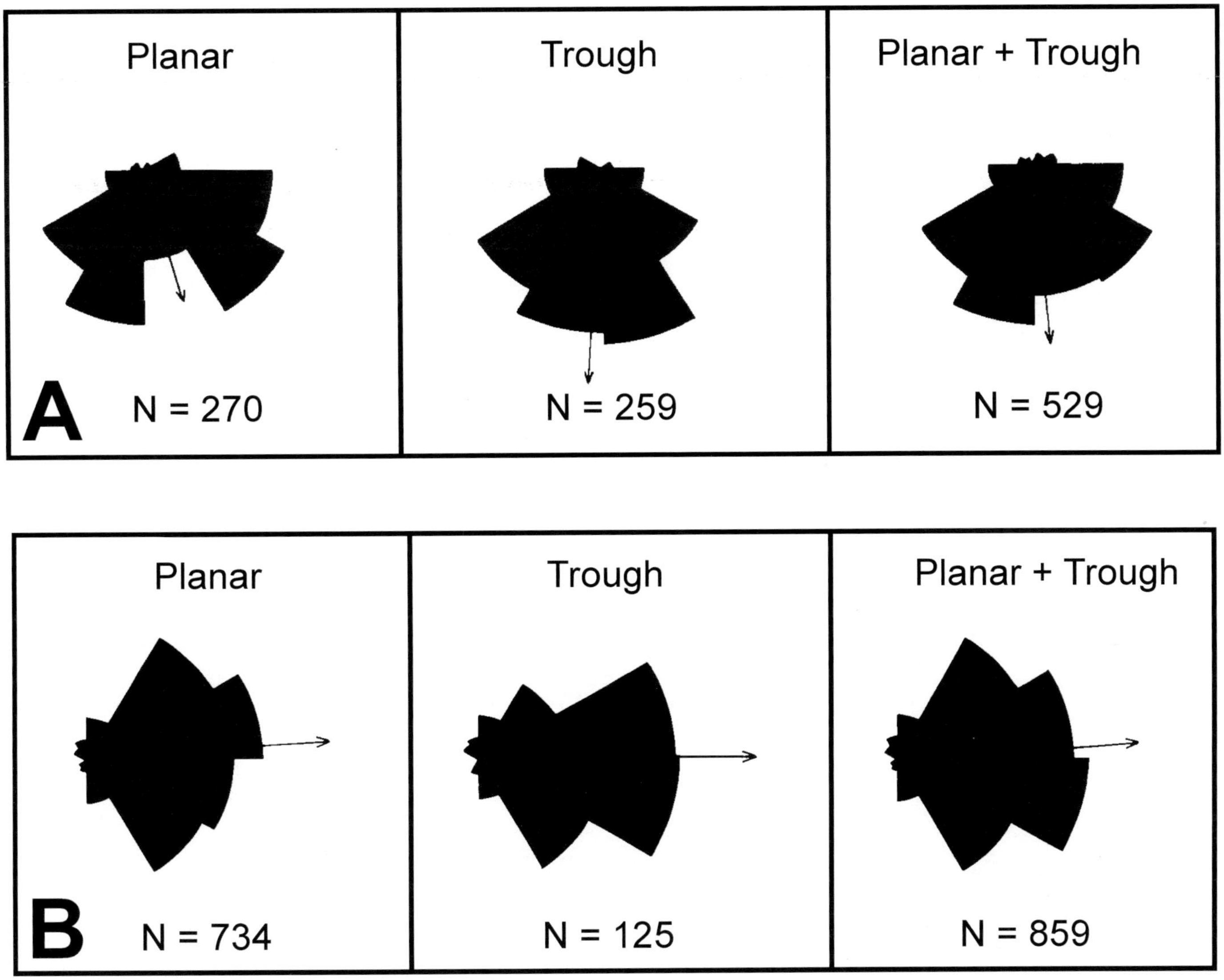

FIGURE 13.—Summary paleocurrent diagrams for A) braid bar and B) point bar in the Holocene Ganga River give rose diagrams for planar, trough, and total measurements. Azimuthal dispersions are bimodal for the planar cross beds, unimodal for troughs, and bimodal with very high values of dispersion for all cross beds in the braid bar. Azimuthal dispersions are unimodal and in the same directions for the point bar (both paleocurrent diagrams are modified from Shukla et al. 1999). Values of azimuthal dispersion are much higher in the braid bar.

uplift of the Rocky Mountains (McMillan et al. 2002). Based on these references there is no evidence to support the contention that the Ogallala has a sheet geometry and inadequate evidence to support a braided-stream interpretation.

The Tuscarora Formation in the central Appalachians (central Pennsylvania and West Virginia) was interpreted as fluvial on the basis of thickness, lithofacies, maximum pebble size, petrographic and fossils data, and cross-bed style and paleocurrent data (Yeakel 1962). Smith (1970) compared the vertical successions of sedimentary structures and grain size to deposits of the Holocene, braided Platte River and concluded a similar environment of deposition. Cotter (1983) reevaluated the Tuscarora and concluded that depositional environments included basal beach and estuary deposits, which are followed upward by cross-bedded, braided fluvial deposits in the east and cross-bedded, shelf sand-wave complexes in the west. Castle (1998, 2001) and Castle and Byrnes (2005) studied the Tuscarora and laterally equivalent units to the west and north in the subsurface and in outcrop. A major conclusion of these studies was the recognition of along-strike variability in stratigraphic patterns that are attributed to the occurrence of structural salients and recesses. The dominate sediment pattern in the salients is coarsening-up aggradational successions. In the recesses the sediment pattern is fining-up incised-valley-fill successions.

In summary the absence of lateral profiles depicting architectural elements or bar forms limits the ability to infer paleochannel form. This, coupled with the inferred lateral variability in sedimentary successions that are related to structural control suggests that a range of paleochannel forms comprise the Tuscarora Formation.

A third example from the subsurface is the Triassic Ivishak Formation (Sadlerochit Gp.), North Slope, Alaska, which is often cited as a type example of a sheet-type hydrocarbon reservoir. The basal Ivishak (Romeo) strata have been interpreted as deltaic by all

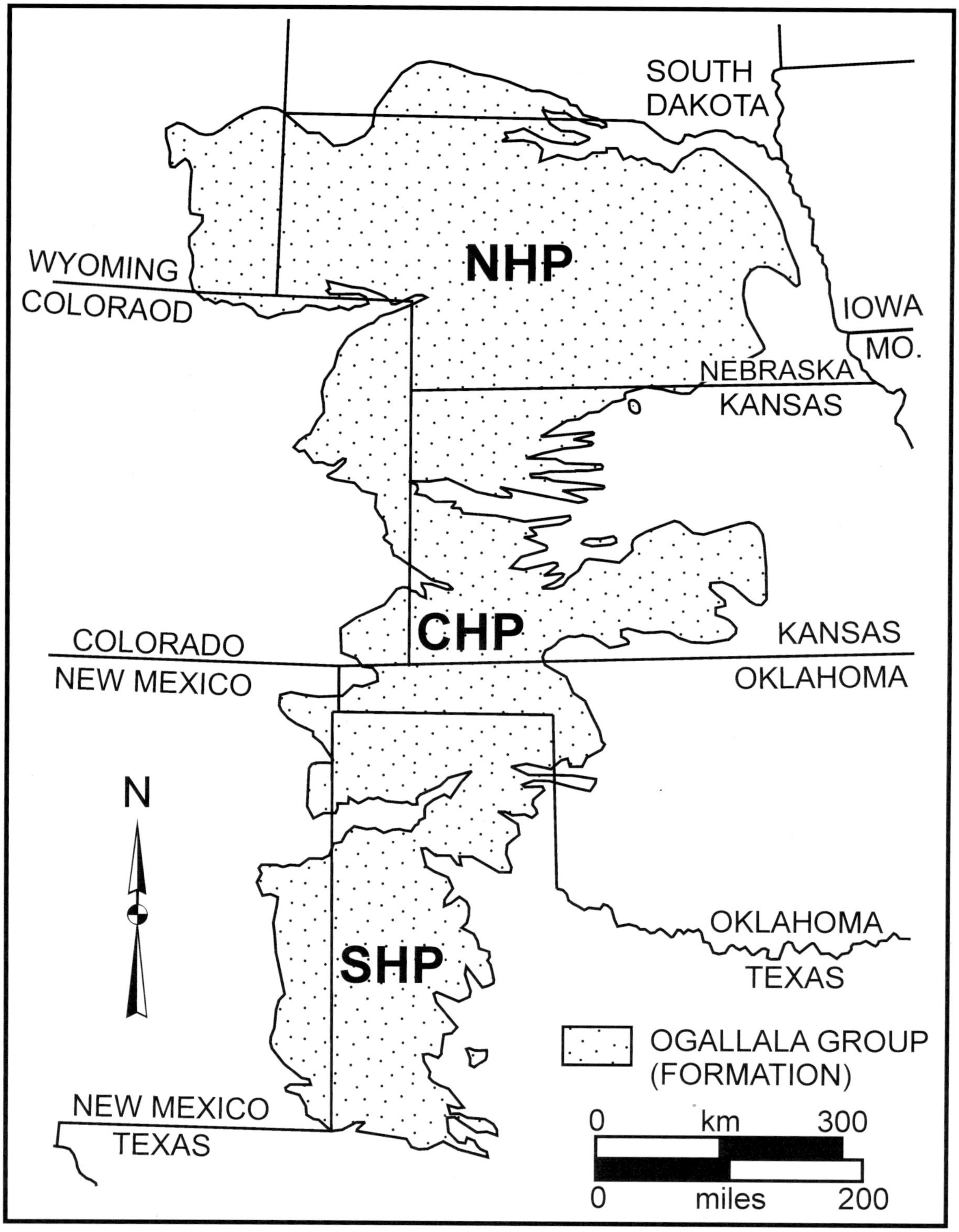

FIGURE 14.—Maximum extent of Ogallala Formation across the southern, central, and northern High Plains of the central USA (modified from Reeves and Reeves 1996).

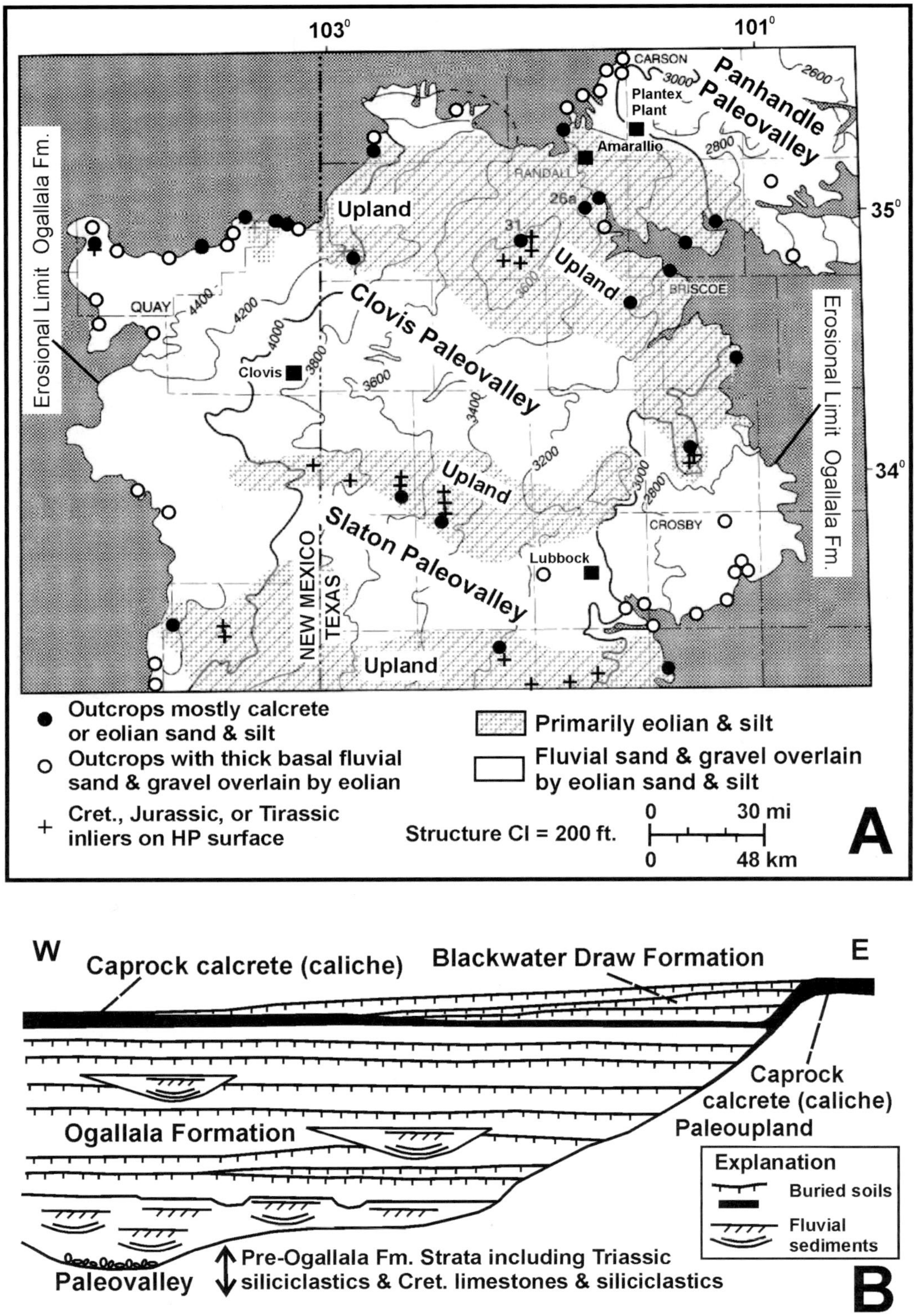

FIGURE 15.—A) Regional structural contour map of the pre-Ogallala erosional surface, illustrating paleotopographic highs or uplands (stippled) and paleo valleys. Paleo valleys are partially filled with fluvial sediments (open circles), and uplands are overlain by eolian deposits (closed circles). Both circle types denote outcrops (redrawn and modified from Gustavson 1996). B) Simplified cross section illustrating the distribution of fluvial sediments and calcic soils in the Tertiary Ogallala and Quaternary Blackwater Draw Formations in the Southern High Plains (modified from Gustavson 1996).

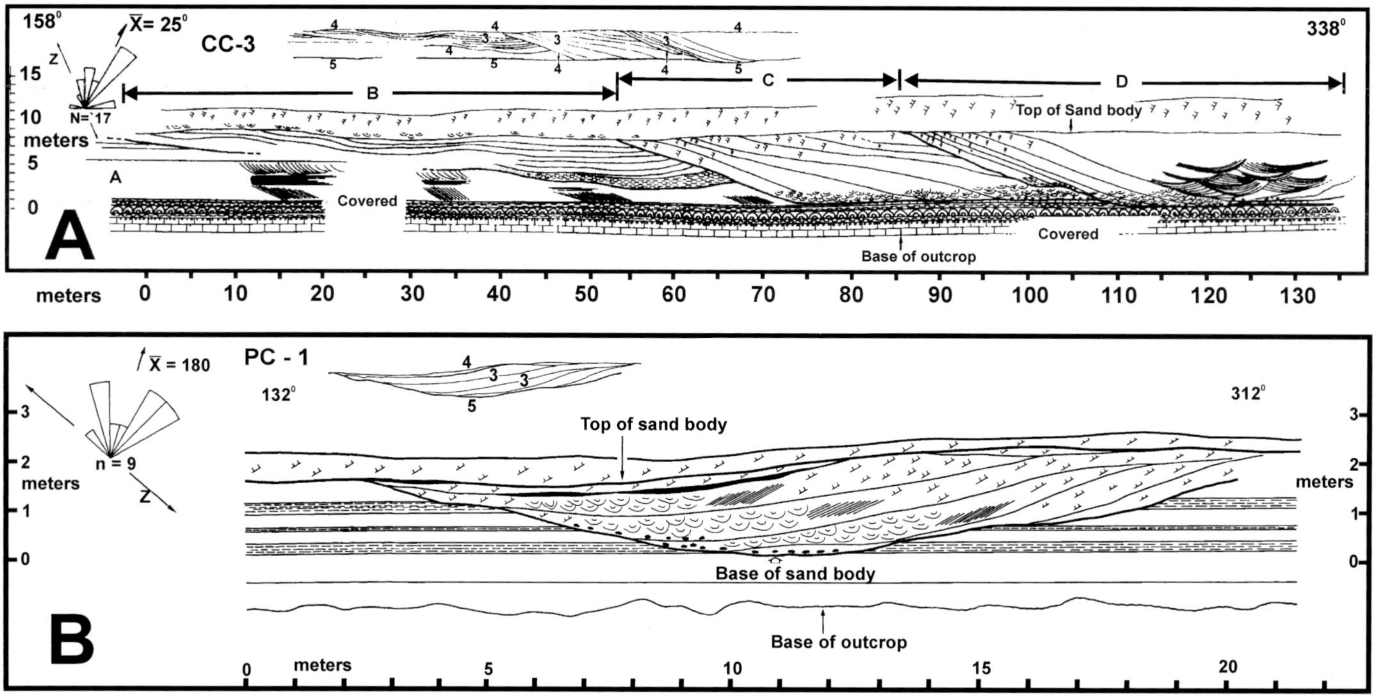

FIGURE 16.—2D profiles of sandstone bodies A) CC2 and B) PC-, Early Eocene, Green River Formation, Nine Mile Canyon, Uinta Basin, Utah (from Szantay 1990). Numbers in upper corners give outcrop orientation. Line Z on paleocurrent rose indicates orientation of outcrop relative to overall paleocurrent direction. Bounding-surface hierarchy is given in small insert of sandstone body after Miall. In profile CC3 note the planar lower bounding surface of the sandstone body that rests directly on carbonate that is truncated by the sandstone body at the right end of the panel and the multiple offlapping lateral-accretion elements. In profile PC-1 note the concave-up nature of the lower bounding surface of the sandstone body that has eroded into mudstone deposits and the single set of lateral accretion elements (modified from Szantay 1990).

researchers (Tye et al. 1999). The bulk of the overlying producing horizons have been interpreted as braided fluvial deposits in most reviews (Miall 1996; Gibling 2006). These fluvial units have been described in continuous cores by Lawton et al. (1987), Atkinson et al. (1988), and Atkinson et al. (1990). Lawton et al. (1987) interpreted some of the sandstone units that displayed fining- and thinning-up successions in the eastern and western parts of the field as meandering fluvial complexes, and the pebbly sandstones and conglomerates as braided-stream deposits. Thin mudstones in this succession were identified as eddy and slough deposits. One modern analog for the fluvial portion of the section is the Canterbury Plains on the south island of New Zealand, which displays gravel deposition in braided streams and sand and gravel deposition in meandering streams. A second is the meandering Kuparuk and braided Sageavnirktok Rivers on the Alaskan North Slope (Lawton et al. 1987).

Atkinson et al. (1988) and Atkinson et al. (1990) organized the various lithofacies (following the scheme of Miall 1985) into ten associations interpreted as fluvial, floodplain, and deltaic. Associations I, II, and III consisting of interbedded conglomerates and sandstones, were interpreted as migrating bars and channels in braided rivers while associations IV and V consisting of gravelly sandstones and sandstones as deposits of either braided or meandering rivers. Their overall interpretation was that of a series of braid deltas on a narrow coastal plain (Atkinson, et al. 1988; Atkinson, et al.1990). Grain size and vertical succession of sedimentary structures were the principal criteria used to infer the braided-stream origin for all of the coarser-grained associations and some of the sandstone units, while other sandstone units were inferred to be deposits of meandering rivers. It unclear why most subsequent writers on the subject equate all of the fluvial deposits in Prudhoe Bay field with braided streams when the original researchers clearly did not. A final note is that the modern rivers of the Canterbury Plain, a claimed analog for the Ivishak deposits, are clearly flowing in incised valleys resulting from incision either from rapid shoreline retreat or uplift of the Southern Alps (Leckie 1994).

Braided-stream deposits might be a component part of the three stratigraphic units discussed above. However, data on large-scale macroforms necessary to suggest an interpretation of plan-view morphology is nonexistent. Widths of these stratigraphic units may not be as great as suggested, and some deposits may be the result of deposition in single, straight or sinuous channels as suggest by several researchers. The fact is that from the available data we simply do not know.

2D and 3D Outcrop Exposures

With 3D exhumed fluvial deposits it is possible to determine paleochannel dimensions and morphology from measurements of parameters such as channel width and depth, radius of curvature, meander wavelength, meander amplitude, sinuosity, horizontal extent of accretion surfaces, and meander-belt width. Nami (1976), Padgett and Ehrlich (1976), and Smith (1987) demonstrated how stream characteristics could be determined from exhumed outcrops of Jurassic meander belts in England, Carboniferous meander-belt deposits in Morocco, and Permian meander-belt deposits in South Africa respectively. These types of exposures are extremely rare and presumably limited to relatively small meander-belt deposits.

Because of the rarity of 3D exhumed fluvial deposits researchers have turned to large-scale 2D outcrops oriented at various directions to

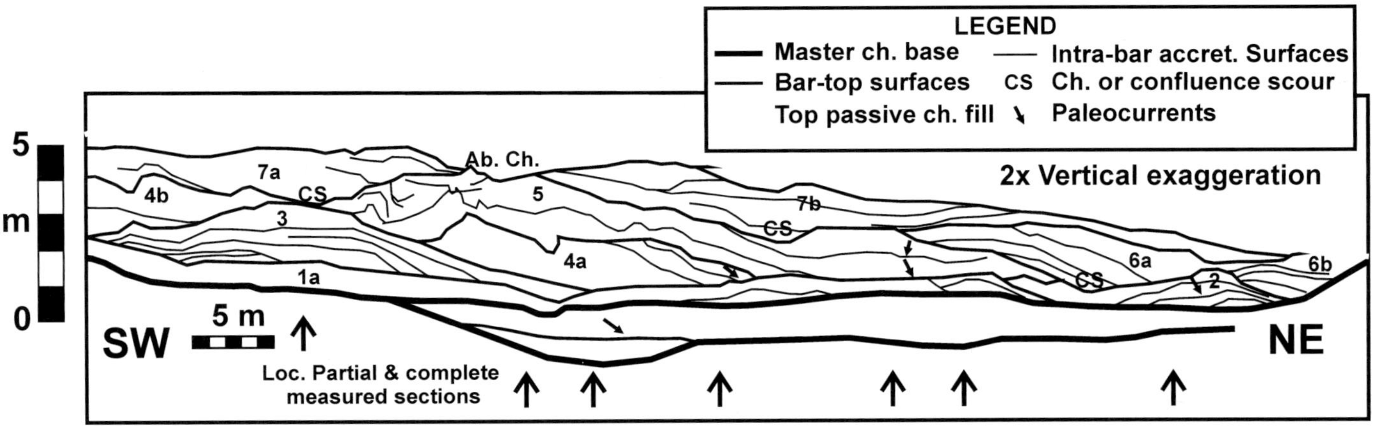

FIGURE 17.—Strike-oriented 2D profile of Upper Cretaceous Blackhawk Formation, central Utah, showing interpreted, stacked bars (numbered 1–7) and channels or confluence scours (CS). Measured vertical sections in original paper are not shown, but locations of these sections are shown at base of profile. Mounded geometries indicate braid bars. Arrows in panel show paleocurrents (redrawn and simplified from Adams and Bhattacharya 2005).

mean paleoflow directions to determine the architecture of large-scale inclined strata (macroforms). This permits assessment of paleochannel form and reliable estimates of paleochannel and channel-belt parameters. The first and most important studies include those of Allen (1983) on the inferred braided-river deposits of the Lower Devonian Brownstones in the Welsh Borders (discussed above) and Bridge and Diemer (1983) on the meandering-river deposits of the Upper Devonian–Lower Carboniferous of Kerry Head, Ireland. The paper by Bridge and Diemer (1983) was followed by a detailed study of meander belt deposits of the Upper Devonian Oneonta Formation (Catskill Magnafacies) in New York (Bridge and Gordon 1985; Gordon and Bridge 1987), and by a comparison of computer simulation results of 3D geometry of point bars with inferred point-bar deposits from the Upper Devonian of New York and the Miocene Siwalik Group of Pakistan (Willis 1993a, 1993b). Sandstone bodies discussed were interpreted as products of laterally migrating and aggrading single-channel, perennial rivers, and all had a reconstructed sinuosity between 1.12 and 1.2. Assuming that the computer simulation models give reasonable results, and given the low estimates of sinuosity for these deposits it seems reasonable to ask if the interpreted bars were true point bars or alternate bars in low-sinuosity paleochannels.

One feature that can aid in the identification of side-bar or point-bar deposits is the common occurrence of channel-form, sandstone bodies with a single compound-bar deposit and adjacent channel fill in large 2D outcrops, oriented normal to paleoflow direction (as illustrated by Bridge, 2006, his Fig. 18C). This architecture is common in 2-D outcrop exposures of the early to middle Eocene Green River Formation in Nine Mile Canyon, southwest flank of the Uinta Basin, Utah (Fig. 16; Szantay 1990; Szantay et al. 1989). It supports an interpretation of deposition in single, low-to-high sinuosity channels (Keighley and Flint 2008). The exposed, channel-form, ribbon sandstone bodies are interbedded with lacustrine mudstones, sandstones, and carbonates and form part of a fluvio-deltaic system along the southern margin of ancient Lake Uinta.

One of the most thoroughly studied stratigraphic units is the middle to lower Miocene Siwalik Group, of the Potwar Plateau of Pakistan (Willis 1993b, 1993c; Willis and Behrensmeyer 1994; Khan et al. 1997; Zaleha 1997a, 1997b). Detail 2D panel diagrams up to 4000 m (2.49 mi.) long at various levels within the stratigraphic succession were made possible from large-scale outcrops exposed along modern rivers in three different study areas. Willis (1993b) interpreted most paleochannels as braided in planform based on convex-up bedset surfaces with limited strike-oriented extent, bidirectional lateral accretion, upstream- and downstream-dipping low-angle bedset surfaces, high number of stories within individual channels, low paleocurrent variance within each story, and coarse-grained channel fills. Zalerha (1997a) interpreted the presence of both single-channel meandering and braided paleochannel plan forms in a different study area. All studies recognized the presence of overbank deposits including minor crevasse channels and splays, floodplain channels, lakes, and paleosols.

Results from these and other studies of the Siwalik Group sedimentary successions across the Gangetic plains are summarized and compared with modern rivers and floodplains in the same area (Jain and Sinha 2003). While not all research reviewed here contains the detail reconstructions summarized above it seems clear that a number of fluvial paleochannel forms are evident in the Siwalik Group, including braided, meandering, straight, and possibly anastomosing patterns (Table 2). A number of authors have proposed the existence of large megafans for the Siwalik Group similar to modern megafans existing in the area today and have suggested that vertical variations in the dimensions and extent of channel sandstone bodies may be related to autocyclic switching of the position of the fans or to extra basinal factors while the entry point or points from the mountain front remains fixed (Willis 1993b; Khan, et al. 1997; Zaleha 1997b). Other authors such as Friend et al. (2001) have moved away from earlier interpretations of fan-like systems to simple river networks for the basal Chinji Formation. The interpretation of megafans, as the dominant large-scale architecture for the Siwalik Group deposits, may be the harbinger of a resurgence of interest in the aggrading distributary fluvial systems (DFS) as an important form of fluvial preservation in continental sedimentary basins (Weissmann et al. 2010; Hartley et al. 2010).

Fluvial sandstone units in the Morrison Formation in the western USA have been the subject of intense study since the 1940s, when it became apparent that they hosted significant uranium deposits. Mullens and Freeman (1957) mapped the Salt Wash Member as a fan-shaped wedge with an apex in south-central Utah, and interpreted it as a distributary stream system. Galloway (1979, 1980) interpreted the Westwater Canyon Member as a humid alluvial fan, based in part on the geometry of this unit and comparison with the Salt Wash geometry (Mullens and Freeman 1957). Interest in these sandstones as outcrop analogs for fluvial sheet sandstones was sparked by Campbell's 1976 paper on the Westwater Canyon Member in New Mexico (discussed

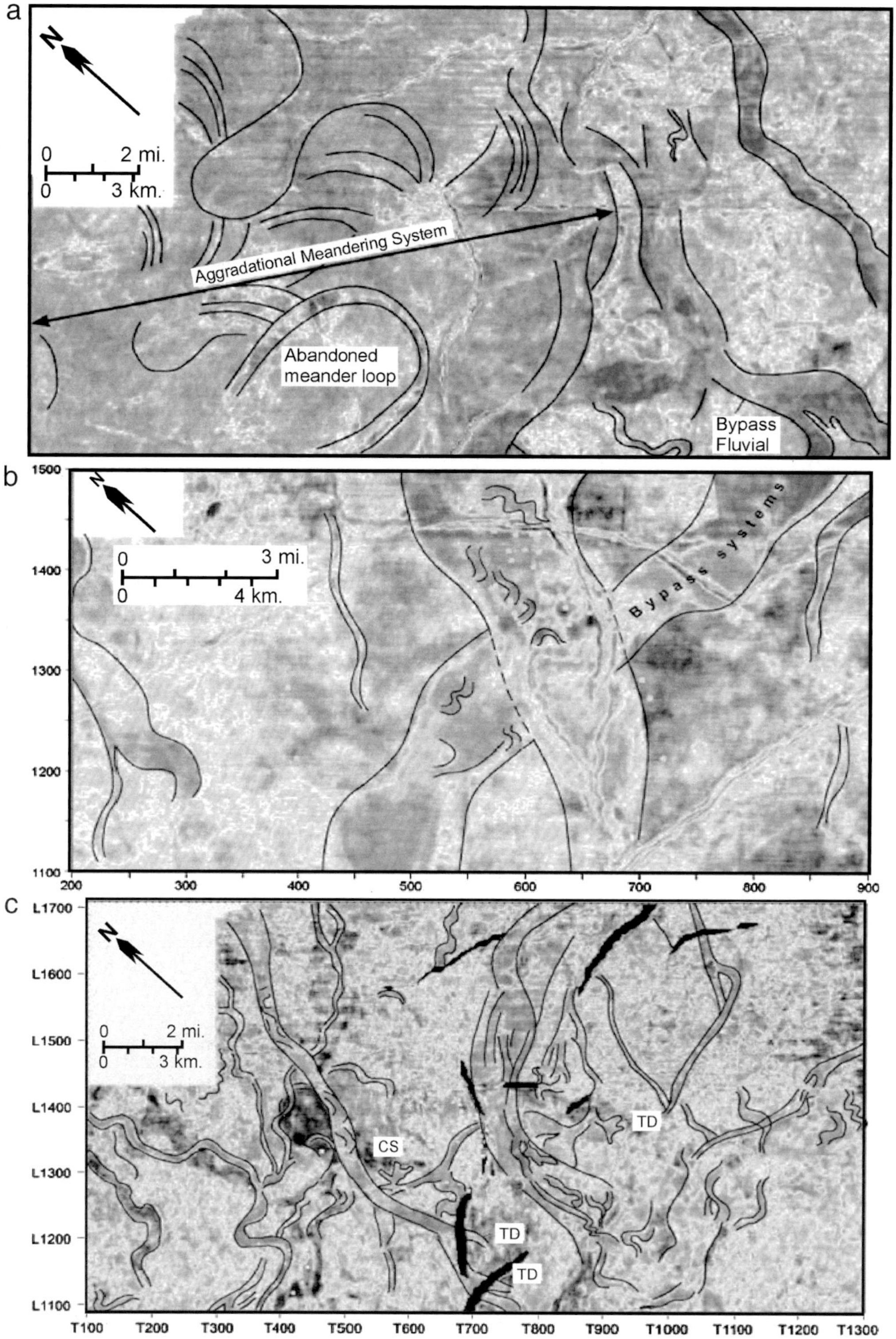

FIGURE 18.—Grayscale images of three proportional amplitude slices through aseismic-amplitude volume illustrating three classes of fluvial systems in Pliocene and Miocene sediments of the Northern Gulf of Mexico (modified from Wood 2007; see electronic supplement for color version of Figures S18A, B, and C). A) Class 1 aggradational meandering fluvial system with low-amplitude abandoned meander loops. This system is impinging on an older Class 2 lowstand bypass fluvial system to the southeast. B) Class 2 bypass valley systems with internal channels several hundreds of meters wide. Small gray dots that are difficult to see in this grayscale image are well locations (see Figure S18B for yellow dots in color slice). Thick dark lines mark major faults. C) Class 3 creeks and distributary channels cutting the older highstand coastal plain, overlain by at least two larger (2–3 km wide) Class 2 systems with multiple internal channels and accretionary architectures.

TABLE 2.—Interpretations from Miocene Siwalik sediments (modified and simplified from Jain and Sinha 2003)

Group	Formation	Potwar Plateau[1]	Jammu and Kashmir Himalaya[2]	Panjab Himalaya[3]	Garhwal & Kumanu Hemilaya[4]	Nepal Himalaya[5]	Sikkim Mimalaya[6]
	Boulder Congl. Pinjor			Gravelly braided rivers in alluvia-fan setting		Debris-flow-dominated gravelly braided rivers	
			broad alluvial plain with ow sinuosity rivers	Braided or anastomosing channels			
Middle Siwalik	Tatrot				Braided rivers	Anastomosing, shallow, sandy, braided rivers	Braided rivers on coalescing alluvial fans
	Dhokpathan	Blue-gray system: large braided rivers spreading over 25 km Buff system: $\leq$ 3 km wide			Braided rivers in alluvial-fan settings	Deep sandy braided; Sandy meandering	
Middle Siwalik	Nagri	Large braided rivers, larger channels than in Chinji		Meandering, entrenching in some cases		Meandering channels	
Lower Siwalik	Chinji	Braided rivers, 2 km wide, megafan setting	Meandering, highly sinuous	Meandering	Gigantic overlapping alluvial fans; meandering rivers	Fine grained meandering	

[1]Willis (1993b), Zaleha (1997a), Khan et al. (1997), Friend, et al. (2001); [2] Sharma et al. (2001), Ranga Rao and Kunt(1987), Visser and Johnson, 1978) [3]Tandon and Narayan (1981), Kumar and Tandon (1985), Bridge and Mackey (1993); [4]Parkash et al. (1974), [5]Nakayama and Ulak (1999); [6]Banerjee and Banerjee (1982)

above). Detailed studies of the fluvial Salt Wash Sandstone Member in the Colorado Plateau were undertaken in the early 1980s (Tyler and Ethridge 1983a, 1983b; Peterson and Tyler 1985). These authors interpreted trunk braided and tributary meandering channels and overbank crevasse splays within the Salt Wash complex, based on vertical lithologic profiles and several lateral profiles, but they failed to delineate macroforms in these deposits. The first delineation of these macroforms came with the work of Miall and Turner-Peterson (1989), Cowan (1991), and Godin (1991) in the Westwater Canyon sandstones of New Mexico. They also inferred the presence of braided and meandering channels and compared the architecture of the Westwater Canyon with that of large Indian subcontinent rivers. These authors discussed the importance of large-scale scour features, related them to confluence scours, and identified the presence of low-amplitude macroforms bounded by flat erosional surfaces. They suggested that these were consistent with multichannel braided systems. They did not identify concave-up features with bidirectional strata sets, but these features appear to be present in some of their strike-oriented panel diagrams.

Despite the presence of extensive and important additional research on the Salt Wash Member of the Morrison Formation in southeastern Utah and the construction of some large-scale panel diagrams (Robinson and McCabe 1997; Kjemperud et al. 2008), the kinds of details needed to identify the internal characteristics of macroforms that scale to channel depth have not been constructed for this member. The detailed macroform architecture for the Westwater Canyon cannot be transferred to the Salt Wash because the two members not only have different source terrains, but also they are separated by a major unconformity (Demko et al. 2004). Thus, the inferred paleochannel form of the Salt Wash remains somewhat speculative.

Some other widespread, dominantly fluvial, stratigraphic units for

←

FIGURE 18 (caption continued).—Several of these channels break into terminal distributary lobes (TD), and a few possible crevasse splays (CS) are identified. Small gray dots that are difficult to sse in this grayscale image are well location (see Figure S18C for yellow dots in color slice). Several thick black lines mark major faults.

which macroforms have been identified in enough detail to suggest channel morphology includes the Upper Cretaceous Castlegate Sandstone, Utah (Miall, 1993), the Cretaceous Chubut Group, Argentina (Bridge, et al. 2000), the Upper Cretaceous Blackhawk Formation, Utah (Adams and Bhattacharya 2005), and the Cretaceous Matasiete Formation (Chubut Group), Argentina (Paredes et al. 2007). Miall (1993) interpreted the Castlegate Sandstone as a braided-stream system similar to the modern-day South Saskatchewan River based on the common occurrence of lateral-accretion, downstream-accretion, and sheet-like elements. The underlying Blackhawk Formation, with isolated channel belts contained within thick mudstone packages, has long been assumed to represent deposits of meandering streams. However, reconstruction of macroforms with bidirectional downlap in strike panels of channel belts (Fig. 17) led Adams and Bhattacharya (2005) to interpret a braided pattern for paleochannels of the Blackhawk, which supports the idea that preservation of thick overbank sediments has little to do with plan-view channel geometry (Friend 1983). The Argentina Cretaceous units studied comprised interbedded sandstones, mudstones, and tuffs. The proportion of channel-belt deposits is generally low, and channel belts are generally not connected. Detailed reconstruction of macroforms revealed the presence of channel-bar deposits with a single adjacent channel-fill deposit and convex-up channel-bar deposits with coeval channel fills (Bridge et al. 2000). Straight, meandering, and braided channel forms were inferred with wide braided channels following pyroclastic supply events (Paredes et al. 2007). Anastomosing channel networks, however, could not be ruled out on the basis of available data.

Detailed reconstruction of macroforms in 3D exposures and 2D profiles support the contention that ancient fluvial successions are more complex than previously thought. Single, straight to meandering channels are not found only in mud-rich sections, and braided channels are not confined to sheet-like sandstone deposits. This is not a new concept, but it has apparently been lost on the majority of geologists who persist in interpreting ancient fluvial deposits on the basis of vertical lithologic profiles and large-scale architecture without obtaining the details necessary to reconstruct paleochannel bars. It is also unreasonable to interpret ancient fluvial deposits in terms of a single end-member planform given the considerable variability documented in modern river channels and the continuum of channel planform types recognized. It is probably more prudent to concentrate on describing channel-bar deposits in outcrop, and from these data and other pertinent information (paleontological and stratigraphic) examine how river systems might have evolved spatially and temporally.

Quantitative Seismic Geomorphology (QSG)

Three-dimensional seismic data enable geoscientists to image the stratigraphic record along selected time, horizon, and stratal (or proportional) slices (Zeng 2007). These slices can provide detailed images of planform geometry of ancient depositional systems and environments that have previously been available only in extremely rare 3D outcrop exposures. The limited literature on reconstructions of fluvial systems using 3D seismic data was reviewed by Ethridge and Schumm (2007), and includes Isa et al. (1992), Hardage et al. (1994), Burnett (1996), Saudi Arabian Oil Company (1996), Posamentier (2001), Miall (2002), Martinez et al. (2004), and Carter (2003). Research in the 1990s was limited in extent to hydrocarbon fields or drilling sites and was designed to show the location and extent of channel systems. Images were not always clear, and some interpretations were not supported by the evidence. Nevertheless they represent important pioneering works in the evaluation of fluvial paleochannels with 3D seismic imaging.

The 2001 paper by Posamentier represents a major step forward in that he recognized both incised-valley systems and un-incised channel systems in Miocene to late Pleistocene fluvial deposits of the Java Shelf. Criteria for the recognition of each system are documented and the implications for distribution of coarse-grained sediments with a sequence stratigraphic framework are discussed. An equally important paper by Miall (2002) examines fluvial systems in Pleistocene fluvial deposits of the Malay Basin. The complex nature of this system is seen in the plethora of channel types present in seismic slices. These include large and small, low- and high-sinuosity channels, equiwidth stable and wide-bend meandering channels, underfit streams, and large braided streams. Both Posamentier and Miall used multiple 3D slices to discuss evolutionary history of these fluvial systems in a sequence stratigraphic framework. Carter (2003) documented the evolution in reservoir geometry of an Oligocene–Miocene meandering fluvial systems at Widuri Field in the Java Sea. Martinez et al. (2004) provided details on the character of the late Miocene unconformity in the Mediterranean and evaluated erosional and depositional processes associated with that major base-level change.

Papers published since the review of Ethridge and Schumm (2007) include Pliocene deposits of the Sunda shelf, offshore Indonesia (Darmadi et al. 2007) and Pliocene and Miocene deposits of the Northern Gulf of Mexico (Wood 2007). Darmadi et al. recognized five fluvial sequences each beginning with an interval of relatively abundant and larger channel-form reflections. In these intervals they recognized a variety of small and large, straight to sinuous channel forms. The largest, sinuous channel, incised into sequence boundary 4, was interpreted to have had a channel width of 3800 m and a depth of 50 m. Wood (2007) recognized three classes of fluvial systems. Class 1 consists of very wide meander belt systems with channels that have relatively high sinuosity, high meander length, and large meander-arc height (Fig. 18A). One or both boundaries of these systems often extended beyond the study area. Class 2 consists of bypass fluvial valley systems up to several kilometers wide, with well-defined meander belts and channels that have low sinuosity, high meander length, and low meander-arc height (Fig. 18B). System 3 consists of distributary channels and creeks with high sinuosity, narrow widths, low meander lengths, and small meander-arc heights (Fig. 18C).

This overview of 3D seismic interpretation of ancient fluvial systems suggests several important conclusions. A wide variety of fluvial morphologies exist, both laterally within the same time or stratal horizons and through time, with the degree of variability increasing with increasing size of the study area. Most channel forms have been identified, with the exception of anastomosing channels, and valleys can usually be differentiated from un-incised channels. Very large meandering rivers are clearly present in the sedimentary record at least as far back as the Miocene. Quantitative characterization of fluvial paleochannels is possible from 3D seismic slices coupled with data from continuous cores, wireline logs, and seismic cross sections. Based on the published literature, 3D seismic interpretations of ancient fluvial systems are, to date, clearly limited to geologically young sedimentary successions at relatively shallow depths. This limitation is related to a reduction in resolution and deterioration in the signal-to-noise ratio with depth (Zeng 2007). In addition reflections from multiple thin beds at depth commonly interfere with each other and form merged events, making problematic the interpretation of thin bedded fluvial deposits, such as those associated with thin, discontinuous channel sandstones in tight gas reservoirs in Wyoming and Colorado.

SUMMARY AND RECOMMENDATIONS

Most classifications of fluvial channels and channel belts are based on plan-view morphology and while these classifications have received their share of criticism, they remain grounded in the literature and in our thought processes regarding river channels. A number of researchers recognize a continuum of channel forms rather than discrete types, while others see abrupt transitions between straight, meandering, and braided channels. The concept of thresholds remains a key factor in analyses of channel response. Studies of modern fluvial channels reveal a high degree of variability in channel morphology along channel lengths and with time, which is related to upstream, local, and downstream controls. This variability is, for the most part, not recognized in ancient fluvial deposits.

Facies models have evolved from vertical lithofacies profiles to those that include lateral profiles illustrating channel-bar architecture. The concentration on channel-bar architecture and internal structure is based on numerous studies over the past two decades using GPR technology. Modern facies models do not provide criteria for differentiating single paleochannels from multiple or anastomosing (anabranching) systems, and this remains a difficult problem in interpreting the morphology of ancient fluvial systems.

A number of misconceptions remain regarding modern rivers and their deposits that continue to cloud our interpretation of ancient fluvial systems in spite of several discussions and reviews. A review of published literature suggests that most inferred plan-view morphologies of ancient paleochannels and channel belts has been based largely on these misconceptions and on vertical lithofacies profiles. Thus many studies contain inadequate data for making inferences regarding the morphology of paleochannels and channel belts. This is especially true for the subsurface, where data have been limited to continuous cores, wireline logs, and 2D seismic. With these data, correlation of inferred channel deposits between wells is often difficult, and vertical lithofacies profiles at best can only provide data to interpret channel depth.

A literature review was undertaken of several well known fluvial deposits that are considered products of deposition by braided streams (Ogallala Group, Tuscorara Formation, and Ivishak Formation). This review reveals that no data exist on the nature of channel-bar deposits in these units, and they are not as widespread as previously suggested.

Detailed reconstructions of preserved bar forms in 3D and 2D outcrops suggest that ancient fluvial successions are much more complex than previously thought. Single, straight to meandering channel deposits are not just confined to mud-rich systems and braided channel deposits to sheet sandstones. Based on these conclusions and the considerable variability found in modern fluvial channels, it is unreasonable to interpret ancient fluvial deposits in terms of a single, end-member planform.

Three-dimensional seismic slices can provide plan-view morphologies of ancient paleochannels in the subsurface, thus permitting reconstruction of channel and channel-belt patterns, distinction between channels and incised valleys, and reconstruction of the evolution of fluvial systems. To date, however, these interpretations are limited to geologically young sequences at relatively shallow depths because of problems associated with decreased resolution with depth and with multiple thin beds.

For modern rivers additional data are needed on width/thickness ratios of river deposits of varying size and complexity, and GPR studies need to concentrate on a variety of aggrading river systems. For ancient outcropping fluvial deposits there is a need for documentation and detailed descriptions of preserved channel bars. Use of these data will aid in determining how fluvial systems evolved spatially and temporally. Many classic fluvial sheet sandstones need to be reexamined, in view of the variability documented in modern fluvial systems, using current facies models.

ACKNOWLEDGMENTS

This paper has benefited greatly from helpful suggestions and comments of earlier drafts by Stan Schumm, reviewers Jan Alexander and Stephen Tooth, and volume editors Stephanie Davidson, Colin North, and Sophie Leleu. To all of these individuals I extend by grateful thanks for suggestions that clarified a number of the main points addressed in this review paper.

REFERENCES

Adams, M.M., and Bhattacharya, J.P., 2005, No change in fluvial style across a sequence boundary, Cretaceous Blackhawk and Castlegate Formations of central Utah, U.S.A: Journal of Sedimentary Research, v. 75, p. 1038–1051.

Alexander, J., and Fielding, C., 1997, Gravel antidunes in the tropical Burdekin River, Queensland, Australia: Sedimentology, v. 44, p. 327–337.

Allen, J.R.L., 1963, Henry Clifton Sorby and the sedimentary structures of sands and sandstones in relation to flow conditions: Sedimentology, v. 42, p. 223–228.

Allen, J.R.L., 1983, Studies in fluviatile sedimentation: bars, bar-complexes and sandstone sheets (low-sinuosity braided streams) in the Brownstones (L. Devonian), Welsh borders: Sedimentary Geology, v. 33, p. 237–293.

Ashworth, P.J., Best, J.L., Roden, J.E., Bristow, C.S., and Klassen, GJ., 2000, Morphological evolution and dynamics of a large, sand braid-bar, Jamuna River, Bangladesh: Sedimentology, v. 47, p. 533–555.

Atkinson, C.D., Trumbly, P.N., and Kremer, M.C., 1988, Sedimentology and depositional environments of the Ivishak Sandstone, Prudhoe Bay field, North Slope, Alaska, *in* Lomando, A.J., and Harris, P.M., eds., Giant Oil and Gas Fields: A Core Workshop v. 2: SEPM (Society for Sedimentary Geology) core workshop 12, p. 561–613.

Atkinson, C.D., McGowen, J.H., Bloch, S., Lundell, L.L., and Trumbly, P.N., 1990, Braidplain and deltaic reservoir, Prudhoe Bay field, Alaska, *in* Barwis, J.H., McPherson, J.G., and Studlick, J.R.J., eds., Sandstone Petroleum Reservoirs: New York, Springer-Verlag, p. 7–29.

Bailey, B.E., 2001, Miocene Stratigraphy of the Niobrara River Valley, Western Cherry County, Nebraska: Unpublished M.S. Thesis, University of Nebraska, Lincoln, NE, 94 p.

Banerjee, I., and Banerjee, S., 1982, A coalescing alluvial fan model of the Siwalik sedimentation–A case history in the eastern Himalaya: Geological Survey of India, Miscellaneous Publication 41(11), p. 1–12.

Bernard, H.A., LeBlanc, R.J., and Major, C.J., 1962, Recent and Pleistocene geology of southeast Texas, *in* Rainwater, E.H., and Zingula, R.P., eds., Geology of the Gulf Coast and central Texas: Geological Society of America, Guidebook for 1962 Annual Meeting, p. 175–224.

Best, J.L., Ashworth, P.J., Bristow, C.S., and Roden, J., 2003, Three-dimensional sedimentary architecture of a large, mid-channel sand braid bar, Jamuna River, Bangladesh: Journal of Sedimentary Research, v. 73, p. 516–530.

Bledsoe, B.P., Hawley, R., and Stein, E.D., 2008, Stream channel classification and mapping systems: Implications for assessing susceptibility to hydromodification effects in southern California, Southern California Coastal Water Research Project, Technical Report 562, 42 p.

Bluck, B.J., 1971, Sedimentation in the meandering River Endrick: Scottish Journal of Geology, v. 7, p. 93–138.

Brice, J.C., 1982, Stream-channel stability assessment: Federal Highway Administration Report No. FHWA/RD-82/021, Washington, DC, 42 p.

Bridge, J.S., 1985, Paleochannel patterns inferred from alluvial deposits: A critical evaluation: Journal of Sedimentary Petrology, v. 55, p. 579–589.

Bridge, J.S., 1993, Description and interpretation of fluvial deposits: a critical perspective: Sedimentology, v. 40, p. 801–810.

Bridge, J.S., 2003, Rivers and Floodplains: Oxford, U.K., Blackwell Publishing, 491 p.

Bridge, J.S., 2006, Fluvial facies models: Recent developments *in* Posamentier, H.W., and Walker, R.G., eds., Facies Models Revisited: SEPM (Society for Sedimentary Geologists), Special Publication 84, p.85–170.

Bridge, J.S., and Diemer, J.A., 1983, Quantitative interpretation of an evolving ancient river system: Sedimentology, v. 30, p. 599–623.

Bridge, J.S., and Gordon, E.A., 1985, Quantitative reconstructions of ancient river systems in the Oneonta Formation, Catskill Magnafacies, *in* Woodrow, D.L., and Sevon, W.D., eds., The Catskill Delta: Geological Society of America, Special Paper 201, p. 163–181.

Bridge, J.S., and Lunt, I.A., 2006, Depositional models of braided rivers, *in* Sambrook Smith, G.H., Best, J.L., Bristow, C.S., and Petts, G.E., eds., Braided Rivers: Process, Deposits, Ecology and Management: International Association of Sedimentologists, Special Publication 36, p. 11–50.

Bridge, J.S. and Mackey, S.D., 1993, A theoretical study of fluvial sandstone body dimensions, *in* Flint, S.S., and Bryant, I.D. eds., Geological Modeling of

Hydrocarbon Reservoirs: International Association of Sedimentologists, Special Publication 15, p. 213–236.

Bridge, J.S., and Tye, R.S., 2000, Interpreting the dimensions of ancient fluvial channel bars, channels, and channel beds from wireline-logs and cores: American Association of Petroleum Geologists, Bulletin, v. 84, p. 1205–1228.

Bridge, J.S., Alexander, J., Collier, R.E.Ll., Gawthorpe, B.L., and Jarvis, J., 1995, Ground-penetrating radar and coring used to study the large-scale structure of point bar deposits in three dimensions: Sedimentology, v. 42, p. 839–852.

Bridge, J.S., Collier, R., and Alexander, J., 1998, Large-scale structure of Calamus River deposits (Nebraska, USA) revealed using ground-penetrating radar: Sedimentology, v. 45, p. 977–988.

Bridge, J.S., Jalfin, G.A., and Georgief, S.M., 2000, Geometry, lithofacies and spatial distribution of Cretaceous fluvial sandstone bodies, San Jorge Basin: Outcrop analog for the hydrocarbon-bearing Chubut Group: Journal of Sedimentary Research, v. 72, p. 341–359.

Brierley, G., 1996, Channel morphology and element assemblages: A constructivist approach to facies modeling *in* Carling, P.A. and Dawson, M.E. eds., Advances in Fluvial Dynamics and Stratigraphy: New York, John Wiley & Sons, p. 263–298.

Brierley, G., and Hickin, 1991, Channel planform as a non-controlling factor in fluvial sedimentology: the case of the Squamish River floodplain, British Columbia: Sedimentary Geology, v. 75, p. 67–83.

Bristow, C.S., 1987, Brahmaputra River: Channel migration and deposition, *in* Ethridge, F.G., Flores, R.M. and Harvey, M.D., Recent Developments in Fluvial Sedimentology: Contributions from the Third International Fluvial Sedimentology Conference: SEPM (Society for Sedimentary Geologists), Special Publication 39, p. 63–74.

Bristow, C.S., 1996, Reconstructing fluvial channel morphology from sedimentary sequences, *in* Carling, P.A., and Dawson, M.R., eds., Advances in Fluvial Dynamics and Stratigraphy: New York, John Wiley & Sons, p. 351–371.

Bristow, C.S., Skelly, R.L., and Ethridge, F.G., 1999, Crevasse splays from the rapidly aggrading, sand-bed, braided Niobrara River, Nebraska: Effect of base-level rise: Sedimentology, v. 46, p. 1029–1047.

Bull, W.B., 1979, Threshold of critical power in streams: Geological Society of America, Bulletin, v. 90, p. 453–464.

Burnett, M., 1996, 3D seismic expression of a shallow fluvial system in west central Texas, *in* Weimer, P., and Davis, T.L., eds., Applications of 3-D Seismic Data to Exploration: American Association of Petroleum Geologists, Studies in Geology 42 and Society for Exploration Geologists, Geophysical Development Series 5, p. 45–56.

Campbell, C.V., 1976, Reservoir geometry of fluvial sheet sandstone: American Association of Petroleum Geologists, Bulletin, v. 60, p. 1009–1020.

Carter, D.C., 2003, 3D seismic geomorphology: Insights into fluvial reservoir deposition and performance, Widuri field, Java Sea: American Association of Petroleum Geologists, Bulletin, p. 87, p. 909–934.

Carson, M.A., 1984, The meandering-braided river threshold: A reappraisal: Journal of Hydrology, v. 73, p. 315–334.

Castle, J.W., 1998, Regional sedimentology and stratal surfaces of a lower Silurian clastic wedge in the Appalachian foreland basin: Journal of Sedimentary Research, v. 68, p. 1201–1211.

Castle, J.W., 2001, Foreland basin sequence response to collisional tectonism: Geological Society of America, Bulletin, v. 113, p. 801–812.

Castle, J.W. and Byrnes, A.P., 2005, Petrophysics of Lower Silurian sandstones and integration with the tectonic-stratigraphic framework, Appalachian basin, United States: American Association of Petroleum Geologists, Bulletin, v. 89, p. 41–60.

Church, M., 1983, Pattern of instability in a wandering gravel-bed channel, *in* Collinson, J.D., and Lewin, J., eds., Modern and Ancient Fluvial Systems: International Association of Sedimentologists, Special Publication 6, p. 160–180.

Church, M., 2006, Bed material transport and the morphology of alluvial river channels: Annual Review of Earth and Planetary Sciences, v. 34, p. 325–354.

Coleman, J.M., 1969, Brahmaputra River, Channel Processes and Sedimentation, Sedimentary Geology, v. 3, p. 129–239.

Collinson, J.D., 1978, Vertical sequence and sand body shape in alluvial sequences, *in* Miall, A.D. ed., Fluvial Sedimentology: Canadian Society of Petroleum Geologists, Memoir 5, p. 577–586.

Cotter, E., 1983, Shelf, paralic, and fluvial environments and eustatic sea-level fluctuations in the origin of the Tuscarora Formation (Lower Silurian) of central Pennsylvania: Journal of Sedimentary Petrology, v. 53, p. 25–49.

Cowan, E.J., 1991, The large-scale architecture of the fluvial Westwater Canyon Member, Morrison Formation (Upper Jurassic), San Juan Basin, New Mexico, *in*, Miall, A.D., and Tyler, N., eds., The Three-Dimensional Facies Architecture of Terrigenous Clastic Sediments and its Implication for Hydrocarbon Discovery and Recovery: SEPM (Society for Sedimentary Geology) Concepts in Sedimentology and Paleontology, v. 3, p. 80–93.

Crews, S., 1983, Internal Structure of a Streamlined Bar, South Platte River, Colorado: Unpublished Report, Fort Collins, Colorado State University, 17 p.

Crowley, K.D., 1983, Large-scale bed configurations (macroforms), Platte River Basin, Colorado and Nebraska: Primary structures and formative processes: Geologic Society of America, Bulletin, v. 94, p. 117–133.

Darmadi, Y., Willis, B.J., and Dorobek, S.L., 2007, Three-dimensional seismic architecture of fluvial sequences on the low-gradient Sunda Shelf, Offshore Indonesia: Journal of Sedimentary Research, v. 77, p. 225–238.

Demko, T. M., Currie, B.S., and Nicoll, K.A., 2004, Regional paleoclimatic and stratigraphic implications of paleosols and fluvial/overbank architecture in the Morrison Formation (Upper Jurassic), Western Interior, U.S.A.: Sedimentary Geology, v. 167, p. 115–135.

Diffendal, R.F., Jr., 1990, The Sidney Gravel and Kimball Formation, supposed parts of the Ogallala Group (Neogene), are not objectively mapable units, *in* Gustavson, T.C., ed., Geological Framework and Regional Hydrology: Upper Cenozoic Blackwater Draw and Ogallala Formation, Great Plains: The University of Texas at Austin, Bureau of Economic Geology, p. 23–38.

Doeglas, D.J., 1962, The structure of sedimentary deposits of braided rivers: Sedimentology, v. 1, p. 167–190.

Eriksson, P.G., Bumby, A.J., Brümer, J.J., and van der Neut, M., 2006, Precambrian fluvial deposits: Enigmatic paleohydrological data from the c. 2–1.9 Ga Waterberg Group, South Africa: Sedimentary Geology, v. 190, p. 25–46.

Ethridge, F.G., and Schumm, S.A., 2007, Fluvial seismic geomorphology: a view from the surface, *in* Davies, R.J., Posamentier, H.W., Wood, L.J., and Cartwright, J.A., eds., Seismic Geomorphology: Applications to Hydrocarbon Exploration and Production: Geological Society of London, special publication 277, p. 205–222.

Ethridge, F.G., Skelly, R.L., and Bristow, C.S., 1999, Avulsion and crevassing in the sandy, braided Niobrara River: Complex response to base-level rise and aggradation, *in* Smith N.D. and Rogers, J.J. eds., Fluvial Sedimentology VI: International Association of Sedimentologists: Special Publication 28, p. 179–191.

Fielding, C.R., and Crane, R.C., 1987, An application of statistical modeling to the prediction of hydrocarbon recovery factors in fluvial reservoir sequences, *in* Ethridge, F.G., Flores, R.M. and Harvey, M.D., eds., Recent Developments in Fluvial Sedimentology: Contributions from the Third International Fluvial Sedimentology Conference: SEPM (Society for Sedimentary Geology), Special Publication 39, p. 321–327.

Fielding, C., Alexander, J., and McDonald, R., 1999, Sedimentology facies from ground-penetrating radar surveys of the modern, upper, Burdekin River of north Queensland, Australia: consequences of extreme discharge fluctuations, *in* Smith, N.D. and Rogers, J., eds., Fluvial Sedimentology VI: International Association of Sedimentologists, Special Publication 28, p. 347–362.

Fielding, C.R., LaGarry, H.E., LaGarry, L.A., Bailey, B.E., and Swinehart, J.B., 2007, Sedimentology of the Whitclay Gravel beds (Ogallala Group) in northwestern Nebraska, USA: Structurally controlled drainage promoted by Early Miocene uplift of the Black Hills Dome: Sedimentary Geology, v. 202, p. 58–71.

Fielding, C.R., Allen, J.P., Alexander, J., and Gibling, M.R., 2009, Facies model for fluvial systems in the seasonal tropics and subtropics: Geology, v. 37, p. 623–626.

Fotherby, L.M., 2009, Valley confinement as a factor of braided river pattern for the Platte River: Geomorphology, v. 103, p. 562–576.

Friend, P.F., 1983, Towards the field classification of alluvial architecture or sequence, *in* Collinson, J.D., and Lewin, J., eds., Modern and Ancient Fluvial

Systems: International Association of Sedimentologists, Special Publication 6, p. 345–354.

Friend, P.F., Mahmood Raza, S., Geehan, G., and Sheikh, K.A., 2001, Intermediate-scale architectural features of the fluvial Chinji Formation (Miocene), Siwalik Group, northern Pakistan: Geological Society of London Journal, v. 158, p. 163–177.

Galloway, W.E., 1979, Morrison Formation–Colorado Plateau, *in* Galloway, W., Kreitler, C.W., and McGowen, J.H., eds., Depositional and Ground-Water Flow Systems in the Exploration for Uranium: A Research Colloquium: The University of Texas at Austin, Bureau of Economic Geology, p. X1-X15.

Galloway, W.E., 1980, Deposition and early hydrologic evolution of Westwater Canyon wet alluvial-fan systems, *in* Rautman, C.A., ed., Geology and Mineral Technology of the Grants Uranium Region 1979: New Mexico Bureau Mines and Mineral Resources, Memoir 38, p. 59–69.

Gawthorpe, R.L., Collier, R.E.Ll., Alexander, J., Leeder, M.R., and Bridge, J.S., 1993, Ground penetrating radar: application to sand body geometry and heterogeneity studies, *in* North, C.P. and Processer, D.J., Eds., Characterization of Fluvial and Aeolian Reservoirs: Geological Society of London, Special Publication 73, p. 421–432.

Gibling, M.R., 2006, Width and thickness of fluvial channel bodies and valley fills in the geological record: A literature compilation and classification: Journal of Sedimentary Research, v. 76, p. 731–770.

Godin, P.D., 1991, Fining-upward cycles in the sandy braided-river deposits of the Westwater Canyon Member (Upper Jurassic), Morrison Formation, New Mexico: Sedimentary Geology, v. 70, p. 61–82.

Goodwin, C.N., 1999, Fluvial classification: Neanderthal necessity or needless normalcy, *in* Olsen, D.S. and Potyondy, J.P., eds., Proceedings Specialty Conference Wildland Hydrology, American Water Resources Association, Technical Publication Series TPS99–3, p. 229–235.

Gordon, E.A., and Bridge, J.S., 1987, Evolution of Catskill (Upper Devonian) river systems: Journal of Sedimentary Petrology, v. 57, p. 234–249.

Gouw, M.J.P., 2008, Alluvial architecture of the Holocene Rhine–Meuse delta (the Netherlands): Sedimentology, v. 55, p. 1487–1516.

Gouw, M.J.P., and Autin, W.J., 2008, Alluvial architecture of the Holocene Lower Mississippi Valley (U.S.A.) and a comparison with the Rhine–Meuse delta (the Netherlands): Sedimentary Geology, v. 204, p. 106–121.

Gustavson, T.C., 1996, Fluvial and Eolian Depositional Systems, Paleosols and Paleoclimate of the Upper Cenozoic Ogallala and Blackwater Draw Formations, Southern High Plains, Texas and New Mexico: The University of Texas at Austin, Bureau of Economic Geology, Report of Investigation 239, 62 p.

Hardage, B.A., Levey, R.A., Pendleton, V., Simmons, J., and Edson, R., 1994, A 3D seismic case history evaluating fluvially deposited thin-bed reservoirs in a gas-producing property: Geophysics, v. 59, p. 1650–1665.

Hartley, A.J., Weissmann, G.S., Nichols, G.J., and Warwick, G.L., 2010, Large distributive fluvial systems: Characteristics, distributions and controls on development: Journal of Sedimentary Research, v. 80, p. 167–183.

Harvey, M.D., Crews, S., Pitlick, J., and Blair, T., 1985, Field Trip 5, Holocene braided streams of eastern Colorado and sedimentological effects of Lawn Lake dam failure, Rocky Mountain National Park, *in* Flores, R.M., and Harvey, M., eds., Field Guidebook to Modern and Ancient Fluvial Systems in the United States: Fort Collins, Colorado Third International Fluvial Sedimentology Conference, p. 87–105.

Heller, P.L., and Paola, C., 1996, Downstream changes alluvial architecture: an exploration of controls on channel-stacking patterns: Journal of Sedimentary Research, v. 66, p. 97–306.

Heller, P.L., Mohrig, D., Vetti, L., Paola, C., and Lynds, R., 2007, An evaluation of recognition criteria for ancient braided vs. meandering streams, abstract: Geological Society of America, Annual Meeting Abstracts with Programs, v. 39, p. 506.

Isa, Z.M., Richards, F.W., and Yunus, H., 1992, Integration of 3D and site survey seismic data in analysis of near-surface hazards to platform location at Dulang field, Malay Basin: Geological Society of Malaysia, Bulletin, v. 32, p. 165–184.

Jackson, R.G., II, 1976, Depositional model of point bars in the lower Wabash River: Journal of Sedimentary Petrology, v. 48, p. 579–594.

Jackson, R.G., II, 1978, Preliminary evaluation of lithofacies models for meandering alluvial streams, *in* Miall, A.D., ed., Fluvial Sedimentology, Canadian Society of Petroleum Geologists Memoir 5, p. 543–576.

Jain, V., and Sinha, R., 2003, River systems in the Gangetic plains and their comparison with the Siwaliks: A review: Current Science, v. 84, p. 1025–1033.

Jorgensen, D.W., 1990, Adjustments of alluvial river morphology and processes to localized active tectonics: Fort Collins, Colorado, unpublished Ph.D. dissertation, Colorado State University, 240 p.

Keighley, D.G., and Flint, S.S., 2008, Fluvial sandstone geometry and connectivity in the Middle Eocene Green River Formation, Nine Mile Canyon, Southwestern Uinta Basin, *in* Longman, M.W., and Morgan, C.D., eds., Hydrocarbon Systems and Production in the Uinta Basin, Utah: Rocky Mountain Association of Geologists and Utah Geological Association, Publication 37, p. 101–119.

Khan, I.A., Bridge, J.S., Kappelman, J., and Wilson, R., 1997, Evolution of Miocene fluvial environments, eastern Potwar Plateau, northern Pakistan: Sedimentology, v. 44, p. 221–251.

Kjemperud, A.V., Schomacker, E.R., and Cross, T.A., 2008, Architecture and stratigraphy of alluvial deposits, Morrison Formation (Upper Jurassic), Utah: American Association of Petroleum Geologists, Bulletin, v. 92, p. 1055–1076.

Knighton, A.D., and Nanson, G.C., 1993, Anastomosis and the continuum of channel pattern: Earth Surface Processes and Landforms, v. 18, p. 613–625.

Kumar, R., and Tandon, S.K., 1985, Sedimentology of Plio-Pleistocene late orogenic deposits associated with intraplate subduction—The upper Siwalik Subgroup of Panjab sub-Himalaya, India: Sedimentary Geology, v. 42, p. 105–158.

Lane, E.W., 1957, A study of the shape of channels formed in natural streams flowing in erodible material: U.S. Army Corps of Engineers Missouri River Division Sediment Series 9, 106 p.

Lawton, T.F., Geehan, G.W., and Voporhees, B.J., 1987, Lithofacies and depositional environments of the Ivshak Formation, Prudhoe Bay Field, *in* Tailleur, L., and Weimer, P. eds., Alaska North Slope Geology: SEPM, Pacific Section, Book 50, p. 61–76.

Leckie, D.A., 1994, Canterbury Plains, New Zealand—Implications for sequence stratigraphic models: American Association of Petroleum Geologists, Bulletin, v. 78, p. 1240–1256.

Leclair, S.F., and Bridge, J.S., 2001, Quantitative interpretation of sedimentary structures formed by river dunes: Journal of Sedimentary Research, v. 71, p. 713–716.

Leclerc, R.F., and Hickin, E.J., 1997, The internal structure of scrolled floodplain deposits based on ground-penetrating radar, North Thompson River, British Columbia: Geomorphology, v. 21, p. 17–38.

Leeder, M.R., 1993, Tectonic controls upon drainage basin development, river channel migration and alluvial architecture: implications for hydrocarbon development and characterization, *in* North, C.P. and Prosser, D.J., eds., Characterization of Fluvial and Aeolian Reservoirs: Geological Society of London, Special Publication 73, p. 7–22.

Leopold, L.B., and Woleman, M.G., 1957, River Channel Patterns: Braided, Meandering and Straight, *in* Physiographic and Hydraulic Studies of Rivers: Washington, D.C., U.S. Geological Survey Professional Paper, 282-B, 85 p.

Lewin, J., and Brewer, P.A., 2001, Predicting channel patterns: Geomorphology, v. 40, p. 329–339.

Lewin, J., and Brewer, P.A., 2003, Reply to Van den Berg and Bledsoe's comment on Lewin and Brewer (2001) "Predicting channel patterns": Geomorphology, v. 53, p. 339–342.

Long, D.G.F., 2002, Aspects of late Paleoproterozoic fluvial style: The Uairen Formation, Roraima, Supergroup, Venezuela, *in* Itermann, W., and Corcoran, P.L., eds., Precambrian Sedimentary Environments: A Modern Approach to Ancient Depositional Systems: International Association of Sedimentologists, Special Publication 33, p. 323–338.

Long, D.G.F., 2004, Alluvial systems, *in* Eriksson, P.G., Altermann, W., Nelson, D.R., Mueller, W.U., and Catuneanu, O., eds., The Precambrian Earth: Tempos and Events: New York, Elsevier, Development in Precambrian Geology 12, p. 599–663.

Lunt, L.A., and Bridge, J.S., 2004, Evolution and deposits of a gravelly braid bar, Sagavanirktok River, Alaska: Sedimentology, v. 51, p. 415–432.

Lunt, L.A., Bridge, J.S., and Tye, R.S., 2004, A quantitative, three-dimensional depositional model of gravelly braided rivers: Sedimentology, v. 51, p. 377–414.

MACKEY, S.D., AND BRIDGE, J.S., 1995, Three-dimensional model of alluvial stratigraphy: Theory and application: Journal of Sedimentary Research, v. B65, p. 7–31.

MAKASKE, B., 2001, Anastomosing rivers: a review of their classification, origin and sedimentary products: Earth-Science Reviews, v. 53, p. 149–196.

MARTINEZ, K.F., CARTWRIGHT, J.A., BURGESS, P.M., AND BRAVO, J.V., 2004, 3-D seismic interpretation of the Messinian unconformity in the Valencia basin. Spain, *in* Davies, R.J., Cartwright, J.A., Stewart, S.A., Lappin, M., and Underhill, J.R., eds., 3-D Seismic Technology: Applications to the Exploration of Sedimentary Basins: The Geological Society of London, Memoir 29, p. 91–100.

MCMILLAN, M.E., ANGEVINE, C.L., AND HELLER, P.L., 2002, Post depositional tilt of the Miocene–Pliocene Ogallala Group on the western Great Plains: Evidence of the late Cenozoic uplift of the Rocky Mountains: Geology, v. 30, p. 63–66.

MERTES, L.A.K., AND DUNNE, T., 2007, Effects of tectonism, climate change, and sea-level change on the form and behavior of the modern Amazon River and its floodplain, *in* Gupta, A., ed., Large Rivers: Geomorphology and Management: New York, John Wiley & Sons, p. 115–144.

MERTES, L.A.K., DUNNE, T., MARTINELLI, L.A., 1996, Channel-floodplain geomorphology along the Solimoes–Amazon River, Brazil: Geological Society of America, Bulletin, v. 108, p. 1089–1107.

MIALL, A.D., 1977, A review of the braided river depositional environment: Earth-Science Reviews, v. 13, p. 1–62.

MIALL, A.D., 1985, Architectural element analysis: A new method of facies analysis applied to fluvial deposits: Earth-Science Reviews, v. 22, p. 261–308.

MIALL, A.D., 1988, Reservoir heterogeneities in fluvial sandstones: Lessons from outcrop studies: American Association of Petroleum Geologists, Bulletin, v. 72, p. 682–697.

MIALL, A.D., 1992, Alluvial deposits, *in* Walker, R.G., and James, N.P. eds., Facies Models: Response to Sea Level Change: St. Johns Newfoundland, Geological Association of Canada, p. 119–142.

MIALL, A.D., 1993, The architecture of fluvial–deltaic sequences in the Upper Mesaverde Group (Upper Cretaceous), Books Cliffs, Utah, *in* Best, J.L. and Bristow, C.S., eds., Braided Rivers: Geological Society of London, Special Publication 75, p. 305–332.

MIALL, A.D., 1995, Description and interpretation of fluvial deposits: a critical perspective (discussion and reply): Sedimentology, v. 42, p. 379–389.

MIALL, A.D., 1996, The Geology of Fluvial Deposits: Sedimentary Facies, Basin Analysis and Petroleum Geology: New York, Springer-Verlag, 582 p.

MIALL, A.D., 1999, In defense of facies classifications and models: Journal of Sedimentary Research, v. 69, p. 2–5.

MIALL, A.D., 2002, Architecture and sequence stratigraphy of Pleistocene fluvial systems in the Malay Basin, based on seismic time-slice analysis: American Association of Petroleum Geologists, Bulletin, v. 86, p. 1201–1216.

MIALL, A.D., 2006, How do we identify big rivers? And how big is big?: Sedimentary Geology, v. 186, p. 39–50.

MIALL, A.D., AND TURNER-PETERSON, C.E., 1989, Variations in fluvial style in the Westwater Canyon Member, Morrison Formation (Jurassic), San Juan Basin, Colorado Plateau: Sedimentary Geology, v. 63, p. 21–60.

MOODY-STEWART, M., 1966, High and low sinuosity stream deposits, with examples from the Devonian of Speitzbergen: Journal of Sedimentary Petrology, v. 36, p. 1102–1117.

MULLENS, T.E., AND FREEMAN, V.L., 1957, Lithofacies of the Salt Wash member of the Morrison Formation, Colorado Plateau: Geological Society of America, Bulletin, v. 88, p. 505–526.

NADLER, C.T., AND SCHUMM, S.A., 1981, Metamorphosis of South Platte and Arkansas Rivers, Eastern Colorado: Physical Geography, v. 2, p. 95–115.

NAKAYAMA, K., AND ULAK, P.D., 1999, Evolution of fluvial style in the Siwalik Group in the foothills of the Nepal Himalaya: Sedimentary Geology, v. 125, p. 205–224.

NANSON, G. C., AND KNIGHTON, A.D., 1996, Anabranching rivers: their cause, character and classification: Earth Surface Processes and Landforms, v. 21, p. 217–239.

NAMI, M., 1976, An exhumed Jurassic meander belt from Yorkshire England: Geological Magazine, v. 113, p. 47–52.

NANSON, G.C., AND GIBLING, M.R., 2003, Rivers and alluvial fans, *in* Middleton, G.V., ed., Encyclopedia of Sediments and Sedimentary Rocks: New York, Springer, p. 568–583.

NEILL, C.R., 1973, Hydraulic and morphologic characteristic of Athabasca River near Fort Assiniboine: Alberta Research Council, Edmonton, Highway and River Engineering Division, Report REH/73/7, 23 p.

NORTH, C., 1996, The prediction of modeling of subsurface fluvial stratigraphy, *in* Carling, P.A., and Dawson, M.R., eds., Advances in Fluvial Dynamics and Stratigraphy: New York, John Wiley & Sons, p. 395–508.

ORE, H.T., 1964, Some criteria for recognition of braided stream deposits: Contributions to Geology, v. 3, p. 1–14.

PADGETT, G., AND EHRLICH, R., 1976, Paleohydrologic analysis of a late Carboniferous fluvial system, southern Morocco: Geological Society of America, Bulletin, v. 86, p. 1101–1104.

PARADES, J.M., FOIX, N., PINOL, F.C., NILLNI, A., ALLARD, J.O., AND MARQUILLAS, R.A., 2007, Volcanic and climatic controls on fluvial style in a high-energy system: the Lower Cretaceous Matasiete Formation, Golfo San Jorge basin, Argentina: Sedimentary Geology, v. 202, p. 96–123.

PARKASH, B., BAJPAI, J.P., AND SAXENA, H.P., 1974, Sedimentary structures and paleocurrents of the Siwaliks exposed between the Yamuna and Gola rivers, U.P. (India): Geological Magazine, v. 111, p. 1–14.

PETERSON, F., AND TYLER, N., 1985, Field guide to the upper Salt Wash alluvial complex, *in* Flores, R.M., and Harvey, M., eds., Field Guidebook to Modern and Ancient Fluvial Systems in the United States: Fort Collins, Colorado, Third International Fluvial Sedimentology Conference, p. 45–64.

POSAMENTIER, H.W., 2001, Lowstand alluvial bypass systems: incised vs. unincised: American Association of Petroleum Geologists, Bulletin, v. 85, p. 1771–1793.

RAINBIRD, R.H., 1992, Anatomy of a large-scale braid-plain quartzarenite from the Neoproterozoic Shaler Group, Victoria Island, Northwest Territories, Canada: Canadian Journal of Earth Sciences, v. 29, p. 2537–2550.

RANGA RAO, A., AND KUNTE, S. V., 1987, Lower Siwaliks of Jammu Hills: Processes of sedimentation: Varanasi, National Seminar on Tertiary Orogeny Proceedings, p. 361–77.

REEVES, C.C., AND REEVES, J.A., 1996, The Ogallala Aquifer (of the Southern High Plains), Volume 1—Geology: Lubbock, Texas, Estacado Books, 360 p.

REYNOLDS, A.D., 1999, Dimensions of paralic sandstone bodies: American Association of Petroleum Geologists, Bulletin, v. 83, p. 211–229.

ROBINSON, J.W. AND MCCABE, P.J., 1997, Sandstone-body and shale-body dimensions in a braided fluvial system: Salt Wash Sandstone Member (Morrison Formation), Garfield County, Utah: American Association of Petroleum Geologists, Bulletin, v. 81, p. 1267–1291.

ROSGEN, D.L., 1994, A classification of natural rivers: Catena, v. 22, p. 169–199.

SAMBROOK SMITH, G.H., ASHWORTH, P.J., BEST, J.L., WOODWARD, J, AND SIMPSON, C.J., 2005, The morphology and facies of sandy braided rivers: Some considerations of scale invariance, *in* Blum, M.D., Marriott, S.B., and Leclair, S.F., eds., Fluvial Sedimentology VII: International Association of Sedimentologists, Special Publication 35, p. 145–158.

SAMBROOK SMITH, G.H., ASHWORTH, P.J., BEST, J.L., WOODWARD, J., AND SIMPSON, C.J., 2006, The sedimentology and alluvial architecture of the sandy braided South Saskatchewan River, Canada: Sedimentology, v. 53, p. 413–434.

SAUDI ARABIAN OIL COMPANY, 1996, Impact of 3-D seismic data on development drilling, Ghinah and Umm Jurf fields, central Saudi Arabia, *in* Weimer, P., and Davis, T.L., eds., Applications of 3-D Seismic Data to Exploration: American Association of Petroleum Geologists, Studies in Geology 42, and Society for Exploration Geophysics, Geophysical Development Series 5, p. 45–56 (compiled by Simms, S.C.).

SCHUMM, S.A., 1977, The Fluvial System: New York, Wiley, 338 p.

SCHUMM, S.A., 2005, River Variability and Complexity: Cambridge, U.K., Cambridge University Press, 220 p.

SCHUMM, S.A., DUMONT, J.F., AND HOLBROOK, J.M., 2000, Active Tectonics and Alluvial Rivers: Cambridge, Cambridge, U.K., University Press, 276 p.

SHARMA, S., SHARMA, M., AND SINGH, I.B., 2001, Facies characteristics and cyclicity of the Lower Siwalik sediments Jammu area: A new perspective: Geological Magazine, v. 138, p. 455–470.

SHUKLA, U.K., SINGH, I.B., SRIVASTAVA, P., AND SINGH, D.S., 1999, Paleocurrent patterns in braid-bar and point-bar deposits: Examples from the Ganga River, India: Journal of Sedimentary Research, v. 69, p. 992-1002.

Simon, A., Doyle, M., Kondolf, M., Shields, F.D., Jr., Rhoads, B., Grant, G., Fitzpatrick, F., Juracek, K., McPhilips, M., and MacBroom, J., 2005, How well do the Rogen classification and associated "Natural Channel Design" methods integrate and quantify fluvial processes and channel response: Anchorage, Alaska, Proceedings of the 2005 World Water and Environmental Resources Congress, p. 584–596.

Singh, I.B., 2007, The Gang River, *in* Gupta, A., ed., Large Rivers: Geomorphology and Management: New York, John Wiley and Sons, p. 347–371.

Skelly, R.L., Bristow, C.S., and Ethridge, F.G., 2003, Architecture of channel-belt deposits in an aggrading shallow sand bed braided river: The lower Niobrara River, northeast Nebraska: Sedimentary Geology, v. 158, p. 249–270.

Smith, N.D., 1970, The braided stream depositional environment: comparison of the Platte River with some Silurian clastic rocks, north-central Appalachians: Geological Society of America, Bulletin, v. 81, p. 2993–3014.

Smith, R.M.H., 1987, Morphology and depositional history of exhumed Permian point bars in the southwestern Karoo, South Africa: Journal of Sedimentary Petrology, v. 57, p. 19–29.

Sønderholm, M., and Tirsgaard, H., 1998, Proterozoic fluvial styles: response to changes in accommodation space (Rivieradal sandstones, eastern North Greenland): Sedimentary Geology, v. 120, p. 257–274.

Sweet, I.P., 1988, Early Proterozoic stream deposits: Braided or meandering—evidence from central Australia: Sedimentary Geology, v. 58, p. 277–293.

Szantay, A.W., 1990, Paleohydrology and paleomorphology of Early Eocene Green River Channel Sandstones, Uinta Basin, Utah: Unpublished M.S. Thesis, Colorado State University, 110 p. plus 2 Appendices and 10 plates.

Szantay, A.W., Ethridge, F.G., and Fouch, T.D., 1989, Paleomorphology and paleohydrology of Eocene channel form sandstones, Unita Basin, abstract: Barcelona, Spain, 4th International Conference on fluvial Sedimentology, Program and Abstracts, p.229.

Tandon, S.K.., and Narayan, D., 1981, Calcrete conglomerate: case-hardened conglomerate and cornstone—A comparative account of pedogenic and non-pedogenic carbonates from the continental Siwalik Group, Punjab, India: Sedimentology, v. 28, p. 353–367.

Tooth, S., and McCarthy, T.S., 2004, Controls on the transition from meandering to straight channels in the wetlands of the Okavango delta, Botswana: Earth Surface Processes and Landforms, v. 29, p. 1627–1649.

Tooth, S., and Nanson, G.C., 2004, Forms and processes of two highly contrasting rivers in arid central Australia, and the implications for channel-pattern discrimination and prediction: Geological Society of America, Bulletin, v. 116, p. 802–816.

Tye, R.S., Bhattacharya, J.P., Lorsong, J.A., Sindelar, S.T., Knock, D.G., Puls, D.D., and Levinson, R.A., 1999, Geology and stratigraphy of fluvio-deltaic deposits in the Ivishak Formation: Applications for development of Prudhoe Bay field, Alaska: American Association of Petroleum Geologists, Bulletin, v. 83, p. 1588–1623.

Tyler, N., and Ethridge, F.G., 1983a, Depositional Setting of the Salt Wash Member of the Morrison Formation, southwest Colorado: Journal of Sedimentary Petrology, v. 53, p. 67–82.

Tyler, Noel and Ethridge, F.G., 1983b, Fluvial architecture of Jurassic uranium bearing sandstones, Colorado Plateau, western United States, *in* Collinson, J.D. and Lewin, J., eds., Modern and Ancient Fluvial Systems: International Association of Sedimentologists, Special Publication 6, p. 533–547.

Van den Berg, J.H., 1995, Prediction of alluvial channel pattern of perennial rivers: Geomorphology, v. 12, p. 259–279.

Van den Berg, J.H., and Bledsoe, B.P, 2003, Comment on Lewin and Brewer (2001): "Predicting channel patterns": Geomorphology, v. 53, p. 333–337.

Vandenberghe, J., and van Overmeeren, R.A., 1999, Ground penetrating radar images of selected fluvial deposits in the Netherlands: Sedimentary Geology, v. 128, p. 245–270.

Visser, C.F., and Johnson, G.D., 1978, Tectonic control of Late Pliocene Molasse sedimentation in a portion of the Jhelum Re-entrant: Geologische Rundschau, v. 67, p. 15–37.

Weissmann, G.S., Hartley, A.J., Nichols, G.J., Scuderi, L.A., Olson, M., Burehler, H., and Bentcah, R., 2010, Fluvial form in modern continental sedimentary basins: Distributive fluvial systems: Geology, v.38, p. 39–42.

Williams, G., 1978, The case of the shrinking channels—the North Platte and Platte Rivers in Nebraska: Geological Survey, Circular 781, 48 p.

Willis, B.J., 1993a, Interpretation of bedding geometry within ancient point-bar deposits, *in* Marzo, M., and Puidefabregas, C., eds., Alluvial Sedimentation: International Association of Sedimentologists, Special Publication 17, p. 101–114.

Willis, B.J., 1993b, Ancient river systems in the Himalayan foredeep, Chinji Village area, northern Pakistan: Sedimentary Geology, v. 88, p. 1–76.

Willis, B.J., 1993c, Evolution of Miocene fluvial systems in the Himalayan foredeep through a two kilometer-thick succession northern Pakistan: Sedimentary Geology, v. 88, p. 77–121.

Willis, B.J., and Behrensmeyer, A.K., 1994, Architecture of Miocene overbank deposits in Northern Pakistan: Journal of Sedimentary Research, v. B64, p. 60–67.

Wood, L.J., 2007, Quantitative seismic geomorphology of Pliocene and Miocene fluvial systems in the northern Gulf of Mexico, U.S.A.: Journal of Sedimentary Research, v. 77, p. 713–730.

Woolf, K.J., and Balzary, J.R., 1996, Fields in the spectrum of channel style: Sedimentology, v. 43, p. 797–805.

Yeakel, L., Jr., 1962, Tuscarora, Juniata, and Bald Eagle paleocurrents and paleogeography in the central Appalachians: Geological Society of America, Bulletin, v. 73, p. 1515–1504.

Zaleha, M.J., 1997a, Fluvial and lacustrine palaeoenvironments of the Miocene Siwalik Group, Khaur area, northern Pakistan: Sedimentology, v. 44, p. 349–368.

Zaleha, M.J., 1997b, Intra- and extrabasinal controls on the fluvial deposition in the Miocene Indo-Gangetic foreland basin northern Pakistan: Sedimentology, v. 44, p. 349–368.

Zeng, H., 2007, Seismic imaging for seismic geomorphology beyond the seabed: potentials and challenges, *in* Davies, R.J., Posamentier, H.W., Wood, L.J., and Cartwright, J.A., eds., Seismic Geomorphology: Applications to Hydrocarbon Exploration and Production: Geological Society of London, Special Publication 277, p. 15–28.

ARCHITECTURE AND DEPOSITIONAL STYLE OF FLUVIAL SYSTEMS BEFORE LAND PLANTS: A COMPARISON OF PRECAMBRIAN, EARLY PALEOZOIC, AND MODERN RIVER DEPOSITS

DARREL G.F. LONG

Department of Earth Sciences, Laurentian University, Sudbury, Ontario, P3E 2C6, Canada

e-mail: dlong@laurentian.ca

Abstract: As rooted vascular plants were not a significant factor in controlling bank stability or surface runoff before the late Silurian, it is not surprising that many modern humid and temperate climate river models are not represented in the Precambrian or early Paleozoic sedimentary record. Many pre-vegetation systems have features that are more closely allied to modern ephemeral and dryland systems, although direct comparison remains elusive. To date, wandering gravel-bed rivers, low-sinuosity sandy braided systems with alternate bars, and fine-grained meandering and anabranching systems have not been positively identified. Precambrian gravel-bed braided systems are common, and have architecture similar to that of modern systems, except that scour hollows are typically absent or difficult to identify. Point-bars in cobble-grade gravel-bed meandering systems are wider and have lower inclination than modern systems. Sandy meandering systems can be identified using the directional relations between foreset orientation and point-bar inclination. Sandy braided systems are dominated by composite barforms with predominantly downstream accretionary elements. Sandy ephemeral channelized upper-flow-regime elements and unconfined sheetflood deposits are common, but they lack many of the associated fine-grained components seen in modern systems.

Most known pre-vegetation fluvial systems were preserved within rifts, along continental margins, or in foreland-basin settings. Nearly all known Precambrian and early Paleozoic river deposits are dominated by channel and sheet-channel facies, and have high lateral continuity and high net-to-gross ratios. Thick overbank and interfluvial deposits are underrepresented, or where identified, are suspect and may represent channel abandonment. Direct comparison with younger systems requires detailed architectural analysis, with specific attention to the relative inclination and directional variability of foresets and first-order to sixth-order surfaces. Unfortunately many studies of modern systems lack this attention to detail, especially of the small-scale morphology of submerged in-channel features, hence direct evaluation of the paleohydraulic characteristics, sinuosity, and slope of older systems remains enigmatic.

Key words: Braided, meandering, ephemeral, architecture, Precambrian

INTRODUCTION

In order to understand how fluvial systems developed before the advent of rooted vegetation it is critical to realize that many of the processes that influenced pre-Devonian fluvial systems were significantly different from those that operate today. Studies of modern sandbars indicate that sand with no plant stabilization is up to 20,000 times easier to erode than sand with only 16% to 18% roots (Smith 1976). Other major differences can be inferred from the lack of sediment binding, baffling, and trapping by plant roots, which would have promoted a tendency for flashy surface runoff, lower bank stability, and faster rates of channel migration than in present-day vegetated areas (Schumm 1968; Cotter 1978; Long 1978, 2004; Fuller 1985; Els 1990, 1998). Even minor vegetation interferes with fluid movement, producing distinctive structures (Rygel et al. 2004). Although anaerobic microbial communities may have been important in Proterozoic soils and groundwater (Horodyski and Knauth 1994; Martini 1994; Ohmoto 1996; Gutzmor and Beukes 1998), they would have had little effect on sediment binding. This is supported by the near absence of algal- and microbial-bound sand chips (Pflüger and Gresse 1996; Schieber 1999; Noffke et al. 2001) in Precambrian fluvial deposits. The presence of compound clay oncoids in fluvial sandstones of the late Paleoproterozoic Athabasca Basin in Canada (Yeo et al. 2002), and microbial traces in the Mesoproterozoic Torridonian sequence in Scotland (Prave 2002), indicates that algal communities could survive in temporary ponds developed in channel thalwegs and on floodplains, but were not sufficiently developed to prevent bed erosion.

In modern systems, sediment supply and characteristics are directly influenced by climate (Schumm 1968, 2005; Blum and Törnqvist 2000; Ashmore and Church 2001). In the Archean, climate zones may have been influenced by the faster rate of the Earth's rotation and differences in the tilt of the rotational axis. Despite lower solar luminosity, climates were predominantly warm (Eriksson et al. 1998). It has been suggested that in the Archean enhanced levels of greenhouse gasses promoted aggressive weathering of labile material (Donaldson and de Kemp 1998), although both the rates of weathering and absolute levels of atmospheric CO_2 remain uncertain (Hessler and Lowe 2006). If the weathering products were not dissolved, or removed by the wind, a greater availability of fines should have promoted mass-flow and hyperconcentrated-flow processes, at least in Archean fan and river systems. By the latest Archean and early Paleoproterozoic the presence of glacial deposits indicates development of a broader range of climatic zones. Warm-humid climates may have prevailed during much of the early Paleoproterozoic (at least to 2.3 Ga), as there is little preserved evidence for arid and hyperarid conditions (Eriksson et al. 1998). Acid-rain effects should have declined in this period, in tandem with decreases in greenhouse gasses, especially after the onset of fully oxygenic conditions after $\sim$ 2.2 Ga (Kasting 2001). The progressive colonization of terrestrial environments by rooted plants in late Silurian to middle Devonian times would have significantly affected microclimate, by modifying albedo and moisture retention, leading to a greater role of organic acids in decomposition of labile components (Jutras et al. 2009).

Most studies of pre-vegetative fluvial systems have relied on direct comparison with idealized models based on small Holocene river systems, predominantly from systems confined to incised valleys in humid-temperate climates (Cotter 1978; Friend 1978; Long 1978; Fuller 1985). This has resulted in the identification of numerous bed-load-dominated, sheetlike, braided alluvial deposits similar to modern

From River To Rock Record: The Preservation Of Fluvial Sediments And Their Subsequent Interpretation
SEPM Special Publication No. 97, Copyright © 2011
SEPM (Society for Sedimentary Geology), ISBN 978-1-56576-305-0, p. 37–61.

sandy (Platte and South Saskatchewan type) and gravel-dominated (Scott and Donjek type) rivers and fans, using models developed by Miall (1977, 1978). Descriptions of high-sinuosity meandering systems are rare from Precambrian successions (Tyler 1978; Sweet 1988). Davies and Gibling (2010a, 2010b) suggest that meandering systems with distinctive inclined heterolithic (muddy/sandy) lateral-accretion elements did not develop before the latest Silurian (Pridolian). Given that many of the preserved deposits of Precambrian fluvial systems are both thick and laterally extensive (Long 1978, 2004), and accumulated in rift, continental-margin, foreland-basin, and successor-basin settings with moderate to high accommodation rates, it may be more appropriate to use models based on modern distributive systems (Tooth 2000; Nichols and Fisher 2007; Weissmann et al. 2009, 2010; Hartley et al. 2010) in similar tectonic settings, although the distal facies of modern and post-Silurian examples are unlike earlier fluvial systems in that they are typically mud rich (e.g., Cain and Mountney 2009)

Major advances have been made in the understanding of Phanerozoic river systems, due to the realization that strata can be divided into genetically related packages, known as architectural elements (Fig. 1), which can be related to specific depositional processes and events (Allen 1983, Miall 1985, 1988, 1996). Several of the element codes used in Figure 1 are not in common usage. DLA (downstream lateral-accretion surfaces), ULA (upstream lateral-accretion surfaces), UA (upstream accretionary elements) were introduced by Long (2006) to provide greater geometrical constraints on elements within larger-scale macroforms. Miall (1985) defined lateral-accretion (LA) surfaces as having the modal orientation of zeroeth-order surfaces (foresets) within $\pm$ 45° of the modal strike of associated 1st or 2nd order bounding surfaces. Long (2006) used $\pm$ 30 to 60° to define DLA components. Similar geometric constraints were used for ULA and UA (Fig. 1, Table S1 in the digital appendix). The element URF (upper-flow-regime elements) was introduced by Fielding (2006) to describe channelized sheet, lens, and scour-fill units, with convex-up geometry. This overlaps with laminated sand sheet (LS) elements of previous authors. Further details of the definition of macroform-scale architectural elements used in this paper are provided in Table S1 (in the digital appendix). Lithofacies codes are summarised in Table 1.

Surfaces that bound elements can be interpreted in terms of a hierarchy of events (Table 2; Miall 1985, 1988, 1996), which may allow recognition of temporal factors that may be of importance in reconstructing channel paleohydraulics (Miall 1996). Although there is no universal acceptance of the hierarchy, numerology, or temporal significance of bounding surfaces proposed by Miall (Bridge 1993), the method provides an objective means of dividing distinct depositional elements, in that strata bounded by lower-order surfaces are typically cut by surfaces of higher order (Miall and Bridge 1995). The range of elements described to date may not be sufficiently comprehensive to describe all modern and ancient fluvial systems, nor is it likely that the temporal range suggested for bounding surfaces can be directly applied to pre-vegetation systems, especially for higher-order surfaces.

Although architectural-element analysis of post-Silurian fluvial strata is now commonplace and has been used to extract much useful information (Miall 1996), far less attention has been paid to the anatomy of fluvial systems prior to the development of rooted land plants (Davies and Gibling 2010b). Only a limited number of studies, including Røe (1987), Rainbird (1992), Bhattacharyya and Morad (1993), Karpeta (1993), Røe and Hermansen (1993), Amireh et al. (1994), Hjellbakk (1997), Sønderholm and Tirsgaard (1998), Tirsgaard and Øxnevad (1998), Chakraborty (1999), and Long (2001, 2002, 2006) provide substantial architectural detail; some (Chakraborty 1999; Van der Neut and Eriksson 1999; Eriksson et al. 2008) have attempted to use this information to reconstruct paleo-flow conditions. The spectrum of five basic river types identified by Miall (1977) has since been expanded to at least sixteen basic models (Miall 1996), founded predominantly on grain size and continuity of flow (Fig. 2). Additional models have been developed for a number of dryland river systems (Tooth 2000; Nanson and Knighton 1996; North et al. 2007; Cain and Mountney 2009), and for seasonal tropical to subtropical river systems (Fielding et al. 2009). Not all of these systems have direct analogues in pre-vegetation fluvial systems.

The object of this paper is to examine what is known about the depositional style of river systems prior to the advent of rooted land plants in the late Silurian, and to show how the scale, architecture, and paleohydraulics of these systems compare and contrast with modern rivers and their deposits. Known pre-vegetation examples are considered below from coarse to fine in terms of the sixteen models (Fig. 2) proposed by Miall (1996).

GRAVEL-DOMINATED SYSTEMS

Miall (1996) recognized five main classes of gravel-bed rivers: these include A) gravel-bed braided rivers with sediment-gravity-flow deposits; B) shallow gravel-bed braided rivers; C) deep gravel-bed rivers; D) gravel-bed wandering systems, and E) gravel-bed meandering systems. Gravel may also dominate proximal parts of distributary fluvial systems in foreland-basins (Cain and Mountney 2009; Hartley et al. 2010). Most documented pre-vegetation gravel-bed rivers appear to resemble the first three types.

A: Gravel-Bed Braided Rivers (and Fans) with Sediment-Gravity-Flow Deposits

Conglomeratic systems containing or dominated by matrix-supported conglomerates (Gms, Gm, SGm, SGh: Fig. 3 A, B, C; see Table 1 for explanation of facies codes) are typically interpreted as Trollheim-type fan deposits with in-channel and possibly lobate debris-flows (SG and GB elements: Räsänen and Mäkelä 1988; Amireh et al. 1994; Corcoran et al. 1999: Cox et al. 2002; Went 2005; Bose et al. 2008; Post and Long 2008; Chakraborty et al. 2009). This interpretation is often supported by evidence of radiating or transverse (fan-apron) paleo-flow, combined with local provenance, and wedge-shaped geometry (cf. Merk and Jirsa 1982; Singh 1984; Williams 2001; Mazumder and Sarkar 2004; Krapez et al. 2008). In rare cases these can be directly associated with rock-fall or scree deposits (Went 2005; Long 2006, 2007; Bose et al. 2008; Chakraborty et al. 2009; Almeida et al. 2009) consisting of local wedge-shaped sheets (1–2 m thick) of angular granule-to boulder-grade clasts of local basement material (Fig. 4 A, B, C).

In modern systems the availability of fine-grained sediment encourages the production of mass-flow and hyperconcentrated-flow deposits, consequently it should be expected that if enhanced weathering of labile components by acidic rainfall occurred in the Archean this should have led to a greater climatic range of debris flow facies, and may even have allowed production of debris-flows in non-fan fluvial systems (Buck and Minter 1985; Krapez 1985). Hyper-concentrated-flows may also have been responsible for deposition of unusual contact framework conglomerates of Cambrian age, containing stacks of imbricated, well-rounded boulders and cobbles, with upstream-directed accretionary elements (UA) produced by plastering of clasts on to the backs of gravel bars during flash-flood events in a valley-confined setting (Pflüger and Seilacher 1991). Hyperconcentrated-flows are also thought to have produced channelized (CH) and sheetlike (CHS) elements in which matrix-supported conglomerates and overlying massive sandstones were deposited by channelized hyperconcentrated-flows, which degraded into normal flows to produce an upper layer of trough cross-stratified sandstone (Dilliard et al.

Element (symbol)	Lithotypes	Precambrian Examples	Geometry and bedforms
Channels **CH** **(CHR, CHS, CV)**	Any combination	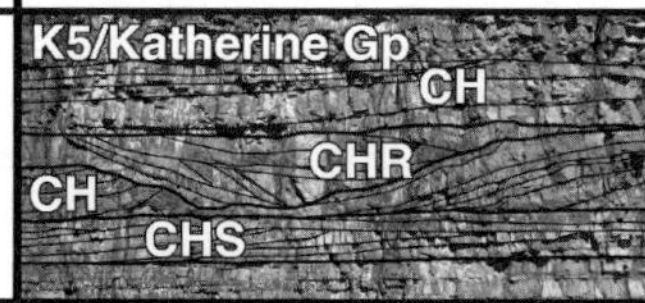	Finger, lens, ribbon (**CHR**) or sheet (**CHS**); concave up, erosional base. **CV:** Crevasse channel (Miall 1985, 1996).
Gravel bars **GB**	Gm, Gp, Gt		Lens or blanket, may be interbedded with SB (Miall 1985, 1996).
Sandy bedforms **SB**	St, Sp, Sh, Sl, Sr, Se, Ss		Lens, sheet, blanket, wedge; occurs in channel-fill, crevasse splays, and minor bars (Miall 1985, 1996). **CB** = Compound bar = amalgamated **SB**.
Sandy sheet-elements **SE**	St, Sp, Sh, Sl		Extensive sheet (Cotter 1987); Width/thickness > 20:1 (Long 2006).
Downstream accretion macroform **DA**	St, Sp, Sh, Sl Sr, Se, Ss (rare Gp, Gt)		Compound stack or lens of SB (or GB), on flat or channeled base, with convex up internal 3rd-order and upper 4th-order surfaces (Miall 1996). Vector mean of 0th-order foresets indicate bedform migration, ± 30° *down-slope* of 3rd-order bounding surfaces.
Upstream accretion macroform **UA**	St, Sp, Sh, Sl Sr, Se, Ss (rare Gm)	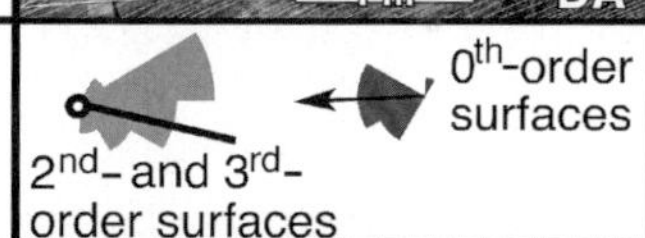	Group of SB (or GB) in which vector mean of 0th-order foresets indicate bedform migration, ± 30° *up-slope* of 3rd-order bounding surfaces.
Lateral accretion macroform **LA**	St, Sp, Sh, Sl, Se (rare Gm, Gt, Gp)	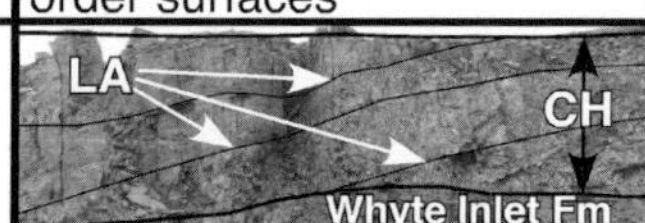	Wedge, sheet, or lobe of SB and/or GB, with inclined 3rd-order bounding surfaces (Miall 1985, 1996).
Downstream lateral accretion macroform **DLA**	St, Sp, Sh, Sl, Se (rare Gm, Gt, Gp)	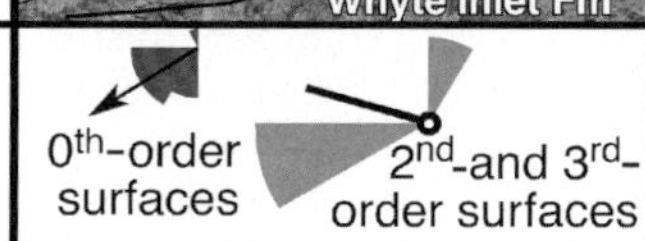	LA element in which vector mean of 0th-order foresets indicate bedform migration, 30 to 60° *down-slope* of 3rd-order bounding surfaces (Long 2006).
Upstream lateral accretion macroform **ULA**	St, Sp, Sh, Sl, Se (rare Gm, Gt, Gp)	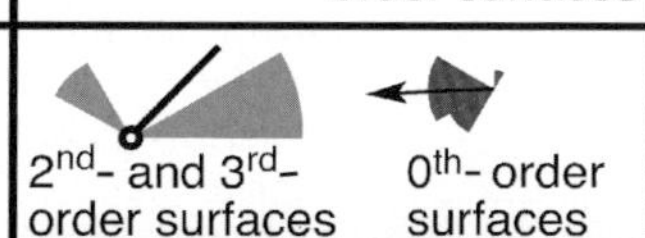	LA element in which vector mean of 0th-order foresets indicate bedform migration, 30 to 60° *up-slope* of 3rd-order bounding surfaces (Long 2006).
Scour hollows **HO**	Gh, Gt, St, Sl, Ss.	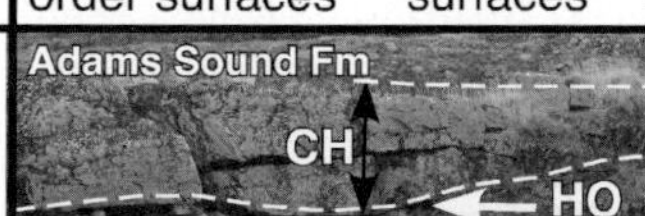	Scoop-shaped hollows with asymmetric fill, and converging paleocurrents (Miall 1996: very rare).
Sediment gravity flows **SG**	Gms, Gm	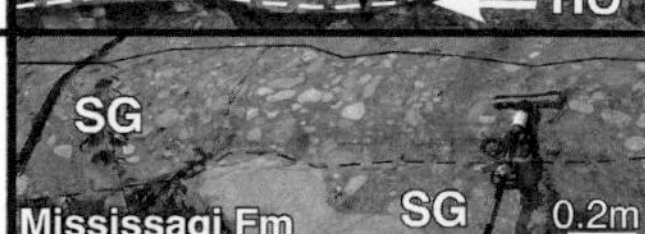	Lobe and sheet, often interbedded with GB (Miall 1985, 1996).
Laminated sand sheet **LS (CS,UFR)**	Sh, Sl, minor Sp, Sr		Sheet, blanket (Miall 1985, 1999); **CS** - local wedge shaped, lobate deposit, formed by crevasse splay (Miall 1996). **UFR** channelized units with convex upper boundary (Fielding 2006).
Overbank **OF**	Fm, Fl		Thin to thick blankets or abandoned channel fills (Miall 1996).

FIGURE 1.—Common architectural elements in early river deposits; codes based on Miall (1985, 1988, 1996), Long (2006), and Fielding (2006). For further details see Table 1.

1999). Direct comparison with modern sandur deposits, associated with catastrophic flood events (jökulhlaup deposits; Maizels 1993; Duller et al. 2008) may provide a better understanding of this depositional style.

Recent studies of fluvial strata in the Paleoproterozoic Mogalakwena Formation (Waterberg Group) in South Africa, indicate that it is dominated by interbedded sheets of conglomerate, granular medium- to coarse-grained sandstone, and matrix-supported conglomerates,

TABLE 1.—Facies codes for fluvial strata

Facies Code	Facies / Sedimentary structures / Source
GRAVEL (G) DOMINATED LITHOFACIES	
Gms	Massive, matrix- supported gravel. No clear structures (1)
Gm	Massive or crudely bedded gravel. Massive, ± imbrication (1)
Gh	Clast-supported, crudely bedded. Horizontal beds ± imbrication (1)
Gt	Gravel, stratified. Trough cross-beds (1)
Gp	Gravel, stratified. Planar cross-beds (1)
MIXED LITHOFACIES, Sandy gravel (SG) and gravelly sand (GS)	
SGm	Sandy-gravel. Massive (2)
SGh	Sandy-gravel. Horizontal bedding (2)
GSm	Gravelly sand. Massive (2)
GSh	Gravelly sand. Horizontal bedding (2)
SAND (S) DOMINATED LITHOFACIES	
St	Sand, fine to very coarse, ± pebbles. Trough cross-beds (1)
Sp	Sand, fine to very coarse, ± pebbles. Planar cross-beds (1)
Sr	Sand, very fine to coarse. Ripple cross-lamination (1)
Sh	Sand, vf. to coarse, ± pebbles. Horizontal lamination, parting lineation (1)
Sl	Sand, very fine to coarse, ± pebbles. Low-angle (<15°) dunes (1)
Ss	Sand, very fine to v. coarse, ± pebbles. Broad, shallow scours (1)
Sm	Sand, fine to coarse. Massive or faint lamination (1)
Se	Erosional scours with intraclasts (1)
MUD (M) DOMINATED FACIES	
Fl	Mud, silt, sand. Fine lamination, small ripples (1)
Fsm	Silt, mud. Massive (3)
Fm	Mud, silt. Massive, desiccation cracks (1)
P	Paleosol, carbonate (calcite/siderite). Pedogenic features (1)

Sources of definitions: 1) Miall 1985; 2) Tucker 2003; 3) Miall 1996.

representing density-modified grainflows, cohesive debris-flows, turbulent channelized hyperconcentrated-flows, unconfined sheetfloods, and minor streamflows (Eriksson et al. 2008). These occur as both sheet elements (SE and CHS: 0.5–1 m thick and hundreds of meters wide) and channel-elements (CH: 2–5 m thick, 20–30 m wide). Note that the term sheetflood is used here in the sense of Hogg 1982 for low-frequency, high-magnitude unconfined overland flows and can involve both upper-flow-regime and lower-flow-regime processes. These are distinct from sheet flows, which are defined as high-frequency, low-magnitude overland flows that occur as continuous sheets, characterized by laminar flow conditions.

Calculations of paleohydraulic parameters for the Mogalakwena Formation, based on maximum (intermediate) clast size, channel dimensions, and cross-bed thickness, indicate that these were deposited on a broad fluvial apron with slopes steeper than the maximum of 0.007 m/m for modern rivers, and less than the minimum slope of modern alluvial fans (0.026 m/m) as defined by Blair and McPherson (1994). Similar paleoslope data were obtained for the Blouberg and Wilgeriver Formations (Van der Neut and Erikkson 1999; Bumby et al. 2004; Eriksson et al. 2006; Eriksson et al. 2008) at the base of the Waterberg Group. All three formations have evidence of sheet-flow, sheetflood, debris-flow, and density-modified-grainflow deposition, all of which are known to occur in arid and semiarid alluvial-fan systems, yet all three systems have consistent unimodal paleocurrent distributions and have been interpreted as braided-river deposits. Studies of humid-region fan systems by Saito and Oguchi (2005) and Hashimoto et al. (2008) indicate that the apparent gap in slopes between fans and rivers noted by Blair and McPherson (1994) may be an artifact of preferential use of data from fans in arid and semiarid regions: they found a continuum of slopes between humid-region fans and floodplain rivers, and noted that some floodplain rivers had higher gradients than the fan systems that fed them. This may indicate that the Paleoproterozoic fluvial systems described by Eriksson et al. (2008) may have been deposited under warm, humid climatic conditions. If this is the case the presence of debris-flow and sheetflood deposits should not automatically be considered as positive evidence for deposition in arid settings.

Matrix-supported cobble and boulder conglomerates of non-fan origin have been recorded by Williams (2005) from the base of the 1.82 Ga King Leopold Sandstone, in the Kimberly district of Western Australia. In this example the conglomerates are interbedded with

TABLE 2.—Hierarchy of fluvial surfaces (after Miall 1985, 1988, 1996).

Order	Time (a)	Process	Example	More information of bounding surfaces
0	10^{-6}	Burst-sweep cycle	Lamina	
1	10^{-5} to 10^{-4}	Bedform migration	Ripple-microform	Set boundary
	10^{-2} to 10^{-1}	Bedform migration	Diurnal dune increment, reactivation surface	
2	10^{-2} to 10^{-1}	Bedform migration	Dune mesoform	Co-set boundary
3	10^{0} to 10^{1}	Seasonal events, 10 a flood	Macroform growth increment, unit bars	Dips at 5–20°
4	10^{2} to 10^{3}	100 a flood, channel and bar migration	Macroform migration (point bar, levee, splay)	Convex up, macroform top, minor channel scours
5	10^{3} to 10^{4}	Avulsion, long-term processes	Channel, delta lobe, mature paleosoils	Flat to concave up, channel base
6	10^{4} to 10^{5}	5th order Milankovich cycles and tectonic pulses	Channel belts, alluvial fan, minor sequence	Flat–regionally extensive
7	10^{5} to 10^{6}	4th order Milankovich cycles and tectonic pulses	Major deposystems, fan tract, sequence	Flat, regionally extensive or base of incised valley
8	10^{6} to 10^{7}	3rd order cycles tectonic and eustatic	Basin-fill complex	Regional disconformity

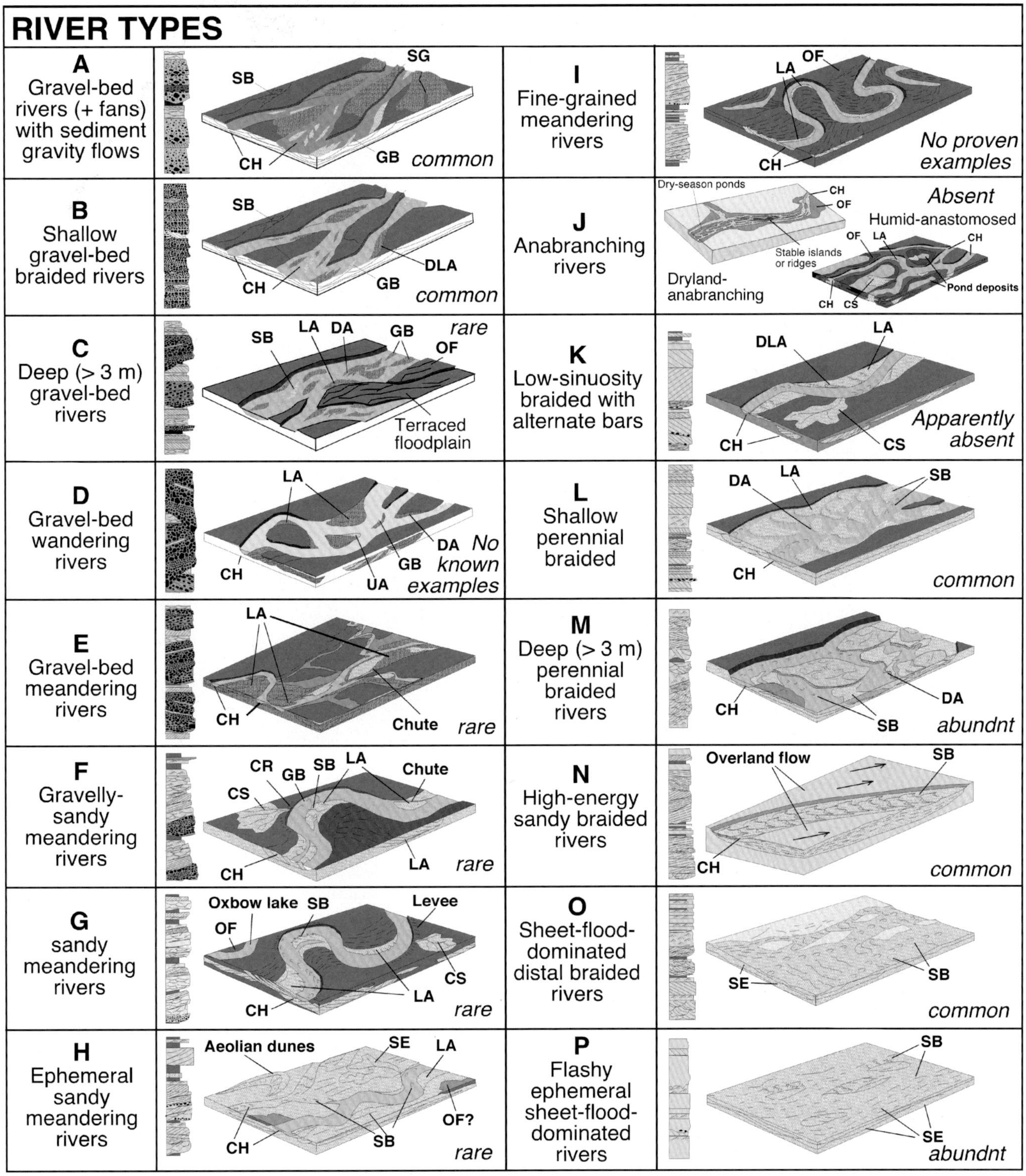

FIGURE 2.—Summary of Miall's (1996) 16 modern river types, with indication of expected elements (see Fig. 1) and idealized profile. Overbank fines (OF), and levee and splay deposits are far less common in pre–late Silurian river systems. Relative abundance of pre-vegetation river types is shown in italics.

FIGURE 3.—Gravel-bed fan systems. A) Deposit of a matrix-supported debris flow, resting directly on Archean granite (below dashed line). Paleoproterozoic Mississagi Fm., NE of Sudbury, Ontario, Canada. B) Matrix-supported (b, c) and framework-supported (a, d, f) large pebble and cobble conglomerates from the Paleoproterozoic Lorrain Fm., near Gowganda, Ontario, Canada (Long and Donaldson 2005). Note preferred alignment of clasts of large pebble to large cobble grade in matrix-supported conglomerate (unit b), suggesting emplacement by hyperconcentrated-flows. Unit e is a planar-laminated, very coarse-grained sandstone, which may represent an overbank deposit (OF element). A fan origin, rather than river origin is suggested for this deposit, as it is of local extent, and is directly associated with a fault. C) Interbedded streamflow (clast-supported) and hyper-concentrated-flow deposits (matrix-supported, with a weak horizontal fabric) in Middle Cambrian strata of the Mount Roosevelt Fm., NE British Columbia (Post and Long 2008).

trough cross-bedded sandstone and granule conglomerate, and rest on an erosional surface marked by linear grooves and local steep-walled sandstone-filled cracks. Williams (2005) interprets these as Nye (subglacial erosional) channels and frost wedges respectively, and interprets the conglomerates as products of deposition in subglacial tunnels. The presence of ventifacts on the surface of these conglomerates suggests that they were subsequently exposed to eolian activity prior to deposition of overlying shallow marine sandstones.

B: Shallow Gravel-Bed Braided Rivers

A number of authors have identified perennial shallow gravel and mixed sandy–gravelly braided-river deposits in pre-Devonian strata (Vos and Tankard 1981; Ethridge et al. 1984; Long 1978, 1987, 2001, 2002; Eriksson and Simpson 1993; Karpeta 1993; Simpson et al. 2002; Borba and Mizusaki 2003; Buatois and Mángano 2003; Hadlari et al. 2006; Krapez et al. 2008). Modern examples (Scott-type rivers) typically have channels less than 1 m deep and are dominated by gravel bars and bedforms, with minor sandy bedforms (GB, SB: Miall 1996; Lunt et al. 2004). GB elements occur as tabular units with abundant internal erosional surfaces, with minor planar cross-stratification developing on the downstream ends of bars (Fig. 5 A, B, C). Channel-fill sequences (CH elements) are common. SB elements typically occur as thin wedges in abandoned channels.

Nonchannelized conglomerate sheets are a common component of many Proterozoic successions (Long 1978; Aspler et al. 2004; Hadlari and Rainbird 2007; Krapez et al. 2008). Some show evidence of thin bedding, enhanced by interbeds of coarse-grained sandstone, as in the lower Cambrian Archer Point Formation on Ellesmere Island, Nunavut (Fig. 5D) which may represent deposition on the top of shallow longitudinal in-channel bars (cf. Fig. 5E). Others (especially in sandy–gravelly systems) show evidence of upwards fining on a scale of 1–5 m (Young 1974; Long 1978;

FIGURE 4.—Talus deposits, at base of the Paleoproterozoic Athabasca Group, Sue Pit, McClean Lake, NE Saskatchewan. A) Poorly sorted, bouldery cobble conglomerate with angular to subangular clasts (talus) directly overlying basement psammites of the Wollaston domain. This talus deposit is onlapped by massive and flat-laminated sandstones of the Read Formation (Long 2007). B) Detail of talus; the presence of both laminated (a) and massive (b) patches of sandy matrix suggests infilling by later sieve mechanisms. C) View of wedge-shaped unit of the talus conglomerate at the south end of the pit (about 20m above A) showing clear evidence of onlap of planar-laminated and massive sandstones of ephemeral-sheetflood origin in the Read Formation.

Krapez et al. 2008). Although some of the latter were originally interpreted to be of meandering-stream origin, a braided-stream origin is more likely, in as much as none contain distinct evidence (LA surfaces and/or systematic upsection changes in paleo-flow direction) of high sinuosity. Hadlari et al. (2006) and Hadlari and Rainbird (2007) suggest that composite fining-upwards sequences in Paleoproterozoic fluvial strata in the Baker Lake Basin could represent either climatically related changes in sediment flux or simply a response to systematic reduction in stream gradient with time.

Shallow ephemeral gravel-bed-river deposits have been described from the Cambrian of Egypt by Khalifa et al. (2006). These include proximal 0.5–1.5m sets of sandy–muddy matrix-supported pebble conglomerate (of possible low-velocity debris-flow origin), interbedded with thin sets of laminated sandstone. These appear to pass down paleoslope to fining-upwards cycles with a basal layer (0.2–0.4 m) of massive to weakly laminated (or rare cross-stratified) granule conglomerate, capped by 0.1–0.2 m thick sets of flat-laminated or low-angle cross-stratified slightly pebbly sandstone with minor intraformational clast layers. These strata appear to have been deposited under transitional to upper-flow-regime conditions. Similar deposits of shallow channelized streamflow or unconfined sheetflood origin have been described by Mazumder and Sarkar (2004) from the Paleoproterozoic Dhanjori Formation in India and by Krapez et al. (2008) in the Mergougil Supersequence, in Australia. Other than thickness, no architectural details are available for either of these two examples, although it is possible that some are channelized on a scale similar to modern examples described by Laronne and Shlomi (2007), which are 50–172 m wide. Sheets of Gm and Gm-Sh in the Mergougil Supersequence can be traced laterally for at least 100 m, and have been interpreted as deposits of within- and between-channel sheet elements (SE) by Krapez et al. (2008), with planar-laminated sand (Sh) units representing deposition on bar surfaces under upper-flow-regime conditions during falling flood stage.

FIGURE 5.—Shallow gravel-bed river systems. A) Moderately poorly sorted, planar cross-stratified large-pebble conglomerate, base of upper member of the late Paleoproterozoic Uairén Formation, Roraima Supergroup, Venezuela (Long 2002), possibly deposited at the downstream end of a longitudinal braid bar. B) Planar cross-stratified pebbly sandstone, Paleoproterozoic Burmside Fm, Wellington Bay, Victoria Island Nunavut (cf. Young 1974). Dispersed pebbles on foreset suggest deposition on a humpback dune. C) Small-scale fining-upwards couplets in Burnside Fm. Pebble conglomerates may reflect channel armoring, with planar and trough cross-stratified sandstones reflecting migration of straight to sinuous-crested sandy bedforms during falling flow conditions. D) Thin planar-bedded, medium- to large-pebble conglomerate, with minor massive sandstone units, Cambrian Archer Fiord Fm, Ellesmere Island, Nunavut. May have been deposited on a bar top, similar to E, from the White River, Yukon Territory.

C: Deep Gravel-Bed Braided Rivers

Modern deep (depth > 3 m) gravel-bed rivers produce channel-fill sequences (CH) thought to be over 3 m thick. Systems like the Donjek River, in northwest Yukon (Fig. 6A, B, E) are currently incising into Plio-Pleistocene deposits, and have multiple terraces, most of which are stabilized by rooted vegetation. As with shallow gravel-bed rivers they are dominated by GB and SB elements, and may locally develop fining-upwards sequences (Miall 1996). Both DA and LA elements are locally preserved, as are decimeter-scale chutes and hollows on bar tops (Fig. 6A, B). Crude horizontal stratification and imbrication may develop locally on longitudinal bars (Fig. 5E).

In most gravel-bed fluvial systems the predominantly sandy units, which cap many modern river bars (Fig. 6B), are selectively removed (both by eolian activity and subsequent channel-bed scour), hence only the deeper parts of the river deposits are preserved. Architectural analysis of early Paleozoic and Precambrian deep gravel-bed rivers is very limited. A preliminary examination of part of the Keweenawan Supergroup near Mamainse Point on Lake Superior (Fig. 6C; Fig. S1 in

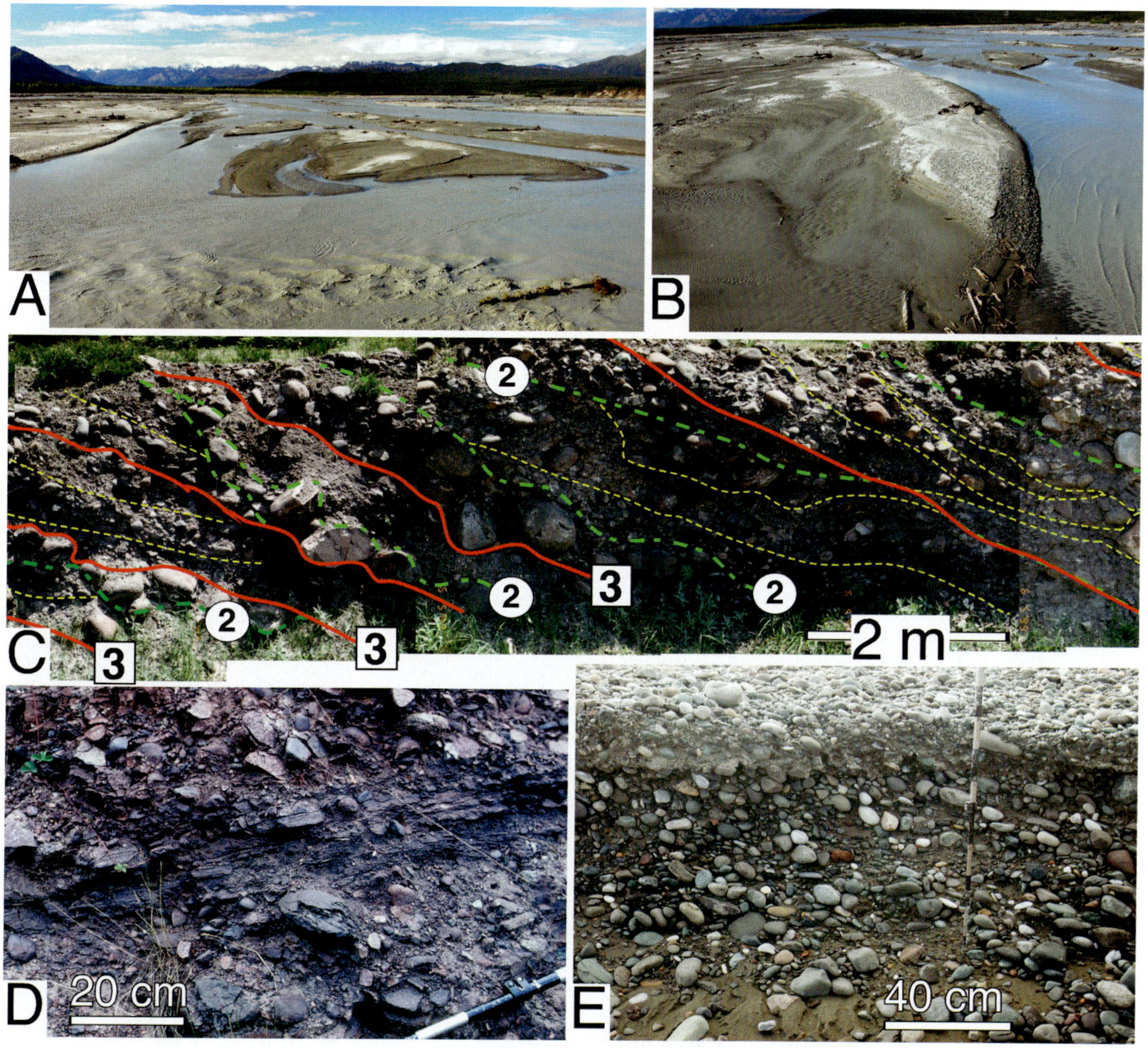

FIGURE 6.—Deep gravel-bed river systems. A) Large-scale gravel bars in the modern Donjek River, Yukon Territory Canada. View upstream from bridge shows secondary and tertiary channels filling chutes on the bar top. Note standing waves in main channel. B) Bar-top hollow, with sand and mud drape, adjacent to tertiary channel, Donjek River. C) Channel-fill boulder and cobble conglomerates of Keweenawan age (~ 1.1 Ga), Mamainse Point, Ontario, Canada. Solid lines (3) indicate third-order (or higher) surfaces defining individual channel complexes; Bold dash (2) indicates LA surfaces; and light dash (1) indicates set boundaries (For more architectural details see Fig. S1). D) Planar-laminated sandstone unit on upper part of LA surface, suggesting deposition under upper-flow-regime conditions (location as C). E) Imbricated large-pebble conglomerate, bar top, Donjek River.

the digital appendix) indicates the presence of narrow irregular channel elements, estimated to be 10–16 m wide (based on 1.5 x LA set width: Leeder 1973) in which flow appears to have been highly variable. These were originally interpreted as inter-flow alluvial-fan deposits by Merk and Jirsa (1982), but are here interpreted as the deposits of a deep gravel-bed river. Maximum grain size is of boulder grade (intermediate diameter: $Di \sim 0.5$ m), suggesting maximum flow velocities in the order of ~ 3.4 m/s (Costa 1983) and bank-full discharge in the order of 117 m^3/s (Schumm 1968). Minor coarse flat-laminated sandstone beds appear to have developed on the top of LA sets during falling flood stage (Fig. 6D) suggesting deposition on bar margins under upper-flow-regime conditions during falling flood stage.

Deposits of deep gravel-bed braided rivers have also been described from the base of the upper member of the Paleoproterozoic Uarién Formation, Roraima Supergroup, in southeast Venezuela. Long (2002) showed that there was distinct evidence for channel incision into underlying trough cross-stratified fine sandstones and silty fine sandstones of the lower member in at least two locations. At one of these locations a sequence of massive and trough cross-stratified cobbly large pebble conglomerates overlies a scoured surface with a local relief of ~ 4 m (Fig. 7A). The steep cutbank appears to be the source of large rafts

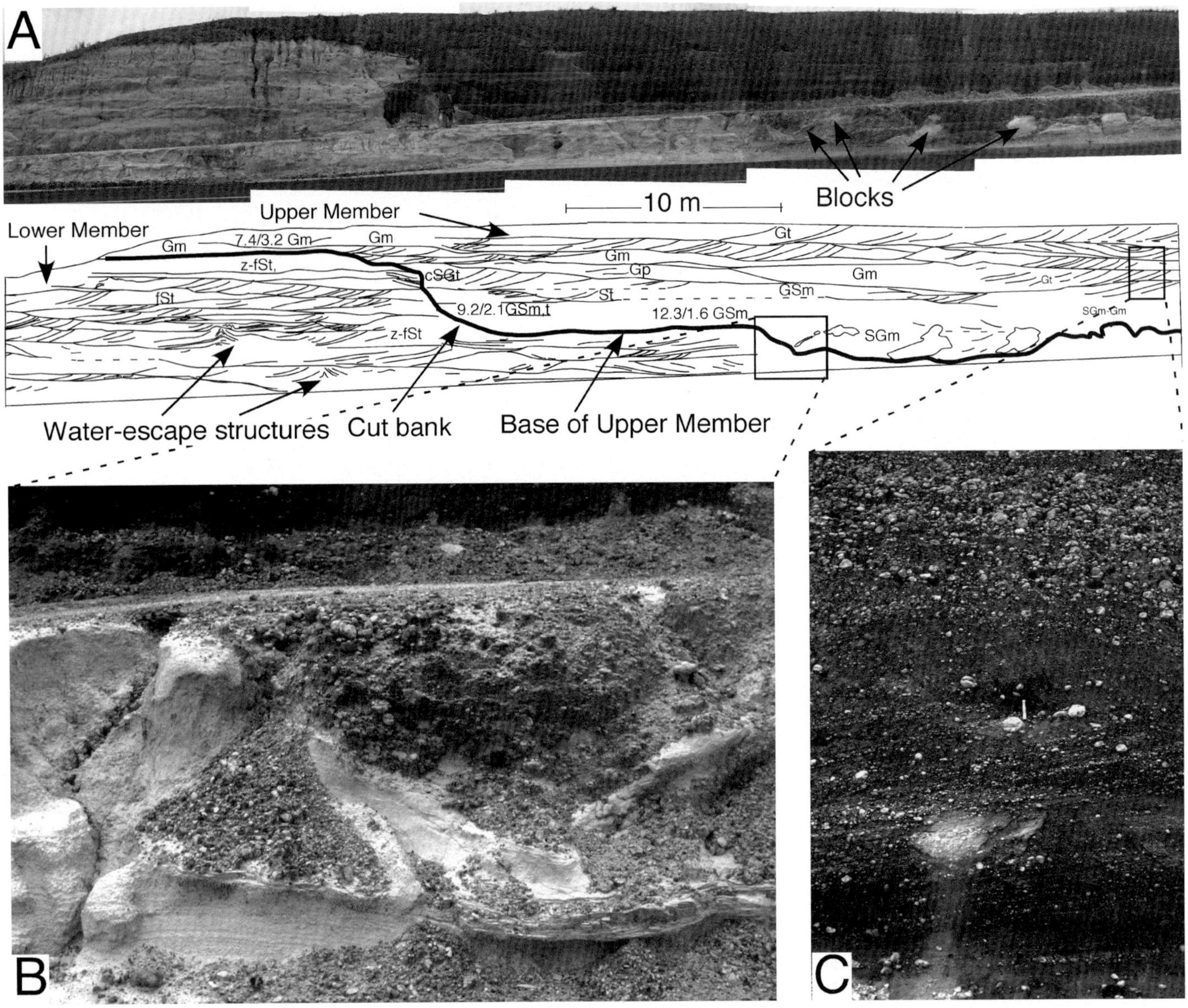

FIGURE 7.—Deep gravel-bed river systems: side-bar development in the late Paleoproterozoic Uairén Formation, Roraima Supergroup, Venezuela (Long 2002). A) Architecture of conglomerates at the base of the upper member. Large blocks along the channel floor (B) were derived by lateral erosion of trough cross-stratified silty fine sandstones (z-fSt) of the lower member. Cosets of cobble and pebble conglomerate are inclined at approximately 3° away from the cutbank, with transverse flow indicated by cross-stratification in the conglomerates (C). Gm = massive conglomerate; Gt = trough cross-stratified conglomerates; SG = sandy conglomerates; GS = gravelly sandstones, St = trough cross-stratified sandstones. 9.2/2.1 = maximum and modal grain size of conglomerates in cm (For more details see Fig. S2).

(boulders) of fine sandstone that are preserved at the channel base (Fig. 7 B). Cross-stratified gravels (Fig. 7A, C) within the channel appear to rest on inclined surfaces, with an apparent dip of ~ 3°, and a lateral extent of 66 m. This may to have developed as a side bar, hence the true width of the channel could not be determined. Estimated maximum velocity (based on $Di = 0.123$ m) is ~ 1.8 m/s (Costa 1983), with an estimated minimum bank-full discharge in excess of ~ 720 m^3/s.

D: Wandering Gravel-Bed Rivers

Wandering gravel-bed river systems (Church 1983; Forbes 1983; Billi et al. 1987; Miall 1996; Roberts et al. 1997; Wooldridge and Hickin 2005) have yet to be positively identified in pre-vegetation fluvial systems. They typically have sinuosities of 1.2 to 1.5 and are characterized by single or local multi-channel systems with large bank-attached point bars, or side bars, with prominent LA elements. Because these systems appear to be transitional into both braided and meandering systems, detailed paleocurrent analysis is required to identify the nature of migration of LA surfaces. In addition there should be clear evidence of channel division, supported by diverging and converging paleo-flow indicators, with evidence of draping of strata around mid-channel compound bars (Billi et al. 1987; Roberts et al. 1997; Long et al. 2000). The apparent absence of this type of river deposit may be a function of limited preservation potential, combined with the limited number of architectural studies of gravel-dominated systems, or may indicate that rooted vegetation is necessary to promote formation of in-channel bars (Nanson and Knighton 1996).

E: Meandering Gravel-Bed Rivers

To date, meandering gravel-bed river systems have not been described from pre-Devonian systems. One possible example is the Ediacaran to early Cambrian Mount Currie conglomerate, exposed in the Olga Mountains (Kata Tjuta) of central Australia (Walter et al. 1995) which had previously been interpreted as an alluvial-fan deposit (Sweet 1993). These boulder and cobble conglomerates occur in laterally extensive sheets (> 500 m) with pronounced LA surfaces (Fig. 8A–D), with primary dips of 7–10°. Average thickness of each sheet is about 10 m (range 7–15 m), with LA surfaces traced laterally for 150–200 m. Sandstone units are rare, and occur both along set boundaries and on LA surfaces (Fig. 8D). Based on a maximum observed grain size of ~ 0.35 m, the average velocity for this system is calculated at ~ 2.9 m/s (Costa 1983), with a bank-full discharge of ~ 5,740 m^3/s (equivalent to one third of the mean annual discharge of the Mississippi River, USA, and 1.6 times the mean annual discharge of the Fraser River, BC, Canada: Orton and Reading 1993). Because Kata Tjuta is a culturally protected site, no direct slope or paleocurrent observations could be made; consequently it is not possible to estimate channel sinuosity. It has been suggested that the lithic arkoses of the Mutitjulu Formation at Uluru (Ayers Rock) are distal equivalents of the Mount Currie Conglomerates at Kata Tjuta (Walter et al. 1995). This is unlikely, in as much as the thin bedded, trough cross-stratified, moderately poorly sorted, granular sandstones (Fig. 8E) at Uluru have a far more restricted detrital zircon assemblage (Camacho et al. 2002).

MIXED GRAVELLY–SANDY, AND SANDY–MUDDY RIVER SYSTEMS

Miall (1996) recognized five distinct categories of mixed river systems. These include F) gravelly–sandy meandering rivers; G) sandy (perennial) meandering rivers; H) ephemeral sandy meandering systems; I) fine-grained meandering systems; and J) anastomosed systems. The last of these models, which is now considered to be a humid organoclastic variety of anabranching rivers, has not been recorded from pre-vegetation systems, nor have their dryland equivalents (Nanson and Knighton 1996; North et al. 2007).

F: Gravelly–Sandy Meandering Rivers

Reliable descriptions of high-sinuosity gravelly–sandy river deposits in the Precambrian are rare (Cotter 1978; Long 1978) and have not been confirmed using architectural analysis: most are based on the assumption that a fining-upwards sequence automatically indicates a meandering fluvial setting. Sweet (1988) described a possible meandering sequence of pebbly (intraclast-rich) to non-pebbly sandstones with marked lateral-accretion surfaces, capped by 3 m of mudstone. Although flow directions indicated by cross-stratification appear to be at a high angle to the LA surfaces (which are shown to be restricted to the upper 4 m of a 9 m-thick story), too few cross-bed directions are recorded to confirm any systematic changes in flow direction up section. The sequence recorded by Sweet (1988) may alternatively represent local lateral-accretion on an in-channel bar within a shallow sandy braided system, with climatically induced reduction in sediment flux producing the fining-upwards sequence, as has been suggested for braided fining-upwards sequences in the Baker Lake Basin by Hadlari et al. (2006) and Hadlari and Rainbird (2007).

G: Sandy (Perennial) Meandering Rivers

Possible meandering-stream deposits have been recorded in two correlative units in northwest Canada: the Neoproterozoic Katherine Group in the Mackenzie Mountains (Long 1978) and the Nelson Head Formation in the Brock Inlier (Long 1978; Rainbird and Young 2009). Both units represent the distal end of a continental-scale fluvial system, with headwaters in the Grenville Orogen, some 5000 km to the southeast (Rainbird et al. 1997; Rainbird and Young 2009). In both units paleocurrents in sets of cross-stratified sandstone are at a high angle to LA surfaces, and change systematically up section. In the Nelson Head Formation (Fig. 9A), LA sets are typically 3–6 m thick, and dip at 0.8° to 4°. My observations indicate that individual LA surfaces can be traced for 350–590 m, indicating that channels may have been as much as 890 m across (cf. Leeder 1973). Mudstone layers are present on some LA surfaces and as channel plugs, suggesting both seasonal flow variations and channel abandonment by upstream avulsion on a broad alluvial plain. Assuming a minimum velocity of about 0.4 m/s, bank-full discharge would have been in the order of 1,426 m^3/s. Estimated sinuosity is 1.47 (Long 1978, using method of Miall 1976).

A smaller meandering-channel element from unit K7 of the Neoproterozoic Katherine Group is illustrated in Figure 9B. This 3.17 m-thick CH element is dominated by medium-grained planar cross-stratified sandstone, with LA surfaces dipping at 6°, with an estimated lateral extent of 40 m. Mean paleo-flow of SB elements is 50° counterclockwise to the orientation of the LA surface. Maximum size of mudstone intraclasts on LA surfaces is 0.02 m, suggesting a stream velocity of ~ 0.78 m/s, and a subsequent bank-full discharge of ~ 99 m^3/s (Schumm 1968). Sinuosity is estimated at 1.65 using the method of Miall (1976), 1.83 using Ferguson (1977), or 2.06 using Ghosh (2000), method 1.

Both examples differ significantly from the classic models for rivers such as the Mississippi and Klarälven, in that they have only minor overbank fines (OF) and lack distinctive crevasse (CR) and splay (CS) elements. The latter may indicate that levees were not well developed before the advent of rooted land plants.

H: Ephemeral Sandy Meandering Rivers

According to Shepherd (1987), ephemeral sandy meandering rivers can be distinguished from sandy meandering systems in that the former are characterized by a predominance of upper-flow-regime bedforms (Sh, Sl) on irregular to stepped LA surfaces. Occasional mudstone intraclasts and mud drapes are present on second-order and third-order sufaces, along with rare wedges of wind-blown sand. Levee and splay facies are rare, and floodplain deposits may be restricted to thin sandstone sheets and minor mudstones (Miall 1996). Neoproterozoic examples of ephemeral sandy meandering systems are present in the Katherine Group (Fig. 10), where they occur as 2.8–5.7 m thick CH elements, interbedded with deposits of perennial sandy meandering systems. In contrast to the perennial examples described above (Fig. 9), LA surfaces (with dips of 1.5 to 16° and widths of up to 240 m) are characterized predominantly by conformable cosets of low-angle cross-laminated and thin planar-laminated medium sandstones (Fig. 10A). In places low-angle sets are overturned (Fig. 10B, C) due to subsequent flow events. As in the Rio Puerco (Shepherd 1987), evidence of multiple scouring of LA surfaces is locally apparent from along-strike observations of set contacts. Mudstone drapes, mudcracks, and mudstone intraclasts are very rare, and wedge-shaped units of impact ripples and pinstripe lamination of eolian origin have yet to be identified. Estimated sinuosity (based on Miall 1976) typically exceeds 1.5.

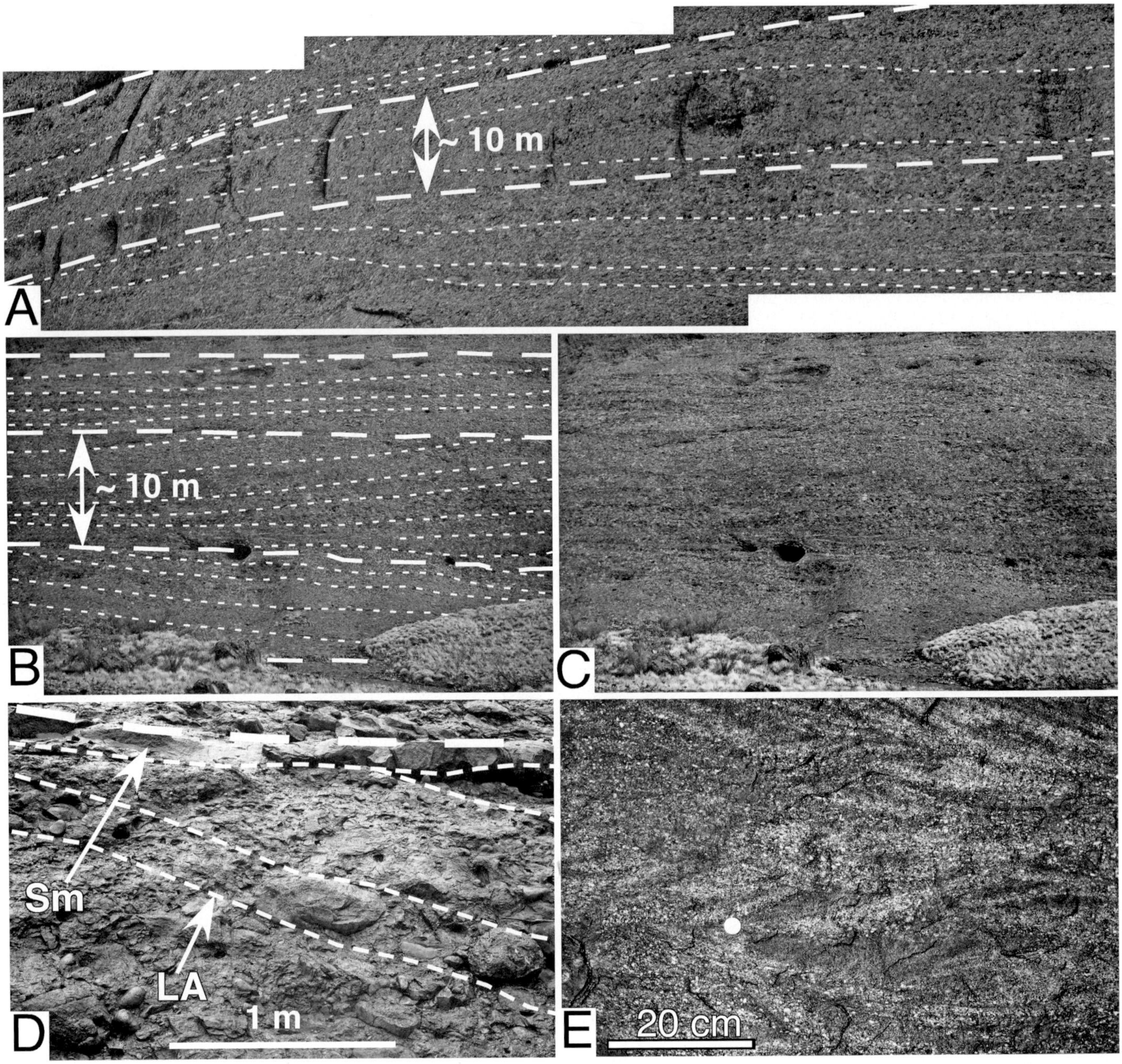

FIGURE 8.—Meandering gravel-bed river deposits, a probable example from the Ediacaran to Early Cambrian Mount Currie conglomerate, Kata Tjuta,. A) 10–15 m thick sheets of boulder and cobble conglomerate (outlined with large dashes) with distinct LA surfaces (small dashes): B) Moderately steeply dipping LA sets. C) As part B, but without lines. D) Detail of upper contact of LA surfaces with overlying channel base. Note thin unit of massive sandstone (Sm) at contact. This may represent an overbank deposit. E) Trough cross-stratified sandstones of shallow sandy braided river origin in the Mutitjulu Formation, at Ulura, previously considered to be lateral equivalents of strata at Kata Tjuta.

I: Fine-Grained Meandering Systems

Morley and Ojakangas (1982), and Ojakangas (2005) suggested that sandstone sets overlain by siltstone and/or mudstone units in the Orienta and Fond du Lac Formations in Minnesota and Wisconsin are the products of meandering river systems. Other similar sequences, including parts of the Mesoproterozoic Solor Church Formation, in Minnesota, USA (Hatch and Morley 1985), and an 80 m-thick, stacked sequence of poorly defined fining-upwards cycles in the late Archean Transvaal Supergroup in South Africa (Tyler 1978), have all been interpreted as meandering-stream deposits, based on vertical sequence analysis. There is no clear evidence for LA surfaces, levee facies, or systematic upsection deviation in paleocurrent flow recorded in any of these examples. The grouped paleocurrents observed by Morley (1967) in outcrops of the Fond du Lac Formation are strongly unidirectional. Observations on sandstone–mudstone cycles in the Fugleberget

FIGURE 9.—Sandy meandering stream deposits. A) Stacked CH elements in the Neoproterozoic Nelson Head Formation at Brock River Canyon, N.W.T. CH and LA elements are enhanced by local mud drapes and rare channel plugs (a). B) Stacked CH elements in Unit K7 of the Neoproterozoic Katherine Group, Dodo Canyon, N.W.T. Section on right shows presence of local intraclast horizons on LA surfaces in one CH element. Gray arrows indicate orientation on individual cross-stratified units (north at top). Black arrows on far right show weighted moving average of sets of 10 cross-beds (cf. Miall 1976), suggesting a progressive clockwise shift in channel orientation up section.

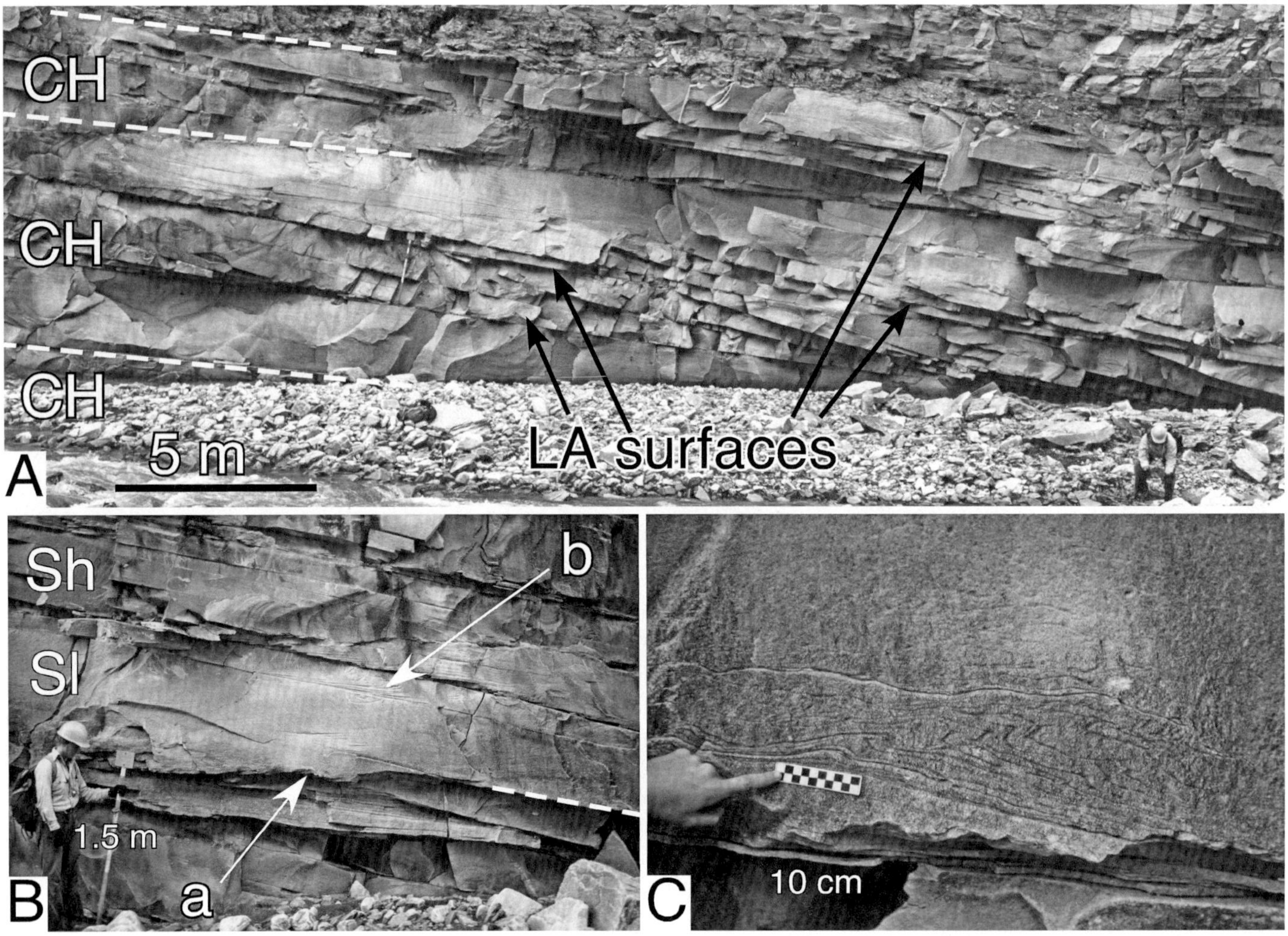

FIGURE 10.—Ephemeral sandy meandering rivers: examples from the Neoproterozoic Katherine Group, N.W.T., Canada. A) LA surfaces in CH units in Loretta Canyon. Note how second-order surfaces, bounding units of Sh, Sl, and minor Sp, taper down the LA surface. B) Lower contact (a) of CH element, showing cosets of Sl and Sh on LA surfaces. Local overturned cross-stratification (b) may be induced by frictional drag from flashy postdepositional flow events. C) Detail of contorted low-angle cross-stratified unit at base of channel.

Formation in Greenland by Sønderholm and Tirsgaard (1998) indicate that in the Neoproterozoic, braiding predominates in distal low-gradient fluvial settings, where meandering would normally be expected to occur. The absence of meanders in this setting may be due to greater fluctuations in fluvial discharge in pre-vegetation times. Given the lack of architectural confirmation, it is more likely that the units described above represent distal braid-plain deposits, analogous to those described as muddy braid-plain deposits by Sønderholm and Tirsgaard (1998).

SANDY FLUVIAL SYSTEMS

Miall (1996) provides models of six types of modern low-sinuosity sand-bed rivers. These include: K) low-sinuosity braided streams with alternate bars; L) shallow perennial braided streams; M) deep perennial braided streams; N) high-energy sandy braided; O) sheetflood distal braided; and P) flashy ephemeral sheetflood systems. All but the first category have been recorded in pre-vegetation systems.

L: Shallow Sandy Braided Rivers

The bulk of recorded pre-vegetation river systems have been interpreted in terms of models of shallow perennial braided rivers. These typically contain stacked sequences of sandy bedforms, which form small-scale depositional sequences with high lateral continuity ("sheet-braided" of Cotter 1978). Width-to-depth ratios may have been significantly greater than in modern systems due to the absence of cohesive banks (20:1 to > 1000:1; Fuller 1985; Els 1990; Rainbird 1992). Clear evidence of channel elements, or even channel margins, are very rare, hence channel depth in many cases can only be approximated from story thickness or average cross-bed thickness. Precambrian sequences dominated by planar cross-stratification tend to be interpreted as products of shallow perennial braided (Platte-type) river deposits (Siedlecka and Edwards 1980; Southwick et al. 1986; Sweet 1988; Eriksson and Simpson 1993; Røe and Hermansen 1993; Long 2002). Eriksson and Simpson (1993) suggest that monotonous sequences of trough cross-bedded sandstone may reflect relatively constant, perennial discharge in pre-vegetation fluvial systems (Fig. 11A, F). Small channelized sets of trough cross stratification may reflect deposition in channel thalwegs during low-flow stage (cf. Rainbird 1992), or small stream channels in distal fluvial settings (Tirsgaard and Øxnevad 1998).

Many examples of deposits of shallow sand-bed braided rivers are best described as sheet elements, in which set boundaries tend to parallel coset boundaries (Long 1978, 2006; Fedo and Prave 1991; Hocking 1991, his FA1; Rainbird 1992; Hjelbakk 1997; Tirsgaard and Øxnevad 1998; Røe and Hermansen 1993; Chakraborty 1999). Many show no marked change in grain size, but may exhibit some thinning of cross strata up section (Fig. 11 B) or superimposed flat-laminated sandstones, deposited under upper-flow-regime conditions on bar tops during falling flood stage (Eriksson and Simpson 1993). Some units pinch out laterally (Røe and Hermansen 1993), suggesting low-relief bank margins, with slopes as low as 7° (Long 2007). Steeper slopes are preserved only where channels cut finer-grained facies, typically with a higher mud or silt content (cf. Tisgaard and Øxnevad 1998). In many deposits there is clear evidence of macroform development, with a predominance of DA and DLA (downstream lateral-accretion) elements, formed by migration of in-channel linguoid or complex bars (Fig. 11C, D, E; Fig. S3 in the digital appendix; Hjelbakk 1997; Chakraborty 1999; Long and Donaldson 2005).

M: Deep Sandy Braided Rivers

Modern deep perennial sand-bed braided rivers, including the South Saskatchewan (Smith et al. 2006) and the Brahmaputra (Best et al. 2003), are typically over 3 m deep and are characterized by the presence of large in-channel macroforms (compound bars) that may have cross-channel and down-channel dimensions from a few hundred meters to several kilometers, and have complex internal accretionary geometries (Miall 1996). Some of these macroforms show evidence of upwards fining and/or decrease in scale of sedimentary structures, but this is not ubiquitous. Because many of the in-channel macroforms have dimensions ranging from a few hundred meters to several kilometers, they tend to concentrate flow in first-order channels. Where these converge they typically produce deep scour hollows (HO), which may be as much as four times the average channel depth (Best et al. 2003). These HO elements have not been observed in pre-vegetation deep sandy river deposits to date.

Large-scale macroforms with complex internal architecture have been recognized in a number of pre-vegetation systems (Hocking 1991, his FA3; Rainbird 1992; Røe and Hermansen 1993; Hjelbakk 1997; Nicholson 1993; Long 2006). Rainbird (1992) noted that almost 90% of the Neoproterozoic Kuujjua Formation on Victoria Island, Nunavut, Canada, consisted of compound cross-bedded cosets of medium- to coarse-grained sandstone in units from 1.5 to 8.5 m thick (some with local mudstone intraclasts at the base), that could be traced laterally, across depositional strike, for up to 20 km, with less than 25% change in thickness. Within these macroforms coset boundaries dipped at an average of 9°, with flow direction of straight to sinuous crested sandy bedforms oriented at 46 to 103° to the slope of the coset boundaries, indicating that these formed as DLA and LA elements. Rainbird (1992) suggested that the Kuujjua Formation was deposited in a large-scale river system on a scale similar to the Brahmaputra. He indicated that the depositional site may have been in an arid-region setting, based on the presence of salt pseudomorphs in associated floodplain, lake, and channel-fill calcareous mudstones.

Similar large-scale composite bedforms, representing DLA elements, have been recorded in the Mesoproterozoic Applecross Formation (Torridon Group) in Scotland by Nicholson (1993). These are characterized by prominent LA surfaces with apparent dips of 8–10°, in channel-fill units up to 9 m thick. Paleocurrents inferred from planar and trough cross-bedded cosets in the two examples illustrated were oriented between 36° and 64° counterclockwise from the orientation of the LA surfaces. Estimated sinuosity of the two examples illustrated by Nicholson (1993) is 1.31 to 1.33, based on cross-bed directional variances of 1764 and 1849, using the method of Ferguson (1977). Nicholson (1993) suggests that deep sand-bed braided rivers in the Applecross Formation had channel widths of 350-1400 m (w/d 50:1–200:1), with bank-full discharge at 1840 to 7350 m^3/s.

On a slightly smaller scale, architectural analysis of part of the Bird Member of the late Paleoproterozoic Manitou Falls Formation in northern Saskatchewan by Long (2006) indicated a strong incremental character of DA elements. These were related to periodic migration of barforms during peak-flood stages, producing unit bars, separated by periods of low flow (Fig. 12A, B; Fig. S4 in the digital appendix). The presence of local transverse lenticular upstream accretionary elements on the top of some CH sets indicates the presence of chute channels (Fig. S4 in the digital appendix), suggesting that at least locally much of the original channel fill was preserved. This is also indicated by local preservation of flat to low-angle inclined granular sandstone layers on one bar top (Fig. 12C). Within this section thin gravel layers accumulated as channel lags and on LA surfaces, with discontinuous mud layers suggesting deposition within temporary ponds in channel thalwegs after major floods. Preserved channel-fill sequences in this section are from 1.8 to 5 m thick. Calculated sinuosity of channels is 1.1 to 1.59 (method of Miall 1976), 1.07 to 1.13 (Ferguson 1977), or 1.14 to 1.43 (Ghosh 2000, method 1). Hollows elements (HO) were not observed.

N: High-Energy Sandy Braided Rivers

Miall's (1996) model of high-energy sandy braided rivers is based on studies of the Jurassic Morrison Formation by Cowan (1991). He suggests that this type of river system should be characterized by deep scours (producing hollows elements; HO) and by abundance of planar lamination (Sh), low-angle cross-stratification (Sl) and sigmoidal cross-stratification characteristic of deposition under transitional and upper-flow-regime conditions (Fig. 13A, B, C). Fielding (2006) and Fielding et al. (2009) have suggested that sheet and lenticular units (their UFR elements), dominated by Sh, Sl, and sigmoidal facies formed under upper-flow-regime conditions, and are indicative of strongly seasonal paleoclimates with pronounced seasonal peaks in precipitation and runoff. Given that runoff would have been more rapid in the absence of rooted vegetation (Schumm 1968), it is expected that this type of river deposit would form in a greater range of climatic settings than at present.

A key to identifying the deposits of pre-vegetation high-energy sandy braided rivers is the presence of sigmoidally bedded sandstone units (Røe 1987; Røe and Hermansen 1993; Hjellbakk 1993, 1997; Sønderholm and Tirsgaard 1998). These typically occur in sheet elements up to 6 m thick, with width-to-depth ratios from 40 to > 100:1, and less commonly in channel fills, 0.1 to 1.3 m thick, with width-to-depth ratios ratios of 15 to 40:1 (Hjellbakk 1997; Røe and Hermansen 1993). Sigmoidally bedded sandstone units are commonly associated with, and may pass laterally into, Sh and Sl units with current lineation. Reactivation surfaces are common (Røe and Hermansen 1993). Beds may show evidence of complex postdepositional deformation, especially beneath reactivation surfaces. This is probably related to liquefaction induced by shear during subsequent floods. Thin siltstones may be preserved locally on the top of sheet elements. Mudstone intraclasts are common, especially along lower set and coset boundaries. Abundant mudstone clasts in these and other pre-vegetation fluvial deposits suggest that fines may have been deposited in channel thalwegs and on floodplain surfaces during falling flood stage, but were eroded by subsequent flows. Massive units (0.4 to 2 m thick, > 50 m wide), interbedded with sigmoidally bedded

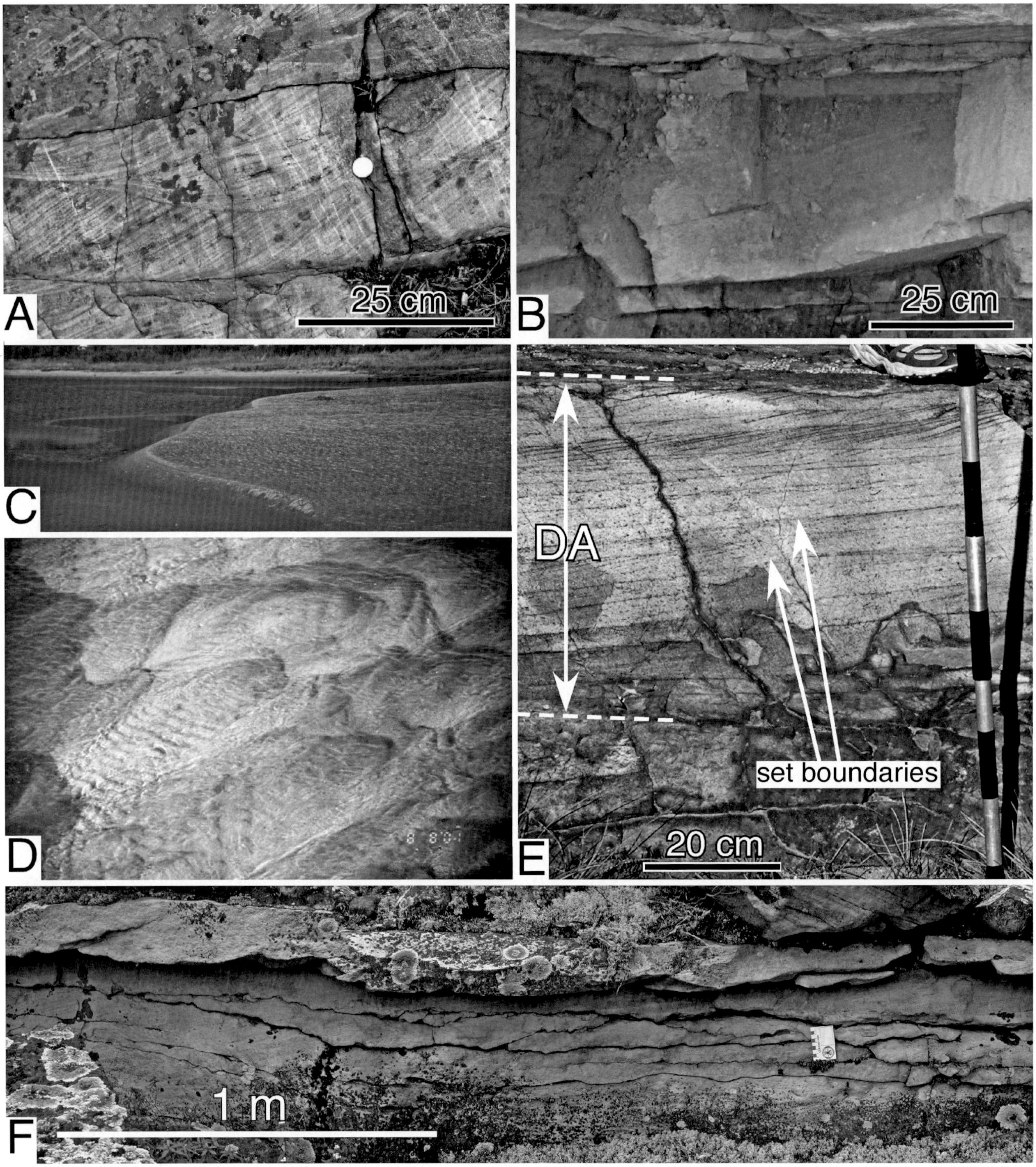

FIGURE 11.—Shallow sandy braided systems. A) Sheet element containing simple bars of trough cross-stratified sandstone, Mesoproterozoic Bonner Formation, Belt Supergroup, near Helena, Montana. B) Large-scale planar cross-stratified sandstone, overlain by thin beds of ripple cross-laminated sandstone, forming a laterally extensive SE element in the late Paleoproterozoic Manitou Falls Formation (Bird Member), Sue Pit, McClean Lake, Saskatchewan. This unit can be traced laterally for over 60 m. C) Planar cross-stratification is produced at the downstream ends of lingoid bars, as in this example from the Platte River, Nebraska. Note ripples on upper surface. D) Trough cross-stratified sandstones can be generated by migration of sinuous-crested dunes, as seen in the main channel of Platte River. E) Small-scale ripples developed within a DA element in the Paleoproterozoic Mississagi Formation, Sudbury, Ontario (For more details see Fig. S3). F) Compound trough cross-stratified sandstone in Dunlop Member of late Paleoproterozoic Manitou Falls Formation, Hughes Lake, Saskatchewan.

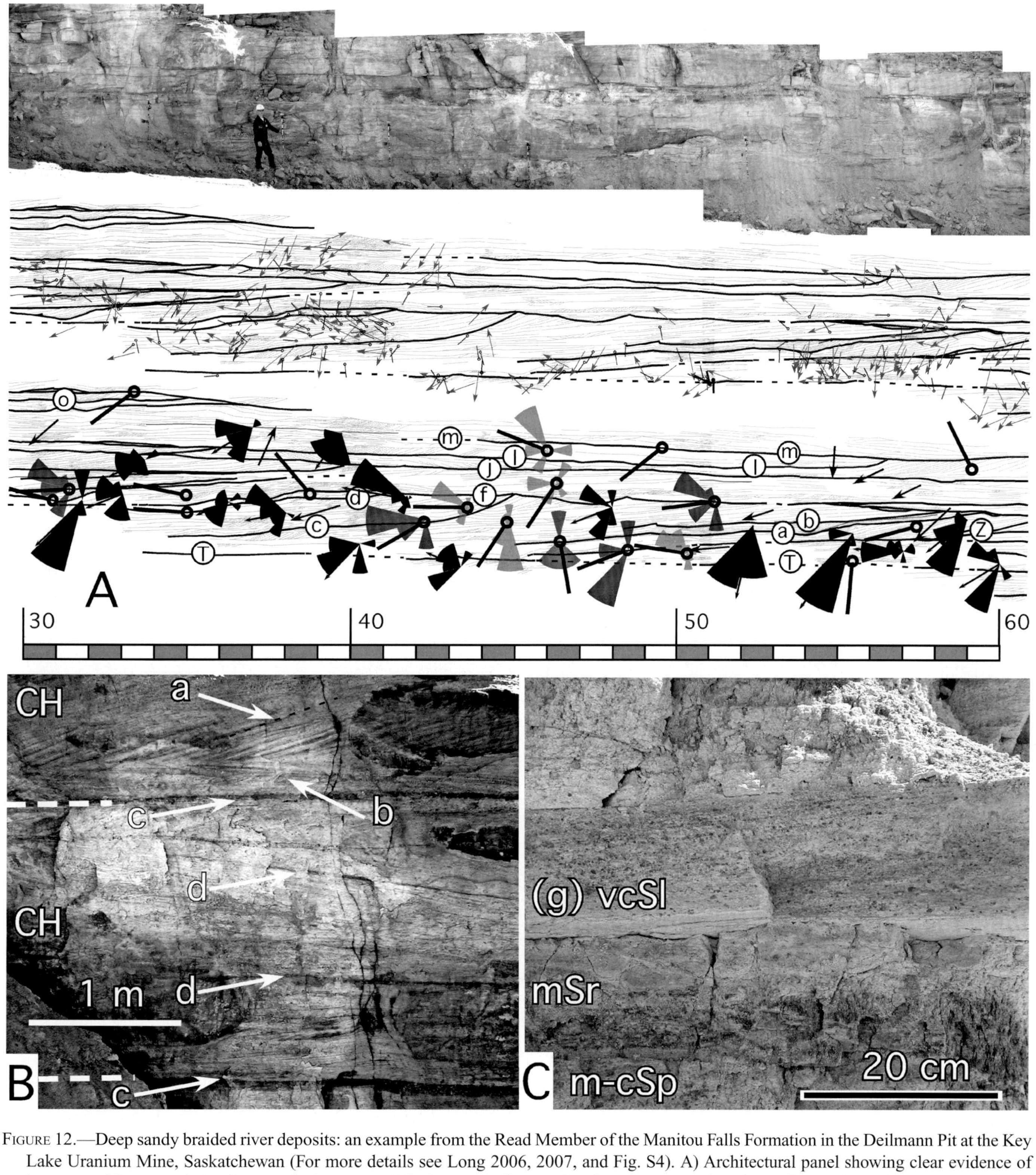

FIGURE 12.—Deep sandy braided river deposits: an example from the Read Member of the Manitou Falls Formation in the Deilmann Pit at the Key Lake Uranium Mine, Saskatchewan (For more details see Long 2006, 2007, and Fig. S4). A) Architectural panel showing clear evidence of incremental bar growth (surfaces a–d). Arrows and black roses indicate paleo-flow from zeroeth-order surfaces; pins and gray roses indicate slopes of second-order and higher order surfaces. B) Detail of two channel elements (a = unit-bar margin; b = ripples in channel thalweg; c = granular sandstone lag on channel base; d = coset boundaries defining growth increments in macroform. C) Granular sandstone unit with low-angle cross-stratification, formed on a bar top under upper-flow-regime conditions.

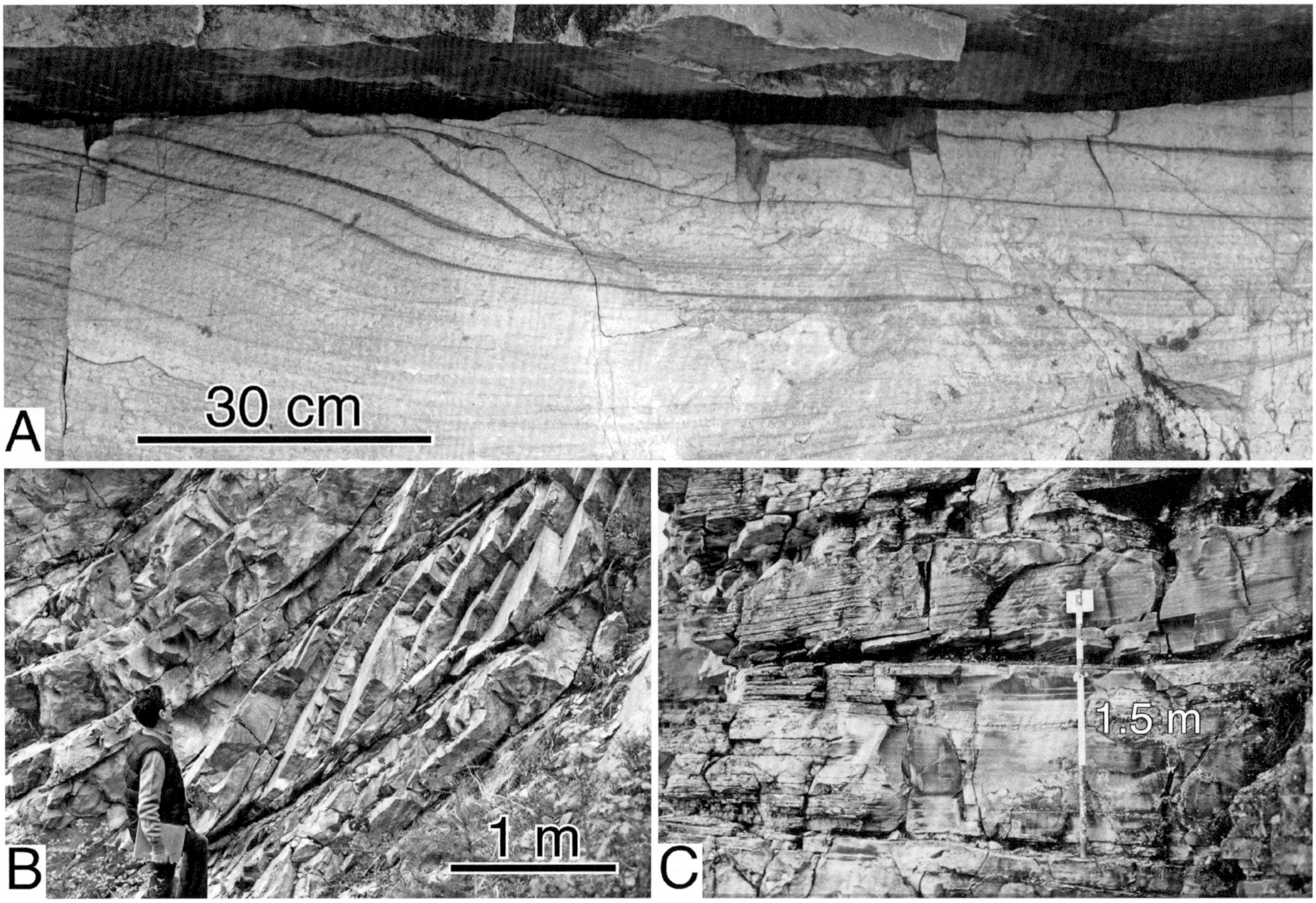

FIGURE 13.—High-energy sandy and sheetflood distal braided river deposits. A) Sinusoidal-bedded medium sandstone, deposited under transitional flow conditions, Unit K5, Neoproterozoic Katherine Group, Mount Eduni, N.W.T., Canada. B) Sigmoidal beds in the Mesoproterozoic Bonner Quartzite, Belt Supergroup, near Helena, Montana, indicate deposition from a sandy ephemeral system. C) SE element with low-angle cross-stratification, Katherine Group, Mackenzie Mountains, N.W.T. These units occur as SE, about a meter thick, and have current lineation which trends at a high angle to the maximum dip of the low-angle cross-stratification.

sandstone sets in the Basnaering Formation may also represent products of rapid deposition from high-energy braided rivers (Hjellbakk 1997).

Hollows elements (HO) appear to be exceptionally rare in pre-vegetation high-energy sandy braided river systems, suggesting that in-channel barforms were not prominent, or if present, were eroded by subsequent flood events before they had any significant influence on flow patterns.

O: Sheetflood Distal Braided Rivers

Miall's (1996) model of sheetflood distal braided rivers is based on observations of terminal river systems by Williams (1971). Miall (1996) suggests that this type of river should be dominated by sheets, lenses, and wedges of SB elements, with local flood cycles up to a few meters thick. Overbank deposits may be rare to absent. In modern dryland systems the presence of sparse vegetation tends to break up overland flows, generating turbulence in rills and small channels, which may eventually evolve into anabranching river systems (Tooth and Nanson 2004). As this would not have happened in pre-vegetation systems (even in semihumid conditions) sheet elements (SE) should predominate.

Pre-Devonian examples of this river type are common (Dam and Andreasen 1990; Bhattacharyya and Morad 1993; Eriksson et al. 1993; Martins-Neto 1994; Long 2002; Borba et al. 2004). Most are dominated by sheets (SE) of flat laminated sandstone, with minor ripple, planar, and trough cross-stratified sandstone. Some pass up section from planar laminated to ripple laminated, and may be capped with a thin layer of mudstone (Martins-Neto 1994; Long 2002), confirming deposition by waning flows.

P: Flashy Ephemeral Sheetflood Systems

Miall (1985) developed a model of flashy ephemeral sheetflood systems based to a large extent on descriptions of surficial deposits found in Bijou Creek, USA (McKee et al. 1967), which Ethridge (cited in Miall 1996) claimed were not representative of deposition at this site. The difference between this and the preceding model is that most deposition occurs predominantly from high-velocity flows under transitional and upper-flow-regime conditions. Possible modern examples include parts of the Finke river in central Australia (Williams

1971), and ephemeral streams in Colorado (Langford and Bracken 1987), Sudan (Abdullatif 1989), and Ethiopia (Billi 2008). In all of these examples thin cosets of Sh are common. Individual flood cycles may be recognizable with scour along the base and, where present, bank margins. Typically, sheet elements are thin (< 1 m), with high width-to-depth ratios. Channels are poorly defined or absent. In the presence of vegetation, high-energy, upper-flow-regime floods can be disrupted by obstacle-induced turbulence as the flow interacts with the stems of pioneer vegetation. This may lead to a dominance of ripples and small dunes on the tops of flood units.

Proterozoic examples (Callaghan et al. 1991; Long 2002, 2006) are typically developed as thin sheet elements, separated by discontinuous mudstone laminae (Fig. 14B) reflecting waning flow conditions. Where isolated pebbles and cobbles are present (Fig. 14 A), they confirm that flood events may have involved hyperconcentrated-flows that could transport material well in excess of the maximum size available (Billi 2008). Lateral continuity of individual SE elements is high (> 40:1 to > 100:1).

FLOODPLAIN DEPOSITS

Levee and splay deposits are rare in pre-vegetation systems, but these are not diagnostic, because they are also relatively uncommon in younger systems. Overbank suspension deposits (OF) are both thin and rare in pre-vegetation systems. Davies and Gibling (2010b) note that thick mudstone intervals form a significant component of fluvial deposits only after the development and spread of advanced rooted vascular plants in the early Silurian. In pre-vegetation systems OF facies may be preserved locally in interfluvial areas and abandoned channels (Rainbird 1992; Mazumder and Sarkar 2004; Rogala et al. 2007; Krapez et al. 2008). In distal fluvial settings, with exceptionally low gradients they may be preserved as extensive sheet elements up to 4 m thick (Knight and Jackson 1994; Sønderholm and Tirsgaard 1998).

FIGURE 14.—A) Thin cosets of thick planar-laminated, medium-grained sandstone in surface exposure of the late Paleoproterozoic Read Formation, Athabasca Group, Saskatchewan (Long 2006, 2007). Note isolated large clasts of quartzite that have been transported by upper-flow-regime flows. B) Thinly laminated sets of planar-laminated medium-grained sandstone (mSh) capped by laminated silty mudstone (zMh), similar to deposits seen at Bijou Creek (Miall 1996), Read Formation. C) Graded couplets of coarse- to fine-grained sandstone of possible levee origin, in Unit K5 of the Katherine Group, near Mt Eduni, N.W.T. Levee deposits dip at a low angle away from the base of overlying CH element. Note large intraclast (near staff) at base of channel. D) Possible overbank (or broad abandoned channel-fill) deposits, Katherine Group. Dark unit consists of poorly stratified silty mudstones (zMh) with thin discontinuous linsen of fine sandstone (fSh, fSr). Stratification is preserved except near the top of the coset, where it is disturbed by sand-filled desiccation cracks.

Although many post–middle Silurian floodplain deposits are dominated by massive (root-bioturbated) mudstones, siltstones, and ripple laminated sandstones (Davies and Gibling 2010b), pre-vegetation systems tend to have better preserved stratification (Fig. 14C, D), with small-scale upwards fining couplets reflecting individual flood events, and minor horizons with sand-filled desiccation cracks. The presence of evaporite casts and pseudomorphs in some deposits may indicate dryland conditions (Rainbird 1992; Knight and Jackson 1994). Well developed pedogenic carbonate horizons indicate stable floodplain conditions which may have persisted for several hundred years (Rogala et al. 2007; Post and Long 2008), and may provide clues to mean annual precipitation (after correction for atmospheric CO_2: Royer 1999; Retallack 2005). Although eolian deposits might be expected to accumulate in floodplain settings prior to the advent of rooted plants, they appear to have a very low preservation potential. Where fine-grained gold particles are present, eolian conditions may be indicated by the presence of flat, discoid grains with flattened margins (toroids of Minter 2006).

CLIMATE

Direct comparison of pre-vegetation fluvial deposits with modern systems may provide some indication of paleoclimate (Leeder et al. 1998), but this is not reliable, because the rates of surface runoff and chemical weathering were significantly different (Schumm 1968). Even the presence of interbedded eolian strata is not in itself a reliable climatic indicator, in as much as it has been suggested that these may have developed in a broader range of climatic settings than at present, and that preservation was selectively biased towards humid systems with rising water tables and high rates of creation of accommodation space (Tirsgaard and Øxnevad 1998; Eriksson et al. 1998). A modern example of this has been recorded in a cold-climate setting on low points on sandur plains in Iceland, where preservation is directly controlled by the position of the groundwater table (Mountney and Russell 2004). Possible climatic conditions may be indicated using paleo-continental reconstructions and other lithological indicators (Maslov et al. 2003). The composition of the conglomerates and sandstones may also provide some indication of climate. For example high quartz content (resistate pebbles) indicate intense weathering typically associated with warm to temperate climates and intense chemical weathering (Medaris et al. 2003), whereas polymict conglomerates (resistate plus labile clasts) indicate slower weathering rates, possibly associated with colder or dryer climatic conditions (Miall 1996). Local presence of caliche nodules, as seen in the Loon Lake Member of the Mesoproterozoic Pass Lake Formation, Sibley Group, northwestern Ontario (Rogala et al. 2007), and the Silurian Holmestrand Formation in Norway (Dam and Andreasen 1990), may indicate arid climatic conditions, whereas abundant mudstone intraclasts suggest rapid erosion and deposition in arid to semiarid climatic conditions as mud clasts are typically broken down within a few hundred meters in perennial rivers (Smith 1972).

An alternative to paleogeographic and lithostratigraphic modeling is to examine the chemical index of alteration (CIA index of Nesbitt and Young 1982) of specific lithotypes (high values in sandstones indicate intense weathering), although this is influenced to some degree by potential inheritance of the chemical characteristics of the provenance area (Maslov et al. 2003).

CONCLUSIONS

Architectural analysis of pre–late Silurian fluvial systems provides the clearest approach to understanding the geometry of early river systems. Using this method it is possible to identify examples of 12 of the 16 river types suggested by Miall (1996). The absence of gravel-bed wandering and low-sinuosity braided systems with alternate bars (Models D and K in Fig. 2) may be a function of the comparative rarity of these fluvial types, or may point to a fundamental difference between ancient and modern systems. The apparent absence of proven examples of fine-grained meandering systems (Model I) supports Sønderholm and Tirsgaard's (1998) contention that these cannot develop in the absence of rooted land plants. This may also explain the apparent underrepresentation of levee and splay deposits that would require plant stabilization for vertical accretion. The absence of bank stabilization by plant roots may also explain the absence of anabranching systems (Model J), The paucity of overbank fines in all fluvial systems developed before the advent of rooted land plants is probably due to a combination of factors, such as low production of primary fines and subsequent removal by eolian deflation.

The depositional products of most pre-vegetation fluvial systems appear to be fairly similar to those of modern braided and ephemeral systems, although individual river systems may have developed in a broader range of climatic settings than modern examples, and HO elements are very rare. Based on the sheetlike geometry of many pre-vegetation fluvial deposits, it is clear that on unconfined braid-plain systems, flood channels were significantly wider, with width-to-depth ratios from 200:1 to more than 1000:1 (Fuller 1985; Els 1990; Rainbird 1992). Most preserved Precambrian fluvial deposits are found as thick, relatively undisturbed successions in basins characterized by moderate rates of creation of accommodation space, such as along passive margins, and in foreland and intracratonic basins, and may have been associated with large river systems or amalgamated braidplains. In contrast most of the existing facies models for modern rivers are based on small Holocene river systems in intermountain and incised valleys, in humid temperate climatic settings associated with uplift and erosion (i.e., thin deposits with negative accommodation). Additional studies of modern rivers are needed in areas with active subsidence to determine if their deposits differ from existing models.

An unusual feature of many Precambrian sandy fluvial systems is the abundance of flat lamination, low-angle cross-stratification, and massive beds, suggestive of upper-flow-regime conditions and hyper-concentrated-flows (Bhattacharyya and Morad 1993; Simpson and Eriksson 1993; Eriksson et al. 1993; Hjellbakk 1993, 1997; Sønderholm and Tirsgaard 1998; Tirsgaard and Øxnevad 1998; Dilliard et al. 1999; Long 2002). Flat lamination is not confined to the tops of larger bedforms, as in Platte- and South Saskatchewan-type rivers: Martins-Neto (1994) suggested that the abundance of planar-laminated sandstone at the bases of SE elements may reflect development of upper-flow-regime conditions at the onset of flooding in ephemeral settings. Røe (1987), Røe and Hermansen (1993), and Hjellbakk (1997) have described extensive sequences dominated by sigmoidal bedded sandstones, passing into flat-laminated and low-angle cross-stratified sandstones. These probably formed at the transition from lower- to upper-flow-regime conditions, and may be a distinctive feature of sheetfloods in pre-vegetation fluvial systems. Abundant mudstone clasts in these and other pre-vegetation fluvial deposits suggest that fines may have been deposited in channel thalwegs and on floodplain surfaces during falling flood stage but were eroded by subsequent flows.

Pre-vegetation meandering systems can be identified by detailed architectural analysis, where systematic upsection changes in flow direction can be seen in SB elements, with flow at a high angle to the slope of associated LA elements. Most documented cases show little to no evidence of fining up section or of fine-grained overbank deposits, leading Sønderholm and Tirsgaard (1998) to suggest that even in low-gradient settings, sandy–muddy systems will tend to braid rather than meander.

Most modern gravelly and sandy braided rivers are characterized by deep scour hollows at flow confluences. These can be as much as two (Parsons et al. 2007) to five (Best 1988; Best and Ashworth 1997; Best et al. 2003) times the average channel depth. Similar results were obtained in flume studies by Gran and Paola (2001) who found that the presence of vegetation increased scour depth at confluences to as much as 6.4 times the average channel depth. The absence of large-scale HO elements in pre-vegetation fluvial systems is enigmatic: it might be related to higher discharge variability and lower stability of in-channel macroforms. Very few Phanerozoic fluvial systems have been identified with clearly defined HO elements (Cowan 1991; Miall and Jones 2003). An alternative explanation may be that these features can be properly identified only when detailed paleoflow observations are made on laterally extensive exposures. This may also explain why they are not recorded more often in younger fluvial deposits, where they may have inadvertently been described as scour features or minor channels.

Detailed architectural analysis is essential for comparison of modern and ancient fluvial systems. Research on modern rivers should include measurement of the inclination and orientation of all channels and macroforms, so that these can be used to provide a better understanding of stream sinuosity, geometry, and hydrology on ancient systems. Studies of distributary river systems in foreland-basin and in craton interiors may provide further insight into the behavior of pre-vegetation fluvial systems. Detailed petrographic and geochemical observations should be made to determine if these could be used to evaluate climate.

ACKNOWLEDGMENTS

Martin Gibling, Nigel Mountney, and Colin North are thanked for their constructive comments on earlier versions of this paper. The National Research Council of Canada is thanked for providing funding through the Discovery Grant Program. Additional financial and logistical support was provided by the Geological Survey of Canada, the Nunavut Geoscience Office, and the Yukon and N.W.T. Geological Surveys.

REFERENCES

ABDULLATIF, O.M., 1989, Channel-fill and sheet-flood facies sequences in the ephemeral terminal River Gash, Kassala, Sudan: Sedimentary Geology, v. 63, p. 171–184.

ALLEN, J.R.L., 1983, Sudies in fluviatile sedimentation: bars, bar complexes and sandstone sheets (low sinuosity braided streams) in the Brownstones (L. Devonian), Welsh Borders: Sedimentary Geology, v. 33, p. 237–293.

ALMEIDA, R.P., JANIKLAN, L., FRAGOSO-CESAR, A.R.S., AND MARCONATO, A., 2009, Evolution of a rift basin dominated by subaerial deposits: the Guaritas Rift, Early Cambrian, southern Brazil: Sedimentary Geology, v. 217, p. 30–51.

AMIREH, B.S., SCHNEIDER, W., AND ABED, A.M., 1994, Evolving fluvial–transitional-marine deposition through the Cambrian sequence of Jordan: Sedimentary Geology, v. 89, p. 65–90.

ASHMORE, P., AND CHURCH, M., 2001, The impact of climate change on rivers and river processes in Canada: Geological Survey of Canada, Bulletin 555, 58 p.

ASPLER, L.B., CHIARENZELLI, J.R., AND COUSENS, B.L., 2004, Fluvial, lacustrine, and volcanic sedimentation in the Angikuni sub-basin, and initiation of ~1.84–1.79 Ga Baker Lake Basin, Western Churchill Province, Nunavut, Canada: Precambrian Research, v. 129, p. 225–250.

BEST, J.L., 1988, Sediment transport and bed morphology at river channel confluences: Sedimentology, v. 35, p. 481–489.

BEST, J.L., AND ASHWORTH, P.J., 1997, Scour in large braided rivers and the recognition of sequence stratigraphic boundaries. Nature, v. 387, p. 275–277.

BEST, J.L., ASHWORTH, P.J., BRISTOW, C.S., AND RODEN, J., 2003, Three dimensional sedimentary architecture of a large, mid-channel bar, Jamuna River, Bangladesh: Journal of Sedimentary Research, v. 73, p. 516–530.

BHATTACHARYYA, A., AND MORAD, S., 1993, Proterozoic braided ephemeral fluvial deposits; an example from the Dhandraul Sandstone Formation of the Kaimur Group, Son Valley, central India: Sedimentary Geology, v. 84, p. 101–114.

BILLI, P., 2008, Bedforms and sediment transport processes in ephemeral streams of Kobo basin, Northern Ethiopia: Catena, v. 75, p. 5–17.

BILLI, P., MAGI, M., AND SAGRI, M., 1987, Coarse-grained low-sinuosity river deposits: example from Plio-Pleistocene Valdarno Basin, Italy, *in* Ethridge, F.G., Flores, R.M., and Harvey, M.D., eds., Recent Developments in Fluvial Sedimentology: SEPM, Special Publication 39, p. 197–203.

BLAIR, T.C., AND MCPHERSON, J.G., 1994, Alluvial fans and their natural distinction from rivers based on morphology, hydraulic processes, sedimentary processes and lithofacies assemblages. Journal of Sedimentary Research 64, p. 450–489.

BLUM, M.D., AND TÖRNQVIST, T.E., 2000, Fluvial responses to climate and sea-level change: a review and look forward: Sedimentology, v. 47 (Supplement 1), p. 2–48.

BORBA, A.W., AND MIZUSAKI, A.M.P., 2003, Santa Bárbara Formation (Caçapava do Sul, southern Brazil): depositional sequences and evolution of an Early Paleozoic postcollisional basin: Journal of South African Earth Sciences, v. 16, p. 365–380.

BORBA, A.W., ANDERSON J.M., AND PIMENTEL, A.M., 2004, Stratigraphic analysis and depositional evolution of the Neoproterozoic Maricá Formation (Southern Brazil): constraints from field data and sandstone petrography: Gondwana Research, v. 7, p. 871–886.

BOSE, P.K., SARKAR, S., MUKHOPADHYAY, S., SAHA, B., AND ERIKSSON, P., 2008, Precambrian basin-margin fan deposits: Mesoproterozoic Bagalkot Group, India: Precambrian Research, v. 162, p. 264–283.

BRIDGE, J.S., 1993, Description and interpretation of fluvial deposits: a critical perspective: Sedimentology, v. 40, p. 801–810.

BUATOIS, L.A., AND MÁNGANO, M.G., 2003, Sedimentary facies, depositional evolution of Upper Cambrian–Lower Ordovician Santa Rosita Formation in northwest Argentina: Journal of South African Earth Sciences, v. 16, p. 343–363.

BUCK, S.G., AND MINTER, W.E.L., 1985, Placer formation by fluvial degradation of an alluvial fan sequence: the Proterozoic Carbon Leader Placer, Witwatersrand Supergroup, South Africa: Geological Society of London, Journal, v. 142, p. 757–764.

BUMBY, A.J., ERIKSSON, P.G., AND VAN DER MEREW, R., 2004, The Early Proterozoic sedimentary record in the Blouberg area, Limpopo Province, South Africa: implications for the timing of the Limpopo orogenic event: Journal of African Earth Sciences, v. 39, p. 123–131.

CAIN, S.A., AND MOUNTNEY, N.P., 2009, Spatial and temporal evolution of a terminal fluvial fan system: the Permian Organ Rock Formation, South-east Utah, USA: Sedimentology, v. 56, p. 1774–1800.

CALLAGHAN, C.C., ERIKSSON, P.G., AND SNYMAN, C.P., 1991, The sedimentology of the Waterberg group in the Transvaal, South Africa: an overview: Journal of African Earth Sciences, v. 13, p. 121–139.

CAMACHO, A., HENSON, B.J., AND ARMSTRONG, R., 2002, Isotopic test of a thermally driven intraplate orogenic model, Australia: Geology, v. 30, p. 887–890.

CHAKRABORTY, P.P., SARKAR, A., DAS, K., AND DAS, P., 2009, Alluvial fan to storm-dominated shelf transition in the Mesoproterozoic Singhora Group, Chattisgarh Supergroup, Central India: Precambrian Research, v. 170, p. 88–106.

CHAKRABORTY, T., 1999, Reconstruction of fluvial bars from the Proterozoic Manchaeral Quartzite, Pranhita–Godavari Valley, India, *in* Smith, N.D., and Rogers, J., eds., Fluvial Sedimentology IV, International Association of Sedimentologists, Special Publication 28, p. 451–466.

CHURCH, M., 1983, Pattern of instability in a wandering gravel-bed channel, *in* Collinson, J.D., and Lewin, J., eds., Modern and Ancient Fluvial Systems: International Association of Sedimentologists, Special Publication 6, p. 169–180.

CORCORAN, P.L., MUELLER, W.U., AND PADGHAM, W.A., 1999, Influence of tectonism and climate on lithofacies distribution and sandstone and conglomerate composition in the Archean Beaulieu Rapids Formation, Northwest Territories, Canada; Precambrian Research, v. 94, p. 175–204.

COSTA, J.E., 1983, Paleohydraulic reconstruction of flash-flood peaks from boulder deposits in the Colorado Front Range: Geological Society of America, Bulletin, v. 94, p. 986-1004.

COTTER, E., 1978, The evolution of fluvial style, with special reference to the central Appalachians Paleozoic, *in* Miall, A.D., ed., Fluvial Sedimentology: Canadian Society of Petroleum Geologists, Memoir 5, p. 361–384.

COWAN, E.J., 1991, The large-scale architecture of the fluvial Westwater Canyon Member, Morrison Formation (Jurassic), San Juan Basin, New Mexico, *in*

Miall, A.D., and Tyler, N., eds., The Three-Dimensional Facies Architecture of Terrigenous Clastic Sediments, and Its Implications for Hydrocarbon Discovery and Recovery: SEPM, Contributions to Sedimentology and Paleontology, v. 3, p. 80–93.

Cox, R., Martin, M.W., Comstock, J.C., Dickerson, L.S., Ekstrom, I.L., and Sammons, J.H., 2002, Sedimentology, stratigraphy, and geochronology of the Proterozoic Mazatzal Group, central Arizona: Geological Society of America, Bulletin, v. 114, p. 1535–1549.

Dam, G., and Andreasen, F., 1990, High-energy ephemeral stream deltas: an example from the Upper Silurian Holmestrand Formation of the Oslo Region, Norway: Sedimentary Geology, v. 66, p. 197–225.

Davies, N.L., and Gibling, M.R., 2010a, Paleozoic vegetation and the Siluro-Devonian rise of fluvial lateral accretion sets: Geology, v. 38, p. 51–54.

Davies, N.L., and Gibling, M.R., 2010b, Cambrian to Devonian evolution of alluvial systems: the sedimentological impact of the earliest land plants: Earth Science Reviews, v. 98, p. 171–200.

Dilliard, K.A., Simpson, E.L., Noto, R.C., and Wizevich, M., 1999, Characterization of fluvial deposits interbedded with flood basalts, Neoproterozoic Catochin Formation, Central Appalachians, USA: Precambrian Research, v. 97, p. 115–134.

Donaldson, J.A., and de Kemp, E.A., 1998, Archaean quartz arenites in the Canadian Shield; examples from the Superior and Churchill provinces: Sedimentary Geology, v. 120, p. 153–176.

Duller, R.A., Mountney, N.P., Russell, A.J. and Cassidy, N.J., 2008, Architectural analysis of a volcaniclastic jökulhlaup deposit, southern Iceland: sedimentary evidence for supercritical flow: Sedimentology, v. 55, 939–964.

Els, B.G., 1990, Determination of some palaeohydraulic parameters for a fluvial Witwatersrand succession: South African Journal of Geology, v. 93, p. 531–537.

Els, B.G., 1998, The auriferous late Archean sedimentation systems of South Africa: unique paleo-environmental conditions: Sedimentary Geology, v. 120, p. 205–224.

Eriksson, P.G., Bumby, A.J., Brümer, J.J., and Van der Neut, M., 2006, Precambrian fluvial deposits: enigmatic paleohydrological data from the c. 2–1.9 Ga Waterberg Group, South Africa: Sedimentary Geology, v. 190, p. 25–46.

Eriksson, P.G., Condie, K.C., Tirsgaard, H., Mueller, W.U., Altermann, W., Miall, A.D., Aspler, L.A., Catuneanu, O., and Chiarenzelli, J.R., 1998, Precambrian clastic sedimentation systems: Sedimentary Geology, v. 120, p. 5–53.

Eriksson, P., Long, D., Bumby, A., Eriksson, K., Simpson, E., Catuneanu, O., Claassen, M., Mtimkulu, M., Mudziri, K., Brümer, J., and Van der Neut, M., 2008, Paleohydrological data from the c 2.0 to 1.8 Ga Waterberg Group, South Africa: a discussion of a possibly unique Paleoproterozoic fluvial style: South African Journal of Geology, v. 111, p. 281–304.

Eriksson, P.G., Schreiber, U.M., Van der Neut, M., Labuschagne, H., Van der Schyff, W., and Potgieter, G., 1993, Alternative marine and fluvial models for the non-fossiliferous quartzitic sandstones of the Early Proterozoic Daspoot Formation, Transvaal Sequence of southern Africa: Journal of African Earth Sciences, v. 16, p. 355–366.

Eriksson, P.G., and Simpson, E.L., 1993, Siliciclastic braided-alluvial sediments interbedded with continental flood basalts in the Early to Middle Proterozoic Mount Isa Inlier, Australia, *in* Marzo, M., and Puigdefábregas, C., eds., Alluvial Sedimentation: International Association of Sedimentologists, Special Publication 17, p. 473–488.

Ethridge, F.G., Tyler, N., and Burns, L.K., 1984, Sedimentology of a Precambrian quartz pebble conglomerate, southwest Colorado, *in* Koster, E.H., and Steel, R.S., eds., Sedimentology of Gravels and Conglomerates. Canadian Society of Petroleum Geologists, Memoir 10, p. 165–174.

Fedo, C.M., and Prave, A.R., 1991, Extensive Cambrian braidplain sedimentation: insights from southwestern U.S.A. Cordillera, *in* Cooper, J.D., and Stevens, C.H., eds., Paleozoic Paleogeography of the Western United States—II: SEPM, Pacific Section, no. 67, p. 227–235.

Ferguson, R.I., 1977, Meander sinuosity and direction variance: Geological Society of America, Bulletin, v. 88, p. 212–214.

Fielding, C.R., 2006, Upper flow regime sheets, lenses and scour fills: extending the range of architectural elements for fluvial sand bodies: Sedimentary Geology, v. 190, 227–240.

Fielding, C.R., Allen, J.P., Alexander, J., and Gibling, M.R., 2009, Facies model for fluvial systems in the seasonal tropics and subtropics: Geology, v. 37, 623–626.

Forbes D.L., 1983, Morphology and sedimentology of a sinuous gravel-bed channel system: Lower Babbage River, Yukon coastal plain, Canada, *in* Collinson, J.D., and Lewin, J., eds., Modern and Ancient Fluvial Systems: International Association of Sedimentologists, Special Publication 6, p. 195–206.

Friend, P.F., 1978, Distinctive features of some ancient river systems, *in* Miall, A.D., ed., Fluvial Sedimentology: Canadian Society of Petroleum Geologists, Memoir 5, p. 531–542.

Fuller, A.O., 1985, A contribution to the conceptual modeling of pre-Devonian fluvial systems: Geological Society of South Africa, Transactions, v. 88, p. 189–194.

Ghosh, P., 2000, Estimation of channel sinuosity from paleocurrent data: a method using fractal geometry: Journal of Sedimentary Research, v. 70, p. 449–455.

Gran, K., and Paola, C., 2001, Riparian vegetation controls on braided stream dynamics: Water Resources Research, v. 37, p 3275–3283.

Gutzmor, J., and Beukes, N.J., 1998, Earliest laterites and possible evidence for terrestrial vegetation in the Early Proterozoic: Geology, v. 26, p. 263–266.

Hadlari, T., and Rainbird, R.H., 2007, Tectonic accommodation and alluvial sequence stratigraphy of a paleoproterozoic continental rift, Baker Lake Basin, Canada: Stratigraphy, v. 3, p. 263–283.

Hadlari, T., Rainbird, R.H., and Donaldson, J.A., 2006, Alluvial, eolian and lacustrine sedimentology of a Paleoproterozoic half-graben, Baker Lake Basin, Nunavut, Canada: Sedimentary Geology, v. 190, p. 47–70.

Hartley, A.J., Weissmann, G.G., Nichols, G.J., and Warwick, G.L., 2010. Large distributive fluvial systems; characteristics, distribution and controls on development: Journal of Sedimentary Research, v. 80, p. 167–183.

Hashimoto, A., Oguchi, T., Hayakawa, Y., Lin, Zhou, Saito, K., and Waskelewicz, T.A., 2008, GIS analysis of depositional slope change at alluvial-fan toes in Japan and the American southwest: Geomorphology, v. 100, p. 120–130.

Hatch, J.R., and Morley, G.B., 1985, Hydrocarbon source rock evaluation of Middle Proterozoic Solor Church Formation, North American Mid-Continental Rift System, Rice County, Minnesota: American Association of Petroleum Geologists, Bulletin, v. 69, p. 1208–1216.

Hessler, A.M., and Lowe, D.R., 2006, Weathering and sediment generation in the Archean: an integrated study of the evolution of siliciclastic sedimentary rocks of the 3.2 Ga Moodies Group, Barberton Greenstone belt, South Africa: Precambrian Research, v. 151, 185–210.

Hjellbakk, A., 1993, A 'flash-flood' dominated braid delta in the upper Proterozoic Næringselva Member, Varanger Peninsula, northern Norway. Norsk Geologisk Tidsskrift, v. 73, p. 63–80.

Hjellbakk, A., 1997, Facies and fluvial architecture of a high-energy braided river: the upper Proterozoic Selodden Member, Varanger Peninsula, northern Norway: Sedimentary Geology, v. 114, p. 131–161.

Hocking, R.M., 1991, The Silurian Tumblagooda sandstone, Western Australia: Geological Survey of Western Australia, Report 27, 104 p.

Horodyski, R.J., and Knauth, L.P., 1994, Life on land in the Precambrian: Science, v. 263, p. 494–498.

Jutras, P.J., Quillan. R.S., and LeForte, M.J., 2009, Evidence from Middle Ordovician paleosols for the predominance of alkaline groundwater at the dawn of land plant radiation: Geology, v. 37, p. 91–94.

Karpeta, W.P., 1993, Sedimentology and gravel bar morphology in an Archean braided river sequence: the Witpan Conglomerate Member (Witwatersrand Supergroup) in the Welkom Goldfield, South Africa, *in* Best, J.L., and Bristow, C.S., eds., Braided Rivers: Geological Society of London, Special Publication 75, p. 369–388.

Kasting, J.F., 2001, The rise of atmospheric oxygen: Science, v. 293, p. 819–820.

Khalifa, M.A., Soliman, H.E., and Wanas, H.A., 2006, The Cambrian Araba Formation in northeastern Egypt: facies and depositional environments: Journal of Asian Earth Sciences, v. 27, p. 873–884.

Knight, R.D., and Jackson, G.D., 1994, Sedimentology and stratigraphy of the

Mesoproterozoic Elwin Subgroup (Aqigilik and Sinasiuvik Formations), uppermost Bylot Supergroup, northern Baffin Island. Geological Survey of Canada, Bulletin 445, 43 p.

Krapez, B., 1985, The Ventersdorp contact placer: a gold–pyrite placer of stream and debris flow origins from the Archean Witwatersrand Basin of South Africa: Sedimentology, v. 32, p. 223–234.

Krapez, B., Barley, M.E., and Brown, S.J.A., 2008, Late Archean synorogenic basins of the Eastern Goldfields superterrane, Yilgarn Craton, Western Australia. Part 1. Kalgoorlie and Gindalbie terranes: Precambrian Research, v. 161, p. 135–153.

Langford, R., and Bracken, B., 1987, Medano Creek, Colorado, a model for upper-flow-regime fluvial deposition: Journal of Sedimentary Petrology, v. 57, p. 863–870.

Laronne, J.B., and Shlomi, Y., 2007, Depositional character and preservation potential of coarse-grained sediments deposited by flood events in hyper-arid braided channels in the Rift Valley, Arava, Israel: Sedimentary Geology, v. 195, p. 21–37.

Leeder, M.R., 1973, Fluviatile fining-upward cycles and magnitude of palaeochannels: Geological Magazine, v. 110, p. 265–276.

Leeder, M.R., Harris, T., and Kirby, M.J., 1998, Sediment supply and climate change: implications for basin stratigraphy: Basin Research, v. 10, p. 7–18.

Long, D.G.F., 1978, Proterozoic stream deposits: some problems of recognition and interpretation of ancient sandy fluvial systems, *in* Miall, A.D., ed., Fluvial Sedimentology: Canadian Society of Petroleum Geologists, Memoir 5, p. 313–341.

Long, D.G.F., 1987, Sedimentary framework of uranium deposits in the southern Cobalt Embayment, Ontario Canada, *in* Pretorius, D., ed., Uranium Deposits in Proterozoic Quartz-Pebble Conglomerates: International Atomic Energy Agency, Vienna, TECDOC-427, p. 155–188.

Long, D.G.F., 2001, Architecture of Late Paleoproterozoic Fluvial Strata in the Manitou Falls Formation at McClean Lake and Key Lake, Eastern Athabasca Basin, Northern Saskatchewan: Saskatchewan Energy and Mines, Geological Survey, Miscellaneous Report 2001–2 (CD).

Long, D.G.F., 2002, Aspects of Late Paleoproterozoic fluvial style: the Uairén Formation. Roraima Supergroup, Venezuela, *in* Alterman, W., and Corcoran, P., eds., Precambrian Sedimentary Environments: A Modern Approach to Ancient Depositional Systems: International Association of Sedimentologists, Special Publication 33, p. 323–338.

Long, D.G.F., 2004, Precambrian Rivers. Chapter 7.8 *in* Eriksson, P., Altermann, W., Nelson, D., Mueller, W., Catuneanu, O., and Strand, K., eds., The Precambrian Earth; Tempos and Events: Elsevier. p. 660–663.

Long, D.G.F., 2006, Architecture of pre-vegetation sandy-braided perennial and ephemeral river deposits in the Paleoproterozoic Athabasca Group, northern Saskatchewan, Canada as indicators of Precambrian fluvial style: Sedimentary Geology, v. 190, p. 71–95.

Long, D.G.F., 2007, Topographic influences on the sedimentology of the Manitou Falls Formation, eastern Athabasca Basin, Saskatchewan, *in* Jefferson, C.W., and Delaney, G., eds., EXTECH IV: Geology and Uranium EXploration TECHnology of the Proterozoic Athabasca Basin, Saskatchewan and Alberta: Geological Survey of Canada, Bulletin 588, p. 267–280.

Long, D.G.F., and Donaldson, J.A., 2005, Modern and ancient clastic sedimentary environments: a collection of field photographs: Geological Association of Canada CSRG-CD 1.

Long, D.G.F., Lowey, G., and Sweet. A.R., 2000, Age and setting of dinosaur track-ways, Ross River, Yukon Territory (NTS 105F/15), *in* Emond, D.D., and Weston, L.H., eds., Yukon Exploration and Geology 2000: Exploration and Geological Services Division, Yukon Region. Indian and Northern Affairs Canada, p. 181–198.

Lunt, I.A., Bridge, J.S., and Tye, R.S., 2004, A quantitative, three-dimensional depositional model of gravelly braided rivers: Sedimentology, v. 51, p. 377–414.

Maizels, J.K., 1993, Lithofacies variations within sandur deposits: the role of runoff regime, flow dynamics and sediment supply characteristics: Sedimentary Geology, v. 85, p. 299–325.

Martini, J.E.J., 1994, A Late Archean–Paleoproterozoic (2.6 Ga) paleosol on ultramafics in the eastern Transvaal, South Africa: Precambrian Research, v. 67, p. 159–180.

Martins-Neto, M.A., 1994, Braidplain sedimentation in a Proterozoic rift basin: the São João da Chapada Formation, southeastern Brazil: Sedimentary Geology, v. 89, p. 219–239.

Maslov, A.V., Krupenin, M.T., and Gareev, E.Z., 2003, Lithological, lithochemical, and geochemical indicators of paleoclimate: evidence from Riphean of the southern Urals: Lithology and Mineral Resources, v. 38, p. 427–446.

Mazumder, R., and Sarkar, S., 2004. Sedimentation history of the Paleoproterozoic Dhanjori Formation, Singhbhum, eastern India. Precambrian Research 130, p. 267–287.

McKee, E.D., Crosby, E.J., and Berryhill, H.L.,Jr., 1967, Flood deposits, Bijou Creek, Colorado, June 1965: Journal of Sedimentary Petrology, v. 37, p. 829–851.

Medaris, L.G., Jr., Singer, B.S., Dott, R.H., Naymark, A., Johnson, C.M., and Schott, R.C., 2003, Late Paleoproterozoic climate, tectonics, and metamorphism in the southern Lake Superior region and proto-North America: evidence from the Baraboo interval quartzites: Journal of Geology, v. 111, p. 243–257.

Merk, G.P., and Jirsa, M., 1982, Provenance and tectonic significance of Keweenawan interlow sedimentary rocks, *in* Wold, R.J., and Hinze, W.J., eds., Geology and tectonics of the Lake Superior basin: Geological Society of America, Memoir 156, p. 97–105.

Miall, A.D., 1976, Palaeocurrent and palaeohyrologic analysis of some vertical profiles through a Cretaceous braided stream deposits, Banks Island, Arctic Canada: Sedimentology, v. 23, p. 459–483.

Miall, A.D., 1977, A review of the braided river depositional environment: Earth-Science Reviews, v. 13, p. 1–62.

Miall, A.D., 1978, Lithofacies types and vertical profile models in braided river deposits: a summary, *in* Miall, A.D., ed., Fluvial Sedimentology: Canadian Society of Petroleum Geologists, Memoir 5, p. 597–604.

Miall, A.D., 1985, Architectural-element analysis; a new method of facies analysis applied to fluvial deposits: Earth-Science Reviews, v. 22, p. 261–308.

Miall, A.D., 1988, Facies architecture in clastic sedimentary basins. Chapter 4, *in* Kleinspehn, K.L., and Paola, C., New Perspectives in Basin Analysis: New York, Springer-Verlag, p. 67–81.

Miall, A.D., 1996, The Geology of Fluvial Deposits: Sedimentary Facies, Basin Analysis and Petroleum Geology: Berlin, Springer-Verlag, 582 p.

Miall, A.D., and Bridge, J.S., 1995, Description and interpretation of fluvial deposits: a critical perspective. Discussion and Reply: Sedimentology, v. 42, 379–389.

Miall, A.D., and Jones, B.G., 2003, Fluvial architecture of the Hawksbury Sandstone (Triassic), near Sydney, Australia: Journal of Sedimentary Research, v. 73, p. 531–545.

Minter, W.E.L., 2006, The sedimentary setting of Witwatersrand placer mineral deposits in an Archean atmosphere, *in* Kesler, S.E., and Ohmoto, H., Evolution of Earth's Atmosphere, Hydrosphere, and Biosphere: Constraints from Ore Deposits: Geological Society of America, Memoir 198, p. 105–119.

Mountney, N.P., and Russell, A.J., 2004, Sedimentology of cold-climate aeolian sandsheet deposits in the Askja region of northeast Iceland: Sedimentary Geology, v. 166, p. 223–244.

Morley, G.B., 1967, Stratigraphy and petrology of the type Fond du Lac Formation, Duluth, Minnesota: Minnesota Geological Survey, Report of Investigations 7, 35 p.

Morley, G.B., and Ojakangas, R.W., 1982, Keweenawan sedimentary rocks of eastern Minnesota and northwestern Wisconsin, *in* Wold, R.J., and Hinze, W.J., eds., Geology and Tectonics of the Lake Superior basin: Geological Society of America, Memoir 156, p. 135–146.

Nanson, G.C., and Knighton, A.D., 1996, Anabranching rivers: their cause, character and classification: Earth Surface Processes and Landforms, v. 21, p. 217–239.

Nesbitt, H.W., and Young, G.M. 1982, Early Proterozoic climates and plate motions inferred from major element chemistry of lutites: Nature, v. 299, p. 715–717.

Nichols, G.J., and Fisher, J.A., 2007, Processes, facies and architecture of fluvial distributary system deposits: Sedimentary Geology, v. 195, p. 75–90.

Nicholson, P.G., 1993, A basin reappraisal of the Proterozoic Torridon Group, northwest Scotland, *in* Frostick, L.E., and Steel, R.J., Tectonic Controlls and

Signatures in Sedimentary Successions: International Association of Sedimentologists, Special Publication 20, p. 183–202.

Noffke, N., Gerdes, G., Klenke, T., and Krumbein, W.E., 2001, Microbially induced sedimentary structures—A new category within the classification of primary sedimentary structures: Journal of Sedimentary Research, v. 71, p. 649–656.

North, C.P., Nanson, G.C., and Fagan, S.D., 2007, Recognition of the sedimentary architecture of dryland anabranching (anastomosing) rivers: Journal of Sedimentary Research, v. 77, p. 925–938.

Ohmoto, H., 1996, Evidence in pre–2.2 Ga paleosols for the early evolution of atmospheric oxygen and terrestrial biota: Geology, v. 24, p. 1135–1138.

Ojakangas, R.W., 2005, Sedimentary fill of the 1100 Ma Midcontinent Rift System (MRS) in the Lake Superior region (abstract): Geological Society of America, Abstracts with Programs, v. 37 (5), p. 73.

Orton, G.J., and Reading, H.G., 1993, Variability of deltaic processes in terms of sediment supply, with particular emphasis on grain size: Sedimentology, v. 40, p. 475–512.

Parsons, D.R., Best, J.L., Lane, S.N., Orfero, O., Hardy, R.J., and Kostaschuk, R., 2007, Form roughness and the absence of secondary flow in a large confluence–diffluence, Rio Paraná, Argentina: Earth Surface Processes and Landforms, v. 32: p. 155–162.

Pflüger, F., and Gresse, P.G., 1996, Microbial sand-chips, a non-actualistic sedimentary structure: Sedimentary Geology, v. 102, p. 263–274.

Pflüger, F., and Seilacher, A., 1991, Flash flood conglomerates, *in* Einsele, G., Ricken, W., and Seilacher, A., eds., Cycles and Events in Stratigraphy: Berlin Springer-Verlag, p. 383–391.

Post, R.T., and Long, D.G.F., 2008, The Middle Cambrian Mount Roosevelt Formation (new) of northeastern British Columbia: Evidence for rifting and development of the Kechicka Graben System: Canadian Journal of Earth Sciences, v. 45, p. 483–498.

Prave, A.R., 2002, Life on land in the Proterozoic: Evidence from the Torridonian rocks of northwest Scotland: Geology, v. 30, p. 811–814.

Rainbird, R.H., 1992, Anatomy of a large-scale braid-plain quartzarenite from the Neoproterozoic Shaler Group, Victoria Island, Northwest Territories, Canada: Canadian Journal of Earth Sciences, v. 29, p. 2537–2550.

Rainbird, R.H., and Young, G.M., 2009, Colossal rivers and massive mountains and supercontinents: was everything bigger in the past?: Earth, v. 54, p. 52–61.

Rainbird, R.H., McNicoll, V.J., Thériault, R.J., Heaman, L.M., Abbot, J.G., Long, D.G.F., and Thorkelson, D.J., 1997, Pan-continental river system draining Grenville Orogen recorded by U-Pb and Sm-Nd geochronology of Neoproterozoic quartzarenites and mudrocks, northwestern Canada: Journal of Geology, v. 105, p. 1–17.

Räsänen, J., and Mäkelä, M., 1988, Early Proterozoic fluvial deposits in the Pyhätunturi area, northern Finland, *in* Laajoki, K., and Paakkola, J., eds., Sedimentology of the Precambrian Formations in Eastern and Northern Finland: Geological Survey of Finland, Special Paper 5, p. 239–254.

Retallack G.J., 2005, Pedogenic carbonate proxies for amount and seasonality of precipitation in plaeosols: Geology, v. 33, 333–336.

Roberts, M.C., Bravard, J-P. and Jol, H.M., 1997, Radar signatures and structure of an avulsed channel: Rhône River, Aoste, France: Journal of Quaternary Science, v. 12, p. 35–42.

Røe, S.-L., 1987, Cross-strata and bedforms of probable dune to upper-stage plane bed origin from a late Precambrian fluvial sandstone, northern Norway: Sedimentology, v. 34, p. 89–101.

Røe, S.-L., and Hermansen, M., 1993, Processes and products of large, Late Precambrian sandy rivers in northern Norway, *in* Marzo, M., and Puigdefábreagas, C., eds., Alluvial Sedimentation: International Association of Sedimentologists, Special Publication 17, p. 151–166.

Rogala, B., Fralick, P.W., Heaman, L.M., and Metsaranta, R., 2007, Lithostratgraphy and chemostratigraphy of the Mesoproterozoic Sibley Group, northwestern Ontario, Canada: Canadian Journal of Earth Sciences, v. 44, p. 1131–1149.

Royer, D.L., 1999, Depth to pedogenic carbonate horizon as a paleoprecipitation indicator?: Geology, v. 27, 1123–1126.

Rygel, M.C., Gibling, M.R., and Calder, J.H., 2004, Vegetation-induced sedimentary structures from fossil forests in the Pennsylvanian Joggins Formation, Nova Scotia. Sedimentology, v. 51, p. 531–552.

Saito, K., and Oguchi, T., 2005, Slope of alluvial fans in humid regions of Japan, Taiwan and the Philippines: Geomorphology, v. 70, p. 147–162.

Schieber, J., 1999, Microbial mats in terrigenous clastics: the challenge of identification in the rock record: Palaios, v. 14, p. 3–12.

Schumm, S.A., 1968, Speculations concerning paleohydrologic controls of terrestrial sedimentation: Geological Society of America, Bulletin , v.79, p. 1573–1588.

Schumm, S.A., 2005, River variability and complexity: Cambridge, Cambridge University Press, 220 p.

Shepherd, R.G., 1987, Lateral accretion surfaces in ephemeral-stream point bars, Rio Puerco, New Mexico, *in* Ethridge, F.G., and Harvey, M.D., eds., Recent Developments in Fluvial Sedimentology: SEPM, Special Publication 39, p. 93–98.

Siedlecka, A., and Edwards, M.B., 1980, Lithostratigraphy and sedimentation of the Riphean Båsnaering Formation, Veranger Peninsula, North Norway: Norges Geologiske Undersokelse, v. 355, p. 27–47.

Simpson, E.L., and Eriksson, K.A., 1993, Thin eolianites interbedded with fluvial and marine succession: Early Proterozoic Whitworth Formation, Mount Isa Inlier, Australia: Sedimentary Geology, v. 87, p. 39–62.

Simpson, E.L., Dilliard, K.A., Rowell, B.F., and Higgins, D., 2002, The fluvial-to-marine transition within post-rift Lower Cambrian Hardyston Formation, Eastern Pennsylvania, USA: Sedimentary Geology, v. 147, p. 127–142.

Singh, S.P., 1984, Fluvial sedimentation of the Proterozoic Alwar Group in the Lalgarh graben, northwestern India: Sedimentary Geology, v. 39, p. 95–119.

Smith, D.G., 1976, Effect of vegetation on lateral migration of anastomosed channels of a glacial meltwater river: Geological Society of America, Bulletin, v. 87, p. 857–860.

Smith, G.H.S., Ashworth, P.J., Best, J.L., Woodwards, J., and Simpson, C.J., 2006, The sedimentology and alluvial architecture of the sandy braided South Saskatchewan River, Canada: Sedimentology, v. 53, p. 413–434.

Smith, N.D., 1972, Flume experiments on the durability of mud clasts: Journal of Sedimentary Petrology, v. 42, p. 378–383.

Sønderholm, M., and Tirsgaard, H., 1998, Proterozoic fluvial styles: response to base-level changes and climate (Riveradal sandstones, eastern North Greenland): Sedimentary Geology, v. 120, p. 257–274.

Southwick, D.L., Morey, G.B., and Mossler, J.H., 1986, Fluvial origin of the lower Proterozoic Sioux Quartzite, southwestern Minnesota: Geological Society of America, Bulletin, v. 97, p. 1432–1441.

Sweet, I.P., 1988, Early Proterozoic stream deposits: braided or meandering—evidence from central Australia: Sedimentary Geology, v. 58, p. 277–293.

Sweet, I.P., 1993, Ayers Rock and the Olgas: Late Proterozoic or Early Cambrian alluvial fan deposits at the southern margin of the Amadeus Basin, central Australia (abstract), *in* Yu, B., and Fielding, C.R., eds., Modern and Ancient Rivers; Their Importance to Mankind: 5th International Conference on Fluvial Sedimentology; Conference Proceedings, Keynote Addresses and Abstracts, p. 129–130.

Tirsgaard, H., and Øxnevad, I.E.I., 1998, Preservation of pre-vegetational mixed fluvio-aeolian deposits in a humid climatic setting: an example from the Middle Proterozoic Eriksfjord Formation, Southwest Greenland: Sedimentary Geology, v. 120, p. 295–317.

Tooth, S., 2000, Process, form and change in dryland rivers: a review of recent research. Earth-Science Reviews, v. 51, p. 67–107.

Tooth S., and Nanson, G.C., 2004, Forms and processes of two highly contrasting rivers in arid central Australia, and the implications for channel-pattern discrimination and prediction: Geological Society of America, Bulletin, v. 116, p. 802–816.

Tucker, M.E., 2003, Sedimentary rocks in the field: Chichester, U.K., J. Wiley and Sons Ltd., 234 p.

Tyler, N., 1978, Early-Proterozoic meandering-stream sedimentation in the Buffalo Springs Group of the Transvaal Supergroup, West-central Transvaal: University of the Witwatersrand, Economic Geology Research Unit, Information Circular, no. 121, 8 p.

Van der Neut, M., and Eriksson, P.G., 1999, Paleohydrological parameters of a Proterozoic braided fluvial system (Wilgerivier Formation, Waterberg Group, South Africa) compared with a Phanerozoic example, *in* Smith, N.D., and Rogers, J., eds., Fluvial Sedimentology IV: International Association of Sedimentologists, Special Publication 28, p. 381–392.

Vos, R.G., and Tankard, A.J., 1981, Braided fluvial sedimentation in the lower

Paleozoic Cape Basin, South Africa: Sedimentary Geology, v. 29, p. 171–193.

Walter, M.R., Veevers, J.J., Calver, C.R., and Grey, K., 1995, Neoproterozoic stratigraphy of the Centralian Superbasin, Australia: Precambrian Research, v. 73, p. 173–195.

Weissmann, G.S., Hartley, A.J., Nichols, G.J., Scuderi, L.A., Olson, M., Buehler, H., and Banteah, R., 2009, The distributary fluvial system (DFS) paradigm: re-evaluating fluvial facies models based on observations from Modern continental sedimentary basins (abstract), *in* Davidson, S., and S Leleu, S., eds., From River to Rock Record, the Preservation of Fluvial Sediments and Their Subsequent Preservation, 12–14 January 2009, Department of Geology and Petroleum Geology, University of Aberdeen, Scotland, Programme and Abstracts, p. 90.

Weissmann, G.S., Hartley, A.J., Nichols, G.J., Scuderi, L.A., Olson, M., Buehler, H., and Banteah, R., 2010, Fluvial form in modern continental sedimentary basins: distributive fluvial systems. Geology, v. 38, p. 39–42.

Went, D.J., 2005, Pre-vegetation alluvial fan facies and processes: an example from the Cambro-Ordovician Rozel Conglomerate Formation, Jersey, Channel Islands: Sedimentology, v. 52, p. 693–713.

Williams, G.E., 1971, Flood deposits of the sand-bed ephemeral streams of central Australia: Sedimentology, v. 17, p. 1–40.

Williams, G.E., 2001, Neoproterozoic (Torridonian) alluvial fan succession, northwest Scotland, and its tectonic setting and provenance: Geological Magazine, v. 138, p. 471–494.

Williams, G.E., 2005, Subglacial meltwater channels and glaciofluvial deposits in the Kimberley Basin, Western Australia: 1.8 Ga low-latitude glaciation coeval with continental assembly: Geological Society of London, Journal, v. 162, p. 111–124.

Wooldridge, C.L., and Hickin, E.J., 2005, Radar architecture and evolution of channel bars in wandering gravel-bed rivers: Fraser and Squamish Rivers, British Columbia: Journal of Sedimentary Research, p.75, p. 844–860.

Yeo, G., Percival, J.B., Jefferson, C.W., Bernier, S., and Hunt, P., 2002, Clay oncoids and crypto-microbial lamination from the Late Paleoproterozoic Manitou Falls Formation, Athabasca Basin, Saskatchewan: Summary of Investigations, Saskatchewan Geological Survey, Saskatchewan Industry and Resources, Miscellaneous Report 2002-4.2. CD-ROM, Paper D9, 9 p.

Young, G.M., 1974, Stratigraphy, paleocurrents and stromatolites of Hadrynian (Upper Precambrian) rocks of Victoria Island, Arctic Archipelago, Canada. Precambrian Research, v. 1, p. 13–41.

DIGITAL APPENDIX

The following supplementary material is available in electronic format either on the DVD accompanying this volume or through the digital archive on the SEPM website at http://www.sepm.org/

Table S1: Macroform-scale architectural elements

Figure S1: Keweenawan (~ 1.1 Ga) channel-fill boulder and cobble conglomerate, Mamainse Point, Ontario. Solid red lines (3) indicate third-order (or higher) surfaces defining individual channel complexes; Bold green dashed lines (2) indicate LA surfaces; and yellow thin dashed lines (1) indicate set boundaries.

Figure S2: Architectural details of deep gravel-bed river deposits in the Uairén Formation. Dark unit is upper member; light unit is lower member. Gm, Gt, Gp = massive, trough, and planar cross-stratified conglomerate; SG = sandy conglomerate; GS = gravelly sandstone; St = trough cross-stratified sandstone.

Figure S3: Architecture of simple downstream (DA) and downstream lateral-accretion (DLA) deposits in a shallow sandy braided river deposit in the Paleoproterozoic Mississagi Formation, based on Long and Donaldson (2005). Arrows indicate paleo-flow from zeroth-order surfaces (foresets), pins indicate slope of set and coset boundaries. Roses on right show how slopes of cross-stratification (black) relate to set (dark gray) and coset (light gray) slopes.

Figure S4: Architecture of deep sand-bed braided river deposits in the Bird Member of the Manitou Falls Formation, based on Long (2006). Top section shows outcrop, surfaces and element orientations, and grouped slope data. Bottom shows individual channel-fill sequences with relative ordering of structures.

RIVER TERRACES IN THE ROCK RECORD: AN OVERLOOKED LANDFORM IN GEOLOGICAL INTERPRETATION?

STUART G. ARCHER
Department of Geology and Petroleum Geology, University of Aberdeen, Kings College, Old Aberdeen, Aberdeen, Scotland, UK, AB24 3UE
e-mail: s.archer@abdn.ac.uk

REYNIR FJALAR REYNISSON
Department of Geology and Petroleum Geology, University of Aberdeen, Kings College, Old Aberdeen, Aberdeen, Scotland, UK, AB24 3UE
Present address: Statoil, Trondheim, Norway

AND

ANNE M. SCHWAB
Department of Geology and Petroleum Geology, University of Aberdeen, Kings College, Old Aberdeen, Aberdeen, Scotland, UK, AB24 3UE
Present address: Marathon International Petroleum (G.B.) Limited, Aberdeen, Scotland, UK, AB15 6FZ

ABSTRACT: Fluvial incision and terracing produces a scale of architectural complexity that is often overlooked in geological interpretations and the construction of subsurface reservoir models. Results of a 2D seismic forward modelling study demonstrate the difficulties and limitations in resolving terraces seismically. We propose that terraced sequence boundaries fall readily into a subsurface data resolution gap between seismic and core to wireline log scale.

Due to limitations in seismic resolution, sequence boundaries are usually interpreted as simple, single surfaces of erosion, for example at the bases of incised-valley fills. However, modern analogues show that sequence boundaries are in fact often compound in nature. The stratigraphic response to pulses of incision and aggradation in an incision-dominant phase can result in flights of stair-stepping terraces. In this paper we recognize two distinct architectural styles: attached and detached terrace flights. Their formation depends on the interplay between the magnitudes of incision versus aggradation.

We suggest that terraced sedimentary architectures, the nature of the terrace-fill lithology, and any associated pedogenesis impacts upon mesoscale to macroscale reservoir permeabilities and therefore have implications for subsurface fluid flow. The existence of terraces, in conjunction with the inherent lack of temporal control in fluvial successions, may help to explain correlation difficulties often encountered and to an extent the compartmentalized production behavior of some fluvial hydrocarbon reservoirs.

KEY WORDS: River terraces, seismic forward modelling, negative accommodation, attached and detached terraces, compound sequence boundaries

INTRODUCTION

Since the work of Gilbert (1879), Penck and Brückner (1909), Mackin (1937), and Fisk (1944), many geomorphologists have recognized river terraces as common, integral parts of fluvial and alluvial landscapes (e.g. Robertson-Rintoul 1986; Stewart et al. 1986; Blum and Valastro 1994; Molnar et al. 1994; Huisink 1997; Antoine et al. 2000; Colombo et al. 2000; Maddy et al. 2001; Anderson et al. 2004; Bridgland et al. 2004; Bridgland and Westaway 2008; Kemp and Rhodes 2010). In contrast however, geologists working with ancient-record subsurface data rarely describe and interpret fluvial terraces (notable exceptions include Arnott et al. 2002 and Zaitlin et al. 2002).

Clearly, the low (geological time scale) preservation potential of erosion-dominated, denuding landscapes and geomorphic features such as terraces explains a large proportion of the discrepancy that exists between the documentation frequency of terraces in the geomorphological and geological literature. However, another potential reason may be that terraces are not easily identifiable in the subsurface and have gone underrecognized. The key question explored in this paper is: "what would it take to observe terraces seismically?" This question is addressed using a 2D seismic forward modelling approach. The results are central to the issue of the subsurface recognition of terraces, revealing why they are often overlooked. The paper goes on to demonstrate the potential importance of terraces to subsurface correlation efforts and illustrates the implications for subsurface fluid flow.

In this paper we present the geomorphic process of terracing in its sequence stratigraphic context, and in doing so we hope to stimulate the continuing debate surrounding the utility and practice of sequence stratigraphy in continental strata (Posamentier 1988; Shanley and McCabe 1994; Blum and Törnqvist 2000).

The Process of Terrace Formation

A generic model demonstrating the incisional and aggradational geomorphic process steps in the formation of a flight of terraces is presented in Figure 1. During fill phase A, no terraces exist—only floodplains and a river channel. As a direct result of incision (cut phase b), two terraces, A and A′ are formed. Floodplains become terraces once the river has incised sufficiently that, even at flood stage, flow does not overtop the terrace. Terraces are abandoned floodplains. Terraces are made up of two distinct morphologic elements, the

From River To Rock Record: The Preservation Of Fluvial Sediments And Their Subsequent Interpretation
SEPM Special Publication No. 97, Copyright © 2011
SEPM (Society for Sedimentary Geology), ISBN 978-1-56576-305-0, p. 63–85.

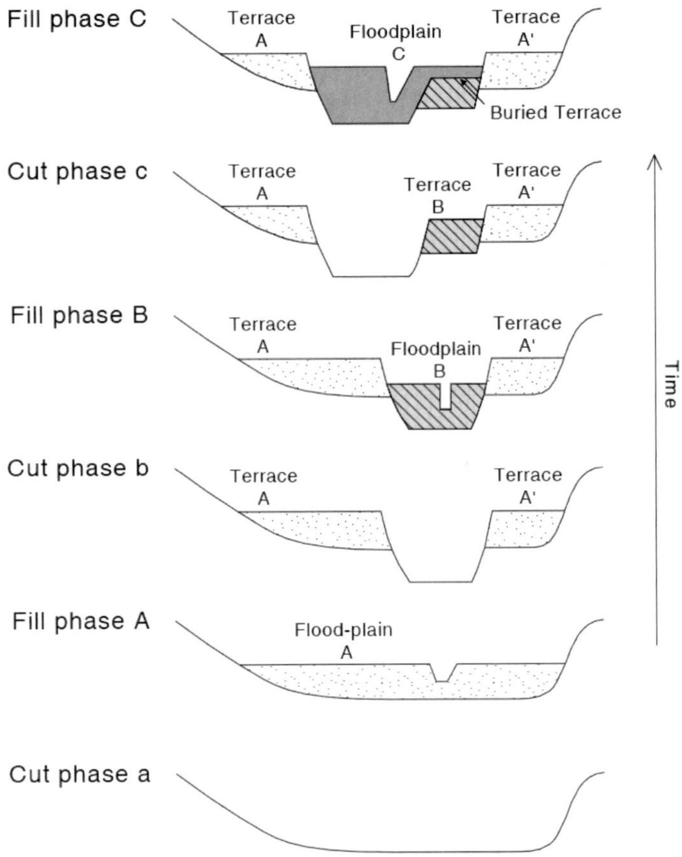

FIGURE 1.—A model of terrace development, modified from Lowe and Walker (1984).

incisional scarp slope and the flat terrace top or stair tread (Leopold, Wolman, and Miller 1964). The term terrace applies to the architectural element as a whole rather than either the scarp or the tread.

In Figure 1, terrace A represents a single aggradational terrace. Terraces A and A′ are paired terraces and terraces such as these, of the same age but physically disconnected in 2D, are defined in this paper as a terrace suite. Multiple terraces of different fill ages are defined here as flights of terraces (terraces A and B on the right of the valley). In contrast to terraces A and A′, terrace B has no cross-valley terrace-suite equivalent, due, at least in part, to the asymmetric erosional focus of cut phase c. Terrace B is then buried during aggradational fill phase C. Fill phase C can be described as overlapping terrace B, whereas fill phase C is inset relative to terrace suite A. Fill phase D overlaps terrace suite A.

From a geological perspective, the process of terrace formation can usefully be considered within a relative chronostratigraphic framework as demonstrated in Figure 2. Figure 2 illustrates that although the law of superposition is never broken, terracing does make for complicated and counterintuitive relationships between stratigraphic ages and height due to the incisional architecture. For example, terrace B is younger, yet topographically lower than terrace A. Floodplain C has a simple stratigraphic and topographic relationship with B—it is higher and younger, but although C is also younger than A, it is inset and lower topographically than A. So although the top of terrace A is closer in stratigraphic position to the top of fill phase C than terrace B, a larger time gap would obviously exist between A and C than between A and B. Added to this there is the sedimentological process complication that floodplain A could be active both during fill phase A and potentially also during the later stages of fill phase C and the early stages of fill phase D.

By providing this relative chronostratigraphic perspective on terrace formation, Figure 2 also demonstrates that the longest hiatus exists between the early part of fill phase C and the deepest eroded and thus oldest part of the original, pre-existing substrate. All other time gaps are of shorter duration.

Terraces and Their Association with Incised Valleys

Incised valleys have received much attention in recent decades, following the advent of sequence stratigraphy as a stratigraphic discipline in the late 1970s and 1980s (e.g., Blum 1993; Dalrymple et al. 1994; Schumm and Ethridge 1994; Shanley and McCabe 1994; Zaitlin et al. 1994; Davies et al. 1999). Although terrace formation is intimately related to the creation of incised valleys (the cut phase) there has been a historical preoccupation with the fill phase (see Figure 3). This is understandable due to the fact that incised-valley fills, and in particular the high net-to-gross, coarse-grained, braided-stream facies that often form the early fill (part A of Figure 3), have been important hydrocarbon reservoir targets (e.g., Dolson et al. 1991).

The criteria for terrace definition introduced earlier are the same as that applied to the formation of an incised valley as defined by Posamentier (2001). Terraces are defined as floodplains that lie above the highest flood stages, in line with Zaitlin et al. (1994), who define incised valleys as being typically larger than a single channel. We believe that the formation of terraces is implicit in the formation of an incised valley.

Compound incised-valley systems form where multiple valley-incision episodes erode into previous incised-valley-fill deposits (Zaitlin et al. 1994). Although compound incised valleys are recognized to exist in the geological record, we focus here on simple, single incised valleys.

Interfluves have often been interpreted at outcrop and in the subsurface (e.g., Wright and Marriott 1993; McCarthy and Plint 2003) and are a common part of sequence stratigraphic models in areas in between incised valleys. We believe that the historical usage of this interpretive term describing paleo-landscape may often be compatible with terraced geometries (or a terraced carapace to the interfluve area).

Controlling Mechanisms for the Formation of Terraces

Terraces ultimately owe their formation to incision, which can be induced by a range of geological factors. This review outlines the role of the three allogenic controls of climate, tectonics, and eustasy as potential triggers of incision since these controls dictate accommodation state, sediment supply, and discharge, and it is the interplay between these factors that governs fluvial stratigraphic architecture. It is within this sequence stratigraphic framework that the possible controlling mechanisms for terrace formation must be viewed.

Eustatic sea-level fall is one of the most commonly cited causal mechanisms to initiate incision on modern and ancient continental margins (Vail et al. 1977; Mitchum and Vail 1977; Van Wagoner et al. 1988; Posamentier and Vail 1988; Van Wagoner et al. 1990). Sequence stratigraphic thinking has evolved greatly in recent decades, and it is of course possible that tectonic uplift can cause relative sea-level fall and thus induce incision, independently of eustatic change (e.g., Hubbard 1988; Bishop 1991; Underhill and Partington 1993). Because rivers can respond to relative sea-level fall by lengthening their long profile and changing their planform character (Miall 1991; Schumm 1993; Wescott 1993), relative sea-level fall does not guarantee an incisive response unless the physiographic setting is favorable (i.e., the shelf is steeper than the coastal plain or the eustatic fall is sufficiently large to expose the shelf edge and steeper gradients associated with the upper slope).

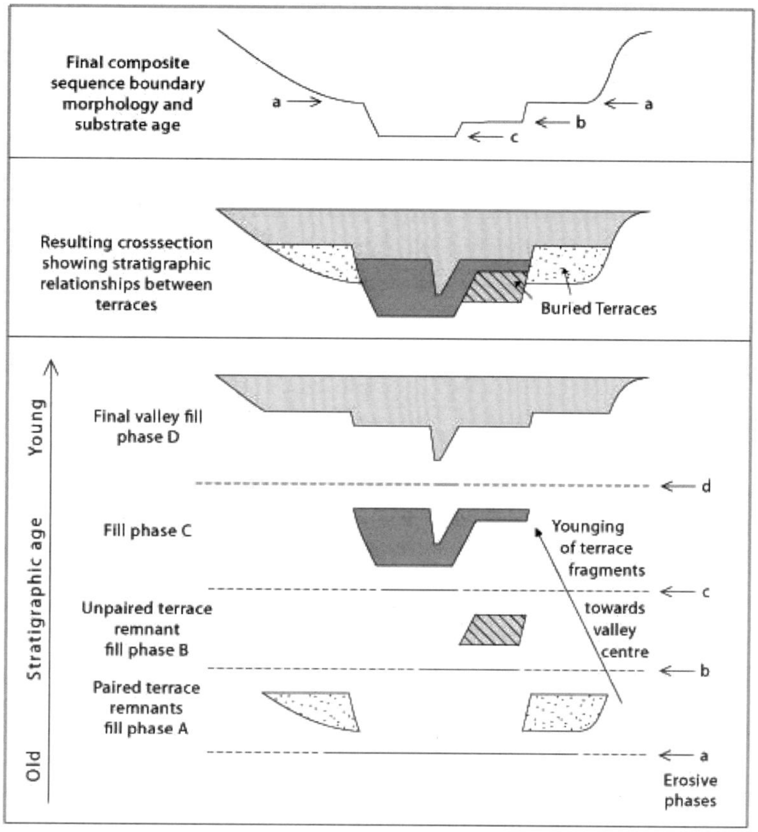

FIGURE 2.—Chronostratigraphic (Wheeler) diagram of the same flight of terraces seen in the model shown in Figure 1.

Up-dip of the effects of eustatic change, the allogenic variables reduce to tectonics and climate (Shanley and McCabe 1994). Examples of tectonic-uplift-induced incision have been documented from Mississippi and Louisiana by Burnett and Schumm (1983) and summarized by Holbrook and Schumm (1999). Where incision and terracing occurs as a function of tectonic uplift there is obviously a requirement for the tectonic deformation state to be temporary if the incisional event is to be preserved in the stratigraphic record. In the longer term, subsidence must eventually ensue and overburden deposits must bury the terraces out of "accumulation space" and into "preservation space" (Kocurek and Havholm 1993; Kocurek 1998).

Although many compelling examples of the incisional response to surface uplift exist in the geomorphological literature (e.g., Molnar et al. 1994), uplift is not a prerequisite for tectonically induced incision and terracing to occur. Differential subsidence rates along a fluvial long profile may be sufficient to leave a particular reach out of equilibrium with its up-dip and down-dip neighboring reaches, thus promoting incision and the potential for terrace formation (Ouchi 1985; Jorgensen 1990). In addition to the reach scale, the tectonic process of differential subsidence, leading to incision along the long profile, also has relevance at the scale of hydraulically linked half grabens or salt-related mini-basins (Archer 2006). Within half grabens or salt-related mini-basins, differential subsidence rates can also lead to lateral ground tilting, incision, and terrace development (e.g., Leeder 1993).

Changes in the balance between discharge and sediment supply can also lead to incision and terrace formation (Lane 1955; Bull 1991). The critical threshold of stream power is an important concept in fluvial geomorphology since degradation ensues if the bedload transport capacity (discharge) exceeds the sediment supply and aggradation occurs if sediment supply exceeds the transporting capacity (see Blum and Törnqvist 2000 or Burbank and Anderson 2001 for fuller summaries). Discharge and sediment supply can be controlled by both climate and tectonics; but purely climate-driven incision and the formation of climatically induced sequence boundaries in the stratigraphic record has recently been attracting significant attention (Jones et al. 1999; Holbrook 2001; Weissmann et al. 2002; Gibling et al. 2005; Mack et al. 2006, Amorosi et al. 2008).

Although highly generalized, there are at least three distinct sedimentary responses to climate change that can be recognized in the geomorphological literature and have relevance for the dynamics of terrace formation:

- Class A sedimentary response—where aggradational pulses are related to cooler, wetter (stadial or glacial) climatic conditions (e.g., Penck and Brückner 1909; Porter et al. 1992; Molnar 1994; Ritter et al. 1995; Berryman et al. 2000; Weissmann et al. 2002),
- Class B sedimentary response—where aggradational pulses are related to a transitional phase of climatic change (e.g., cooler to warmer; Bull 1991; McDonald et al. 2003; Pan et al. 2003).
- Class C sedimentary response—where aggradational pulses are related to warmer, drier (interstadial or interglacial) conditions (e.g., Reheis et al. 1996).

Within the category of climate related incision we include here the impact of lake-level fall as a potential mechanism to drive the creation of terraces. An important consideration where terracing is associated with lacustrine base-level fall is that there can be a very direct coupling

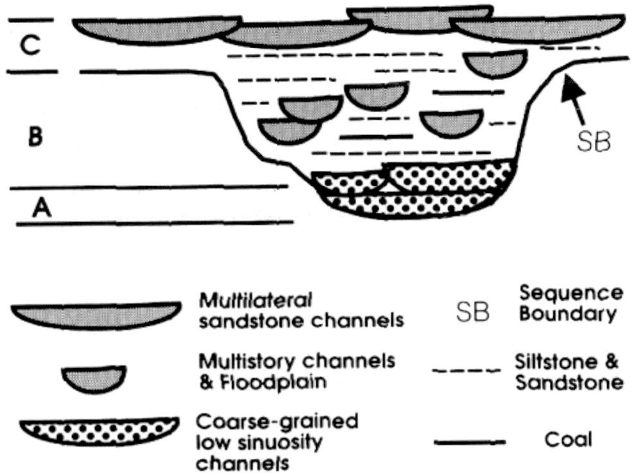

FIGURE 3.—A model for the formation and filling of an incised-valley, from Ethridge et al. (1998). Note that terrace formation is implied by the flat portion of the sequence-bounding surface on the left-hand side of the valley. However, multiple incisional events and aggradational-terrace-fill lithologies are not included explicitly in this model.

between lake-level fall and a decrease in fluvial discharge and sediment delivery if both the lake and the fluvial system exist within the same climatic regime (Perlmutter and Matthews 1990; Bohacs et al. 2000).

Finally, terracing can also be initiated by autogenic, complex response mechanisms (Schumm and Parker 1973) where degradation and aggradation are intrinsic to the river rather than externally forced by a change in allogenic variables.

Controlling Mechanisms in the Context of Fluvial Long Profiles: Figure 4 shows that there is some degree of predictability to incisional geometries in that the magnitude of the erosion should increase towards the driving mechanism. This simple observation may have utility in the subsurface where erosion levels can be mapped to increase in a certain direction, but the causal allogenic mechanism is uncertain.

Figure 4A suggests that in the case of eustatic or lacustrine base level fall, erosion is at a maximum distally and reduces up the long profile to the point where the incision is zero (distal intersection point or knickpoint). In a longitudinal sense it is clear, however, that the pre-existing physiography of the coastal prism (or piedmont), the shelf, the shelf edge and the upper slope strongly influences the location of incision and the resulting morphology of the incised valley (Zaitlin et al. 1994; Talling 1998).

In the case of purely climate-driven incision observed on alluvial or fluvial fans, for example (Ritter et al. 1995; Weissmann et al. 2002), erosion tends to increase proximally, towards the range front (Fig. 4B), from which discharge and sediment issue, and diminishes down the long profile to the point where the incision is zero (proximal intersection point). The geometry seen in example 4B can also be driven by tectonic uplift and tilting (e.g., Stokes and Mather 2000).

As stated previously, differential subsidence rates along a long profile can also lead to erosion due to the formation of perched half grabens or salt-related mini-basins (see mini-basin Cii in Figure 4C). These basins are tectonically perched due to slower rates of subsidence and can be regarded as out of equilibrium with respect to the long profile and thus prone to incision and terracing.

Figure 4D reveals more complex erosive trends associated with tectonic uplift, where erosion increases in both up-dip and down-dip senses towards the center of the uplift. Examples of tectonic uplift include magmatic or salt-related doming, flexural bulging in a foreland basin, structural inversion, and isostatic responses to deglaciation or water or sediment removal. Intrabasinal faulting can also cause incision on the uplifting footwall block (4E), as previously shown by Holbrook and Schumm (1999).

The geometric patterns of incision outlined above may well interact depending on the paleogeography. The interaction between up-dip climate and down-dip eustatic base-level controls has been highlighted in previous work by Blum (1993), Shanley and McCabe (1994), and Blum and Törnqvist (2000); however, the stratigraphic signal resulting from interfering allogenic controls is notoriously difficult to deconvolve in ancient-record fluvial deposits. The unifying relationship that provides a framework in which to understand these complexities is the interplay between accommodation state and sediment supply.

An "Accommodation State" Model for Terrace Formation

In order to illustrate the range of potential cross-valley terrace architecture as a function of varying accommodation states, we outline a model in Figure 5 that considers terrace formation during a hypothetical base-level transit cycle where aggradation during terrace fill is held constant. The ratio of degradation to aggradation associated with any single terrace cut-and-fill cycle must be seen in the longer-term context of net-negative-accommodation, which is required for the excavation of a valley to proceed. Once the base-level transit cycle turns around and the incised valley starts to fill, the system enters the net-positive-accommodation stage of its evolution.

In this paper we define "negative-accommodation terraces" as those formed by short-duration aggradational pulses during a long-term incisional episode and "positive-accommodation terraces" as those formed by short-duration degradational pulses during a long-term basin-filling episode. In precise quantitative terms, the ratio of degradation to aggradation, expressed as a number greater than 1 during incisional episodes, becomes a ratio less than 1 during predominantly aggradational time periods (Fig. 5). Due to the potential confusion that surrounds the definition of strath (or bedrock or veneer) terraces, and the potential complexity of imposing process-time definitions, we have chosen not to define these two categories as degradational and aggradational terraces.

Terrace-forming erosion surfaces, developed during both base-level fall and base-level rise in this model, are challenging to differentiate in the ancient record. The key evidence for distinguishing the two is that the surfaces developed during base-level rise are less likely to be re-eroded and thus may be more continuous than the terrace remnants developed against valley walls during base-level fall. Furthermore, the erosion surfaces developed during base-level rise may exhibit less mature paleosol profiles due to shorter terrace exposure time in a predominantly aggradational setting (e.g., Bestland 1997). Negative-accommodation terraces tend to form inset terrace flights that young towards the center of the incised valley. In contrast, positive-accommodation terraces often develop during a phase of reducing confinement, where the valley form shallows and widens, encouraging the terraces to overlap and young out (see overlapping positive-accommodation terrace geometries in Figure 5).

Attached versus Detached Terrace Architectures: Terraces developed during the state of negative accommodation can usefully be described as attached or detached depending on the ratio of incision to aggradation (Figs. 5, 6). The definition of attached terraces is where the substrate and incised-valley fill are not in physical connection because the terrace-fill materials overlap. In contrast, detached terraces allow the juxtaposition of the substrate and valley-fill materials. The generation of detached terraces requires that the incisional cut be

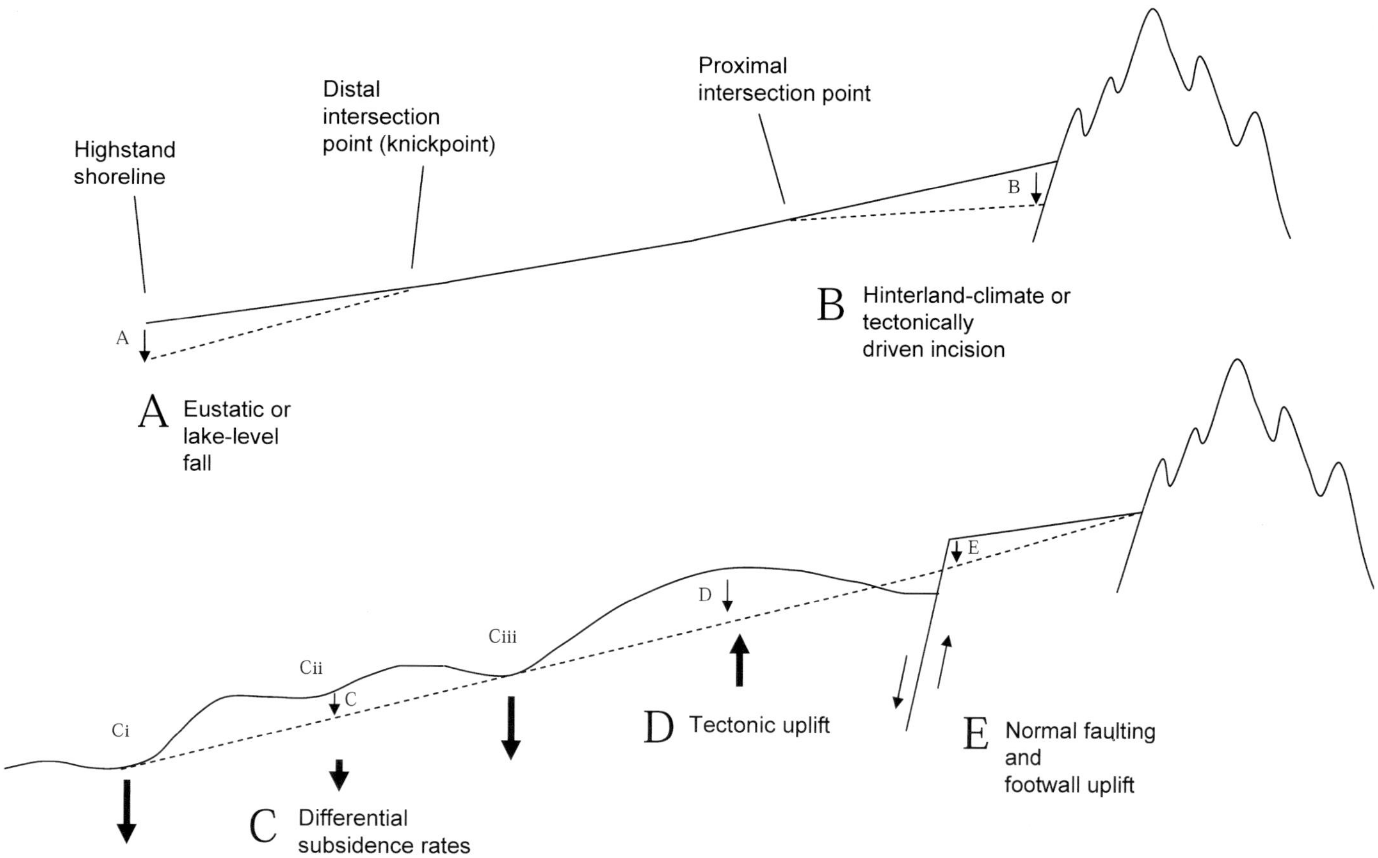

FIGURE 4.—Spatially variable incision patterns where sequence-bounding erosion levels increase towards the causal allogenic mechanism. A) Incision and terracing is base-level driven (including, marine or lake level fall); B) incision and terracing is climate driven from the hinterland catchment area or tectonically driven; C) incision and terracing is driven by differential tectonic subsidence rates; D) incision and terracing is driven by tectonic uplift; E) incision and terracing is driven by normal faulting.

deeper than the thickness of the next aggradational pulse. The degradation-to-aggradation ratio, for each individual terrace cut-and-fill cycle, must be less than 2:1 to develop "negative-accommodation attached terrace flights" and greater than 2:1 for "negative-accommodation detached terrace flights" to exist (Figs. 5, 6).

There is a tendency to encourage the development of attached terraces (Fig. 6A) where rates of generation of negative accommodation are low and the development of detached terraces (Fig. 6B) where rates of generation of negative accommodation are high (assuming that sediment supply variations and discharge are steady state).

Quaternary Analogues

In recent years, major advances in the understanding of how allogenic forcing factors combine to control stratigraphic development in fluvial systems have come from studies of the Quaternary record. Examples of these types of studies include: the Colorado River (Blum 1993; Blum and Price 1998), the Trinity and Sabine Rivers (Thomas and Anderson 1994), the Rhine–Muese (Törnqvist et al. 2003; Wallinga et al. 2004), the Seine and Somme (Antoine et al. 2000), the Thames (Maddy et al. 2001) the Severn-Avon (Bridgland 2000) and the Gangetic Plains (Gibling et al. 2005). Even though the long-term preservation potential of the terrace records of most of the European river systems in this list of examples may be compromised due to their tectonic setting, there are still valuable recent process-based lessons to be learned. These recent advances, made through a better understanding of process, are being fed back and used to condition interpretations of the ancient stratigraphic record. A key message is that climate is a more important factor than was previously recognized in many of the early sequence stratigraphic models (Cecil 2003; Bridgland and Westaway 2008).

Observations surrounding the formation of sequence boundaries in modern landscapes and in the Quaternary record demonstrate that erosion histories are complex and that the resulting sequence bounding surfaces are in fact often compound in nature.

We synthesize these concepts by introducing a simple hypothetical example of terrace formation, placed in a Quaternary record, sequence stratigraphic context (Fig. 7). The terrace flight in Figure 7 exhibits a 100 ka incisional record of base-level fall which has superimposed on it a terrace architecture created by cyclic pulses of aggradation, which are modulated by climate.

Long-timescale terrace records such as that in Figure 7 appear to be best developed due to the interaction of two or more allogenic variables, and it is with incision (through eustatic fall or surface uplift or both) already set as the background canvas that the clearest observations of well-preserved terrace flights are made. Terrace formation and the resulting geometry depends on the interplay between the rate of generation of accommodation and the timing and magnitude of pulses of sediment flux relative to discharge (which controls the individual terrace-scale degree of aggradation or degradation).

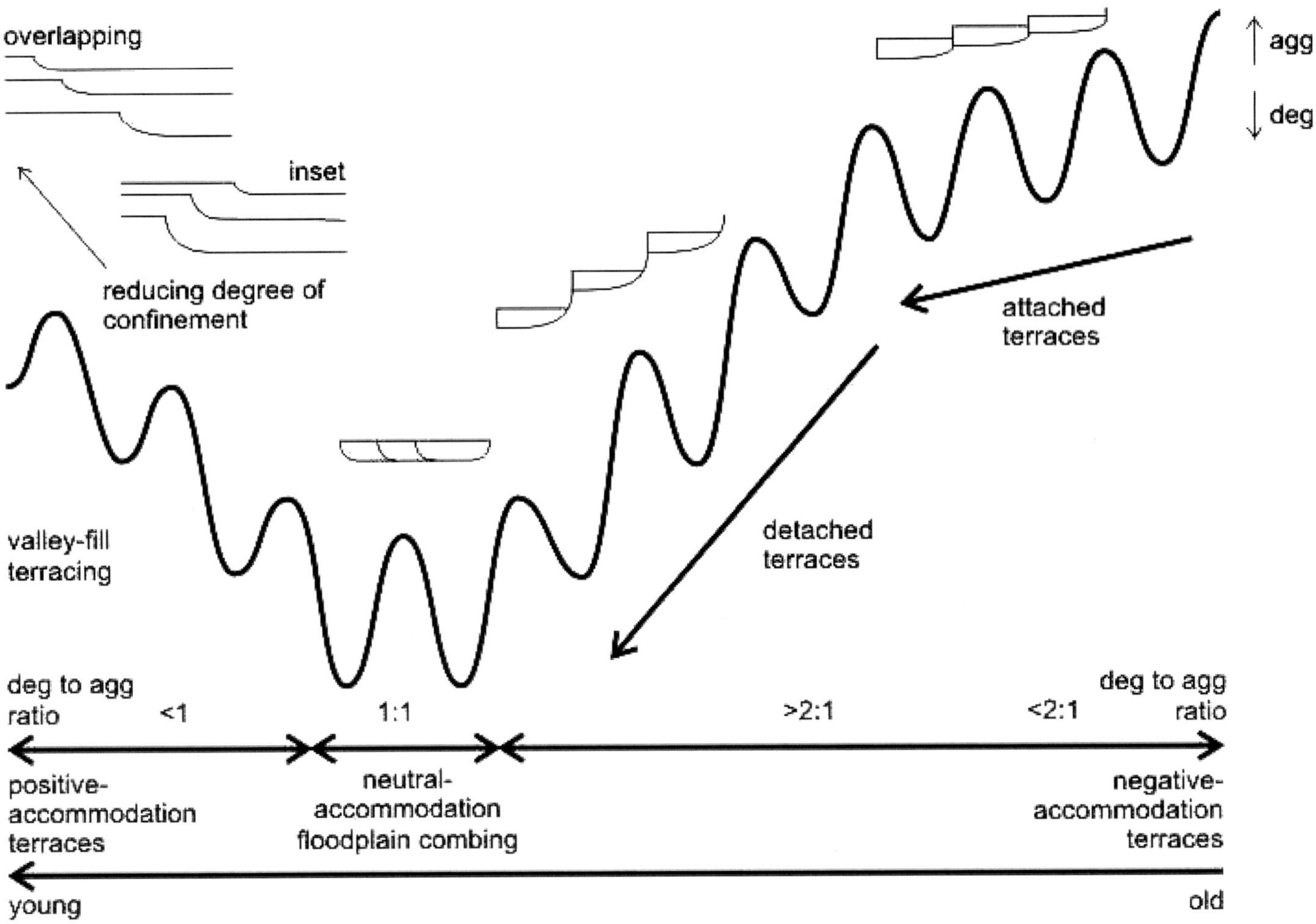

FIGURE 5.—A model illustrating the development of different terrace styles during changing accommodation states.

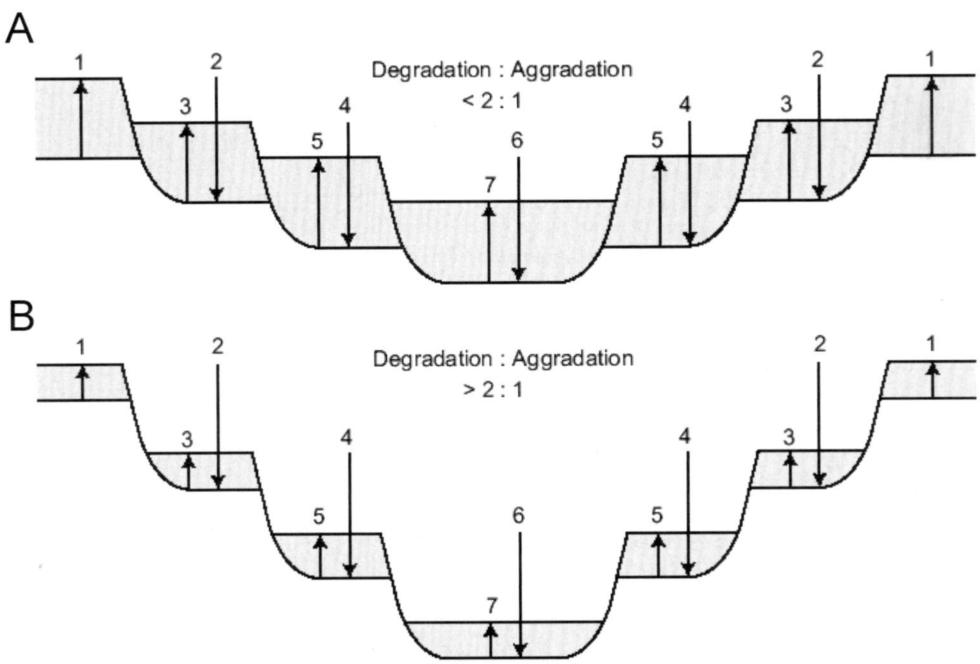

FIGURE 6.—Attached versus detached terrace flight architectures. A) Attached paired terraces. High sediment supply and low negative accommodation scenario. B) Detached paired terraces. Low sediment supply and high negative accommodation scenario.

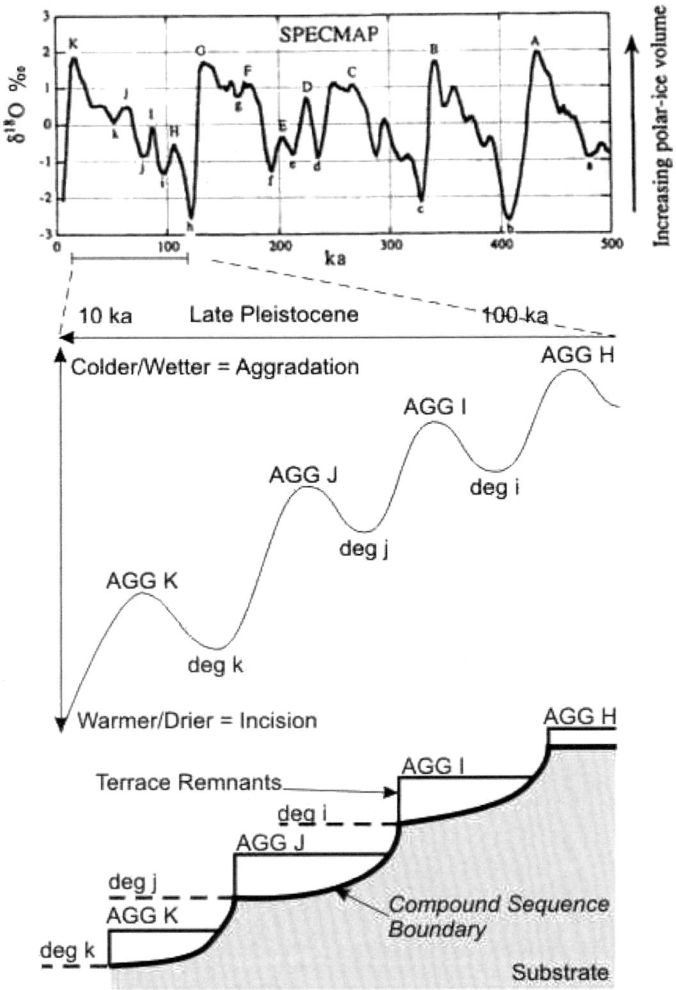

FIGURE 7.—An example of terrace development in a Quaternary sequence stratigraphic context. The source of the oxygen isotope curve is Imbrie et al. 1984.

In Figure 7, stadial and interstadial climatic fluctuations are shown to manifest themselves in the stratigraphic record as aggradational and erosional phases respectively (stadials in upper case, interstadials in lower case). This sedimentary response (Class A) would appear to be the dominant manner in which fluvial systems have responded to climatic change in the recent geological past (100 ka).

Numerous phases of degradation and aggradation, in an incision-dominant phase, result in the pulsed superimposition of stair-stepping terraces upon the sequence boundary. The realization that climate can drive the cyclic processes of aggradation and degradation leads us to the conclusion that many incised valleys may have been cut episodically rather than by a single event. The multiphase terraced carapace in Figure 7 is bounded on its underside by a compound sequence-bounding master surface that is stepped in appearance as opposed to smooth.

METHODOLOGY

A 2D seismic forward-modelling project was undertaken in order to explore the question: "what would it take to observe terraces seismically?" Seismic forward modelling is a valuable experimental method commonly employed where subsurface imaging is challenging. It allows the generation of a synthetic seismic response by convolving an acoustic-impedance model with a given seismic wavelet (Mellman and Kunzinger, 1992; Schwab and Pince, 1996; Schwab et al. 2007). In this project, the various modelled synthetic seismic sections test the seismic response to three key variables:

1. Terrace thickness and architecture,
2. Acoustic impedance contrasts at terrace tops and bases,
3. Seismic frequency.

GXT-EarthWave 1.2 was the modelling program used to construct the synthetic seismic sections. The input ranges of the three key variables used in the models are discussed in detail below.

Terrace Thickness and Architecture: The range of terrace thicknesses used in the modelling were acquired through a review of the geomorphological literature (Antoine et al. 2000; Bridgland 2000; Maddy and Bridgland 2000; Wisniewski and Pazzaglia 2002). Terrace thickness was defined as the thickness of a single aggradational terrace. Data on 38 individual terraces were gathered. Although this represents a small proportion of the available literature on modern and Quaternary terraces globally, the data set is felt to be sufficiently representative for this forward modelling exercise to be valid. The average terrace thickness was 6 m, and ranged between 1.7 and 17 m. These data were used in the construction of the earth models and are shown in histogram form in Figure 8. In Models 1–3, terrace thickness was increased artificially from left to right to show the impact of varying terrace thickness on the seismic response, while Models 4 and 5 have a constant terrace thickness of 8 m. Although we are aware of the existence of strath terraces (which incise into bedrock) and purely incisional (degradational) terraces in alluvium which have no or very thin aggradational components (Burbank and Anderson 2001), we have chosen to use models of terraces that included an aggradational element.

The incised-valley widths and depths used in the modelling have been selected as a representative range from the literature (Schumm and Ethridge 1994; Throne 1994; Davies et al. 1999). The depth range is from 90 m (Models 1 and 2) to 22 m (Models 4 and 5).

Acoustic-Impedance Contrasts at Terrace Tops and Bases: Constant-layer acoustic-impedance data were input for three architectural elements within the models, the original valley substrate, the terrace fill, and the incised-valley fill (Fig. 9). In some cases the valley fill and the substrate was multilayered. Acoustic-impedance properties in the models were calculated using average velocities from a range of common lithologies from Rider (2002) and Gardner's equation (Gardner et al. 1974) was used to obtain density from the velocity. The terrace material itself was kept constant as gravel with a relatively slow velocity (2500 m/s), and we varied the nature of the substrate and the fill material to provide a range of acoustic-impedance contrasts. High and low and positive and negative acoustic-impedance contrast models were developed to test the sensitivity of the seismic response at terrace tops and bases to a wide range of impedance contrasts.

Seismic Frequency: Three seismic wavelet frequencies were used as a proxy for ancient-record burial depth. Synthetic seismic was constructed by convolving wavelets of 30, 50, and 70 Hz peak frequency to represent a range of data resolutions representing deep to shallow buried terraces. A zero-phase Ricker wavelet of normal polarity was used in all models (i.e., a peak is an increase in acoustic impedance). All models include shallow and deep control layers to provide consistent test markers (top and bottom layers in Figure 9).

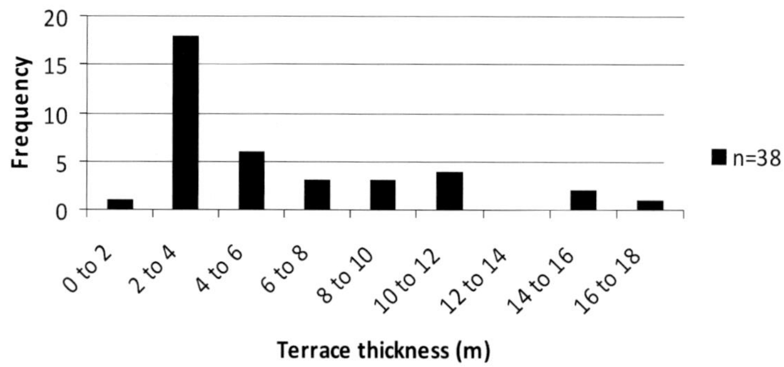

FIGURE 8.—A histogram of terrace thickness obtained from geomorphic literature ($n = 38$, average = 6 m). Sources: Antoine et al. (2000); Bridgland (2000); Maddy and Bridgland (2000); Wisniewski and Pazzaglia (2002).

MODEL DESCRIPTIONS AND SYNTHETIC SEISMIC RESULTS

Five models have been selected to illustrate various permutations of terrace thickness, acoustic-impedance contrasts, and seismic resolution.

Model 1

Model 1 (Fig. 10) was designed to represent the best possible subsurface scenario modelled, incorporating thick detached terraces and high acoustic-impedance contrasts (Fig. 10B). The model includes an incised valley cut by a flight of three terrace suites and incorporates five separate detached terraces. Terrace thickness ranges from 10 to 30 m and increases from left to right in increments of 5 m. The model is 8 km wide, and the total incised-valley depth is 90 m (Fig. 10A).

The geological scenario modelled incorporates large acoustic-impedance contrasts where the relatively thick terraces contain low-velocity gravel relative to the sandy valley-fill material and the shale-prone substrate. The velocity of the valley fill (IVF) is 3200 m/s, the velocity of the terrace fill is 2500 m/s and the velocity of the substrate is 4200 m/s. These velocities result in the top of the terraces having a negative reflection coefficient ($R = -0.15$), the bases having a stronger positive reflection coefficient ($R = 0.32$) and the interfluve areas at the edges of the model, where terraces are absent, having a lower positive reflection coefficient ($R = 0.17$). The resulting seismic reflections show

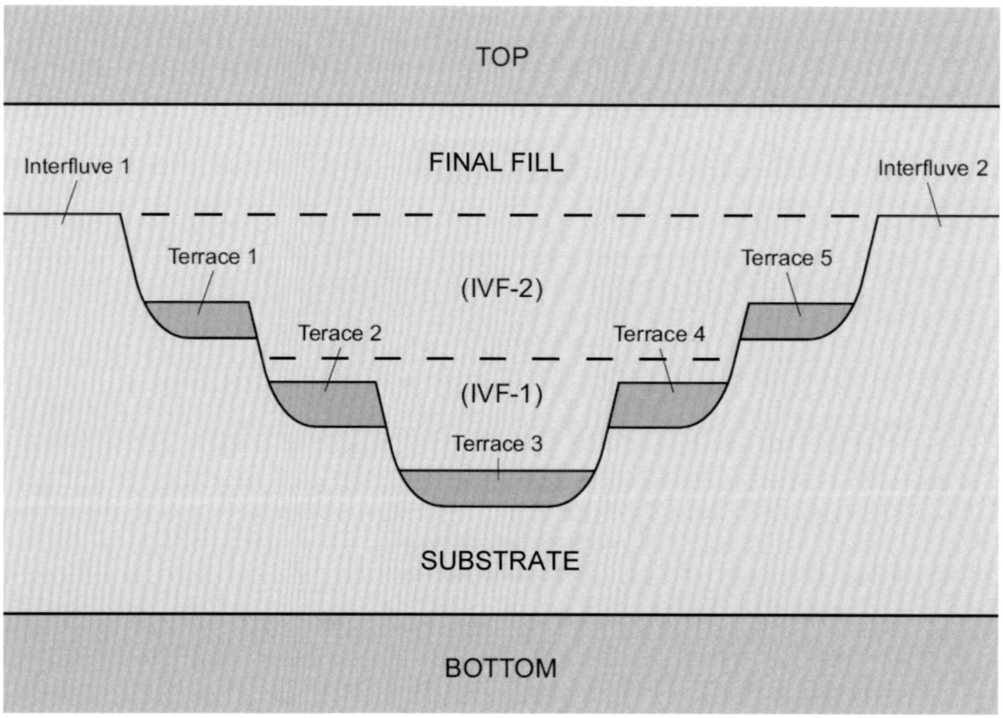

FIGURE 9.—Input earth model to explain the names of layers and numbering of terraces. Layer names are in capital letters and dashed lines indicate layers only in multi-layer-fill-models (MLF), where incised valley fills (IVF) are split into IFV-1 and IVF-2. The numbering of interfluves and terraces are those used in the synthetic seismic section descriptions of Models 1 and 2. Models 3, 4, and 5 have seven terraces, and they are numbered in the same fashion, increasing from left to right.

a high-amplitude white trough with black peak side lobes at the tops of the terraces, and a high-amplitude black peak with white trough side lobes at the base of the terraces, laterally changing to a weaker-amplitude black peak and white trough side lobes on the interfluves at the edges of the model.

On the 70 Hz peak frequency synthetic seismic section (Fig. 10) the model-layer tuning thickness ranges from 9 to 15 m, depending on the layer-interval velocity, with the terrace tuning thickness at 9 m. Terrace 1 is 10 m thick and is characterized by a strong trough at the top of the terrace and a strong peak at the base. Because this terrace is just above tuning thickness, the lower side lobe of the terrace top constructively interferes with the main peak of the terrace base to create an enhanced high-amplitude peak. Positive interference of the central trough of the upper reflection with the upper side lobe of the lower reflection also increases the amplitude of the trough and results in a strong trough overlying a strong peak. As the terraces become thicker, the side lobes begin to destructively interfere with each other rather than the central troughs and peaks, and this interaction between the strong trough and underlying peak can be observed as a "bumpy" lower-frequency trough (Terrace 2 in Figure 10B). On terrace 3, which is 20 m thick, a new peak emerges from between the two troughs which represents the lower side lobe of the upper terrace reflection. Terraces 4 and 5, where thicknesses are 25 and 30 m, have fully resolved top and bottom reflections. A slight "push-down" effect can be seen with increasing terrace thickness on the bottom reflection (compare terrace 1 to terrace 5).

The layer tuning thickness on the 50 Hz peak-frequency synthetic seismic section (Fig. 10C) ranges from 13 to 21 m, with the terrace tuning thickness at 13 m. The lower side lobes of the terraces tops interfere with the upper side lobes of the terrace bases on terraces 1–3, and only terraces 4 and 5 exhibit a middle peak which is the lower side lobe of the upper reflection. However, since the amplitude of this lower side lobe never reaches the full amplitude seen in the upper side lobe, we conclude that at 50 Hz no terrace reflections are "fully" resolved. Slight "push-down" effects can still be seen with increasing terrace thickness on the bottom reflection.

The layer tuning thickness on the 30 Hz peak-frequency synthetic seismic section (Fig. 10D) ranges from 21 to 35 m, with the terrace tuning thickness at 21 m. Terrace 3 (20 m) shows the strongest trough and peak reflections due to the constructive interference of the side lobes, since it is nearly at terrace tuning thickness. The two thinner terraces (1 and 2) have no side lobes and the trough and peak reflections are of weaker amplitude due to destructive interference. With increasing thickness, the upper side lobe peak of the top terrace reflection exhibits clear amplitude increase and the frequency of the central trough decreases due to side-lobe interference, but the terraces

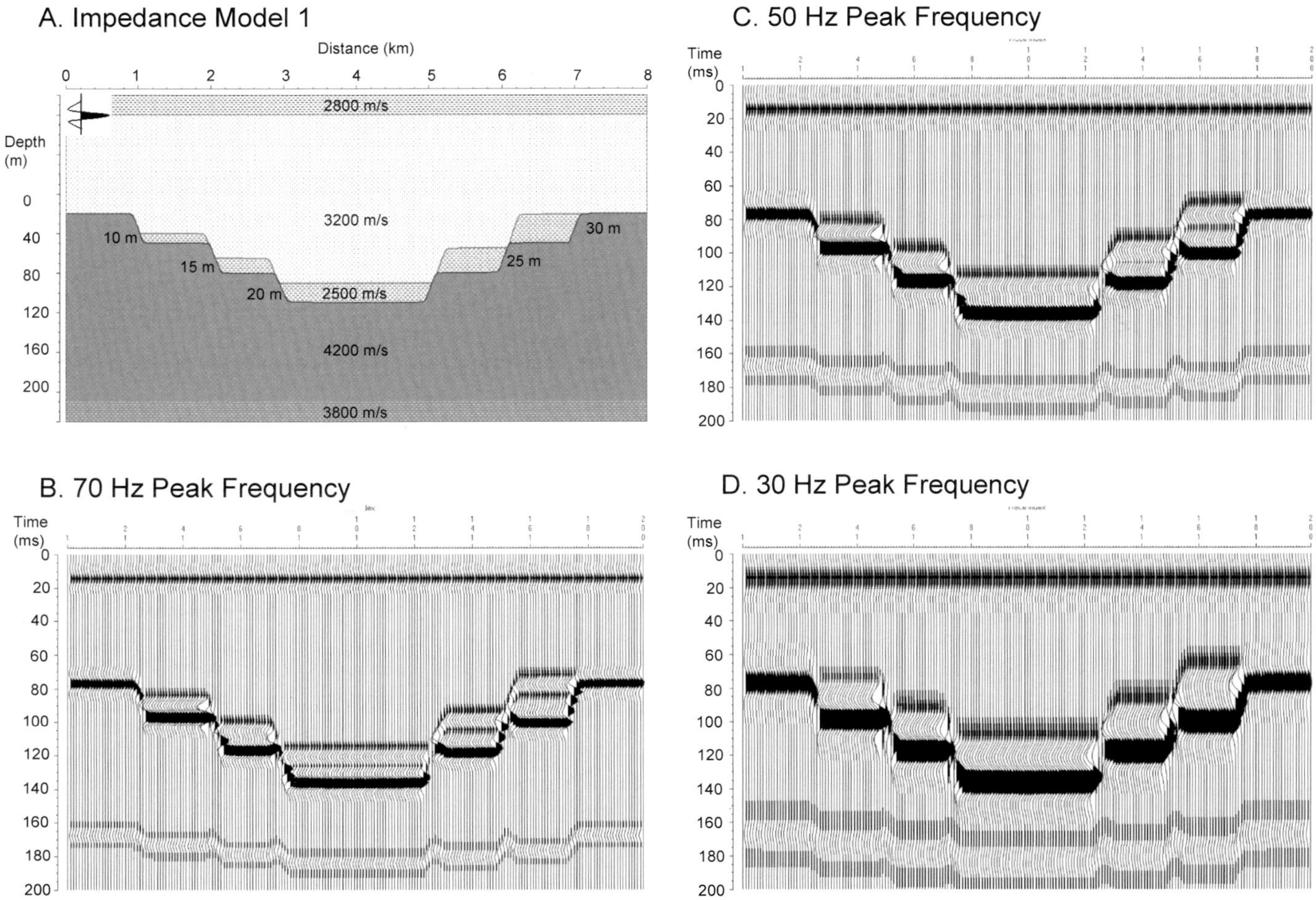

FIGURE 10.—Results of 2D seismic forward Model 1 (best possible scenario for terrace recognition). Terrace thickness ranges from 10 to 30 m, total incised-valley depth is 90 m and the model is 8 km wide. A) Acoustic impedance input model; B) synthetic seismic at 70 Hz peak frequency; C) synthetic seismic at 50 Hz peak frequency; D) synthetic seismic at 30 Hz peak frequency.

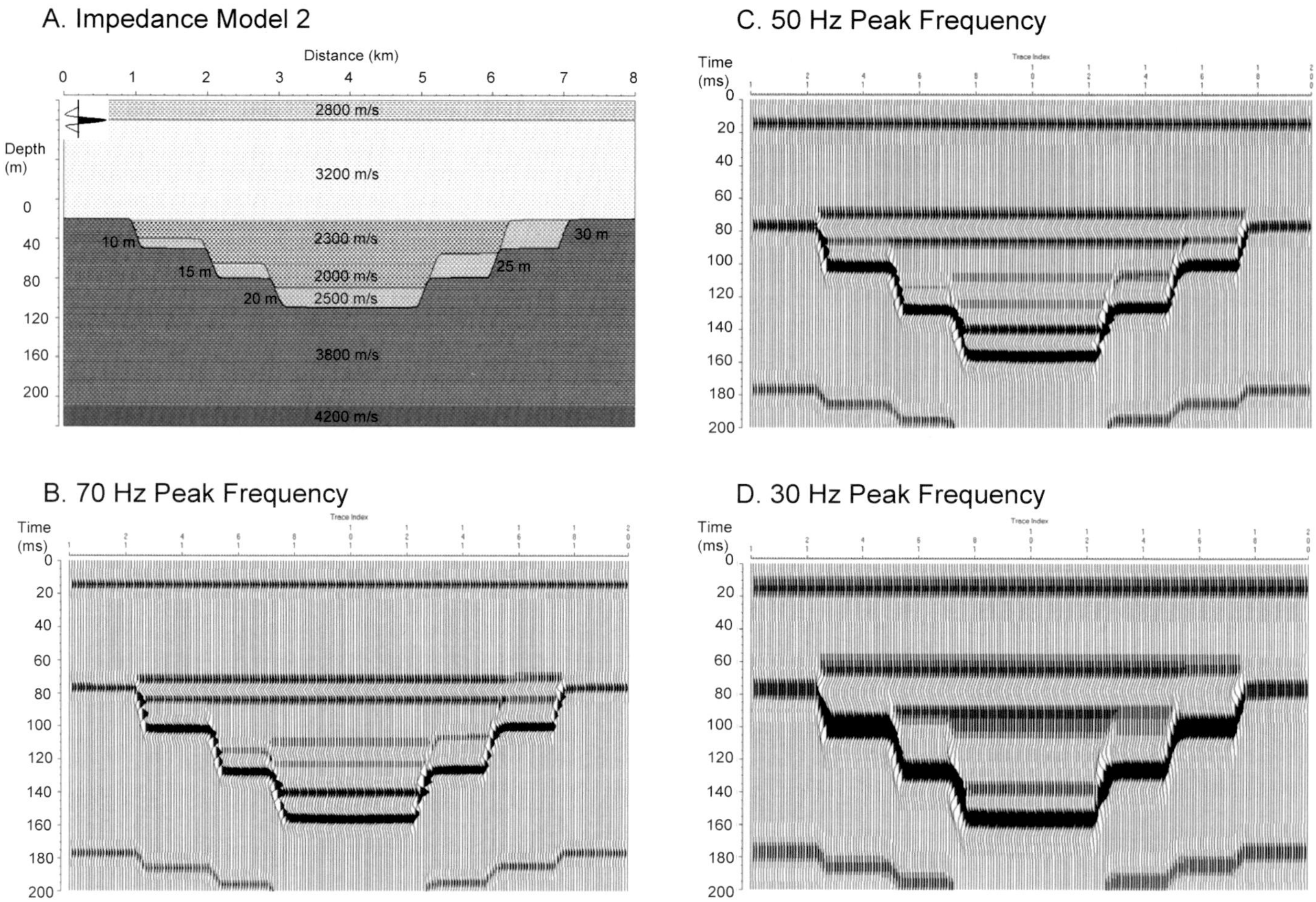

FIGURE 11.—Results of 2D seismic forward Model 2 (includes a multilayer valley-fill). Terrace thickness ranges from 10 to 30 m, total incised-valley depth is 90 m, and the model is 8 km wide. A) Acoustic impedance input model; B) synthetic seismic at 70 Hz peak frequency; C) synthetic seismic at 50 Hz peak frequency; D) synthetic seismic at 30 Hz peak frequency.

do not reach sufficient thickness to fully resolve the reflections at the tops and bases of the terraces at 30 Hz peak frequency. A slight "push-down" effect can just be seen with increasing terrace thickness on the bottom reflection.

On the synthetic seismic of this model, a central trough reflection images the terrace tops and a central peak reflection images the terrace bases. Only the thickest terraces (> 20 m for 70 Hz, and > 25 m for 50 Hz) image the side lobes, while all other thinner terraces have increasing degrees of side-lobe interference. A stepped appearance to the reflection at the base of the valley is clearly observable on all synthetic seismic sections.

Model 2

Model 2 (Fig. 11) was designed to represent a more complex variant on the best possible subsurface scenario of Model 1, incorporating thick detached terraces, a multilayered fill, and relatively high acoustic-impedance contrasts (Fig. 11A). Model 2 incorporates an incised valley cut by a flight of three terrace suites and consists of five separate detached terraces. Terrace thickness ranges from 10 to 30 m and increases from left to right in increments of 5 m. The model is 8 km wide, and the total incised-valley depth is 90 m (Fig. 11A).

The geological scenario in Model 2 is largely based on Model 1, but in order to introduce more complexity, the incised-valley fill is heterogeneous and consists of two layers with different velocities, designed to simulate the upward fining of the incised-valley-fill materials. The model again incorporates an incised valley whose relatively thick terraces contain low-velocity gravels. Additionally, the velocity contrast between the terrace-fill gravels and the shaly substrate was diminished in comparison to Model 1. The velocity of the final fill is 3200 m/s, the velocity of IVF-2 is 2300 m/s, and 2000 m/s for IVF-1. The terrace fill is 2500 m/s, and the velocity of the substrate is 3800 m/s (Fig. 11A).

The top surface of the terraces create a moderately strong positive reflection coefficient ($R = 0.14$) where they is buried by IVF-1 (Terrace 3), but this reflectivity diminishes towards the margins of the valley, where velocity contrasts between the terrace fill and IVF-2 are lower (indeed Terrace 5 has a negative refection coefficient due to the higher velocity of the overlying final fill). The velocity contrast across the bottom surfaces of the terraces has a positive reflection coefficient ($R = 0.26$), but this diminishes markedly on to the interfluves ($R = 0.11$), producing weaker peak reflections with trough side lobes. There are also weak negative reflection coefficients produced between the final fill and IVF-2 ($R = -0.21$), and IVF-2 and IVF-1 ($R = -0.09$), producing trough reflections with peak side lobes.

On the 70 Hz peak frequency synthetic seismic section (Fig. 11) the layer tuning thickness ranges from 7 to 15 m depending on the layer

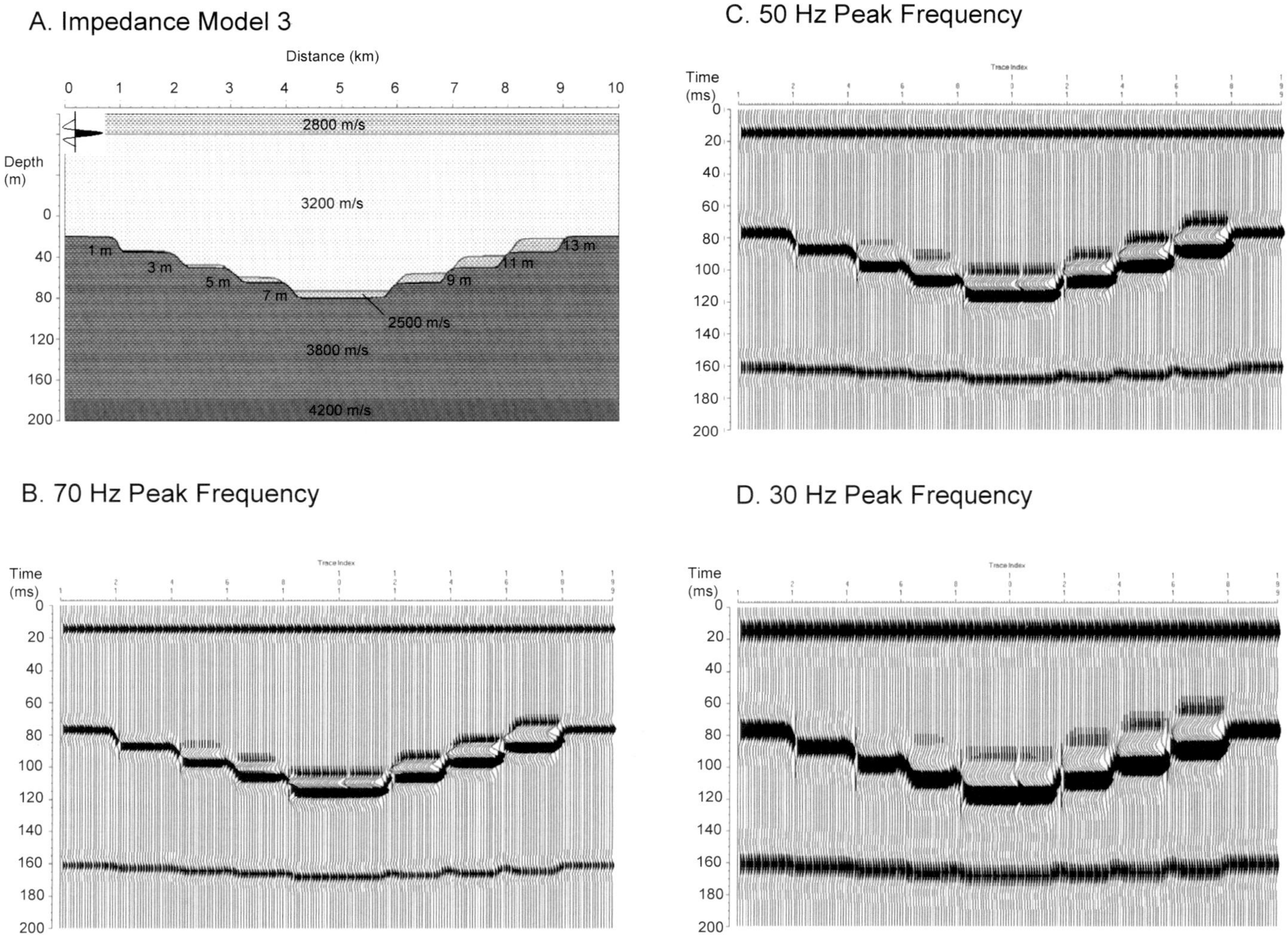

FIGURE 12.—Results of 2D seismic forward Model 3 (thinnest-terrace scenario). Terrace thickness ranges from 1 to 13 m, total incised-valley depth is 60 m, and the model is 10 km wide. A) Acoustic impedance input model; B) synthetic seismic at 70 Hz peak frequency; C) synthetic seismic at 50 Hz peak frequency; D) synthetic seismic at 30 Hz peak frequency.

interval velocity, with the terrace tuning thickness at 9 m. As in Model 1, terraces 2–5 all have resolvable central peaks (and nearly fully resolvable trough side lobes) that correspond to the tops and bases of the terraces. The complex low-velocity valley fill reduces the clarity of the top terrace peak reflection, except on terrace 3, where the anomalously low-velocity fill creates a distinctly stronger terrace-top peak reflection. Above terrace 3, the complex fill produces a trough between IVF-2 and IVF-1. The contrast between the final fill and IVF-2 creates an obvious continuous trough that exhibits an amplitude decrease at the top surface of terrace 5.

The layer tuning thickness on the 50 Hz peak frequency synthetic seismic section (Fig. 11C) ranges from 10 to 21 m, with the terrace tuning thickness at 13 m. The layer tuning thickness on the 30 Hz peak frequency synthetic seismic section (Fig. 11D) ranges from 17 to 35 m, with the terrace tuning thickness at 21 m. With decreasing frequency, the interference between valley-fill reflections and terrace-top reflections increases, such that it is difficult to distinguish the two reflections at 30 Hz due to interference (Fig. 11D). The peaks associated with the bases of terraces 1 and 5 are enhanced in both amplitude and frequency due to constructive interference with the base final fill reflection in the case of terrace 5 and the base final fill reflection and the top terrace reflection in the case of terrace 1 (Fig. 11D).

In this model the impact of the added complexity of the low-velocity valley-fill is that terrace tops can only be subtly imaged on higher-frequency data with thicker terraces. Although weakly developed, both the central peaks and trough side lobes are resolved for the terraces thicker than 20 m. The most clearly imaged terrace tops are terrace 3, where the contrast is moderately large ($R = 0.14$) and the thickest 30 m terrace on 70 Hz data, where the uniquely negative terrace-top impedance contrast is large between the terrace and the final fill. The seismic reflections from the tops and bases of the terraces interfere with the internal valley-fill reflections, for thinner terraces at 70 Hz and all terrace thicknesses at 50 and 30 Hz. Although the velocity contrast between the terrace fill and the substrate is lower than in Model 1, and thus the resulting amplitude lower, a stepped reflection (that shows particularly variable amplitude at 30 Hz) occurs along the base of the valley on all synthetic seismic sections. The reflection generated from the base of the model shows an even more distinct "push down" due to the effect of the overlying low-velocity valley fill.

Model 3

Model 3 (Fig. 12) was designed to represent a challenging subsurface scenario where thin, detached terraces reveal the difficulties of resolving the presence of terraces seismically. Model 3 incorporates an incised-valley cut by a flight of four terrace suite levels and seven separate, detached terraces. Terrace thickness ranges between 1 and 13 m and increases from left to right in increments of 2 m. The model is 10 km wide, and the total incised-valley depth is 60 m (Fig. 12A).

The geological scenario modelled represents an incised valley that is not as deep as Models 1 and 2 and incorporates relatively thin terraces containing low-velocity gravels. Both the valley-fill materials (sandstone) and the substrate (shale) are homogeneous (Fig. 12A). The velocity of the valley fill in this model is 3200 m/s, the velocity of the terrace fill is 2500 m/s, and the velocity of the substrate is 3800 m/s. These input velocities result in the tops of the terraces having a moderately strong negative reflection coefficient ($R = -0.15$), producing a central trough reflection with peak side lobes, and the bottom has a stronger positive reflection coefficient ($R = 0.26$), producing a central peak reflection with trough side lobes.

On the 70 Hz peak frequency synthetic seismic section (Fig. 12) the layer tuning thickness ranges from 9 to 15 m depending on the layer interval velocity, with the terrace tuning thickness at 9 m. The amplitude of the main peak reflection and overlying trough increases from left to right with increasing terrace thickness due to constructive interference between the reflections at the tops and bases of the terraces. None of the terraces in this model are thick enough to produce a complete reflection for the terrace top or base. On terrace 6 there is a subtle increase in the amplitude and frequency of the main trough due to interference. The main indication that two separate reflections exist is the increase in amplitude of the upper side lobe peak of the top terrace reflection. "Push-down" effects are barely noticeable at the base model reflection.

The layer tuning thickness on the 50 Hz frequency model (Fig. 12C) is 13 to 21 m depending on the layer interval velocity, with the terrace tuning thickness at 13 m. The layer tuning thickness on the low-frequency data of 30 Hz (Fig. 12D) is 21 to 35 m, with the terrace tuning thickness at 21 m. With decreasing frequency, none of the terraces have fully resolvable top and base reflections. At 30 Hz, terrace thicknesses of 5 m or less show little or no perceptible clues to the presence of a terrace-top reflection.

In this model, terrace tops and bases are never both fully resolvable, but the presence of terraces can still be inferred from the development of a weak-amplitude upper side lobe peak from the upper reflection and the stepped appearance of the erosion surface central peak reflection, which is still observable at all frequencies.

Model 4

Model 4 (Fig. 13) was designed to represent a challenging seismic environment, where a multilayered substrate adds complexity to the combination of 8-m-thick attached terraces. Model 4 incorporates an incised-valley cut by four terrace suite levels and seven individual but attached terraces. The model is 10 km wide, and the total incised valley depth is 22 m. The synthetic seismic was constructed to more closely resemble actual seismic in that noise and gain were added to the synthetic seismic signal.

The geological scenario modelled incorporates a shallow incised valley lined with attached terraces that contain low-velocity gravels. The valley-fill materials are homogeneous sandstones, but the valley is cut into a complex heterogeneous substrate consisting of interbedded lithologies. The velocity of the valley fill in this model is 3200 m/s, the velocity of the terrace fill is 2500 m/s, and the velocity of the substrate alternates between 3200 m/s for sandstone and 2800 m/s for shale. The top surfaces of the terraces result in a moderately strong negative reflection coefficient ($R = -0.15$), which produces a trough with peak side lobes, and the terrace bottom produces a variable (but all positive) reflection coefficient due to the interbedded substrate ($R = 0.26$ for terraces on shale and $R = 0.32$ for terraces on sandstone), which produces a varying-amplitude peak with trough side lobes.

On the 70 Hz peak frequency synthetic seismic section (Fig. 13B), the terrace tuning thickness is 9 m, which is close to the modelled terraces thickness of 8 m. Therefore, it is not possible to resolve the top and bottom terrace reflections fully. Both the central trough of the upper reflection and the central peak of the terrace base reflection have been enhanced in amplitude due to constructive interference with the upper side lobe of the lower reflection and lower side lobe of the upper reflection respectively.

Terraces 2 and 6 exhibit a more distinct terrace-like seismic response, due to the stronger acoustic-impedance contrasts between the substrate and the terraces. In previous models, a clue to the presence of terraces has come from the topography and flighted architecture of the valley-bottom reflection; however, the stepped appearance of the base reflection in Model 4 is less clear than on the previous models. This is because of the alternating lower reflection coefficients along the base terrace due to the complex substrate, the fact that the terraces display attached geometries, and the fact that all the terrace thicknesses are just below the tuning thickness. At 70 Hz peak frequency (Fig. 13B) the stepped appearance is only just observable and is sufficiently ambiguous that the steps could be interpreted as faults instead of terrace flights.

At 30 Hz (Fig. 13C) the stepped breaks between terraces are not at all clear, and the high-amplitude peak response at the base of the terraces amalgamates with the trough at the top of the terraces to form one continuous reflection. This amalgamated continuous reflection appears as a concave-up parallel trough-peak set, which could be interpreted as the base of a nonterraced incised valley.

Due to the terrace tuning thickness, it is not possible to fully resolve top and bottom terrace reflections. The observation of the stepped base is difficult to make at low frequency due to a combination of attached terrace geometries, without large incisional cut phases, and the complex substrate, which induces amplitude variations in the basal reflection.

Model 5

Model 5 (Fig. 14) was designed to represent a highly challenging seismic environment, where a multi-layered substrate and fill sequence adds complexity to the combination of 8-m-thick attached terraces. Model 5 incorporates an incised-valley cut by a flight of four terrace suite levels and seven individual but attached terraces. The model is 10 km wide, and the total incised valley depth is 22 m. The synthetic seismic was constructed to more closely resemble "actual" seismic in that noise and gain were added to the synthetic seismic signal.

The geological scenario modelled incorporates an incised valley lined by terraces containing gravel and cut into a substrate of interbedded sandstone and shale, as in Model 4, but it has the additional complication of a multilayer fill sequence of interbedded sandstone and shale. The velocity of the valley fill and the substrate in this model alternates between 3200 m/s for sandstone and 2800 m/s for shale. The velocity of the terrace fill is 2500 m/s. The tops of the terraces result in a moderately strong to weak negative reflection coefficients ($R = -0.15$ for sandstone and $R = -0.05$ for shale), which produce a variable-amplitude trough with peak side lobes, and the terrace bottom is a moderate to weak (but all positive) reflection coefficient due to the interbedded substrate ($R = 0.26$ for terraces on shale and $R = 0.32$ for terraces on sandstone), which produces a variable-amplitude peak with trough side lobes.

On the 70 Hz peak frequency synthetic seismic section (Fig. 14B), the terrace tuning thickness is 9 m, which is close to the modelled terraces thickness of 8 m. Due to the terrace tuning thickness, is not possible to fully resolve top and bottom terrace reflections. With this

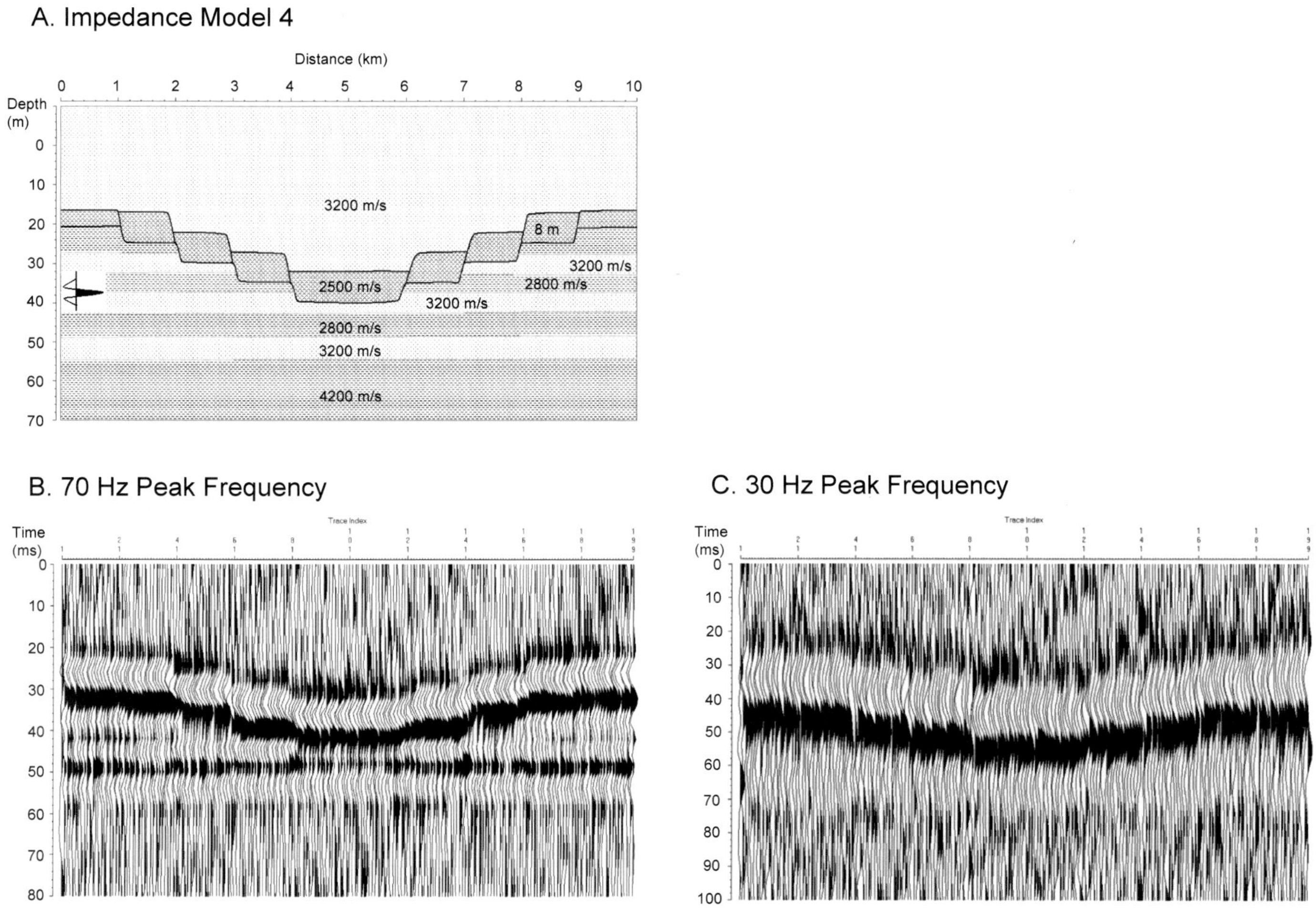

FIGURE 13.—Results of 2D seismic forward Model 4 (approximation of actual seismic and multilayer substrate). Terrace thicknesses are all 8 m, total incised-valley depth is 22 m, and the model is 10 km wide. A) Acoustic impedance input model; B) synthetic seismic at 70 Hz peak frequency; C) synthetic seismic at 30 Hz peak frequency.

level of substrate and fill complexity, the reflectivity across the section at the top and base terrace surfaces is variable. Terraces 2, 4, and 6 exhibit the most distinct terrace base reflection, which is due to a stronger acoustic-impedance contrast between the terrace and the substrate (gravel on sandstone). Terraces 1, 3, 4, 5, and 7 exhibit the most distinct terrace-top reflection, which is due to a stronger acoustic-impedance contrast between the overlying valley fill and the terraces (sandstones on gravels).

An indication that terraces exist can be gleaned only from the flighted architecture of the valley-bottom peak reflection at 70 Hz peak frequency, but the stepped appearance of the top terrace reflection (Fig. 14B) is not as obvious as in Model 4 (Fig. 14B) due to interference from the variable valley fill in Model 5. One result of the horizontal reflections related to the fill is that the stepped breaks along the basal reflection are less likely to be interpreted as faults since the overlying reflection is not similarly stepped. Lastly, the interpretation of fluvial-channel lag deposits that was possible in Model 4 is less feasible in Model 5 due to the presence of the horizontally bedded multilayered fill.

In the 30 Hz section (Fig. 14C), neither individual terrace top and base reflections, nor even a stepped appearance to the composite basal erosion surface reflection, are observable.

Summary of the Results of the Seismic Forward Modelling

The results of this work show that:

- We should not expect to commonly observe terraces directly on most actual seismic data.
- Only where conditions are favorable regarding terrace thickness, acoustic-impedance contrasts between terraces, fill, and substrate, and ancient-record burial depth which optimizes the seismic resolution and reflection tuning thickness, can terraces be directly identified seismically.
- However, even if terrace top and base reflections cannot be individually discerned, pulsed incision and terracing as the valley is cut can result in a stepped appearance to the sequence bounding surface reflection.
- On occasion, amplitude changes across this basal bounding surface reflection can allude to the presence of terraces (e.g., Model 1).
- This stepped reflection character is best observed where the incisional cut phase is deeper than the subsequent aggradational fill is thick (e.g., Models 1–3 show detached terraces, and Models 4 and 5 show attached terraces; see Figures 5 and 6 for definitions).

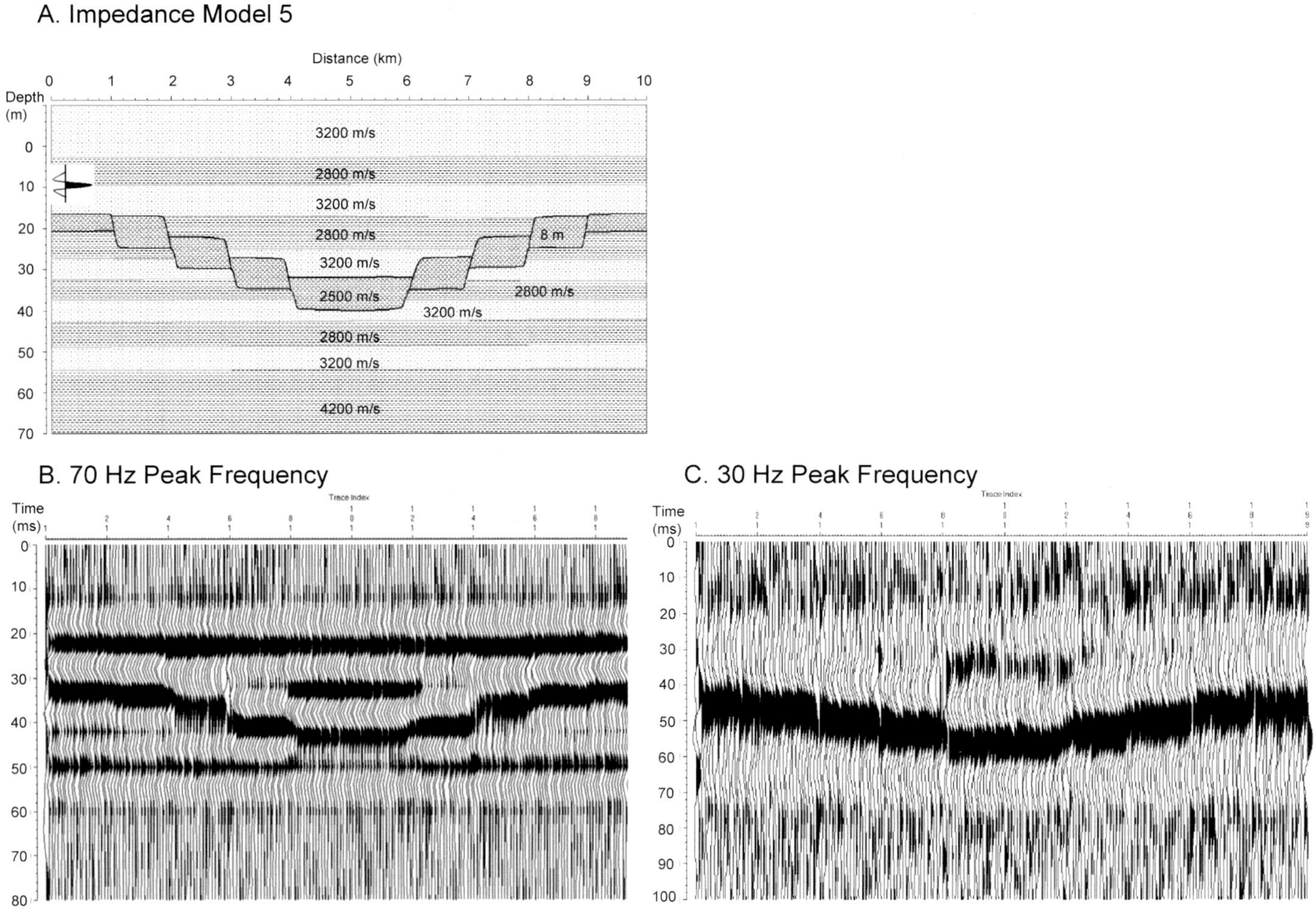

FIGURE 14.—Results of 2D seismic forward Model 5 (approximation of actual seismic with multilayer substrate and fill). Terrace thicknesses are all 8 m, total incised valley depth is 22 m, and the model is 10 km wide. A) Acoustic impedance input model; B) synthetic seismic at 70 Hz peak frequency; C) synthetic seismic at 30 Hz peak frequency.

- Where observed, this stepped profile to the valley-bottom reflection might in some circumstances be confused with small-scale normal faulting.
- The models presented assume that terrace fills are made up of porous, low-velocity gravels, which create in most models (except Model 2) a negative acoustic-impedance contrast at the terrace tops. The presence of hard, floodplain-shale-filled terraces, where the incised-valley fill is a porous sandstone, would create a positive acoustic-impedance contrast at the terrace top.
- The presence of paleosols could alter the acoustic-impedance character of terraces. For example, mature, arid or semiarid (calcrete, dolocrete or silcrete) soil profiles, if thick enough, could enhance the acoustic-impedance contrast at the terrace top, improving the seismic response and making this surface a higher-amplitude reflection than the basal erosion surface.
- Multilayered substrates and multilayered valley-fill successions (Models 2, 4, and 5) can make the observation of terraces even more problematic.
- Just because terraces cannot be readily observed on seismic does not mean they are not present in the subsurface at a sub-seismic scale.

Seismic Stratigraphy

The poorly resolved multiple surfaces of incision and terracing seen in Figures 13 and 14 illustrate why seismic stratigraphers often interpret sequence boundaries as simple, single surfaces, for example at the bases of incised-valley fills (see Figure 6 for an example of a classic model). When an incision surface is picked seismically as a single surface of erosion (e.g., Suter and Berryhill 1985; Stevenson and McMillan 2004) then assumptions are easily made about the valley having been excavated in a short period of geological time, as one erosional event. The interpretation of a single erosional surface would overlook the time transgressive, cyclic processes of cut and fill, the existence of the 8-m-thick terrace fills, and the formation of a compound sequence-bounding master surface that are often inherent to the process of incision (see Figures 13 and 14)..

Although we recognize that simple, flat, sequence-bounding surfaces of erosion exist and have been interpreted at outcrop (e.g., Holbrook 1996), the results of our seismic forward modelling results demonstrate that seismically observable erosion surfaces that appear as simple, convex-up, single surfaces may hide the true complexity that exists at a sub-seismic scale.

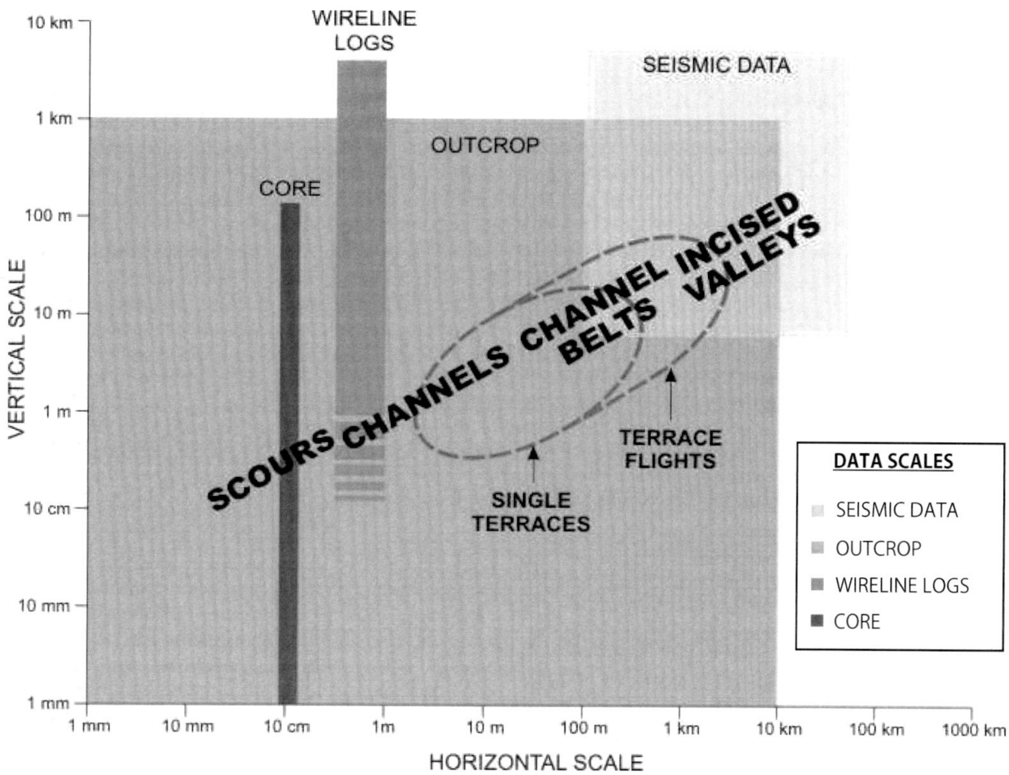

FIGURE 15.—Terraces fall into a subsurface data-resolution gap. Modified from Morris and Normark (2000).

Sequence Stratigraphy

Although we acknowledge that incised valleys do not only form close to the coast, in response to eustatic sea-level fall (see Figure 4), this next section integrates some of the observations made in this paper with traditional paralic sequence stratigraphic concepts.

Hunt and Tucker (1992) were among the first to draw attention to processes ongoing during the regressive part of a relative sea-level cycle. The sequence stratigraphic example shown in Figure 7 and the models in Figures 10–14 reveal flights of terraces whose deposits can be viewed in the context of the falling-stage systems tract (Plint and Nummedal 2000). Incision and terracing in the fluvial realm coincides with the onset of the forced regression, which ends when aggradation of the valley-fill sequence starts. The basal, master erosion surface is a compound sequence boundary that therefore develops diachronously. There has been much discussion in the literature concerning the positioning of the sequence-bounding stratal surface (summarized in Catuneanu 2006 and Catuneanu et al. 2009), but rather than being positioned as instantaneous events either at the onset of base-level fall or the end of base-level fall we suggest that the compound surface evolves and can be re-sculpted by continuing incision and terracing during the entire duration of the falling-stage systems tract.

The upper terrace surfaces are not strictly speaking erosional surfaces since they owe their origin simply to a switch from aggradation to degradation. However, because incision and terracing may be continuing elsewhere (as may be suggested by the pedification of the terrace top), the surface could therefore be designated as a terrace-top sequence boundary since it represents an exposed and potentially erosional surface during continuing base-level fall. Although these flat upper terrace surfaces represent time gaps and are thus stratigraphic surfaces of importance, they are commonly not extensive due to the continuation of incision. It may therefore be more appropriate to define this surface sequence stratigraphically as a flooding surface since it also represents a back-stepping of facies belts and often demarcates the boundary between fluvial terrace materials and estuarine incised-valley-fill materials (see outer and middle incised-valley segments closer to the coast in Zaitlin et al. 1994). The deposits above this terrace-top flooding surface mostly represent the transgressive systems tract since deposits related to the lowstand, if preserved at all, are only present in the very lowest part of the valley feature.

An obvious facies variant exists up-dip of the limit of marine sedimentation where the terrace-top surface is first buried by fluvial deposits rather than marine deposits. In this case, in the absence of a distinct lithological change between the terrace fill and the incised-valley fill or evidence for a transgressive lag, or transgressive surface of erosion, the terrace-top surface may not be easy to distinguish (Shanley and McCabe 1993). A pedified terrace top, if present, is the best indicator as to the presence of the hiatus associated with the terrace-top surface.

Some interpreters may wish to view terraces as negative-accommodation parasequences since each terrace is composed of a discrete aggradational pulse.

DISCUSSION

River Terraces—Why an Overlooked Landform in Geological Interpretation?

Geomorphologists have for many years recognized that terraces are a common, integral part of many landscapes. By comparison, however, geologists working in the subsurface rarely describe and interpret terraces. Confirmation of this imbalance comes from Gibling (2006), who states that although terraces are widely represented in modern valleys, they are rarely identified in datasets. So, why does an apparent disparity exist between river and rock record?

Terraces Fall into a Subsurface Data-Resolution Gap: The primary reason why terraces go unrecognized in the subsurface is that they fall into a data-resolution gap. Put simply, mesoscale architectural elements such as terraces are generally too small to be observed on seismic and too big for core and wireline log data (Fig. 15).

This study explored the question: "what would it take to observe fluvial terraces seismically?", and through forward seismic modelling the results have illustrated the geophysical challenges associated with resolving terraces. The forward models demonstrate that it would take a highly favorable combination of terrace thickness, acoustic contrasts, and data frequency to observe terraces seismically, so it is clear from our results that terraces are not readily identifiable on seismic in most subsurface environments.

High-resolution, shallow 3D seismic data and the geophysical techniques of time and horizon slicing offer the best opportunity to potentially image terraces (Ethridge and Schumm 2007). The practice of high-resolution seismic geomorphology in shallowly buried alluvial strata of Quaternary age has facilitated the imaging of terraces in the Java Sea Shelf (Posamentier 2001) and in the Gulf of Mexico (Simms et al. 2007).

Other geophysical methods, including ground-penetrating radar and shallow resistivity surveys, may be of sufficiently high resolution to detect terraces in the shallow subsurface; however, a proper evaluation of the utility of these techniques for the detection of shallow-burial terraces and their deposits is outside the scope of this paper.

The results of the low-frequency forward-modelling experiments in this study may be more analogous to deeper seismic records—for example, the deeply buried fluvial successions of the Carboniferous of the Southern North Sea, where we speculate that fluvial formations could include sub-seismic-scale terraced incision surfaces in association with incised valleys.

At the small-scale end of the data-resolution gap, the scale of the architectural complexity that terraces represent is simply too large to resolve from a single core or wireline log. Core descriptions of fluvial successions often include interpretations of "floodplain" depositional environments. Although no doubt correctly interpreted in terms of environment, sedimentologists would rarely interpret "*terraced floodplain*" architectures from conventional core, several centimeters wide. We suggest that a proportion of what is typically interpreted as floodplain facies in core has the architectural expression of a terrace.

Although terraces are likely to have been more commonly sampled than traditional single-well facies interpretations would suggest, it is the essentially 1D nature of core and wireline data that precludes the interpretation of terraces directly from single-well studies. Multiple wells with close well spacing, chronostratigraphically overlapping core coverage, and the integration of core, wireline, seismic, and pressure and production data are all required to enable the interpretation of what have previously been termed "valley and terrace" or "valley cut and fill" deposits in the subsurface (e.g., Arnott et al. 2002; Potocki et al. 1997; Zaitlin et al. 2002).

Terraces Underinterpreted in Outcrop: If, as we advocate in this paper, terraces fall into a subsurface data-resolution gap between seismic and core and wireline data, then the optimally scaled environment to observe terraces in the rock record is surely outcrop. Intriguingly, though, outcrop studies of fluvial successions do not include many documented examples of terraces.

Bestland (1997) provides one of the few examples of paleo-terraces that have been recognized in outcrop. The Painted Hills study area in Oregon, U.S.A., exposes early Oligocene rocks sufficiently well to allow the observation of a truncation surface with a stepped geometry where nearly horizontal sections (treads) are separated by steep sections (risers) with relief of 4–5 m. The "cut and fill" episodes are interpreted by Bestland (1997) to be in response to changes in climate, hinterland vegetation, and therefore sediment supply.

The dearth of interpreted outcrop examples found in the literature suggests that there may be limitations and difficulties surrounding the identification of terraces in outcrop. One geometrical issue is that 2D outcrop sections are often not perfectly oriented as a cross-valley profile, suggesting that oblique outcrop observations may be a limiting factor in their recognition. Furthermore, the lateral extent of 2D sections would have to be large and favorably oriented to expose both sides of an incised valley which would be required to identify paired terrace suites. The available field of view strongly conditions the scale of observations we are able to make at outcrop, and mesoscale architectural features such as terraces require large outcrops for them to be recognized in their full stratigraphic context.

At outcrop, a terrace appears, for most of its extent, as part of a simple succession. This is because the greatest proportion of the extent of any terrace simply possesses a conformable upper tread surface and a sharp erosional contact with the substrate below. Only a very small volumetric percentage of a terraces 3D geometry is actually characterized by an angular erosional scarp, bounding the terrace edge (either associated with the first cut phase or a subsequent incisional event). Many terrace edges or "scalloped scarps" may have been mistakenly described at outcrop as channel margins due to the limited horizontal and vertical scales of investigation that are required to provide stratigraphic context for a terrace interpretation to be confidently made.

Since the term terrace describes purely architecture and does not carry a lithological connotation, terrace fills can be made up of either floodplain or channel facies and the channel facies may or may not contain a basal lag. Terraces may or may not be pedified, depending on the length of exposure time and the rate of soil formation in the given climatic setting (Kraus 1999). If environmental conditions and exposure times are conducive to soil formation then the pedification of the upper surface of the terrace remnant, in close association with (a terrace-fill thickness above) a sequence boundary may be the strongest lithologic evidence for terracing.

Detailed criteria for the identification of terraces in outcrop need to be outlined in future work, but here we speculate that some of what has previously been interpreted as floodplain or interfluve may actually possess a terraced architecture if the full scale (vertical and horizontal extent) of the whole mesoscale architectural element could have been properly observed.

Terraces May Have Low Preservation Potential: The possibility exists that terraces may be entirely or partially eroded by a range of postdepositional processes. Even if terraces do survive the preservational passage into the rock record, they are not likely to appear in the pristine geomorphic form that they are often depicted in text-books or as we currently observe them in underfilled, degradational landscapes.

Flume-tank experiments undertaken by Strong and Paola (2006) nicely demonstrate the difficulty of preserving falling-stage deposits such as terraces. Valley-widening processes, active during the lowest part of the base-level transit cycle, may well undercut, erode, and rework the terraces that were deposited during base-level fall. Catastrophic floodplain stripping has been documented by Nanson (1986), and whole-scale channel-belt scour of this nature would obviously pose a problem for the preservation of terraces. The possibility also exists that terraces can be eroded into by tributaries, which would decrease their longitudinal continuity. The degradation of terraces by interfluve-sourced drainage during continuing base-level fall has been previously recognized as an important process (Leeder and Stewart 1996; Posamentier 2001; Gibling et al. 2005) that will partially determine the preservation potential of terraces. In certain climatic settings eolian process may act to deflate and rework terraces, depending on their consolidation and vegetation cover. Depending on the slope of the valley, climatic setting, and lithological makeup of the valley sides, hillslope processes and mass movements may reduce terrace preservation potential. Colluvial deposits have been recorded in association with terracing by Bestland (1997).

As confinement decreases during base-level rise, valley expansion due to valley-wall erosion and lateral wandering by the fluvial system have also been proven experimentally by Strong and Paola (2006) to reduce the preservation potential of terraces. Where base-level rise is accompanied by a marine transgressive surface of erosion, the terraces may be truncated or entirely eroded. These ravinement surfaces, capable of eroding into older terraced strata, can be related to tidal or wave processes (Zaitlin et al. 1994). However, where base-level rise constitutes a fast, non-erosive transgressive event as interpreted using shallow seismic in the Java Sea by Posamentier (2001), then a better chance of preserving terraces exists.

Terraces are Most Common in Actively Denuding Landscapes: Many of the terraces and terrace flights described in the geomorphological literature are from upland areas undergoing long-term erosion and are not considered part of actively subsiding sedimentary basins. These regions, although they can provide helpful analogues to better understand the formation of terraces, rarely become part of the stratigraphic record and as such must be used as subsurface analogues with extreme caution (Weissmann et al. 2010). Although terraces may be more common in eroding catchment areas than in the middle reaches of fluvial systems, we contest that terraces can also form in actively subsiding sedimentary basins and therefore cannot be ruled out of rock-record interpretations. It is not correct to categorize terraces as purely hinterland erosional landforms, since their long-term preservation potential, and ability to be locked in as part of the geological record, is proven in ancient-record outcrops (Bestland 1997), in Quaternary depositional systems observed in the field (Blum 1993), and on high-resolution 3D seismic data (Posamentier 2001; Simms et al. 2007).

Terracing in Other Depositional Settings: Terraces are known to form in many other depositional settings including tidal channels, shallow marine shorefaces (Walker and Eyles 1991), deepwater canyons and deepwater channel complexes, revealing that the results of the seismic forward modelling in this paper are farther reaching than just the fluvial realm. Terraces ought to be an integral part of depositional records where cut-and-fill processes are in action. However, because the term terraces was first generated as a descriptive geomorphological term with particular reference to rivers, there appears to have been a reluctance to apply the term in nonfluvial depositional environments.

Implications for the Allocation of Geological Time in Fluvial Strata

Wheeler (1964) noted that "many inadequacies of concept and practice stem from the popular notion that stratigraphy is the science of past sedimentation, to the exclusion of degradation. . .". However, the identification and quantification of missing time in fluvial successions is challenging (Dunay and Hailwood 1995) and therefore many efforts to model fluvial reservoirs simply assume, in the absence of evidence to the contrary, that aggradation was constant. This assumption, although often encouraged by limitations of seismic data, poor chronostratigraphic control, and lack of core coverage, obviously precludes the possibility that time gaps could be present and would overlook the possibility that incisional, terraced sequence boundaries could exist. These shortcomings bring into question the utility of calculations of long-time-averaged sedimentation rate (or subsidence rate) that consider preserved thickness and time-duration estimates (after accounting for compaction).

Key evidence for the existence of time gaps and potential incision in a fluvial succession surrounds the recognition of paleosols (Wright 1989; Hanneman and Wideman 1991; Tandon and Gibling 1994; Kraus 1999). Wright and Marriott (1993) developed an elegant model linking paleosol maturity to changes in base level. Bestland (1997) recognized associations between weakly developed paleosols and rapid depositional rates and contrasted these against well developed colluvially reworked paleosols and incision and terrace formation. An appreciation of the geological time locked into strongly developed mature paleosol formation should trigger at least the consideration of the processes of incision and terracing.

The time locked into some of the oldest and highest terraces developed during the creation of incised valleys and their fill represents the time it takes to both cut and fill the valley. The development of the sequence-bounding interfluve paleosol stops only when the deposition related to the final stages of the valley-fill succession ensues. Leeder and Stewart (1996) postulated that, as the water table drops in response to base-level fall, the paleosols should record a change from relatively wet to relatively dry to relatively wet as the valley fills again.

The process of pedogenesis and the potential for terrace formation during sequence-boundary formation should be central to the choice of key stratal surfaces to be used in the subdivision and correlation of fluvial strata. The fact that the vertical resolution of wireline logs is often not sufficient to "see" paleosols emphasizes the need to acquire core data or high-resolution image logs. The recognition of pedified exposure surfaces with chronostratigraphic significance, and the development of Wheeler diagrams such as Figure 2, are critical to ensuring a more sophisticated allocation of geological time in fluvial stratigraphic models.

Implications for the Correlation of Fluvial Reservoirs

The process of incision and terrace formation has implications for the correlation of fluvial successions. Sequence boundaries are often cited as important surfaces for correlation (e.g., Van Wagoner et al. 1990). The formation of terraces can add stratigraphic complexity to a sequence boundary. Correlation of sequence boundaries may not be straightforward where it can be demonstrated that the erosion surface is compound and time transgressive in nature.

Reservoir-characterization efforts often use intrareservoir floodplain shales as correlation markers, but floodplain shales could represent discontinuous terrace fragments of the same or different age, even though log and core character are similar. Even in the rare examples where the same-age shale can be unequivocally correlated, there is no guarantee that it has not been eroded out between wells by a later phase of incision and terracing acting at the interwell scale (for example, the correlation of terrace A with terrace A′ following cut phase b in Figure 2). This simple example provides a cautionary reminder that correlation between wells does not necessarily guarantee physical connection.

In the case of a simple conglomeratic terrace fill, terraces could wrongly be correlated in wells as a continuous, basal valley-fill lag deposit, when in reality the conglomeratic terraces might not be chronostratigraphically equivalent or even physically connected (Fig. 6B).

As was revealed in Figure 2, there are counterintuitive age–height relationships that exist in terrace complexes as the terraces young and become deeper towards the valley center. These age–height relationships could add a currently underappreciated level of complexity to the correlation of fluvial deposits in the subsurface.

The complex labyrinth-like reservoir architecture that is commonly assigned to fluvial reservoirs may, in part, be due to the often unseen

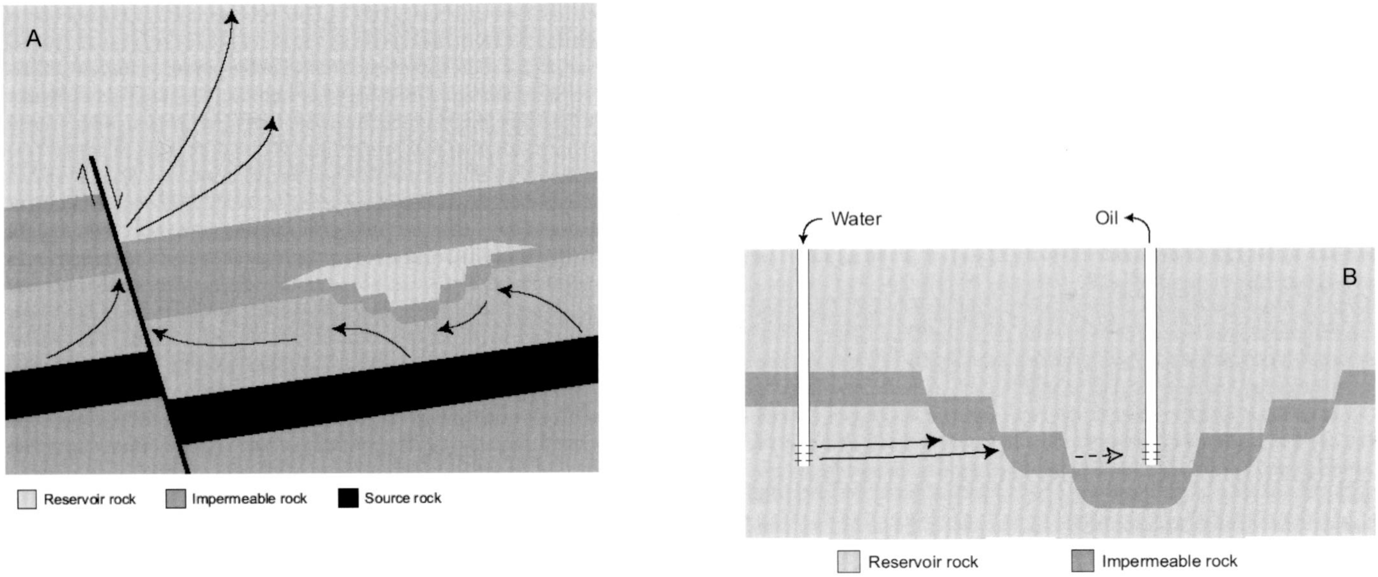

FIGURE 16.—Implications of terraces for subsurface fluid flow. The potential impact of fluvial terraces is shown in two potential scenarios. A) Exploration scale. Impermeable fluvial terraces line the base of an incised-valley fill and encase the sand fill such that it cannot be charged by migrating hydrocarbons. B) Production scale. Impermeable fluvial terraces can add a level of architectural heterogeneity that is often overlooked in the design of waterfloods.

architectural complications of terracing. Channels are often implicated as the primary source of complexity in fluvial reservoirs, but although the formation of elevated floodplains and terracing is implicit in the process of deep channelization, terraces still remain largely a geomorphic term and have not yet become an integral part of many interpretations of the ancient rock record.

One subsurface example where incision and terracing has caused correlation challenges is the Lower Cretaceous Basal Quartz system of incised valleys in Alberta (Arnott et al. 2002). This low-accommodation foreland-basin setting results in a relatively condensed fill succession, with little space between unconformity-bounded sequences, a general lack of floodplain fines, and the repeated cannibalization of older deposits by younger. A complex arrangement of old paleovalley remnants (termed terraces and mesas in Arnott et al. 2002) preserved between younger valley fills leads to rapid changes in lithology and therefore reservoir quality over short distances. Korus et al. (2008) also recognized stratigraphic complexity associated with the development of compound paleovalley fills in the Carboniferous of West Virginia, USA.

Implications for Geological Modelling

Geological modelling efforts endeavor to capture the critical geological factors that control fluid flow in the subsurface, since geomodels are routinely used as inputs into reservoir simulators for the purpose of history matching and ultimately production forecasting. Incision and terracing produces a mesoscale of architectural complexity that should be considered during the construction of subsurface geological models. It is critical that static and dynamic data be integrated where possible and that core, wireline logs, seismic, pressure, and production data where available be included in any reservoir-characterization effort. Zaitlin et al. (2002) point out that although the valley and terrace deposits of the Lower Cretaceous Basal Quartz are 40 m thick, they are difficult to map in the absence of core and production data. Potocki et al. (1997) was able to recognize valley and terrace incision through the integration of both static and dynamic data.

Should sufficient static or dynamic evidence exist to support the inclusion of terraces into subsurface models, possible geological modelling solutions include the deterministic addition of individual terrace geobodies, in association with key hiatal surfaces. In order to closely represent the terraced architectures observed in modern analogues, an object-based modelling approach with sufficiently high vertical resolution to include individual terraces would be most appropriate. If attached terraces are believed to exist, these could also be incorporated using amalgamated geobodies that represent terrace flights rather than individual terraces. Alternatively, if the detailed geometry and architecture of the terraces is poorly understood, or the geomodel pixel based, then the vertical permeability could simply be altered manually in the simulator by the application of a transmissibility multiplier to cells adjacent to a sequence-bounding stratal surface.

Implications for Subsurface Fluid Flow and Compartmentalization

The main economic application of these observations is that complex terraced architectures and their fills have implications for fluid flow. This is because terraces can impact mesoscale and macroscale vertical and horizontal permeability. Industrial and societal applications that could benefit from a greater appreciation of incision and terracing include oil and gas exploration and production, extraction from and recharge of ground-water aquifers, coal mining, and the prospecting of alluvial placers for ore deposits (e.g., uranium or gold).

Figure 16 illustrates attached terrace fills acting as barriers and baffles to fluid flow and negatively impacting the exploration for and production of hydrocarbons. Figure 16A reveals the potential difficulty of charging incised-valley fills during hydrocarbon migration and Figure 16B illustrates a scenario that has obvious implications for water-injection schemes due to the reduction of lateral connectivity.

Two critical parameters control the potential impact of terracing on subsurface fluid flow: attached versus detached architectural status, and the nature of the fill material. The impact that terraces have on subsurface fluid flow and communication depends greatly on the material that fills the terraces, and it is worth pointing out that terrace fills can be characterized by single or multiple lithologies both within and between terraces (Fig. 17). Zaitlin et al. (2002) provides an example of the use of detailed petrography to differentiate subtle mineralogical differences between units in incised-valley and terrace deposits, and Ratcliffe et al. (2004), building on this work, was able to demonstrate how chemostratigraphy and heavy-mineral analysis were employed to differentiate stratigraphic units from a slightly younger part of the same Basal Quartz succession.

The nature of the fill material becomes more important in the case of attached terraces where the substrate and valley fill are not in direct communication. The maximum sealing potential of terraces, as shown in Figure 6A and 17, is the combination of attached terraces (developed during low rates of incision or high sediment supply) and fine-grained floodplain fills.

Although we have emphasized the potential pitfalls that terraces may bring to the issue of subsurface fluid flow, there are also potential benefits. For example, terraces could be involved in the development of stratigraphic traps (e.g., Wood and Hopkins 1992) associated with incised valleys. Within a trap, in the case of a coarse-grained terrace-fill material that simply cannibalizes and recycles material into younger terraces to form an attached flight, then the terraces could create a flow unit of its own. The potential exists that this flow unit could act as a high-performance reservoir (or equally a thief zone during secondary recovery). High-permeability, coarse-grained valley-fill lag deposits could theoretically connect back to permeable, older (and higher) terraces. Significantly, this hydraulically connected flow unit would crosscut stratigraphic boundaries and timelines.

In addition to the primary lithology of the fill, secondary modifications due to pedogenesis are also important to the macro-permeability structure of terrace flights. A depositional hiatus on a floodplain leads to pedification, and in dryland environments, lengthy exposure times can lead to the formation of aridisols. Aridisol soil profiles have added importance since, if sufficiently mature and laminar in structure, they can be impermeable and act as barriers to subsurface fluid flow. As shown in Figure 17, paleosol maturity decreases from the oldest to youngest terraces as a function of exposure time (Wright 1989). Less mature paleosols tend to occupy the centers of incised valleys, and this may explain subtle geographic differences in production behavior in incised-valley reservoirs, where vertical communication is known to be better in the valley center than at the valley margins.

As was mentioned in the correlation section, a key difference between reservoir-characterization efforts in purely aggradational systems versus those in which incision and terracing exist is the potential that both reservoir rocks or non-reservoir seal rocks may be eroded into and could therefore be discontinuous. Physical detachment of flow units due to incision can increase the degree of compartmentalization and may require the drilling of more wells to access resources.

In summary, from the standpoint of hydrocarbon exploration and production, the existence of terraces can have positive and negative implications for charging, trap formation, and connectivity in the subsurface. If terraces are attached, fine-grained, pedified, and extensive in the depositional dip direction then they could represent an underrecognised, yet important, form of stratigraphic compartmentalisation.

CONCLUSIONS

This paper addresses why there is a disparity between how commonly fluvial terraces are observed by the geomorphological community and the relative lack of recognition of terraces in the ancient record by geologists. Clearly the largest proportion of the disparity is due to the large volume of literature regarding terrace formation in uplifting hinterlands. In addition to this factor, two key data-resolution and scale-of-investigation issues have combined to hamper the consideration and inclusion of terraces into interpretations of ancient fluvial and alluvial successions.

- The existence of an inherent subsurface data-resolution gap that exists between seismic and core and wireline data.
- Vertical and horizontal scale limitations and the lack of true three-dimensionality at outcrop.

Another reason why we often fail to interpret terraces in the rock record is that there is currently a lack of criteria for the recognition of terraces.

The results of a simple seismic forward-modelling experiment demonstrate that terraces, if present, may be observed on seismic only in very specific circumstances. Architectural (terrace thickness), lithological (acoustic-impedance contrasts), and geophysical (seismic frequency) characteristics must interact positively in order for terraces to be imaged directly. In many cases, a stepped appearance to the composite sequence-bounding surface may be the only evidence indicative of terracing.

Because terraces are not readily observable on seismic data, does not mean that they do not exist in subsurface fluvial successions (absence of evidence is not evidence of absence). We emphasize that although erosive pulses and the formation of terraces may be easily overlooked during seismic interpretation, it is the multidisciplinary integration of geological, geophysical, and engineering data (static and dynamic if available) across a spectrum of scales that is required to reveal the presence of incision and terracing in the subsurface.

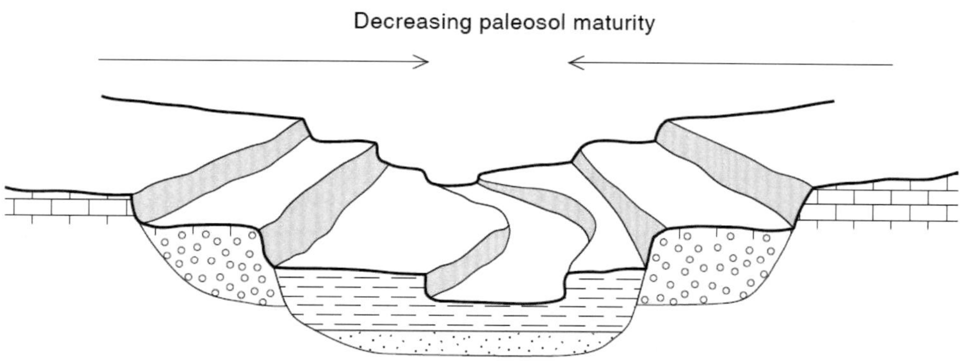

FIGURE 17.—Illustration of the range of possible lithological fill types within a flight of attached terraces.

The existence of terraces can have important implications for subsurface fluid flow at both exploration and production scales. As this paper demonstrates, this is especially true if the terraces are attached, fine-grained and have the potential to form a compound stratigraphic seal. The potential for stratigraphic compartmentalization suggests that terraces may play an influential role in the creation of stratigraphic traps and in the production behavior of some fluvial hydrocarbon reservoirs. The omission of this intermediate scale of stratigraphic architectural complexity from interpretations of erosion surfaces, in conjunction with the common problem of lack of chronostratigraphic control, may partially explain the correlation difficulties so often experienced in ancient fluvial successions.

In conclusion, there is a strong physical process-based case for the formation and preservation of terraces in the rock record, and they must be incorporated into geological interpretations on a more frequent basis if the present is *truly* the key to the past.

ACKNOWLEDGMENTS

The seismic forward-modelling work was undertaken as part of an M.Sc. project by Fjalar Reynisson at the University of Aberdeen (Reynisson 2005), under the supervision of Stuart Archer and Anne Schwab. Some of the ideas in this paper were previously presented by the lead author at a joint BGRG–BSRG International Conference on Drylands held at the Geological Society of London in 2005 (Archer et al. 2005). AUDRI Phase I sponsors (Anadarko, BP, ConocoPhillips, Shell, and Total) are thanked for funding part of this research. The authors are extremely grateful to the reviewers Brian Zaitlin and Frank Ethridge, and editors Stephanie Davidson and Colin North, for their suggested improvements to the manuscript. Barry Fulton is thanked kindly for draughting many of the figures.

REFERENCES

Amorosi, A., Pavesi, M., Lucci, M.R., Sarti, G., and Piccin, A., 2008, Climatic signature of cyclic fluvial architecture from the Quaternary of the central Po Plain, Italy: Sedimentary Geology, v. 209, p. 58–68.

Anderson, E., Harrison, S., Passmore, D.G., Mighall, T.M., and Wathan, S., 2004, Late Quaternary river terrace development in the Macgillycuddy's Reeks, southwest Ireland: Quaternary Science Reviews, v. 23, p. 1785–1801.

Antoine, P., Lautridou, J.P., and Laurent, M., 2000, Long-term fluvial archives in NW France: response of the Seine and Somme rivers to tectonic movements, climatic variations and sea-level changes: Geomorphology, v. 33, p. 183–207.

Archer, S.G., 2006, The sequence stratigraphy and correlation of dryland fluvial systems: a geomorphic perspective from the Basin and Range Province, SW USA: unpublished PhD thesis, University of Aberdeen, Aberdeen, UK, 486 p.

Archer, S.G., Fordham, A.M., Hartley, A.J., North, C.P., and Warwick, G.L., 2005, River terraces—a forgotten landform in subsurface modelling? (abstract), *in* Drylands—Linking Landscape Processes to Sedimentary Environments: BGRG/BSRG International Conference, Geological Society of London.

Arnott, R.W.C., Zaitlin, B.A., and Potocki, D., 2002, Stratigraphic response to sedimentation in an accommodation-limited setting, Lower Cretaceous Basal Quartz, south-central Alberta, Canada: Bulletin of Canadian Petroleum Geology, v. 50, p. 92 –104.

Berryman, K., Marden, M., Eden, D., Mazengarb, C., Ota, Y., and Moriya, I., 2000, Tectonic and paleoclimatic significance of Quaternary river terraces of the Waipaoa River, east coast, North Island, New Zealand: New Zealand Journal of Geology and Geophysics, v. 43, p. 229–245.

Bestland, E.A., 1997, Alluvial terraces and paleosols as indicators of early Oligocene climate change (John Day Formation, Oregon): Journal of Sedimentary Research, v. 67, p. 840–855.

Bishop, D.G., 1991, High-level marine terraces in western and southern New Zealand: Indicators of the tectonic tempo of an active continental margin, *in* Macdonald, D.I.M. ed, Sedimentation, Tectonics and Eustasy: International Association Sedimentologists, Special Publication 12, p. 67–78.

Bohacs, K.M., Carroll, A.R., Neal, J.E., and Mankiewicz, P.J., 2000, Lake-basin type, source potential, and hydrocarbon character: an integrated-sequence-stratigraphic–geochemical framework, *in* Gierlowski-Kordesch, E.H., and Kelts, K.R., eds., Lake Basins through Space and Time: American Association of Petroleum Geologists, Studies in Geology 46, p. 3–34.

Blum, M.D., 1993, Genesis and architecture of incised-valley fill sequences: a Late Quaternary example from the Colorado River, Gulf Coastal Plain of Texas, *in* Weimer, P., and Posamentier H. W., eds., Siliciclastic Sequence Stratigraphy: Recent Developments and Applications: The American Association of Petroleum Geologists, Memoir 58, p. 259–283.

Blum, M.D., and Price, D.M., 1998, Quaternary alluvial plain construction in response to interacting glacio-eustatic and climatic controls, Texas Gulf Coastal Plain, *in* Shanley, K.W., and McCabe, P.J. eds., Relative role of eustacy, climate and tectonism in continental rocks: SEPM, Special Publication 59, p. 31–48.

Blum, M.D., and Törnqvist, T.E., 2000, Fluvial responses to climate and sea-level change: a review and look forward: Sedimentology, v. 47, p. 2–48.

Blum, M.D., and Valastro, S. Jr., 1994, Late Quaternary sedimentation, Lower Colorado River, Gulf Coastal Plain of Texas: Geological Society of America, Bulletin, v. 106, p. 1002–1016.

Bridgland, D.R., 2000, River terrace systems in north-west Europe: an archive of environmental change, uplift and early human occupation: Quaternary Science Reviews, v. 19, p. 1293–1303.

Bridgland, D., and Westaway, R., 2008, Climatically controlled river terrace staircases: a worldwide Quaternary phenomenon: Geomorphology, v. 98, p. 285–315.

Bridgland, D., Maddy, D., and Bates, M., 2004, River terrace sequences: templates for Quaternary geochronology and marine-terrestrial correlation: Journal of Quaternary Science, v. 19, p. 203–218.

Bull, W.B., 1991, Geomorphic responses to climate change: New York, Oxford University Press, 326 p.

Burbank, D.W., and Anderson, R.S., 2001, Tectonic Geomorphology: Malden, Blackwell, 274 p.

Burnett, A.W., and Schumm, S.A., 1983, Alluvial river response to neotectonic deformation in Louisiana and Mississippi: Science, v. 222, p. 49–50.

Catuneanu, O., 2006, Principles of Sequence Stratigraphy: Amsterdam, Elsevier, 375 p.

Catuneanu, O., Abreub, V., Bhattacharya, J.P., Blum, M.D., Dalrymple, R.W., Eriksson, P.G., Fielding, C.R., Fisher, W.L., Galloway, W.E., Gibling, M.R., Giles, K.A., Holbrook, J.M., Jordan, R., Kendall, C.G. St.C., Macurda, B., Martinsen, O.J., Miall, A.D., Neal, J.E., Nummedal, D., Pomar, L., Posamentier, H.W., Pratt, B.R., Sarg, J.F., Shanley, K.W., Steel, R.J., Strasser, A., Tucker, M.E., and Winker, C., 2009, Towards the standardization of sequence stratigraphy: Earth-Science Reviews, 92, p. 1–33.

Cecil, C.B., 2003 The concept of autocyclicity and allocyclicity controls on sedimentation and stratigraphy, emphasizing the climatic variable, *in* Cecil, C.B., and Edgar, N.T., eds., Climate Controls on Stratigraphy: SEPM, Special Publication 77, p. 13–20.

Colombo, F., Busquets, P., Ramos, E., Vergés, J., and Ragona, D., 2000, Quaternary alluvial terraces in an active tectonic region: the San Juan River Valley, Andean Ranges, San Juan Province, Argentine: Journal of South American Earth Sciences, v. 13, p. 611–626.

Dalrymple, R.W., Boyd, R., and Zaitlin, B.A., 1994, History of research, types and internal organisation of incised-valley systems: introduction to the volume, *in* Dalrymple, R.W., Boyd, R., and Zaitlin, B.A., eds., Incised-Valley Systems: Origin and Sedimentary Sequences: SEPM, Special Publication 51, p. 3–10.

Davies, S., Hampson, G., Flint, S., and Elliott, T., 1999, Continental-scale sequence stratigraphy of the Namurian, Upper Carboniferous and its applications to reservoir prediction, *in* Fleet, A.J., and Boldy, S.A.R., eds., Petroleum Geology of Northwest Europe: Proceedings of the 5th Conference: Geological Society of London, p. 757–770.

Dolson, J., Muller, D., Evetts, M.J., and Stein, J.A., 1991, Regional paleotopographic trends and production, muddy sandstone (Lower Cretaceous), Central and Northern Rocky Mountains: American Association of Petroleum Geologists, Bulletin, v. 75, p. 409–435.

DUNAY, R.E., AND HAILWOOD, E.A., 1995, Non-biostratigraphical methods of dating and correlation: an introduction, *in* Dunay, R.E., and Hailwood, E.A., eds., Non-biostratigraphical Methods of Dating and Correlation: Geological Society of London, Special Publication 89, p. 1–2.

ETHRIDGE, F.G., AND SCHUMM, S.A., 2007, Fluvial seismic geomorphology: a view from the surface, *in* Davies, R.J. Posamentier, H.W., Wood, L.J., and Cartwright, J.A., eds., Seismic Geomorphology: Applications to Hydrocarbon Exploration and Production: Geological Society of London, Special Publication 277, p. 205–223.

ETHRIDGE, F.G., WOOD, L.J., AND SCHUMM, S.A., 1998, Cyclic variables controlling fluvial sequence development: problems and perspectives, *in* Shanley, K.W., and McCabe, P.J., eds., Relative role of Eustacy, Climate and Tectonism in Continental Rocks: SEPM, Special Publication 59, p. 17–29.

FISK, H.N., 1944, Geological investigation of the alluvial valley of the lower Mississippi River: U.S. Army Corps of Engineers, Mississippi River Commission, Vickburg, 78 p.

GARDNER, G.H.F., GARDNER, L.W., GREGORY, A.R., 1974, Formation velocity and density: the diagnostic basics for stratigraphic traps: Geophysics, v. 39, p. 770–780.

GILBERT, G.K., 1879, Geology of the Henry Mountains: U.S. Government Printing Office, Washington DC, 170 p.

GIBLING, M.R., 2006, Width and thickness of fluvial channel bodies and valley fills in the geological record: a literature compilation and classification: Journal of Sedimentary Research, v. 76, p. 731–770.

GIBLING, M.R., TANDON, S.K., SINHA, R., AND JAIN, M., 2005, Discontinuity-bounded alluvial sequences of the southern Gangetic Plains, India: aggradation and degradation in response to monsoonal strength: Journal of Sedimentary Research, v. 75, p. 369–385.

HANNEMAN, D.L., AND WIDEMAN, C.J., 1991, Sequence stratigraphy of Cenozoic continental rocks, southwestern Montana: Geological Society of America, Bulletin, v. 103, p. 1335–1345.

HOLBROOK, J., 1996, Complex fluvial response to low gradients at maximum regression: a genetic link between smooth sequence boundary morphology and architecture of overlying sheet sandstone: Journal of Sedimentary Research, v. 66, p. 713–722.

HOLBROOK, J., 2001, Origin, genetic interrelationships, and stratigraphy over the continuum of fluvial channel-form bounding surfaces: an illustration from middle Cretaceous strata, southeastern Colorado: Sedimentary Geology, v. 144, p. 179–222.

HOLBROOK, J., AND SCHUMM, S.A., 1999, Geomorphic and sedimentary response of rivers to tectonic deformation: a brief review and critique of a tool for recognizing subtle epeirogenic deformation in modern and ancient settings: Tectonophysics, v. 305, p. 287–306.

HUBBARD, R.J., 1988, Age and significance sequence boundaries on Jurassic and Early Cretaceous rifted continental margins: American Association of Petroleum Geologists, Bulletin, v. 72, p. 49–72.

HUISINK, M., 1997, Late-glacial sedimentological and morphological changes in a lowland river in response to climatic change: the Maas, southern Netherlands: Journal of Quaternary Science, v. 12, p. 209–223.

HUNT, D., AND TUCKER, M.E., 1992, Stranded parasequences and the forced regressive wedge systems tract: deposition during base level fall: Sedimentary Geology, v. 81, p.1–9.

IMBRIE, J., HAYS, J.D., MARTINSON, D.G., MCINTYRE, A., MIX, A.C., MORLEY, J.J., PISIAS, N.G., PRELL, W.L., AND SHACKLETON, N.J., 1984, The orbital theory of Pleistocene climate: support from a revised chronology of the maritime O^{18} record, *in* Berger, A., Imbrie, J., Hays, J., Kukla, G., and Saltaman, B., eds., Milankovitch and Climate: Boston, D. Reidel Publishing Co., p. 269–305.

JONES, S.J., FROSTICK, L.E., AND ASTIN, T.R., 1999, Climatic and tectonic controls on fluvial incision and aggradation in the Spanish Pyrenees: Geological Society of London Journal, v. 156, p. 761–769.

JORGENSEN, D. W., 1990, Adjustment of alluvial river morphology and process to localised active tectonics: Unpublished Ph.D. thesis, Colorado State University, 240 p.

KEMP, J., AND RHODES, E.J., 2010, Episodic fluvial activity of inland rivers in southeastern Australia: Palaeochannel systems and terraces of the Lachlan River: Quaternary Science Reviews, v. 29, p. 732–752.

KOCUREK, G., 1998, Aeolian system response to external forcing factors—a sequence stratigraphic view of the Saharan region, *in* Alsharhan, A.S., Glennie, K., Whittle, G.L. and Kendall, C.G. St.C. Quaternary Deserts and Climate Change: Rotterdam, Balkema Press, p. 327–337.

KOCUREK, G., AND HALVHOLM, K.G., 1993, Eolian sequence stratigraphy: A conceptual framework, *in* Weimer, P., and Posamentier, H.W., eds., Siliciclastic Sequence Stratigraphy: Recent Developments and Applications: Amercian Association of Petroleum Geologists, Memoir 58, p. 393–410.

KORUS, J.T., KVALE, E.P., ERIKSSON, K.A., AND JOECKEL, R.M., 2008, Compound paleovalley fills in the lower Pennsylvanian New River Formation, West Virginia, USA: Sedimentary Geology, v. 208, p. 15–26.

KRAUS, M., 1999, Paleosols in clastic sedimentary rocks: their geologic applications: Earth-Science Reviews, v. 47, p. 41–70.

LANE, E.W., 1955, The importance of fluvial morphology in hydraulic engineering: American Society of Civil Engineers, Proceedings, Journal of the Hydraulics Division, v. 81, no. 745, p. 1–17.

LEEDER, M.R., 1993, Tectonic controls upon drainage basin development, river channel migration and alluvial architecture: implications for hydrocarbon reservoir development and charcterisation, *in* North, C.P. and Prosser, D.J., eds., Characterisation of Aeolian and Fluvial Reservoirs: Geological Society of London, Special Publication 73, p. 7–23.

LEEDER, M.R., AND STEWART, M.D., 1996, Fluvial incision and sequence stratigraphy: alluvial responses to relative sea-level fall and their detection in the geological record, *in* Hesselbo, S.P., and Parkinson, D.N., eds., Sequence Stratigraphy in British Geology: Geological Society of London, Special Publication 103, p. 25–39.

LEOPOLD, L.B., WOLMAN, M.G., AND MILLER, J.P., 1964, Fluvial Processes in Geomorphology: San Francisco, W.H. Freeman and Company, 522 p.

LOWE, J.J., AND WALKER, M.J.C., 1984, Reconstructing Quaternary Environments: Essex, U.K., Longman Scientific and Technical, 389 p.

MACK, G.H., SEAGER, W.R., LEEDER, M.R., PÉREZ-ARLUCEA, M. AND SALYARDS, S.L., 2006, Pliocene and Quaternary history of the Rio Grande, the axial river of the southern Rio Grande rift, New Mexico, USA. Earth-Science Reviews, v. 79, p. 141–162.

MACKIN, J.H., 1937, Erosional history of the Big Horn Basin, Wyoming: Geological Society of America, Bulletin, v. 48, p. 813–893.

MADDY, D., AND BRIDGLAND, D.R., 2000, Accelerated uplift resulting from Anglian glacioisostatic rebound in the Middle Thames Valley, UK: evidence from the river terrace record. Quaternary Science Reviews, v. 19, p. 1581–1588.

MADDY, D., BRIDGLAND, D., AND WESTAWAY, R., 2001, Uplift-driven valley incision and climate controlled river terrace development in the Thames Valley, UK: Quaternary International, v. 79, p. 22–36.

MCCARTHY, P.J., AND PLINT, A.G., 2003, Spatial variability of palaeosols across Cretaceous interfluves in the Dunvegan Formation, NE British Columbia, Canada: paleohydrological, paleogeomorphical and stratigraphic implications: Sedimentology, v. 50, p. 1187–1220.

MCDONALD, E.V., MCFADDEN, L.D., AND WELLS, S.G., 2003, Regional response of alluvial fans to the Pleistocene–Holocene climatic transition, Mojave Desert, California, *in* Enzel, Y., Wells, S.G., and Lancaster, N., eds., Paleoenvironments and Paleohydrology of the Mojave and Southern Great Basin Deserts: Geological Society of America, Special Paper 368, p. 189–205.

MELLMAN, G., AND KUNZINGER, P.A., 1992, Forward modeling of seismic data, *in* Morton-Thompson, D., and Woods, A.M., eds., Development Geology Reference Manual: American Association of Petroleum Geologists, Methods in Exploration Series, v. 10, p. 392–394.

MIALL, A.D., 1991, Stratigraphic sequences and their chronostratigraphic correlation: Journal of Sedimentary Petrology, v. 61, p. 497–505.

MITCHUM, R.M. JR., AND VAIL, P.R., 1977, Seismic stratigraphy and global changes of sea level, part 7: stratigraphic interpretation of seismic reflection patterns in depositional sequences, *in* Payton C.E., ed., Seismic Stratigraphy—Applications to Hydrocarbon Exploration: American Association of Petroleum Geologists, Memoir 26, p. 135–144.

MOLNAR, P., BROWN, E.T., BURCHFIEL, B.C., DENG, Q.D., FENG, X.Y., LI, J., RAISBECK, G.M., SHI, J.B., WU, Z.M., YIOU, F., AND YOU, H.C., 1994, Quaternary climate change and the formation of river terraces across growing anticlines on the North Flank of Tien Shan, China: Journal of Geology, v. 102, p. 583–602.

Morris, W.R., and Normark, W.R., 2000, Sedimentologic and geometric criteria for comparing modern and ancient sandy turbidite elements, *in* Weimer, P., Slatt, R.M., Coleman, J., Rosen, N.C., Nelson, H., Bouma, A.H., Styzen, M.J., and Lawrence, D.T., eds., Deep Water Reservoirs of the World, Gulf Coast Section SEPM Foundation, p. 606–628.

Nanson, G.C., 1986, Episodes of vertical accretion and catastrophic stripping—a model of disequilibrium floodplain development: Geological Society of America, Bulletin, v. 97, p. 1467–1475.

Ouchi, S., 1985, Response of alluvial rivers to slow active tectonic movement: Geological Society of America, Bulletin, v. 96, p. 504–515.

Pan, B., Burbank, D., Wang, Y., Wu, G., Li, J., and Guan, Q., 2003, A 900 k.y. record of strath terrace formation during glacial–interglacial transitions in northwest China: Geological Society of America, Bulletin, v. 31, p. 957–960.

Penck, A., and Brückner, E., 1909, Die Alpen im Eiszeitalter: Leipzig, Verlag Tauchnitz, 1396 p.

Perlmutter, M.A., and Matthews, M.D., 1990, Global cyclostratigraphy—a model, *in* Cross, T.A., ed, Quantitative Dynamic Stratigraphy: Englewood Cliffs, New Jersey, Prentice Hall, p. 233–260.

Plint, A.G., and Nummedal, D., 2000, The falling stage systems tract: recognition and importance in sequence stratigraphic analysis, *in* Hunt, D., and Gawthorpe, R.L., eds., Sedimentary Responses to Forced Regressions: Geological Society of London, Special Publication 172, p. 1–17.

Porter, S.C., An, Z. S., and Zheng, H., 1992, Cyclic Quaternary alluviation and terracing in a nonglaciated drainage basin on the north flank of the Qinling Shan, Central China: Quaternary Research, v. 38, p. 157–169.

Posamentier, H.W., 1988, Fluvial deposition in a sequence stratigraphic framework, *in* James, D.P., and Leckie, D.A., eds., Sequences, Stratigraphy and Sedimentology: Surface and Subsurface, Canadian Society of Petroleum Geologists, Memoir, 15, p. 582–583

Posamentier, H.W., 2001, Lowstand alluvial bypass systems: incised vs. unincised: American Association of Petroleum Geologists, Bulletin, v. 85, p. 1771–1793.

Posamentier, H.W., and Vail, P.R., 1988, Eustatic controls on clastic deposition II—sequence and systems tract models, *in* Wilgus, C.K., Hastings, B.S., Kendall, C.G. St.C., Posamentier, H.W., Ross, C.A., and Van Wagoner, J.C., eds., Sea Level Changes: An Integrated Approach: SEPM, Special Publication 42, p. 125–154.

Potocki, D., Raychaudhuri, I., Thorburn, L., Galas, C., and King, H., 1997, Integrated reservoir characterization of a heterogeneous channel sandstone: the Duchess Lower Mannville X pool: Journal of Canadian Petroleum Technology, v. 38, p. 53–60.

Ratcliffe, K.T., Wright, A.M., Hallsworth, C., Morton, A., Zaitlin, B.A., Potocki, D., and Wray, D.S., 2004, Chemostratigraphic characterization of fine grained facies in a low accommodation setting: an example from the (Lower Cretaceous) Basal Quartz, Southern Alberta: American Association of Petroleum Geologists, Bulletin, v. 88, p. 1419–1432.

Reheis, M.C., Slate, J.L., Throckmorton, C.K., Mcgeehin, J.P., Sarnawojcicki, A.M., and Dengler, L., 1996, Late Quaternary sedimentation on the Leidy Creek fan, Nevada–California: Geomorphic responses to climate change: Basin Research, v. 8, p. 279–299.

Rider, M., 2002, The Geological Interpretation of Well Logs: Rider-French Consulting Ltd., Sutherland, Scotland, 280 p.

Ritter, J.B., Miller, J.R., Enzel, Y., and Wells, S.G., 1995, Reconciling the roles of tectonism and climate in Quaternary alluvial-fan evolution: Geology, v. 23, p. 245–248.

Robertson-Rintoul, M.S.E., 1986, A quantitative soil-stratigraphic approach to the correlation and dating of postglacial river terraces in Glen Feshie, Southwest Cairngorms: Earth Surface Processes and Landforms, v. 11, p. 605–617.

Reynisson, F.R., 2005, The seismic recognition of fluvial terraces: Unpublished M. Sc. thesis, University of Aberdeen, Aberdeen, UK, 105 p.

Schumm, S.A., 1993, River response to baselevel change: Implications for sequence stratigraphy: Journal of Geology, v. 101, p. 279–294.

Schumm, S.A., and Ethridge, F.G., 1994, Origin, evolution and morphology of fluvial valleys, *in* Dalrymple, R.W., Boyd, R., and Zaitlin, B.A., eds., Incised-Valley Systems: Origin and Sedimentary Sequences: SEPM, Special Publication 51, p. 11–27.

Schumm, S.A., and Parker, R.S., 1973, Implications of complex response of drainage systems for Quaternary alluvial stratigraphy: Nature, v. 243, p. 99–100.

Schwab, A.M., and Pince, J-M., 1996, Oligocene–Miocene shallow carbonates in offshore Tunisia: a composite case study for a stratigraphic seismic forward model: Société Géologique de France, Bulletin, v. 167, p. 717–727.

Schwab, A.M., Cronin, B.T., and Ferreira, H., 2007, Seismic expression of channel outcrops: Offset stacked versus amalgamated channel systems: Marine and Petroleum Geology, v. 24, p. 504–514.

Shanley, K.W., and McCabe, P.J., 1993, Alluvial architecture in a sequence stratigraphic framework: a case history from the upper Cretaceous of southern Utah, U.S.A. *in* Flint, S.S., and Bryant, I.D., eds., Quantitative Modeling of Clastic Hydrocarbon Reservoirs and Outcrop Analogues: International Association of Sedimentologists, Special Publication 15, p. 21–55.

Shanley, K.W., and McCabe, P.J., 1994, Perspectives on the sequence stratigraphy of continental strata: American Association of Petroleum Geologists, Bulletin, v. 79, p. 544–568.

Simms, A.R., Anderson, J.B., Milliken, K.T., Taha, Z.P., and Wellner, J.S., 2007, Geomorphology and age of the Oxygen isotope stage 2 (last lowstand) sequence boundary on the northwestern Gulf of Mexico continental shelf, *in* Davies, R.J., Posamentier, H.W., Wood, L.J., and Cartwright, J.A., eds., Seismic Geomorphology: Applications to Hydrocarbon Exploration and Production: Geological Society of London, Special Publication 277, p. 29–46.

Stevenson, I.R., and McMillan, I.K., 2004, Incised valley fill stratigraphy of the Upper Cretaceous succession, proximal Orange Basin, Atlantic margin of southern Africa: Geological Society of London, Journal, v. 161, p. 185–208.

Stewart, A.J., Blake, D.H., and Ollier, C.D., 1986, Cambrian river terraces and ridgetops in central Australia: oldest persisting landforms?: Science, v. 233, p. 758–761.

Stokes, M., and Mather, A.E., 2000., Response of Plio-Pleistocene alluvial systems to tectonically induced base-level changes, Vera Basin, SE Spain: Geological Society of London, Journal, v. 157, p. 303–316.

Strong, N., and Paola, C., 2006, Fluvial landscapes and stratigraphy in a flume: The Sedimentary Record, v. 4, No 2, p. 4–8.

Suter, J.R., and Berryhill, H.L., Jr., 1985, Late Quaternary shelf-margin deltas, Northwest Gulf of Mexico: American Association of Petroleum Geologists, Bulletin, v. 69, p. 77–91.

Talling, P.J., 1998, How and where do incised valeeys form if sea level remains above the shelf edge?: Geology, v. 26, p. 87–90.

Tandon, S.K., and Gibling, M.R., 1994, Calcrete and coal in late Carboniferous cyclothems of Nova Scotia, Canada: climate and sea-level changes linked: Geology, v. 22, p. 755–758.

Thomas, M.A., and Anderson, J.B., 1994, Sea-level control on the facies architecture of the Tertiary/Sabine incised-valley system, Texas Continental Shelf, *in* Dalrymple, R.W., Boyd, R., and Zaitlin, B.A., eds., Incised-Valley Systems: Origin and Sedimentary Sequences: SEPM, Special Publication 51, p. 63–82.

Throne, J., 1994, Constraints on riverine valley incision and the response to sea level change based on fluid mechanics, *in* Dalrymple, R.W., Boyd, R., and Zaitlin, B.A., eds., Incised-Valley Systems: Origin and Sedimentary Sequences: SEPM, Special Publications 51, p. 29–43.

Törnqvist, T.E., Wallinga, J., and Busschers, F.S., 2003, Timing of the last sequence boundary in a fluvial setting near the highstand shoreline—insights from optical dating: Geological Society of America, Bulletin, v. 31, p. 279–282.

Underhill, J., and Partington, M., 1993, Jurassic thermal doming and deflation in the North Sea: implications of the sequence stratigraphic evidence, *in* Parker, J.R., ed., Petroleum Geology of Northwest Europe: Proceedings of the 4th Conference: Geological Society of London, p. 337–346.

Vail, P.R., Mitchum, R.M. Jr., and Thompson, S. III., 1977, Seismic stratigraphy and global changes of sea level, part 4: global cycles of relative changes of sea level, *in* Payton, C.E., ed. Seismic Stratigraphy—Applications to Hydrocarbon Exploration: American Association of Petroleum Geologists, Memoir 26, p. 83–97.

Van Wagoner, J.C., Posamentier, H.W., Mitchum, R.M. Jr., Vail, P.R., Sarg,

J.F., Loutit, T.S., and Hardenbol, J., 1988, An overview of the fundamentals of sequence stratigraphy and key definitions, *in* Wilgus, C.K., Hastings, B.S., Kendall, C.G. St.C., Posamentier, H.W., Ross, C.A., and Van Wagoner, J.C., eds., Sea Level Changes: An Integrated Approach: SEPM, Special Publication 42, p. 39–45.

Van Wagoner, J.C., Mitchum Jr., R.M., Campion, K.M., and Rahmanian, V.D., 1990, Siliciclastic Sequence Stratigraphy in Well Logs, Cores and Outcrop: Concepts for High Resolution Correlation of Time and Facies: American Association of Petroleum Geologists, Methods in Exploration Series, 7, 55 p.

Walker, R.G., and Eyles, C.H., 1991, Topography and significance of basin-wide sequence bounding erosion surfaces in the Cretaceous Cardium Formation, Alberta, Canada: Journal of Sedimentary Petrology, v. 61, p 473–496.

Wallinga, J., Törnqvist, T.E., Busschers, F.S., and Weerts, H.J.T., 2004, Allogenic forcing of the late Quaternary Rhine–Meuse fluvial record: the interplay of sea-level change, climate change and crustal movements: Basin Research, v.16, p.535–547.

Weissmann, G.S., Mount, J.F., and Fogg, G.E., 2002, Glacially driven cycles in accumulation space and sequence stratigraphy of a stream-dominated alluvial fan, San Joaquin valley, California, USA: Journal of Sedimentary Research, v. 72, p. 240–251.

Weissmann, G.S., Hartley, A.J., Nichols, G.J., Scuderi, L.A., Olsen, M., Buehler, H., and Banteah, R., 2010, Fluvial form in modern continental sedimentary basins: Distributive fluvial systems: Geological Society of America, Bulletin, v. 38, p. 39–42.

Wescott, W.A., 1993, Geomorphic thresholds and complex response of fluvial systems—some implications for sequence stratigraphy: American Association of Petroleum Geologists, Bulletin, v.77, p. 1208–1218.

Wheeler, H.E., 1964, Baselevel, lithosphere surface, and time stratigraphy: Geological Society of America, Bulletin, v. 75, p. 599–610.

Wisniewski, P.A. and Pazzaglia, F.J., 2002, Epeirogenic controls on Canadian river incision and landscape evolution, Great Plains of Northeastern New Mexico: Journal of Geology, v. 110, p. 437–456.

Wood, J.M., and Hopkins, J.C., 1992, Traps associated with paleovalleys and interfluves in an unconformity bounded sequence: Lower Cretaceous glaucinitic member, Southern Alberta, Canada: American Association of Petroleum Geologists, Bulletin, v. 76, p. 904–926.

Wright, V.P., 1989, Geomorphic and stratigraphic relationships of alluvial soils: a guide for interpreting ancient palaeosol-bearing sequences, *in* Allen, J.R.L., and Wright, V.P., eds., Palaeosols in Siliciclastic Sequences: University of Reading, UK, Postgraduate Research Institute for Sedimentology, p. 26–48.

Wright, V.P., and Marriott, S.B., 1993, The sequence stratigraphy of fluvial depositional systems: the role of floodplain storage: Sedimentary Geology, v. 86, p. 203–210.

Zaitlin, B.A, Dalrymple, R.W., and Boyd, R., 1994, The stratigraphic organization of incised valley systems associated with relative sea-level change, *in* Dalrymple, R.W., Boyd, R. and Zaitlin, B.A., eds., Incised-Valley Systems: Origin and Sedimentary Sequences: SEPM, Special Publication 51, p. 45–60.

Zaitlin, B.A., Potocki, D., Warren, M.J., Rosenthal, L., and Boyd, R., 2002, Depositional styles in a low accommodation foreland basin setting: an example from the Basal Quartz (Lower Cretaceous), southern Alberta: Bulletin of Canadian Petroleum Geology, v. 50, p. 31–72.

Short Term: Autogenically Dominated?

FLUVIAL SYSTEMS AND THEIR DEPOSITS IN HOT, SEASONAL SEMIARID AND SUBHUMID SETTINGS: MODERN AND ANCIENT EXAMPLES

CHRISTOPHER R. FIELDING

Department of Earth & Atmospheric Sciences, 214 Bessey Hall, University of Nebraska–Lincoln, Nebraska 68588-0340, USA

Email: cfielding2@unl.edu

JONATHAN P. ALLEN

Department of Earth & Atmospheric Sciences, 214 Bessey Hall, University of Nebraska–Lincoln, Nebraska 68588-0340, USA

Present address: Chevron North America Exploration & Production, 9525 Camino Media, Bakersfield, CA 93311, USA

JAN ALEXANDER

School of Environmental Sciences, University of East Anglia, Norwich, Norfolk, NR4 7TJ, UK

MARTIN R. GIBLING

Department of Earth Sciences, Dalhousie University, Halifax, Nova Scotia B3H 3J5, Canada

MICHAEL C. RYGEL

Department of Geology, 44 Pierrepont Avenue, State University of New York at Potsdam, Potsdam, NY 13676, USA

AND

JOHN H. CALDER

Nova Scotia Department of Natural Resources, P.O. Box 698, Halifax, Nova Scotia B3J 2T9, Canada

ABSTRACT: Observations and data from modern streams and recent deposits demonstrate that river systems in subhumid and semiarid seasonal settings in tropical and subtropical latitudes can have deposit characteristics very different from those predicted from previously published facies models based on other climatic settings. The differences result primarily from the extremely variable discharges that are typical of these climatic settings and contribute to a distinct fluvial style. Five critically important characteristics of the deposits are illustrated by data from NE Australia. The deposits with these characteristics occur in streams with variable discharge and are independent of channel size. (1) The channel-fill lithosomes are erosionally based and exhibit complex lateral facies changes. (2) Within the channel-fill lithosomes there are abundant mud partings, some of which are pedogenically modified. (3) In most cases the complex internal architecture lacks the macroform elements, such as lateral-accretion cross bedding, typical of other fluvial sediment bodies. This is the case even within point bars. (4) Sedimentary structures formed under high flow stage are abundant and frequently preserved in the lithosomes. (5) Trees and other vegetation adapted to occasional inundation by fast-flowing water may colonize channel floors (not only on banks), and these both generate organic sediment and influence the flow and resulting sediment deposition patterns. In addition to the characteristics of the channel-fill deposits, the overbank deposits may have diagnostic character and the sand petrography may differ from that in other settings. The same features are identified in the rock record and a detailed case study from the Pennsylvanian of the Maritimes Basin Complex of Atlantic Canada illustrates the distinction between the seasonal tropical and other fluvial styles. The upper Namurian (Yeadonian) Boss Point Formation shows a transition from perennial to subhumid to semiarid, seasonal tropical fluvial styles within a succession of sheet-like braidplain channel bodies. The overlying, basal Westphalian (Langsettian), Little River Formation is composed entirely of more lensoid channel bodies displaying the subhumid to semiarid, seasonal tropical fluvial style, and associated overbank deposits. The overlying coal-bearing Joggins Formation shows a gradual return to a perennial fluvial style with channel body geometries similar to the underlying unit. A clear signal of paleoclimate change can be deconvolved from variations in accommodation regime, providing a hitherto unavailable discriminant for interpreting fluvial successions. Recognition of the strongly seasonal, tropical to subtropical fluvial style in the rock record, and of changes in character through vertical successions, will aid paleoclimate interpretation and subsurface reservoir analysis in fluvial successions.

KEY WORDS: seasonal tropics, subhumid, semiarid, rivers, paleoclimate analysis, alluvial stratigraphy, Pennsylvanian

INTRODUCTION

The interpretation of rocks that are fluvial in origin depends inherently on an understanding of the processes and products of modern rivers. Whilst it may be easy to describe and quantify an ancient fluvial succession from outcrop or core, using these data to predict sedimentary facies in unsampled areas still requires considerable intuition. Data from an analogous modern river or rivers might be expected to aid interpretation of the fluvial succession. However, for an analogue to be useful in the interpretation of palaeoenvironment or for predicting reservoir geology requires that it have a good match to the sedimentary processes operating at physical and temporal scales at

From River to Rock Record: The Preservation of Fluvial Sediments and Their Subsequent Interpretation
SEPM Special Publication No. 97, Copyright © 2011
SEPM (Society for Sedimentary Geology), ISBN 978-1-56576-305-0, p. 89–111.

which the comparisons are made. A major disadvantage of using a direct comparison to improve qualitative understanding of an ancient deposit is that the analogue is likely to be misleading or incorrect if the comparison is uncritically stretched beyond the aspects of the system for which it was proposed. Modern rivers are highly variable (in space and time; e.g., Ethridge, this volume), and their documentation is also variable. To select one river or a group of rivers as an appropriate analogue for an ancient river that deposited sediments about which we have little data, in conditions that we know little of, is at best difficult. An additional problem is that although many modern rivers have been documented, relatively few recent deposits have been studied in sufficient detail to be of great use for the interpretation of ancient counterparts from core or rock face, and variation in the preservation of deposits adds to the variability of facies and architecture. Attempting to allocate modern rivers directly as analogues also does not allow for the variation in conditions controlling sedimentary and preservation processes through geological time (e.g., with evolution of different plants). The only acceptable approach therefore is to look for groups of characteristics of the sedimentary facies that can be compared with groups of characteristics in similar modern fluvial systems. This is the basis of the facies-association approach and of this paper. This is not to support the notion that a new facies model is defined for every river deposit described, but rather for groups of rivers with similar characteristics that are distinctly different from those of previously established facies-association models. This approach also recognizes that rivers will exist with deposit characteristics intermediate between facies association models.

The fluvial facies models embedded in the perception of most sedimentologists, and in the literature, represent only a limited range of possible climatic settings: predominantly humid and to a lesser extent arid environments (e.g., Collinson 1996; Galloway and Hobday 1996; Bridge 2003; 2006). By comparison, deposits of fluvial systems in settings with widely varying discharge are underrepresented and those of the hot semiarid and subhumid climates have been largely ignored, particularly in settings where runoff is both strongly seasonal and very variable from year to year. Large seasonal variation in discharge (by orders of magnitude) occurs in cold climates associated with ice and snow melt and warmer climates associated with monsoon and cyclone activity (hurricanes and typhoons). The seasonal discharge variability can be expressed by the ratio between mean annual flood discharge and mean discharge, but this is not a complete measure of discharge variability, as it excludes inter-annual variability, which is hugely important in areas of monsoon and tropical cyclone activity (particularly in settings where most of the annual rain can fall in a matter of hours or days).

In areas where, for most of the time, there is a considerable excess of evaporation over precipitation, river discharge is low or very low most of the time. When precipitation occurs, either it is too light to produce significant runoff or it may be short-duration, high-intensity rainfall that leads to rapid and efficient runoff. The impact of these short-duration rainfall events may be increased by sparse vegetation cover (related to the generally low near-surface water availability) and dry surfaces (reducing infiltration). In these conditions, stream flow may be minimal or non-existent for long periods, allowing trees to colonize the beds of rivers, where they exploit the water within the channel deposits. The tree species have adapted to this unusual ecological niche where they are intermittently inundated by deep and fast-flowing water. The dearth of facies models for fluvial sediments of hot-climate rivers with highly variable discharge (i.e., variable-discharge rivers where the variability is not related to ice/snow melt) has persisted even though modern rivers of this type drain large portions of Australia, Asia, India, SW North America, South America, and Africa.

Comparison of modern climatic belts with paleogeographic reconstructions suggests that, through much of geological history, large land areas must have been hot or warm and experienced strongly seasonal and erratic rainfall. Consequently it is likely that fluvial deposits formed in these settings must be relatively common in the stratigraphic record. Few ancient fluvial deposits have been attributed to these settings. Deposits with the distinctive fluvial style documented herein have been recognized in the literature (e.g., Rust and Gibling 1990; Browne and Plint 1994; Plint and Browne 1994; Keighley and Pickerill 1996) but typically are regarded as an eccentric variation on known fluvial styles. We submit that the rivers of the hot semiarid and subhumid settings preserve a distinctive type of alluvial deposit, recognition of which may assist in stratigraphic analysis, paleoclimatic interpretation, and reservoir description. In this paper, we summarize the facies characteristics of large modern rivers draining the subhumid tropics of NE Australia in order to illustrate this fluvial style. We then provide a detailed example of this style from the rock record that represents the activity of similar, fluvial

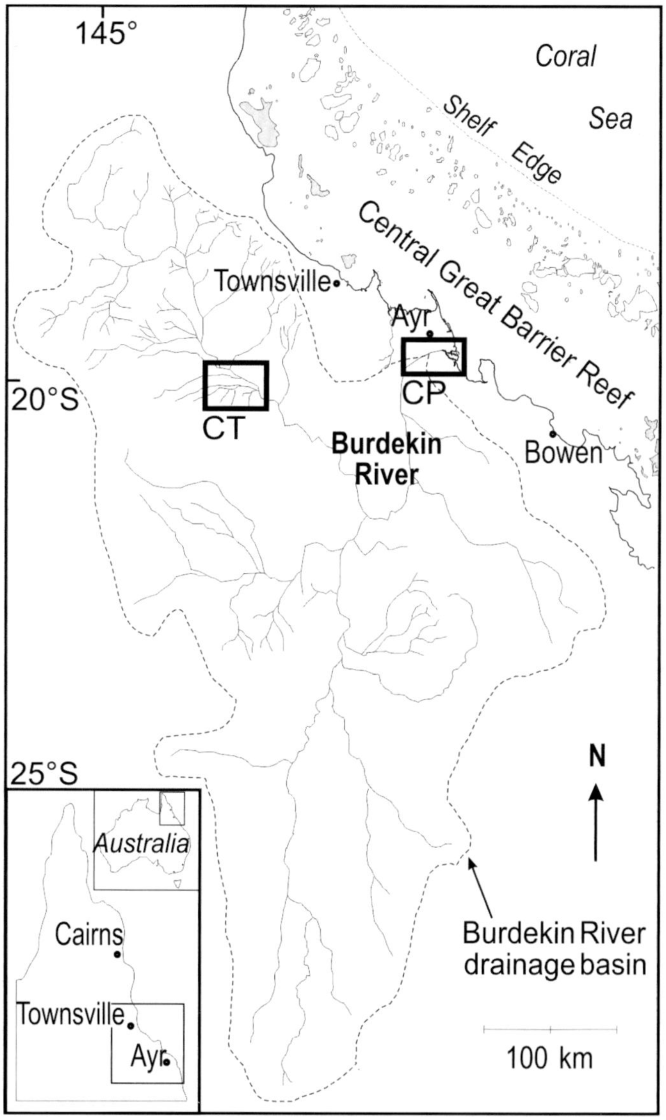

FIGURE 1.—Location of the Burdekin River drainage basin in NE Australia, and of the study sites (outlined) on the upper (Charters Towers, CT) and lower river (Coastal Plain, CP) mentioned in the text.

systems in the paleotropical Pennsylvanian Maritimes Basin complex of Atlantic Canada. We use this ancient example to illustrate applications of this new facies model to resource exploration and development, and to paleoclimatic analysis. This paper expands upon the brief description of the facies model and its applications given in Fielding et al. (2009).

MODERN RIVERS IN THE SUBHUMID TO SEMIARID REGIONS OF QUEENSLAND, NE AUSTRALIA

The subhumid to semiarid regions of NE Australia are drained by a number of rivers that have similar hydrological and sedimentological characteristics. The most studied of these is the Burdekin River System (Fig. 1), which drains an area of 129,000 km^2 and discharges onto the Great Barrier Reef shelf (Fielding and Alexander 1996; Alexander et al. 1999a). The mean annual rainfall is 500–700 mm over most of the catchment, although the coastal strip and eastern headwaters have a more humid climate. The main control on rainfall is the passage of tropical cyclones and monsoonal depressions, which together make the precipitation strongly seasonal, and extremely variable from year to year. Most of the annual precipitation occurs in short-duration intense rainfall events that are associated with the cyclones and depressions, and these generate high-magnitude runoff events. In lower reaches of the Burdekin River for example, discharge is well below 100 m^{3}s^{-1} for most of the time but with the passage of a cyclone or depression through the catchment discharge may rise by several orders of magnitude for just a few days. Discharge events with peak flow of 20,000–30,000 m^{3}s^{-1} have a 10–16 yr return period in the lower reaches while in some years discharge never rises above 3000–5000 m^{3}s^{-1} (Alexander et al. 1999a). Because of the erratic rainfall occurrence, long droughts (of several years) may be broken by sudden intense rainfall events. Consequently vegetation cover may be low, runoff efficient, and local erosion rapid, leading to high suspended-sediment concentrations. High flow competence and sediment availability also tends to allow movement of gravel-grade sediment throughout the system. Amos et al. (2004) found that sediment transport during moderate flow events on the lower Burdekin River was supply-limited, rather than discharge-limited, and this is likely to be more pronounced with larger discharge events.

Event hydrographs for the Burdekin River and other rivers of NE Australia rise extremely rapidly to a sharp peak and decline almost as rapidly (Alexander et al. 1999a). Riverbed macroforms and mesoforms formed under Froude-supercritical flow conditions, and large-scale dunes formed in deep subcritical flow, form in the high-discharge events and are frequently preserved because there is insufficient time for them to adapt to falling-stage conditions. This is in strong contrast to the situation that arises in streams characterized by more gradual changes in discharge, where the river bed is typically gradually modified during falling stage, covered by dunes of various sizes and the subsurface dominated by cross-bedding (Jones 1977).

The hydrology of these Queensland rivers has similarities to that of other strongly seasonal climatic settings, and there is little in the character of discharge that is diagnostic of latitude. However, there are other differences resulting from the warm climate setting that, along with the sedimentary structures, are likely to be diagnostic. Firstly, the combination of strongly variable discharge and river-bed vegetation growth (a consequence of water stress combined with high temperatures and high rates of evapotranspiration) gives rise to the distinctive characteristics that we document herein. Secondly, the weathering products of a hot seasonal climate produce a different petrographic and microfossil record.

In most parts of the drainage basin, the river channels scale with the peak flow of the major flow events (Fielding et al. 1999). Thus flow events with 10–16 yr (or more) return intervals are the channel-forming discharges—unlike most perennial systems, where channels scale with the 1–2 yr events (Williams 1978). It is these major flow events that do the significant work of redistributing sediment and forming the bulk of the alluvium and during which the major channel bars form. During most of any year, however, the channel bed is exposed, with little or no low-stage flow (Fig. 2A–C) and very little sediment movement. In mid reaches of the Burdekin River for example, the channel is as much as 25 m deep and typically 100–500 m wide (increasing to 800 m and more in lower reaches) whereas the flow is mostly less than 1 m deep and perhaps 5–50 m wide, giving the appearance through most of any year of an "underfit" system. This apparent underfit, however, is misleading because water overspills the channel banks during major flow events (Fig. 2D).

The channel is typically of low to moderate sinuosity (average 1.3 in one detailed study area near Charters Towers, Queensland). A variety of sandy barforms is evident on most reaches (Fig. 2); these have the external form of point bars and lateral bars, and are covered by dunes as much as 3 m in height (Fig. 2C), plane beds (Fig. 2E), antidunes (Fig. 2E) (Alexander and Fielding 1997), and gravel patches (Fig. 2C).

Stable arborescent vegetation on the channel floor and on bars is (in the Burdekin, principally the paperbark tree *Melaleuca argentea*) mainly in flow-parallel, linear groves (Fig. 2B; Fielding and Alexander 1996). These linear groves engineer their environment by trapping sediment to form linear, flow-parallel bars (Fielding et al. 1997). Tree seeds are dispersed during major discharge events, deposited in the channel floor and bar deposits, and germinate and grow rapidly in the wet sand during prolonged periods between major flow events. Individual trees (from sapling stage) and shrubs growing on the river bed interact with the flow and transported sediment during subsequent flow events to form obstacle marks, with frontal scour pits, centroclinal cross-bedding (Underwood and Lambert 1974), and current crescents on the downstream side (Fig. 2B; Nakayama et al. 2002). These trees, which adopt a prostrate, downstream-facing stance and have upstream-extending roots that give stability, are often buried within the channel deposits and some accumulate considerable logjams that can also become buried (Fielding et al. 1997; Alexander et al. 1999b; Nakayama et al. 2002). These features can collectively be referred to as vegetation-induced sedimentary structures ("VISS" of Rygel et al. 2004). Detrital plant material is abundant in a variety of within-channel settings (Fig. 2H; Alexander et al. 1999b).

Although the bulk of the sediment deposited within the channel is sand and gravel, mud layers are formed during the waning stage of some large flow events and during smaller discharge events. Mud layers formed in the late stages of large events tend to drape the topography and thicken into swales (Fig. 2F). Mud layers deposited during smaller events form in the deeper parts of channels and on bar edges, whereas bar tops are generally sandy. These mud layers may be buried rapidly in later discharge events (Fig. 2H), and the rapid inundation of the mud preserves desiccation cracks, root tubules, and some color mottling (indicative of Entisol development).

The channel lithosome consists predominantly of well-sorted but compositionally immature sand, with subordinate gravel (Fig. 2G), and partings of plant-debris rich and root-penetrated, color-mottled mud (Fig. 2H). Macroform architectural features that are typical of temperate perennial rivers (e.g., point-bar lateral-accretion sets) are present in some subsurface imagery, but they are typically cryptic or absent from most ground-penetrating radar (GPR) data from the study area (Fig. 3). This is because of the frequent complete reworking of the alluvium of whole bar complexes during major flow events (Fielding et al. 1999) such that they do not record gradual and sequential growth over long periods. GPR surveys across the same cross-stream transect at Dalrymple Bend in two successive years (1996, 1997) demonstrated total rearrangement of the internal alluvial architecture over the intervening period (Fielding et al. 1999). A characteristic of the Burdekin River alluvium is its complex internal heterogeneity (Fig. 3).

A
B
C
D
E
F
G
H

←

FIGURE 2.—Illustrations of the upper Burdekin River and its sedimentology near Charters Towers, Queensland. A) Aerial view of the river, showing its low-sinuosity, low-stage thalweg channel and exposed, sand-covered bed. Note the occurrence of groves of trees on the river bed. River is ~ 250 m wide, flow towards the viewer. B) View of a linear grove of young *Melaleuca argentea* trees (< 5 years) at Brigalow Bend, showing downstream trailing habit (to left) and concentration of coarse-grained sediment downstream of the flow obstacle formed by the trees. C) View of Dalrymple Bend, near Charters Towers, showing development of 3-m-high, sinuous-crested dunes and gravel patches on the lower point-bar surface (geologists for scale). View is down the point bar surface towards the low-stage thalweg, flow from left to right. D) View of Brigalow Bend during a high-stage flow event on 25 March 1997 (18 hours after peak flow), showing complete inundation of riverbed vegetation (approximately same field of view as Part B, flow from left to right). E) surface of Brigalow Bend showing typically plane-bedded character with gravel streamlines (background) and a sinuous-crested gravel antidune (foreground). Flow towards the viewer. F) Surface of the lower point bar on Big Bend, showing mud layer desiccating as water level drops after a flow event. Notebook for scale. G) Cutbank exposure at Dalrymple Bend of cross-bedded sand with thin gravel layers, near the top of the point bar. Notebook is 20 cm long, flow from right to left. H) Cutbank exposure of interbedded sand and mud with abundant detrital plant debris on the upper point bar at Big Bend, near Charters Towers. Shovel for scale, flow from right to left. Detrital plant material from this site yielded a conventional radiocarbon age of 110 ± 50 years, suggesting that it may record the devastating flood of 1870.

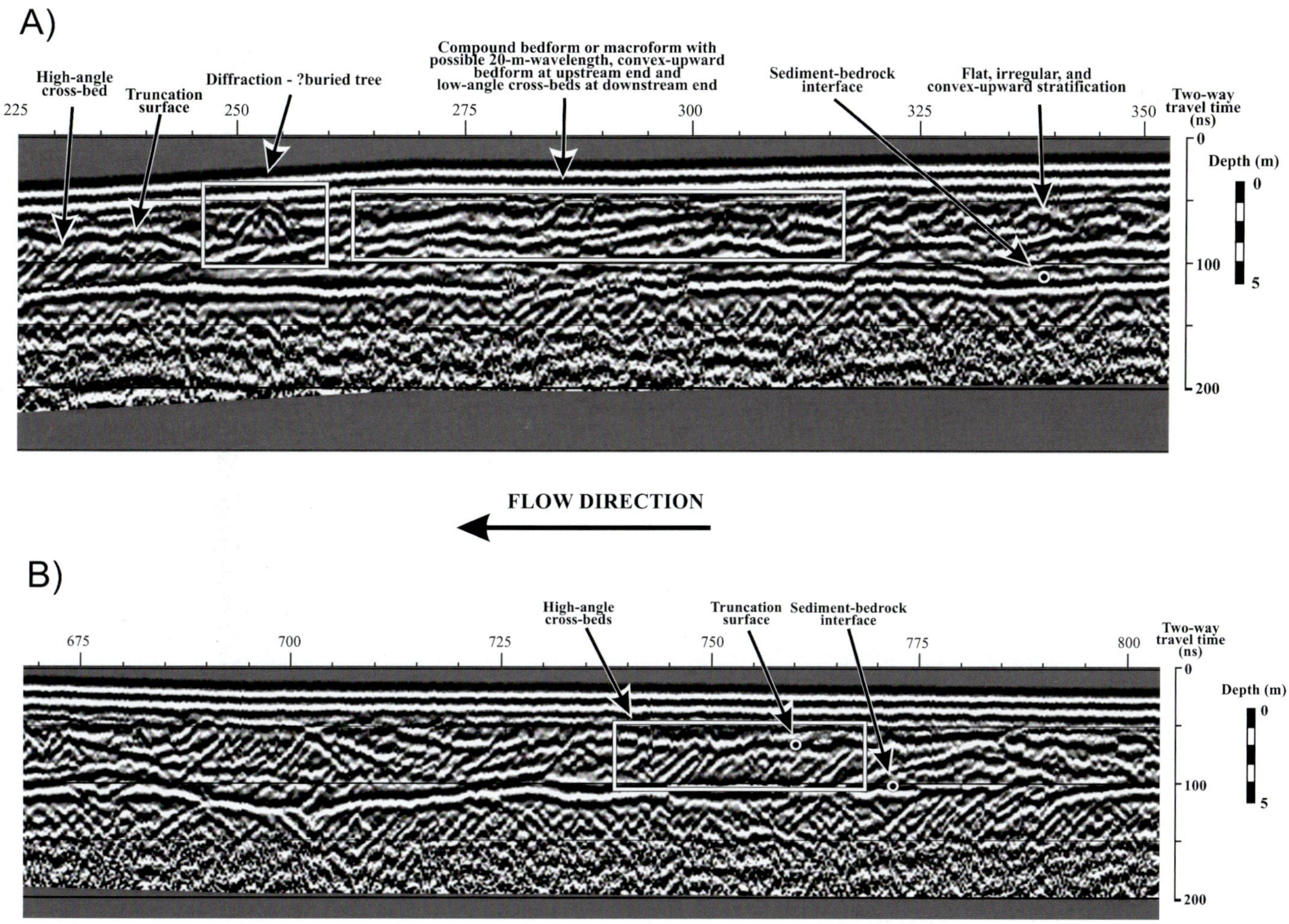

FIGURE 3.—Ground-penetrating radar (GPR) data from a flow-parallel transect across the Dalrymple Bend point bar on the upper Burdekin River, near Charters Towers. A) Data from the upstream portion of the point bar, showing a dominance of flat, irregular, undulating and convex-upward stratification, with local high-angle cross-bedding at the downstream end. B) Data from the downstream end of the point bar, showing a greater dominance of high-angle cross-bedding, with abundant scour surfaces. Flow from right to left in both images. Modified from Fielding et al. (1999).

Parts of the extant channel-fill body are dominated by large-scale (as much as 3 m) sets of cross-bedded sand (Fig. 3B), which is most commonly though not exclusively developed in the lower portion of complex bars. In places these sets are associated with smaller-scale cross-bedding (Fig. 2G), but they have also been observed with surface expression on dry areas of the river bed. High-angle cross-beds are produced by downstream migration of dunes, sediment waves, and less commonly lower-stage barforms. The lower portion of the alluvial fill also preserves relatively narrow, deeply scoured trough forms that in flow-parallel sections show large-scale (< 3 m) cross-bedding of limited lateral extent (< 10 m: Fig. 4). These features are interpreted as macroturbulent scours formed during high flow stage, and filled by slipface progradation due to the flow-separation effect over the scour. GPR data through the sandbodies also show common erosional surfaces, typically overlain by smaller-scale cross-bedded and plane-bedded sand (Fig. 3). Zones of cross-bedding pass cross-stream and down-stream into areas dominated by flat, undulating, low-angle and irregular stratification (Fig. 3). These stratification styles are interpreted as a record of high-flow-stage conditions, ranging from the plane-bed to antidune and possibly chute-and-pool stability fields (these proposed flow conditions are consistent with qualitative observations by us of the water surface during peak flow events). Mud layers up to 0.2 m thick, some pedogenically modified and desiccated, are evident both on the river bed and in the subsurface, as observed in channel cutbank exposures (Fig. 2H) and indirectly from GPR surveys. Most mud layers appear to be continuous over distances of up to a few tens of meters. Detrital plant material is abundant within the channel fill as dispersed debris, log jams associated with *in situ* trees and buried stumps, and concentrations of litter in the lee of cross-stratification (Fig. 2H; Alexander et al. 1999b).

In summary, the channel lithosome produced by the upper Burdekin River (and by other, less well-documented rivers in similar environmental conditions) comprises an erosionally based sand body with discontinuous mud partings, some gravel lenses, and abrupt lateral facies changes. Architectural features such as laterally inclined bedding in point bars, typical of perennially flowing rivers in temperate and other climatic settings, are not well developed. Sandbodies tend to have a greater thickness relative to width than arid-setting sandbodies, are more architecturally complex, and contain an abundance of structures formed around the dune–plane bed transition and under supercritical flow. In addition they preserve *in situ* evidence for plant colonization of the active river bed that is distinctive of this fluvial style.

A variation on this theme is illustrated by the coastal-plain reaches of the Burdekin River (Fielding et al. 2005). Here the river is straighter, broader (< 1000 m), and not as deeply entrenched, but is otherwise similar in form and in its discharge characteristics (Fig. 5; Alexander et al. 1999a). At low stage, the river bed is again largely exposed, with a more substantial low-stage channel reflecting stronger base discharge, and with large braid bars, point bars, and alternate bars visible on the surface (Fig. 5). As in inland sites, riverbed vegetation is prominent on the modern river bed (Fig. 5). At flood stage, the entire channel (and much of the coastal plain) is inundated by fast-flowing water (Alexander et al. 1999a; Amos et al. 2004). GPR data from these reaches (Fig. 6, and Fig. 7 of Fielding et al. 2005) show a much greater degree of preservation of cross-bedding, with sets as much as 4 m thick, and in flow-transverse section, steep-sided scour fills with elements of convex-upward stratification that are comparable to those found in the upper river (Fig. 4A). As at other sites, little evidence of significant macroform structure is visible in these data (Fig. 6).

On the coastal plain, the channel is flanked by a low-relief levee. Deposition on levees and floodplains occurs during very large discharge events and relatively infrequently when compared to perennial-river levees and floodplains. In these large flood events, significant sand is carried in suspension and floodplains may be inundated under 1–2 m of water. This has the consequence of generating floodplain successions with relatively high sand content compared with perennial-discharge settings (Alexander and Fielding 2006). Such deposits comprise sand beds that show little vertical grain-size variation, interbedded with lesser mud and gravel beds. Except on the very distal coastal plain, individual beds within the levee and floodplain deposits have limited lateral extent and cannot be readily correlated over distances of tens of meters (Alexander and Fielding 2006). Thus, the coastal-plain reaches of the Burdekin River show sedimentological characteristics that are comparable to those of the inland river, but at a larger scale and with some subtle differences.

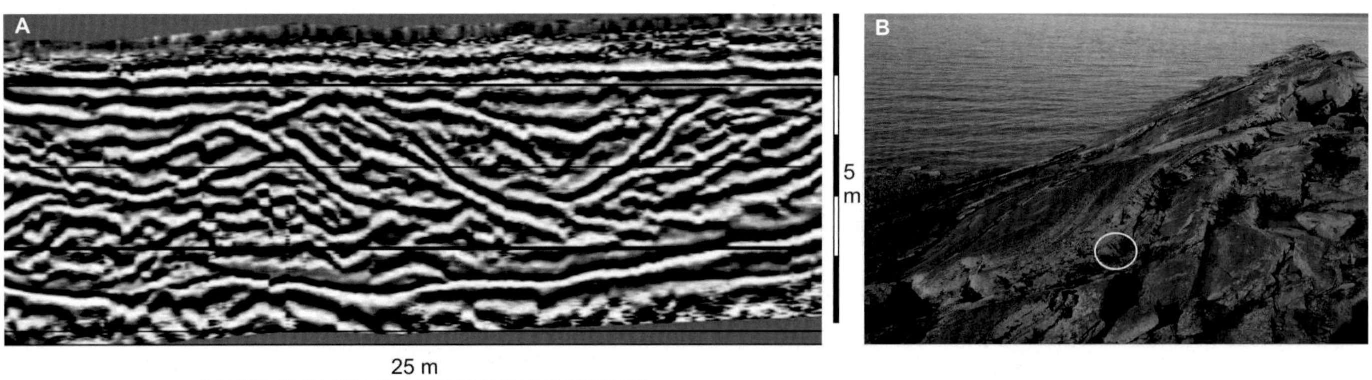

FIGURE 4.—GPR data from the subsurface of the upper Burdekin River, illustrating the stratal architecture typical of the tropical seasonal fluvial style. A) GPR image of part of a cross-stream transect from beneath the low-stage thalweg at Dalrymple Bend on the upper Burdekin River, showing large trough-like scours with scour infill continuous laterally into convex-upward bedding structure. The water table is close to the surface in this image, and the continuous event near the base of the image is the sediment–bedrock interface. The structure is interpreted as the fill of macroturbulent scours, possibly formed under Froude-critical conditions. B) A possible ancient analogue for the structures shown in Part A, from the Pennsylvanian Little River Formation (at 280 m in Fig. 9) in the type section at Lower Cove (Calder et al. 2005). The bedding structure has both scour fill and convex-upward drape elements. Paleoflow was away from the viewer. Hammer (circled) for scale. Images modified after Fielding et al. (1999).

FIGURE 5.—Illustrations of the lower Burdekin River and its sedimentology near Ayr, Queensland. A) Aerial view of the Burdekin River at The Rocks, showing development of a large point bar similar to those in the upper river. Note the larger area of the low-stage channel, reflecting greater perennial discharge. Flow towards the viewer, river is ~ 800 m wide at the bend apex. B) Aerial view of the river bed downstream of the Inkerman Bridge, showing complex bar morphology in a crossover zone. Note also the presence of living trees on the alluvial surface. River is ~ 800 m wide, flow from right to left. C) Aerial view of the river bed upstream of the Inkerman Bridge, showing the generally planar nature of the alluvial surface, with scattered living trees. River is ~ 800 m wide, flow from right to left. D) Ground view of the river bed upstream of the diffluence point with The Anabranch (first distributary), showing the generally planar sand bed with scattered gravel patches. Field of view ~ 1000 m wide, flow towards the lower left. GPR image of Figure 6B runs across this field of view.

A FACIES MODEL FOR FLUVIAL CHANNEL BODIES OF THE SUBHUMID TROPICS AND SUBTROPICS

Fielding et al. (2009) proposed a facies model for fluvial channel bodies of the seasonal tropics and subtropics. The essential features of this model are summarized in Figure 7. The model is based on studies of modern rivers in NE Australia, notably the Burdekin as described above, but it is informed by studies of analogous ancient successions (notably many formations preserved in the Carboniferous Maritimes Basin complex of Atlantic Canada: Gibling et al., 2008). Distinctive characteristics of these rivers include: (1) erosionally based channel-fill lithosomes that exhibit complex lateral facies changes, with (2) abundant, pedogenically modified mud partings, (3) complex internal architecture that may lack the macroform elements typical of other fluvial sediment bodies, (4) an abundance of sedimentary structures formed under high flow stage (here termed Froude-transcritical and supercritical structures: FTSS), and (5) an abundance of *in situ* trees that colonize channel floors and are adapted to inundation by fast-flowing water. The climatic conditions that generate the highly variable discharge responsible for the distinctive characteristics of these rivers are not strictly limited to the tropics, but rather the distribution of the fluvial style depends on factors such as continental distribution and mean sea-surface temperatures. Accordingly, the facies model can be considered to apply to the seasonal tropics and some areas of the subtropics. Flashy-discharge streams that exist in other climates might show some of the physical characteristics described above, but since temperatures are lower, it is expected that the full suite of features will not be developed. Deposits of rivers that display the subhumid to semiarid, seasonal tropical to subtropical characteristics summarized above contrast with (but are transitional into) those characterized by more perennial discharge, which are expected to be more dominated by cross-bedding, and those of dryland environments, that show evidence of flashy discharge but are typically strongly reddened and typically lacking in macroform architectural elements.

An additional element of this model is that channel lithosomes from strongly seasonal discharge regimes might be expected to preserve channel-margin "wings" (distally thinning extensions of the channel margin) to a considerably greater extent than channel bodies from other climatic and runoff regimes. This is because of the ability of rivers in strongly seasonal tropical climates to inundate the floodplain with fast-

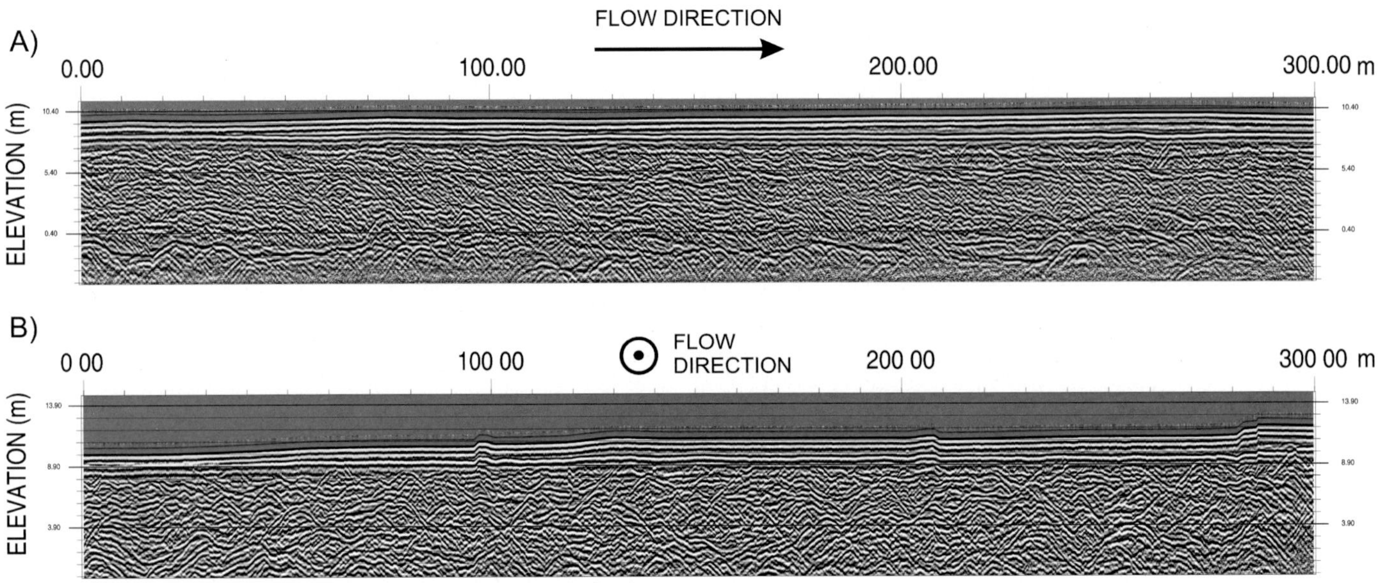

FIGURE 6.—GPR data from the subsurface of the lower Burdekin River, showing a stratal architecture more consistent with perennially flowing rivers. A) Flow-parallel transect near the Inkerman Bridge on the upper delta plain, showing the dominance of irregular scour surfaces and trough cross-bedding on various scales. Individual sets are traceable in a downstream direction for up to a few tens of meters. Local point diffractions (e.g., at 265 m, near surface) suggest buried trees or large boulders. The prominent, continuous event near the base of the image is the interface between Holocene alluvium and Pleistocene deposits. Note the undulating nature of the basal erosion surface. B) Flow-transverse section 1 km upstream from the first distributary on the upper delta plain, showing a dominance of trough-like scour fills that correspond to cross-beds in the flow-parallel orientation. The Pleistocene–Holocene interface is below the base of the section at this locality. Some of the trough forms are steep-sided and preserve convex-upward bedding elements similar to those shown in Figure 4A, suggesting a similar origin. Water table in both images is close to the surface. Two-way time in nanoseconds has been converted to meters depth based on a velocity for water-saturated sand of 0.06 m/ns.

flowing water during high-stage flow events, and consequently the ability of these systems to transmit coarse-grained sediment out onto the floodplain (in contrast to the typically lower-velocity overbank flows of more temperate climate systems). Presence of overbank "wings" to channel bodies has been noted in a number of studies of ancient successions, and could be a further useful discriminant of the strongly seasonal fluvial style. For example, such features have been extensively documented from Cenozoic fluvial deposits of the Pyrenean foreland basins (e.g., Friend et al. 1986; Marzo et al. 1988; Hirst 1991), which are thought to have experienced a seasonal, subhumid to semiarid, subtropical climate.

Since the strongly seasonal fluvial style produces sediment bodies that are distinctively different from other fluvial milieux, recognition of this style might be of value in exploration for subsurface fluids and their deposits (hydrocarbons, water, uranium). Furthermore, if it is possible to recognize and correlate intervals of strata characterized by this fluvial style, the model may have applications to stratigraphic and paleoclimatic analysis. In the following section, we document an unusually stratigraphically complete section of early Pennsylvanian age and coastal-alluvial-plain origin from the Maritimes Basin complex of Atlantic Canada. We show how the strongly seasonal fluvial style can be recognized within a discrete interval, enclosed by intervals characterized by more typical fluvial style.

RIVER TO ROCK: PRESERVATION OF ANCIENT SEASONAL TROPICAL FLUVIAL DEPOSITS IN THE PENNSYLVANIAN OF ATLANTIC CANADA

The Maritimes Basin of Atlantic Canada (Fig. 8A) was in low latitudes during the Pennsylvanian (Scotese and McKerrow 1990; Ziegler et al. 2002), analogous to modern NE Australia. Numerous Pennsylvanian formations contain fluvial channel bodies that show remarkable analogy with the modern river sediments described above. Stratigraphic intervals that preserve a record of strongly seasonal discharge, tens to hundreds of meters thick, are enclosed by similarly thick intervals characterized by conventional fluvial style. These intervals can be correlated across the Maritimes Basin complex (Allen 2009), suggesting that they represent long-term variations in environmental conditions that were of regional to interregional extent.

In order to illustrate the features of the seasonal fluvial style, we provide details of a discrete stratigraphic interval from the Cumberland Basin of Nova Scotia (Fig. 8B), of early Pennsylvanian (lower Bashkirian, Yeadonian to Langsettian) age. This interval encompasses, from base to top, the upper Boss Point Formation, the entire Little River Formation, and the lower Joggins Formation (Ryan and Boehner 1994; Davies et al. 2005; Calder et al. 2005) and is 1120 m thick in total (Fig. 9). This section highlights the contrast between fluvial styles, which can be recognized irrespective of evidence for other external forcing factors such as changes in the accommodation/sediment supply ratio (A/S) and tectonic processes. This example also illustrates how recognition of these contrasts in fluvial style can yield additional insights into basin evolution and environmental change.

Location and Stratigraphy

The section of interest is superbly exposed in the now World Heritage–listed coastal cliffs and intertidal platforms of the eastern shore of Chignecto Bay, mainland Nova Scotia, around the village of Joggins (Fig. 8C). The section, which in aggregate exposes some 4 km of upper Mississippian to lower Pennsylvanian strata almost contin-

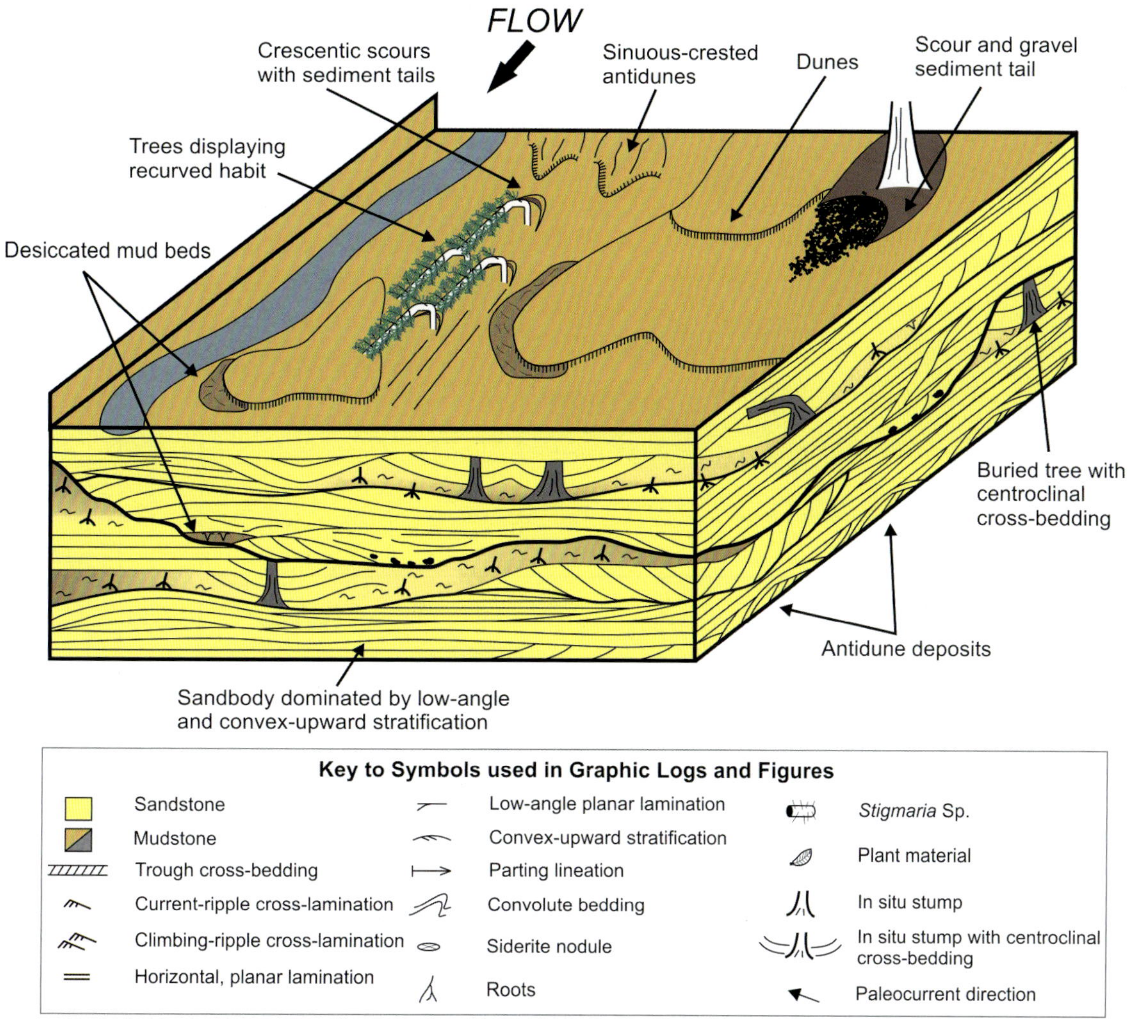

FIGURE 7.—Three-dimensional block model summarizing the essential elements of the facies model for strongly seasonal tropical and subtropical fluvial systems (from Fielding et al. 2009).

uously, was formed under a subsidence regime more rapid than virtually anywhere else in the Euramerican Pennsylvanian province (Calder 1998; Davies and Gibling 2003). Perhaps the principal reason for the extraordinary early Pennsylvanian subsidence rates experienced by this area was subsurface withdrawal of > 1.5 km of Mississippian Windsor Group salt deposits at this time (Waldron and Rygel 2005). The major consequence of this is that the Bashkirian succession is extremely thick relative to most other areas even within the Maritimes Basin complex, and it is likely to be unusually stratigraphically complete inasmuch as it preserves no known major unconformities.

The Boss Point Formation comprises c. 800 m of alternating sheetlike sandstone bodies and mudrock-dominated units (Ryan and Boehner 1994; Browne and Plint 1994; Plint and Browne 1994). The above authors interpreted these strata as the record of repeatedly alternating braidplain and lacustrine environments of deposition, the changes driven by downdrop along basin-bounding faults. Some of the mudrock intervals in the Boss Point Formation at Boss Point (the type section, and the lower part of the section of interest here; Fig. 8C), are strongly colored (brown, purple, locally brick red) and root-penetrated, and they were interpreted as the product of pedogenic processes acting on floodplain and shallow floodbasin deposits (Plint and Browne 1994). The contact between the Boss Point Formation and the overlying Little River Formation (Calder et al. 2005) is defined at the top of the uppermost thick, apparently sheetlike sandstone body ("South Reef") on the shore platform at Lower Cove (Fig. 10).

The overlying Little River Formation (as defined by Calder et al. 2005) comprises 636 m of complexly interbedded sandstone bodies, interbedded sandstone–mudrock units and thick intervals of mainly pedogenically modified mudrocks. Sandstone bodies are smaller in cross-sectional area than those of the underlying Boss Point Formation, and they appear randomly distributed throughout the section, although most are of limited lateral extent (Fig. 10). The contact with the overlying Joggins Formation has been taken at the lowermost coal bed in the section, within what appears to be a gradual transition in stratigraphic style. The lower part of the Joggins Formation (916 m in total) comprises a similar array of channelized sandstone bodies of mainly limited dimensions (Rygel and Gibling 2006), gray and more reddened mudrock intervals, with minor sheet-like sandstones < 2 m thick. The Joggins Formation is interpreted in terms of alternations among shallow, open-water coastal environments, poorly drained alluvial settings, and well-drained alluvial settings (Davies and Gibling 2003; Davies et al. 2005).

The overall environmental interpretation that can be made from the entire Boss Point to Joggins Formation interval from previously

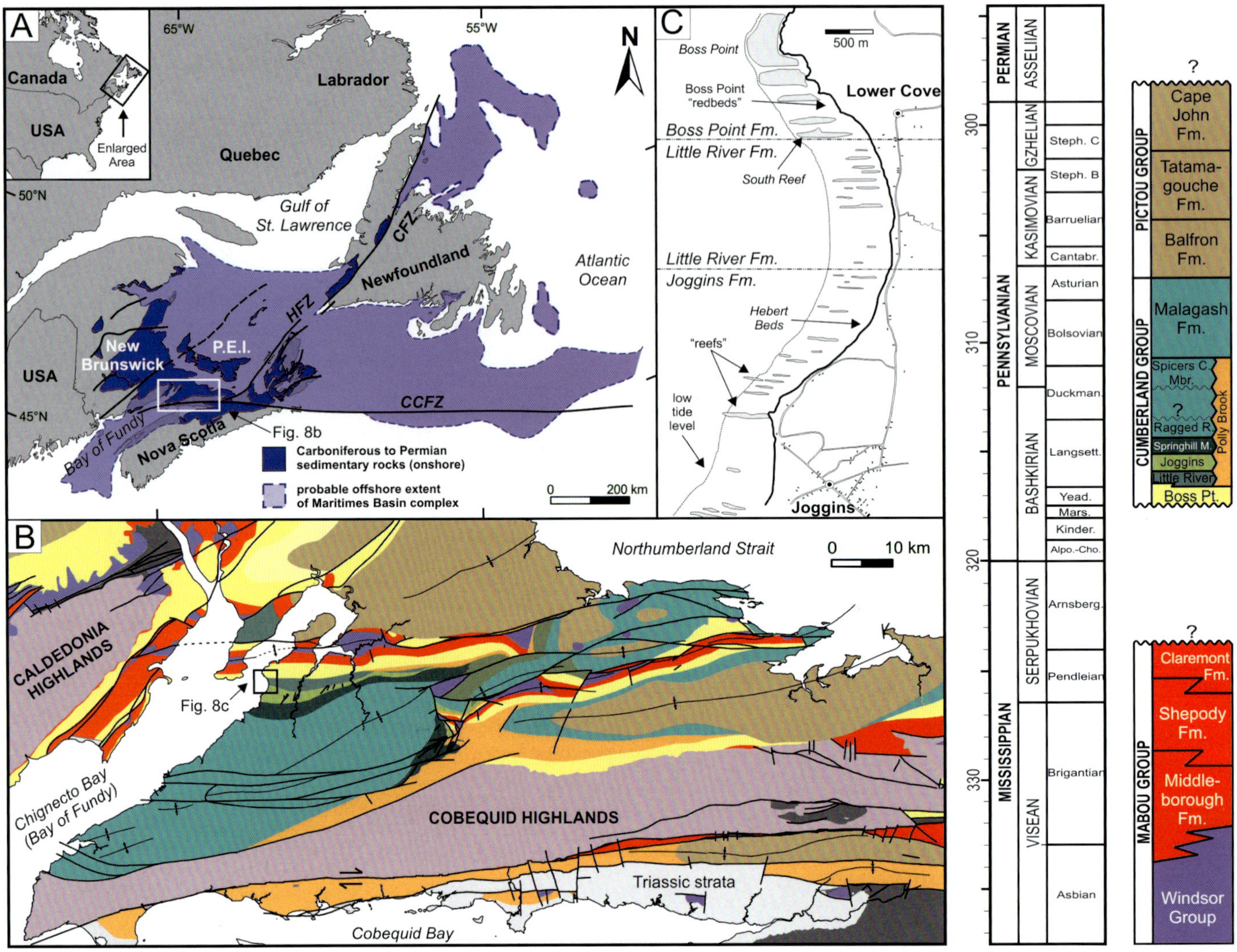

FIGURE 8.—Maps of A) the Carboniferous Maritimes Basin Complex in Atlantic Canada, and B) the Cumberland Basin in mainland Nova Scotia, C) study sites, and D) generalized Pennsylvanian stratigraphy of the Cumberland Basin. CCFZ, Cobequid-Chedabucto Fault Zone; HFZ, Hollow Fault Zone; CFZ, Cabot Fault Zone; Arnsberg., Arnsbergian; Alpo., Alportian; Cho., Chokierian; Kinder., Kinderscoutian; Mars., Marsdenian; Yead., Yeadonian; Langsett., Langsettian; Duckman., Duckmantian; Cantabr., Cantabrian; Steph., Stephanian.

published work is that it represents a coastal alluvial plain that experienced a series of changes through time in degree of drainage and in fluvial style. The dominant direction of sediment transport was northeastward along the axis or structural grain of the basin complex (Fig. 8), although some data suggest that transverse drainage, perhaps from fluvial megafans flanking border highlands, was an important contributor at certain times (Gibling et al., 1992). The details of changes in fluvial style have not been previously interpreted in terms of controlling mechanisms. These characteristics and temporal shifts are explored in more detail below.

Sedimentology

Boss Point Formation: The sedimentology of the Boss Point Formation has been described in detail by Browne and Plint (1994) and Plint and Browne (1994). In their facies analyses, these authors recognized that, whereas most of the apparently sheetlike channel sandstone bodies of the Boss Point Formation were dominated by cross-bedding, with lesser ripple cross-lamination and planar lamination, a minor proportion of these sandstones were composed dominantly of planar, low-angle, and convex-upward lamination (referred to by the authors as hummocky cross-stratification), with subordinate cross-bedding. They interpreted these structures as arising from flow at and around Froude-critical conditions during the final stages of channel filling. We concur with these observations, and add that the proportion of individual sandstone bodies containing Froude-transcritical and -supercritical structures (FTSS), and sandstone bodies dominated by these structures, increases gradually upward through the formation. The log of the type section of the Boss Point Formation by Browne and Plint (1994) and Plint and Browne (1994) does not cover the uppermost 160 m of the unit, however, nor the contact with the overlying Little River Formation (previously the base of the Joggins Formation, prior to the definition of the Little River Formation by Calder et al. 2005).

The uppermost 160 m of the Boss Point (Fig. 9 and the more detailed version of this figure in the digital appendix) shows a progressive

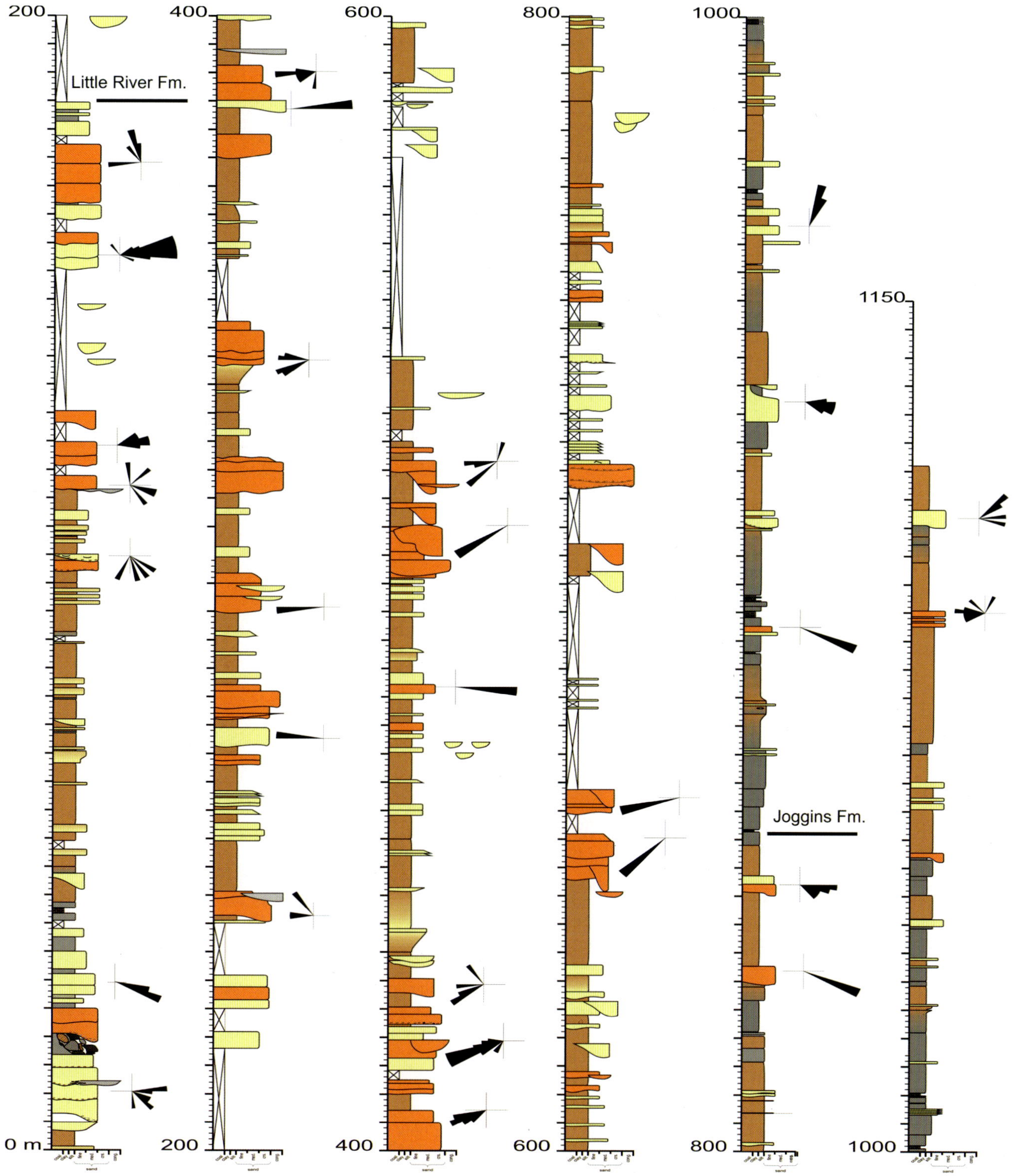

FIGURE 9.—Detailed graphic log of a measured section through the uppermost Boss Point Formation, the entire Little River Formation, and the lower Joggins Formation at Lower Cove, Nova Scotia, and key to symbols and colors used. [An expanded version of this figure, showing additional detail, is available in the digital appendix.]

FIGURE 10.—Annotated aerial photograph showing exposure of the uppermost Boss Point Formation and the Little River Formation (Calder et al. 2005). Sandstones stand out on the shore platform as white areas. Note the pronounced change in fluvial style from the apparently sheetlike sandstones of the Boss Point Formation into the less laterally extensive channel bodies of the Little River Formation. Some of the latter appear to comprise lateral and vertical amalgamations of small, lensoid bodies. Note the large-scale, inclined macroform bedding in the uppermost Boss Point Formation sandstone bodies south of North Reef and at South Reef, which, from relationships with the orientation of small-scale structures, can be interpreted as lateral-accretion bedding. The measured section of Figure 9 runs along the shore platform over the length of the aerial photograph.

increase in the proportion of mudrocks and thinly interbedded mudrocks and sandstones, and an increase in the proportion of reddened mudrocks relative to drab facies (Figs. 10, 11). Within this section is a bed of bivalve and ostracod fossil–bearing, nonmarine limestone. This interval also shows a more irregular decrease in the proportion of channel sandstones in the vertical section, as well as in their thickness. This pattern persists until the uppermost 85 m of the formation, which reverts to a more typically "Boss Point" lithology in preserving three thick, apparently sheetlike channel sandstone bodies that stand out on the shore platform as "reefs" (Fig. 10). Each of these bodies shows paleocurrent directions (east to south) that are consistent with the bulk of the underlying Boss Point Formation (Fig. 9; Browne and Plint 1994). The top of the Boss Point Formation is taken at the top of the uppermost of these composite, sheet sandstone bodies (Fig. 10).

Comparing the three sandstone bodies, the first is 3 m thick and single-story, the second 14 m thick and multistory, and the third is 22 m thick and multistory. All three contain a combination of both cross-bedding and flat, low-angle, and convex-upward stratification (FTSS), along with sparse bioturbation referable to *Planolites*. Many occurrences of flat and convex-upward stratification show preservation of trains of low-amplitude, discordant laminae, preserved as form sets on bedding planes (Fig. 11F). These features are distinct from truncated ripples in

FIGURE 11.—Illustrations of the upper Boss Point Formation along the Boss Point to Lower Cove shore section. A) Typical Boss Point Formation lithologies: alternating thick mudrock intervals and thick, composite sheetlike sandstone bodies. The sandstone body shown here is internally dominated by flat and low-angle lamination. Approximately 40 m of section is exposed in the field of view. B) Close-up view of thick, sheetlike sandstone body at the top of the Boss Point Formation at South Reef (175 m in Figure 9), showing large, sigmoidal cross-set (1.0 m thick) underlain and overlain by flat and low-angle stratification. Beds dip at c. 30 degrees towards the right. C) View of lensoid channel sandstone body in the uppermost Boss Point Formation encased in somewhat reddened mudrock facies. D) Close-up view of color-mottled and root-penetrated mudrocks with thin-bedded sandstones in the uppermost Boss Point Formation. Hammer 0.28 m for scale. E) View of long-wavelength convex-upward bedding structure, interpreted as the deposit of a large antidune, in the upper Boss Point Formation. The bedform had a wavelength of ~ 24 m. F) Bedding-plane view of plane-bedded sandstone with parting lineation and preserving trains of low-amplitude bed waves as a form set. These small bedforms are < 5 mm in height. Coin (toonie) is 3 cm in diameter.

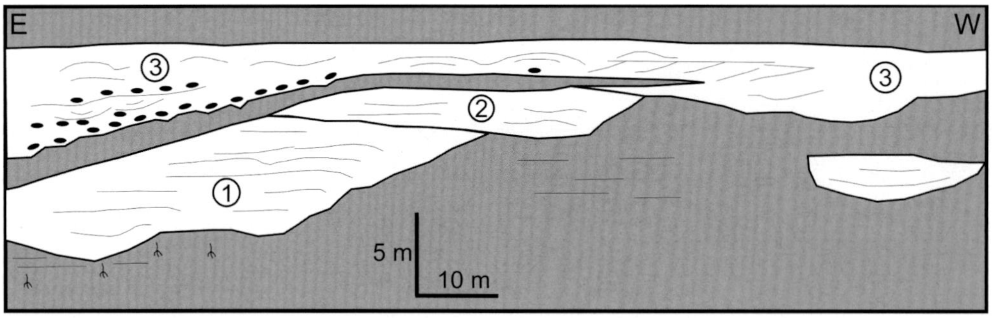

FIGURE 12.—Line drawing of a composite, multistory channel sandstone in the lower Little River Formation (at 280 m in Figure 9), showing the geometry of component stories and distribution of major internal bedding features.

being of considerably lower amplitude. The mainly thin-bedded siltstone–sandstone interval overlying the first of these sandstones is typically pale red/brown, containing dispersed carbonate glaebules near the base and plant roots throughout. The bases of sandstone stories are strongly erosional and in places lined by intraformational breccia layers. The uppermost sandstone body at South Reef (Fig. 10) displays a wide variety of bedding geometries, including sigmoidal and humpback cross-bedding forms (Fig. 11B) and large, undular forms with wavelengths of 10–24 m (cf. Fig. 11E). The uppermost story of this complex is the first to show a change in paleocurrent direction (west to north) that is characteristic of the overlying Little River Formation (Fig. 9). It is also one of two bodies in the logged interval to show dipping macroform bedding structures (Fig. 10).

The thick, laterally extensive sandstone bodies of the Boss Point Formation (e.g., Fig. 11A) have been interpreted as the products of sand deposition in large, braided fluvial channels (Browne and Plint 1994; Plint and Browne 1994). The dominance of cross-bedding in the lower half of the formation suggests a relatively even discharge regime that routinely facilitated preservation of cross-bedding via the modification of higher-stage bedforms and macroforms during the falling stage of runoff events (cf. Jones 1977). The increased abundance of FTSS in the upper half of the formation suggests a progressive change in the discharge regime towards a strongly seasonal character, in which short, high-discharge flow events characterized by rapid changes in flow stage facilitated the preservation of high-stage flow structures. Applying published fluid dynamic equations (Kennedy 1963, Middleton 1965, Allen 1968, and see Duller et al. 2008), the largest convex-upward forms can be estimated to have formed in ~ 4 m of water under flows of ~ 6 m/s. Trains of small, discordant laminae on some bedding planes (Fig. 11F) are interpreted as the deposits of low-amplitude bedwaves active under transcritical and supercritical conditions documented experimentally by Bridge and Best (1988), Paola et al. (1989), Cheel (1990), and Best and Bridge (1992).

The changes in abundance and thickness of channel sandstones upward through the uppermost Boss Point Formation (Figs. 9, 11A, C) suggest that fluvial style changed progressively if nonlinearly through the Yeadonian into the Langsettian. Thick, laterally extensive sheet sandstones interpreted as braidplain deposits by Plint and Browne (1994) become less abundant upward, making a final appearance at the top of the formation. The internal structure of those bodies also changes progressively upward, with proportionally more FTSS preserved towards the top of the formation. A change in sediment dispersal direction from eastward to broadly northwestward is also noted at the top of the Boss Point Formation (Fig. 9). Furthermore, evidence of waterlogging and standing water (gray mudrocks, carbonaceous plant debris and rootlets, freshwater limestones, and carbonaceous shales) gradually diminishes in abundance upward through this interval (Fig. 9). The upward decrease in the abundance and thickness of channel bodies can in a simple sense be interpreted as resulting from an increase in the ratio of accommodation to sediment supply (A/S). But the evidence for increasingly well-drained alluvial surfaces argues against an increase in accommodation.

Little River Formation: The sedimentology of the Little River Formation has been documented by Calder et al. (2005). The formation contains a number of fine- to coarse-grained sandstone bodies < 11 m thick and at least 534 m wide (Calder et al., 2005). Some of these bodies are single story, but others are composite, multistory bodies composed of two or more, vertically or laterally amalgamated, lensoid sandstone stories (e.g., Fig. 12). In general, individual stories show a simple channel-form, lensoid geometry in cross section, with the only evidence of macroform bedding structure on the ground or in aerial photographs being local, subtle, low-angle discordances between bedsets. Some bodies show lateral thinning into extensive "wings" that ultimately pinch out. Bases of many bodies, and of component stories within multistory bodies, are strongly erosional. Intraformational-clast breccias and conglomerate layers are common in these bodies, as are concentrations of plant debris, particularly at story bases. Concentrations of feldspathic granules were noted locally in the formation by Calder et al. (2005). Some bodies contain siltstone partings, which are difficult to trace over distances of more than a few meters.

Thick sandstone bodies in the Little River Formation are internally dominated by flat and low-angle lamination, convex-upward (undular) bedding on various scales with wavelengths < 15 m and amplitude 0.1–0.4 m, local backset cross-bedding, and rare, supercritically climbing, dune-scale, nearly symmetrical aggradational structures (Fig. 13). Numerous examples were noted of nested, low-angle, truncated trough forms (Fig. 13D) in paleoflow-parallel sections similar to those produced experimentally by Alexander et al. (2001) in Froude-supercritical flow conditions. Similarly to the Boss Point Formation, several occurrences of flat and convex-upward stratification show preservation of trains of low-amplitude, discordant laminae, preserved as form sets on bedding planes. Several bodies preserve large, undular bedding structures (amplitude 0.2–0.4 m, wavelength 10–15 m) with convex-upward and concave-upward elements, the angles of dip of which are greater than in other, similar structures (Fig. 4B). Backset cross-beds typically fill listric, concave-upward scours several meters in length (cf. Duller et al. 2008). The scale of undular structures does not vary systematically upward through most bodies, with some of the largest structures preserved at the tops of sandstone bodies. Some erosional scours show an undulating erosional topography of < 0.5 m with a regular wavelength. Bases of some beds show sole structures, including gutter casts, and some bodies preserve *in situ* tree stumps, in at least one case inclined in the direction of paleoflow with upstream trailing roots (Fig. 13E, F). Centroclinal cross-bedding (Underwood and Lambert 1974) is locally preserved around upright, *in situ* tree stumps ("vegetation-induced sedimentary structures", or "VISS" of Rygel et al. 2004). Subordinate cross-bedding and ripple cross-lamination is also preserved. Many cross-sets show a sigmoidal cross section, with preservation of topset, lower than angle-of-repose foreset, and asymptotic bottomset elements (Fig. 13B). Another conspicuous feature of these sandstone bodies is the common preservation of ripple form sets on extensive bedding planes, in some bodies several successive bedding planes, where each set shows somewhat different ripple morphology, ranging from straight-crested, through sinuous-crested, to linguoid, locally with interference patterns (Fig. 13H). Palaeocurrent data from thick sandstone bodies in the Little River Formation show a consistently westward orientation in nearly all exposures (Fig. 9).

Thick sandstone bodies are separated by thin-bedded sandstone–siltstone and mudrock intervals that together make up a larger component of the formation than in the underlying interval. Mudrocks are characterized by a dominantly pale red/brown color, and some show color mottling to green or gray (Fig. 13G). All are associated with drab root compressions, and some show concentrations of calcareous glaebules and rhizocretions. Carbonate glaebules become particularly abundant near the top of the formation, where mudrocks are also more conspicuously brick-red in color (higher Chroma values).

The thicker sandstone bodies (2 m+) in the Little River Formation are interpreted as channel deposits of a coastal alluvial plain, separated in time and space by the mudrock-dominated deposits of mainly well-drained floodplain and intermittent floodbasin environments. Channels are inferred to have had more modest dimensions than those that deposited the thick, sheetlike sandstones of the Boss Point Formation, and to have been more variable in planform. Within channel deposits, planar lamination with parting lineation (with or without low-angle discordant bedding) is interpreted as upper-plane-bed deposits. Nested, low-angle trough forms in paleoflow-parallel sections are interpreted as the deposits of small antidunes (cf. Alexander et al. 2001). Convex-upward stratification and undular structures are interpreted as the products of formation and migration of larger antidunes, and rare backset bedding filling listric scours may be the product of transient chute–pool development (cf. Fielding 2006; Duller et al. 2008). Given the model of Duller et al. (2008, their Figure 14) and hydraulic equations for antidune stability (Kennedy 1963; Middleton 1965; Allen 1968; Duller et al. 2008) and resulting bedding geometry, many of the larger structures may record large antidunes forming and migrating under < 2.5 m of water at flow velocities of < 5 m/s, at Fr $\sim$ 1. The supercritically climbing, nearly symmetrical structures noted above are interpreted as the deposits of rapidly aggrading and downpaleocurrent-migrating small antidunes. Trains of low-angle discordant laminae on some bedding planes are interpreted as low-amplitude bedwaves, as suggested above for the Boss Point Formation. One additional, pertinent observation here is that convex-upward, antidune stratification seems not to be common either in experimental studies or in other case histories of natural sediments (e.g., Alexander et al. 2001; Duller et al. 2008), yet is abundant in these fluvial deposits. The reason for this is not yet clear, but it may be related to very rapid falls in flow stage in the formative rivers.

The suite of sedimentary structures preserved in the Little River Formation is collectively referred to herein as Froude-transcritical and supercritical structures (FTSS), and is directly comparable to the suite of bedforms and structures preserved on the bed of the modern Burdekin River of NE Australia. Sigmoidal cross-bedding has been shown to be associated with the transition from the dune stability field to the upper-plane-bed stability field (Saunderson and Lockett 1983; Bridge and Best 1988; Hand and Bartberger 1988), and is here interpreted in the same manner. The preservation of so many FTSS, along with abundant plane bedding, transitional dune–plane bed phase structures, and well-preserved form sets of ripples indicates a discharge regime characterized by highly peaked flow hydrographs, with falling limbs so rapid that bedforms were unable to equilibrate to changing flow conditions and thus were preserved intact.

The nonchannel facies are interpreted as the deposits of mainly well-drained floodplains, from their red color (the brightness of which increases (higher Chroma values) upward through the formation) and evidence of pedogenic modification. Local preservation of carbonate glaebules and nodules (Stages II and III of Machette 1985) implies formation of immature aridisols (calcisols of Mack et al. 1993), whereas intervals of drab color imply some degree of waterlogging of the alluvial surface. Calder et al. (2005) characterized the paleosols of the Little River Formation as mainly well-drained entisols (protosols) with local and intermittent evidence for gleying, and local aridisols. Thin, sheetlike sandstones are interpreted as overbank sheetflow and splay deposits. Overall, the impression is gained of a generally well-drained floodplain under a strongly seasonal, hot climate that intermittently inundated the surface with floodwaters.

The Little River Formation preserves a complete contrast in fluvial style from the underlying Boss Point Formation, a change heralded by the uppermost 160 m of the latter as documented above. In the Little River Formation channel sandstone bodies, FTSS dominate over cross-bedding, and in general channel sandstones show all the hallmarks of the strongly seasonal fluvial style documented from modern NE Australia. The assemblage of physical and biogenic sedimentary structures points towards a strongly seasonal, hot environment in which flow events were characterized by rapid rises and falls of flow stage, allowing widespread preservation of FTSS. The presence of *in situ* tree fossils attests to the growth of trees within channel floors, and their adaptation to this habitat. Extra-channel facies also argue for a predominantly well-drained alluvial landscape that was periodically inundated by floodwaters but generally experienced a significant excess of evaporation over precipitation. Plant fossils recovered from the Little River Formation support the subhumid to semiarid landscape interpretation, being mainly taxa associated with well-drained environments (Calder et al. 2005). Lycopsids (the floral group associated with wetland settings in the paleotropical Pennsylvanian) appear for the first time in the middle of the unit, and become more prevalent upward in the overlying Joggins Formation (Calder et al. 2005).

Joggins Formation: The base of the Joggins Formation is taken arbitrarily at the base of the lowermost coal bed in the stratigraphic succession (Fig. 14A). Other than this marker, there are only subtle changes in the lithofacies assemblage from the upper Little River Formation into the lower Joggins Formation (which explains the historical inclusion of the former into the latter; Calder et al. 2005). The lower Joggins Formation comprises an array of gray and variably reddened mudrocks, thin coals, carbonaceous shales and carbonaceous limestones, interbedded mudrocks and thin sandstones, tabular, sheet-like sandstone bodies, and thicker, erosionally based sandstone bodies. The sedimentology of these rocks has been addressed by Davies and Gibling (2003), Davies et al. (2005), and Rygel and Gibling (2006), who defined three facies associations. These are (1) an open-water facies assemblage, in which limestones and laminated gray mudrocks represent maximum flooding events and tabular sheet-like sandstones represent small-scale deltas prograding into shallow, open water, (2) a poorly-drained floodplain assemblage, in which coals and carbonaceous shales record the early phases of flooding events and gray mudrocks record permanently waterlogged alluvial surfaces, and (3) a well-drained floodplain assemblage, in which variously reddened mudrocks and sandstones record a well-drained alluvial surface. Erosionally based sandstone bodies of interpreted alluvial channel origin are preserved in both assemblages 2 and 3 and were interpreted to record natural geomorphic variability, rather than any external forcing mechanism, by Rygel and Gibling (2006). The three facies assemblages occur in intervals of typically tens of meters thickness, irregularly inter-

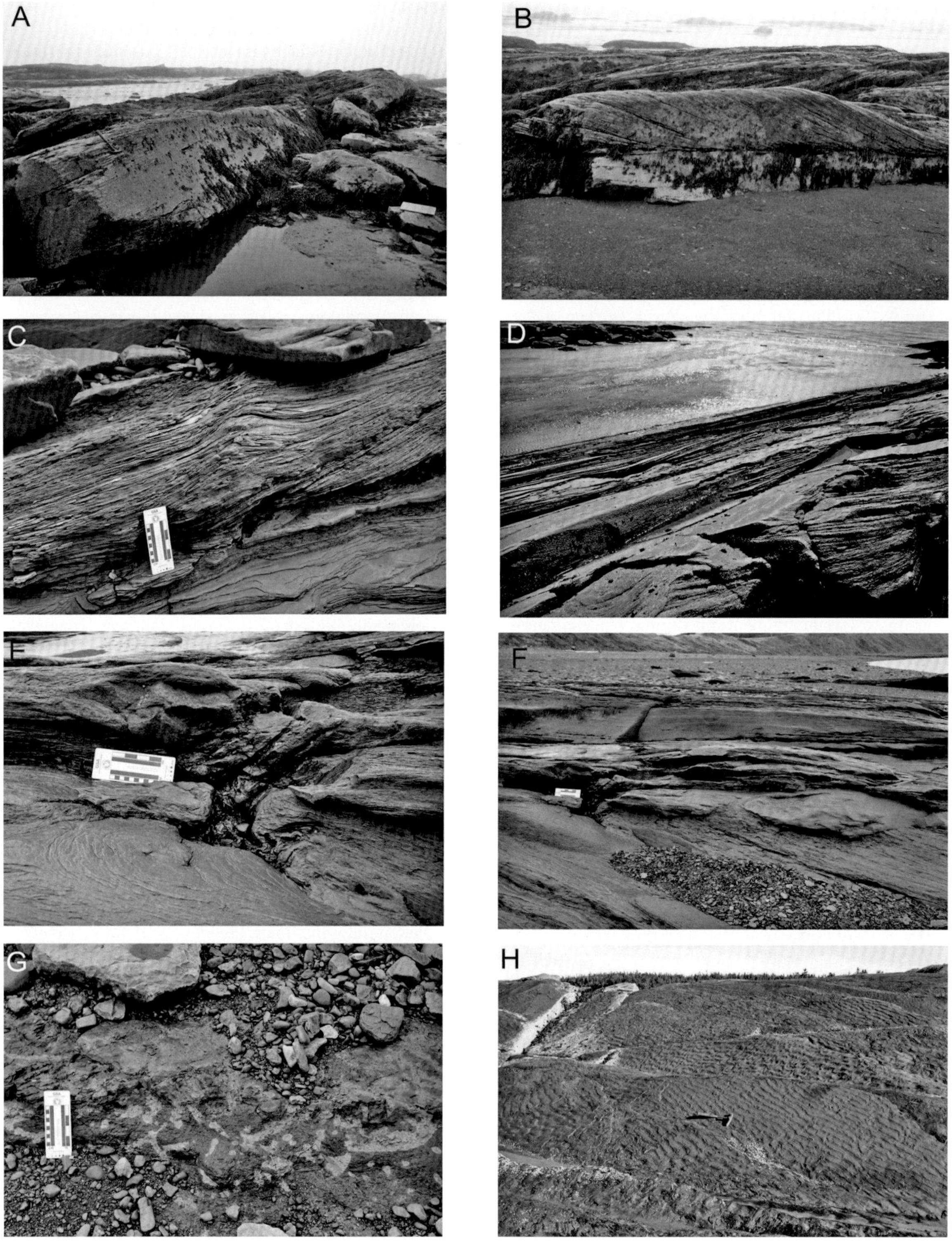
A
B
C
D
E
F
G
H

←

FIGURE 13.—Illustrations of the Little River Formation in the Lower Cove shore section. A) Typical Little River Formation channel sandstone body, showing a dominance of flat and low-angle stratification. Hammer for scale, 0.28 m. B) Cross-set showing sigmoidal cross-sectional geometry within a channel sandstone body. Hammer for scale, 0.28 m. C) Supercritically climbing, nearly symmetrical aggradational structure (at 269 m in Figure 9), interpreted as the record of a climbing and downstream-migrating small antidune. Scale card is 0.15 m. D) Low-angle discordant stratification, including small, low-angle trough fills interpreted as antidunal bedding. Hammer for scale, 0.28 m. E) *In situ*, inclined small fossil tree within a channel sandstone body (at 650 m in Figure 9), showing upstream-trailing roots. Scale card is 0.15 m. F) Context of Part E, showing *in situ* tree fossil encased in flat and low-angle laminated sandstone with minor cross-bedding indicating paleoflow towards the right, similar to the direction of tilting of the tree fossil. Scale card is 0.15 m. G) Little River Formation mudrock, showing bright red color characteristic of the uppermost part on the formation, with drab green-gray mottling along presumed root channels. H) Ripple form sets on bedding planes close to the top of a channel sandstone body (at 281 m in Figure 9). Note the varying ripple planform in successive form sets, and local preservation of interference patterns.

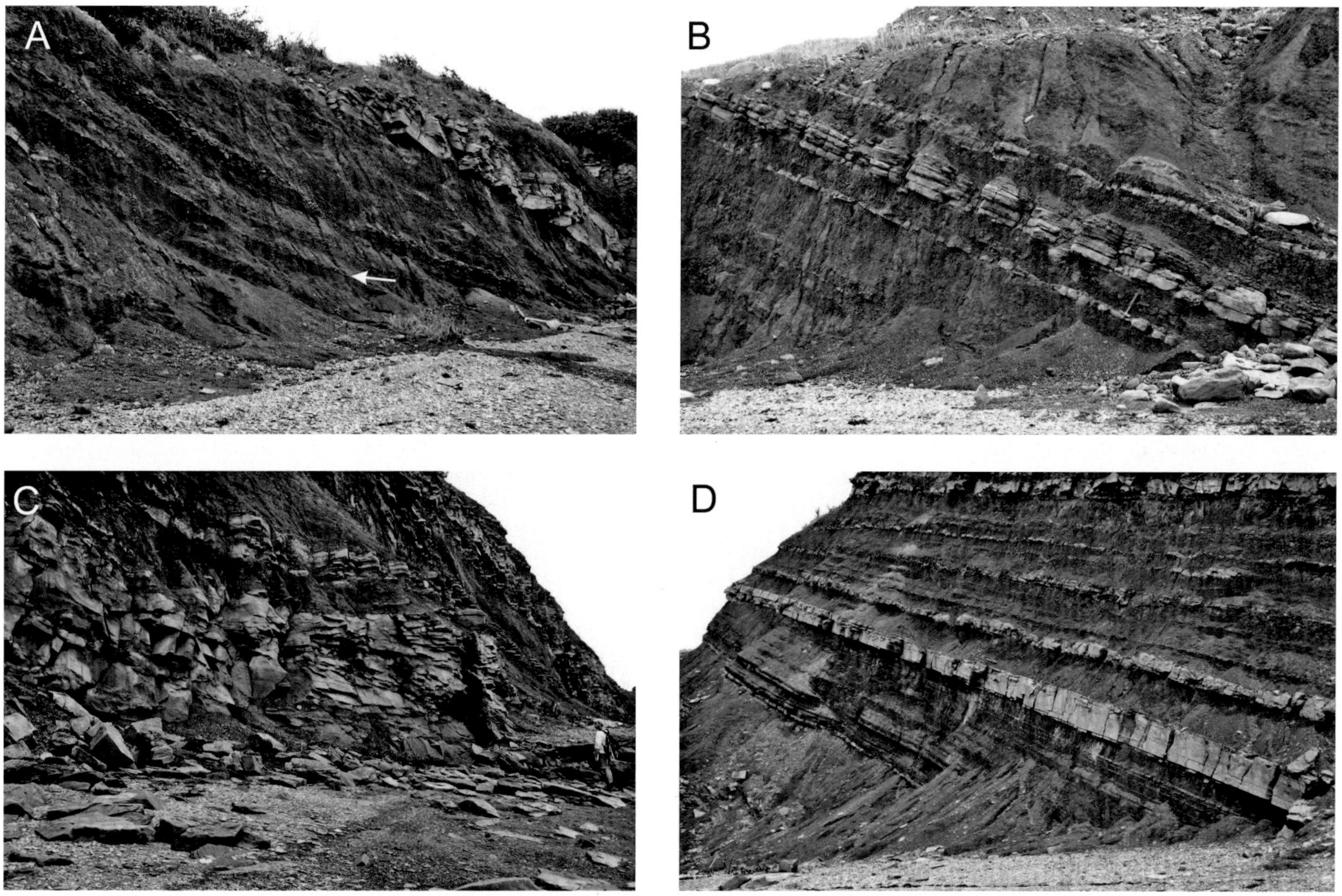

FIGURE 14.—Illustrations of the lower Joggins Formation in the Lower Cove to Joggins shore section. A) Cliff face exposing the contact between the Little River Formation and overlying Joggins Formation at a thin coal seam (white arrow). Note the lack of immediate change in lithofacies from one unit to the next. 25 m section exposed. B) Example of typical mudrocks with thin-bedded floodplain or floodbasin sandstones in the lower Joggins Formation. Note the persistence of brown and red colors in floodplain mudrocks. 4 m section exposed. C) Discontinuous channel sandstone in the lower Joggins Formation (at 930 m in Figure 9), showing a dominance of cross-bedding with minor flat and low-angle stratification. 20 m section exposed. D) Example of alternating well-drained (reddened) and poorly drained to open-water (gray) facies associations in the lower Joggins Formation, with distinctive sheet sandstones interpreted by Davies and Gibling (2003) as coastal-plain, shallow-water minor delta deposits. 20 m of section exposed.

spersed throughout the formation. Preserved paleosols are mainly entisols (protosols) and alfisols (argillisols) with minor, poorly developed aridisols (calcisols) near the base of the formation (Davies and Gibling 2003).

The main difference between the lower Joggins Formation and the underlying Little River Formation, other than the local preservation of coal and carbonaceous shale, is that there is once again preservation of drab, gray mudrocks and carbonaceous material in plant-fossil adpressions. Mudrocks also contain more abundant plant fossils, mainly referable to lycopsids, and nonmarine bivalves are common. The proportion of sandstone in the section decreases somewhat, and the few thick sandstone bodies preserve a combination of FTSS and cross-bedding, similar to the uppermost Boss Point Formation (Fig. 9). Paleocurrent data from these sandstones are directed eastward once again.

The facies assemblage of the lower Joggins Formation could be interpreted as recording a further increase in the A/S ratio. Here, the changes in facies noted above could be interpreted in terms of an increase in accommodation, the most likely cause being salt withdrawal alluded to above. However, the findings of Waldron and Rygel (2005) imply that salt withdrawal from the subsurface occurred throughout the time represented by the Little River Formation up to the Malagash Formation (Fig. 8). Accordingly, a climatic change is considered the most likely explanation of these patterns. The lower Joggins Formation is herein interpreted as recording a gradual return to a more equable climate, perhaps transitional between the strongly seasonal climate of the Little River Formation and the more pluvial climate of the Boss Point Formation. This regime may have included periods of strongly seasonal climate, in that there are many similarities between the facies assemblage of the lower Joggins and Little River Formations. For example, one sandstone interval (the "Hebert Beds" at 1095 m in Fig. 9) has been studied in detail by Falcon-Lang et al. (2004) and interpreted as the deposit of a dryland waterhole (shallow channel fill) analogous to those in the "Channel Country" desert of central Australia. Given the combination of FTSS and cross-bedding preserved in this interval, and preservation of plant and invertebrate animal fossils, a strongly seasonal, tropical fluvial setting is here considered a more appropriate analogy. Interestingly, this body is unique in the lower Joggins Formation in showing westward paleocurrent directions, similar to the underlying Little River Formation.

Alternating with intervals of strongly seasonal climate were times of more humid climate, as evidenced by the preservation of thin coals, carbonaceous shales, and nonmarine limestones. Farther up into the Joggins Formation, coal seams are significantly thicker and have been mined extensively in the area. Accordingly, the lower Joggins Formation can be regarded as recording a climatic regime that was transitional from the strongly seasonal, semiarid tropical system of the Little River Formation into the more humid climate characteristic of much of the Joggins Formation (Davies and Gibling 2003; Davies et al. 2005; Rygel and Gibling 2006).

Significance of the Studied Interval

The interval documented herein, from the uppermost Boss Point Formation, through the Little River Formation, into the lower Joggins Formation, is interpreted to record a cycle of progressive change in fluvial style (Fig. 15). These changes are in turn interpreted to record a complete cycle of paleotropical climate change from humid to strongly seasonal and subhumid or semiarid to again relatively humid, over a stratigraphic interval of ~ 1100 m recording a timeframe of ~ 1 million years (Fig. 16). This is comparable to interpreted climatic cycles recognized in the Pennsylvanian "cyclothem" stratigraphy of midcontinent USA by Feldman et al. (2005). That such a coherent record of long-term climate change is preserved in the present study area is likely due to the extraordinary rates of tectonic accommodation in the Cumberland Basin (Waldron and Rygel 2005; Gibling et al. 2008). Nonetheless, Allen (2009) shows that stratal intervals in which characteristics of the strongly seasonal fluvial style dominate can be recognized across the Maritimes Basin complex (~ 1000 km), and correlate remarkably well across the region. This signal can be deconvolved from probable tectonic effects on the stratigraphy by the use of the facies characteristics, many of which are independent of accommodation. This suggests that the long-term changes in paleoclimate documented herein may be significant and related to global events occurring at this time, specifically the late Paleozoic Ice Age. It is emphasized that these changes are long-term trends, and that shorter-term cycles or changes in paleoclimate might be superimposed on such a pattern (as indeed has been recognized in coal-bearing rocks of the overlying Springhill Mines Formation by Calder 1994, and in the nearby Sydney Basin by Gibling et al. 2004).

The timing of the strongly seasonal climate interval documented here coincides with the transition from one of several Carboniferous glacial epochs in eastern Australia recognized by Fielding et al. (2008; their C3) into a succeeding nonglacial epoch. Interestingly, a further interval of strongly seasonal paleoclimate is documented by Fielding et al. (2009) and Allen (2009) from higher in the Pennsylvanian succession and correlated regionally: this interval also coincides with a transition from the nonglacial epoch noted above into the next eastern Australian glacial epoch (C4 of Fielding et al. 2008). It can be concluded, then, that recognition of regionally extensive stratigraphic intervals characterized by the strongly seasonal paleotropical fluvial style can inform paleoclimatic analysis of the pre-Quaternary rock record. Recognition of such changes in fluvial styles may also contribute to a more complete understanding of the controls on the A/S ratio in alluvial and coastal-plain successions. Many studies invoke changes in subsidence regime to explain variations in facies stacking in fluvial successions (e.g., Martinsen et al. 1999; Anderson and Cross 2001; Catuneanu and Elango 2001). From the data presented, however, it is evident that a change in paleoclimate may cause facies changes without any change in A/S ratio, even given that changes in sediment supply might be expected to accompany climatic changes.

A further aspect of the stratigraphic interval documented above that deserves mention is the evidence for a reversal of sediment dispersal direction from the Boss Point Formation into the Little River Formation (east-directed to west-directed) and back again into the Joggins Formation (with occasional short-term reversals) (Fig. 15). That the changes in sediment dispersal direction, and hence paleoslope orientation, broadly coincide with changes in fluvial style suggests that these environmental changes had profound effects on the nature of the drainage basin (or vice versa). The overwhelmingly dominant direction of sediment dispersal in the Maritimes Basin complex throughout the Pennsylvanian was eastward to northeastward (Gibling et al. 1992; Gibling et al. 2008), which means that the seasonal tropical Little River Formation stands out as a significant anomaly. The reasons for this reversal are, however, unclear. It may be that when the dominant regional drainage to the northeast (Gibling et al. 1992) was curtailed during such semiarid times, the motif of flashy flow events reflecting perhaps megafans, feeding from the Cobequid Highlands to the south of the Cumberland Basin, was recorded in the basin. It is also possible that tectonic processes changed climate states in the hinterland by modifying physiography, and that these events somehow coincided with changes in atmospheric circulation associated with global climate change. Clearly, resolution of this enigma requires further research and is beyond the scope of the present paper.

DISCUSSION: THE ROLE OF FLUVIAL STYLE IN STRATIGRAPHIC AND RESERVOIR INTERPRETATION

The above case study demonstrates the usefulness of recognizing a further variant on fluvial style that records hot, strongly seasonal,

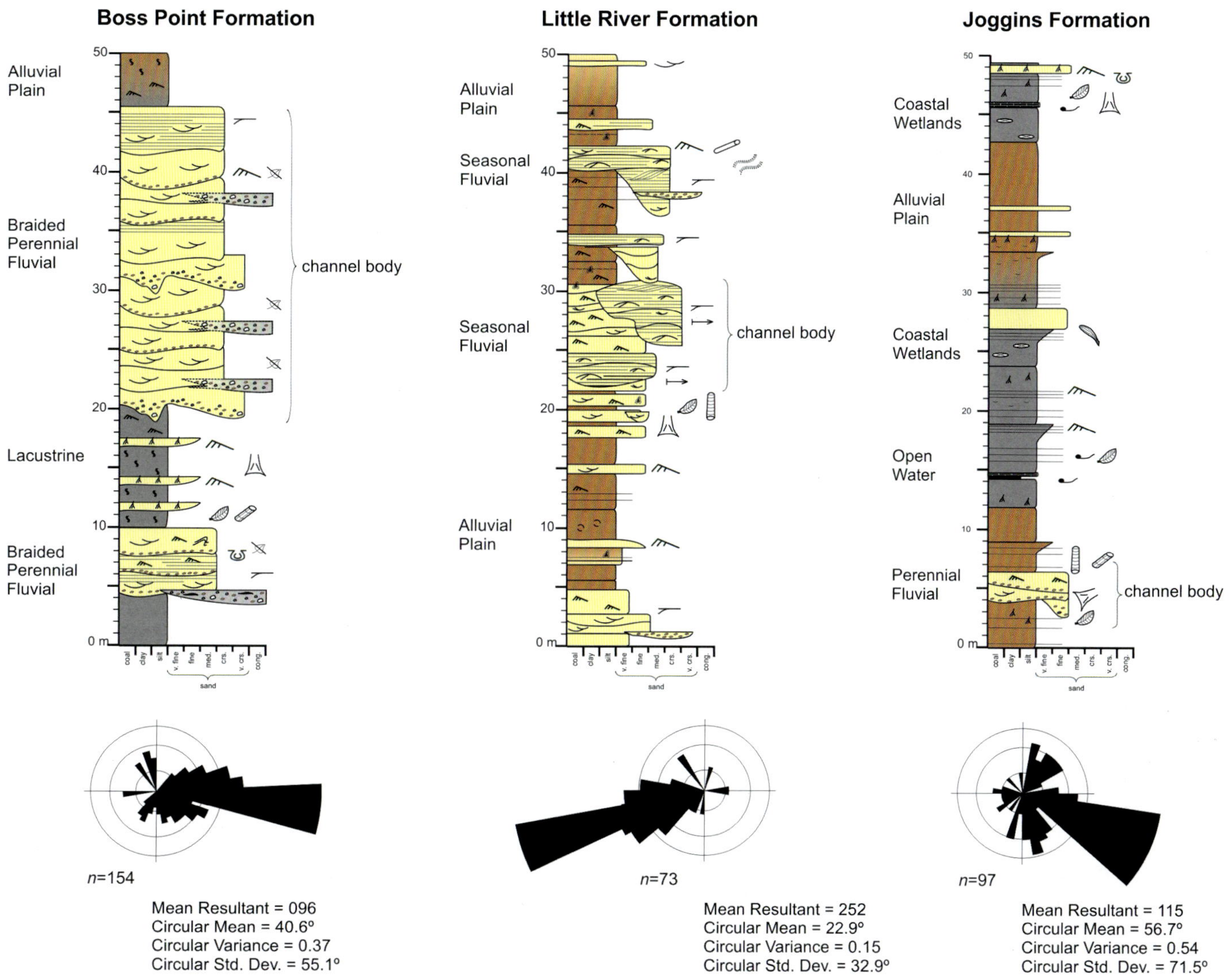

FIGURE 15.—Suite of three schematic graphic logs summarizing the sedimentological characteristics of the upper Boss Point Formation, Little River Formation, and the lower Joggins Formation. Also shown are all paleocurrent data collected from the three units, to highlight the polar reversal in sediment dispersal direction recorded in the Little River Formation.

subhumid to semiarid climates. In addition to its basic utility in characterizing the paleoclimate of a target succession, the recognition of correlatable long-term changes in paleoclimate may contribute towards understanding of basin evolution and environmental change through a period of Earth history. A succession of discrete intervals each characterized by a certain fluvial style in the Pennsylvanian succession of the Maritimes Basin complex of Atlantic Canada (Allen 2009) has been interpreted in terms of long-term paleoclimate evolution possibly teleconnected to glacial events in the paleo–southern hemisphere. The implication of this is that stratigraphic changes in fluvial architecture recorded in the paleo-tropics may have significance for understanding global climate change in Earth history. Since it has been established from the modern that changes from one fluvial style to another can occur simultaneously in different sections of the same river, however, care must be taken not to overinterpret occurrences of the seasonal tropical fluvial style within, say, a single channel body, in the rock record. The patterns documented herein occur progressively over significant vertical intervals, and can be confidently interpreted as a true record of long-term environmental change.

The recognition of the subhumid to semiarid, seasonal tropical to subtropical fluvial style documented herein and in Fielding et al. (2009) may also prove useful in exploration for and production of fluids from subsurface reservoirs. Firstly, since the fluvial channel bodies characteristic of the seasonal tropical fluvial style are typically complex, with mudrock partings, a composite channel body might be mistaken on wireline logs for an array of thin channel sands and overbank deposits, thus downgrading reservoir potential (or the net/gross ratio). In fact, despite the complex internal architecture of the seasonal tropical fluvial channel bodies, mudrock partings are typically laterally discontinuous, which means that the net/gross ratio would likely be higher than anticipated, and all sandstone layers within a body would likely be in fluid communication with each other. Recognition of the seasonal fluvial style from wireline logs alone is considered impossible, as is the recognition of any fluvial style from such data. However, we submit that the deposits of fluvial channel facies from

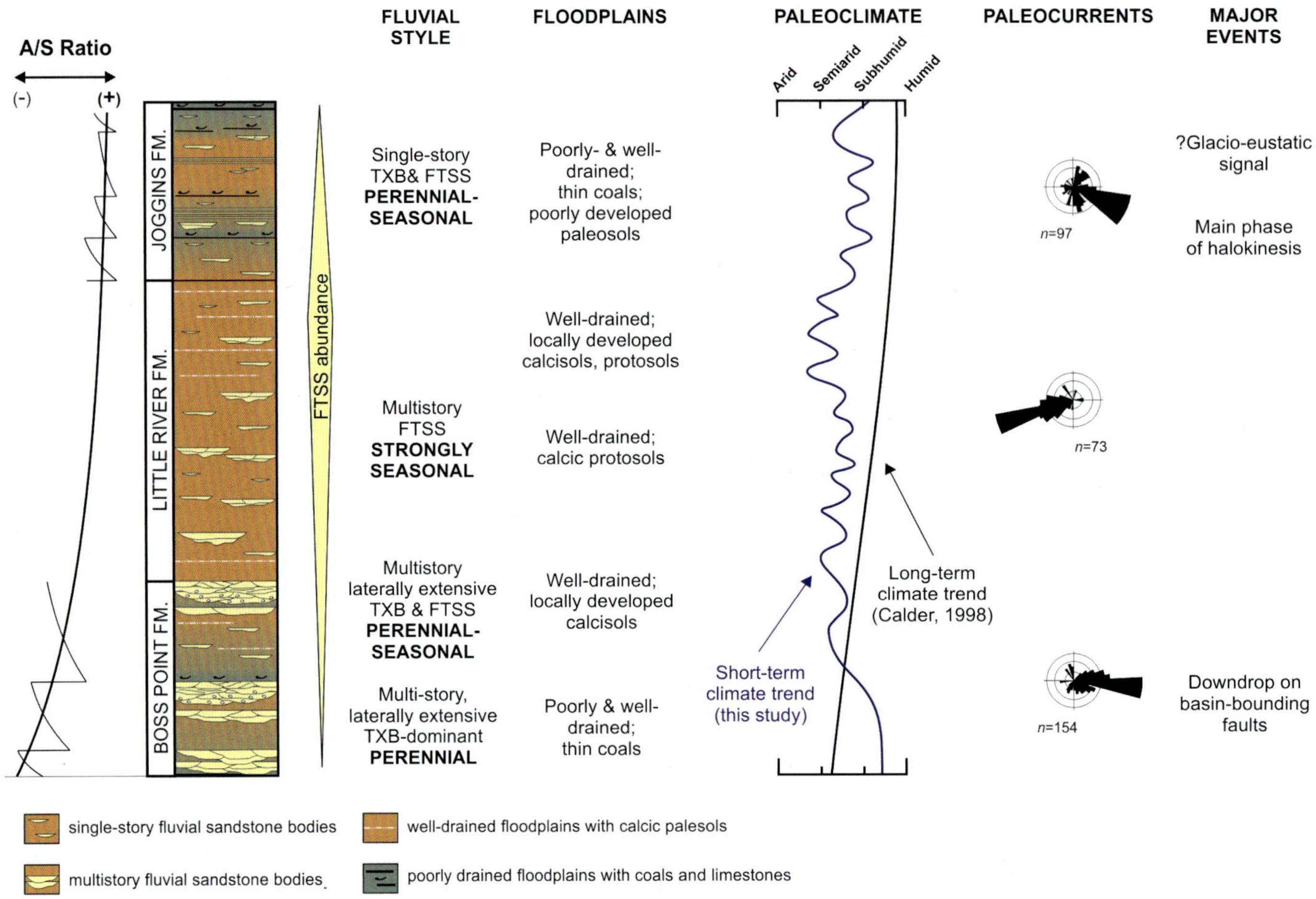

FIGURE 16.—Synthesis diagram summarizing all main aspects of the interval and their interpretation. Note the gradual increase and subsequent decrease in the abundance of Froude-transcritical and supercritical structures (FTSS) through the section, which vary independently of accommodation. This, together with other criteria noted on the diagram, allows the deconvolution of climatic from tectonic effects on the stratigraphy. TXB, trough cross-bedding.

seasonal tropical settings should be recognizable from core data. Erosionally based sandstone bodies with common mudrock partings are dominated internally by flat and low-angle discordant lamination, with subordinate cross-bedding and ripple cross-lamination. Some sandstones might be confused with the products of shoreface settings, since shoreface sands also preserve abundant flat and low-angle lamination, but the association with *in situ* plant growth, pedogenic fabrics, and other features should allow confident discrimination of a fluvial origin. By analogy with the modern example, deltaic systems composed of clean, well sorted sand in proximal parts might be predicted downdip from such fluvial deposits, similar to the Burdekin River Delta of NE Australia (Fielding et al. 2005).

A further aspect of potential reservoirs of seasonal tropical fluvial origin that is worthy of mention is their texture and mineralogy. Sands from the alluvium of the modern Burdekin River show good sorting and a general lack of fines (Fielding and Alexander 1996), suggesting that subsurface analogues might possess high initial porosity and permeability. Furthermore, these sands are mineralogically immature (Fig. 17), showing a wide variety of labile grain types that would in other tropical climates be eliminated by chemical weathering processes at the surface. Thus, if sands deposited in seasonal tropical fluvial environments could be removed from the surface weathering environment by rapid subsidence or aggradation, then there is the potential for the production of secondary porosity in such sands by dissolution of framework grains in the subsurface (cf. Schmidt and McDonald 1979). Sandstones from the Little River Formation, accumulated during a period of very rapid subsidence, show a feldspathic composition and preservation of aggregates of feldspathic granules locally (Calder et al. 2005), consistent with this idea. We therefore suggest that the channel bodies of seasonal tropical fluvial paleoenvironments might be an underappreciated potential reservoir facies in the subsurface.

CONCLUSIONS

This paper highlights the distinctive nature of rivers that characterize the seasonal tropics and subtropics, and their resulting sedimentary deposits. The characteristics of these deposits in combination constitute a distinct fluvial style that is readily recognizable in the rock record. Fluvial deposits displaying the distinctive seasonal tropical fluvial style occur within a discrete portion of the Pennsylvanian stratigraphy in the Cumberland Basin of Atlantic Canada. Analysis of this interval shows how interpretations of climate change can be deconvolved from tectonic effects, thereby providing a new criterion for use in stratigraphic analysis of fluvial successions. Analysis of these changes also informs the broader interpretation of geological history of the basin.

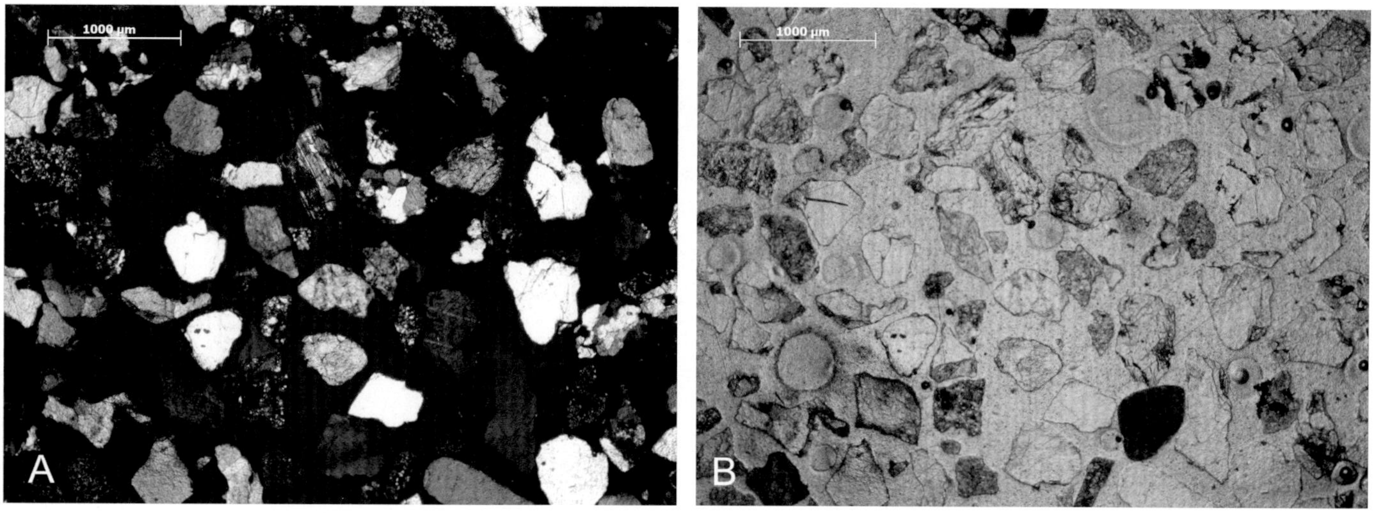

FIGURE 17.—Photomicrographs of Burdekin River sand. A) View of a sand sample from the river bed near Charters Towers (Fig. 1) in cross-polarized light, B) Same view in plane-polarized light. Scale bar is 1 mm. Note the mineralogical immaturity of the sand (high feldspar and lithic content), but textural maturity (good sorting, no fines). This character is typical of sands from the Burdekin River of NE Australia.

This paper emphasizes the imprint of climate change on alluvial stratigraphy. The vertical stratigraphic changes identified in the Pennsylvanian case study (Fig. 16) could be interpreted in terms of changes in the accommodation/sediment supply (A/S) ratio. Knowing the tectonic history of the basin, most of the changes in alluvial architecture observed can be attributed to climatic change that might have varied sediment supply (specifically, a change to a strongly seasonal, semiarid to subhumid climate and concomitant decrease in sediment yield, and vice versa). However, interpreting the change from the predominantly red alluvial deposits of the Little River Formation into the more coal and lacustrine limestone–rich, overlying Joggins Formation as the consequence of a change in A/S ratio would be flawed. This is because all evidence suggests that the accommodation regime for the two formations was the same, and the sand/mud ratio suggests that sediment supply was comparable. The difference in lithology can, rather, best be interpreted in terms of increasing runoff (via climate change) leading to a more frequently or permanently waterlogged alluvial surface, without any change in accommodation or sediment supply. This paper provides a new criterion by which to recognize the effects of climate change in the paleotropical realm, and we hope that this will stimulate further research on the effects of climate change on alluvial stratigraphy.

ACKNOWLEDGMENTS

The Coffman Endowment in Sedimentary Geology at UNL supported fieldwork in the Maritimes Basin. JPA also received financial support from the Geological Society of America and American Association of Petroleum Geologists. MRG was supported by a Discovery Grant from the Natural Sciences and Engineering Research Council of Canada. We thank the reviewers Russell Dubiel and Paul McCarthy, the Editors, and copy-editor John Southard for their constructive reviews of the submitted manuscript.

REFERENCES

ALEXANDER, J., AND FIELDING, C.R., 1997, Gravel antidunes in the tropical Burdekin River, Queensland, Australia: Sedimentology, v. 44, p. 327–337.

ALEXANDER, J., AND FIELDING, C.R., 2006, Coarse-grained floodplain deposits in the seasonal tropics: towards a better facies model: Journal of Sedimentary Research, v. 76, p. 539–556.

ALEXANDER, J., FIELDING, C.R., AND POCOCK, G.D., 1999a, Flood behaviour of the Burdekin River, tropical north Queensland, Australia, *in* Marriott, S.B. and Alexander, J., eds., Floodplains: Interdisciplinary Approaches: Geological Society of London Special Publication 163, p. 27–40.

ALEXANDER, J., FIELDING, C.R., AND JENKINS, G., 1999b, Plant-material deposition in the tropical Burdekin River, Australia: implications for ancient fluvial sediments: Palaeogeography, Palaeoclimatology, Palaeoecology, v. 153, p. 105–125, doi: 10.1016/S0031–0182(99)00073–5.

ALEXANDER, J., BRIDGE, J.S., CHEEL, R.J., AND LECLAIR, S.F., 2001, Bedforms and associated sedimentary structures formed under supercritical water flows over aggrading sand beds: Sedimentology, v. 48, p. 133–152.

ALLEN, J.P., 2009, The Carboniferous evolution of the Maritimes Basin Complex, Atlantic Canada: unpublished Doctoral Thesis, Lincoln, University of Nebraska–Lincoln, 547 p.

ALLEN, J.R.L., 1968, The nature and origin of bed-form hierarchies: Sedimentology, v. 10, 161–182.

AMOS, K.J., ALEXANDER, J., HORN, A., POCOCK, G.D., AND FIELDING, C.R., 2004, Supply limited sediment transport in a high-discharge event of the tropical Burdekin River, North Queensland, Australia: Sedimentology, v. 51, p. 145–162.

ANDERSON, D.S., AND CROSS, T.A., 2001, Large-scale architecture in continental strata, Hornelen Basin (Devonian), Norway: Journal of Sedimentary Research, v. 71, p. 255–271.

BEST, J.L., AND BRIDGE, J.S., 1992, The morphology and dynamics of low amplitude bedwaves upon upper stage plane beds and the preservation of planar laminae: Sedimentology, v. 39, p. 737–752.

BRIDGE, J.S., 2003, Rivers and Floodplains; Forms, Processes and Sedimentary Record: Oxford, U.K.,Blackwell Science Ltd, 491 p.

BRIDGE, J.S., 2006, Fluvial facies models: recent developments, *in* Posamentier, H.W., and Walker, R.G., eds., Facies Models Revisited: SEPM, Special Publication 84, p. 85–170.

BRIDGE, J.S., AND BEST, J.L., 1988, Flow, sediment transport and bedform dynamics over the transition from dunes to upper-stage plane beds: implications for the formation of planar laminae: Sedimentology, v. 35, p. 753–763, doi: 10.1111/j.1365–3091.1988.tb01249.x.

BROWNE, G.H., AND PLINT, A.G., 1994, Alternating braidplain and lacustrine deposition in a strike-slip setting: the Pennsylvanian Boss Point Formation of the Cumberland Basin, Maritime Canada: Journal of Sedimentary Research, v. B64, p. 40–59.

Calder, J.H., 1994, The impact of climate change, tectonism and hydrology on the formation of Carboniferous tropical intermontane mires: the Springhill coalfield, Cumberland Basin, Nova Scotia: Palaeogeography, Palaeoclimatology, Palaeoecology, v. 106, p. 323–351.

Calder, J.H., 1998, The Carboniferous evolution of Nova Scotia, *in* Blundell, D.J., and Scott, A.C., eds., Lyell: The Past is the Key to the Present: Geological Society of London, Special Publication 143, p. 261–302.

Calder, J.H., Rygel, M.C., Ryan, R.J., Falcon-Lang, H.J., and Hebert, B.L., 2005, Stratigraphy and sedimentology of early Pennsylvanian red beds at Lower Cove, Nova Scotia, Canada: the Little River Formation with redefinition of the Joggins Formation: Atlantic Geology, v. 41, p. 143–167.

Catuneanu, O., and Elango, H.N., 2001, Tectonic control on fluvial styles: the Balfour Formation of the Karoo Basin, South Africa, Sedimentary Geology, v. 140, p. 291–313.

Cheel, R.J., 1990, Horizontal lamination and the sequence of bed phases and stratification under upper-flow-regime conditions: Sedimentology, v. 37, p. 517–529, doi: 10.1111/j.1365-3091.1990.tb00151.x.

Collinson, J.D., 1996, Alluvial sediments, *in* Reading, H.G., ed., Sedimentary environments: Processes, Facies and Stratigraphy: Oxford, U.K., Blackwell Science, p. 37–82.

Davies, S.J., and Gibling, M.R., 2003, Architecture of coastal and alluvial deposits in an extensional basin: the Carboniferous Joggins Formation of eastern Canada: Sedimentology, v. 50, p. 415–439.

Davies, S.J., Gibling, M.R., Rygel, M.C., Calder, J.H., and Skilliter, D.M., 2005, The Pennsylvanian Joggins Formation of Nova Scotia: sedimentological log and stratigraphic framework of the historic fossil cliffs: Atlantic Geology, v. 41, p. 115–142.

Duller, R.A., Mountney, N.P., Russell, A.J., and Cassidy, N.C., 2008, Architectural analysis of a volcaniclastic jökulhlaup deposit, southern Iceland: sedimentary evidence for supercritical flow: Sedimentology, v. 55, p. 939–964.

Falcon-Lang, H.J., Rygel, M.C., Calder, J.H., and Gibling, M.R., 2004, An early Pennsylvanian waterhole deposit and its fossil biota in a dryland alluvial plain setting, Joggins, Nova Scotia: Geological Society of London Journal, v. 161, p. 209–222.

Feldman, H.R., Franseen, E.K., Joeckel, R.M., and Heckel, P.H., 2005, Impact of longer-term modest climate shifts on architecture of high-frequency sequences (cyclothems), Pennsylvanian of midcontinent USA: Journal of Sedimentary Research, v. 75, p. 350–368.

Fielding, C.R., 2006, Upper flow regime sheets, lenses and scour fills: extending the range of architectural elements for fluvial sediment bodies: Sedimentary Geology, v. 190, p. 227–240, doi: 10.1016/j.sedgeo.2006.05.009.

Fielding, C.R., and Alexander, J., 1996, Sedimentology of the upper Burdekin River of north Queensland, Australia—an example of a tropical, variable discharge river: Terra Nova, v. 8, p. 447–457, doi: 10.1111/j.1365-3121.1996.tb00770.x.

Fielding, C.R., Alexander, J., and Newman-Sutherland, E., 1997, Preservation of in situ vegetation in fluvial channel deposits—data from the modern Burdekin River of North Queensland, Australia: Palaeogeography, Palaeoclimatology, Palaeoecology, v. 135, p. 123–144, doi: 10.1016/S0031-0182(97)00022-9.

Fielding, C.R., Alexander, J., and McDonald, R., 1999, Sedimentary facies from GPR surveys of the modern, upper Burdekin River of north Queensland, Australia: consequences of extreme discharge fluctuations, *in* Smith, N.D., and Rogers, J., eds., Fluvial Sedimentology VI: International Association of Sedimentologists Special Publication 28, p. 347–362.

Fielding, C.R., Trueman, J.D., and Alexander, J., 2005, Sedimentology of the modern and Holocene Burdekin River Delta of north Queensland, Australia—controlled by river output, not waves and tides, *in* Giosan, L., and Bhattacharya, J.P., eds., River Deltas: Concepts, Models, and Examples: SEPM, Special Publication 83, p. 476–496.

Fielding, C.R., Frank, T.D., Birgenheier, L.P., Rygel, M.C., Jones, A.T., and Roberts, J., 2008, Alternating glacial and non-glacial intervals characterize the late Paleozoic Ice Age: stratigraphic evidence from eastern Australia: Geological Society of London Journal, v. 165, p. 129–140.

Fielding, C.R., Allen, J.P., Alexander, J., and Gibling, M.R., 2009, A facies model for fluvial systems in the seasonal tropics and subtropics: Geology, v. 37, p. 623–626.

Friend, P.F.F., Hirst, J.P.P., and Nichols, G.J., 1986, Sandstone-body structure and river process in the Ebro Basin of Aragon, Spain: Cuadernos Geología Ibérica, v. 10, p. 9–30.

Galloway, W.E., and Hobday, D.K., 1996, Terrigenous Clastic Depositional Systems, Second Edition: Berlin, Springer-Verlag, 489 p.

Gibling, M.R., Calder, J.H., Ryan, R., van de Poll, H.W., and Yeo, G.M., 1992, Late Carboniferous and Early Permian drainage patterns in Atlantic Canada: Canadian Journal of Earth Sciences, v. 29, p. 338–352.

Gibling, M.R., Saunders, K.I., Tibert, N.E., and White, J.A., 2004, Sequence sets, high-accommodation events, and the coal window in the Carboniferous Sydney Coalfield, Atlantic Canada, *in* Pashin, J.C., and Gastaldo, R.A., eds., Sequence Stratigraphy, Paleoclimate, and Tectonics of Coal-Bearing Strata: American Association of Petroleum Geologists, Studies in Geology Series, no. 51, p. 169–197.

Gibling, M.R., Culshaw, N., Rygel, M.C., and Pascucci, V., 2008, The Maritimes Basin of Atlantic Canada: basin creation and destruction in the collisional zone of Pangea, *in* Hsü, K.J., ed., Sedimentary Basins of the World, Volume 5, The Sedimentary Basins of the United States and Canada, A.D. Miall, ed.: Amsterdam, Elsevier Science, p. 211–244.

Hand, B.M., and Bartberger, C.E., 1988, Leeside sediment fallout patterns and the stability of angular bedforms: Journal of Sedimentary Petrology, v. 58, p. 33–43.

Hirst, J.P.P., 1991, Variations in alluvial architecture across the Oligo-Miocene Huesca fluvial system, Ebro Basin, Spain, *in* Miall, A.D., and Tyler, N., eds., The Three-Dimensional Facies Architecture of Terrigenous Clastic Sediments and Its Implications for Hydrocarbon Discovery and Recovery: SEPM, Concepts in Sedimentology and Paleontology Series, no. 3, p. 111–121.

Jones, C.M., 1977, Effects of varying discharge regimes on bed-form sedimentary structures in modern rivers: Geology, v. 5, p. 567–570.

Keighley, D.G., and Pickerill, R.K., 1996, The evolution of fluvial systems in the Port Hood Formation (Upper Carboniferous), western Cape Breton Island, eastern Canada: Sedimentary Geology, v. 106, p. 97–144.

Kennedy, J.F., 1963, The mechanics of dunes and antidunes in erodible-bed channels: Journal of Fluid Mechanics, v. 16, p. 521–544.

Machette, M.N., 1985, Calcic soils of the southwestern United States, *in* Weide, D.L., and Faber, M.L., eds., Soils and Quaternary Geology of the Southwestern United States: Geological Society of America, Special Paper 203, p. 1–21.

Mack, G.H., James, W.C., and Monger, H.C., 1993, Classification of paleosols: Geological Society of America, Bulletin, v. 105, p. 129–136.

Martinsen, O.J., Ryseth, A., Helland-Hansen, W., Flesche, H., Torkildsen, G., and Idil, S., 1999, Stratigraphic base level and fluvial architecture: Ericson Sandstone (Campanian), Rock Springs Uplift, SW Wyoming, U.S.A: Sedimentology, v. 46, p. 235–259.

Marzo, M., Nijman, W., and Puigdefabregas, C., 1988, Architecture of the Castissent fluvial sheet sandstones, Eocene, South Pyrenees, Spain: Sedimentology, v. 35, p. 719–738.

Middleton, G.V., 1965, Antidune cross-bedding in a large flume: Journal of Sedimentary Petrology, v. 35, p. 922–927.

Nakayama, K., Fielding, C.R., and Alexander, J., 2002, Variations in character and preservation potential of vegetation-induced obstacle marks in the variable discharge Burdekin River of north Queensland, Australia: Sedimentary Geology, v. 149, p. 199–218, doi: 10.1016/S0037-0738(01)00173-7.

Paola, C., Wiele, S.M., and Reinhart, M.A., 1989, Upper-regime parallel lamination as the result of turbulent sediment transport and low-amplitude bedforms: Sedimentology, v. 36, p. 47–60.

Plint, A.G., and Browne, G.H., 1994, Tectonic event stratigraphy in a fluvio-lacustrine, strike-slip setting: the Boss Point Formation (Westphalian A), Cumberland Basin, Maritime Canada: Journal of Sedimentary Research, v. B64, p. 341–364.

Rust, B.R., and Gibling, M.R., 1990, Braidplain evolution in the Pennsylvanian South Bar Formation, Sydney Basin, Nova Scotia, Canada: Journal of Sedimentary Petrology, v. 60, p. 59–72.

Ryan, R.J., and Boehner, R.C., 1994, Geology of the Cumberland Basin, Cumberland, Colchester and Pictou Counties, Nova Scotia: Halifax, Nova Scotia Department of Natural Resources, Memoir 10, 222 p.

Rygel, M.C., and Gibling, M.R., 2006, Natural geomorphic variability recorded in a high-accommodation setting: fluvial architecture of the Pennsylvanian Joggins Formation of Atlantic Canada: Journal of Sedimentary Research, v. 76, p. 1230–1251.

Rygel, M.C., Gibling, M.R., and Calder, J.H., 2004, Vegetation-induced sedimentary structures from fossil forests in the Pennsylvanian Joggins Formation, Nova Scotia: Sedimentology, v. 51, p. 531–552.

Saunderson, H.C., and Lockett, F.P.J., 1983, Flume experiments and structures at the dune–plane bed transition, *in* Collinson, J.D., and Lewin, J., eds., Modern and Ancient Fluvial Systems: International Association of Sedimentologists, Special Publication 6, p. 49–58.

Schmidt, V., and McDonald, D.A., 1979, The role of secondary porosity in the course of sandstone diagenesis, *in* Scholle, P.A., and Schluger, P.R., eds., Aspects of Diagenesis: SEPM, Special Publication 26, p. 175–207.

Scotese, C.R., and McKerrow, W.S., 1990, Revised world maps and introduction, *in* McKerrow, W.S., and Scotese, C.R., eds., Palaeozoic Palaeogeography and Biogeography: Geological Society of London, Memoir 12, p. 1–21.

Underwood, J.R., Jr., and Lambert, W., 1974, Centroclinal cross strata. A distinctive sedimentary structure: Journal of Sedimentary Petrology, v. 44, p. 1111–1113.

Waldron, J.W.F., and Rygel, M.C., 2005, Role of evaporite withdrawal in the preservation of a unique coal-bearing succession: Pennsylvanian Joggins Formation, Nova Scotia: Geology, v. 33, p. 337–340.

Williams, G.P., 1978, Bank-full discharge of rivers: Water Resources Research, v. 14, p. 1141–1154, doi: 10.1029/WR014i006p01141.

Ziegler, A.M., Rees, P.M., and Naugolnykh, S.V., 2002, The Early Permian floras of Prince Edward Island, Canada: differentiating global from local effects of climate change: Canadian Journal of Earth Sciences, v. 39, p. 223–238.

DIGITAL APPENDIX

The following supplementary material is available in electronic format either on the DVD accompanying this volume or through the digital archive on the SEPM website at http://www.sepm.org/

Larger version of Figure 9, with additional detail: a measured section through the uppermost Boss Point Formation, the entire Little River Formation, and the lower Joggins Formation at Lower Cove, Nova Scotia.

INTERPRETING FLUVIAL HYDROMORPHOLOGY FROM THE ROCK RECORD: LARGE-RIVER PEAK FLOWS LEAVE NO CLEAR SIGNATURE

SUZANNE F. LECLAIR

Department of Earth and Environmental Sciences, Tulane University, New Orleans, LA 70118, U.S.A.
Presently an Independent Researcher in Montreal, QC, Canada
e-mail: suzanneleclair@sympatico.ca

Abstract: Most recent studies on dune in large rivers focus on bedform development and sediment transport, and little consideration is usually given to the preservation potential of dune deposits. In this study, the Leclair-Bridge (2001) and the Paola-Borgman (1991) models are combined to estimate the thicknesses of bedform-scale deposits within channel-scale deposits in a large river at the end of annual high flows. Modeling relies on high-resolution bed elevations measured on 15-km-long 2-D lines during five successive surveys in the Lower Mississippi River.

Results from this study challenge our common assumptions: peak-flow deposits are not preferentially preserved, and dune height does not scale with flow depth even on a single survey. The model predicts that only few peak-flow deposits would be preserved in the deep parts of the channel (and would be from small dunes); instead, synchronous large-dune deposits would be found at elevation up to 10 m above the bend-scour level. Therefore, even full preservation of peak-flow dune deposits would lead to incorrect paleo-hydraulic reconstruction. Moreover, peak-flow deposits would hardly be distinguished from other deposits due to short-term partial erosion and overlap in thickness distributions of stacked channel-scale cosets and dune sets. If such dune deposits would be seen in an outcrop, the estimated mean dune height, based on measurements from several vertical sections where mean set thickness varies only by ~ 30%, would represent average seasonal flow conditions, not peak flows (methodology based on analysis of natural and experimental deposits presented in this paper), and the large size of the system would not be recognized.

Our interpretation problems are not methodological, but fundamental. Future research should integrate fluvial morphodynamics and sedimentology, in rivers of various sizes and patterns. Datasets are presented as supplementary material for teaching, eventual comparative studies, and for developing stratigraphic models.

Key words: river dune, large river, peak flow, experimental sedimentology, bedform migration, cross-set, Mississippi, Loire

INTRODUCTION

Dunes are very common bedforms in sandy-gravelly rivers, and hence stratification formed by the migration of dunes (Fig.1) is generally the most common sedimentary structure in fluvial deposits. Over the last few decades, our understanding of bedform dynamics and related deposits has greatly improved; however, the interpretation of fluvial deposits is still challenging at times—e.g., if outcrop exposures show no large sedimentary structures, would it mean that large dunes never occurred or they were not preserved? Are deposits from a specific stage of the hydrograph preferentially preserved? Answering such questions is important because deposits from large dunes formed during bankfull conditions are used as indicators of channel scale, flow regime, climate, etc. This paper investigates short-term controls on dune stratigraphy, and specifically targets the week-to-month time scale over annual peak discharge. It also provides guidance for interpreting dune deposits as observed in cores and outcrops by addressing shorter-time-scale controls such as dune migration patterns (in supplementary material, Appendix S1).

The objective of this paper is to understand better the preservation potential of deposits from river dunes occurring at different times of the hydrograph, and hence to help improving our interpretation of the fluvial sedimentary record. The quantitative interpretation of sedimentary structures is useful for, e.g., comparing fluvial patterns of ancient rock formations (Adams and Battacharya 2005) or for potentially estimating catchment size and paleohydrology of ancient rivers by relying on scaling relationships between dune geometry and channel depth (cf. Yalin 1992 or Bridge 2003; more on this later) and geomorphological regional curves (Davidson and North 2009). The study site is the Lower Mississippi River, and hence this paper also brings new insights on large river systems, which are still a challenge to identify in the rock record (Miall 2006). The Leclair-Bridge (2001) and the Paola-Borgman (1991) models are combined herein to estimate the thicknesses of bedform-scale deposits within channel-scale deposits. A review of these models is provided in Appendix S1, but briefly, they go like this: the Paola-Borgman theory (1991) proposes a functional relationship between random topography and the preserved stratification that can be summarized as

$$Tk_m = 16.445(\sigma^2/x_m) \quad (1)$$

where Tk_m is a general term for mean thickness of bedform set or channel deposit, σ^2 the variance of bed elevation or scour-depth distribution, and x_m the mean value. Leclair and Bridge (2001) slightly modified the Paola-Borgman model in order to consider the natural variability of dune height and scour depth elevation below mean bed level as dunes migrate (based on experimental results), and they proposed a simple method for estimating mean dune height (h_m) from mean set thickness (s_m):

$$h_m = 2.9(\pm 0.7)s_m \quad (2)$$

In modern environments, h_m can be measured or estimated from the probability distribution of bed elevation (Leclair and Blom 2005; Leclair and Miller 2006), and hence mean set thickness (s_m) can be predicted based on Equation 2.

In addition to data from the Mississippi River, deposits from the Quaternary Loire River and experimental sediment peels are analyzed with the aim of setting criteria for the proper use of the Leclair-Bridge model.

From River To Rock Record: The Preservation Of Fluvial Sediments And Their Subsequent Interpretation
SEPM Special Publication No. 97, Copyright © 2011
SEPM (Society for Sedimentary Geology), ISBN 978-1-56576-305-0, p. 113–123.

VARIATION IN BED TOPOGRAPHY DURING ANNUAL PEAK FLOWS AND EXPECTED PRESERVED DUNE STRATIGRAPHY IN A MODERN RIVER

Methodological Approach

Study Site: The Lower Mississippi River in Louisiana, U.S.A., was surveyed during the 2005 spring high-discharge conditions. This study is not meant to be a regional study (cf. Saucier 1994, Ritternour et al. 2005), but is an attempt to understand processes leading to preserved stratigraphy in large rivers in general. Most previous dune studies on large rivers (e.g., Jackson 1975, 1976; Shugar et al. 2010) or specifically on the Mississippi River (Harbor 1998; Nittrouer et al. 2008), aimed at understanding bedform development and/or sediment transport, without relating these to the characteristics of the deposits. Conversely, this study focuses on potential stratigraphy due to changes in river bed morphology while keeping the complex hydrosedimentary dynamics beyond the scope of this paper.

Flow data are those available from the U.S. Army Corps of Engineers (USACE) and provide the general background conditions during the surveys. The sampling period was one of relatively high river stages—it ranged from 6.4 m to 11 m (Fig. 2), compared to the long-term (1935–2010) bankfull stage of ~ 9 m at Baton Rouge gauging station (~ 11 km downstream from study site). The historical maximum and minimum river stage at this station are 13.7 m and ~ 0.6 m (USACE stage hydrographs for user-defined gauging station and time period are available online, see References). Discharge comes for the nearest upstream gauging station (Talbert Landing, ~ 120 km from the study site), and there is no significant tributary downstream from this point. During the survey period, discharge varied from 28,357 m^3/s (twice the value of the long-term mean discharge) right after the highest flow peak, to 13,471 m^3/s at the end of the second flow peak (Table 1).

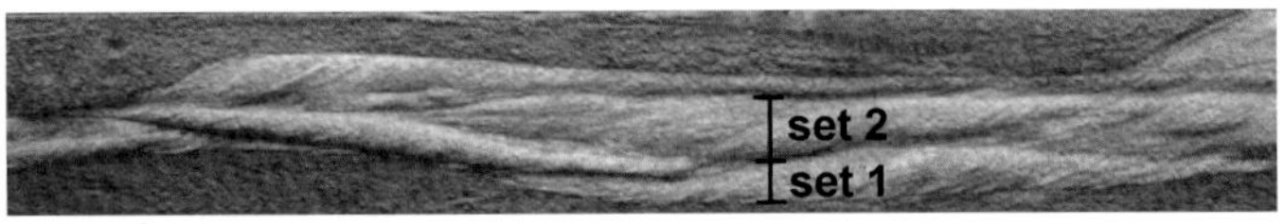

FIGURE 1.—Definition figure using a sediment peel showing preserved sets from a non-aggrading experiment (Leclair 2000); a coset is a group of sets; see Allen (1982) for definitions.

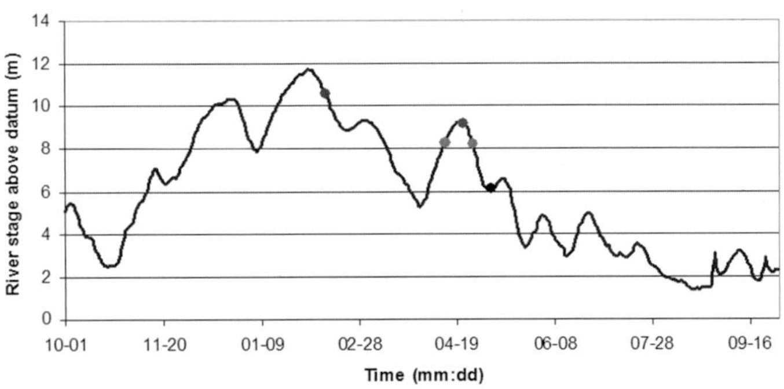

FIGURE 2.—Stage hydrograph for the survey period (source: USACE online data for the Baton Rouge gauging station). Dots indicate time of surveys.

Sampling Scheme and Instrumentation: In the sampling reach, variation in bed elevation is solely due to fluvial processes: the site is located at a distance of ~ 370 km from the river mouth, in a zone away from tidal effects, and where there is no cargo transportation—only barge traffic, therefore sand dredging is not required. The banks in the study area are covered with vegetation and the man-made levees stand on the natural levees, and hence the flow is not as constrained as in the lower reaches of the river, where they are built on the river banks (Wolman et al. 1990). However, the submerged banks are protected with revetments, and therefore the alongstream survey line was located at a distance of ~ 150 m from the (left) bank in order to avoid these structures. The survey line stretches over straight and sinuous reaches of the river, and includes sharp and moderately curved meander bends (Fig. 3). The sampling scheme was planned in order to record the variation in elevation due to bend scours and bar tops, that is, the widest range of bed elevation possible in a meandering river. The two first surveys consisted in recording high-resolution (variable, yet often sub-meter) bed elevation over a 40-km-long line (with the boat going upstream), then, the first 15 km of this line were surveyed three more times.

The instruments used were a 1 MHz single narrow (1.7°) beam transducer and a high-resolution ECHOTRAC echo sounder (at least ± 0.03 m accuracy in the range of flow depths reported here), coupled with a DGPS and a motion sensor compensating for roll, pitch, and heave. The HYPACK software was used to input (USACE) map of the area, pre-determine the location of the target survey line, and to process the bed elevations relative to water level. HYPACK also provides a function for assessing the accuracy of the boat track during data acquisition; most of the time, successive survey lines more or less overlap, and occasional deviation from the target survey line is usually less than ± 15 m (data from some ship tracks is provided in Appendix S2). The river width at the study site is about 1 km or more, and it is assumed that recorded data are representative of the overall bed conditions along the target survey line. The objective of this paper is not to accurately assess the deposit thickness at a specific location, but to estimate the trends in deposition–erosion patterns at the channel scale. Yet, in cases when boat could not be kept on track in the strong

TABLE 1.—Flow conditions during the 2005 Mississippi River surveys. Local range of flow depth due to reach topography is indicated).

Date	Mean discharge, Q (m^3/s)	Q/annual Q	River stage (m) Max = 13.7	Flow depth at selected reaches– mean (range) (m)			
				0–5 km	7–8 km	8–10 km	14–15 km
11 February	28357	2.1	11	18 (10–27)	23 (19–27)	24 (18–29)	26 (25–28)
13 April	20150	1.5	8.8	15 (10–21)	21 (17–26)	22 (17–28)	24 (22–25)
22 April	22753	1.7	9.4	16 (10–23)	21 (16–26)	24 (23–27)	25 (23–27)
27 April	17914	1.3	8.5	14 (8–20)	20 (16–25)	23 (17–28)	24 (22–25)
6 May	13471	1.0	6.4	12 (8–18)	18 (15–22)	20 (15–25)	21 (20–23)

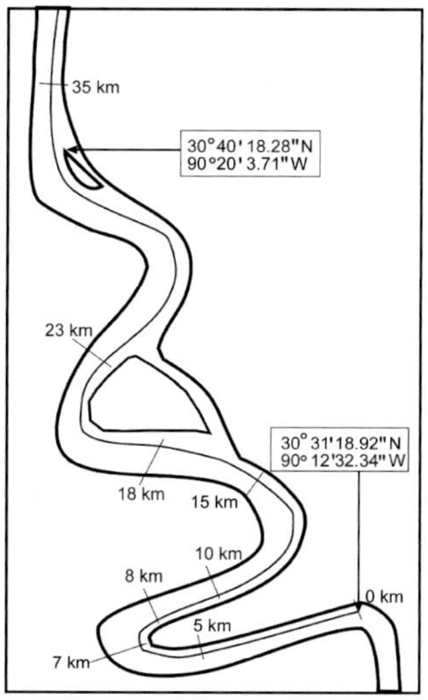

FIGURE 3.—Schematic Mississippi River planform at study site, with target survey line; coordinates are indicated for points of reference; short lines are markers related to material presented in other figures.

currents (essentially at the downstream sharp river bend), data between surveys were not compared for this part of the line.

Analytical Approach: The present approach, previously used in morphodynamics studies (Parker et al. 2000; Leclair 2005) considers the statistics of distributions of bed elevation. This approach is preferred to the traditional description of distributions of dune height, length, etc. because the dataset allows encompassing all multi-scale, superimposed bedforms; moreover, it reflects the original concept of Paola and Borgman (1991), which was not restricted to bedforms but also included channel-scale bed topography.

Recorded elevation data were corrected for alongstream changes in river stage in order to get bed elevations relative to a constant datum (NGVT 1929; correction based on water-surface slope computed with water level from the nearest gauging stations downstream and upstream from the study site at the time of the survey). Visual comparative analysis of the successive 40-km-long bed profiles was done in order to put into context the quantitative analysis of variation in bed elevation of the selected 15-km-long river reach.

Visual Observations: Preserved Channel-Scale Deposits Eight Weeks after Annual Peak Flow

The variation in bed elevation over the ~ 40-km-long survey line shows an alternation of high and low reaches corresponding to point bars and scour bends, as expected in a meandering river (Fig. 4; data in Appendix S2). The range of bed elevations is about 20 m, and the transition between low and higher bed heights can be quite sharp. Between the two first surveys, the overall bed topography remained about the same, with the macro-bedforms keeping about their same position. However, most deep scours were deeper during the second than the first survey—and other deep-water locations were also at lower elevations in April than in February (Fig. 4, e.g., at distance near 10,000 m), leading to little preservation of high-discharge deposits in topographic lows. Deposits from the highest spring peak flow appear to be preserved along the point bars, yet not at the upstream bar tops that were systematically eroded between the two first surveys (Fig. 4).

Some small dunes (with height about 1–2 m and length of 30–40 m) are preserved on bar tops (Fig. 4, at distance > 30,000 m, and Fig. 5A), but similar or smaller dunes are also well preserved 14 m deeper (Fig. 5B). Moreover, these small-dune deposits at distance ~ 15,000 m (Fig. 4) would stand at elevations about 10 m below synchronous strata from large dunes (most often 3–4 m high, but up to 5–7 m, and about 185 m long) preserved at other reaches of the river. This is contrary to our common assumption that large dunes occur, and are preserved, in the deepest parts of channels. The potential for short-term preservation of the large dunes from the peak discharge varies spatially: over a distance of approximately 8 km (middle of survey line), some dunes may have been completely preserved (Fig. 5C) while others were mostly eroded

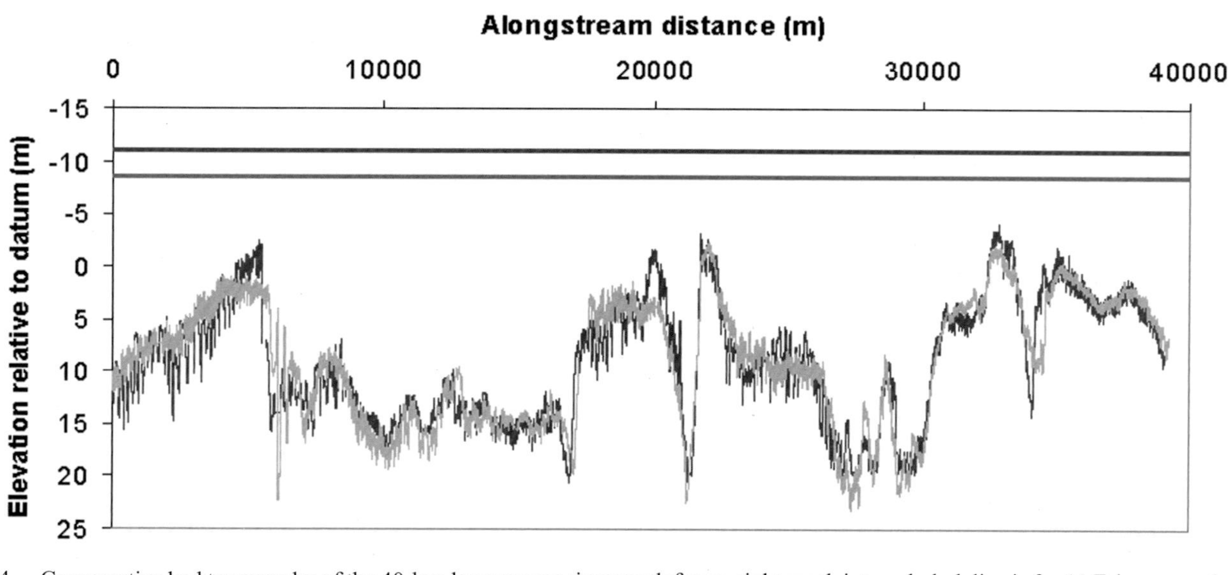

FIGURE 4.—Comparative bed topography of the 40-km-long survey river reach for an eight-week interval: dark line is for 11 February and pale line is for 13 April. See Appendix S2 for Excel files (data and graph with user-defined axis scale).

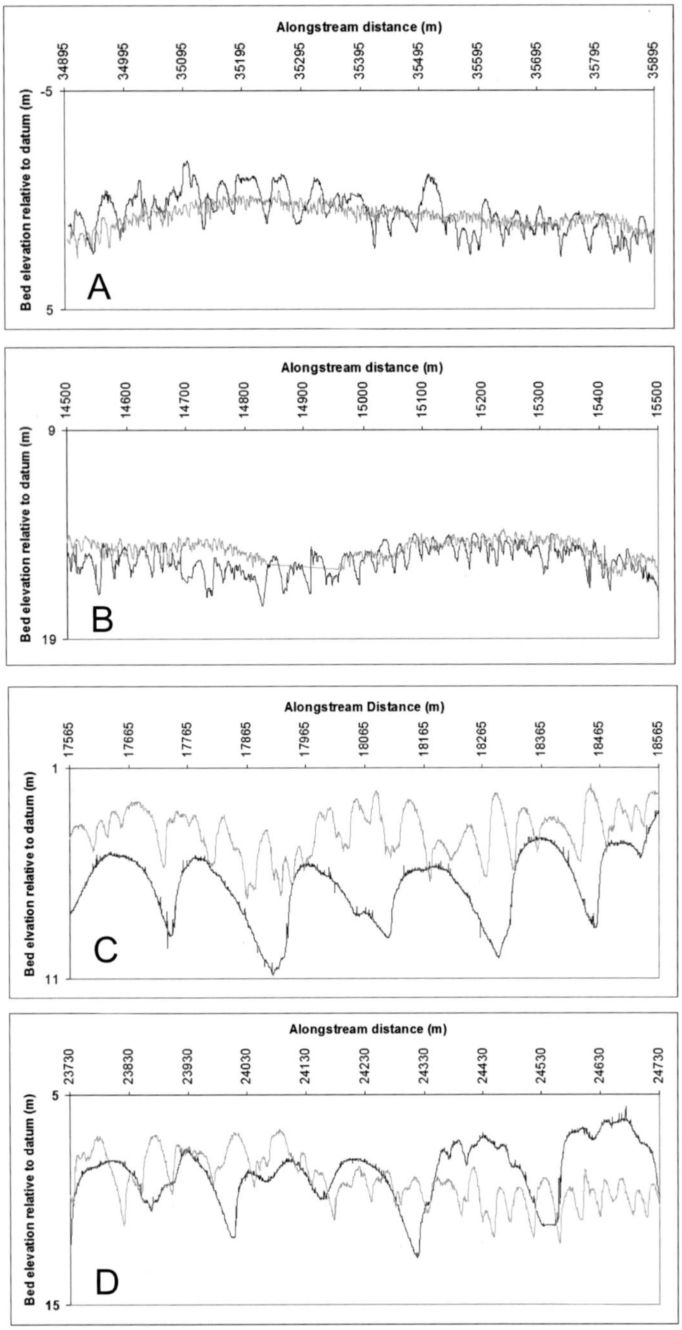

FIGURE 5.—Comparative bed topography of selected river reaches reach for an eight-week interval: dark line is for 11 February and pale line is for 13 April: A) small dunes at high bed elevation; B) small dunes in deep waters; C, D) large dunes from peak flow completely reworked between surveys. Scale is the same for all graphs.

(Fig. 5D at ~ 24,330–24,730 m), or partially preserved (e.g., Fig. 5D at ~ 24,230 m). The comparative bed topography shows that after eight weeks, there was no such thing as large remnant dunes with superimposed smaller dunes (another common assumption): the bed was entirely reworked.

Modeling Preserved Channel-Scale Stratigraphy at the End of the Spring Peak Flows (Twelve Weeks)

The Paola-Borgman theory requires the assumption of randomness in bed-elevation variation; therefore because, at the present time scale of interest, the point bars in this meander do not randomly become bend scours, this analysis considers subsamples from river reaches where bed elevation varies within a specific morphological feature. The selected reaches include: the 0–5 km straight reach (Fig. 6A); the lee side of a macro-bedform upstream from the sharp bend (Fig. 6B); the stoss side of this same bedform (Fig. 6C); and the upstream bend scour (Fig. 6D; see Figure 3 for location and Appendix S2 for original 0–15 km bed-profile Excel graph). It is assumed that the stratigraphy of the river bed (point bars, macro-bedforms, etc.) is essentially composed of dune stratification, based on observations in other large rivers where this sedimentary structure was dominant (Ashworth et al. 2000; Sambrook-Smith et al. 2009). No overall aggradation occurred in the river by the end of the survey period: on average, the final bed elevation on comparative bed profiles (Fig. 6, black line) is at, or below, the level of dune crests from the first peak flow (brown line).

Model Description: Considering the large data files, a simplified analytical method was used to minimize data processing: (1) the trend due to the overall bed slope in the river reach was removed in order to retain variation relative to the local mean bed level; (2) selected percentiles larger than the 50th of the distribution were computed, given that only deep elevations contribute to deposit preservation (Fig. 7; Leclair 2000) and that the theory is not affected by the number of occurrences (Paola and Borgman 1991); (3) plots of comparative cumulative distributions were made (Fig. 8); (4) the selected data were merged two by two with those from the successive surveys; (5) the mean and variance of each combined distribution were computed; and (6) the mean channel deposit preserved between surveys was computed (Equation 1). The method could be refined, using all individual elevation data, analysis of type of distributions, etc., but a simple approach suffices here to shed light on general trends in the potential preservation of dune deposits.

The graphs showing the comparative distributions of bed elevation allows determining if there was deposition or erosion between surveys: when a line plots on the right-hand side of the line from the previous survey (Fig. 8), it represents higher bed elevations, and hence a net deposit; conversely, when a line plots to the left of the line from the previous survey, the bed was eroded (e.g., Fig. 8A, 22 April versus 13 April, and Fig. 8D, 27 April versus 22 April).

In the analysis of experimental (Leclair and Blom 2005) or field data (for reaches slightly downstream of the present site; Leclair and Miller 2006), it was observed that the upper end of the percentile distribution of bed heights (e.g., Fig. 7) always corresponded to the measured mean dune height. The reasons for this are not clear, but this relationship allows an estimate of approximate mean dune height for the different surveys and selected reaches, and hence to predict the mean set thickness (using Equation 2) that would be expected to have been produced by such dunes. Combined models, at the dune scale within the channel scale, can then be used to predict the characteristics of the final deposit.

Results

Predicted Mean Dune Height for Individual Reaches and Surveys: Apart from the straight reach (0–5 km), where the local largest dunes occur during the first survey (highest discharge), predicted values for mean dune height in the other selected reaches are higher during the second flow peak than at other times (Table 2). For the stoss side of the macro-bedform (8–10 km), where flow depth was 23–27 m during the first survey, predicted mean dune height is

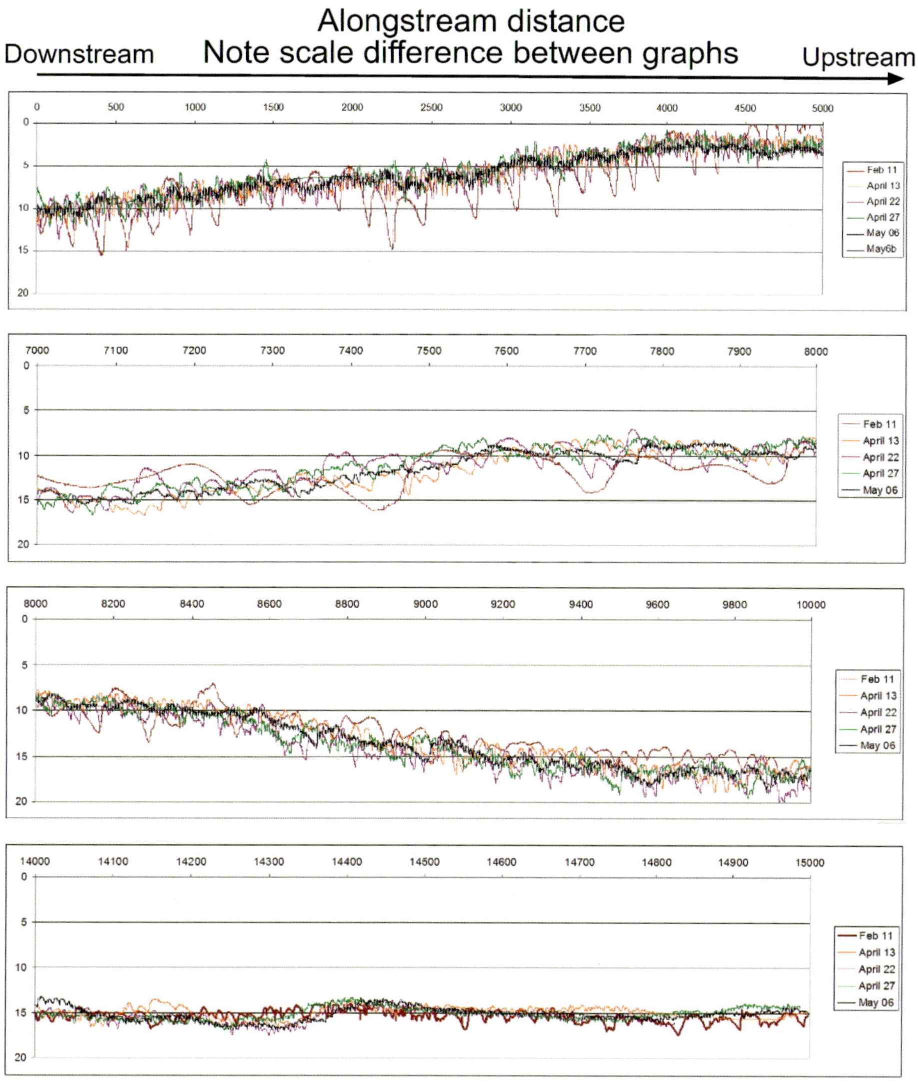

FIGURE 6.—Comparative bed topography of selected river reaches over a twelve-week period for: A) the straight reach; B) the lee side of a macro bedform upstream from the sharp bend; C) the stoss side of this same bedform; and D) the upstream bend scour. See Appendix S2 for Excel graph (with user-defined axis scale).

only 0.8 m, reaches 3.5 m by the time of the second peak flow, and decreases to 1.6 m by the end of the survey period (Table 2). For the upstream scour bend, predicted mean dune height is generally lower than for other reaches. Otherwise, the range of predicted mean dune height for the various reaches at various survey times is somehow narrow, with many values between 1.7 and 2.4 m (Table 2).

The ratio of observed local flow depth to predicted mean dune height is lower in the straight reach (ranging from 5 to 8) than in the upstream bend (ranging from 11 to 19); over the macro-bedform, this ratio ranges from 7 to 13, except for the first survey, where it shows the highest value (30 at reach 8–10 km; Table 2).

Predicted Coset and Set Thickness at the End of the Survey Period: Erosion or deposition between successive surveys is not systematic for all selected reaches, except during receding flow (Table 3). The models provide average values, and variability can occur at a given reach : for example, Figure 8B shows that deposits from the first flow peak exist at depth between ~ 0–1 m below mean bed level (brown line most to the left) while deposits from receding flows are at 2–4 m below mean bed level. Predicted average thickness for reach-scale deposits formed in the first eight-week interval is on average 3.1 m in the straight channel, only 0.5 m in the 8–10 km reach, and nonexistent in the meander bend. No deposit from the second flow peak is predicted elsewhere than at the 0–5 km reach (Table 3; see Figure 8B showing 27 April receding-flow line at quite a distance to the left from the 22 April line, i.e., erosion in just a few days). By the end of the receding flows, most previously deposited cosets are only partially preserved.

Results from the combined models (bedform-scale within channel-scale) indicate that the final deposits from the selected reaches are a mostly an assemblage of cosets 1.3–1.5 m thick (on average) containing sets with mean thickness ranging 0.5 m to 0.7 m (Table 3). Successive channel-scale deposits (cosets) are slightly thinning upward (Table 3), and showing altogether a thinning-upstream mean set thickness (given dune heights of 3.4 m to 1.7 m, mean set thickness is about 1 m to 0.5 m from the lowest coset to the top). However, apart

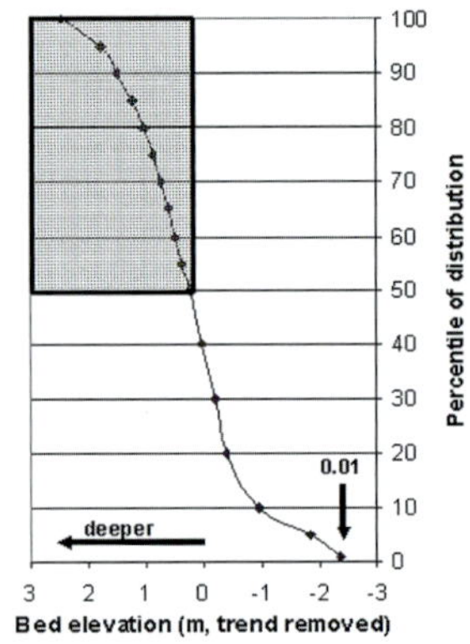

FIGURE 7.—Description graph for analytical methods of percentiles of bed-elevation distributions: 0.01 percentile is used as a proxy for mean dune height; percentiles in shadowed box (deeper than mean bed level) are used for estimating the thickness of preserved channel deposits.

from the sets formed by the largest dunes, sets in other cosets have about the same mean thickness (0.4–0.6 m).

Implications for Interpretation

This study is based on a limited number of surveys, yet all under flow conditions at or above average long-term discharge, and along a single along-stream survey line. This last point should not be a constraint because experimental results on meander dynamics (Peakall et al. 2007) showed that point-bar growth is controlled mainly by alongstream transport and deposition on the same side of the channel. Therefore, the present results provide potentially important information for the interpretation of fluvial deposits in general, and specifically those from large rivers.

On short time scales the preservation potential of channel-scale deposits from peak flow varies spatially, in both the alongstream and the vertical (bed elevation) directions, and is not particularly high. Very little depositional evidence of the high flows was preserved in the deep parts of the channel; instead, deposits from dunes of various sizes, including large ones, were relatively well preserved at higher bed elevations (up to about 10 m from the base of the bend scour). If these deposits were exposed in a kilometer-long outcrop in the far future, it

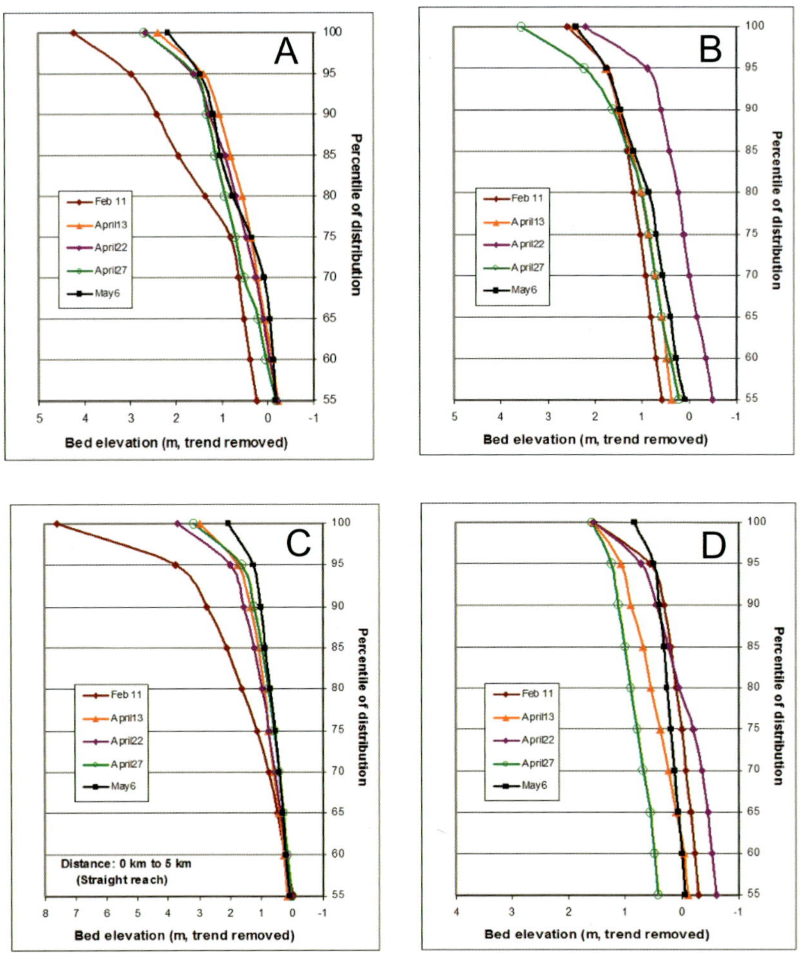

FIGURE 8.—Comparative cumulative distributions of the largest percentiles of the bed elevation (deeper than the mean bed level) for the selected reaches: A) the lee side of a macro-bedform upstream from the sharp bend. B) the stoss side of this same bedform; C) the straight reach; and D) the upstream bend scour.

TABLE 2.—Predicted mean dune height (based on first percentile of detrended distribution of river-reach bed elevation) and observed ratio of mean flow depth to predicted mean dune height.

River reach:	0–5 km	7–8 km	8–10 km	14–15 km
Date	Dune height, m (mean flow depth/ mean dune height)			
11 February	3.4 (5)	2.3 (10)	0.8 (30)	1.4 (19)
13 April	1.9 (8)	2.3 (9)	2.4 (9)	1.5 (16)
22 April	2.1 (7)	2.7 (8)	3.5 (7)	2.2 (11)
27 April	2.0 (7)	2.3 (8)	2.4 (9)	1.3 (18)
6 May	1.7 (7)	2.0 (9)	1.6 (13)	1.2 (18)

TABLE 3.—Reach-scale predicted mean thickness for preserved coset (plain script) and related mean set thickness (in parentheses) between surveys. Last-survey bed material is not considered a coset and hence only mean set thickness is indicated. Note that bed erosion (mean thickness of eroded material in italics) will decrease the final thickness of previous coset. Survey date is reversed to show vertical succession of cosets.

River reach:		0–5 km	7–8 km	8–10 km	14–15 km
		Preserved coset or *eroded material* thickness, m (mean set thickness, m)			
6 May	From second peak flow	(0.5)	(0.6)	(0.5)	(0.4)
27 April to 6 May	From receding flow	1.1 (0.6)	1.4 (0.7)	1.3 (0.7)	0.6 (0.4)
22 April to 27 April	From second peak flow	1.5 (0.6)	*1.4*	*2.0*	*2.7*
13 April to 22 April	From rising flow	*1.3*	*1.5*	1.5 (0.7)	2.2 (0.5)
11 February to 13 April	From first peak flow	3.1 (1)	2.0 (0.7)	0.5 (02)	*1.4*

would be quite difficult to relate them to a single "event". Based on our current knowledge of scaling relationships between dune geometry and flow depth, these results show that even if the deposits from the spring highest flow are completely preserved, their interpretation would probably not lead to a correct reconstruction of the formative hydraulic conditions, because the ratio of flow depth to dune height for the 11 February varies from 5 to 30 between reaches. This ratio also varies between surveys, although all values except one fall in the 3–20 range of numerous empirical data from various river patterns and flow conditions (cf. Bridge 2003).

Deposits from dunes migrating during rising discharge do not seem to have a high potential for preservation; this is because at several reaches of the river, the bed surface rises during rising stage and is eroded afterward (see Fig. 6). Previous researchers (Hickin 1995; Paige and Hickin 2000) also observed that annual peak discharge and bed elevation were out of phase, and maybe more pronounced in large rivers.

The most continuous deposits seems to be formed by dunes migrating during receding flows (Table 3), and the peak flow signature can be seen in isolated deposits. Data are lacking to describe dune geometry during the first falling limb of the hydrograph and demonstrate that the area with the deep scours in Figure 6A could be occupied by deposits from dunes smaller than 3 m high. The present results are potentially critical for a wide range of sedimentological issues because fluvial models generally require determination of a representative-event input (Wolinsky 2010), and caution should be used before choosing peak-flow conditions as model input.

Predicted mean set thicknesses for all surveys do not show a wide range, and hence the various distributions of set thickness probably overlap from one coset to the other. If such sets were observed in a core or an outcrop, it might be difficult to distinguish deposits formed during different flow conditions, because they could look similar. The next section presents a methodological approach for the analysis of natural dune deposits.

FROM RIVER TO QUATERNARY OUTCROP

Context and Study Site

The Quaternary deposits from the Loire River, in France, were previously studied with the aim of understanding the fluvial response to climatic changes (Blum and Törnqvist 2000; Straffin and Blum 2002; Leclair et al. 2003). The outcrop example (Fig. 9A; high-resolution color pictures provided in Appendix S2) is in a quarry near the city of Orléans, France). This outcrop was chosen because the large cross-sets visible in Fig. 9A are the only ones that could be observed in probably all the accessible deposits of the Loire valley (given that exposures depend on land use, e.g., vineyards versus quarries). Thin cross-sets seem to be more frequent than larger ones, just as predicted for the modern Mississippi River (in fact, these field observations triggered the questions suggested in the introduction of this paper). In the ~ 3-m-high portion of the trowel-refreshed outcrop (Fig. 9B) are stacked at least five different deposits formed by dunes, with the more continuous ones being less than 1 m thick (Fig. 9C). The objective of this section is to develop criterion for the applicability of the Leclair-Bridge model to partially preserved deposits. The criterion is based on the fact that for experimental results, mean set thickness at a point is about equivalent to the value computed at any other point or over a section, which is not the case for poorly preserved cosets (see Appendix S1 for details).

Analytical Approach

Outlines of cosets and sets were drawn (Fig. 9C), and set thicknesses were measured for cosets C, D, and E. Distinct cosets were determined from differences in strata type and flow direction suggested by the sedimentary structures (flow from right to left in C, and in across-stream direction in D). For each coset, the mean set thickness was

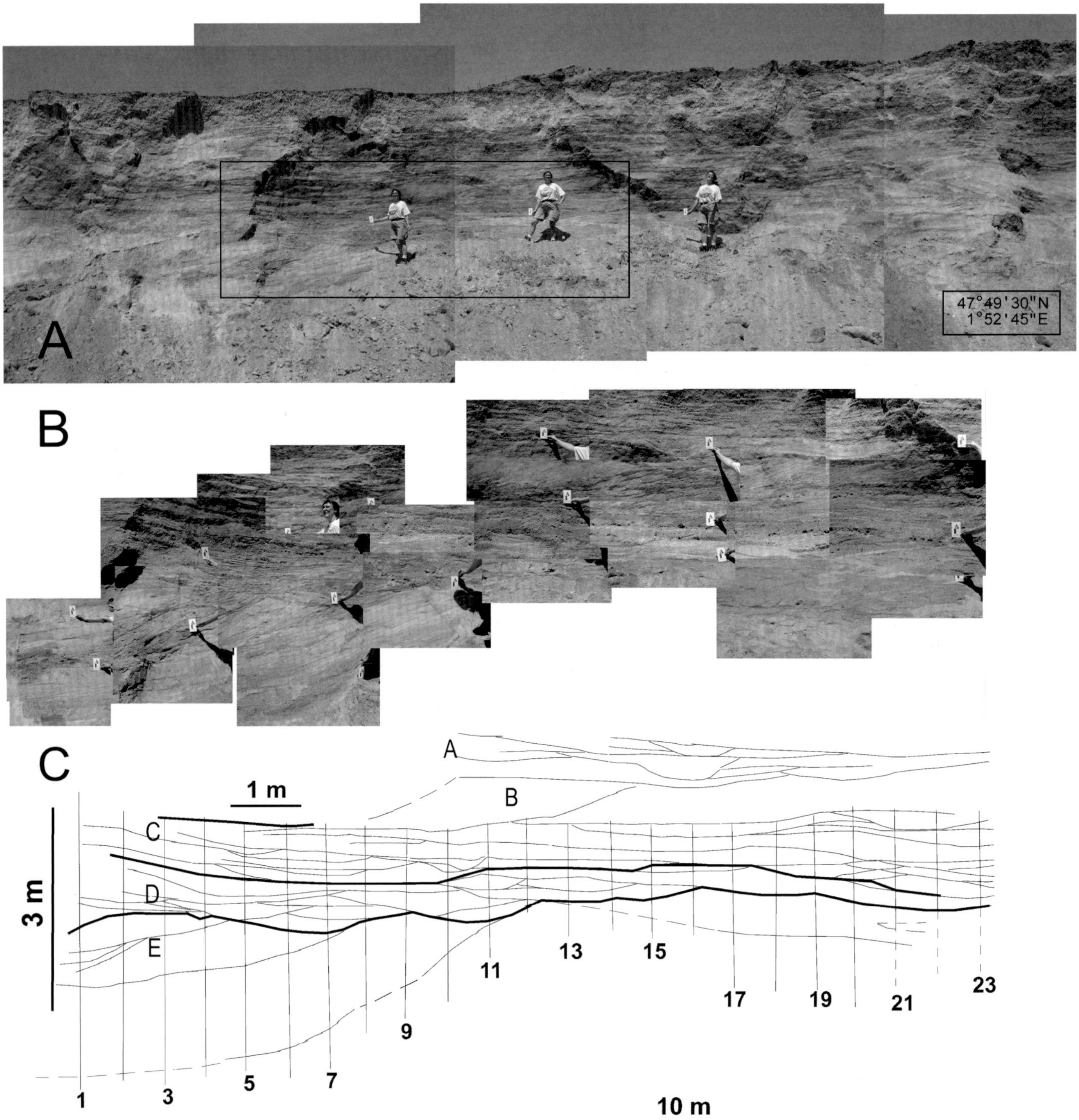

FIGURE 9.—A) View of an outcrop of the Quaternary Loire River near Orléans, France; coordinates are indicated for information; scale card is 15 cm high; box indicates approximate location of photo mosaic in Part B) showing dune sets in details, and Part C) coset and set outlines. See Appendix S2 for high-resolution color pictures.

computed for each vertical section, and for all the vertical sections. The reliability of the estimate was computed as the percent difference between measurements from individual sections and the overall mean value for the entire coset, which is the best (and only) estimate available.

Results

Mean set thickness measured at vertical sections for coset C averages 0.14 m and ranges from 0.09 m to 0.19 m (Table 4). The difference between at-a-section mean thickness and the overall set

TABLE 4.—Mean set thickness(s_m) from all data for a given coset from a Quaternary Loire River outcrop (see Fig. 9), compared to range of values when the estimate is computed from individual vertical sections (with percentage difference (*diff*) with mean value from all data).

	All data	Data from individual vertical sections				
Coset	s_m (m)	s_{min} (m)	s_{max} (m)	diff_{min} %	diff_{max} %	Notes
C	0.14	0.09	0.19	−31	44	All sections
D_a	0.14	0.08	0.29	−38	123	All sections
D_b	0.12	0.1	0.15	−25	15	Only sections with > 2 sets
E	0.53	0.07	1.06	−87	100	All sections

thickness for coset C varies from −31% to 44%, but in most cases is ± 23%. Coset D shows a mean value similar to C but a wider range if all vertical sections are considered (D_a, Table 4); if the sections presenting a single set (on the right-hand side of the outcrop, Fig. 9C) are removed from the distribution, then the range of at-a-section mean set thickness is only 0.1–0.15 m, and the difference from the mean ranges from −25% to 15% (D_b, Table 4). Results from coset E show a wide range of values and difference from the mean varies from −87% to 100% (Table 4).

Implications for Interpretation

In most cases, the variability in the mean set thickness from the Loire cosets C and D is no more than the natural variability itself observed in the 34 flume experiments of Leclair (2001; 2.9 ± 24%). Based on Equation 2, the overall mean set thickness from cosets C and D_b would indicate that the height of the formative dunes was about 0.3–0.5 m and 0.26–0.4 m, respectively. The method should not be used when there is less than a third left from an original deposit in outcrop exposures (e.g., coset E; see Appendix S1 for explanation), although this situation is not obviously recognized in the first place. Comparing at-a-section mean set thicknesses with the overall mean value can indicate low preservation or coset amalgamation in cases where the percent difference is large, as it might be the case in deposit D_a where some vertical sections—ones that show suspected outranged values—may not belong to the determined coset. A natural deposit can be considered quite representative of the original deposit when the mean set thickness from several vertical sections show less than 30% variability. A consequence of this analysis is that single at-a-point incomplete distributions of sets, as can happen in cores, are not reliable for the quantitative interpretation of dune deposits (see also Appendix S1).

SUMMARY DISCUSSION

Characteristics of natural deposits and modeling results indicate that our observations of dune set thickness in fluvial systems will likely be based on data from amalgamated cosets. In the Loire River example, it is not known if the selected cosets were formed on a short time scale or not, yet mean dune height estimated from several vertical sections showing a narrow range of mean set thickness is the best available result. If the Mississippi River predicted deposits were observed in an outcrop, cosets would also be merged erroneously because the distributions of set thickness overlap, and yet, because of this narrow range in mean set thickness, the result would be considered valid. Such a result would represent average dune height for average seasonal flow conditions, not peak flows, and hence the size of the fluvial system would be underestimated.

In natural deposits, sets from large dunes can be seen at various elevation (Fig. 9A, in selected area and on the right-hand side of the picture), and are typically interpreted as the lower part of different channels. Data from the Mississippi River indicate that the base of large-dune cosets cannot be interpreted systematically as a channel thalweg because large dunes were observed at about 10 m higher than small dunes in the bend scour. This unexpected dune geometry may be due to dissimilar bend curvatures that affect the development of velocity-magnitude zones and related bedform height (Jackson 1975, 1976). Some dunes at high bed elevations in the most upstream part of the 40-km-long survey lines are indeed small, and hence there are some reaches where the flow-velocity field produces expected bedform geometry. Our knowledge of flow-velocity fields over dunes in large rivers has improved in recent years (e.g., Parson et al. 2005; Shugar et al. 2010), but more research is definitively needed. In particular, the relationships between dune geometry and sediment transport conditions still required more investigation because dune height does not relate only to flow depth (cf. Gill 1971; Yalin and Karahan 1979; Leclair and Blom 2005). On long time scales, dune stratigraphy alone will provide clues for interpretation of the rock record, and therefore, morphodynamics studies will need to integrate dune sedimentological characteristics.

CONCLUSION

Observations and modeling results from a study on variation in bed elevation during annual peak flows in a modern large river strongly suggest that, even on short time scales, the preserved deposits would bear no clear signature for the correct interpretation of flow regime or channel size.

Predicted dune deposits preserved shortly (12 weeks) after the annual high flows were not preferentially from peak flow or systematically from any other specific flow conditions, although a large part may represent receding discharge. Dune preservation potential varies spatially, and the final deposits would likely be an amalgam of dune sets from various discharge conditions, yet showing a narrow range of thicknesses; estimated dune height from such deposits would provide an average value for unsteady seasonal conditions, not bankfull flows.

At the 40 km alongstream study site, large (3–7 m high) dunes occurred at elevation ~ 10 m higher than synchronous small (1–2 m high) dunes, and hence neither the thalweg elevation nor the maximum channel depth could be reconstructed correctly from dune sets, even if they were fully preserved. The ratio of observed mean flow depth to predicted mean dune height ranges from 5 to 30, and varies between river reaches and surveys, making it very difficult to reconstruct the formative flow depth.

The above results have important implications for our interpretation of fluvial deposits and the recognition of large rivers in the rock record. The problems with which sedimentologists are confronted are not methodological, but more fundamental. The Paola-Borgman model is a brilliant concept and the Leclair-Bridge model is valid for estimating dune height even if there is only ~ 30% preserved from the original deposit, at least in the case of outcrop exposure where little variation in mean set thickness is observed between several vertical sections (limited exposures in core are not reliable). It is our difficulties in properly distinguishing cosets in the sedimentary record, and mostly, our lack of knowledge on the relationships between fluvial morphodynamics and sedimentology, that hinders a correct interpretation of flow depth from preserved set characteristics and predicted bedform geometry.

ACKNOWLEDGMENTS

First, warmful thanks to Stephanie Davidson, Sophie Leleu, and Colin North, for inviting me to the "*From River to Rock Record*" conference, and getting me motivated for writing this paper, and thanks to *Environnement Illimité Inc.*, in Montreal, for their support. Thanks to Chris Paola and Leon Borgman for their inspiring paper. The Mississippi-dune program was funded by Tulane University and the Louisiana Board of Regents Support Fund, Research and Development, Research Competitiveness Subprogram during 2003–2006. Thanks to Richard Byrd's team at Odom Hydrographics in Baton Rouge, Louisiana—especially David Andrews and Marvin Story, for their help with instrumentation and navigation. Thanks to Zac Miller, Tiffany Dawn Jobe, and especially Nathan Bradbury (who measured hundreds of partially preserved cross-set thicknesses), Tulane University undergraduate student assistants at the time of the author's appointment at this institution. Loire River data were collected in collaboration with Eric Straffin (Edinboro University of Pennsylvania), as part of Michael Blum's NSF project while the author was a postdoctoral researcher at the University of Nebraska Lincoln. Flumes studies were part of Leclair's Ph.D. program at Binghamton University NY, funded and by *Le Fonds pour la Formation de Chercheurs et l'Aide à la Recherche* (FCAR, Québec, Canada) and by a grant to John S. Bridge by the Petroleum Research Fund of the American Chemical Society. Finally, sincere thanks to reviewers Phil Ashworth and Bill Arnott for their thoughtful comments on an earlier version of this paper.

REFERENCES

Allen, J.R.L., 1982, Sedimentary Structures; Their Character and Physical Basis: Amsterdam, Elsevier, 663 p.

Adams M.M., and Bhattacharya, J.P., 2005, No change in fluvial style across a sequence boundary, Cretaceous Blackhawk and Castlegate Formations of central Utah, U.S.A.: Journal of Sedimentary Research, v. 75, p. 1038 - 1051.

Ashworth P.J., Best, J.L., Roden, J.E., Bristow, C.S., and Klaassen, G.J., 2000, Morphological evolution and dynamics of a large, sand braid-bar, Jamuna River, Bangladesh: Sedimentology, v. 47, p 533–555.

Blum M.D., and Törnqvist, T.E., 2000, Fluvial responses to climate and sea-level change: a review and look forward: Sedimentology, v. 47 (suppl. 1), p. 2–48.

Bridge, J.S., 2003, Rivers and Floodplains: Forms, Processes, and Sedimentary Record: Oxford, U.K., Blackwell Publishing, 491 p.

Bridge, J.S., and Best, J.L., 1997, Preservation of planar laminae due to migration of low-relief bedforms over aggrading upper-stage plane beds: comparison of experimental data with theory: Sedimentology, v. 44, p. 253–262.

Bridge, J.S., and Demicco, R.V., 2008, Earth Surface Processes, Landforms and Sediment Deposits: Cambridge, U.K., Cambridge University Press, 815 p.

Collinson, J.D., 1970, Bedforms of the Tana River, Norway: Geografiska Annaler, A, v.52, p. 31–56.

Davidson, S.K., and North, C.P., 2009, Geomorphological regional curves for prediction of drainage area and screening modern analogues for rivers in the rock record: Journal of Sedimentary Reasearch, v. 79, p. 773–792.

Gabel, S.L., 1993, Geometry and kinematics of dunes during steady and unsteady flows in the calamus river, Nebraska, USA: Sedimentology, v. 40, p. 237–269.

Gill, M.A., 1971, Height of sand dunes in open channel flows: American Society of Civil Engineers, Proceedings, Journal of the Hydraulics Division, v. 97, p. 2067–2073.

Harbor, D.J., 1998, Dynamics of bedforms in the Lower Mississippi River: Journal of Sedimentary Research, v. 68, p. 750–762.

Hickin, E.J., 1995, Hydraulic geometry and channel scour, Fraser River, British Columbia, Canada, *in* Hickin, E.J., ed., River Geomorphology, New York, John Wiley & Sons, p. 155–167.

Holbrook, J., Scott., R.W., and Oboh-Ikuenobe, F.E., 2006, Base-level buffers and buttresses: a model for upstream versus downstream control on fluvial geometry and architecture within sequences: Journal of Sedimentary Research, v. 76, p. 162–174.

Jackson, R.G., II, 1975, Velocity–bed-form–texture patterns of meander bends in the lower Wabash River of Illinois and Indiana: Geological Society of America, Bulletin, v. 86, p. 1511–1522.

Jackson, R.G., II, 1976, Depositional model of point bars in the Lower Wabash River: Journal of Sedimentary Petrology, v. 46, p. 579–594.

Leclair, S.F., 2000, Preservation of cross-strata due to migration of subaqueous dunes: Unpublished Ph.D. thesis, Binghamton University, New York, 422 p.

Leclair, S.F., 2002, Preservation of cross-strata due to migration of subaqueous dunes: an experimental investigation: Sedimentology, v. 49, p. 1157–1180.

Leclair, S.F., 2005, Up-scaling dune morphodynamic models for the study of bedload transport in large rivers (abstracts): EOS, Transactions, American Geophysical Union, 86(18), Joint Assembly Supplement, Abstract h24a-06.

Leclair, S.F., and Blom, A., 2005, A qualitative analysis of the distribution of bed-surface elevation and the characteristics of associated deposits for subaqueous dunes, *in* Blum, M.D., Marriott, S.B., and Leclair, S.F., eds., Fluvial Sedimentology VII: International Association of Sedimentologists, Special Publication 35, p. 121–134.

Leclair, S.F., 2006, New pieces to the puzzle of reconstructing paleofluxes from river dune deposits: Geology, v. 34, p. 401–404.

Leclair, S.F., and Bridge, J.S., 2001, Quantitative interpretation of sedimentary structures formed by river dunes: Journal of Sedimentary Research, v. 71, p. 713–716.

Leclair, S.F., and Miller, J.Z., 2006, Time variation of probability distributions of dune-bed elevation in a large river, *in* Parker, G., and Garcia, M., eds., River, Coastal and Estuarine Morphodynamics: RCEM 2005: London, Taylor and Francis Group, v. 2, p. 802–812.

Leclair, S.F., Straffin, E.C., and Blum, M.D., 2003, Quantitative interpretation of fluvial deposits: an application to the Quaternary Loire River, France (abstract): Geological Society of America Annual Meeting Abstracts with Programs, v. 35, no. 6, p. 23–24.

Miall, A.D., 2006, How do we identify big rivers? And how big is big?: Sedimentary Geology, v. 186, p. 39–50.

Nittrouer, J.A., Allison, M.A., and Campanella, R., 2008, Bedform transport rates for the Lowermost Mississippi River: Journal of Geophysical Research, v. 113, F03004, 16 p., doi: 10.1029/2007JF000795.

Nordin, C.F., 1971, Statistical Properties of Dune Profiles: U.S. Geological Survey, Professional Paper 562-F, 41 p.

Paige, A.D., and Hickin, E.J., 2000, Annual bed-elevation regime in the alluvial channel of Squamish River, Southwestern British Columbia, Canada: Earth Surface Processes and Landforms, v. 25, p. 991-1009.

Paola, C., and Borgman, L., 1991, Reconstructing random topography from preserved stratification: Sedimentology, v. 38, p. 553–565.

Parker G., Paola, C., and Leclair, S.F., 2000, Probabilistic Exner sediment continuity equation for mixture with no active layer: Journal of Hydraulic Engineering, v. 126, p. 818–826.

Parson, D.R., Best, J.L., Orfeo, O., Hardy, R.J., Kostaschuck, R., and Lane, S.N., 2005, Morphology and flow fields of three-dimensional dunes, Rio Paraná, Argentina: Results from simulataneous multibeam echo sounding and acoustic Doppler current profiling: Journal of Geophysical Research, v. 110, F04S03, 9 p., doi: 10.1029/2004JF000231.

Peakall, J., Ashworth, P.J., and Best, J.L., 2007, Meander-bend evolution, alluvial architecture, and the role of cohesion in sinuous river channels: a flume study: Journal of Sedimentary Research, v. 77, p. 197–212.

Pontén, A., and Plink-Björklund, P., 2007, Depositional environments in an extensive tide-influenced delta plain, middle Devonian Gauja formation, Devonian Baltic basin: Sedimentology, v. 54, p. 969-1006.

Rittenour, T.M., Goble, R.J., and Blum, M.D., 2005, Development of an OSL chronology for Late Pleistocene channel belts in the lower Mississippi valley, USA: Quaternary Science Reviews, v. 24, p. 2539–2554.

Rubin, D.M., and Hunter, R.E., 1982, Bedform climbing in theory and nature: Sedimentology, v. 29, p. 121–138.

Sambrook-Smith, Ashworth, P.J., Best, J.L., Lunt, I.A., Orfeo, O., and Parsons, D.R., 2009, The sedimentology and alluvial architecture of a large braid bar, Rio Paraná, Argentina: Journal of Sedimentary Research, v. 79, p. 629–642.

Saucier, R.T., 1994, Geomorphology and Quaternary geologic history of the Lower Mississippi: The Mississippi River Commission, Vicksburg, Miss.: U.S. Army Corps of Engineers, Waterways Experiment Station, 2 vols.

Straffin, E.C., and Blum, M.D., 2002, Holocene fluvial response to climate change and human activities, Burgundy, France: Netherlands Journal of Geosciences / Geologie en Mijnbouw, v. 81, p. 417–430.

Shugar, D.H., Kostaschuk, R., Best, J.L., Parsons, D.R., Lane, S.N., Orfeo, O., And. Hardy, R.J. , 2010, On the relationship between flow and suspended sediment transport over the crest of a sand dune, Río Paraná, Argentina: Sedimentology, v. 57, p. 252–272.

USACE New Orleans district, Hydraulics branch, URL last accessed on 16 August 2010: http://www.mvn.usace.army.mil/eng/edhd/watercon.htm (general) and http://www.mvn.usace.army.mil/cgi-bin/watercontrol.pl?01160 (Baton Rouge)

Wolinsky, M.A., 2010, Upscaling sedimentary processes: From bed to basin: American Association of Petroleum Geologists, Search and Discover Article #90104©2010, American Association of Petroleum Geologists Annual Convention and Exhibition-11–14 April 2010.

Wolman, M.G., Church, M., Newbury, R., Lapointe, M., Frenette, M., Andrews, E.D., Lisle, T.E., Buchanan. J.P., Schumm, S.A., and Winkley, B.R., 1990, The riverscape, *in* Wolman, M.G., and Riggs, H.C., eds., Surface Water Hydrology: Boulder, Colorado: Geological Society of America, The Geology of North America, v. O-1, p. 281–328.

Yalin, M.S., 1992, River Mechanics: Oxford, UK, Pergamon Press, 220 p.

Yalin, M.S., and Karahan, E., 1979, Steepness of sedimentary dunes: Journal of Hydraulics Division, v. 105, p. 381–392.

DIGITAL APPENDIX

The following supplementary material is available in electronic format either on the DVD accompanying this volume or through the digital archive on the SEPM website at http://www.sepm.org/

This digital appendix contains previously unpublished data from Leclair's experimental investigation on dune deposit preservation, in addition to results from a new analysis of selected data from Leclair (2002).

File "Leclair_theory_review": (in Microsoft Word and in Adobe Acrobat PDF formats) presents: (1) a brief review of the Paola-Borgman theory; (2) new testing of the applicability of the Leclair-Bridge model to the interpretation of partially preserved cosets; and, (3) a description of the effects of dune migration patterns on cross-set boundaries, with examples of sedimentary structures solely due to each of these patterns and guidance (or warning) for identifying cosets and cross-sets in outcrops and cores.

Files named: "Video_S...": four video files (in Quicktime .MOV format) on dune migration patterns, as seen through the flume side wall, from Leclair's experiments at Binghamton University, NY, U.S.A.

Files named "Pict_S...": picture files (JPEG format) for five 1-m long experimental sediment peels, from Leclair's experiments at Binghamton University, NY, U.S.A.

Four Microsoft Excel files with: 1) some graphs of ship tracks of survey lines; 2) data (distance, bed elevation) for two 40-km long survey lines on the Mississippi River; 3) a graph file of these data; 4) a graph file of bed profiles for five ~15-km long survey lines of this river. In this last case, although the raw data are not provided, having the Excel file allows one to modify the graph scales in order better to see the detail.

Files named "Loire_9...": color pictures (JPEG format) of the Loire River deposit described in this article.

PEDOGENIC AND SEDIMENTOLOGIC CRITERIA FOR RECOGNITION OF OVERBANK SUB-ENVIRONMENTS IN A TRIASSIC ANABRANCHING-RIVER DEPOSIT

PARTHASARATHI GHOSH AND SOUMEN SARKAR
Geological Studies Unit, Indian Statistical Institute, 203 B.T. Road, Kolkata 700108, India
email: pghosh@isical.ac.in

ABSTRACT: A variety of sub-environments associated with geomorphologic features like splay systems, floodplain ponds/lakes, and abandoned alluvial ridges commonly occur in the overbank areas of modern river valleys. Though the overbank areas are generally characterized by lower flow strength, supply of finer-grained detritus and increased landform stability compared to the channels, the subtle differences among different overbank sub-environments may at times impart discernible lithologic and pedologic characteristics to their deposits.

Based upon the sedimentology of the channel-fill bodies, the upper part of the Triassic Denwa Formation of a central Indian Gondwana basin has been interpreted as the deposit of an anabranching fluvial system comprising a number of virtually coexistent channels of different orders and characters that operated in a wide, low-gradient plain. Earlier studies had identified the preservation of very thick deposits of pedogenically modified fluvial overbank sediments in this succession. Most of the paleosols preserved within the overbank deposits are similar to modern Vertisols and calcic Vertisols, suggesting a seasonal semiarid to subhumid climatic setting. Though in terms of carbonate accumulation most of these paleosols are immature to at best submature, they display a wide variation in the organization of vertic features.

This case study attempts to delineate the deposits of different overbank sub-environments within the upper Denwa Formation with the help of paleosol profile characters, stacking pattern of the profiles, and the nature of the host sediments. The sedimentologic and pedogenic characters of extra-channel deposits, taken together, led to the identification of three distinct associations representing three overbank sub-environments. The association comprising alternations of sheet bodies of heteroliths and mudstones, with simple or compound, paleosol profiles, with weak accumulation of carbonate, is interpreted as the deposits of splay complexes. The second association that comprises thicker mudstone intervals, virtually devoid of sand, and have cumulate to composite stack of vertic features without carbonate concretions, has been interpreted as deposits of distal floodplain depressions. The third association comprising mudstones with slightly mature paleosols that contain up to small-pebble-size calcareous concretions has been interpreted as deposits of sediment-starved elevated sites within the distal floodplain.

The deposits of Denwa fluvial system show marked differences with the existing humid anabraching models. Again, in spite of having a semiarid to subhumid climate this system differs from the Cooper Creek, Australia, in having a mixed sediment load, resulting in the development of channel–levee–floodplain geomorphology and contrast between the in-channel and overbank lithologies. On the other hand, Denwa does not show a consistent trend in paleosol maturity away from the channel areas as suggested from the study of Willwood Formation. This deviation can be attributed to the multichannel character of the fluvial system, which allowed distal floodplain sites to receive sediment input from multiple sources. This caused a rather continuous pattern of sedimentation in distal sites and inhibited the widespread development of mature paleosols. The variability observed between the distal floodplain paleosols reveals the influence of the factors like local floodplain topography, sediment accretion pattern, and drainage conditions.

KEY WORDS: Paleosols, Overbank, Anabranching fluvial system, Triassic, India

INTRODUCTION

A river system consists of both channel and overbank environments. The coarser detritus is transported and deposited within the channels, while finer sediments (silt and clay) are deposited in the floodplain. The flow characteristics, geomorphology, and depositional mechanisms of modern floodplains have not been studied in as much detail as that for fluvial channels (Bridge 1984, 2003). Similar bias can be noted in literature dealing with ancient fluvial successions. One possible reason for this may be that, in case of an ancient fluvial deposit, it is easier to study the sedimentary features in coarser-grained sediments than those in muddy sediments using inexpensive and traditional methods. Moreover, in some ancient fluvial deposits the fines are not preserved in significant proportions due to autocyclic processes of those systems. These may account for better documentation of channel fill and its proximal facies in the fluvial literature compared to that of floodplain facies.

This bias is also translated into the classification of fluvial systems and deposits. The internal organization of the preserved channel fills and the way they are stacked (cf. Miall 1985, 1996) forms a major basis for the fluvial facies models while the characteristics of finer-grained facies are less emphasized. However, presently a few studies, based on the observations of the near-surface deposits, have addressed the issues of floodplain facies and stratigraphy (cf. Rust 1981, Nanson and Croke 1992, Aslan and Blum 1999, Brooks 2003, Page et al. 2003, and others).

In modern river valleys where wide overbank areas coexist with the channels and channel belts, a variety of overbank sub-environments exists. They include splay systems, floodplain ponds and lakes, abandoned alluvial ridges, etc. The deposits of these sub-environments are likely to be incorporated in the rock record unless a major part of the overbank deposits are removed by autocyclic process prior to their preservation. It can be noted that there are modern river valleys where overbank cover much more extensive areas than the channel belts and where channel belts do not sweep over the entire expanse of overbank areas quite frequently (Gibling et al. 1998). In such a situation thick, fine-grain, extra-channel deposits along with the coarser grained channel-fill deposits are expected to be preserved in the resulting succession. Again, there are reports of quite a number of ancient fluvial successions where significant amount of fines are preserved. Such successions are found all around the globe and from different intervals of the Phanerozoic Era. The examples include Lower Old Red Sandstones, United Kingdom (Allen 1974), the Eocene of United States of America (Willwood Formation; Bown and Kraus 1987, Kraus

From River To Rock Record: The Preservation Of Fluvial Sediments And Their Subsequent Interpretation
SEPM Special Publication No. 97, Copyright © 2011
SEPM (Society for Sedimentary Geology), ISBN 978-1-56576-305-0, p. 125–142.

1997, Kraus and Gwinn 1997), the Miocene of Pakistan (Willis and Behrensmeyer 1994), the Tertiary of Spain (Puigdefabregas and van Vliet 1978), the Permian of South Africa (Stear 1983), the Permian of India (Ray and Chakraborty 2002), the Upper Cretaceous of India (Ghosh 1997), and the Lower Cretaceous of United Kingdom (Stewart 1983).

Therefore, fluvial successions, with thick extra-channel fines preserved, provide with the scope to study the various components of ancient overbanks in more details. Bown and Kraus (1987), for example, identified proximal and distal overbank deposits in a mud-dominated fluvial succession (Willwood Formation) with the help of paleosol maturity.

The upper part of the Triassic Denwa Formation, the topic of the present discussion, is a red-mudstone-dominated fluvial succession. The details of the channel-fill deposits, channel-proximal sandbodies, and stratigraphic architecture of this succession were discussed elsewhere and the succession was identified as a deposit of muddy anabranching river system (Ghosh et al. 2006). In the present discussion we have used the term anabranching, as a "low-energy multiple-channel system in which stable or slowly migrating anabranches having resistant banks are separated by wide interfluves containing both depressed and elevated areas". Modern anabranching rivers are found in a wide range of climates from cool temperate to tropical, from humid to semiarid. Although the driving force for the development of anabranches is essentially the same for all (Huang and Nanson, 2007), North et al. (2007) suggested that the anabranching rivers of humid regions and dryland regions produce distinctly different fluvial geomorphology and sedimentary successions. But anabranching systems described so far are limited, and the full range of variation is yet to be recognized. Several ancient successions like the Denwa succession, as discussed later, may display characters intermediate between the humid and dryland end members. In this work we demonstrate how an integrated analysis of the sedimentologic and pedogenic characters of the upper part of the Denwa succession in the Satpura Gondwana basin, central India, can be used to identify the overbank sub-environments of an anabranching river system that formed under a subhumid to semiarid climatic setting.

GEOLOGICAL BACKGROUND

The Satpura basin is the westernmost of the series of discrete Gondwana basins occurring along the Narmada-Son-Damodar valley (Fig. 1A). The basin is rhomb-shaped, approx. 200 km long and ~ 60 km wide. Its longer sides are marked by the ENE–WSW-trending Son-Narmada south fault and Tapti north fault. Chakraborty et al. (2003) and Chakraborty and Ghosh (2005) proposed that the Satpura basin is of pull-apart origin and developed due to the extensions related to strike-slip movements along these two faults.

The Satpura Gondwana basin is filled with a siliciclastic succession about 4 km thick, ranging in age from Permian to the Late Cretaceous (Fig. 1A; Crookshank 1936, Pascoe 1959). The regional strike of the basin fill is ENE–WSW, and the regional formation dip (~ 5°) is to the north.

The Triassic succession of the Satpura Basin comprises the Pachmarhi and the overlying Denwa formations (Fig. 1B). The Pachmarhi Formation overlies the Upper Permian Bijori Formation with an erosional disconformity (Crookshank 1936, Chakraborty and Sarkar 2005) and passes gradationally upwards to the Denwa Formation. The latter is overlain with an angular unconformity by the conglomerate-dominated Jurassic (?) Bagra Formation (Casshyap et al. 1993; Maulik et al. 2000). The Pachmarhi–Denwa succession, therefore, constitutes a sedimentary package bounded by unconformity surfaces. This Triassic sediment package is suggested to have been deposited under a single tectonic regime having a monotonic increase in aggradation rate (Maulik et al. 2000).

The Denwa Formation is a red-mudstone-dominated fluvial succession replete with vertebrate fossils. On the basis of faunal content, an early Middle Triassic age has been assigned to the Denwa Formation (Bandyopadhyay and Sengupta 1999). The sand-mud ratio in the lower part (~ 120 m thick) of the Denwa succession is 9:11, and that decreases to about 1:9 in its upper part (~ 380 m). Based on contrasting lithofacies, sand-mud ratio, and mesoscale and macroscale architectural elements Maulik et al. (2000) informally subdivided the Denwa Formation (300 to 600 m thick) into a lower and an upper unit with a gradational contact.

Sedimentology of the Denwa Formation

The lower unit of the Denwa Formation comprises an alternation of sheetlike sandstone (medium- to fine-grained) bodies, 3–15 m thick, interleaved with red mudstone intervals (1.5–10 m). The mudstone intervals contain decimeter–centimeter-scale fine sandstone interlayers. Maulik et al. (2000) interpreted the lower part of the Denwa succession as deposits of sandy braided channel belts and their muddy floodplains.

The upper unit of the Denwa Formation is muddier (sand-mud ratio 1:9) than the lower unit and mainly comprises ribbon-shaped (1 to 5 m thick) and sheetlike bodies (1–3 m thick) of sandstones and sand-mud heteroliths along with some small lenticular bodies of calcirudite / calcarenite (~ 1 m thick) that are encased in pedogenically modified red mudstones (1 to 10 m thick). In the upper Denwa succession there is no perceptible vertical or lateral bias in the lithology and internal organization of sand and heterolithic bodies. The upper Denwa sandstones are quartzose with subordinate amount of plagioclase feldspars (4–17%) and rock fragments (0–4%). These are medium- to fine-grained and moderately sorted with commonly occurring carbonate cement.

The common architectural elements in the upper part of the Denwa Formation include (I) single-story and multistory sandy ribbons , (II) sandy tabular bodies, (III) ribbon bodies of inclined heterolithic strata (IHS) , (IV) mudstone-dominated ribbon bodies with or without IHS, (V) minor ribbons of intraformational conglomerate dominated by caliche-derived calcareous peloids, (VI) sandy to heterolithic sheet bodies, (VII) tabular bodies of complexly organized sand-mud heteroliths, and (VIII) red mudstone with variable degree of pedogenic modification.

The single-story sandy ribbons (element I) are 3 to 4.5 m thick and extend laterally for 100 to 200 meters. These bodies, unlike other large ribbon bodies of the Denwa succession, are dominated by cosets of large (dune scale), wedge-shaped planar cross-strata, are rich in siliciclastic sand deposits, and typically lack sand-mud heterolithic lateral-accretion deposits. This is the only channel-fill type that is characterized by the presence of thick planar cross-strata that suggest migration of 2-D bedforms under a north flowing current in relatively straight channels.

In multistory ribbon bodies (7–8 m thick and ~ 100 m wide), a sub-variety of element I, individual stories (2–3 m thick and 30–40 m wide) are characterized by concavo-planar cross-sectional profile and are internally composed of decimeter-thick trough cross-strata, low-angle planar lamination, and ripple cross-lamination. Each story displays a fining- and thinning-upward trend. These bodies, though rare, indicate reoccupation of older channel belts by newer channels. Both of these kinds of bodies belonging to element I that are sand-dominated and lack lateral accretion deposits possibly represent the deposits of laterally stable, sandy low-sinuosity channels.

The sandy tabular bodies (element II; 7–8 m thick and ~ 100 m wide) are characterized by sand-dominated IHS overlying basal sandy trough cross-stratified lithosomes that indicate flow direction at a high angle to direction of inclination of IHS. These bodies are interpreted as

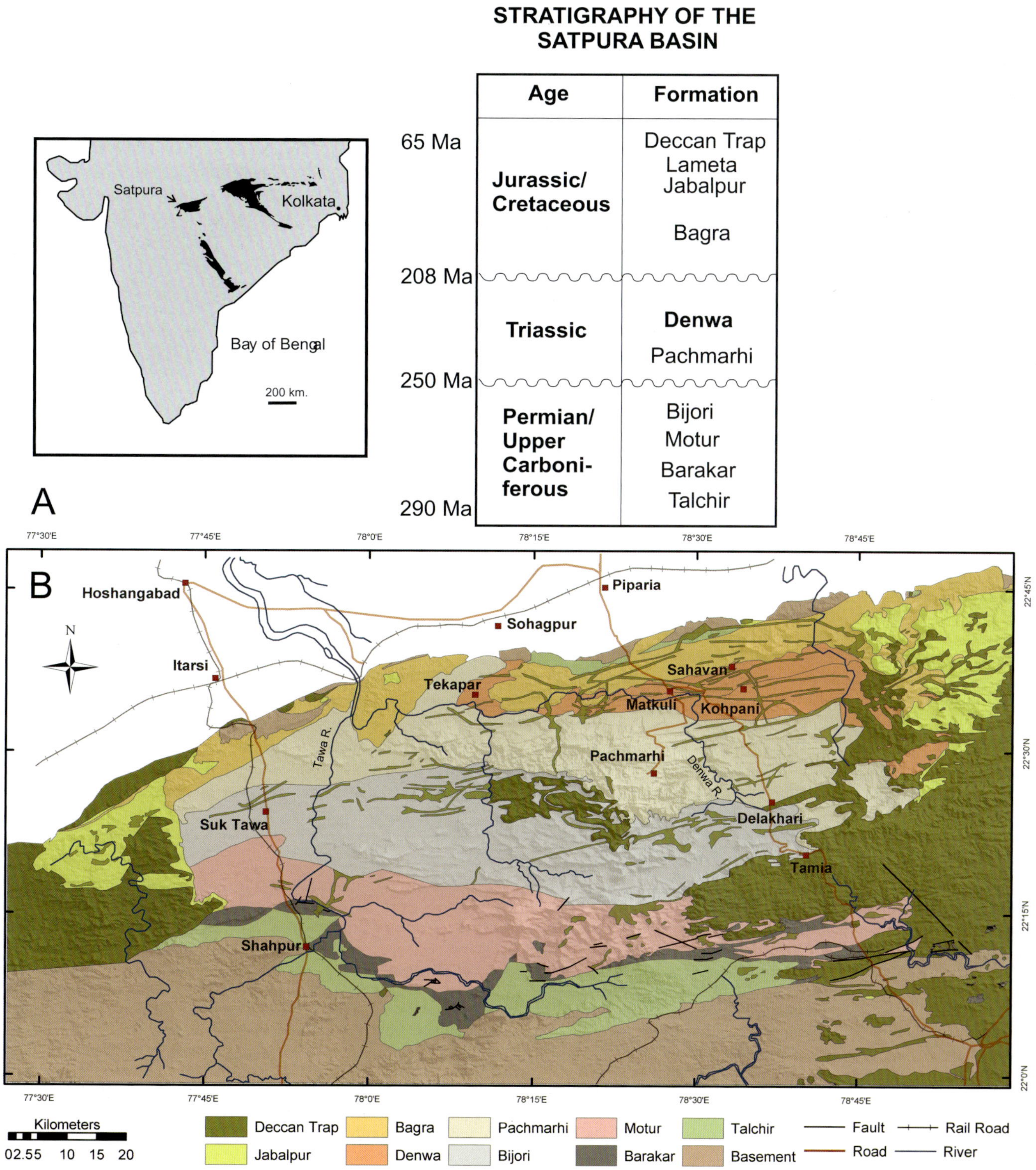

FIGURE 1.—A) Outcrops of the Gondwana basins in peninsular India and the stratigraphy of the Satpura basin fill; B) geological map of the Satpura Gondwana basin showing the exposure of the Dewna Formation.

fills of mixed-load, sinuous, laterally migrating channels with laterally accreting point bars.

Ribbon bodies (element III; ~ 5 m thick and ~ 100 m wide), dominated by laterally stacked, co-directional lensoid sets of IHS separated by curved discontinuity surfaces (Fig. 2A), represent deposits of mixed-load to suspended-load, sinuous channels that were filled by obliquely accreting fine-grained point bars or river banks.

Also common in the upper Denwa succession are mudstone-

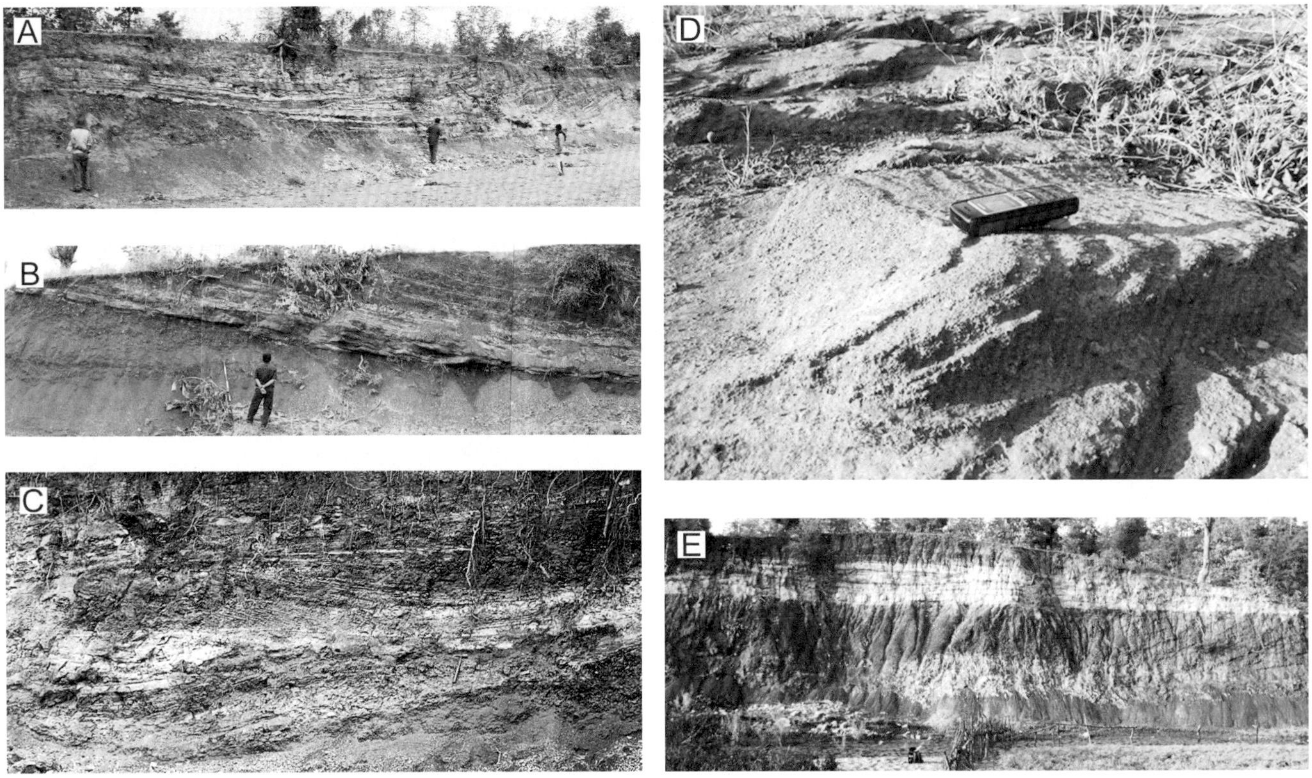

FIGURE 2.—A) A ribbon-shaped channel-fill body (element II) encased in red mudstone. Note that channel fill comprises several laterally stacked sets of inclined heterolithic strata (IHS). Paleoflow is towards the observer. B) Part of a ribbon-shaped, mudstone-dominated channel-fill body (element IV) encased in mudstone. The projected thicker beds in the basal part of the channel fill comprise calcirudite and calcarenite. C) Part of a ribbon-shaped channel-fill body (element IV) comprising mainly nested sets of mud-dominated heterolithic strata. Note that in the lower part the heteroliths fill the channel scour conformably. Hammer for scale. D) Cross-stratification in a small channel-fill body (element V) comprising calcirudite. E) A sand-dominated heterolithic sheetlike body (element VI) in the upper part of the exposure. Note the sharp planar basal contact with mudstone showing well-developed inclined ped-bounding surfaces (in the right-hand side of the photo). Also note the presence of another similar body (in the lower half of the exposure) that has been obliterated and disrupted by desiccation. Backpack in the foreground for scale.

dominated ribbonlike bodies (element IV; 5–7 m thick and ~ 150 m wide) encased within red mudstones. These are filled with either muddy IHS overlying a thin basal lithosome dominated by calcirudite and calcarenite (Fig. 2B) or concentrically and asymmetrically arranged muddy heterolithic strata (Fig. 2C). The former subvariety of ribbon-shaped deposits has been interpreted as the deposits of suspended-load channels that were dominantly filled by oblique aggradation of muddy banks and benches and by passive plugging upon abandonment. The later subvariety, smaller in size, possibly represent short-lived laterally stable, suspended-load channels.

From the foregoing description, it is apparent that a wide variation in geometry, dimension, mesoscale architecture, and lithology is the hallmark of the channel-fill bodies of the upper part of the Denwa Formation. The ribbon-shaped channel fills with laterally accreting IHS are comparable with modern mixed-load and suspended-load channels as well as their ancient analogues (Taylor and Woodyer 1978, Woodyer et al. 1979, Stewert 1983, Thomas et al. 1987, Brooks 2003). The overwhelmingly mud-dominated ribbon-shaped deposits of the Upper Denwa Formation are comparable with abandoned to partially abandoned mud-filled channels described from ancient successions (Stewert 1983, Muñoz et al. 1992, Kirschbaum and McCabe 1992, Newell et al.1999) and are akin to those interpreted to represent "avulsion splay channel" deposits by Kraus and Davies-Vollum (2004). The low width/depth ratio (13 to 30) of most of the upper Denwa channel-fill deposits is attributable to the cohesive nature of the muddy banks, which allowed only limited lateral migration of the channels.

Plano-concave bodies (element V: ~ 1 m thick and 5 to 20 m wide) made up of calcirudite and calcarenite along with minor amount of silt- to fine-sand-size siliciclastic grains occur encased in red mudstone. They have sharp and erosional boundaries with their host red mudstone. The constituent detrital glaebules of these bodies internally display pedogenic microfabrics like clotted micrite, corroded detrital quartz, and calcite- and barite-filled radial and circumgranular cracks (Ghosh et al. 2006). Internally these bodies are crudely bedded and /or trough cross-stratified (Fig. 2D). The dominance of caliche-derived calcareous peloids along with the virtual absence of siliciclastics suggest derivation of detritus mostly from local sources and implies that the depositional sites were denied of siliciclastic influx. These bodies possibly represent deposits of short-lived, minor floodplain channels that formed in response to sporadic rainfall and drained local floodplain areas. Poorly developed or incipient caliches were reworked from the uppermost layer of the floodplain sediments, and these calcite peloids were deposited in small channels wandering on the Denwa floodplain. Allen and Williams (1979) and Mariott and Wright (1996) interpreted similar deposits (their Type B conglomerates) from the

Siluro-Devonian of Wales as representative of interfluvial drainage system that captured only the precipitation falling locally on the alluvial plain and perhaps the water of the main rivers during severe floods. Sarkar (1988, his Type II) and Gomes-Graz and Alonso- Zarza (2003, their Type 1) also interpreted similar bodies as small channel or scour fills in the floodplain.

The sheet bodies of the upper Denwa Formation (element VI; 1–8 m thick and 10 m to few hundreds of meters wide) comprise centimetre- to decimetre-scale flat-bedded (with or without parting lineations) sandy or muddy heteroliths. The basal bounding surface is flat and erosional. Often these bodies overlie sand-filled desiccation cracks present in the underlying mudstone unit. The bedding is parallel or gently inclined to the basal bounding surface. The dimensions and lithology of the sheet bodies show wide variations. Some of the thicker sand beds in the lower part of the sheet bodies display ripple-scale to small-dune-scale planar and trough cross-strata. The muddy heterolithic strata in places are characterized by the presence of articulated bivalve shell of unionids. In one instance a mixed skeletal assemblage of amphibians and unio and dipnoan fishes occur in this heterolithic package. At places small (30–50 cm thick), ribbonlike bodies with sandy heterolithic fills are found encased within thicker sheet bodies. Also present are plano-convex lenticular lithosomes with wavy stratification in the basal part of some of the sheet bodies. The sheet bodies which generally lack any channel-shaped basal scour surface and dominated by, flat, parallel to low-angle lamination suggest deposition from sheet flow. The typical assemblage of fossil amphibians, fishes, and shells of unionids along with wave-influenced lamination attest to the presence of at least ephemeral floodbasin ponds. However, the absence of carbonaceous material in the upper Denwa succession indicates the absence of persistent lakes or bogs, similar to those reported from humid fluvial settings (Aslan and Blum 1999, Morozova and Smith 1999, 2000).

The channel-fill bodies described earlier show a mean paleocurrent direction towards the northwest, whereas the sheet bodies display a bipolar, bimodal distribution with two prominent modes symmetrically disposed around the mean direction shown by the channel-fill deposits. These bodies are interpreted as crevasse-splay deposits and were possibly emplaced by simple diversion of flow into the subdued overbank areas, a style akin to progradational avulsion of Morozova and Smith (1999, 2000) and aggradational avulsion of Mohrig et al. (2000).

The tabular bodies (element VII; 3–5 m thick and tens of meters wide) comprise complexly organized sandstone, mudstone, and heterolithic strata and are characterized by sharp, planar to slightly undulating lower boundaries, absence of any vertical trend in primary structures or grain size, and absence of large-scale bedforms, and thus appear to have been deposited by weakly channelized, low-energy flows and have been built up mainly through intermittent vertical rather than lateral aggradation. Rapid flow fluctuation and ephemerality are reflected by sharp vertical changes from sandstone to mudstone and presence of desiccation cracks in mudstone strata.

The fine-grained units (Element VIII) comprise mudstones and siltstones that are mainly grayish red (5R 4/2), dark red brown (10R 3/4), and purple (5RP 2/2) in color. Granulometry shows that the mudstones are composed of 25–35% of sand, 31–63% silt, and 12–23% clay. Montmorillonite occur as a dominant constituent of these mudstones (< 2 μm fraction). Under microscope, microscale networks of fractures separating blocky peds and floating siliciclastic grains (fine sand) are discernible in the mudstones. The mudstones are mostly massive. In places strata of fine to very fine sandstones and calcarenites, a few millimeters to a centimeter thick, are found interlayered with mudstones. These coarser-grained strata internally show parallel stratification and current-ripple cross-laminations. The fine-grained intervals and sheet bodies of heteroliths show varying degree of pedogenesis.

The process of suspension settling of fines from overbank flooding alone cannot account for this huge thickness of fines in the upper Denwa succession. Conversely, as suggested by many others (e.g., Slingerland and Smith, 2004) avulsions provide a longer-term mechanism for continuous supply of fine sediment to the floodplains. Repeated episodes of flow diversion into the floodbasin accompanied by rapid semicontinuous floodplain aggradation possibly produced massive mudstone intervals, several meter thick, alternating with crevasse-splay sands and heteroliths in the upper Denwa succession. In general a large accumulation of fines in a fluvial setting requires a suitable source lithology, a conducive climate (Reid and Frostick, 1985), and the existence of a geomorphic setting that traps the fine-grained detritus (Mariott and Wright, 1996).

The preservation of large proportion of silt- and clay-size detritus in many of the channel-fill bodies of the Denwa formation and the rare occurrence of dune-scale cross-strata might indicate that the overall flow strength of the Denwa fluvial system was low. In case of Denwa, the channel banks were reinforced by a high proportion of clay in the overbank sediments. This in turn facilitated reduction of channel width/depth ratios (cf. Nanson and Knighton 1996). Together with low flow strength, resistant banks, constricted channels along with a seasonally fluctuating and/or ephemeral flow regime possibly gave rise to the development of a multichannel fluvial system.

Coexistence of channel-fill bodies of varying dimensions, lithology, and internal organization in restricted stratigraphic interval of the upper Denwa succession was one of the reasons that led us to suggest (Ghosh et al. 2006) a multichannel system having penecontemporaneous channels with different fill histories. Flow in adjacent anabranches at bankfull is laterally independent (Schumm 1977, Bridge 1993), enabling the channels and islands they bound to develop complex and variable forms. It is, however, impossible to demonstrate simultaneous flow in the channels. The channel-fill character of the Denwa succession suggests that not all the channels were equally important: there were one or two important, larger anabranches that dominated the efficiency of the overall system (see also Huang and Nanson 2007).

However, North et al. (2007) have argued that a succession comprising ribbon-shaped bodies encased in mudstones can be built up by an avulsing single-thread meandering channel. If this had been true then we would have expected a greater homogeneity in lithology and internal organization of channel fills. The lack of homogeneity between different channel-fill bodies occurring in the upper Denwa succession does not favor such an interpretation. Though lateral-accretion deposits are present in some of the channel fills, these are absent in others (e.g., element I). On the other hand, at any particular stratigraphic level the Denwa channel-fill bodies are laterally separated by 1 to 2 km of floodplain fines and other extra-channel deposits (Ghosh et al. 2006).

Taking into consideration the tectonic strike of the Denwa succession (ENE–WSW) and a mean flow direction towards northwest, as derived from the channel bodies, it can be suggested that the interchannel distances were in the order of hundreds of meters (800 m to 1600 m). Considering the large lateral extent of the Denwa alluvial plain, the building up of this succession through nodal avulsion of a single-thread channel would have left large floodplain areas starved of sediment for a period long enough to develop regionally extensive mature paleosols. The occurrence of localized rather than regionally extensive paleosols in the upper Denwa indicates that the sites of pedogenesis were probably controlled by the spatial distribution of the anabranches of the system.

Two modern anabranching river systems are well documented to date: the Cooper Creek system in Australia, and the Saskatchewan River in Canada, representing the dryland system and the humid system respectively. The characters of the Denwa system show some similarities and some differences with both of these.

TABLE 1.—Pedogenic features of the paleosols in the upper part of the Denwa Formation.

Pedogenic features	Description	Horizon	Remarks & Interpretations
P1) Desiccation cracks	Numerous, narrow, subvertical, downward tapering wedge-shaped cracks filled with fine-grained sandstone, siltstone, and / or granule-size detrital pedogenic glaebules.	Occur as the uppermost horizon (60 cm to 2 m) of the paleosol profiles.	Dehydration of subaerially exposed sediment surface.
	1–80 cm in length, a few mm to 4 cm wide. In rare instances, cracks as wide as 7 cm and as long as 2 m are present.	Few large cracks cut across entire profile.	
	Most wedges bifurcate downward. Distinct from the rhizocretions by noncylindrical 3D geometry, sharp and jagged sidewalls, and poorly calcareous, sandy fills.	Comparable to Bss horizon of modern vertisols.	
P2) Wedge-shaped peds	Wedge-shaped mudstone units, 5 cm to 2 m thick and 10 cm to 3 m wide, bounded by inclined (5° to 20°) and mutually intersecting, gently curved surfaces that are a few cm to ~ 3 m long. The surfaces dip in random directions. The surfaces are polished and show slickenside lineation (Fig. 3F). Some of the larger surfaces are encrusted with mm-thick carbonate. Their mutually intersecting character gives rise to pseudo-anticlines.	Occur either as the topmost or middle horizon (1–5 m thick) of the profiles.	Disruptions by shrink and swell of clay minerals within the muddy substrate due to repeated hydration and dehydration.
	In some paleosols wedge-shaped peds occur in two diverse scales. Smaller peds are overprinted by larger peds. Smaller peds are 10 to 30 cm wide and 5–15 cm thick. Larger peds are 2–3 m wide and 1–2 m thick.	Comparable to Bss horizon of modern Vertisols.	
P3) Rhizocretions	Vertical or subvertical, cylindrical calcareous concretions (rhizocretions) that taper downward. These are 2–40 cm long and 0.5–5 cm in diameter (Fig. 3A). The degree of cementation by carbonate varies. The well indurated rhizocretions are characterized by micritic cylinders with a centrally located axial zone that is filled with either calcite spar or silt. In some cases a faint concentric zonation can be observed in transverse sections. The outer surface is smooth and at places shows a mm-thick coat of iron cement.	10–50 cm thick, and in rare instances they are 2 m thick. Such a horizon occurs either in isolation or as the lowermost horizon of the profile.	Accumulation of carbonate in the rhizosphere mediated by the respiration of living plant roots. Some of the larger cylindrical concretions may be comparable to the concretions formed around tree stumps (cf. Kraus and Gwinn 1997)
	The less-indurated variety show a reticulate network of calcium carbonate around a central tubular carbonate-enriched zone (centimeter wide). The network holds in it small (mm to cm) angular blocks of red mudstone. The outer surface is typically rough and irregular.	Similar to Bk horizon of modern calcic Vertisols.	
P4) Glaebules	Isolated, subspherical or platelike calcareous concretions (glaebules: Fig. 3C). Small to large pebble size. The platy glaebules show honeycomb surface texture. Surface of the spherical glaebules is smooth.	The horizons are generally 30–50 cm thick. Occur in association with rhizocretions.	Precipitation of carbonate from meteoric water upon supersaturation of the dissolved ions due to evaporation. The spherical glaebules were probably precipitated from the pore waters, and the platy glaebules with honeycomb surface texture were possibly formed in small cracks in otherwise impervious mudstone.

TABLE 1.—*Continued.*

Pedogenic features	Description	Horizon	Remarks & Interpretations
	Composed mostly of micrite and minor microspar (Fig. 3D) with a few silt to sand size detrital quartz grains floating in the carbonate groundmass. Microscopic features include clotted micrite and corroded detrital quartz grains. Some glaebules show radial fractures filled with blocky spar. Spar-filled, circum-granular cracks are also present in some of the glaebules.	Similar to Bk horizon of modern calcic vertisols.	
P5) Gleys	Roughly circular (diameter: 1–5 cm) colored mottles in shades of white and yellow (N8, 5Y 7/2, 10Y 6/2, 5YR 3/2). The gleys typically occur in the mudstone intervals that lack interbeds of detritus coarser than silt.	Occurs either all through the paleosol profiles or in the basal 30–50 cm of the profiles. In places associated with rhizocretions. Similar to Bg and Cg horizon of modern Vertisols.	Leaching of more mobile oxides of iron due to local and temporary reducing soil conditions.

The upper Denwa succession shows similarities with humid or organoclastic anabranching river deposits in having ribbon-shaped, sandy and heterolithic channel-fill bodies encased in mudstones, numerous sheet bodies resembling splay deposits, and a strong contrast between channel and overbank suites of lithologies. It differs from the humid subvariety in lacking any organic or lacustrine marsh deposits. At the same time the Denwa displays some similarities with the Cooper Creek model by the presence of significant amounts of mud in the channel fills (elements III and IV), and rarity of vertical channel amalgamation. However, the Denwa sheet bodies have sharp and erosional basal contact and they often overlie mudstone units showing evidence of desiccation suggesting emplacement over a dry floodplain surface.

As also emphasized by Nanson and Knighton (1996, p. 236) anabranching systems described so far are limited and undoubtedly do not represent the full range that will eventually be recognized. As discussed by North et al. (2007), in humid and dryland regions, different classes of anabranching rivers exist and they produce distinctly different sedimentary successions. Even in a dryland setting there may be several subvarieties. We believe that the currently well studied Cooper Creek system may not define the full range of possibilities in some important aspects. Apart from climate (amount of precipitation, seasonality of precipitation and temperature variation, and the presence of vegetation) and relief, the nature of sediment supplied to a river basin is also largely controlled by the composition, texture, and structure of the source rocks (Bridge, 2003, p. 12). So, it is expected not all rivers in the dryland are necessarily characterized by a sediment load that is either entirely muddy or entirely sandy. A particular combination of semiarid to subhumid climate and source rocks may produce a mixed sand-mud load unlike the mud-dominated or sand-dominated examples of dryland rivers cited by North et al. (2007). Several ancient successions like the presently discussed Denwa succession may just be showing characters intermediate between these two. The presence of both mud-dominated and sand-dominated anabranching rivers from the present day dryland of Australia indicates that the overall sediment load was not influenced by prevailing climate but by some other factors, possibly source rocks. In spite of the dominance of mud the presence of sand in the sediment load possibly produced the channel–levee–floodplain morphology in the Denwa system due to hydrodynamic segregation of different grain-size populations.

PALEOSOL PROFILES

Description

The channel-fill deposits of the Denwa succession, regardless of their size and internal organization, do not show evidence of alteration by the pedogenic process. The paleosols are found in the overbank deposits comprising red mudstone and sheet bodies of sand-mud heterolith. The paleosols extend laterally for about 20–100 m. Typically none of them could be recognized as a regionally extensive stratigraphic unit. The paleosol profiles are generally 30 cm to 1.5 m thick and only at places they attain a thickness of about 5 meters in case of cumulate soils. The profiles comprise two or more pedohorizons described in Table 1 (Figs. 3 and 4A). The contact between the horizons is diffused and often overlapping.

A horizon of wedge-shaped peds (Table 1, Fig. 3E) bounded by inclined, slightly curved, polished surfaces with slickenside lineations (Fig. 3F) and developed on a red mudstone host is a conspicuous feature of most of the Denwa paleosols. The horizon comprising wedge-shaped peds is usually 1 m to 1.5 m thick. However, at places it can attain a thickness up to 5 m. In many profiles, desiccation cracks that are intimately associated with the wedge-shaped peds form a distinct horizon (60 cm–2 m thick) above the horizon of wedge-shaped peds. In some of the profiles a horizon (usually 10–50 cm thick) of calcareous glaebules and rhizocretions occurs below the horizon of wedge-shaped peds. Gleys occur mainly in the profiles developed upon a predominantly mudstone host material. They are usually present in the lower part of the horizon of wedge-shaped peds and also in the horizon of poorly indurated rhizocretions (P3, Table 1). Not all the horizons are present in every profile. However, all types of the profiles are apparently equally abundant all through the stratigraphic column of the upper Denwa Formation. Again, there is no appreciable lateral bias in their mode of occurrence.

The presence of isolated simple paleosol profiles (sensu Marriott and Wright 1993) encased in unaltered host sediments is not very common within the succession. Generally they occur as compound sets (sensu Marriott and Wright 1993) (Fig. 4A) comprising two to three

FIGURE 3.—Pedogenic features of the upper part of the Denwa Formation. A) A large, weakly calcareous rhizocretion. Scale 15 cm long. B) A thin body of cross-stratified calcirudite erosionally overlying pedogenically modified massive mudstone containing calcareous rhizocretion (arrow). C) Spherical and subspherical calcareous glaebules found in the mudstone shown in Figure 2B. Scale divisions are in centimeters. D) Photomicrograph of a calcarenite showing detrital calcareous glaebules. Note virtual absence of sand-size siliciclastic grains. E) Profusely developed pedogenic slickenside surfaces bounding wedge-shaped peds. The surfaces are coated with thin films of carbonate. Note well developed large gleys. Hammer for scale (arrow). F) Polished surface of pedogenic slickenside showing slickenside lineations. Scale 15 cm long.

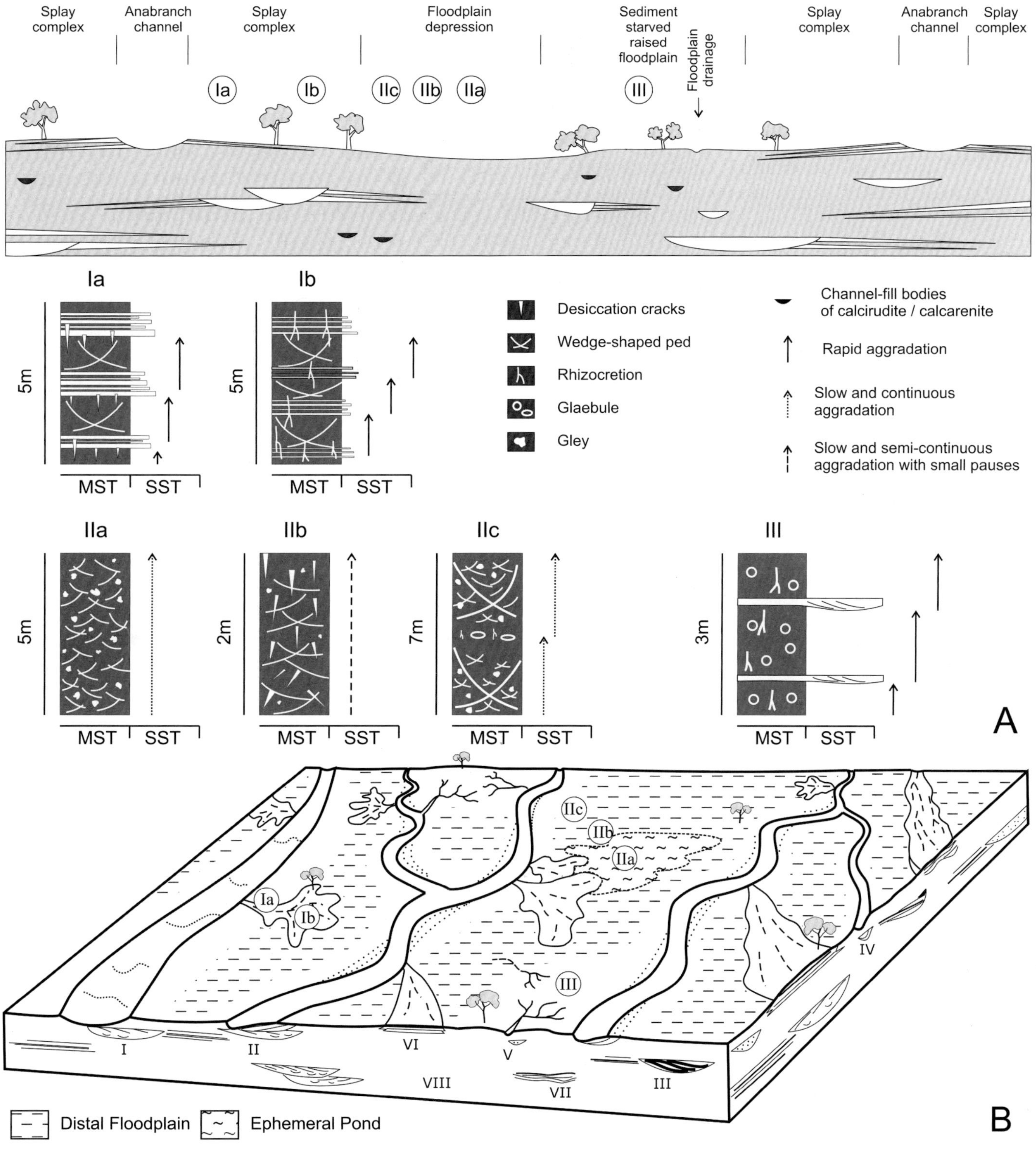

FIGURE 4.—A) Schematic lithologs of different types of associations (1 to 3) discussed in the text (lower half of the diagram). The lateral separations between arrows suggest pauses in aggradation. A schematic transverse profile (upper half) shows the various sub-environments of the upper Denwa fluvial system. The probable sites for development of different litho-pedogenic associations (discussed in the text) are indicated. B) A schematic block diagram showing the spatial relationship between the lithofacies–pedofacies associations and the various components of the fluvial system. The architectural elements of the upper Denwa succession are shown by Roman numerals, whereas the identifiers of different litho-pedogenic associations are circled.

profiles vertically separated by thinner unaltered sediments. In places the horizons of wedge-shaped peds (P2, Table 1) and rhizocretions (P3) of the overlying profile extends down to the upper part of the underlying profile (Fig. 4A) producing a composite set (sensu Marriott and Wright 1993). Some other profiles are cumulate in nature (Fig. 4A). In that case, intervals of mudstones up to 5 m thick show the presence of small wedge-shaped peds (individual peds about 10–70 cm thick) over its entire thickness. The peds are often gleyed or are associated with small cracks (3–15 cm long) that are deformed along the ped boundaries.

Interpretation

The widespread occurrence of well developed wedge-shaped peds, bounded by polished surfaces showing slickensidic lineation in a montmorillonite-rich mudstone host, indicates that shrink and swell of the solum was a major pedologic process in the Denwa soils. The horizon of wedge-shaped peds (P2) is comparable with the Bss subsurface horizon noted in modern Vertisols (cf. Soil Survey Staff 1975). The presence of an overlying horizon of cracks makes the Denwa paleosols very similar to the modern Vertisols, which are characterized by large surface cracks that remain open for the dry part of the year and where the cracks are underlain by a disrupted, festoon-shaped subsurface horizon produced due to the shrinking and swelling of the smectitic clays present in the host (cf. Retallack 2001). Some of the Denwa paleosols show the presence of carbonate glaebules (P4) and films on the ped surfaces along with the vertic features (P2), and they are best compared with the modern calcic Vertisols and their ancient analogues (Allen 1973, Blodgett et al 1993, Driese and Mora 1993).

Modern Vertisols occur in regions with subhumid to semiarid climates (Retallack 2001). Climatic conditions are the main criteria for distinguishing the suborders of modern Vertisols. However, for that purpose, the particular information needed about how many days per year the cracks remain open is apparently impossible to ascertain for paleosols. The absence of horizons containing carbonaceous organic material might suggest that the Denwa soils lacked organic surface or subsurface horizons like those of soils formed under a humid climate. The Denwa paleosols also lack horizons that are similar to leached surface horizon, fragipan, spodic horizon, etc., typical of humid soils. The accumulation of authigenic carbonate in the profile in the form of macroscopic nodules and films on the ped indicate moisture deficiency at least for a significant part of the year. The presence of montmorillonite, plagioclase feldspar, and sulfates (barite) in the Denwa sediments also corroborates this climatic setting.

On the other hand, the paleosols of the Denwa Formation are not readily comparable with the typical soils of arid regions (e.g., Aridisols) because in this case the carbonate nodules are small in size and do not coalesce with each other to form petrocalcic horizons. Moreover, the horizon of calcareous glaebules (P4) usually lies below the horizon of wedge-shaped peds (P2), and therefore it is likely that most of the carbonate in these profiles accumulated at a depth of more than 1 m. However, the intraformational caliche-derived detritus preserved in the upper Denwa was possibly produced by stripping of the pedogenically modified upper floodplain surface. Therefore, it is likely that some of the Denwa soils also had dispersed carbonate nodules at shallower depths, resembling a weakly developed shallow calcic horizon. Though it cannot be ruled out that the erosion never reached the deeper soil horizons, the contribution from the shallower horizons was possibly volumetrically more significant.

Though most of the Denwa paleosols are similar to modern Vertisols and calcic Vertisols, some of the profiles that have poorly developed horizons can best be equated with modern Inceptisols. Again, the profiles that lack a well-developed horizon of wedge-shaped peds, and have an upper horizon of rhizocretion (P3) and a horizon of glaebules (P4) below it could be similar to the modern Alfisols. However, a light-colored, organic-poor epipedon and pronounced clay coatings on skeletal grains, which are typical of Alfisols are not present in the Denwa paleosols.

Considering all the available pedologic and lithologic evidence it appears that the calcic Vertisols of the Denwa Formation were formed under a subhumid to semiarid climatic setting. The presence of large and well developed wedge-shaped peds in most of the paleosols also suggests that the climate was strongly seasonal, possibly with more dry months than wet ones. Poor preservation of organic matter and sparse development of rhizocretions indicate a poorly vegetated landscape.

The size and concentration of carbonate concretions are generally considered to represent the maturity of calcic soils (Leeder 1975, Allen 1986, Gile et al., 1966). Considering the small dimension of glaebules in the some of the Denwa paleosols and the fact that they do not coalesce with each other, it can be suggested that they represent immature to at best moderately mature soils that are comparable with Type A and B profiles of Allen (1974, 1986) and stages I to II calcretes of Gile et al. (1966). A period of about one thousand years is usually required for the appearance of macroscopic carbonate concretions in soil profiles (Leeder 1975). In contrast, vertic features develop rather quickly, and Vertisols can form in only a few hundreds of years on claystone and shales of smectitic composition (Retallack 2001, Mariott and Wright 1993, Wright and Mariott 1996). Therefore, in case of the Denwa, the variations in the types of horizons present, the nature of profile stacking, and the degree of development and dimension of vertic features reflect local topography, sediment aggradation rate, and water-table condition rather than the maturity of the soils. For example, where sedimentation was rapid and unsteady, compound sets of weakly developed profiles separated by unaltered host sediments formed (cf. Kraus 1999) (Fig. 4A). Vertically successive profiles overlap to produce composite sets (Fig. 4A) when and where the rate of pedogenesis exceeded the rate of deposition in a regime of unsteady sedimentation. On the other hand, a situation of steady but slow sedimentation led to the development of cumulative sets (cf. Kraus 1999) (Fig 4A).

However, within this short time span required for the development of vertic features, it is expected that the formation of a meter-thick wedge-shaped ped would require a longer residence time compared to what is required for development of similar features on a centimeter scale while other factors are held constant. In this study, the degree of soil development is assessed with the help of size of the wedge-shaped peds, amount of carbonate accumulation in the rhizocretions, and the size and concentration of the glaebules. The profiles in the upper Denwa succession that comprise wedge-shaped peds associated with tens- of- centimeters- scale cracks and lack macroscopic expression of carbonate accumulation are considered to represent poorly developed soils whereas the profiles containing rhizocretions, well indurated by carbonates, and numerous glaebules (about 2 cm in diameter) are considered to be better-developed soils. The profiles containing only a few small (a few millimeters in diameter), sparsely dispersed glaebules, poorly cemented rhizocretions, and vertic features or profiles that have composite set of vertic features of different scales are considered to represent moderately developed soils.

ASSOCIATION OF LITHOLOGIC AND PEDOGENIC FEATURES

Based on the combination of the pedogenic features and the lithofacies of the host sediment, the following associations could be identified.

Type 1 Association

Description: The Type 1 association (Table 2) comprises pedogenically modified sandy to muddy heterolithic sheetlike bodies (elements VI) intercalated with red-mudstone-dominated units having laminae of

TABLE 2.—Associations of lithologic and pedogenic features in the upper part of the Denwa Formation.

Associations	Lithofacies	Paleosols	Nature of sedimentation	Location with respect to the fluvial system
Type 1a	Sheetlike bodies of sand-dominated heteroliths, 50 cm to 3 m thick, interbedded with red mudstone-dominated units, 1 m to 1.5 m thick. Mudstone-dominated units contain mm- to cm-thick siltstone and fine-sandstone interbeds.	Paleosol profiles (75 cm to 2 m thick) comprise upper horizon of sand-filled cracks overlying a horizon of well developed wedge-shaped peds. Gleys, rhizocretions, and glaebules are generally absent. The profiles are stacked vertically to produce a compound set where the vertically adjacent profiles are separated by unaltered host sediments.	Fast but highly episodic.	Proximal parts of splay complex.
Type 1b	Mud-dominated, sheetlike heterolithic bodies, 30 cm to 2 m thick, interbedded with red mudstone units 2 to 3m thick.	Paleosol profiles (~ 30 cm to 1 m thick) comprise an upper unit of wedge-shaped peds overlying a horizon of rhizocretions. Rhizocretions are weakly cemented with carbonate. Gleys and carbonate glaebules are either absent or are not well developed. The profiles are vertically stacked to produce compound or composite sets.	Moderately fast and episodic. Smaller pauses in sedimentation compared to Type 1a.	Distal parts of splay complex.
Type 2a	Massive mudstone units, 7 to 10 m thick. Drab purple in color (5RP). Interbeds of heteroliths, calcirudite or calcarenite absent.	Palaeosols comprise a cumulate stack of wedge-shaped peds, 1 to 5 m thick. Individual peds are 30 cm to ~ 1 m wide and 10 cm to 70 cm thick. Profuse, large gleys are present all throughout. Sand or silt filled desiccation cracks are absent.	Slow and continuous sedimentation.	Deeper part of flood-basin depressions.
Type 2b	-do-	Paleosols comprise a cumulate stack, 3 m to 5 m thick, of wedge-shaped peds and associated silt filled cracks. Individual peds are 30–50 cm wide and 10–30 cm thick. The cracks (3 to 15 cm in length) are deformed and rotated along the ped boundaries. Gleys are small and sparsely distributed.	Slow and semicontinuous. Small pauses between individual sedimentation events.	Intermediate between Types 2a and 2c.
Type 2c	-do-	Paleosols comprise a composite stack, ~ 5 m thick, of very large (2–3 m wide and 1–2 m thick) and smaller (10–30 cm wide and 5–15 cm thick) wedge-shaped peds. The bounding surfaces of larger peds are lined with carbonate films. Gleys are small and sparsely distributed. Small calcareous rhizocretions and glaebules are present. Vertically adjacent paleosols are separated by ~ 50 cm of massive mudstones (contain fine-sandstone laminae, 1 to 5 mm thick) forming compound sets.	Slow and continuous but with longer (compared to Type 2b) pauses between individual sedimentation events.	Shallower parts of flood-basin depressions.
Type 3	Massive red mudstone units containing sparse ribbon bodies (~ 1 m thick; 5 to 20 m wide) of trough cross-stratified calcirudite and/or calcarenite. Sheet bodies of heteroliths are absent.	Paleosol profiles comprise a horizon (10 to 20 cm) of rhizocretions, overlying a horizon (30 to 50 cm) of numerous dispersed, calcareous glaebules. The rhizocretions (< 10 cm in length) and glaebules (up to 3 cm in diameter) are well cemented with carbonate. Simple paleosol profiles occur in isolation.	Individual events of sedimentation is rapid but are separated by periods of nondeposition and /or erosion.	Slightly raised, abandoned floodplain areas.

siltstone and fine sandstone, 1 mm to < 1 cm thick. The paleosols are characterized by a horizon of wedge-shaped peds (P2), overlain by a horizon of desiccation cracks (P1) and underlain by a horizon of rhizocretions with weak to moderate carbonate cements (P3).

Based on the relative thickness of the heterolithic bodies and the intervening mudstone-dominated intervals as well as on the characters of the paleosol profiles, this association can be classified into two subtypes (Types 1a and 1b).

The Type 1a association (Table 2, Figs. 4, 5) comprises sheet-like bodies of sand-dominated heteroliths, 50 cm to 3 m thick (element VI), interbedded with red mudstone-dominated units 1 m to 1.5 m thick. A number of such couplets, stacked vertically are found in some of the

FIGURE 5.—Field photograph showing the Type 1a association. Note well developed desiccation cracks at the top and large inclined boundaries of wedge-shaped peds below (arrow). Scale bar is 50 cm long.

exposures. The sheetbodies extend laterally for 50 m to ~ 150 m, and beyond that they gradually thin out. The lower and upper bounding surfaces of the heterolithic bodies are generally sharp and planar. Internally the heterolithic bodies comprise a sandy (fine sand) member up to 10 cm thick (sensu Thomas et al. 1987) alternating with a mudstone member 1 to 5 cm thick. They are flat bedded, and the sandy bed surfaces show parting lineation.

Only the upper part of the couplet comprising a heterolithic body and its overlying mudstone show development of pedogenic features. The stratifications and primary sedimentary structures in the basal part of the couplets are preserved. The paleosol profiles (75 cm to 2 m thick) show weak horizonation with mainly two pedohorizons: upper horizon of desiccation cracks (P1) and a lower horizon of wedge-shaped peds (P2). These paleosol profiles occur either as an isolated simple set vertically separated by non-pedogenized host sediments or form a stack of compound sets (sensu Marriott and Wright 1993).

The Type 1b association (Table 2, Figs. 4, 6) comprises red mudstones 2 to 3 meters thick interbedded with mud-dominated heterolithic sheet bodies 30cm to 2m thick (element VI). These heterolithic bodies are in places found to be the lateral variants of Type 1a heterolithic sheetbodies. Internally heteroliths comprise a sandy (fine to very fine sand) member 1 to 3 cm thick that alternate with a muddy member 10 to 15 cm thick. The sandy members can be massive, parallel laminated, or current ripple cross-laminated. The mudstone members are typically massive.

Individual paleosol profiles are developed on a couplet comprising a sheetbody of muddy heterolith overlain by a thicker mudstone unit. The profiles comprise an upper unit of wedge-shaped peds (P2) overlying a horizon of rhizocretions (P3). The upper horizon of wedge-shaped peds occurs mainly in the mudstones, and the lower horizon of large (10–40 cm long) rhizocretions (P3) occurs mainly in the upper part of the muddy heterolithic body. Most of the rhizocretions are weakly calcareous, barring a few moderately calcareous ones. Sand- or silt-filled desiccation cracks are generally absent but wherever present are less than 10 cm in length and occur as the uppermost horizon of the profile. Unlike Type 1a, in this association unaltered host sediments separating the vertically adjacent profiles are absent and vertical stacking of profiles produces compound sets. In many places, however, the rhizocretions of the overlying profiles continue down to the upper part of the underlying profile, forming composite sets.

Interpretation: The intervals comprising sheetbodies of heteroliths interbedded with thick red mudstones have been identified as deposits of splay-avulsion complexes (cf. Ghosh et al. 2006). The weakly developed horizonation, dominance of vertic features, and absence of large, well cemented calcareous concretions characterize the associated paleosols as immature ones. The occurrence of simple paleosol profiles vertically separated by units of unmodified host sediments together with the immature character of the paleosols suggests that pedogenesis was brief and was punctuated by the events of sedimentation. This corroborates well with the splay origin of the host sediments, where events of rapid and episodic sedimentation is expected.

FIGURE 6.—Field photograph of the Type 1b association. Note large rhizocretions in muddy heterolithic body (arrow) and large inclined pedogenic slickenside surfaces in the overlying and underlying muddy units. Scale bar 1.5 m long.

The lateral transitions from thick and numerous sandy heterolithic sheetbodies in mudstones (Type 1a) to thinner and sparse muddy heteroliths (Type 1b) possibly represent a proximal-to-distal transition in a splay complex. The presence of only vertic features (P1 and P2) in Type 1a paleosols and the development of rhizocretions (P3) in that of Type 1b suggest that the later was better developed than the former. This in turn implies increased pauses between the individual events of sediment aggradation. That the floodplain sites represented by Type 1b underwent repeated pedogenesis is also highlighted by the presence of composite profiles in Type 1b. Though both the Type 1a and Type 1b sites (Fig. 4) experienced rapid episodic sedimentation, the presence of isolated paleosol sets in Type1a and composite profiles in Type 1b demonstrate a proximal-to-distal decrease in net sediment aggradation rate.

Type 2 Association

Description: The Type 2 association (Table 2, Figs. 4, 7, 8, 9) is represented by very thick (7 to 10 m) pedogenically modified mudstone intervals that generally lack interbeds of heteroliths or calcirudite and calcarenite. The pedogenically modified mudstones of this association, in contrast to those of Types 1 and 3, are drab and have a purple hue (5RP). Vertically stacked sets of wedge-shaped peds and a variable degree of development of gleys are the typical pedogenic features. Sand-filled desiccation cracks are in general absent, though in cases small silt-filled desiccation cracks can be noted. In rare instances small calcareous rhizocretions and glaebules are found in the paleosol profiles of this association. The organization of pedogenic horizons of the paleosol profiles shows considerable variation.

Based on the organization of wedge-shaped peds, the intensity of development of gleys, and the presence of calcareous glaebules, this association can be further classified into three subtypes (Table 2). Though these subtypes are not distinct by the nature of their host sediments, the associated paleosols show perceptible differences.

The Type 2a association (Figs. 4, 7) comprises cumulate stack, 1–5 m thick, of wedge-shaped peds (individually 30 cm to ~ 1 m wide and 10 cm to 70 cm thick) and well-developed large gleys. Cracks, filled with either sand or silt, are typically absent.

The Type 2b association (Figs. 4, 8) is typically found in mudstone units underlying major channel-fill sandstone bodies. The paleosol profiles are 3–5 m thick. They have silt-filled cracks, 3 cm to 15 cm, long in their upper part and wedge-shaped peds (30–50 cm wide and 10–30 cm thick) below it. The profiles are typically stacked in a cumulative fashion. The desiccation cracks of the older profiles are deformed and rotated along the surfaces bounding the wedge-shaped peds of younger profiles. Gleys are small and sparse.

The paleosol profiles of the Type 2c association (Table 2), like Type 2a, comprise mainly wedge-shaped peds (Figs. 4, 9). However, unlike Type 2a, the wedge-shaped peds in Type 2c form composite stacks where the peds occur in two diverse scales. The smaller-scale peds (10 to 30 cm wide and 5 to 15 cm thick) are overprinted by larger peds (2–3 m wide and 1–2 m thick). Bounding surfaces of the peds are often coated with carbonate films, 1 to 3 mm thick. These paleosol profiles are in places found to be vertically separated by massive red mudstone ~ 50 cm thick (containing sheets of very fine sandstone, 1 to 5 mm thick), showing the presence of rhizocretions and platy glaebules. The rhizocretions are small (< 5 cm in length) and are poorly calcareous. The glaebules are paper thin, showing honeycomb texture on their upper and lower surfaces. They are found to be disposed along bedding

FIGURE 7.—Field photograph of the Type 2a association. Note thick cumulate stack of wedge-shaped peds and profuse small gleys (lighter coloured). The scale is about 1 m in length.

parallel planes. Gleys occur in patches. Desiccation cracks filled with sandy or silty material are not noted.

Interpretation: The presence of very thick mudstone intervals that virtually lack any interbedding of sediments having grain size coarser than silt suggests that the Type 2 association represents deposits of a distal floodplain. The depositional site was possibly situated beyond the reach of the splays emanating from the channels and was thus deprived of coarser detritus. The drab color of these sediments, their purple hue, and the presence of gleys indicate that the sites of Type 2 was poorly drained compared to those of the other two types. The evidence for persistent lakes, ponds, or bogs, which are common in the floodplains of modern anabranching systems under a humid climatic setting (cf. Aslan and Blum 1999, Morozova and Smith 1999, 2000) are absent in the Denwa Formation. However, ephemeral lakes and ponds probably existed in the floodplain depressions during wet seasons. The sediments of the Type 2 association probably accumulated close to such distal floodplain depressions where the water table was closer to the surface than that at the sites of the other two types (Types 1 and 3). Anabranches surrounding these sites fed them with water and sediments. Wash load generated from the distal margin of the splay systems during rain might also have contributed to these sites.

The cumulate stacking of vertic features suggests slow but virtually continuous sedimentation of fines at these sites. The overbank sedimentation rate commonly lies between 1 and 10 mm/year (Kraus 1999). The theoretical model of Wright and Marriott (1996) predicts that for a residence time of 10^2–10^3 years a sediment aggradation rate of less than 2 mm/year is suitable for the formation of cumulate profiles having a horizon thickness of 0.5 m. However, cumulate soils that have thicker soils horizons might form even at a slightly higher value of aggradation rate. The absence of calcareous concretions in the two subtypes (Types 2a and b) of this association and their weak development in Type 2c could be a collective reflection of low residence time and a reducing soil environment. Though the sedimentation was generally continuous it was punctuated by minor pauses in cases of Types 2b and c. This is evidenced by the presence of small silt-filled cracks in Type 2b and carbonate coats on the ped bounding surfaces and small calcareous concretions between the vertically adjacent profiles in Type 2c. Again the composite nature of stacking of wedge-shaped peds in Type 2c suggests that the net aggradation rate, though generally low, fluctuated with time. While the smaller peds were produced during a phase of slow and continuous sedimentation, their overprinting by larger peds indicates that at times sedimentation slowed down further. Among the subtypes of Type 2, the sedimentation was possibly least punctuated for Type 2a.

Expansion and contraction of smectitic clays upon hydration and dehydration produce wedge-shaped peds (Yaalon and Kalmar 1978). However, at depth exceeding $\sim$ 2 m the superincumbent pressure overcomes the force of clay expansion and gradually inhibits ped development (Yaalon and Kalmar 1978). Within this limit, if a homogeneous column of sediment containing expandable clays is hydrated and dehydrated over its entire thickness it is expected that the thickness of the wedge-shaped peds will be similar to that of the sediment column. In contrast, if the zone of hydration–dehydration remains restricted to the upper part of the column, the resulting peds are likely to be much thinner. Therefore the size of wedge-shaped peds can potentially be linked with the depth of seasonal wetting or with the nature of water-table fluctuation.

The presence of numerous gleys in Type 2a on the one hand indicates poorly drained sites having a shallow water table. On the other hand, the smaller size of the associated wedge-shaped peds might indicate a smaller range of water-table fluctuation. The weaker development of gleys in Type 2b suggests better drained conditions. However, the overprinting of smaller peds by larger peds in Type 2c might suggest an increase in the range of water-table fluctuation at a later stage of pedogenesis.

The distinctions between subtypes 2a and 2c that indicate decreased continuity in sedimentation, increased variability in net aggradation, increased variation of the depth of hydration–dehydration, and the change from poorly drained to a better drained soil condition may together suggest that Types 2a and 2c represent deeper and marginal parts of the floodplain depressions respectively, whereas Type 2b represents deposits of an intermediate location (Fig. 4).

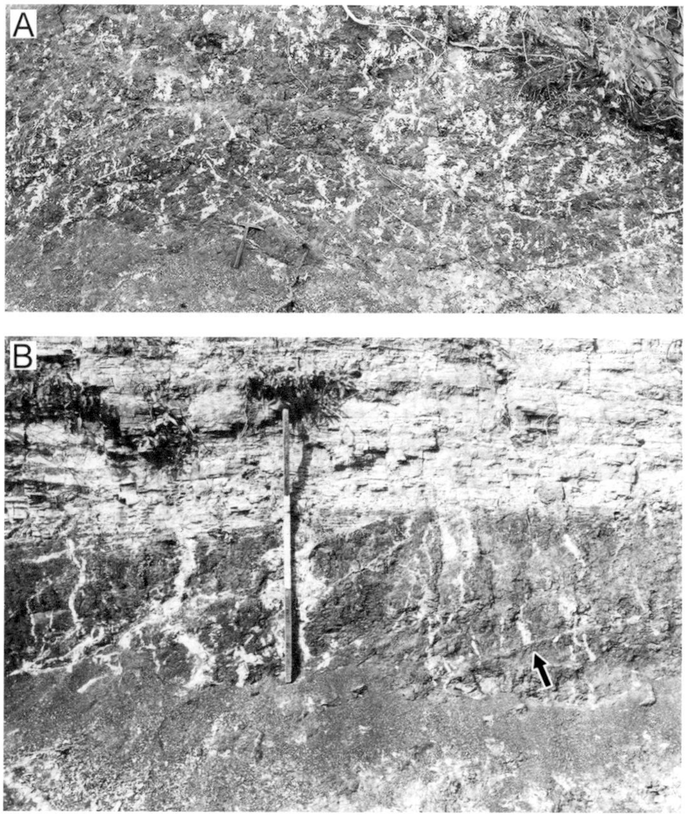

FIGURE 8.—A) Field photograph showing the Type 2b association. Note innumerable light-colored, narrow, wedge-shaped, silt-filled desiccation cracks deformed and disrupted by curvilinear to rectilinear traces of diversely oriented, inclined boundaries of wedge-shaped peds. Hammer for scale. B) Close-up view of the Type 2b association showing displacement and truncation (arrow) of desiccation cracks by the inclined traces of ped bounding surfaces. Scale bar is 1.5 m long.

FIGURE 9.—Field photograph of the Type 2c association. Note thick mudstone interval showing composite stack of traces of surfaces bounding wedge-shaped peds of various scales. The larger surfaces in the lower part of the photograph are coated with carbonate films. One of the smaller traces in the upper part of the photograph is marked by arrow. Also note patchy development of gleys (light-colored circular to semicircular blotches). Scale length 1 m.

Type 3 Association

Description: The association (Table 2) comprises massive red mudstone units with numerous, sparse, well developed, calcareous glaebules (up to 3 cm in diameter) and rhizocretions that are less than 10 cm in length (Figs. 3B, C). The horizon of rhizocretions generally overlies the horizon of glaebules with a gradational or overlapping contact.

Small (~ 1 m thick and 5 to 20 m in lateral extent) ribbon bodies (element V) of trough cross-stratified calcirudite and calcarenite are found to be encased within the calcareous glaebule-bearing mudstones. The calcirudites and calcarenites comprise mainly medium-sand- to small-pebble-size calcareous peloids that internally show pedogenic microfabrics (Fig. 3D; cf. Ghosh et al. 2001, Ghosh et al. 2006).

Interpretation: The presence of centimeter-scale calcareous glaebules and rhizocretions suggests prolonged pedogenesis of these mudstones (compared to the other types, i.e., Types 1 and 2) that produced better developed paleosols comparable to Stage II of Gile et al. (1966) and Type B of Allen (1974, 1986). These paleosols are somewhat more similar to modern Alfisols than to calcic-Vertisols.

Allen and Williams (1979), Allen (1986), Ekes (1993), Marriott and Wright (1996), Singh (1996), and Gómez-Gras and Alonso-Zarza (2003) interpreted similar calcirudite and calcarenite bodies as deposits of short-lived, small channels within the floodplain. These minor channels possibly drained local floodplain areas and derived detritus by stripping away the upper, pedogenically modified part of the overbank sediments. The framework grains of the calicrudite and calcarenite, which are nothing but caliche-derived calcareous detritus, provide additional evidence for existence of calcareous glaebule- and rhizocretion-bearing paleosols in the associated mudstones. It also implies that the glaebules developed in the mudstones were not coalesced together to produce a mature caliche profile, resistant to erosion by minor floodplain channels. The predominance of detrital glaebules in the calcirudites and calcarenites bodies and the paucity of coarse siliciclastics imply that the glaebules were derived from local sources where influx of sand-grade siliciclastic was minimal.

The presence of better-developed paleosols in Type 3 suggests the existence of sites within the distal Denwa floodplain that were deprived of sediment for a considerably long period and remained subaerially exposed, leading to the development of calcisols. It might be argued that these sites were slightly raised compared to the rest of the distal floodplain and thus escaped sedimentation for a longer period than at other sites (e.g., Type 2). Similar preferential accumulation of carbonates in slightly raised sites has also been noted by Singh (1996) in the modern Ganga River floodplain. The small but positive relief possibly enhanced the rate of evaporation at these sites as compared to the sites with lower relief. The enhanced evaporation led to faster attainment of supersaturation of calcium carbonate precipitating ions at these sites. Again, during rainfall these areas of positive relief were more prone to preferential erosion. The eroded paleosol material was possibly distributed locally within the floodplain through small ephemeral floodplain channels and was finally deposited nearby as small calcirudite and/or calcarenite bodies. The absence of vertic features in the mudstones of this association, which are common in other associations, might indicate faster deposition of detritus derived from erosion by sheetwash.

DISCUSSION

One of the characteristic features of the upper Denwa succession is that all the paleosols are locally developed and do not occur as regionally extensive units. Again, there is no perceptible bias in the type of paleosols either laterally along strike or vertically across the succession. The same is true for the fluvial architecture and lithology. Again, most of the Denwa paleosol profiles are dominated by different kinds of vertic features and show weak accumulation of pedogenic carbonate. In terms of paleosol maturity, which is assessed with the help of the degree of pedogenic carbonate accumulation, most of the Denwa paleosols can be compared to immature to at best submature soils. It is likely that individual soil profiles had less than a thousand years to develop. Therefore, the variations observed in the Denwa paleosols possibly do not reflect any allogenic control induced by tectonic and climatic changes and large-scale autogenic controls arising out of changes in fluvial system (e.g., avulsion) that operate over a longer time scale.

In the present situation the degree and nature of soil formation was possibly governed by differential rate of sediment aggradation and drainage conditions in different parts of the Denwa overbank areas. The overbank geomorphology (Fig. 4) and processes possibly exerted a dominant control over the variations observed in the upper Denwa paleosols. These variations are comparable to the "mesoscale' variations of Kraus and Aslan (1999) and Kraus (2002). At this spatio-temporal scale two factors together define the soil–landscape relationships, i.e., catenary and pedofacies (see Bown and Kraus 1987, Kraus 1997).

For the Eocene Willwood Formation the increase in soil maturity has been linked with the decreasing aggradation rates away from the channel (Bown and Kraus 1987). In the Denwa, the absence of carbonate accumulation in the paleosols developed on proximal splay deposits (Type 1a) and their weak accumulation in the rhizocretions of the paleosols of distal splay deposits (Type 1b) as well as the presence of calcareous concretions in the paleosols developed on sediment-starved floodplain surfaces (Type 3) may apparently indicate a trend, though subdued, in increasing maturity similar to those proposed by Bown and Kraus (1987). However, most of the paleosols developed on the distal floodbasin sites are virtually devoid of carbonates.

This difference might have arisen due to two reasons. Firstly, Bown and Kraus's (1987) fluvial model essentially considers overbank flanking a single channel where the floodplain aggradation rate decreases and soil maturity increases with increasing distance from the channel. The Denwa succession, in contrast, represents deposits of a network of coexisting channels separated by wide interfluve areas (Fig. 4; cf. Ghosh et al. 2006). In such a situation a particular floodplain site is likely to be fed by more than one channel. This may result in more continuous sediment aggradation in those sites, resulting in reduced residence time. In other words, most floodplain sites are unlikely to remain sediment starved for a long period. This might have favored development of cumulate profiles of vertic features and gleys and hindered widespread development of paleosols containing calcareous glaebules. The development of comparatively mature soils was restricted only to those parts of the floodplains that were slightly elevated (Fig. 4) and thus remained sediment starved for a longer period. Secondly, a shallow groundwater table could have prevented the precipitation of carbonates at the distal sites.

The degree and nature of development of vertic features in the paleosols developed in the Denwa floodplain sediments reflects differences in short-term local aggradation rates, floodplain geomorphology, and local drainage conditions. The variations in sedimentation pattern and hydromorphy related to the deeper to shallower part of the floodplain depressions are reflected in the paleosol characters of the subtypes of the Type 2 association (Fig. 4).

The observed pattern of paleosol characters and their association with lithofacies may help in identifying deposits of different overbank sub-environments in similar other mud-dominated fluvial successions. However, relationships presented here are pertinent only for muddy, ephemeral, anabranching fluvial systems having very wide, low-gradient overbank areas under a subhumid to semiarid climate. The paleosol–lihofacies associations are likely to be different for a single-strand meander system and for a different climatic setting.

CONCLUSIONS

Integrated analyses of paleosol profile characters and the stacking pattern of the profiles in the context of host lithofacies and depositional environments, as demonstrated in this study, may help in better identification of different overbank sub-environments in otherwise monotonous extra-channel deposits. In the upper part of the Denwa Formation the intervals comprising alternations of sheet bodies of heteroliths and mudstones, with simple to compound, immature to submature Vertisol profiles, represent the deposits of splay complexes. Thick mudstone intervals having immature Vertisols that show cumulate and composite stacking patterns of vertic features represent the deposits of distal floodplain depressions. The mudstones containing relatively mature paleosols (calcic Vertisols) in close association with caliche-derived calcirudite and calcarenites represent elevated areas of distal floodplain.

The splay deposits of the upper Denwa Formation show a proximal to distal trend of increasing carbonate accumulation in the soil profiles. This trend is partly analogous to the proximal–distal changes in soil maturity reported from the Eocene Willwood Formation. This trend is commensurate with the decrease in sediment aggradation rate away from a channel. Despite this trend the paleosols developed in further distal areas of the Denwa floodplain are less well developed than expected. The monotonic proximal–distal trend with respect to a particular channel was perturbed by contribution of sediment and water from more than one coexisting channel. Variations in sediment aggradation rate, overbank geomorphology, and drainage condition controlled the paleosol character as well as their degree of development in overbank sites.

The spectrum of variations in lithofacies and paleosol association discussed here is expected in deposits of a muddy anabranching fluvial system that has very wide low-gradient overbank areas and ephemeral discharge characters and operates under a subhumid to semiarid climatic setting. The lithofacies and pedofacies associations for other fluvial types and for a different climatic setting are likely to be different from those of the upper part of the Denwa Formation.

ACKNOWLEDGMENTS

This work is a part of integrated research initiative of Geological Studies Unit in the Satpura Gondwana basin and is funded by the Indian Statistical Institute. We thank Ms. Sampa Ghosh for her help in petrographic studies under microscope and Mr. Ajay Das for drafting the illustrations. We are grateful to the reviewers, Brenda Buck and Susan Marriott, and the editors for their constructive suggestions.

REFERENCES

Allen, J.R.L., 1973, Compressional structures (Pattern Ground) in Devonian pedogenic limestones: Nature, Physical Science, v. 243, p. 84–86.

Allen, J.R.L., 1974, Studies in fluviatile sedimentation: implications of pedogenic carbonate units, lower Old Red Sandstone, Anglo-Welsh outcrop: Geological Journal, v. 9, p. 181–208.

Allen, J.R.L., 1986, Pedogenic calcretes in the Old Red Sandstone facies (Late Silurian–Early Carboniferous) of the Anglo-Welsh area, southern Britain, in Wright, V.P., ed., Paleosols: Their Recognition and Interpretation: Oxford, U.K., Blackwell Scientific Publications, p. 58–86.

Allen, J.R.L., and Williams, B.P.J., 1979, Interfluve drainage on Silurian–Devonian alluvial plains in Wales and the Welsh Borders: Geological Society of London, Journal, v. 136, p. 361–366.

Aslan, A., and Blum, M.D., 1999, Contrasting styles of Holocene avulsion, Texas Gulf Coastal Plain, U.S.A., in Smith, N.D., and Rogers, J., ed., Fluvial

Sedimentology, VI: International Association Sedimentologists, Special Publication 28, p. 193–209.

BANDYOPADHYAY, S., AND SENGUPTA, D.P., 1999, Middle Triassic vertebrates of India: Journal of African Earth Science, v. 29A, p. 233–241.

BLODGETT, R.H., CRABAUGH, J.P., AND MCBRIDE, E.F., 1993, The color of red beds—a geologic perspective, in Bigham, J.M., and Ciolkosz, E.J., eds., Soil Color: Soil Science Society of America, Special Publication 31, p. 127–158.

BOWN, T.M., AND KRAUS, M.J., 1987, Integration of channel and floodplain suites, I. Developmental sequence and lateral relations of alluvial paleosols: Journal of Sedimentary Petrology, v. 37, p. 587–601.

BRIDGE, J.S., 1984, Large-scale facies sequences in alluvial overbank environments: Journal of Sedimentary Petrology, v. 54, p. 583–588.

BRIDGE, J.S., 1993, The interaction between channel geometry, water flow, sediment transport and deposition in braided rivers, in Best, J.L., and Bristow, C.S., eds., Braided Rivers: Geological Society of London, Special Publication 75, p. 13–71.

BRIDGE, J.S., 2003, Rivers and Floodplains; Forms, Processes and Sedimentary Record: Oxford, U.K., Blackwell Science Ltd., 504 p.

BROOKS, G.R., 2003, Alluvial deposits of a mud-dominated stream: the Red River, Manitoba, Canada: Sedimentology, v. 50, p. 441–458.

CASSHYAP, S.M., TEWARI, R.C., AND KHAN, A., 1993, Alluvial fan origin for Bagra Formation (Mesozoic Gondwana) and tectono-stratigraphic implications: Geological Society of India, Journal, v. 42, p. 262–279.

CHAKRABORTY, C., MANDAL, N., AND GHOSH, S.K., 2003, Kinematics of the Gondwana basins of peninsular India: Tectonophysics, v. 377, p. 299–324.

CHAKRABORTY, C., AND GHOSH, S.K., 2005, Pull-apart origin of the Satpura Gondwana basin, central India: Journal of Earth System Science, v. 114(3), p. 259–273.

CHAKRABORTY, T., AND SARKAR, S., 2005, Evidence of lacustrine sedimentation in the Upper Permian Bijori Formation, Satpura Gondwana basin: Palaeogeographical and tectonic implications: Journal of Earth System Science, v. 114 (3), p. 303–323.

CROOKSHANK, H., 1936, Geology of the northern slopes of the Satpura between the Morand and Sher rivers: Geological Survey of India, Memoirs, v. 66(2), p. 173–381.

DRIESE, S.G., AND MORA, C.I., 1993, Physico-chemical environment of pedogenic carbonate formation in Devonian vertic paleosols, Central Appalachians, USA: Sedimentology, v. 40, p. 199–216.

EKES, C., 1993, Bedload transported pedogenic mud aggregates in the Lower Old Red Sandstone in southwest Wales: Geological Society of London, Journal, v. 150, p. 469–472.

GHOSH, P., 1997, Geomorphology and paleoclimatology of some Upper Cretaceous paleosols in central India: Sedimentary Geology, v. 110, p. 25–49.

GHOSH, P., GHOSH, P., AND BHATTACHARYA, S.K., 2001, CO_2 in the Late Paleozoic and Mesozoic atmosphere from soil carbonate and organic matter, Satpura basin, central India: Palaeogeography, Palaeoclimatology, Palaeoecology, v. 170, p. 219–270.

GHOSH, P., SARKAR, S., AND MAULIK, P., 2006, Sedimentology of a muddy alluvial deposit: Triassic Denwa Formation, India: Sedimentary Geology, v. 191, p. 3–36.

GIBLING, M.R., NANSON, G.C., AND MAROULIS, J.C., 1998, Anastomosing river sedimentation in the Channel Country of central Australia: Sedimentology, v. 45, p. 595–619.

GILE, L.H., PETERSON, F.F., AND GROSSMAN, R.B., 1966, Morphological and genetic sequences of carbonate accumulation in desert soils: Soil Science, v. 101, p. 347–360.

GÓMEZ-GRAS, D., AND ALONSO-ZARZA, A.M., 2003, Reworked calcretes: their significance in the reconstruction of alluvial sequences (Permian and Triassic, Minorca, Balearic Islands, Spain): Sedimentary Geology, v. 158, p. 299 319.

HUANG, H.Q., AND NANSON, G.C., 2007, Why some alluvial rivers develop an anabranching pattern: Water Resources Research, v. 43, no. W07741, 10.1029/2006WR005223.

KIRSCHBAUM, M.A., AND MCCABE, P.J., 1992, Controls on the accumulation of coal and on the development of anastomosed fluvial systems in the Cretaceous Dakota Formation of southern Utah: Sedimentology, v. 39, p. 581–598.

KRAUS, M.J., 1997, Lower Eocene alluvial paleosols: Pedogenic development, stratigraphic relationships, and paleosol/landscape associations: Palaeogeography, Palaeoclimatology, Palaeoecology, v. 129, p. 387–406.

KRAUS, M.J., 1999, Paleosols in clastic sedimentary rocks: their geologic applications: Earth-Science Reviews, v. 47, p. 41–70.

KRAUS, M.J., 2002, Basin-scale changes in floodplain paleosols: Implications for interpreting alluvial architecture: Journal of Sedimentary Research, v. 72, p. 500–509.

KRAUS, M.J., AND ASLAN, A, 1999, Palaeosol sequences in floodplain environments: a hierarchical approach, in Smith, N. D., and Rogers, J., eds., Fluvial Sedimentology VI: International Association Sedimentologists, Special Publication, 28, p. 303–321.

KRAUS, M.J., AND DAVIES-VOLLUM, K.S., 2004, Mudrock-dominated fills formed in avulsion splay channels: examples from the Willwood Formation, Wyoming: Sedimentology, v. 51, p. 1127–1144.

KRAUS, M.J., AND GWINN, B., 1997, Facies and facies architecture of Paleogene floodplain deposits, Willwood Formation, Bighorn Basin, Wyoming, USA: Sedimentary Geology, v. 114, p. 33–54.

LEEDER, M.R., 1975, Pedogenic carbonates and flood sediment accretion rates: A quantitative model for alluvial arid-zone lithofacies: Geological Magazine, v. 112, p. 257–270.

MARRIOTT, S.B., AND WRIGHT, V.P., 1993, Palaeosols as indicators of geomorphic stability in two Old Red Sandstone alluvial suites, South Wales: Geological Society of London, Journal, v. 150, p. 1109–1120.

MARRIOTT, S.B., AND WRIGHT, V.P., 1996, Sediment recycling on Silurian–Devonian floodplains: Geological Scociety of London, Journal, v. 153, p. 661–664.

MAULIK, P.K., CHAKRABORTY, C., GHOSH, P., AND RUDRA, D., 2000, Meso and macro scale architecture of a Triassic fluvial succession: Denwa formation, Satpura Gondwana basin, M. P.: Geological Society of India, Journal, v. 56, p. 489–504.

MIALL, A.D., 1985, Architectural-element analysis: a new method of facies analysis applied to fluvial deposits: Earth-Science Reviews, v. 21, p. 261–308.

MIALL, A.D., 1996, The Geology of Fluvial Deposits: Berlin, Springer, 582 p.

MOHRIG, D., HELLER, P.L., PAOLA, C., AND LYONS, W.J., 2000, Interpreting avulsion process from ancient alluvial sequences: Guadalope–Matarranya system (northern Spain) and Wasatch Formation (western Colorado): Geological Society of America, Bulletin, v. 112, p. 1787–1803.

MOROZOVA, G.S., AND SMITH, N.D., 1999, Holocene avulsion history of the lower Saskatchewan fluvial system, Cumberland Marshes, Saskatchewan–Manitoba, Canada, in Smith, N.D., and Rogers, J., eds., Fluvial Sedimentology VI: International Association of Sedimentologists, Special Publication 28, p. 231–249.

MOROZOVA, G.S., AND SMITH, N.D., 2000, Holocene avulsion styles and sedimentation patterns of the Saskatchewan River, Cumberland Marshes, Canada: Sedimentary Geology, v. 130, p. 81–105.

MUÑOZ, A., RAMOS, A., SÁNCHEZ-MOYA, AND Y., SOPEÑA, A., 1992, Evolving fluvial architecture during a marine transgression: Upper Buntsandstein, Triassic, central Spain: Sedimentary Geology, v. 75, p. 257–281.

NANSON, G.C., AND CROKE, J.C., 1992, A genetic classification of floodplains: Geomorphology, v. 4, p. 459–486.

NANSON, G.C., AND KNIGHTON, A.D., 1996, Anabranching rivers: Their cause, character and classification: Earth Surface Processes and Landforms, v. 21, p. 217–239.

NEWELL, A.J., TVERDOKHLEBOV, V.P., AND BENTON, M.J., 1999, Interplay of tectonics and climate on a transverse fluvial system, Upper Permian, southern Uralian foreland basin, Russia: Sedimentary Geology, v. 127, p. 11–29.

NORTH, C.P., NANSON, G.C., AND FAGAN, S.D., 2007, Recognition of the sedimentary architecture of dryland anabranching (anastomosing) rivers: Journal of Sedimentary Research, v. 77, p. 925–938.

PAGE, K.J., NANSON, G.C., AND FRAZIER, P.S., 2003, Floodplain formation and sediment stratigraphy resulting from oblique accretion on the Murrumbigee River, Australia: Journal of Sedimentary Research, v. 73, p. 5–14.

PASCOE, E. H., 1959, A Manual of the Geology of India and Burma, Third Edition: Government of India Press, Calcutta, 2130 p.

PIUGDEFABREGAS, C., AND VAN VLIET, A., 1978, Meandering stream deposits from the Tertiary of the southern Pyrenees, in Miall, A.D., ed., Fluvial

Sedimentology: Canadian Society of Petroleum Geologists, Memoir 5, p. 469–485.

Ray, S., and Chakraborty, T., 2002, Lower Gondwana fluvial succession of the Pench–Kanhan valley, India: stratigraphic architecture and depositional controls: Sedimentary Geology, v. 151, p. 243–271.

Reid, I., and Frostick, L.E., 1985, Towards a better understanding of bedload transport, in Ethridge, F.G., Flores, M., and Harvey, M. D., eds., Recent Developments in Fluvial Sedimentology: SEPM, Special Publication 39, p. 13–19.

Retallack, G.J., 2001, Soils of the Past; An Introduction to Paleopedology: Oxford, U.K., Blackwell Science Ltd., 404 p.

Rust, B.R., 1981, Sedimentation in an arid-zone anastomosing fluvial system: Cooper's Creek, Central Australia: Journal of Sedimentary Petrology, v. 51, p. 745–755.

Sarkar, S., 1988, Petrology of caliche-derived peloidal calcirudite/calcarenite in the Late Triassic Maleri Formation of the Pranhita–Godavari valley, south India: Sedimentary Geology, v. 55, p. 263–282.

Schumm, S.A., 1977, The Fluvial System: New York, Wiley, 338 p.

Singh, I.B., 1996, Geological evolution of Ganga plain—an overview: Palaeontological Society of India, Journal, v. 41, p. 99–137.

Slingerland, R., and Smith, N.D., 2004, River avulsions and their deposits: Annual Review of Earth Planetary Sciences, v. 32, p. 257–285.

Soil Survey Staff, 1975, Soil taxonomy: a basic system of soil classification for making and interpreting soil surveys: Soil Conservation Survey Handbook, v. 436, U.S. Department of Agriculture, Washington DC.

Stear, W., 1983, Morphological characteristics of ephemeral stream channel and overbank splay sandstone bodies in the Permian Lower Beaufort Group, Karoo Basin, South Africa: International Association of Sedimentologists, Special Publication 6, p. 405–420.

Stewart, D.J., 1983, Possible suspended-load channel deposits from Wealden Group (Lower Cretaceous) of Southern England: International Association of Sedimentologists, Special Publication 6, p. 369–384.

Taylor, G., and Woodyer, K.D., 1978, Bank deposition in suspended-load streams, in Miall, A.D., ed., Fluvial Sedimentology: Canadian Society of Petroleum Geologists, Memoir 5, p. 257–275.

Thomas, R.G., Smith, D.G., Wood, J.M., Visser, J., Calverley-Range, E.A., and Koster, E.H., 1987, Inclined heterolithic stratification—terminology, description, interpretation and significance: Sedimentary Geology, v. 53, p. 123–179.

Willis, B.J., and Behrensmeyer, A.K., 1994, Architecture of Miocene overbank deposits in northern Pakistan: Journal of Sedimentary Research, v. B64, p. 60–67.

Woodyer, K.D., Taylor, G., and Crook, K.A.W., 1979, Depositional processes along a very low-gradient, suspension–load stream: the Barwon River, New South Wales: Sedimentary Geology, v. 22, p. 97–120.

Wright, V.P., and Marriott, S.B., 1996, A quantitative approach to soil occurrence in alluvial deposits and its application to the Old Red Sandstone of Britain: Geological Society of London, Journal, v. 153, p. 907–913.

Yaalon, D.H., and Kalmar, D., 1978, Dynamics of cracking and swelling clay soils: Displacement of skeletal grains, optimum depth of slickensides, and rate of intra-pedonic turbation: Earth Surface Processes, v. 3, p. 31–42.

STRATIGRAPHY OF COUNTER-POINT-BAR AND EDDY-ACCRETION DEPOSITS IN LOW-ENERGY MEANDER BELTS OF THE PEACE-ATHABASCA DELTA, NORTHEAST ALBERTA, CANADA

DERALD G. SMITH
Department of Geography, University of Calgary, Calgary, AB, T2N 1N4, Canada
Email: dgsmit@ucalgary.ca

STEPHEN M. HUBBARD
Department of Geoscience, University of Calgary, Calgary, AB, T2N 1N4, Canada

JASON R. LAVIGNE
Bergshrund Integrated Sedimentologic Solutions, 2104 3rd Avenue NW, Calgary, AB, T2N 0K4, Canada

DALE A. LECKIE, AND MILOVAN FUSTIC
801-7th Ave. SW, Nexen Inc., Calgary, AB, T2P 3P7, Canada

ABSTRACT: Previously termed concave bank-bench deposits, point-bar-tail deposits, and distal point-bar deposits, counter-point-bar deposits have received little research attention, and their stratigraphy and sedimentology is poorly understood in modern and ancient river meander belts. The stratigraphy and lateral continuity of counter-point-bar deposits were studied in the modern Peace River Delta, northeast Alberta, Canada. Morphology of counter-point-bar deposits was identified as having concave-shaped scroll-bar and vegetation patterns, and they are always located immediately downriver from sandy point bars. Stratigraphy of counter-point-bar deposits, composed of 55 to 86% silt, were studied using a vibracorer to depths of 6.5 m and channel-bottom grab samples to depths of 20 m over a river distance of 1800 m. This contrasts with 95 to 100% sand in adjacent point-bar deposits. Like point bars, counter point bars are also lateral-accretion deposits, but are composed of inclined heterolithic strata. Counter-point-bar deposits can form in short-radius meanders, as well as in wide-radius broad meanders. Counter-point-bar deposits formed where channels impinge at low angles (10 to 40°) against bedrock valley sides or other erosion resistant sediment.

Within the Athabasca River Delta, a different feature called an eddy-accretion deposit, was studied and compared to counter-point-bar deposits. Eddy-accretion deposits formed as short-radius, pronounced concave scroll-bar and vegetation patterns that notably arc up-valley. In contrast to the counter-point-bar deposits analyzed, these deposits are thicker, sandier, and not as laterally extensive for a channel of a given size. Eddy-accretion deposits formed where river flow impinges more directly against (40°-140°) a valley side or channel margin composed of bedrock or resistant sediment such as an oxbow-lake fill, abandoned-channel fill or, as in this case, a Holocene prodelta deposit. The eddy-accretion deposit studied is up to 500 m wide and 24 m thick, although probably closer to 34 m thick during near-bankfull discharge. It is composed of up to 80% sand.

KEY WORDS: counter-point-bar deposits, point-bar deposits, eddy-accretion deposits, meander belt stratigraphy

INTRODUCTION

This paper builds on an ongoing research program focused on understanding counter-point-bar deposits (CPBD) (Smith et al., 2009). Smith et al. (2009) examined the stratigraphy, sedimentology, and spatial relationship of CPBD to point-bar deposits (PBD) in a river reach with sharp meander bends (Fig. 1). That study area was located 210 km up the Peace River from the Peace River Delta where the research for this paper was carried out. The primary objective in this paper is to determine if a similar CPBD-PBD relationship occurs within a low arcing gentle meander bend (Fig. 2). A secondary objective was to examine an eddy-accretion deposit (EAD) in a large low-gradient meander bend in the lower Athabasca River, located within the Athabasca River Delta, and compare it morphologically and sedimentologically with the CPBD studied in the nearby Peace Delta (Figs. 1, 2). Both the Peace and Athabasca deltas have merged to become one enormous channel and wetland complex, and when combined is called the Peace-Athabasca Delta (PAD), one of the largest lacustrine deltaic complexes in the world (Fig. 1).

In large, low-gradient, meandering rivers, CPBDs are different from PBDs (Fig. 2). In planview, point bars have convex-shaped scroll-bar morphologies, in contrast to counter point bars, which form concave-shaped patterns. The latter are always located immediately downriver from and adjacent to point bars (Fig. 2). CPBDs are particularly apparent in meandering rivers that tend to migrate down regional dip or down-valley (down-valley translation), rather than those meanders that migrate towards the valley sides (lateral translation, Daniel, 1971). Also, counter point bars tend to be slightly lower (1–2 m) in elevation than adjacent point bars and consist of finer-grained sediment (Nanson and Page, 1983; Hickin, 1986; Makaske and Weerts, 2005; Smith et al., 2009).

There has been a slow evolution of thought about CPBDs. The term concave-bank bench was initially used to describe concave scroll-bar morphology located at sharp bends in modern meandering rivers (Taylor, 1971; Woodyer *et al.*, 1979; Hickin, 1979, 1986; Page and Nanson, 1982; Page, 1983; Nanson and Page, 1983; Nanson and Croke, 1992), and thought to be caused by a hydraulic separation zone consisting of a train of small reverse eddy currents moving upstream along the deposit during bankfull flow (Taylor *et al.*, 1971). The term concave bank-bench deposit was difficult to conceptualize for most academic, industry, and government people, and thus its extent and significance was never fully understood, recognized, nor accepted.

From River To Rock Record: The Preservation Of Fluvial Sediments And Their Subsequent Interpretation
SEPM Special Publication No. 97, Copyright © 2011
SEPM (Society for Sedimentary Geology), ISBN 978-1-56576-305-0, p. 143–152.

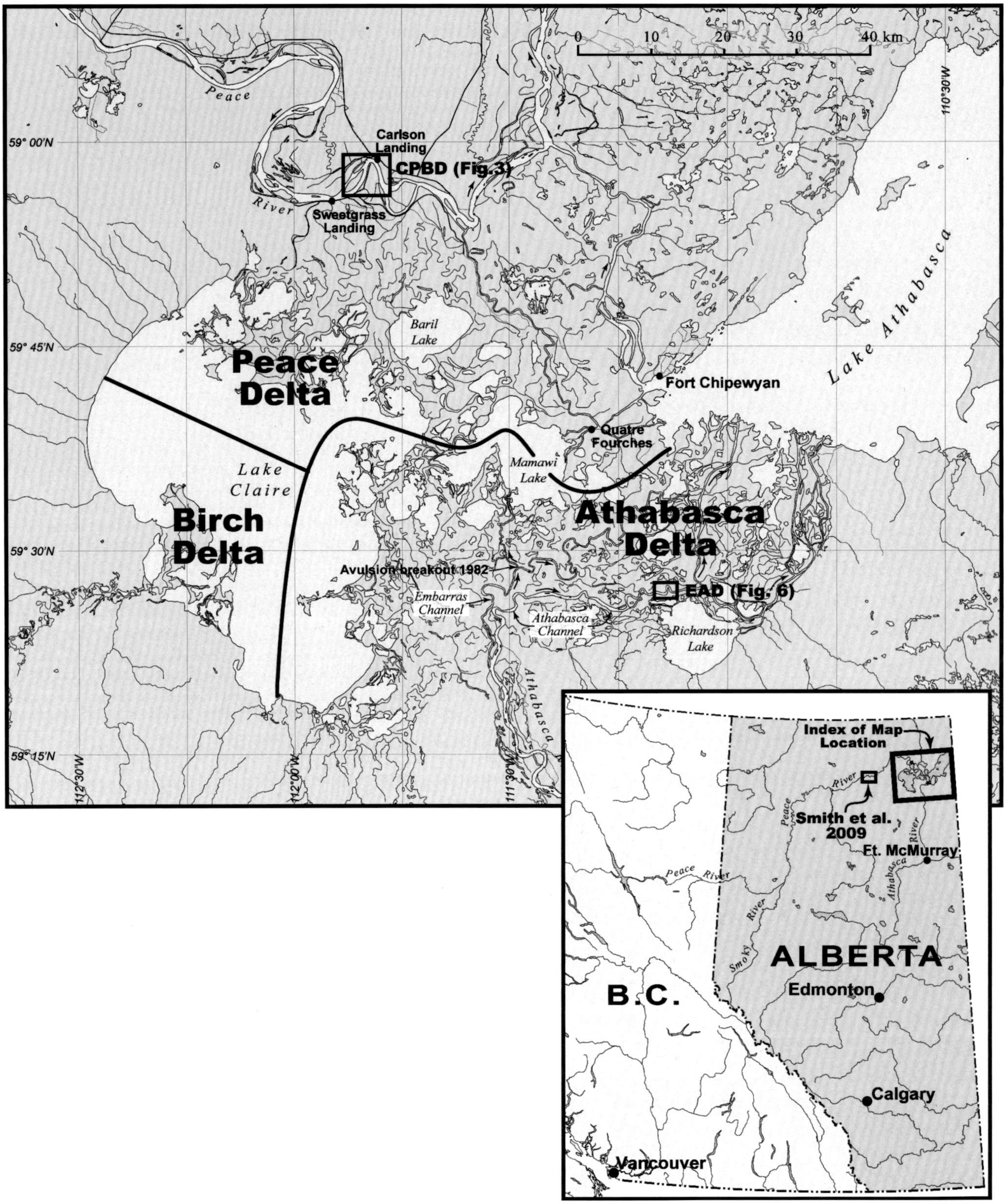

FIGURE 1.—Map of the Peace-Athabasca Delta (PAD) and the locations of the two meander sites studied, one exhibiting a counter point bar deposit on the Peace River and the other an eddy accretion deposit on the Athabasca River, NE Alberta, Canada.

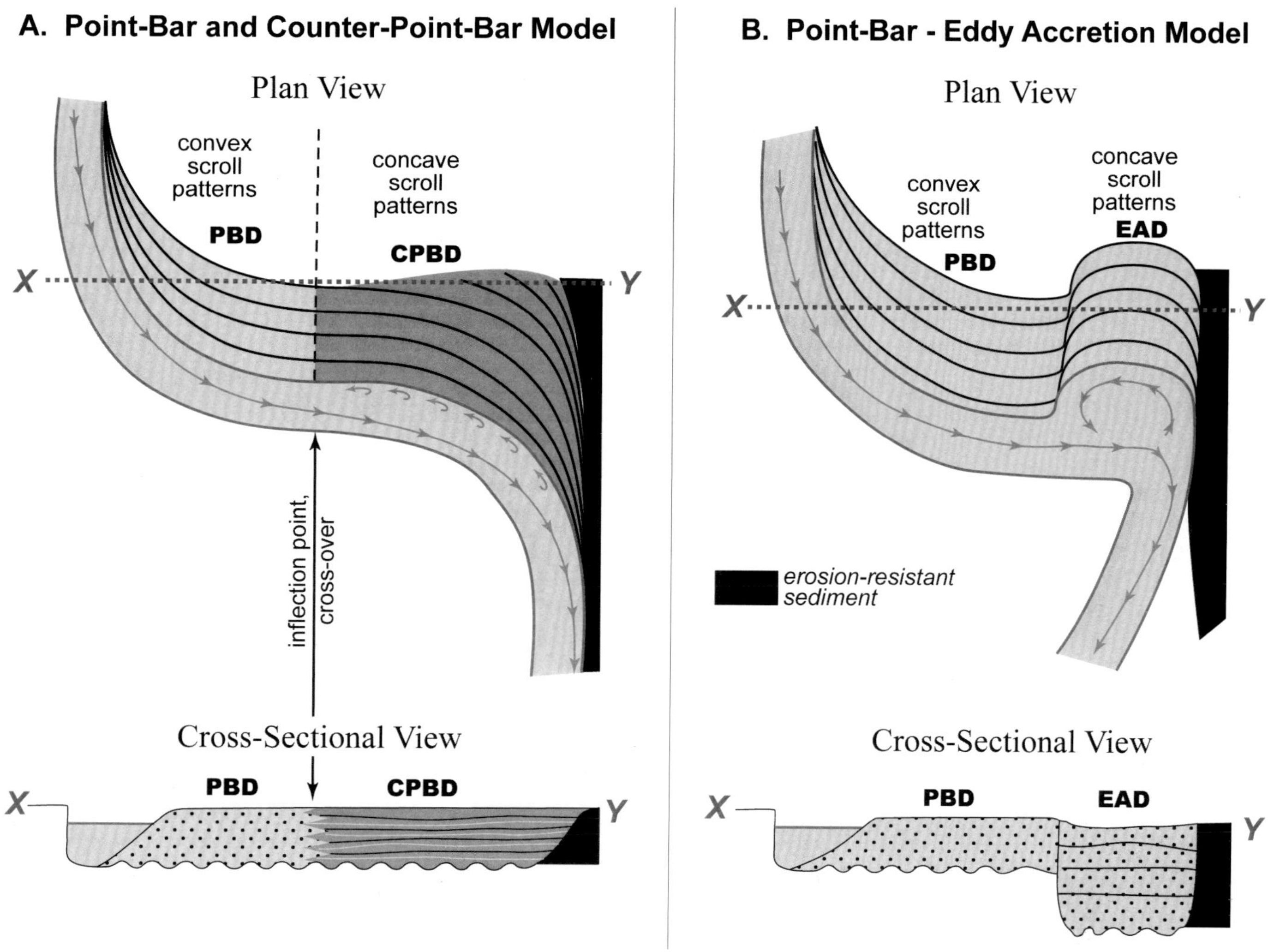

FIGURE 2.—Schematic plan-form view comparing the scroll bar morphology of a counter point bar deposit with an eddy accretion deposit and how both are laterally restricted by erosion resistant finer-grained sediment.

Another term, counterpoint deposits, was introduced by Lewin (1983). Later, counterpoint bar and counterpoint accretion terms were used to describe lateral-accretion morphology and processes (Nanson and Croke, 1992; Makaske and Weerts, 2005).

An eddy-accretion deposit (EAD) (Carey, 1969; Burge and Smith, 1999) shares morphologic characteristics with counter point bars, but is stratigraphically thicker. EADs form where river flow impinges more directly (90°+ 50°) against a valley side composed of resistant sediment such as bedrock, an oxbow-lake fill, abandoned-mud-channel fill, and deltaic deposits consisting of clay and silt. In the Athabasca River of the PAD, river impingement into Holocene deltaic silt and clay cause the EAD that was studied. When river flow is forced make an abrupt bend, about a third of the flow separates to form a large reverse eddy current and scour pool on the upstream side of the obstacle (Carey, 1969; Burge and Smith, 1999). As the meander and EAD slowly migrate down-valley, a sharp concave bend or scroll pattern is left as the surface expression of the process (Fig. 2). More importantly, an unusually thick sand, up to twice as thick as nearby point-bar sand, infills the scour pool and migrates down-valley (Burge and Smith, 1999). Morphologically, counter point bars and eddy accretions look similar in plan view (Fig. 2), but compositionally they are quite different, with CPBDs dominated by silt and EADs dominated by sand. Both counter point bars and eddy accretions are influenced by river impingement on resistant sediment along channel margins, at a low angle of 10 to 40° between the channel and resistant sediment body, and a relatively high angle of 40 to 140°, respectively.

STUDY AREAS

Sampling was carried out in two areas of the PAD in Wood Buffalo National Park, located in northeast Alberta, Canada. The first field site was selected along a 3.4 km reach immediately upriver from Carlson's Landing of the lower Peace River in the PAD (Fig. 1). The second PAD site was located on the main distributary channel of the lower Athabasca River, 50 km upriver from Lake Athabasca (Fig. 1). Both study sites were examined in the first two weeks of July 2008.

The single-channel Peace River system was discharging 1900 $m^3 s^{-1}$ (Environment Canada, 2008). On the basis of the 10-year return period of discharge (commonly used in Alberta to approximate bankfull discharge), bankfull flow was estimated to be 12,000 $m^3 s^{-1}$ (Kellerhals

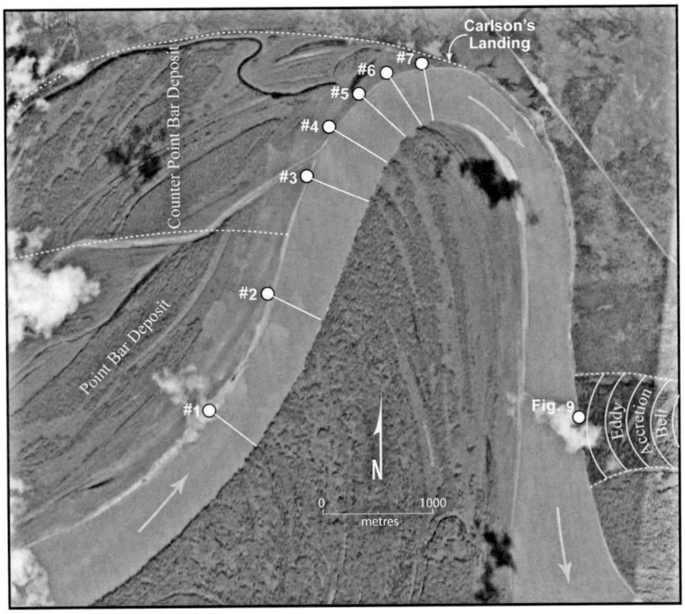

FIGURE 3.—Satellite image from Google Earth™ showing the channel reach of counter point bar deposits studied along a gently arcing meander bend of the Peace River in the Peace River Delta, NE Alberta. Locations of vibracores, channel cross-sections and channel bed sediment sampling sites of the counter point bar deposit are indicated.

et al., 1972). Channel-bed and point-bar sediment consisted of sand, commonly in the fine to medium range with occasional coarse grains. Suspended sediment was generally high, with a maximum measured load of 14,600 metric tonnes/day (Environment Canada, 1980). Seasonal humid, brief warm summers and long cold winters influence discharge. With its headwaters on the eastern slope of the Canadian Rocky Mountains, the Peace River watershed, upriver from the study area, covers 300,000 km^3 (Environment Canada, 1991). The major source of peak flow comes from spring snowmelt-induced ice-jam floods and summer rainfall floods from the Smoky River tributary, with headwaters located in northwestern Alberta (Smith, 2003). Ice-jam floods have been shown to cause the highest river stage, which in turn causes unusually large amounts of overbank sedimentation in the PAD (Smith, 2003). Water from the Peace eventually enters the Arctic Ocean via the Slave and Mackenzie rivers.

Channel width is rather uniform in the study reach, varying from 432 to 482 m, and maximum bankfull depth is 20 m. However, some channel depths may erode deeper during bankfull discharge and floods. River gradient is estimated at 0.00004 (4 cm/km, Kellerhals et al., 1972), a value that may be an overestimate since it was measured 500 km upriver from the PAD, and a value closer to 0.00002 may be more realistic.

The Athabasca River watershed, upstream of the delta, covers 60,000 km^2, about 20% of that of the Peace River watershed. The Athabasca River study site is located approximately midway between the delta front and the delta apex, 50 km in each direction on the main branch (Athabasca) of one of two major distributary channels. In the second week of July 2008, flow discharge in all of the Athabasca River distributaries was 1090 m^3s^{-1} (Environment Canada, 2008). The main stem of the Athabasca carried an estimated 60% (654 m^3s^{-1}) of the total discharge. In Alberta bankfull discharge closely approximates the 10-year flood, estimated at 4060 m^3 s^{-1} (Kellarhals et al., 1972). Thus, bankfull discharge on the main stem of the Athabasca was estimated at 2436 m^3s^{-1}.

River slope in the Athabasca River distributary study reach is 0.00003 (3 cm/km, Kellerhalls et al., 1972). During the study (July 12 and 13, 2008), channel width varied from 145 to 205 m, maximum depth measured 24 m below bankfull discharge, and the water surface was 2 m below bankfull. The studied meander bend was attached at each end to a nearly straight channel, with a sinuosity index of about 1.3.

METHODS

Vibracoring was used to recover sediment cores from the subsurface of the CPBD and EAD (Smith, 1984, 1992), following the same research design as Smith et al. (2009). Coring sites on the Peace River were 5 to 10 m from the channel edge on surfaces at least 1 m above the river level (Fig. 3). Some sites could not be accessed due to insufficient bearing strength of the saturated silt. On the Athabasca River, cores were taken mainly from an island surface 30 cm above the river level. Once coring tubes were hoisted from the subsurface, sediment cores were vibrated out of the tubes (7.5 cm diameter aluminum irrigation pipes) into plastic-pipe receiving troughs. Cores in the troughs were cut in half and logged stratigraphically for grain size. Initially, an American/Canadian Stratigraphic grain size chart and 20-power field microscope were used to compare sand grain-size. The presence of clay was determined in the field by using the "rolling into ribbons test"; however, no significant amounts were present. In the laboratory, grain size analysis of silt was carried out on a Mastersizer laser particle size analyzer (manufactured by Malvern Instruments) to confirm field identification of silt in core and channel bottom samples.

Channel-bottom sediments were sampled and river depths measured from a boat using a Lane bottom drag sampler and an Eagle Optima (Eagle Electronics, Catossa, Oklahoma, USA) non-recording echo sounder. Six to eight sites along each cross section of the channel bottom were sampled. Horizontal distances from the shore to the riverbed sampling points were measured using an Impulse Laser Tech XL lidar range finder.

At each selected CPBD and EAD sample site, cross-channel distances were surveyed using the lidar range finder. Channel cross-sectional profiles along the CPBD reach were spaced 175 to 790 m apart, and the EAD sample sites were 80 to 125 m apart. Vertical measurements were taken with the lidar range finder in vertical mode to record the core sites relative to bankfull heights. Latitude and longitude of core points was determined with a global positioning system (GPS).

The regional base maps for the study were modified from landsat imagery downloaded from Google Earth™. Data on channel cross-section location data were transferred onto copied satellite images (courtesy Google Earth™)). Channel cross-sectional profiles were plotted to show channel width, depth, locations of sediment sample sites, and sediment grain size.

RESULTS FROM COUNTER-POINT-BAR DEPOSITS (CPBDs)

Along the selected meander-bend research site in the Peace River Delta, 210 km downriver from the site studied by Smith et al. (2009), seven vibracores were recovered (Figs. 3, 4). The two shallowest cores varied from 2.7 to 4.2 m deep and were taken from the adjacent PBD (Fig. 3). Five vibracores to depths varying between 6.4 and 4.8 m were extracted from the CPBD.

A sand-dominated lithofacies was present in core #1, taken from the point bar. Though only 2.7 m deep, coarse sand dominated from 1.4 to 2.7 m, suggesting that coarse sand extends through the full PBD

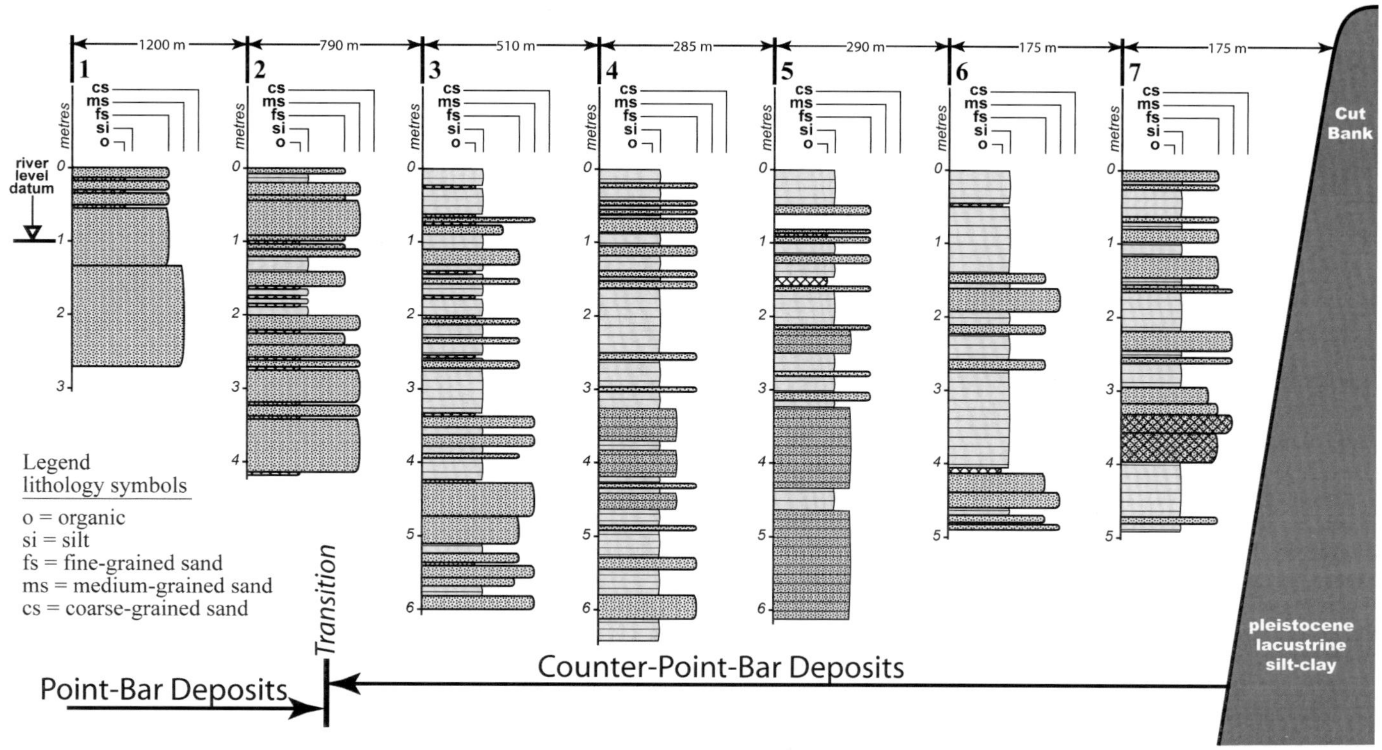

FIGURE 4.—Lithostratigraphic profiles of 7-vibracores taken from point bar and counter point bar deposits along the left bank of the Peace River (see Fig. 3 for locations). Note that the grain size trend fines in the downstream direction from the point bar deposits to the counter point bar deposit.

stratigraphy to a probable depth of 17 to 18 m (Figs. 4, 5). The coarse-grained sand was probably rafted into the reach by river ice during spring breakup. In core #2, located 1200 m farther downriver, medium-grained sand extends from 2.4 to 4.2 m, and probably extends to a depth of at least 14 to 15 m or probably greater (Fig. 5). Sediment in core #2 is interpreted to also represent a sandy PBD. Somewhere in the 790 m channel distance between cores #2 and #3 is a transition from PBD to CPBD (Figs. 3, 4). Whether or not the lithofacies transition is gradual or rapid is unknown and needs further research attention.

The silt-dominated CPBD covers a distance of at least 1435 m from between cores #2 and #3 to beyond core #7 to the valley-side cut-bank. By extending the CPBD to midway between cores # 2 and 3, an additional 400 m, it is possible that it covers a distance of 1835 m (Figs. 3, 4). Within the five silt-dominated cores (3 to 7), the percentage of net-sand-to-gross stratigraphy (N/G) varies from 14 to 45%. This CPBD sequence consists of irregular silt beds, up to 1.4 m thick, with occasional thin lenses of fine and medium sand, often of millimeter thickness, but locally up to 30 cm thick. Sometimes thin laminae of organic detritus up to 1.0 mm in thickness are present. Sedimentary structures within the silt and sand beds consist of nearly horizontal laminations (possibly dipping towards the channel) and occasional sequences of current ripples. From grain-size analysis, very fine and fine sand mixed within the silt beds amounted to 5 to 10 % of the total sediment volume. This proportion of silt to fine sand was typical of all samples taken from all the five CPBD cores. This contrasts with ripple and trough cross-stratified medium and coarse-grained sand-dominated lithofacies within PBDs, which consist of N/G of 95 to 100% (estimates from trench samples, vibracores, and channel-bottom grab samples; Fig. 4). In these cores of limited length (2.4 to 4.2 m), it would appear the PBD lacked the classic fining-up succession with a silt/clay overbank cap. However, future lateral accretion will likely produce a fining-up trend. Clay-size particles were absent, probably reflecting insufficient settling velocities and time in flowing water, while most medium sand grains and all coarse sand grains were too heavy to stay in suspension and were transported only as bed load.

Though not proven in this study, sand interbeds or lenses within CPBD silt-dominant facies are probably dipping towards the channel at about the same angles as are shown in the underwater slopes, averaging 8.5°, ranging from 6.5 to 10.8° (Fig. 5). On the basis of silt present in grab samples taken from the channel bed, it is likely that silt extends to the deepest parts of the channel, as shown in cross-sectional profiles 5 and 6 (Fig. 5). The highest velocity flow must occur on the opposite side of the channel from the CPBD during deposition.

DISCUSSION OF CPBDs

In Figures 4 and 5, a sandy PBD is shown in vibracore logs 1 and 2, with a transition to CPBD present between cores 2 and 3, and a corresponding transition to inclined heterolithic stratification (IHS; Thomas et al., 1987). IHS present in vibracores 3 through 7 from the CPBD reach extends over a horizontal distance of at least 1435 m, and possibly an additional 400 m to half way between cores 2 and 3, totaling 1835 m. IHS within the CPBD likely extends to depths of nearly 22 m, tentatively confirmed by channel cross-sectional profile # 7 and channel-bottom silt sediment samples (Fig. 5). All vibracores were taken from a surface approximately one meter above river level. Most river depths within the CPBD zone, varied from 8.5 to 17 m, not including the one meter of elevation above river level from which vibracores were taken (Fig. 5). The data presented in Figure 5 show the river deepening as it moves from proximal to distal along the CPBD.

The subaqueous slope angle of the CPBD in this study varies from 6.5° to 10.8°, and averaging 8.5°, which should be approximately equivalent to dip angles present in the IHS of the CPBD (Fig. 5). Smith et al. (2009) reported a steepening trend downriver along a CPBD within sharp bends, and a similar trend was noted in this study, with the exception of cross section 7. Within the PBD zone, the underwater slope angle was 3.5°, much less than expected. Compared to underwater CPBD slope angles of up to 22° in the tightly spaced meanders (low-radius bends) in the study by Smith et al. (2009), underwater slope angles the PAD were much lower. This raises the possibility that lower-angle IHS dips in ancient rock sequences may be associated with broad meander bends and steeper dips with tighter meander bends.

Lithostratigraphy in the Peace River CPBD consists of IHS, which until recently was demonstrated only in meandering tidal-estuary channel point-bar deposits (Smith 1988), though its presence in meandering rivers was suggested by Jackson (1981) and Calverley (1984). In vibracores 3 to 7 (Fig. 4), IHS varies between silt and medium sand, but mostly fine sand. The sand percentage or N/G varied considerably, with values of 45%, 40%, 14%, 26% and 45%, from cores 3,4,5,6, and 7, respectively. These numbers do not include the 5 to 10% sand mixed within silt beds. The data mimic the CPBD results studied earlier by Smith et al. (2009) from the site 210 km upriver.

It seems that in large, low-gradient rivers, CPBDs are characterized by silt-dominated IHS that contrasts with the more homolithic sand-dominant facies present in adjacent PBD. The likely 22 m (cross section #7 of Fig. 5) of IHS will subsequently be capped by up to 5 m of overbank silt and clay, deposited mostly by ice-jam floods (Hugenholtz et al., 2009; Smith et al., 2009). Overbank deposits from ice-jam floods tend to be several times thicker than those deposited from non-ice-jam-induced overbank flooding.

A final question is: can CPBDs be caused by flow impinging on an older CPBD, acting as an erosion- resistant barrier? The proportion of a CPBD in a low-energy meander belt may increase with meander-belt age, because CPBDs tend to be preferentially preserved at the expense of PBDs. Although not proven in this study, this implication perhaps deserves further research.

RESULTS FROM EDDY-ACCRETION DEPOSITS (EADs)

Less common than CPBDs, EADs located in a meandering belt in the Athabasca River in the Athabasca Delta sector of the PAD were also investigated (Figs. 1, 6). The surface morphology of EAD exhibits a more pronounced concave scroll pattern and a much-reduced radius of arc than are present in CPBDs (Fig. 2).

In terms of thickness, EADs are often up to twice that of adjacent PBDs, which contrasts with silt-dominated CPBDs which have a thickness approximate to that of PBDs. The overly thick sand in EADs results from unusually deep scour holes in the channel bed, caused by river flow impinging almost directly (40 to 140°) against resistant sediment on the side of a meander belt, rather than at much lesser angles as with CPBDs (Carey, 1969; Burge and Smith, 1999). The direct head-on force of the flow against the resistant sediment causes about a third of the flow to form a large reverse eddy or whirlpool on the upstream side, while the remainder of the flow continues on it downriver course (Fig. 2). At the same time macro turbulent eddies erode the channel bed (Burge and Smith, 1999). The large reverse eddy and channel-bed scour are likely to be most effective during near-bankfull or flood-flow conditions. Width of the concave morphology represents the size of the reverse eddy during high-flow conditions. EADs studied previously suggest that scour holes partially infill following near-flood conditions. In the Mississippi River, Carey (1969)

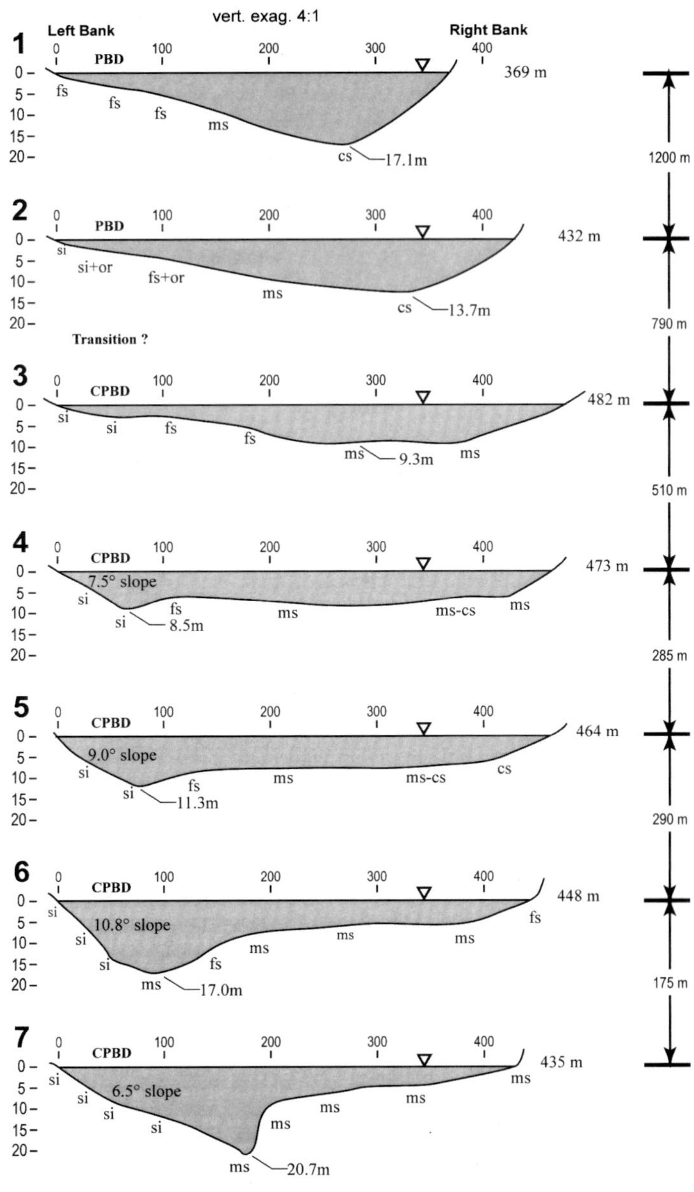

FIGURE 5.—Channel cross-sectional profiles from the 7-coring locations along the Peace River showing width, depth and sediment grain size from grab samples taken from the channel bed (see Fig. 3 for locations).

measured PBD thicknesses of 30 m while adjacent EAD measured 60 m. Similarly, in the smaller Kooteney River in southeastern British Columbia, Canada, Burge and Smith (1999) measured PBD thickness of 9 m and EAD at 18 m.

From lithostratigraphic vibracore logs, sediment samples taken from the channel bed, and measured cross-channel profiles, the character of EADs along the Athabasca River in the Athabasca Delta was established. In Figure 6 and 7 stratigraphic log #1, with a N/G of 48%, was taken from the PBD located 625 m upriver from the proximal EAD. Cores #2 to 4 from within the EAD were dominated by sand with N/G percentages of 70, 85, and 85, respectively (Fig. 7). Core #4 penetrated 9.8 m of EAD strata, and probably extends to a total depth of at least 22 m, based on channel depth soundings (Fig. 8). Core #5

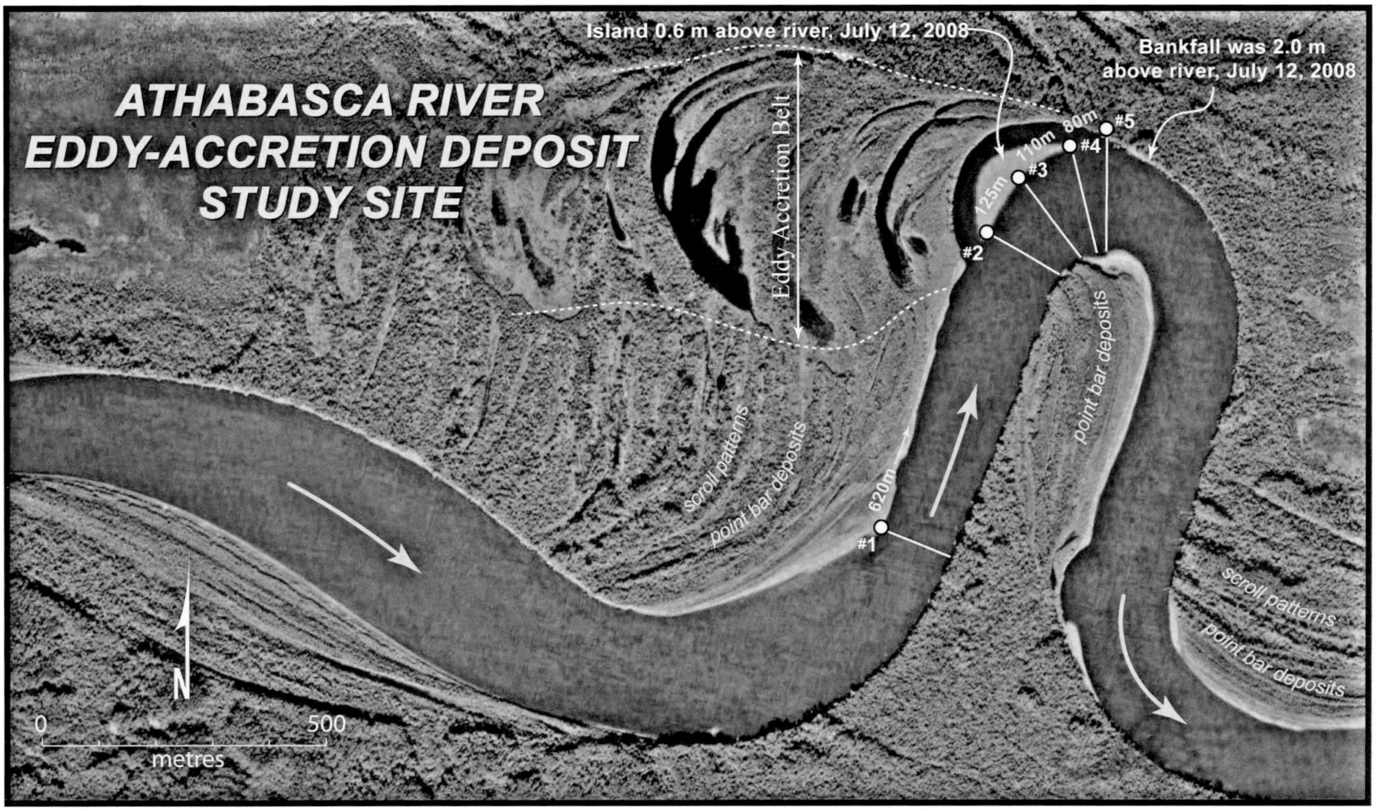

FIGURE 6.—Satellite image from Google Earth™ showing the morphology of an eddy accretion belt and locations of vibracores and channel cross-sectional profiles from the Athabasca River in the Athabasca River Delta, NE Alberta (see Fig. 1 for location).

was taken from the erosion-resistant sediment barrier on the channel margin, and is dominated by prodelta silt-clay mud. The first of the two sand beds in core #5 is interpreted to represent a 30-cm-thick distributary-mouth-bar sand from the former delta channel mouth (Gorecki, 1990), and the upper 60 cm sand is probably an overbank deposit from recent floods. Note that this mud-dominated strata (silt-clay) probably extends continuously to 24 m, based on channel bed depths. Of that 24 m, only 0.9 m likely consists of sand.

Sediment samples taken from lateral-accretion bedding on the channel bottom of the EAD suggest that sand dominates to a depth of between 19 and 22 m over a width of about 300 m (Fig. 6). However, perhaps prior to 1982 the width may have been as great as 500 m (reasoning to be discussed later in discussion section). On the lateral-accretion surfaces fine sand dominates the subaqueous slope (Fig. 8). In channel cross-sectional profiles #2 and 4, stiff burrowed clay was sampled near the deepest parts of the channel bed. This clay is interpreted to represent offshore lacustrine deposits, similar to those sediments present in core #5 (Fig. 7).

The subaqueous slopes of the EAD, are 14°, 8.7°, and 16° for channel cross-sectional profiles 2, 3, and 4, respectively. Most of the sediment on the subaqueous portions of the lateral-accretion slopes consists of fine sand (Fig. 8), in contrast to silt on CPBDs in the Peace River Delta.

A fourth lithostratigraphic log of EADs was obtained from vibracore and cut-bank exposure from the left bank of the Peace River, located 2000 m downriver from core # 7 (Fig. 3). The log was taken 200 m downriver from the paleo cut-bank of the EAD, and shows a nearly continuous sand stratatigraphy from 2.4 m to the end of the core at 8.3 m (Fig. 9). Considering the total strata cored, sand comprises 70%, but if one excludes the upper 2.4 m of silty overbank is excluded, the EAD consists of 97% sand. Given the fact that the PBD associated with this EAD was likely 17 m thick and the ratio thickness of PBD to EAD is usually 1:2, then the EAD is estimated at 34 m thick (Burge and Smith, 1999).

DISCUSSION OF EADs

The depth of the channel scour pool and adjacent lateral-accretion fill of EADs has important implications as a modern analogue for ancient fluvial meander belts. In the Athabasca River, thickness of EAD sediment is at least 24 m relative to bank-full flows, but could be as thick as 34 m. The latter figure is based on the approximate 1:2 thickness ratios of PBD to EAD, determined from borehole data taken from the Mississippi River and the Kootenay River, British Columbia (Carey, 1969; Burge and Smith, 1999). The thickness of the PBD on the Athabasca was measured at 17 m (Fig. 8, cross section #1). However, a lesser ratio of 2:3 of PBD to EAD was determined for Beaver River in Alberta (Burge and Smith 1999). Using the above ratios, the EAD on the Athabasca could be as great as 34 m, or as little as 25 m relative to bank-full. Using the same ratios as discussed above, the EAD (Fig. 9) on the Peace River would range from 34 m to at least 23 m below July 2008 river levels. In this comparative discussion one unknown is whether the low river gradients of the Peace and Athabasca rivers play a significant role in the depth of scour of the channel bed at sites with EAD, and is an area of future research.

Lithostratigraphic results of the EAD were consistent with previously studied examples (Burge and Smith, 1999). The Kootenay and Beaver rivers in British Columbia and Alberta, respectively, were

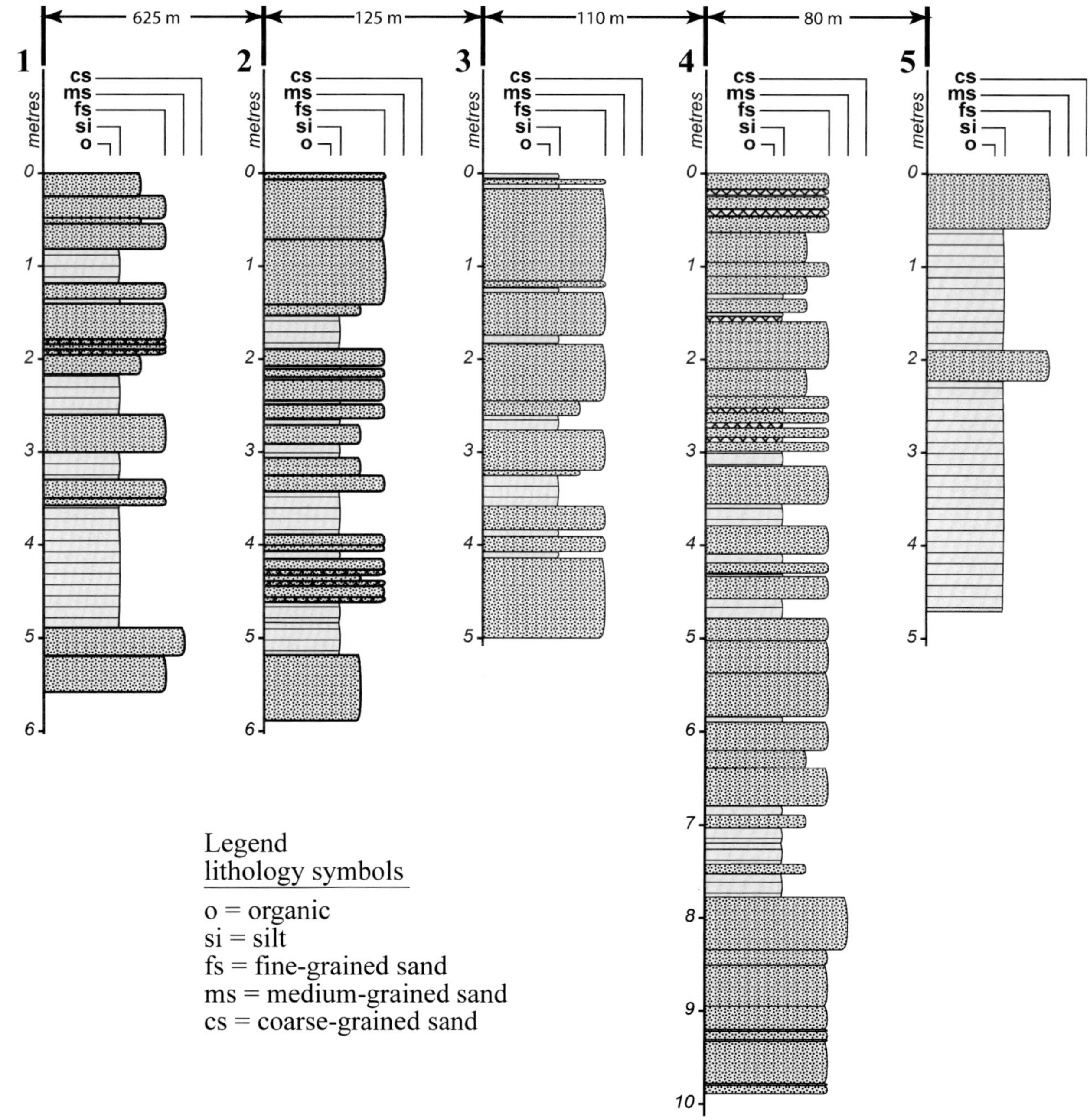

FIGURE 7.—Lithostratigraphic profiles of 5-vibracores taken from along the left bank of the main distributary of the Athabasca River (see Fig. 6 for locations).

characterized by greater than 90 percent sand on average (Burge and Smith (1999). In the three vibracores (#2, 3, 4) extracted from the Athabasca EAD, sand percentages were 70, 85 and 85 percent, respectively.

Note that the width of the EAD belt on the Athabasca River has decreased from 500 to 300 m with time (Fig. 6). This decrease might be attributed to the fact that flow discharge and channel width of the main distributary of the Athabasca has been waning over the past 26 years, since the avulsion or breakout of flow discharge from the Embarras to Mamawi Lake, just upstream of the study reach, beginning in 1982 (Calverley, 1984). Over the past 26 years larger volumes of river flow have been captured by this enlarging avulsion channel, which has led to an increase in the width and discharge of the Embarras feeder channel. This subsequently reduced discharge and channel width in the main Athabasca distributary, resulting in the reduced width of the EAD belt.

Can an EAD evolve into a CPBD, or CPBD evolve into an EAD? It seems to be a reasonable hypothesis depending on the angle of channel approach to a laterally resistant sediment body along the margin of a meander. This question might be answered with an examination of the morphologic deposits from aerial photographs and satellite imagery. At

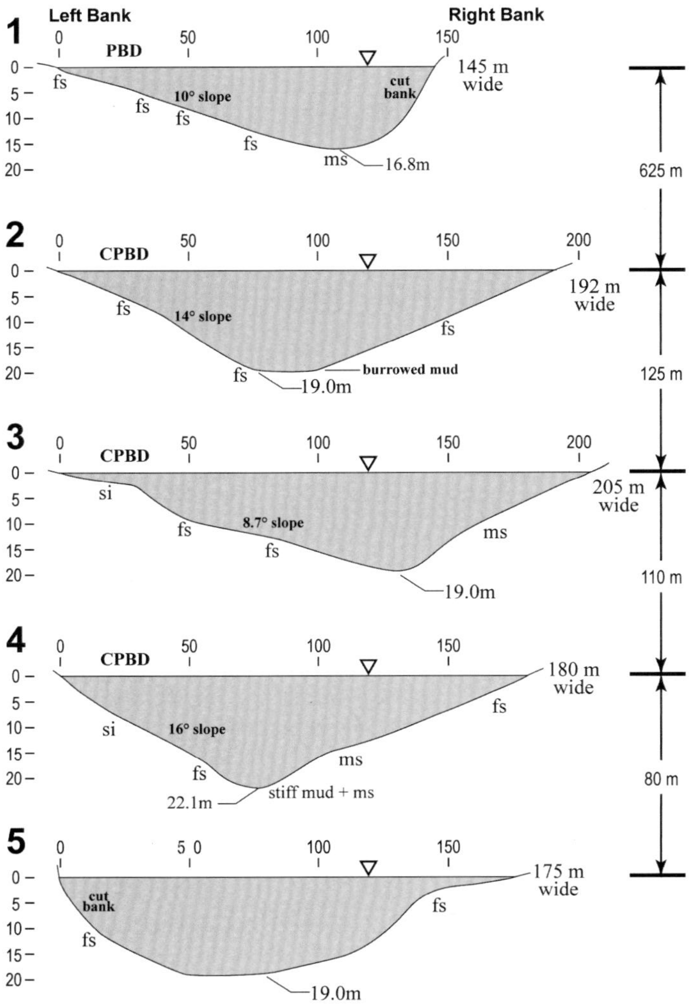

FIGURE 8.—Channel cross-sectional profiles from 5-locations along the Athabasca River showing width, depth and sediment grain size from grab samples taken from the channel bed (see Fig. 6 for locations).

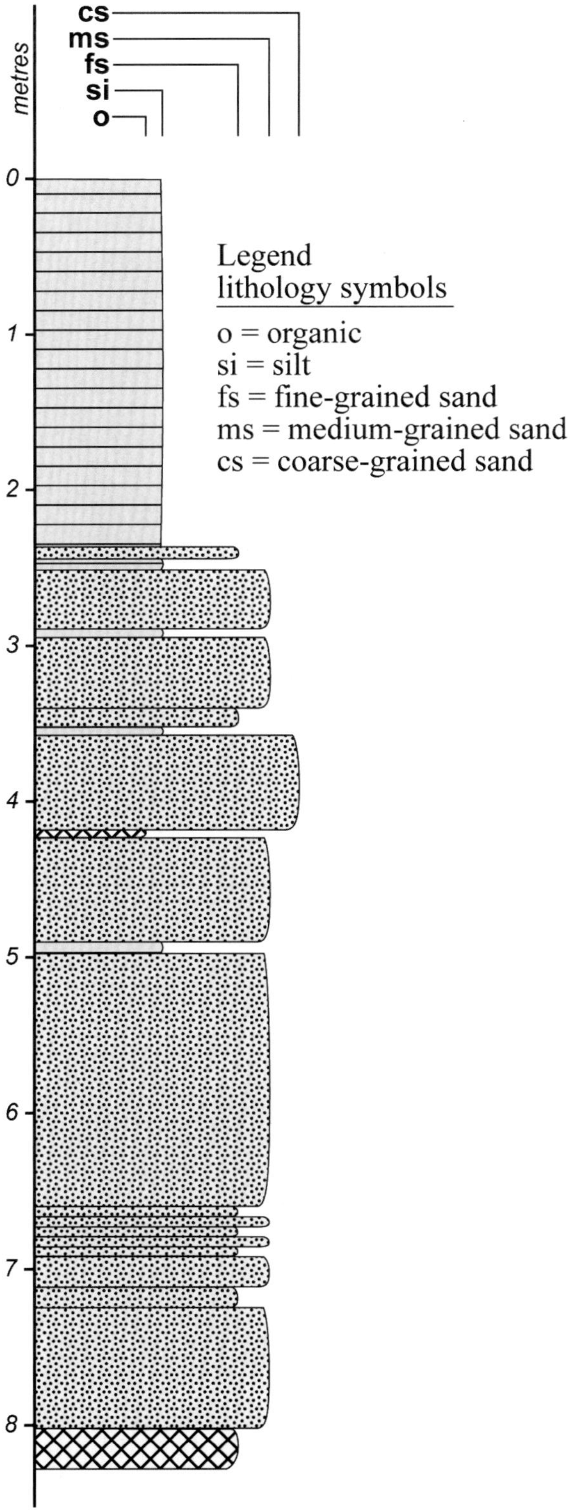

FIGURE 9.—Lithostratigraphic log of an eddy accretion deposit from the Peace River (see Fig. 3 for location).

this point in the research the evolution from one depositional style to another remains speculative, but it seems plausible. A more interesting question is: would such a depositional change be detectable in an ancient-rock meander belt? If present, one should expect a transition from a siltstone-dominant CPBD to a thicker (up to 2 times) nearly homolithic sandstone deposit. Conversely, a trend from thick sandstone to half-thickness CPBD might be expected across the shift from EAD to CPBD.

The presence of EADs in both the Athabasca and Peace River tributaries of the PAD is significant because the laterally extensive deltaic plains into which eddy accretions deeply eroded were resistant clay-silt deposits. This study is the first EAD reported from a deltaic setting. It stands to reason that EADs should be present in ancient delta plains, including deposits of important hydrocarbon-producing basins across the globe.

CONCLUSIONS

1. The surface morphology of counter point bar consists of concave scroll-bar patterns, which always form downriver from point bars that exhibit convex scroll-bar morphology.

2. Counter-point-bar deposits can form in low-gradient (0.00004) gentle arcing meander bends, as shown in this study, as well as in short-radius meander bends, as was determined from previous work by Smith et al. (2009) carried out 210 km up the Peace River from the current study site.

3. In the broad gently arcing meander studied on the Peace River, inclined heterolithic strata of silt dominated by alternating sand and silt beds in a counter-point-bar deposit extended over a channel distance of 1800 m with a probable thickness varying between 9.5 and 22 m.

4. From the five vibracores taken from the counter-point-bar deposits to depths of 6.5 m, percentages of silt varied from 55 to 86%, while sand in point-bar deposits varied from 95 to 100%.

5. In contrast to counter-point-bar deposits, the surface morphology of eddy-accretion deposits shows more pronounced concave scroll patterns.

6. Eddy-accretion deposits consist of unusually thick sand-dominant accumulations that infill deep scour trenches in the channel bed caused by macro turbulence at an abrupt channel bend consisting of erosion-resistant silt-clay, in the case of the Athabasca and Peace river reaches studied.

7. Eddy-accretion deposits can form in low-gradient rivers (0.00003) associated with deltaic plains, as well as in higher-gradient meander belts such as the Mississippi, U.S.A., Kootenay, British Columbia, Canada. and Beaver, Alberta, Canada (Carey, 1969; Burge and Smith, 1999).

ACKNOWLEDGMENTS

Financial support was kindly provided by Nexen Inc. and the Natural Sciences and Engineering Research Council (NSERC) of Canada. Drafting of diagrams was provided by Shawn Meuller and Robin Poitras from the Department of Geography, University of Calgary. The authors are grateful to reviewers Esther Stouthamer and Rick Donselaar for their suggestions to improve this manuscript. Parks Canada provided permission to carry out this research in Wood Buffalo National Park.

REFERENCES

Burge, L.M., and Smith, D.G., 1999, Confined meandering river eddy accretions: sedimentology, channel geometry and depositional processes: *in* Smith, N.D., and Rogers, J., eds., Fluvial Sedimentology VI: International Association of Sedimentology, Special Publication 28, p. 113–130.

Calverley, A., 1984, Sedimentology of lateral accretion channel deposits with epsilon mud interbeds in the upper Athabasca River Delta Plain, N.E. Alberta: Unpublished M.Sc. thesis, University of Calgary, Canada, 116 p.

Carey, W.C., 1969, Formation of floodplain lands: American Society of Civil Engineers Proceedings., Journal of the Hydraulics Division, v. 95, p. 981–994.

Daniel, J.F., 1971, Channel movement of meandering Indiana streams: United States Geological Survey, Professional Paper 732-A, 18 p.

Environment Canada, 1980, Historical Sediment Data Summary 1978: Water Survey of Canada, Ottawa, Canada, 121 p.

Environment Canada, 1991, Historical Streamflow Summary 1990: Water Survey of Canada, Ottawa, Canada, 629 p.

Environment Canada, 2008, Historical Streamflow Summary 2009: Water Survey of Canada, Ottawa, Canada.

Gorecki, R.C., 1990, Geomorphology and sedimentology of the lower delta plain and subaqueous delta plain of the Athabasca river Delta, NE, Alberta: Unpublished M.Sc. thesis, University of Calgary, Alberta, Canada, 207 p.

Hickin, E.J., 1979, Concave-bank benches on the Squamish River, British Columbia, Canada: Canadian Journal of Earth Sciences, v. 16, p. 200–203.

Hickin, E.J., 1986, Concave-bank benches in the floodplains of Muskwa and Fort Nelson Rivers, British Columbia: Canadian Geographer, v. 30, p. 111–122.

Hugenholtz, C.H., Smith, D.G., and Livingston J.M., 2009, Application of Floodplain stratigraphy to determine the recurrence of ice-jam flooding along the lower Peace and Athabasca Rivers, Alberta: Canadian Water Resources Journal, v. 34, p. 1–16.

Jackson, R.G., 1981, Sedimentology of muddy fine-grained channel deposits in meandering streams of the American Middle West: Journal of Sedimentary Petrology, v. 51, p. 1169–1192.

Lewin, J., 1983, Changes in channel pattern and floodplain, *in* Gregory, K.J., ed. Background to Paleohydrology: Chichester, U.K., Wiley, p. 303–319.

Kellerhals, R., Neill, C.R., and Bray, D.I., 1972, Hydraulic and Geomorphic Characteristics of Rivers in Alberta: Research Council of Alberta: River Engineering and Surface Hydrology, Report, 72–1, 52 p.

Makaske, B. and Weerts, H.J.T., 2005, Muddy lateral accretion and low stream power in a sub-recent confined channel belt, Rhine-Meuse delta, central Netherlands: Sedimentology, v. 52, p. 651–668.

Nanson, G.C., and Croke, J.C., 1992, A genetic classification of floodplains: Geomorphology, v. 2, p. 459–486.

Nanson, G.C., and Page, K.J., 1983, Lateral accretion of fine-grained concave benches on meandering rivers, *in* Collinson, J., and Lewin, J., eds, Modern and Ancient Fluvial systems: International Association of Sedimentologists, Special Publication 6, p. 133–143.

Page, K., 1983, Concave bench evolution and sedimentation on the Manawatu River, New Zealand: New Zealand Geographer, v. 39, p. 59–63.

Page, K., and Nanson, G., 1982, Concave-bank benches and associated floodplain formation: Earth Surface Processes, v. 7, p. 529–543.

Smith, D.G., 1984, Vibracoring fluvial and deltaic sediments, tips on improving penetration and recovery: Journal of Sedimentary Petrology, v. 54, p. 660–663.

Smith, D.G., 1988, Modern Point Bar Deposits Analogous to the Athabasca Oil Sands, Alberta, Canada. *in* de Boer, P.L., van Gelder, H. and Nio, S.D., eds., Tidal Environments and Facies, Reidel Publishing, Dordrecht, The Netherlands, p. 417–432.

Smith, D.G., 1992, Vibracoring: recent innovations: Journal of Paleolimnology, v. 7, p. 137–143.

Smith, D.G., 2003, 1100 years of ice-jam flooding in the Peace River Delta interpreted from flood bed sediments and ice-scarred trees, *in* Hicks, F., ed., 12th Workshop on the Hydraulics of Ice Covered Rivers: Canadian Geophysical Union, Hydrology Section, p. 241–260.

Smith, D.G, Hubbard, S.M., Leckie, D.A., and Fustic, M., 2009, Counter point bar deposits: lithofacies and reservoir significance in meandering modern Peace River and ancient McMurray Formation, Alberta, Canada: Sedimentology, v. 56, p. 1655–1669.

Taylor, G., Crook, K.A.W., and Woodyer, K.D., 1971, Upstream-dipping foreset cross-stratification: origin and implication for paleoslope analyses: Journal of Sedimentary Petrology, v. 41, p. 578–581.

Thomas, R.G., Smith, D.G., Wood, J.M., Visser, J., Calverley-Range, E.A., and Koster, E.H., 1987, Inclined heterolithic stratification—terminology, description, interpretation and significance: Sedimentary Geology, v. 53, p. 123–179.

Woodyer, K.D., Taylor, G., and Crook, K.A.W., 1979, Sedimentation and benches in a very low gradient suspended load stream, the Barwon River, New South Wales: Sedimentary Geology, v. 22, p. 97–120.

FLUVIAL PALEOTRANSPORT DERIVED FROM TROUGH CROSS-BEDDING: EXAMPLE FROM THE LOWER CLAIR GROUP, WEST OF SHETLAND, USING ORIENTED WHOLE-CORE IMAGES

ANDREW SMALLEY
Task Geoscience Inc., 10375 Richmond Avenue, Millenium Tower, Suite 1300, Houston, Texas 77042, U.S.A.
e-mail: andy.smalley@taskgeoscience.com

ABSTRACT: The use of cross-bedding to determine sediment dispersal patterns in ancient river systems has long been established. From well data, such measurements can be made but require a degree of inference and conceptualization due to the limited coverage of the wellbore and the quality of the data available. This study presents results derived from whole-core image data acquired in a horizontal well. The orientation of this well is approximately perpendicular to sediment transport direction in an ancient braided river system, from the Devonian Lower Clair Group. This provides a unique borehole view of fluvial sandstones, where trough cross-bedding is sampled in transverse section. Where whole trough sets are sampled, the orientation of the hydraulic dune axes which formed them can be determined. Although these data are invaluable, all borehole measurements are subject to the effects of borehole bias, which in this case, is not significant in terms of bedding dip but has a significant effect in terms of bedding strike. Essentially dunes with axes which trend oblique to the well path, which is likely to occur given the direction of the well and the expected braided sinuosity of channel-ribbons, will be poorly sampled. Despite this, trough axes form an orthogonal set when plotted together on a stereonet, which is the geometry expected as channel-ribbons converge at bar fronts, where hydraulic dunes are most likely to be preserved within a river system. Based on this assumption, the bisector of this trend could represent the channel braid axis of the preserved river system, which trends to the SSE. However, given that data were acquired from a horizontal well oriented approximately perpendicular to the paleotransport direction, it will more likely reflect lateral variation within the river system. Based on this assumption, and inferences from elongate-bar geometries, an investigation of lateral variation in braid axis orientation was carried out. This revealed divergent patterns in channel braid axes, reminiscent of terminal fan geometries. From these patterns, at least two terminal fan systems of more than 100 m in width and with axes to the SSE and SSW have been identified, which overlap at their margins.

KEY WORDS: Clair Field, fluvial sediment dispersal, braided rivers, whole-core image analysis, borehole bias

INTRODUCTION

Paleotransport analysis using cross-bedding is a well established method for inferring channel orientation in ancient river systems. In fluvial sandstones two principal types of cross-bedding are preserved: planar (formed by ridge dunes) and trough (formed by sinuous-crested dunes). For paleotransport analysis a different approach is required for these features, where planar cross-bedding forms consistent sets, whereas trough cross-bedding forms scattered datasets reflecting their curvature. In this example only trough cross-bedding has been recognized so distinguishing between these features is not an issue. Measuring such features from borehole data can be complicated due to borehole bias: in vertical wells, data sampled from trough cross-bedding tends to form a scattered distribution as each trough set is sampled from a particular part of a trough. In deviated wells drilled parallel to paleotransport direction, borehole bias can have a very significant effect to the point of apparent absence of data statistically (anomalous trends are over sampled). In this example the borehole was drilled near horizontal, approximately perpendicular to paleotransport direction, which allows whole trough sets to be sampled. Such sampling is not usually possible from well data.

One uncertainty for the Clair depositional system was the large-scale regional orientation of fluvial channel axes, which is fundamental for reservoir characterization. Previous studies have suggested a source area generally to the north (Allen and Mange Rajetzky, 1992, Nichols, 2005) based on clast composition and heavy-mineral assemblages. However, paleotransport, based on rock-fabric measurements, has not been presented so far for the Clair field. This study presents measurements of sediment dispersal based on actual rock-fabric trends from Clair Unit V (Table 1). From these trends, inferences are made for various fluvial geometrical elements based on a conceptual depositional model. Although this study represents data from a single well and therefore has limited regional application, the physical measurements from oriented whole-core images are of very high resolution and quantitative. This provides the basis for realistic inferences to be made in the vicinity of the wellbore, within the constraints of the conceptual depositional model. A consideration of the effects of borehole bias on these measurements is also taken into account, a very important consideration when interpreting rock-fabric measurements from well data.

BACKGROUND TO THE CLAIR FIELD AND THE STUDY WELL

Clair is a giant oilfield 75 km west of the Shetland Islands (Fig. 1). Discovered in 1977, an extended appraisal campaign delayed development. Production finally commenced in 2005. During appraisal it was established that production rates were controlled by the complexity of the reservoir, in particular the presence of natural fractures, and also by matrix heterogeneity. 206/8-A02, the first horizontal development well, had an extensive data acquisition package designed to fully characterize the reservoir section (see Witt et al., 2009 for further details on field appraisal and start-up). Data acquisition included borehole images and core from the principal reservoir sections.

Stratigraphy and Geological Setting

The Clair Field is situated in the extensional Faroe–Shetland Basin, formed during Devonian to Cretaceous times. It is a structurally complex field, affected by several phases and styles of deformation (Barr et al, 2007), the age and relative timing of which is uncertain. The reservoir interval comprises the Devonian Old Red Sandstones of the Lower Clair Group. Lower Clair Group stratigraphy represents a fluvial

From River To Rock Record: The Preservation Of Fluvial Sediments And Their Subsequent Interpretation
SEPM Special Publication No. 97, Copyright © 2011
SEPM (Society for Sedimentary Geology), ISBN 978-1-56576-305-0, p. 153–165.

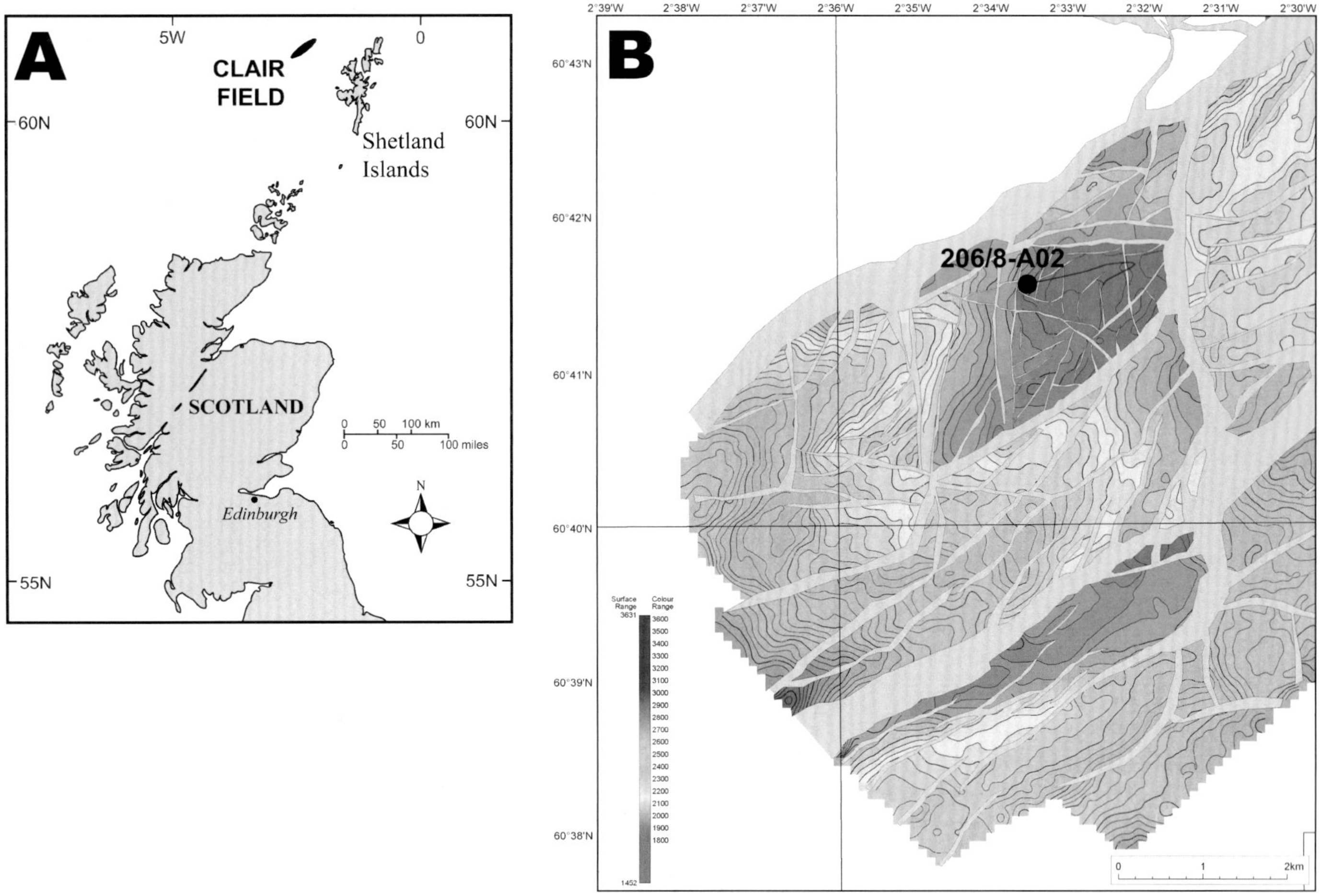

FIGURE 1.—(A) Location of the Clair Field, to the west of the Shetland Islands. (B) Depth map of top of reservoir (red = shallow, blue = deep), showing location of the study well (206/8-A02) and distribution of the major faults (grey areas with no contours).

succession (Nichols, 2005) divided into six reservoir units of varying quality (Units I to VI; Allen and Mange Rajetzky, 1992). This well targeted the homogeneous Unit V, the best-quality unit in the field. Table 1 presents the Clair stratigraphic units in the Lower Clair Group and their interpreted depositional environments.

The Lower Clair Group shows a range of non-marine facies associations. These range from locally derived talus and alluvial fan conglomerates, through poorly confined and channelized fluvial systems, to lakes and wind influenced sandy floodplains (Allen and Mange Rajetzky, 1992; McKie and Garden, 1996; Nichols 2005).

This study involves core from Unit V. The main facies (and their interpretations) in the cored interval, also typical of Unit V in general in the field, are in descending abundance: trough cross-bedded sandstone (in-channel), parallel laminated sandstone (channel margin and overbank), and wavy or irregularly laminated sandstone (wind-influenced damp floodplain). Mudstone is a very minor component. Unit V has been interpreted as having been deposited by braided rivers, with varying degrees of eolian reworking, mainly affecting the floodplain elements (Allen and Mange Rajetzky, 1992; McKie and Garden, 1996; Nichols, 2005).

TABLE 1.—Stratigraphic units of the Lower Clair Group and their interpreted depositional environments. After Nichols (2005).

Clair Unit	Depositional environment
VI	Fluvial sheet flood, lacustrine and lake margin.
V	Fluvial channel and sheet flood, lacustrine and lake margin.
IV	Fluvial channel and overbank.
III	Eolian sand flat
II	Fluvial channel.
I	Lacustrine and lake fan delta.

DATASET USED

The following dataset was used in this study:

- Whole-core and slabbed-core digitally acquired images. Whole-core images are acquired by photographing the outer surface of the core; slabbed-core images are captured once the core has been slabbed and represent a cross section of the core (see Figures 2 and 3 for a comparison). Both sets of images are useful for interpreting geological features. Both whole-core and slabbed-core images are presented in Enclosures 1a and 1b.

- OBMI (Oil-Base MicroImager, Schlumberger, 2001) and UBI (UltraSonic Borehole Imager, Schlumberger, 2002) image logs for orientation reference (not presented in this study).
- Open-hole log suite including gamma-ray, caliper, compensated neutron, density and array induction resistivity logs.
- Graphic core log descriptions for facies calibration (not available for publication).
- Depth structure maps with well location (Fig. 1).

METHODOLOGY

Data Preparation

After quality checking the orientation measurements from the borehole images (using the procedure given in Appendix 1), the whole-core images were depth shifted and oriented to the borehole images using obvious features, recognizable in both datasets (Fig. 4). After applying various photo edits to the core images (in oil-bearing sandstones, core photographs can appear very dark and of low contrast, so the brightness and contrast needs to be adjusted to bring out features and make them visible to the naked eye. This is a relatively straightforward process with digital photographs using most digital image-editing software), they can be interpreted in the same way as borehole images (see Harker et al, 1990; Lofts et al., 1997; Pavlovic et al., 2004; Donselaar & Schmidt, 2005; Garcıa-Carballido et al, 2009 for further details on methodologies for borehole image interpretation). Geological features can be classified and oriented by interactively fitting sine curves to them. The advantage of this approach, for the purpose of this study, is that different bedding features can be classified geologically (such as planar-laminae and cross-bedding; see Figures 3 and 4). This allows such features to be isolated within the dataset in order that different types of trend analysis can be performed (e.g., structure, paleotransport etc).

Correction for Structural Tilt

In order to analyze paleotransport from sedimentary successions, sedimentary fabrics need to be identified, and ultimately sedimentary features that were formed by the transport of sediment in the depositional system: cross-bedding in this case. By measuring the orientation of such features, paleotransport can be quantified. However, no real attempt to quantify paleotransport direction can be made without first determining the structure of the formation, within which the host sediments occur. This process allows a structural de-rotation to be applied to the sedimentary data, returning them to the paleohorizontal, at the time of deposition.

Structural dip is best defined by well-bedded intervals that are assumed to be originally horizontally stratified (i.e., mudstone beds). However, the section of the Clair Group encountered in the study interval in this well consists almost entirely of sandstone lithologies, with only localized muddy intervals. Consequently it was necessary to use other bedding types for determining structural dip that are less suitable for this purpose. In this study, parallel laminae within sandstones were used for determining structural dip, which has implications for confidence in the results.

Where possible, relatively flat sandstone bedding features (parallel laminae) were used to reduce the influence of sedimentary trends on the dataset (steeper sedimentary features such as cross-bedding reflect sediment transport. Parallel laminae may reflect an element of depositional dip in that they are deposited on bar tops and in floodplain settings, which are, none the less, closer to the original structure of the formation). Even where sandstone laminae could be distinguished from cross-bedding, an element of structural uncertainty remains using such features as sandstone laminae because they are influenced by depositional dip. Mean dip trends derived from sandstone laminae therefore represent structural dip ± depositional dip, where depositional dip represents a minor but uncertain variable. Determination of structural dip using parallel laminae in sandstones therefore could at best be considered a near-approximation of true structural trends. In the

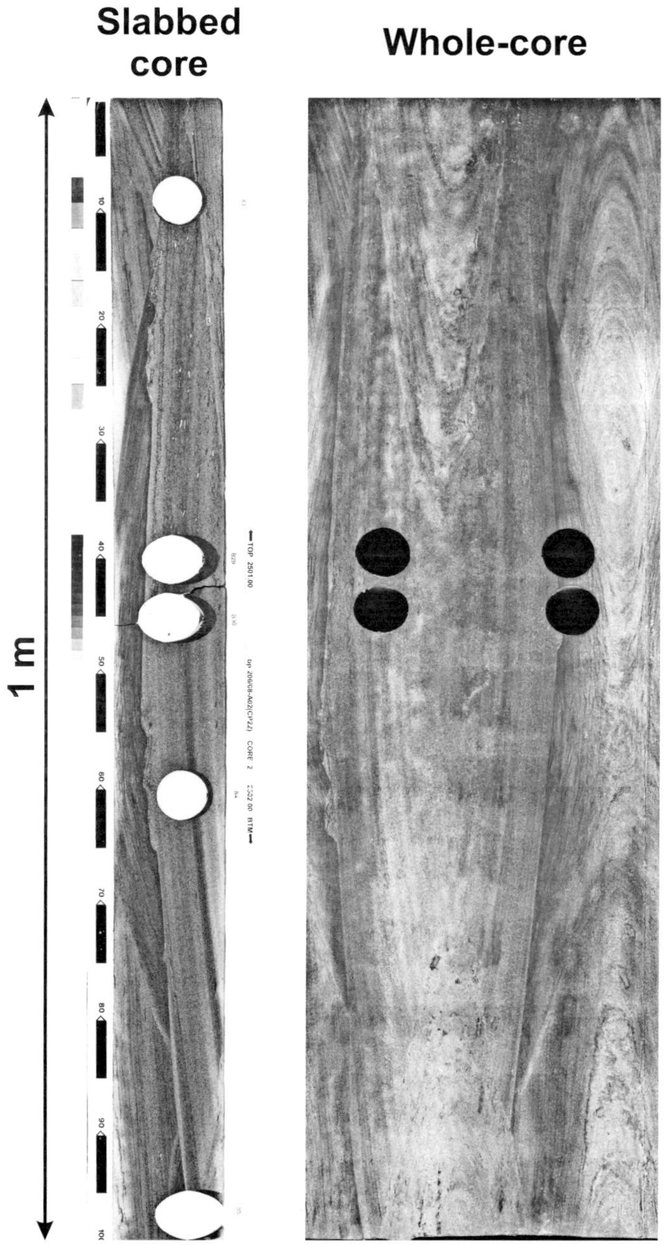

FIGURE 2.—Comparison of slabbed-core and whole-core images. Slabbed-core provides a cross-sectional view, which allows the viewer to see the relationships between features in the borehole section. Whole-core images capture the outer surface of the core before it has been slabbed. As with borehole images, planar features appear as sinusoids in whole-core images, which can be picked as true dip features once the whole-core images have been oriented.

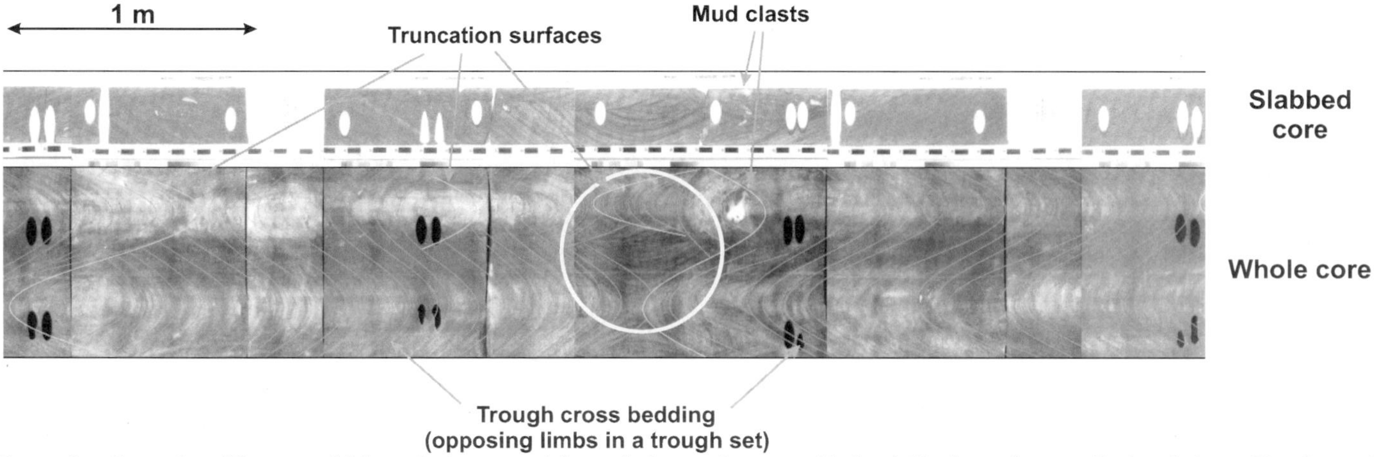

FIGURE 3.—Examples of features which can be interpreted from whole-core images, with the slabbed-core images displayed above. The shape of the trough set limbs are clearly seen in the slabbed-core images. Note that cross-bedding features in the axial region of the trough set are not sampled (circled). This demonstrates borehole bias as the axial part of the troughs strike parallel to the plane of the borehole. Also, the complexity of features at the trough axis means that it is not possible to pick features here.

absence of more reliable data, this approach to determining structural dip is the only route available (seismic data are not always reliable in the Clair Field and, given the influence of sub-seismic and sometimes very subtle faulting, it is important to understand local structure on a well-by-well basis). On top of the uncertainty in depositional dip, other structural uncertainties influence the final dataset, given the extent of fracturing and small-scale faulting encountered in the cored interval. Localized, irresolvable fault drag may introduce anomalous trends into the paleotransport dataset.

Paleotransport Analysis

The cored interval was sub divided into "sand packets", which are defined as intervals of consistency in cross-bedding trend, or where whole trough sets have been sampled. In a large part of the cored interval, overlapping sets of cross-bedded sandstones (trough sets) are encountered, usually separated by truncation surfaces (e.g., Fig. 5). In order to isolate each trough set, the structurally de-rotated cross-bedding was systematically reclassified, within each neighboring set, as "Cross-bedding 1", "Cross-bedding 2" and so forth. Reclassified and structurally de-rotated cross-bedding dip data were then imported into stereoplots, for advanced stereographic analysis. Stereographic analysis of the cross-bedding dip data aims to determine the plunge of trough cross-bedding axes. This assumes that each trough set reflects a simple sinuous-crested dune structure, with foreset and toeset components, formed parallel to fluvial channel-ribbon axes. Because the well is oriented approximately orthogonal to fluvial channel axes, the core samples trough sets in transverse section (Fig. 6). (For further discussion on determining paleotransport direction using trough cross-bedding see Collinson and Thompson, 1982; Ricci Lucchi, 1995; Luthi and Banavar, 1998; Luthi, 2001; Glover and Borman, 2007; Donselaar and Schmidt, 2005). Foreset dips from each trough set are treated in the same way as folded bedding in stereographic projections, in order to determine their axial plunge, using eigenvector systems. This was achieved by fitting a great circle between the centers of clusters on corresponding limbs of each trough set. The axial plunge of each trough set should equate to the dip of the hydraulic dune axis (the direction of dune migration; see Fig. 6). An axial plunge was determined for each trough set where foresets have been sampled on both trough limbs. In this case, data tend to form clusters which connect on a great circle, representing corresponding limbs of a trough

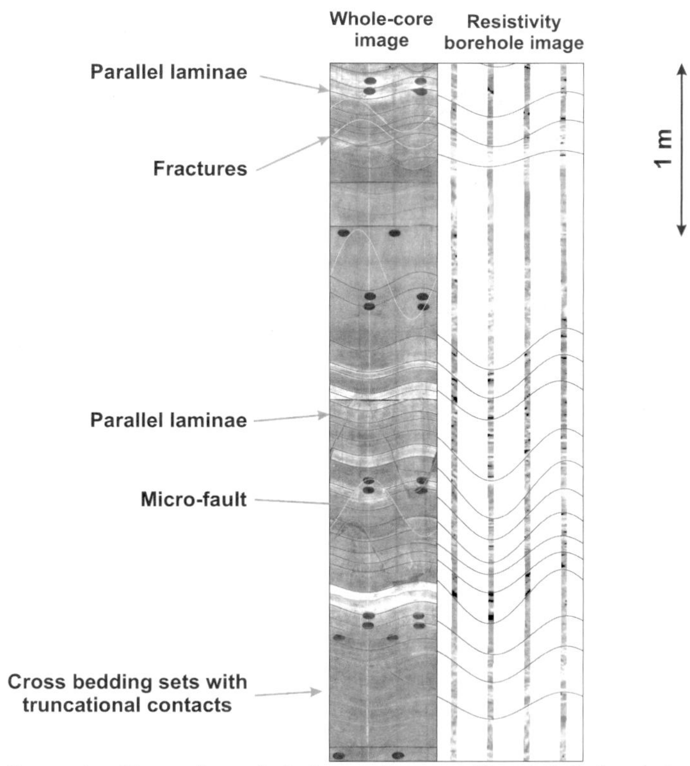

FIGURE 4.—Comparison of whole-core images with resistivity borehole images (OBMI). Features can be picked from both sets of images by interactively fitting sine curves to them. Apparent dip is calculated from the amplitude of the sinusoid and the borehole width, trigonometrically. The azimuth of sinusoid features are taken as their azimuth at maximum dip. Apparent dip values are converted to true dip using the borehole survey gathered by the borehole imaging tool, which is measured in relation to magnetic north, using magnetometers. Whole-core images are oriented to the borehole image by rotating them azimuthally and matching sinusoidal features. Data not from this well.

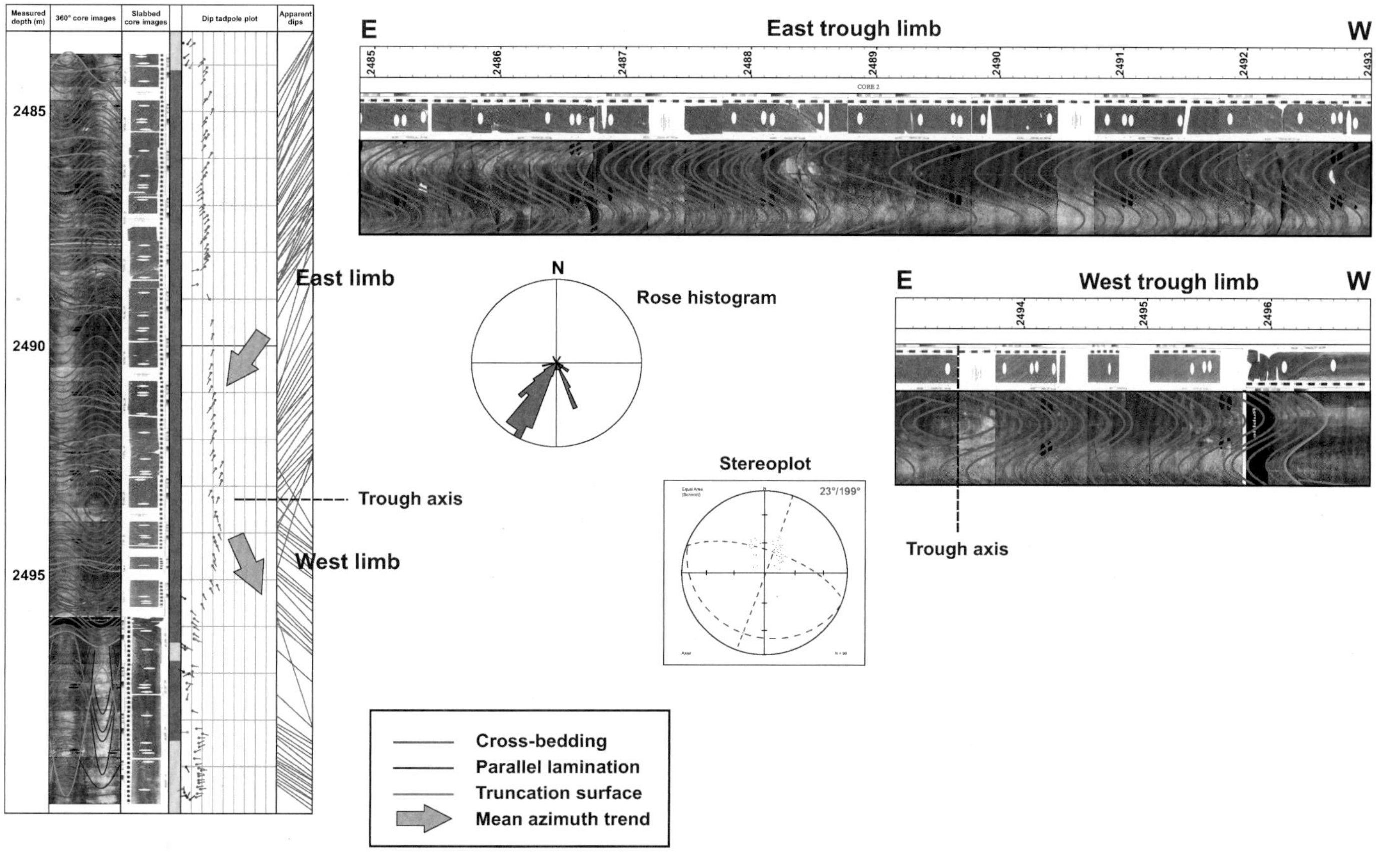

FIGURE 5.—Example of fluvial trough cross-bedding set from well 206/8-A02 (sand packet 4 in Figure 8). In this example structurally de-rotated trough cross-bedding dips are consistent either side of the trough axis. Here, condition "A" in Figure 9B is satisfied, hence data are sampled from both trough limbs. However, this is complicated by the geometry of the preserved sand body hosting the cross-bedding, which results in more data being sampled from the E limb of the trough set. This demonstrates sampling bias related to borehole width.

set. However, some of these clusters form secondary trends (i.e., they may sit on their own great circle; see for example sand packet 4), which represent curvature of the plunge axis, or curvature of the dune crest (essentially a doubly plunging fold; see Luthi, 2001).

Stereographic analysis of the structurally de-rotated cross-bedding was carried out for each "sand packet". "Sand packets" 1, 4, 6, 16, 17, 18, 20, 23, 26, 31, 33a, and 34a sample cross-bedding foresets within trough sets from both limbs and therefore allow determination of trough axial plunge. Cross-bedding sets in other "sand packets" sample only one limb, which is not sufficient to determine axial plunge. These results are presented as a well trajectory plot in Figure 8 and are summarized in Table 2.

Modeling the curvature of trough sets where only one limb has been sampled has been attempted in other fluvial studies (Luthi, 2001; Donselaar and Schmidt, 2005). These techniques are beyond the scope of this study but may be of interest for future reference.

Borehole Sampling Bias

Sampling of geological features from a borehole is unavoidably affected by bias (Terzaghi, 1965). This reflects the fact that borehole sampling is limited by borehole width (which can result in apparent trend consistency when sampling trough cross-bedding, demonstrated in Donselaar and Schmidt, 2005), and, more relevant to this case, borehole orientation in relation to feature orientation (relative dip). Essentially features which strike or dip parallel to the plane of the borehole will be under-sampled due to borehole bias. Figures 8A and 8B demonstrate the effects of borehole bias on sampling trough cross-bedding in this well. Figure 8A represents a schematic cross section of the wellbore cut parallel to the well path (W–E). Trough sets encountered along the well profile will be well sampled in their limbs but poorly sampled in their axial regions. Cross-bedding picks from such features should form relatively steep, consistent orthogonal sets (as seen in the example illustrated in Figure 5). In terms of bedding dip, the effects of borehole bias is not significant in this well section.

Figure 8B represents a schematic map view of well 206/8-A02, in relation to expected trough set orientations in Clair Unit V, with conditions A, B, and C representing possible trough orientations deposited from rivers sourced generally from the north. Condition A represents a trough set with an axis trending orthogonal to the well path. Conditions B and C represent trough sets whose axes trend oblique to the well path in opposite directions (essentially end members of channel-ribbon orientation). For the purpose of determining trough axes, borehole bias effects that result in undersampling from one of the trough limbs is more important as both limbs need to be sampled. In condition A, both limbs are sampled and therefore borehole bias is not an issue. However, in conditions B and C data are sampled from only one trough limb, and so bias becomes an issue. In fact, trough sets that

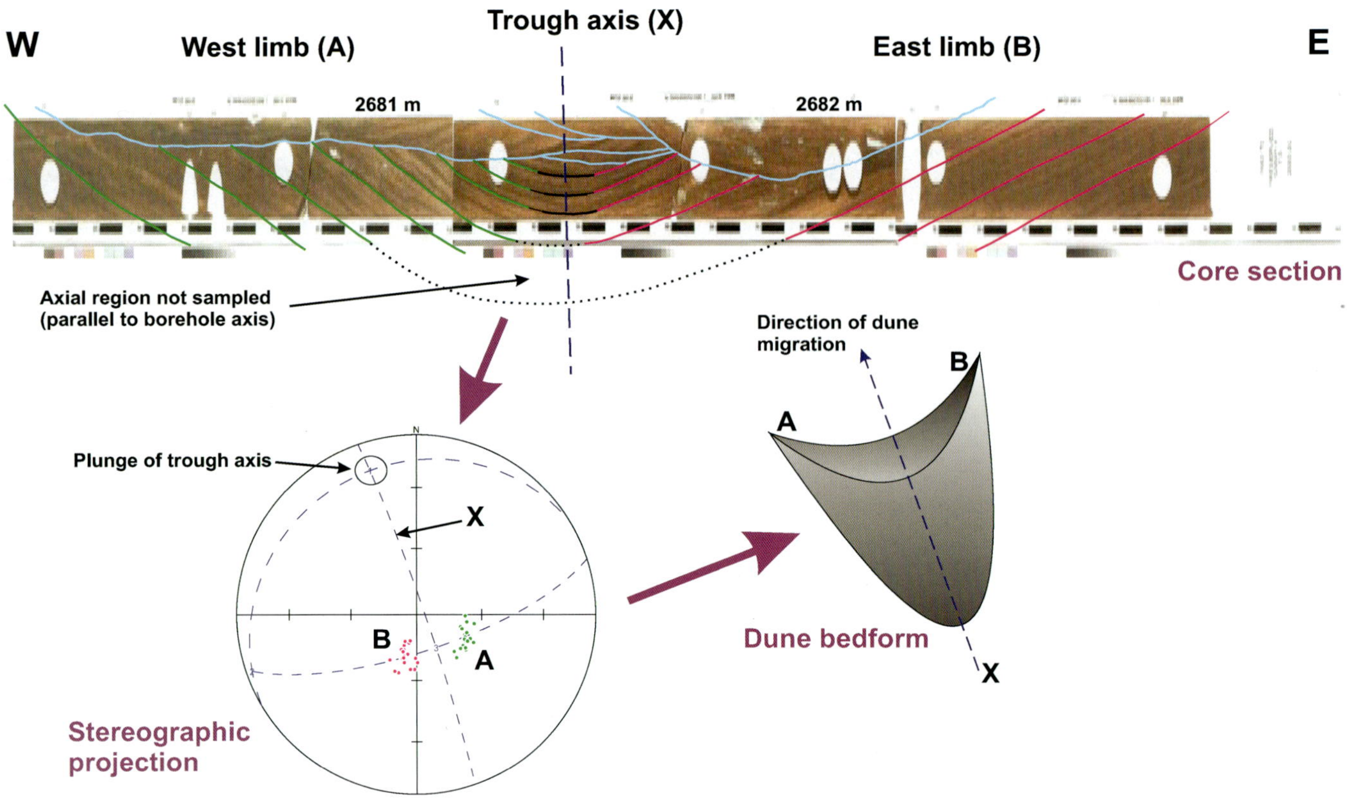

FIGURE 6.—Explanation of how trough cross-bedding is used to determine paleotransport direction, well 206/8-A02. Note the stereoplot is upper-hemisphere, depicting poles-to-planes, hence data points appear in opposite quadrants.

satisfy conditions B and C will be excluded from the dataset, as only whole trough sets were used. The effects of borehole bias on cross-bedding data sampling results in the full spread of data not being sampled or, to put it another way, the full extent of channel-ribbon sinuosity is not captured, due to channel-ribbon end members occurring oblique to the wellbore. Inferences for fluvial geometry are still valid, though statistical confidence may be reduced.

Figures 9A and 9B demonstrate borehole bias in terms of borehole width limiting the perspective and sampling of features. These two examples clearly demonstrate the dilemma of unlocking the puzzle of architectural elements and their relationships from sedimentary successions (Miall, 1989), where data are limited by the borehole view. In order to define architectural elements accurately, their cross-cutting relationships, lateral extent, and geometries need to be assessed. Within a single borehole view such relationships are not easily identified. Lamination types and macroforms may be recognized but their ordering will not be immediately apparent.

PALEOTRANSPORT ANALYSIS

In the cored interval cross-bedded units were encountered, which tend to occur as extensive intervals with consistent azimuth trends (Fig. 5). More complex units also occur, which comprise several overlapping cross-bedding sets, separated by truncation surfaces (Fig. 10). This apparent variation in complexity most probably reflects the way in which the well is sampling trough sets from these units rather than depositional variation. Cross-bedding occurs in the form of trough-shaped sets, though trough axes are only occasionally seen in core. The sub horizontal trajectory of this well follows a direction that is approximately orthogonal to the Clair Unit V fluvial channel axes (well deviation to the west; see Figure 1 and Figure 7). Consequently, the cored section in this well provides a unique dataset, where whole trough-cross-bedded units have been sampled in transverse section. Such transverse sampling is not usually available, particularly in vertical wells where trough cross-bedding tends to be sampled from one limb at a time. Trough-cross-bedding dip data, derived from the cored interval in this well, can thus be analyzed to an advanced level in order to determine paleotransport much more accurately than is normally possible in vertical wells. Figure 6 demonstrates this approach graphically, using a real example from the cored section of this well.

In terms of gross sediment dispersal the cored interval can be subdivided into two zones, above and below the fault contact at 2653 m (Fig. 7). Above this contact, trough axial trends vary from SE, SSE, SW, SSW, and ESE. Below the 2653 m fault contact, however, strikingly different trough axial plunge trends are evident, indicating orthogonal WSW and NNW paleotransport directions. These trends are not consistent with the rest of the cored interval, suggesting that they are anomalous and could reflect uncertainty in structural resolution, related to foot-wall deformation associated with the fault contact at 2653 m. Without further structural resolution, the data in the faulted section were excluded from paleotransport analysis.

Focusing on the section above 2653 m, where the data are reliable, ten measurements of trough axes were made, which give a sub-orthogonal distribution (Fig. 11). Such a distribution would be expected for channel-ribbon orientation in a braided fluvial system.

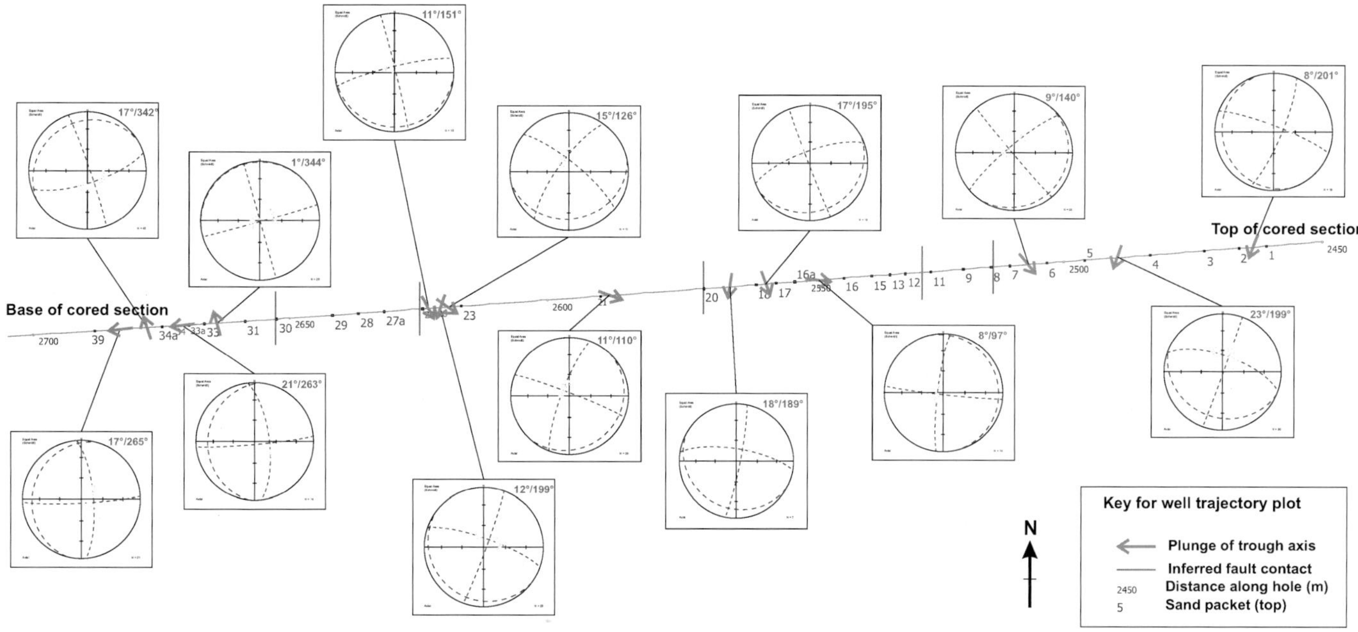

FIGURE 7.—Well trajectory plot in map view for well 206/8-A02, within the cored interval. Cyan arrows represent the plunge of trough cross-bedding axes, derived from stereographic projections of the data (inset). The inset stereoplots are Schmidt upper-hemisphere projections of poles to trough cross-bedding foresets. Data points therefore show the opposite dip/azimuth sense, whereas the plunge derived from these data points display true dip/azimuth sense. Plunges were derived using eigenvectors. Fault contacts are merely representative of where faults occur within the cored interval and do not display true orientation of fault surfaces. Only high-confidence data are presented (i.e., with data sampled from both trough limbs). The SSW, SSE, SE, ESE, and SW trends reflect trends of braided channel-ribbons, within a broad channel belt trending to the SSE (see Figure 11). Below the fault at 2653 m, strong orthogonal NNW and WSW trends occur, which are probably influenced by faulting.

However, the significance of such trends cannot be fully appreciated without relating them to a valid conceptual depositional model, which is discussed below.

CONCEPTUAL DEPOSITIONAL MODEL AND INFERRED FLUVIAL GEOMETRIES

Based on studies of present day braided rivers (Cant and Walker, 1978; Bridge, 1993, Bristow, 1993) and examination of satellite images (available on Google Earth), a conceptual depositional model was constructed (Fig. 11) for Clair Unit V, assuming braided fluvial morphology suggested by previous studies on the Clair Field (Allen and Mange Rajetzky, 1992; McKie and Garden, 1996; Nichols, 2005). This conceptual depositional model assumes that hydraulic dunes are generally preserved on the fronts of lensoid bars, by downstream accretion. Figure 12 expands on the geometry of trends expected in a braided fluvial system and how they relate to each other in a hierarchical sense. Cross-bedding measurements from corresponding limbs of trough sets can be used to determine trough axes (axes of hydraulic dunes). As hydraulic dunes form on bar fronts, where channel-ribbons converge, they should align systematically with this convergent geometry, hence they form sub-orthogonal sets within each preserved barform. The bisector of this convergence, or of convergent trough/dune axes, can therefore be used to infer channel braid axis. On this basis it can be assumed that when trough axes are plotted on a stereonet they should form a sub-orthogonal distribution, the bisector of which gives the channel braid axis for the fluvial system responsible for depositing Clair Unit V. To demonstrate this, a stereonet of trough axes is included in Figure 11, giving an inferred channel braid axis to the SSE for Clair Unit V. However, given that these data are derived from a near-horizontal well, they are more likely to reflect lateral relationships in the braid system, or even different flow events in the river system, so an overall channel braid axis, inferred from laterally distributed sample points, may not be entirely valid.

An investigation of the lateral variation in orientation of the channel braid axis was carried out, based on the geometrical criteria demonstrated in Figure 12. Based on the well trajectory plot illustrated in Figure 8, elongate bars were inferred from trough axes, reflecting convergent channel-ribbon geometries at bar fronts (Fig. 13). From these, local channel braid axes were inferred. The inferred channel braid axes suggest divergent azimuth patterns reminiscent of terminal fan geometries presented in Nichols (2005). On this basis, at least two terminal fan systems are suggested, which apparently overlap at their margins. Such terminal fans would be at least 100 m wide in the vicinity of the well, with axes trending SSE and SSW. Similar divergent and laterally overlapping fans were mapped from Middle Devonian outcrop studies on Orkney (Astin, 1985), using sediment dispersal and grain-size distribution.

CONCLUSIONS

Whole-core images provided a high-resolution dataset from a braided fluvial sandstone succession. From these data, bedding fabrics were easily recognized and oriented. The orientation of the well allowed sampling of trough cross-bedding in a unique way because the well is oriented approximately perpendicular to paleochannel axes.

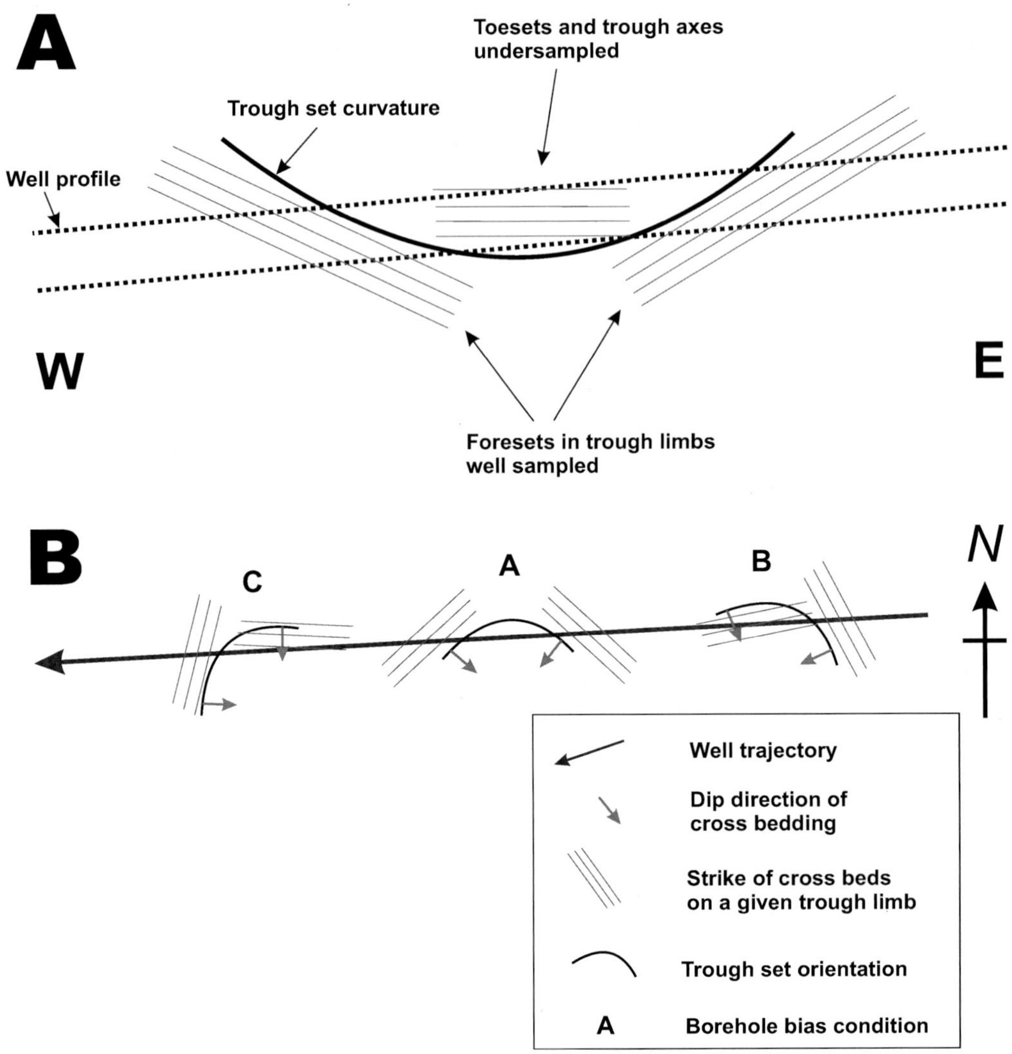

FIGURE 8.—(A) Schematic cross section of a borehole profile of well 206/8-A02. This example considers the influence of bedding dip on borehole bias. Essentially bedding features that dip approximately parallel to the wellbore will be under sampled. In this case, bedding features within the axial region of each trough set and toesets will be under sampled. (B) Schematic plan view of well 206/8-A02, with sketched theoretical trough sets which may be encountered given the observed data spread within the study section. Essentially planes encountered along the well path that strike parallel to the well will be under sampled. Borehole bias conditions A, B, and C represent possible orientations for trough sets within Clair Unit V, where A is most favorable for paleotransport analysis, with cross-bedding sampled from both limbs. Conditions B and C will be under sampled in one trough limb.

This allowed whole trough sets to be sampled in cross section. Considering the effects of borehole bias from this well, it can be demonstrated that effects related to dip angle are not significant. However, borehole bias has a considerable effect in terms of feature strike and well orientation. Essentially, trough cross-bedding sets with an axis falling roughly perpendicular to the well orientation will be best sampled; those which trend oblique to the wellbore will be under sampled. Such an oblique trough set orientation is considered a real possibility given a braided-channel sinuosity. Borehole bias effects therefore have the effect of reducing the spread of results, hence the full sinuosity of channel-ribbons within Clair Unit V are not captured.

Where whole trough sets were sampled, the orientation of hydraulic dunes which formed such troughs could be determined. From this, measurement of paleochannel-ribbon orientation in the Clair fluvial system during the deposition of Clair Unit V can be made. On this basis trough axes to the SE, SSE, SW, SSW, and ESE were determined in the studied section, with anomalous trends to the WSW and NNW in the lower section, probably reflecting fault disruption. Ignoring the anomalous faulted section, trough axes form orthogonal sets when plotted together on a stereonet, will reflect channel-ribbon sinuosity where they converge on bar fronts. The bisector of this convergence can be used to determine the orientation of the channel braid axis, which trends to the SSE. However, given that the study well is sub-horizontal it is sampling lateral variation in the Clair Unit V fluvial system. This lateral variation was investigated in order to infer local channel braid axes in the vicinity of the well, based on assumed

TABLE 2.—Paleotransport trends by sand packet, well 206/8-A02, Clair Unit V.

Sand packet	Axial plunge	Source	Sample (n)	Confidence	Comments
1 2461.70–2466.30 m	8°/201°	Cross-bedding 1	19	Moderate	Troughs display a SW-trending axial plunge.
4 2483.50–2496.29 m	23°/199°	Cross-bedding 1	90	High	SW-trending axial plunge.
6 2503.50–2510.80 m	9°/140°	Cross-bedding 1	23	Moderate	SE axial plunge trend.
16 2542.82–2552.40 m	8°/97°	Cross-bedding 1	14	Moderate	ESE-trending axial plunge. Data are poorly sampled from the N limb.
17 2556.00–2559.87 m	17°/195°	Cross-bedding 1	19	Moderate	Troughs display a SSE-trending axial plunge.
18 2542.82–2552.40 m	18°/189°	Cross-bedding 1	7	Low	SSW axial plunge trend. Data have been poorly sampled from the W limb.
20 2570.00–2590.00 m	11°/110°	Cross-bedding 1, Cross-bedding 2	28	Moderate	ESE axial plunge trend. Cross-bedding 1 and Cross-bedding 2 were combined as they sit on the same great circle.
23 2617.00–2622.20 m	15°/126°, 12°/199°	Cross-bedding 1, Cross-bedding 5	11, 29	Moderate, high	Cross-bedding 1: SE-plunging trough axis. Cross-bedding 5: SSW-trending trough axis. Cross-bedding 5 samples data from both limbs in equal proportions.
26 2622.20–2624.55 m	11°/151°	Cross-bedding 1	15	Moderate	SS- trending axial plunge.
31 2659.00–2667.00 m	1°/344°	Cross-bedding 2	29	Low	NNW axial plunge trend. Data are well sampled from both trough limbs, with a secondary plunge trend discernible on the NNE limb. Strong dataset, though axial trend is probably influenced by localized faulting.
33a 2670.00–2672.91 m	21°/263°	Cross-bedding 1	14	Low	WSW trough axial plunge trend. Both limbs have been sampled in approximately equal proportions. May be influenced by localized faulting.
34a 2675.20–2688.30 m	17°/342°, 17°/265°	Cross-bedding 2, Cross-bedding 5	43, 21	Low	Cross-bedding 2: NNW axial plunge, with data sampled on both limbs in comparable proportions. Cross-bedding 5: WSW-trending axial plunge, with data well sampled in both limbs. Both trends may be influenced by localized faulting.

elongate, lensoid bar geometry. From this approach, a systematic lateral divergence in local channel braid axes was observed. These divergent patterns are reminiscent of terminal fan geometries, of which at least two are encountered in the study section, which overlap at their margins. Such terminal fans would be at least 100 m wide with axes to the SSE and SSW in the vicinity of the well and are reminiscent of fan geometries mapped from Middle Devonian outcrop studies.

ACKNOWLEDGMENTS

The author would like to thank: BP for releasing the data used in this study; Stuart Buck (Task Geoscience) for assistance in the initial sediment dispersal analysis, Doug Boyd (Integrated Sedimentology) for input on the sedimentology of the Lower Clair Group, and Andy Witt (BP) for contributing to the introductory section on the Clair Field.

REFERENCES

ALLEN, P.A., AND MANGE RAJETZKY, M.A., 1992, Devonian-Carboniferous sedimentary evolution of the Clair area, Offshore North-Western UK: Impact of changing provenance: Marine and Petroleum Geology, v. 9, p. 29–52.

ASTIN, T.R., 1985, The palaeogeography of the Middle Devonian Lower Eday Sandstone, Orkney: Scottish Journal of Geology, v. 21, p. 353–375.

BARR, D., SAVORY, K.E., FOWLER, S.R., ARMAN, K. AND MCGARRITY, J.P., 2007, Pre-development fracture modelling in the Clair field, west of Shetland, *in* Lonergan, L., Jolly, R.J.H., Rawnsley, K., and Sanderson, D.J., eds, Fractured Reservoirs, Geological Society of London, Special Publication, 270, p. 205–225.

BRIDGE, J.S., 1993, The interaction between channel geometry, water flow, sediment transport and deposition in braided rivers, *in* Best, J.L., and Bristow, C.S., eds, Braided Rivers: Geological Society of London, Special Publication, 75, p. 13–71.

BRISTOW, C.S., 1993, Sedimentary structures exposed in bar tops in the

FIGURE 9.—(A) Outcrop of Lower Devonian fluvial sandstones at Whiting Ness, Arbroath, Angus, NE Scotland. This figure demonstrates the difference in perspective from an outcrop and approximate borehole view, which might be seen in core from a hypothetical vertical well taken through the same section (inset). (B) Outcrop of Upper Devonian alluvial-fan sandstones and conglomerates at Whiting Ness, Arbroath, Angus, NE Scotland. This figure demonstrates the difference in perspective from an outcrop and approximate borehole view, which might be seen in core from a hypothetical horizontal well taken through the same section (inset).

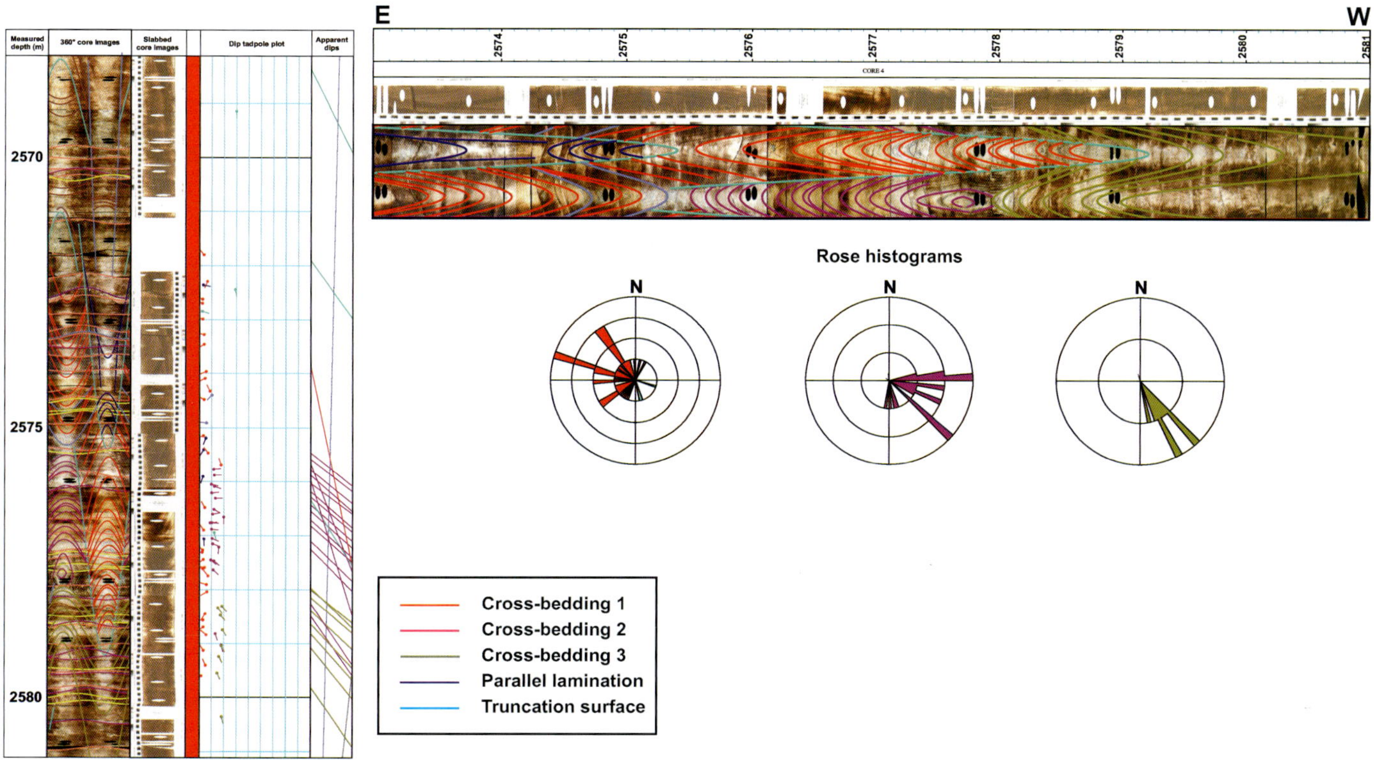

FIGURE 10.—Example of complex fluvial trough cross-bedding sets from well 206/8-A02. In this example the well encounters several sand packets, each comprising individual trough cross-bedding sets, separated by truncation surfaces. Consequently, dip tadpole patterns are complex. Rose histograms display varied azimuth trends for each cross-bedding set, with a combined data spread of approximately 180 degrees. In this example the borehole is probably sampling the thin edges of several sand packets, which stack within the scale of the borehole width.

Brahmaputra River, Bangladesh, *in* Best, J.L., and Bristow, C.S., eds, Braided Rivers: Geological Society of London, Special Publication, 75, p. 277–289.

CANT, D.J. AND WALKER, R.G., 1978, Fluvial processes and facies sequences in the South Saskatchewan River, Canada: Sedimentology, v. 25, p. 625–648.

COLLINSON, J.D. AND THOMPSON, D.B., 1982, Sedimentary Structures: London, George, Allen & Unwin, p. 83–84.

DONSELAAR, M.E. AND SCHMIDT, J.M., 2005, Integration of outcrop and borehole image logs for high-resolution facies interpretation: example from a fluvial fan in the Ebro Basin, Spain: Sedimentology, v. 52, p, 1021–1042.

GARCIA-CARBALLIDO, C., STYLES, A., AND POPPELREITER, M., 2010, Reduction of gross rock volume uncertainty in a salt flank structure using dipmeter and image-log data: Poppelreiter, M. ed., Dipmeter and Borehole Image Log Technology: Tulsa, American Association of Petroleum Geologists, Memoir 92.

GLOVER, P.W.J. AND BORMAN, P., 2007, The characterisation of trough and planar cross-bedding from borehole image logs: Journal of Applied Geophysics, v. 62, p. 178–191.

HARKER, S.D., MCGANN, G.J., BOURKE, L.T. AND ADAMS, J.T., 1990. Methodology of Formation Micro-Scanner image interpretation in Claymore and Scapa fields (North Sea), *in* Hurst, A., Lovell, M.A., and Morton, A.C., eds., Geological Applications of Wireline Logs, Geological Society of London, Special Publication, 48, p. 11–25.

LOFTS, J.C., BEDFORD, J., BOULTON, H. VAN DOORN, J.A. AND JEFFREYS, P., 1997, Feature recognition and the interpretation of images from horizontal wellbores, *in* Lovell, M.A. & Harvey, P.K., eds., Developments in Petrophysics: Geological Society of London Special Publication 122, p. 345–365.

LUTHI, S.M., 2001, Geological Well Logs; their use in reservoir modelling: Berlin, Springer-Verlag, p. 247–258.

LUTHI, S.M., AND BANAVAR, J.R., 1998, Application of Borehole Images to Three-Dimensional Geometric Modeling of Eolian Sandstone Reservoirs, Permian Rotliegende, North Sea: American Association of Petroleum Geologists Bulletin, v. 72, p. 1074–1089.

MCKIE, T. AND GARDEN, I.R., 1996, Hierarchical stratigraphic cycles in the non-marine Clair Group (Devonian) UKCS, *in* Howell, J.A. & Aitken, J.F., eds., High Resolution Sequence Stratigraphy: Innovations and Applications, Geological Society of London Special Publication 104, p. 139–157.

MIALL, A.D., 1989, Architectural elements and bounding surfaces in channelized clastic deposits: notes on comparisons between fluvial and turbidite systems, *in* Tiara, A. and Masuda, F., eds., Sedimentary Facies in the Active Plate Margin, Tokyo, Terra Scientific Publishing, p. 3–15.

NICHOLS, G.J., 2005, Sedimentary evolution of the Lower Clair Group, Devonian, West of Shetland: climate and sediment supply controls on fluvial, eolian and lacustrine deposition, *in* Doré, A.G. and Vining, B.A., eds., Petroleum Geology: North-West Europe and Global Perspectives—Proceedings of the 6th Petroleum Geology Conference, p. 957–967.

PAVLOVIC, M., SMALLEY, A., BORGHI, M., IBRAHIM, M., BALLOSINO, P. AND DYMMOCK, S., 2004, Structural interpretation of borehole image data in wells drilled with non-conductive mud systems: an example from the Adriatic Coast, Italy—including integration with 3-D reservoir model. Presented at the American Association of Petroleum Geologists European Region Conference held in Prague, Czech Republic, 10–13 October, 2004.

RICCI LUCCHI, F., 1995, Sedimentographica: A Photographic Atlas of Sedimentary Structures: New York, Columbia University Press, p. 63–64.

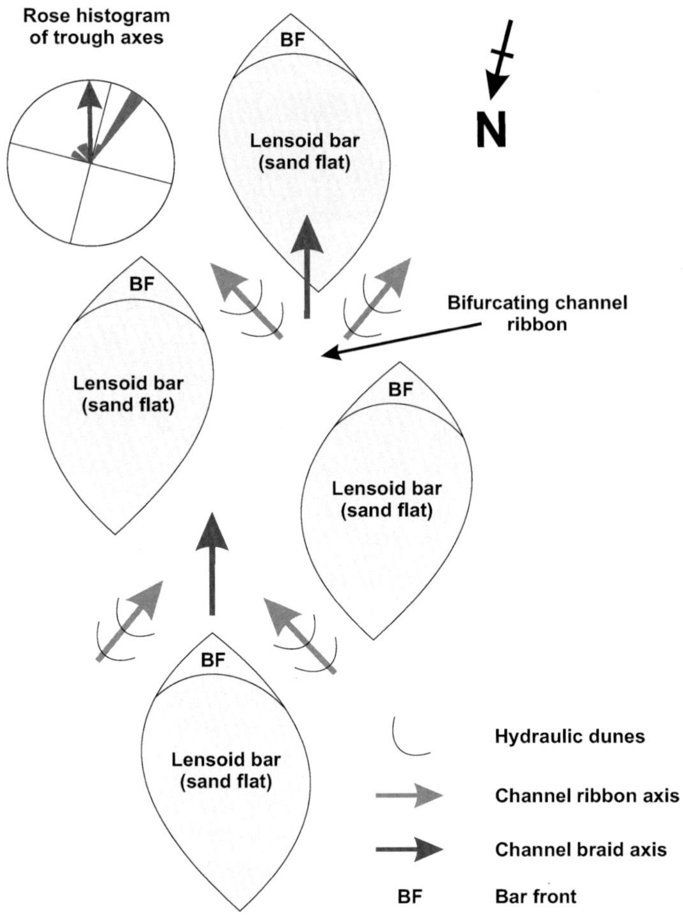

FIGURE 11.—Conceptual depositional model of braided fluvial channel elements and their relationship to channel-belt orientation, after Cant and Walker (1978) and observations of satellite images from the Brahmaputra river, Bangladesh (available on Google Earth). In braided rivers, hydraulic dunes are generally preserved on bar fronts by downstream accretion.

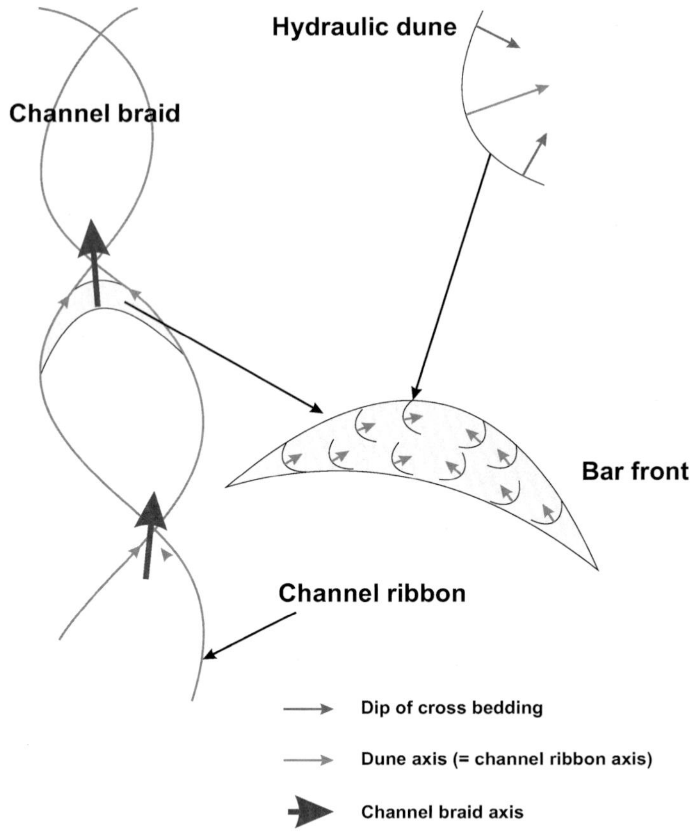

FIGURE 12.—Explanation of various geometrical elements within a braided channel system, used to construct conceptual models in this study. Hydraulic dunes are preserved on bar fronts by downstream accretion of lensoid bars. The axial traces of these dunes can be measured from trough cross-bedding (trough axes). In this model, trough axes align with channel-ribbons where they coalesce at bar fronts. The bisector of channel-ribbons gives the channel braid axis at this point. It is not likely that the complete geometry of bar fronts will be preserved in the rock record, due to erosion and avulsion. However, given that they display systematic geometrical relationships, channel braid axes can be inferred from those elements that are preserved.

SCHLUMBERGER, 2002, UBI. Advanced Borehole Imaging Independent of Mud Type. Brochure, SMP 5871.

SCHLUMBERGER, 2001, OBMI. Borehole Imaging in Oil-base Mud. Brochure, SMP 5806.

TERZAGHI, R.D, 1965, Sources of error in joint surveys: Geotechnique, v. 15, p. 287–304.

WITT, A.J., FOWLER, S.R., KJELSTADLI, R.M., DRAPER, L.F., BARR, D. AND MCGARRITY, J.P., 2009, Managing the start-up of a fractured oil reservoir: development of the Clair field, West of Shetland. Presented at the 7th Petroleum Geology Conference: From Mature Basins to New Frontiers, QEII Centre, London, 30 March – 02 April, 2009.

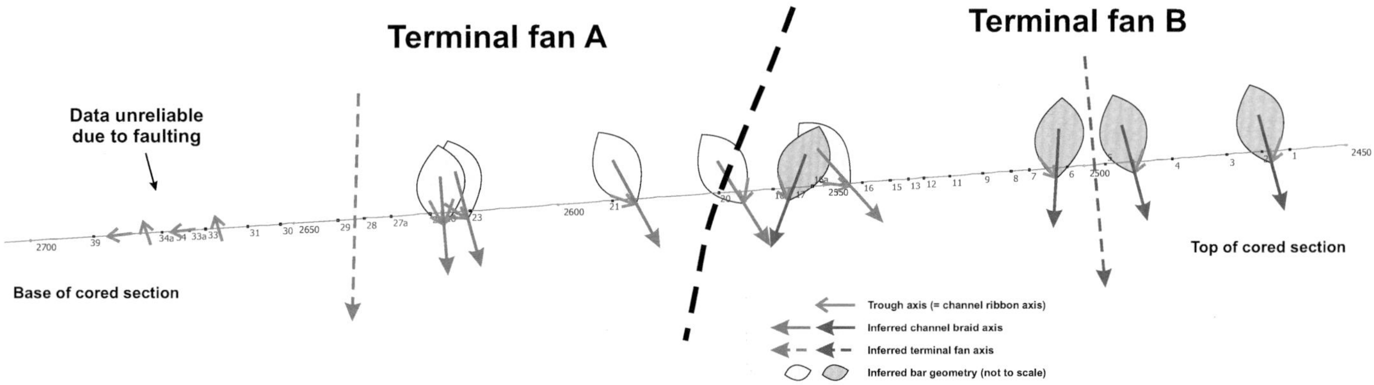

FIGURE 13.—Well trajectory plot in map view presented in Figure 7, with superimposed inferred braided fluvial geometries. From the trough axis trends simple longitudinal bars have been inferred, based on the conceptual depositional model illustrated in Figure 11 and the geometrical relations shown in Figure 12. The inferred channel braid axes display systematic divergent azimuth swings, suggesting fan distributions. Based on models for terminal fans (Nichols, 2005), fan axes can be inferred as shown. From this distribution at least two terminal fans are suggested, which overlap at their margins. This would suggest fan widths of at least 100 m in the vicinity of this well.

DIGITAL APPENDIX

The following supplementary material is available in electronic format either on the DVD accompanying this volume or through the digital archive on the SEPM website at http://www.sepm.org/
Figure S1: Full, interpreted image log, part 1: 2461 m to 2581 m
Figure S1: Full, interpreted image log, part 2: 2581 m to 2700 m

APPENDIX 1: BOREHOLE IMAGE ORIENTATION QUALITY CONTROL (QC) PROCEDURES

In order to check the reliability of borehole image orientation measurements, a QC check of the data which are used to record orientation is performed. This involves:

Comparing the hole orientation survey with an independent survey.

With any borehole imaging tool, a hole geometry survey is measured from a given point in the tool string. This measuring sonde (most commonly referred to as the GPIT (Schlumberger) or ORIT (Baker Hughes) for wireline tools) uses magnetometers and accelerometers, which measure borehole position and shape in three dimensions. The magnetometers record data in relation to magnetic north and the accelerometers record tool movement in 3-D space. In deviated wells, the accelerometers record the position of borehole highside for the tool, as this generally reflects gravity. Borehole highside is the point in the borehole from which the tool is referenced. The position of borehole highside in 3-D space, in relation to magnetic north, is measured by the magnetometers. After applying a correction for true north, the borehole highside measurement represents the trajectory of the wellbore, or the borehole survey, from which the tool is referenced. An important quality check for borehole image data is to compare this survey with an independent reference. Ideally a comparison is made with a gyroscopic borehole survey, which is made after the borehole has been completed. A gyroscopic survey is more reliable as it does not use magnetometers, which can be affected by conditions in the well. More commonly, a driller's directional survey is all that is available. A directional survey is measured while the hole is being drilled by the drilling engineer who is steering the well. By comparing these surveys discrepancies can be flagged and corrected where possible.

Raw orientation data check.

As explained above, magnetometers and accelerometers from the imaging tool orientation cartridge are fundamental for orienting the tool. Any measurement relating to the orientation of features derived from an imaging tool therefore depend on the reliability of their measurements. A simple and standard way of checking these data is to generate cross-plots. Three measurements are made by the orientation sonde by both the accelerometers and magnetometers in directions x, y and z, where the z direction is parallel to the tool axis. Magnetometers and accelerometers measuring in the x-y plane (perpendicular to tool axis) are more useful here, as they record both the movement and positioning of the tool measurement along the borehole. When plotted on a symmetric scale, where the tool has rotated the magnetometer and accelerometer data should form an arc which sits on a circle centered at "0:0" on the plot. If the data are not centered on zero then there is a problem with the measurement, which can be corrected accordingly. If there is noise in the data (often there is magnetic interference with the magnetometers from the formation or magnetized metal fragments in the drilling mud) then this should also be visible as scatter within the plotted arc.

DOWNSTREAM CHANGES AND ASSOCIATED FLUVIAL-EOLIAN INTERACTIONS IN AN ANCIENT TERMINAL FLUVIAL SYSTEM: THE PERMIAN ORGAN ROCK FORMATION, SE UTAH, U.S.A.

STEPHEN A. CAIN

Earth Sciences, School of Physical and Geographical Sciences, Keele University, Staffordshire, ST5 5BG, U.K.
Present address: Lukoil Overseas UK Ltd, Charles House, 5-11 Regent Street, London, SW1Y 4LR, UK.
e-mail: scain@lukoiloverseasuk.com

AND

NIGEL P. MOUNTNEY

Earth Sciences, School of Earth and Environment, University of Leeds, Leeds, LS2 9JT, UK.
e-mail: n.mountney@see.leeds.ac.uk

ABSTRACT: The fluvial Organ Rock Formation, which forms part of the Pennsylvanian–Permian Cutler Group of the Paradox foreland basin, is exposed across much of SE Utah and adjoining parts of northern Arizona and represents a wedge of coarse-grained fluvial strata that progressively fines southwestwards (distally) away from its source area, the Uncompahgre Uplift. By the time of onset of Organ Rock deposition (Leonardian-Artinskian), the Paradox basin was in an overfilled state, resulting in the progradation of a 100-m-thick wedge of fluvial strata across a wide part of the basin floor. These deposits record a downstream transition from a proximal fluvial system that was dominated by in-channel, high-energy, bed-load sedimentation, through a medial zone in which channels lay within belts that were subject to a variety of lateral accretion, avulsion, and anabranching processes, to a distal zone where evidence for in-channel sedimentation is less abundant and in which sheet flood and eolian dune elements are dominant.

A suite of high-resolution architectural panels detailing stratigraphic relationships across a range of outcrop sections totaling 5 km in length were integrated with a set of broader regional stratigraphic panels based on correlations between 84 vertical sedimentary logs. These were together used to propose two possible paleogeographic models that account for the observed systematic downstream variability in fluvial architecture and associated style of fluvial-eolian interactions across the region. The first paleogeographic model represents the Organ Rock fluvial system as a channel network that is continuous from its source (Uncompahgre Highlands) to its point of termination on a low-relief desert plain in the distal part of the basin. This model is comparable, in part, to the processes responsible for downstream changes in channel morphology predicted by current models of terminal fluvial systems. By contrast, the second paleogeographic model represents the Organ Rock fluvial system as two separate channel networks that occupied more proximal and more distal parts of the basin, respectively, and which were separated by an intermediate floodout zone in which channel competency and physical expression was lost. This latter model, which is supported by field observations, does not conform to current models of terminal fluvial systems and has implications for the hydrological controls on the evolution of such systems and for the transportation and distribution of sediment within such systems.

KEY WORDS: Fluvial; eolian; distributary; ephemeral; Cutler Group; Utah

INTRODUCTION

Terminal fluvial systems represent an unusual style of fluvial drainage network in which fluvial discharge does not drain to a significant standing body of water (e.g., the sea or a lake), but instead terminates on an alluvial plain or in marginal eolian dune-field settings. Such a definition can be applied to fluvial systems in both fan and non-fan settings. Current depositional models of terminal fluvial systems, commonly referred to as "terminal fans", developed in dryland settings (e.g., Friend, 1978; Fisher et al., 2008; Lang et al., 2004; Olsen, 1987; Kelly and Olsen, 1993; Hampton and Horton, 2007) imply a fan setting for such systems in which channelized fluvial flow moves progressively from proximal to distal parts of the system through a continuous series of channels that display a pattern of downstream bifurcation. The downstream reduction in the size of individual channels but the corresponding increase in the number of active channels has been widely cited to exert a primary control on the style of resultant sedimentation in these systems, whereby systematic downstream changes in preserved fluvial architecture are typically attributed to a decrease in discharge and flow competency (Friend, 1978; Kelly and Olsen, 1993). The applicability of such models has, however, been questioned (see North and Warwick, 2007, and Cain and Mountney, 2009, for more detailed discussions), in particular because of their apparent inability to accurately predict downstream changes in fluvial architecture and facies distributions in ancient terminal fluvial systems.

It has long been known from studies of modern dryland fluvial systems that discharge is not ubiquitously sourced from upland catchment areas but is instead commonly additionally sourced from intrabasinal catchments where localized convective storms act to affect only small parts of the drainage basin during any one event (e.g., Cooke et al., 1993; Tooth 2000b; North and Warwick, 2007). As such, only part of the fluvial system may experience active fluvial flow in the aftermath of one such event, and subsequent storm events elsewhere in the basin will likely induce flow through other parts of the channel network. A consequence of this, and one that is largely overlooked by current depositional models for terminal fluvial systems, is that fluvial channel networks in different parts of the fluvial system may evolve, to some degree, independently of each other and therefore preserved downstream changes in fluvial architecture may not necessarily be closely related to gradual downstream reductions in discharge and flow competency. Furthermore, as observed in modern dryland settings (e.g., Tooth and Nanson, 1999; Tooth, 2000a; North and Warwick, 2007), terminal fluvial systems do not ubiquitously develop an overall fan-shaped morphology or repeatedly bifurcate downstream as is predicted by current terminal fan models (e.g., Kelly and Olsen, 1993; Hampton and Horton, 2007). These models also offer little insight into

From River To Rock Record: The Preservation Of Fluvial Sediments And Their Subsequent Interpretation
SEPM Special Publication No. 97, Copyright © 2011
SEPM (Society for Sedimentary Geology), ISBN 978-1-56576-305-0, p. 167–185.

the nature and extent of fluvial–eolian interactions, which are widespread and exhibit a varied character in both modern (e.g., Langford, 1989; Bullard and Livingstone, 2002) and ancient (e.g., Langford and Chan, 1989; Mountney et al., 1999; Mountney and Jagger, 2004) semiarid and arid fluvial systems. Such relationships have important implications when, for example, considering hydrocarbon reservoir architecture and heterogeneity (North, 1985; Glennie, 1998).

Given that dryland fluvial systems are highly susceptible to rapid and significant geomorphological change in response to the impact of repeated high-magnitude flood events (Graf, 1983, 1988), the morphology of dryland rivers and their preserved sedimentology are more likely to be shaped by low-frequency, high-magnitude floods that eradicate the record of earlier smaller flood events (Reid and Frostick, 1997; Billi, 2007). As such, most dryland river systems are likely to undergo dramatic changes in form and sedimentary architecture during the next substantial discharge event. For these reasons, depositional models based solely on evidence from the flow behavior of modern systems cannot be used to account for the observed sedimentology. Furthermore, the sedimentology of ancient successions preferentially records the signature of larger but low-frequency flood events, and the deposits are unlikely to be representative of "normal" flow conditions.

For clarification purposes, the term "terminal fan" in this study is used only in reference to previously published models to which this terminology has been applied. The term "terminal fluvial system" is used herein as a generic descriptor that can be applied to any fluvial system, in a fan or non-fan setting, in which fluvial flow terminates before a significant standing body of water is reached, whereas the term "fluvial fan" is used to describe a large-scale fan-shaped package of fluvial sediment developed by the migration and aggradation of river systems that may not necessarily be dissimilar to rivers in non-fan settings. The Organ Rock Formation has been interpreted as representing a large-scale terminal fluvial fan system (see Cain and Mountney, 2009) in that it represents a fluvial fan system in which fluvial flow terminated before a significant standing body of water was reached. Although the Organ Rock Formation lacks a "classic" fan-shaped morphology, owing partly to its large scale, the succession represents the southwestward (basinward) extension of the long-lived semiarid fluvial fan system represented by the undifferentiated part of the Cutler Group (cf. Mack and Rasmussen, 1984; Barbeau, 2003), in which fluvial sediments were dispersed laterally over a large area in response to sediment delivery to the overfilled Paradox foreland basin.

The aim of this paper is to relate preserved fluvial and eolian sedimentary architecture observed in the Organ Rock Formation to processes known to operate in modern dryland fluvial settings and to investigate the processes responsible for the downstream sedimentological and architectural trends observed in outcrop. The specific objectives of this study seek to build on the detailed sedimentological and stratigraphic characterization of the Organ Rock Formation presented by Cain and Mountney (2009) and include: (1) documentation of the downstream variation in preserved fluvial architecture and sedimentary style via intermediate-resolution and high-resolution architectural element and paleocurrent analyses; (2) investigation of the complex nature of fluvial–eolian interactions, which are known to be inherent to many semiarid fluvial systems and successions; and (3) to propose two basin-scale depositional models that depict alternative paleogeographic scenarios that may account for the downstream variations in preserved fluvial architecture.

BACKGROUND AND GEOLOGICAL SETTING

The fluvial Organ Rock Formation, which forms part of the Pennsylvanian–Permian Cutler Group of the Paradox foreland basin, is exposed across much of southeastern Utah and adjoining parts of northern Arizona and represents a wedge of coarse-grained fluvial strata that progressively fines southwestwards (distally) away from its source area, the Uncompahgre Uplift (Stanesco et al., 2000; Cain and Mountney, 2009). In the subsurface, the unit extends into southwestern Colorado, northwestern New Mexico, and northern Arizona (Fig. 1; Condon, 1997). Over much of the outcrop area, the Organ Rock Formation conformably overlies the predominantly eolian Cedar Mesa Sandstone (Permian) and a range of styles of interaction demonstrate the gradational nature of the transition between the two successions (Baars, 1975; Blakey, 1996; Condon, 1997; Stanesco et al., 2000). The Organ Rock Formation is conformably overlain by either the Permian White Rim or De Chelly sandstones (both eolian), in areas to the west of the Colorado River and around Bluff, respectively (Dubiel et al., 1996; Condon, 1997; Stanesco et al., 2000). However, in much of the Canyonlands and Hite areas and across the Monument Upwarp, the Triassic Moenkopi Formation rests unconformably on the Organ Rock Formation (Baars, 1975; Condon, 1997; Stanesco et al., 2000). A detailed review of the sedimentology and stratigraphy of the Organ Rock Formation is provided by Cain and Mountney (2009), with earlier, more generalized discussions presented by Baars (1962, 1975), Nuccio and Condon (1996), Condon (1997), and Stanesco et al. (2000).

As is the case for current models of ancient terminal fluvial systems, the deposits of the Organ Rock Formation also exhibit pronounced downstream variations in fluvial sedimentology that have been attributed to a progressive decrease in fluvial discharge along the fluvial reach (Cain and Mountney, 2009). However, further detailed analyses of fluvial architecture observed in proximal, medial, and distal parts of the succession (this study) suggests that preserved fluvial architecture may be significantly more complex than that recognized previously and which is depicted by existing models derived from other similar fluvial systems (e.g., Friend, 1978; Kelly and Olsen, 1993; Hampton and Horton, 2007). Observations introduced in this study include an overall converging rather than diverging plan-view fluvial architecture, an overall downstream decrease in channel size and frequency, and an apparent nonsystematic relationship in preserved fluvial architecture between medial and distal parts of the system.

METHODOLOGY AND DATA

Geometric relationships of fluvial and eolian facies bodies and the juxtaposition of these bodies observed in the Organ Rock Formation were constrained using a suite of oriented architectural element panels and basin-scale fence diagrams. These data, presented in this study, depict the detailed facies architecture throughout proximal, medial, and distal parts of the succession and were integrated with observations made by Cain and Mountney (2009), who present a full lithofacies scheme, an architectural-element classification, and representative sedimentary log data for the Organ Rock Formation. Two-dimensional architectural panel analysis records 5 km of bed-scale architectural relationships observed in proximal, medial, and distal parts of the succession, whereas the pseudo-three-dimensional architectural-panel analysis highlights basin-scale variations in stratigraphic and sedimentary architecture in medial and distal parts of the succession. Based on the architectural-element analysis presented in this study a classification for fluvial–eolian interactions is herein introduced based on primary observations made from the Organ Rock Formation. By combining these data with paleocurrent data obtained from fluvial and eolian cross-bed foreset azimuths (cf. Cain and Mountney, 2009) and by drawing comparisons from present-day dryland fluvial systems, such as those observed across the northern plains of central Australia (e.g. Tooth and Nanson, 1999; North et al., 2007), two possible paleogeographic scenarios are proposed, each of which is employed to account for the downstream variability in preserved fluvial architecture and the nature of associated fluvial–eolian interactions.

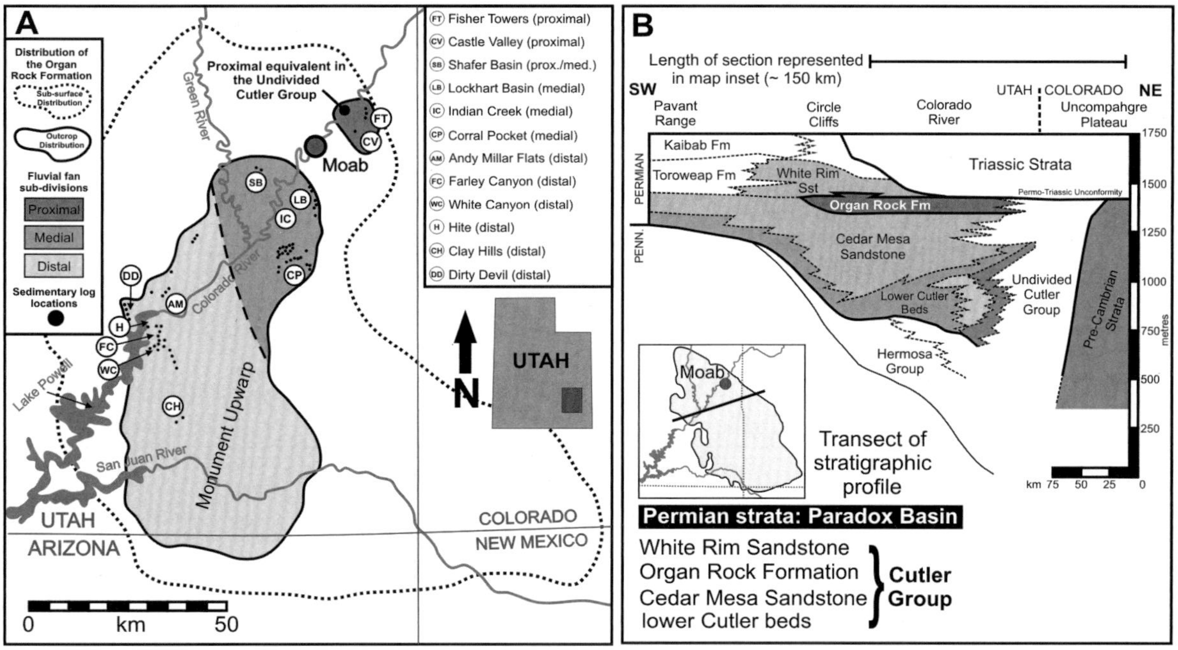

FIGURE 1.—A) Regional distribution of the Organ Rock Formation and its proximal equivalent in the undifferentiated Cutler Group in the subsurface and at outcrop. Locations of sedimentary logs are denoted by black dots. B) Regional stratigraphy of the Paradox Basin in South East Utah.

SEDIMENTARY ARCHITECTURE

The regional distribution of facies and their respective associations throughout the succession has enabled the Organ Rock Formation to be informally subdivided into proximal, medial, and distal fluvial regions (Fig. 1). A summary of the principal fluvial and eolian facies and architectural elements encountered in these regions is shown in Figure 2; detailed descriptions are given by Cain and Mountney (2009). The medial part of the succession was further subdivided into two vertical parts, informally referred to as the "lower" medial Organ Rock Formation and the "upper" medial Organ Rock Formation. Of these, the upper-medial part is interpreted to record the backstepping of more distal facies over more medial facies during an overall episode of fluvial-system retreat (Cain and Mountney, 2009). Therefore, to provide a more accurate assessment of proximal-to-distal architectural variations, the upper-medial part of the succession is not considered in detail in this architectural-element analysis study.

Proximal Organ Rock Formation

In the most proximal parts of the basin, to the northeast of the town of Moab, the Organ Rock Formation passes into the age-equivalent undivided Cutler Group and in this region the type section for the succession is the area around Fisher Towers, approximately 30 km northeast of Moab (Fig. 1). The true stratigraphic thickness of the succession is difficult to determine in this proximal part of the basin because the lower part of the unit is indistinct from underlying fluvial deposits that are contemporaneous with the Cedar Mesa Sandstone in more distal parts of the basin. In this region, where the Organ Rock Formation merges into the undivided Cutler Group, fluvial deposits are represented by an 1800-m-thick wedge of arkosic sandstones and conglomerates (Condon, 1997) that was deposited throughout the development of the Paradox Basin (Barbeau, 2003). An upper boundary, however, is discernible and represents the Permian–Triassic boundary where the Triassic Moenkopi Formation unconformably overlies the lower Permian undifferentiated Cutler Group (Baars, 1975; Condon, 1997; Stanesco et al., 2000).

Proximal Facies Architecture: In proximal regions, facies diversity is relatively low, with channelized fluvial facies (CH) accounting for 89% and nonchannelized fluvial facies (UC) accounting for 9% of the total succession (Fig. 3). Only 2% of the proximal succession represents eolian sandsheet and related facies (Fig. 3). Fluvial channel-fill facies are arranged into laterally and vertically amalgamated channel-fill complexes (F1) (Fig. 4) that tend to form laterally extensive units displaying overall sheet-like geometries, some examples of which extend for distances in excess of 5 km perpendicular to paleoflow (NW-SE) and which routinely can be traced for over 15 km parallel to paleoflow (NE-SW) (Cain and Mountney, 2009).

Paleocurrent analyses of the proximal fluvial deposits utilized dip-azimuth measurements obtained exclusively from preserved downstream and laterally migrating fluvial barforms recorded either in log sections or on architectural-element panels. Measurements reveal a resultant mean fluvial paleoflow direction of 236° (southwest) and reflect fluvial drainage away from the Uncompahgre Uplift (Fig. 4).

The fluvial style in proximal parts of the basin is interpreted as representing a high-energy, bedload-dominated braided fluvial reach (Cain and Mountney, 2009). A radial dispersion of paleocurrents in this region, as derived from foreset dip-azimuth analysis of trough and planar cross-bedded fluvial sandstones, indicates the presence of both downstream and laterally migrating barforms. The quality of the exposure enables lateral-accretion sets and large bars to be observed in plan view at several localities, and the curved form of the preserved foresets partly accounts for the broad dispersion in the paleocurrent data. The absence of significant proportions of preserved overbank deposits is testament to the lateral mobility and cannibalistic nature of these channel networks, though it may also reflect a lack of appreciable

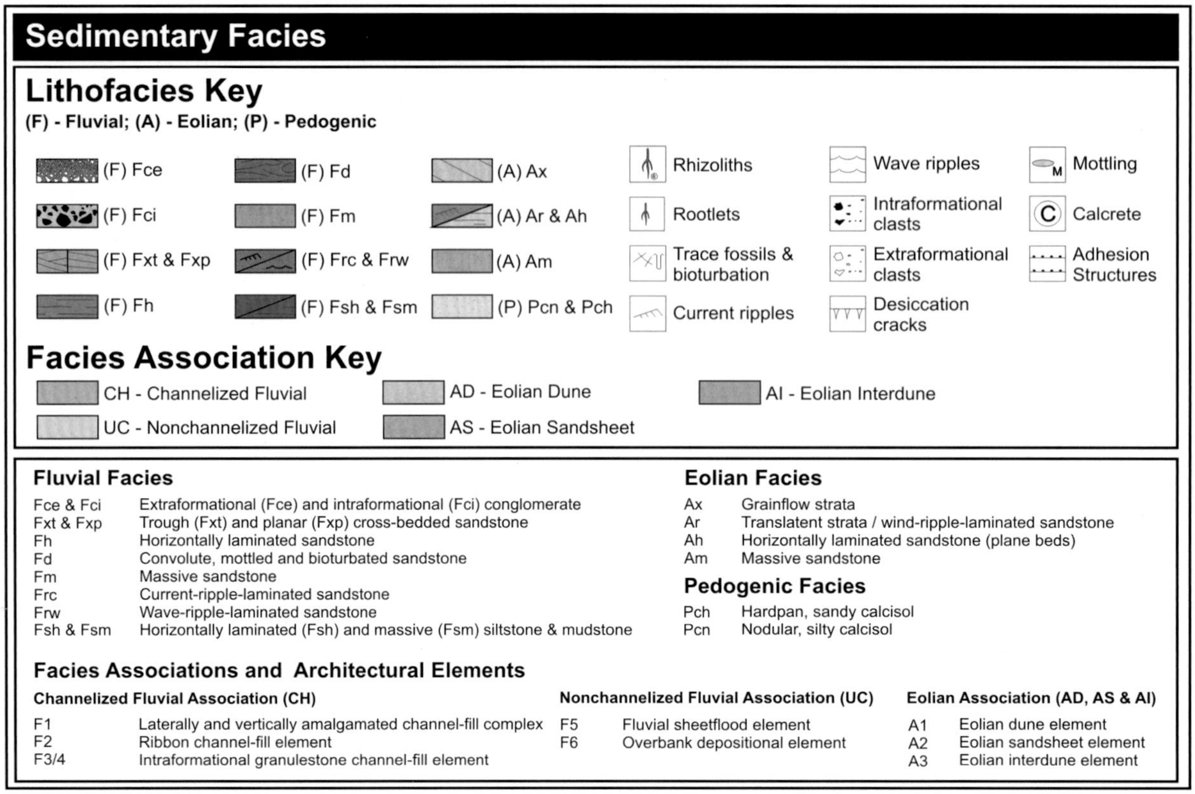

FIGURE 2.—Key to standard colors, symbols, and abbreviations used throughout this paper. Summary of sedimentary facies and architectural elements identified in the outcrop belt of the Organ Rock Formation. See Cain and Mountney (2009) for more detailed descriptions.

overbank sedimentation as a result of a relatively high fluvial system gradient.

Medial Organ Rock Formation

The type locality for the medial Organ Rock Formation is represented by the area in and around Corral Pocket 45 km south of Moab (Fig. 1) from where a framework of 46 sedimentary logs was erected. This location, and that around the neighboring Indian Creek (Fig. 1), has been interpreted previously (e.g., Condon, 1997; Stanesco et al., 2000, Cain and Mountney, 2009) to represent a predominantly fluvial environment of deposition, characterized by medium- to coarse-grained, arkosic fluvial channel-fill sandstone elements (F1 and F2), interbedded with thick and tabular floodplain elements composed mostly of fine- to medium-grained sandstone with lesser amounts of mudstone (F5 and F6). Eolian dune (A1) and sandsheet (A2) elements are also recognized.

A suite of outcrop-scale correlation panels arranged in a pseudo-three-dimensional framework highlights the gross-scale architecture of the Organ Rock Formation in medial parts of the basin (Fig. 5). The boundary between lower-medial and upper-medial parts of the succession is distinct, and is interpreted as a regional, time-significant surface (Cain, 2010), representing a significant change in fluvial regime. The lower-medial section is characterized by channelized (CH) and nonchannelized (UC) fluvial elements, and by eolian dune (AD, A1) and sandsheet (AS, A2) elements. The upper-medial section is characterized almost exclusively by nonchannelized fluvial sheetflood elements (F5). The following description and analyses of the medial Organ Rock Formation focus exclusively on the lower part of the succession.

Medial Facies Architecture: The "lower" Organ Rock Formation in its medial reaches is dominated by channelized fluvial deposits (CH) that account for up to 63% of the succession (Fig. 3), and which exhibit a vertical change in gross-scale channel architecture (Figs. 5, 6). The lowermost 10 to 15 m of the Organ Rock Formation in this region is characterized by laterally extensive multilateral and multistory sheets of fluvial channel-fill sandstones (F1), individual examples of which can be traced continuously over an area of 50 km^2. In places these interfinger at a local scale with cross-bedded eolian sandstones (A1) of the underlying Cedar Mesa Sandstone (Fig. 5). This relationship resulted in a complex range of styles of fluvial–eolian interaction (Fig. 5; see later analyses). Fluvial channel complexes (F1) are typically 10 to 15 m thick, with variations attributed to varying topography and fluvial incision at the boundary between the Organ Rock Formation and the underlying Cedar Mesa Sandstone (Fig. 6). Nonchannelized fluvial elements (F5 and F6) are only rarely developed in this lower part of the succession, and those that are present are typically laterally discontinuous, having apparently been cut out by subsequent fluvial incision (Figs. 5, 6).

The 30–40 m-thick interval above, which immediately overlies the lowermost 10 to 15 m of the medial Organ Rock Formation, is represented by more discrete channel-fill elements (F2) that are characterized by a decrease in lateral and vertical connectivity compared that seen with the lower fluvial channel sheets (Figs. 5, 6). This decrease in channel-fill connectivity is coincident with an increase in preservation of nonchannelized fluvial deposits (F5 and F6), which appear thick-bedded and laterally extensive (traceable for distances in excess of 250 m). Thick-bedded eolian sandsheet elements (A2) are also widespread, and these are interbedded with, and grade laterally

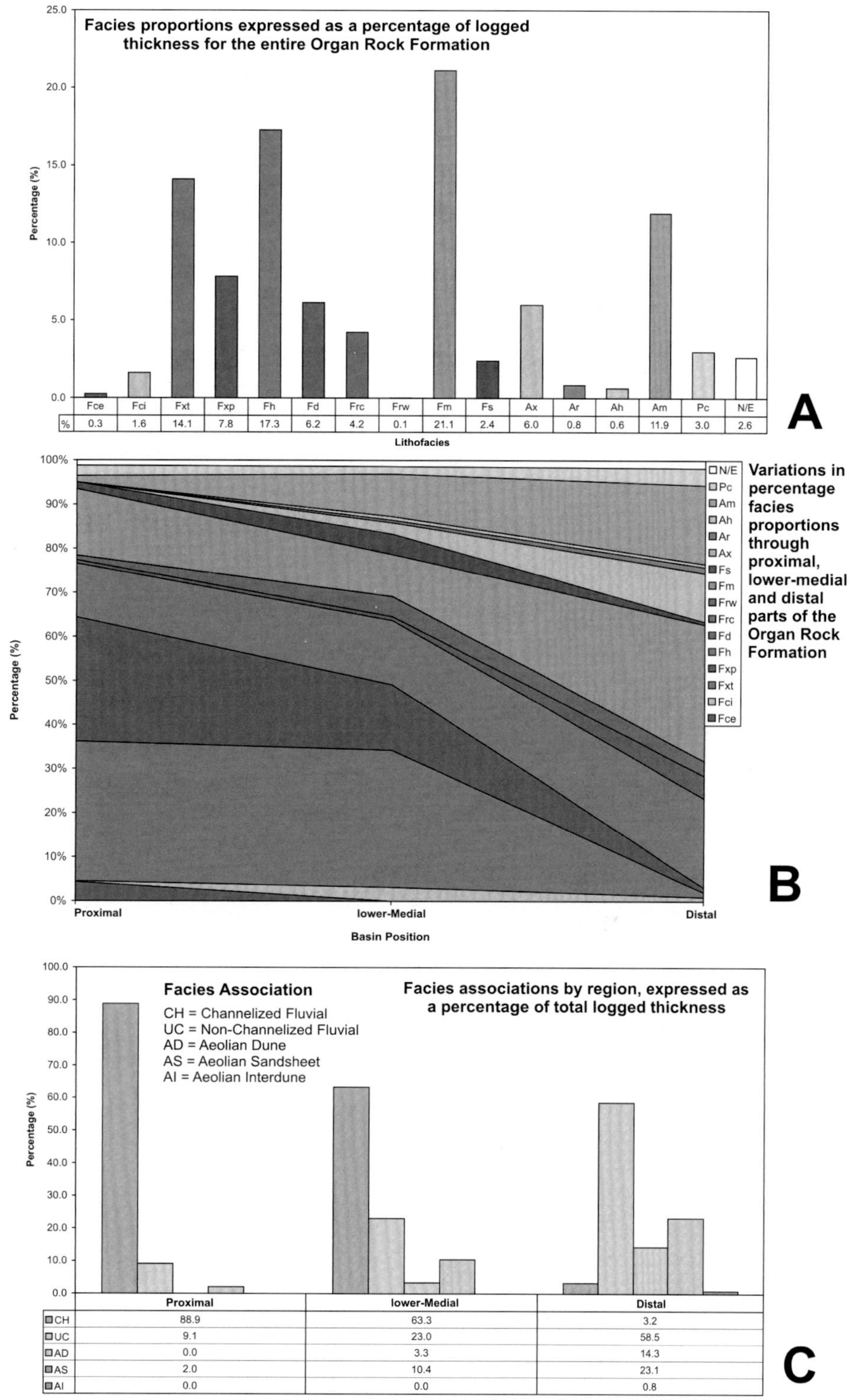

FIGURE 3.—A) Percentage facies proportions for the whole Organ Rock Formation. B) Regional variations in percentage facies proportions throughout the Organ Rock Formation. C) Comparison of percentage facies associations between proximal, lower-medial and distal Organ Rock Formation. See Figure 2 for explanation of facies and architectural-element codes.

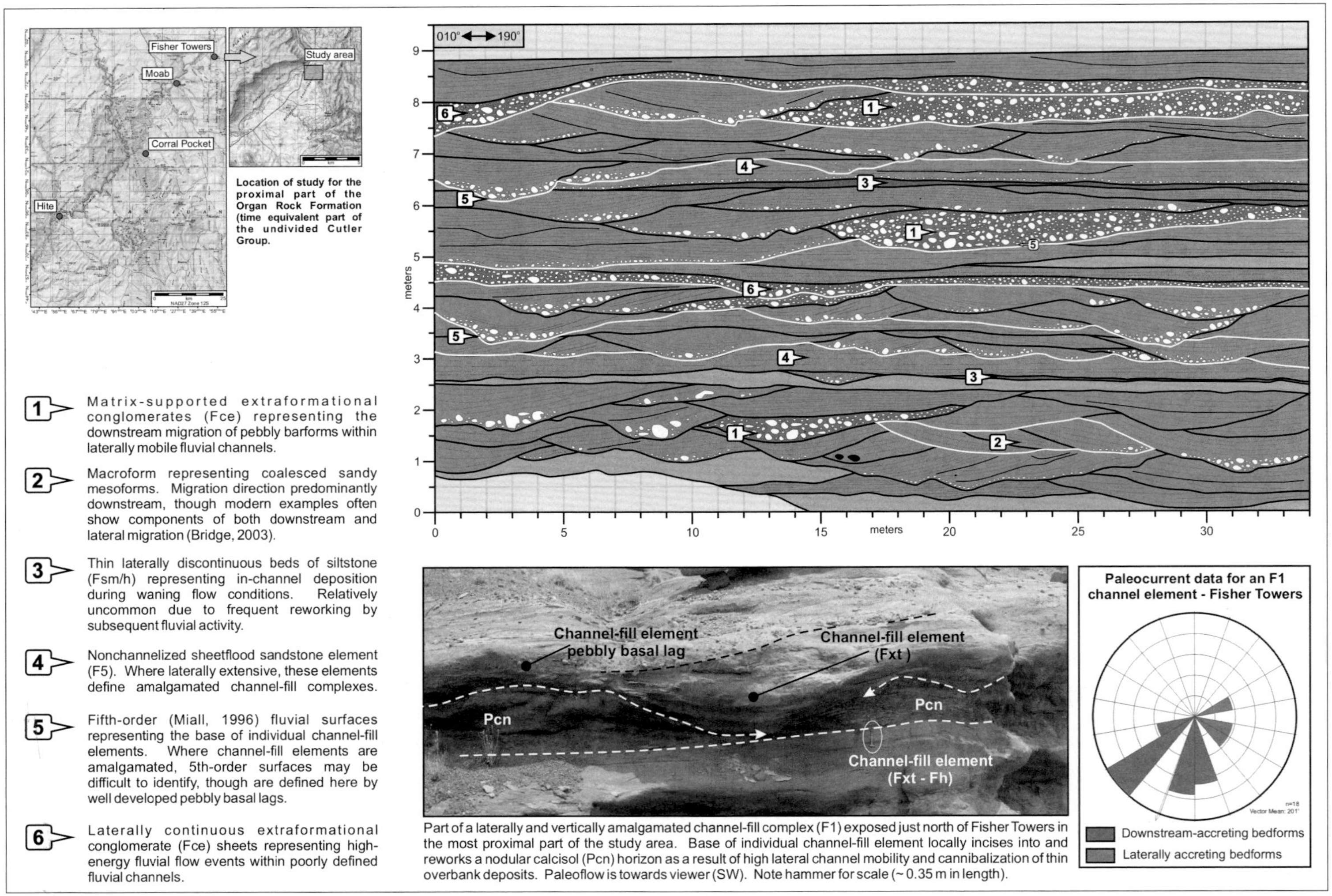

FIGURE 4.—Two-dimensional architectural panel depicting typical fluvial architecture characteristic of proximal parts of the Organ Rock Formation (time equivalent Undivided Cutler Group). Panel location is approximately 5.5 km north of Fisher Towers). Refer to lithofacies key in Figure 2.

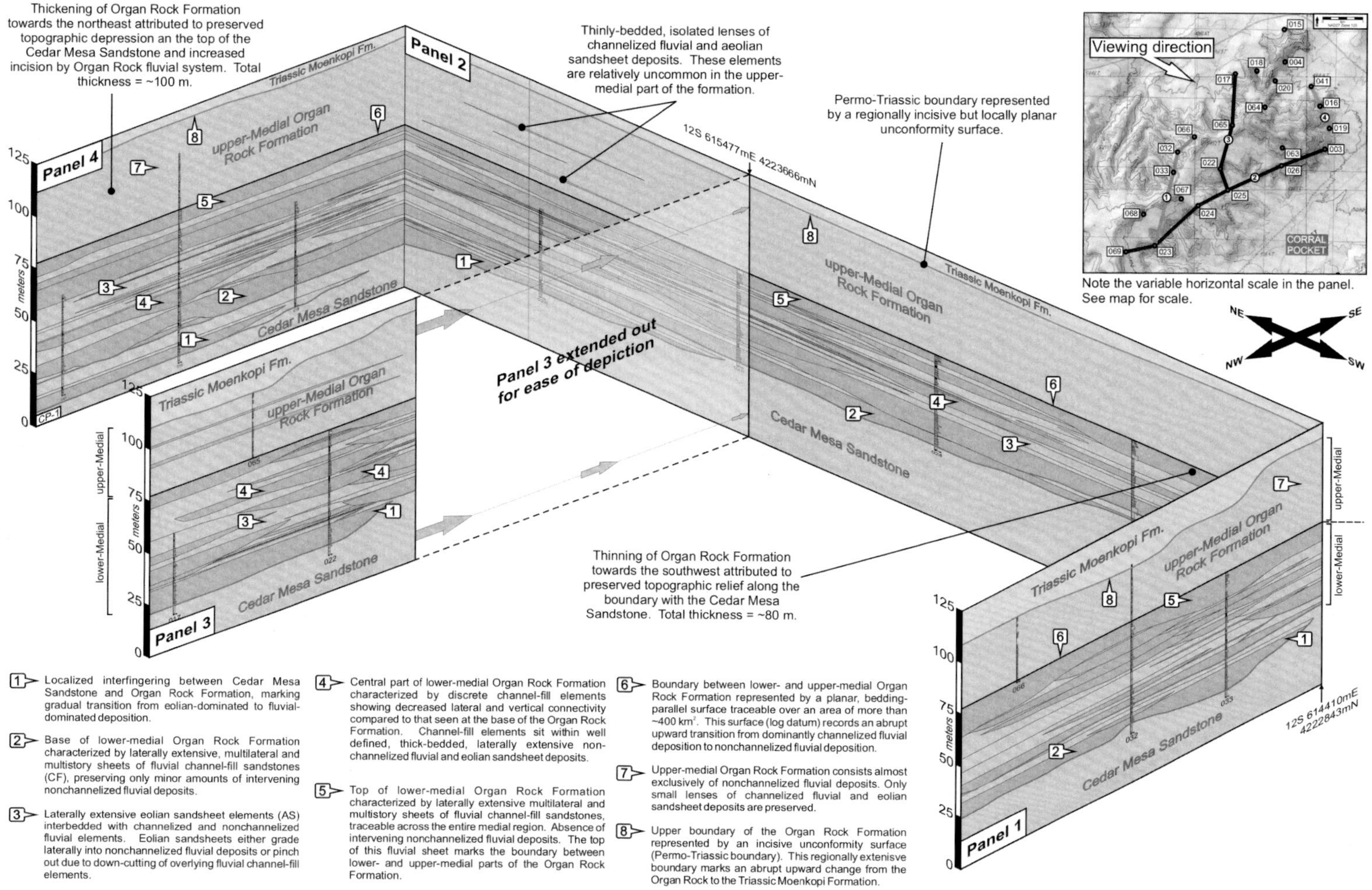

FIGURE 5.—Intermediate-scale, pseudo-three-dimensional reconstruction of the gross-scale sedimentary architecture that typifies medial parts of the Organ Rock Formation. See map inset for viewing direction and panel positions. Also see Figure 1 for the regional location of Corral Pocket and Figure 2 for the key to facies associations.

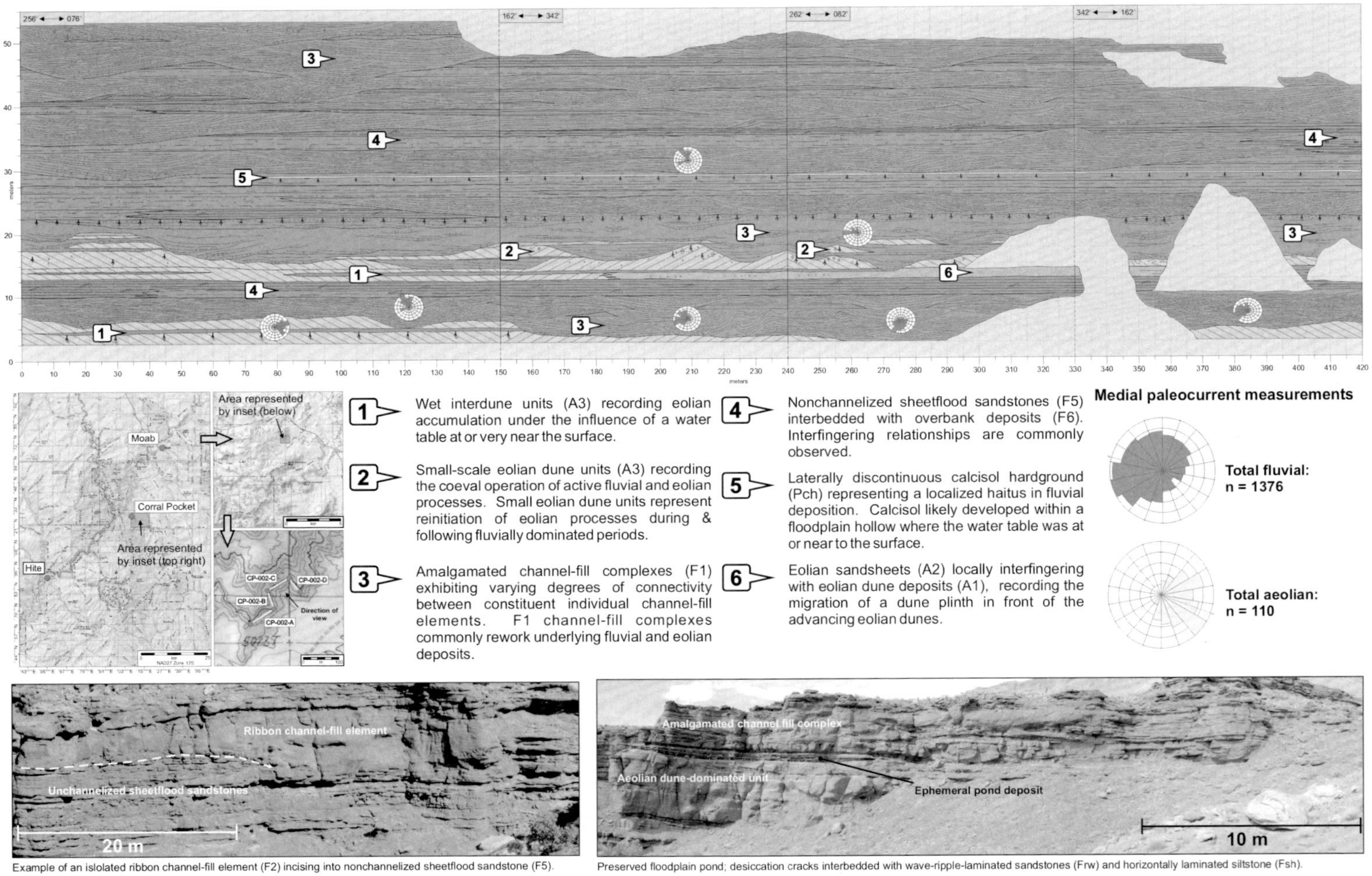

FIGURE 6.—Composite diagram of architectural panels depicting the high-resolution fluvial and eolian architecture characteristic of lower medial parts of the Organ Rock Formation. The photographs depicted are from a part of Corral Pocket not shown on the panel but are representative of the architecture in the medial part of the succession. Refer to lithofacies key in Figure 2.

into, nonchannelized fluvial deposits. Overlying channelized fluvial sandstones typically erode down into these deposits.

The uppermost 15 to 20 m of the "lower" Organ Rock Formation is represented by a change from discrete fluvial channel-fill elements (F2) back to multilateral and multistory fluvial channel-fill elements (F1), which are arranged into laterally extensive units with sheet-like architecture. Such units can be traced over an area of at least 40 km^2. This upper channel sheet varies in thickness from 5 m to 20 m and in places has minor nonchannelized fluvial elements enveloped within it. The top of this upper channel sheet marks the boundary between the lower-medial and upper-medial Organ Rock Formation.

Paleocurrent measurements representative of the medial part of the Organ Rock fluvial system were primarily derived from both the Corral Pocket area and from the Lockhart Basin area, 15 km to the north (Fig 1). Foreset azimuths were recorded from both fluvial and eolian bedsets.

In the Lockhart Basin region, 173 fluvial paleocurrent readings from channel-fill facies show a wide radial pattern of dispersion, with the resultant mean direction being northwest (309°) and a modal direction between 000° and 020°. In the Corral Pocket region, 1203 fluvial paleocurrent readings predominantly from amalgamated channel-fill complexes show a wider radial pattern of dispersion indicative of a predominantly southwest to west-directed paleoflow with a resultant mean direction to 282° and a modal direction between 240° and 260°.

By combining data from both the Corral Pocket and Lockhart Basin regions and by treating all data equally, a resultant mean fluvial paleocurrent direction of west-northwest (285°) is discerned, but the modal paleocurrent direction is west-southwest (240° to 260°; Fig. 6). The radial dispersion of fluvial paleocurrent data from the lower-medial part of the succession therefore could indicate that the data set likely contains measurements from both downstream-migrating and laterally-migrating fluvial barforms, particularly where radial variations are observed within a single channel complex. However, a modal paleoflow direction between 000° and 020° observed in the Lockhart Basin region contradicts more regional trends (cf. Stanesco et al., 2000). It is not possible with the current data set to determine if this modal value is significant or whether it is somehow biased due to an inadequate distribution of measurements and must therefore be interpreted with caution. It is possible that these measurements could result from bedforms that were migrating laterally across a channel belt, in directions oblique or even perpendicular to the regional paleoflow given by the switch from proximal to medial to distal facies toward the southwest. Alternatively, they may represent some degree of channel sinuosity. Eolian paleocurrent data across the medial region (Lockhart Basin to Corral Pocket) reveal a modest radial pattern of dispersion, with a resultant mean direction to the southeast (110°) and a modal direction between 120° and 140° (Fig. 6), a pattern that is similar to that recorded from the underlying Cedar Mesa Sandstone (Loope, 1984; Mountney and Jagger, 2004; Mountney 2006a).

Distal Organ Rock Formation

The type locality for the distal part of the Organ Rock Formation in the southwest of the study area is the area around Hite Marina at Lake Powell (Fig 1). Distal parts of the succession have previously been interpreted as representing a low-relief coastal plain dominated by streams and subject to marine reworking (Baars, 1962, 1975; Condon, 1997; Stanesco et al., 2000). There is, however, no observed evidence from which to infer proximity to an adjacent shoreline in the Hite Marina area, and no evidence for marine incursions and reworking in the Organ Rock Formation was noted in this study. Instead, the distal part of the succession is dominated by interbedded and interfingering fluvial and eolian sediments, which are indicative of a sheetflood-dominated distal fluvial plain that grades into eolian sandsheets and marginal areas of adjoining eolian dune fields (Cain and Mountney, 2009).

The gross-scale architecture of the distal Organ Rock Formation, as determined by lateral tracing of surfaces and the construction of pseudo-three-dimensional fence-panel diagrams (Figs. 7, 8), typically lacks the channelized fluvial architectures characteristic of the proximal and lower-medial parts of the succession. Instead, distal regions are dominated by tabular, sheetflood elements (F5) similar to those observed in the upper-medial part of the succession. In contrast to the upper-medial Organ Rock Formation, however, the distal part of the succession, though consisting predominantly of nonchannelized fluvial deposits, is additionally characterized by a significant proportion of eolian dune (A1) and eolian sandsheet (A2) elements, which represent 14% and 23% of the preserved succession, respectively (Fig. 3).

Distal Facies Architecture: Preserved fluvial architecture in the distal part the Organ Rock Formation is dominated by laterally extensive, tabular, nonchannelized fluvial sheetflood deposits (F5), which account for 59% of the succession, and which have small and isolated channelized fluvial elements (3% of the succession) nested within them (Fig. 3). The fluvial sheets are separated vertically by eolian dune (A1) and sandsheet (A2) elements, 1–15 m thick, that represent 14% and 23% of the succession, respectively. Where channelized fluvial deposits are observed, they typically occur as intraformational, granule-stone-filled elements (F3) and heterolithic (fine sandstone to siltstone) interbedded channel-fill elements (F4), though unlike those observed in the proximal and lower-medial parts of the succession, they are typically isolated, with little to no lateral or vertical connectivity. Fluvial sediments are typically interbedded with abundant eolian sandsheet (A2) deposits, which form relatively thin, laterally extensive sheets that exhibit both localized and more regional pinch-outs (Fig. 7). A characteristic feature of the distal Organ Rock Formation is a prominent eolian dune–dominated interval that can be traced throughout much of the region (Fig. 7), and which covers an observed area of at least 500 km^2, with a substantially greater extent almost certainly present in the subsurface (Stanesco et al., 2000). The high-resolution architecture that is typical of the distal part of the succession is depicted in Figure 8, which represents a 135-m-long section that demonstrates the two-dimensional arrangement of fluvially-related architectural elements. Observable in this panel is the laterally extensive nature of the sheetflood sandstones and the rarely preserved channelized fluvial architecture, which is characterized by relatively thin (< 5 m thick) and narrow (< 100 m in width) channel-fill elements when compared to those preserved in medial parts of the basin. Though this section is relatively thin overall, it does reveal the rare preservation of internal sandbody geometries and is considered representative of distal fluvial sandbodies as a whole.

The base of the prominent eolian unit is sharp and flat and appears to represent a paleo-horizontal surface (Stanesco et al., 2000). It is therefore considered a reliable datum from which to hang log sections measured in the area. The eolian dune–dominated interval attains a maximum thickness of 15 m in the Farley Canyon area before gradually thinning and pinching out to the south and to the northeast. The top of this prominent eolian unit is irregular, showing evidence of localized incision and reworking by overlying fluvial sediments (Fig. 7).

Above the prominent eolian unit, further evidence for channelized fluvial deposition is difficult to discern, and where examples do occur, deposits cannot be traced for more than a few tens of meters. In a style similar to that encountered below the prominent eolian unit, widespread nonchannelized fluvial deposits (F5) in the upper part of the succession are sheet-like and laterally continuous, and are interbedded with regionally extensive eolian sandsheet and more localized eolian dune deposits, which form extensive sheets and localized lenses, respectively (Fig. 7).

A total of 19 fluvial and 69 eolian paleocurrent measurements were obtained from the Hite area (Fig. 8). Fluvial paleocurrent data are

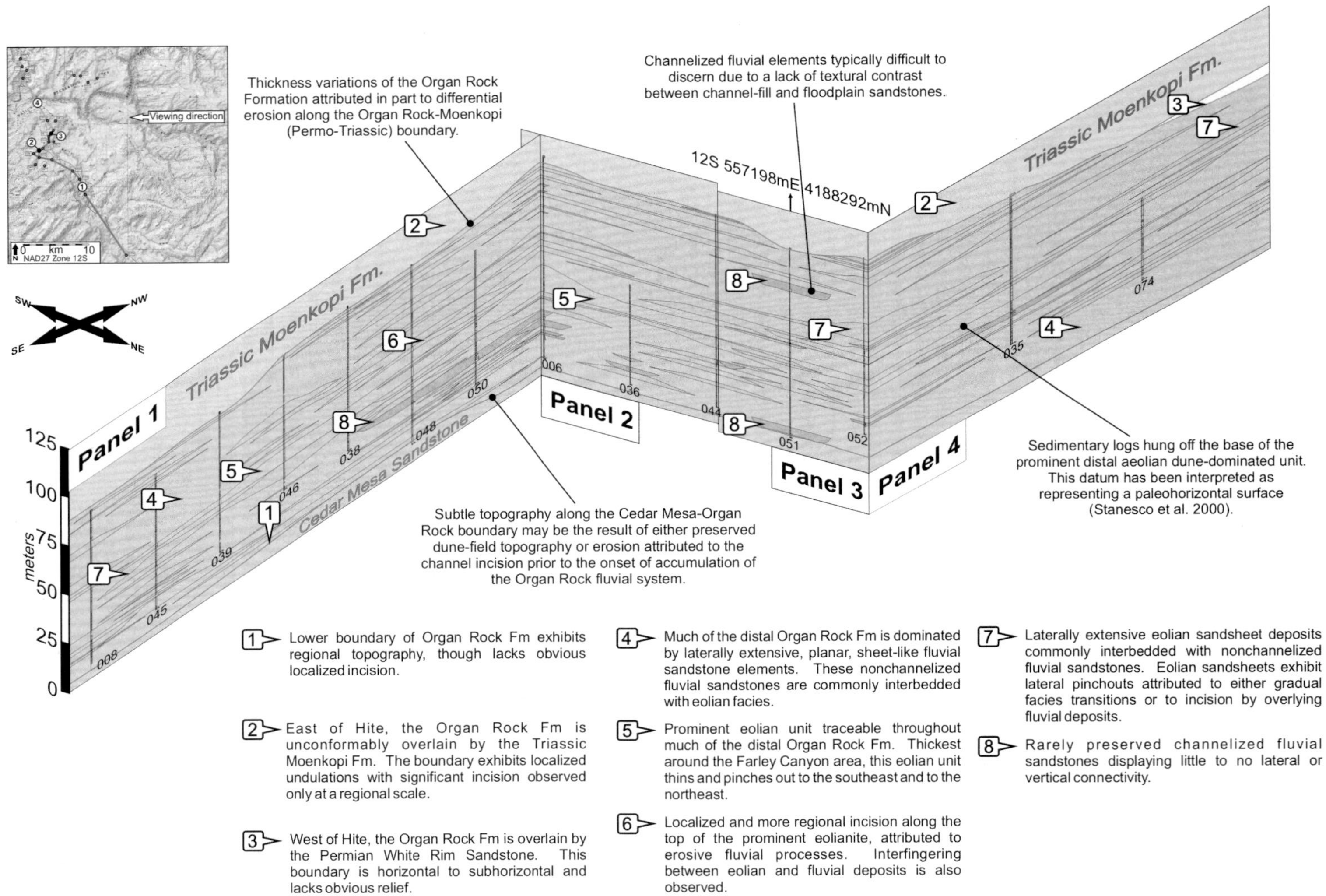

FIGURE 7.—Intermediate-scale, pseudo-three-dimensional reconstruction of the gross-scale sedimentary architecture that typifies distal parts of the Organ Rock Formation. See map inset for viewing direction and panel positions. Also see Figure 1 for the regional location of Hite. Refer to facies association key in Figure 2.

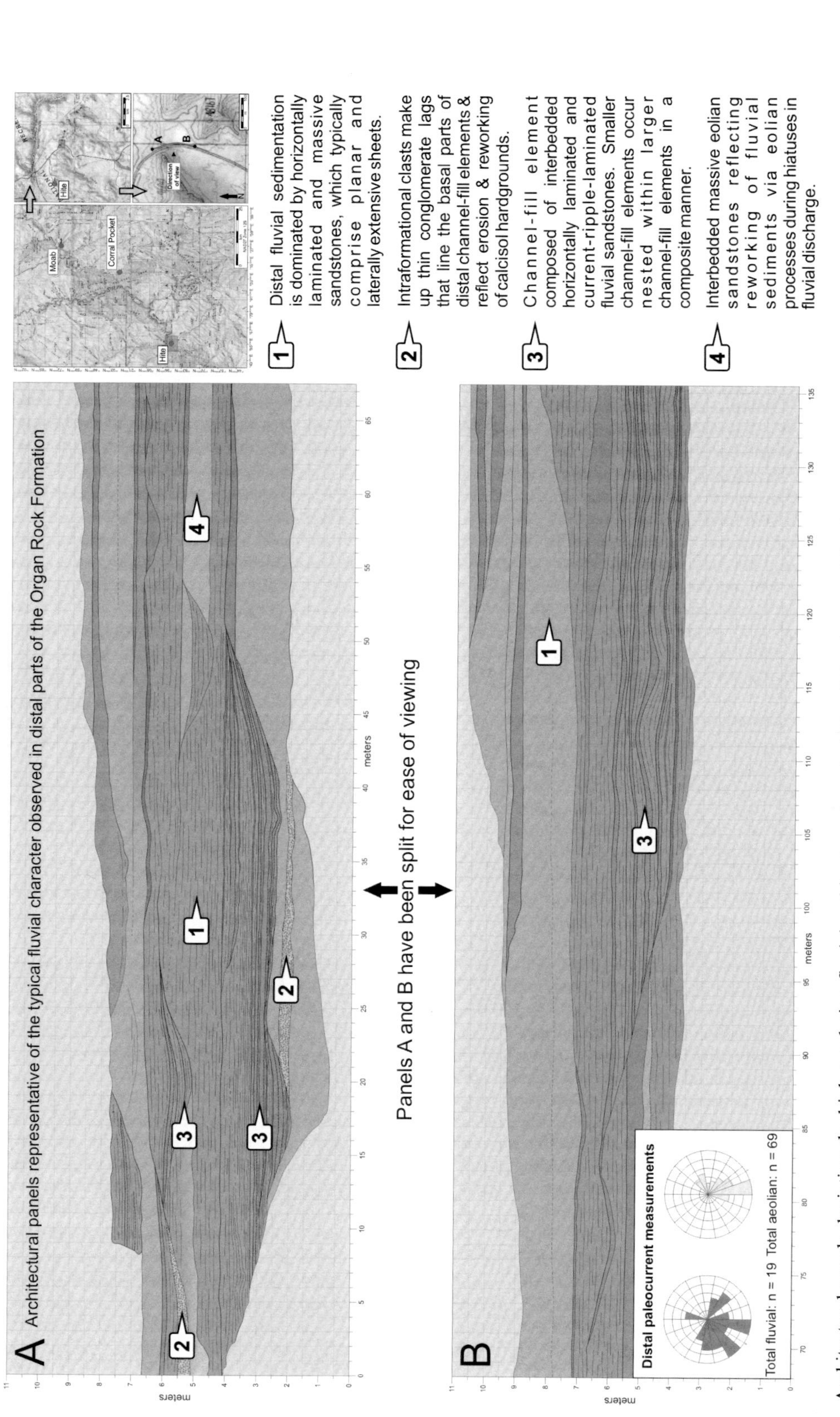

FIGURE 8.—Architectural panels depicting the high-resolution fluvial architecture characteristic of distal parts of the Organ Rock Formation. Refer to lithofacies key in Figure 2.

limited, due to a general paucity of suitable facies and structures that yield paleocurrent information, notably an absence of cross bedding and ripple cross lamination. However, where present, paleocurrent indicators conform to the regional trends identified from the proximal and medial parts of the Organ Rock Formation. At Andy Millar Flats (see Fig. 1 for location), measurements reveal a southwest-directed (219°; $n = 9$) fluvial paleocurrent direction, whereas at Farley Canyon (see Fig. 1 for location) a mean fluvial paleocurrent direction of south-southeast (167°; $n = 7$) is observed. Approximately 10 km to the southeast of Farley Canyon, an average fluvial paleocurrent direction of southwest (236°; $n = 3$) is again observed. Combining these data provides a regional average paleoflow direction to the southwest (212°; Fig. 8). Eolian paleocurrent data from dune cross bedding indicate a mean dune migration direction to the southeast (Fig. 8), with mean directions at Andy Millar flats (n = 8) and White Canyon ($n = 61$) toward 163° and 146°, respectively.

FLUVIAL–EOLIAN INTERACTIONS

Within the Organ Rock Formation interactions between fluvial and eolian processes are most evident in the lower-medial and distal parts of the succession. In proximal parts of the succession, rare eolian sandsheet elements partly infill the uppermost parts of fluvial channel-fill elements, though they are typically less than a few tens of meters in lateral extent because they are cut out by neighboring fluvial channels.

Four styles of fluvial–eolian interaction were identified in the medial and distal parts of the Organ Rock Formation, and these occur at a variety of scales, from regional scale down to the grain scale, and include: (1) interdune flooding, (2) local-scale dune breakthrough, (3) major fluvial incursion, and (4) fine-scale interlaminations of fluvial sheetwash and sand-drift sediment (Figs. 9, 10). The first three styles represent distinct stages in a sequential transition from eolian-dominated deposition in situations where the water table and its capillary fringe lay significantly below the depositional surface and did not influence eolian accumulation, to fluvial-dominated deposition in situations where the water table lay at or above the depositional surface. Examples of these four styles of fluvial–eolian interaction are most common in the lower-medial and distal parts of the succession, where they occur at a variety of stratigraphic levels.

Interdune Flooding

Evidence for interdune flooding includes the presence of lenses of red-brown siltstone, brown parallel-laminated and climbing-ripple-laminated fine-grained sandstone (Fig. 9A), orange structureless (homogenized) sandstone, and thin beds of red silty mudstone between yellow and orange sets of cross-bedded sandstone of eolian dune origin (with foresets dominated by laminae of grainflow strata 30 to 50 mm thick). These lenses, which are 10 to 100 m wide and 0.05 to 3 m thick, can routinely be traced along their length over distances of several hundred meters to points where they pinch out against eolian dune strata. Bioturbation and root mottling is common in the deposits of these lenses but tends to be rare or absent from the adjoining eolian dune strata.

These lenses represent interdune elements in which sedimentation occurred in response to fluvial incursion and / or the localized ponding of rainwater following localized storm events. The red-brown siltstone and brown sandstone indicate the introduction of sediment derived from outside the immediate vicinity of the dune field, whereas the orange homogenized sandstone records the localized slumping of dune flanks at the margin of the interdunes, possibly during rainfall events. The red silty mudstone records the accumulation of windblown dust within spatially enclosed, water table–controlled damp or wet interdune hollows. Similar deposits have been described from elsewhere in the Cutler Group (Langford and Chan, 1988, 1989; Mountney and Jagger, 2004).

Examples of interdune flooding observed in lower-medial and distal parts of the Organ Rock Formation record either the rise of the water table to a position at or above the level of the accumulation surface or the onset of fluvial incursions into an area following an episode of eolian accumulation (Fig. 9A). Wet interdune sedimentation in the form of massive, slumped and locally water-reworked sandstone of former eolian origin arises due either to a temporary ponding of water in dune fields or to a longer-term rise in water-table level to or above the level of the accumulation surface, thereby recording the episodic or continual flooding of the interdune areas (Mountney, 2006b).

Flooding attributed to short-term intra-dune-field flash flooding can occur without the need for a rise in water table level either by fluvial incursion of water flowing from beyond the dune-field margin or by extra-erg rainfall events that temporarily exceed the infiltration capacity of the substrate (Langford, 1989; Langford and Chan, 1989). The effects of interdune floods arising from externally sourced flood waters are to introduce sediment from outside the dune field, and, as is the case of the Organ Rock Formation, resultant flood deposits can contrast markedly in color and/or texture with intervening eolian dune sediments (Fig. 9A). Interdune floods arising from intra-dune-field rainfall events, by contrast, introduce little or no external sediment and instead result in the localized slumping and homogenization of sand from the flanks of dunes that bound the interdune.

Longer-term controls on wet interdune sedimentation are dictated by variations in ground-water levels and reflect changes in the balance between ongoing sedimentation, subsidence, and regional climatic variations (Kocurek and Havholm, 1993). During episodes of elevated water table, wet interdune units typically expand at the expense of adjoining dunes, which decrease in size (Mountney, 2006b). Where fully enclosed by surrounding dunes, sedimentation in persistently wet interdune occurs chiefly by the trapping of wind-blown mud, silt, and sand at the wet surface. Deposits in the Organ Rock Formation interpreted to result from this process are typically darker in color than surrounding dune sandstones, they are poorly sorted, have a finer mean grain size, and are characterized by aqueous ripple strata, bioturbation (cf. Ahlbrandt and Fryberger, 1979), rhizolith and paleosol development (cf. Langford and Chan, 1988, 1989), soft-sediment deformation, homogenization and dewatering structures (cf. Mountney and Jagger, 2004). Localized chert (silcrete) and freshwater limestone pond deposits are also recognized (cf. Driese, 1985; Loope, 1984).

The resultant architecture of flooded interdune deposits is controlled partly by the size and shape of the original interdune depression, which itself is dictated by the morphology of the surrounding dunes but also by the longevity of the flooded interdune and its persistence as it migrates in front of advancing dunes (cf. Langford and Chan, 1988). The ratio between the rate of water-table rise and the rate of downwind bedform migration therefore exerts a primary control on preserved architecture (Mountney and Thompson, 2002). Wet interdune elements in the Organ Rock Formation form vertically isolated lenses that typically extend laterally for only a few hundred meters, indicating that the wet interdune areas were only transient features across the depositional surface. At certain stratigraphic horizons, wet interdune elements decrease in size, number, and frequency, suggesting that, at times, the wet interdune ponds were replaced by expanding eolian bedforms during periods of relative water-table fall (e.g., Mountney and Jagger, 2004, their Fig. 17).

Localized Dune Breakthrough by Flood Water

Examples of localized eolian dune breakthrough are recognized in the uppermost parts of eolian dune–dominated units in the lower-medial part of the Organ Rock Formation (Fig. 9B). Channel-like forms from 5 to 30 m wide with erosional bases up to 5 m (very rarely 10 m) deep cut down into underlying eolian dune strata at a number of

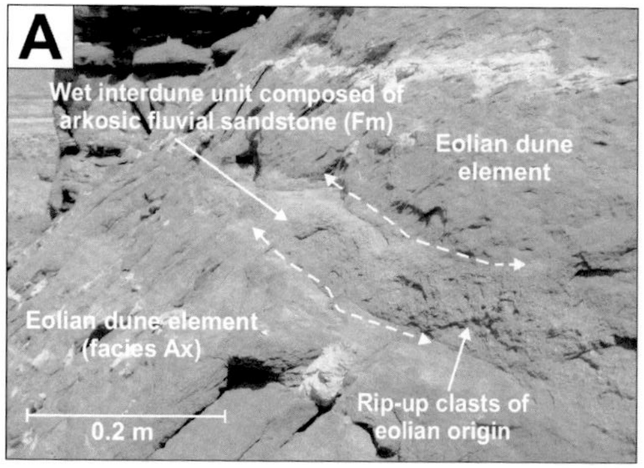

Externally, fluvially derived arkosic fluvial sandstone deposited within interdune hollow during dune-field flooding.

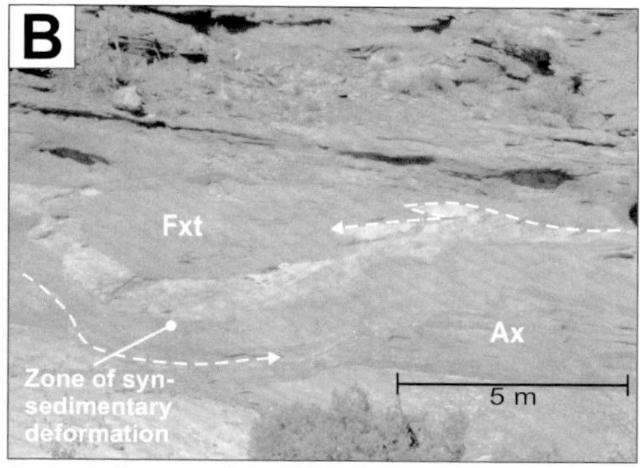

Preserved expression of localized high-energy dune breakthrough. Interdune flood waters carved out trench-like channels through dune trains.

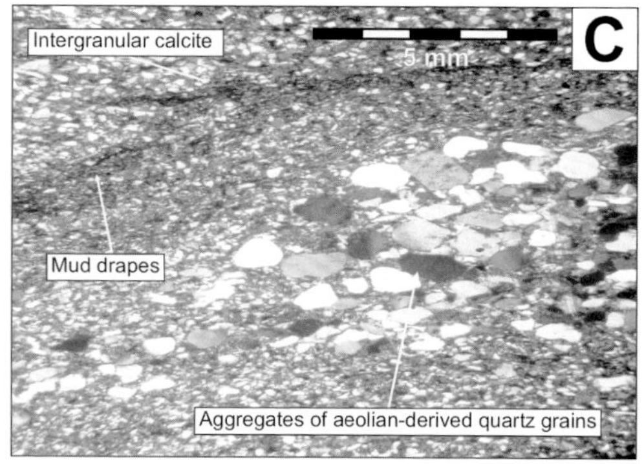

Thin-section from a deposit of fluvial overbank sandstone showing small-scale reworking of aggregates of quartz originally deposited via eolian processes.

Major fluvial incursion representing a major flooding of the underlying aeolian dune field (Cedar Mesa Sandstone). Records the preserved expression of a major change from eolian-dominated deposition (Cedar Mesa Sandstone) to fluvial-dominated deposition (Organ Rock Formation). Person circled for scale is 1.8-m tall.

FIGURE 9.—Outcrop expression of fluvial–eolian interaction observed through medial and distal parts of the Organ Rock Formation.

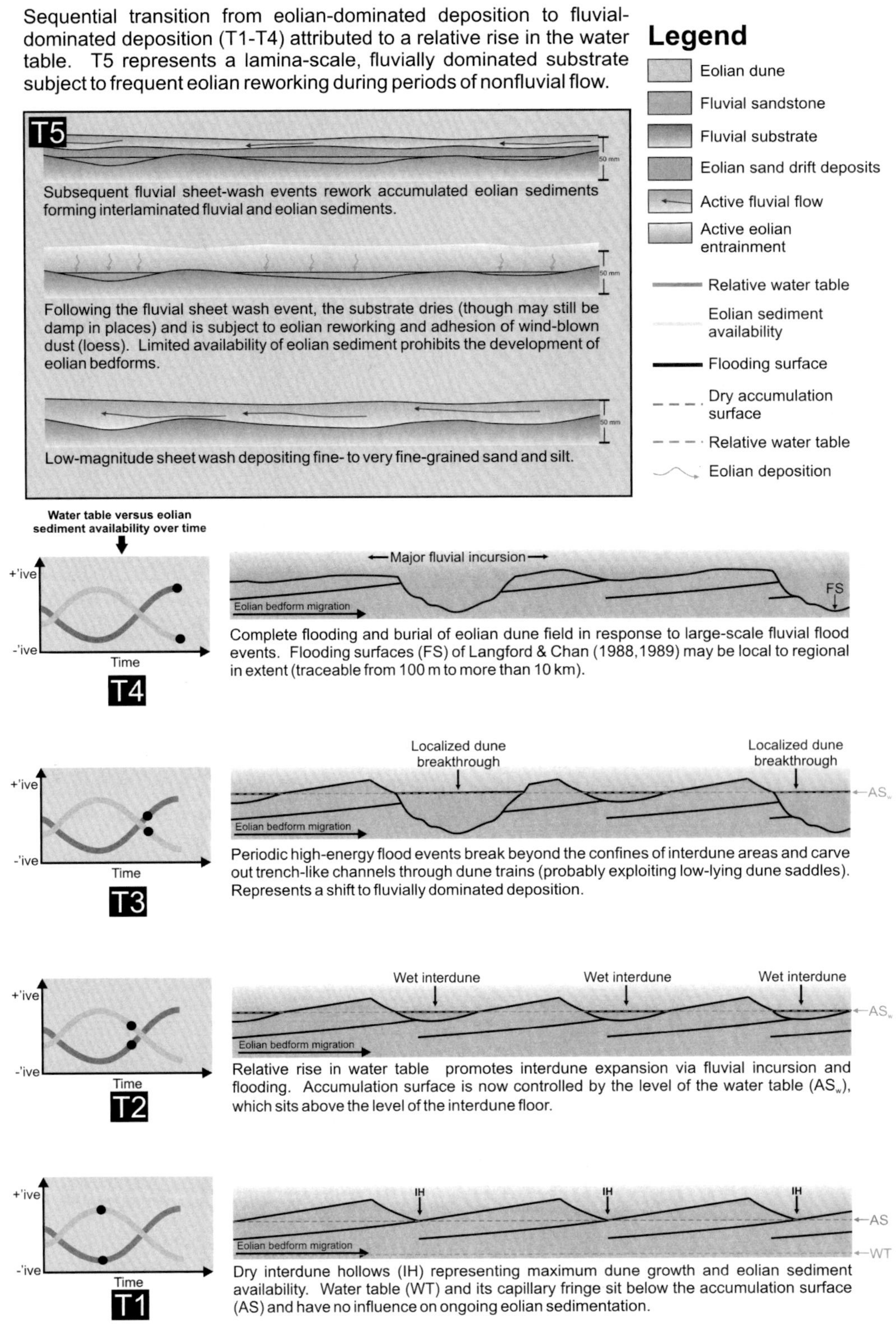

FIGURE 10.—Schematic model depicting paleoenvironmental conditions responsible for preserved architectural relationships between fluvial and eolian facies as inferred from fluvial–eolian interactions observed in the Organ Rock Formation. See Figure 9 for photo examples.

places in the lower-medial part of the Organ Rock Formation. These channels are filled with red-brown sandstone that is, for the most part, structureless but which rarely has a weak planar lamination developed within it. Margins of the channels are steep and, in places, are almost vertical. The eolian sandstones into which the channels downcut commonly exhibit convolute lamination and other signs of rapid dewatering such as homogenization within zones up to 10 m either side of the channels. Elsewhere brittle deformation in the form of small (0.3

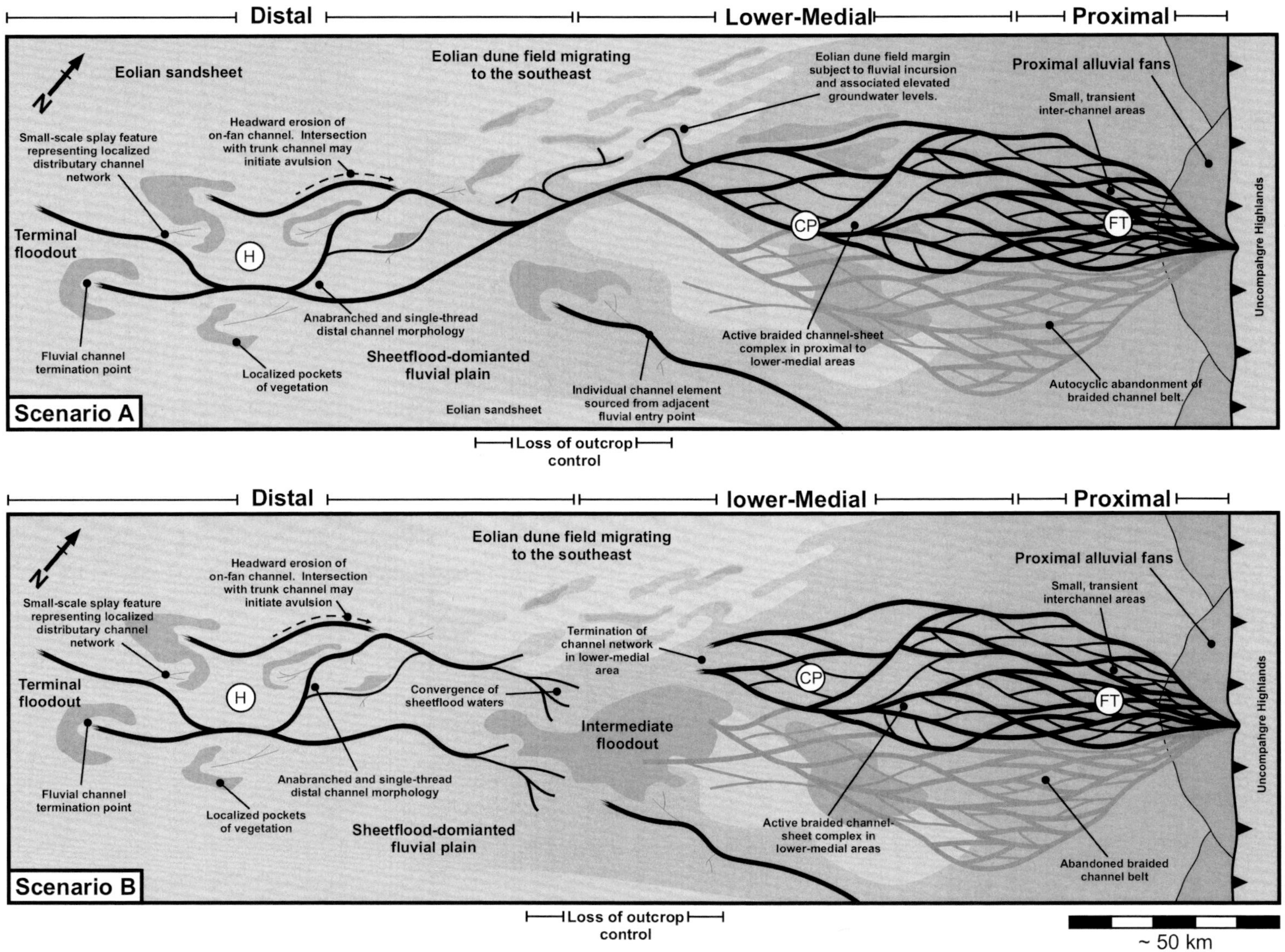

FIGURE 11.—Alternative models for downstream changes in fluvial architecture observed from proximal to distal parts of the Organ Rock Formation. A) continuous channel network flowing from its source (Uncompahgre Highlands) to its termination on a low-relief desert plain in distal parts of the basin. B) two separate channel networks occupying proximal-medial and distal parts of the basin, respectively, and which are separated by an intermediate floodout. FT is Fisher Towers; CP is Corral Pocket; H is Hite. These areas are representative of the proximal, medial, and distal parts of the Organ Rock Formation, respectively.

m long) growth faults and sandstone breccia intraclasts are present in and adjacent to the channels. By contrast, at the samestratigraphic level but away from the channels, the eolian sandstones are cross-bedded and arranged are into sets 1 to 4 m thick.

This style of fluvial–eolian interaction represents the depositional product of episodic and often catastrophic flood events, which have sufficient energy to break beyond the confines of interdune areas and carve out trench-like channels through adjoining dunes. Unlike wet interdune deposits, these trench-like channels do not conform to a specific architecture that is dictated by the morphology of surrounding eolian bedforms. Instead, examples of channels associated with localized dune breakthrough are typically very narrow (width < 10 m; W/D < 2) and show evidence for high-energy reworking of dune bedforms (Fig. 9B).

Rapid incision and emplacement of associated fluvial sediments (Fig. 10) is implied by load-induced, brittle and ductile deformation of underlying eolian dune sediments that were both semi-lithified (indurated) and partially saturated (cf. McKee et al., 1971; Doe and Dott, 1980; Horowitz, 1982). Partial induration, together with associated evidence of stabilization by vegetation, and saturation of eolian sediments in the immediate vicinity of the break-through channels, is indicative of high ground water levels prior to break-through. This style of fluvial–eolian interaction signifies higher magnitude fluvial events compared to those associated with interdune flooding and a close association with overlying, laterally extensive fluvial sheets suggests that dune break-through features are likely to be linked to an overall change in depositional regime, and mark a style of transition from eolian-dominated to fluvial-dominated deposition.

Major Fluvial Incursion

Major fluvial incursions into eolian dune fields are represented by laterally extensive fluvial channel complexes that are underlain by localized to regional-scale eolian flooding surfaces similar to those described by Langford and Chan (1988, 1989). At one well exposed locality in Corral Pocket (Fig. 9D), multiple channels (at least five but probably more), each filled with brown sandstone of fluvial origin, are observed to be developed at one common stratigraphic level (Fig. 9D). These channels are spaced between 20 and 200 m apart, and each is of a similar size (5 m wide and 3 m deep). They represent the preserved product of a single flood event that breached the dunes in the eolian dune field at a series of low points—probably via saddles or interdunes between dunes in a train. The single horizon represented by these channels represents a flood surface and records an episode of major fluvial incursion into a dune field (cf. Langford and Chan, 1988, 1989; Mountney and Jagger, 2004).

At a regional scale, flooding surfaces observed in the Organ Rock Formation exhibit modest incision of up to 10 m (though typically < 5 m); however, these are commonly directly underlain by dune breakthrough features that themselves exhibit abrupt incision of 2 to 5 m. Flooding surfaces mark a significant shutdown of the eolian depositional system in favor of an exclusively fluvial depositional regime, a transition that is likely related to and driven by relatively long-term variations in climate and associated long-term discharge trends.

Interlaminated Fluvial Sheet-Wash and Eolian Sand-Drift Deposits

This style of small-scale fluvial–eolian interaction is evident in sheet-like accumulations of fine-grained sandstone that are characterized by millimeter-scale interlamination of fluvial sheet-wash and eolian sand-drift deposits (Gilbert Kelling, personal communication; Fig. 9C). In thin section, aggregates of medium-grained, well-rounded quartz grains of eolian origin form discrete lenses that are interlaminated with very fine-grained quartz, mica, and clay particles that appear to be of fluvial sheet-wash origin based on their textural and petrographic characteristics. In these thin fluvial deposits, a distinctive fabric, which is best observed in thin section, is present and is characterized by the alignment of mica and clay minerals (Fig. 9C), which constitute mud drapes formed when suspended sediment is deposited upon cessation of fluvial flow (cf. Reid and Frostick, 1997).

This intimate association of fluvial and eolian sediments reflects the cyclical occurrence of low-magnitude unconfined fluvial sheet-wash events followed by drying of the substrate and the liberation of fluvially derived sediments for eolian entrainment and reworking (Fig. 10). Elsewhere, parts of the substrate remained damp and acted to trap eolian grains via adhesion. If the flooding was widespread then the limited availability of eolian sediment for transport would have likely inhibited the formation of eolian bedforms across these surfaces, though limited volumes of loose sand and silt were able to blow across and adhere to the damp surface and accumulate in localized hollows. Alternatively, the flooding could have been more localized and may have had little impact on ongoing regional dune field construction. Subsequent sheet-wash events repeatedly reworked surface sediments forming lamina-scale intercalations of mixed fluvio- eolian sediments.

DISCUSSION

Comparisons of channelized fluvial characteristics between proximal and medial parts of the Organ Rock Formation show a clear relationship in both channel-fill composition (extrabasinally-sourced arkosic sandstone) and preserved channel-fill architecture (Cain and Mountney, 2009). Systematic decreases in overall grain size and increases in the preservation of nonchannelized deposits reflect downstream changes that would be expected in a fluvial system that was becoming progressively more distal from its source. However, the relationship between channelized fluvial architecture observed in the lower-medial and distal parts of the succession is far more ambiguous; distal fluvial channel architecture is apparently unrelated to that observed in lower-medial regions.

To determine the relationship between preserved fluvial architecture observed in lower-medial and distal parts of the Organ Rock Formation, two paleogeographic scenarios were developed, which are constrained by a combination of primary outcrop data coupled with observations made from potential modern analogues (Fig. 11).

Paleogeography Scenario A

Paleogeography Scenario A represents the Organ Rock fluvial system as a single, continuous fluvial channel network that is continuous from its source to its termination in distal parts of the basin, and in which discharge progressively decreases from proximal to distal parts of the succession in response to transmission losses and evaporation. As such, with sufficient discharge and survivability, a flood event may flow, within the confines of a fluvial channel, for the full length of the fluvial system before ultimately terminating in a floodout. It is this assumption of continuous channelized fluvial flow from source to termination that is key in current models of ancient terminal fluvial systems and their prediction of a systematic downstream change in fluvial style and morphology (e.g. Friend, 1978; Olsen, 1987; Kelly and Olsen, 1993; Hampton and Horton, 2007). These models predict that, with a progressive downstream decrease in discharge, fluvial channels exhibit a systematic decrease in cross-sectional area, attributed to repeated, large-scale river bifurcation and the splitting of flow between coevally flowing channels. Recently, the role of downstream bifurcation in dryland fluvial systems vhas been contested by North and Warwick, (2007), who argue that the significance of distributary networks has been overstated. Downstream reductions in channel cross-sectional area are identified in distal regions of many present-day dryland fluvial systems (e.g., Tooth, 1999a, 1999b; 2000a, 2000b), though whether such reductions are typical of most such systems is unclear (Billi, 2007).

Paleogeography Scenario A conforms to the principle of a progressive and continuous downstream decrease in discharge; however, the resultant downstream change in channel morphology is quite different from that predicted by current models, which predict a distributary-style fluvial network (e.g., Friend, 1978; Kelly and Olsen, 1993; Nichols and Fisher, 2007). Instead, the Organ Rock fluvial system shows no evidence for the downstream bifurcation of fluvial channels (cf. Kelly and Olsen, 1993), but rather the opposite. Indeed, distal parts of the succession show evidence for a dramatic reduction in the number of preserved channel-fill elements, suggesting that the Organ Rock fluvial system exhibited overall converging channel system plan-view morphology.

Converging plan-view channel system morphology was likely enabled by the gradual, downstream convergence of the extensive, braided channel sheets, observed in proximal and lower-medial parts of the succession, into a single-thread trunk channel as a result of decreasing discharge. It is known from modern dryland environments (e.g., Cooper Creek, Australia) that some dryland fluvial systems exhibit a compound plan-view morphology, in which larger floods occupy extensive floodplain complexes, reverting to flow in single thread or anabranching channels during smaller non-flood events (e.g., Rust, 1981). In these studies, this relationship between flood magnitude and system morphology is attributed to temporal changes in fluvial regime; however, it could alternatively be possible for the same relationship to apply to spatial variations in flow magnitude, whereby in cases where a fluvial system loses energy downstream, it reverts from a very broad multithread to a single-thread or relatively narrow anabranching morphology. This also seems to be the case for the Organ Rock Formation, judging by the observed downstream decrease in both the size and the number of channels from the medial area to the distal area.

A progressive decrease in fluvial energy as envisaged by Paleogeography Scenario A in this study also highlights a paradox, in that anabranching rivers that are associated with decreased flood energy commonly exhibit a coarse-grained bedload (Tooth and Nanson, 1999). Indeed, intraformational granulestone channel-fill elements (F3) observed in the distal Organ Rock Formation are interpreted to represent segments of an anabranching channel network (Cain and Mountney, 2009) and are composed predominantly of granulestone intraclasts derived from the local reworking of calcrete paleosol horizons. Nanson and Knighton (1996), however, state that although a decrease in overall flow strength may lead to anabranching, via channel choking and subsequent avulsion, the act of splitting flow between two or more narrower anabranches acts to concentrate the flow, thereby maintaining flow velocity and bed shear stress. Sediment-transport efficiency can therefore be maintained or even increased without the need for an increase in slope gradient or discharge (Nanson and Huang, 1999).

Paleogeography Scenario B

Paleogeography Scenario B differs from traditional terminal fluvial models in that it does not represent a continuous channel network from source to termination, but instead represents two spatially disconnected fluvial channel networks separated by a zone of intermediate floodout. As such, each channel network could be effectively independent of the other, with changes in one not necessarily affecting the other. Minor flood events might only occupy more proximal parts of the system, whereas major flood events might pass through a zone of floodout to reconverge and continue downstream to the distal part of the system. Unlike current models of terminal fluvial systems, this model does not assume progressive and continuous downstream reductions in discharge, and therefore does not predict systematic downstream changes in channel morphology. Flow termination could potentially occur at any point in the system (either progressively or abruptly), depending on the magnitude of the flood event.

In Paleogeography Scenario B, the more proximal channel network is represented by extensive, braided channel complexes that characterize proximal and lower-medial parts of the Organ Rock Formation, and which are sourced from the Uncompahgre Highlands. However, unlike in Paleogeography Scenario A, where these high-energy braided channel complexes converge into a single-thread channel before flowing to distal parts of the system, in Paleogeography Scenario B the braided channel complexes terminate in an intermediate floodout zone with formerly channelized waters being dispersed as unconfined sheetfloods (F5) (cf. Tooth, 1999a, 1999b; Figs. 7, 8). The second, distal, channel network is located downstream of the intermediate floodout zone and is not directly sourced by flow emanating from the Uncompahgre Highlands. Instead, the distal channel network is activated by the convergence of sheetflood waters, sourced from either the intermediate floodout zone or from localized convective storms, which produce intense flash flood events (cf. Tooth, 2000b) that are associated with high runoff coefficients and a dominance of Hortonian overland flow (Graf, 1988; Reid and Frostick, 1997). It is therefore possible that in Paleogeography Scenario B, the distal channel network can be active and evolving while the more proximal channel network remains inactive. The potential for independent channel network evolution may therefore account for the distinct contrast in channel morphology and sediment character between lower-medial and distal parts of the Organ Rock Formation. This hypothesis does not preclude the possibility of the distal channel network being sourced directly from flood waters emanating from the highlands at the basin margin during high-discharge events that crossed the intermediate floodout zone.

The relatively sudden loss of channelized fluvial architecture south and west of Corral Pocket in the medial part of the basin is best explained by the model for Paleogeography Scenario B, which accounts for the abrupt transition from dominantly channelized fluvial deposition to nonchannelized fluvial deposition in which sheetflood process predominate. Rare occurrences of channelized fluvial sandstones in distal parts of the basin lack the externally sourced arkosic sandstones that characterize medial channel-fill elements, and are instead composed of internally sourced, fluvial and eolian sandstones and reworked hardground paleosol material. The preserved contrast in sedimentary characteristics and architecture between proximal to medial parts and distal parts of the succession is suggestive of the presence of an intermediate floodout to the south and west of the medial part of the fluvial system, downstream from which, and for sustained periods of time, the distal part of the system evolved independently via the influence of flood events sourced from localized rainfall events.

The roll of contemporaneous eolian processes also exerted a significant influence on fluvial drainage and the preserved architecture of fluvial deposits. For significant periods of time, the basin floor was dominated by extensive eolian dune fields that migrated overall to the southeast. Architectural-element analysis and paleocurrent data indicate that the fluvial and eolian systems interacted in a complex manner at multiple scales. During periods of eolian-dominated deposition, fluvial incursions into the dune-field margins were confined to and controlled by the geometries of the eolian interdune areas. However, as deposition became increasingly influenced by fluvial processes, the influence that interdune areas exerted on fluvial deposition decreased with fluvial drainage become more pervasive until complete flooding of the eolian dune field occurred.

CONCLUSIONS

The downstream variability in preserved fluvial architecture observed in the Organ Rock Formation is indicative of a fluvial system that traversed a semiarid desert plain (Cain and Mountney, 2009) on which both fluvial and eolian processes were contemporaneously active. Previous depositional models of such systems (e.g., Friend, 1978; Kelly and Olsen, 1993; Hampton and Horton, 2007; Nichols and Fisher, 2007) imply that channelized fluvial flow moves progressively from proximal to distal parts of the system, which results in systematic downstream changes in preserved fluvial architecture attributed to a decrease in discharge and flow competency. However, observations made from the Organ Rock Formation suggest that the relationship between preserved fluvial architecture and the governing processes in semiarid dryland systems may in fact be more complex in this case study. Due to the dominant processes known to shape modern dryland systems (i.e., infrequent but high-magnitude flood events) most dryland river systems are considered to be out of equilibrium with "normal" flow conditions owing principally to their long recovery periods following large flood events (Knighton and Nanson, 1997; Gibling, 2006). It is therefore difficult to produce predictive models for what are essentially unpredictable fluvial systems.

Paleogeography scenarios A and B presented in this study represent two generalized models that attempt to relate processes known to operate in modern dryland fluvial systems to the preserved fluvial architecture observed in the Organ Rock Formation. Paleogeography Scenario A draws comparisons with current published models of terminal fluvial systems in that it represents a continuous fluvial reach from the Uncompahgre Highlands in the northeast to the to the point of termination to the southwest of the basin. However, it differs from several previous models in that it does not assume a distributary channel network, but instead one that converges from a broad, braided channel belt to a single-thread or anabranching channel network. A relationship between discharge and channel morphology may relate to spatial changes in discharge whereby as discharge progressively decreases downstream, the fluvial system reverts from a braided channel network to one that is characterized by single thread. Paleogeography Scenario B modifies this model further by suggesting that the abrupt decrease in channel size and frequency between medial and distal parts of the Organ Rock succession can be explained by the presence of a zone of intermediate floodout that separates two different styles of channel network. This proposed model has broad implications for the interpretation of other ancient dryland fluvial successions, including those with economic resource implications such as those acting as hydrocarbon reservoirs in the Permian of the Southern North Sea Basin and the Triassic of the East Irish Sea Basin and North Sea Central Central Graben.

ACKNOWLEDGMENTS

We a grateful to Oliver Wakefield (Keele University) for assistance with field data collection and to Oliver Jordan (Statoil) for valuable discussions about the geology of the Paradox Basin. The US National Park Service and the staff of the ranger service of Canyonlands National Park are thanked for granting permission to conduct research in the park. SAC is grateful for funding support from Keele University and for fieldwork grants from the British Sedimentological Research Group (BSRG) and International Association of Sedimentologists (IAS). Kathryn Amos and Richard Langford are thanked for their helpful and constructive reviews, as are the editors of this volume, who provided valuable advice.

REFERENCES

Ahlbrant, T.S., And Fryberger, S.G., 1979, Sedimentary features and significance of interdune deposits, *in* Ethridge, F.G., and Flores R.M., eds., Recent and Ancient Non-Marine Depositional Environments: Models for Exploration: Society of Economic Paleontologists and Mineralogists, Special Publication 31, p. 293–314.

Baars, D.L., 1962, Permian system of the Colorado Plateau: American Association of Petroleum Geologists, Bulletin, v. 46, p. 149–218.

Baars, D.L., 1975, The Permian system of Canyonlands country, *in* Fasset, J.E., and Wengerd, S.A., eds., Canyonlands Country: Four Corners Geological Society, Eighth Field Conference Guidebook, p. 123–127.

Barbeau, D.L., 2003, A flexural model for the Paradox basin: implications for the tectonics of the ancestral Rocky Mountains: Basin Research, v. 15, p. 97–115.

Billi, P., 2007, Morphology and sediment dynamics of ephemeral stream terminal distributary systems in the Kobo Basin (northern Welo, Ethiopia): Geomorphology, v. 85, p. 98–113.

Blakey, R.C., 1996, Permian eolian deposits, sequences and sequence boundaries, Colorado Plateau, *in* Longman, M.W., and Sonnenfield, S.D., eds., Paleozoic Systems of the Rocky Mountain Region: SEPM, Rocky Mountain Section, [Society of Economic Paleontologists and Mineralogists], p. 405–426.

Bullard, J.E., and Livingston, I., 2002, Interactions between aeolian and fluvial systems in dryland environments: Area, v. 34, p. 8–16.

Cain, S.A., 2010, Sedimentology and stratigraphy of a terminal fluvial fan system: the Permian Organ Rock Formation, South East Utah: unpublished Ph.D. thesis, University of Keele, England, 341 p.

Cain, S.A., and Mountney, N.P., 2009, Spatial and temporal evolution of a terminal fluvial fan system: the Permian Organ Rock Formation, south east Utah, USA: Sedimentology, v. 56, p. 1774–1800.

Condon, S.M., 1997, Geology of the Pennsylvanian and Permian Cutler Group and Permian Kaibab Limestone in the Paradox Basin, Southeastern Utah and Southwestern Colorado. U.S. Geological Survey, Bulletin 2000-P, 46 p.

Cooke, R.U., Warren, A., and Goudie, A.S., 1993, Desert Geomorphology: University College London, CL Press, London, 534 p.

Doe, T.W., and Dott, R.H., Jr., 1980, Genetic significance of deformed cross bedding—with examples from the Navajo and Weber Sandstones of Utah: Journal of Sedimentary Petrology, v. 50, p. 793–811.

Driese, S.G., 1985, Interdune pond carbonates, Weber Sandstone (Pennsylvanian–Permian), northern Utah and Colorado: Journal of Sedimentary Petrology, v. 55, p. 187–195.

Dubiel, R.F., Huntoon, J.E., Condon, S.M., and Stanesco, J.D., 1996, Permian deposystems, paleogeography and paleoclimate of the Paradox Basin and vicinity, *in* Longman, M.W., and Sonnenfield, S.D., eds., Paleozoic Systems of the Rocky Mountain Region: SEPM, Rocky Mountain Section, [Society of Economic Paleontologists and Mineralogists], p. 427–443.

Friend, P.F., 1978, Distinctive features of some ancient river systems, *in* Miall, A.D., ed., Fluvial Sedimentology: Canadian Society of Petroleum Geologists Memoir, v. 5, p. 531–542.

Gibling, M.R., 2006, Width and thickness of fluvial channel bodies and valley fills in the geological record: a literature compilation and classification: Journal of Sedimentary Research, v. 76, p. 731–770.

Glennie, K.W., 1998, Petroleum Geology of the North Sea: Basic concepts and recent advances: Oxford, U.K., Blackwell Sciences Ltd, 635 p.

Graf, W.L., 1983, Downstream changes in stream power in the Henry Mountains, Utah: Annals of the Association of American Geographers, v.73, p.373–387.

Graf, W.L., 1988, Fluvial Processes in Dryland Rivers: Berlin, Springer-Verlag, 143 p.

Hampton, B.A., and Horton, B.K., 2007, Sheetflow fluvial processes in a rapidly subsiding basin, Altiplano plateau, Bolivia: Sedimentology, v. 54, p. 1121–1147.

Horowitz, D.H., 1982, Geometry and origin of large-scale deformation structures in some ancient wind-blown sand deposits: Sedimentology, v. 29, p. 155–180.

Kelly, S.B., and Olsen, H., 1993, Terminal Fans—a review with reference to Devonian examples: Sedimentary Geology, v. 85, p. 339–374.

Knighton, A.D., and Nanson, G.C., 1997, Distinctiveness, diversity and uniqueness in arid zone river systems, *in* Thomas, D.S.G., ed., Arid Zone Geomorphology: Process, Form and change in Drylands, Second Edition: New York, John Wiley & Sons Ltd, p. 185–203.

Langford, R.P., 1989, Fluvial–aeolian interactions: part 1, modern systems: Sedimentology, v. 36, p. 1023–1035.

Langford, R.P., and Chan, M.A., 1988, Flood surfaces and deflation surfaces within the Cutler Formation and Cedar Mesa Sandstone (Permian), southeastern Utah: Geological Society of America, Bulletin, v. 100, p. 1541–1549.

Langford, R.P., and Chan, M.A., 1989, Fluvial–aeolian interactions: part 2, ancient systems: Sedimentology, v. 36, p. 1037–1051.

Loope, D.B., 1984, Eolian origin of Upper Palaeozoic sandstones, southeastern Utah: Journal of Sedimentary Petrology, v. 54, p. 563–580.

McKee, E.D., Douglass, J.R., and Rittenhouse, S., 1971, Deformation of lee side laminae in eolian dunes: Geological Society of America, Bulletin, v. 82, p. 359–378.

Mountney, N.P., 2006a, Periodic accumulation and destruction of aeolian erg sequences in the Permian Cedar Mesa Sandstone, White Canyon, southern Utah: Sedimentology, v. 53, p. 789–823.

Mountney, N.P., 2006b, Eolian Facies Models, *in* Posamentier, H., and Walker, R.G., eds., Facies Models Revisited: SEPM [Society of Economic Paleontologists and Mineralogists], Special Publication 84, p. 19–83.

Mountney, N.P., and Jagger, A., 2004, Stratigraphic evolution of an aeolian erg margin system: the Permian Cedar Mesa Sandstone, SE Utah, USA: Sedimentology, v. 51, p. 713–743.

Mountney, N.P., and Thompson, D.B., 2002, Stratigraphic evolution of an aeolian erg margin system: the Permian Cedar Mesa Sandstone, SE Utah, USA: Sedimentology, v. 49, p. 805–833.

Mountney, N.P., Howell, J., Flint, S., and Jerram, D., 1999, Climate, sediment supply and tectonics as controls on the deposition and preservation of the aeolian-fluvial Etjo Sandstone Formation, Namibia: Geological Society of London Journal, v. 156, p. 771–777.

Nanson, G.C., and Huang, H.Q., 1999, Anabranching rivers: divided efficiency leading to fluvial diversity, *in* Millar, A.J., and Gupta, A., eds., Varieties of Fluvial Form: Chichester U.K., Wiley, p. 477–494.

Nanson, G.C., and Knighton, D.A., 1996, Anabranching rivers: their cause, character and classification: Earth Surface Processes and Landforms, v. 21, p. 217–239.

Nichols, G.J., and Fisher, J.A., 2007, Processes, facies and architecture of fluvial distributary system deposits: Sedimentary Geology, v. 195, p. 75–90.

North, F.K., 1985, Petroleum Geology: London, Allen & Unwin, 607 p.

North, C.P., and Warwick, G.L., 2007, Fluvial fans: myths, misconceptions, and the end of the terminal-fan model: Journal of Sedimentary Research, v. 77, p. 693–701.

North, C.P., Nanson, G.C., and Fagan, S.D., 2007, Recognition of the sedimentary architecture of dryland anabranching (anastomising) rivers: Journal of Sedimentary Research, v. 77, p. 925–938.

Nuccio, V.F., and Condon, S.M., 1996, Burial and Thermal History of the Paradox Basin, Utah and Colorado, and the Petroleum Potential of the Middle Pennsylvanian Paradox Formation: U.S. Geological Survey, Bulletin 2000-O, 41 p.

Olsen, H., 1987, Ancient ephemeral stream deposits: a local terminal fan model from the Bunter Sandstone Formation (L. Triassic) in the Tønder-3, -4 and -5 wells, Denmark, *in* Frostick, L. and Reid, I., eds., Desert Sediments: Ancient and Modern: Geological Society of London, Special Publication 35, p. 69–86.

Reid, I., and Frostick, L.E., 1997, Channel form, flows and sediments in deserts, *in* Thomas, D.S.G., ed., Arid Zone Geomorphology: Process, Form and Change in Drylands, Second Edition: Chichester U.K., John Wiley & Sons Ltd, p. 205–229.

Rust, B.R., 1981, Sedimentation in an arid-zone anastomosing fluvial system: Cooper's Creek, Central Australia: Journal of Sedimentary Petrology, v. 51, p. 745–755.

Stanesco, J.D., Dubiel, R.F., and Huntoon, J.E., 2000, Depositional environments and paleotectonics of the Organ Rock Formation of the Permian Cutler Group, southeastern Utah, *in* Sprinkel, D.A., Chidsey, T.C., Jr., and Anderson, P.B., eds., Geology of Utah's Parks and Monuments: Utah Geological Association, Publication 28, p. 591–605.

Tooth, S., 1999a, Floodouts in central Australia, *in* Miller, A.J. and Gupta, A., eds., Varieties of Fluvial Form: New York, John Wiley & Sons, p. 219–247.

Tooth, S., 1999b, Downstream changes in floodplain character on the Northern Plains of arid central Australia, *in* Smith, N.D., and Rogers, J., eds., Fluvial Sedimentology VI: International Association of Sedimentologists Special Publication 28, p. 93–112.

Tooth, S., 2000a, Downstream changes in dryland river channels: the Northern Plains of arid central Australia: Geomorphology, v. 34, p. 33–54.

Tooth, S., 2000b, Process, form and change in dryland rivers: a review of recent research: Earth-Science Reviews, v. 51, p. 67–107.

Tooth, S., and Nanson, G.C., 1999, Anabranching rivers on the Northern Plains of arid central Australia: Geomorphology, v. 29, p. 211–233.

Tooth, S., and Nanson, G.C., 2004, Forms and processes of two highly contrasting rivers in arid central Australia, and the implications for channel-pattern discrimination and prediction: Geological Society of America, Bulletin, v. 116, p. 802–816.

Medium Term: Complex Control Interactions

ARCHITECTURE AND BEHAVIOR OF DRYLAND FLUVIAL RESERVOIRS, TRIASSIC SKAGERRAK FORMATION, CENTRAL NORTH SEA

TOM McKIE
Shell UK Exploration and Production, 1 Altens Farm Road, Nigg, Aberdeen AB12 3FY, UK
E-mail Tom.Mckie@Shell.com

ABSTRACT: Fluvial reservoirs are inherently heterogeneous. They typically have a complex connectivity between sandbodies of limited predictability and have highly variable reservoir properties at a range of scales. Attention is typically focused on the connectivity of fluvial channels because this primarily determines the feasibility of hydrocarbon recovery. However, in sand-rich fluvial reservoirs, where connectivity is less of an issue, the internal heterogeneity related to deposition and preservation of the fluvial deposit still results in uneven fluid movement and presents a challenge for prediction of reservoir behavior. The Triassic Skagerrak Formation provides an example of a sand-rich, dryland fluvial reservoir that would, prior to production, be regarded as having no critical issues relating to depositional architecture. The Skagerrak was deposited as widespread, coarsening-upward sheets by terminal fluvial systems extending several hundred kilometers from the basin margins. These sheets are typically playa and splay dominated in the lower part and become increasingly channel dominated upwards. Multistory channel-belt packages at the top of coarsening-upward sheets form the main producing intervals, and were likely to have been the product of mobile, avulsive, multiple channel systems which generated kilometer-scale channel belts. Well-test and pressure data indicate that these channel belts constitute a dual-permeability system consisting of a network of higher-permeability bodies that are variably distributed within a lower-permeability matrix. This behavior is a product of the channel belts comprising discontinuous, coarse-grained thalweg and lower bar bodies embedded in a finer-grained matrix of upper bar and splay sands. Intervals where these coarse-grained facies are well connected form high-permeability pathways which dominate early production, and these tend to be located at the bases of multistory units. However, long-term production rates are ultimately determined by inflow to these depleting pathways from the finer-grained, lower-permeability upper bar and splay facies. Well tests typically encounter boundaries, which reflect the common presence of internal barriers and baffles within the channel belts. These boundaries are composed of abandonment plugs, bar-draping fines, mud-chip conglomerates, and cemented calcrete-clast lags. However, despite the apparent abundance of flow baffles, long-term production indicates that the channel belts are fully connected laterally and that such features are likely to be discontinuous. Whilst the high sand–shale ratio of the Skagerrak would suggest that there should be few problems related to the connectivity of the fluvial sandbodies, the vertical connectivity is considerably reduced as a result of compartmentalizing shales. The origin of these shales is variable. Predictable shale packages of semi-regional extent mark intervals of terminal fluvial contraction, resulting in regional interfingering of fluvial sand sheets and floodbasin fines. In addition, more localized bar-top and floodplain shale remnants which scale with the areal extent of a field introduce more random flow barriers which are less predictable. Despite developments in the conceptual understanding of the facies architecture of such fluvial reservoirs, prediction of reservoir behavior is hindered by a paucity of quantitative and qualitative data on the geometry of preserved fluvial lithosomes with which to construct 3D models of complex bar architectures at a sub-channel-belt scale. This is particularly important when such belts are larger than the field extent, and the heterogeneities which need to be modelled are therefore at a finer hierarchical scale. The prediction of effective permeability of the reservoir requires information on the grain-size architecture within these lithosomes, together with the geometry and distribution of flow-baffling fines drapes and mud-chip lags. However, there are currently insufficient data from good-quality outcrops to be able to fully capture the potential impact of the natural variability of this architecture and construct predictive, quantitative models with a range of realistic geometries and properties.

KEY WORDS: dryland, fluvial, reservoir, Skagerrak, North Sea

INTRODUCTION

Fluvial deposits typically form more heterogeneous reservoirs than most other depositional settings. The preserved sandbodies are generally discontinuous, with variable lateral and vertical connectivity; the internal heterogeneity of the sandbodies can be complicated by lateral variations in grain size and sorting influencing the permeability structure; and internal baffles and barriers resulting from mud drapes and differential cementation can be common (cf. Brayshaw et al., 1996). Whilst there has been much focus on the connectivity of fluvial sandbodies in reservoirs deposited in mud-prone alluvial settings, more sand-prone systems are commonly assumed to be well connected with relatively few production issues. In the early stages of field development a primary concern is how well connected the reservoir sandbodies are, and how this connectivity will dictate the number and trajectory of wells required to access economic hydrocarbon volumes. However, there remains the significant challenge of predicting production behavior over the lifetime of a field. In all fields late-life production, as reservoir pressure is depleted, or the effects of uneven sweep of hydrocarbons becomes apparent, is more problematic than production in early field life, and as a result fluvial reservoirs can present particular reservoir management challenges in the later stages of their depletion.

Critical to understanding the behavior of fluvial reservoirs is the integration of knowledge of the alluvial architecture, in particular the likely spatial distribution of permeability barriers, baffles, and high-permeability elements, with the dynamic behavior of the reservoir over time, in order to reduce the geological uncertainty and refine prediction of future behavior. Research has been undertaken to understand the likely behavior of fluvial reservoirs using computer simulation of outcrop data (e.g., North and Taylor, 1996; Pranter et al., 2007), but such models are commonly two dimensional or restricted in areal extent, and do not necessarily capture the three-dimensional movement of fluids over the scale of a field or the typical interwell distances encountered in offshore developments. Feedback from the dynamic behavior of fluvial reservoirs over time is also required to constrain what controls flow in these reservoirs and guide the collection of appropriate analogue data from outcrops. Commonly such information resides in reservoir engineering databases and not in the sedimentological literature (Martin, 1993), but ultimately the dynamic behavior of the reservoir during relatively short-duration well tests and long-term production provides the clues as to what are the critical fluvial

From River To Rock Record: The Preservation Of Fluvial Sediments And Their Subsequent Interpretation
SEPM Special Publication No. 97, Copyright © 2011
SEPM (Society for Sedimentary Geology), ISBN 978-1-56576-305-0, p. 189–214.

architectural elements controlling fluid movement and pressure transmission in the subsurface. This understanding can provide the framework for effective reservoir management, identification of new in-field opportunities, and guide the collection of quantitative outcrop analogue data for input into numerical simulation models.

Compared to paralic settings, dryland reservoirs are underrepresented as a fluvial reservoir type. Greater focus has been applied to incised-valley-fill, coastal-plain, delta-distributary, and estuarine channel-fill reservoirs, where relative sea-level changes exert an overarching control on depositional architecture (Shanley and McCabe, 1994). Whilst there have been attempts to apply an allocyclic approach to fluvial successions based on the interaction of tectonics, climate and sediment supply (e.g., Steel and Ryseth, 1990; Legarretta and Uliana, 1998; Labourdette and Jones, 2007) these are typically hampered by the lack of a high-resolution chronostratigraphic framework and an understanding of the cause-and-effect relationships between catchment and river system in the geological record (Blum and Törnqvist, 2000; Miall, 2006).

This paper therefore focuses on the fluvial architecture and reservoir behavior of dryland fluvial reservoirs of the Triassic Skagerrak Formation in the Central North Sea, using data from multiple fields across the region to distil common elements of their dynamic behavior which can be linked to the fluvial architecture and heterogeneity. These sandstones are the principal reservoir in a number of North Sea fields, as well as forming a secondary reservoir to many Jurassic accumulations (Goldsmith et al., 2003). Although they are less prolific than the widespread paralic and shallow marine Jurassic, and deep marine Tertiary reservoirs of the North Sea, they can be productive, particularly under the high pressure conditions encountered at depth in the Central North Sea area (Fig. 1), where very high initial flow rates are achievable in good-quality reservoir, and there is sufficient reservoir energy to drive hydrocarbons through poorer quality reservoir which would not be considered economic at shallower depths.

TRIASSIC FLUVIAL SYSTEMS IN THE CENTRAL NORTH SEA

The Triassic basins across northwest Europe were characterized by prevailing dryland conditions, with widespread terminal fluvial systems, common playa and eolian facies (Hounslow and Ruffell, 2006; McKie and Williams, 2009), and episodes of evaporite-prone marine incursion from Tethys (Geluk, 2005). In the Central North Sea the succession is dominated by fluvial, floodplain, and playa deposits, which are broadly divided into an early Triassic, mud-rich, playa section up to 500 m thick (Smith Bank Formation,) and a middle to late Triassic, sand-rich, fluvial section (Skagerrak Formation) in excess of 500 m thick (cf. Fisher and Mudge, 1998). The Skagerrak Formation has been further subdivided into alternating sand-prone and mud-prone members (Fig. 2, Goldsmith et al., 1995, 2003), with the more sand-prone Anisian to Carnian, Judy and Joanne Members forming the main reservoir intervals across the region.

Regional Tectonostratigraphy

Early Triassic rifting, followed by middle to late Triassic post-rift thermal subsidence and subdued polyphase extension (Steel and Ryseth, 1990; Geluk, 2005) dominated the filling style of these basins. Rifting propagated southward from the Arctic, via the Northern and Central North Sea regions (Coward, 1995; Tomasso et al., 2002), and ultimately penetrated the Southern North Sea (Fig. 1A). In this area residual thermal subsidence following Permian igneous activity (van Wees et al., 2000) had generated a geomorphic depression vulnerable to episodic marine flooding from the Tethys Sea (Aigner and Bachmann, 1992; Geluk, 2005). In the Central North Sea the early, syn-rift basin fill is composed of playa, sabkha and lacustrine mudrocks (Smith Bank Formation) punctuated by localized fluvial-eolian Bunter fluvial deposits (Fig. 2A). The section rests on a

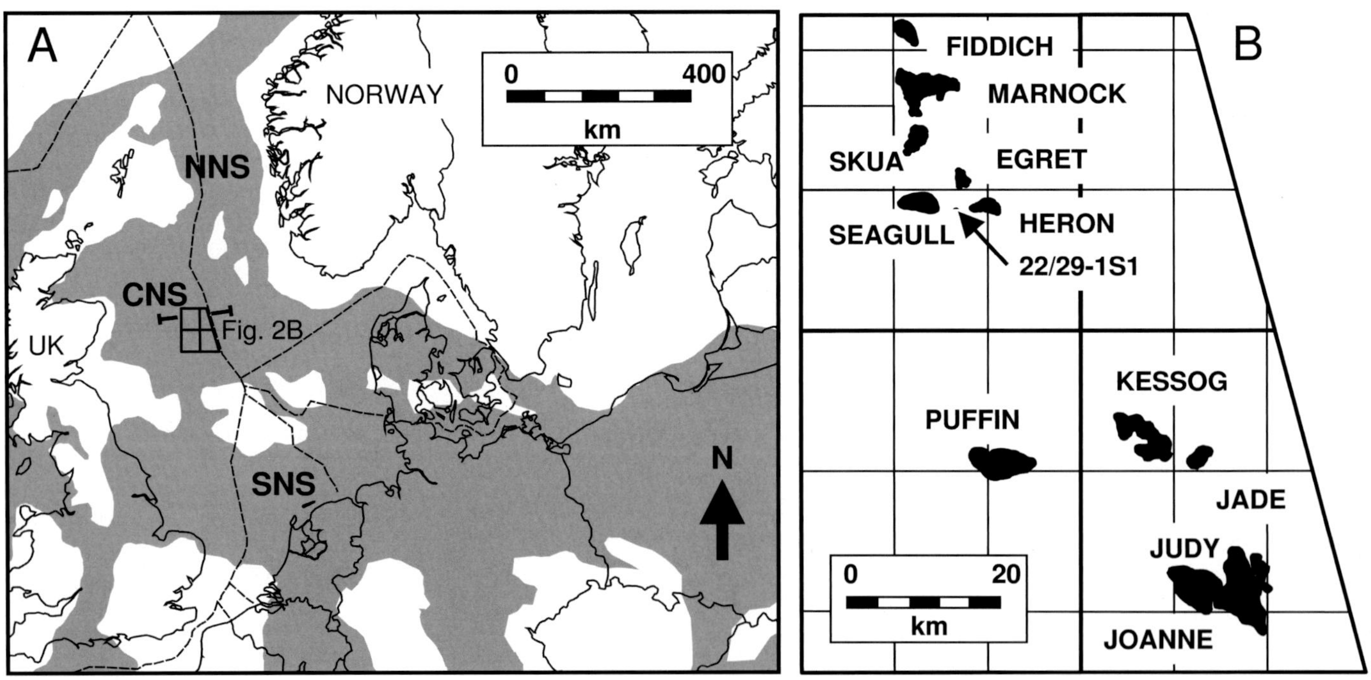

FIGURE 1.—Central North Sea location, fields and basin-fill. (A) Regional distribution of the Triassic basin-fill (shaded) in the North Sea region. NNS Northern North Sea; CNS Central North Sea; SNS Southern North Sea. (B) Location of fields producing from Triassic fluvial reservoirs in the Central North Sea.

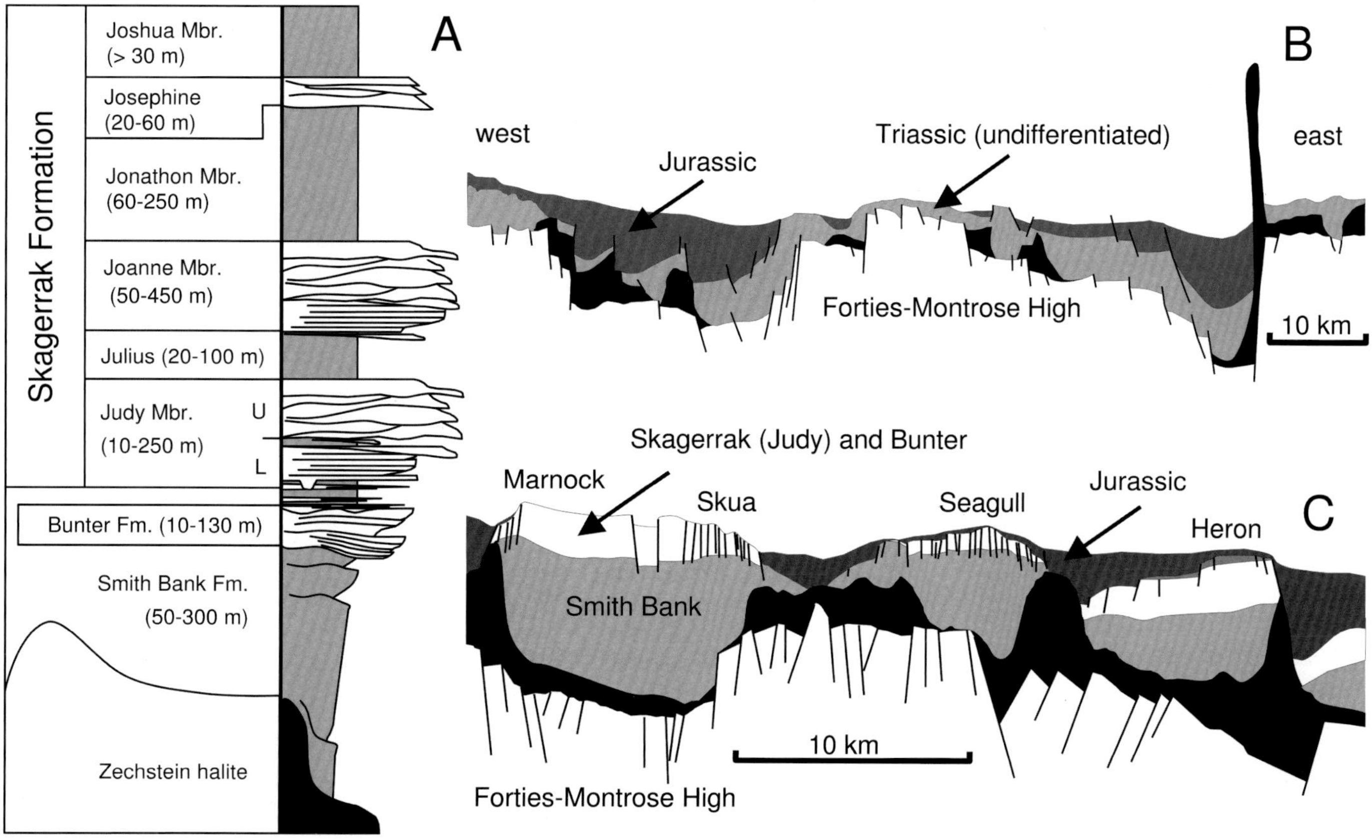

FIGURE 2.—Triassic stratigraphy and tectonic setting. (A) Schematic stratigraphic column (based on Goldsmith et al., 2003) illustrating the Triassic succession in the Central North Sea. (B) Geoseismic section across the Central North Sea illustrating the overall tectonic setting of the Triassic and setting of the Forties-Montrose High region where the Marnock, Skua, Seagull, and Heron fields are located (after Erratt et al., 1999). (C) Geoseismic section across the Forties-Montrose High area showing the detailed setting of the Marnock, Skua, Seagull and Heron fields (after McKie et al., 2010).

significant thickness of Permian halite (Zechstein), and this was mobilized during extension, resulting in dramatic thickness variations (Figs 2B, C, 3) and the formation of minibasins (Stewart and Clark, 1999). However, over much of the area the post-rift Skagerrak shows more subdued thickness variations, and on seismic has a more tabular, layered fabric on a kilometer scale, suggesting that salt movement had slowed as the instability induced by extensional faulting ceased. Although halokinesis was more subdued, the differential subsidence induced by sediment loading locally influenced the stacking of channel belts and the relative proportion of channel versus floodplain deposits, but the basin-wide dispersal of Skagerrak sands would indicate that any surficial expression of this subsidence was unable to maintain areas of sustained lacustrine ponding and divert or block river courses (cf. Matthews et al., 2007).

Sediment Dispersal

Paleocurrent data derived from borchole image logs and oriented core show an overall dominance of flow to the south towards the Southern North Sea (Fig. 4). When the data are isolated over narrower stratigraphic intervals, this bulk pattern shows the local tendency of flow to the southeast and southwest, reflecting episodes of both axial flow and transverse flow from the Fennoscandian margin. Heavy-mineral provenance data from the Skagerrak Formation indicate that these sediments were largely deposited from fluvial systems draining off the Scottish Highlands to the west and Fennoscandia to the east (Fig. 4A, cf. Mange Rajetzky, 1995), with mixed and end-member compositions suggesting that the transverse Fennoscandian drainage expanded and contracted across the basin, and was reworked by, or formed tributary systems, to the axial drainage. Ultimately this fluvial drainage flowed southward towards the Southern North Sea, where carbonates and halites (Muschelkalk) accumulated in a restricted marine basin (McKie and Williams, 2009). The Central North Sea was connected to this basin, but the Skagerrak shows no evidence of marine penetration. The transition from terrestrial clastics into marine carbonates in this region has been described by Pöppelreiter (2002), Borkhataria et al. (2005), and Borkhataria et al. (2006) as consisting of interfingered lagoonal micrites and pedogenically modified, siliciclastic mudstones and marls, with no significant coarse-grained detritus. This fine-grained, mixed-lithology, lagoonal and paleosol facies is likely to be representative of the terminal nature of the Skagerrak fluvial systems, which do not appear to have deposited coarse-grained terrestrial or deltaic facies in any location marginal to the Muschelkalk.

Climate

Across northwest Europe the early to middle Triassic saw a transition from the arid, eolian and evaporitic depositional regime of the Permian into a more fluvially dominated setting with less evidence of aridity. Eolian deposition persisted in small dune fields located

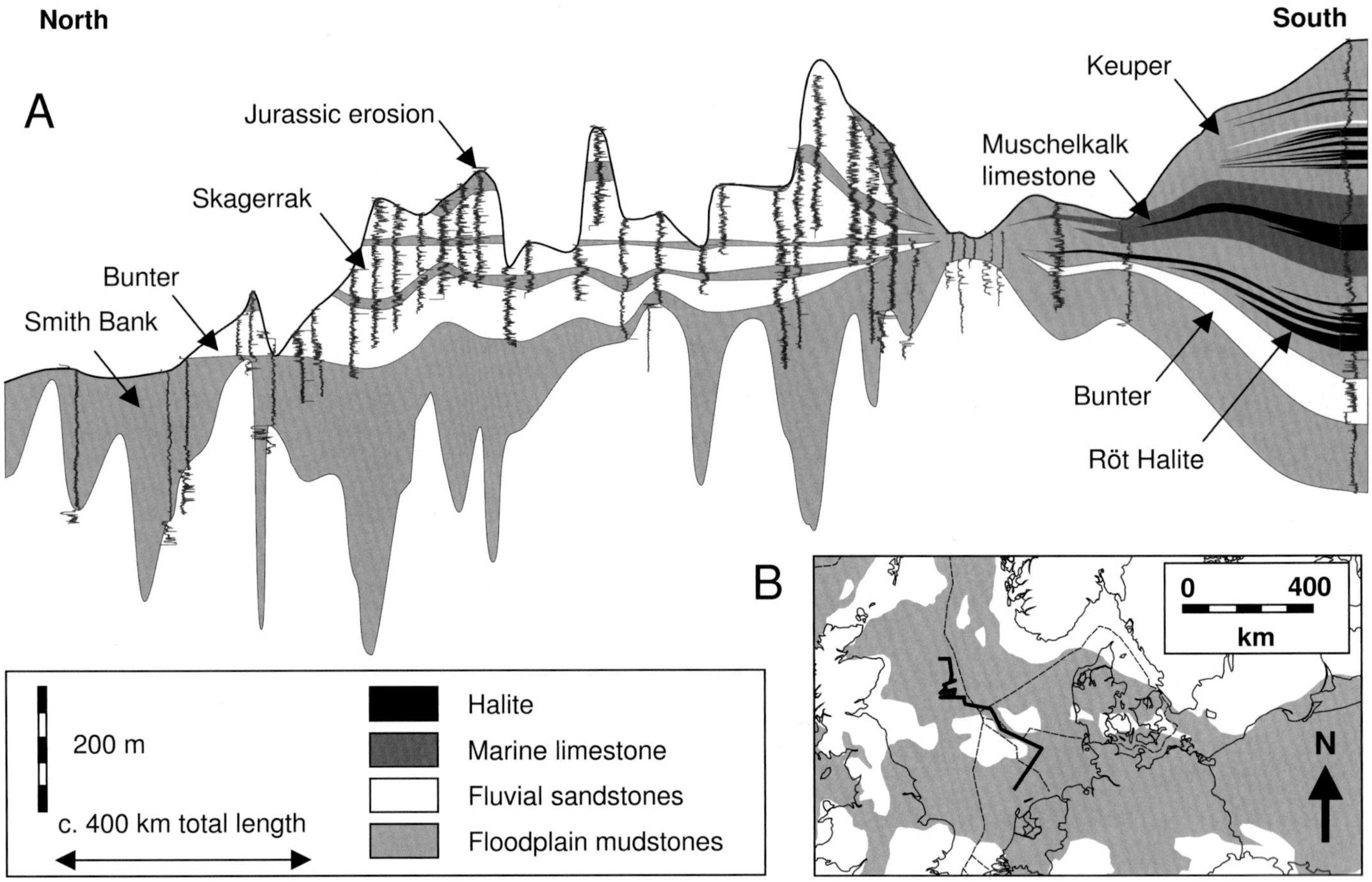

FIGURE 3.—(A) Regional correlation of the Triassic stratigraphy from the Central North Sea into the adjacent Southern North Sea Basin. The correlation is based on limited biostratigraphic data, magnetostratigraphy, heavy-mineral and well-log correlation, and regional megamerge seismic interpretation of the major shale-member reflections and seismic facies contrast between the Smith Bank and Skagerrak. Thickness changes in the Smith Bank Formation below the well penetrations are based on seismic reflections at the Smith Bank Shale–Zechstein halite contact at the well locations and serve to illustrate the true preserved thickness of the entire Triassic succession. (B) Triassic basin-fill distribution, with line of correlation highlighted. Wells, from north to south, comprise 22/10-1, 22/11-1, 22/11-3S1, 22/12a-2, 22/12b-4, 22/14a-2, 22/14b-4, 22/19a-1, 22/24a-1, 22/24b-5ST1, 22/29b-2, 22/29-1S1, 22/30a-6, 22/30a-2, 22/23b-5, 29/3b-4, 29/6a-3, 29/5b-3FZ, 29/8-3, 29/10-3S1, 29/5a-7, 29/5a-3, 30/7a-4, 30/7a-7, 30/7a-8, 30/12b-2, 30/12b-3, 30/17-4, 31/21-1, 31/21-3, 31/21-6, 31/21-5, 31/21-4, West Lulu-1, Ravn-1, L02-01.

along the fringes of terminal fluvial systems (Mader, 1982; Uličný, 2004; Geluk, 2005; Hounslow and Ruffell, 2006; Meadows, 2006; Brookfield, 2008), and evaporite deposition, in the form of stratigraphically restricted but areally widespread halite and anhydrite, appears to have been the product of episodic marine penetration into hydrologically closed sub-basins (Thompson and Meadows, 1997; Kovalevych et al., 2002; Geluk, 2005; McKie and Williams, 2009). Paleoclimate simulations of the early Triassic suggest that the North Sea region experienced a monsoonal climatic regime originating in Tethys (Péron et al., 2005), which may have been replaced by a more maritime climate by the middle Triassic following flooding by the Muschelkalk Sea (van der Zwan and Spaak, 1992). In the Central North Sea the early Triassic is dominated by playa deposits characterized by a virtual absence of root traces, burrows, and paleosols, but by the middle Triassic the section shows evidence of widespread soil formation and pervasive root traces and trace fossils (Porter, 2006). Goldsmith et al. (1995) document a concomitant transition from a lower Triassic section barren of spores and pollen into a middle Triassic section with relatively good recovery of gymnosperms, pteridophytes, bryophytes, and arborescent lycopods indicative of an upland vegetation of conifers, widespread shrubland vegetation, and episodic marsh conditions. The presence of *Taeniaesporites* is also potentially indicative of a dryland setting with a maritime influence (Hubbard and Boulter, 2000). Whilst localized eolian and evaporite conditions persisted across northwest Europe and east Greenland through the middle Triassic (Hounslow and Ruffell, 2006; McKie and Williams, 2009), the Central North Sea basin appears to have experienced a more equable, Mediterranean-like climate as a result of proximity to the Muschelkalk Sea.

The upper Triassic section is poorly preserved in the Central North Sea, but regionally shows the development of widespread, mud-rich fluvial and playa systems with sands restricted to the basin margins (Ruffell and Shelton, 1999). Perennial lacustrine facies are sporadically distributed throughout the region (O'Donnell, 1993; Milroy and Wright, 2000; Porter and Gallois, 2008), and palynological data are indicative of episodically more humid conditions (Simms and Ruffell, 1990; van der Zwan and Spaak, 1992). In the Central North Sea the section shows widespread paleosol disruption with common, large

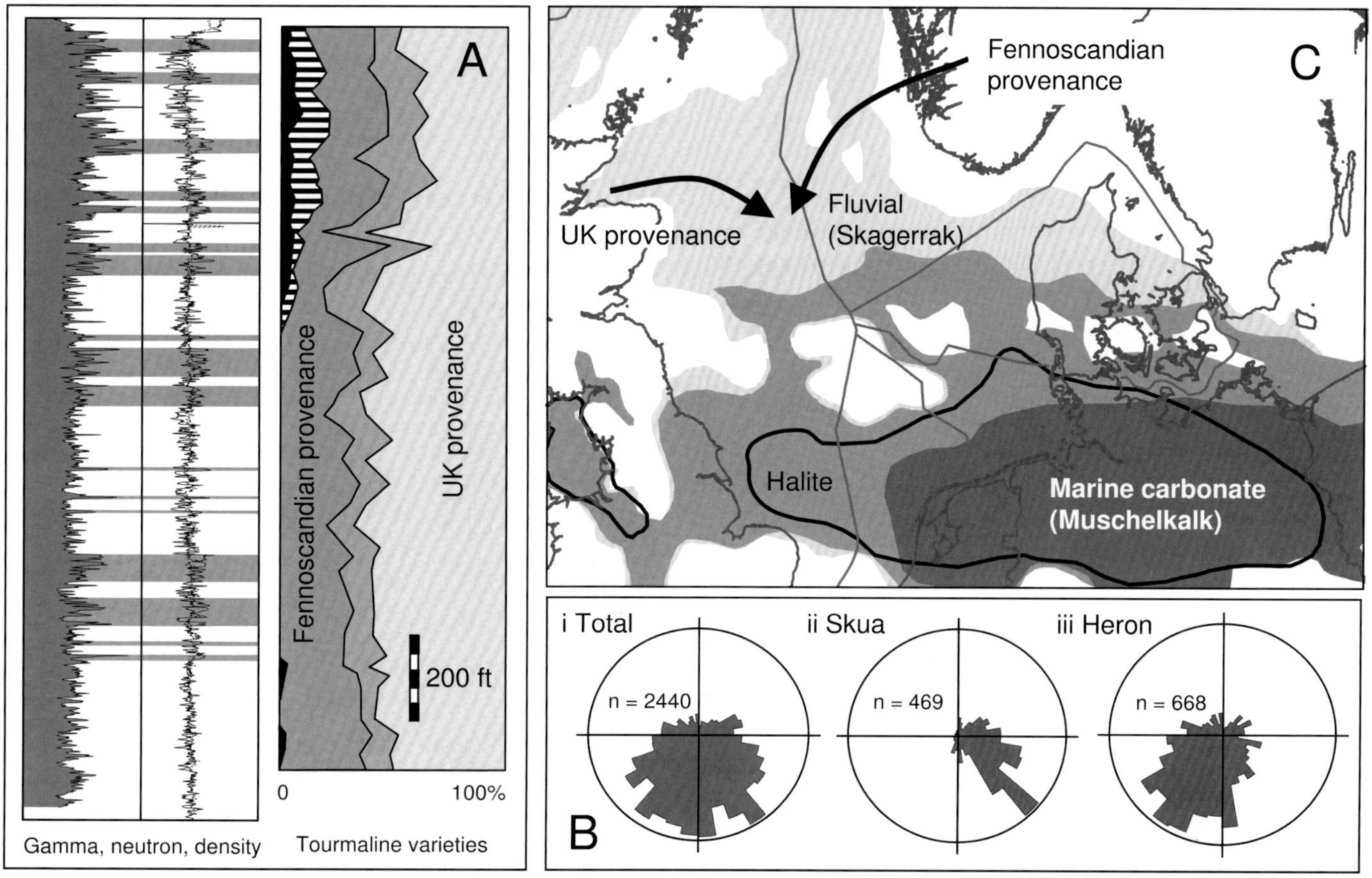

FIGURE 4.—Triassic fluvial drainage. (A) Stratigraphic section from Marnock well 22/24b-5S1 showing vertical variations in percentage of tourmaline species based on shape, color, internal structure, and optical properties (modified after Mange-Rajetzky, 1995) which are used to differentiate Fennoscandian and UK provenance. (B) Middle Triassic paleocurrent data for the Skagerrak Formation in the Forties-Montrose High region based on borehole image and oriented core data. (i) Total data for cross-bedding from wells 22/14b-4, 22/24b-9 (Skua), 22/24d-10 (Egret), 22/29-2S1, and 22/29-3 (Seagull), 22/29-5RE and 22/30a-6 (Heron). (ii) Data from cross-bedding in the basal upper Judy section of 22/24b-9. (iii) Data from cross-bedding the upper Judy section of Heron field. (C) Middle Triassic paleogeography with inferred fluvial drainage pattern based on provenance and paleocurrent data.

rhizoliths indicative of a well-established vegetation cover. However, halite occurrences in the Southern North Sea also indicate that arid conditions were generally pervasive.

CENTRAL NORTH SEA FLUVIAL SYSTEMS

The dryland fluvial systems of the Skagerrak Formation are organized into large-scale coarsening- and cleaning-upward cycles several hundred meters thick. These cycles are bounded by regionally correlatable shale members representing episodes of reduced sand dispersal into the basin. The cycles are typically composed of a basal section of playa or floodplain heterolithics and sand-prone, generally unconfined fluvial facies, which pass upward into increasingly sand-prone and channel confined facies (cf. Jones et al., 2005; McKie et al., 2010). This vertical succession has been interpreted to reflect the repeated expansion and contraction of the fluvial systems in response to tectonics and base-level changes (cf. Steel and Ryseth, 1990), although it is also likely that fluctuations in sediment supply from the main catchments also played a significant role (McKie and Williams, 2009). This paper concentrates on the middle Triassic Judy Member, which consists of a large-scale coarsening-upward cycle, the uppermost part of which defines the main reservoir interval in the Heron, Egret, Skua, and Marnock fields (Fig. 1B). The Judy Member in this area can be further divided into two subsidiary cycles of subtly differing provenance and sedimentological character: a lower interval largely derived from the Scottish highlands which generally lacks paleosols and biogenic traces, and an upper, channel-dominated section containing sediments of mixed Fennoscandian and Scottish provenance (Mange-Rajetzky, 1995) with evidence of more widespread plant and animal colonization.

Skagerrak Facies

Although the Skagerrak in the Central Graben has long been recognized as a dryland alluvial succession (Hodgson et al., 1992; Fisher and Mudge, 1998), the interpretation of the depositional system has evolved from a simple braidplain model (Hodgson et al., 1992) to a more diverse suite of dryland depositional styles (McKie et al., 2010). The facies architecture in the area of interest (Fig. 1) can be divided into three broad associations comprising channel-belt and floodplain associations defining the upper section, and dryland terminal splay complexes defining the lower section.

Channel-Belt Facies Association: These sand-prone deposits are typically expressed by sharp-based, fining-upward (or locally trendless) cycles 1–5 m thick composed of very coarse- to very fine-grained sandstone (dominantly fine- to medium-grained). The cycles are stacked into multistory intervals up to 20 m thick. The basal surfaces of these cycles are erosional and mantled by intraformational clasts of blocky and platy mud clasts, reworked calcic paleosol, and rhizolith material. Granule- and pebble-grade extraformational grains are rare. The upper parts of the fining-upward cycles commonly grade upwards into pedogenically modified mudstones and very fine-grained heterolithic facies, with common adhesive meniscate bioturbation (cf. Smith et al., 2008), root traces, and calcic rhizoliths penetrating into the underlying sands. Mud curls and desiccation cracks are locally present, but not common in these sections. Three end-member channel-belt deposits can be differentiated:

- Low-angle accretionary deposits dominate this facies, and are composed of fine- to medium-grained, low-angle cross-stratified beds up to 1 m thick, organized into bed sets up to 4 m thick (Fig. 5). In general this low-angle stratification is difficult to discern in core without the presence of angular truncations (although these are relatively common), but borehole image data corrected for tectonic dip indicate the dominance of low-angle dips of 5–20° in this pervasive facies. Isolated cross-bed sets occur internally within these deposits, typically immediately above the basal lag.
- Cross-stratified fills are less common (< 10% of this facies association) but are broadly similar in thickness, lithology, and clast composition to the low angle accretionary fills. Internally these deposits comprise both trough and planar cross-stratification throughout the lower section, with current-ripple and horizontal to low-angle lamination present in the upper section of the sandbody as this grades into overlying siltstones and mudstones. The orientations of the bounding surfaces to cross-bed sets from borehole image data suggest that many set boundaries are inclined at c. 5–10° and commonly at an angle of up to 90° to the enclosed cross-beds, although bulk data for cross-bedding and bounding surfaces broadly show similar trends.
- Mud-prone fills are characterized by up to 4 m of heterolithic siltstones and very fine-grained sandstones (Fig. 6), or more rarely by uniform successions of planar-laminated and current-ripple-laminated siltstones and mudstones. These heterolithic beds show variable proportions of root-trace disruption and bioturbation by adhesive meniscate burrows (Fig. 7), either descending from the upper part of the fill or from colonization surfaces within the heterolithic units. Soft-sediment deformation is locally present in the form of small-scale slump structures. These deposits typically grade upwards from sharp based lag intervals (c. 1 m thick) of intraclast-bearing, fine- to medium-grained sandstone identical to the basal lags of the fining-upward cycles described above. Mud-prone fills tend to be common in the south of the study area towards Puffin

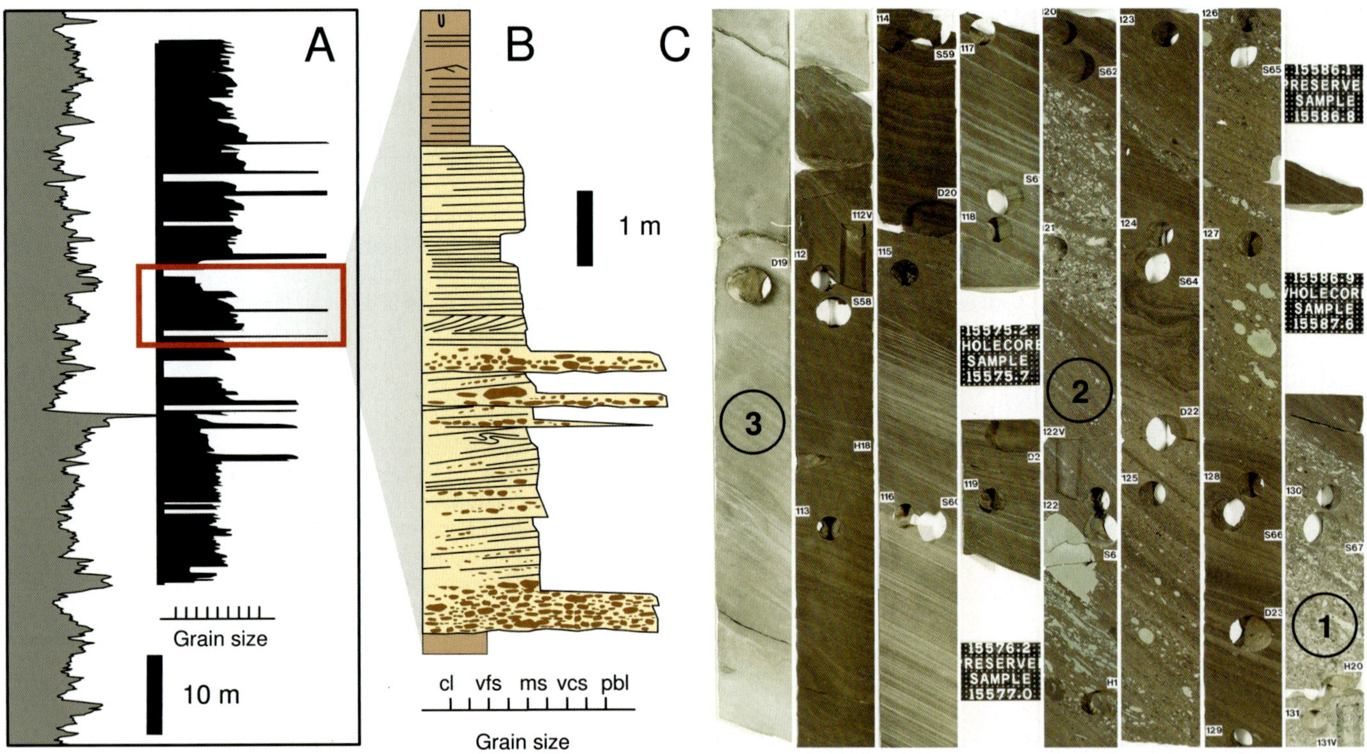

FIGURE 5.—Skagerrak fluvial deposits composed of low-angle accretionary channel-fills dominate the succession. (A) Illustrative gamma-log motif and outline grain-size profile through a multistory channel-fill succession in well 22/29-5RE (Heron). (B) Graphic log from selected section in Part A of typical Skagerrak channel-fill comprising low-angle planar lamination with scattered intraclasts of reworked calcrete material and mud clasts. (C) Core photographs of low-angle accretionary deposits in Part B, showing a basal section containing abundant mud and calcrete intraclasts (1), low angle cross-stratification (2), and an upper section (3) of plane-bedded and current-ripple-laminated siltstones and very fine-grained sandstones (N.B. bedding is inclined at c. 40° relative to core due to well deviation). The top of the section is upper left and base is lower right. Core sections are 1 m long.

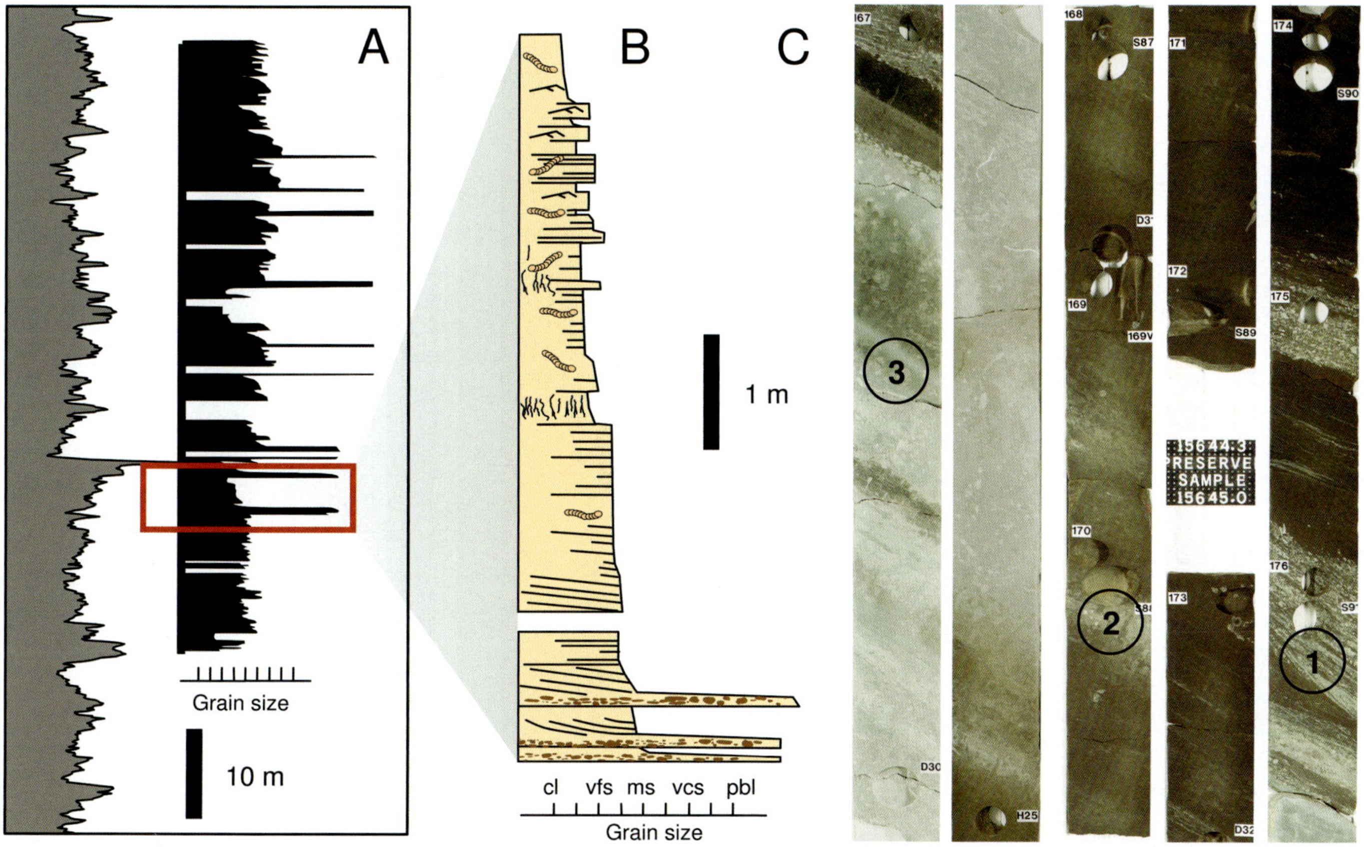

FIGURE 6.—Mud-prone channel-fills composed of paleosol-disrupted heterolithic deposits. (A) Illustrative gamma-log motif and outline grain-size profile through a sand-prone to heterolithic channel-fill succession in well 22/29-5RE (Heron). (B) Graphic log of selected section from 22/29-5RE and corresponding core photographs (C), illustrating a basal lag of reworked calcrete and mud intraclasts (1), sand-prone medial section with remnant lamination (2), passing upward into root-trace-disrupted (3) heterolithic deposits (N.B. bedding is inclined at c. 40° relative to core due to well deviation). The top of the section is upper left and base is lower right. Core sections are 1 m long.

(Fig. 1), and in stratigraphic intervals transitional into the regional shale members.

Paleocurrent data, derived from borehole images and oriented core (corrected for tectonic dip), are widely scattered but, as described above, show a dominance of flow to the south with the local dominance of south-southwest or southeast modes reflecting drainage from the main catchments (Fig. 4B, C). At a finer scale, cross-bedding and low-angle accretionary lamination from individual multistory channel intervals tend to show a similar, unimodal distribution (Fig. 8) with variable degrees of scatter. The upper, heterolithic parts of the component fining-upward cycles are commonly gently inclined at 5–10°, and show a wider range of orientations, although these are still oriented in a direction broadly similar to the underlying low-angle lamination and cross-bedding (Fig. 8).

These coarse-grained, erosionally based and upward-fining intervals show widespread evidence of floodplain soil erosion, and are interpreted as the products of channel migration across the Skagerrak alluvial plain. Such facies are widely distributed and dominate the upper parts of the Judy succession, with the lack of evidence for areas of sustained floodplain deposition possibly indicating a dominance of mobile, potentially multiple channel systems and frequent avulsion leading to a uniformly channel-dominated architecture over wide areas (up to 10 km). The predominance of low-angle planar lamination and low-angle cross-bedding suggests that bar accretion took place under shallow, high-velocity flows transitional between lower and upper flow regimes, possibly analogous to the examples described by Allen (1982). The more cross-stratified intervals may record more complex bar accretion of compound bedforms, or record dune migration along channel thalwegs in cases where only the lower part of the fining-upward cycle is cross-stratified. The muddy upper sections of the fining-upward cycles represent accretion of the upper parts of bars, and these were commonly subaerially exposed and sparsely vegetated, and supported a restricted fauna of burrowing insects. This may reflect a variable discharge regime, with dry-season colonization or desiccation of upper bar surfaces and wet-season, bank-full conditions reworking desiccated mud drapes and soil material. Although the abundant, reworked calcrete is also likely to be a product of thalweg scouring of buried soil profiles, the mud intraclasts tend to be composed of laminated clay which contrasts with the more silt-prone, biogenically disturbed floodplain and bar-top fines. The clay intraclasts may therefore be largely derived from reworking of in-channel, slack-water drapes and plugs rather than the more persistently subaerial bar-top and floodplain material.

The more heterolithic and mud-prone deposits, with evidence of repeated burrowing and rooting of the upper-bar section, are indicative of channels which experienced infrequent bank-full conditions and sustained periods of reduced or limited flow (albeit with sufficient year-round water availability to maintain riparian vegetation and

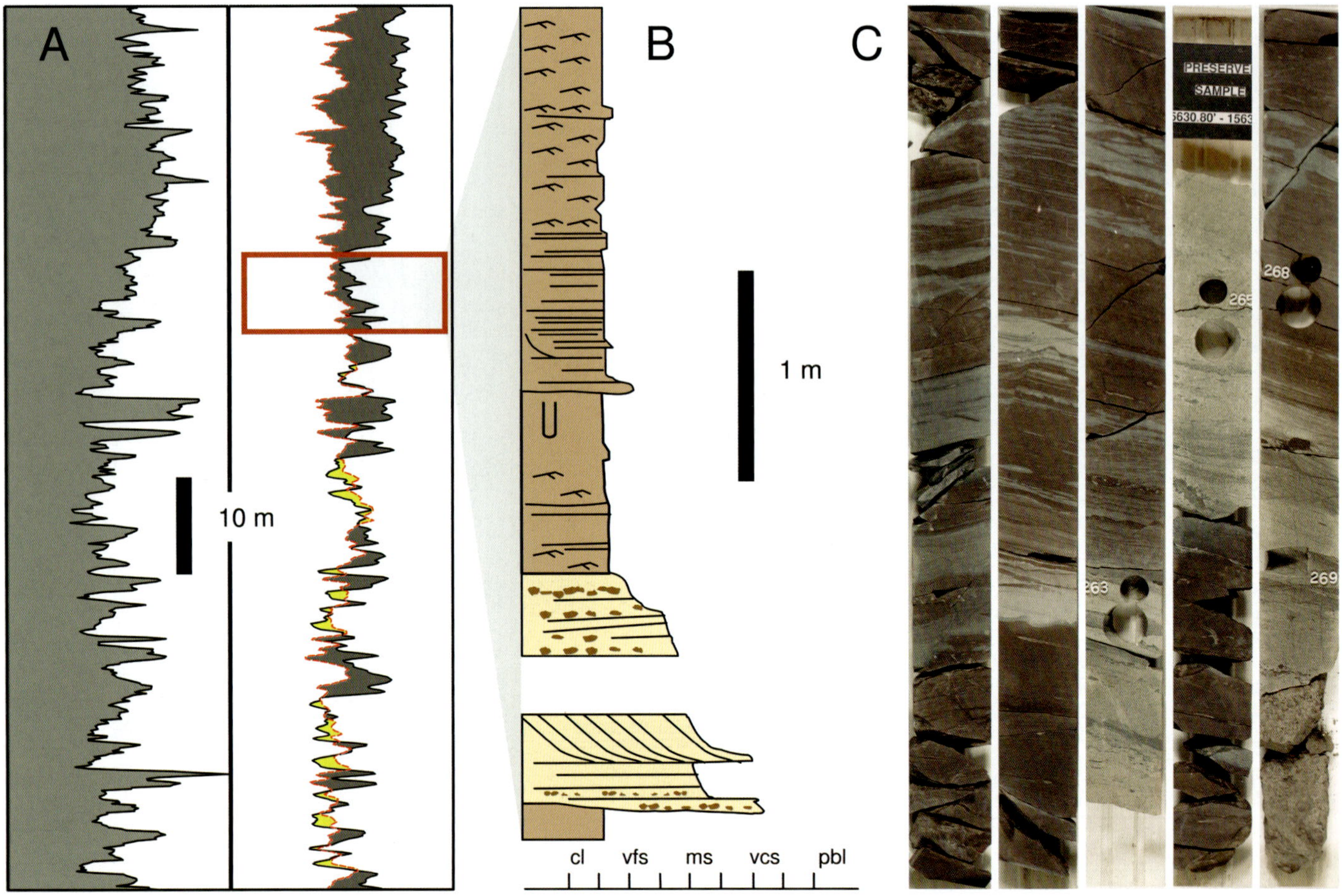

FIGURE 7.—Example of a mud-rich channel-fill from well 22/29-1S1. (A) Representative gamma, neutron, and density profile of an interval composed of mud-rich channel-fill deposits (yellow neutron/density cross over intervals record sandstones, gray equates to shale-prone intervals). (B) Graphic log of a mud-prone fill distinguished by an erosionally-based, intraclast-bearing sandstone overlain by current-ripple-laminated and planar-laminated siltstones. (C) Core photographs of interval in Part B. The top of the core section is upper left and base is lower right. Core sections are 1 m long.

burrowing insects). The increased frequency of such deposits in a down-paleoflow direction (over c. 50 km) may be indicative of the reduced frequency of bank-full discharges in the lower reaches of the Skagerrak terminal system. Mud-prone abandonment deposits are rarely encountered in core, and comprise laminated siltstones overlying thalweg deposits, indicative of an abrupt loss of stream power and progressive infilling of a permanently subaqueous, abandoned channel.

The paleocurrent data show that both the sand-prone, lower-bar cross-bedding and inclined laminae, and mud-prone upper-bar deposits tend to dip in broadly similar orientations (albeit with variable scatter), which could be interpreted to reflect deposition in relatively straight channels containing downstream-accreting, transverse bars. However, the common presence of vegetation and burrowing on these upper-bar sections indicates that a predominance of subaqueous, transverse bars is unlikely. Alternatively the data may reflect deposition within channel systems dominated by the downstream translation of lateral, medial, or point bars, with a preservational bias towards the downstream-accreted part of the bar paleoflow record (cf. Willis, 1989). If the channels were sinuous it could be argued that the basal cross-bedding data collected from thalweg scours should preserve a widely scattered record of channel sinuosity compared to the more biased record of bar accretion, and that the overall consistency of the Skagerrak data favors relatively straight channel reaches with downstream accretion of medial and lateral bars. Given the broadly semi-arid, terminal fluvial setting of the Skagerrak it would be tempting to argue for a highly erratic hydrology for the fluvial systems. Fielding et al. (1999), and Fielding et al. (2009) have argued that such streams should be characterized by a variable bedform architecture with abundant, erosional cut-and-fill fabrics, common upper-flow-regime structures, and in-situ vegetation which colonized dry channel floors. The general absence of such features, combined with the apparent consistency of the Skagerrak orientation data, could point to a depositional record dictated by sustained seasonal flooding and steady downstream bar migration rather than episodic and erratic flood events. In contrast, the channel-fills in the terminal splay association (see below) do show abundant core-scale erosional features (e.g., mud-clast-mantled gutter casts), which may point to a more erratic hydrology for this biogenically impoverished facies.

Floodplain Facies Association: These deposits form heterolithic sections 1–30 m thick which are ubiquitously pedogenically modified and burrowed (Fig. 9). They tend to occur as discrete packages vertically separating multistory channel-belt deposits or less commonly preserved on the margins between belts, and increase in thickness and proportion in a down-paleoflow direction over distances of several

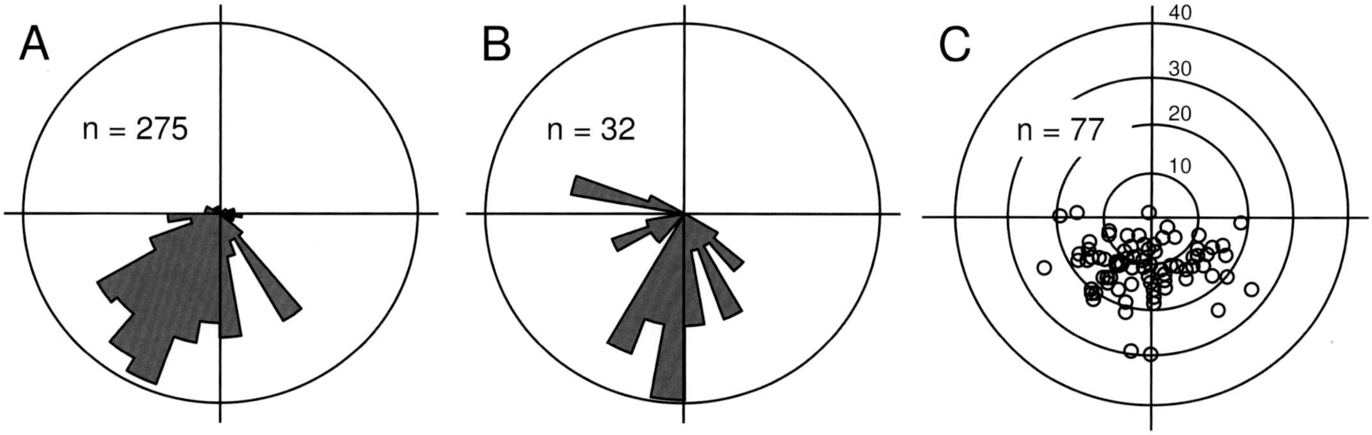

FIGURE 8.—Paleocurrent data from oriented core in well 22/29-5RE (Heron) comparing the A) gently inclined lower-bar sandstone cross-bedding and B) upper-bar heterolithic unit. Both are oriented southwest, albeit with greater scatter in the heterolithic unit, suggesting a dominance of downstream accretion. (C) Wulff upper-hemisphere stereoplot of poles to inclined cross-bedding and low-angle accretionary surfaces for a single c. 30 m multistory unit in well 22/29-2S1 from borehole image logs. A comparable pattern of similarly oriented high-angle cross-bedding and low-angle accretionary surfaces is present, but with a greater scatter compared to the oriented core data.

kilometers to tens of kilometers. The contact with underlying channel-belt deposits is commonly transitional as a result of the biogenic disruption. The paleosols in these successions are variable in maturity, showing weak to pervasive disruption by root traces and adhesive meniscate burrows, variable clay illuviation, calcic paleosol maturity, and level of oxidation or reduction. Two end-member facies are recognized defining these heterolithic intervals.

Splay Deposits. These commonly have a structureless to disrupted appearance with abundant burrow mottling by adhesive meniscate burrows and *Skolithos,* and common root traces (Fig. 9A, D). They comprise poorly sorted, very fine- to fine-grained sandstone containing a high proportion of intraformational clay and carbonate material. Sharp grain-size breaks, lags of intraformational mud and calcrete, and truncated paleosol profiles, rather than clearly defined erosional surfaces, commonly mark the bases of units on a scale of 0.2–1 m. Typically these stack into fining- and coarsening-upward bedsets < 10 m thick. Indistinct cross-lamination and plane bedding are locally discernible.

These sands are interpreted to represent weakly confined to unconfined splay deposition which was subject to pervasive, post-depositional paleosol modification. The sharp bases and localized lags suggest some scouring and preservation of small-scale bedforms, but in general these deposits were probably deposited as poorly defined sheets. The association with channel facies, either within the same vertical section or as down-paleoflow equivalents, suggests an origin as lateral and terminal splays to the channels (cf. Tooth, 2005; Fisher et al., 2007). The evidence for shallow scouring could either be attributable to a network of shallow channels, or alternatively could be the product of erosion around shrubland vegetation, e.g., as shadow bars and scours (Rygel et al., 2004). It is also possible that some of these sands could be a record of levee accretion, but there are no diagnostic criteria which could be used with the core dataset to determine whether these occur (cf. Brierley et al., 1997). Variations in paleosol maturity and drainage could provide clues to floodplain elevation, but there are no systematic patterns in the Skagerrak which would suggest that levees occur as a discrete end-member facies which can be differentiated from equally bioturbated splay deposits under ubiquitously well-drained conditions.

Floodplain Fines. These deposits are dominated by very fine-grained sandstones, siltstones, and mudstones, forming intervals from 1–30 m thick which are again pervasively disrupted by paleosols. The paleosols are variable in character, ranging from weakly developed entisols, through aridisols with relatively immature calcrete profiles (stages I to III, Machette, 1985) and alfisols with abundant illuviated clay (Fig. 9B, C, D). The color of these paleosols varies from red to green, gray and black, although caution must be exercised in interpreting the significance of Skagerrak soil color because of widespread iron reduction by fluid movement associated with the hydrocarbon charge, and bitumen staining following in-situ hydrocarbon cracking under high temperature conditions. Ped fabrics are common in mud-rich intervals, and soil slickensides are locally common.

These mudrocks record vegetated floodplain conditions. The common presence of calcrete, ped fabrics, and slickensides suggest generally well-drained conditions, but with common water-table fluctuations. The overprinting by pervasive adhesive meniscate burrows suggests episodically higher soil-moisture conditions favorable for insect larvae (cf. Smith and Hasiotis, 2008; Smith et al., 2008) and indicates the former presence of abundant organic detritus derived from plant material. The palynological data described previously suggest a mix of drought-resistant shrubland and wetland environments, possibly with a maritime influence (cf. Kelber and van Konijnenburg-van Cittert, 1998; Hubbard and Boulter, 2000). Overall the evidence suggests that floodplain drainage was variable, with episodic flooding leading to repeated clay infiltration and triggering insect burrowing activity, alternating with episodes of well-drained floodplain conditions and calcrete formation. The core data are insufficient to establish whether variations in floodplain drainage can be related to channel-belt proximity or position on alluvial ridges, or to a seasonal pattern of discharge. In the Marnock field, paleosol types appear to be correlatable across the field (c. 5 km), but whilst this could be used to favor temporal changes in floodplain drainage there is insufficient areal coverage to establish that the observed changes are not due to kilometer-scale channel movement.

Marsh and Palustrine Deposits. This relatively uncommon facies forms intervals c. 1–10 m thick composed of black mudrocks and heterolithic beds with localized limestone (micrite) horizons < 1 m thick and laminar carbonate cementation. The heterolithic sections

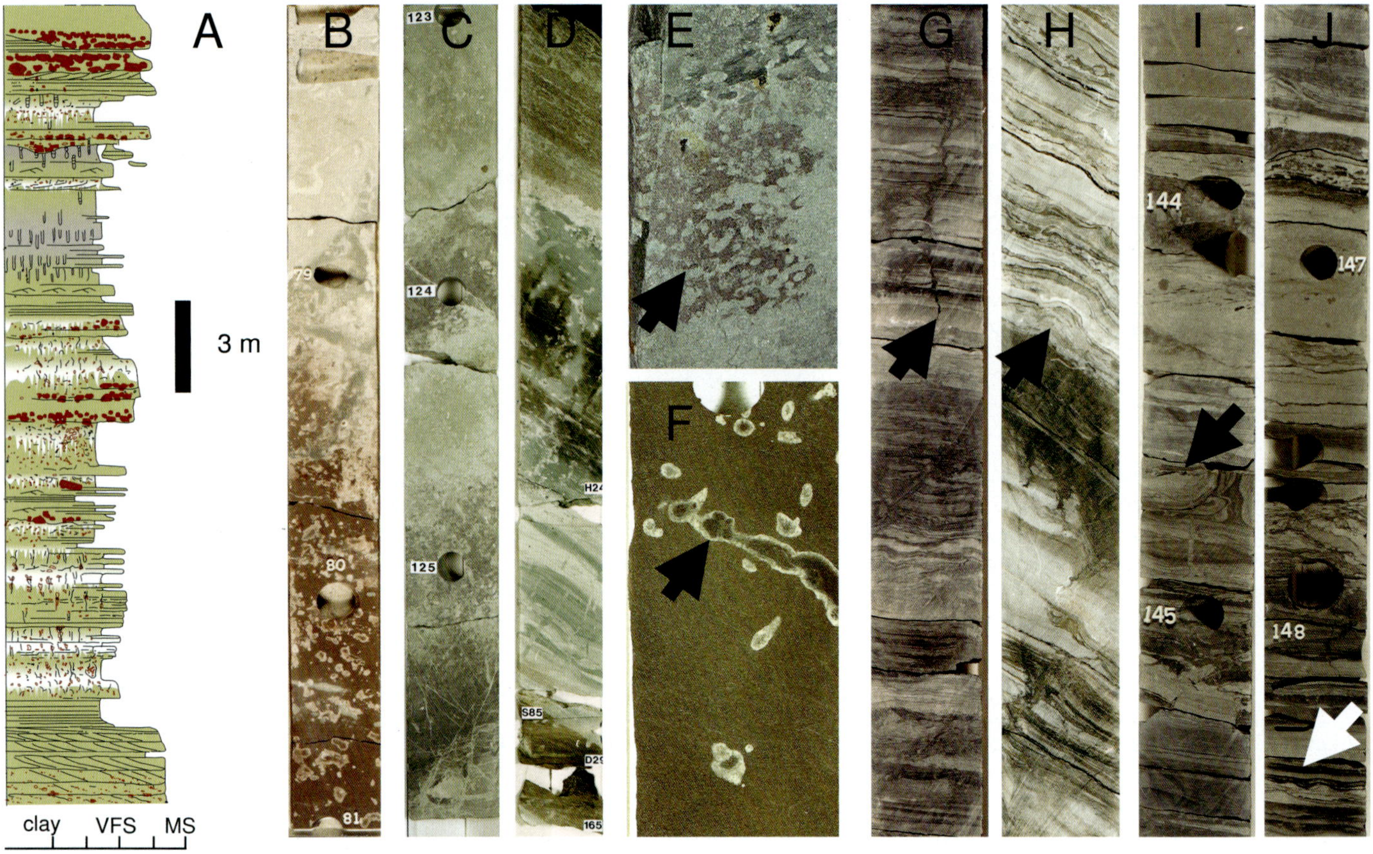

FIGURE 9.—Representative core photographs of the floodplain facies association. (A) Graphic log from well 22/24b-5ST1 (Marnock) showing the interbedding of sand-prone and mud-prone splay and floodplain intervals. (B) Core photograph of calcic paleosol profile, Heron 22/30b-2. (C) Rooted and burrowed paleosol with pervasive illuviated clay, Puffin 29/10-3. (D) Lightly rooted, immature paleosol with drab haloes around root traces, Heron 22/29-5RE. (E) Pervasive disruption by adhesive meniscate burrows, Puffin 29/10-3. (F) Calcic rhizoliths (arrowed), Heron 22/29-5RE. (G) Rooted marsh heterolithic deposits, Heron 22/30b-6. (H) Heterolithic marsh deposits with irregular and domal algal fabrics (arrowed), Heron 22/29-5RE. (I, J) Interbedded hyperpycnite sandstones and lacustrine mudstones with dewatering fabrics and wave-ripple laminae (arrowed), Heron 22/30b-6. All core sections 0.1 m wide.

show the development of wave-ripple lamination, indeterminate bioturbation, algal laminites, flat-pebble conglomerates, and root-trace disruption (Fig. 9G, H). Deeply penetrating (up to 0.3 m) desiccation cracks are common. Soft-sediment deformation, in the form of clastic dikes and fissures, are locally abundant. Palynological analysis of this facies indicates a dominance of pteridophytes, bryophytes, and arborescent lycopods (Goldsmith et al., 1995).

The black mudrocks record subaqueous conditions in relatively persistent water bodies, either as floodplain lakes or small ponds. In the heterolithic sections more marginal conditions predominated. The plant rooting, presence of algal structures, and early carbonate cementation are indicative of palustrine and marsh conditions, with sustained groundwater carbonate cementation combined with the relatively clear-water conditions favorable for algal mats.

Hyperpycnite Facies. These relatively uncommon heterolithic deposits comprise 1–5 m of interbedded, normally graded, very fine- to fine-grained sandstones and mudrocks on a scale of 0.01–0.1 m (Fig. 9I, J). There is no systematic pattern to the stacking of these beds, although subtle muddying-upward trends are locally discernible. The sandstone beds are typically erosionally based and structureless, although wave- and current-ripple laminae are present, together with rare humpback antidune lamination (Fielding, 2006), and soft-sediment deformation and dewatering fabrics. The interbedded mudrocks are typically black, laminated, and identical to those associated with the marsh facies.

These heterolithic deposits record episodic density-current deposition, either as rapidly deposited massive and dewatered sands, or via tractional deposition from more dilute suspensions. Dewatering and slumping may suggest some instability, which could indicate that these heterolithics were deposited on a sloping surface, e.g., a lacustrine delta front, but the general absence of such structures (and lack of coarsening-upward, progradational patterns) could suggest that they represent hyperpycnite flow of rare flood events directly into floodplain lakes and ponds (cf. Sáez et al., 2007).

Dryland Terminal Splay Association: This is the dominant association in the lower Judy section. In contrast to the upper section, the lower interval (Fig. 2A) is almost entirely devoid of burrows and root traces, yields little or no palynological recovery (Goldsmith et al., 1995), shows limited paleosol development, and appears to have a provenance restricted to the Scottish Highlands without any significant Fennoscandian input (Mange-Rajetzky, 1995). The section broadly comprises three end-member facies:

Channel and Scour-Fill Deposits. These are most common at the top of the lower Judy, but also occur throughout the succession in association with the splay deposits described below. They comprise up to 3 m thick, sharply or erosionally based, fining-upward or trendless packages of fine- to medium-grained sandstones with abundant mud-flake lags (Fig. 10). Mudclast-filled gutter casts and scours are common throughout. However, unlike the upper Judy, reworked calcrete is absent from this interval. The sandbodies have an internal fabric varying from structureless to indistinctly dewatered, or dominated by horizontal to gently inclined planar lamination (with abundant mica concentrations defining laminae). Cross-bedding is uncommon except in the uppermost part of the succession.

These deposits record deposition under a variety of weakly channel-confined conditions. The abundance of mud-flake lags may either be indicative of channel reworking of overbank muds, or local reworking of desiccated mud drapes. The structureless to dewatered sandstones suggest catastrophic deposition, either through bank collapse or rapid deposition from waning, sediment-laden flows, whilst the low-angle and horizontal plane-bedded intervals, in association with common scours and gutter casts, indicate shallow, high-velocity flows with common scouring and erosion (cf. Fielding, 2006) and a dominance of flood processes. The vertically isolated nature of these sand bodies and evidence of rapid cut and fill suggest that these systems did not form established channel belts, and instead comprised discrete ribbons prone to sediment plugging and avulsion. Their close association with weakly confined splay and playa facies indicates that floodwaters expanded and shallowed into lateral and terminal splays on the margins of playa, but under drier conditions than the splays encountered in the upper section described above. The appearance of cross-stratified deposits towards the top of the section is apparently coincident with the appearance of Fennoscandian-derived tourmaline in 22/25b-5 (Fig. 4A), and is likely to mark a change in fluvial style related to the increased water availability, which subsequently allowed significant biogenic activity to occur in the upper section.

Terminal Splay Complexes. These consist of micaceous, very fine- to fine-grained sandstones (Fig. 10) organized into coarsening-upward cycles up to 10 m thick (in turn grouped into composite cycles up to 40 m thick). At the bases of these cycles are thin shales (typically < 0.5 m thick), which are well laminated or massive, commonly desiccated, and very rarely burrowed. Wave-ripple lamination is commonly developed, and rare, delicately interlaminated shales and siltstones are locally present. The contact with the overlying sandstones is either sharp (with mud clasts or desiccation cracks), or gradational via planar-laminated and current-ripple-laminated siltstones and very fine-grained sandstones. The 0.3–1 m thick beds forming these cycles are dominated by planar-lamination and climbing-current-ripple lamination, with dewatering fabrics locally developed. The tops of the cycles are either

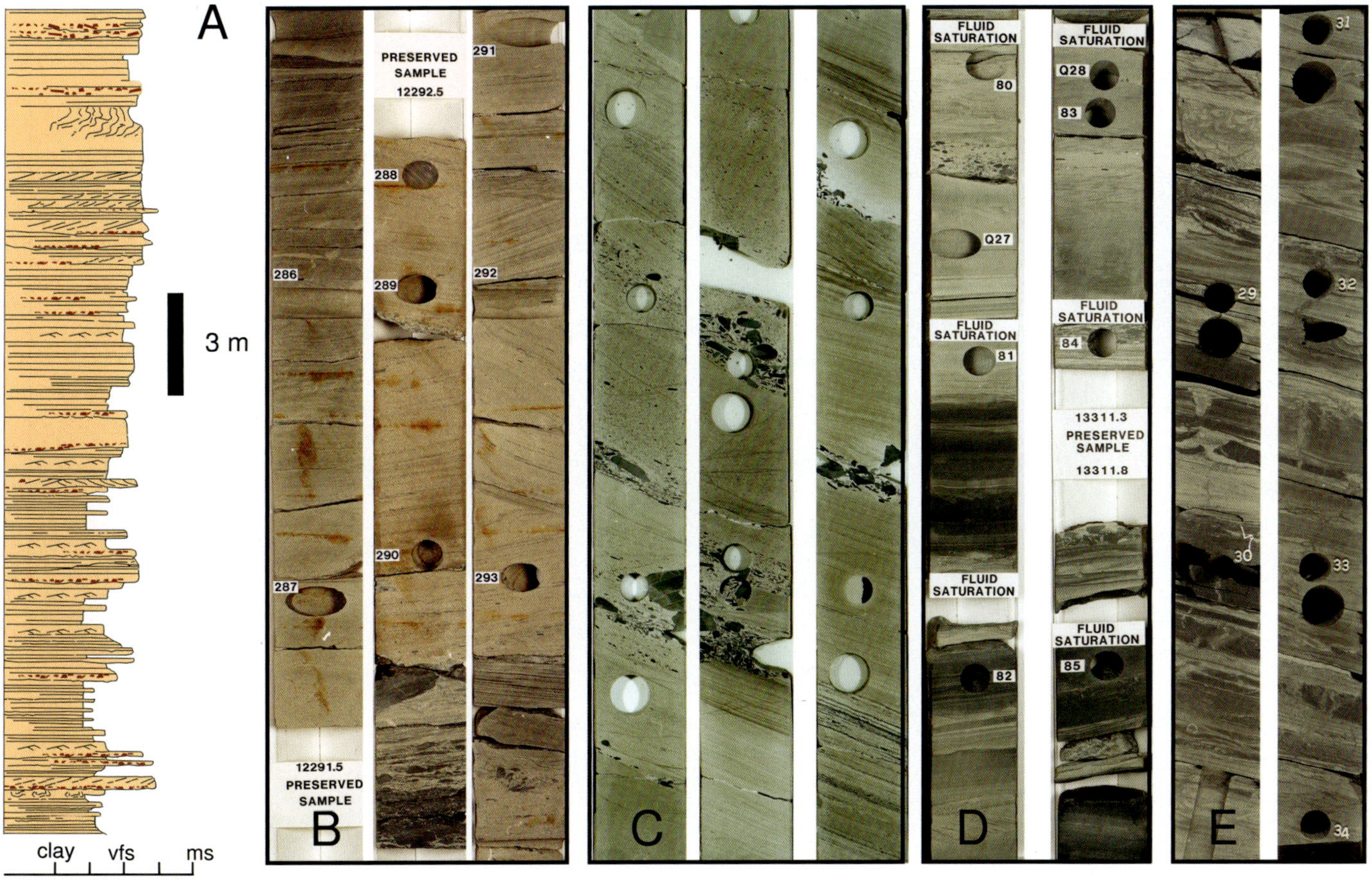

FIGURE 10.—Lower Judy facies. (A) Illustrative core-log section of terminal splay sandstones coarsening upward into cross-bedded and planar-laminated channel-fills, Skua 22/24b-9. (B) Cross-bedded fining-upward package from 22/24b-9 with large mud-clast lag. (C) Weakly confined, largely planar-laminated sandstones with local mudclast concentrations, 21/18-5. (D) Splay deposits coarsening upward from laminated mudstones into planar-laminated and current-ripple-laminated sandstones, 29/8-3. (E) Heterolithic playa deposits, 21/30-18.

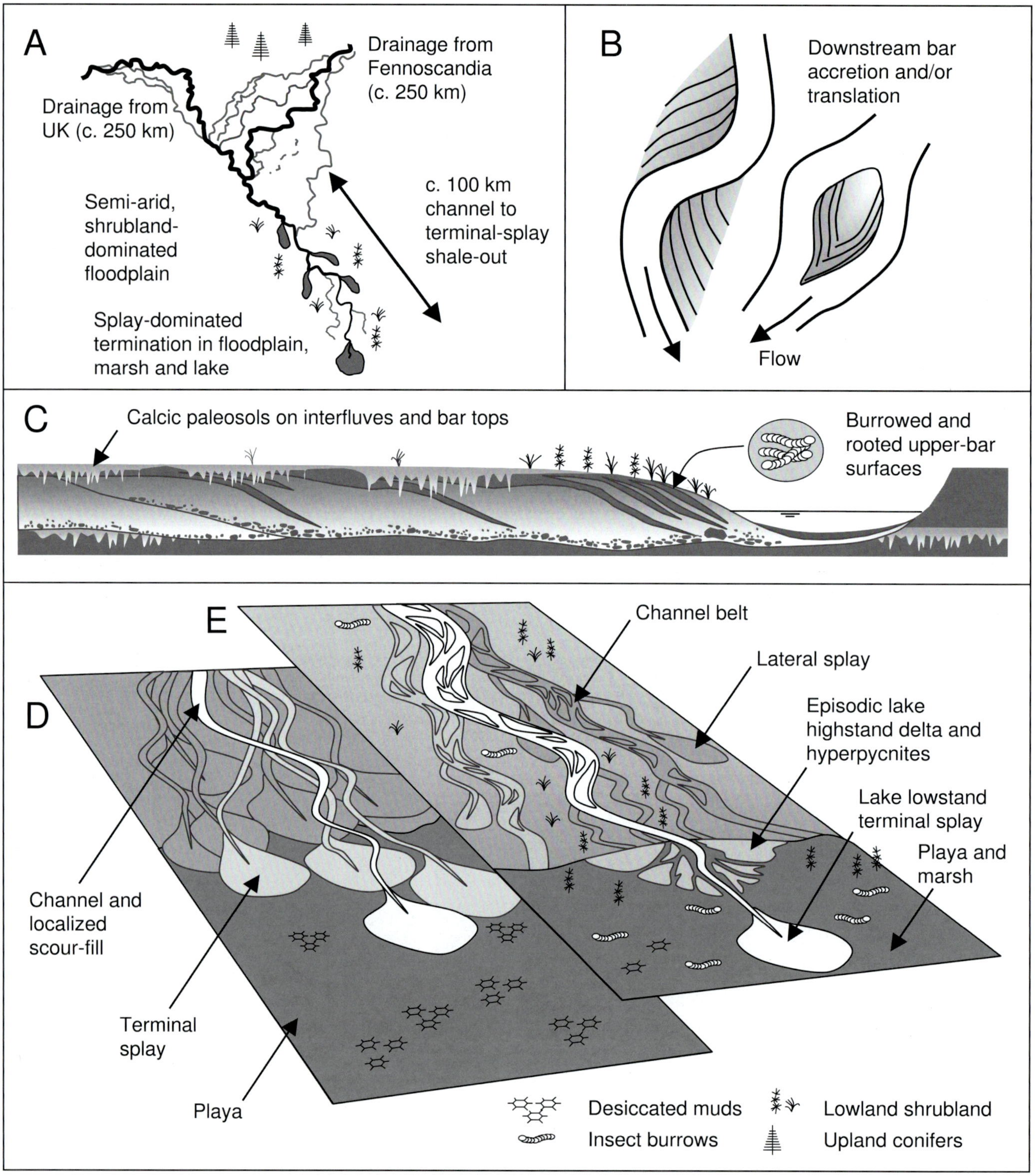
A
Drainage from Fennoscandia (c. 250 km)
Drainage from UK (c. 250 km)
c. 100 km channel to terminal-splay shale-out
Semi-arid, shrubland-dominated floodplain
Splay-dominated termination in floodplain, marsh and lake
B
Downstream bar accretion and/or translation
Flow
C
Calcic paleosols on interfluves and bar tops
Burrowed and rooted upper-bar surfaces
E
Channel belt
Lateral splay
Episodic lake highstand delta and hyperpycnites
Lake lowstand terminal splay
Playa and marsh
D
Channel and localized scour-fill
Terminal splay
Playa
Desiccated muds
Insect burrows
Lowland shrubland
Upland conifers

truncated by channel-fill deposits or sharply overlain by the mudstone of a succeeding cycle.

The basal shales to these cycles record subaqueous deposition, and the paucity of biogenic activity and common desiccation suggest that these were ephemeral water bodies (although rare intervals of delicately interlaminated shales and siltstones indicate the persistence of water bodies through multiple flood events). The overlying plane-bedded and climbing-ripple-laminated sandstones represent the product of deposition from shallow, unconfined flows, and their arrangement into coarsening-upward cycles is suggestive of progradation. The variably sharp (with evidence of desiccation) or gradational contacts suggest that these progradational sand systems alternated between entering standing water bodies or covering subareally exposed surfaces. The overall setting appears to have been of a flood-dominated system characterized by rapid deposition into ephemeral water bodies or across dry lake floors, with the channel and scour-fill deposits described above representing the associated master channels which supplied these terminal splay complexes.

Playa Deposits. These heterolithic deposits form intervals up to c. 20 m thick, but commonly they are less than 1 m thick. They are most common in the lowermost parts of the Judy section, and form the dominant facies in the transition between the Skagerrak and Smith Bank Formations. The heteroliths are composed of millimeter- to centimeter-scale alternations of very fine- to fine-grained sandstone and mudstone, organised into meter-scale cleaning- and coarsening-upward cycles. Wave-ripple and current-ripple lamination, and desiccation fabrics, are abundant (Fig. 10), with rare, irregular adhesion laminae. Burrows and root traces are present but extremely rare. Dewatering fabrics are present in thicker sandstone intervals.

These mud-prone deposits represent low-energy deposition in a setting characterized by episodic sand deposition and erratic water-table conditions that fluctuated between subaqueous, damp, and desiccated in response to ephemeral flood events. No significant thicknesses of fully subaqueous, lacustrine muds are present, and it is likely that these represent extensive playa rather than the fluctuating margins to perennial lakes.

In conclusion, the Skagerrak alluvial plain would appear to have been traversed by dryland, terminal fluvial systems with a drainage length in excess of several hundred kilometers (Fig. 11A) and under two end-member climatic regimes (reflected in the upper and lower stratigraphic subdivisions). In the lower part of the section the level of late Scythian–Anisian aridity (Price, 1999) limited biogenic activity, whereas in the upper, Anisian–Ladinian section larger streams traversed a well-drained alluvial plain colonized by drought-resistant shrubs, and sufficient water was retained within the channels to maintain plants and burrowing insects on upper bar surfaces during dry seasons (Fig. 11B, C). Seasonal discharge variations were likely influenced by the Tethyan monsoon, and the increased vegetation cover recorded in the upper section may have been a result of marine flooding of the adjacent Southern North Sea increasing local precipitation (McKie and Williams, 2009).

In the lower part of the Judy section the drier, flashier discharge conditions resulted in fluvial systems characterized by rapid cut and fill, which (in the study area) fed terminal splay complexes on the margins of ephemeral lakes (Fig. 11D). The terminal splay complexes on the margin of the Lake Eyre Basin may offer appropriate modern process analogues (Lang et al., 2004; Fisher et al., 2008), although the Lake Eyre systems are fed by incised feeder channels which may have resulted in a more pronounced "fan" morphology than would have existed in the more strongly aggradational Skagerrak systems described here (cf. North and Warwick, 2007). In the upper Judy section, more established channel belts terminated in vegetated floodbasin muds with locally developed marshes and lakes (Fig. 11D), in a gross setting possibly analogous to the Mesopotamian marshes (Baltzer and Purser, 1990; Baeteman et al., 2004) or possibly Lake Eyre during pluvial phases (Croke et al., 1996). In detail the Mesopotamian marshes have been extensively modified by human activity, and form only a broad depositional analogue in terms of arid setting, proximity to a marine basin, and maintenance of localized wetland conditions by runoff from distant catchments via axial and transverse drainage. The fluvial systems remained terminal in character, but waning flood waters terminated in minor channels and splays which were extensively modified by soil processes and colonized by shrubland vegetation. Locally, possibly during wetter climatic conditions, lacustrine water bodies developed with hyperpycnite-dominated deltas. Orientation data from the channel-fills in the upper section suggest that the streams in the study area were of a low sinuosity, with the preserved record dominated by the downstream accretion or translation of lateral and medial bars. The overall dominance of channel facies over wide areas may have been facilitated by a tendency towards avulsive, multiple channel systems, but the limited subsurface data are inadequate to resolve such details of channel form. Downstream facies changes are detectable over distances of several tens of kilometers; in the north of the study area the vertical channel proportion in the upper section is c. 50–60%, but reduces to c. 25%, and the fills become increasingly mud-prone and pedogenically overprinted in the area of the Puffin field c. 50 km down regional paleoflow to the south.

←

FIGURE 11.—Summary of the gross depositional setting of the main fluvial reservoir facies in the Skagerrak Formation. (A) The overall paleogeographic setting was of terminal fluvial systems draining transversely off the basin margins and axially towards playa and marsh settings landward of a carbonate-dominated inland marine basin. In the upper Judy these systems traversed a shrubland alluvial plain maintained by perennial water supply. In the lower Judy more arid conditions prevailed. (B) The fluvial systems appear to have been of low sinuosity, and dominated by either downstream bar translation or accretion. (C) The channel belts show extensive soil modification, and variable fills ranging from sand-prone with low-angle accretionary surfaces, to heterolithic and siltstone fills with variable levels of rooting and burrowing. (D) illustrative cartoon showing the overall interpretation of depositional relationships between playa, splay, and channel facies in the lower Judy dryland terminal-splay facies association. (E) illustrative cartoon showing the interpreted depositional arrangement of playa, marsh, lacustrine, channel-belt, floodplain, and splay facies in the more vegetated upper Judy section. The basinal climate was likely to have been relatively arid throughout, with the upper section reflecting increased runoff from the main catchments in Fennoscandia.

Facies Architecture

The sand-prone members of the Skagerrak Formation record discrete episodes of terminal fluvial expansion into the Central North Sea basin, punctuated by episodes of fluvial contraction and the widespread encroachment of floodbasin muds, either as playa and sabkha during the drier Bunter and lower Judy episodes, or as shrubland, marshes, and less common lakes during the upper Judy (and younger) sections (Fig. 12). Each sand-prone fluvial cycle is represented by a lower interval dominated by splay facies, which passes upwards into more channel-confined facies. Further subdivision of these large-scale cycles is possible. For example, in the area of the Heron, Egret, Marnock, and Skua fields (Figs. 1, 12) the intra-Judy

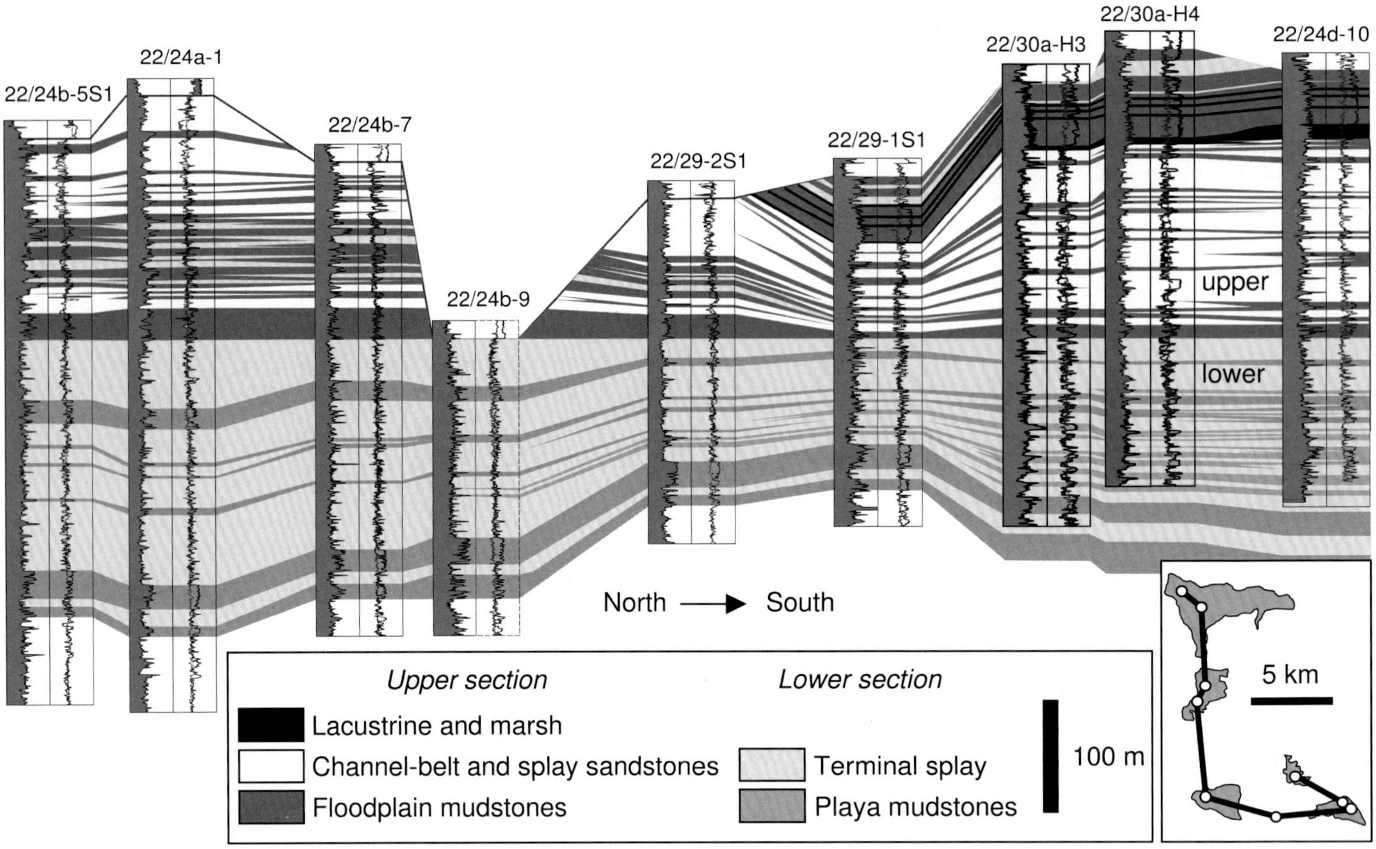

FIGURE 12.—Correlation of the Judy architecture between Marnock (22/24b-5S1, 22/24a-1), Skua (22/24b-7, 22/24b-9), Seagull (22/29-2S1), 22/29-1S1, Heron (22/30a-H3, H4), and Egret (22/24d-10) fields. The broad thickness changes can be attributed to differential subsidence induced by subdued salt movement. However, the overall architecture is comparable over the c. 25 km line of correlation and shows the continuity of the main floodplain shale intervals which punctuate the fluvial stratigraphy. Gamma log to left, neutron and density logs to right.

subdivision into upper and lower sections can be made along a shale-prone package which separates dryland splay complexes derived from the Scottish Highlands from a section of generally more channel-confined sediments derived from both Fennoscandia and Scotland (Fig. 13). This package marks a change in provenance, level of water supply, and facies architecture in this part of the basin. Higher-resolution internal subdivisions of the lower and upper cycles are possible based on semi-regional shale packages which appear to correlate more locally across the area:

- In the lower Judy the meter-scale playa-shale intervals separate up to 10 m thick, coarsening-upward packages of stacked terminal splay complexes. These commonly show successively more evidence of channel scouring in the upper part of the succession as the interval coarsens and becomes more sand-rich upward. These packages can be correlated over distances of tens of kilometers (Figs. 12, 13), and apparently have a similar, potentially correlatable expression over distances of up to 50 km (cf. McKie and Audretsch, 2005), although without a high-resolution chronostratigraphic framework correlation over these distances is increasingly speculative.
- In the upper Judy section the floodplain-shale packages separate upward-fining, multistory channel-belt deposits up to 35 m thick. These are apparently correlatable on a kilometer scale, but they are less laterally persistent than the sandbodies seen in the lower Judy section. The multistory channel packages (Fig. 14) enclosed by these shales are typically floored by up to one meter of bedded conglomerate (cf. McLaurin and Steel, 2007) composed of reworked calcrete and mud clasts, and medium- to coarse-grained, cross-stratified sandstone which represent thalweg lags. These are commonly sites of extensive post-depositional calcite cementation which nucleated around the reworked calcrete material. The overlying channel section fines upward through amalgamated fills in the lower part, but typically shows increased preservation of burrowed and pedogenically modified bar-top facies in the upper part, ultimately grading upward into a succeeding floodplain package.

Although the Skagerrak appears to comprise an alluvial stratigraphy punctuated by widespread shales, the origin of these cycles is equivocal. The shale package separating the upper and lower Judy sections records a fundamental shift in the dominance of the two basin-margin catchments, and the availability of year-round moisture for plant and animal colonization. The current dataset is unable to resolve which mechanisms were the cause behind the observed change in fluvial style, which could be attributed to regional climatic wettening following Anisian aridity (cf. Price, 1999), increased precipitation following Muschelkalk flooding of the Southern North Sea, or expansion of a Fennoscandian drainage which had its headwaters in a region of greater year-round precipitation (cf. Sellwood and Valdes, 2006) via a more remote climatic change. The change in the level of biogenic activity appears to mark a permanent change in the

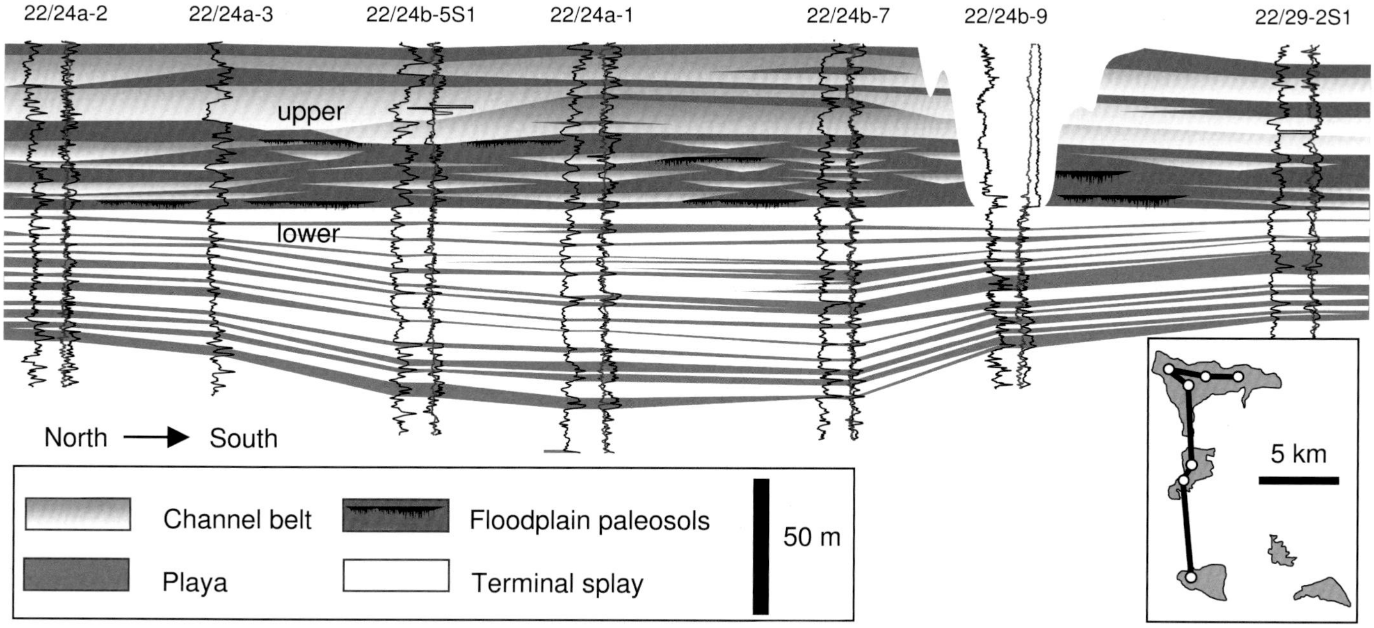

FIGURE 13.—Correlation of the transitional section between the lower and upper Judy sections across Marnock, Skua, and Seagull Fields illustrating the change from sheet-like terminal splay packages into mud-prone floodplain with isolated sandbodies, passing abruptly upwards into laterally widespread channel-belt deposits. Gamma log to left, neutron and density logs to right.

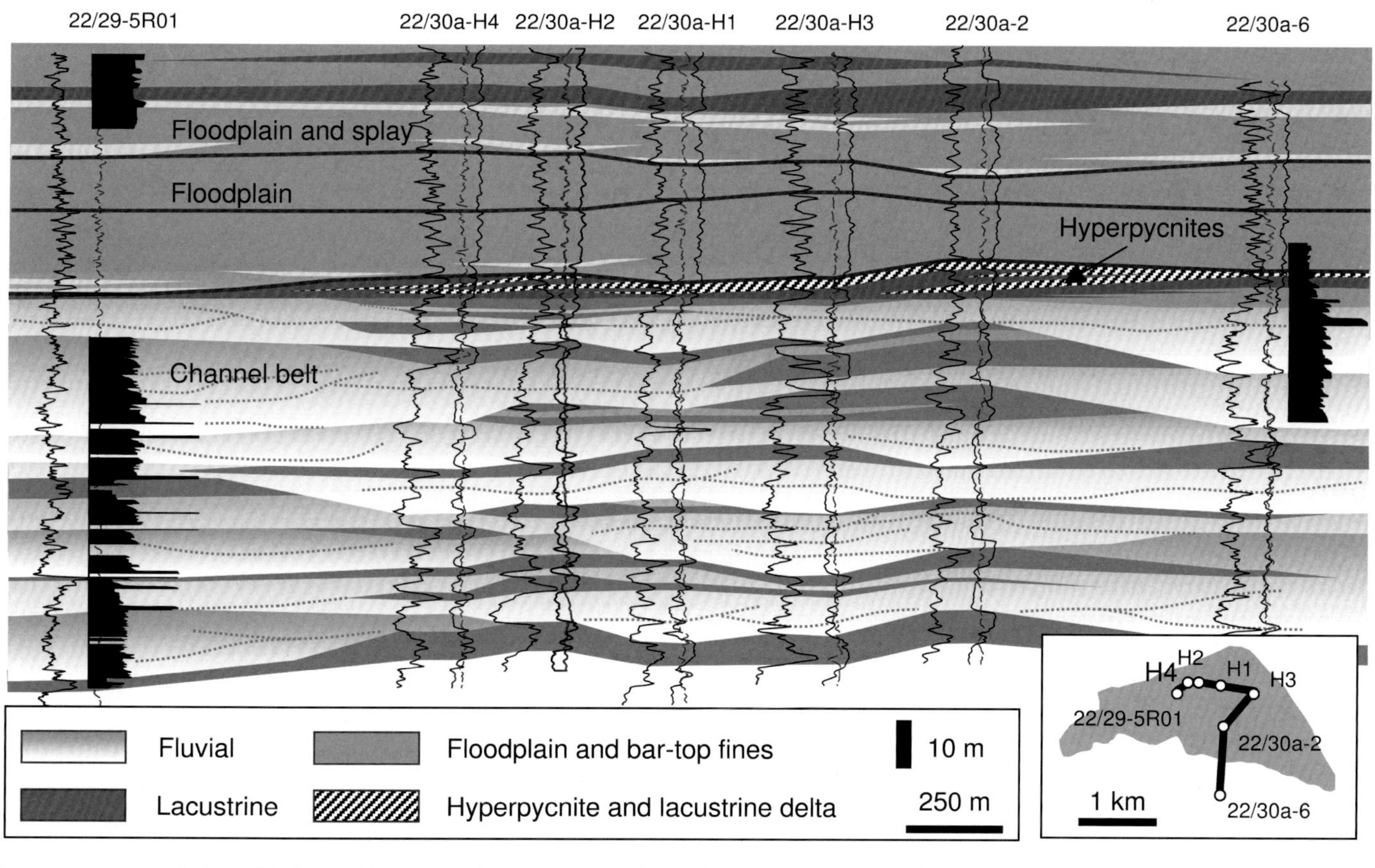

FIGURE 14.—Correlation of facies architecture in the uppermost Judy section across the Heron Field showing the architecture of the channel-belt deposits extending across the field, passing upward into a hyperpycnite and floodplain-dominated section.

depositional character of the Skagerrak across the region. This vegetated alluvial plain with scattered marshes and lakes stands in marked contrast to the more arid and evaporitic tendency of the Triassic elsewhere across northwest Europe (Hounslow and Ruffell, 2006; Ruffell et al., 2002), which may indicate a change to perennial runoff from Fennoscandia which was capable of maintaining year-round soil-moisture and shrubland conditions (McKie and Williams, 2009).

The shales bounding the smaller scale (10–40 m) cycles within the Judy subdivisions also may have multiple origins. Although occurring in different depositional settings, both the lower and upper Judy examples could be a record of random avulsion over lateral distances comparable to, or greater than, the spacing between individual fields. Alternatively, these alternating sand-prone and shale-prone packages could record the downstream expansion and contraction the terminal fluvial system, again over distances of tens of kilometers, in response to changes in sediment supply or runoff, or adjustment to local changes in base level in response to subdued salt movement at depth and the development of small playa or lakes. In the more channel-prone sections fluvial aggradation and incision as result of changes in discharge, sediment supply, or base level could also generate multistory cycles recording systematic upward-muddying motifs, and provided that the incisional phases did not completely remove the underlying floodplain, would result in widespread shale preservation. However, not all shale packages are correlatable between fields, particularly in the upper, channel-prone section. In most cases the shales preserved towards the tops of individual multistory channel packages are composed of bar-top fines, or erosional remnants of floodplain muds, which cannot be correlated across individual fields. They appear to have kilometer-scale extents, which may be a reflection of the typical width of a Skagerrak channel belt (which can be correlated over c. 200 m to 3 km within individual fields where well densities permit).

There are no obvious associations of such shale-bounded cycles with changes in paleosol type or provenance, suggesting that if they were a product of changes in sediment supply or discharge driven by climate change, this was not expressed in the basinal climate record. The absence of any significant changes in biogenic activity and soil type therefore favors avulsion, or the downstream transmission of changes in runoff or sediment yield modulated by climate changes in the headwaters of the system, which were not manifested as comparable changes within the basin. Up to 5 m thick, weakly pedogenically modified, heterolithic avulsion deposits (sensu Kraus and Wells, 1999) are locally preserved in areas of greater subsidence, but in areas of reduced accommodation such deposits are increasingly truncated by erosionally amalgamated channel-belt deposits, suggesting that avulsion processes were abrupt (cf. Jones and Hajek, 2007), and that deep channel erosion following avulsion commonly reworked any sediments deposited during the avulsion process.

RESERVOIR BEHAVIOR

The principal fields producing from the Skagerrak in the UK Central Graben area are located in two main areas: the Forties-Montrose High fields of Heron, Egret, Skua, and Marnock (Pooler and Amory, 1999), together with the undeveloped Culzean, Seagull, and Fiddich discoveries; and the fields to the south in the J-Ridge area comprising Jade (Jones at al., 2005), and Judy, together with the nearby, undeveloped Puffin, Jasmine and Kessog accumulations (Fig. 1). In addition to these fields there are numerous Jurassic reservoirs that have additional hydrocarbons deeper in the Triassic which have yet to be exploited.

This discussion of the dynamic behavior of the Skagerrak reservoirs largely focuses on the producing fields of the Heron Cluster area (Heron, Egret, Skua), with additional data drawn from other fields to compare static data (e.g., porosity and permeability relationships) and well test results. The Heron Cluster fields are located in salt-withdrawal structures (Hodgson et al., 1992; Smith et al., 1993) which have been deeply truncated by middle Jurassic erosion during regional thermal uplift (Underhill and Partington, 1993) and subsequently dissected by late Jurassic extensional faulting (Helgeson, 1999; Husmo et al., 2002). As such they are remnants of the formerly extensive Triassic stratigraphy, which have survived repeated erosional episodes through salt-controlled subsidence conveying the section below the regional level of truncation. The fields show variable degrees of structural deformation, but all contain similar reservoir fluids (volatile oil) and are produced through natural pressure depletion. As a result the geological controls on the dynamic behavior of these fields can be compared. The effects of faulting on well tests and production are detectable as persistent no-flow boundaries, large-scale pressure compartments and boundaries to 4-D seismic anomalies, and these can therefore be discounted in order to identify the contribution of the fluvial architecture to the production behavior. Overall these reservoirs exhibit behavioral patterns which are likely to be common to fluvial reservoirs in general, with some details which may be of particular issue in terminal systems.

Vertical Barriers to Fluid Flow

These Skagerrak reservoirs ubiquitously show evidence of stratigraphic compartmentalization by shales. The shales in the Skagerrak Formation have a variety of origins, ranging from playa mudstones associated with terminal splay complexes, through bar-top and floodplain shales with variable levels of paleosol maturity and drainage, to marsh and lacustrine facies. The shale-prone intervals bounding the progradation of terminal splay complexes, and some examples bounding multistory channel-belt sections, appear to be correlatable over a field and inter-field scale of many tens of kilometers, and have the potential to form vertical flow barriers in the subsurface. The evidence to suggest that many (but not all) of these form stratigraphic compartments is based on geochemical data indicative of fluid compartmentalization and dynamic data indicating that reservoir depletion is stratigraphically restricted:

- The geochemical composition of the oils in these Skagerrak reservoirs commonly shows vertical compositional differences across correlatable shale intervals (Fig. 15), indicating that over geological time scales the fluids were unable to mix across these barriers (cf. Smalley et al., 2004). In the case of the Skua and Heron fields, production wells perforated or drilled above such barriers failed to produce compositionally distinct oils encountered immediately below, despite close proximity (70 m) and large pressure differentials of $>$ 6000 psi (McKie et al., 2010) which should have induced flow over these short distances.
- The rate of pressure decline in the naturally depleting reservoirs is consistent with the volume of the shale-bounded stratigraphic interval which has been perforated. Re-perforation of wells across shale barriers show renewed pressure as previously unaccessed stratigraphy contributes to production (Fig. 16).
- The $^{87/86}$Sr isotope data from residual formation water salts locally indicate that as well as the hydrocarbons being physically isolated, the original formation water also did not mix across the shale barriers.

The occurrence of stratigraphic barriers in fluvial reservoirs has been noted (e.g., Hamilton et al., 1998). However, the repetitive occurrence of shale barriers may be an inherent property of reservoirs deposited in terminal fluvial systems. Similar stratigraphically compartmentalized reservoirs, where shales separate different hydrocarbons or support multiple fluid contacts, have been described from Alwyn (Harker et al., 2003), Wytch Farm (McKie et al., 1997, 1998;

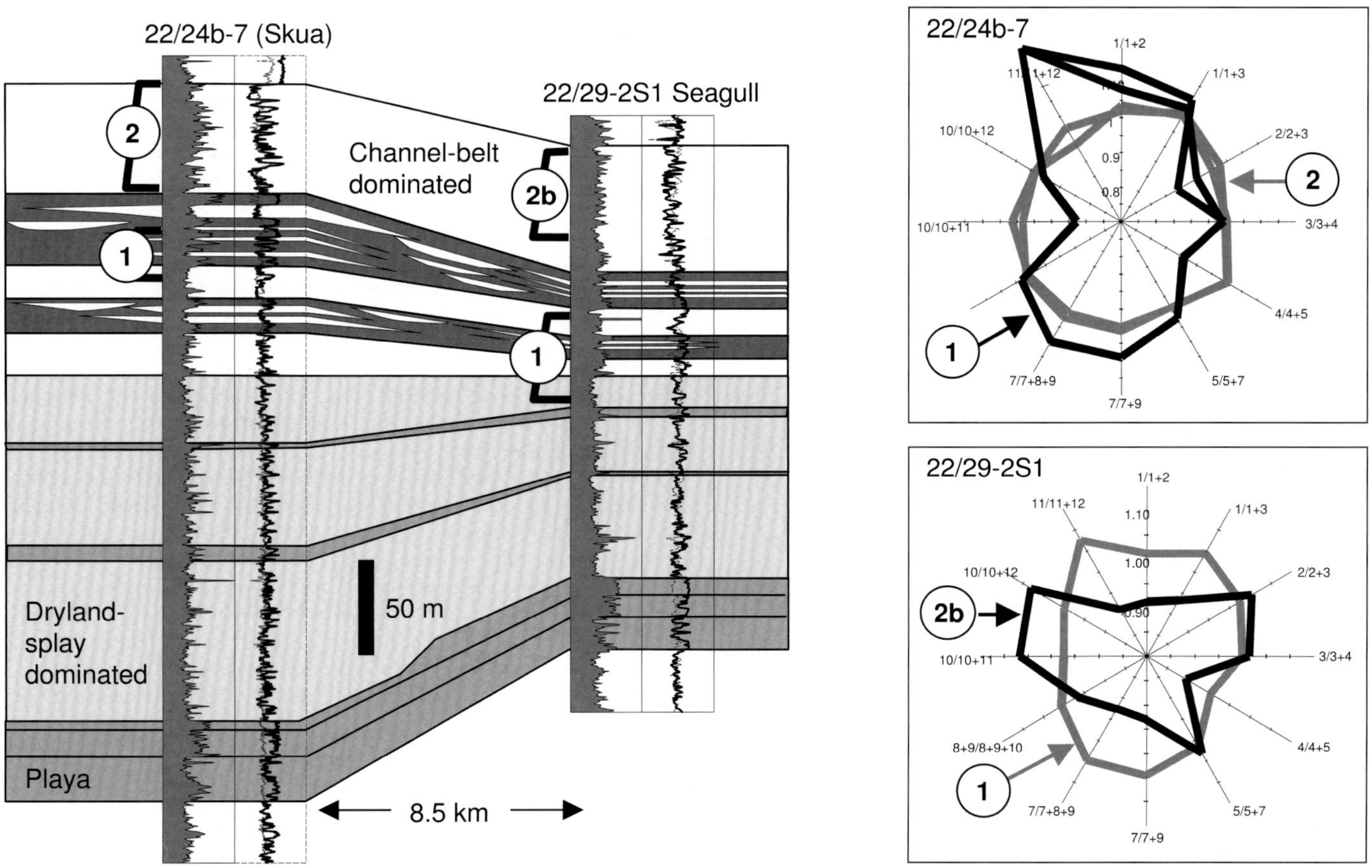

FIGURE 15.—Shale barriers in the Skagerrak are composed of correlatable packages defining changes in depositional character. The shale-prone package marking a change in depositional style between the lower terminal-splay-dominated section and the more vegetated, channel-dominated upper section (with corresponding provenance change), can be correlated between the fields in the area (Fig. 13) and also marks the boundary to fluid compartments. In the Skua (22/24b-7) and Seagull (22/29-2S1) fields this interval shows a change in the geochemical composition of the oils in these reservoirs (illustrated as star plots of multidimensional gas-chromatography data, plotted as ratios of the aromatic compounds in the C9–C10 alkane range).

Hogg et al., 1999), Nevis (Farquharson and Gibson, 2005), Beryl (O'Donnell, 1993), Jade (Jones et al., 2005), and Chaunoy (Eschard et al., 1998). In these cases playa, floodplain and bar-top shales have kilometer-scale dimensions greater than, or comparable to, the structures hosting the hydrocarbon accumulation. The preservation of shale packages on this scale could be the result of lateral stream avulsion over distances greater than the typical field size, or the natural interfingering of terminal fluvial sandstones with floodbasin fines. The broadly contemporaneous, terminal fluvial Stubensandstein Formation in Germany (Hornung and Aigner, 1999, 2002a, 2002b) may provide evidence that the latter is more likely. The architecture of this system comprises sharp-based sand packages c. 40 m thick which extend several hundred kilometers into the basin, and are bounded by equally widespread playa shale packages (Hornung and Aigner, 2002a, 2002b). In the subsurface this succession would exhibit a scale of compartmentalization similar to that seen in the Skagerrak reservoirs. These examples are exclusively from the Triassic of northwest Europe, but outcrop data from farther afield suggest that this could be a general pattern. The Stubensandstein example may have been deposited in a basin with subdued extensional tectonics which may have favored basinward extension of the fluvial system rather than trapping sediment in areas of higher subsidence, but in more tectonically active basins such floodbasin fines also appear to blanket terminal fluvial sand systems over distances of several tens of kilometers as a result of contraction or avulsion of the system (Nichols and Fisher, 2007; Sáez et al., 2007), and shale packages on such a scale would be equally likely to compartmentalize fields of the aerial extent seen in the Central North Sea.

However, not all apparently significant and correlatable shale packages show evidence of being barriers. Figure 17 shows an example from the Heron field where a potentially widespread shale package shows a gradational change in the $^{87/86}$Sr profile across this interval. If the shale had functioned as an effective barrier the isotopic composition would have undergone a step change in composition. In this case there is gradational change, indicating that whilst not an absolute barrier the shale functioned as a significant baffle to fluid mixing (Webb and Kuhn, 2003) but not as a continuous physical barrier. In core this shale package is composed of weakly burrowed, bar-top heterolithic units which would not necessarily be expected to have field-wide continuity. However, it does indicate that the bar-top shales bounding multistory channel cycles are potentially capable of scaling to a size similar to the fields in the area, and are therefore capable of forming significant baffles and barriers. It is also likely that a range of bar-top shale dimensions are present and that only the larger examples are capable of forming full barriers (Robinson and McCabe, 1997).

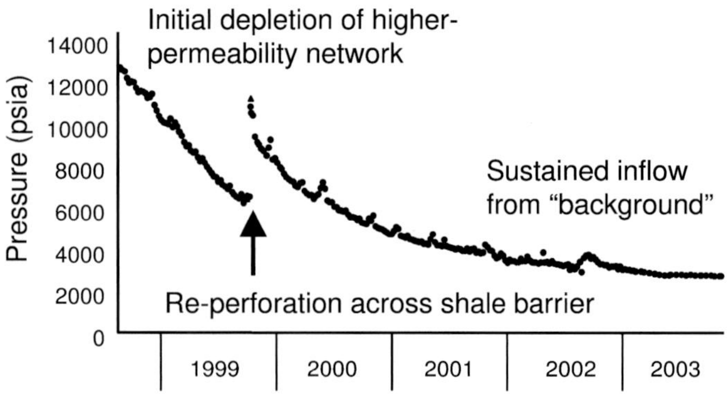

FIGURE 16.—Pressure data from the Heron field. Re-perforation of Heron well H3 in 1999 across a floodplain shale package accessed undepleted reservoir, indicating that the floodplain package acted as a field-wide vertical flow barrier. The long-term rate of pressure decline is indicative of a heterogeneous reservoir in which high initial flow rates and pressure decline are from a connected network of high-permeability reservoir, followed by sustained lower rates as slow inflow from poorer-quality reservoir feed into the depleted high-permeability network. The higher-permeability network is likely to comprise coarse-grained thalweg and lower-bar facies (Figs 18, 19), with the poorer matrix inflow from upper-bar and splay facies.

Permeability Structure

The principal control on the pore-scale porosity and permeability of the Skagerrak reservoir is the primary depositional texture of the sandstone (dominated by grain size, sorting, and clay content). Burial diagenesis has degraded these sandstones to varying degrees, but it has not destroyed the relative relationships between the main textural groupings, and as a result the permeability structure of the reservoir is intimately linked to the facies architecture. An example of core-plug porosity and permeability data, coded by facies, from the Marnock and Skua fields is shown in Figure 18, which illustrates that the higher permeability values occur exclusively in the coarser-grained sandstones occurring in the basal parts of the channel-belt facies, as thalweg and lower-bar deposits that were not subsequently cemented by groundwater calcrete (cf. Hall et al., 2004). However, although core-plug data provide an indication of the facies control on porosity and permeability, the latter is also scale dependent, and the *effective* permeability of a sandbody over a particular length scale is controlled by the permeability heterogeneity (essentially grain-size distribution), and the distribution of any flow baffles and barriers. In fluvial reservoirs the effective permeability is most likely dominated by the bedform-to bar-scale grain-size distribution and the spatial arrangement of flow baffles (e.g., mud drapes) over this scale. Beyond this scale the *connectivity* of the various channel belt elements within the fine-grained floodplain matrix becomes increasingly important. The dimensions, internal heterogeneity, and spatial arrangement of the thalweg and lower-bar deposits (combined with the presence or absence of cemented calcrete lags) are therefore likely to define the

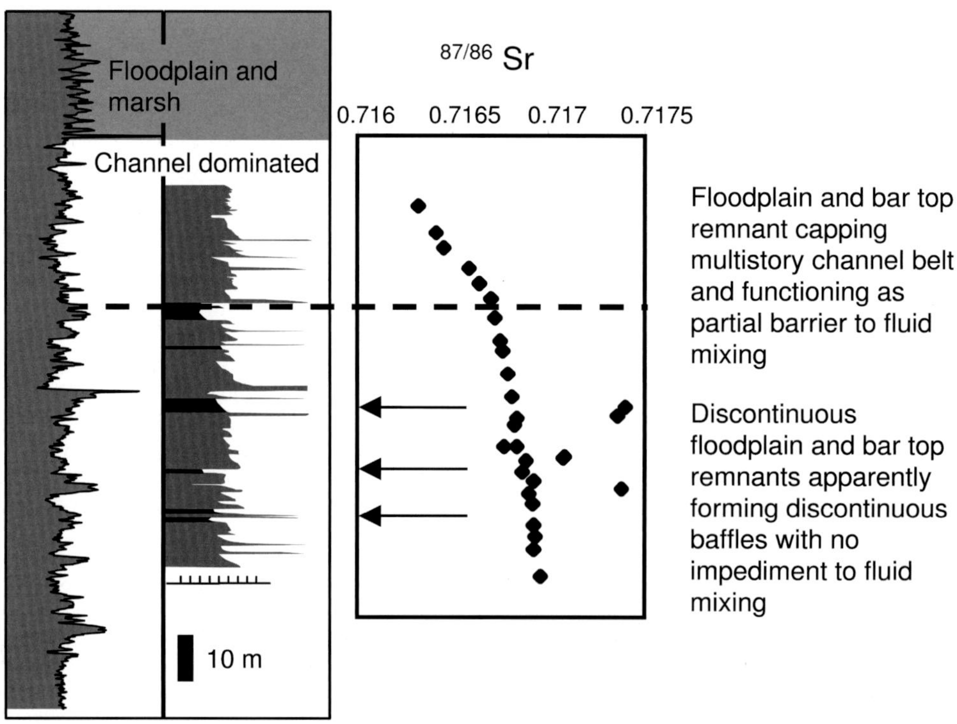

FIGURE 17.—Plot of $^{87/86}$Sr data from Heron well 22/29b-5RE showing an inflection in the data trend across a shale-prone package defining the top of a multistory channel-belt section. The data indicate that the shale package acted as a significant boundary inhibiting formation water mixing, but the absence of a step change in $^{87/86}$Sr values show that this was not a complete barrier. In this example the bar top and floodplain package either represents a local avulsion, with fluvial connectivity maintained within the area of the field, or the package has been locally eroded by the overlying channel base, or there is a limited connection pathway via minor channel and splay facies through the mud-rich succession.

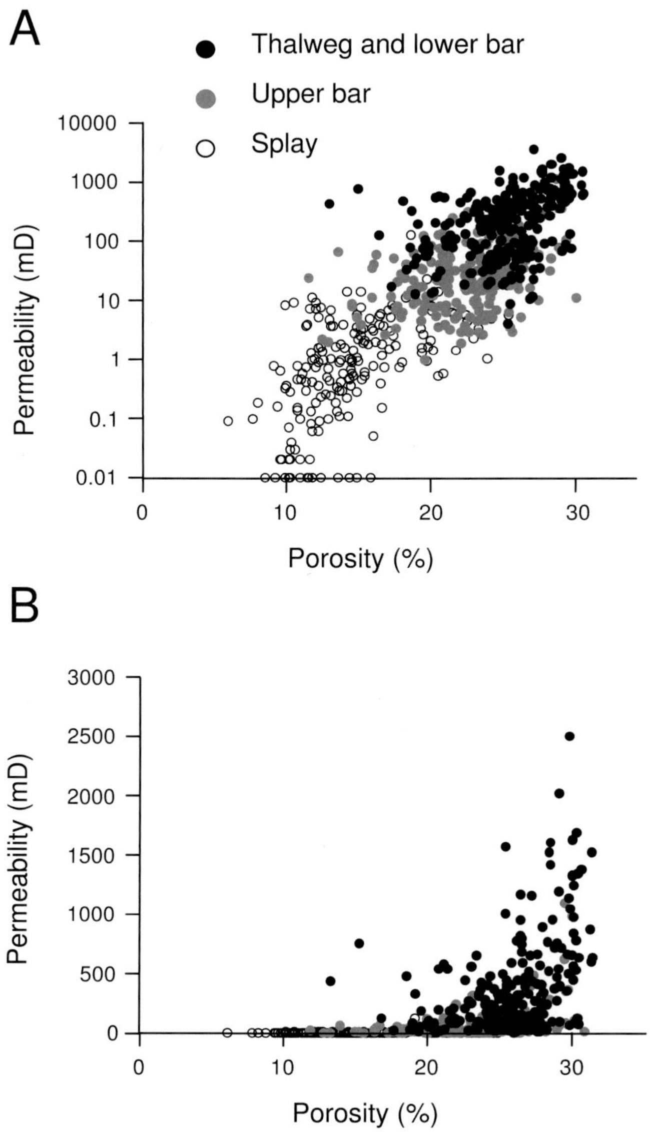

FIGURE 18.—Skagerrak porosity and permeability data from the Marnock and Skua fields. These data are from an area which experienced a common diagenetic and burial history, and the differences are entirely attributable to primary depositional texture (and early carbonate cementation). The plot in Part A has permeability scaled logarithmically, and shows the overall dominance of the higher porosity and permeability distribution by coarse-grained channel thalweg and lower-bar deposits. The upper-bar and splay sandstones have moderate properties, with thin, fine-grained distal splay sands embedded in floodplain fines having limited properties. Linear scaling of permeability (Part B) gives a more realistic representation of the relative importance of the thalweg and lower-bar facies to flow, with the finer-grained splay and floodplain facies forming an overall background of poor-quality reservoir.

location and effectiveness of any of high-permeability pathways in the reservoir, with the dominance of such pathways accentuated by their permeability contrast with the "background matrix" sands. Simulation experiments by Jones et al. (1994) have indicated that a permeability contrast by a factor of 50 marks an approximate threshold where higher-permeability pathways begin to dominate flow. In addition to the architecture of the higher-permeability facies the distribution of bar-top and floodplain muds, together with bar-draping fines, function as flow baffles that degrade the effective horizontal and vertical permeability of the sandbodies. The contribution of these facies-controlled permeability groupings is illustrated by flow-meter data from Egret Field (Fig. 1), which shows the short-term measurement of flow rates during well testing (Fig. 19). Flow measurements from this well illustrate that the finer-grained, lower-permeability terminal splay complexes in the lower part of the succession do not make a significant contribution to flow. A greater proportion of flow is contributed by the coarser-grained, higher-permeability channel-belt succession in the upper part, and in particular the bases of individual multistory units where step changes in cumulative flow occur. The basal parts of these multistory units are composed of amalgamated, coarse-grained thalweg and lower-bar deposits, with upper-bar deposits largely eroded. The dominance of these intervals is accentuated by the permeability contrast between the coarse-grained lags and the finer-grained "background matrix" of the rest of the channel-belt succession, which can differ by several orders of magnitude (Fig. 18).

Whilst the coarser-grained channel-base elements may dominate flow during the short duration of a well test (typically days), the duration with which these dominate flow over production timescales depends on the dimensions of the original coarse-grained deposit (elliptical scour bodies or extensive sheets resulting from lateral migration), and the degree to which successive bodies were amalgamated and connected during the superposition of successive channel belts. In the fields discussed here the pattern of pressure depletion is consistent with a broadly sheet-like architecture in which the reservoirs behave as a dual-permeability system characterized by very high initial flow rates and rapid pressure decline, followed by sustained lower flow rates and reduced rates of pressure decline (Fig. 16). The initial flow and depletion rates are more rapid than would be predicted for a more homogeneous reservoir lacking shales and pronounced grain-size contrasts (e.g., many upper-shoreface reservoirs), and record flow from a connected, high-permeability network (rather than isolated scour fills), with the sustained longer-term flow at reduced pressures the product of slow inflow from the finer-grained upper-bar and splay architecture (and their connection to more isolated thalweg bodies) into the depleted thalweg and lower bar network.

Lateral Connectivity

Despite the repeated evidence for *vertical* stratigraphic barriers in Skagerrak reservoirs the long-term performance of these fields suggests that *lateral* connectivity is generally good. Material-balance calculations suggest that the rate of pressure decline is consistent with depletion of the reservoir volume of the stratigraphic layers which have been accessed by well perforations. However, well tests during field appraisal consistently show the presence of flow boundaries, which tend to give a pessimistic impression of Skagerrak connectivity that is not borne out by the long-term field performance. In some cases these boundaries can be attributed to faults in the reservoir, but the persistent detection of boundaries in Skagerrak well tests suggests that these are commonly encountering internal flow baffles that are an integral part of the Skagerrak facies architecture.

A compilation of the calculated dimensions of connected drainage areas from Skagerrak well tests is shown in Figure 20, using data drawn from exploration, appraisal, and production wells across the region. The data have been filtered to exclude tests that encountered faults (either visible on seismic or inferred via successive tests which encountered a recurring boundary along the well bore) and tests which covered an excessive vertical interval where no meaningful inference could be made about any single channel or multistory unit. It should be

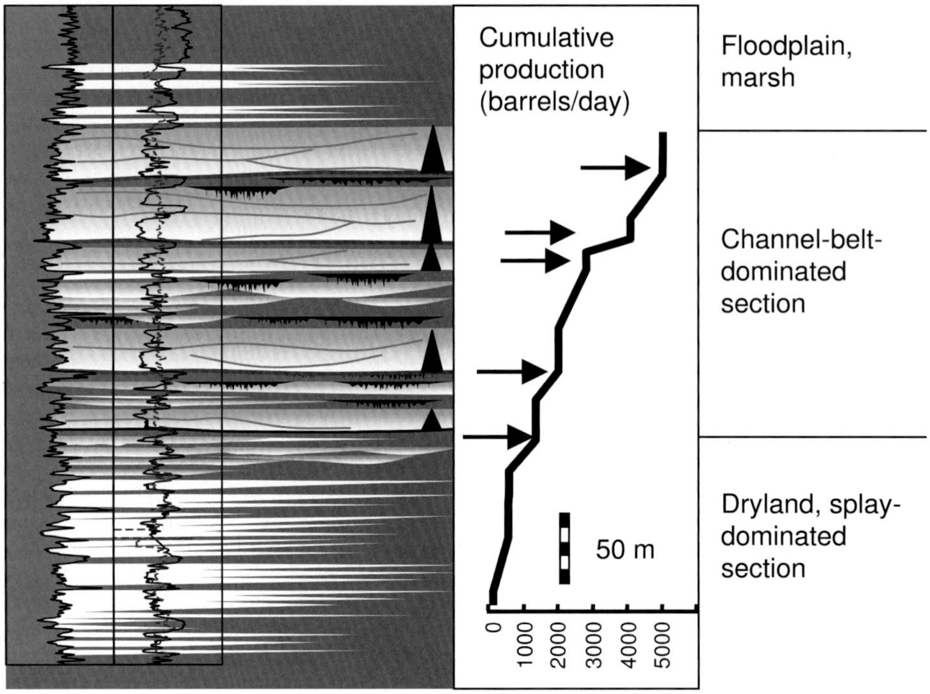

FIGURE 19.—Flowmeter data from Heron well 22/30a-2 showing the overall facies architecture with a plot of cumulative flow rate during well testing. The lower section of splay facies shows limited contribution to flow, with increased flow in the upper, more channel-dominated sections. The largest contributions come from the basal parts of multistory channel intervals (denoted by black triangles), within sections composed of amalgamated thalweg and lower-bar deposits.

noted that the distance to flow boundaries calculated from well tests is critically dependent on assumptions made regarding the permeability of the reservoir (which is used with the timing of changes in the rate of pressure decline to estimate distance), and that these calculations are inherently imprecise. However, regardless of the precision of the measurements the calculated dimensions are indicative of common flow baffles on a lateral scale of tens to a few hundreds of meters, with few tests showing either no boundaries (infinite radial flow) or boundaries approaching kilometer-scale distances. In the Skagerrak channel belts there are a range of candidate flow baffles which could form boundaries during well tests:

- The flow boundaries could reflect the presence of permeability baffles at the margins to channel belts, and that the test dimensions are a reflection of the lateral scale of the Skagerrak channel belts. Such boundaries could include abandonment plugs, channel-margin shale drapes (Lynds and Hajek, 2006), or groundwater carbonate cementation creating "tight" flow barriers.
- Alternatively, these baffles could be the result of internal heterogeneities within the Skagerrak channel belts at the bar and bedset scale, e.g., bar-draping fines, mud-chip conglomerates, or cemented calcrete-clast concentrations.

Although there is no direct means of confirming whether one, or both, models are correct, comparison with analogous outcrop channel-belt dimensions (e.g., Gibling, 2006) would suggest that when the scale of the Skagerrak fluvial system is considered, the typical range of channel-body thickness versus the lateral scale of the boundaries seen in well tests would generate unrealistically narrow and deep channel bodies for such a large system. Estimates of the scale of the channel belts, based on empirical relationships between cross-bedding, flow depth, and channel-belt width (e.g., Bridge and Tye, 2000; Bridge, 2003), suggest widths of 0.5–3 km and would appear to confirm this, although the Skagerrak fills suffer from insufficient stacked cross-bedded intervals to have any confidence in the results. The test data fall within the range of dimensions possible for crevasse and meandering stream deposits presented by Gibling (2006), but would still represent streams at the lower end of the global size distribution. If the channels belts were of such a small scale one might anticipate that the overall sand:shale ratio of the reservoir would be lower, that the connectivity encountered during long-term production would be limited, and that sandbody correlation over well spacings of a few hundred meters would be more problematic (cf. Fig. 14). It would seem more likely that the majority of boundaries encountered during well testing are inclined internal heterogeneities within the channel belt rather than narrow channel belts. The available core data from the Skagerrak Formation would suggest that the most likely candidates to form inclined baffles in the channel-belt facies would be bar-draping fines (Miall, 1988), mud-chip conglomerates (Doyle and Sweet, 1995; Taylor and Ritts, 2004), groundwater calcrete-cemented bar surfaces, and cemented calcrete-clast conglomerates (Newell, 2006). These are commonly observed in core, but there are limited data on their orientation and inclination. Oriented core and borehole image data suggest that they generally dip at up to ten degrees, but there are insufficient examples to be able to infer their likely relationship to the more data-rich cross-bed and upper-bar orientations.

DISCUSSION

The Skagerrak is an example of a sand-rich fluvial reservoir that would, prior to production, be regarded as having no critical issues relating to depositional architecture. Whilst the sand–shale ratio of the section would suggest that there should be few problems related to the lateral and vertical connectivity of the fluvial sandbodies, the vertical

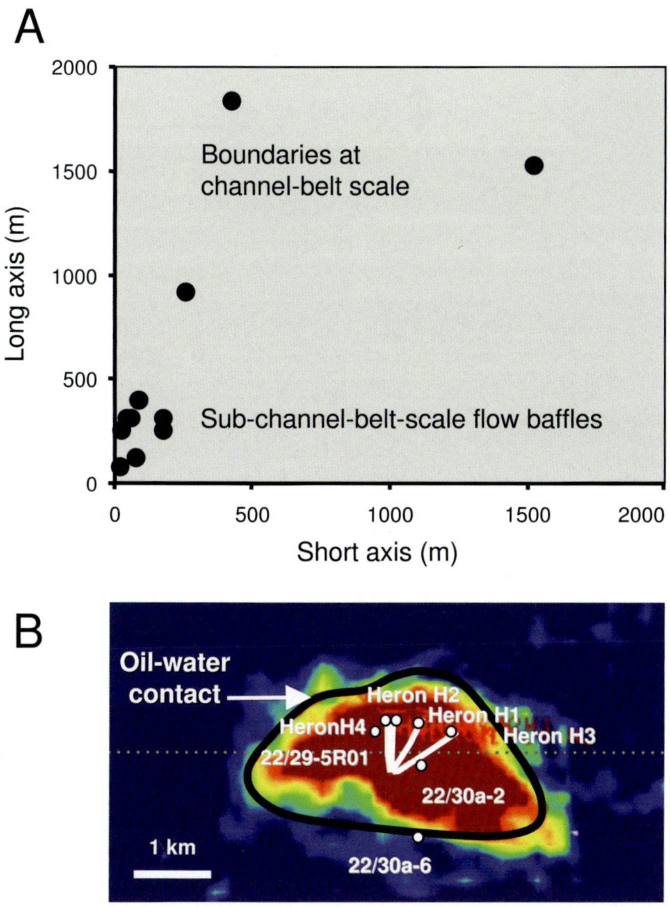

FIGURE 20.—Lateral connectivity of the Skagerrak reservoir. (A) Well test drainage areas. These data are based on calculations of the distance from the well bore to flow boundaries seen during well tests. These are calculated from changes in the rate of pressure depletion and using reasonable assumptions of the permeability away from the well bore. The test data are selected based on elimination of tests close to seismic-scale faults, tests which show successive boundaries indicative of subseismic faults, and tests which cover intervals greater than several stacked fills. The calculated distances have no precision (particularly where the boundaries are calculated from a test covering more than a single fill) but serve to illustrate that the Skagerrak channel-belt facies have common flow baffles on a scale of tens to hundreds of meters. (B) 4-D seismic response over the Heron Field (with warmer colors indicating greater difference between surveys and corresponding to areas of depletion-induced stress changes). The response shows a field-wide anomaly, indicating that the entire area of the field is being depleted. Despite the presence of flow baffles during well testing, long-term production shows that provided wells are perforated across any stratigraphic barriers then the lateral connectivity of the Skagerrak channel belts is sufficient to drain a typical field area. In more structurally deformed fields the faulting provides additional flow baffling, but generally does not act as flow barriers (McKie et al., 2010).

connectivity of the reservoirs is considerably reduced as a result of areally widespread shales. This was not anticipated during development of the Heron field and required a series of well interventions to re-perforate unaccessed stratigraphy (McKie and Audretsch, 2005). The compartmentalizing shale barriers appear to be related to the repeated expansion, contraction, and avulsion of the terminal fluvial system, which left a depositional record of field-scale, mud-prone, floodbasin facies bounding sand-rich channel-belt and splay reservoir. These mud-prone layers appear to be correlatable over a scale of kilometers to tens of kilometers, and most are of greater arial extent than all but the largest fields in the region. This pattern of stratigraphic compartmentalization appears to be a characteristic of many reservoirs deposited as terminal fluvial systems and may be a characteristic of terminal fluvial reservoirs in general. The expression of these compartmentalizing shales is variable; shale packages of semi-regional extent record major changes in fluvial style or define the boundaries to major cycles of terminal fluvial expansion. However, some stratigraphic barriers appear to have no distinguishing sedimentological characteristics (but are detectable by differences in fluid geochemistry) and may record random avulsive behavior of the fluvial system leaving field-scale floodplain remnants capable of acting as reservoir flow barriers, or bar-top fines acting as barriers or baffles.

Within the individual reservoir layers defined by these bounding shales the production data indicate that the sandbodies are laterally well connected (albeit with common baffles detected during transient well tests). The level of connectivity is in all likelihood partly a result of the typical well spacing (or small field) being of a comparable scale to most Skagerrak channel-belt or splay complexes (c. 0.25–5 km). However, whilst sandbody connectivity is not generally an issue in these reservoirs, the channel-belt permeability structure, and its impact on flow behavior, is still a major unknown. Knowledge of this structure, or a range of different but equally possible structures given the limitations of subsurface data to resolve this scale of architecture, is essential to understand the likely effective permeability of the reservoir over interwell distances and to be able to construct predictive models of flow rates and long-term reservoir performance. Although there has been a concerted effort to model the connectivity of fluvial sandbodies in more mud-prone settings (e.g., Shanley, 2004; Larue and Hovadik, 2006; de Jager et al., 2009; Pranter et al., 2009) there have been fewer studies on the effects of internal fluvial heterogeneities on flow behavior. Detailed studies of bar-scale architecture have been undertaken (e.g., Allen, 1982; Haszeldine, 1983; Kirk, 1983; Corbeanu et al., 2005), but there are still very limited data available from good outcrop analogues on the internal permeability structure of fluvial sandbodies (e.g., Hartkamp-Bakker and Donselaar, 1993; Ringrose et al., 1999) although this has a critical impact on the prediction of reservoir performance (Larue and Friedman, 2005).

Reservoir behavior is typically investigated using numerical models of reservoir properties (e.g. porosity, permeability, fluid saturation) which are constructed from grid cells representing reservoir volumes tens to hundreds of meters in length and width, and of a meter scale in thickness (Keogh et al., 2007). Any heterogeneity in the properties being modelled below this scale is not explicitly captured, despite being critical to controlling effective permeability (Jones et al., 1993; Toro-Rivera et al., 1994; Ringrose et al., 2008). The grid cells are commonly populated using numerically averaged, vertical well data (with permeability estimated from the well-log porosity curve using an empirical relationship between core-plug porosity and permeability), and extrapolated in the horizontal plane. Given that permeability is scale dependent, and in general fluvial reservoirs can be very heterogeneous at the scale of most model grid cells, it is likely that the effective permeability assigned to such grids using this method is erroneous and most likely to be overestimated. There are currently insufficient data from good-quality outcrops to be able to fully understand and capture the impact of bar-scale architecture, and how

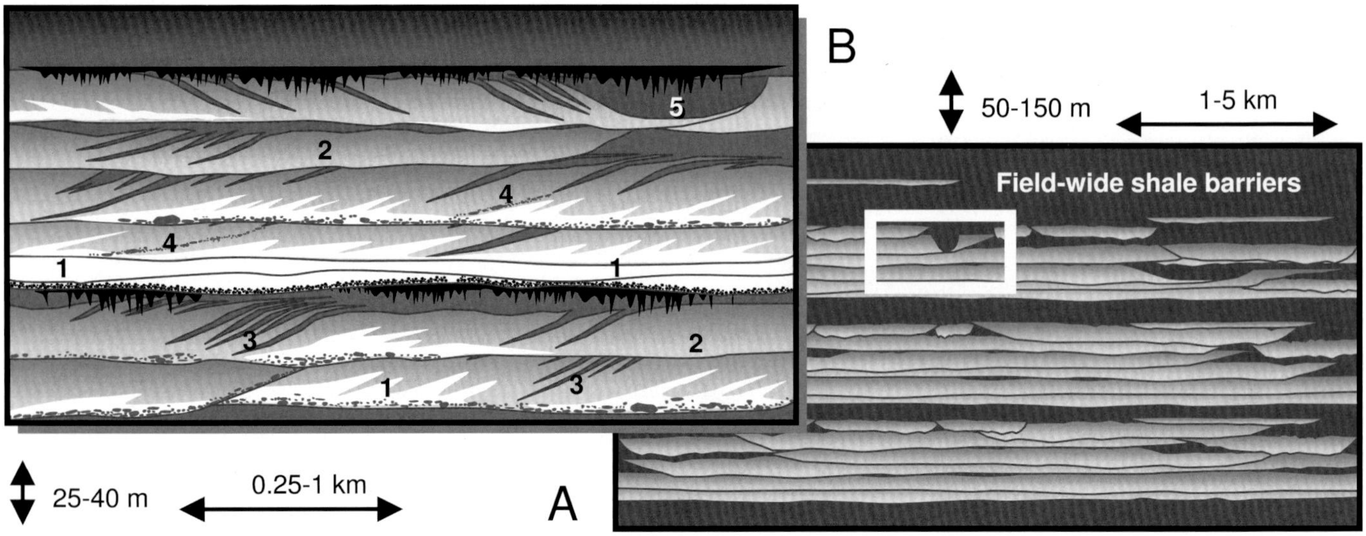

FIGURE 21.—Conceptual architectural model for the Skagerrak fluvial reservoirs. Reservoir layers are composed of channel-belt deposits, vertically compartmentalised by field-wide floodplain and bar-top remnants (Part A). Within individual channel-belt sandbodies (Part B) initial flow is dominated by the coarse-grained, higher-permeability, thalweg and lower-bar deposits (1). Longer-term flow is sustained by inflow from finer-grained, poorer-quality upper-bar and splay sandbodies (2). Internal baffling by bar-draping fines (3) mud-chip lags (4) and abandonment plugs (5) reduce the effective horizontal permeability, and are detected as flow barriers during transient well tests. However, long-term production indicates that the channel belts are fully connected.

this information can be captured in reservoir models (Zheng et al., 2000, Zheng et al., 2007). Some attempts to build simulation models of outcrops at this scale have been made (e.g., Høimyr et al., 1992; North and Taylor, 1996; Ringrose et al., 1999; Pranter et al., 2007), but without further studies of this kind the full range of possibilities needed to model uncertainty in the subsurface cannot be captured. In particular, although there has been attention focussed on the distribution of flow barriers, e.g., shale drapes, the 3-D permeability distribution (fundamentally a function of grain size) has not been examined in any detail. Some studies have used minipermeameter data to derive semi-variograms of permeability in fluvial deposits (e.g., Dreyer et al., 1990; Davis et al., 1993; Liu et al., 1996; North and Taylor, 1996), but these are limited to short horizontal transects which do not reveal the full 2-D or 3-D variability. These studies do however indicate that permeability in fluvial sandbodies varies very rapidly, with semi-variogram ranges of the order of meters, but do not give an indication of the scale of "representative elementary volume" (Nordahl and Ringrose, 2008; Ringrose et al., 2008) required to be able to describe and capture the variation of fluvial reservoirs in numerical models. What is clear however is that averaging vertical well measurements through a fluvial succession is highly unlikely to realistically describe the lateral permeability. As the Skagerrak examples show, the continuity and connectivity of the higher-permeability thalweg deposits (and also the distribution of calcrete-clast conglomerates and mud drapes) are critical to understanding which intervals dominate early production (Martin, 1993; Cortez and Corbett, 2005), and provide the internal "plumbing" which drains the reservoir away from the well bore.

There is a continuing lack of good outcrop descriptions of the true geometry of preserved fluvial sandbodies available to calibrate subsurface models. Outcrop work clearly demonstrates how variable the geometry of the preserved lithosomes can be (e.g., Donselaar and Overeem, 2008), and how divergent these are from the simple ribbon geometries offered by most modelling packages for replicating the architecture and connectivity of fluvial sandbodies. Most investigations of connectivity in fluvial reservoirs have used a simple template of ribbon-shaped sandbodies of various widths and sinuosities, but recent work (e.g., Patterson et al., 2006; Donselaar and Overeem, 2008) has shown that the connectivity and continuity of bar-scale elements are more important for predicting fluid flow, and that these do not necessarily correspond to such geometries. In addition, the well-test data from the Skagerrak indicate that it is flow baffles at or below the bar scale which influence short-term flow, and that a realistic estimation of permeability requires an understanding of the permeability structure at this scale. Some attempts have been made to understand well-test responses using geological templates (e.g., Corbett et al., 1998; de Rooij et al., 2002), but again these are limited by our understanding of what are realistic geometries based on outcrop analogues, and it is debatable whether the fluvial styles used (braided vs. meandering) are a realistic description of the sandbody styles likely to be encountered in the subsurface.

CONCLUSIONS

The Skagerrak Formation is an example of a sand-prone fluvial reservoir deposited in a dryland, terminal fluvial setting. The fluvial systems were of low sinuosity, dominated by downstream bar accretion or translation and, although prone to avulsion, ultimately generated a stratigraphy of multistory channel sheets with a broadly similar character over lateral distances of tens of kilometers. Despite the relatively high sand:shale ratio of the Skagerrak Formation the connectivity of the reservoir is not straightforward. The section is vertically compartmentalized, internal baffles are common, and fluid flow and pressure depletion reflect a heterogeneous reservoir (Fig. 21). This behavior is intimately linked to the alluvial architecture of the fluvial system:

- The formation comprises a series of upward-coarsening cycles composed of fine-grained terminal splay facies, which pass upward into more channel-confined facies. The upper, channel-dominated sections form effective reservoir, whilst the splay facies are typically fine-grained and mud-prone, and lack the permeability to flow at economic rates.

- The channel belts in the upper parts of these cycles are composed of upward-fining, multistory sheets. The basal parts of these units are composed of coarse-grained, amalgamated thalweg and lower-bar deposits which form a connected permeable network that dominates initial production. Longer-term flow is dependent on the rate of recharge of this network by a poorer-quality matrix of upper-bar and splay sands. The channel belts are vertically compartmentalized by floodplain shales recording major changes in depositional regime, and more locally by floodplain and bar-top remnants which scale with the arial extent of the fields.
- Whilst lateral connectivity appears to be good within individual multistory units, this is commonly impeded by baffles composed of bar-draping fines, mud-chip lags, and cemented calcrete-clast conglomerates.
- The stratigraphic compartmentalization common to all these fields may be a characteristic of dryland fluvial reservoirs, although this is not unknown from more humid systems.

The behavior of the Skagerrak reservoir is consistent across the region. Capturing this behavior in numerical simulation models requires a detailed understanding of a range of potential alluvial architectures and heterogeneity structures which are not supported by knowledge from outcrop analogues. The prediction of the permeability structure of the reservoir, and the effect of internal baffles and barriers, requires a knowledge of bar-scale heterogeneity and grain-size distribution from a range of fluvial systems in order to define the uncertainty range possible in subsurface data sets.

ACKNOWLEDGMENTS

Andres Aslan and Brian Jones are thanked for their detailed and insightful comments. Colin North, Stephanie Davidson, Sophie Leleu, and John Southard are also thanked for their editorial input.

REFERENCES

Aigner, T., and Bachmann, G.H., 1992, Sequence stratigraphic framework of the German Triassic: Sedimentary Geology, v. 80, p. 115–135.

Allen, J.R.L., 1982, Studies in fluviatile sedimentation: bars, bar-complexes and sandstone sheets (low-sinuosity braided steams) in the Brownstones (L. Devonian), Welsh Borders: Sedimentary Geology, v. 33, p. 237–293.

Baeteman, C., Dupin, L., and Heyvaert, V., 2004, The Persian Gulf shorelines and the Karkeh, Kurun and Jarrahi Rivers: a geoarchaeological approach: Akkadica, v. 125, p. 155–215.

Baltzer, F., and Purser, B.H., 1990, Modern alluvial fan and deltaic sedimentation in a foreland tectonic setting: the Lower Mesopotamian Plain and the Arabian Gulf: Sedimentary Geology, v. 67, p. 175–197.

Blum, M.D., and Törnqvist, T.E., 2000, Fluvial responses to climate and sea level change: a review and look forward: Sedimentology, v. 47, p. 2–48.

Borkhataria, R., Aigner, T., Pöppelreiter, M.C., and Pipping, J.C.P., 2005, Characterisation of epeiric "layer-cake" carbonate reservoirs: Upper Muschelkalk (Middle Triassic), the Netherlands: Journal of Petroleum Geology, v. 28, p. 119–146.

Borkhataria, R., Aigner, T., and Pipping, K.J.C.P., 2006, An unusual, muddy, epeiric carbonate reservoir: the Lower Muschelkalk (Middle Triassic) of the Netherlands: American Association of Petroleum Geologists, Bulletin, v. 90, p. 61–89.

Brayshaw, A.C., Davies, G.W., and Corbett, P.W.M., 1996, Depositional controls on primary permeability and porosity at the bedform scale in fluvial reservoir sandstones, *in* Carling, P.A., and Dawson, M., eds., Advances in Fluvial Dynamics and Stratigraphy: Chichester, U.K., John Wiley & Sons, p. 373–394.

Bridge, J.S., 2003. Rivers and Floodplains- Forms, Processes and Sedimentary Record: Oxford, U.K., Blackwell, 504 p.

Bridge, J.S., and Tye, R.S., 2000, Interpreting the dimensions of ancient fluvial channel bars, channels, and channel belts from wireline-logs and cores: American Association of Petroleum Geologists Bulletin, v. 84, p. 1205–1228.

Brierley, G.J., Ferguson, R.J., and Woolfe, K.J., 1997, What is a fluvial levee?: Sedimentary Geology, v. 114, p. 1–9.

Brookfield, M.E., 2008, Palaeoenvironments and palaeotectonics of the arid to hyperarid intracontinental latest Permian- late Triassic Solway basin (U.K.): Sedimentary Geology, v. 210, p. 27–47.

Coward, M.P., 1995, Structural and tectonic setting of the Permo-Triassic basins of northwest Europe, *in* Boldy, S.A.R., ed., Permian and Triassic Rifting in Northwest Europe: Geological Society of London, Special Publication 91, p. 7–39.

Corbeanu, R.M., Wizevich, M.C., Bhattacharya, J.P., Zeng, X., and McMechan, G.A., 2005. Three-dimensional architecture of ancient lower delta-plain point bars using ground-penetrating radar, Cretaceous Ferron Sandstone, Utah, *in* Chidsey, T.C. Jr., Adams, R.D., and Morris, T.H., eds., Analog for fluvial–deltaic reservoir modeling: Ferron Sandstone of Utah: American Association of Petroleum Geologists, Studies in Geology 50, p. 427–450.

Corbett, P., Zheng, S-Y, Pinisetti, M., Mesmari, A., and Stewart, G., 1998, The integration of geology and well testing for improved fluvial reservoir characterisation: Society of Petroleum Engineers 48880, p. 471–488.

Cortez, O.D., and Corbett, P.W.M., 2005, Time lapse production logging and the concept of flowing units: Society of Petroleum Engineers 94436, p. 1–12.

Croke, J., Magee, J., and Price, D., 1996, Major episodes of Quaternary activity in the lower Neales River, northwest of Lake Eyre, central Australia: Palaeogeography, Palaeoclimatology, Palaeoecology, v. 124, p. 1–15.

Davis, J.M, Lohmann, R.C., and Phillips, F.M., 1993, Architecture of the Sierra Ladrones Formation, central New Mexico: depositional controls on the permeability correlation structure: Geological Society of America, Bulletin, v. 105, p. 998-1007.

de Jager, G., van Doren, J.F.M., Jansen, J.D., and Luthi, S.M., 2009, An evaluation of relevant geological parameters for predicting the flow behaviour of channelized reservoirs: Petroleum Geoscience, v. 15, p. 345–354.

de Rooij, M., Corbett, P.W.M., and Barens, L., 2002, Point bar geometry, connectivity and well test signatures: First Break, v. 20, p. 755–763.

Donselaar, M., and Overeem, I., 2008, Connectivity of fluvial point-bar deposits: an example from the Miocene Huesca fluvial fan, Ebro Basin, Spain: American Association of Petroleum Geologists, Bulletin v. 92, p. 1109–1129.

Doyle, J.D., and Sweet, M.L., 1995, Three-dimensional distribution of lithofacies, bounding surfaces, porosity, and permeability in a fluvial sandstone—Gypsy Sandstone of Northern Oklahoma: American Association of Petroleum Geologists, Bulletin, v. 79, p. 70–96.

Dreyer, T., Scheie, Å., and Walderhaug, O., 1990, Minipermeameter-based study of permeability trends in channel sand bodies: American Association of Petroleum Geologists, Bulletin, v. 74, p. 359–374.

Eschard, R., Lemouzy, P., Désaubliaux, G., Bacchiana, C., and Smart, B., 1998, Combining sequence stratigraphy, geostatistical simulations, and production data for modeling a fluvial reservoir in the Chaunoy Field (Triassic, France): American Association of Petroleum Geologists, Bulletin, v. 82, p. 545–568.

Erratt, D., Thomas, G.M., and Wall, G.R.T., 1999, The evolution of the central North Sea Rift, *in* Fleet, A.J. and Boldy, S.A.R., eds., Petroleum Geology of Northwest Europe; Proceedings of the 5th Conference: Geological Society of London, p. 63–82.

Farquharson, G.W., and Gibson, J.P.C., 2005, A significant satellite: the Nevis South Field in Block 9/13, *in* Doré, A.G. and Vining, B., eds., Petroleum Geology: North-West Europe and Global Perspectives- Proceedings of the 6th Petroleum Geology Conference: Geological Society of London, p. 389–403.

Fielding, C.R., 2006, Upper flow regime sheets, lenses and scour fills: extending the range of architectural elements for fluvial sediment bodies: Sedimentary Geology, v. 190, p. 227–240.

Fielding, C.R., Alexander, J., and McDonald, R., 1999, Sedimentary facies from GPR surveys of the modern, upper Burdekin River of north Queensland, Australia: consequences of extreme discharge fluctuations, *in* Smith, N.D., and Rogers, J., eds., Current Research in Fluvial Sedimentology: International Association of Sedimentologists, Special Publication 28, p. 347–362.

Fielding, C.R., Allen, J.P., Alexander, J., and Gibling, M.R., 2009, Facies model for fluvial systems in the seasonal tropics and subtropics: Geology, v. 37, p. 623–626.

Fisher, J.A., Krapf, C.B.E., Lang, S.C., Nichols, G.J., and Payenberg, T.D., 2008, Sedimentology and architecture of the Douglas Creek terminal splay, Lake Eyre, central Australia: Sedimentology, v. 55, p. 1915–1930.

Fisher, J.A., Nichols, G.J., and Waltham, D.A., 2007, Unconfined flow deposits in distal sectors of fluvial distributary systems: examples from the Miocene Luna and Huesca Systems, northern Spain: Sedimentary Geology, v. 195, p. 55–73.

Fisher, M.J., and Mudge, D.C., 1998, Triassic, *in* Glennie, K.W., ed., Petroleum Geology of the North Sea: Oxford, U.K., Blackwell, p. 212–244.

Geluk, M.C., 2005, Stratigraphy and tectonics of Permo-Triassic basins in the Netherlands and surrounding areas: unpublished Ph.D. thesis, University of Utrecht, 171 p.

Gibling, M.R., 2006, Width and thickness of fluvial channel bodies and valley fills in the geological record: a literature compilation and classification: Journal of Sedimentary Research, v. 76, p. 731–770.

Goldsmith, P.J., Rich, B., and Standring, J., 1995, Triassic correlation and stratigraphy in the South Central Graben, UK North Sea, *in* Boldy, S.A.R., ed., Permian and Triassic Rifting in Northwest Europe: Geological Society of London, Special Publication 91, p. 123–143.

Goldsmith, P.J., Hudson, G., and Van Veen, P., 2003, Triassic, *in* Evans, D., Graham, C., Armour, A., and Bathurst, P., eds, The Millenium Atlas: Petroleum Geology of the Central and Northern North Sea: Geological Society of London, p. 105–127.

Hamilton, D.S., Holtz, M.H., Ryles, P., Lonergan, T., and Hillyer, M., 1998, Approaches to identifying reservoir heterogeneity and reserve growth opportunities in a continental-scale bed-load fluvial system: Hutton Sandstone, Jackson Field, Australia: American Association of Petroleum Geologists, Bulletin, v. 82, p. 2192–2219.

Haszeldine, R.S., 1983, Fluvial bars reconstructed from a deep, straight channel, upper Carboniferous coalfield of northeast England: Journal of Sedimentary Petrology v. 53, p. 1233–1247.

Hall, J.S., Mozley, P., Davis, J.M., and Roy, N.D., 2004, Environments of formation and controls on spatial distribution of calcite cementation in Plio-Pleistocene fluvial deposits, New Mexico, U.S.A: Journal of Sedimentary Research, v. 74, p. 643–653.

Harker, S.D., Richardson, G., Sides, L.E., and Cooper, R., 2003, Alwyn North Triassic main gas-condensate: drilling deeper promotes production: Petroleum Geoscience, v. 9, p. 133–143.

Hartkamp-Bakker, C.A., and Donselaar, M.E., 1993, Permeability patterns in point bar deposits: Tertiary Loranca Basin, central Spain, *in* Flint, S.S., and Bryant, I.D., eds., The Geological Modelling of Hydrocarbon Reservoirs and Outcrop Analogues: International Association of Sedimentologists, Special Publication 15, p. 157–168.

Helgeson, D.E., 1999, Structural development and trap formation in the Central North Sea HPHT play, *in* Fleet, A.J., and Boldy, S.A.R., eds., Petroleum Geology of Northwest Europe; Proceedings of the 5th Conference: Geological Society of London, p. 1029–1034.

Hodgson, N.A., Farnsworth, J., and Fraser, A.J., 1992, Salt-related tectonics, sedimentation and hydrocarbon plays in the Central Graben, North Sea, UKCS, *in* Hardman, R.F.P., ed., Exploration Britain; Geological Insights for the Next Decade: Geological Society of London, Special Publication 67, p. 31–63.

Hogg, A.J.C., Evans, I.J., Harrison, P.F., Meling, T., Smith, G.S., Thompson, S.D. and Watts, G.F.T., 1999, Reservoir management of the Wytch Farm oil field, Dorset, UK: providing options for growth into later field life, *in* Fleet, A.J. and Boldy, S.A.R., eds., Petroleum Geology of Northwest Europe, Proceedings of the 5th Conference: Geological Society of London, p. 1157–1172.

Høimyr, Ø., Kleppe, A., and Nystuen, J.P., 1992, Effects of heterogeneities in a braided stream channel sandbody on the simulation of oil recovery: a case study from the Lower Jurassic Statfjord Formation, Snorre Field, North Sea, *in* Ashton, M., ed., Advances in Reservoir Geology: Geological Society of London, Special Publication 69, p. 105–134.

Hornung, J., and Aigner, T., 1999, Reservoir and aquifer characterization of fluvial architectural elements: Stubensandstein, Upper Triassic, southwest Germany: Sedimentary Geology, v. 129, p. 215–280.

Hornung, J., and Aigner, T., 2002a, Reservoir architecture in a terminal alluvial plain; an outcrop analogue study (Upper Triassic, southern Germany); Part 1, Sedimentology and petrophysics: Journal of Petroleum Geology, v. 25, p. 3–29.

Hornung, J., and Aigner, T., 2002b, Reservoir architecture in a terminal alluvial plain; an outcrop analogue study (Upper Triassic, southern Germany); Part II, Cyclicity, controls and models: Journal of Petroleum Geology, v. 25, p. 151–178.

Hounslow, M.W., and Ruffell, A.H., 2006, Triassic: seasonal rivers, dusty deserts and saline lakes, *in* Brenchley, P.J., and Rawson, P.F., eds., Geology of England and Wales: Geological Society of London, p. 295–324.

Hubbard, N.L.B., and Boulter, M.C., 2000, Phytogeography and paleoecology in Western Europe and Eastern Greenland near the Triassic–Jurassic boundary: Palaios, v. 15, p. 120–131.

Husmo, T., Hamar, G.P., Hoiland, O., Romuld, A., Spencer, A.M., and Titterton, R., 2002, Lower and Middle Jurassic, *in* Evans, D., Graham, C., Armour, A., and Bathurst, P., eds., The Millenium Atlas: Petroleum Geology of the Central and Northern North Sea: Geological Society of London, p. 129–155.

Jones, A.D., Auld, H.A., Carpenter, T.J., Fetkovich, E., Palmer, I.A., Rigatos, E.N., and Thompson, M.W., 2005, Jade Field: an innovative approach to high-pressure, high-temperature field development, *in* Doré, A.G., and Vining, B., eds., Petroleum Geology: North-West Europe and Global Perspectives-Proceedings of the 6th Petroleum Geology Conference: Geological Society of London, p. 269–284.

Jones, A., Doyle, J., Jacobsen, T., and Kjønsvik, D., 1993, Which sub-seismic heterogeneities influence waterflood performance? A case study of a low net-to-gross fluvial reservoir, *in* De Haan, H.J., ed., New Developments in Improved Oil Recovery: Geological Society of London, Special Publication 84, p. 5–18.

Jones, A.D.W., Verly, G.W., and Williams, J.K., 1994, What reservoir characterisation is required for predicting waterflood performance in a high net-to-gross fluvial environment? *in* Aasen, J.O., Berg, E., Buller, A.T., Hjelmeland, O., Holt, R.M., Kleppe, J., and Torsæter, O., eds., Third North Sea Oil and Gas Reservoirs Conference Proceedings: Norwegian Institute of Technology, p. 223–232.

Jones, H.L., and Hajek, E.A., 2007, Characterizing avulsion stratigraphy in ancient alluvial deposits: Sedimentary Geology, v. 202, p. 124–137.

Kelber, K.-P., and van Konijnenburg-van Cittert, J.H.A., 1998. *Equisetites arenaceus* from the Upper Triassic of Germany with evidence for reproductive strategies: Review of Palaeobotany and Palynology, v. 100, p. 1–26.

Keogh, K.J., Martinius, A.W., and Osland, R., 2007, The development of fluvial stochastic modelling in the Norwegian oil industry: a historical review, subsurface implementation and future directions: Sedimentary Geology v. 202, p. 249–268.

Kirk, M., 1983, Bar development in a fluvial sandstone (Westphalian 'A'), Scotland: Sedimentology v. 30, p. 727–742.

Kovalevych, V., Peryt, T.M., Beer, W., Geluk, M., and Halas, S., 2002, Geochemistry of early Triassic seawater as indicated by study of the Röt halite in the Netherlands, Germany, and Poland: Chemical Geology, v. 182, p. 549–563.

Kraus, M.J., and Wells, T.M., 1999, Recognizing avulsion deposits in the ancient stratigraphical record, *in* Smith, N.D. and Rogers, J., eds., Fluvial Sedimentology VI: International Association of Sedimentologists, Special Publication 28, p. 251–268.

Labourdette, R., and Jones, R.R., 2007, Characterization of fluvial architectural elements using a three-dimensional outcrop data set: Escanilla braided system, South-Central Pyrenees, Spain: Geosphere, v. 3, p. 422–434.

Lang, S.C., Payenberg, T.H.D., Reilly, M.R.W., Kassan, J., Hicks, T., and Benson, J., 2004, Modern analogues for dryland sandy fluvial–lacustrine deltas and terminal splay reservoirs: Australian Petroleum Production and Exploration Association, Journal, v. 44, p. 329–356.

Larue, D.K., and Friedman, F., 2005, The controversy concerning stratigraphic architecture of channelised reservoirs and recovery by waterflooding: Petroleum Geoscience, v. 11, p. 131–146.

LARUE, D.K., AND HOVADIK, J.M., 2006, Connectivity of channelized reservoirs: a modelling approach: Petroleum Geoscience, v. 12, p. 291–308.

LARUE, D.K., AND HOVADIK, J.M., 2008, Why is reservoir architecture an insignificant uncertainty in many appraisal and development studies of clastic channelized reservoirs?: Journal of Petroleum Geology, v. 31, p. 337–366.

LEGARRETA L., AND ULIANA M.A., 1998, Anatomy of hinterland depositional sequences; Upper Cretaceous fluvial strata, Neuquen Basin, west-central Argentina, *in* Shanley, K.W., and McCabe, P.J., eds., Relative Role of Eustasy, Climate, and Tectonism in Continental Rocks: SEPM Special Publication 59, p. 83–92.

LIU, K., BOULT, P., PAINTER, S., AND PATERSON, L., 1996, Outcrop analog for sandy braided stream reservoirs: permeability patterns in the Triassic Hawkesbury Sandstone, Sydney Basin, Australia: American Association of Petroleum Geologists, Bulletin, v. 80, p. 1850–1866.

LYNDS, R., AND HAJEK, E., 2006, Conceptual model for predicting mudstone dimensions in sandy braided-river reservoirs: American Association of Petroleum Geologists, Bulletin, v. 90, p. 1273–1288.

MACHETTE, M.N., 1985, Calcic soils of the southwestern United States, *in* Weide, D.L., ed., Soils and Quaternary Geology of the Southwestern United States: Geological Society of America, Special Paper 203, p. 1–21.

MADER, D., 1982, Aeolian sands in continental red beds of the middle Buntsandstein (Lower Triassic) at the western margin of the German Basin: Sedimentary Geology, v. 31, p. 191–230.

MANGE-RAJETZKY, M., 1995, Subdivision and correlation of monotonous sandstone sequences using high-resolution heavy mineral analysis, a case study: the Triassic of the Central Graben, *in* Dunay, R.E., and Hailwood, E.A., eds., Non-Biostratigraphical Methods of Dating and Correlation: Geological Society of London, Special Publication 89, p. 23–30.

MARTIN, J.H., 1993, A review of braided fluvial hydrocarbon reservoirs: the petroleum engineer's perspective, *in* Best, J.L., and Bristow, C.S., eds., Braided Rivers: Geological Society of London, Special Publication 75, p. 333–367.

MATTHEWS, W.J., HAMPSON, G.J., TRUDGILL, B.D., AND UNDERHILL, J.R., 2007, Controls on fluvio lacustrine reservoir distribution and architecture in passive salt-diapir provinces: insights from outcrop analogs: American Association of Petroleum Geologists, Bulletin, v. 91, p. 1367–1403.

MCKIE, T., AND AUDRETSCH, P., 2005, Depositional and structural controls on Triassic reservoir performance in the Heron Cluster, ETAP, Central North Sea, *in* Doré, A.G., and Vining, B., eds., Petroleum Geology: North-West Europe and Global Perspectives; Proceedings of the 6th Petroleum Geology Conference: Geological Society of London, p. 285–298.

MCKIE, T., AND WILLIAMS, B., 2009, Triassic palaeogeography and fluvial dispersal across the northwest European Basins: Geological Journal v. 44, p. 711–741.

MCKIE, T., AGGETT, J., AND HOGG, A.J.C., 1997, Sequence architecture of a Triassic semi-arid, fluvio-lacustrine reservoir, Wytch Farm Field, southern England, *in* Shanley, K.W., and Perkins, B.F., eds., Shallow Marine and Nonmarine Reservoirs: Sequence Stratigraphy, Reservoir Architecture and Production Characteristics: 18th Annual SEPM Gulf Coast Section Research Conference, Proceedings, p. 197–207.

MCKIE, T., AGGETT, J., AND HOGG, A.J.C., 1998, Reservoir architecture of the upper Sherwood Sandstone, Wytch Farm Field, southern England, *in* Underhill, J.R., ed., The Development and Evolution of the Wessex Basin and Adjacent Areas: Geological Society of London, Special Publication 133, p. 399–406.

MCKIE, T., JOLLEY, S., AND KRISTENSEN, M.B., 2010, Stratigraphic compartmentalisation in dryland fluvial reservoirs—examples from the Triassic Skagerrak Formation, UKCS, *in* Jolley, S.J., Fisher, Q.J., Ainsworth, A.B., Vrolijk, P., and Delisle, S., eds., Reservoir Compartmentalization: Geological Society of London, Special Publication, **347**, p. 165–198.

MCLAURIN, B.T., AND STEEL, R.J., 2007, Architecture and origin of an amalgamated fluvial sheet sand, lower Castlegate Formation, Book Cliffs, Utah: Sedimentary Geology, v. 197, p. 291–311.

MEADOWS, N.S., 2006, The correlation and sequence architecture of the Ormskirk Sandstone Formation in the Triassic Sherwood Sandstone Group of the East Irish Sea Basin, NW England: Geological Journal, v. 41, p. 93–122.

MIALL, A.D., 1988, Reservoir heterogeneities in fluvial sandstones: lessons from outcrop studies: American Association of Petroleum Geologists, Bulletin, v. 72, p. 682–697.

MIALL, A.D., 2006, Reconstructing the architecture and sequence stratigraphy of the preserved fluvial record as a tool for reservoir development: a reality check: American Association of Petroleum Geologists, Bulletin, v. 90, p. 989-1002.

MILROY, P.G., AND WRIGHT, V.P., 2000, A highstand oolitic sequence and associated facies from a Late Triassic lake basin, south-west England: Sedimentology, v. 47, p. 187–209.

NEWELL, A.J., 2006, Calcrete as a source of heterogeneity in Triassic fluvial sandstone aquifers (Otter Sandstone Formation, SW England), *in* Barber, R.D., and Tellam, J.H., eds., Fluid Flow and Solute Movement in Sandstones: The Onshore UK Permo-Triassic Red Bed Sequence: Geological Society of London, Special Publication 263, p. 119–127.

NICHOLS, G.J., AND FISHER, J.A., 2007, Processes, facies and architecture of fluvial distributary system deposits: Sedimentary Geology, v. 195, p. 75–90.

NORDAHL, K., AND RINGROSE, P.S., 2008, Identifying the representative elementary volume for permeability in heterolithic deposits using numerical rock models: Mathematics and Geosciences, v. 40, p. 753–771.

NORTH, C.P., AND TAYLOR, K.S., 1996, Ephemeral-fluvial deposits: integrated outcrop and simulation studies reveal complexity: American Association of Petroleum Geologists Bulletin, v. 80, p. 811–830.

NORTH, C.P., AND WARWICK, G.L., 2007, Fluvial fans: myths, misconceptions, and the end of the terminal-fan model: Journal of Sedimentary Research, v. 77, p. 693–701.

O'DONNELL, D., 1993, Enhancing the oil potential of secondary Triassic reservoirs in the Beryl A Field, UK North Sea, *in* Spencer, A.M., ed., Generation, Accumulation and Production of Europe's Hydrocarbons: European Association of Petroleum Geoscientists, Special Publication 3, p. 37–44.

PATTERSON, P.E., JONES, T.A., DONOVAN, A.D., OTTMANN, J.D., AND DONOFRIO, C.J., 2006, Hierarchical description and geologic modeling of a braided stream depositional system, *in* Coburn, T.C., Yarus, J.M., and Chambers, R.L., eds., Stochastic Modeling and Geostatistics: Principles, Methods, and Case Studies, Volume II: American Association of Petroleum Geologists, Computer Applications in Geology 5, p. 221–236.

PÉRON, S., BOURQUIN, S., FLUTEAU, F. AND GUILLOCHEAU, F., 2005, Paleoenvironment reconstructions and climate simulations of the Early Triassic: impact of the water and sediment supply on the preservation of fluvial system: Geodinamica Acta, v. 18, p. 431–446.

POOLER, J., AND AMORY, M., 1999, A subsurface perspective on ETAP—an integrated development of seven North Sea fields, *in* Fleet, A.J., and Boldy, S.A.R., eds., Petroleum Geology of Northwest Europe; Proceedings of the 5th Conference: Geological Society of London, p. 993–1006.

PÖPPELREITER, M., 2002, Facies, cyclicity and reservoir properties of the Lower Muschelkalk (Middle Triassic) in the NE Netherlands: Facies, v. 46, p. 119–132.

PORTER, R., 2006, Ichnology and sedimentology of Triassic continental sequences: onshore and offshore UK: unpublished Ph.D. thesis, University of Bristol, 218 p.

PORTER, R.J., AND GALOIS, R.W., 2008, Identifying fluvio–lacustrine intervals in thick playa-lake successions: an integrated sedimentology and ichnology of arenaceous members in the mid–late Triassic Mercia Mudstone Group of south-west England, UK: Palaeogeography, Palaeoclimatology, Palaeoecology, v. 270, p. 381–398.

PRANTER, M.J., ELLISON, A.I., COLE, R.D., AND PATTERSON, P.E., 2007, Analysis and modeling of intermediate-scale reservoir heterogeneity based on a fluvial point-bar outcrop analog, Williams Fork Formation, Piceance Basin, Colorado: American Association of Petroleum Geologists, Bulletin, v. 91, p. 1025–1051.

PRANTER, M.J., COLE, R.D., PANJAITAN, H., AND SOMMER, N.K., 2009, Sandstone-body dimensions in a lower coastal-plain depositional setting: Lower Williams Fork Formation, Coal Canyon, Piceance Basin, Colorado: American Association of Petroleum Geologists, Bulletin, v. 93, p. 1379–1401.

PRICE, G.D., 1999, The evidence and implications of polar ice during the Mesozoic: Earth-Science Reviews, v. 48, p. 183–210.

RINGROSE, P., PICKUP, G., JENSEN, J., AND FORRESTER, M., 1999, The Ardross

Reservoir gridblock analog: sedimentology, statistical representivity, and flow upscaling, *in* Schatzinger, R., and Jordan, J., eds., Reservoir Characterization - Recent Advances: American Association of Petroleum Geologists, Memoir 71, p. 265–276.

Ringrose, P.S., Martinius, A.W., and Alvestad, J., 2008, Multiscale geological reservoir modelling in practice, *in* Robinson, A., Griffiths, P., Price, S., Hegre, J., and Muggeridge, A., eds., The Future of Geological Modelling in Hydrocarbon Development: Geological Society of London, Special Publication 309, p. 123–134.

Robinson, J.W., and McCabe, P.J., 1997, Sandstone-body and shale-body dimensions in a braided fluvial system: Salt Wash Sandstone Member (Morrison Formation), Garfield County, Utah: American Association of Petroleum Geologists, Bulletin, v. 81, p. 1267–1291.

Ruffell, A., and Shelton, R., 1999, The control of sedimentary facies by climate during phases of crustal extension: examples from the Triassic of onshore and offshore England and Northern Ireland: Geological Society London, Journal, v. 156, p. 779–789.

Ruffell, A., McKinley, J.M., and Worden, R.H., 2002, Comparison of clay mineral stratigraphy to other proxy indicators in the Mesozoic of NW Europe: Royal Society of London, Philosophical Transactions, Mathematical, Physical and Engineering Sciences, v. 360, p. 675–693.

Rygel, M.C., Gibling, M.R., and Calder, J.H., 2004, Vegetation-induced sedimentary structures from fossil forests in the Pennsylvanian Joggins Formation, Nova Scotia: Sedimentology, v. 51, p. 531–552.

Sáez, A., Anadón, P., Herrero, M.J. and Moscariello, A., 2007, Variable style of transition between Palaeogene fluvial fan and lacustrine systems, southern Pyrenean foreland, NE Spain: Sedimentology, v. 54, p. 367–390.

Sellwood, B.W., and Valdes, P.J., 2006, Mesozoic climates: general circulation models and rock record: Sedimentary Geology, v. 190, p. 269–287.

Shanley, K.W., 2004, Fluvial reservoir description for a giant, low-permeability gas field: Jonah Field, Green River Basin, Wyoming, U.S.A., *in* Robinson, J.W., and Shanley, K.W., eds., Jonah Field: Case Study of a Giant Tight-Gas Fluvial Reservoir: American Association of Petroleum Geologists, Studies in Geology 52, and Rocky Mountain Association of Geologists 2004 Guidebook, p. 159–182.

Shanley, K.W., and McCabe, P.J., 1994, Perspectives on the sequence stratigraphy of continental strata: report of a working group at the 1991 NUNA Conference on high resolution sequence stratigraphy: American Association of Petroleum Geologists, Bulletin, v. 74, p. 544–568.

Simms, M.J., and Ruffell, A.H., 1990, Climatic and biotic change in the late Triassic: Geological Society of London, Journal, v. 147, p. 321–327.

Smalley, C., England, W.A., Muggeridge, A., Abacioglu, Y., and Cawley, S., 2004, Rates of reservoir fluid mixing: implications for interpretation of fluid data, *in* Cubitt, J.M., England, W.A., and Larter, S., eds., Understanding Petroleum Reservoirs: Towards an Integrated Reservoir Engineering and Geochemical Approach: Geological Society of London, Special Publication 237, p. 99–113.

Smith, J.J., and Hasiotis, S.T., 2008, Traces and burrowing behaviors of the cicada nymph *Cicadetta calliope:* neoichnology and paleoecological significance of extant soil-dwelling insects: Palaios, v. 23, p. 503–513.

Smith, J.J., Hasiotis, S.T., Kraus, M.J., and Woddy, D.T., 2008, *Naktodemasis bowni:* new ichnogenus and ichnospecies for adhesive meniscate burrows (amb), and paleoenvironmental implications, Paleogene Willwood Formation, Bighorn Basin, Wyoming: Journal of Paleontology, v. 82, p. 267–278.

Smith, R.I., Hodgson, N., and Fulton, M., 1993, Salt control on Triassic reservoir distribution, UKCS central North Sea, *in* Parker, J.R., ed., Petroleum Geology of Northwest Europe; Proceedings of the 4th Conference: Geological Society of London, p. 547–557.

Steel, R. and Ryseth A., 1990, The Triassic–early Jurassic succession in the northern North Sea: megasequence stratigraphy and intra-Triassic tectonics, *in* Hardman, R.F.P., and Brookes, J., eds., Tectonic Events Responsible for Britain's Oil and Gas Reserves: Geological Society of London, Special Publication 55, p. 139–168.

Stewart, S.A., and Clark, J.A., 1999, Impact of salt on the structure of the Central North Sea hydrocarbon fairways, *in* Fleet, A.J., and Boldy, S.A.R., eds., Petroleum Geology of Northwest Europe: Proceedings of the 5th Conference: Geological Society of London, p. 179–200.

Taylor, A.W., and Ritts, B.D., 2004, Mesoscale heterogeneity of fluvial–lacustrine reservoir analogues: examples from the Eocene Green River and Colton Formations, Uinta Basin, Utah, USA: Journal of Petroleum Geology, v. 27, p. 3–26.

Thompson, J., and Meadows, N.S., 1997, Clastic sabkhas and diachroneity at the top of the Sherwood Sandstone Group: East Irish Sea Basin, *in* Meadows, N.S., Trueblood, S., Cowan, G., and Hardman, M., eds., The Petroleum Geology of the East Irish Sea and Adjacent Areas: Geological Society of London, Special Publication 124, p. 237–251.

Tomasso, M., Underhill, J.R., Hodgkinson, R.A., and Young, M.J., 2002, Structural styles and depositional architecture in the Triassic of the Ninian and Alwyn North fields: implications for basin development and prospectivity in the Northern North Sea: Marine and Petroleum Geology, v. 25, p. 588–605.

Tooth, S., 2005, Splay formation along the lower reaches of ephemeral rivers on the Northern Plains of arid central Australia: Journal of Sedimentary Research, v. 75, p. 636–649.

Toro-Rivera, M.L.E., Corbett, P.W.M., and Stewart, G., 1994, Well test interpretation in a heterogeneous braided fluvial reservoir: Society of Petroleum Engineers 28828, p. 223–238.

Uličný, D., 2004, A drying-upward aeolian system of the Bohdašín Formation (early Triassic), Sudetes of NE Czech Republic: record of seasonality and long-term palaeoclimate change: Sedimentary Geology, v. 167, p. 17–39.

Underhill, J.R., and Partington, M.A., 1993, Jurassic thermal doming and deflation in the North Sea: implications of the sequence stratigraphical evidence, *in* Parker, J.R., ed., Petroleum Geology of Northwest Europe; Proceedings of the 4th Conference: Geological Society of London, p. 337–346.

van der Zwan, C.J., and Spaak, P., 1992, Lower to Middle Triassic sequence stratigraphy and climatology of the Netherlands, a model: Palaeogeography, Palaeoclimatology, Palaeoecology, v. 91, p. 277–290.

van Wees, J.-D., Stephenson, R.A., Ziegler, P.A., Bayer, U., McCann, T., Dadlez, R., Gaupp, R., Narkiewicz, M., Bitzer, F., and Scheck, M., 2000, On the origin of the Southern Permian Basin, Central Europe: Marine and Petroleum Geology, v. 17, p. 43–59.

Webb, P.J., and Kuhn, O., 2003, Enhanced scale management through the application of inorganic geochemistry and statistics: Society of Petroleum Engineers 87458, p. 1–15.

Willis, B.J., 1989, Palaeochannel reconstructions from point bar deposits: a three-dimensional perspective: Sedimentology, v. 36, p. 757–766.

Zheng, S-Y., Corbett, P.W.M., Ryseth, A., and Stewart, G., 2000, Uncertainty in well test and core permeability analysis: a case study in fluvial channel reservoirs, northern North Sea, Norway: American Association of Petroleum Geologists, Bulletin, v. 84, p. 1929–1954.

Zheng, S-Y., Legrand, V.M., and Corbett, P.W.M., 2007, Geological model evaluation through well test simulation: a case study from the Wytch Farm oilfield, southern England: Journal of Petroleum Geology, v. 30, p. 41–58.

AVULSION AND ITS IMPLICATIONS FOR FLUVIAL-DELTAIC ARCHITECTURE: INSIGHTS FROM THE HOLOCENE RHINE–MEUSE DELTA

ESTHER STOUTHAMER

Department of Physical Geography, Faculty of Geosciences, Utrecht University, P.O. Box 80.115, 3508 TC Utrecht
e-mail: e.stouthamer@geo.uu.nl

KIM M. COHEN

Department of Physical Geography, Faculty of Geosciences, Utrecht University, P.O. Box 80.115, 3508 TC Utrecht.
and Deltares | Applied Geology and Geophysics, Princetonlaan 6, Utrecht

AND

MARC J.P. GOUW

Department of Physical Geography, Faculty of Geosciences, Utrecht University, P.O. Box 80.115, 3508 TC Utrecht

ABSTRACT: Avulsion is a principal process in the formation of fluvial-deltaic successions and a primary control on deltaic architecture. It determines the distribution of sediment and water and hence influences which location in the delta receives clastic sedimentation in what amounts for what time. It also determines channel network configuration and recurrence interval of channel belts in the sedimentary succession. Starting from an overview of the depositional history and (quantified) boundary conditions of the Holocene Rhine–Meuse delta, The Netherlands, this paper discusses the implications of avulsions for fluvial-deltaic architecture based on extensive field data, highlighting the geometric properties, channel-deposit proportion (CDP), channel-belt connectedness (CR) and channel-belt width to thickness ratios (W/T).

Our study area stretches over 150 km downstream distance and covers fluvial, lagoonal, and estuarine reaches of the Holocene delta. Delta width and thickness increase considerably in downstream direction, from 15 km x 4 m at the apex to 60 km x 20 m at the river mouth.

Overall, the architecture and avulsion history is governed by relative base-level rise until 3000 years ago and by human-increased sediment delivery thereafter. Channel-belt CDP, CR, and W/T ratios decrease downstream within the wedge-shaped fluviodeltaic sequence. Their spatial variation is due to differences in provided accommodation space and inherited floodplain geometry, and due to distribution of stream power over branches. Initial high avulsion frequencies decreased with dropping rates of sea-level rise since 9000 years until ~ 3000 years ago. Thereafter avulsion frequency increased again due to increased delivery of fine sediment. This multi-millennial trend in avulsion frequency is overprinted with an ~ 600 year periodicity, attributed to intrinsic avulsive behavior occurring in the distributary network.

The highest avulsion frequencies occurred in the areas with highest CDP. Differential tectonics between the upstream and downstream delta interrupt overall downstream architectural trends, with a local minimum in W/T and a maximum in CDP and CR over the relative upthrown block. This downstream change favored nodal avulsion immediately downstream of the block.

Amalgamated channel belts formed along the sandy margins of the wide backfilled paleo-valley. In contrast, channels of an anastomosed river system in the western–central part of the delta formed isolated channel belts, encased in cohesive floodplain sediments until it was abandoned ~ 4000 years ago. In the last 3000 years, when the situation in the delta changed dramatically, major channel belts became established due to major avulsions in the apex region, leading to a substantial increase in channel-deposit proportion.

Seven sequences of avulsions, shifting upstream along a feeder branch, have been recognized. Depending on the mechanism creating these sequences, the sequences may influence fluvial-deltaic architecture. Based on the observations in the Rhine–Meuse delta, these sequences seem to be caused mainly by the backwater effect after an avulsion occurred downstream and not by continued growth of alluvial ridges as was suggested by Mackey and Bridge (1995). Therefore, the high CDP and CR in the upper Rhine–Meuse delta is not related to a particular avulsion-sequence-driving mechanism being dominant.

KEY WORDS: avulsion, fluvial-deltaic architecture, avulsion sequence, delta evolution, Rhine–Meuse delta

INTRODUCTION

Avulsion, the shifting of a channel belt to another location on the floodplain, is a key process in the development of fluvially dominated deltaic successions. Avulsion frequencies determine the areal distribution of depositional (sub)environments that control deltaic architecture. Avulsion determines the channel recurrence interval at a given location on the floodplain, and the resulting channel densities and distributary network. Within economically valuable deltaic formations, former avulsions determined channel-belt density and interconnectedness of channel-belt sands, both in ancient settings (reservoir geology) and in subrecent delta deposits (hydrology, engineering geology).

To be able to predict and understand the architecture and facies distribution in deltaic deposits with confidence, processes over multiple time and spatial scales need to be considered.

The Rhine–Meuse delta in the Netherlands offers a unique opportunity to study delta development and resulting deltaic successions. Deposits from the last glacial-interglacial half cycle in this area are well preserved, mapped, and dated. Deltaic deposition succeeded valley deposition some 9000 years BP (all ages are in cal yr). The deltaic deposits have been studied extensively, resulting in detailed reconstructions of the Holocene paleogeographic development (Berendsen and Stouthamer 2000; Berendsen and Stouthamer 2001), avulsion history (Stouthamer and Berendsen 2000; Stouthamer 2001), growth and occupancy of accommodation space (Cohen 2005; Erkens et al. 2006; Erkens 2009) and alluvial architecture (Gouw and Erkens 2007; Gouw 2008; Hijma et al. 2009; Hijma 2009). Due to the availability of these studies, it is possible to study the relationships between avulsion and deltaic architecture over the full length of the Holocene and over the full area where fluviodeltaic deposits were created in the last 9000 years.

From River To Rock Record: The Preservation Of Fluvial Sediments And Their Subsequent Interpretation
SEPM Special Publication No. 97, Copyright © 2011
SEPM (Society for Sedimentary Geology), ISBN 978-1-56576-305-0, p. 215–231.

The objectives of this paper are (1) to update our overview of the depositional history and (quantified) boundary conditions of the Holocene Rhine–Meuse delta, integrating the newest results and insights, and (2) to reveal and discuss implications of avulsions for fluvial-deltaic architecture, with particular focus on channel-deposit proportion (CDP), channel-belt connectedness (CR) and channel-belt width to thickness ratios (W/T).

GEOLOGICAL SETTING

The Holocene Rhine–Meuse delta is located in the southern part of the North Sea Basin. Over the last 2 My, the rivers Rhine and Meuse have shifted their valleys repeatedly in the basin hinge zone and formed deltas at various locations (Fig. 1). Below the Holocene coastal plain, the sedimentary record alternates between fluvial-valley, fluviodeltaic, and shallow marine barrier and tidal-inlet deposits (Törnqvist et al. 2003; Westerhoff 2008). Their formation was governed by sea-level oscillation affecting base level, by climatic change affecting sediment supply and fluvial style, and by the local tectonic setting of basin subsidence affecting sequence preservation (e.g., Zagwijn 1974). Continuous subsidence, abundant sediment availability, and repeated repositioning of valleys in the hinge zone area promoted preservation of Quaternary deposits. In addition, the vicinity of the area to former Scandinavian ice sheets has co-controlled the positioning, architectural expression, and state of preservation of the Late Pleistocene sequence directly underlying the Holocene delta (Cohen et al. 2002; Busschers et al. 2007; Busschers et al. 2008) and the ice-marginal topography from the penultimate glaciation (~ 150 ka, Saalian) that bounds the Last Glacial valley and Holocene delta.

During the Holocene, the rivers Rhine and Meuse jointly formed a delta (Figs. 2, 3). The upstream part of the delta occupies a nearly stable block in the hinge zone of the North Sea basin. In contrast, the downstream part occupies an area of marked subsidence, separated by a neotectonically active fault zone (Geluk et al. 1994; Cohen et al. 2002; Cohen et al. 2005). Around the time of the Last Glacial Maximum and the initial deglaciation (25–15 ka) the Rhine and Meuse were confined to a braided-river incised valley of up to 25 km active

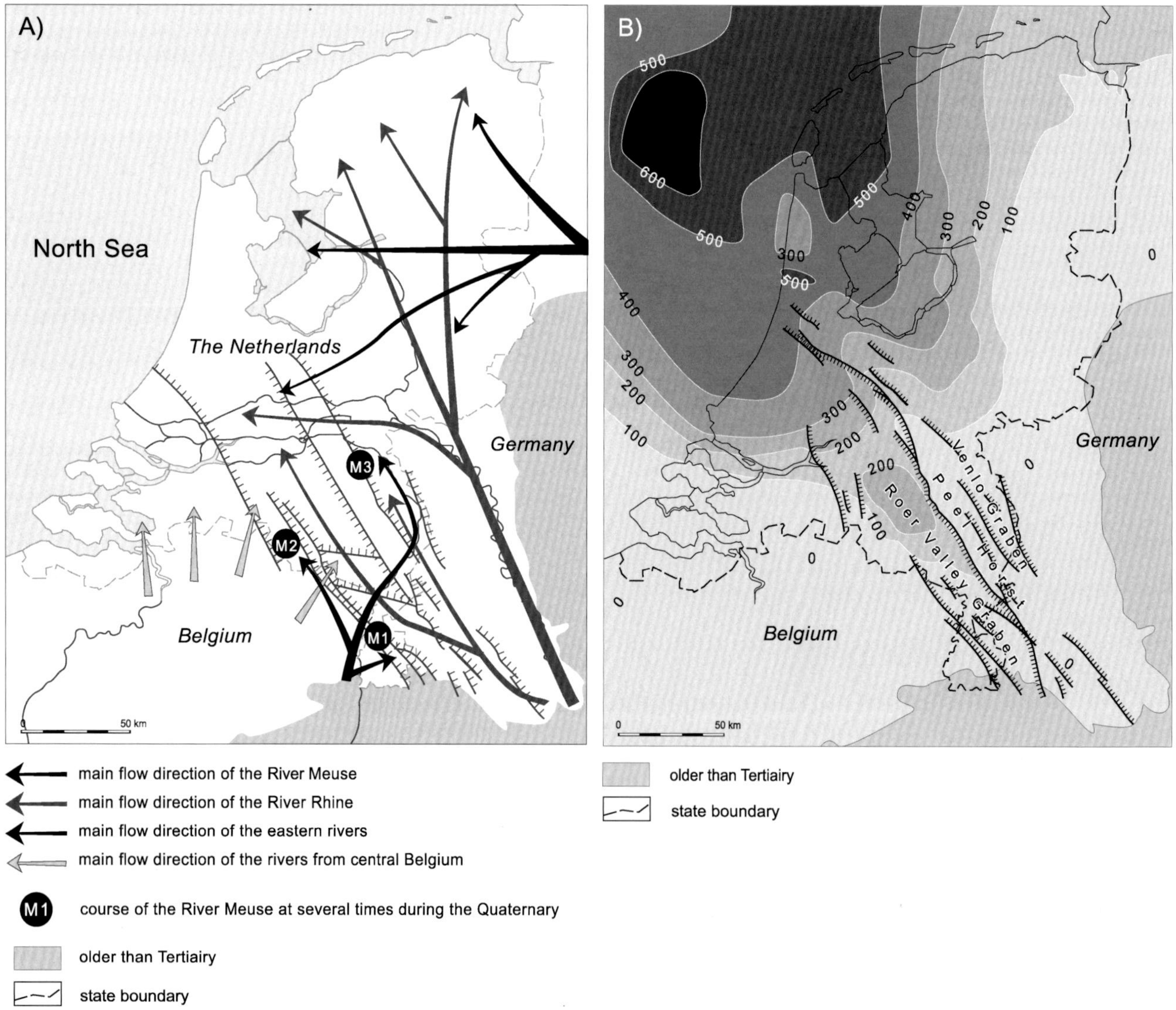

FIGURE 1.—A) Shifting of the Rhine and Meuse rivers over the last 2 My, forming deltas at various locations in the North Sea Basin hinge zone. B) Thickness of the Quaternary deposits in the subsiding North Sea Basin in the Netherlands (after Berendsen 2008).

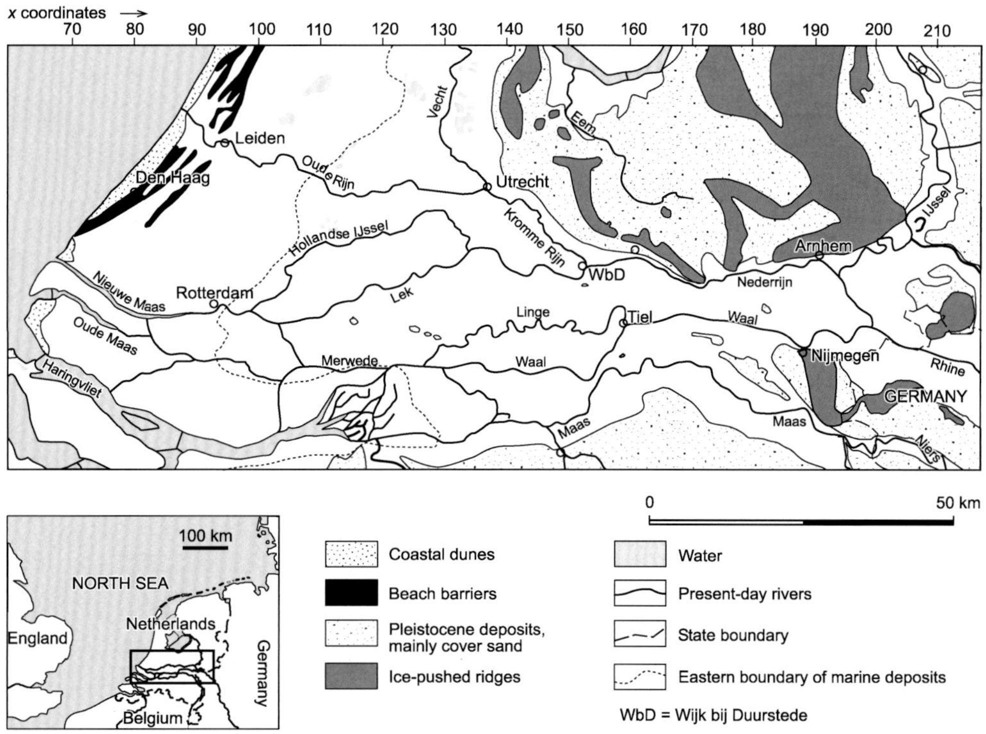

FIGURE 2.—The Holocene Rhine–Meuse delta in the central part of the Netherlands. Within the Rhine–Meuse delta, rivers run E–W. X-coordinates according to the Dutch map coordinate system (in kilometers, after Stouthamer and Berendsen 2001).

width. Similarly to continuations upstream of the delta, this braidplain was progressively abandoned in late glacial times (15–10 ka), leaving two distinct floodplain levels, now buried below the delta, dipping at ~ 20 cm/km (Pons 1957; Berendsen et al. 1995; Huisink 1997; Busschers et al. 2007; Hijma et al. 2009).

Outside channel belts, the thickness of the Holocene sequence is approximately 2 m near the delta apex and just over 20 m at the present river mouth and coastal barrier. The upstream part of the delta is bordered by higher topography of sandy deposits of Pleistocene age and relatively narrow (15–25 km wide). The downstream part of the delta grades laterally into tidal-estuary and lagoonal Holocene coastal-plain environments, and widens considerably to ~ 60 km (Fig. 2). The distance between the modern apex of the delta and the river mouth is ~ 150 km, of which the last ~ 50 km has a nearly zero gradient.

EVOLUTION OF THE HOLOCENE DELTA

The transformation from incised-valley to deltaic and back-barrier coastal environments commenced 9000 years BP at the location of the present river mouth (near Rotterdam; Fig. 2). The Early Holocene transition from valley to estuary occurred fast and was associated with rapid aggradation backfilling the paleovalley, to widening of the area of fluvial deposition and to transformation of seasonally inundated floodplains to permanently wet floodbasin environments (Hijma et al. 2009; Hijma 2009). These systems buried the network of channels of the preceding valley and spread out over surrounding inherited topography (Fig. 3a). By 7000 yr BP, fluviodeltaic backfilling had reached the fault zone separating the lower and upper delta (Fig. 3b). At the same time the coastal barrier had become established at the modern position and the transgressed valley was transformed from a broad, shallow estuary to a more protected lagoon. Fluviodeltaic environments became established at the inland side of the lagoon. Tidal inlets prevailed closer to the barrier, some of which received the Rhine and Meuse distributaries and functioned as estuaries. Coastal peat land prevailed between the fluvial channels. By 6000 yr BP the lower delta had reached more or less its present size (e.g., Berendsen and Stouthamer 2000) (Fig. 3).

An important aspect of the transition from valley to delta is the reduction in floodplain gradient, i.e., watertable slope (Van Dijk et al. 1991; Berendsen and Stouthamer 2000; Cohen 2005), influencing fluvial style, channel-belt geometry, and avulsion potential. For the last 7000 years, the inland part of the delta has maintained a 10–15 cm/km slope with a modest decrease downstream, whereas the downstream-most tidal-lagoonal part has a nearly zero slope. In contrast, the valley slope immediately upstream of the delta apex and its continuation below the delta is ~ 20 cm/km, with modest neotectonic variation and downstream slightly decreasing trend (Fig. 4) (Berendsen et al. 1995; Cohen et al. 2002, Cohen et al. 2005; Hijma et al. 2009).

Accommodation, Sediment Supply and Deposition

Progressive inland backfilling and floodbasin groundwater-level rise have been reconstructed at many sites by dating peats at the base of the Holocene deposits throughout the central and lower delta (Van Dijk et al. 1991; Berendsen and Stouthamer 2000; Cohen 2005; Hijma and Cohen 2010). These dates provide a quantitative history of accommodation in time and space (lateral and longitudinal), parallel to sea-level rise as observed at the river mouth (Jelgersma 1961; Van de Plassche 1995). Due to the westward-sloping braided-river deposits in the Late Pleistocene incised river valley, valley filling started in the low-lying western part of the valley at the coast and progressed to the east with time, resulting in onlapping Holocene deposits (Fig. 4).

In the lower delta ongoing land subsidence kept creating accommodation space, even during the Late Holocene, when eustatic sea-level rise had tailed out. Especially between 6000 and 3000 yr BP

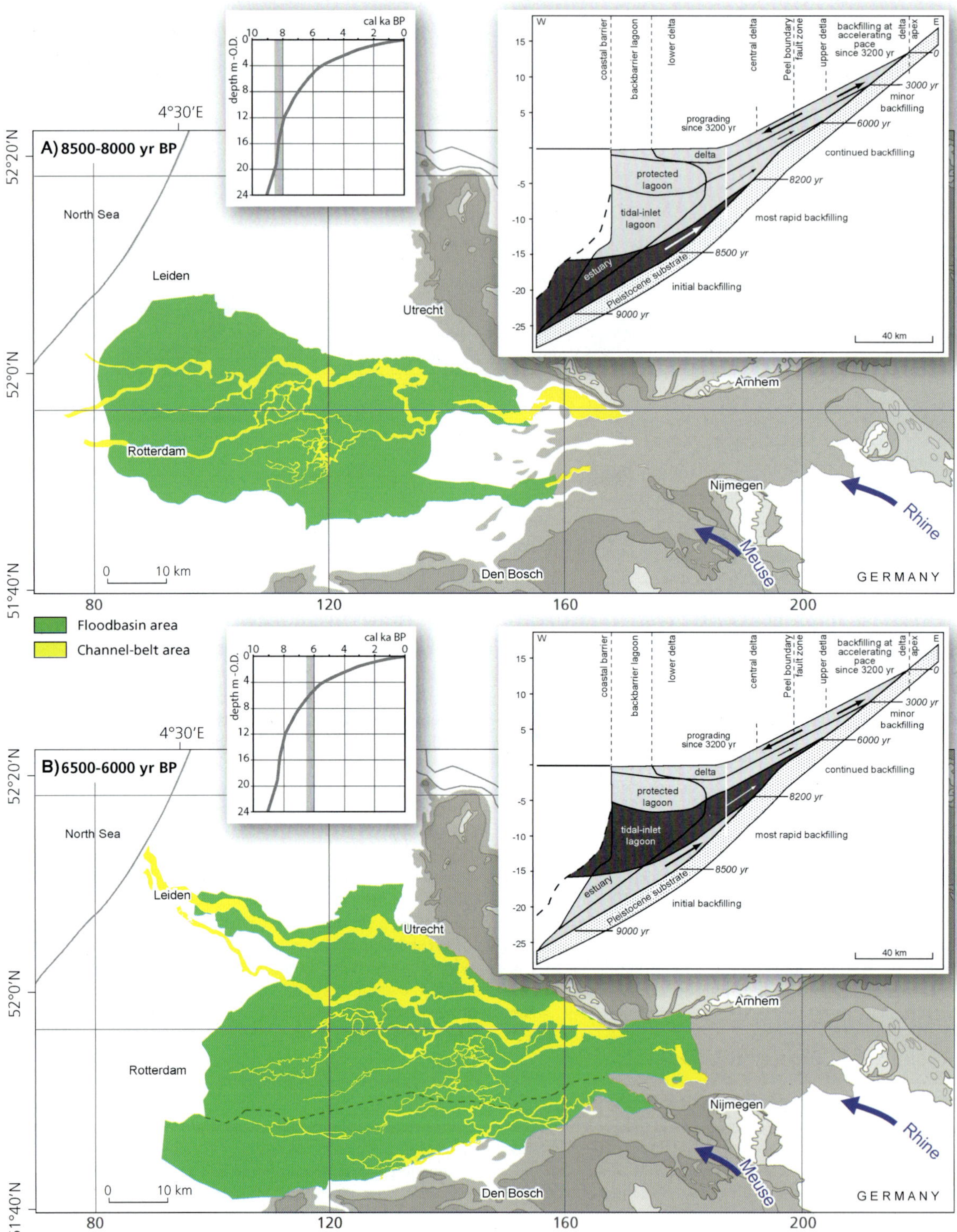

FIGURE 3.—Paleogeographic maps of the Rhine–Meuse delta for the time periods A) 8500–8000 yr BP, B) 6500–6000 yr BP, C) 4000–3500 yr BP, and D) 2500–2000 yr BP (after Erkens 2009). The maps show the extension of the Holocene fluvial delta with time. For every time period Holocene sea-level, sea-level rise history (after Hijma and Cohen 2010), and a schematic representation of the backfilling and delta progradation history of the delta is shown.

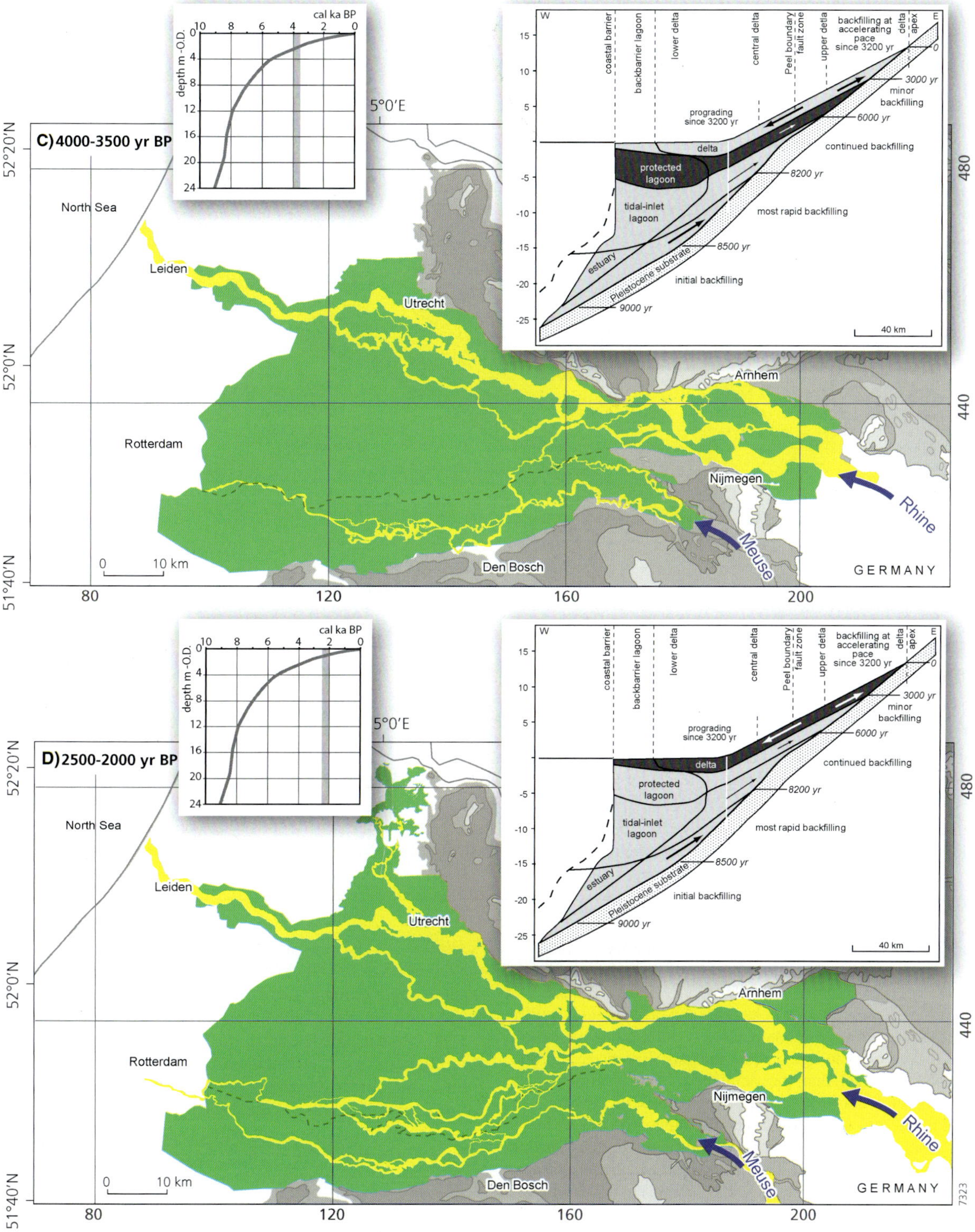

FIGURE 3.—*Continued*

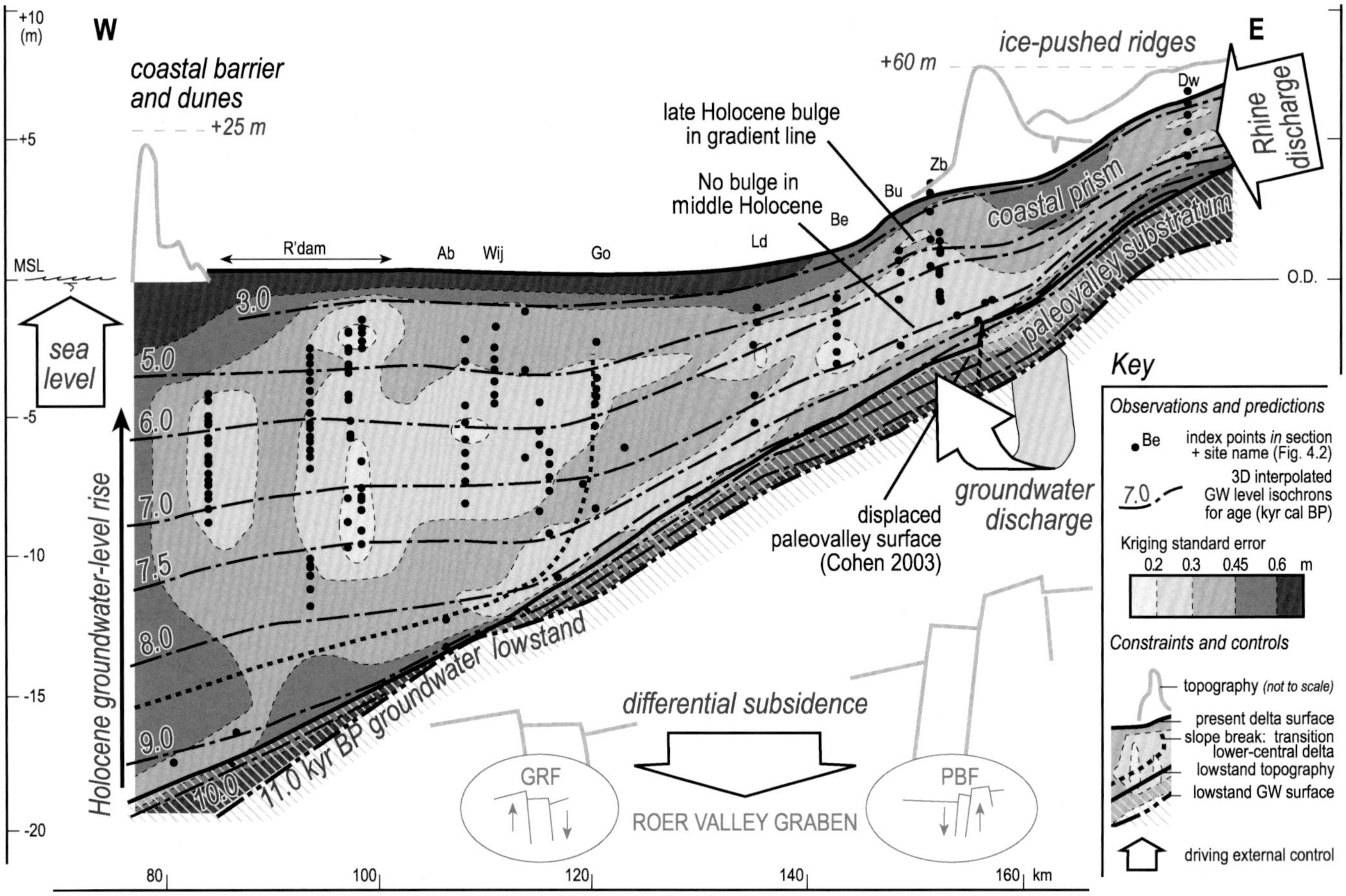

FIGURE 4.—Floodbasin groundwater-level gradients for an axial section through the Rhine–Meuse delta (from Cohen 2005). Based on 3D geostatistical interpolation of basal peat index point between upper and lower envelope surface (interpolation error increases away from data points). For the last ~ 7000 years, a stable zonation was in effect, with nearly zero gradient, 1 cm/km in the lower delta (downstream of $x = 120$) and gentle gradient, 10 cm/km in the central delta upstream. This marks distances of inland tidal control and seaward river-discharge control, in the situation after completion of postglacial eustatic sea-level rise. Note superimposed local secondary effects due to neotectonic structure and ice-pushed ridge fed groundwater discharge.

land subsidence is considered to have been the dominant contributor to accommodation space and spatial distribution of aggradation (Gouw and Erkens 2007; Erkens 2009). After 3000 yr BP, aggradation and backfilling in the apex region of the delta markedly accelerated, leading to the re-annexation of former Rhine pathways draining to the central Netherlands lagoon. From storage budget analysis (Erkens et al. 2006; Erkens 2009) it is evident that the delivery of fine sediments had increased notably and was the driving control of the major Late Holocene geological and geomorphological changes observed in the delta. It marks delta response to enhanced erosion in the German hinterland, and altered routing of fine sediment through the drainage basin, due to progressive forest clearings by prehistoric humans (Hoffman et al. 2007; Erkens et al. 2009). The excess fine sediment arriving in the delta was stored in two ways. In the upper delta, the apex region aggraded and widened and proceeded to backfill the valley and enlarged the fluviodeltaic environment in upstream direction. In the lower delta, notable deltaic progradation took place (Cohen 2005; Gouw and Erkens 2007), resulting in burial of the peaty floodbasins with clastic overbank deposits, major avulsive reorganization of the distributary network, and the formation of channels of increased meander wave length (Stouthamer and Berendsen 2001).

Intra-Holocene climate change is not considered a major control on sediment supply, peak discharge, and depositional style in the delta or valley (Erkens 2009). In the Holocene, climate was relatively stable and no major climatic shifts are recorded. The width of the lower Rhine valley and the size of the catchment has buffered the impact of minor variations that paleoclimatic studies in the hinterland have identified for subparts of the catchment. Hydrological changes related to hinterland forest clearing (reduced evapotranspiration, increased runoff) may have affected the discharge regime of the Rhine. Discharge variation of the Rhine is estimated to have been within 10%, which is minor compared to the ~ 160% increase in the load of fine sediments reconstructed for the last millennia (Erkens 2009).

Between 900 and 700 years ago (1100–1300 AD) all rivers in the delta were embanked, stopping floodbasins from receiving sediments. Flooding and channel-migration activity has since remained restricted to the embanked flood plains. Consequently, our reconstructions of natural delta behavior end at 850 yr BP.

Boundary Conditions

For modeling deltaic architecture for, e.g., scientific or reservoir engineering purposes, quantitative information for the input parameters and boundary conditions of delta formation are needed. This information should be deduced from field data as much as possible (e.g., Paola 2000). Here, we have summarized values quantified for the Holocene Rhine–Meuse delta (Table 1). See Gouw (2008) for a more extensive overview.

AVULSION HISTORY

In the Holocene Rhine–Meuse delta various kinds of avulsions occurred: (1) avulsions of the main channels of the Rhine and Meuse, (2) avulsions of secondary channels, which are relatively small channels (usually < 250 m wide) of crevasse splays and the straight channels of the anastomosing river system that was active in parts of the central delta between 8000 and 4000 yr BP, fed from the trunk channel system along the northern delta rim, and (3) avulsion associated to bay-head delta formation at distributary mouths into tidal estuaries (e.g., Hijma et al. 2009). The avulsion history and delta buildup seem to be fully controlled by avulsions of the main channels, which are predominantly upstream driven. Therefore, here we present the avulsion history related to channel-belt avulsions in the fluvial-deltaic plain based on the detailed reconstruction of Stouthamer and Berendsen (2000, 2001).

The oldest documented avulsions of the Rhine–Meuse delta occurred 9000–8500 yr BP, in the Rotterdam area. From ~ 8500 to ~ 4500 yr BP (~ 3700 ^{14}C yr BP), avulsion sites shifted upstream with sea-level rise, as part of deltaic backfilling (Figs. 5, 6). Thereafter, inland shifting halted. From ~ 4500 to ~ 3000 yr BP, avulsion sites were controlled by the Peel Boundary Fault zone tectonic feature, which hosts four out of six of the *nodal* avulsion sites in the delta (i.e., sites showing two or more avulsion events triggered at the same location; Stouthamer and Berendsen, 2000) (Fig. 6). After ~ 3000 yr BP, the avulsion frequency increased significantly, and avulsion sites occurred all over the delta (Fig. 6). Between 3000 and 2000 yr BP, the number of coeval channels increased, after which it remained constant. This was the first time that a new main trunk distributary formed across the Peel Horst.

Avulsion Frequency

Avulsion frequency, the number of avulsions per time interval, in the Rhine–Meuse delta was reconstructed by Stouthamer and Berendsen (2001). They calculated frequencies including all identified 91 deltaic avulsion events inland of Rotterdam, including partial avulsions creating secondary distributaries and full avulsion rerouting the main delta channels, by plotting the ages of these avulsion events in a ^{14}C intensity histogram as a probability density function (for detailed description of this method, see Stouthamer and Berendsen 2001) (Fig. 6). During initial rapid backfilling of the downstream part of the delta

TABLE 1.—Summary of boundary conditions for the formation of the Holocene Rhine-Meuse delta. Where feasible, temporal variation is accounted for and separate values are given for the lower, central and upper delta (x coordinates 70–110; 110–160; 160–210 km, respectively).

		Reference(s)
Climate	Relatively stable climate in the temperate zone. Minor climatic variability within the Holocene. Effects on variation of discharge and sediment regime minor compared to human-induced catchment impacts.	Erkens (2009)
Discharge	Rhine: 2200 m^3/s mean annual flow, 12.600 m^3/s flood of record (1926 AD) Meuse: 260 m^3/s mean annual flow, 3.000 m^3/s flood of record	20th century measurements by Rijkswaterstaat
Sediment supply	$\geq 2.7 \pm 0.5\ 10^9$ kg/yr bed load: suspended load breakdown: 1:1 (Holocene-averaged value) 1:2 (average last 2000 years) 1:3.4 (modern value)	Erkens et al. (2006) Erkens and Cohen (2009)
Basin subsidence rate	Downstream of PBF (central, lower delta): 0.3 mm/yr Upstream of PBF (upper delta): 0.1 mm/yr	Cohen (2003), Gouw (2008)
Floodbasin aggradation rate	Lower delta: 3–0.3 mm/yr (7500–7000 yr BP: 3 mm/yr; 2500–1800 yr BP 0.3 mm/yr) Upper delta: 1–0.3 mm/yr (5000–3000 yr BP: 1 mm/yr; 2500–1800 yr BP: 0.3 mm/yr)	Cohen (2005), Stouthamer and Berendsen (2007)
Channel-belt aggradation rate	Lower delta: < 0.4–4.2 mm/yr Upper delta: < 0.3–2.8 mm/yr	Stouthamer (2001, 2005), Erkens and Cohen (2009)
Floodplain width	Lower delta: 60 km Upper delta: 15 km	
Floodplain length	Apex to river mouth: 150 km	
Holocene floodplain gradient	Lower delta: 2–5 cm/km Central delta: 10–15 cm/km Upper delta: 15–20 cm/km	Van Dijk et al. (1991), Cohen (2005)
Paleo-valley gradient	Lower delta: 20–30 cm/km Central delta: 15–25 cm/km Upper delta: 20–30 cm/km	Verbraeck (1984), Törnqvist (1998), Stouthamer and Berendsen (2000), Cohen et al. (2002), Hijma et al. 2009

PBF = Peel Boundary Fault zone.

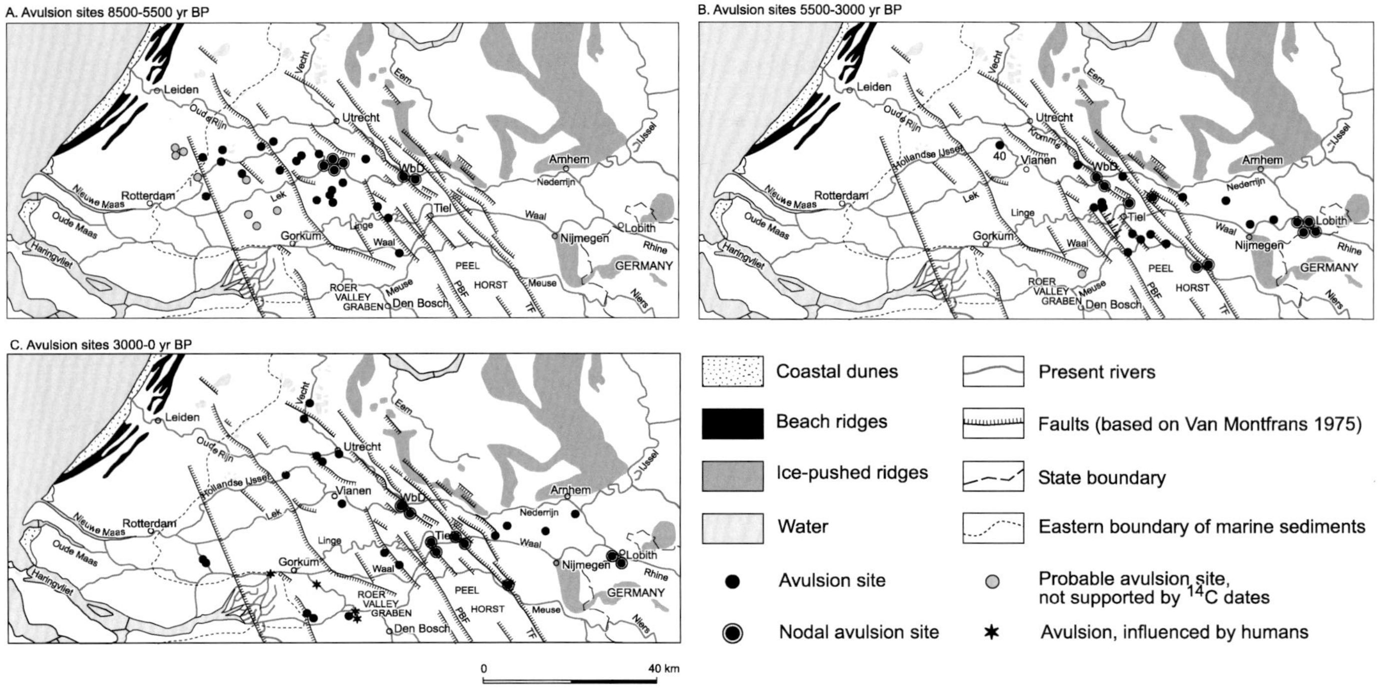

FIGURE 5.—Location of avulsion sites in three time intervals during the Holocene. A) ~ 8500–5000 yr BP: avulsion sites shifted eastward due to relative sea-level rise, B) 5500–3000 yr BP: many avulsions occurred on faults bordering the Peel Block, and near the apex of the delta, and C) 3000–1000 yr BP: avulsion sites occurred all over the delta due to increased within-channel sedimentation and slightly increased discharge. Between 1000 yr BP and the present, only few avulsions occurred, which were strongly influenced by humans. The position of the apex–avulsion node coincides with the long-term tectonic hinge line and occurs in a relatively confined reach of the lower valley (dissected ice-marginal topography). Avulsion that occurred upstream of the Peel Boundary Fault zone in many cases only created secondary channels that rejoined the main distributary instead of crossing the Peel Horst independently. Notably, the largest nodal site (WbD 6 avulsions in 3200 years; Stouthamer 2001) was fed by the Rhine trunk channel in trapped position between the Peel Horst highest block and the ice-marginal topography (Berendsen and Stouthamer 2000; Cohen 2003). TF = Tegelen Fault, PBF = Peel Boundary Fault, and WbD = Wijk bij Duurstede (after Stouthamer and Berendsen 2000).

(9000–7300 yr BP), the first channel-belt avulsions seem to lag the onset of floodbasin aggradation by some 800 years. A relatively high avulsion frequency occurred in the downstream delta between 8500 to 7300 yr BP. Thereafter, avulsion frequency was low from 7300 to 3200 yr BP. The maximum avulsion frequency observed between 8000 and 7300 yr BP relates to the rapidly growing lateral and vertical accommodation space at that time (see above). The relatively low frequency from 7300 to 3200 yr BP may be related to the steadily dropping rate of accommodation growth under decreasing rates of sea-level rise (Figs. 4, 6) and associated development of the coastal barrier. At this time the backfilling process continued but at a relatively slow rate. Since 6000 yr BP the backbarrier delta was relatively stable and can be considered more or less in equilibrium (transgression ends, highstand begins). Abundant erosion-resistant peats (notably from woody *Alnus* swamps) resulted in the fixation of channels invading the back-barrier area (Stouthamer and Berendsen 2000; Van Asselen et al. 2009). Especially, the landward side of the lagoonal delta had become unsuitable for the formation of new channels because of the occurrence of extensive backswamps. This situation contrasts with the earlier transgressive stage of delta formation, when the environment was wetter and open reed marsh occurred (*Phragmites* peats, floodbasin lakes; Bos et al. 2009), encouraging avulsion. During the Late Holocene, the swamps became partially cleared, and then clay covered during the last peaks in avulsion frequency and ultimately embanked, successfully stopping further avulsion. The peaty flood basins also induced low cross-valley gradients, unfavorable for successful avulsion. Peat formation is an autogenic control, in this case, stabilizing the avulsion frequency.

The increase in avulsion frequency between 3200 and 1400 yr BP (reaching a maximum around 3000 yr BP) coincides with the increased delivery of fine sediment. It caused general rerouting of Rhine discharge from the northern rim to the south–central part of the delta. It is important to note that this maximum in avulsion frequency, unlike the earlier maximum at 8500–7200 yr BP, occurred totally independent of accommodation space creation by sea-level rise (which by 3000 yr ago had become nearly zero). The first peak, during transgression, was triggered downstream; the second peak, well into the highstand, was triggered upstream.

The avulsion frequency shows a periodicity of ~ 600 years during the Holocene, which has been suggested to be an autogenic characteristic of the system (Stouthamer and Berendsen 2007). A clue to the internal delta processes that cause avulsions to alternate in frequency at submillenial scale might be found in the recognition of avulsion sequences. An avulsion sequence in its strict definition groups a series of avulsion sites that shift progressively upstream along a main channel, with a simultaneous *decreasing* interavulsion period (Mackey and Bridge 1995). Seven avulsion sequences may have occurred in the Holocene Rhine–Meuse delta, suggesting a periodicity of ~ 600 yr (Fig. 6). Each peak in the avulsion frequency coincides with an avulsion sequence. When deltas indeed generate

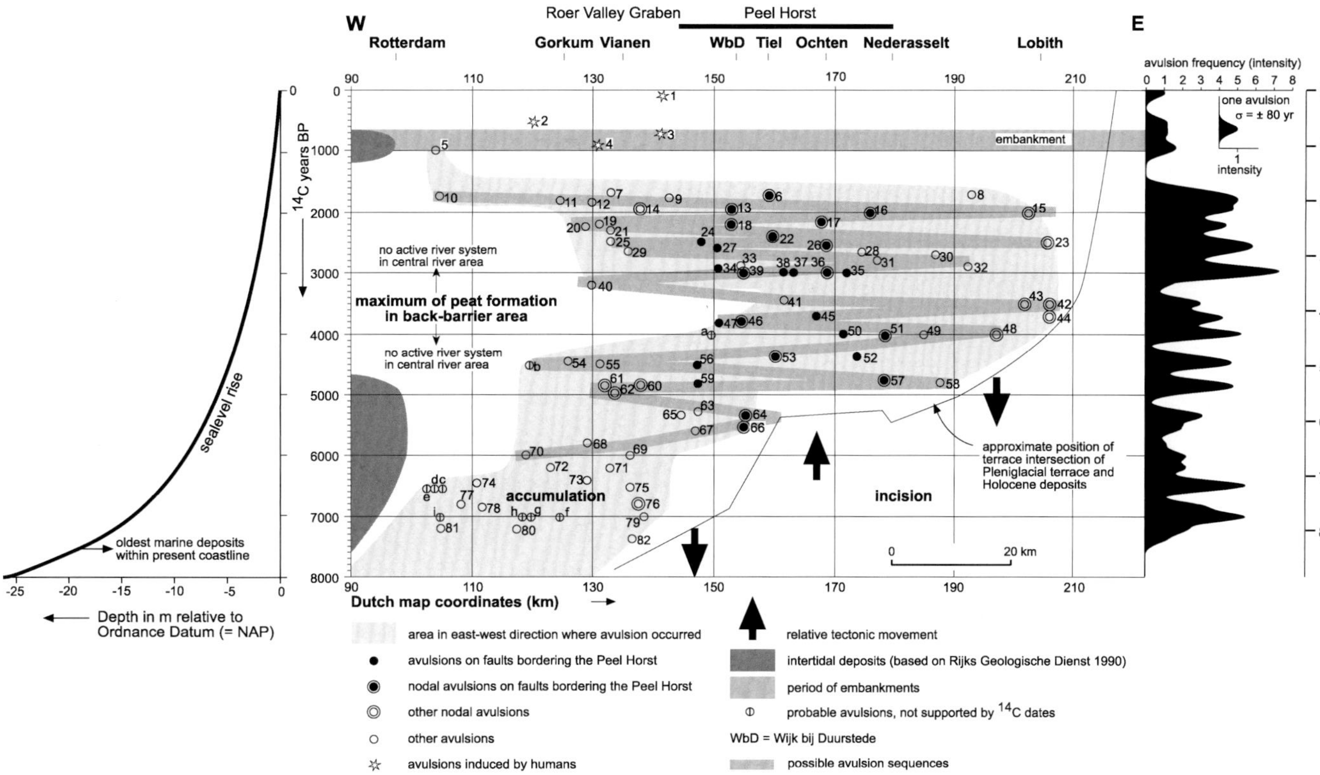

FIGURE 6.—Time-space diagram of avulsion sites in the Holocene Rhine–Meuse delta, showing the general upstream shift of avulsion sites as a result of sea-level rise. After 5500 yr BP many avulsions occurred on or near faults bordering the upthrown Peel Horst. Left: Holocene relative sea-level rise in The Netherlands. Right: avulsion frequency. Possible avulsion sequences seemed to be independent of the allogenic factors and coincide with peaks in the avulsion frequency (from Stouthamer and Berendsen 2007). The numbers refer to avulsions listed by Berendsen and Stouthamer (2001, their Table 10.1).

avulsions in sequences, this theoretically predicts relatively few avulsions to occur in downstream delta parts and more in upstream parts.

In conclusion, in the avulsion frequency, trends at different time scales can be observed: (1) a long-term decreasing trend that is attributed to sea-level rise and creation of accommodation space in the older millennia, replaced in the youngest ~ 3000 years BP to increased delivery of fine sediment and altered discharge regimes; (2) an overprinted autogenic cyclicity due to intrinsic avulsive behavior of individual distributary channels at the submillenial time scale (~ 600 yr periodicity).

Interavulsion Period

The length of time between successive avulsions (interavulsion period) increased from 500 yr (8000 yr BP) to 1500 yr (2800 yr BP) and then decreased to 1000 yr (1500 yr BP) (Fig. 7). Note that interavulsion period and avulsion frequency are not necessarily directly related, because the number of channels varies due to the occurrence of partial avulsions, in which case temporarily two active channels exist and the interavulsion period is determined along every feeding channel whereas the avulsion frequency is determined for the entire delta. Since 3200 yr BP, variation in the interavulsion period increased due to the coming into existence of large meandering channel belts, in combination with a decreased vertical aggradation rate. After 2800 yr BP the interavulsion period decreased, coinciding with an increase in avulsion frequency in all cases (Stouthamer and Berendsen 2007).

DESCRIPTIVE PARAMETERS OF FLUVIAL-DELTAIC ARCHITECTURE

Quantifying Alluvial Architecture

Alluvial architecture is defined as the geometry, proportion, and spatial distribution of different types of fluvial deposits in alluvial successions (cf. Leeder 1978; Allen 1978). The alluvial architecture of the Holocene Rhine–Meuse delta has been quantified by Gouw (2008) conform the method of Mackey and Bridge (1995) based on five detailed N–S oriented cross sections through the fluvial part of the delta by the following parameters: sand–body width/thickness ratio (SBW/SBT), channel-belt deposit proportion (CDP), overbank-deposit proportion (ODP), and organic proportion (OP), connectedness ratio (CR), and areal connectedness ratio (ACR).

Definition of Alluvial-Architecture Parameters: A sand body is defined as a sediment body principally composed of sand(stone). A sand body may be deposited by a single channel (isolated sand body) or by successive channels (amalgamated sand body). Sand-body width is defined as the maximum horizontal distance between the edges of the sand body in cross section. Sand-body thickness is the difference in elevation between the base and the top of the sand body. The channel-belt deposit proportion is the area of channel-belt deposits divided by the total area of the Holocene alluvial succession in cross section. The overbank-deposit proportion and the organics proportion are defined in a similar manner. The connectedness ratio is defined as the ratio between

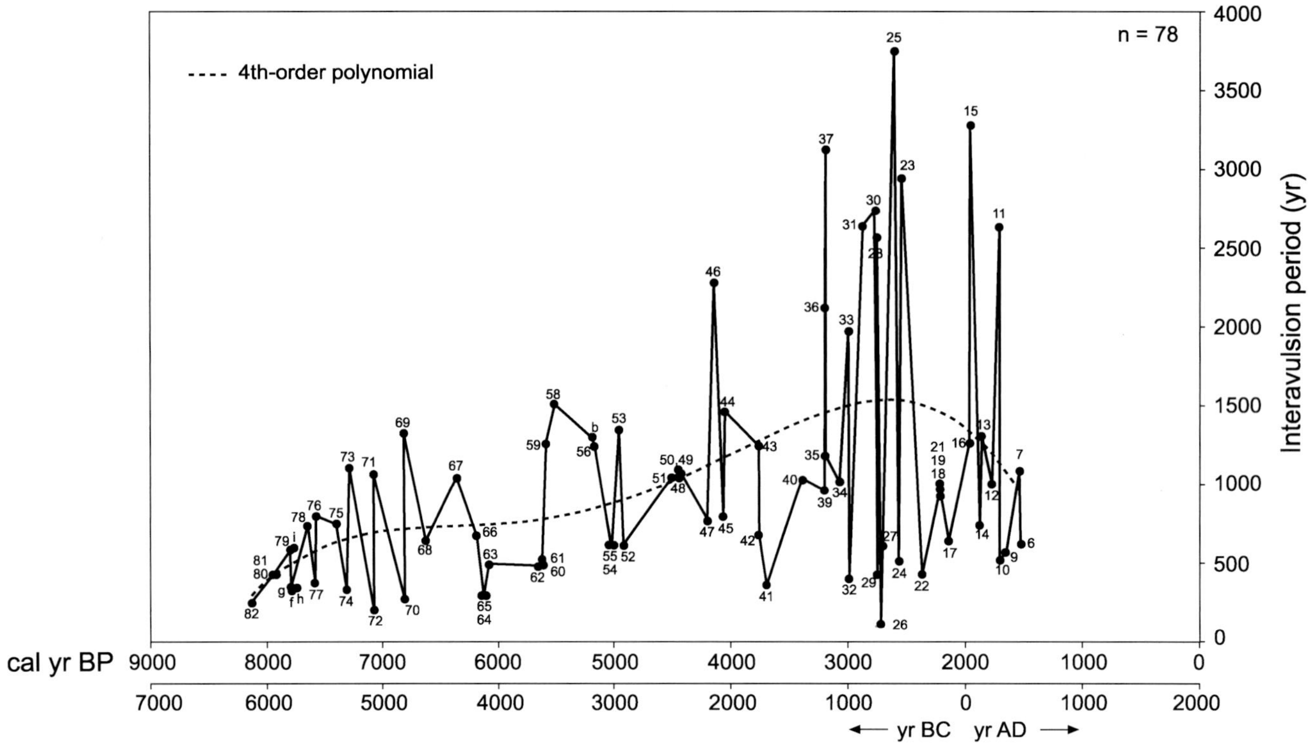

FIGURE 7.—The interavulsion period of channel belts in the Rhine–Meuse delta (from Stouthamer and Berendsen, 2007). Interavulsion period increased from 500 yr (8000 yr BP) to 1500 yr (2800 yr BP) and then decreased to 1000 yr (1500 yr BP). Since 3200 yr BP variation in the interavulsion period increased. The numbers refer to avulsions listed by Berendsen and Stouthamer (2001, their Table 10.1).

the length of horizontal contact between connected channel belts and the summed width of all channel belts in cross section. Areal connectedness ratio is the area of contact between connected channel belts divided by the total area of all channel belts in cross section (Fig. 8).

Architectural Configuration

The main river channels in the Rhine–Meuse delta for a long time since the start of deltaic aggradation tended to adhere to the sandy margins of the wide backfilled paleo-valley. The presence of noncohesive banks at the margins of the delta resulted in meandering distributaries, with high channel sinuosities and relatively wide channel belts. Locally, these channels formed isolated channel belts, but over a considerable length they formed amalgamated channel-belt complexes (Gouw 2008). In contrast, channels in the western–central part of the delta formed isolated channel belts, encased in cohesive floodplain sediments. This multistage anastomosed river system is characterized by relatively small (< 250 m) straight channels and widespread crevassing until it was abandoned ~ 4000 yr BP. In the last 3000 years, when the situation in the central delta changed dramatically, major channel belts became established, due to major avulsions > 60 km upstream in the apex region and at the central fault-zone nodal sites. As these channels ("Krimpen system"; cf. Berendsen 1982) took over discharge from the former trunk channel along the northern delta rim ("Utrecht system"; cf. Berendsen 1982), they formed relatively wide, confined-meandering channel belts (Makaske 1998; Hesselink et al. 2003; Gouw 2008). This altered the alluvial architecture drastically, raising CDP from < 0.5 before 3000 yr BP to > 0.5 after 3000 yr BP (Gouw 2008).

In the eastern, confined part of the delta, meandering channel belts existed already since the onset of aggradation due to relative sea-level rise. This is due to the relative abundance of easily erodible paleovalley sand at shallow depth and the relatively confined delta floodplain.

In the lowermost part of the delta, fluviodeltaic channels postdating transgression of the area are mostly lacking. Tidal-inlet channel systems and estuaries prevail (Hijma et al. 2009), with one major exception in the very north of the delta, which is the fluviotidal "Oude Rijn" outlet channel belt (Fig. 2). Channels of the Rhine and the Meuse merged into broad tidal estuaries at a considerable distance (~ 30 km) inland of the North Sea coast line.

Distinct spatial and temporal trends have been recognized in the alluvial architecture (Gouw 2007; Gouw 2008). Width/thickness (W/T) ratios decrease by a factor of ~ 4 in the downstream direction. Channel-deposit proportion (CDP) decreases from ~ 0.7 upstream to ~ 0.3 downstream. Organic-matter proportion (OP) increases from less than 0.05 in the upstream part of the delta to more than 0.25 in the downstream delta. Overbank-deposit proportion (ODP) is approximately constant (0.4). The connectedness ratio (CR) is ~ 0.25 upstream, which is approximately two times larger than in the downstream part of the delta. These trends are attributed to variations in available accommodation space, floodplain geometry, and channel-belt size (Gouw 2008). For example, channel belts tend to narrow in a downstream direction, which reduces W/T, CDP, and CR (Fig. 9).

AVULSIONS CONTROLLING FLUVIAL-DELTAIC ARCHITECTURE

Avulsion Frequency and Channel-Deposit Proportion

In the Rhine–Meuse delta the avulsion frequency appears to be positively correlated with CDP and CR (Gouw 2008). Before 3200 yr

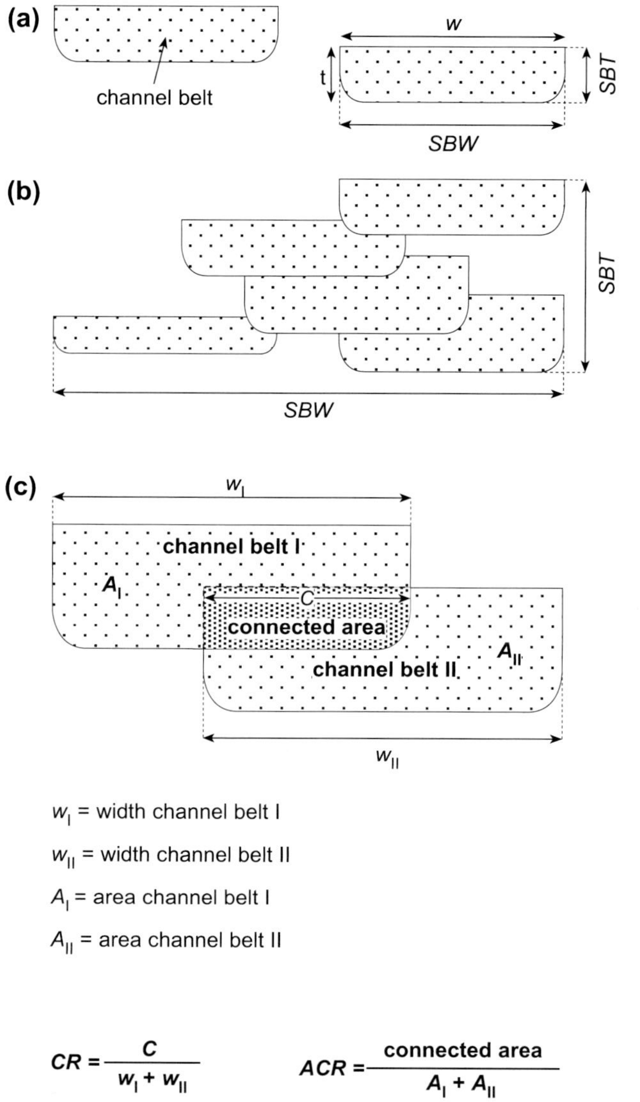

FIGURE 8.—Definition diagram of alluvial architecture parameters. A) In case of unconnected channel belts, sand-body width (*SBW*) equals channel-belt width (*w*) and sand-body thickness (*SBT*) equals channel-belt thickness (*t*). B) When two or more channel belts are connected, *SBW* is defined as the maximum horizontal distance between the edges of the sand body. *SBT* is the maximum vertical distance within the sand body. C) The connectedness ratio (*CR*) is defined as the ratio between the length of horizontal contact between connected channel belts and the summed width of all channel belts in cross section. The areal-connectedness ratio (*ACR*) is the area of contact between connected channel belts divided by the total area of all channel belts in cross section (after Gouw 2008).

BP avulsion frequency and CDP were low; after 3200 yr BP avulsion frequency and CDP were high. Low CDP and CR occurred during high rates of creation of accommodation space in combination with a relatively low avulsion frequency after 7300 yr BP and more or less constant discharge and sediment supply. In this period, large flood basins were formed in which peat developed, echoed by downstream-increasing trends of OP in the delta. The increasing CDP and CR after 3200 yr BP is the result of the then higher avulsion frequency and is explained by low accommodation space, importantly increased sediment supply, overbank aggradation, and within-channel sedimentation (Stouthamer and Berendsen 2001; Erkens et al. 2006) and, possibly, slightly increased discharge. In other deltas CR and CDP may *not* be positively related. For example, in the case of decreasing channel-belt widths, CDP can increase whereas CR decreases.

The evolution of architecture has also a spatial component. CDP and CR are higher in the confined eastern part of the delta and lower in the unconfined, wider, western part of the delta. We portray the post-3200 yr BP developments in the central delta as "upstream architectural style" prograding across "downstream architectural style", due to changing upstream boundary conditions (see above). The strongest indicator for this upstream drive is that while the central delta experienced progradation, in the apex region backfilling occurred at accelerated pace (Cohen 2005; Erkens 2009).

Levee Buildup Versus Floodbasin Filling: A Control on Avulsion and Architecture

The high avulsion frequency observed after 3200 yr BP marks the start of a phase of accelerated levee buildup throughout the delta (Stouthamer 2005). We use the ratio between proximal levee aggradation ("local" overbank sedimentation) and more widespread distal aggradation in flood basins ("regional" aggradation or accommodation) to express differences in overbank sedimentation conditions in the delta in time and space (Fig. 10). Initially (Stouthamer 2005), we quantified local aggradation rates for individual river systems (dividing levee thickness by channel longevity) and compared it to floodbasin aggradation rates derived from reconstructions of groundwater level rise (interpolations of inland basal peat paleo-waterlevel markers; Cohen 2005). Recently completed sediment budgetting analysis (Erkens 2009) has provided numbers on the amounts of distal and proximal overbank deposition trapped in the subregions of the delta, as a time series spanning the Holocene with 500 yr increments (Erkens and Cohen 2009). With this new dataset, we calculated "regionally averaged" values for "local aggradation" per time slice (dividing the volume of levee and crevasse facies by their representative areal extent). In the central delta, the results confirm the originally identified break in local/regional aggradation ratio to occur at ~ 3200 yr BP. The sediment budgetting analysis also delivered values for floodbasin aggradation rates in the upper delta, where the groundwater-level-rise method could not do this before. Figure 10 thus shows increased spatial coverage and increased temporal resolution, now allowing to explore the impact of the local/regional aggradation ratio in the apex region and the central part of the delta coevally.

The increase in local/regional aggradation rate ratio at 3200 yr BP is attributed to two combined developments: (1) regional aggradation rates had steadily decreased to a background value creating accommodation space at a modest rate, owing to land subsidence; (2) critically increased loads of fine sediment changing the rates and heights at which levees and crevasse splays could develop, owing to the prehistoric deforestation of the German hinterland.

DISCUSSION: ASPECTS OF AVULSION AND FLUVIODELTAIC ARCHITECTURE

Local/Regional Aggradation Rate Ratio and Consequences for Architecture

Local/regional aggradation rate ratio and CDP are positively related. The balance between local and regional aggradation rates determines cross-valley gradients, which is generally regarded an important prerequisite for avulsion. In this view, the ratio of local to regional aggradation rate is a proxy for the probability of avulsion. This was

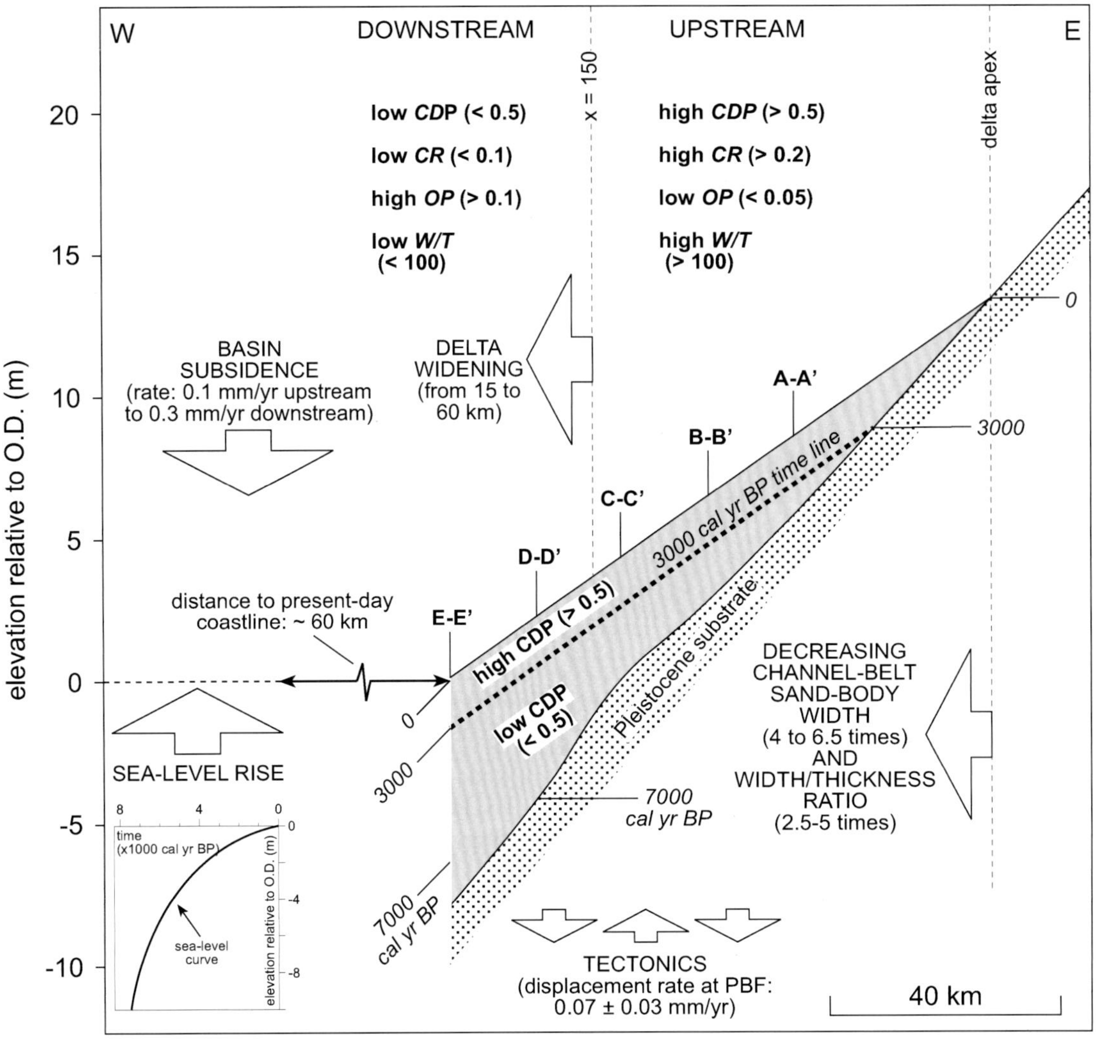

FIGURE 9.—Summary diagram showing the spatio-temporal trends in the alluvial-architecture parameters and their controlling factors for the fluviodeltaic part of the Rhine–Meuse delta. In the upstream delta CDP, CR, and W/T are high, whereas OP is low. Downstream, CDP, CR, and W/T are low and OP is high. The CDP in the succession formed before 3000 yr BP is lower than that of the succession formed after 3000 yr BP. The observed trends are attributed to variations in available accommodation space, floodplain geometry, and channel-belt size. Fault tectonics cause local deviations in the general trends in fluvial-deltaic architecture (from Gouw 2008).

also suggested by, e.g., Bryant et al. (1995), Heller and Paola (1996), and Mohrig et al. (2000), who found that avulsion is caused by superelevation of the channel or the channel belt above the floodplain and the associated increase in cross-valley slope relative to either the downchannel or the downvalley slope (e.g., Mackey and Bridge 1995; Jones and Schumm 1999; Törnqvist and Bridge 2002).

Local/regional aggradation rate ratios decrease from the delta apex to the distal delta from 1–2 mm/yr to 0–1 mm/yr respectively. The highest local aggradation rates occurred just west of the Peel Horst. In time, the ratios increased from 1.4 to 2.0 mm/yr at the delta apex and from 0.5 to 0.7 mm/yr in the downstream part. The longitudinal variation can be attributed to differences in subsidence (greatest between *x coordinate* = 110 and *x coordinate* = 155, causing increased regional aggradation), delta-plain width (narrow upstream, *x coordinate* = 160, increasing regional aggradation due to floodbasin sedimentation) and elevation above sea level (inland part of the valley or delta does not sense sea-level rise until it is within a few meters; Cohen 2005). Temporal variation is due to deceleration of base-level rise, affecting regional rates (flood-basin aggradation) and increased loads of fine sediment in the Late Holocene, affecting local rates (levee overbank sedimentation). These trends match the trends that are observed in the alluvial architecture: a high CDP and CR in the upper delta and a low CDP and CR in the downstream delta.

The high avulsion frequency between 8000 and 7300 yr BP is not reflected in the local/regional aggradation rate ratios. An explanation for this is that channel belts at the time of relatively fast transgression were more unstable than later in the Holocene. These avulsions occurred in the western part of the delta, which by that time was drowning. Regional aggradation rate was high, almost as high as the local aggradation rate (Erkens and Cohen 2009). Therefore the local/regional aggradation rate ratio, in this case, cannot have been the mechanism driving avulsion. These avulsions probably occurred due to the formation of mouth bars. This is reflected in the relatively short interavulsion periods.

In the central part of the delta the differences in local/regional aggradation rate ratio are relatively small, especially compared to the upper delta, where the impact of overbank aggradation in the last ~

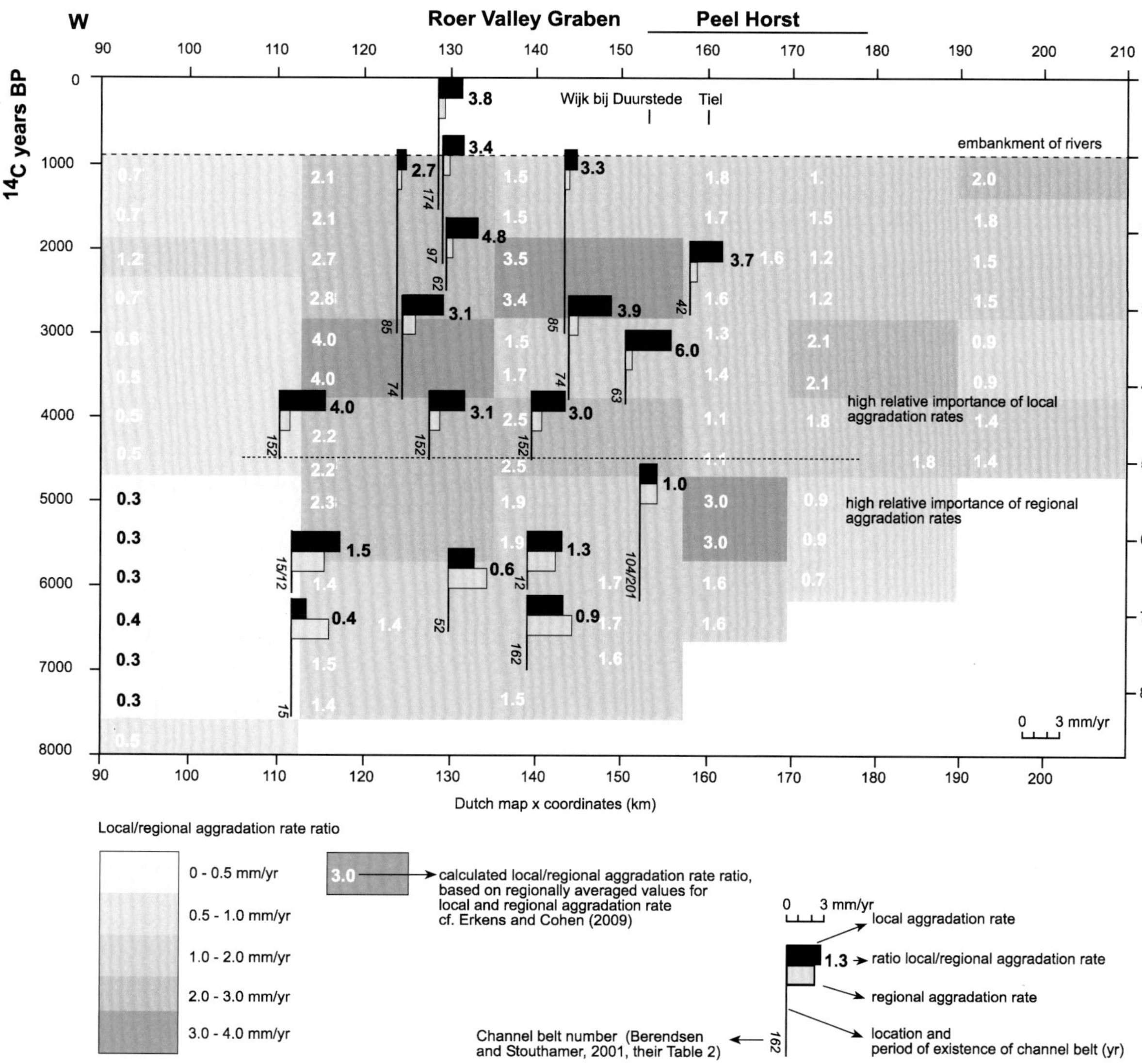

FIGURE 10.—Space-time diagram of locally and regionally averaged aggradation rates and their ratio plotted for the axis of the delta. Based on Stouthamer (2005), updated with data from Erkens and Cohen (2009); see text for details. Note that the active tectonic faults cross the section from SE to NW (Fig. 5).

3000 years is stronger. But, considering the alluvial architecture, most differences occur in the central part of the delta due to tectonics locally disturbing the general trends in the alluvial-architecture parameters. Although channel belts crossing the Peel Block are relatively narrow (Gouw and Berendsen, 2007) CDP, CR, and ACR are a bit higher there. This could be related to the high concentration of avulsion sites that is attributed to neotectonics (Stouthamer and Berendsen 2000) at the Peel Block and surrounding fault zones. Mackey and Bridge (1995) showed that CDP and connectedness may be high in areas directly downstream from avulsion sites. In conclusion, local disturbances in general (at delta scale) trends in the alluvial architecture can be attributed to local tectonics.

Relative Importance of Avulsions and Architecture

Some avulsion events in the Rhine–Meuse delta were more important for the evolution and resulting architecture than others. The most important avulsions occurred in the upstream part of the delta, in the area with the highest aggradation rates leading to absolute elevation of the surface. In the central and downstream delta, aggradation leads to compensation of subsidence but not to substantial surface elevation. Crevassing may have happened as many times in the upper and in the central and downstream delta, but the chance for succesfull avulsion is much higher in the upper delta due to the higher local aggradation rates creating superelevation. This zone of relatively high local aggradation rates was since 5500 yr BP located at the Peel Boundary Fault zone, forming the apex of the delta by that time, and shifted upstream to the present delta apex since 3200 yr BP.

In the western-central part of the delta many partial avulsions occurred between 8000 and 4000 yr BP. This sustained the anastomosed river system of secondary channels that functioned parallel to their source trunk channel along the northern delta fringe (Fig. 3b). The high number of active channels (some larger, many smaller) and repeated avulsions resulted in a diffusive distribution of sediment instead of concentration in a single corridor only. If only full avulsions (spawning channels that take over the entire discharge and cause complete abandonment of the old channel) are considered, the frequency of avulsion is much reduced (Stouthamer and Berendsen 2001). This reduction occurred under stable supply of fines and bedload (Erkens and Cohen 2009), delivered by the Rhine from a densely vegetated drainage basin (e.g., Erkens et al. 2009). Eventually, such a single-corridor situation came into existence around 4000 yr BP, but it was to be sustained only briefly and did not survive the 3000 yr BP changes.

Style of Avulsion and Architecture

Alluvial architecture in the Rhine–Meuse delta can be thought of as being the result of avulsion style. Examples of differing avulsion style are (1) avulsion by diversion into the flood basin, creating mainly single channel belts; (2) avulsion by reoccupation of an abandoned channel belt, creating stacked channel belts and; (3) avulsion by annexation of a new flood basin, formerly outside the delta, creating a new delta. Aslan and Blum (1999) related avulsion style to the stage of valley filling, based on field data of the Trinity, Nueces, and Colorado Rivers, Texas, USA. They concluded that channel reoccupation is most prevalent during early and late stages of valley filling when aggradation rates are low, due to low sediment supply or limited accommodation space. On the contrary, avulsion by diversion into the floodbasin and hence the formation of new channel belts occurs during periods of rapid accommodation-space creation and valley filling. This relationship was also found by Morozova and Smith (2000) for the Saskatchewan River, Canada. In the Rhine–Meuse delta this also seems to apply; the number of reoccupations was low during rapid sea-level rise prior to 5000 yr BP and high during slow sea-level rise (Stouthamer 2005).

High avulsion frequencies do not necessarily lead to high reoccupation frequencies. In the initial stages of delta development, avulsion frequency was high but reoccupation frequency was low. Excessive reoccupation occurred only where main channel belts fixed themselves along the delta rims. In that situation, channel belts were either rejuvenated soon after abandonment or not at all. These regions are characterized stratigraphically by the nearby coexistence of isolated and amalgamated stacked channel belts, tied to either natural levee deposits at multiple levels or exceptionally thick (> 3 m) natural-levee deposits, whereas regions of other avulsion style have not (Stouthamer 2005).

Avulsion Sequences and Architecture

The occurrence of avulsion sequences that are observed in the Rhine–Meuse delta, but were also found in the Po, Kosi, Yellow (Mackey and Bridge 1995), and Burdekin river deltas (Fielding et al. 2006) and in experiments (Hoyal and Sheets 2009), can be explained in three ways. The first way (1) is by continuous growth of alluvial ridges, which steadily increases cross-valley slopes, and predicts chances for avulsion to be largest along matured sections of a channel belt. The newly formed channel downstream of a first avulsion node is less likely to avulse than the upstream feeding channel, because it has immature, low natural levees, far below the avulsion threshold. This favors next avulsions to occur upstream past avulsions, leading to an avulsion sequence moving upstream towards the apex of the delta, where it terminates. A new sequence is then triggered by a next "first" avulsion in the downstream part of the delta (Mackey and Bridge 1995). This mechanism inevitably leads to decreasing interavulsion periods during a sequence. An alternative explanation (2) attributes the hydraulic backwater effect, and morphological adaption by sedimentation effecting response to it, an upstream avulsion-triggering role (e.g., Kleinhans et al. 2010). When an avulsion is ongoing, the flow velocity in the channel upstream of the active bifurcation is reduced and the river raises its water level (adaptation length = channel depth/water level slope). This hydraulic backwater effect affects a river of the size of the Rhine over some ten to twenty kilometers of upstream distance, somewhat helped by the bed-level adaption that it triggers. Along this reach it increases the frequency of levee overtopping and, hence, avulsion. This favours next avulsions to occur upstream past avulsions, leading to an avulsion sequence moving upstream towards the apex of the delta, where it terminates. A third idea (3) is that avulsion sequences are driven by allogenic processes rather than the autogenic mechanisms of explanation 1 and 2. During the transgression (9000–6000 yr BP) it is obvious from the replacement of fluvial by intertidal environments that avulsion locations were pushed upstream by external downstream forcing. During the human-overprinted youngest 3000 years, increased sediment load accelerated natural levee buildup, especially in the relatively confined upstream part of a delta. It has increased avulsion frequency and decreased interavulsion period in this area, and the upstream shifting of the delta apex can be seen as the culmination of an external-upstream forced avulsion sequence.

The extensive dataset of the Rhine–Meuse delta, covering 9000 years of delta formation, allows to balance the various hypotheses. Of the autogenic explanations, which need not be mutually exclusive, explanation 2 seems to apply to all avulsion sequences in the Rhine–Meuse delta. Explanation 1 implies a decreasing interavulsion period during an avulsion sequence, but that was the case in half of the avulsion sequences only. In the other half, interavulsion period did not decrease, but even increased (Stouthamer and Berendsen 2007). Explanation 3 involves major autogenic changes, and can explain particularities and timing of specific avulsion sequences, but certainly not all sequences are driven by the same forcing. Avulsion sequences in the Rhine–Meuse delta occur with a more or less constant periodicity of ~ 600 years and are recognized to have occurred since 6800 yr BP (Fig. 6).

The occurrence of avulsion sequences driven by mechanism 1 in the past has been used to explain the relatively high CDP and CR in the upper part of, e.g., the Yellow River delta (Mackey and Bridge 1995). We think that the high CDP and CR in the upper Rhine–Meuse delta is not related to a particular avulsion-sequence-driving mechanism being dominant. The inherited confined geometry of this part of the delta and the Late Holocene allogenically increased suspended load that it trapped are more important to the Rhine delta architecture and the youngest part of its contained avulsion sequences.

SUMMARY AND CONCLUSIONS

The most important observations and thoughts regarding the Holocene evolution of the Rhine–Meuse delta and the relationship between avulsion and fluviodeltaic architecture are summarized below.

- The formation of the present-day delta occurred in three phases: (1) backfilling of the incised river valley due to eustasy and subsidence (eustasy + subsidence, ~ 9000–4500 yr BP; subsidence, ~ 4500–3000 yr BP), (2) delta progradation and apex-region expansion, at accelerating rate due to increase in sediment supply (3000–1000 yr BP), and (3) Human-controlled embanked stage (last 1000 years) without sedimentation in the cultivated floodbasins.

- In the avulsion frequency, trends at different time scales occurred: (1) a long-term decreasing trend, attributed to sea-level rise and creation of accommodation space in the Early and Mid Holocene, counteracted in the youngest ~ 3000 years by increased delivery of fine sediment and slightly altered discharge regimes; (2) an overprinted autogenic cyclicity due to intrinsic avulsive behavior of individual distributary channels at the submillennial time scale (~ 600 yr periodicity).

- It is important to note that the first peak in the avulsion frequency, during transgression, was triggered downstream and the second peak, well into the high stand, was triggered upstream.

- Antecedent topography and fault-zone neotectonics together caused local overprinted effects in the avulsion history and hence delta formation and architecture.
- W/T ratios, CDP, and CR decrease in a downstream direction, whereas OP increases and ODP is approximately constant. These trends are attributed to variations in available accommodation space, floodplain geometry, and channel-belt size. At the sandy margins of the backfilled paleovalley the main river channels tended to adhere, resulting in relatively wide, high-sinuosity meandering channel belts forming amalgamated channel-belt complexes. In the western-central part of the delta isolated, relatively small channel belts, encased in cohesive floodplain sediments, formed. During the Late Holocene, after 3000 yr BP, major avulsions occurred in the apex region and at the central fault-zone nodal sites, resulting in the formation of major wide, confined-meandering channel belts, inreasing CDP drastically.
- Avulsion frequency, CDP, and CR are positively related. Before 3200 yr BP avulsion frequency, CDP, and CR were low, whereas after 3200 yr BP avulsion frequency, CDP, and CR were high. The low CDP and CR occurred during high rates of creation of accommodation space in combination with a relatively low avulsion frequency after 7300 yr BP and constant discharge and sediment supply. The increasing CDP and CR after 3200 yr BP is the result of the increased avulsion frequency in combination with limited accommodation space, importantly increased sediment supply, overbank aggradation, and within-channel sedimentation, and, possibly, slightly increased discharge. Spatially, CDP and CR are higher in the confined eastern part of the delta and lower in the unconfined, wider, western part of the delta.
- Local/regional aggradation rate ratio, determining cross-valley gradients, and CDP are positively related. Local/regional aggradation rate ratios decrease from the delta apex to the distal delta, with the highest local aggradation rates occurring just west of the Peel Horst. The longitudinal variation can be attributed to differences in subsidence, delta-plain width, and elevation above sea level, and temporal variation is due to deceleration of base-level rise affecting regional flood-basin aggradation rates and increased loads of fine sediment in the Late Holocene affecting local levee overbank sedimentation rates. These trends match the trends that are observed in the alluvial architecture—a high CDP and CR in the upper delta and a low CDP and CR in the lower delta—because there is a direct relationship between local/regional aggradation rate ratio, avulsion frequency, and CDP and CR.
- The most important avulsions of the main channels of the Rhine and Meuse determining delta formation occurred in the upstream part of the delta, the area with the highest aggradation rates leading to absolute elevation of the surface. In the central and downstream delta, aggradation leads to compensation of subsidence but not to substantial surface elevation. The chance for succesfull avulsion therefore is much higher in the upper delta. The zone of relatively high local aggradation rates was since 5500 yr BP located at the Peel Boundary Fault zone, forming the apex of the delta by that time, and shifted upstream to the present delta apex since 3200 yr BP. Downstream of the fault zone, the high number of many secondary small channels and repeated partial avulsions resulted in a diffusive distribution of sediment instead of concentration in a single corridor.
- Avulsion style changed between stages of valley filling. Avulsion by diversion into the flood basin created mainly single-channel belts during the transgression phase (9000–6000 yr BP), avulsion by reoccupation of abandoned channel belts created stacked channel belts during highstand (6000–3000 yr BP), and avulsion by annexation of new flood basin during the last 3000 years, under human-increased fine sediment loads, dramatically shifted the Rhine discharge from NW to SW, completely changing the distributary network and affecting the entire delta.
- Avulsion sequences, of which seven were recognized in the Rhine–Meuse delta, seem to be caused mainly by the backwater effect and not by increasing cross-valley slopes upstream of avulsion locations, the mechanism by which the relatively high CDP and CR in the upper part of, e.g., the Yellow River delta was explained (Mackey and Bridge 1995). We think that the high CDP and CR in the upper Rhine–Meuse delta is not related to a particular avulsion-sequence-driving mechanism being dominant. The inherited confined geometry of this part of the delta and the Late Holocene allogenically increased suspended load that it trapped are more important to the Rhine delta architecture and the youngest part of its contained avulsion sequences.

ACKNOWLEDGMENTS

This paper benefited from thorough and constructive reviews by John Swenson and Brian Willis, and from discussions with co-workers Gilles Erkens and Maarten Kleinhans.

REFERENCES

Allen, J.R.L., 1978, Studies in fluviatile sedimentation: an exploratory quantitative model for the architecture of avulsion-controlled alluvial suites: Sedimentary Geology, v. 21, p. 129–147.

Aslan, A., and Blum, M.D., 1999, Contrasting styles of Holocene avulsion, Texas Gulf Coastal Plain, USA, *in* Smith, N.D., and Rogers, J., eds., Fluvial Sedimentology VI: International Association of Sedimentologists, Special Publication 28, p. 193–209.

Berendsen, H.J.A., 1982, De genese van het landschap in het zuiden van de provincie Utrecht: een fysisch-geografische studie: Published Ph.D. Thesis, Utrecht University, Utrechtse Geografische Studies 25, 255 p.

Berendsen, H.J.A., 2008, De vorming van het land: Inleiding in de geologie en de geomorfologie: Assen, Van Gorcum, 410 p.

Berendsen, H.J.A., and Stouthamer, E., 2000, Late Weichselian and Holocene palaeogeographic evolution of the Rhine–Meuse delta, The Netherlands: Palaeogeography, Palaeoclimatology, Palaeoecology, v. 161, p. 311–335.

Berendsen, H.J.A., and Stouthamer, E., 2001, Palaeogeographic Development of the Rhine–Meuse Delta, The Netherlands: Assen, Van Gorcum, 270 p.

Berendsen, H.J.A., Hoek, W.Z., and Schorn, E.A., 1995, Late Weichselian and Holocene river channel changes of the Rivers Rhine and Meuse in the Central Netherlands (Land van Maas en Waal), *in* Frenzel, B., ed., European River Activity and Climate Change during the Lateglacial and Early Holocene: European Science Foundation Project European Paläoklimaforschung / Paleoclimate Research 14, special issue, p. 151–171.

Bos, I.J., Feiken H., Bunnik, F., and Schokker, J., 2009, Influence of organics and clastic lake fills on distributary channel processes in the distal Rhine–Meuse delta (The Netherlands): Palaeogeography, Palaeoclimatology, Palaeoecology, v. 284, p. 355–374.

Bryant, M., Falk, P., and Paola, C., 1995, Experimental study of avulsion frequency and rate of deposition: Geology, v. 23, p. 365–368.

Busschers, F.S., Kasse, C., Van Balen, R.T., Vandenberghe, J., Cohen, K.M., Weerts, H.J.T., Wallinga, J., Johns, C., Cleveringa, P., and Bunnik, F.P.M., 2007, Late Pleistocene evolution of the Rhine–Meuse system in the southern North Sea basin. Imprints of climate change, sea-level oscillation and glacio-isostasy: Quaternary Science Reviews, v. 26, p. 3216–3248.

Busschers, F.S., Van Balen, R.T., Cohen, K.M., Kasse, C., Weerts, H.J.T., Wallinga, J., and Bunnik, F.P.M., 2008, Response of the Rhine–Meuse fluvial system to Saalian ice-sheet dynamics: Boreas, v. 37, p. 377–398.

Cohen, K.M., 2003, Differential subsidence within a coastal prism: Published Ph.D. thesis, KNAG/Faculty of Geographical Sciences, Utrecht University, Netherlands Geographical Studies 316, 172 p.

Cohen, K.M., 2005, 3D geostatistical interpolation and geological interpolation of palaeo-groundwater rise within the coastal prism in the Netherlands, *in*

Giosan, L., and Bhattacharya, J.P., eds., River Deltas: Concepts, Models, and Examples: SEPM, Special Publication 83, p. 341–364.

Cohen, K.M., Stouthamer, E., and Berendsen, H.J.A., 2002, Fluvial deposits as a record for neotectonic activity in the Rhine Meuse delta, the Netherlands: Netherlands Journal of Geosciences/Geologie en Mijnbouw, v. 81, p. 389–405.

Cohen, K.M., Gouw, M.J.P., and Holten, J.P., 2005, Fluvio-deltaic floodbasin deposits recording differential subsidence within a coastal prisma (central Rhine–Meuse delta, the Netherlands), *in* Blum, M.D., Marriott, S.B., and Leclair, S.F., eds., Fluvial Sedimentology VII: International Association of Sedimentologists, Special Publication 35, p. 295–320.

Erkens, G., 2009, Sediment dynamics in the Rhine catchment. Quantification of fluvial response to climate change and human impact: Published Ph.D. thesis, KNAG/Faculty of Geographical Sciences, Utrecht University, Netherlands Geographical Studies 388, 278 p.

Erkens, G., and Cohen, K.M., 2009, Quantification of intra-Holocene sedimentation in the Rhine–Meuse delta: a record of variable sediment delivery, *in* Erkens, G., Sediment dynamics in the Rhine catchment. Quantification of fluvial response to climate change and human impact: Published Ph.D. Thesis, KNAG/Faculty of Geosciences, Utrecht University, Netherlands Geographical Studies 388, 278 p.

Erkens, G., Cohen, K.M., Gouw, M.J.P., Middelkoop, H., and Hoek, W.Z., 2006, Holocene sediment budgets of the Rhine Delta (The Netherlands): a record of changing sediment delivery, *in* Rowan, J.S., Duck, R.W., Werritty, A., eds., Sediment Dynamics and the Hydromorphology of Fluvial Systems: International Association of Hydrological Sciences, Special Publication 306, p. 406–415.

Erkens, G., Dambeck, R., Volleberg, K.P., Bouman, M.T.I.J., Bos, J.A.A., Cohen, K.M., Wallinga, J., and Hoek, W.Z., 2009, Fluvial terrace formation in the northern Upper Rhine Graben during the last 20 000 years as a result of allogenic controls and autogenic evolution: Geomorphology, v. 103, p. 476–495.

Fielding, C.R., Trueman, J.D., and Alexander, J., 2006. Holocene depositional history of the Burdekin River delta of Northeastern Australia: a model for a low-accommodation, highstand delta: Journal of Sedimentary Research, v. 76, p. 411–428.

Geluk, M.C., Duin, E.J.Th., Dusar, M., Rijkers, R.H.B., Van den Berg, M.W., and Van Rooijen, P., 1994, Stratigraphy and tectonics of the Roer Valley Graben: Geologie en Mijnbouw, v. 73, no. 2–4, special publication: Seismotectonics and Seismic Hazard in the Roer Valley Graben; with Emphasis on the Roermond Earthquake of April 13, 1992, p. 129–141.

Gouw, M.J.P., 2007, Alluvial architecture of the Holocene Rhine–Meuse delta (The Netherlands) and the Lower Mississippi Valley (U.S.A.): Published Ph.D. thesis. KNAG/Faculty of Geographical Sciences, Utrecht University, Netherlands Geographical Studies 364, 192 p.

Gouw, M.J.P., 2008, Alluvial architecture of the Holocene Rhine–Meuse delta (the Netherlands): Sedimentology, v. 55, p. 1487–1516.

Gouw, M.J.P., and Berendsen, H.J.A., 2007, Variability of fluvial sand body dimensions and the consequences for alluvial architecture: Observations from the Holocene Rhine–Meuse delta (The Netherlands) and Lower Mississippi Valley: Journal of Sedimentary Research, v. 77, p. 124–138.

Gouw, M.J.P., and Erkens, G., 2007, Architecture of the Holocene Rhine–Meuse delta (The Netherlands): a result of changing external controls, *in* Stouthamer, E., and Ten Brinke, W., eds., Fluvial Sedimentology: Netherlands Journal of Geosciences/Geologie en Mijnbouw, v. 86, p. 23–54.

Heller, P.L., and Paola, C., 1996, Downstream changes in alluvial architecture: an exploration of controls on channel-stacking patterns: Journal of Sedimentary Research, v. 66, p. 297–306.

Hesselink, A.W., Weerts, H.J.T, and Berendsen, H.J.A., 2003, Alluvial architecture of the human-influenced River Rhine, The Netherlands: Sedimentary Geology, v. 161, p. 229–248.

Hijma, M.P., 2009, From river valley to estuary. The early–mid Holocene transgression of the Rhine–Meuse valley, The Netherlands: Published Ph.D. thesis, KNAG/Faculty of Geosciences, Utrecht University, Netherlands Geographical Studies 389, 192 p.

Hijma, M.P., and Cohen, K.M., 2010, Timing and magnitude of the sea-level jump preluding the 8200 yr event: Geology, v. 38, p. 275–278, doi: 10.1130/G30439.1.

Hijma, M.P., Cohen, K.M., Hoffmann, G., Van der Spek, A.J.F., and Stouthamer, E., 2009, From river valley to estuary: early to middle Holocene evolution of the lower Rhine–Meuse delta, the Netherlands: Netherlands Journal of Geosciences/Geologie en Mijnbouw 88, p. 13–53.

Hoffman, T., Erkens, G., Cohen, K.M., Houben, P., Seidel, J., and Dikau, R., 2007, Holocene floodplain sediment storage and hillslope erosion within the Rhine catchment: The Holocene, v. 17, p. 105–118.

Hoyal, D., and Sheets, B., 2009, Morphodynamic evolution of experimental cohesive deltas: Journal of Geophysical Research, v. 114, F02009, doi:10.1029/2007JF000882.

Huisink, M., 1997, Lateglacial sedimentological and morphological changes in a lowland river in response to climate change: the Maas, southern Netherlands: Journal of Quaternary Science, v. 12, p. 209–223.

Jelgersma, S., 1961, Holocene sea-level changes in the Netherlands: Thesis Leiden, Mededelingen Geologische Stichting, CVI-7.

Jones, L.S., and Schumm, S.A., 1999, Causes of avulsion: an overview, *in* Smith, N.D., and Rogers, J., eds., Fluvial Sedimentology VI: International Association of Sedimentologists, Special Publication, v. 28, p. 171–178.

Kleinhans, M.G., Weerts, H.J.T., and Cohen, K.M., 2010, Avulsion in action: Reconstruction and modelling sedimentation pace and upstream flood water levels following a Medieval tidal-river diversion catastrophe (Biesbosch, The Netherlands, 1421–1750 AD): Geomorphology, v. 118, doi:10.1016/j.geomorph.2009.12.009.

Leeder, M.R., 1978, A quantitative stratigraphic model for alluvium with special reference to channel deposit density and interconnectedness, *in* Miall, A.D., ed., Fluvial Sedimentology: Canadian Society of Petroleum Geologists, Memoir 5, p. 587–596.

Mackey, S.D., and Bridge, J.S., 1995, Three-dimensional model of alluvial stratigraphy: theory and application: Journal of Sedimentary Research, v. B65, p. 7–31.

Makaske, B., 1998, Anastomosing rivers: forms, processes and sediments: Published Ph.D. Thesis, Utrecht University, Netherlands Geographical Studies 249, 287 p.

Mohrig, D., Heller, P.L., Paola, C., and Lyons, W.J., 2000, Interpreting avulsion process from ancient alluvial sequences: Guadaloupe Matarranya system (northern Spain) and Wasatch Formation (western Colorado): Geological Society of America, Bulletin, v. 112, p. 1787–1803.

Morozova, G.S., and Smith, N.D., 2000, Holocene avulsion styles and sedimentation patterns of the Saskatchewan River, Cumberland Marshes, Canada: Sedimentary Geology, v. 130, p. 81–105.

Paola, C., 2000, Quantative models of sedimentary basin filling: Sedimentology, v. 47 (Suppl. 1), p. 121–178.

Pons, L.J., 1957, De geologie, de bodemvorming en de waterstaatkundige ontwikkeling van het Land van Maas en Waal en een gedeelte van het Rijk van Nijmegen: 's-Gravenhage. Verslagen van Landbouwkundige Onderzoekingen 63.11, Ph.D. Thesis Wageningen, Bodemkundige Studies 3.

Rijks Geologische Dienst, 1990, Vereenvoudigde geologische kaart van Rotterdam en omgeving: Haarlem, Rijks Geologische Dienst.

Stouthamer, E., 2001, Holocene avulsion history of the Rhine–Meuse delta, The Netherlands: Published Ph.D. thesis. KNAG/Faculty of Geographical Sciences, Utrecht University, Netherlands Geographical Studies 283, 224 p.

Stouthamer, E., 2005, Reoccupation of channel belts and its influence on alluvial architecture in the Holocene Rhine–Meuse delta, The Netherlands, *in* Giosan, L., and Bhattacharya, J.P., eds., 2005, River Deltas: Concepts, Models, and Examples: SEPM, special publication 83, p. 319–339.

Stouthamer, E., and Berendsen, H.J.A., 2000, Factors controlling the Holocene avulsion history of the Holocene Rhine–Meuse delta (The Netherlands): Journal of Sedimentary Research, v. 70, p. 1051–1064.

Stouthamer, E., and Berendsen, H.J.A., 2001, Avulsion frequency, avulsion duration and interavulsion period of Holocene channel belts in the Rhine–Meuse delta, The Netherlands: Journal of Sedimentary Research, v. 71, p. 588–597.

Stouthamer, E., and Berendsen, H.J.A., 2007, The relative roles of autogenic and allogenic processes: Sedimentary Geology, v. 198, p. 309–325.

Törnqvist, T.E., 1998, Longitudinal profile evolution of the Rhine–Meuse system during the last deglaciation: interplay of climate change and glacio-isostasy?: Terra Nova 10, p. 11–15.

Törnqvist, T.E., and Bridge, J.S., 2002, Spatial variation of overbank aggradation rate and its influence on avulsion frequency: Sedimentology, v. 49, p. 891–905.

Törnqvist, T.E., Wallinga, J., and Busschers, F.S., 2003, Timing of the last sequence boundary in a fluvial setting near the highstand shoreline – Insights from optical dating: Geology, v. 31, p. 279–282.

Van Asselen, S., Stouthamer, E., and Van Asch, Th.W.J., 2009, Effects of peat compaction on delta evolution: A review on processes, responses, measuring and modeling: Earth–Science Reviews, v. 92, p. 35–51.

Van de Plassche, O., 1995, Periodic clay deposition in a fringing peat swamp in the lower Rhine–Meuse river area, 5400–3400 cal BC: Journal of Coastal Research, special issue v. 17, p. 95–102.

Van Dijk, G.J., Berendsen, H.J.A., and Roeleveld, W., 1991, Holocene water level changes in the Rhine–Meuse delta (The Netherlands), and implications for sea level reconstruction: Geologie en Mijnbouw, v. 70, p. 311–326.

Van Montfrans, H.M., 1975, Toelichting bij de ondiepe breukenkaart met diepteligging van de Formatie van Maassluis, schaal 1:600.000, *in* Zagwijn, W.H. and Van Staalduinen, C.J., eds., 1975, Toelichting bij de geologische overzichtskaarten van Nederland: Rijks Geologische Dienst Haarlem, p. 103–109.

Verbraeck, A., 1984, Toelichtingen bij de Geologische Kaart van Nederland, schaal 1:50.000, blad Gorinchem Oost (38O): Rijks Geologische Dienst, Haarlem, 335 p.

Westerhoff, W.E., 2008, Stratigraphy and sedimentary evolution: the lower Rhine–Meuse system during the Late Pliocene and Early Pleistocene (southern North Sea Basin): Ph.D. thesis. Vrije Universiteit Amsterdam, TNO B&O Geological Survey of the Netherlands: Utrecht, 168 p.

Zagwijn, W.H., 1974, The palaeogeographic evolution of The Netherlands during the Quaternary: Geologie en Mijnbouw, v. 53, p. 369–385.

A TIDALLY INFLUENCED, HIGH-LATITUDE COASTAL-PLAIN: THE UPPER CRETACEOUS (MAASTRICHTIAN) PRINCE CREEK FORMATION, NORTH SLOPE, ALASKA

PETER P. FLAIG

Dept. of Geology and Geophysics, and Geophysical Institute, University of Alaska Fairbanks, P.O. Box 755780, Fairbanks, Alaska 99775, U.S.A.

Current address: University of Texas at Austin, Bureau of Economic Geology, Quantitative Clastics Laboratory, 10100 Burnett Rd, Austin, Texas 78758, U.S.A.

E-mail: peter.flaig@beg.utexas.edu

PAUL J. McCARTHY

Dept. of Geology and Geophysics, and Geophysical Institute, University of Alaska Fairbanks, P.O. Box 755780, Fairbanks, Alaska 99775, U.S.A.

E-mail: mccarthy@gi.alaska.edu

AND

ANTHONY R. FIORILLO

Museum of Nature and Science, P.O. Box 151469, Dallas, Texas 75315, U.S.A.

E-mail: tfiorillo@natureandscience.org

Abstract: The Prince Creek Formation is an Upper Cretaceous, dinosaur-bearing, high-latitude alluvial succession deposited on an ancient coastal-plain that crops out in bluffs along the Colville, Kogosukruk, and Kikiakrorak Rivers of northern Alaska. Studies that document the complex stratigraphy and architecture of high-latitude alluvial systems deposited under greenhouse conditions are extremely rare. It is exceptionally uncommon to find extensive, accessible outcrops that also contain numerous Arctic dinosaur fossils; hence the Prince Creek Formation is of great significance not only to sedimentologists but also to paleontologists involved in reconstructing high-latitude dinosaur habitats.

Maastrichtian strata of the Prince Creek Formation record deposition on a tidally influenced high-latitude coastal-plain in (i) first-order meandering trunk channels, (ii) second-order meandering distributary channels, (iii) third-order fixed (anastomosed?) distributary channels, and (iv) on floodplains. Conglomerate and medium- to coarse-grained multistory sandbodies are found exclusively in regionally restricted 13–17-m-thick fining-upward successions (FUSs) that display inclined heterolithic stratification (IHS) capped by finer-grained, organic-rich facies. These relatively thick FUSs are interpreted as first-order meandering trunk channels. Thinner (2–6-m thick), single-story, heterolithic sheet sandbodies composed predominantly of IHS and including abundant mud-filled channel plugs are the most frequently encountered channel form. Trough cross-lamination at the base of the IHS records paleoflow at high angles relative to the dip of the inclined beds, indicating that lateral accretion of point-bars was the principal depositional mechanism. These single-story sandbodies are interpreted as second-order meandering distributary channels. Fine-grained, 1.5–3.0-m-thick, ripple cross-laminated ribbon sandbodies deposited mainly by vertical accretion above an arcuate erosion surface and containing only minor IHS are interpreted as third-order fixed (anastomosed?) distributary channels. Thinner (0.2–1.0-m-thick) current-rippled sheet sands and silts are interpreted as small-scale crevasse splays and levees. Organic-rich siltstone and mudstone, carbonaceous shale, coal, bentonite, and tuff are interpreted as deposits of lakes, ponds, swamps, marshes, mires, paleosols, and ashfall on floodplains.

Heterolithic sheet sandstones deposited by small, sinuous meandering distributary channels typically appear lenticular along strike, commonly incise into pre-existing distributary channels, and interfinger with and incise into organic-rich floodplain facies. Fixed, ribbon-form (anastomosed?) distributaries incise either into meandering distributaries or into floodplain facies, with numerous ribbons typically preserved in tiers at the same stratigraphic level. Spatial relationships between channel types, and between channels and floodplain facies, indicate that the bulk of deposition took place on crevasse-splay complexes adjacent to trunk channels. Crevasse-splay complexes were constructed by the lateral migration of sinuous meandering distributaries and the vertical filling of fixed (anastomosed?) distributaries, with splay complexes separated from each other by organic floodplain facies. Flow in meandering distributaries and fixed (anastomosed?) distributaries may have been contemporaneous. Alternatively, fixed (anastomosed?) distributaries may record the initial or waning stages of flow during splay-complex formation or abandonment.

IHS composed of rhythmically repeating, coarse-to-fine couplets of current-rippled sandstone and siltstone or mudstone is found in all three types of channels. The rhythmic and repetitive nature of these couplets together with relatively thick, muddy fine-grained members in couplets suggest that flow in channels was likely influenced by tidal effects.

Drab colors in fine-grained sediments, abundant carbonaceous plant material, and common siderite nodules and jarosite suggest widespread reducing conditions on poorly drained floodplains influenced, in more distal areas, by marine waters. However, carbonaceous root traces found ubiquitously in all distributary channels and most floodplain facies along with common Fe-oxide mottles indicate that the alluvial system likely experienced flashy, seasonal, or ephemeral flow, and a fluctuating water table. The flashy nature of the alluvial system may have been driven by recurring episodes of vigorous seasonal snowmelt in the Brooks Range orogenic belt as a consequence of the high paleolatitude of northern Alaska in the Late Cretaceous.

Key words: Cretaceous, Delta, Meandering, Anastomosed, Crevasse-splay Complex, Inclined Heterolithic Stratification (IHS).

From River to Rock Record: The Preservation of Fluvial Sediments and Their Subsequent Interpretation
SEPM Special Publication No. 97, Copyright © 2011
SEPM (Society for Sedimentary Geology), ISBN 978-1-56576-305-0, p. 233–264.

INTRODUCTION

The Upper Cretaceous (Maastrichtian) Prince Creek Formation (Fm.) is exposed semi-continuously in 20–50 m bluffs along the Colville, Kogosukruk, and Kikiakrorak Rivers in northern Alaska (Figs. 1, 2). Outcrop exposures of the Prince Creek Fm. have, to date, been primarily the focus of paleontological studies to examine and collect dinosaur specimens because the formation contains the richest concentration of dinosaur bones of any high-latitude location in the world (Rich et al. 1997, 2002). Although rigorous documentation of vertebrate and plant fossils of the Prince Creek Fm. has resulted in ground-breaking discoveries regarding Alaskan dinosaurs and biota (e.g., Parrish et al. 1987; Parrish and Spicer 1988; Clemens and Nelms 1993; Fiorillo and Gangloff 2000; Gangloff et al. 2005; Fiorillo et al. 2009; Fiorillo et al. 2010), and although a number of local sedimentologic studies have focused on areas near these vertebrate bonebeds (Phillips 1989, 2003; Brandlen 2008; Fiorillo et al. 2010), no comprehensive regional outcrop study had yet been undertaken. A comprehensive depositional model for the Prince Creek Fm. is required to: (1) place current and future paleontological discoveries in a regional paleoenvironmental and geomorphologic context within which a high-latitude dinosaur ecosystem thrived during the Late Cretaceous, and (2) improve our understanding of facies relationships, sediment-body geometries, alluvial architecture, and vertical and lateral variations within a high-latitude coastal-plain succession.

The purpose of this paper, therefore, is to examine the facies and alluvial architecture of the Prince Creek Fm. in order to (1) document the various depositional environments, (2) determine the type of alluvial system(s) present, and (3) develop a depositional model for the Prince Creek Fm. in the Colville River region during the Late Cretaceous.

GEOLOGIC SETTING AND PREVIOUS WORK

The Colville Basin is a foreland basin on the North Slope of Alaska (Fig. 1). Jurassic to Early Cretaceous subduction led to northward emplacement of allochthons and uplift of the Brooks Range orogenic belt (Moore et al. 1994). Additional crustal loading in the Aptian to Albian, driven by continued convergence, initiated lithospheric flexure and subsidence to the north of the Brooks Range, producing the east–west-trending Colville Basin (Moore et al. 1994; Cole et al. 1997). Uplift and erosion of the evolving mountain belt drove sediments from the west into the foredeep, resulting in a predominantly axial basin fill (Molenaar 1985; Mull 1985; Moore et al. 1994; Mull et al. 2003; Decker 2007). Additional sediments entered the basin transversely from the south (Moore et al. 1994). The Colville Basin continued to fill throughout the Late Cretaceous and Tertiary, preserving in excess of 4,000 m of sediment at some locations (Molenaar et al. 1987).

The Prince Creek Fm. (Fig. 3) is a coastal-plain succession composed of conglomerate, sandstone, siltstone, mudstone, carbonaceous shale, coal, bentonite, and tuff (Roehler 1987, Mull et al. 2003; Flores et al. 2007a; Flores et al. 2007b; Fiorillo et al. 2009, Fiorillo et al. 2010). Deposits of the Prince Creek Fm. are the most proximal in a predominantly progradational, Late Cretaceous to Paleocene continental to marine succession that also includes shallow to marginal marine facies of the Schrader Bluff Fm., slope to deepwater facies of the Canning Fm., and the distal Hue Shale (Fig. 3; Mull et al. 2003; Decker 2007). The Prince Creek Fm. was originally subdivided into two clastic tongues: the Turonian–Coniacian Tuluvak Tongue, and the

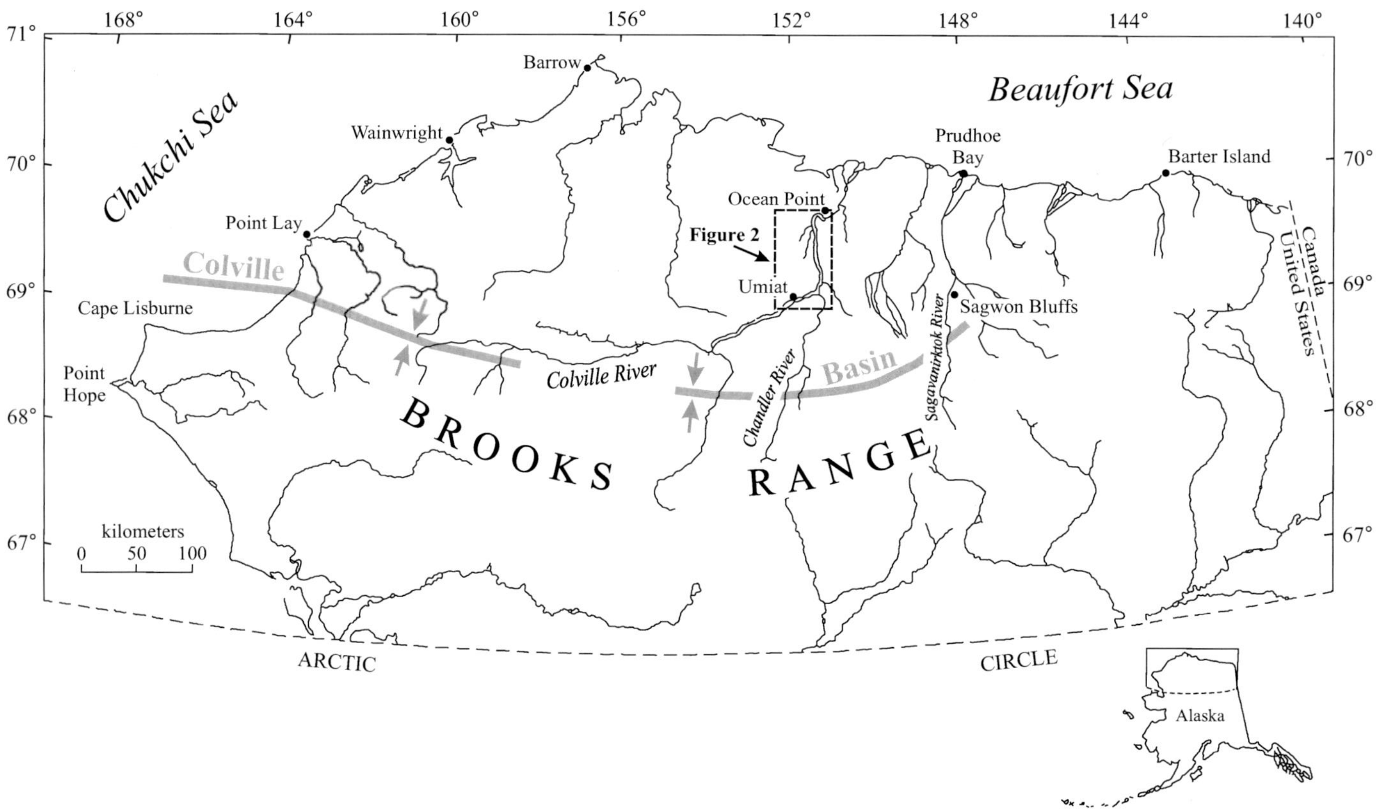

FIGURE 1.—Northern Alaska above the Arctic Circle, including the Brooks Range, the Colville Basin, and the location of the study area depicted in Figure 2 (Modified from Fiorillo and Gangloff 2000, and Bird 2001).

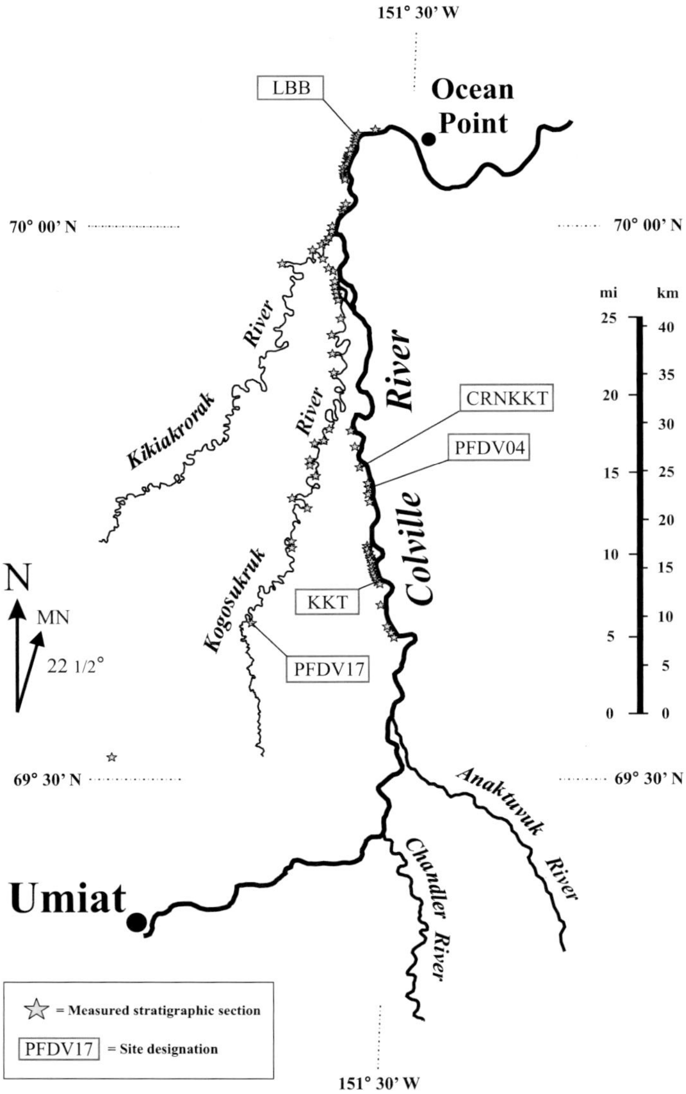

FIGURE 2.—Study area including rivers, measured stratigraphic sections, and the location of sites discussed in the text. MN = magnetic north.

Campanian–Maastrichtian Kogosukruk Tongue (Gryc et al. 1951). Mull et al. (2003) revised this nomenclature by removing the Tuluvak Tongue from the Prince Creek Fm. and raising the Tuluvak to formation status. The Prince Creek Fm. is now redefined to include only the former Kogosukruk Tongue and Paleocene strata originally assigned to the Sagwon Member of the Sagavanirktok Fm. at Sagwon Bluffs (Figs 1, 3).

No surface exposure containing a complete succession of the Prince Creek Fm. exists (Mull et al. 2003), and the total thickness along the Colville River can only be estimated from outcrop studies at ~ 450 m (Detterman et al. 1963; Brosge et al. 1966). Strata along the Colville River have a gentle structural dip (3° or less) to the northeast and locally contain minor faults and folds. Estimates for depth of burial in the study area from downhole vitrinite reflectance profiles and sonic-porosity logs range from ~ 2000 to 6000 ft. (~ 600-1800 m) (Burns et al. 2005).

The entire formation generally fines upward from a conglomeratic base (Flores et al. 2007b) to muddier, finer-grained facies at the top. In surface exposures along the Colville River near the confluence of the Chandler and Anaktuvuk Rivers (Fig 2), the base of the Prince Creek Fm. is erosional, with fluvial channel sandstones and conglomerates incised into shallow-marine sandstones of the lower Schrader Bluff Fm. (Mull et al. 2003; Flores et al. 2007b). Locally, this contact is interpreted as an unconformity beneath an incised paleovalley (Flores et al. 2007a; Flores et al. 2007b) and correlates regionally with a submarine mass-wasting and erosion surface (mid-Campanian unconformity of Decker 2007). At Ocean Point (Fig. 2), the contact between the top of the Prince Creek Fm. and the overlying upper Schrader Bluff Fm. is transgressive between alluvial and interdistributary-bay facies of the Prince Creek Fm. and shallow marine shoreface deposits of the upper Schrader Bluff Fm. (Phillips 2003). In our study area, the Prince Creek Fm. is truncated by Pliocene to Pleistocene shoreface sands of the shallow marine Gubik Fm. (Black 1964) or the Holocene unconformity.

Palynomorphs (Frederiksen et al. 1986, 1988, Frederiksen et al. 2002; Frederiksen 1991; Frederiksen and McIntyre 2000; Flores et al. 2007b; Brandlen, 2008; Fiorillo et al. 2010, unpublished data), floral and faunal evidence (Parrish and Spicer 1988; Brouwers and de Deckker 1993), and K/Ar and $^{40}Ar/^{39}Ar$ analyses (Conrad et al. 1990; unpublished data) indicate that the age of the Prince Creek Fm. ranges from Campanian to Paleocene; however, all of the deposits in our study area are Early Maastrichtian (Flores et al. 2007b; Fiorillo et al. 2010, unpublished data).

There is no modern analogue for an ecosystem that experienced relatively warm temperatures under a polar-light regime (Spicer 2003). The Late Cretaceous Arctic was nearly devoid of polar ice caps and supported luxuriant high-latitude forests (Spicer et al. 1992; Spicer 2003). Although the present latitude of strata in the study area is ~ 70° N, earlier work suggests that the Prince Creek Fm. may have been deposited at paleolatitudes as high as 82°–85° N (Brouwers et al. 1987; Witte et al. 1987; Besse and Courtillot 1991; Rich et al. 2002, Spicer and Herman 2010). The Cretaceous was a time of global warmth (Parrish et al. 1987; Zakharov et al. 1999; Spicer et al. 1992; Deconto et al. 2000; Spicer 2003) during a greenhouse period in Earth's history (Fischer 1981; Nordt et al. 2003; Gallagher et al. 2008). Temperature estimates for the North Slope of Alaska during the Late Cretaceous (Maastrichtian) based on megaflora suggest that the warmest month mean was 10–12° C with a coldest month mean of 2–4° C (Brouwers et al. 1987). Overall, mean annual temperatures for the North Slope were likely ~ 5° C (Spicer and Parrish 1987; Parrish and Spicer 1988; Spicer 2003; Fiorillo et al. 2009; Fiorillo et al. 2010). Mean annual precipitation is estimated at between 500 mm a^{-1} and 1500 mm a^{-1} (Spicer and Parrish 1990; Brandlen 2008). These estimates are in contrast to modern records for the North Slope of Alaska (Haugen 1982; Spicer 2003) indicating a mean annual temperature of −12° C at Prudhoe Bay and mean annual precipitation of 140 mm a^{-1} at Sagwon Bluffs (Fig. 1).

Characteristics of paleosols reveal that the coastal-plain was wet with periods of (seasonal?) drying (Brandlen 2008; Fiorillo et al. 2009; Fiorillo et al. 2010). Vegetation consisted mainly of deciduous conifers (*Taxodiaceae*), pine (*Pinaceae*), horsetails (*Equisetites*), and ferns (*Omundaceae, Polipodiaceae, Cyantheaceae*) (Spicer and Parrish 1987; Spicer 2003; Brandlen 2008; unpublished data). Abundant charcoal indicates that forest fires were common during periodic dry periods (Spicer and Parrish 1987; Spicer 2003). No evidence of cryoturbation or ground ice has been found in Late Cretaceous sediments or paleosols of the Prince Creek Fm.; however, growth ring patterns and false rings in fossilized wood suggest that limited seasonal freezes probably occurred (Spicer et al. 1992; Spicer 2003). Elevations in the Brooks Range likely exceeded 1500 m, with mean annual temperatures below 0° C at high elevations (Spicer 2003) that likely resulted in at least seasonal snow and ice (Spicer and Parrish 1990; Fiorillo et al. 2009).

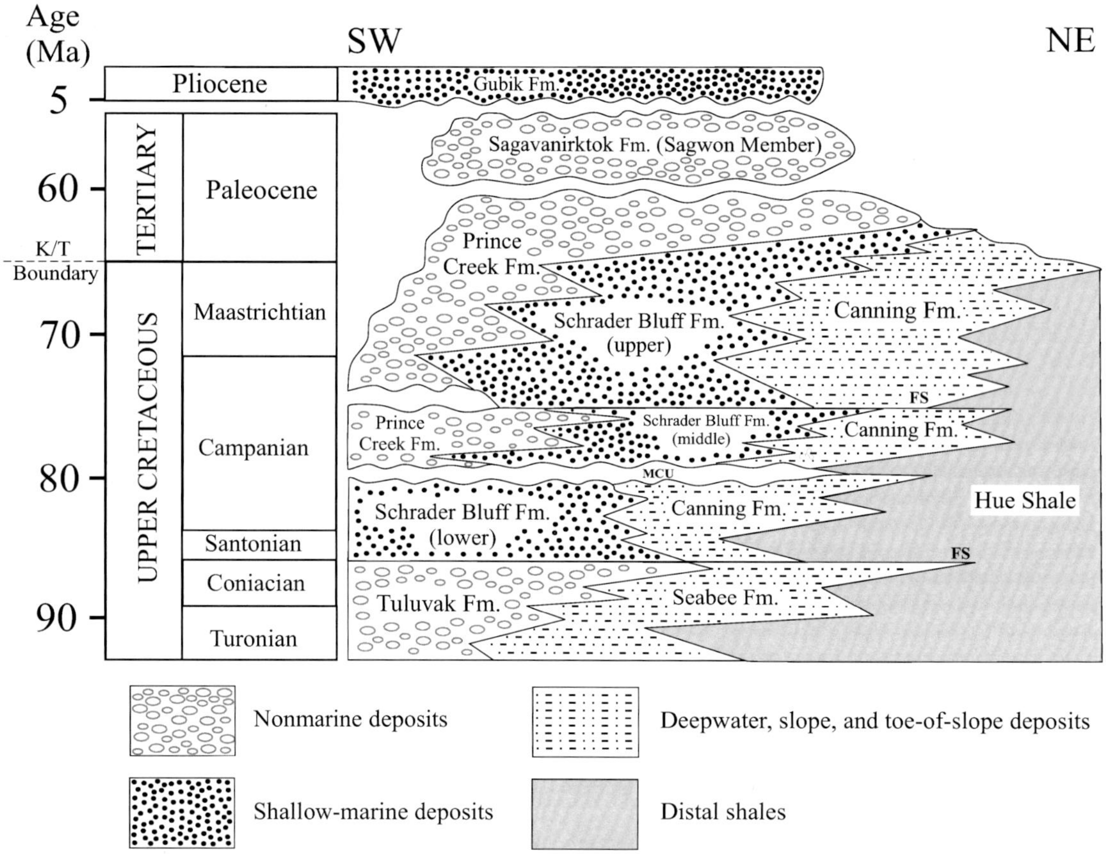

FIGURE 3.—Generalized chronostratigraphic column for the central North Slope of Alaska. Abbreviations include: FS (flooding surface), MCU (mid-Campanian unconformity), and K/T, (Cretaceous/Tertiary boundary). Revised from Mull et al. (2003), Garrity et al. (2005), and Decker et al. (in press).

METHODS

Fieldwork was conducted along the Colville, Kogosukruk, and Kikiakrorak rivers (Figs. 1, 2) during July and August 2005–2007. Site selection was limited to exposures from immediately south of the Kikak–Tegoseak dinosaur quarry to northeast of the Liscomb dinosaur quarry (KKT and LBB in Fig. 2) in order to avoid duplication of previous work on the Prince Creek Fm. (Phillips 2003; Flores et al. 2007b).

Sites were selected in order to obtain a representative sample of the regional stratigraphy and on the basis of accessibility of outcrop, complexity of alluvial architecture, observable flora and fauna, and regional geographic and stratigraphic relationships. 75 measured sections, including detailed descriptions of grain size, lithology, contacts, sedimentary structures, flora, fauna, and vertical and lateral facies relationships form the data set for this paper (Fig. 2). All exposures were photographed to generate photomosaics that were modified into interpretive line drawings of alluvial architecture. Interpreted photomosaics were compared with measured sections, paleoflow orientations, and hand samples for consistency during architectural analysis and paleoenvironmental interpretation.

FACIES AND FACIES ASSOCIATIONS

Thirteen facies are defined on the basis of grain size, thickness, sedimentary structures, the nature of upper and lower contacts, and flora and fauna (Table 1). These thirteen facies define four facies associations (FA) on the basis of facies relationships and stacking patterns (Table 2, Fig. 4). Facies associations record sediment deposition in three distinct types of channels (FA-I, FA-II, FA-III) and on floodplains (FA-IV).

Facies Association I: Large Sinuous Channels

FA-I is an erosionally based, 13–17-m-thick succession that fines upward from a basal conglomerate and medium-grained multistory sandbody through an interval of IHS into organic-rich siltstone, carbonaceous shale, coal, and mudstone at the top (Fig. 5A). FA-I consists of a combination of facies that may include F-1, F-2a, F-3, F-6a, F-6b, F-7a, F-7b, F8, and F-9 (Tables 1, 2; Fig. 4).

At the base of each FUS in FA-I is a 0.2–0.5-m-thick conglomerate lag (F-1; Fig. 5B). The basal surface of the conglomerate is erosional with relief of up to 13 m or more. The succession fines upward from F-1 into a 7–9-m-thick, medium-grained, multistory sandbody (F-2a, Fig. 5A) containing cosets of trough cross-lamination with individual sets averaging 0.5 m thick (Fig. 5C). Sandbodies of FA-I are conspicuously devoid of root traces and invertebrate fossils but contain convolute bedding (Fig. 5D), scour-and-fill structures, spherical mudstone balls, and coalified wood (Fig. 5E). F-2a grades upward into 2–5 m of IHS (F-3; Fig. 5A). Inclined beds in the IHS dip at angles of 5–10° and tangentially onlap the top of F-2a (Fig. 5A). Paleoflow recorded from trough cross-lamination in the IHS is typically at a high angle relative to the dip direction of the inclined beds. The heterolithic succession passes gradationally upward into a 3–4-m-thick succession of finer-grained facies that typically includes ripple cross-laminated to massive rooted siltstone (F-6b), blocky rooted mudstone (F-7b), carbonaceous shale (F-8), and coal (F-9).

Interpretation: The FUS of FA-I is interpreted as the channel thalweg and point-bar deposits of large, sinuous, suspended-load meandering channels (*sensu* Allen 1964, 1965a, 1965b, 1970; Schumm 1968, 1985; Leeder 1973; Elliott 1976; Smith 1987, 1988). Erosionally based, fining-upward successions containing basal lags, trough cross-laminated sandbodies, lateral-accretion surfaces, and fine-grained upper facies are typical of meandering systems (Allen 1965a, 1970; Elliott 1976; Walker 1979; Nanson 1980; Ethridge et al. 1981; Smith 1987; Choi et al. 2004). Conglomerate lags on scoured surfaces in FUSs are commonly interpreted as channel thalweg deposits (Allen 1964, 1965b, 1970; Nanson 1980; Plint 1983). Sets of trough cross-lamination record migration of 3-D dunes along the bed of the channel (Allen 1963, 1965b; Elliott 1976; Plint 1983; Choi et al. 2004). The multistory nature of the channel fill reflects scouring within the channel during flooding events, and bar-form migration (Elliott 1976, Gibling 2006, McLaurin and Steel 2007). IHS dipping normal or at high angles to paleoflow is interpreted as lateral-accretion surfaces on point bars that record the migration of the channel as it wandered across the floodplain (Allen 1963, 1965a; Elliott 1976; Collinson 1978; Nanson 1980; Thomas et al. 1987). Convolute bedding in sandbodies indicates high sedimentation rates with shearing of sediments by current drag while the sediments were still liquefied (Banks 1973). Scour-and-fill structures are interpreted as chute channels on point bars (Plint 1983). Spherical mudstone balls form in channels with high suspended-sediment concentrations by the erosion of both mud drapes on point bars and fine-grained channel margins (Baker et al. 1995; Dalrymple and Choi 2007). Large sinuous channels are interpreted as suspended-load channels on the basis of relatively high concentrations of mud in the upper portions of FUSs and thick mud drapes on point-bar surfaces (Schumm 1968; Stewart 1983). Drifted logs, leaves, wood, and plant fragments are common in point-bar sands (Allen 1965b). Finer-grained facies (F-6b, F-7a, F-7b, F-8, F-9) record vertical accretion at the tops of point-bars at shallow depths and low bed shear stresses (Allen 1965a, 1970; Elliott 1976; Nanson 1980). The absence of root traces in sandbodies and in the IHS of FA-I suggest that these larger channels experienced continuous flow conditions which did not allow vegetation to root in these relatively deep channels.

Facies Association II: Small Sinuous Channels

FA-II is a 2.0–6.0-m-thick FUS composed predominantly of a single-story, 1.5–5.0-m-thick sandbody dominated by IHS (Fig. 6A). FA-II contains a combination of facies which may include F-2b, F-3, F-6b, F-7a, F-7b, F-8, and F-9 (Tables 1, 2; Fig. 4).

The base of FA-II is erosional with relief of up to 6.0 m (Figs. 6A, B, D), and there is no conglomerate above the basal erosion surface. Instead, fine-grained, trough cross-laminated sandstone (F-2b), 0.3–1.0-m-thick, commonly containing coalified wood and mudstone clasts, is typically found at the base (Figs. 6A, 6B, 6C). Crossbed set thicknesses average 0.2 m. The thickness of this cross-bedded sand is highly variable, and it may be locally absent; however, where present it is thinner and finer grained than the medium-grained sandbody (F2a) found near the bases of large sinuous channels (FA-I, Fig. 5A).

Lying gradationally above this sand or directly above the basal erosion surface is a 1.5–5.0-m-thick FUS mostly containing IHS composed of alternating fine-grained sand and silt or mud (F-3; Figs. 6A, D). The inclined beds that make up the IHS generally appear to dip at higher angles (max. 20°) than the IHS in large sinuous channels (FA-I) (max. 10°). Both the coarse and fine members of IHS couplets in FA-II contain abundant vertical to subhorizontal root traces up to 30 cm long and 2.0 cm in diameter (Figs. 6E, F). The IHS typically passes gradationally upward into a thin (< 0.5 m) succession of rooted mudstone (F-7b), carbonaceous shale (F-8), and coal (F-9). In many instances the entire IHS succession grades laterally into an asymmetrical, trough-shaped feature filled with rooted mudstone and siltstone overlying a concave-up erosion surface (Fig. 6D).

Interpretation: The FUS of FA-II is interpreted as point-bar deposits of small, sinuous meandering channels. High mud contents in these channels suggest that they carried most of their sediment in suspension and are best classified as suspended-load streams (Schumm 1968). Meandering is evidenced by prominent IHS with paleoflow orientations at high angles relative to the dip of inclined beds, indicating lateral accretion of channels (Allen 1963; Collinson 1978; Thomas et al. 1987). Fine-grained facies at the top of FA-II record vertical accretion at the tops of point bars (F-6b, F-7a, F-7b, F-8, F-9). Asymmetrical, trough-shaped features filled with rooted and current-rippled organic-rich siltstone and mudstone are similar in geometry, lithology, and facies relationships to mud-filled abandoned channels formed during neck cutoff (Fisk 1944; Collinson 1978; Chevren 1978; Ethridge et al. 1981; Plint 1983). Mud-filled abandoned channels and laterally extensive point-bar deposits are common in sinuous meandering systems (Allen 1965b; Schumm 1985; Miall 1985).

Facies Association III: Small Low-Sinuosity Channels

FA-III is composed of a FUS that is thinner than the majority of FUSs of small sinuous channels (FA-II) and considerably thinner than those of large sinuous channels (FA-I). Small low-sinuosity channels of FA-III are dominated by a current-rippled sandbody overlying an arcuate, concave-up erosional base. FA-III contains a limited number of facies including F-2b (extremely rare), F-3, F-4, F-7a, F-7b, and F-8 (Tables 1, 2; Fig. 4).

FA-III is composed predominantly of a single-story, 1.5–3.0-m-thick, very fine- to fine-grained sandbody (F-4) that sits atop a concave-up basal erosion surface with relief of up to 3 m (Figs. 7A, B). Ripple cross-lamination is the principal sedimentary structure, although, rarely, trough-cross-laminated sand (F-2b) can be found between the current-rippled sand (F-4) and the basal erosion surface. These sandbodies are ubiquitously rooted. IHS (F-3) is largely absent from the FUS and, where present, is found only at the top of the channel fill (Fig. 7C). The sandbody of FA-III typically fines upward into a thin layer of rooted, organic-rich mudstone (F-7b) and carbonaceous shale (F-8). At some locations "wings" composed of horizontally bedded, very fine-grained, ripple cross-laminated sandstone or siltstone < 0.5 m thick extend laterally from the uppermost portion of the channel fill (Fig. 7A). Rarely, small low-sinuosity channels of FA-III may be devoid of sand and, instead, are filled with rooted and ripple cross-laminated mudstone and tuff. In one case, two smaller channels were found to have scoured into the top of a larger channel-fill succession of FA-III (Fig. 7D).

Interpretation: FA-III is interpreted as a laterally stable, ribbon sandbody (*sensu* Friend et al. 1979; Friend 1983) similar in facies and geometry to anastomosed channels (Smith 1976; Smith and Smith 1980; Smith and Putnam 1980; Knighton and Nanson 1993; Eberth and Miall 1991; Nadon 1994; Stouthamer 2001; Makaske 2001). Many of these channels are interpreted as (anastomosed?) crevasse channels that form part of larger splay complexes (Smith 1983; Fiorillo et al. 2010).

Most of FA-III is composed of a single, vertically accreted sandbody that lacks predominant lateral-accretion surfaces. The only evidence of lateral accretion is found near the top of the channel-fill sequence (Fig. 7C). Minor lateral-accretion surfaces have been reported in channels interpreted as anastomosed (Kirschbaum and McCabe 1992; Törnqvist et al. 1993; Smith and Pérez-Arlucea 1994; Makaske 2001; Abbado et

TABLE 1.—Description and interpretation of the thirteen facies identified in the Prince Creek Formation, North Slope, Alaska

Facies		Thickness	Contacts	Physical Sedimentary Structures & Minerals
F-1—Conglomerate		< 0.5 m	Basal Erosional Upper Gradational	Common: Quartz and chert pebbles, cobbles, and boulders up to 30 cm long along longest axis, trough cross-lamination, mud and coal rip-up clasts
F-2—Trough cross-laminated Medium- to fine-grained sandstone	F-2a—Medium-grained	2.0–9.0 m	Basal Erosional to gradational Upper Gradational	Common: Multistory trough cross-lamination, ripple cross-lamination, mud and coal rip-up clasts, mud balls, siderite nodules, convolute bedding Rare: Cut-and-fill structures. Tuffaceous concretions, quartz and chert pebbles
	F-2b—Fine-grained	0.3–1.0 m	Basal Erosional to gradational Upper Gradational	Similar to F-2a but Trough cross-laminations are single-storey
F-3—Very fine-to fine-grained heterolithic ripple-cross-laminated sandstone, siltstone, and mudstone		1.5–6.0 m	Basal Erosional Upper Sharp to gradational	Common: Sand-mud or silt-mud couplets (IHS), ripples, reddish-orange mottles, siderite concretions, jarosite. Paleoflow indicators at high angles to dip of inclined bed Rare: Mud and coal rip-up clasts. Tuffaceous concretions
F-4—Very fine- to fine-grained ripple cross-laminated to massive sandstone		0.05–3.0 m	Basal Erosional to gradational or deformed Upper Sharp to gradational or deformed	Common: Ripples, reddish-orange mottles, siderite nodules, mud drapes on ripples Rare: Soft-sediment deformation, rip-up clasts, carbonate concretions, tuffaceous concretions, jarosite, burrows (<15 mm diam.)
F-5—Very fine- to fine-grained plane parallel-laminated sandstone and siltstone		0.2–1.5 m	Basal Sharp to gradational or deformed Upper Sharp to gradational or deformed	Common: Plane-parallel stratification. Primary current and parting lineation. Rare: Ripples, jarosite
F-6—Siltstone	F-6a—Extremely rare to absent root traces	0.6–5.0 m	Basal Sharp to gradational Upper Sharp to gradational	Common: Ripples Rare: Siderite nodules, burrows (<15 mm diam.), jarosite, vivianite
	F-6b—Common root traces	0.1–8.0 m	Basal Sharp to gradational or deformed Upper Sharp to gradational or deformed	Common: Reddish-orange mottles, siderite nodules, jarosite, vivianite Rare: Ripples, tuffaceous concretions
F-7—Mudstone	F-7a—Extremely rare to absent root traces	0.5–2.0 m	Basal Sharp to gradational or deformed Upper Sharp to gradational or deformed	Common: Reddish-orange to yellow mottles (to 10 cm diameter), jarosite, vivianite
	F-7b—Aggregated (blocky or platy), with common roots	0.5–4.0 m	Basal Sharp to gradational, or deformed Upper Sharp to gradational or deformed	Common: Reddish-orange to yellow mottles, siderite nodules, carbonate concretions Rare: Tuffaceous concretions, jarosite
F-8—Carbonaceous Shale		0.1–2.0 m	Basal Sharp to gradational or deformed Upper Sharp to gradational or deformed	Common: Jarosite
F-9—Coal		0.1–0.8 m	Basal Sharp to gradational Upper Sharp to gradational	Common: Mud interbeds, jarosite
F-10—Tuff/Bentonite		0.05–0.3 m	Basal Sharp or deformed Upper Sharp or deformed	Lacks sedimentary structures. Basal contact is frequently deformed.

TABLE 1.—*Extended.*

Flora / Fauna	Grain Size & Sediment Texture	Depositional Environment	Occurrence
Common: Carbonized or silicified wood fragments and plant fragments, wood impressions, silicified vertebrate bone fragments, coalified logs.	Clast to matrix-supported. Well-rounded pebble to boulder-size clasts of quartz, chert, and minor quartzite rock fragments. Matrix of medium-to coarse-grained subrounded to subangular quartz and chert rich sand.	Channel thalweg highest-energy bedload-transport deposit.	Found only at base of FA-I
Common: Silicified or carbonized wood and plant fragments, wood impressions, silicified vertebrate bone fragments. Rare: Silicified or coalified logs, tree stumps (max. diam. 18 cm). Vertebrate bone up to 0.5 m long	Medium-grained, moderately well-sorted subrounded to subangular quartz and chert rich sand.	3-D dunes deposited in channel thalweg on point-bar toes.	FA-I above F-1
Same as F-2a except no large vertebrate bones, logs, or stumps	Moderately well-sorted, subrounded to subangular quartz and chert rich fine-grained sand.	Same as F-2a but found in smaller channels	At base of FA-II and very rarely FA III
Common: Roots (range from 5 to 30 cm-long and .02 to 2 cm diameter). Carbonized wood and plant fragments, coalified or silicified logs (max. diam. 20 cm).	Alternating cm-to dm-scale couplets of fine-grained sand and silt or mud	Laterally accreting point bars in meandering channel-fill successions	Above F-2 in both FA-I and FA II and at the top of FA-III
Common: Carbonized roots up to 30 cm long, Carbonized wood and plant fragments. Rare: Silicified vertebrate bone and bone fragments, logs, clams, gastropods	Typically composed of homolithic very fine-grained sand. May contain glass shards and may weather white.	Anastomosed channel fill and small-scale crevasse splays	At base of FA-III and in FA-IV
Common: Carbonized wood and plant fragments	Alternating poorly sorted very fine-grained sand and organic silt	Channel-proximal levees and sandy lakes	FA-IV
Common: Carbonized or silicified wood and plant fragments, coalified logs Rare: Clams, gastropods	Predominantly silt with interspersed clay-size particles and organic material	Deepest floodplain lakes (e.g., oxbow lakes) and large channel abandonment (point-bar)	FA-IV and upper FA-I
Common: Carbonaceous root traces, carbonized or silicified wood and plant fragments. Rare: Silicified vertebrate bones and bone fragments, pelecypods, gastropods, amber, burrows	Similar to F-6a but includes both macroscopic and microscopic vertically oriented, branching carbonaceous root traces	Distal levees, distal crevasse splays, lake margins, paleosols	FA-IV and near top of FUS in FA-I and FA-II
Common: Carbonized plant fragments Rare: Clams, gastropods, burrows, bioturbation	Massive mudstone with interspersed organics	Shallow floodplain lakes and ponds, and some channel abandonment successions (point-bar)	FA-IV and near top of FUS in FA-I, FA-II, and FA-III
Common: Roots (range from 1 to 20 cm long and .01 to 0.5cm thick), carbonized plant fragments	Mudstone with a blocky to platy structure and interspersed organics	Paleosols	FA-IV and top of FUS in FA-I and FA-II
Common: Carbonized wood and plant fragments, coalified logs. Rare: amber	Clay with shale partings.	Mires	FA-IV and near top of FUS in FA-I and FA II
Common: Carbonized wood and plant fragments, coalified logs	Composed entirely of carbonized plant fragments. Brittle with common vitreous luster.	Backswamps	FA-IV and near top of FUS in FA-I and FA-II
Large roots up to 30 cm long and 5 cm diameter. May contain embedded carbonaceous plant or wood fragments	Massive. Composed of a mix of unaltered and altered glass shards and clay. Weathers white or light green.	Ashfall deposits	FA-IV

TABLE 2.—Description and interpretation of the four facies associations identified in the Prince Creek Formation.

Facies Association	Paleo-environment	Inclusive Facies	Diagnostic Features	Occurrence
FA-I	Large Sinuous Channel	F-1,F-2a, F-3, F-6a, F-6b, F-7a, F-7b, F8, F-9	13–17-m-thick fining-upward succession capped by organic siltstone, carbonaceous shale, coal, or paleosol	Regionally restricted FUS deposited in larger meandering trunk channels
FA-II	Small Sinuous Channel	F-2b, F-3, F-6b, F-7a, F-7b, F-8, F-9	2–7-m-thick fining-upward succession. Lenticular sandbody containing inclined heterolithic stratification (IHS) indicative of lateral accretion. Abundant mud-filled abandoned channels. Often capped by thin (< 1 m thick) organic shale, coal, or paleosol.	Most common type of channel deposit. Deposited in smaller meandering distributary channels. Often associated with FA-III. Interdigitates with FA-IV.
A-III	Small Low- Sinuosity Channel	F-2b, F-3, F-4, F-7a, F-7b, F-8	1–3-m-thick concave-up ribbon-form channel. IHS may occur in top 0.5 m. Deposits indicate primary vertical accretion and minor lateral accretion.	Frequent deposit of small, fixed (anastomosed?) distributary channels. May incise into FA-II or FA-IV.
FA-IV	Interbedded Floodplain Sand, Silt, and Mud	F-4, F-5, F-6a, F-6b, F-7a, F-7b, F-8, F-9, F-10	Variable thickness. Contains the bulk of the fine-grained sediments in the study area. Often found as laterally extensive organic-rich successions encasing FA-I, FA-II, and FA-III	Extremely common. Includes all overbank sedimentation, abandoned channel fill, and soil-forming environments

al. 2005), suggesting that some lateral migration of anastomosed channels is possible (Thomas et al. 1987), however vertical accretion dominates over lateral accretion in anastomosing systems (Smith and Smith 1980; Smith 1983; Eberth and Miall 1991; Nadon 1994; Makaske 2001). The "wings", which extend laterally outward from the tops of some of these ribbon-form channels, are commonly described from channels interpreted as anastomosed (Smith and Smith 1980; Smith 1983; Eberth and Miall 1991; Kirschbaum and McCabe 1992; Nadon 1994; Kraus 1996). These "wings" are interpreted as levees which help to stabilize channels and prevent lateral migration (Kraus 1996; Nadon 1993, 1994; Makaske 2001; Heritage et al. 2001). Anastomosed channels remain largely "fixed" in position on the floodplain owing to low stream power, low gradients, and bank stability resulting from abundant silt, mud, and vegetation (Smith 1976; Kirschbaum and McCabe 1992; Törnqvist 1993; Törnqvist et al. 1993; Nadon 1994; Abbado et al. 2005; Guow and Berendsen 2007). Anastomosis is common in low-gradient systems containing abundant fine-grained material and high concentrations of suspended load (Smith 1983; Smith et al. 1989).

Facies Association IV: Interbedded Sandstone, Siltstone, and Mudstone

FA-IV contains finer-grained facies which may include F-4, F-5, F-6a, F-6b, F-7a, F-7b, F-8, F-9, and F-10 (Tables 1, 2; Fig. 4). FA-IV lacks the coarser-grained facies of large sinuous channels (FA-I) and small sinuous channels (FA-II) and sedimentary structures indicative of high-energy environments. Sandbodies are typically < 2 m thick, nonchannelized, sheet-like, ripple-cross-laminated or planar-laminated, and coarsen upward or fine upward or both (Figs. 8A, 8B). Contacts with underlying strata are erosional-to-gradational or undulating-to-deformed, and lack significant relief. Burrows up to 15 mm in diameter are rare. Organic-rich sandstones and siltstones of FA-IV are found in a number of dinosaur quarries along the Colville River and contain abundant vertebrate bones and bone fragments (Phillips 2003; Fiorillo et al. 2009; Fiorillo et al. 2010). Many of these sandstones and siltstones overlie finer-grained, organic-rich facies (Fig. 8B). Some sheet-like sandbodies, when traced laterally, are attached to the margins

→

FIGURE 4.

Key

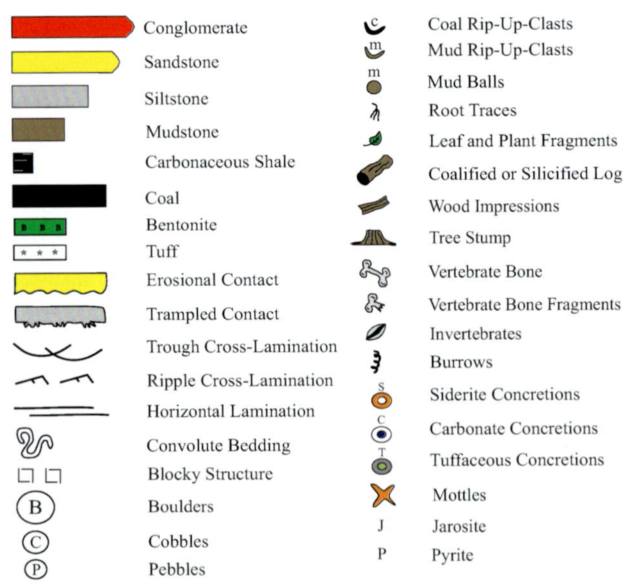

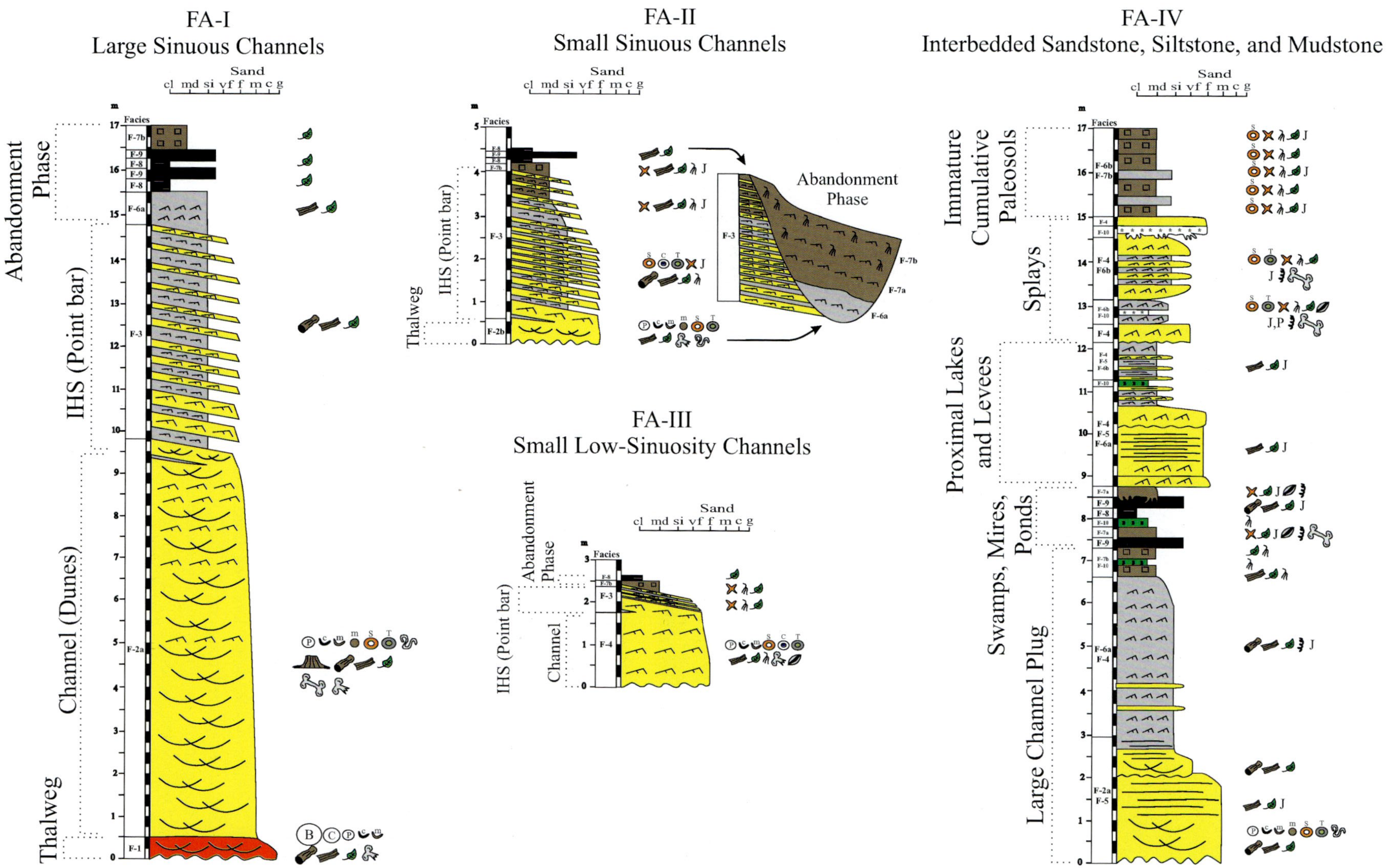

FIGURE 4.—Idealized measured sections for the four facies associations of the Prince Creek Formation found in the Colville River region. Facies Association I (FA-I), large sinuous channels; FA-II, small sinuous channels; FA-III, small low-sinuositycChannels; and FA-IV, interbedded sandstone, siltstone, and mudstone. Detailed descriptions of individual facies and facies associations are given in Tables 1 and 2.

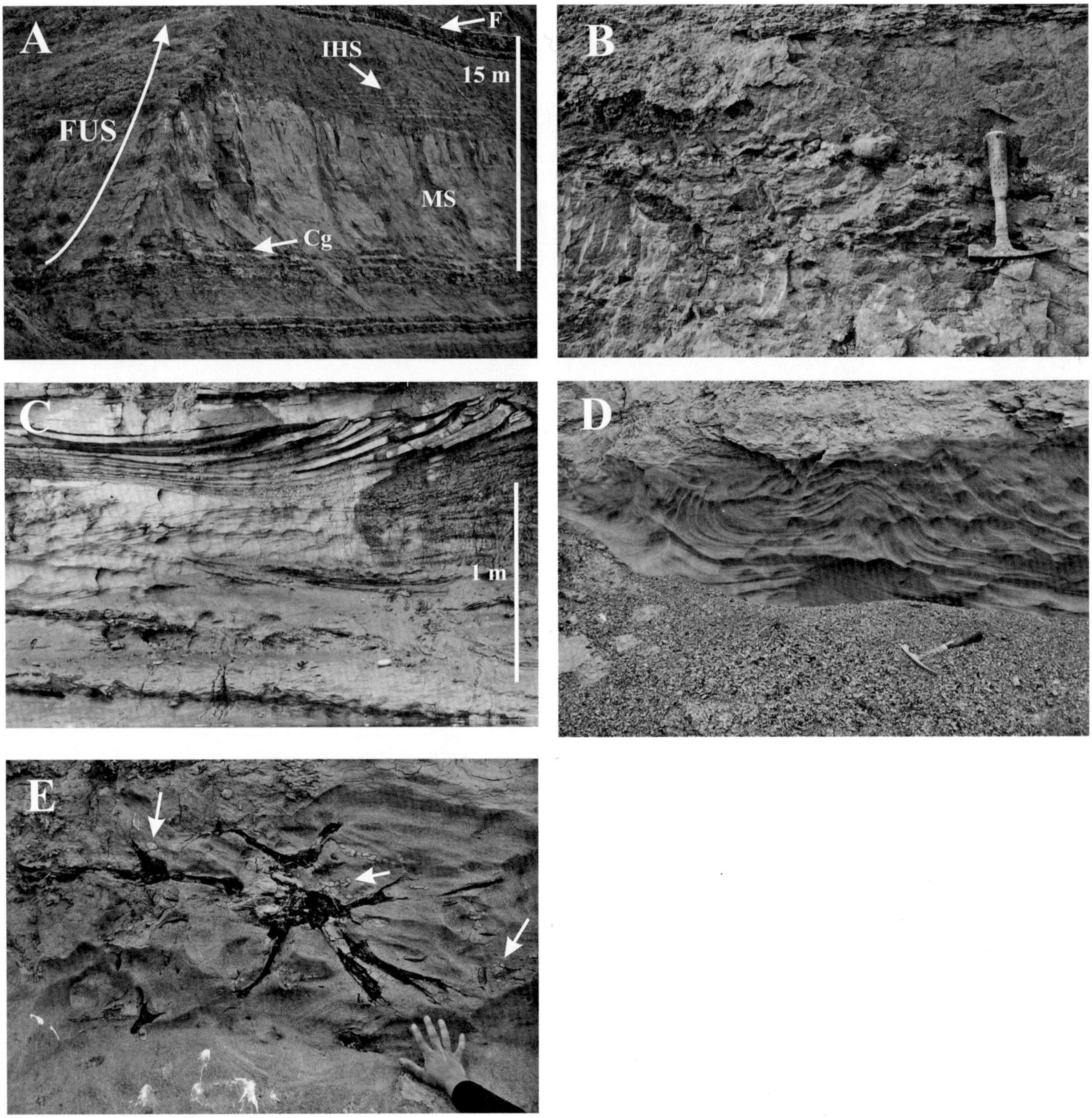

FIGURE 5.—Facies, physical sedimentary structures, flora, and fauna typical of FA-I (large sinuous channels). A) Complete fining-upward succession (FUS) including coarse-grained conglomeratic base (Cg), multistory medium-grained sandbody (MS), inclined heterolithic stratification (IHS), and fine-grained top (F); B) pebble-to-cobble conglomerate; C) cross-bed coset composed of 0.5-m-thick sets of trough cross-lamination; D) convolute bedding; E) carbonized tree (18 cm diam.) with roots and (arrows) spherical mudstone balls Hammer is 30 cm long.

of channels (Fig. 8C). Mudstones (F7; Figs. 8B, D, E) commonly overlie and underlie sheet sandbodies of FA-IV (F4, F5, Fig. 8A). At some locations siltstones and mudstones (F6, F7) contain clams, gastropods, snails, and rare burrows up to 15 mm in diameter (Fig. 8F). Jarosite halos on roots, jarosite mottles, and bluish vivianite mottles become more common up-section. Organic-rich shales (F8; Fig. 8B) are common, but true coals (always < 1 m thick) are relatively rare (F9) (Fig. 8H). Undulating-to-deformed basal contacts (Fig. 8H) and iron-

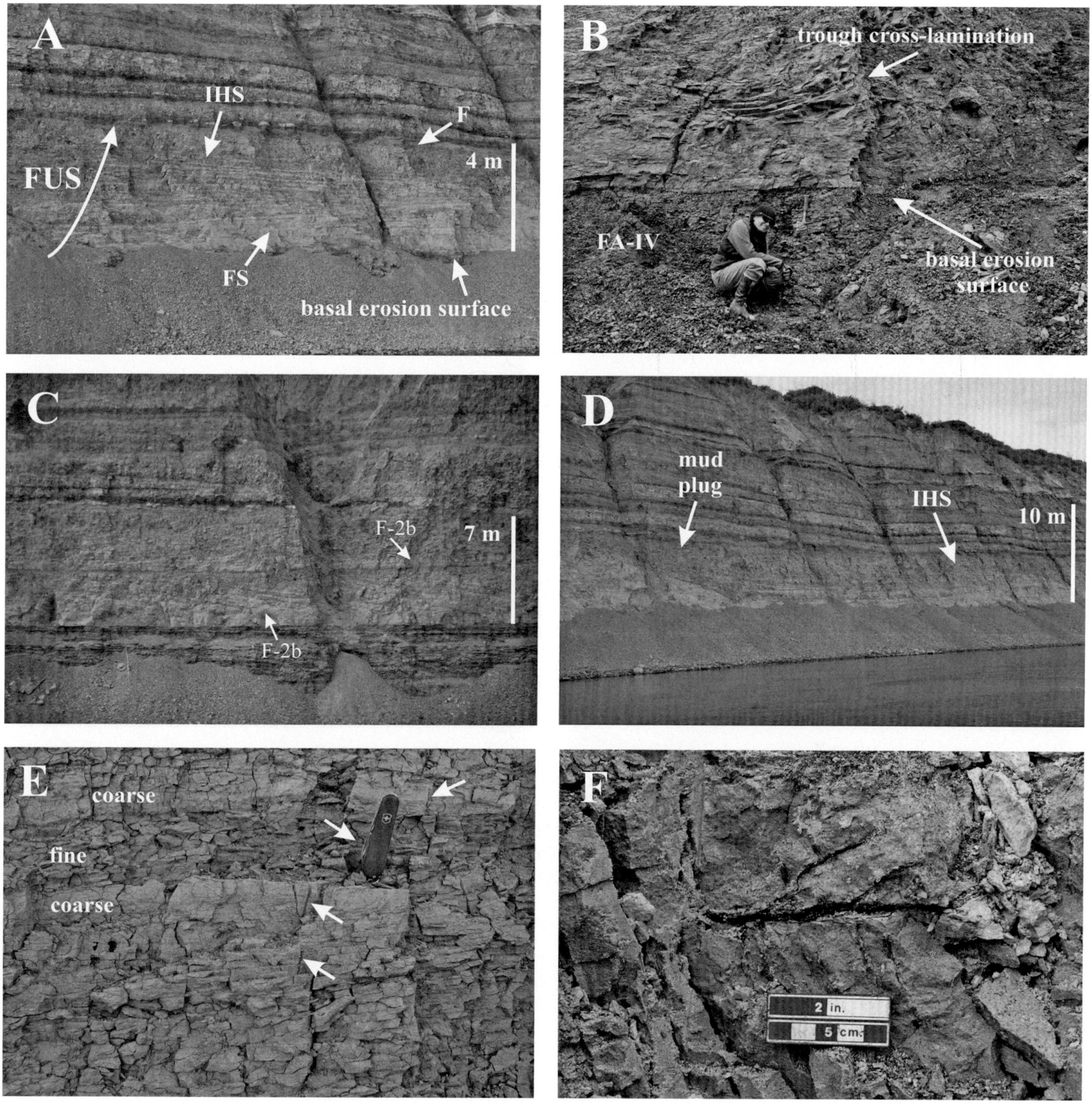

FIGURE 6.—Facies, physical sedimentary structures, and flora typical of FA-II (small sinuous channels). A) Complete fining-upward succession (FUS) including basal erosion surface, fine-grained sand base (FS), inclined heterolithic stratification (IHS), and fine-grained top (F); B) trough cross-lamination in a fine-grained sandbody in erosional contact with floodplain deposits of FA-IV; C) stacked channels with apparent dips in opposing directions; the stratigraphically lower channel contains coarse members of inclined heterolithic stratification (IHS) that thicken down-dip and merge with trough cross-laminated sand (F-2b); D) IHS grading laterally into mud-filled channel plug; E) alternating coarse and fine members of IHS couplets containing vertical carbonized roots (arrows); F) carbonized root 15 cm long and 0.5 cm in diameter exhibiting horizontal and vertical growth as well as branching. Pocket knife is 9 cm long.

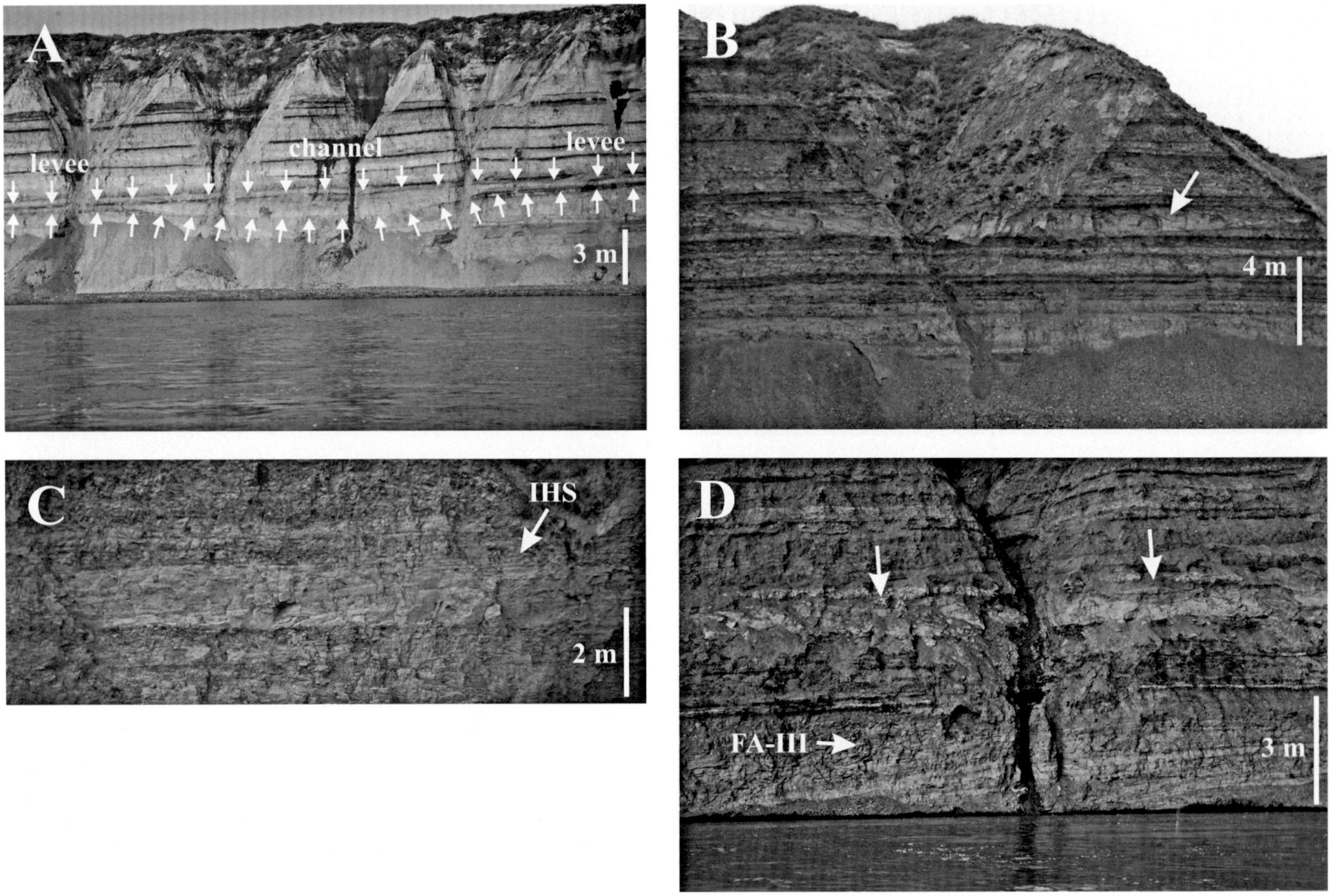

FIGURE 7.—Facies and physical sedimentary structures typical of FA-III (small low-sinuosity channels) A) channel-form (outlined by arrows) displaying arcuate shape and "wings" interpreted as channel-attached levees; B) channel (arrow) encased in floodplain facies of FA-IV; C) small low-sinuosity channel (FA-III) including minor inclined heterolithic stratification (IHS) at the top of the channel-fill sequence incised into a small sinuous channel of FA-II; D) small low-sinuosity channel (FA-III) containing two smaller channels (arrows) incised into the top of the channel-fill sequence; image also includes a small sinuous channel (FA-II).

oxide mottles are found in nearly all fine-grained facies. Thin (< 0.5 m) tuffs and bentonites (F10) may be locally interbedded with the other facies (Fig. 8G).

Interpretation: FA-IV is interpreted to represent floodplain deposition. Laterally extensive, thin, sheet-like, ripple cross-laminated or planar-laminated sandstones and siltstones with erosional-to-gradational or undulating-to-deformed basal contacts, abundant root traces, Fe-oxide mottles, and comminuted plant material overlying finer-grained organic facies such as carbonaceous shale, organic-rich mudstones or coal are interpreted as small-scale crevasse splays and channel-proximal levees (cf. Coleman et al. 1964; Elliott 1974; Bridge 1984; Fielding 1984; Kraus 1987; Mjøs et al. 1993; Smith and Pérez-Arlucea 1994; Changsong et al. 1995; Rajchl and Uličný 2005; Martín-Closas and Galtier 2005). Sandstones represent proximal splay deposits and higher flow velocities while siltstones represent distal splay deposits and waning flow (e.g., Pérez-Arlucea and Smith 1999; Rajchl and Uličný 2005). Similar deposits have also been attributed to levees (Allen 1965b; Fielding 1986; Pérez-Arlucea and Smith 1999; Stouthamer 2001; Michaelsen et al. 2000), although a levee interpretation is more convincing when thin, sheet-like sandstones and siltstones can be seen to thicken and grade laterally into channel deposits (e.g., Fig. 8C). Undulating-to-deformed basal contacts of thin sheet sandstones (Fig. 8H) suggest trampling by dinosaurs (e.g., La Porte and Behrensmeyer 1980; Loope 1986; Avanzini et al. 1997).

Lacustrine intervals in alluvial successions commonly contain fine-grained sandstone, siltstone, or mudstone that is typically planar-laminated or ripple cross-laminated (e.g., Fielding 1984; Astin 1990; Besley and Collinson 1991, Smith and Pérez-Arlucea, 1994; McCarthy et al. 1999; Tanner 2000). FUSs composed predominantly of 3–4 m of planar-laminated or current-rippled siltstone (F-6) that lacks evidence of roots but contains carbonized plant fragments, carbonized wood, and unusually abundant coaly logs and overlies 3–5 m of fine to medium-grained, planar-laminated (F-5) and trough-cross-laminated (F-2a) sandstone (Figs. 4, 8I) are interpreted as channel plugs of large sinuous channels (FA-I) formed after neck cutoff (e.g., Lewis and Lewin 1983; Miall 1996; Pérez-Arlucea and Smith 1999). Thinner packages of ripple cross-laminated and rarely rooted or burrowed siltstone and mudstone (typically < 1 m thick) that may contain brackish-water clams (*Nucula* aff. *N. percrassa* Conrad; Fig. 8F) along with indeterminate gastropods with spiral ornaments and indeterminate snails (personal communication, Robert Blodgett) are interpreted as small lakes and ponds that occupied topographic lows on the floodplain

(Pérez-Arlucea and Smith 1999). Often these siltstones and mudstones also exhibit undulating-to-deformed (trampled) basal contacts.

Organic-rich mudstone, carbonaceous shale, and coal were deposited in swamps, marshes, and mires (cf. Coleman et al. 1964; Tatsch, 1980; Smith and Smith 1980; Smith 1983; Flores 1984; Fielding 1984; Smith and Pérez-Arlucea 1994; Fielding 1987a; Pérez-Arlucea and Smith 1999). The presence of jarosite and rare vivianite indicate an oxygen-reduced, organic-rich environment (Postma 1977, Davies-Vollum 1999, Driese and Ober 2005) and may signal proximity to sulfate-rich marine or tidal waters (Wright 1986, Ufnar et al. 2001).

Carbonaceous root traces are ubiquitous in facies of FA-IV, with the exception of the rippled siltstone (F-6) found in channel plugs, indicating copious vegetation on the floodplain. The abundance of preserved carbonaceous plant material along with common siderite concretions suggests that many environments experienced anoxic or reducing conditions (Retallack 2001; Ashley et al. 2004; Johnson and Graham 2004).

Drab-colored, rooted siltstones and mudstones with common Fe-oxide mottles are interpreted as poorly drained paleosols (Fiorillo et al. 2010). Individual paleosol profiles are difficult to discern in the field (owing to frequent clastic influx and cumulative pedogenesis) and are similar to modern Inceptisols and Entisols (Fiorillo et al. 2009). Although drab colors, siderite concretions, and preserved organic matter suggest poorly drained conditions on floodplains, abundant rooted horizons and ferruginous features indicate periodically better-drained, or possibly seasonally dry conditions (Smith et al. 1989; Nadon 1994; Ashley et al. 2004; Johnson and Graham 2004; Brandlen 2008; Fiorillo et al. 2009).

PALEOFLOW

Paleoflow orientations recorded from channel fill sucessions are useful for determining basin geometries and sources for ancient river systems (e.g. Barrett 1970; Vavra 1984; Isbell 1990; Collinson et al. 1994; Flaig 2005), and their dispersion can also assist in distinguishing the type of fluvial system responsible for deposition (Miall 1976; Collinson 1978; Bridge 1985).

A total of 68 paleoflow orientations were recorded from channel-fill successions at 20 locations in the study area (Fig. 9). Measurements used to calculate mean paleoflow were restricted to trough cross-lamination on dune foresets in fine to coarse-grained sandbodies. Paleocurrent orientations indicate that regional paleoflow in the Prince Creek Fm. varied from northwest to southeast. The overall vector mean for all paleoflow in the study area is 060°. This value is consistent with previously published data indicating regional paleoflow to the northeast (Phillips 2003; Mull et al. 2003), although our data also indicate a high degree of variability in flow direction. No convincing evidence for bidirectional flow was found in any facies.

ALLUVIAL ARCHITECTURE

Architectural analysis (*sensu* Miall 1985) is a technique that is used to decipher the sequence of events responsible for the formation of alluvial successions and the nature of the system(s) responsible for their deposition (Eberth and Miall 1991; Wood 1989; Roberts 2007; McLaurin and Steel 2007). In this architectural analysis we integrate measured stratigraphic sections with interpreted line drawings on photomosaics. We also identify genetic lateral and vertical relationships between sandbodies and between sandbodies and floodplain facies. Architectural components used in the architectural analysis include Facies Associations I, II, III, and IV. The analysis is divided into three sections, with each section focusing on a specific channel type (FA I, II, and III). Many photomosaics include channels of more than one type. The architecture of FA-IV (floodplains) is intimately linked to the other facies associations (channels) and is, therefore, incorporated into each section. Due to the vertical nature and inaccessibility of many of the bluffs, the dip direction of IHS discussed here is apparent dip rather than true dip. Outcrop trend and paleoflow are incorporated into each photomosaic to assist in interpretation.

Large Sinuous Channels (FA-I)

The alluvial architecture of large sinuous channels (FA-I) is best observed near measured section PFDV-17 (Figs. 2, 10, 11). PFDV-17 was chosen for architectural analysis because both the basal conglomerate and the finer-grained organic-rich top of the FUS of FA-I are visible in outcrop. Section PFDV-17 also contains the largest clasts found in the study area (quartz boulders, longest axis = 30 cm).

Figure 10 is a 125-m-wide composite image showing a 28-m-thick vertical succession containing one 13-m-thick FUS of FA-I (large sinuous channels), two laterally extensive 2–4-m-thick FUSs of FA-II (small sinuous channels), and interdigitating facies of FA-IV. Section PFDV-17 (Fig. 11) was measured while walking up-section along the ridge on the left of the image in Figure 10; hence the entire FUS of FA-I, only one of the two FUSs of FA-II (between 6.25–8.0 m), and a sampling of the interdigitating facies of FA-IV appears in the measured section. The erosion surface at the base of the conglomerate in the large sinuous channel (FA-I), which exhibits 13 m of relief, is incised into an interval of stacked facies of FA-IV. Measurements recorded from trough cross-laminations in the medium-grained sandstone near the base of FA-I indicate that paleoflow in the large channel was directed obliquely out of the plane of the image at 343°, which is approximately 060° to the northwest relative to the strike of the outcrop (045°/225°). The medium-grained sandbody grades upward into an interval of IHS, capped by an interval of siltstone, carbonaceous shale, coal, and rooted mudstone. The IHS is composed of decimeter-scale, coarse-to-fine couplets of sandstone and mudstone, with the muddy component at least equal in thickness, and sometimes thicker, than the sandy component. IHS containing sand and mud couplets of approximately equal thickness is also visible in the two small sinuous channels (FA-II) in Figure 10. Depositional environments of FA-IV (interbedded sandstone, siltstone, and mudstone) in PFDV-17 include levees, crevasse splays, shallow lakes, swamps, mires, soils, and ashfall deposits. Thick mud drapes on point-bar surfaces (IHS) in large sinuous channels (FA-I) and small sinuous channels (FA-II) and the thick succession of fine-grained facies of FA-IV found stratigraphically above, below, and interfingering with channels suggest that these were suspended-load channels (Schumm 1981; Stewart 1983; Orton and Reading 1993).

At PFDV-17 initial deposition took place on the floodplain. A small sinuous channel (FA-II) migrated into the area over the floodplain sediments, depositing both levee and within-channel facies. The channel migrated out of the area and floodplain facies (FA-IV) were deposited above channel deposits. Paleoenvironments included crevasse splays, swamps, soils, and a 0.4-m-thick ashfall deposit preserved in a small pond or mire. A second small sinuous channel (FA-II) migrated through the region followed by another interval of floodplain deposition. The initial indication of a large sinuous channel (FA-I) in the vicinity is evidenced by the appearance of a relatively thick levee or splay sandstone deposited on top of the floodplain deposits. This sandstone is visible in the photomosaic (Fig. 10) but was not measured in section PFDV-17 (Fig. 11). Large channel incision into previously deposited levee and floodplain facies followed. Lateral migration of the large channel is indicated by point-bar deposits preserved as prominent IHS. Levees, crevasse splays, and organic-rich floodplain facies were deposited along the flanks of the channel. The entire succession was capped by a final episode of floodplain deposition after the large channel migrated out of the area.

A
30 m
Thin Sheet Sands
Siltstone
Carbonaceous shale and coal
B
sandstone
CUS
siltstone
mudstone
shale
Jarosite
C
Levee
15 m
FA-II
IHS
D
E

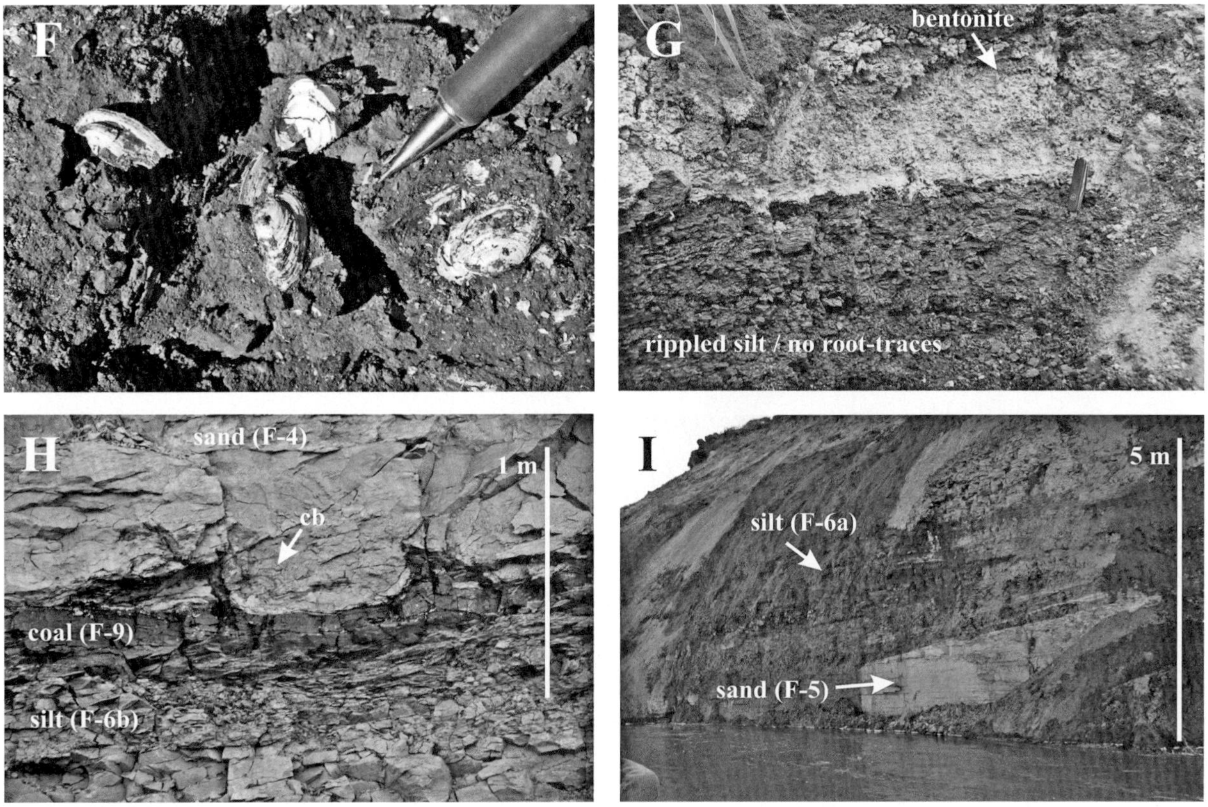

FIGURE 8.—*Continued*

Small Sinuous Channels (FA-II)

The alluvial architecture of small sinuous channels (FA-II) is visible at many locations in the study area; however, diagnostic attributes of FA-II are exceptionally well exposed near measured section PFDV-04 and location CRNKKT (Figs. 2, 12, 13). Bluffs near PFDV-04 (Fig. 12) were chosen for architectural analysis because: (1) they are composed of a series of vertically stacked small sinuous channels (FA-II) containing IHS with apparent dips which trend in opposing directions, (2) numerous mud-filled abandoned channels are visible in the bluffs, (3) the exposure also contains a small low-sinuosity channel (FA-III) incised into a small sinuous channel (FA-II), and (4) a small drainage allowed access to the north for the measurement of section PFDV-04 (Fig. 11). An additional photomosaic of bluffs at CRNKKT (Fig. 13) is included to illustrate more extensive lateral relationships between FA-II and FA-IV over a distance of 1.6 kilometers, approximately along strike of paleoflow.

Section PFDV-04: Figure 12 is a composite image that covers an area 400 m wide by 30 m high. A series of at least eight vertically stacked small sinuous channels of FA-II, many containing mud-filled channel plugs, are visible in the photomosaic. Small sinuous channels (FA-II) interfinger with interbedded sandstone, siltstone, and mudstone (FA-IV). Floodplain facies were deposited on crevasse splays and levees and in lakes, ponds, marshes, and swamps; and they include immature cumulative soils. Small sinuous channels (FA-II) in PFDV-04 are single-story and vertically separated by 0.2–5.0-m-thick packages of floodplain facies (FA-IV) unless the stratigraphically higher channel erodes into the underlying channel. Also included on the right side of the image (Fig. 12) and in measured section PFDV-04 (Fig. 11) is a single small low-sinuosity channel (FA-III) incised into a small sinuous

←

FIGURE 8.—Facies and physical sedimentary structures typical of FA-IV (interbedded sandstone, siltstone, and mudstone). A) Thin sheet sandstones interbedded with siltstone, carbonaceous shale, and coal; B) coarsening-upward succession (CUS) showing sheet sandstone with abundant carbonaceous roots with jarosite halos (arrows) gradationally overlying interbedded siltstone, mudstone, and carbonaceous shale containing jarosite; C) sheet sandstone interpreted as a levee grading laterally into a small sinuous channel (FA-II) containing inclined heterolithic stratification (IHS); D) drab-colored, cumulative, brown, gray, and black paleosols; E) carbonaceous root traces (arrows) in paleosols. Hammer is 30 cm long; pocket knife is 9 cm long; F) clams (*Nucula* aff. *N. percrassa* Conrad) in ripple cross-laminated organic siltstone; G) olive-green to yellow bentonite layer above a rippled siltstone lacking root traces; H) undulating-to-deformed "punch-down" contact at the base of a sandstone containing convolute bedding (cb) and peat (now coal) "injected" upward into the sandbody, likely as the result of trampling by dinosaurs; I) fine-grained large sinuous channel (FA-I) plug containing rippled siltstone overlying planar-laminated fine-grained sandstone. Pocket knife is 9 cm long, pencil is 15 cm long.

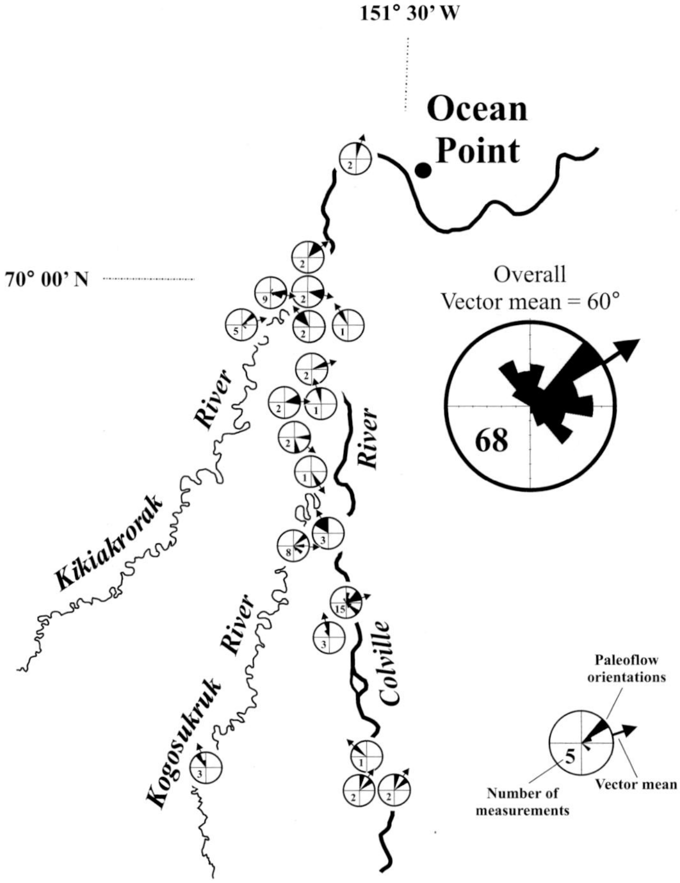

FIGURE 9.—Paleocurrent orientations recorded from trough cross-laminations in channel thalwegs at 20 separate locations in the Prince Creek Formation along the Colville, Kogosukruk, and Kikiakrorak Rivers. Included in the diagram are paleoflow vector rose diagrams that include the number of measurements recorded and mean paleoflow at each location. The diagram also displays the overall vector mean of paleoflow in the study area.

channel (FA-II). An additional mud- and tuff-filled small low-sinuosity channel (FA-III) located at the same stratigraphic level as the small low-sinuosity channel in Figure 12 can be found 125 m to the north, out of the field of view.

The bluffs at PFDV-04 strike at 165°/345°. Paleocurrent orientations recorded from trough cross-lamination at the base of three separate small sinuous channels (FA-II) in PFDV-04 (channels #1, #2, and #3, Fig. 12) indicate that mean paleoflow varied between 360°, 058°, and 136° at angles ranging from 15 to 75° relative to the apparent dip of the IHS and the strike of the bluffs. The mean resultant vector for paleoflow at PFDV-04 is 072°. Flow at high angles relative to the apparent dip of inclined surfaces in the IHS supports an interpretation of FA-II as remnants of laterally accreting point-bars. The sinuous nature of these channels is evidenced by the opposing apparent dip of IHS in stacked channels in this succession (Gradzinski 1970; Stewart 1983, Smith 1987; Flores et al. 2009b) along with the abundance of mud-filled abandoned channels (Collinson 1978; Miall 1996). The sand-filled small low-sinuosity channel (FA-III) at the right of the image (~ 15–17 m in section PFDV-04, Fig. 11) was filled mainly by vertical accretion, although minor IHS visible near the top of the channel (Fig. 7C) indicates that limited lateral migration of the channel likely occurred.

At PFDV-04, point-bars deposited by lateral migration of small sinuous channels (FA-II) alternate with highly organic floodplain facies (FA-IV). The migration direction of these sinuous channels was highly variable and abandonment of channels was common. Small sinuous channels (FA-II) incise either into pre-existing point-bar deposits of other small sinuous channels or into organic-rich floodplain deposits. The sand-filled small low-sinuosity channel (FA-III) incises into point-bar deposits of a small sinuous channel (FA-II, channel #3, Fig. 12) and is overlain by abandoned small sinuous channels (FA-II, channels #6 and #7, Fig. 12), suggesting that the smaller, fixed channel was active after point bar deposition by a small sinuous channel (FA-II, channel #3, Fig. 12) but prior to new small-sinuous-channel (FA-II) and floodplain (FA-IV) deposition.

Section CRNKKT: Figure 13 is a composite image of bluffs along the Colville River 2 KM to the north of PFDV-04 at location CRNKKT (Fig. 2) and covers an area 1600 m wide by 35 m high. The bluffs at CRNKKT strike at 155°/335°, similar to the strike of bluffs at PFDV-04. Figure 13 also includes a series of at least eight vertically stacked small sinuous channels (FA-II). Mud-filled channel plugs are present, but they are omitted in the interpreted line drawing for clarity. No paleoflow data were recorded at CRNKKT; however, apparent dips of IHS and the close proximity of CRNKKT to PFDV-04 suggest that paleoflow was likely toward the east-northeast, similar to that recorded in channels at PFDV-04.

Small sinuous channels (FA-II) in Figure 13 grade laterally from channelized deposits with erosional bases and prominent IHS into levees or splays with sharp to gradational basal contacts. Laterally extensive sandbodies assume an overall lenticular shape. The small sinuous channel (FA-II) designated "channel #2" is divided into two separate lobes by thinner levee deposits which are overlain by carbonaceous shale and coal. This geometry suggests that two channels either were contemporaneous, and separated by thin, interconnected levees, or they were successively deposited in close proximity to each other.

At CRNKKT, point bars of small sinuous channels (FA-II) and associated levees and splays form lenticular deposits in a range of orientations oblique to channel axes. Small sinuous channels (FA-II) alternate with and pinch out into floodplain facies that typically both overlie and underlie these channels. Small sinuous channels (FA-II) commonly erode into other, pre-existing small sinuous channels (FA-II) and levee, splay, or organic-rich deposits of FA-IV. Geometries suggest that multiple channels may have been active at the same time in close proximity to each other on the floodplain.

Small Low-Sinuosity Channels (FA-III)

The alluvial architecture of small low-sinuosity channels (FA-III) is best viewed at a location stratigraphically below and immediately to the south of the Kikak–Tegoseak dinosaur quarry (KKT in Fig. 2). This location was chosen for architectural analysis of FA-III because three, small, low-sinuosity ribbon-form sandbodies (FA-III) are exposed at the same stratigraphic level in bluffs (Fig. 14). The bluffs in Figure 14 are vertical, and no stratigraphic succession was measured there. However, measured section KKT, located 200 m to the north of the bluffs with an interval from ~ 4.5–6.0 m containing a small low-sinuosity channel (FA-III), is included for reference (Fig. 11). The interval from 15 to 17 m in measured section PFDV-04 (Fig. 11) provides an additional example of a small low-sinuosity channel (FA-III). Details of depositional environments near the Kikak–Tegoseak dinosaur quarry can be found in Brandlen (2008) and Fiorillo et al. (2010).

Figure 16 covers an area 900 m wide by 40 m high. The bluffs near

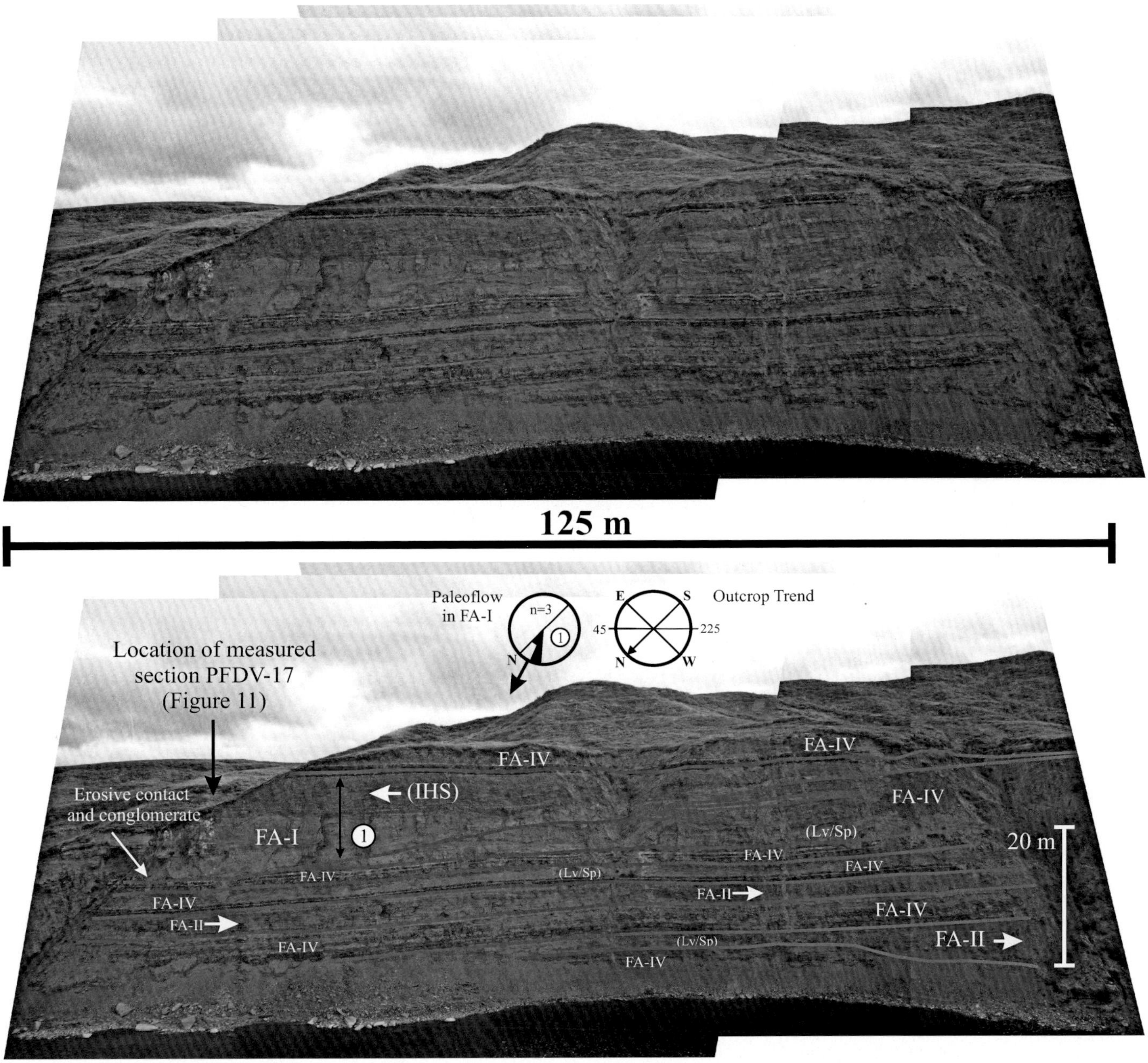

FIGURE 10.—Photomosaic and interpreted line drawing at location PFDV-17 including outcrop trend, paleoflow orientations, and location of measured section PFDV-17 (see Fig. 11). Image contains one complete fining-upward succession (FUS) of FA-I (large sinuous channel), two FUS of FA-II (small sinuous channels), and interbedded floodplain deposits of FA-IV (interbedded sandstone, siltstone, and mudstone). Abbreviations include: FA, facies association; IHS, inclined heterolithic stratification; and Lv/Sp, levee or splay. Arrows on image show apparent dip direction within IHS (point bars) and apparent direction of channel migration. See text for detailed interpretation. [A larger version of this figure can be found in the digital appendix.]

KKT visible in the photomosaic strike at 165°/355°. Paleoflow orientations taken from Brandlen (2008) ($n = 16$) show that the mean direction of paleoflow near the Kikak–Tegoseak dinosaur quarry is 298°, 057° to the west of the strike of the outcrop.

Most of the succession shown in Figure 14 is composed of interbedded sandstone, siltstone, and mudstone (FA-IV). The image includes two small sinuous channels (FA-II), one near the base of the succession and one near the top. The stratigraphically lower of the two channels grades laterally to the south into a levee or splay that ultimately pinches out into organic-rich floodplain deposits. Also included in Figure14 are three ribbon-form small low-sinuosity channels (FA-III). All three ribbon-form channels are sand-filled, located at the same stratigraphic level in the bluffs, and incise into floodplain deposits (FA-IV).

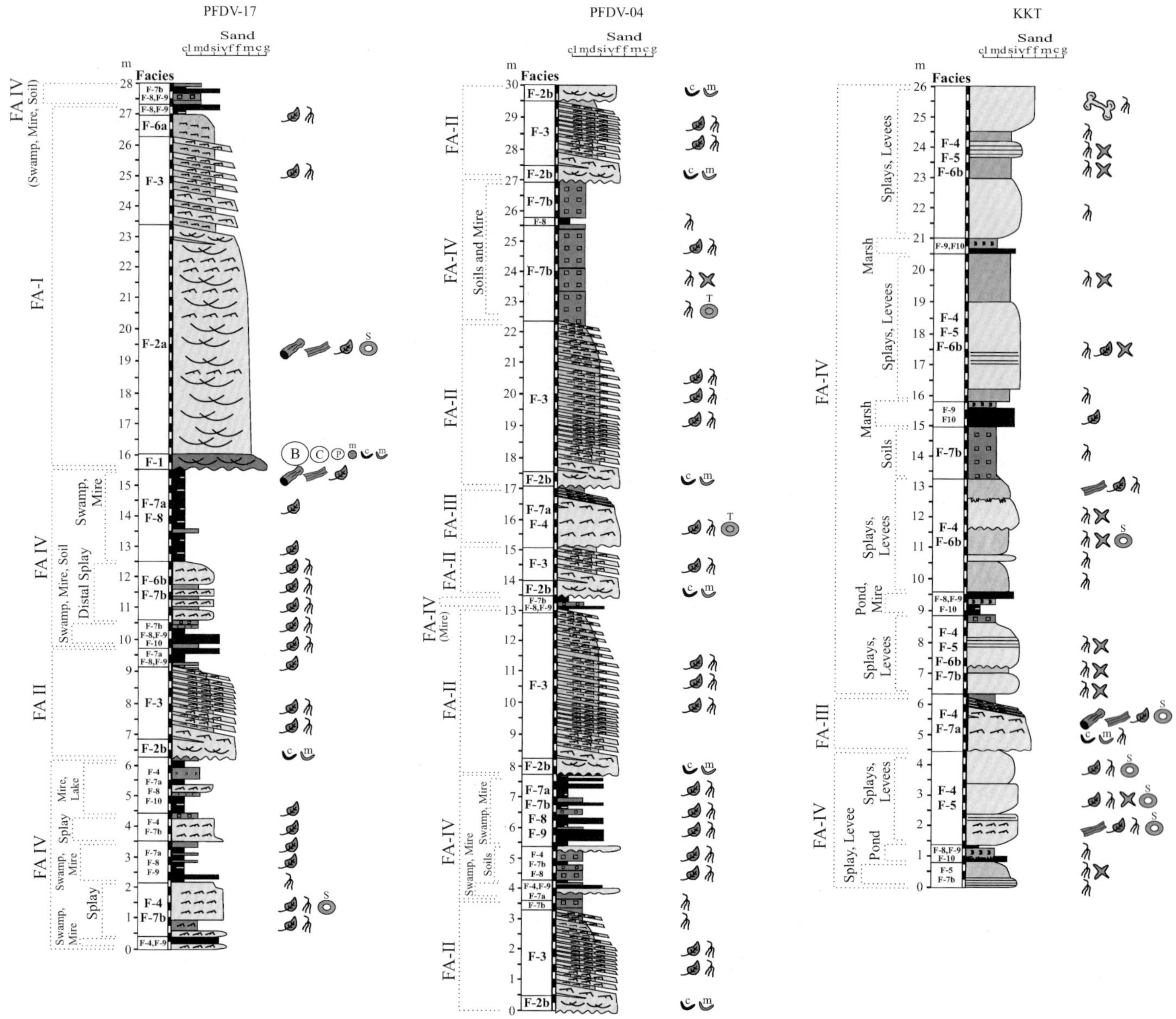

FIGURE 11.—Measured stratigraphic sections PFDV-17, PFDV-04, and KKT. Diagrams include paleoenvironmental interpretations. See Figure 4 for key to symbols. Detailed descriptions of individual facies and facies associations are given in Tables 1 and 2.

Near KKT, most deposition occurred on floodplains. Small sinuous channels (FA-II) and associated levees and splays interfinger with and pinch out into floodplain facies. Small low-sinuosity channels (FA-III) near KKT incise into floodplain deposits (FA-IV). This is in contrast to the small low-sinuosity channel (FA-III) at PFDV-04 in Figure 12, which incises into a small sinuous channel (FA-II).

CHANNEL CHARACTERISTICS

Although a 2-D architectural analysis is useful in establishing typical genetic relationships between channels and floodplain facies, it is also helpful to extract sandbody geometries from alluvial successions. However, incomplete or inaccessible exposures of the Prince Creek Fm. did not always allow quantitative estimates of channel geometries in the field; therefore, we use a series of methods (Schumm 1963, Ethridge and Schumm 1978, Gardner 1983, McLaurin and Steel 2007) to estimate channel depth, channel width, and sinuosity for each of the three channel types in the Prince Creek Fm.

Channel Depth

Channel depths for large sinuous channels (FA-I), small sinuous channels (FA-II), and small low-sinuosity channels (FA-III) in the Prince Creek Formation are estimated using two techniques. Estimates of bankfull depth provided are minimum estimates due to compaction of sediments during burial (~ 2000–6000 ft. (~ 600-1800 m), Burns et al. 2005). Other values (e.g., bankfull width) that incorporate channel depth into calculations should also be considered minimum values.

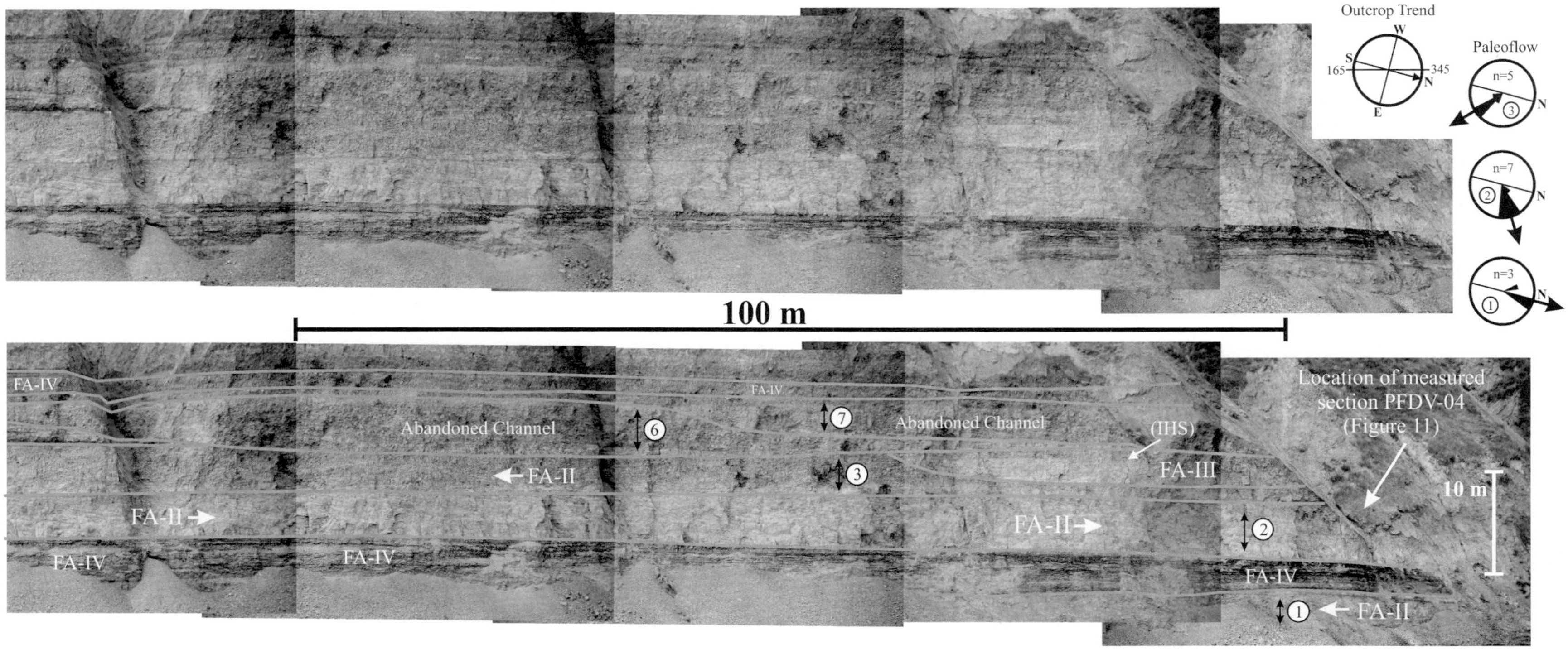

FIGURE 12.—Photomosaic and interpreted line drawing at location PFDV-04 (Fig. 2) including outcrop trend, paleoflow orientations, and location of measured section PFDV-04 (Fig. 11). Image contains eight small sinuous channels (FA-II) along with associated mud-filled abandoned channels, one example of a small low-sinuosity channel (FA-III), and interbedded sandstone, siltstone, and mudstone (FA-IV). Abbreviations include: FA, facies association; IHS, inclined heterolithic stratification; and Lv/Sp, levee or splay. Arrows on image show apparent dip direction in IHS (point bars) and apparent direction of channel migration See text for detailed interpretation. [A larger version of this figure can be found in the digital appendix.]

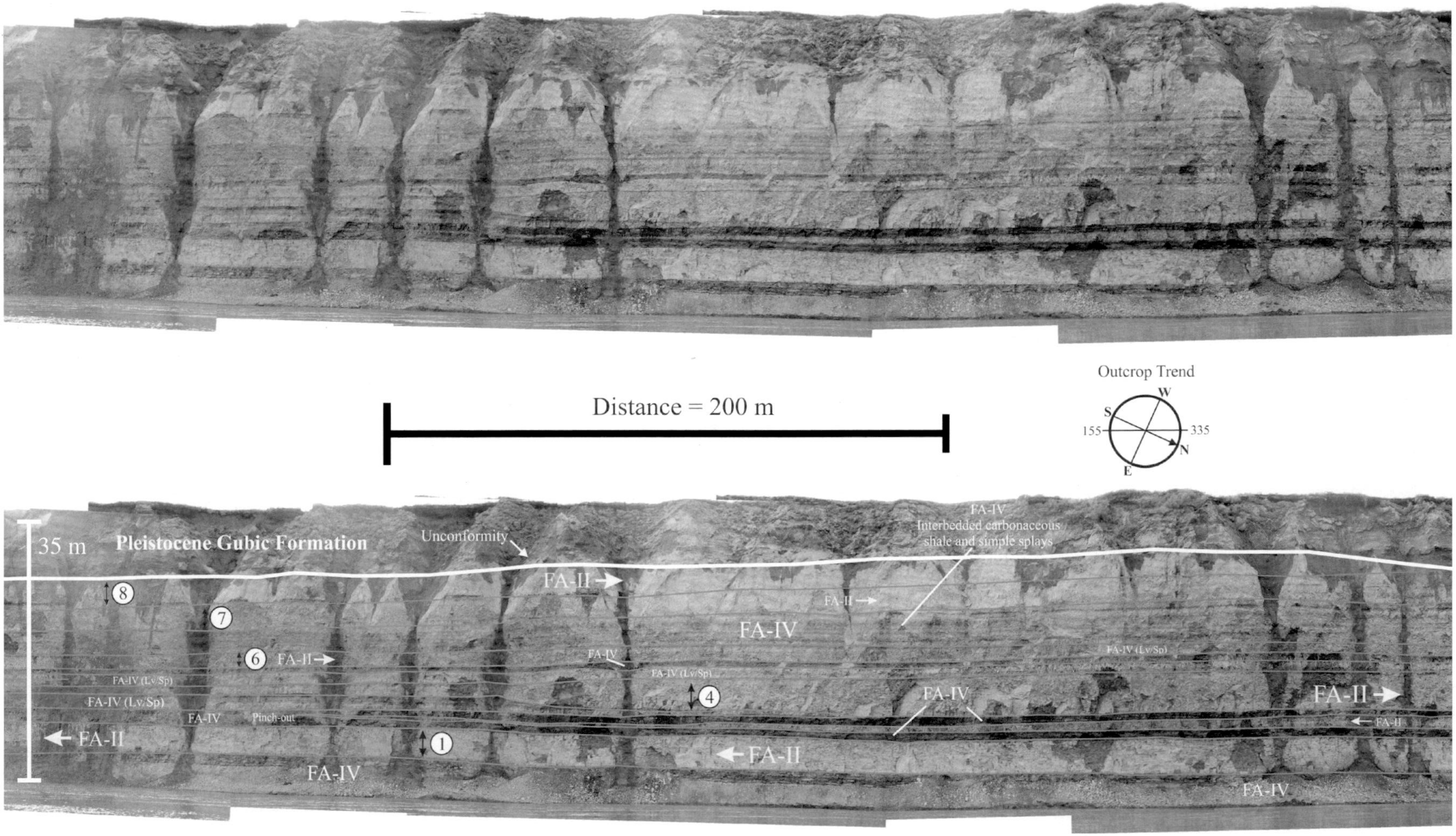

FIGURE 13.—Photomosaic and interpreted line drawing at location CRNKKT (Fig. 2) including outcrop trend. Overall paleoflow is inferred to be similar to Figure 12. Image contains eight small sinuous channels (FA-II) along with interbedded sandstone, siltstone, and mudstone (FA-IV). Abbreviations include: FA, facies association; IHS, inclined heterolithic stratification; and Lv/Sp, levee or splay. Arrows on image show apparent dip direction in IHS (point bars) and apparent direction of channel migration. See text for detailed interpretation. [A larger version of this figure can be found in the digital appendix.]

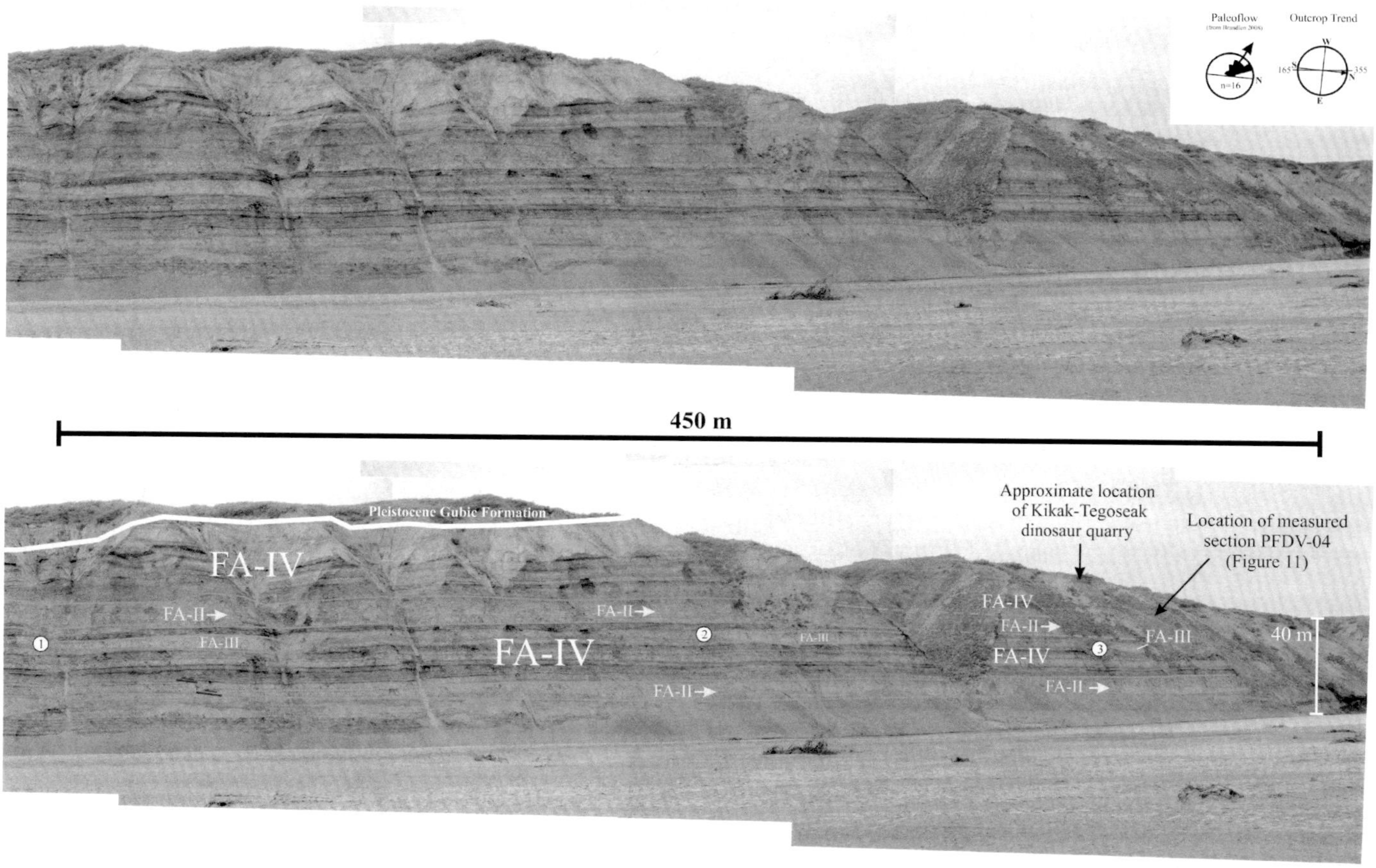

FIGURE 14.—Photomosaic and interpreted line drawing at location KKT (Fig. 2) including outcrop trend, paleoflow orientations, and location of measured section KKT (Fig. 11). Image contains three small low-sinuosity channels (FA-III), two small sinuous channels (FA-II), and interbedded sandstone, siltstone, and mudstone (FA-IV). Abbreviations include: FA, facies association; IHS, inclined heterolithic stratification; and Lv/Sp, levee or splay. Arrows on image show apparent dip direction in IHS (point bars) and apparent direction of channel migration. See text for detailed interpretation. [A larger version of this figure can be found in the digital appendix.]

FIGURE 15.—Rhythmically repeating centimeter- to decimeter-scale sandstone–mudstone couplets on point-bar deposits (IHS) common to the Prince Creek Formation. Sand-rich beds in image appear more resistant while clay-rich beds are more easily eroded.

Initial calculations employ a series of field measurements taken from a number of different channels of each type. In complete FUSs, bankfull depth was assumed to be the thickness of the succession from the base of the thalweg to the top of the point-bar surface or channel-fill succession (Ethridge and Schumm 1978, Gardner 1983). Outcrop measurements indicate that minimum flow depths for large sinuous channels (FA-I, $n = 2$) range from 13 to 17 m with an average depth of 15 m. Flow depths for small sinuous channels (FA-II, $n = 36$) range from 2 to 6 m with an average depth of 4 m, while flow depths for small low-sinuosity channels (FA-III, $n = 5$) range from 1.5 to 3.0 m with an average depth of 2.5 m. These results were compared to values obtained using the method of Leclair and Bridge (2001) in which the average crossbed set thickness serves as a proxy for flow depth by assuming that crossbed set thickness is equal to one-third dune height and that flow depth is six to ten times dune height. Using an average crossbed set thickness of 0.5 m for large sinuous channels (FA-I) (which allows for the inclusion of measurements from incomplete FUSs throughout the study area) a range of flow depths between 9 and 15 m was calculated. An average crossbed set thickness of 0.2 m for small sinuous channels (FA-II) gives a range of flow depths between 3.5 and 6.0 m. Crossbeds are extremely rare in small low-sinuosity channels (FA-III), and the thickness of crossbed sets was not recorded in these channels. Results for large sinuous channels (FA-I) and small

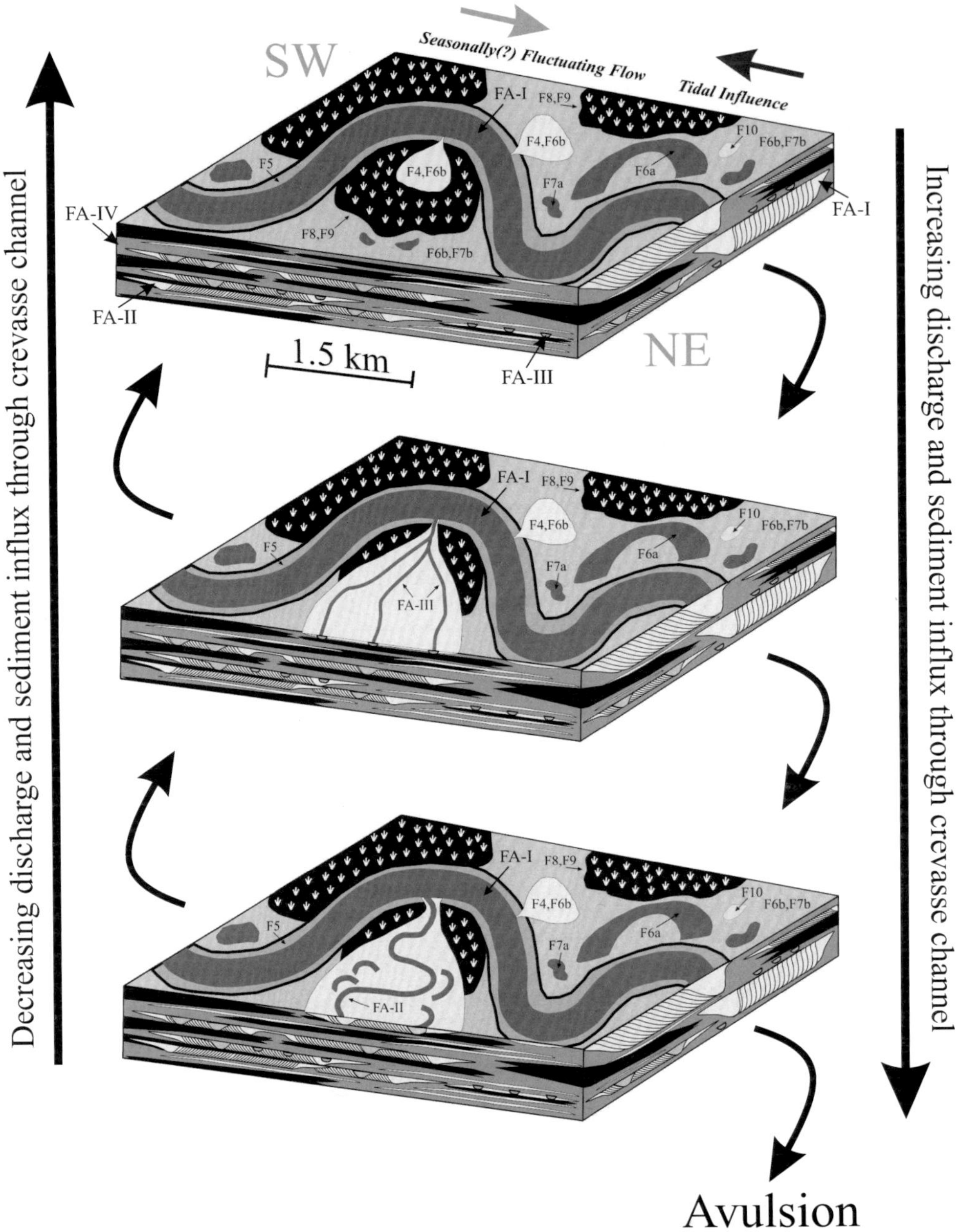

FIGURE 16.—Depositional model for the Prince Creek Formation in the Colville River region of Northern Alaska. Abbreviations and descriptions of facies and facies associations are included in Table 1 and Table 2.

sinuous channels (FA-II) from both methods are comparable. Therefore, ranges for average maximum bankfull depths of 9–17 m for large sinuous channels (FA-I), 2–6 m for small sinuous channels (FA-II), and 1.5–3.0 m for small low-sinuosity channels (FA-III) were used in the following calculations to estimate channel width.

Channel Width

Outcrop exposures containing complete FUSs of large sinuous channels (FA-I) are rare, and no exposure containing both channel margins was found. Because of the scarcity of laterally extensive exposures, widths for large sinuous channels (FA-I) must be estimated. Additionally, most smaller channels and channel plugs in the study area are in vertical bluffs that are either inaccessible or make quantitative measurement of channel width challenging. Because of difficulties involved in direct measurement and because of incomplete exposures, we use the empirical regression equation of Bridge and Mackey (1993) to estimate the width of channels:

$$w_c = 8.88 d_m^{1.82} \qquad (1)$$

where w_c is equal to the bankfull channel width and d_m is equal to the mean bankfull depth. Mean bankfull depth is approximately equal to one-half the maximum bankfull depth (Bridge and Mackey 1993). Incorporating half the range of our calculated maximum bankfull depths into Equation 1 we obtain estimates for bankfull channel widths of 140–435 m for large sinuous channels (FA-I), 10–65 m for small sinuous channels (FA-II), and 5–19 m for small low-sinuosity channels (FA-III). Using these results for channel width and maximum bankfull, width-to-thickness ratios (*w/t*) are estimated at: 15:1 to 25:1 for large sinuous channels (FA-I), 4:1 to 11:1 for small sinuous channels (FA-II), and 3:1 to 6:1 for small low-sinuosity channels (FA-III). Following Gibling (2006), large sinuous channels (FA-I) are classified as narrow sheets (range = 15:1 to 100:1), small sinuous channels (FA-II) are classified as broad ribbons (5:1–15:1), and small low-sinuosity channels (FA-III) are classified as narrow ribbons (< 5:1).

Sinuosity

Schumm (1963) provides an empirical equation that can be used to estimate sinuosity based on the relationship between sinuosity and the ratio of channel width to channel depth:

$$sn = 35(W_b/D_{max})^{-0.27} \quad (2)$$

where W_b is equal to bankfull width and D_{max} is equal to maximum bankfull depth. By incorporating our calculated range of bankfull widths and bankfull depths into Equation 2, Schumm's equation yields sinuosities for large sinuous channels (FA-I) ranging from 1.23 to 1.98 with an average sinuosity of 1.60, and sinuosities for small sinuous channels (FA-II) from 1.37 to 3.0 with an average value of 2.19.

Channel sinuosity can also be estimated by examining overall paleoflow variation, where low dispersion of paleoflow suggests low sinuosity and a high dispersion suggests high sinuosity. Bridge et al. (2000) suggest that a relationship exists between the range of paleoflow orientations in a channel system and the sinuosity of the system. They provide two sinuosity equations that assume bends within channels can be represented as circular arcs or sine-generated curves:

$$sn = \Phi/\sin\Phi(\text{circular} - \text{arc method}) \quad (3)$$

$$sn = 484/(4.84 - \Phi^2)(\text{sine} - \text{generated} - \text{curve method}) \quad (4)$$

where sn is equal to the sinuosity of the channel and Φ is equal to one-half of the maximum paleocurrent range in radians. In order to obtain more accurate estimates for the sinuosity of individual channel types in the Prince Creek Fm., large sinuous channels (FA-I) were separated from small sinuous channels (FA-II). In addition, because of the wide dispersion of paleocurrents for small sinuous channels (FA-II), we limit the range to paleocurrents collected from the three stacked channels at PFDV-04. This minimizes the error that would occur by inadvertently including paleoflow from distributaries on both sides of a main channel. Instead, we limit the estimated sinuosity to a series of apparently sequential small sinuous channels (FA-II) at a single location.

The range of paleocurrents recorded from trough cross-lamination in the thalwegs of large sinuous channels (FA-I, $n = 21$) is 190°. Taking one-half of the range (95°) in radians and incorporating it into Equation 3 yields a sinuosity value of 1.66. Substitution in Equation 4 yields a higher value of 2.31. The range of paleocurrents ($n = 15$) for small sinuous channels (FA-II) at PFDV-04 is 180°. Taking one-half of the range and incorporating it into Equation 3 yields a sinuosity value of 1.57. Using Equation 4 yields a higher value of 2.03.

All results indicate that both large sinuous channels (FA-I) and small sinuous channels (FA-II) should be classified as highly sinuous channels (*sensu* Miall 1996). The relative lack of IHS (point-bars) in small low-sinuosity channels (FA-III), and their consistent arcuate shape in outcrop suggests that they were most likely low-sinuosity channels.

DISCUSSION

Channel Hierarchy in the Prince Creek Formation

Channel-fill deposits of the Prince Creek Fm. are identified as FUSs that include a combination of two or more of the following attributes: (1) a basal erosion surface, (2) IHS with flow directions at high angles relative to the dip of inclined surfaces indicating lateral accretion, and (3) a concave-up, arcuate geometry. Channels are ranked as *first-order*, *second-order*, and *third-order* using a hierarchy based on (1) relative thicknesses of FUSs, (2) facies character, (3) grain size, (4) sedimentary structures, (5) geometries, and (6) regional spatial relationships. Frequency relationships are inherent to a hierarchical classification (Catuneanu et al. 2009). First-order channels are expected to be the least frequent, second-order channels of a higher frequency, and third-order channels the most frequent. However, a difficulty arises when adapting a hierarchical scheme to ancient alluvial successions because preservation potential becomes a factor. This is in contrast to modern alluvial systems, where channel geometries, channel order, and channel frequency are evident in plan view (e.g., Fisk 1944; Morisawa 1985; Morisawa and Montgomery 1985; Roberts 1997; Morozova and Smith 2000; Olariu and Bhattacharya 2006). In the Prince Creek Fm., first-order channels appear less frequently than second- or third-order channels. However, second-order channels appear to be the most frequent while third-order channels are of intermediate frequency. This is likely an artifact of the relatively poor preservation potential of third-order channels (see below), and we therefore choose to downplay apparent frequency in our hierarchical scheme.

First-Order Channels: Large sinuous channels (FA-I) are the least frequently encountered in outcrop, exhibit FUSs that are the thickest overall, contain the coarsest grain sizes in the region, and include sedimentary structures suggesting the highest flow velocities (Ethridge et al. 1981; Morisawa 1985; Morisawa and Montgomery 1985; Olariu and Bhattacharya 2006). Large sinuous channels (FA-I) of the Prince Creek Fm. are therefore interpreted as *first-order* meandering trunk channels. First-order trunk channels are composed of FUSs that are typically > 13 m thick; at least twice the thickness of FUSs found in second-order channels. First-order channels also contain a basal pebble-to-boulder conglomerate and medium-grained sandbodies that exhibit the highest frequency of trough cross-lamination and convolute bedding. Sandbodies in first-order channels lack carbonaceous root traces, suggesting greater depths and continuous flow. Conglomerate is noticeably absent in second- and third-order channels, and no other channel type in the Prince Creek Fm. consistently contains grain sizes larger than fine-grained sand. Multistory sandbodies are also found exclusively in first-order channels. Fill successions interpreted as large sinuous channels (FA-I) do not meet the requirements for incised-valley fills (Gibling 2006), in that their basal erosion surfaces are local phenomena and roughly scale to the depth of individual FUSs. Also FUSs of large sinuous channels (FA-I) are less than an order of magnitude greater than FUSs of small sinuous channels (FA-II).

Second-Order Channels: Small sinuous channels (FA-II) have intermediate FUS thicknesses and intermediate grain sizes, contain sedimentary structures suggesting intermediate or mixed flow velocities, and are therefore interpreted as *second-order* meandering distributary channels. FUSs in second-order channels range in thickness from 1.5 to 5.0 m, are thinner than first-order channels, and are generally thicker than third-order channels. Second-order channels also lack the conglomerate found at the base of first-order channels, and all sandbodies are fine-grained and single-story. Sedimentary structures include both trough cross-lamination and ripple cross-lamination; however ripple cross-lamination is most common. All second-order channels contain carbonaceous root traces throughout, suggesting shallower depths than first-order channels and indicating possible fluctuating flow velocities and water depths in these channels.

Third-Order Channels: Small low-sinuosity channels (FA-III) have the thinnest average FUSs, the finest grain sizes, and sedimentary structures indicating the lowest average flow velocities of all channels, and are interpreted as *third-order* fixed (anastomosed?) distributary channels. Third-order channel FUSs range in thickness from 1.5 to 3.0 m and are, on average, the thinnest of all channel types. Third-order channels also lack the conglomerate of first-order channels, and all sandbodies are single-story and ubiquitously rooted. Ripple cross-lamination is the dominant sedimentary structure. Trough cross-lamination is extremely rare. Third-order channels consistently lack the thick packages of IHS that dominate the deposits of first-order and second-order channels, suggesting that flow velocities were generally lower than required for lateral channel migration.

Floodplains

Floodplains (FA-IV) of the Prince Creek Fm. were predominantly poorly drained and sporadically inundated with sediment. Large abandoned channels were present near trunk channels. Facies deposited in smaller abandoned channels, lakes, ponds, swamps, marshes, and mires are found interbedded with distributary channels, suggesting close spatial relationships between these environments. Crevasse splays frequently covered low areas with silt and sand. Bentonites and tuffs are preferentially preserved in lakes and ponds and are occasionally reworked into crevasse splays or preserved in fixed channels. Undulating-to-deformed basal contacts are common in floodplain deposits, suggesting that dinosaurs occupied areas adjacent to major rivers and distributaries. Abundant root traces indicate that vegetation grew in most environments on the floodplain.

Paleosols of the Prince Creek Fm. are cumulative and poorly developed. Detailed descriptions, geochemistry, and mineralogy of paleosols near measured section KKT (Fig. 2) are presented elsewhere (Brandlen 2008; Fiorillo et al. 2009). Drab-colored fine-grained sediments, well-preserved carbonaceous plant material, and abundant siderite nodules suggest poorly drained conditions and shallow standing water typical of wetlands (Bown and Kraus 1987; Kraus 1987; Kraus and Bown 1988; Retallack 2001; Ashley et al. 2004; Rajchl and Uličný 2005). Jarosite, an oxidation product of pyrite (Kraus 1998, Davies-Vollum 1999, Kraus and Hasiotis 2006), becomes more common up-section in the Prince Creek Fm. as strata approach the contact with the shallow-marine shoreface successions of the upper Schrader Bluff Fm. The presence of jarosite may signal the intermixing of marine or tidal waters with pore fluids in floodplain sediments (Wright 1986, Ufnar et al. 2001, McCarthy and Plint 2003). Micromorphological features of paleosols include microscopic, ovoid elliptical burrows (up to 0.5 mm in diameter) attributed to insects and earthworms (McCarthy and Plint 2003), ferruginous void and grain coatings and nodules, mottled soil aggregates and depletion features, as well as evidence of illuvial clay coatings in voids. These features suggest that floodplains were subjected to fluctuating drainage conditions, or perhaps seasonally dry conditions, with significant portions of the floodplain drying out for a period of time (McCarthy et al. 1998; Brandlen 2008; Fiorillo et al. 2009; Fiorillo et al. 2010).

The relatively thin coals in the study area may be the result of repeated episodes of sediment influx onto the floodplain during floods, retarding prolonged peat production (Flores 1984, Fielding 1987b; Spicer 2003; Fiorillo et al. 2010). Additionally, fluctuating groundwater-levels and periodic drying out of floodplain sediments (Spicer 2003; Fiorillo et al. 2009; Fiorillo et al. 2010) may have also controlled peat production and preservation (Diessel 1992; van Asselen et al. 2009).

Crevasse Splays and Crevasse-Splay Complexes

Crevasse splays are defined as sheet sandbodies that are lobate or elongate in plan view and lenticular or lens-shaped along strike,which commonly interdigitate with finer-grained floodplain deposits (Fisk 1944; Coleman and Gagliano1964; Allen 1965b; Elliott 1974; Collinson 1978; Bridge 1984; Smith et al. 1989; Mjøs et al. 1993; Jorgensen and Fielding 1996; Morozova and Smith 2000). Some workers separate crevasse splays into two categories: (1) "small-scale" or simple crevasse splays and (2) large-scale composite crevasse splays or "crevasse-splay complexes" (Elliott 1974; Fielding 1987a; Smith et al. 1989; Mjøs et al. 1993; Smith and Pérez-Arlucea 1994). Smith et al. (1989) in their study of crevasse splays in the Cumberland Marshes point out that the development of splays on an alluvial plain is typically not an event, but rather a complex evolutionary process. They identify three stages in the formation of splays adjacent to trunk channels. Stage I splays are the smallest, record initial crevassing of the trunk channel, and commonly consist of a relatively thin, lobate, lenticular (along strike) sheet sandbody that lies directly on older floodplain sediments. In contrast, Stage II splays may be best described as "splay complexes" in that they are larger, evolve from Stage I splays, and are composed of a network of active and abandoned channels. Stage III splays evolve from Stage II splays and exhibit lower channel densities than Stage II splays. With additional flow velocity redirected through the crevasse channel, Stage III splays may capture the main flow of a channel during channel avulsion.

The Prince Creek Fm. contains large sinuous channels (FA-I) which we classify as first-order meandering trunk channels (Figs. 6A, 12) and thin, laterally extensive, ripple cross-laminated sheet-like sandbodies (small-scale splays of FA-IV, Figs. 9A, 15) that we interpret as Stage I splays (Smith et al. 1989). In addition to these smaller Stage I or "small-scale" splays, second-order meandering distributaries and third-order fixed (anastomosed?) distributaries of the Prince Creek Fm. construct larger "splay complexes" adjacent to trunk channels, similar to Stage II splays described by Smith et al. (1989). Evidence in the Prince Creek Fm. for construction of these splay complexes includes: (1) meandering distributaries (FA-II) and fixed (anastomosed?) distributaries (FA-III) eroded into and encased in organic-rich floodplain deposits throughout the succession (Figs. 10, 12, 13, 14) indicating deposition of coarser-grained facies onto finer-grained floodplains; (2) distributary channels deposited as laterally extensive sheets that commonly exhibit lenticular or lens-shaped geometries when viewed in orientations oblique to strike (Fig. 13), which are similar to geometries expected for crevasse splay lobes in cross section; (3) distributaries incised into thinner sheet sandstones of FA-IV (Fig. 14) suggesting a common, genetic relationship between distributary channels and small-scale Stage I splays; (4) distributaries incised into pre-existing distributary channels (Figs. 7A, 7C, 12), which most likely record the reworking of splay complexes by distributary channels; and (5) tiers of fixed, ribbon-form distributaries (FA-III) in close proximity to each other at the same stratigraphic level (Fig. 14) and incised into either meandering distributaries or floodplain facies, suggestive of anastomosed channels on splay complexes and floodplains.

The geometries of sandbodies constructed by distributary channels of the Prince Creek Formation are similar to those found in numerous descriptions of crevasse-splay complexes (Fisk 1944; Coleman and Gagliano1964; Fielding 1984; Fielding 1986; Smith et al. 1989; Jorgensen and Fielding 1996; Bristow et al. 1999; Pérez-Arlucea and Smith 1999; Morozova and Smith 2000; Stouthamer 2001; Tooth 2005). Crevasse splays of the Prince Creek Fm. likely began as Stage I splays similar to those described by Smith et al. (1989). With increased discharge through the crevasse channel, Stage I splays evolved into Stage II crevasse-splay complexes through the formation of a series of fixed channels that incised into the Stage I splay in areas proximal to the trunk channel, and into floodplain deposits at more distal locations. The flow in these channels either did not achieve velocities sufficient for the channels to begin to wander on the fine-grained floodplain, or lateral migration was inhibited by abundant mud and vegetation such as peat (Smith 1976; Kirschbaum and McCabe 1992; Törnqvist 1993; Nadon 1994). However, as additional flow was diverted into the crevasse channel, and as flow velocities and sedimentation increased, channels on the splay complex began to wander, ultimately evolving into meandering channels. This change in channel form may have been driven, in part, by peat compaction (Smith and Pérez-Arlucea 2004; Rajchl and Uličný 2005; van Asselen et al. 2009). When subjected to a load, peat compacts (Tatsch 1980; Elliott 1985; Diessel 1992; Allen 1999; van Asselen et al. 2009). It has been suggested that peat compaction can play a role in shaping stratal geometries by controlling accommodation space and inhibiting channel incision (Diessel 1992; Michaelsen et al. 2000; Smith and Pérez-Arlucea 2004; Rajchl and Uličný 2005; van Asselen et al. 2009). As flow velocities and sedimentation rates in distributary channels on the crevasse-splay complex increase, peat is compacted below the channel fills and, ultimately, peat compaction ceases, which limits accommodation (Elliott 1985; van Asselen et al. 2009). This compacted peat layer inhibits vertical scouring at the base of the channel and promotes lateral migration of the channel (Smith and Pérez-Arlucea 2004; van Asselen et al. 2009). Accretion in these channels passes from predominantly vertical to principally lateral, producing sheet-like sandbodies with point bars and high width-to-depth ratios (Smith and Pérez-Arlucea 2004; Rajchl and Uličný 2005; van Asselen et al. 2009). Lateral migration of the channel continues until the peat layer below the thalweg is breached or the channel is abandoned (Smith and Pérez-Arlucea 2004). During this stage in the evolution of the crevasse-splay complex both fixed and meandering distributary channels may have been active at different locations depending on flow velocities in distributary channels, local substrate conditions (i.e., sand-rich or mud-rich), and local accommodation and the depth to the peat layer. With additional discharge diverted through the crevasse channel, an avulsion might have occurred (Morozova and Smith 2000; Stouthamer 2001). Alternatively, the main crevasse channel could have become plugged by sediment, forcing abandonment of the splay complex. In the event of rapid abandonment, meandering distributaries and fixed distributaries would fill with fine-grained sediment and floodplain facies would be deposited above the abandoned splay complex (e.g., Coleman and Gagliano 1964; Smith and Pérez-Arlucea 1994, 2004; Pérez-Arlucea and Smith 1999; Morozova and Smith 2000). In the event of a slow abandonment, the cycle of deposition would reverse itself, driving the system back through the fixed (anastomosed?) splay complex, Stage I splay formation, and ultimately deposition of floodplain facies. In either scenario, the preservation potential for fixed (anastomosed?) distributaries may be lower than the preservation potential for meandering distributaries, based on (1) their stratigraphic position below, or possibly adjacent to, higher-energy meandering distributaries on the splay complex that could capture the bulk of the flow (Pérez-Arlucea and Smith 1999), (2) their relatively narrow cross section and lack of lateral accretion (Smith and Pérez-Arlucea 2004), or (3) in the case of a slow abandonment, their relatively shallow depths and increased height relative to the thalweg of the main channel and base level (Reading 1996; Bridge 2003; Kjemperud et al. 2008). The preservation potential for fixed (anastomosed?) distributaries is decreased further in the event of the reactivation of a formerly active crevasse channel with flow velocities and sedimentation rates sufficient to cause lateral migration of the channel, and scouring of the splay complex or avulsion (Smith et al. 1989; Morozova and Smith 2000; Stouthamer 2001; Smith and Pérez-Arlucea 2004).

The exact reason for the crevassing of trunk channels in the Prince Creek Fm. is unknown; however, high concentrations of suspended load along with abundant mud and volcanic ash in channels may have been a contributing factor. Extreme seasonality brought about by the high-latitude (paleo-Arctic) setting (Spicer et al. 1992; Spicer 2003) may have driven seasonal snowmelt in the Brooks Range and triggered spring floods which subsequently carved splay channels.

Inclined Heterolithic Stratification in the Prince Creek Formation

IHS is defined as "large-scale, waterlain [*sic*], lithologically heterogeneous siliciclastic sedimentary sequences whose constituent strata are inclined at an original depositional angle to the horizontal or paleohorizontal" (Thomas et al. 1987, p. 125). Typically, IHS is composed of a series of repeating sand–silt, sand–mud, or silt–mud couplets inclined relative to a basal, horizontal erosion surface. Recognizing IHS in a sedimentary succession is important because the physical properties of the IHS can be used to identify the likely depositional environment. For example, paleoflow at high angles to the dip of the inclined beds indicates lateral accretion in point bars while paleoflow at low angles relative to the dip of inclined beds indicates downstream accretion in mid-channel bars or Gilbert-type deltas (Allen 1963; Stanley and Surdam 1978). IHS successions in the Prince Creek Fm. fine upward (e.g., Figs. 5A, 6A), and paleoflow is consistently at high angles to the dip of the inclined beds, suggesting that IHS records lateral migration of point-bars (Allen 1965a; Stanley and Surdam 1978; Thomas et al. 1987).

Although IHS has been recognized in purely fluvial environments (Stewart 1981; Jackson 1981; Arche 1983; Page et al. 2003), most IHS is attributed to tidally influenced settings (Thomas et al. 1987; Smith 1987, 1988; Nio and Yang 1991; Nakajo 1998; Rogers 1998; Changsong et al. 1995; Choi et al. 2004; Joeckel et al 2005; Yang et al. 2007; Dalrymple and Choi 2007; Hovikoski et al. 2008). In fact the recognition of IHS has played a key role in the reinterpretation of some coastal-plain successions as tidally influenced (e.g., the McMurray Formation of Alberta: Mossop and Flach 1983; Flach and Mossop 1985; Crerar 2003; Crerar and Arnott 2007). On low-gradient coastal plains the tidal zone may reach tens to hundreds of kilometers inland, with highly sinuous meandering channels being common, especially in the landward reaches of tidal influence, where tidal current speeds decrease (Dalrymple et al. 1992; Orton and Reading 1993; Eberth 1996; Crerar 2003; Dalrymple and Choi 2007). High levels of suspended sediment are common on delta-plains where high suspended-sediment concentrations and thick mud drapes on point bars are driven in part by a low gradient and river–tidal interaction in the transition between the terrestrial and marine zones (Schumm 1968; Schumm and Kahn 1972; Smith 1987, 1988; Baker et al. 1995; Dalrymple and Choi 2007). Mud drapes on lateral-accretion surfaces have been attributed to tidal influence in delta distributary channels (Smith 1987, 1988; Wood et al. 1988; Gastaldo et al. 1995; Changsong et al. 1995; Nakajo 1998; Rogers 1998; Makaske and Weerts 2005; Joeckel et al. 2005; Hovikoski et al. 2008). Although mud drapes do occur on point-bar surfaces in purely fluvial deposits (e.g., Jackson 1981), they are not as abundant or as thick as those deposited during slackwater periods in tidally influenced deposits (Thomas et al. 1987; Wood et al. 1988; Dalrymple and Choi 2007). Rounded mud balls and fluid muds are also common in this environment (Baker et al. 1995; Dalrymple and Choi 2007; Bhattacharya and MacEachern 2009). The bulk of the channels of the Prince Creek Fm. are classified as highly sinuous, transported a large percentage of their sediment as suspended load, and contain thick mud drapes on point-bar surfaces (Fig. 15) and spherical mudstone balls (Fig. 5E), similar to tidally influenced channels.

IHS containing rhythmic-style repetition of coarse-to-fine couplets is present at some stratigraphic level in all three orders of channels

(Figs. 5A, 6A, 7C,) indicating that all three types of channels (including small low-sinuosity channels of FA-III) migrated laterally during some period in their evolution (Smith 1987, 1988; Dalrymple and Choi 2007) and likely experienced tidal effects that affected sediment deposition in channels (Thomas et al. 1987; Smith 1987, 1988; Gastaldo et al. 1995; Makaske and Weerts 2005).

Faunal assemblages of the Prince Creek Fm., when encountered, are extremely high dominance and low diversity, suggesting biologically stressed brackish-water conditions (personal communication, Robert Blodgett). Bioturbation of sediments is often low in tidally influenced channels due to stressed environments characteristic of turbid, brackish-water conditions (Dalrymple and Choi 2007). Bivalves and gastropods are most tolerant of these stressed conditions (MacEachern et al. 2005; Dalrymple and Choi 2007). Bioturbation is rare in channel deposits of the Prince Creek Fm., and bivalves and gastropods, although rare, are the most common invertebrates and are most likely responsible for bioturbation in channel sediments.

Although we recognize that the thickness of mud drapes and other fine-grained laminae should be demonstrably cyclic in order to conclusively infer tidal influence on a system (Dalrymple and Choi 2007), we consider that the large relative thickness of mud drapes on point-bar surfaces and the rhythmically repetitive nature of the coarse-to-fine couplets in the IHS is such that tidal influence is the best explanation for prominent IHS in the Prince Creek Fm.

Depositional Model for the Prince Creek Formation

A depositional model for the Prince Creek Fm. in the Colville River region is presented in Figure 16. The figure includes a large, sinuous meandering trunk channel (FA-I) along with small-scale crevasse splays, crevasse-splay complexes, and associated floodplain deposits. Flow is from the southwest to the northeast and fluctuated seasonally(?). Tidal effects enter the system from the northeast. Levees (F-5) extend outward from the main channel. Abandoned channels, smaller lakes, ponds, mires, swamps, and paleosols (FA-IV) are widespread on the floodplain (F-6a, F-6b, F-7a, F-7b, F-8, F-9, and F-10). Initially, a small crevasse channel is open and crevassing occurs as a small-scale crevasse splay (F-4, F-6b) or Stage I splay of Smith et al. (1989). As discharge and sediment influx through the crevasse channel increases, fixed (anastomosed?) channels (FA-III) form on the splay complex and extend onto the distal floodplain. Floodplain facies are deposited adjacent to splay complexes. With additional discharge and sediment influx, peat compacts below fixed (anastomosed?) distributary channels until maximum compaction is achieved, limiting accommodation space. Distributary channels begin to wander, and a meandering distributary channel (FA-II) replaces the fixed (anastomosed?) distributaries (FA-III) on the splay complex. With an additional increase in discharge through the crevasse channel an avulsion of the main channel would occur. Alternatively, if the crevasse channel is blocked and discharge decreases, fixed (anastomosed?) distributaries (FA-III) reoccupy the splay complex and incise into previously deposited meandering distributaries (FA-II). With a further decrease in discharge a Stage I splay (small-scale splay of FA-IV) reoccupies a portion of the splay complex. If the crevasse channel is blocked completely, overbank sedimentation on floodplains becomes the dominant mode of deposition.

Although the meandering distributary channels (FA-II) of the Prince Creek Fm. are classified as broad ribbons (Gibling 2006), in actuality point-bar deposits of these systems create laterally extensive, asymmetrical, heterolithic sheets that appear lenticular along strike and pinch out into floodplain deposits. These lenticular sheets, composed principally of point-bars and channel-plugs, may contain incised, fixed (anastomosed?) distributaries (FA-III) and are separated from each other by organic-rich floodplain facies (FA-IV), which may also be incised by tiers of fixed (anastomosed?) distributaries. Deposits of larger meandering trunk channels (FA-I) are rare, thicker, regionally restricted, and alternate with all other deposits in the area.

This depositional model for the Prince Creek Fm. is compatible with the findings of Phillips (2003) who suggests that the stratigraphically higher, uppermost facies of the Prince Creek Fm. to the northeast of our study area at Ocean Point (Figs. 1, 2) were deposited in meandering distributaries and in interdistributary bays transgressed by shallow-marine deposits of the Schrader Bluff Fm. Interdistributary-bay deposits increase in abundance up-section in the Prince Creek Fm. with brackish-water indicators such as clams, gastropods, jarosite, and vivianite becoming more common. The entire Prince Creek Fm. generally fines upward from conglomeratic sands above incised valleys at the base (Flores et al. 2007a; Flores et al. 2007b) to finer-grained, interdistributary facies with brackish-water indicators at the top (Brouwers and de Deckker 1993; Phillips 2003). Tidally influenced sedimentation would not be unexpected on this high-latitude, low-gradient, muddy Cretaceous coastal plain during a period of base-level rise culminating in an apparent marine transgression.

CONCLUSIONS

The Upper Cretaceous (Maastrichtian) Prince Creek Formation in the Colville River region of northern Alaska is a coastal-plain succession containing (i) larger, first-order, sinuous meandering trunk channels, (ii) smaller, second-order, sinuous meandering distributary channels, (iii) third-order, fixed (anastomosed?) distributary channels, and (iv) organic-rich floodplains.

First-order meandering trunk channels are regionally restricted, exhibit the thickest FUSs, contain the largest clasts and coarsest grain sizes, exhibit an interval of IHS indicative of lateral accretion, display thick mud drapes on point-bars, and include sedimentary structures that indicate the highest flow velocities in the region. Sandbodies in trunk channels are multistory and lack root traces, also suggesting high flow velocities in relatively deep suspended-load channels.

Second-order meandering distributary channels exhibit smaller FUSs, lack conglomerates, and include sedimentary structures indicative of intermediate flow velocities. Meandering distributaries are single-story and composed predominantly of IHS with thick mud drapes (point bars) along with mud-filled abandoned channels common to sinuous, suspended-load channels.

Fixed (anastomosed?) distributaries display the smallest FUSs, include sedimentary structures indicative of the lowest flow velocities, and lack prominent IHS, demonstrating that they were filled predominantly by vertical rather than lateral accretion.

Both meandering and fixed (anastomosed?) distributaries contain carbonaceous root traces throughout the channel-fill succession, suggesting shallower water levels and fluctuating flow.

Abundant, thin sheet sandbodies are interpreted as small-scale crevasse splays and levees. Organic-rich siltstone and mudstone, carbonaceous shale, coal, bentonite, and tuff interpreted as abandoned channels, small lakes, ponds, swamps, marshes, mires, paleosols, and ashfall deposits.

Meandering distributaries typically appear lenticular along strike, erode into pre-existing distributary deposits, and interfinger with and incise into organic-rich floodplain facies. Fixed (anastomosed?) distributaries either incise into meandering distributary deposits or into organic-rich floodplain facies. Multiple fixed (anastomosed?) distributaries may be preserved in tiers at the same stratigraphic level. Spatial relationships between channels and floodplains indicate that the bulk of deposition in the Prince Creek Fm. occurred on crevasse–splay complexes adjacent to trunk channels. Splay complexes were constructed above organic-rich floodplain deposits by the lateral migration of meandering distributaries and the vertical filling of anastomosed distributaries. Flow in meandering and fixed (anastomosed?) distributaries may have been contemporaneous or, alternatively, fluctuating discharge through the main crevasse channel and peat compaction may have controlled the type of system dominating the splay complex.

IHS is found in all three orders of channels. Thick mud drapes on point bars and the rhythmic and repetitive nature of the coarse-to-fine couplets in the IHS suggest that flow in channels was likely influenced by tidal effects. Common jarosite and vivianite also suggest an influx of marine waters at the distal end of the system.

Drab colors in paleosols and fine-grained sediments, abundant carbonized plant material, and common siderite suggest that floodplains were wet and reducing conditions were widespread. However, carbonaceous root traces throughout most fine-grained facies and in all distributary channels along with Fe-oxide mottles and illuvial clay coatings in paleosols indicate that the alluvial system likely experienced flashy, seasonal, or ephemeral flow and a fluctuating water table. The flashy nature of this Arctic alluvial system may have been driven by seasonal snowmelt in the evolving Brooks Range orogenic belt as a consequence of the high paleolatitude of the Prince Creek Fm. near the end of the Cretaceous Period.

ACKNOWLEDGMENTS

This research is part of the doctoral dissertation of Peter P. Flaig. The project was funded through the National Science Foundation Office of Polar Programs grants OPP-425636 to McCarthy and OPP-424594 to Fiorillo. Additional funding was provided by the Geist Fund at the University of Alaska Museum of the North, the Alaska Geologic Society, the University of Alaska–Fairbanks Graduate School Fellowship Program, the Geological Society of America, the Evolving Earth Foundation, and BP America. Logistical support was provided by the Barrow Arctic Science consortium (BASC), CH2M Hill (formerly Veco Polar Resources), Wrights Air, Air Logistics, Alaska Air Taxi, Evergreen Helicopters, and the support staff at Umiat, Alaska. We thank Dolores van der Kolk, Thomas Adams, David Norton, Roland Gangloff, Douglas Hissom, Susi Tomsich, and Jason Addison for their diligent assistance in the field. We also thank David Houseknecht, Paul Decker, Marwan Wartes, and Charles "Gil" Mull for enlightening discussions regarding the evolution of the Colville Basin and the Cretaceous stratigraphy of Alaska. This manuscript benefitted greatly from insightful and helpful reviews by Guy Plint and Gary Hampson.

REFERENCES

ABBADO, D., SLINGERLAND, R., AND SMITH, N.D., 2005, Origin of anastomosis in the upper Columbia River, British Columbia, *in* Blum, M. D., Mariott, S. B., and Leclaire, S.F., eds., Fluvial Sedimentology VII: International Association of Sedimentologists, Special Publication 35, p. 3–15.

ALLEN, J.R.L., 1963, The classification of cross-stratified units with notes on their origin: Sedimentology, v. 2, p. 93–114.

ALLEN, J.R.L., 1964, Studies in fluvitile sedimentation: Six cyclothems from the lower Old Red Sandstone, Anglo-Welsh Basin: Sedimentology, v. 3, p. 163–198.

ALLEN, J.R.L., 1965a, Fining upward cycles in alluvial successions: Geological Journal, v. 4, p. 229–246.

ALLEN, J.R.L., 1965b, A review of the origin and characteristics of recent alluvial sediments: Sedimentology, v. 5, p. 89–191.

ALLEN, J.R.L., 1970, Studies in fluvitile sedimentation; a comparison of fining-upwards cyclothems, with special reference to coarse-member composition and interpretation: Journal of Sedimentary Petrology, v. 40, p. 299–323.

ALLEN, J.R.L., 1999, Geological impacts on coastal wetland landscapes: some general effects of sediment autocompaction in the Holocene of northwest Europe: The Holocene, v. 9, no. 1, p. 1–12.

ARCHE, A., 1983, Coarse-grained meander lobe deposits in the Jarma River, Madrid, Spain, *in* Collinson, J.D., and Lewin, J., eds., Modern and Ancient Fluvial Systems: International Association of Sedimentologists, Special Publication 6, p. 313–321.

ASHLEY, G.M., MAITIMA MWORIA, J., MUSYA, A.M., OWENS, R.B., DRIESE, S.G., HOVER, V.C., RENAUT, R.W., GOWAN, M.F., MATHAI, S., AND BLATT, S.H., 2004, Sedimentation and recent history of a freshwater wetland in a semi-arid environment: Lobi Swamp, Kenya, East Africa: Sedimentology, v. 55, p. 1301–1321.

ASTIN, T.R., 1990, The Devonian lacustrine sediments of Orkney, Scotland; Implications for climate cyclicity, basin structure, and maturation history: Geological Society of London, Journal, v. 147, p. 141–151.

AVANZINI, M., FRISA, S., VAN DEN DRIESSCHE, K., AND KEPPENS, E., 1997, A dinosaur tracksite in an Early Liassic tidal flat in Northern Italy: Paleoenvironmental reconstruction from sedimentology and geochemistry: Palaios, v. 12, p. 538–551.

BAKER, E.K., HARRIS, P.T., KEENE, J.B., AND SHORT, S.A., 1995, Patterns of sedimentation in the macrotidal Fly River delta, Papua New Guinea, *in* Flemming, B.W., and Bartholomä, A., eds., Tidal Signatures in Modern and Ancient Sediments: International Association of Sedimentologists Special Publication 24, p. 193–211.

BANKS, N.L., 1973, The origin and significance of some downcurrent-dipping cross-stratified sets: Journal of Sedimentary Petrology, v. 43, p. 423–427.

BARRETT, P.J., 1970, Paleocurrent analysis of the mainly fluvitile Permian and Triassic Beacon rocks, Beardmore Glacier area, Antarctica: Journal of Sedimentary Petrology, v. 40, p. 395–411.

BESLY, B.M., AND COLLINSON, J.D., 1991, Volcanic and tectonic controls of lacustrine and alluvial sedimentation in the Stephanian coal-bearing sequence of the Malpas–Sort basin, Catalonian Pyrenees: Sedimentology, v. 38, p. 3–26.

BESSE, J., AND COURTILLOT, V., 1991, Revised and synthetic apparent polar wander paths of the African, Eurasian, North American and Indian plates, and true polar wander since 200 Ma: Journal of Geophysical Research, v. 96, p. 4029–4050.

BHATTACHARYA, J.P., AND MACEACHERN, J.A., 2009, Hyperpycnal rivers and prodeltaic shelves in the Cretaceous Seaway of North America: Journal of Sedimentary Research, v. 79, p. 184–209.

BIRD, K.J., 2001, Alaska—A twenty-first-century petroleum province, *in* Downey, M.W., Threet, J.C., and Morgan, W.A., eds., Petroleum Provinces of the Twenty-First Century: American Association of Petroleum Geologists, Memoir 74, p. 137–165.

BLACK, R.F., 1964, Gubik Formation of Quaternary age in northern Alaska: U.S. Geological Survey, Professional Paper 302-C, p. 59–91.

BOWN, T.M., AND KRAUS, M.J., 1987, Integration of channel and floodplain suites I. Developmental sequence and lateral relations of alluvial paleosols: Journal of Sedimentary Petrology, v. 57, p. 587–601.

BRANDLEN, E., 2008, Paleoenvironmental reconstruction of the Late Cretaceous (Maastrichtian) Prince Creek Formation, near the Kikak–Tegoseak dinosaur quarry, North Slope, Alaska: Unpublished Master's Thesis, University of Alaska–Fairbanks, Fairbanks, Alaska, 225 p.

BRIDGE, J.S., 1984, Large scale facies sequences in alluvial overbank environments: Journal of Sedimentary Petrology, v. 54, p. 583–588.

BRIDGE, J.S., 1985, Paleochannel patterns inferred from alluvial deposits; a critical evaluation: Journal of Sedimentary Petrology, v. 55, p. 579–589.

BRIDGE, J.S., 2003, Rivers and Floodplains; Forms, Processes, and Sedimentary Record: Malden, Massachusetts, Blackwell Science Ltd., 491 p.

BRIDGE, J.S., AND MACKEY, S.D., 1993, A theoretical study of fluvial sandbody dimensions, *in* Flint, S.S., and Bryant, I.D. eds., The Geologic Modeling of Hydrocarbon Reservoirs and Outcrop Analogues: International Association of Sedimentologists, Special Publication 15, p. 213–236.

BRIDGE, J.S., GUILLERMO, A.J., AND GEORGIEFF, S.M., 2000, Geometry, lithofacies, and spatial distribution of Cretaceous fluvial sandstone bodies, San Jorge Basin, Argentina: Outcrop analog for the hydrocarbon-bearing Chubut Group: Journal of Sedimentary Research, v. 70, p. 341–359.

BRISTOW, C.S., SKELLY, R.L., AND ETHRIDGE, F.G., 1999, Crevasse splays from the rapidly aggrading, sand bed, braided, Niobrara River, Nebraska: effect of base level rise: Sedimentology, v. 46, p. 1029–1047.

BROSGE, W.P., WHIMINGTON, C.L., AND MORRIS, R.H., 1966, Geology of the Umiat–Maybe Creek region, Alaska: U.S. Geological Survey, Professional Paper 303-H, p. 548–570.

BROUWERS, E.M., AND DE DECKKER, P., 1993, Late Maastrichtian and Danian ostracod faunas from northern Alaska: Reconstructions of environment and paleogeography: Palios, v. 8, p. 140–154.

Brouwers, E.M., Clemens, W.A., Spicer, R.A., Ager, T.A., Carter, L.D., and Sliter, W.V., 1987, Dinosaurs on the North Slope, Alaska: High latitude, latest Cretaceous environments: Science, v. 25, p. 1608–1610.

Burns, W. M., Hayba, D. O., Rowan, E. L., and Houseknecht, D. W., 2005, Estimating the amount of eroded section in a partially exhumed basin from geophysical well logs: an example from the North Slope, United States Geological Survey Special Paper 1732-D, p. 1–18.

Catuneanu, O., Abreu, V., Bhattacharya, J.P., Blum, M.D., Dalrymple, R.W., Eriksson, P.G., Fielding, C.R., Fisher, W.L., Galloway, W.E., Gibling, M.R., Giles, K.A., Holbrook, J.M., Jordan, R., Kendall, C.G.St.C., Macurda, B., Martinsen, O.J., Miall, A.D., Neal, J.E., Nummendal, D., Pomar, L., Posamentier, H.W., Pratt, B.R., Sarg, J.F., Shanley, K.W., Steel, R.J., Strasser, A., Tucker, M.E., and Winker, C., 2009, Towards the standardization of sequence stratigraphy: Earth-Science Reviews, v. 92, p. 1–33.

Changsong, L., Sitian, L., and Zhen, L., 1995, Facies architecture, stratigraphic sequences, and coal occurrences in the Late Carboniferous and Early Permian delta complexes of the North Huabei Basin, China, *in* Oti, M.N., and Potsma, G.P., eds, Geology of Deltas: Rotterdam, Netherlands, A.A. Balkema, p. 125–138

Chevren, V.B., 1978, Fluvial and deltaic facies in the Sentinel Butte Formation, central Williston Basin: Journal of Sedimentary Petrology, v. 48, p.159–170.

Choi, K.S., Dalrymple, R.W., Seung, S.C., and Seong-Pil, K., 2004, Sedimentology of modern inclined heterolithic stratification (IHS) in the macrotidal Han River Delta, Korea: Journal of Sedimentary Research, v. 74, p. 677–689.

Clemens, W.A., and Nelms, L.G., 1993, Paleoecological implications of Alaskan terrestrial vertebrate fauna in latest Cretaceous time at high paleolatitudes: Geology, v. 21, p. 503–506.

Cole, F., Bird, K.J., Toro, J., Roure, F., O'Sullivan, P.B., Pawlewicz, M., and Howell, D.G., 1997, An integrated model for the tectonic development of the frontal Brooks Range and Colville Basin 250 km west of the Trans-Alaska Crustal Transect: Journal of Geophysical Research, v. 102, p. 20,685–20,708.

Coleman, J.M., and Gagliano, S.M., 1964, Cyclic sedimentation in the Mississippi River deltaic plain: Gulf Coast Association of Geological Societies, Transactions, v. 14, p. 67–80.

Coleman, J.M., Gagliano, S.M., and Webb, G.E., 1964, Minor sedimentary structures in a prograding distributary: Marine Geology, v. 1, p. 240–258.

Collinson, J.D., 1978, Vertical sequence and sand body shape in alluvial sequences, *in* Miall, A.D., ed., Fluvial Sedimentology: Canadian Society of Petroleum Geologists, Memoir 5, p. 577–587.

Collinson, J.W., Isbell, J.L., Elliot, D.H., Miller, M.F., and Miller, J.M.G., 1994, Permian–Triassic Transantarctic basin, *in* Veevers, J.J., and Powell, C.M., eds., Permian–Triassic Pangean Basins and Foldbelts along the Panthalassan Margin of Gondwanaland: Geological Society of America, Memoir 184, p. 173–222.

Conrad, J.E., McKee, E.H., and Turrin, B.D., 1990, Age of tephra beds at the Ocean Point Dinosaur Locality, North Slope, Alaska, based on K-Ar and $^{40}Ar/^{39}Ar$ Analyses: U.S. Geological Survey, Bulletin 1990-C, p. 1–12.

Crerar, E.E., 2003, Sedimentology and stratigraphic evolution of a tidally-influenced marginal-marine complex: the Lower Cretaceous McMurray Formation, Athabasca oil sands deposit, northeastern Alberta: Unpublished Masters Thesis, University of Ottawa, Ottawa, Ontario, Canada, 309 p.

Crerar, E.E., and Arnott, R.W.C., 2007, Facies distribution and stratigraphic architecture of the Lower Cretaceous McMurray Formation, Lewis Property, northeastern Alberta: Bulletin of Canadian Petroleum Geology, v. 55, p. 99–124.

Dalrymple, R.W., and Choi, K., 2007, Morphologic and facies trends through the fluvial–marine transition in tide-dominated depositional systems: A schematic framework for environmental and sequence-stratigraphic interpretation: Earth-Science Reviews, v. 81, p. 135–174.

Dalrymple, R.W., Zaitlin, B.A., and Boyd, R., 1992, Estuarine facies models: Conceptual, basis, and stratigraphic applications: Journal of Sedimentary Petrology, v. 62, p. 1130–1146.

Davies-Vollum, K.S., 1999, The formation of beds underlying carbonaceous shales as Aquic paleosols: examples from the Bighorn Basin of Wyoming: International Journal of Coal Geology, v. 41, p. 239–255.

Decker, P. L., 2007, Brookian Sequence stratigraphic correlations, Umiat Field to Milne Point Field, west-central North Slope, Alaska: Preliminary Interpretive Report 2007-2, Alaska Department of Natural Resources, Fairbanks, Alaska, 21 p., 1 map.

Decker, P.L., LePain, D.L., Wartes, M.A., Gillis, R.J., Mongrain, J.R., Kirkham, R.A., and Schellenbaum, D.P., in press, Sedimentology, stratigraphy, and subsurface expression of Upper Cretaceous strata in the Sagavanirktok River area, east central North Slope, Alaska, *in* Wartes, M.W., and Decker, P.L., eds., Preliminary Results of Recent Geologic Field Investigations in the Brooks Range Foothills and North Slope, Alsaka: Alaska Division of Geologic & Geophysical Surveys, Preliminary Interpretive Report 2009-1C, 3 sheets.

Deconto, R.M., Brady, E.C., Bergengren, J., and Hay, W.W., 2000, Late Cretaceous climate, vegetation, and ocean interactions, *in* Huber, B.T., MacLeod, K.G., and Wing, S.L., 2000, Warm Climates in Earth History: Cambridge, UK, Cambridge University Press, p. 275–296.

Detterman, R.L., Bickel, R.S., and Gryk, G., 1963, Geology of the Chandler River region, Alaska: U.S. Geological Society, Special Paper 303-E, p. 223–324.

Diessel, C.F.K., 1992, Coal-Bearing Depositional Systems: Berlin, Springer-Verlag, 721 p.

Driese, S.G., and Ober, E.G., 2005, Paleopedologic and paleohydrologic records of precipitation seasonality from Early Pennsylvanian "underclay" paleosols, U.S.A: Journal of Sedimentary Research, v. 75, p. 997-1010.

Eberth, D.A., 1996, Origin and significance of mud-filled incised valleys (Upper Cretaceous) in southern Alberta, Canada: Sedimentology, v. 43, p. 459–477.

Eberth, D.A., and Miall, A.D., 1991, Stratigraphy, sedimentology, and evolution of a vertebrate-bearing braided to anastomosed fluvial system, Cutler Formation (Permian–Pennsylvanian), north-central New Mexico: Sedimentary Geology, v. 72, p. 225–252.

Elliott, T., 1974, Interdistributary bay sequences and their genesis: Sedimentology, v. 21, p. 611–622.

Elliott, T., 1976, The morphology, magnitude, and regime of a Carboniferous fluvial-distributary channel: Journal of Sedimentary Petrology, v. 46, p. 70–76.

Elliott, R.E., 1985, Quantification of peat to coal compaction stages, based especially on the phenomena in the East Pennine Coalfield, England: Yorkshire Geological Society, Proceedings, v. 45, part 3, p. 163–172.

Ethridge, F.G., and Schumm, S. A., 1978, Reconstructing paleochannel morphologic and flow characteristics; methodology, limitations and assessment, *in* Miall, A.D. ed., Fluvial Sedimentology, Canadian Society of Petroleum Geologists, Memoir, p.703–721.

Ethridge, F.G., Jackson, T.J., and Youngberg, A.D., 1981, Floodbasin sequence of a fine-grained meander belt subsystem: The coal-bearing Lower Wasatch and Upper Fort Union Formations, southern Powder River Basin, Wyoming, *in* Ethridge, F.G., and Flores, R.M. eds., Recent and Ancient Nonmarine Depositional Environments; Models for Exploration: Society of Economic Paleontologists and Mineralogists, Special Publication 31, p. 191–209.

Fielding, C.R., 1984, Upper delta plain lacustrine and fluviolacustrine facies from the Westphalian of the Durham coalfield, NE England: Sedimentology, v. 31, p. 547–567.

Fielding, C.R., 1986, Fluvial channel and overbank deposits from the Westphalian of the Durham coalfield, NE England: Sedimentology, v. 33, p. 119–140.

Fielding, C.R., 1987a, Lower delta-plain interdistributary deposits; an example from the Westphalian of the Lancashire Coalfield, northwestern England: Geological Journal, v. 22, p. 151–162.

Fielding, C.R., 1987b, Coal deposition models for deltaic and alluvial plain sequences: Geology, v. 15, p. 661–664.

Fiorillo, A.R., and Gangloff, R.A, 2000, Theropod teeth from the Prince Creek Formation (Cretaceous) of northern Alaska, with speculations on arctic dinosaur paleoecology: Journal of Vertebrate Paleontology, v. 20, p 675–682.

Fiorillo, A.R., Tykoski, R.S., Currie, P.J., McCarthy, P.J., and Flaig, P.P., 2009, Description of two Troodon partial braincases from the Prince Creek

Formation (Upper Cretaceous), North Slope, Alaska: Journal of Vertebrate Paleontology, v. 29, p. 178–187.

Fiorillo, A.R., McCarthy, P.J., Flaig, P.P., Brandlen, E., Norton, D., Jacobs, L., Zippi, P., and Gangloff, R.A., 2010, Paleontology and paleoenvironmental interpretation of the Kikak–Tegoseak dinosaur quarry, (Prince Creek Formation: Late Cretaceous), northern Alaska: A multi-disciplinary study of an ancient high-latitude, ceratopsian dinosaur bonebed, *in* Ryan, M.J., Chinner-Algeier, B.J., and Eberth, D.A., eds., New Perspectives on Horned Dinosaurs: The Royall Tyrell Museum Ceratopsian Syposeum: Bloomington, Indiana, Indiana University Press, p. 456–477.

Fischer, A.G., 1981, Climatic oscillations in the biosphere, *in* Nitecki, M.H., ed., Biotic Crisis in Ecological and Evolutionary Time: New York, Academic Press, p. 103–132.

Fisk, H.N., 1944, Geological investigation of the alluvial valley of the lower Mississippi River: Tulsa Geological Society, Digest, v. 15, 78 p.

Flach, P.D., and Mossop, G.D., 1985, Depositional environments of the Lower Cretaceous McMurray Formation, Athabasca Oil Sands, Alberta: American Association of Petroleum Geologists, Bulletin, v. 69, p. 1195–1207.

Flaig, P.P., 2005, Changing fluvial style across the Permian–Triassic boundary: Beardmore Glacier Region, Central Transantarctic Mountains, Antarctica: Unpublished Master's Thesis, University of Wisconsin–Milwaukee, Milwaukee, Wisconsin, 288 p.

Flores, R.M., 1984, Comparative analysis of coal accumulation in Cretaceous alluvial deposits, southern United States Rocky Mountain basins, *in* Stott, D.F., and Glass, D.J., eds., The Mesozoic of Middle North America: Canadian Society of Petroleum Geologists, Memoir 9, p. 373–385.

Flores, R.M., Stricker, G.D., Decker, P.L., and Myers, M.D., 2007a, Sentinel Hill Core Test 1: facies descriptions and stratigraphic reinterpretations of the Prince Creek and Schrader Bluff Formations, North Slope, Alaska: U.S. Geological Survey, Professional Paper 1747, 31 p.

Flores, R.M., Myers, M.D., Houseknecht, D.W., Stricker, G.D., Brizzolara, D.W., Ryherd, T.J., and Takahashi, K.I., 2007b, Stratigraphy and facies of Cretaceous Schrader Bluff and Prince Creek Formations in Colville River Bluffs, North Slope, Alaska: U.S. Geological Survey, Professional Paper 1748, 52 p.

Frederiksen, N.O, 1991, Pollen zonation and correlation of Maastrichtian marine beds and associated strata, Ocean Point dinosaur locality, North Slope, Alaska: U.S. Geological Survey, Bulletin 1990-E, 24 p.

Frederiksen, N.O., and McIntyre, D.J., 2000, Palynomorph biostratigraphy of mid(?) Campanian to upper Maastrichtian strata along the Colville River, North Slope of Alaska: U.S. Geological Survey, Open-File Report 00–493, 36 p.

Frederiksen, N.O., Ager, T.A., and Edwards, L.E., 1986, Comment on "Early Tertiary marine fossils from northern Alaska: Implications for Arctic Ocean paleogeography and faunal evolution": Geology, v. 14, p. 802–803.

Frederiksen, N.O., Ager, T.A., and Edwards, L.E., 1988, Palynology of Maastrichtian and Paleocene rocks, lower Colville River region, North Slope, Alaska: Canadian Journal of Earth Sciences, v. 25, p. 512–527.

Frederiksen, N.O., McIntyre, D.J., and Sheehan, T.P., 2002, Palynological dating of some Upper Cretaceous to Eocene outcrop and well samples from the region extending from the easternmost part of NPRA in Alaska to the western part of ANWR, North Slope of Alaska: U.S. Geological Survey, Open-File Report 02-405, 37 p.

Friend, P.F., 1983, Towards the field classification of alluvial architecture or sequence, *in* Collinson, J.D., and Lewin, J., Modern and Ancient Fluvial Systems: International Association of Sedimentologists, Special Publication 6, p. 345–354.

Friend, P.F., Slater, M.J., and Willimas, R.C., 1979, Vertical and lateral buildings of river sandstone bodies, Ebro Basin, Spain: Geological Society of London, Journal, v. 136, p. 39–46.

Gallagher, S.J., Wagstaff, B.E., Baird, J.G., Wallace, M.W., and Li, C.L., 2008, Southern high latitude climate variability in the Late Cretaceous greenhouse world: Global and Planetary Change, v. 60, p. 351–364.

Gangloff, R.A., Fiorillo, A.R., and Norton, D.W., 2005, The first Pachycephalosaurine (Dinosauria) from the Paleo-Arctic and its paleogeographic implications: Journal of Paleontology, v. 79, p. 997-1001.

Gardner, T.W., 1983, Paleohydrology and paleogeomorphology of a Carboniferous, meandering, fluvial sandstone: Journal of Sedimentary Petrology, v. 53, p. 991-1005.

Gastaldo, R.A., Allen, G.P., and Huc, A.-Y., 1995, The tidal character of fluvial sediments of the modern Markham River Delta, Kalimantan, Indonesia, *in* Flemming, B.W., and Bartholomä, A., eds., Tidal Signatures in Modern and Ancient Sediments: International Association of Sedimentologists, Special Publication 24, p. 171–181.

Garrity, C.P., Houseknecht, D.W., Bird, K.J., Potter, C.J., Moore, T.E., Nelson, P.H., and Schenk, C.J., 2005, U.S. Geological Survey 2005 oil and gas resource assessment of the central North Slope, Alaska: Play maps and results: U.S. Geological Survey, Open-File Report 2005-1182, 29 p.

Gibling, M.R., 2006, Width and thickness of fluvial channel bodies and valley fills in the geological record: a literature compilation and classification: Journal of Sedimentary Research, v. 76, p.731–770.

Gradziński, R., 1970, Sedimentation of dinosaur-bearing Upper Cretaceous deposits of the Nemegt Basin, Gobi Desert: Palaeontologica Polonica, v. 21, p. 147–229.

Gryc, G., Patton, W.W., and Payne, T.G., 1951, Present stratigraphic nomenclature of northern Alaska: Washington Academy of Sciences, Journal, v. 41, p. 159–167.

Guow, M.J.P., and Berendsen, H.J.A., 2007, Variability of channel-belt dimensions and the consequences for alluvial architecture: observations from the Holocene Rhine–Meuse Delta (the Netherlands) and Lower Mississippi Valley (U.S.A.): Journal of Sedimentary Research, v. 77, p. 124–138.

Haugen, R.K., 1982, Climate of remote areas in north-central Alaska, 1975–1979 summary, *in* US Army Cold Regions Research and Engineering Laboratory, Report 82-35, 110 p.

Heritage, G.L, Charlton, M.E., and Regan, S.O., 2001, Morphological classification of fluvial environments: An investigation of the continuum of channel types: Journal of Geology, v. 109, p. 21–33.

Hovikoski, J., Räsänen, M., Gingras, M., Ranzi, A., and Melo, J., 2008, Tidal and seasonal controls in the formation of Late Miocene inclined heterolithic stratification deposits, western Amazonian foreland basin: Sedimentology, v. 55, p. 499–530.

Isbell, J.L., 1990. Fluvial sedimentology and basin analyses of the Permian Fairchild and Buckley formations, Beardmore Glacier region, and the Weller Coal Measures, Southern Victoria Land, Antarctica [Ph.D. thesis]: Department of Geological Sciences, The Ohio State University, Columbus, Ohio, 347 p.

Jackson, R.G., 1981, Sedimentology of muddy fine-grained channel deposits in meandering streams of the American middle west: Journal of Sedimentary Petrology, v. 51, p. 1169–1192.

Joeckel, R.M., Kudvigson, G.A., Witzke, B.J., Kvale, E.P., Phillips, P.L., Brenner, R.L., Thomas, S.G., and Howard, L.M., 2005, Paleogeography and fluvial to estuarine architecture of the Dakota Formation, (Cretaceous, Albian) eastern Nebraska, USA, *in* Blum, M.D., Marriott, S.B., and Leclair, S.F., eds., Fluvial Sedimentology VII: International Association of Sedimentologists, Special Publication 35, p. 453–480.

Johnson, C.L., and Graham, S.A., 2004, Cycles in perilacustrine facies of Late Mesozoic rift basin, southeastern Mongolia: Journal of Sedimentary Research, v.74, p. 786–804.

Jorgensen, P.J., and Fielding, C.R., 1996, Facies architecture of alluvial floodbasin deposits: Three-dimensional data from the Upper Triassic Callide Coal Measures of east–central Queensland, Australia: Sedimentology, v. 43, p. 479–495.

Kirschbaum, M.A., and McCabe, P.J., 1992, Controls on the accumulation of coal and on the development of anastomosed fluvial systems in the Cretaceous Dakota Formation of southern Utah: Sedimentology, v. 39, p. 581–598.

Kjemperud, A.V., Schomacker, E.R., and Cross, T.A., 2008, Architecture and stratigraphy of alluvial deposits, Morrison Formation (Upper Jurassic), Utah: American Association of Petroleum Geologists, Bulletin, v. 92, p. 1055–1076.

Knighton, A.D., and Nanson, G.C., 1993, Anastomosis and the continuum of channel pattern: Earth Surface Processes and Landforms, v. 18, p. 613–625.

Kraus, M.J., 1987, Integration of channel and floodplain suites, II. Vertical relations of alluvial Paleosols: Journal of Sedimentary Petrology, v. 57, p. 602–617.

Kraus, M.J., 1996, Avulsion deposits in lower Eocene alluvial rocks, Bighorn Basin, Wyoming: Journal of Sedimentary Research, v. 66, p. 354–363.

Kraus, M.J., 1998, Development of potential acid sulfate paleosols in Paleocene floodplains, Bighorn Basin, Wyoming, USA: Palaeogeography, Palaeoclimatology, Palaeoecology, v. 144, p. 203–224.

Kraus, M.J., and Bown, T.M., 1988, Pedofacies analysis: a new approach to reconstructing ancient fluvial sequences, *in* Reinhardt, J., and Sigleo, W.R., eds., Paleosols and Weathering Through Geologic Time; Principles and Applications: Geological Society of America, Special Paper 216, p. 143–152.

Kraus, M.J., and Hasiotis, S.T., 2006, Significance of different methods of rhizolith preservation to interpreting paleoenvironmental and paleohydrologic settings: Examples from Paleogene paleosols, Bighorn Basin, Wyoming, U.S.A.: Journal of Sedimentary Research, v. 76, p. 633–646.

La Porte, L.F., and Behrensmeyer, A.K., 1980, Tracks and substrate reworking by terrestrial vertebrates in quaternary sediment of Kenya: Journal of Sedimentary Petrology, v. 50, p. 1337–1346.

Leclair, S.F., and Bridge, J.S., 2001, Quantitative interpretation of sedimentary structures formed by river dunes: Journal of Sedimentary Research, v. 71, p. 713–716.

Leeder, M.R., 1973, Fluviatile fining-upwards cycles and the magnitude of paleochannels: Geological Magazine, v. 110, p. 265–276.

Lewis, G.W., and Lewin, J., 1983, Alluvial cutoffs in Wales and the Borderlands, *in* Collinson, J.D., and Lewin, J., eds., Modern and Ancient Fluvial Systems: International Association of Sedimentologists, Special Publication 6, p. 145–154.

Loope, D.B., 1986, Recognizing and utilizing vertebrate tracks in cross section: Cenezoic hoofprints from Nebraska, Palaios, v. 1, p. 141–151.

MacEachern, J.A., Pemberton, S.G., Bann, K.L., and Gingras, M.K., 2005, Departures from the archetypal ichnofacies: Effective recognition of physico-chemical stresses in the rock record, *in* MacEachern, J.A., Bann, K.L., Pemberton, S.G., and Gingras, M.K. eds., Applied Ichnology: SEPM, Short Course Notes, no. 52, p. 65–93.

Makaske, B., 2001, Anastomosing rivers: A review of their classification, origin and sedimentary products: Earth-Science Reviews, v. 53, p. 149–196.

Makaske, B., and Weerts, H.J.T., 2005, Muddy lateral accretion and low stream power in a sub-recent confined channel belt, Rhine–Meuse delta, central Netherlands: Sedimentology, v. 52, p. 651–668.

Martín-Closas, C., and Galtier, J., 2005, Plant taphonomy and paleoecology of Late Pennsylvanian intramontane wetlands in the Graissessac–Lode've Basin (Languedoc, France): Palaios, v. 20, p. 249–265.

McCarthy, P.J., Martini, I.P., and Leckie, D.A., 1998, Use of micromorphology for interpretation of complex alluvial paleosols: examples from the Mill Creek Formation (Albian), southwestern Alberta, Canada: Palaeogeography, Palaeoclimatology, Palaeoecology, v. 143, p. 87–110.

McCarthy, P.J., Faccini, U.F., and Plint, G., 1999, Evolution of an ancient coastal plain: palaeosols, interfluves and alluvial architecture in a sequence stratigraphic framework, Cenomanian Dunvegan Formation, NE British Columbia, Canada: Sedimentology, v. 46, p. 861–891.

McCarthy, P.J., and Plint, A.G., 2003, Spatial variability of paleosols across Cretaceous interfluves in the Dunvegan Formation, NE British Columbia, Canada: paleohydrological, paleogeomorphological, and stratigraphic implications: Sedimentology, v. 50, p. 1187–1220.

McLaurin, B.T., and Steel, R.J., 2007, Architecture and origin of an amalgamated fluvial sheet sand, lower Castlegate Formation, Book Cliffs, Utah: Sedimentary Geology, v. 197, p. 291–311.

Miall, A.D., 1976, Paleocurrent and paleohydrologic analysis of some vertical profiles through a Cretaceous braided stream deposit, Banks Island, Arctic Canada: Sedimentology, v. 23, p. 459–483.

Miall, A.D., 1985, Architectural-element analysis: A new method of facies analysis applied to fluvial deposits: Earth-Science Reviews, v. 22, p. 261–308.

Miall, A.D., 1996, The Geology of Fluvial Deposits; Sedimentary Facies, Basin Analysis, and Petroleum Geology: Berlin, Springer-Verlag, 582 p.

Michaelsen, P., Henderson, R.A., Crosdale, P.J., and Mikkelsen, S.O., 2000, Facies architecture and depositional dynamics of the Upper Permian Rangal Coal Measures, Bowen Basin, Australia: Journal of Sedimentary Research, v. 70, p. 879–895.

Mjøs, R., Walderhaug, O., and Prestholm, E., 1993, Crevasse splay sandstone geometries in the Middle Jurassic Ravenscar Group of Yorkshire, UK, *in* Marzo, M., and Puigdefabregas, C., eds., Alluvial Sedimentation: International Association of Sedimentologists, Special Publication 17, p. 167–184.

Molenaar, C.M., 1985, Subsurface correlations and depositional history of the Nanushuk Group and related strata, North Slope, Alaska, *in* Huffman, A.C., ed., Geology of the Nanushuk Group and Related Rocks, North Slope, Alaska: U.S. Geological Survey, Bulletin 1614, p. 37–60.

Molenaar, C.M., Bird, K.J., and Kirk, A.R., 1987, Cretaceous and Tertiary stratigraphy of northeastern Alaska, *in* Tailleur, I.L., and Weimer, P., eds., Alaskan North Slope Geology: SEPM, Pacific Section, Book 50, v. 1, p. 513–528.

Moore, T.E., Wallace, W.K., Bird, K.J., Karl, S.M., Mull, C.G., and Dillon, J.T., 1994, Geology of northern Alaska, *in* Plafker, G, and Berg, H.C., eds., The Geology of Alaska: Boulder, Colorado, The Geological Society of America, The Geology of North America, v. G-1, p. 49–140.

Morisawa, M., 1985, Topologic properties of delta distributary networks, *in* Michael, W.J., ed., Models in Geomorphology: Winchester, Massachusetts, Allen & Unwin, p. 239–268.

Morisawa, M., and Montgomery, W., 1985, Delta distributary networks: some quantitative aspects, *in* Kumar, A., ed., Facets of Geomorphology; Allahabad, India, Thinker's Library, p. 106–121.

Morozova, G.S., and Smith, N.D., 2000, Holocene avulsion styles and sedimentation patterns of the Saskatchewan River, Cumberland Marshes, Canada: Sedimentary Geology, v. 130, p. 81–105.

Mossop, G.D., and Flach, P.D., 1983, Deep channel sedimentation in the Lower Cretaceous McMurray Formation, Athabasca oil sands, Alberta: Sedimentology, v. 30, p. 493–509.

Mull, C.G., 1985, Cretaceous tectonics, depositional cycles, and the Nanushuk Group, Brooks Range and Arctic Slope, Alaska, *in* Huffman, A.C. Jr., ed., Geology of the Nanushuk Group and Related Rocks, North Slope, Alaska: U.S. Geological Survey, Bulletin 1614, p. 7–36.

Mull, C.G., Houseknecht, D.W., and Bird, K.J., 2003, Revised Cretaceous and Tertiary stratigraphic nomenclature in the Colville Basin, northern Alaska: U.S. Geological Survey, Professional Paper 1673, 51 p.

Nadon, G.C., 1993, The association of anastomosed fluvial deposits and dinosaur tracks, eggs, and nests: Implications for the interpretation of floodplain environments and a possible survival strategy for Ornithopods: Palaios, v. 8, p. 31–44.

Nadon, G.C., 1994, The genesis and recognition of anastomosed fluvial deposits: Data from the St. Mary River Formation, southwestern Alberta, Canada: Journal of Sedimentary Research, v. B64, p. 451–463.

Nakajo, T., 1998, Tidal influences on distributary-channel sedimentation of the Tertiary delta in the Taishu Group, Tsushima Islands, southwestern Japan: Journal of Geosciences, Osaka City University, v. 41, p. 37–46.

Nanson, G.C., 1980, Point bar and floodplain formation of the meandering Beatton River, northeastern British Columbia: Sedimentology, v. 27, p. 3–29.

Nio, S.D., and Yang, C.S., 1991, Diagnostic attributes of clastic tidal deposits: a review, *in* Smith D.G., Reinson, G.E., Zaitlin, B.A., and Rahmani, R.A., eds., Clastic Tidal Sedimentology: Canadian Society of Petroleum Geologists, Memoir 16, p. 3–28.

Nordt, L., Atchley, S., and Dworkin, S., 2003, Terrestrial evidence for two greenhouse events in the latest Cretaceous: GSA Today, v. 13, no. 12, p. 4–9.

Olariu, C., and Bhattacharya, J.P., 2006, Terminal distributary channels and delta front architecture of river-dominated delta systems: Journal of Sedimentary Research, v. 76, p. 212–233.

Orton, G.J., and Reading, H.G., 1993, Variability in deltaic processes in terms of sediment supply with particular emphasis on grain size: Sedimentology, v. 40, p. 475–512.

Page, K.J., Nanson, G.C., and Frazier, P.S., 2003, Floodplain formation and sediment stratigraphy resulting from oblique accretion on the Murrumbidgee River, Australia: Journal of Sedimentary Research, v. 73, p. 5–14.

Parrish, J.T., and Spicer, R.A., 1988, Late Cretaceous terrestrial vegetation: A near-polar temperature curve: Geology, v. 16, p. 22–25.

Parrish, J.M., Parrish, J.T., Hutchison, J.H., and Spicer, R.A., 1987, Late

Cretaceous vertebrate fossils from the North Slope of Alaska and implications for dinosaur ecology: Palaios, v. 2, p. 377–389.

Pérez-Arlucea, M., and Smith, N. D., 1999, Depositional pattern following the 1870's avulsion of the Saskatchewan River (Cumberland Marshes, Saskatchewan, Canada): Journal of Sedimentary Research, v. 69, p. 62–73.

Phillips, R.L., 1989, Summary of late Cretaceous environments near Ocean Point, North Slope, Alaska: Geological studies in Alaska by the U.S. Geological Survey: U.S. Government Printing Office, Washington D.C., p. 101–106.

Phillips, R.L., 2003, Depositional environments and processes in Upper Cretaceous nonmarine and marine sediments, Ocean Point dinosaur locality, North Slope, Alaska: Cretaceous Research, v. 24, p. 499–523.

Plint, A.G., 1983, Sandy fluvial point-bar sediments from the Middle Eocene of Dorset, England, *in* Collinson, J.D., and Lewin, J., eds., Modern and Ancient Fluvial Systems: International Association of Sedimentologists, Special Publication 6, p. 355–368.

Postma, D., 1977, The occurrence and chemical composition of recent Fe-rich mixed carbonates in a river bog: Journal of Sedimentary Petrology, v. 47, p. 1089–1098.

Rajchl, M., and Uličný, D., 2005, Depositional record of an avulsive fluvial system controlled by peat compaction (Neogene, Most Basin, Czech Republic): Sedimentology, v. 52, p. 601–625.

Reading, H.G., 1996, Sedimentary Environments: Processes, Facies and Stratigraphy: Malden, Massachusetts, Blackwell Science Ltd., 688 p.

Retallack, G.J., 2001, Soils of the Past; An Introduction to Paleopedology: London, Blackwell Science, 404 p.

Rich, T.H., Gangloff, R.A., and Hammer, W.R., 1997, Polar dinosaurs, *in* Currie, P.J., and Padian, K., eds., Encyclopedia of Dinosaurs: San Diego, Academic Press, p. 562–573.

Rich, T.H., Gangloff, R.A., and Hammer, W.H., 2002, Polar Dinosaurs: Science, v. 295, p. 979–980.

Roberts, H.H., 1997, Dynamic changes of the Holocene Mississippi River delta plain: the delta cycle: Journal of Coastal Research, v. 13, p. 605–627.

Roberts, E.M., 2007, Facies architecture and depositional environments of the Upper Cretaceous Kaiparowits Formation, southern Utah: Sedimentary Geology, v. 197, p. 207–233.

Roehler, H.W., 1987, Depositional environments of the coal-bearing and associated formations of Cretaceous age in the National Petroleum Reserve in Alaska: U.S. Geological Society, Bulletin 1575, 16 p.

Rogers, R.R., 1998, Sequence analysis of the Upper Cretaceous Two Medicine and Judith River Formations, Montana: Nonmarine response to the Claggett and Bearpaw marine cycles: Journal of Sedimentary Research, v. 68, p. 615–631.

Schumm, S.A., 1963, Sinuosity of alluvial rivers on the Great Plains: Geological Society of America, Bulletin, v. 74, p. 1089–1100.

Schumm, S.A., 1968, Speculations concerning paleohydrologic controls of terrestrial sedimentation: Geological Society of America, Bulletin, v. 79, p. 1573–1588.

Schumm, S.A., 1981, Evolution and response of the fluvial system, sedimentologic implications, *in* Ethridge, F.G. and Flores, R.M., eds., Recent and Ancient Nonmarine Depositional Environments: SEPM, Special Publication 31, p. 19–29.

Schumm, S.A., 1985, Patterns of alluvial rivers: Annual Review of Earth and Planetary Sciences, v. 13, p. 5–27.

Schumm, S.A., and Kahn, H.R., 1972, Experimental study of channel patterns: Geological Society of America, Bulletin, v. 83, p. 1755–1770.

Smith, D.G., 1976, Effect of vegetation on lateral migration of anastomosed channels of a glacial meltwater river: Geological Society of America, Bulletin, v. 87, p. 857–860.

Smith, D.G., 1983, Anastomosed fluvial deposits: modern examples from Western Canada, *in* Collinson, J.D., and Lewin, J., eds., Modern and Ancient Fluvial Systems: International Association of Sedimentologists, Special Publication 6, p. 155–168.

Smith, D.G., 1987, Meandering river point bar lithofacies models: modern and ancient examples compared, *in* Ethridge, F.G., Flores, R.M., Harvey, M.D., and Weaver, J.N., eds., Recent Developments in Fluvial Sedimentology: SEPM, Special Publication.39, p. 83–91.

Smith, D.G., 1988, Modern point bar deposits analogous to the Athabasca Oil Sands, Alberta, Canada, *in* de Boer, P.L., van Gelder, A., and Nio, S.D., eds., Tide-Influenced Sedimentary Environments and Facies: Dordrecht, D. Reidel, p. 417–432.

Smith, D.G., and Putnam, P.E., 1980, Anastomosed river deposits: Modern and ancient examples in Alberta, Canada: Canadian Journal of Earth Sciences, v. 17, p. 1396–1406.

Smith, D.G., and Smith, N.D., 1980, Sedimentation in anastomosed river systems: examples from alluvial valleys near Banff, Alberta: Journal of Sedimentary Petrology, v. 50, p. 157–164.

Smith, N.D., and Pérez-Arlucea, M., 1994, Fine-grained splay deposition in the avulsion belt of the Lower Saskachewan River, Canada: Journal of Sedimentary Research, v. B64, p. 159–168.

Smith, N.D., and Pérez-Arlucea, M., 2004, Effects of peat on the shapes of alluvial channels: examples from the Cumberland Marshes, Saskatchewan, Canada: Geomorphology, v. 61, p. 323–335.

Smith, N.D., Cross, T.A., Dufficy, J.P., and Clough, S.R., 1989, Anatomy of an avulsion: Sedimentology, v. 36, p. 1–23.

Spicer, R.A., 2003, Changing climate and biota, *in* Skelton, P., ed., The Cretaceous World: Cambridge, U.K., Cambridge University Press, p. 85–162.

Spicer, R.A., and Parrish, J.T., 1987, Plant megafossils, vertebrate remains, and paleoclimate of the Kogosukruk Tongue (Late Cretaceous), North Slope, Alaska: U.S. Geological Survey, Circular C 0998, p. 47–48.

Spicer, R.A., and Parrish, J., 1990, Late Cretaceous–early Tertiary paleoclimates of northern high latitudes: a quantitative view: Geological Society of London, Journal, v. 147, p. 329–341.

Spicer, R.A., and Herman, A.B., 2010, the Late Cretaceous environment of the Arctic: A quantitative reassessment based on plant fossils: Palaeogeography, Palaeoclimatology, Palaeoecology, doi: 10.1016/j.palaeo.2010.02.025

Spicer, R.A., Parrish, J.T., and Grant, P.R., 1992, Evolution of vegetation and coal-forming environments in the Late Cretaceous of the North Slope of Alaska, *in* McCabe, P.J., and Parrish, J.T., eds, Controls on the Distribution and Quality of Creataceous Coals: Geological Society of America, Special Paper 267, p. 177–192.

Stanley, K.O., and Surdam, R.C., 1978, Sedimentation on the front of Eocene Gilbert-type deltas, Washakie Basin, Wyoming: Journal of Sedimentary Petrology, v. 48, p. 557–573.

Stewart, D.J., 1981, A meander-belt sandstone of the Lower Cretaceous of southern England: Sedimentology, v. 28, p. 1–20.

Stewart, D.J., 1983, Possible suspended-load channel deposits from the Wealden Group (Lower Cretaceous) of southern England, *in* Collinson, J.D., and Lewin, J., eds., Modern and Ancient Fluvial Systems: International Association of Sedimentologists, Special Publication 6, p. 369–384.

Stouthamer, E., 2001, Sedimentary products of avulsions in the Holocene Rhine–Meuse Delta, The Netherlands: Sedimentary Geology, v. 145, p. 73–92.

Tanner, L.H., 2000, Palustrine–lacustrine and alluvial facies of the (Norian) Owl Rock Formation, (Chinile Group), Four Corners region, southwestern U.S.A.; Implications for Late Triassic paleoclimate: Journal of Sedimentary Research, v. 70, p. 1280–1289.

Tatsch, J.H., 1980, Coal Deposits: Sudsbury, Massachusetts, Tatsch Associates, 590 p.

Thomas, R.G., Smith, D.G., Wood, J.M., Visser, J., Calverley-Range, E.A., and Koster, E.H., 1987, Inclined heterolithic stratification—Terminology, description, interpretation, and significance: Sedimentary Geology, v. 53, p. 123–179.

Tooth, S., 2005, Splay formation along the lower reaches of ephemeral rivers on the northern plains of arid central Australia: Journal of Sedimentary Research, v. 75, p. 636–649.

Törnqvist, T.E., 1993, Holocene alternation of meandering and anastomosing fluvial systems in the Rhine–Meuse Delta (central Netherlands) controlled by sea level rise and subsoil erodibility: Journal of Sedimentary Petrology, v. 63, p. 683–693.

Törnqvist, T.E., van Ree, M.H.M, and Faessen, E.L.J.H., 1993, Longitudinal facies architectural changes of a Middle Holocene anastomosing distributary system (Rhine–Meuse Delta, central Netherlands), *in* Fielding, C.R., ed., Current Research in Fluvial Sedimentology: Sedimentary Geology, v. 85, p. 203–219.

Ufnar, D.F., González, L.A., Ludvigson, G.A., Brenner, R.L., and Witzke,

B.J., 2001, Stratigraphic implications of meteoric sphaerosiderite $\delta^{18}O$ values in paleosols of the Cretaceous (Albian) Boulder Creek Formation, N.E. British Columbia foothills, Canada: Journal of Sedimentary Research, v. 71, p. 1017–1028.

van Asselen, S., Stouthammer, E., and van Asch, Th.W.J., 2009, Effects of peat compaction on delta evolution: a review on processes, responses, measuring, and modeling: Earth-Science Reviews, v. 92, p. 35–51.

Vavra, C.L., 1984, Provenance and alteration of the Triassic Fremow and Falla Formations, Central Transantarctic Mountains, Antarctica: Columbus, Ohio, Ohio State University, Institute of Polar Studies, Report v. 87, 98 p.

Walker, R.G., ed., 1979, Facies Models: Geological Association of Canada, 211 p.

Witte, K.W., Stone, D.B., and Mull, C.G., 1987, Paleomagnetism, paleobotany, and paleogeography of the Cretaceous, North Slope, Alaska, *in* Tailleur, I., and Weimer, P., eds., Alaska North Slope Geology: SEPM Pacific Section and Alaska Geological Society, v. 1, p. 571–579.

Wood, J.M., 1989, Alluvial architecture of the Upper Cretaceous Judith River Formation, Dinosaur Provincial Park, Alberta, Canada: Bulletin of Canadian Petroleum Geology, v. 37, p. 169–181.

Wood, J.M., Thomas, R.G., and Visser, J., 1988, Fluvial processes and vertebrate taphonomy: The Upper Creatacous Judith River Formation, south-central Dinosaur Provincial Park, Alberta, Canada: Paleogeography, Paleoclimatology, Paleoecology v. 66, p. 127–143.

Wright, V.P., 1986, Pyrite formation and the drowning of a paleosol: Geological Journal, v. 21, p. 139–149.

Yang, B., Dalrymple, R.W., Gingras, M.K., Seungshoo, C., and Lee, H., 2007, Up-estuary variation of sedimentary facies and ichnocoenosis in an open-mouthed, macrotidal, mixed energy estuary, Gomso Bay, Korea: Journal of Sedimentary Research, v. 77, p. 757–771.

Zakharov, Y.D., Boriskina, N.G., Ignatyev, A.V., Tanabe, K., Shigeta, Y., Popov, A.M., Afanasyeva, T.B., and Maeda, H., 1999, Paleotemperature curve for the Late Cretaceous of the northwestern circum-Pacific: Cretaceous Research, v. 20, p. 685–697.

DIGITAL APPENDIX

The following supplementary material is available in electronic format either on the DVD accompanying this volume or through the digital archive on the SEPM website at http://www.sepm.org/ Larger versions of the photomosaic and interpreted line drawing pairs shown in Figs 10, 12, 13 and 14.

ROLE OF SCOURING AND BASE-LEVEL CHANGE IN PRODUCING ANOMALOUSLY THICK FLUVIAL SUCCESSIONS: AN EXAMPLE FROM THE TANA RIVER, NORTHERN NORWAY

RAYMOND S. EILERTSEN
Geological Survey of Norway, Polarmiljøsenteret, N-9296 Tromsø, Norway
e-mail: raymond.eilertsen@ngu.no

AND

GEOFFREY D. CORNER
Department of Geology, Faculty of Science and Technology, University of Tromsø, N-9037 Tromsø, Norway

ABSTRACT: An anomalously thick (> 30 m), multistory, Holocene fluvial succession exposed in a 600-m-long, postglacially raised river-terrace section at Masjoksletta, in the lower Tana River in northern Norway, is more than five times thicker than the "normal" single-story fluvial unit (3–5 m) found elsewhere in the lower part of the valley. We describe the facies composition, depositional architecture, and basal relief of the succession based on studies of exposures and ground-penetrating-radar data, and reconstructs the shoreline trajectory in order to evaluate the role of base-level change and deep scouring during deposition.

The succession at Masjoksletta comprises three, stacked, sandy, fluvial units erosionally overlying deltaic sediments. The lowermost unit (I) is 20 m thick and composed of relatively thick cross-beds of uniform grain size, in contrast to the overlying units (units II and III), which are 8–8.5 m thick and characterized by fining-upward grain size and decreasing-upward bed thickness. Unit I formed following rapid base-level fall but is interpreted as a scour-fill unit based on the following properties: (1) anomalous thickness (> 20 m) compared with a local norm of 3–5 m, (2) erosion and deposition up to 28 m below contemporaneous sea level, compared with modern channels depths of < 10 m, and (3) thick beds and uniform grain size compared with overlying fining-upward units. The middle (II) and uppermost (III) units are interpreted as having formed in response to a stillstand in relative sea-level related to the regional early-to-mid-Holocene Tapes transgression, which, combined with rapid delta progradation, caused the river bed to aggrade.

A comparison of channel flow depths for units I–III, derived from cross-bed set thickness and our regional data, shows that the methodology of using cross-bed thickness to derive flow depth may give a useful approximation but underestimates channel depth for scours and possibly also other high-aggradation settings. We suggest that the characteristics reported here for scour-and-fill sediments (unit I) at Tana, such as great thickness, uniform bed thickness, and uniform grain size, can be used to identify scour-fill deposits in other areas and for distinguishing between autogenic, localized channel scouring and allogenically controlled incision.

KEY WORDS: River terraces, fluvial facies, scour, base-level change, Norway

INTRODUCTION

Thick successions of stacked fluvial channel deposits are commonly interpreted to result from aggradation due to a rise in base level. In contrast, deeply eroded bases of fluvial sand bodies are commonly linked to sequence boundaries, interpreted as having formed during base-level fall (Salter 1993). However, deeply incised erosion surfaces may result from scouring, and such processes may occur at any time during a relative sea-level cycle. It is therefore important to distinguish between the deposits of channel fills formed by autogenic channel-base scouring and subsequent infill, and lithostratigraphic units reflecting regional allogenic processes such as base-level change (Friend 1978; Weimer 1984; Collinson 1986; Salter 1993; Best and Ashworth 1997; Miall 2006). Also, scours and their fills, comprising fluvial sand bodies of varying geometries, may represent primary targets for petroleum exploration (Ardies et al. 2002). In this paper we follow the definition of Salter (1993) whereby scour refers to the processes of erosion that are intrinsic to the fluvial channel and of local extent, whereas incision refers to episodes of erosion stimulated by extrinsic or allocyclic (Beerbower 1964) controls affecting the river over considerably larger distances.

The role of scour and fill in producing thick fluvial sequences is rarely mentioned as an alternative interpretation to base-level rise and aggradation, even though the deepest sediments in a river have the greatest preservation potential. There is, however, growing awareness based on an examination of modern rivers that scouring may produce thick sediment bodies, especially within braided systems and at channel confluences (Salter 1993; Best and Ashworth 1997; Fielding 2007). Several authors have also suggested that short-lived scour-and-fill events, which are effectively instantaneous in geological terms, may be responsible for depositing a disproportionately large amount of sediment (Smith et al. 1989; Ashworth et al. 1999; Ashworth et al. 2007; Holbrook 2001; Sheets et al. 2002). However, few examples of such phenomena have been described from the rock record, suggesting that they may have been overlooked or misinterpreted. Scouring is an important factor in controlling sandbody thickness, as it promotes vertical thickening. Lateral channel migration or in-channel scour migration has the potential to rework a large proportion of the sandbody base (Salter 1993; Bristow et al. 1993). Best and Ashworth (1997) suggested that scour depths greater than five times the mean depth of the upstream channel(s) should be documented before dismissing autocyclic processes as the cause of the scouring.

We report here results from an investigation of anomalously thick fluvial sediments exposed in a Holocene river terrace section at Masjoksletta, in the lower Tana valley, northern Norway. The succession is more than five times thicker than the normal fluvial succession in the area (cf. Sivertsen 1997; Corner 2006). We evaluate the role played by autocyclic (scouring) and allocyclic (changes in base level) processes in producing such a succession. The study site is particularly interesting because the depositional pattern can be related

From River To Rock Record: The Preservation Of Fluvial Sediments And Their Subsequent Interpretation
SEPM Special Publication No. 97, Copyright © 2011
SEPM (Society for Sedimentary Geology), ISBN 978-1-56576-305-0, p. 265–279.

to relatively well-established sea-level changes analyzed with reference to the known pattern of regional sea-level change, information not readily obtainable in ancient records. The present river, moreover, serves as an analogue for the sediments found in the section.

SETTING

The study area is a terrace section 20 m high and 600 m long at Masjoksletta on the Tana River, located approximately 35 km from the river mouth, and 3 km south of the Masjok tributary river, in Finnmark, northern Norway (Fig. 1). The terrace top has an elevation of 26–28 m a.s.l., and mean river-level is c. 6–7 m a.s.l. The terrace rises to 31 m a.s.l. close to the valley side to the west, and is incised by lower, younger terraces to the north and south (Fig. 1). The present climate is subarctic with a mean annual precipitation and temperature of c. 940 mm and −0.7 °C, respectively (Norwegian Meteorological Institute http://met.no/index.shtml). Bedrock in the area consists predominantly of Late Precambrian sedimentary rocks belonging to the Vestertana Group, mostly sandstones and shales (Siedlecka and Roberts 1992).

The Tana River is a bedload-dominated, sandy-to-gravel-bed river with an intermediate-transitional, meandering–braided (wandering) pattern (cf. Collinson 1970; Schumm 1981). The present river has a gradient of ~ 0.18 m km^{-1} between the study site and the river mouth. The lower 10–15 km is tidally influenced. Channel depths exceeding 5 m occur in places adjacent to the locality at present (Lotsari et al. 2010), and a maximum depth of 10 m has been recorded farther downstream (Corner et al. 1995). The width of the active bed ranges from c. 1 km at Masjoksletta to 2 km at Rustefjelbma (Fig. 1). The present river channel contains three major types of macroforms: (1) linguoid dunes, (2) bars, and (3) stable vegetated islands (Collinson 1970; Corner et al. 1995). Coalescence of the linguoid dunes results in the formation of diagonal bars (Røe and Corner 1994), which in turn build up compound side bars by oblique-upstream accretion (Corner et al. 1995). The Tana River displays a three-order hierarchy of channels following the definition by Bristow (1987).

The river drains an area of 16,380 km^2 and is 360 km long. It has a markedly seasonal, "mountain regime" discharge pattern, with high discharge ($>$ 500 m^3 s^{-1}) during a brief period of snowmelt flood in spring (May–June) and low discharge ($<$ 70 m^3 s^{-1}) during winter, when the river, and periodically also the delta plain, are ice-covered. Annual mean discharge at Polmak, 25 km upstream from the section, between 1962 and 1999 was 168 m^3 s^{-1}. Annual mean flood discharge is 1600 m^3 s^{-1}, with a maximum recorded discharge value of 3843 m^3 s^{-1} in May 1920 (http://www.nve.no). Commonly, spring ice breakup commences five days before maximum spring floods, and ice may play an important geomorphic role in the fluvial environment (Collinson 1971).

The lower Tana valley was deglaciated between 13,000 and 10,000 radiocarbon years ago as shown by marginal moraines and regional dating correlation (Sollid et al. 1973, Andersen 1979). Below the postglacial marine limit (defined as the maximum elevation of marine inundation following deglaciation), the valley contains a thick succession of glacigenic clays, sands and gravels, overlain by postglacial deltaic and fluvial sands which occur in a series of terraces (Corner 1975; Corner et al. 1995, Corner and Eilertsen 2008). These terraces lie below the marine limit, situated at a maximum of 32–34 m a.s.l. at Tana Bru, descending down valley through 30–31 m a.s.l. at Masjoksletta, and 1–2 m a.s.l. at the present delta (Fig. 1B). The terraces formed during a predominant relative sea-level fall caused by glacio-isostatic uplift. Differential uplift following deglaciation, with rates increasing towards the southeast, has caused tilting of the raised terraces (see later discussion of base level). The present rate of uplift is c. 1–2 mm/yr (Sørensen et al. 1987; Dehls et al. 2000), promoting fluvial incision and enhanced sediment supply, and gradually raising active parts of the delta and river bed beyond the reach of delta and river-forming processes. A staircase of fluvial terraces capping deltaic deposits can be mapped within the valley, and the associated fluvial sediments are mostly between 3 and 5 m thick (Sivertsen 1997). Available sediment along the river banks is mainly of sand and gravel size. Thick, slide-prone, glaciomarine and marine clay deposits occur locally farther upstream and intermittently provide pulses of fine-grained sediment to the river.

METHODS

Methods employed for studying the deposits included surface mapping, sedimentary logging, photomosaics, and ground-penetrating radar (GPR) profiling. Ages based on radiocarbon dating are given as calibrated ages in thousands of years (ka) before present.

Surficial deposits and associated terraces at different levels, mapped by Corner et al. (1995), provide a morphostratigraphic framework for correlating lithological successions and reconstructing delta progradation and fluvial terrace formation in relation to falling base level (Fig. 1B). Deltaic (primary) terraces can be easily distinguished from fluvial (secondary) terraces based on relative elevation and morphology (Corner 2006, Eilertsen et al. 2006).

Nineteen logs were measured and photographed along the section at Masjoksletta in order to document grain size, bedding characteristics, and type of bounding surfaces (Figs. 2, 3). Sediment samples were sieved, dried, and weighted and classified according to the grain-size nomenclature based on the Udden-Wentworth (phi) grade scale (Wentworth 1922). Photomosaics were used to identify and measure distinct boundaries.

Mauring and Rønning (1993) and Mauring et al. (1995) measured a series of GPR profiles in the Tana Valley. The profile used in this study was measured along the river bank between position 3200 and 2870 m (Fig. 2), using a PulseEkkoIV GPR system with 100 MHz antennas. The profile was measured in step mode with 0.5 m step intervals. The depth scale was determined by CMP (common mid-point) measurement (Sensors & Software, 1996) for converting two-way travel time to depth in meters, giving a velocity of 0.06 m/ns.

RESULTS

Seven facies types and three fluvial depositional units (I–III) are recognized in the section. The depositional units are distinguished based on their stratigraphic position, boundaries, and facies composition. The lowermost unit lies unconformably above deltaic deposits. The units are described and interpreted in stratigraphical order. The underlying deltaic deposits and overlying younger fluvial deposits are also briefly described. Table 1 shows the facies characteristics and their interpretations.

Underlying Deltaic Sediments

The deltaic deposits are more than 6 m thick in exposure above river level but extend to much greater depth below river level (Corner et al. 1995). They consist of parallel-bedded sand, silt, and clayey silt with intermittent thin silty clay to clay layers. Individually accreted sets show a very regular progressive vertical thickening and thinning, and can be followed laterally for several tens of meters with an overall dip (3–14°) towards the north (Fig. 4A, B). These deposits are of tidal origin (Eilertsen et al. 2005). Similar facies are found both downstream and upstream from the study site (Hestem 1997; Corner et al. 1995) and are interpreted as delta foreset–bottomset sediments.

Lower Fluvial Unit (I)

Unit I is between 3 and 20 m thick in the main part of the section, and is documented in logs 1–4, 6, 9–14, and 17–19 (Fig. 3). It

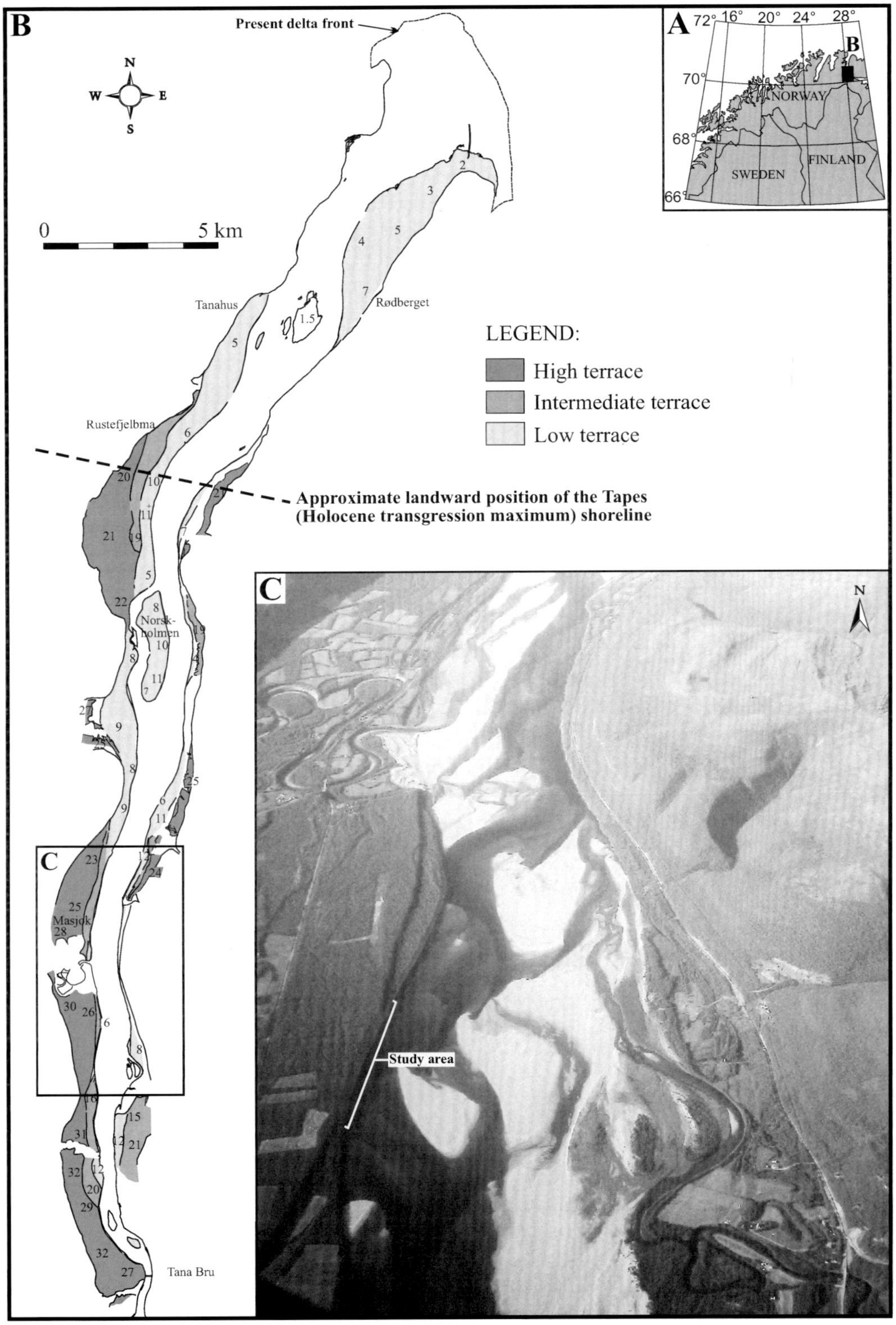

FIGURE 1.—Location maps of the study area. A) Index map. B) Distribution of terraces (simplified from Corner et al., 1995). Altitude in meters. Higher terraces are generally older than lower ones. C) Aerial photograph of the Tana River at Masjoksletta. Note the well-defined terrace scarps.

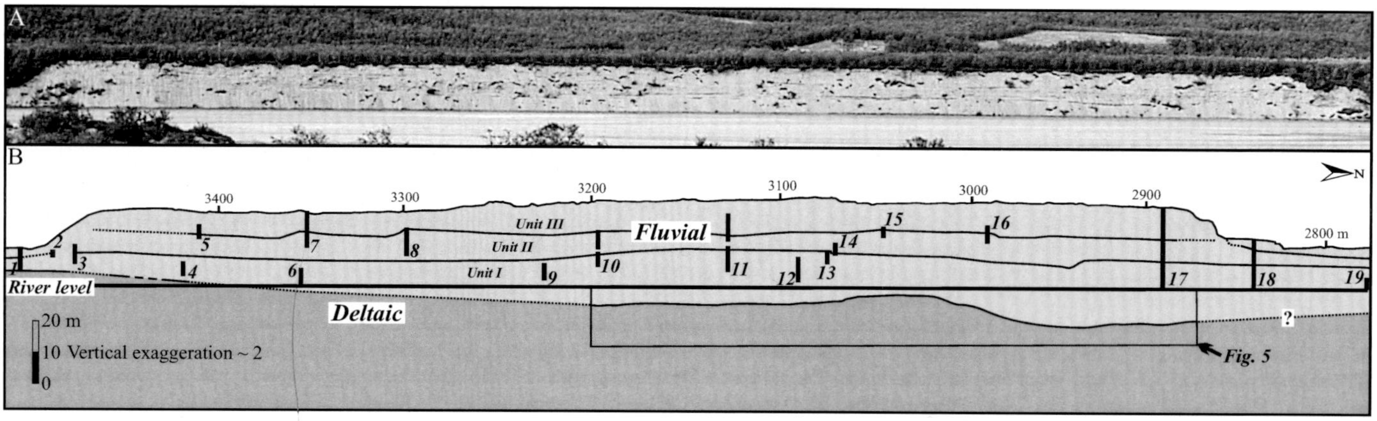

FIGURE 2.—Terrace section at Masjoksletta. A) Photograph showing the river terrace section. B) Sketch of the section showing the location of sedimentological logs, GPR profile (Fig. 5), and boundaries of the stacked fluvial succession (units I, II, and III). The succession overlies deltaic sediments and is truncated at either end of the section by younger, incised, fluvial sediments.

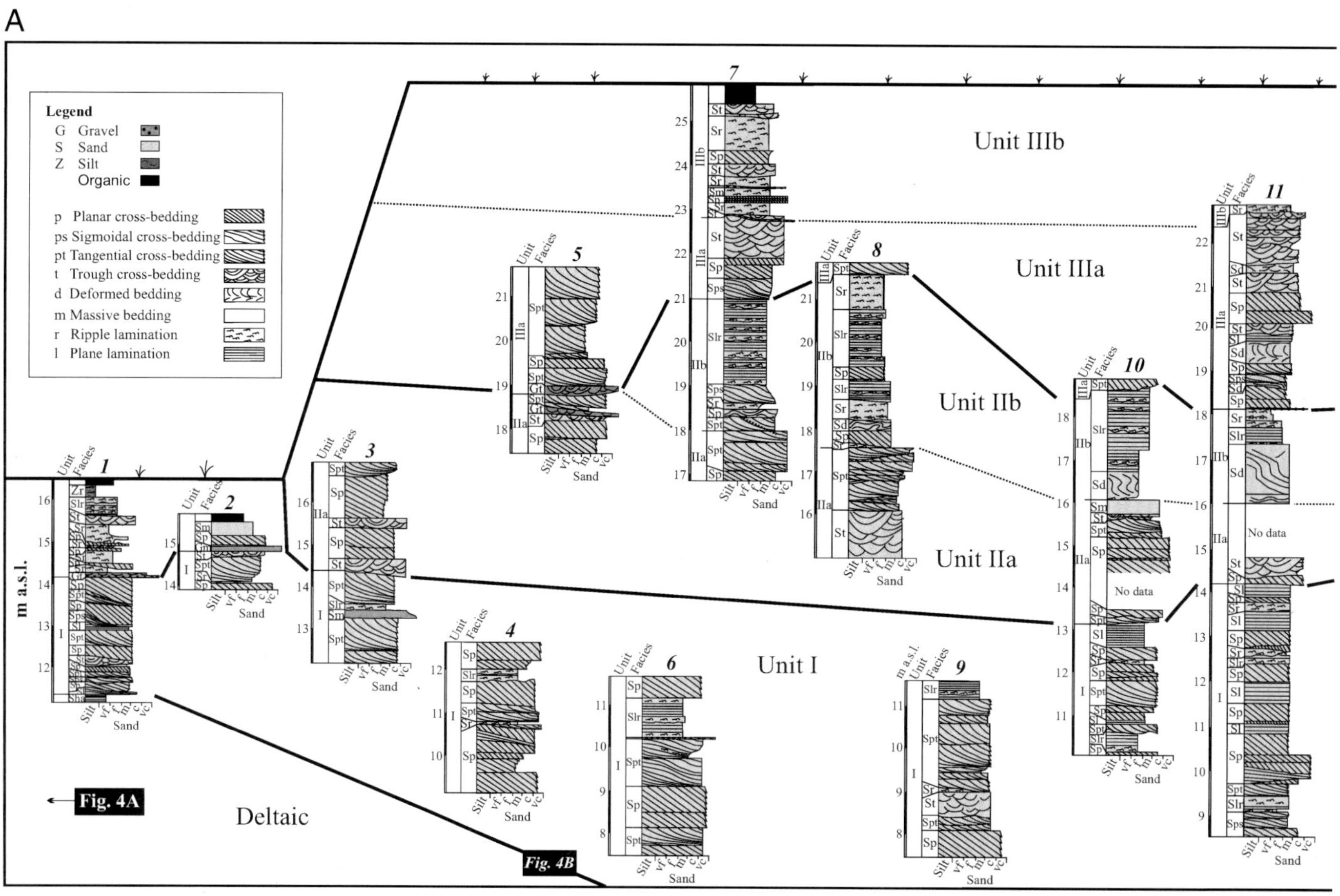

FIGURE 3.—Sedimentological logs from the terrace section at Masjoksletta, arranged according to their relative position along the transect (see Figure 2 for precise location). A) southern (upstream) part of the transect. B) northern (downstream part of the transect). The vertical scale is in meters above sea level.

erosionally overlies the deltaic unit. Its lower boundary is exposed up to 6 m above river level in the southern part of the section and descends below river level north of position 3350 (Figs. 2, 3). A GPR profile taken along the river bank between position 3200 and 2870 reveals that the lower boundary descends to a maximum depth of 13 m below river level, i.e., 7 m below present sea level (Fig. 5). The boundary thus has a maximum relief of 17 m along the investigated 600 m length of the section. It is particularly steep in the northern part of the profile, where it descends 12 m over a distance of 70 m (between position 3010 and 2940). Internal GPR reflections within unit I show several lens-shaped, transparent zones bounded by strong reflections. They are up to 3 m thick and are stacked in an offset manner in relation to each other. Sediments in the exposed part of the section comprise predominantly planar cross-bedded medium to coarse sand in sets up to 2 m thick, with a mean thickness of 46 cm (Fig. 4C). Beds of coarse and fine gravel also occur. Sigmoidal and tangential cross bedding, and plane lamination is also common. Plane-laminated beds can commonly be traced laterally (downstream) for 20–30 m into sigmoidal cross-bed sets (Fig. 4D). The abundance of plane lamination increases upward in the unit. Trough cross-bedding is relatively rare, and ripple cross-lamination occurs only in alternation with plane lamination. Overall, set thickness decreases upwards.

The relatively great thickness of unit I and the high relief of its lower boundary, the abundance of sigmoidal cross-bedding and relatively thick bedsets, the relatively uniform grain size compared with overlying units, and the stacking pattern of dune beds observed in the GPR profile, together suggest that unit I was deposited rapidly. The preservation of flat-laminated topsets implies formation under conditions of rising flow power from dune to upper-plane-bed stability conditions (Røe 1987; Fielding 2006). The thickness of unit I greatly exceeds the present normal depth of the river channel and the thickness of the uppermost, terrace-capping, fluvial unit (3–5 m; Corner et al. 1995; Sivertsen 1997).

Middle Fluvial Unit (II)

Unit II is between 4 and 8.5 m thick and is documented in logs 3, 5, 7–8, 10–11, and 14–18 (Fig. 3). It is overlain by unit III in the main part of the section and is truncated and erosionally overlain by younger fluvial sediments at either end of the terrace section. The lower boundary is erosional and defined by a sharp increase in grain size, in places with gravel lag at the base. The unit can be divided into two subunits based on grain size and facies within an overall fining-upwards trend. *Subunit IIa* consists of coarse to very coarse, sandy planar cross-beds with an an average set thickness of 31 cm, often with tangential foresets typically overlain by thinner sets of trough cross-beds. Individual beds can be followed laterally for several tens of meters. For example, two distinct tangential cross-bed sets at the bottom of the unit in logs 18 and 19 can be followed laterally for more than 210 m. The boundary between subunit IIa and IIb is very distinct

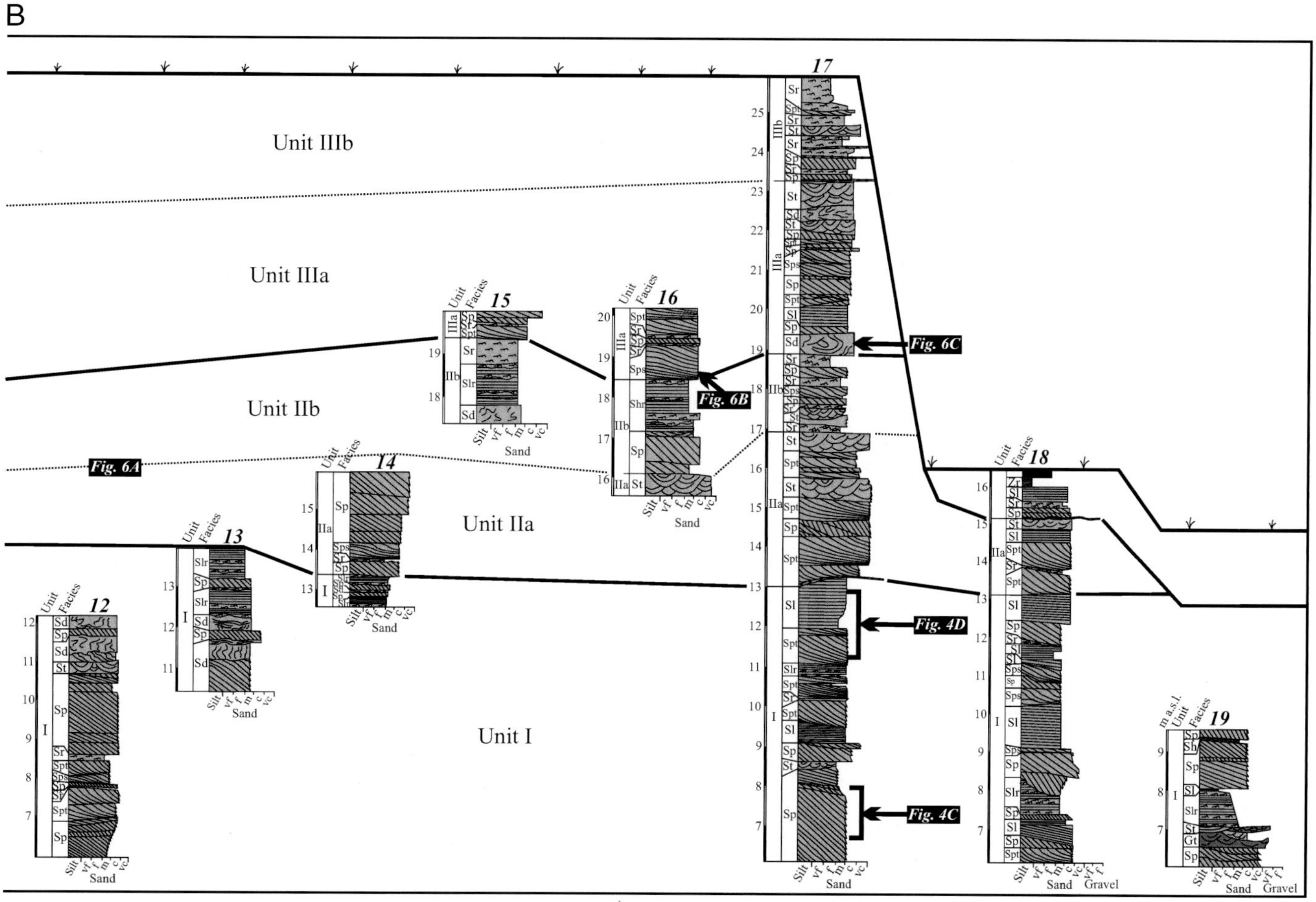

FIGURE 3.—Continued.

TABLE 1.—Facies characteristics and their interpretation.

Lithofacies (*facies codes*)	Description	Occurrence	Interpretation (depositional mechanism)
1. Rhythmic bedding (S*ha*/Z*hb*) (Fig. 4A-B)	Alternating beds of fine to medium sand grading into very fine sand (*Sha*), occasional silt laminae (*Zhb*); 1 mm to > 8 cm in thickness, thickness varying in a rhythmic manner with 14–28 couplets within each interval; occasional thin medium sand bed separated by two silt to very fine sand beds, forming double mud drapes. Groups of 5 to 8 of these cycles goes from thin to thick to thin, and are bordered by several-cm-thick, mud-rich layers in which the couplets are no longer visible. Ripples are often found above the mud-rich layer, followed by a new cycle.	Abundant	Suspension controlled by tidal currents ("tidal drawdown" *sensu* Smith et al. 1990; De Boer et al. 1989; Kvale et al. 1989). The thick-thin cyclicity involving 14–28 couplets is interpreted as semi-monthly (spring-neap) tidal cycles. Neap-spring-neap cycles reflect the half-synodic lunar orbit around the earth. The presence of 28 couplets indicates preservation of nearly every ebb tidal event during two weeks of deposition. Deposition during the spring and summer months, quiescent during winter (Eilertsen et al. 2005).
2. Planar cross bedding (S*p* – angular; S*ps* – sigmoidal, S*pt* – tangential; Fig. 4C)	Fine to medium sand, occasionally coarse sand to gravel beds; 15 cm to 2 m thick, < 1 m most common; thickness of individual cross laminations may vary. Individual beds may be followed for more than 200 m.	Abundant	Lower flow regime, traction currents. Linguoid, transverse and diagonal 2D dunes (Collinson 1970; Collinson and Thompson 1989; Røe and Corner 1994). Angular, tangential and sigmoidal forms reflects an increase from relatively low to relatively high sedimentation rates (Jopling 1965; Collinson and Thompson 1989), possibly from lower-stage to upper-stage flow regime.
3. Trough cross bedding (S*t*/G*t*)	Medium to coarse sand, occasional gravelly; 10–40 cm thick; cosets up to 1.2 m thick.	Common	Lower flow regime, traction currents. Represents deposition of 3D dunes. Collinson (1970) described these forms as commonly deposited on top of larger 2D dunes in the Tana River.
4. Plane lamination (S*l*; Fig. 6B)	Very fine to medium/coarse sand; few cm to 1 m thick; occasionally dipping laterally into planar cross beds; often interbedded with ripples; individual beds may be more than 200 m long.	Common	Upper flow regime, traction currents. Interbedding with ripples suggests fluctuating flow, possibly generated by tidal modulation of river flow, seasonal variations in discharge, or fluctuations in flow direction due to bar migration (or a combination of these).
5. Ripple bedding *(Sr)*	Silty very sand to coarse sand; commonly with relief of between 3–5 cm and 10–20 cm; often interbedded with plane-laminated beds; occasional climbing ripples are found migrating up dune foresets; rare silt-draped ripples also found.	Common	Lower flow regime, traction currents. Climbing ripples up dune foreset slopes suggest flow separation at the lee side causing backflow of the current (Hunter 1985; Collinson and Thompson 1989).
6. Deformed beds (S*d*; Fig. 6C)	Fine to very coarse sand beds; up to 75 cm thick; folds, overturned foresets, mud flames, dish structures, intraclasts; individual beds may be followed more than 210 m.	Rare/ common	Liquefaction structures, overturned foresets caused by shear stress of water saturated bed (Mckee et al. 1962; Coleman 1969; Reineck and Singh 1980). Suggest rapid deposition (Collinson and Thompson 1989).
7. Massive beds (S*m*/G*m*)	Massive, medium sand to medium gravel; 10–40 cm thick.	Rare	Rapid deposition, coarse clasts or deformation makes it impossible to detect internal structures.

due to an abrupt decrease in grain size. It descends ~2 m between positions 3350 and 3190, before flattening out (Fig. 3A). *Subunit IIb* consists of medium to fine sand and contains abundant deformed beds and alternating planar-laminated and ripple-laminated beds. The latter are up to 2 m thick and can be followed laterally for several hundred meters. The deformed beds consist of overturned dune foreset lamination and liquefaction structures (Fig. 6C), and individual beds can also be followed laterally for almost two hundred meters. Generally, bed thickness decreases upwards.

Coarser-grained *subunit IIa* is interpreted as channel thalweg

FIGURE 4.—Examples of lithofacies types found in the section at Masjoksletta (see Figure 3 for location). A) Rhythmic bedding (tidal rhythmites) from the underlying deltaic sediments. Note thickening and thinning of individual laminae corresponding to spring and neap tidal cycles, and person for scale. B) Boundary between deltaic sediments and unit I representing the basal scour surface. Note erosional contact truncating laminae in the deltaic deposits. C) Planar cross-bedding close to river level in unit I. D) Planar tangential cross-bedding overlain by plane lamination.

deposits in which planar cross beds represent 2D (mainly linguoid) dunes, on which smaller-scale 3D dunes are superimposed. This is a typical depositional pattern described by Collinson (1970) for the present river. The finer-grained deposits of overlying subunit IIb suggest deposition in the final stage of channel fill, possibly as part of a sidebar or compound bar. The thick beds of alternating ripple bedding and plane lamination at the top of this unit suggest fluctuating flow, from upper to lower flow regime, at shallower water depths than for underlying unit IIa. One possible mechanism causing such alternating facies may be tidal modulation of river flow as the river current strength decreases during flood tide and increases during ebb tide (Eilertsen et al. 2005). Other mechanisms could be seasonal discharge fluctuations or simply bar migration causing redirection of stream flow.

Upper Fluvial Unit (III)

Unit III is found at the top of the section and is erosionally truncated at the northern and southern ends of the section (Figs. 2, 3). It has a sheetlike geometry and a thickness of between 5 and 8 m. It is documented in logs 5, 7–8, 10–11, and 15–17 (Fig. 3). The boundary between units II and III is sharp and erosional, slightly undulating, and marked by an increase in grain size (Fig. 6B). The unit can be divided into two subunits based on grain size and facies within an overall fining-upwards trend. *Subunit IIIa* consists predominantly of coarse to very coarse sand, occasionally medium sand. Planar cross bedding having an average set thickness of 24 cm predominates in the lower parts of the subunit. Trough cross beds and rare ripple cross-lamination and plane lamination occur in the upper part of the subunit. Bed thickness varies but is generally less than 50 cm. *Subunit IIIb* is characterized by medium to fine sandy ripple beds, overlain by medium, occasionally coarse sandy, trough and planar cross bedding.

Coarser-grained *subunit IIIa* is interpreted as channel thalweg deposits in which the planar and trough cross beds represent 2D (mainly linguoid) dunes with smaller, superimposed 3D dunes. The subunit is similar to unit IIa below, but with a slightly higher incidence of trough cross bedding. The finer-grained deposits of overlying *subunit IIIb* suggest deposition in the final stage of a channel fill, probably as part of a sidebar, compound bar, and/or overbank deposits.

Overlying Fluvial Sediments

Younger fluvial deposits erosionally overlie or truncate units I–III at the northern and southern end of the high Masjoksletta section, in an intermediate-level terrace (terrace top is 16 m a.s.l.; Fig. 2). This

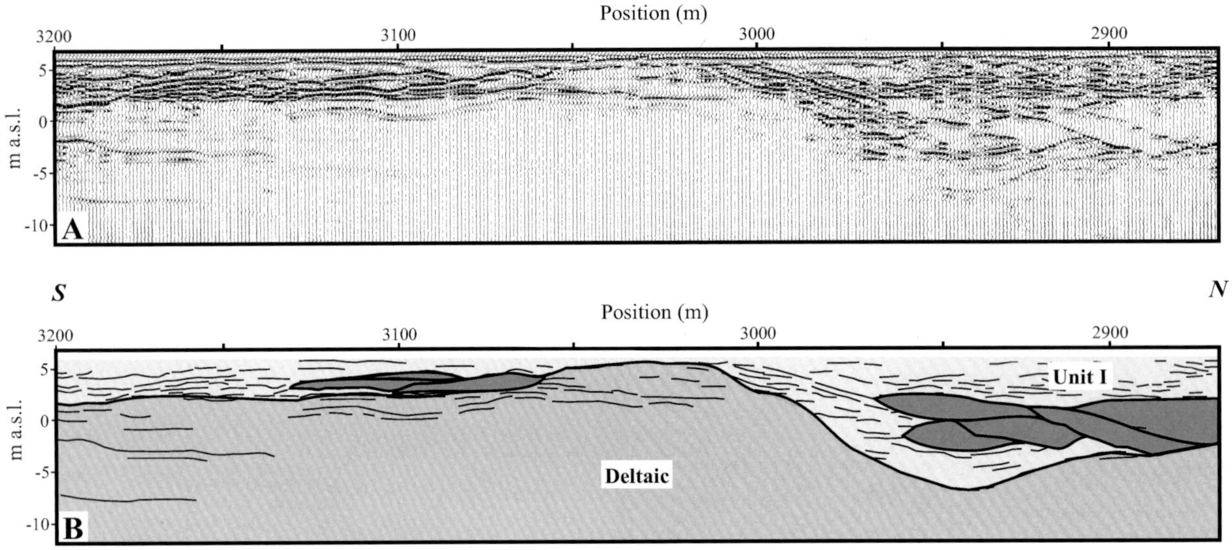

FIGURE 5.—Ground-penetrating-radar profile (100 MHz) between position 2880 and 3200 m measured parallel to the section at river level (6 m a.s.l.), illustrating fluvial scour-infill sediments (unit I) erosionally overlying deltaic sediments. A) Raw data, B) interpretation. Note the ~ 12-m-deep trough at 2950 m. Lens-shaped forms, indicated in dark gray, possibly represents stacking of dunes (see text for further explanations). See Figure 2 for location.

terrace unit is 2.5–7 m thick and increases in thickness towards the north. Its lower boundary is marked in most places by a gravel lag, overlain by sandy to gravelly planar cross bedding and fine to medium sand ripple beds. The latter may be organized into bed sets more than 1 m thick towards the top of the unit. The unit is interpreted as a fluvial channel fill, formed as the Tana River eroded into the larger terrace and subsequently migrated towards the east.

DISCUSSION

Figure 7 illustrates the characteristics of each of the three main fluvial units described above. The lowermost unit (Unit I) is characterized by great thickness (maximum 20 m), uniform grain size (coarse sand), and relatively thick bed sets. In contrast, overlying units II and III are thinner (maximum 8–8.5 m), fining upwards, generally with thinner beds and decreasing bed thickness upwards, and with fine-grained sediments (fine to medium sand) largely comprising ripple cross-lamination in their upper parts. Uppermost unit (III) represents a nearly complete infill of a channel, whereas units I and II are truncated at the top and only partly preserved (see below). Together, the three units form a succession of fluvial deposits more than 30 m thick, in strong contrast to the norm of 3–5 m found elsewhere in the lower Tana valley (Sivertsen 1997).

We discuss below two factors that can explain the exceptional thickness of the stacked fluvial succession at Masjoksletta: (1) base-level control of accommodation space, and (2) deep scour and subsequent fill. We also attempt to reconstruct channel depth based on the method of Bridge and Tye (2000) and LeClair and Bridge (2001) using cross-set thickness distributions to estimate dune height and corresponding flow depth. Although prediction of flow depth by this method is imprecise, our data provide a useful opportunity to compare results obtained by this method with our independent estimates obtained from a comparison with the modern river and a reconstruction of base level at the time of deposition. We end with a brief discussion of criteria for distinguishing between autogenic scour and allocyclic incision due to base-level fall.

Base-Level Change

Base-level elevation during deposition of the fluvial sediments in the Masjoksletta section, and their approximate age, can be deduced from the regional pattern of postglacial relative sea-level change combined with terrace-level correlation.

Figure 8A shows postglacial relative sea-level curves for two positions in the lower Tana valley (Masjoksletta and Rustefjelbma) derived from a comparison of sea-level curves from five areas between Varangerfjorden and the Kola Peninsula farther east (Corner et al. 2001). The curves were constructed by interpolation, using the elevation of the regionally distinct, tilted, mid-Holocene (c. 7 ka), Tapes transgression maximum shoreline (Sollid et al. 1973) as a guide. The Tapes shoreline has an elevation of between 19 and 21 m a.s.l. in the lower Tana valley below Rustefjelbma, and a theoretical level of between 23 and 24 m a.s.l. at Masjoksletta when extrapolated upvalley (at a gradient of 0.22 m/km). The amplitude of the Tapes sea-level rise (c. 9–7 ka) decreases inland because of differential uplift, hence the difference in shape between the two curves shown in Fig. 8A (cf. Corner 2006).

The highest fluviodeltaic terraces in the lower Tana valley descend gradually, over a distance of 20 km, from approximately 33–40 m a.s.l at Tana Bru, through 28–31 m a.s.l. at Masjoksletta, to c. 21 m a.s.l. at Rustefjelbma, beyond which they descend more abruptly, over a distance of c. 19 km, to present sea level (Fig. 1; Corner et al. 1995; Sivertsen 1997; Corner and Eilertsen 2008). Deltaic terraces at Rustefjelbma, about 16 km downstream from Masjoksletta, correspond approximately to the level of the Tapes shoreline, which means that higher deltaic terraces upstream are older; their elevation just a few meters above the level of the extrapolated Tapes shoreline shows that they formed immediately prior to the Tapes sea-level rise, i.e., c. 10–9.5 ka. The deltaic sediments underlying fluvial sediments at Masjoksletta were formed at this time. The top of the studied fluvial section at Masjoksletta (26–28 m a.s.l.) lies c. 3–5 m above the level of the extrapolated Tapes shoreline (23–24 m a.s.l.). The mean gradient of a level joining this terrace with the 21 m terrace at Rustefjelbma, adjusted for postglacial tilt, is only slightly greater (c. 0.25 m/km) than

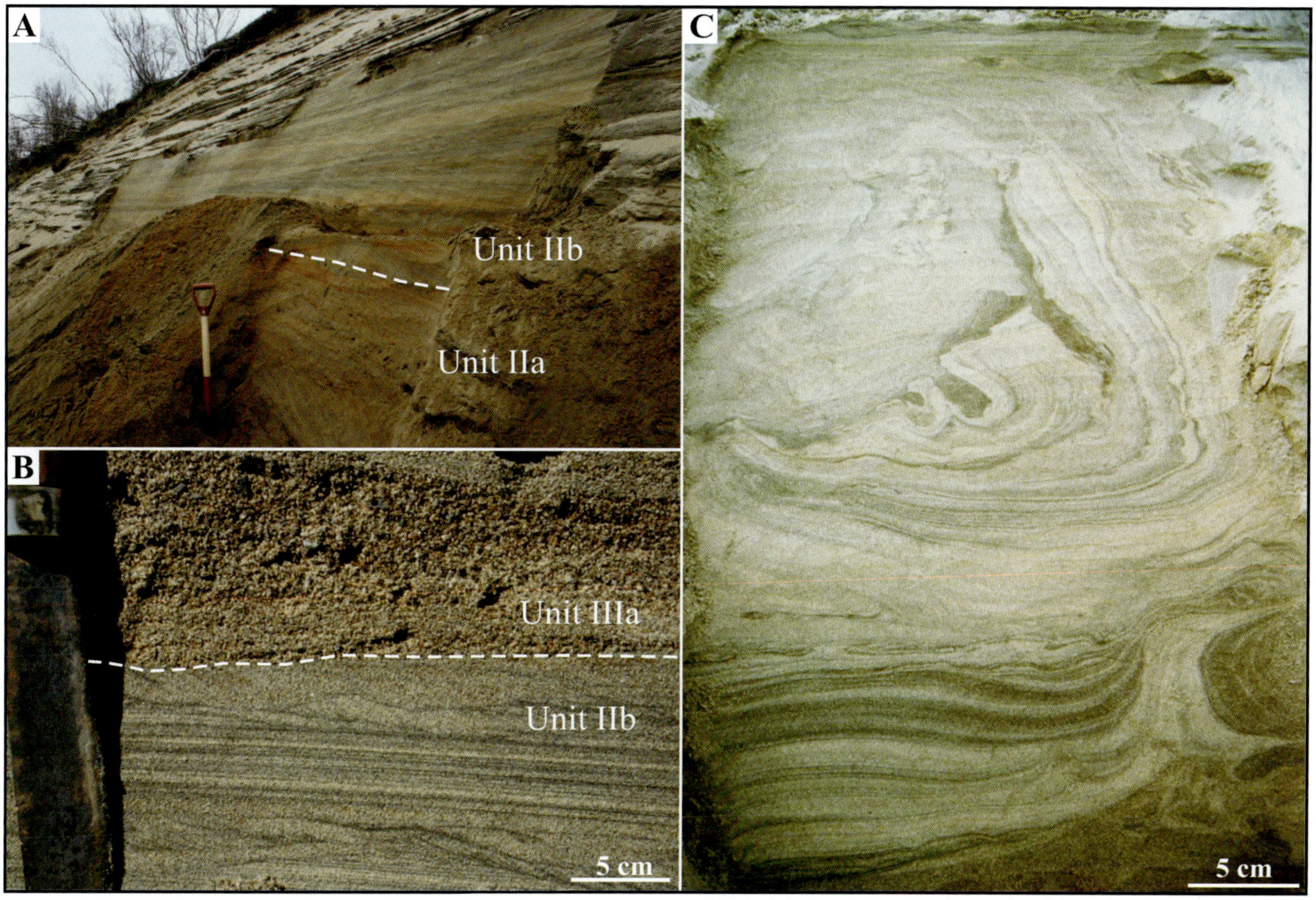

FIGURE 6.—Illustration of unit boundaries and facies in the section at Masjoksletta (see Figure 3 for location). A) Boundary between coarse-grained thalweg channel sediments of *subunit IIa* and overlying finer-grained sigmoidal crossbeds and plane-laminated beds of *subunit IIb*. Spade is ~ 1 m long. B) Close-up of boundary between *subunits IIb* and *IIIa*. C) Deformation (flame) structures in *subunit IIIa*.

the gradient of the present river (c. 0.24 m/km). This suggests that the top of the fluvial terrace and unit III at Masjoksletta were graded to the Tapes shoreline and formed during the Tapes highstand (c. 7 ka). Units I and II consequently have an age that is intermediate between the Tapes highstand and the age of the underlying deltaic sediments (i.e., between 7 and 10 ka), and were probably formed during the pre-Tapes lowstand and subsequent Tapes sea-level rise (c. 9–7 ka).

The predicted level of the pre-Tapes lowstand at Masjoksletta, derived from the sea-level curves, is c. 21 m a. present sea level, and has approximately the same level as the Tapes highstand elevation at the isostatically less uplifted locality at Rustefjelbma farther down the valley, which means that the shoreline trajectory during deposition of units I–III was essentially horizontal (Fig. 8B). It appears, therefore, that the regional Tapes transgression promoted the development of the thick, stacked succession of fluvial sediments at Masjoksletta by providing a two-thousand-year-long virtual stillstand in base-level elevation during progradation of the fluviodeltaic system at this location. This occurred within what was otherwise a falling-stage systems tract (cf. Corner 2006).

Stratigraphic Development

Figure 9 shows a reconstruction of the stratigraphic development of the stacked fluvial unit at Masjoksletta in relation to base-level change and terrace aggradation and progradation. As discussed above, base level during deposition of units I–III lay c. 21 m a.s.l., over which time (c. 9–7 ka) the fluviodeltaic system prograded c. 16 km. The base of unit I reach a maximum depth of 7 m below present sea level, which is approximately 28 m below base level at this time. Erosion to such a depth, which is many times greater than normal channel depth, must be related to a deep scour (discussed below). The uppermost part of unit I is truncated at 14 m a.s.l., c. 7 m below contemporaneous base level, which suggests that approximately 7 m of sediment were removed from unit I when unit II was emplaced.

Unit II formed when the river mouth was located several kilometers downstream from Masjoksletta (Fig. 9C). The shoreline trajectory shown in Figure 8, combined with a river gradient of 0.25 m/km, indicates that the top of unit II was deposited close to 22–23 m a.s.l. This suggests that 3–4 m of the top of unit II has been removed by erosion prior to deposition of overlying unit III (Fig. 9). We interpret the presence of interbedded plane lamination and ripple cross-lamination (facies Slr) in units I and II, and their absence in unit III, as possible indicators of reduced tidal modulation of river flow as the shoreline regressed.

The preservation of a nearly complete channel-infill unit (unit II) overlain by another channel-infill unit (unit III) can be achieved only if there is a significant accommodation (preservation space) available. The stillstand in relative sea level following the pre-Tapes lowstand and

Unit	Thickness (m)	Facies (listed in order of abundance)	Composite log	Characteristics
IIIb **IIIa**	5 - 8	Sr, St, Sp, Spt ………… Spt, Sp, Sps, St, Sd, Sl, Sr, Slr	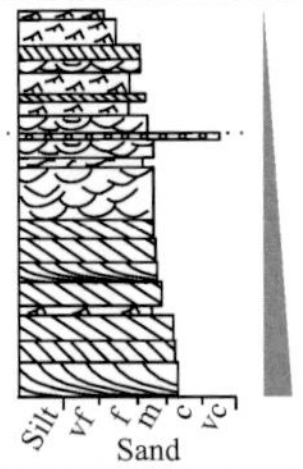	Fining-upwards unit. Lower subunit deposited as thalweg deposits, overlain by finer-grained ripple dominated sediments. Indicates shallowing and subsequent infilling of river channel.
IIb **IIa**	4 - 8.5	Slr, Sr, Sd, Sp, Sps, St ………… Spt, Sp, St, Sps, Sl, Sm, Sr, Sd	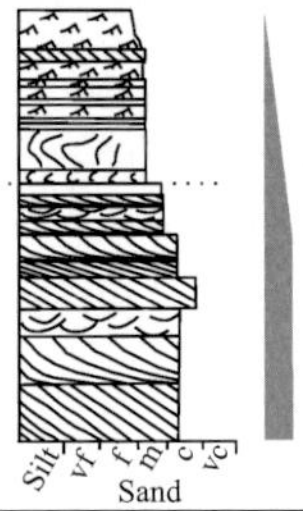	Fining-upwards unit. Lower subunit deposited as thalweg deposits and dominated by planar cross bedding, overlain by finer-grained, interbedded plane and ripple beds. Decreasing bed thickness upwards.
I	3 - 20	Sp, Spt, Slr, Sl, Sps, St, Sd, Sr, Sm, Gt	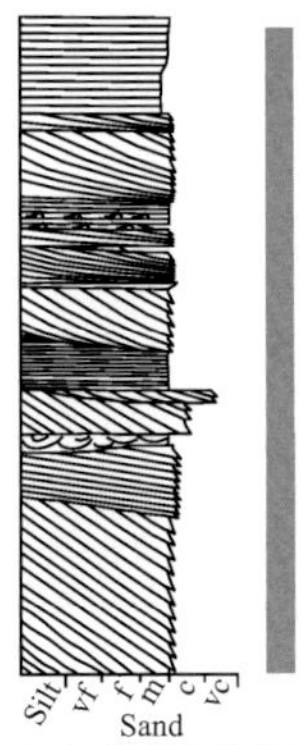	Scouring followed by infill. Relatively uniform grain size and thick beds. Dominance of planar cross bedding. Rapid deposition.

FIGURE 7.—Summary diagram showing the characteristics of the fluvial units at Masjoksletta.

related to the Tapes transgression is one possible mechanism that may have caused the river bed to aggrade, and which could account for the superposition of channel deposits. The process of aggradation associated with a stillstand or slight rise in sea level is most significant close to the shoreline, and its effect diminishes with distance upstream and as overall gradient decreases (Blum 1993; Shanley and McCabe 1994; Guccione 1994; Törnqvist 1998). The study site, situated at a maximum distance of 15–16 km from the shoreline, is well within the range of such base-level effects (Blum and Törnqvist 2000). Moreover, the river had to migrate laterally in order to preserve most of unit II, before returning westwards to deposit unit III. The lower boundary of unit II at 14 m a.s.l. suggests that the river must have scoured up to 6 m below contemporaneous sea level during deposition of the unit (Fig. 9). Using an estimated elevation of 22–23 m a.s.l. for the paleosurface of unit II, and 26–28 m a.s.l. for the top of unit III (terrace top), suggests an aggradation of ~ 4–5 m between their formation.

Role of Scouring

Reconstruction of base-level elevation during deposition of unit I indicates that the base of this unit was formed c. 27 m below base level, corresponding to a channel depth slightly greater than 27 m below river level. This is five times thicker than the fluvial succession found elsewhere in the lower Tana valley and more than five times deeper than the mean channel thickness today (Corner et al. 1995; Sivertsen 1997; Lotsari et al. 2010). The cause is clearly related to scour processes.

Deep scours may form under various conditions where flow acceleration, secondary currents, and/or vortices occur at river channel features such as confluences, obstructions, bends, and abrupt changes in flow direction (Galay et al. 1987). Among these sites, channel confluences have the greatest potential for creating deepest scour (Galay et al. 1987), and scour depths up to five times the mean depth of the upstream channel have been recorded (Ashmore and Parker 1983; Klaassen and Vermeer 1988; Best 1988; Best and Ashworth 1997). By using an average depth of 5 m for the Tana River based on the thickness of raised fluvial depositional units (Sivertsen 1997) and known present day bathymetry (Corner et al. 1995), the scour at the study site was ~ 5 times deeper than the average channel and thus in agreement with previous studies of channel confluences. Channel-bend scours at an eroding bend are known to scour 2.5 times the incoming channel (Galay et al. 1987), and although Eilertsen and Hansen (2008) reported channel bend scours more than 3 times deeper, it is unlikely that the scour at Masjoksletta was the result of scouring in a channel bend. The presence of a constriction or obstruction may result in scouring to depths up to five times the incoming channels (Galay et al. 1987). Eilertsen and Hansen (2008) reported a scour 23 m deep in the Lake Øyeren Delta distributaries caused by the combination of a major flood

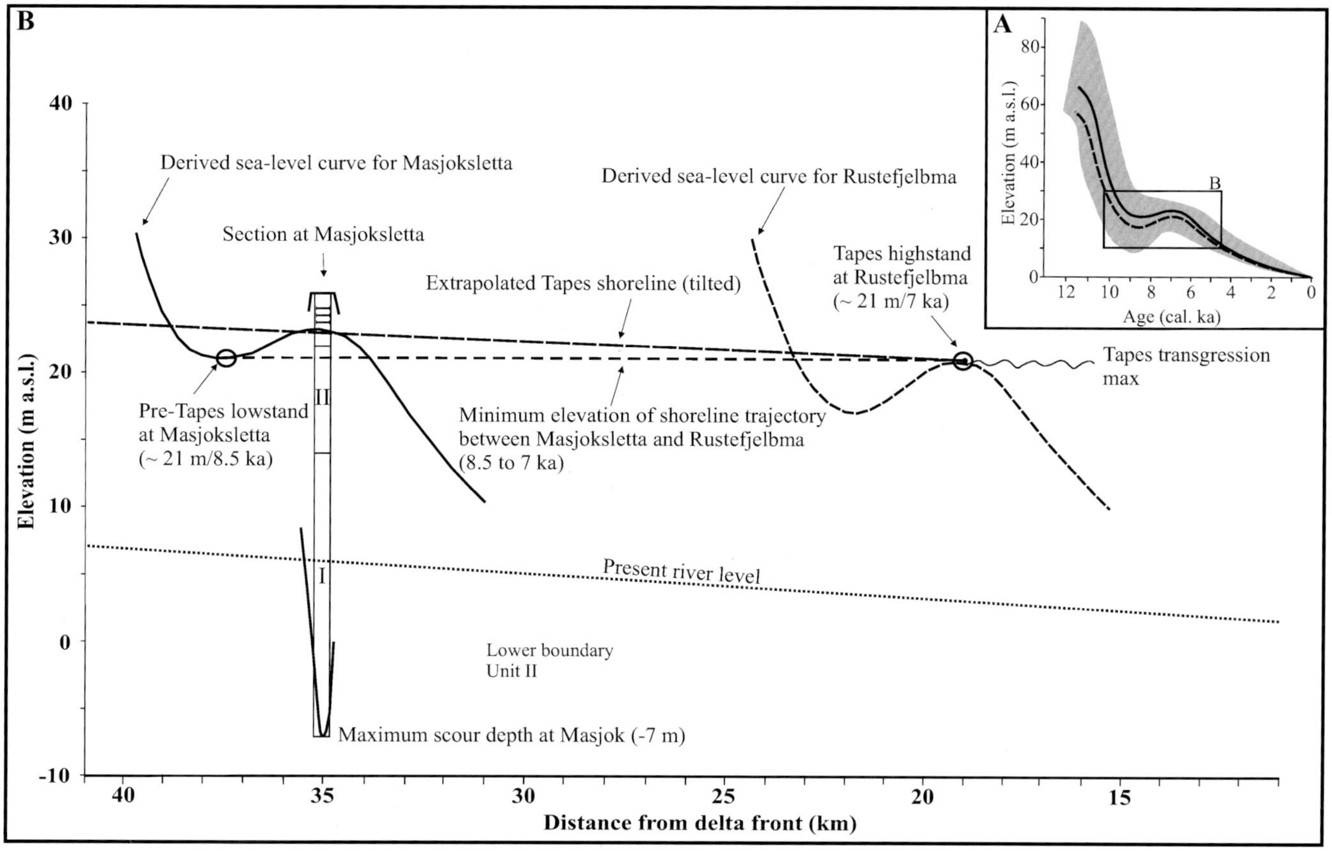

FIGURE 8.—Reconstruction of base-level changes during deposition of the Masjoksletta section based on regional relative sea-level-change data and local terrace correlation. (A) Postglacial relative sea-level curves for Masjoksletta (continuous line) and Rustefjelbma (dashed line), in the lower Tana valley, derived from a comparison of sea-level curves from five areas east of Tana (gray-shaded envelope; cf. Corner et al. 2001) and using Tapes shoreline isobases as a reference level for the elevation of the mid-Holocene transgression (cf. Sollid et al. 1973). (B) Correlation of fluvial terrace deposits at Masjoksletta with shore levels and sea-level change. The fluvial terrace top at Masjoksletta is correlated with deltaic terraces at Rustefjelbma and the Tapes transgression maximum shoreline beyond. Older fluvial sediments in the Masjoksletta section were deposited before or during the Tapes transgression (8.5–7 ka); the minimum elevation of the shoreline trajectory during their deposition is indicated.

and the gradual build up of a mid-channel bar resulting in channel constriction and subsequent increase in velocity and turbulence. A similar mechanism is possible here. Also, the presence of river ice (or ice jams) may induce scouring both as a result of scouring around the ice jam blocks itself but also due to increased flow velocity caused by channel constriction. Smith and Pearce (2002) reported small-scale scour holes beneath and between grounded ice jam blocks from the Milk River in Canada and USA. However, we do not know how common river ice was 6,000–8,000 years ago.

Relative few studies of infill patterns in scour holes have been published. Bristow et al. (1993) published a facies model for channel confluences, and postulated that scour fill most likely is the result from the migration of mouth bars into the confluence scour, giving very large sets of cross-stratification relative to others within the channel. The largest sets in Tana are found in unit I, and as such fit with this description. Cowan (1991) described the infill of scours at channel confluences (their HO (hollow) architectural element; see also Miall and Jones 2003) as composed of cosets of low-angle cross-bedding. The sediments of unit I are composed of several facies types, and fit more with the range of elements forming the fifth-order channel architectural element (CH of Miall and Jones 2003) than the HO element. Unit I can be distinguished from overlying units by the preservation and abundance of sigmoidal cross-bedding, its relatively thick beds, stacking pattern of dunes, and relatively uniform grain size. All of these characteristics suggest rapid deposition, maybe in the range of one or a few seasons rather than over many years.

Once a deep scour is formed, it commonly appears to be self-maintained and relatively stable, though migrating within the channel itself (Moore 1970; Lapointe 1985; Fassnacht and Conly 2000; Eilertsen and Hansen 2008). The scour at the Lake Øyeren Delta distributaries (Eilertsen and Hansen 2008), mentioned above, migrates at a rate of 13 m yr^{-1}, accompanied by at the same time by upstream filling. Best and Ashworth (1997) showed that the deepest part of large scours in the Jamuna River migrated with a rate of up to 1.8 km yr^{-1} with deposition of up to 17 m in the same time period. This shows that erosion and deposition in scour regions may be very rapid.

The noncohesive, fine-sandy lithology of the underlying deltaic sediments would have been favorable for scouring. Furthermore, it is conceivable that the relatively fine-grained sediments of the basal part of unit I, compared with the basal subunits of unit II and III, could reflect the local provenance of the scour fill during rapid erosion and deposition.

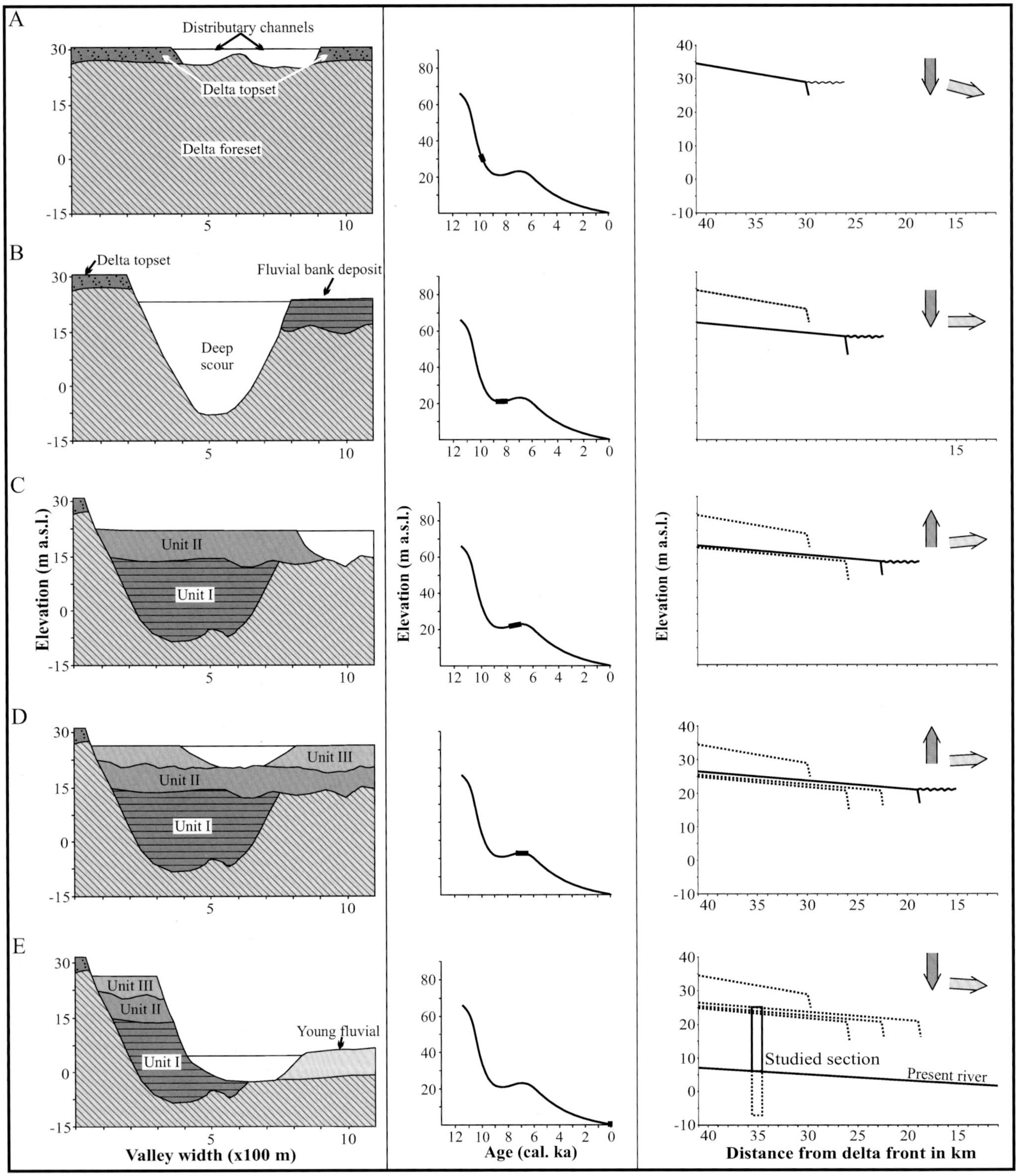

FIGURE 9.—Schematic illustration of the depositional history of the succession at Masjoksletta. Note that during the Tapes transgression, sediment supply was probably sufficient to promote progradation and aggradation. E illustrates the present situation. Horizontal and subhorizontal arrows show shoreline trajectory. Arrows pointing up and down show fluvial aggradation and incision, respectively.

TABLE 2.—Data table with measured and estimated thicknesses, and inferred maximum water depth during deposition of the units. Mean cross-set thickness, dune height, and thickness/height calculations are according to Bridge and Tye (2000) and Leclair and Bridge (2001). * Depth range calculated for the values 3 and 20 as most dunes fall within this range (with a normal range between 6 and 10) according to Leclair and Bridge (2001). See text for further discussion. All values are in meters.

Unit	Measured unit thickness (m)	Estimated max unit thickness (m)	Inferred bankfull water depth (m)	Mean cross-set thickness (s_m)	Mean dune height (h_m= 2.9s_m)	Depth range* (normal range)	Flow depth/dune height ratio (d/h_m)
I	3–20	< 28	~ 28	0.46	1.33	4–26.6 (8–13.3)	~ 20
II	4–8.5	12	> ~ 12	0.31	0,9	2.7–18 (5.4–9)	~ 13
III	5–8	8	> ~ 8	0.24	0,7	2.1–14 (4.2–7)	~ 11.5

Reconstruction of Flow Depth

Bridge and Tye (2000) and Leclair and Bridge (2001) used cross-set thickness distributions to estimate dune heights and corresponding flow depths. Mean dune height (h_m) is approximately 2.9 ($\pm$ 0.7) x mean cross-set thickness (s_m) (Leclair and Bridge 2001), and the former can be related to the water depth existing at the time of dune migration (Yalin 1964; Allen 1982). Mean dune height generally increases with flow depth (d), although the scatter is large (Yalin 1964; Allen 1982; Bridge and Tye 2000). Most dunes fall within the range of $3 < d/h_m < 20$ and are commonly between 6 and 10, although the values are based on measurements of dunes that were not necessarily in equilibrium with the prevailing flow conditions (Leclair and Bridge 2001).

The mean cross-set thickness (with standard deviation) of units I, II, and III is 46 cm (34), 31 cm (18), and 24 cm (19), respectively, giving a mean dune height of 133 cm (unit I), 90 cm (unit II), and 70 cm (unit III) (Table 2). If the ratio of flow depth to mean dune height is between 3 and 20 (Yalin 1964; Allen 1982; Bridge and Tye 2000), the flow depth should have been 4–26.6 m for unit I, 2.7–18 m for unit II, and 2.1–14 m for unit III. Maximum channel depth for unit I is 28 m, which is slightly higher than the maximum scale range (Table 2). The values for units II and III are slightly higher than the common range given by Leclair and Bridge (2001). The reasons for this can be that the height of dunes relative to their flow depth increases from the lower boundary of their stability fields to a maximum in the middle before decreasing close to the upper boundary at the transition to upper-stage plane beds (Leclair and Bridge 2001). It is possible that the dunes preserved in unit I were formed closer to the latter, explaining the large value of d/h_m. Also, it is not certain that the dunes were in equilibrium with the prevailing flow conditions. Formation of the scour surface could be related to the maximum flow conditions, with subsequent infill during waning flow. However, the 3-m-thick dunes recorded in the deepest part of unit I on the GPR profile (Fig. 5) would suggest a d/h_m ratio of 9, which is well within the range given by Leclair and Bridge (2001).

Our analysis suggests that the methodology of using cross-set thickness distributions to estimate formative flow depth can give a useful approximation, but should be used with caution in relation to scour infill, and possibly also to other high-aggradation settings.

Distinguishing Between Autocyclic Scour and Allocyclic Incision

Interpretation of base-level fall, erosional "sequence" boundaries, and the accumulation of sediments in incised valleys is commonly assessed using a number of diagnostic criteria (Van Wagoner et al. 1990; Hampson et al. 1996): (1) the basal erosion surface must be of regional (basin-wide) extent, (2) facies associations overlying the basal erosion surface should be markedly different from those below the surface, (3) the erosional unconformity should remove underlying strata, (4) erosional relief on the sequence boundary is significantly greater than the mean fluvial channel depth, and (5) incised-valley fills have a distinctive internal architecture that is commonly multistory, and that records the progressive rise in base level through the filling of the valley (cf. Fielding 2007). The first of these criteria is the only one that is truly diagnostic (Fielding 2007), although it is very difficult to satisfy unequivocally. As shown in this study, diagnostic criteria 2, 3, and 4 above are met for the scour surface in Tana; the scour infill sediments directly overlie distinctly different deltaic sediments, a considerable thickness (> 30 m) of the underlying strata has been removed, and the erosional relief of the scour surface is greater than the mean fluvial channel depth. However, the scour surface at Tana also coincides with what is potentially a sequence boundary associated with a falling-stage systems tract (Corner 2006), and the cited criteria do not assist in identifying its deep-scour origin as distinct from a sequence boundary. Fielding et al. (2005) and Fielding et al. (2006) reported from the modern Burdekin River delta plain that local erosion has removed the entire Holocene transgressive and older highstand systems tract. Another example is from the Jamuna–Ganges confluence, where a deep scour cuts well into underlying deposits (Umitsu 1993). Best and Ashworth (1997) have suggested that the presence of a basal erosion surface greater than five times the mean channel depth, and ideally a correlation of this surface between basins is the best criterion with which to confirm an incised-valley-fill interpretation. Our study concurs with this conclusion. We would also suggest that the characteristic deposits described for unit I at Tana, such as great thickness (up to five times the normal thickness of fluvial sediments in the area), uniform bed thickness, and uniform grain size, may serve as an additional diagnostic tool for identifying scour-fill deposits and hence for distinguishing between autogenic, localized channel scouring and allogenically controlled incision. However, few studies of scour infill have been made and more work is needed in order to verify this.

CONCLUSIONS

A 30-m-thick fluvial succession at Masjoksletta in the Tana valley, northern Norway, comprises three stacked units which, together, are more than five times thicker than the "normal" fluvial unit found elsewhere in the valley. The anomalous thickness can be related to a combination of deep scour and fill (deposition of unit I) and subsequent two-phase aggradation (deposition of units II and III) during rapid progradation of the fluviodeltaic system. The succession formed during a glacio-isostatically controlled local stillstand in relative sea-level fall causing an essentially horizontal shoreline trajectory at an elevation of c. 21 m a.s.l., corresponding to the regional Tapes transgression between 9 and 7 ka. Deposition occurred at an estimated distance of 5–15 km from the river mouth.

The lowermost, thickest (< 20 m) unit (unit I) is characterized by relatively thick beds and uniform grain size, sigmoidal cross-bedding, and stacking of dunes, all suggesting rapid deposition, possibly within one season. This anomaly in thickness, together with the sedimentological characteristics, is ascribed to localized scouring and subsequent rapid infill. Approximately 7 m of sediment were removed from unit I by erosion when overlying unit II was emplaced. Units II and III are 8–8.5 m thick and characterized by upward-fining grain size and upward-decreasing bed thickness, comprising sandy planar cross-beds in the lower parts, often with tangential foresets, and thinner sets of trough cross-beds, deformed beds, and alternating planar-laminated and ripple-laminated beds in the upper parts. Approximately 3–4 m of unit II has been removed by erosion prior to deposition of overlying unit III. Units II and III formed in response to a stillstand in relative sea level related to the regional mid-Holocene Tapes transgression, which, combined with rapid delta progradation, caused the river bed to aggrade.

Based on our results, we suggest that characteristics such as great thickness (up to five times the normal thickness of fluvial sediments in the area), uniform bed thickness, and uniform grain size may serve as additional diagnostic tools for identifying scour-fill deposits and hence for distinguishing between autogenic, localized channel scouring and allogenically controlled incision. Our results also show that the use of cross-set thickness distributions to calculate dune heights and flow depth should be used with caution when applied to scour-and-fill sediments, and possibly also to other high-aggradation settings.

ACKNOWLEDGMENTS

The work is based on the first author's Master's thesis, carried out as part of a Tana delta project financed by Norsk Hydro and the University of Tromsø in 1992–1994, in collaboration with the Geological Survey of Norway. We thank the editors Stephanie K. Davidson, Colin P. North, and Sophie Leleu, and reviewers Chris Fielding and Trevor Hoey for the constructive reviews that helped improve the manuscript.

REFERENCES

Allen, J.R.L., 1982, Sedimentary Structures; Their Character and Physical Basis, vol. 1: Amsterdam, Elsevier, 663 p.

Andersen, B.G., 1979, The deglaciaton of Norway 15,000–10,000 B.P.: Boreas, v. 8, p. 79–87.

Ardies, G.W., Dalrymple, R.W., and Zaitlin, B.A., 2002, Controls on the geometry of incised valleys in the Basal Quartz Unit (Lower Cretaceous), Western Canada Sedimentary Basin: Journal of Sedimentary Research, v. 72, p. 602–618.

Ashmore, P., and Parker, G., 1983, Confluence scour in coarse braided streams: Water Resources Research, v. 19, p. 392–402.

Ashworth, P.J., Best, J.L., Peakall, J., and Lorsong, J.A., 1999, The influence of aggradation rate on braided alluvial architecture: field study and physical scale modeling of the Ashburton River gravels, Canterbury Plains, New Zealand, *in* Smith, N.D., and Rogers, J., eds., Fluvial Sedimentology VI: International Association of Sedimentologists, Special Publication 28, p. 333–346.

Ashworth, P.J., Best, J.L., and Jones, M.A., 2007, The relationship between channel avulsion, flow occupancy and aggradation in braided rivers: insights from an experimental model: Sedimentology, v. 54, 3, p. 497–513.

Beerbower, J.A., 1964, Cyclothems and cyclic depositional mechanisms in alluvial plain sediments: Kansas Geological Survey, Bulletin 169, p. 31–42.

Best, J.L., 1988, Sediment transport and bed morphology at river channel confluences: Sedimentology, v. 35, p. 481–498.

Best, J.L., and Ashworth, P.J., 1997, Scour in large braided rivers and the recognition of sequence stratigraphic boundaries: Nature, v. 387, p. 275–277.

Blum, M.D., 1993, Genesis and architecture of incised valley fill sequences: a late Quaternary example from the Colorado River, Gulf coastal plain of Texas, *in* Weimer, P., and Posamentier, H.W., eds., Siliciclastic Sequence Stratigraphy: Recent Developments and Applications: American Association of Petroleum Geologists, Memoir 58, p. 259–283.

Blum, M.D., and Törnqvist, T.E, 2000, Fluvial response to climate and sea level: A review and look forward: Sedimentology, v. 47, p. 1–48.

Bridge, J.S., and Tye, R.S., 2000, Interpretating the dimensions of ancient fluvial channel bars, channels, and channel belts from wireline-logs and cores: American Association of Petroleum Geologists, Bulletin, v. 84, p. 1205–1228.

Bristow, C.S., 1987, Brahmaputra River: channel migration and deposition, *in* Ethridge, F.G., Flores, R.M., and Harvey, M.D., eds., Recent Developments in Fluvial Sedimentology: SEPM, Special Publication 39, p. 63–74.

Bristow, C.S., Best, J.L., and Roy, A.G., 1993, Morphology and facies models of channel confluences: International Association of Sedimentologist, Special Publication 17, p. 91–100.

Coleman, J.M., 1969, Brahmaputra River: channel processes and sedimentation: Sedimentary Geology, v. 3, p. 129–239.

Collinson, J.D., 1970, Bedforms of the Tana River, Norway: Geografiska Annaler, v. 52A, p. 31–56.

Collinson, J.D., 1971, Some effects of ice on a river bed: Journal of Sedimentary Petrology, v. 41, 2, p. 557–564.

Collinson, J.D., 1986, Alluvial sediments, *in* Reading, H.G., ed., Sedimentary Environments and Facies, Second Edition: Oxford, U.K., Blackwell, p. 20–62.

Collinson, J.D., and Thompson, D.B., 1989, Sedimentary Structures, Second Edition: London, Unwin Hyman, 207 p.

Corner, G.D., 2006, A transgressive–regressive model of fjord-valley fill: stratigraphy, facies and depositional controls, *in* Dalrymple, R.W., Leckie, D.A., and Tillman, R.W., eds., Incised Valleys in Time and Space: SEPM (Society for Sedimentary Geology), Special Publication 85, p. 161–178.

Corner, G.D., and Eilertsen, R.S., 2008, Raised beaches, falling-stage deltas, river terraces and postglacial fjord-valley fill, arctic Norway: 33 International Geological Congress, excursion no. 40, August 1–4, 2008, Excursion guide, 87 p.

Corner, G.D., Andreassen, K., Rønning, J.S., Mauring, E., and Kristoffersen, Y., 1995, Geology of the Tana delta: a morphological, sedimentological and geophysical study of a regressive, sandy Holocene fjord-delta: University of Tromsø, report to Norsk Hydro A.S., 145 p.

Corner, G.D., Kolka, V.V., Yevzerov, V.Y., and Møller J.J., 2001, Postglacial relative sea-level change and stratigraphy of raised coastal basins on Kola Peninsula, northwest Russia: Global and Planetary Change, v. 31 (1–4), p. 153–175.

Corner, R., 1975, The Tana valley terraces, a study of the morphology and sedimentology of the lower Tana valley terraces between Mansholmen and Maskejokka, Finnmark, Norway: Department of Physical Geography, University of Uppsala, Ungi Report, 38, 293 p.

Cowan, E.J., 1991, The large-scale architecture of the fluvial Westwater Canyon Member, Morrison Formation (Jurassic), San Juan Basin, New Mexico, *in* Miall, A.D., and Tyler, N., eds., The Three-Dimensional Facies Architecture of Terrigenous Clastic Sediments, and Its Implications for Hydrocarbon Discovery and Recovery: SEPM, Concepts in Sedimentology and Paleontology, no. 3, p. 80–93.

De Boer, P.L., Oost, A.P., and Visser, M.J., 1989, The diurnal inequality of the tide as a parameter for recognizing tidal influences: Journal of Sedimentary Petrology, v. 59, 6, p. 912–921.

Dehls, J.F., Olesen, O., Bungum, H., Hicks, E.C., Lindholm, C.D., and Riis, F., 2000, Neotectonic map: Norway and adjacent areas, 1:3 mill. Geological Survey of Norway.

Eilertsen, R.S., and Hansen, L., 2008, Morphology of river bed scours on a delta plain revealed by interferometric sonar: Geomorphology, v. 94, p. 58–68.

Eilertsen, R., Corner, G.D., Kvale, E., and Jensen, M., 2005, Quaternary tidal rhythmites as a proxy for paleoprecipitation (abstract): Geological Society of America, Annual Meeting, Salt Lake City, USA, 16–19 October, 2005, Abstracts with Programs, v. 37, no. 7, p. 364.

Eilertsen, R., Corner, G.D., Aasheim, O., Ystborg, H., Andreassen, K., and Kristoffersen, Y., 2006, Valley-fill stratigraphy and evolution of the Målselv fjord-valley, northern Norway, *in* Dalrymple, R.W., Leckie, D.A., and Tillman, R.W., eds., Incised Valleys in Time and Space: SEPM (Society for Sedimentary Geology), Special Publication 85, p. 179–195.

Fassnacht, S.R., and Conly, M., 2000, Persistence of a scour hole on the East Channel of the Mackenzie Delta, N.W.T.: Canadian Journal of Civil Engineering, v. 27, p. 798–804.

Fielding, C.R., 2006, Upper flow regime sheets, lenses and scour fills:

Extending the range of architecture elements for fluvial sediment bodies: Sedimentary Geology, v. 190, p. 227–240.

Fielding, C.R., 2007, Sedimentology and stratigraphy of large river deposits: recognition in the ancient record, and distinction from 'incised valley fills', *in* Gupta, A., ed., Large Rivers: Geomorphology and Management: New York, John Wiley & Sons, p. 97–113.

Fielding, C.R., Trueman, J.D., Dickens, G.R., and Page, M., 2005, Geomorphology and internal architecture of the ancestral Burdekin River across the Great Barrier Reef shelf, north-east Australia, *in* Blum, M.D., Marriott, S.B., and Leclair, S.F., eds., Fluvial Sedimentology VII: International Association of Sedimentologists, Special Publication 35, p. 321–347.

Fielding, C.R., Trueman, J.D., and Alexander, J., 2006, Holocene depositional history of the Burdekin River delta of northeastern Australia: a model for a low-accommodation, highstand delta: Journal of Sedimentary Research, v. 76, p. 411–428.

Friend, P.F., 1978, Distinctive features of some ancient river systems, *in* Miall, A.D., ed., Fluvial Sedimentology: Canadian Society of Petroleum Geologists, Memoir 5, p. 531–542.

Galay, V.J., Yaremko, E.K., and Quazi, M.E., 1987, River bed scour and construction of stone riprap protection, *in* Thorne, C.R., Bathurst, J.C., and Hey, R.D., eds., Sediment Transport in Gravel-bed Rivers: Chichester, United Kingdom, John Wiley & Sons, p. 353–379.

Guccione, M.J., 1994, Indirect response of the Peace River, Florida, to episodic sea-level change: Journal of Coastal Research, v. 11, p. 637–650.

Hampson, G.J., Elliott, T., and Flint, S.S., 1996, Critical application of high resolution sequence stratigraphic concepts to the Rough Rock Group (Upper Carboniferous) of northern England, *in* Howell, J.A., and Aitken, J.F., eds., High Resolution Sequence Stratigraphy: Innovations and Applications: Geological Society of London, Special Publication 104, p. 221–246.

Hestem, N., 1997, Sedimentological examination of fluviodeltaic sediments, lower Tana River, Finnmark (in Norwegian): Master's thesis, University of Tromsø, Norway, 123 p.

Holbrook, J.M., 2001, Origin, genetic interrelationships, and stratigraphy over the continuum of fluvial channel-form bounding surfaces: An illustration from middle Cretaceous strata, southeastern Colorado: Sedimentary Geology, v. 124, p. 202–246.

Hunter, R.E., 1985, Subaqueous sand-flow cross-strata: Journal of Sedimentary Petrology, v.55, 6, p. 886–894.

Jopling, A.V., 1965, Hydraulic factors controlling the shape of laminae in laboratory deltas: Journal of Sedimantary Petrology, v. 35, p. 777–791.

Klaassen, G.J., and Vermeer, K., 1988, Confluence scour in large braided rivers with fine bed material: International Conference on Fluvial Hydraulics, Vituki, Budapest, Proceedings, p. 395–408.

Kvale, E.P., Archer, A.W., and Johnson, H.R., 1989, Daily, monthly, and yearly tidal cycles within laminated siltstones of the Mansfield Formation (Pennsylvanian) of Indiana: Geology, v. 17, p. 365–368.

Lapointe, M.F., 1985, Aspects of channel bathymetry and migration patterns in the Mackenzie Delta, N.W.T.: National Hydrology Research Institute, Cold Regions Section, Environment Canada, Ottawa, Ontario.

Leclair, S.F., and Bridge, J.S., 2001, Quantitative interpretation of sedimentary structures formed by river dunes: Journal of Sedimentary Research, v. 71, p. 713–716.

Lotsari, E., Veijalainen, N., Alho, P., and Käyhkö, J., 2010, Impact of climate change on future discharges and flow character of the Tana River, subarctic northern Fennoscandia: Geografiska Annaler, v. 92, p. 263–284.

Mauring, E., and Rønning, J.S., 1993, Geophysical measurements in Tanadalen municipality, Finnmark (in Norwegian): Geological Survey of Norway, Report 93.122, 33 p.

Mauring, E., Koziel, J., Lauritsen, T., Rønning, J.S., and Tønnesen, J.F., 1995, Measurements with georadar. Theory, use, techniques and examples (in Norwegian): Geological Survey of Norway, Report, 95.127, 19 p.

Mckee, E.D., Reynolds, M.A., and Baker, C.H., 1962, Laboratory studies on deformation in unconsolidated sediments: U.S. Geological Survey, Professional Paper 450D, p. D151-D155.

Miall, A.D., 2006, How do we identify big rivers? And how big is big?: Sedimentary Geology, v. 186, p. 39–50.

Miall, A.D., and Jones, B.G., 2003, Fluvial architecture of the Hawkesbury Sandstone (Triassic), near Sydney, Australia: Journal of Sedimentary Research, v. 73, p. 531–545.

Moore, B.R., 1970, Scour and fill processes in a deep river hole, Ohio River, Louisville, Kentucky: Journal of Sedimentary Petrology, v. 40, p. 449–456.

Reineck, H-.E., and Singh, I.B., 1980, Depositional Sedimentary Environments, Second Edition: New York, Springer-Verlag, 551 p.

Røe, S-.L., 1987, Cross-strata and bedforms of probable transitional dune to upper-stage plane-bed origin from a Late Precambrian fluvial sandstone, northern Norway: Sedimentology, v. 34, p. 89–101.

Røe, S-.L., and Corner, G.D., 1994, Macroform characteristics in the Tana River: a degrading fluvial system in arctic Norway (abstract): Alluvial Basins: Past and Present Environments, Netherlands, 10–15 September, 1994.

Salter, T., 1993, Fluvial scour and incision: models for their influence on the development of realistic reservoir geometries, *in* North, C.P., and Prosser, D.J., eds., Characterization of Fluvial and Aeolian Reservoirs: Geological Society of London, Special Publication 73, p. 33–51.

Schumm, S.A., 1981, Evolution and response of the fluvial system: sedimentologic implications, *in* Ethridge, F.G., and Flores, R., eds., Recent and Ancient Nonmarine Depositional Environments: Models for Exploration: SEPM, Special Publication 31, p. 19–29.

Sensors & Software, 1996, PulseEKKO IV User's guide: Sensors & Software Inc.: Mississauga, Ontario, 59 p.

Shanley, K.W., and McCabe, P.J., 1994, Perspectives on the sequence stratigraphy of continental strata: American Association of Petroleum Geologists, Bulletin, v. 78, p. 544–568.

Sheets, B.A., Hickson, T.A., and Paola, C., 2002, Assembling the stratigraphic record: depositional patterns and timescales in an experimental alluvial basin: Basin Research, v. 14, p. 287–303.

Siedlecka, A., and Roberts, D., 1992, The bedrock geology of Varanger Peninsula, Finnmark, North Norway: an excursion guide: Geological Survey of Norway, Special Publication 5, p. 1–45.

Sivertsen, F., 1997, Fluvial lithofacies in terraces along the lower Tana Valley north of Tana Bru, Finnmark (in Norwegian): Master's thesis, University of Tromsø, Norway, 185 p.

Smith, D.G., and Pearce, C.M., 2002, Ice jam–caused gullies and scour holes on northern river flood plains: Geomorphology, v. 42, p. 85–95.

Smith, N.D., Cross, T.A., Dufficy, J.P., and Clough, S.R., 1989, Anatomy of an avulsion: Sedimentology, v. 36, p. 1–24.

Smith, N.D., Phillips, A.C., and Powell, R.D., 1990, Tidal drawdown: a mechanism for producing cyclic sedimentation laminations in glaciomarine deltas: Geology, v. 18, p. 10–13.

Sollid, J.L., Andersen, S., Hamre, N. Kjeldsen, O., Salvigsen, O., Sturød, S., Tveitå, T., and Wilhelmsen, A., 1973, Deglaciation of Finnmark, North Norway: Norsk Geografisk Tidsskrift, v. 27, p. 233–325.

Sørensen, R., Bakkelid, S., and Torp, B., 1987, Nasjonalatlas for Norge, kartblad 2.3.3., landhevning (land uplift): Statens Kartverk.

Törnqvist, T.E., 1998, Longitudinal profile evolution of the Quaternary Rhine–Meuse river system during the last glaciation: Interplay of climate change and glacioeustasy?: Terra Nova, v. 10, p. 11–15.

Umitsu, M., 1993, Late Quaternary sedimentary environments and landforms in the Ganges delta: Sedimentary Geology, v. 83, p. 177–186.

Van Wagoner, J.C., Mitchum, R.M., Campion, K.M., and Rahmanian, V.D., 1990, Silisiclastic sequence stratigraphy in well logs, cores and outcrops: concepts for high resolution correlation of time and facies: American Association of Petroleum Geologists, Methods in Exploration, v. 7., 55 p.

Weimer, R.J., 1984, Relation of unconformities, tectonics and sea-level changes, Cretaceous of Western Interior, U.S.A., *in* Schee, J.S., ed., Interregional Unconformities and Hydrocarbon Accumulation: American Association of Petroleum Geologists, Special Publication 36, p. 7–35.

Wentworth, C.K., 1922, A scale of grade and class terms for clastic sediments: Journal of Sedimentological Petrology, v. 30, p. 377–392.

Yalin, M.S., 1964, Geometrical properties of sand waves: American Society of Civil Engineers (Proceedings), Journal of the Hydraulics Division, v. 90, no. HY5, Part 1, p. 105–119.

DEPOSITIONAL DYNAMICS AND PRESERVATION POTENTIAL IN A PROGRADATIONAL LACUSTRINE FLUVIO-DELTAIC SETTING: IMPLICATIONS FOR HIGH-RESOLUTION SEQUENCE STRATIGRAPHY (UPPER TRIASSIC, NORTHWESTERN CHINA)

JENS HORNUNG AND MATTHIAS HINDERER

Technische Universität Darmstadt, Institut für Angewandte Geowissenschaften, Schnittspahnstrasse 9, 64287 Darmstadt, Germany

e-mail: hornung@geo.tu-darmstadt.de

ABSTRACT: Lacustrine deltaic systems differ in several ways from marine ones, and classic sequence stratigraphical concepts need to be modified. This study uses a quantitative outcrop analogue approach with an exceptionally high-resolution record of a lacustrine delta complex (centimeter to decimeter) of the Upper Triassic at the southern margin of the Junggar Basin (Xinjiang, northwestern China). Sedimentological logging of lithofacies and architectural elements were combined with gamma-ray measurements and two-dimensional maps of outcrop faces. The data are analyzed and interpreted in terms of depositional dynamics, cyclicity, stacking pattern, accommodation vs. sediment supply, and preservation potential. The sedimentary inventory comprises various types of gravelly channel bodies, sheetlike sandy and clayey units, as well as ferrocrete horizons and coal seams organized in four depositional environments: Delta slope, delta front, delta top and distal alluvial plain. In contrast to existing fan-delta models, weakly incised, coarse-grained channel fills reaching far beyond the delta front are intercalated into fine-grained delta-slope deposits. Thus, we propose a new type of a mixed-load fan delta, which might be common in other very large lake systems. A fourfold cycle hierarchy with systematic superposition of cycles was identified. It shows a closely linked control of sediment transport, depositional processes, and accommodation space typical for lake systems. According to preservation of cycles and regional geodynamic data, tectonic rates did not change markedly in the Late Triassic. However, a decrease of bisaccate pollen and biodiversity but an increase of spores suggests an increase in humidity, but also in climate "extremes" towards the Jurassic–Triassic boundary. Preservation potential and cycle symmetry are related to the accommodation space by the sediment supply (A/S) ratio, which shows an overall decrease during Norian times with a subsequent increase during the Rhaetian. Gamma-ray measurements (GR) also show a distinct cyclicity, however frequently independent of grain sizes. Channelized coarse-grained sediments show increased natural gamma-ray emission (total counts) due to ^{40}K in claystone clasts. The U/Th ratio is often higher in the delta-front environment due to interaction of fresh river water and lake water.

KEY WORDS: Lacustrine delta, depositional cycles, cycle symmetry, coal, preservation potential, sequence stratigraphy, stratigraphic base-level, reservoir potential, facies prediction, lithofacies, architectural-elements, spectral gamma-ray, U/Th-ratio, uranium, thorium, Junggar Basin, China, sedimentology.

INTRODUCTION

Deltas are of specific interest, because they are sensitive recorders of continent–ocean interactions, bear huge resources of gas and oil, and are often densely inhabited, which directly link delta dynamics with environmental and engineering issues. While marine deltaic systems have attained much attention in the international research community, lacustrine deltas are still underrepresented, although they may be as important as marine ones and contain valuable information on environmental, climatic, and tectonic signals of their hinterland. Results from research on marine deltas are not directly transferable to lacustrine systems, because these are much more sensitive to fluctuations in climate-driven sediment and water supply (Bohacs et al., 2000; Bohacs et al., 2003; Bohacs et al., 2007; Pla-Pueyo et al., 2009; Shanley and McCabe, 1994; Keighley et al., 2003; Hamilton and Tadros, 1994; Einsele, 2000; Reading, 1996), which may result in fundamentally different depositional patterns. This has been documented by seismic investigations of large lacustrine deltas in rift and foreland basins (e.g., Wood, 1994; Scholz and Finney, 1994; Moore et al., 1997; Back et al., 1999; Escalona and Mann, 2006; for a summary, see Scholz and Hutchinson, 2000). Additional new insight into the architecture and dynamics of lacustrine deltas also come from high-resolution analogue studies of modern and fossil systems partly combined with well-log analysis (e.g., Ayers et al. 1986; Horton and Schmitt, 1996; Overeem et al., 2003; Bohacs et al., 2007). Both together have stimulated nonmarine sequence stratigraphical concepts for lake basins (e.g., Aigner et al., 1996; Aigner et al., 1998; Aigner et al., 1999; Csato et al., 1997; Carroll and Bohacs, 1999; Strecker and Steidtmann, 1999; Diessel et al., 2000; Farrell, 2001; Bartov et al., 2002; Keighley et al., 2003).

According to these studies, the following differences between marine and lacustrine deltas are most obvious: (1) water-level fluctuations operate at different scales, usually showing much higher frequencies and magnitudes in lakes; (2) basin morphologies and differential subsidence patterns are often more accentuated, especially in rift settings; (3) suspended-sediment dispersal may strongly vary with time due to temporal stratification and changing salinity; and (4) tides play no prominent role, whereas wave and storm action may be an important distributary process (Dam and Surlyk, 1992). Although, in general, lacustrine systems demonstrate a high potential to enter the geological record, the nature and preservation of specific architectural elements may be significantly different, e.g., higher rates and frequencies of lake-level fluctuations directly affect the available accommodation space and thus the preservation potential of architectural elements. Forced regression and complete subaerial exposure of deltas are quite common phenomena in lacustrine systems and even may remove highstand deposits, which often dominate marine systems (e.g., Overeem et al., 2003; Keighley et al., 2003). This favors the formation and preservation of thick transgressive systems tracts.

We carried out a high-resolution sedimentological study in an extremely well-exposed Upper Triassic fluvio-deltaic sequence at the southern edge of the Junggar Basin (China) to unravel its architecture

From River To Rock Record: The Preservation Of Fluvial Sediments And Their Subsequent Interpretation
SEPM Special Publication No. 97, Copyright © 2011
SEPM (Society for Sedimentary Geology), ISBN 978-1-56576-305-0, p. 281–310.

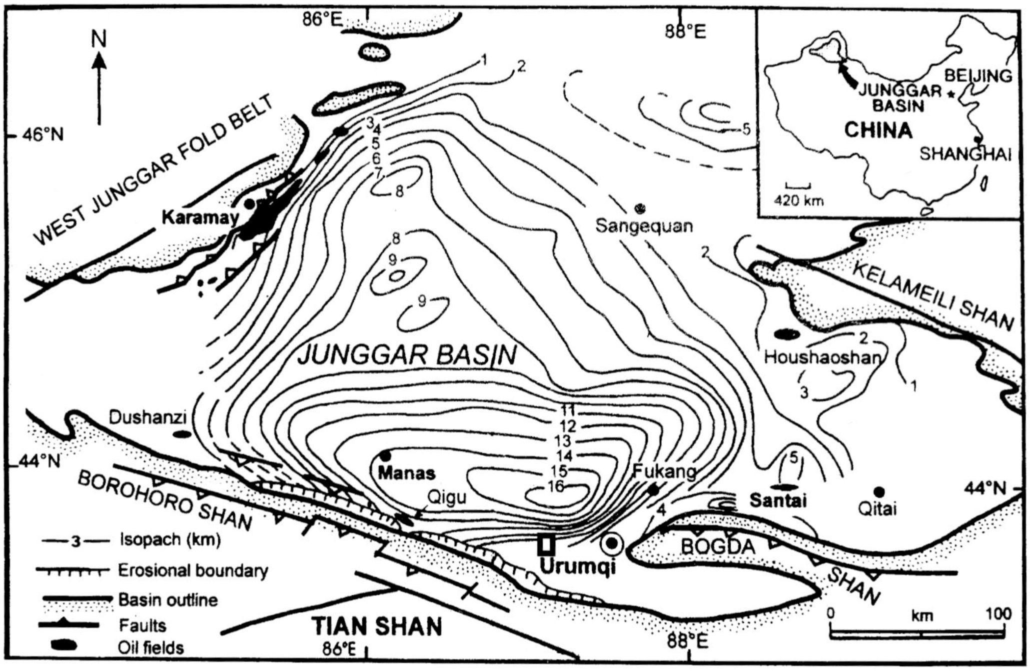

FIGURE 1.—The Junggar Basin with geographic position and total sediment thickness from Permian to Tertiary and major oil fields (in solid black). After Tang et al. (1997). Box to the west of Urumqi marks the study area of the Haojiagou Valley section.

and depositional control. At this time, the Junggar Basin formed an asymmetric inner-cratonic foreland basin, which was flooded by a huge inland lake since the uppermost Permian (Carroll and Warts, 2003). The adjacent thrust belt of the ancestral Tian Shan in the south supplied large amounts of clastic sediments and fed large rivers. According to the proximity of the hinterland, we expect significant shorter river lengths and smaller catchments than in standard marine settings. Local controls on sediment discharge like uplift, thrusting, rock erodibility, or climate patterns interact with regional but high-frequency controls of accommodation space like lake-level variations and subsidence and can be well addressed by a source-to-sink approach.

In this paper, we focus on the classification of preservation potentials and cycle stacking patterns of regional relevance in order to contribute to a basin-wide high-resolution sequence-stratigraphical framework for the Upper Triassic and Lower Jurassic of the Junggar and Turpan Basins. For detailed lithofacies descriptions and environmental interpretations, the reader is referred to Hornung et al. (2011, in press). Furthermore, the Upper Triassic to Lower Jurassic sequences in southern Junggar basin offer a well-exposed example of a large lacustrine delta complex from which general conclusions on the sequence stratigraphy of nonmarine deltas can be drawn. Specific emphasis is put on the cyclic hierarchies operating in different sedimentary environments to distinguish autocyclic, local and regional controls and their possible relationships.

GEOLOGICAL AND DEPOSITIONAL SETTING

The Junggar Basin is almost completely bordered by fold belts: in the northwest the West Junggar Fold Belt, in the northeast the Kelameili Shan and Altai Shan, and in the south the Tian Shan and Bogda Shan mountain ranges (Fig. 1). Two other prominent basins exist south and southeast of the Tian Shan and Bogda Shan, the much bigger Tarim basin and the smaller Turpan basin. The Junggar Basin displays the shape and tectonic structures of a broken foreland basin (Hendrix, 2000) with a long multiphase history that is closely related to the general tectonic evolution in Eurasia. An overview of tectonic evolution at larger scale is given, e.g., by Zhang et al. (1984), Watson et al. (1987), Coleman (1989), Lawrence (1990), Sengör et al. (1993), Otto (1997), Ritts (1998), and Greene et al. (2004). The Junggar Basin is an example of a long-lasting polyphase basin dating back to the late Proterozoic. During the Paleozoic, several kilometers of marine and epicontinental sediments were deposited. Two orogenic pulses in the early and late Permian converted the Junggar Basin into a foreland basin, which received much detritus from the rising ancestral Tian Shan (Windley et al., 1990; Allen et al., 1991). Further collisional events in southern Asia during the Mesozoic and Cenozoic caused repeated tectonic pulses in the Tian Shan orogenic belt, which reactivated the foreland basin and led to repeated progradation of clastic tongues and wedges along the southern margin, which interfinger with basinal sediments. The total sediment thickness reaches about 16 km in the depocenter northwest of Urumqi, of which 6 km are Mesozoic sediments (Tang et al., 1997; Bian et al., 2010). The Junggar Basin contains economically important oil and gas reservoirs.

Most authors (e.g., Allen et al., 1993; Allen et al., 1995; Carroll et al., 1990; Carroll et al., 1995; Zhou, 1997; Tang et al., 1997) follow the hypothesis that the Junggar Basin is separated from the ocean through the uplift of the ancestral Tian Shan–Bogda Shan orogenic systems since late Permian times (Fig. 2). Huang (1978) favors a later separation of the Junggar and Tarim basins in Mesozoic times, which would imply that marine conditions remained significantly longer. However, this hypothesis has not been widely accepted, because of strong evidence for the Paleozoic uplift of the ancestral Tian Shan (e.g., Bally et al., 1986; Hendrix et al., 1992; Hendrix, 2000). From our sedimentological study, we also discard a marine control in the Late Triassic, thus following the majority of authors who favors a landlocked sedimentary environment with a mostly lacustrine record during the Mesozoic (e.g., Lee, 1985; Graham et al, 1990; Allen et al., 1991). The paleolatitude during the Late Triassic was around 45° N, similar to nowadays (e.g., Smith and Briden, 1977; Sharps, 1990; Golonka et al., 1994; Cogne et al. 1995, Allen et al., 1995).

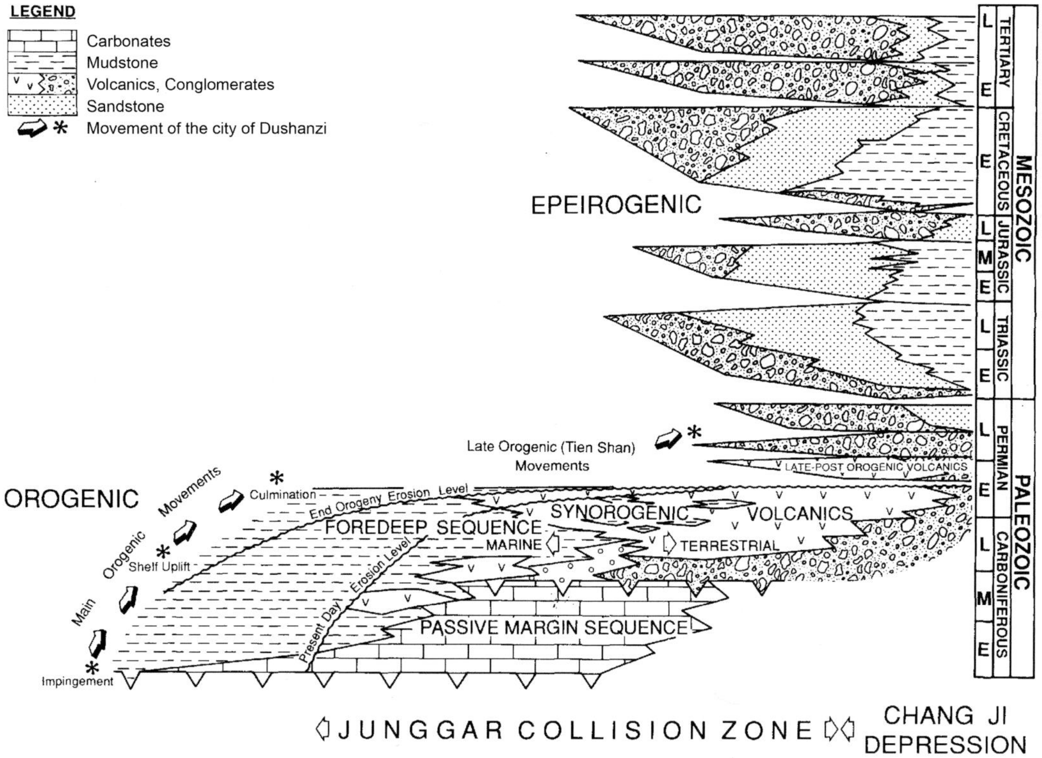

FIGURE 2.—Compilation of tectonic movements and sediment distribution in the ancestral Tian Shan and south Junggar Basin based on seismic and literature data. Note that during the Triassic no uplift event is recognized. From Lawrence (1990) with legend added.

At the southern margin of the Junggar Basin repeated marginal thrusting uplifted Permian to Tertiary deposits, which are well exposed and can be traced along largescale folds. Qualitative and nonscaled descriptions of Triassic to Jurassic depositional environments from these sequences are provided by Qi et al. (1994) and Schneider (1996). A summary of the regional development of depositional environments and basin evolution is found in Bian et al. (2010). Ashraf et al. (1999) and Ashraf et al. (2010) performed detailed palynostratigraphical work along several stratigraphical sections in the southern Junggar Basin, which yield a solid biostratigraphical subdivision and age estimation and make up the stratigraphical basis of our sections.

We investigated the Rhaetian Haojiagou and Norian Huangshanjie formations in an outcrop section located 40 km west of Urumqi in the Haojiagou valley (see box in Figs. 1 and 3). Coordinates (UTM) of the study area center are: 45 T 517729 4833682. The complete Mesozoic section exposed in this area is at least 3.2 km thick (Fig. 1) and was lithologically logged first by Ashraf et al. (1999). The investigated formations comprise sediments around 600 m thick, dipping 10–30° to the north. Using palynomorphs, Ashraf et al. (1999) and Ashraf and Wang (2001) dated the Upper Triassic to middle Jurassic succession by 360 samples and found an overall trend of decreasing bisaccate pollen and an increase of spores (Fig. 3). As bisaccate pollen are wind driven and can fly only under dry conditions, he stated that the decrease in bisaccate pollen and increase in spores can be interpreted as decreasing aridity and increasing humidity, respectively. In addition, he found a significant reduction of biodiversity, which he interpreted as an increase of climate variability. The recorded plant species point to a warm, subtropical climate. All studies were carried out in the framework of the Sino-German Joint Group on Prehistoric Life and Geology in Xinjiang. Field work was done in 2001 with some additions in 2007.

METHODOLOGY AND DATABASE

To achieve the goals of this study one-dimensional and two-dimensional quantitative data with a high resolution were necessary. During a two-months field campaign in 2001, we logged stratigraphical sections with sedimentological and geophysical methods to characterize lithofacies types and facies associations making up architectural elements (Dreyer, 1993; Miall, 1996). Distribution of architectural elements, depositional architecture, and cycle stacking patterns were investigated using 2-D photomosaic mapping. By integrating all data, we set up a sequence stratigraphic model for the investigated outcrop area.

1-D Logging

We sedimentologically logged a 505-m-thick continuous section at a centimeter to decimeter scale through the complete Upper Triassic, closely following the section surveyed by Ashraf et al. (1999). In addition, parallel sections with a total thickness of 220 m were logged to verify lateral heterogeneities. Along with the sedimentological logging, we measured total counts of natural (passive) gamma-ray radiation in spacing of about 5–10 cm on clean outcrop surfaces. This yielded in total 4500 measurement points. We used a total counts probe from G.B.H.-Elektronik and a spectral device from Exploranium Ltd. (GR 130); both are hand held and have a 2.5″ NaJ-Detector and a photomultiplier. The company calibrated the devices with test pads (Grasty et al., 1991), and in the field peak stabilization was done in a two-hour interval or when temperature changed with a ^{137}Cs source. The stabilization interval was automatically controlled and displayed by the device itself. However, the device was lead-protected so that

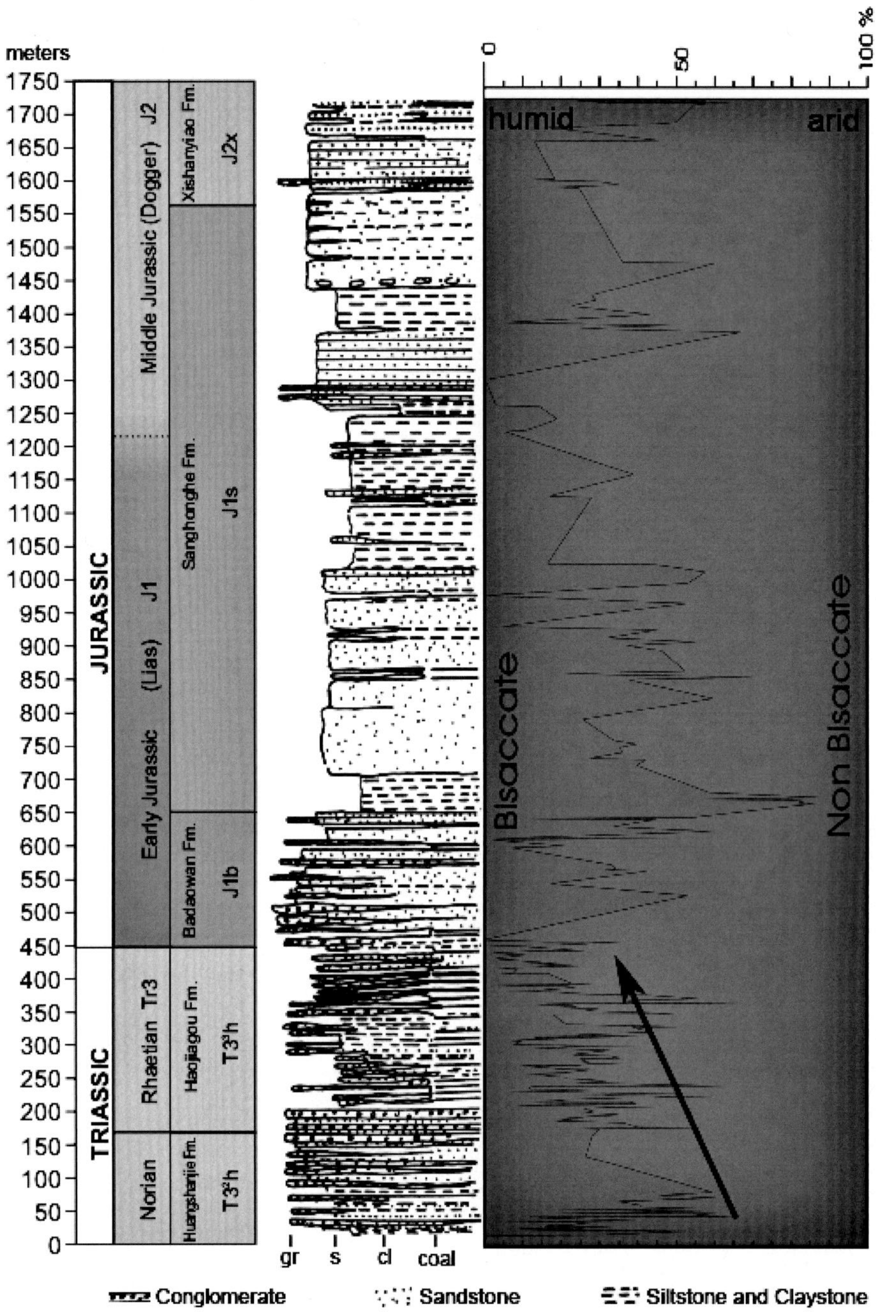

FIGURE 3.—Lithostratigraphic column and palynostratigraphic overview of the Upper Triassic to Middle Jurassic in the investigation area based on data from Ashraf et al. (1999, 2001, 2004, and 2010). To the right the ratio of bisaccate to other pollen grains give paleoclimatic implications. As plants spreading wind-driven bisaccate pollen are privileged in drier conditions, an increase of bisaccate pollen may be a proxy for the expansion of drier habitats. Note the trend towards more humid conditions during the Late Triassic (arrow), ignoring pollen transport mechanisms.

radiation could only enter from the rock face. This reduces incursion of cosmic rays by a factor of two (compared to air measurements). Therefore, measuring time in total counts mode could be reduced to 10 s and with the spectral mode to 60 s. Anyway, in all figures we plotted raw count rates to avoid problems with the transformation of counts into radionuclide concentrations.

To ensure that these relatively short measuring times reproduce reliable spectral gamma-ray data, which are above the inherent statistical error, we carried out time-series measurements up to 600 s. These series show that from a time window greater than 180 s onwards data are very stable. However, not until below 60 s, data become seriously unstable, and so this reflects a best value considering the signal-to-noise ratio and the large amount of individual measurements undertaken. The same was done for total counts measurements.

FIGURE 4.—Upper part of Haojiagou Formation as an example of a typical outcrop in the study area. Strata dip here at 25–30° towards the northeast; resistant units form ridges used to construct scaled photomosaics. Note the scale bar at lower left corner.

2-D Logging

The outcrop section is very suitable for a ground-based photomosaic surveying technique. Because the strata dip at 10–30° to the north, each depositional cycle is exposed as a ridge with an outcrop face. Hence, outcrop faces could be scanned from one ridge to another, yielding a photomosaic of each ridge (Figs. 4, 5). To ensure a minimum of optical distortions, telephoto lenses were used and a GPS-surveyed scale (fitting points) was sprayed on the section. With the help of the scale and the stratigraphical dip angles, the real thickness could be calculated. GPS positions and dip angles showed the original location in the stratigraphical section. In total, an area of 600,000 m^2 was covered.

Depositional Cycles and Sequence Stratigraphy

We use the sequence stratigraphic concept and nomenclature (Fig. 6) according to Wheeler (1964), Jervey (1988), Homewood et al. (1992), Homewood et al. (2000), Cross et al. (1993), Martinsen et al. (1999), Schlager (1993), Shanley and McCabe (1994), Cross and Lessenger (1997), Cross and Lessenger (1998), and Anderson and Cross (2001), just to name a few. These authors define the stratigraphic-base-level as

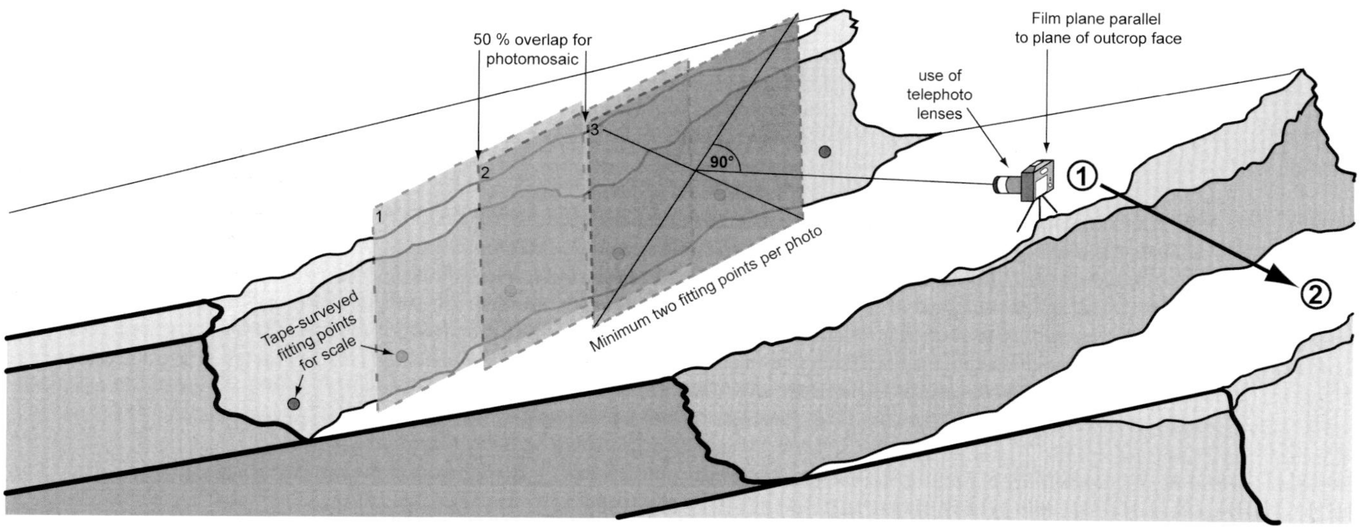

FIGURE 5.—Photomosaic surveying technique as applied in this study. Control nodes at the outcrop face are marked with a tape and surveyed with GPS. By patching all mosaics together (comprising 600,000 m^2), the distribution of architectural elements, sedimentary architecture, and cycle stacking patterns can be mapped quantitatively. For details, see text.

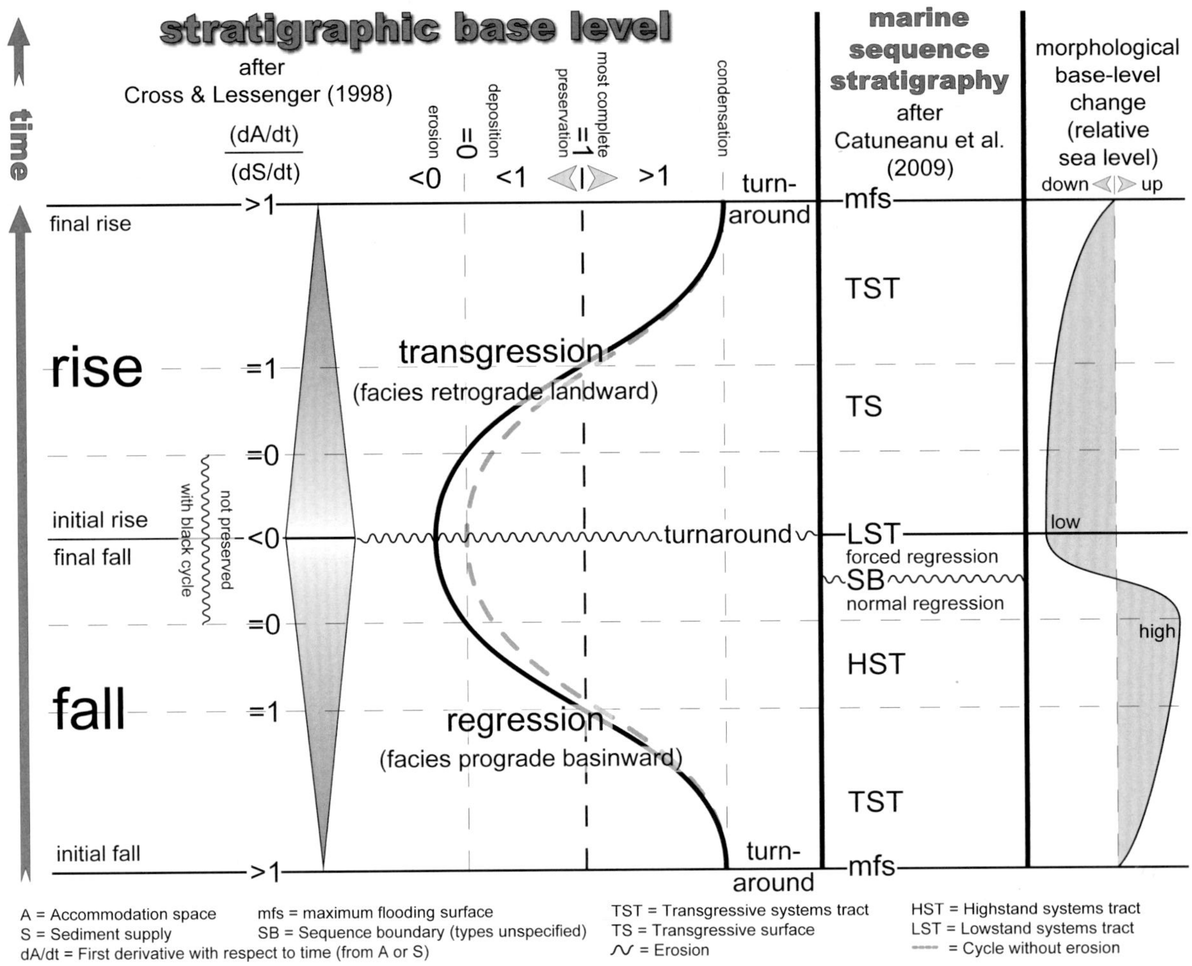

FIGURE 6.—Comparison between the concept of stratigraphic base level and the marine sequence stratigraphy. Note that the A/S ratio changes along downdip sections and parts of the displayed stratigraphic-base-level cycle will not be preserved. For detail see text. After Hornung (1999).

the change of the ratio of sediment supply (S) to accommodation space (A), which means the first derivative with respect to time from the A/S ratio: *d/dt* (A/S) (see also Hornung, 1999; Hornung and Aigner, 2002b). This implies that a retreat of fluvial systems towards the basin margin indicates a stratigraphic-base-level rise (decreasing supply and/or increasing accommodation space) and a progradation towards the basin center a fall (increasing supply and/or decreasing accommodation space). As a common basis most authors agree that a stratigraphic-base-level fall is reflected by progradational and a stratigraphic-base-level rise by retrogradational depositional patterns, respectively (Catuneanu et al., 2009). In turn, the inflection point from rise to fall, the so-called "turnaround" corresponds to the highest rate of lake-level rise (in marine sequence stratigraphy the maximum flooding) and the fall to rise turnaround approximately the lowstand systems tract in marine sequence stratigraphic terminology.

For all figures, a top-up triangle indicates a stratigraphic-base-level rise, a top-down triangle symbolizes the stratigraphic-base-level fall (Fig. 6). The point of touching triangles is the turnaround point. It is considered to have a regional relevance for correlation of the sections, inasmuch as it is independent of lithofacies and controlled from outside the depositional environment, such as climate, lake level, sediment supply, and subsidence (Bohacs et al., 2000). In continental settings, autocyclic processes mask this depositional cyclicity, such as channel migration and avulsion. Therefore, this study used kilometer-wide 2-D sections to eliminate local autogenetic effects on deposition and to figure out distinct cycle hierarchies and relationships such as symmetry and stacking patterns.

DEPOSITIONAL ENVIRONMENTS AND LITHOFACIES MAKEUP

We identified and characterized 19 lithofacies types (see Table 1). Full sedimentological descriptions and environmental interpretations can be found in Hornung et al. (2011, in press). Fig. 7 provides a brief summary of some major lithofacies types and architectural elements. Together with architectural-element analyses, four different depositional environments could be distinguished:

- Distal alluvial plain,
- Delta top,
- Delta front, and
- Delta slope.

Typical distal alluvial-plain facies associations show channels

TABLE 1A.—The lithofacies classification of dominantly gravel-size clastic sediments used at the Haojiagou Valley section. The first capital letter symbolizes the dominant grain size, the second lowercase letter the bedding type, and the third lower-case letter the framework characteristics. With lower-case letters, matrix grain sizes could be added in front of this main code, and after the main code additional information separated by a comma. Table contains only main features; for a detailed description, see Hornung et al. (2011).

Classification of lithofacies

G, Gravel (all dominantly gravel-size clastic sediments)

General description: commonly elongated, around 2–3 cm diameter, poorly sorted, subrounded to rounded

General Interpretation: high stream power, > 10 cm/s

Code	Type	Framework	Main features	Interpretation
Gm	Gravel, massive	m, matrix supported g, grain supported o, open framework	• Components randomly distributed	Rapid sedimentation due to sudden decrease in stream power and/or sediment overload.
Gp	Gravel, planar cross-bedded		• Planar crossbedded with • dip angles 15 to 30° • Internal and laterally graded foresets • set thickness 10 to 100 cm	Approx. set thickness: < 20 cm: 2-D bedforms; > 20 cm: scour fill, mid-channel bars or transverse bars with straight crests.
Gt	Gravel, trough cross-bedded		• trough cross-bedded • Internal and laterally graded foresets • set thickness 10 to 100 cm	Approx. set thickness: < 20 cm: 3-D bedforms; > 20 cm: scour fill, mid channel bars or transversal bars with curved crests.
Gh	Gravel horizontally bedded		• sometimes graded • sets of 3 to 15 cm thickness • occasionally imbrications	Roll over or gravel carpet deposition.

comprising crossbedded sandstone and conglomeratic units (Fig. 7a, C and Fig. 7b, A), occasionally with a massive appearance. Horizontally bedded and massive siltstone and claystone sheets represent overbank deposits. In addition, thin coal seams and ferruginous pedogenetic crusts are intercalated. The iron hydroxide forming these crusts precipitated in the pore space of sandstones and increase in thickness and frequency towards the upper Haojiagou Formation. They might be used as a proxy for changing redox conditions remobilizing the iron. Temporarily lower groundwater tables may favor cementation in pore space.

The delta-top depositional environment is characterized by coaly, silty and clayey lithofacies types, which show dominantly horizontal stratification up to massive bedding (Fig. 7a, A). Conglomeratic channels slightly incise and bundled crevasse sand sheets intercalate these deposits (Fig. 7a, B). Similar facies associations and depositional environments may reoccur in the coal-bearing Jurassic and are described in Schneider (1996).

Structureless conglomeratic channels with randomly deposited clasts, beach (Fig. 7a, D) and foreshore sands, and some coastal siltflats become dominant at the delta-front depositional environment. The conglomeratic channels incise into the other depositional elements.

Finally, the delta-slope depositional environment comprises massive and horizontally stratified fine sand and siltstone. They are frequently normally graded, laterally highly persistent, and show a sheetlike geometry with constant thickness (Fig. 7b, C). Occasionally massive claystones intercalate. Slightly incised, conglomeratic channels are a prominent feature showing a massive appearance, randomly deposited clasts, and a low (1:5–1:45) w/t relationship (Fig. 7b, B). Another type of channel shows sheetlike (1:150–1:400) w/t relationships and is almost not incised.

The lithologic composition of the section comprises neither carbonates nor marly beds, but exclusively siliciclastic facies types. This applies also for calcite or calcite cements, which are completely absent in the section. Sedimentary structures indicate unidirectional flow and some wave action. Remarkably, almost no slumping was observed.

Whereas plant debris is quite common for the whole Upper Triassic succession in the Haojiagou valley (Sun et al., 2001) neither bioturbated horizons nor bivalves or any other shell accumulations were found. The section is sterile also in terms of other typical lacustrine biota, e.g., ostracods, algae, microbialites, gastropods, fish, turtles, crocodiles, or even ichnofossils that are most common in the Jurassic (Maisch et al., 2001; Maisch et al., 2004; Pfretzschner et al. 2004; Wings et al, 2007; Schellhorn et al., 2007). Furthermore, we did not find any indications that shells or concretions might be dissolved during diagenesis.

Another unexpected observation was the lack of clear signatures of subaerial exposure and emergence like desiccation cracks, evaporates, red beds, or pedogenetic features like concretions or rooting (except three minor ones) even for units which must have been deposited under supra-lacustrine or subaerial conditions (the Haojiagou formation; e.g., the coal deposits). Probably, the iron crusts are pedogenetically influenced and might be formed by evaporitic pumping causing a simple impregnation of pore space in permeable units on top of mudstone beds. In some horizons few, but very big (> 1 m) groundwater-induced ferrogenetic concretions were found.

More detailed lithological descriptions, facies models, and interpretation of depositional environments for the Upper Triassic are provided in Hornung et al. (2011, in press).

Even though the Haojiagou valley section exhibits perfect outcrops, the third dimension (downgradient) is missing. Therefore, it is not clear if the different depositional environments are in a continuous lateral relationship after Walther's law or more likely, if a complete reorganization of the sedimentary environment took place around the transition zone of the distinguished depositional environments.

TABLE 1B.—The lithofacies classification of dominantly sand-size clastic sediments used at the Haojiagou valley section. Table contains only main features; for a detailed description, see Hornung et al. (2011).

Classification of lithofacies

S, Sand (all dominantly sand-size clastic sediments)

General description:
fine sand to very coarse sand
yellow to greyish quartz arenites

General Interpretation:
Intermediate stream power

Code	Type	Main features	Interpretation
Sc	Sand, chaotic	• lithic graywacke, poorly sorted • plant chaff, extraclasts and mudclasts • mostly convolute bedded or structureless, • rarely indistinct trough-crossbeds • 0.1–1 m thick	Rapidly deposited mainly as scour fill
Sm	Sand, massive	• medium- to very coarse-grained • sometimes extraclast and mudclasts • structureless, mostly ungraded • 0.05–2 m thick	Dewatered or primarily structureless
St	Sand, trough cross-bedded	• medium- to very coarse-grained • extraclasts at base and on some foresets • graded foresets • sets 0.1–0.5 m thick	Approx. set thickness: < 30 cm: 3D bedforms (dunes) > 30 cm: bars (curved crests)
	Sand, planar cross-bedded	• fine- to coarse-grained • dip angles of 15–35° • sets 0.1–0.5 m thick	Approx. set thickness: < 30 cm: 2D bedforms > 30 cm: bars (straight crests)
Sl	Sand, low-angle cross-bedded	• fine- to medium-grained, well sorted • < 10° inclined planar foresets • 0.1–1 m thick	Upper flow regime due to wave action
Sh	Sand, horizontally bedded	1. fine-grained, silty and clayey, laminated 2. fine- to medium-grained quartz arenite, 3. like 2. + very well sorted and rounded, parting lineation, lamina thickness approx. 0,5 cm.	1. subaqueous suspension fallout 2. plane bed (alluvial) 3. wave action on shoreline
Sr	Sand, ripple cross-laminated	• fine- to medium-grained • small-scale trough or planar cross-bedded	2D and 3D current and wave ripples

TABLE 1C.—The lithofacies classification of dominantly muddy clastic sediments used at the Haojiagou valley section (40 km west of Urumqi, Junggar Basin, NW China). Table contains only main features; for a detailed description, see Hornung et al. (2011).

Classification of lithofacies

F, Fines (all dominantly fine-grained or muddy clastic sediments)

General description: clay to coarse silt (< 125 μm)

General Interpretation: Weak or no stream power

Code	Type	Main features	Interpretation
Fl	Fines, laminated	• Muddy very fine sandstones, siltstones, and claystones • Horizontally bedded or laminated • Often showing normal grading	Distal suspension fallout and hyperpycnal flows
Fm	Fines, massive	• Siltstones and claystones • No primary sedimentary structures • Can show weak normal or reverse grading	Proximal suspension fallout
Fc	Fines, clast-bearing	• Mudstones or siltstones • Widely dispersed very coarse sand or gravel clasts • structureless, not graded	Mudflow or density flow

TABLE 1D.—The classification of coal, organic, and pedogenetic facies types. For a detailed description, see Hornung et al. (2011).

Classification of lithofacies			
C, Coal			
General description: Often clayey Interfered by ripple cross-stratified shaly coal and coaly siltstones			General Interpretation: Backswamps and lake marshes
Code	**Type**	**Main features**	**Interpretation**
Cm	Coalbed, massive	• structureless • thickness < 10 cm	Coal swamp without clastic input
Cl	Coalbed, laminated	• often clayey • platy, laminae 0.1–2 mm thick	Frequent flooding of coal swamp (distal)
Cp	Coalbed, peaty	• immature browncoal • intercalated with clastics (< sand) • load casts, mottling, convolute bedding • thickness < 10 cm	Coal swamp with frequent alluvial flooding (proximal)
O, Organic sediments			
Op	Organics, peat	• organic matter, laminated, not charred • thickness 5 to < 1 mm	Change of source plants, water content, humic soils, etc.
P, Pedogenic			
Pf	Pedogenic ferrocrete	• ferruginous conglomerate, sand or silt sheet • also as large isolated concretions	Evaporation of rising pore fluids (marsh ironstones).

SEQUENCE STRATIGRAPHICAL RELEVANCE OF ARCHITECTURAL ELEMENTS

Architectural elements were distinguished sensu Miall (1996) and Dreyer (1993). Most architectural elements of this study are part of two or even three of the four depositional environments. Two of them (channels and levees) occur in subaerial and subaqueous settings, but then show properties and relationships different from the surrounding architectural elements. Therefore, we described them separately and distinguished them with the index [sa] for subaqueous. Table 2 gives an overview of the classified architectural elements, which are dominant for each depositional environment and Fig. 8 provides a general legend for the following figures of sections. For detailed descriptions and depositional models see Hornung et al. (2011, in press).

Distal-Alluvial-Plain Depositional Environment

The floodplain-dominated distal alluvial plain is displayed by the upper Haojiagou Formation (Fig. 9). It shows river channels, levees, crevasse splays, and mainly floodplain sedimentation as well as some coal mires in interchannel areas. The river style can be classified as mixed-load comprising a broad spectrum of grain sizes. The bed-load component (coarse sand to gravel) is confined to distinct, very low-sinuosity channels and show internal braid patterns (multilateral amalgamation) as well as frequent avulsions, but no significant lateral migration.

Channel (CH): Incised channels indicate a negative accommodation space, which can be local due to amalgamation. All deposits in a channel (except of lags) indicate already a rise in base-level due to increasing accommodation space (Cross and Lessenger, 1998). To distinguish between regional allogenetically and local autogenetically driven incision processes, the degree of preservation of the underlying overbank deposits is most helpful. If downcutting of the channel is limited to its natural extent, levees or coarsening-upward bundled crevasse-splay complexes are preserved below the channel base. This is indicates regionally available accommodation space. If accommodation space decreases or become negative on a regional scale, channel incision would remove large parts of the underlying overbanks (Ramon and Cross, 1997). In these cases, a base-level fall to rise turnaround is assumed at the channel base. When channels are deposited in a generally high-accommodation setting, incision rates cannot be used to detect A/S conditions. Then the base-level fall to rise turnaround point was set at the biggest grain size observed in the succession, because it represents the highest depositional energy.

To exclude misinterpretations from autocyclic processes, it is essential to observe the channel stacking patterns at a regional scale, best in 2-D sections or several parallel logs. In such cases, decreasing or increasing channel spacing was used to determine turnaround points. In a low-accommodation setting, channels stack or amalgamate with others, because only the basal parts of a fluvial cycle are preserved. With increasing accommodation space, channel spacing increases (Ramon and Cross, 1997; Einsele, 2000). Again, if this signal is not well developed, the point of highest or lowest depositional energy was used.

Crevasse Splay (CS): Crevasse splays are part of the overbank fluvial system and are preserved only under moderate A/S conditions. They indicate the presence of a channel nearby. Overall fining-upward trends in crevasse-splay successions indicate a diverging channel and vice versa a coarsening-upward trend a converging channel.

Archaeological data of the Mississippi floodplain suggest that changes in hurricane frequency in the order of 10^3 years can enter the (short-term) geological record (Kidder, 2003). In principle, several authors go along that long-term climatic changes are reflected in a change of fluvial style (Kasse et al., 2003; Blum and Törnqvist, 2000;

FIGURE 7A.—Some major lithofaciestypes types and architectural elements observed in the Haojiagou Valley section. Small scale is 1 cm/dm; large scale is 1 m. A) Massive siltstones (Fm) in the upper part; lower part shows horizontally bedded (partially laminated) siltstones (Fl). B) Laminated and massive coal (Cl and Cm) in a Swamp (SW) architectural element intercalated by sand and mud from bundled crevasse splays (CS). C) Grain-supported, planar crossbedded gravel (Gpg) in an alluvial channel (CH). D) Very well sorted and well rounded, low-angle crossbedded, fine sand interpreted to be formed by swash and backswash wave action in a beach (BH) architectural element. Surface is partly covered by incrustations on a crack plane.

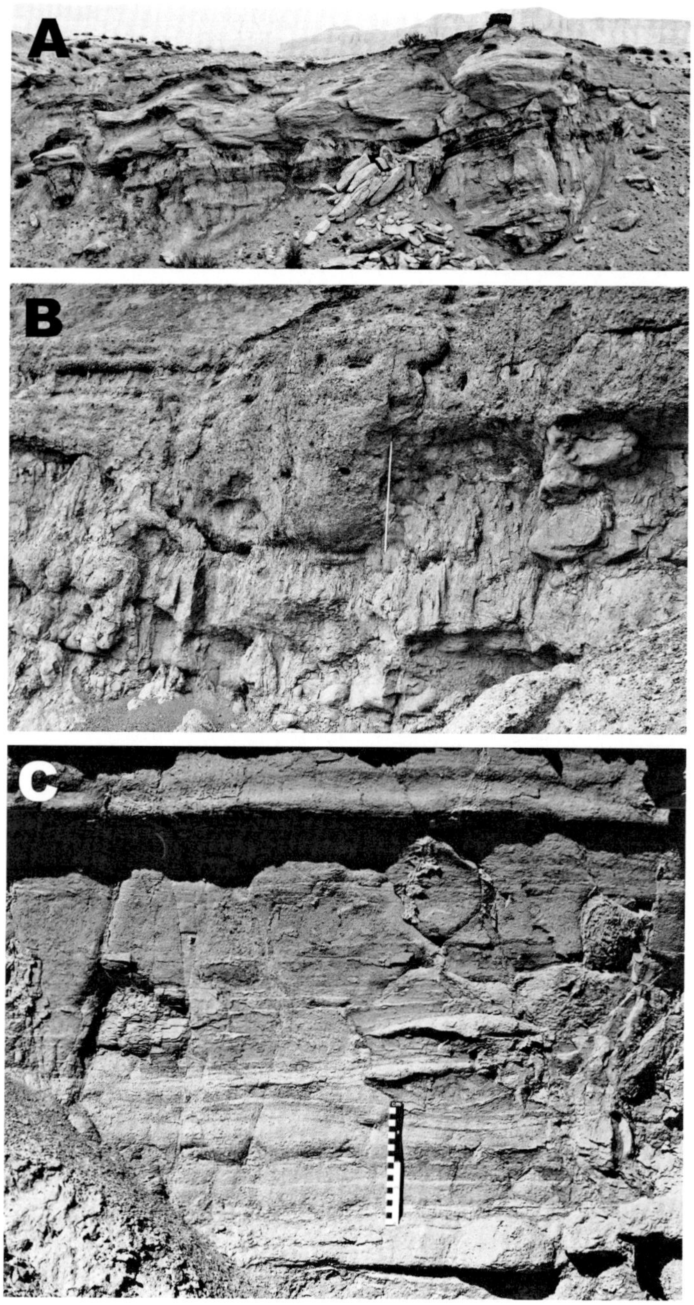

FIGURE 7B.—Three major architectural elements of the Haojiagou Valley section. Small scale is 1 cm/dm; large scale is 1 m. A) Clearly organized, fluvial channel (CH) in the upper Haojiagou Formation sitting on top of levee (LV), floodplain (FF), and crevasse-splay (CS) overbank architectural elements. B) Subaqueous channel ($CH_{(sa)}$) showing disorganization and massive, randomly distributed and partially matrix-rich gravels. Central part is incised and upper part is tailing out. C) Subaqueous fan (SF) showing individual turbidite events with normal graded bedding, slightly erosional bases, and an overall fining-upward trend. Note some upstream-accreted ripples in the central part of the picture. Parts B and C belong to the lower Huangshanjie Formation.

Hinderer, 1999; among many others). In this study, the long-term geological record of crevasse splays is preserved. However, higher frequencies or increase in thickness of crevasse splays may indicate that the fluvial system is affected by external controls, e.g., increased seasonality or decline of vegetation, mostly coinciding with an increase of sediment supply (Einsele, 2000). Hence, at a large or overall scale the frequency, thickness, or depositional energy of observed crevasse splays can be diagnostic of changes in discharge and of the stability of an alluvial system, and with caution systematic changes might be used as a climate proxy within the same depositional environment and the same degree of preservation. Only in the latter case (and if better suitable architectural elements are absent) it was used to set a base-level turnaround point.

However, by just comparing the preserved crevasse splays throughout stratigraphy, they get significantly bigger just after the Triassic–Jurassic boundary, forming a cuesta in the present landscape. Already during the Rhaetian, we observe an increase of floodplain development (frequency and thickness) as well as an increase in thickness (but not frequency) of crevasse-splay units, which can occur in bundles. With respect to the high degree of preservation throughout the upper Haojiagou formation, we speculate that this increased crevasse activity may indicate a reorganization of discharge patterns and/or river types. With great caution, this might be used as a proxy for increased seasonal discharge.

Floodplain Fines (FF): Floodplain intervals mark an increase of preservation potential, as they are easily erodible by channel shifting processes in times when less accommodation space is available. At the same time they suffer from a lack of sediment supply as only suspended-load is deposited. Hence, floodplains indicate a high A/S ratio and a high stratigraphic-base-level in a fluvial system.

Levee (LV): Levees in a distal alluvial plain are preserved only when accommodation space is available at maximum because this overbank architectural element is the topographically highest in a fluvial system and therefore first eroded when channels migrate. Since maximum accommodation space may also indicate a maximum A/S rate, the base-level rise turnaround point should be set here. In other fluvial systems where fewer levees are available, such as, e.g., braided rivers, the purity of floodplain claystones might be used.

Mudflow (MF): Mudflows from the hinterland may have climatic, weathering, and/or tectonic causes (Blair and McPherson, 2008). In this case, they might have a sequence stratigraphical relevance as they indicate trends of regional base-level. If so, they accumulate in distinct levels in the stratigraphical record and then can be used for paleoenvironmental interpretation and for setting turnaround points. If they are formed by local processes, e.g., at waterlogged levee slopes, they are randomly distributed and just reflect autocyclic or chaotic processes.

Delta-Top and Delta-Front Depositional Environment

The delta-top paleoenvironment is obvious in the lower Haojiagou Formation by a dominance of coal seams (Figs. 10, 11). Because coal seams are interbedded with small channels and crevasse-splay sediments, we interpret this part as coastal backswamps (SW) of a delta system. In the upper Huangshanjie Formation, a coastal environment (delta front) becomes dominant. This unit is characterized by lithofacies types and architectural elements, which indicate a shallow lake, lakeshore, and river-mouth depositional environment. Delta-top and delta-front environments are characterized by the interaction of fluvial and shallow lacustrine depositional processes. Plenty of accommodation space is available, and sediment delivery is highly variable from almost none to oversupply. Therefore, preserva-

TABLE 2.—Compilation of the observed architectural elements in the lower Huangshanjie Formation (Haojiagou section, Junggar Basin, NW China). A sequence stratigraphic interpretation is provided in the text.

Classification of architectural elements

Code	Architectural element-type	Lithology and depositional pattern	Genetic Interpretation
Distal alluvial plain			
CH	Channel	• Gpg, Ghm, Gmm, Gtm, Gpm, Gmo, Gmg, • Sc, St, Sh, Sp • 1–2 m incision • Multilateral • Commonly overall fining-upward • 1–4 m thick, w/t 1:5 – 1:45	Main river bed
CS	Crevasse splay	• Consist of Sm, Sh, Sc, St, Sr, Fl, Fm • Poorly sorted, clear fining-upward trend • May form bundles with Fm, Fl • Up to 2 m thick	Levee breakthrough and deposition on the near-channel floodplain
FF	Floodplain fines	• Intercalations of Fl, Fm • Commonly siltstone-mudstone couplets • Fining-upwards and coaling-upwards	Interchannel areas with frequent flooding
LV	Levee	• Sheets, lenses, channel-like forms of Sc, Sr, Sp, St, Sm, Sh, Gtm, Gmm, Ghm alternating with Fm, Fl, Fc • Reactivation surfaces, convolute bedding, overall coarsening-up, inhomogeneous	Overbank deposition next to channels during flood events
MF	Mudflow	• Fc • Randomly distributed coarser grains • No trends • Up to 0.5 m thick	Transport as cohesive suspension and "freezing" due to dewatering or decrease in relief.
Delta-top			
SW	Swamp	• Cl, Cm, Cp, Op alternating with Fl, Fm • Cleaning upward trends • Pure beds 0.5–10 cm thick, overall up to 4 m	Backswamp and lakeshore marsh
Delta-front			
MB	Mouth bar	• Ghm, Gmm, Gpm, St, Sm, Sp, Sr, Sh, Fm, Gpg • Extremely heterogeneous, strong grain-size and lithofacies variations between layers. • Individual layers weakly sorted with high extraclast content.	Subaqueous foreshore continuation of fluvial channels where sediment overload is deposited rapidly.
BH	Beach	• Sl, Sh, Sm, Sr • Very well sorted, very well rounded • No trends • 20 cm thick, amalgamated up to 1 m	Beach sands in swash zone, shoreline
RS	Rippled sand sheet	• Sh, Sr, St • Well sorted, well preserved wave-Sr • No trends • 10 to 150 cm thick	Coastal sand bars/sand waves, shallow water
CM	Coastal mudflat	• Fm, Sh, Sr, Fl • Lenticular- and flaser-bedded very small fine-grained channels • Bioturbation and proportion of organic carbon are increasing upwards, overall fining-up • Units up to 2 m thick	Delta-interchannel area
Delta-slope			
$CH_{[sa]}$	Channel subaqueous	• Mainly Gmm, Ghm possible • Very poor sorting, no trends, homogeneous appearance • Shows wings on both margins, drapes on top • Central part 1–2 m steeply incised, scoured base • w/t 1:5 – 1:45	Feeder channel of turbidity current or foreshore continuation of river channels

TABLE 2.—*Continued.*

Classification of architectural elements			
Delta slope (continued)			
Code	**Architectural element-type**	**Lithology and depositional pattern**	**Genetic Interpretation**
$LV_{[sa]}$	Levée	• Sm, Sh, Sr, Fl, Fc, Fm • Very homogeneous, weakly inclined layers • Internal normal grading, overall coarsening-upward	Deposition along subaqueous channels by spillover
DF	Delta foreset	• Fl, Sm • Coarsening and thickening-upwards, rarely slumps • Beds 0.1–0.5 m, units 0.5–2 m	Prograding Gilbert-type delta front foresets and suspension fallout from water column.
SF	Subaqueous fan	• Fl, Fm, Fc with high proportion of organic carbon; Sr, Sh, Sm, plant debris • Erosional base, fining-upward, homogenous; overall coarsening- or fining-upward • 0.05–1 m, units 0.5–15 m thick, w/t 1:150 – 1:400	Stacked turbidites from distal to prodelta position

tion potential helps to distinguish the individual role of A and S in the stratigraphic-base-level. For setting turnaround points, also the neighboring architectural elements have to be considered to eliminate autogenetic processes and to get representative data for a succession. Furthermore, the position of architectural elements in the depositional cycle and its relationship to A/S conditions was used. Those reflecting weakest depositional energy (may be equivalent to biggest lake extent) are diagnostic of a rise stratigraphic-base-level turnaround point. The other way round, those representing the highest depositional energy in a cycle mark a fall-to-rise turnaround point.

Swamp (SW): This architectural element, composed of coal deposits up to 3 m thick (Fig. 7a, B) typically occurs at the delta top or at coastal-plain positions. However, there are only a few pure coal beds. Dominantly the coal seams contain high amounts of clay, are thinly laminated, homogeneous, and consist of fine clarain. These are interpreted to represent sapropelic coals. On the other hand, some coal seams consists of durain and vitrain, while fusain (fiber coal) is particularly common, which is formed almost exclusively by forest fires (Ligouis, 2001). Indication of in situ deposition (rooting) is present in coal seams but lacking in other sediments.

A coal swamp or mire needs a slightly submerging plain and a low flux of clastic input, so that vegetation is able to catch-up with the lake level or compensate for a switching of the depositional system (Hamilton and Tadros, 1994; Diessel et al., 2000). In turn, pure and thick coal beds may indicate a weak or distant clastic source that may not be able to enter the system and an elevation around the lake level or ground-water level. However, in this study swamps or mires are interpreted to indicate the retrogradation of a fluvial system either laterally or up-dip. Therefore, regionally distributed coal swamps are diagnostic of shallow flooding of the delta top or coastal plain. However, regionally extensive coals can exhibit the essential attributes of sequence boundaries (Hamilton and Tadros, 1994). In case of a lacustrine clayey top, the turnaround point has to be set here (drowning) in order to consider lake-level highstands. In case of an alluvial top (emergence, and/or progradation, and/or poisoning by clastic input), the end of the rise has to be set at the top of the coal seam. To exclude local autogenetic processes like temporal plain floodings, this is applied only at mesoscale where coal seams accumulate and show strong lateral persistence.

Mouth Bar (MB): In a lake system, mouth bars represent the furthermost progradation of an alluvial system into a lake (fall to rise turnaround) or vice versa in a sequence of pure alluvial deposits, mouth bars mark the maximum rise turnaround. At the same time, preservation potential of mouth bars is very low, because they are

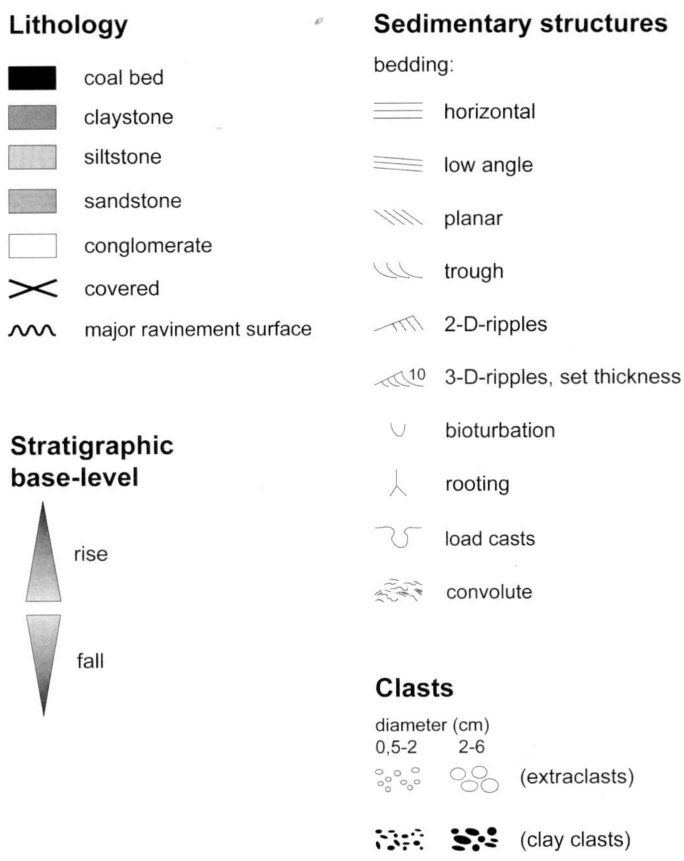

FIGURE 8.—General legend for this study.

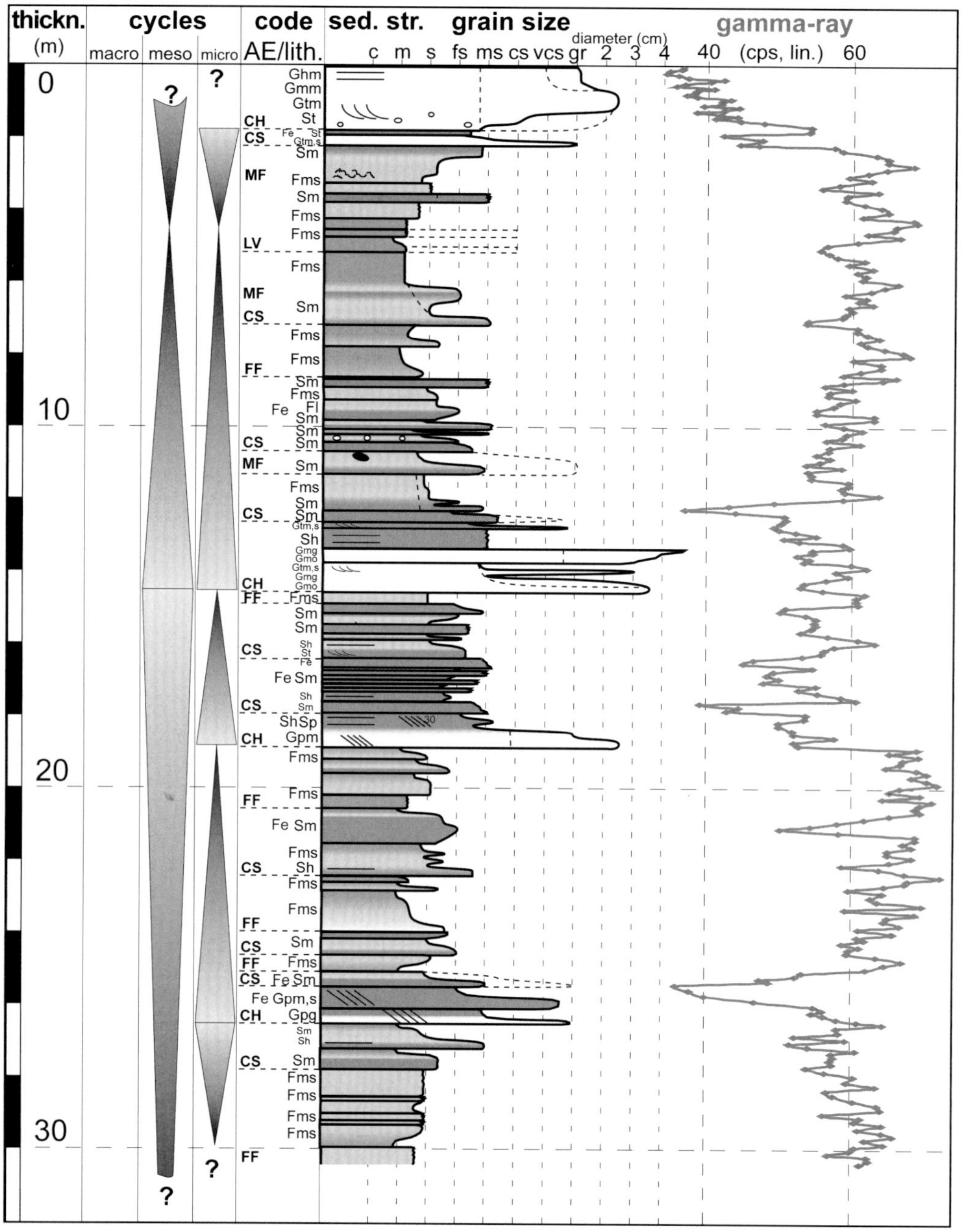

FIGURE 9.—Typical example of the upper Haojiagou Formation. Lithological section with observed depositional cycles, architectural elements (AE), and lithofacies code (lith.). The litholog shows sedimentary structures and grain sizes (see Fig. 8 general legend). The passive gamma-ray log is shown in red where measurement locations are indicated by dots. Note the abnormal behavior of the gamma-ray log around the channel bodies. Details are provided in the text.

easily removed by migration of the river mouth. Hence, if mouth bars are preserved, they indicate plenty of accommodation space or frequent channel switching.

Rippled Sand Sheets (RS): RS elements demonstrate clastic input into the lacustrine foreshore or offshore depositional environment, where the fine-grained fraction is removed via suspension. Dominantly wave ripples are observed, indicating a foreshore position, whereas current ripples rarely occur, which would indicate offshore deposits below the storm wave base. The RS element takes a transitional position from and into lacustrine architectural elements, hence, it can be placed near the fall turnaround with higher clastic input and restricted accommodation space due to wave action or currents. Transitional to fluvial architectural elements, it represents higher accommodation space and smaller sediment supply, so that it has to be placed near the rise turnaround.

Here, the storm wave base governs preservation, so that the portion of preserved ripple sets and bed thickness may be a good proxy to

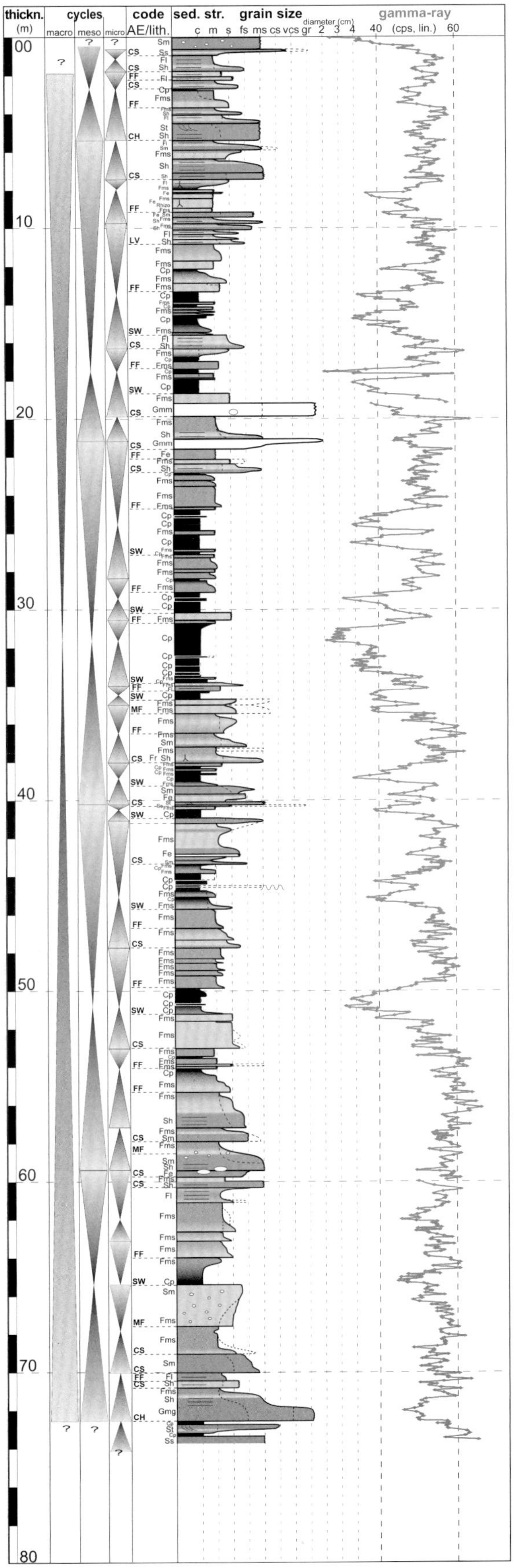

estimate A/S conditions. However, thick beds with low preservation of ripple structures may indicate much accommodation space but dominance of sediment supply (low A/S ratio), whereas complete preservation means a dominance of accommodation space over sediment supply (high A/S ratio).

Beach (BH): The beach architectural element directly indicates a paleo–lake level and the highest depositional energy due to wave action. A turnaround point can be set here related to the context of the succession. Coming from fluvial conditions it represents a base-level rise, coming from a deltaic or lacustrine environment, a fall.

Coastal Mudflat (CM): This unit may indicate the maximum lateral or up-dip retrogradation of the alluvial system due to the absence of direct sediment input, e.g., from a feeder channel. At the same time, it represents weak depositional energy in an almost standing shallow water body, meaning the deepest submergence in the coastal depositional environment. Therefore, it indicates a dominance of accommodation space over sediment supply. Preservation potential and importance for sequence stratigraphic interpretation is comparable to rippled-sand-sheet elements (RS).

Delta-Slope Depositional Environment

The prodelta and delta-slope environment is represented by the lower Huangshanjie Formation (lowermost unit, Fig. 12). Here, laterally persistent lithofacies types and sheetlike architectural elements dominate. They may be formed by frequent floods producing underflow currents and event layers (Hornung et al., 2011 in press). Depositional cycles can be detected mainly just by means of grain-size variations, which reflects stream velocities and distance to the sediment supply system. Finest sediment in a succession represents the base-level-rise turnaround, the coarsest the fall turnaround. However, to average local autocyclic processes such as individual flood events, observations at a regional scale are required for further interpretation.

Delta Foreset (DF): The delta-foreset architectural element consists of several individual beds, showing coarsening-upward and fining-upward trends. The bases and tops of these bundled foresets give hint to proximity and the grain-size spectrum of the clastic supply. The most fine-grained part of a delta foreset element may indicate a "maximum flooding surface" (base-level rise to fall turnaround) or maximum retrogradation of the clastic input. On the other hand, the largest grain size may indicate a progradation of the clastic source into the lake (base-level fall to rise turnaround). Processes might be analogous to Gilbert-type deltas (Gilbert, 1885).

Subaqueous Fan (SF): This architectural elements (Fig. 7b, C) is dominated by repeated mass-flow deposits. Most are mud turbidites. Individual beds show normal grading, due to decreasing stream power during the turbidite event. SF elements may occur also in bundles with overall fining-upward and coarsening-upward trends. Due to its proximity to its source at the upper delta slope, the most fine-grained

←
FIGURE 10.—Representative succession showing almost half of the lower Haojiagou Formation. Lithological section displays observed depositional cycles, architectural elements (AE), and lithofacies code (lith.). The litholog shows sedimentary structures and grain sizes (see general legend Fig. 8). The passive gamma-ray log (red) was measured directly at the cleaned outcrop section; the measurement locations are indicated by dots.

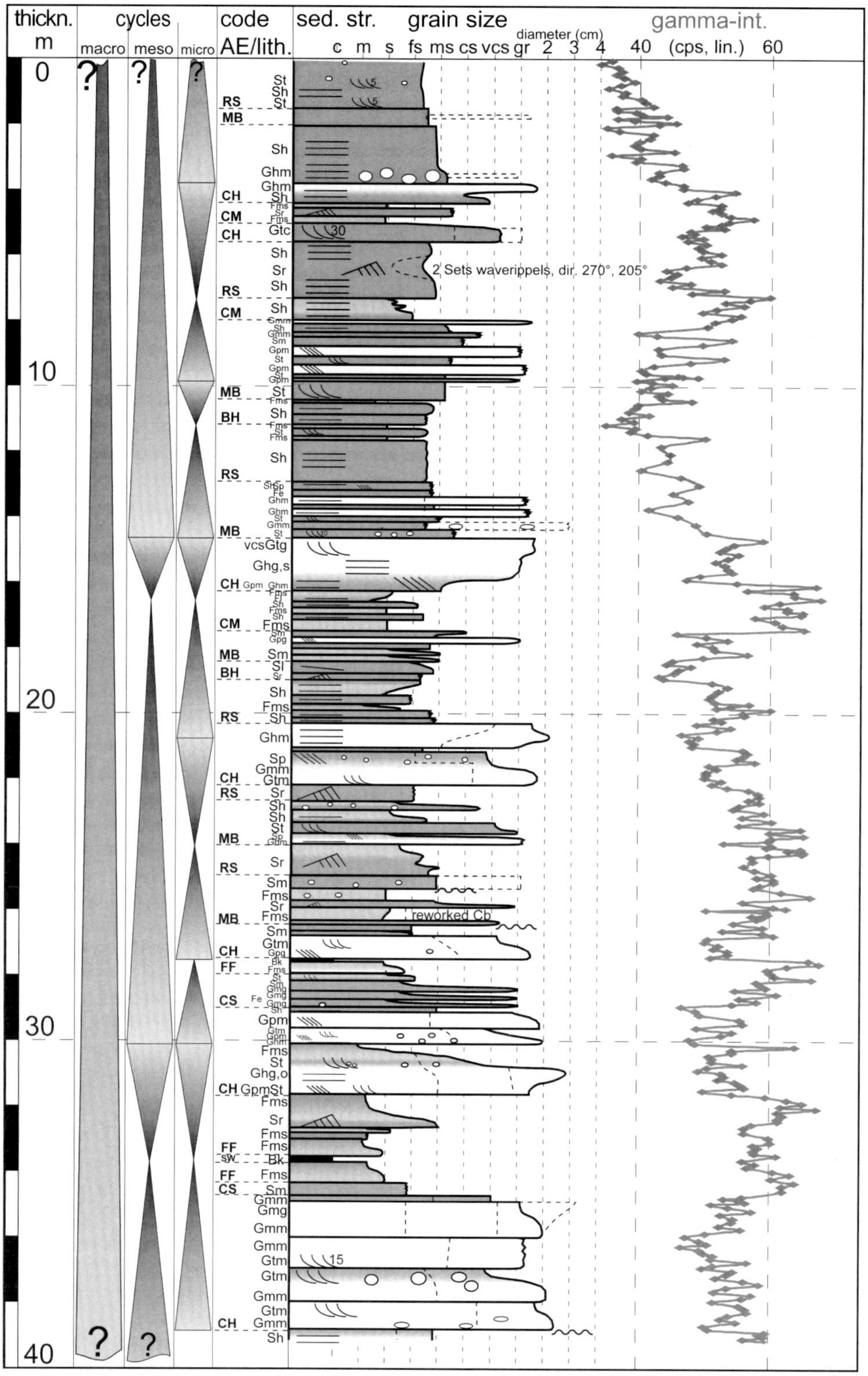

FIGURE 11.—Typical example of the upper Huangshanjie Formation. Lithological section with observed depositional cycles, architectural elements (AE), and lithofacies code (lith.). The litholog shows sedimentary structures and grain sizes (see general legend Fig. 8). The passive gamma-ray log (red) was measured directly at the cleaned outcrop section; measurement locations are indicated by dots.

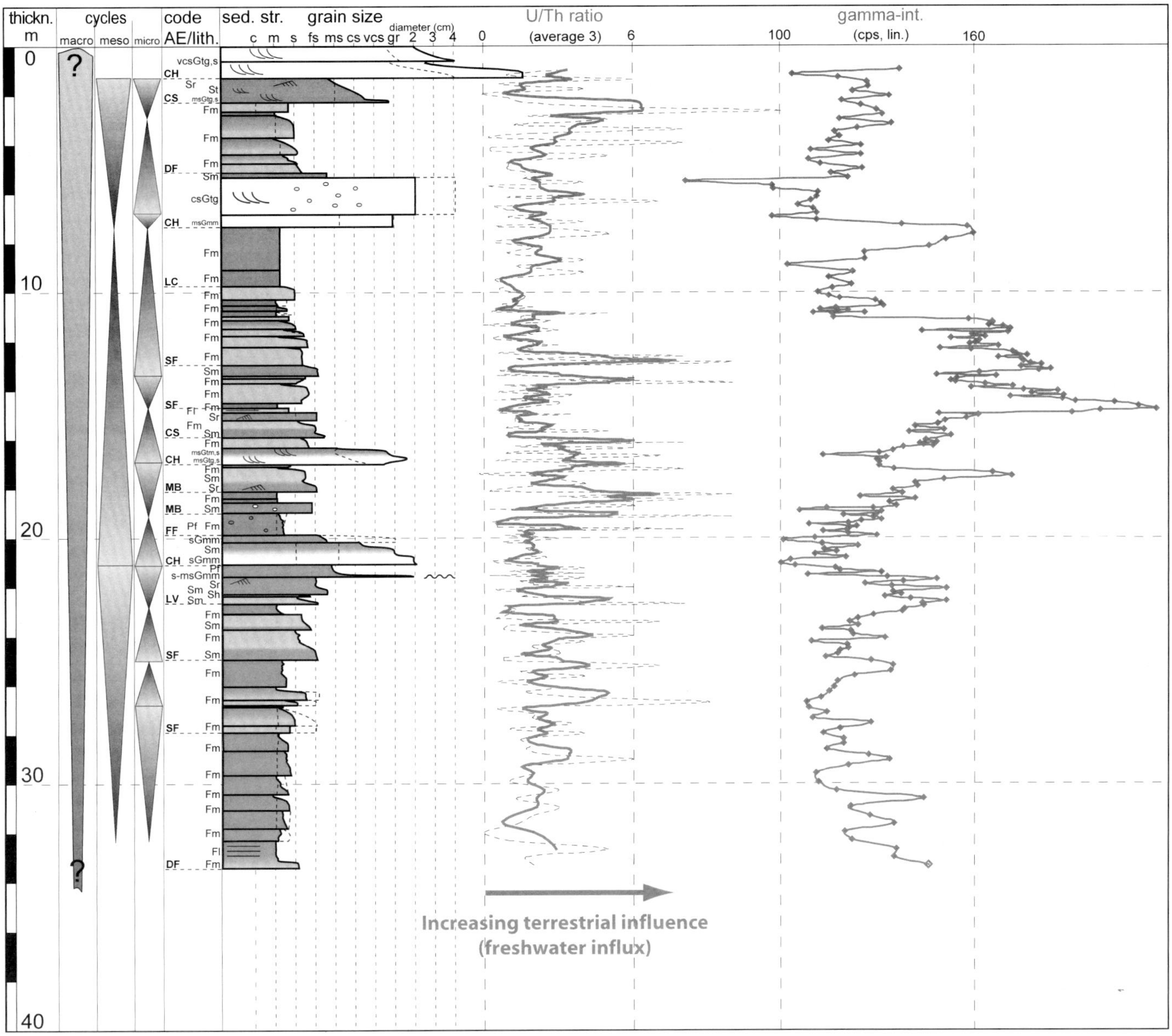

FIGURE 12.—Lithological section with observed depositional cycles, architectural elements (AE), and lithofacies code (lith.) of the Lower Huangshanjie Formation. The litholog shows sedimentary structures and grain sizes (see general legend, Fig. 8). The passive gamma-ray log (red) shows the highest values of the entire Haojiagou succession and was measured directly at the cleaned outcrop section (locations are indicated by dots). Here, spectral gamma-ray data were recorded and used to plot the uranium-to-thorium ratio, which may be an indicator of lacustrine (hypersaline?) versus freshwater input of rivers. For detail see text. The green log represents a moving average of three values; the black thin dotted line represents the original data. Note again the almost independent behavior of both logs concerning grain-size distributions (details see text).

part of a fan complex may indicate a "maximum flooding surface" (stratigraphic-base-level turnaround from rise to fall), e.g., the base of an overall coarsening-upward succession. In turn, the coarsest part represents the base-level fall to rise turnaround.

Channel Subaqueous ($CH_{[sa]}$): Subaqueous channel fills represent a dramatic increase in sediment supply compared to the surrounding fine-grained deposits. From the channel fill only the coarse-grained part (gravel) is preserved, which means that most of the sediment bypassed (sand, silt, clay). In terms of A/S conditions this means high S and low A, thus a low stratigraphic base-level. These coarse-grained subaqueous channel fills are interpreted as a regional signal of increased clastic sediment supply. Therefore, we set a fall turnaround here.

Levee Subaqueous ($LV_{[sa]}$): It indicates the presence of a channel nearby. If no channel was exposed, this element was used as its deputy.

DEPOSITIONAL CYCLICITY

We can distinguish for each depositional environment specific patterns of stratigraphic-base-level cycles in the studied Haojiagou valley section. They are characterized by architectural-element makeup, preservation potential, symmetry relationships, lateral persistence, and thickness. We could establish four different scales of depositional cyclicity in the Upper Triassic proportion of the Haojiagou valley section:

- Microscale cycles, 2–10 m thick (average 5 m), of local lateral extent and therefore considered as autogenetic.
- Mesoscale cycles, 10–30 m thick (average 15 m). Likely, they are partly autogenetic or masked by autogenetic processes, but because they are regionally traceable, they might represent the local response to small changes of the stratigraphic base level.
- Macroscale cycles, 40–100 m thick (average 60 m) and mostly regionally traceable and thus considered to be externally controlled and suitable for regional correlations using the stratigraphic-base-level concept.
- Overallscale cycles comprise 190–250 m of stratigraphy and have the highest potential to correlate sections by sequence stratigraphical concepts.

Distal-Alluvial-Plain Sedimentary System

The smallest detectable sedimentary cycles of this depositional environment are at microscale (Fig. 9). They mostly display a fining-upward, occasionally a coarsening-upward trend and comprise up to five architectural elements making up compartments of stacked fluvial cycles.

Depositional cyclicity at mesoscale contain several of these microscale fluvial cycles. Thickness is about 10–20 m, revealing a prominent overall fining-upward trend and frequently towards the top a coarsening-upward trend. The architectural elements in the cycle are stacked and form a succession that starts with channels (CH) containing gravel bars, followed by sandy bedforms, crevasse-splay bundles (CS) that interfinger with silty or clayey floodplains (FF), and occasionally backswamp (SW) deposits. At the top, frequently a crevasse-splay (CS) or levee (LV) complex is found forming the coarsening-upward trend.

As these mesocycles are composed of microscale cycles and show a significant lateral extent, they are considered to be regionally controlled, at least far from pure autogenetic processes with randomly distributed architectural elements. The preservation potential of individual architectural elements is closely related to their position in a stratigraphic-base-level cycle. As expected during a base-level fall, a slight incision of channels takes place (negative accommodation space), and channel fill starts not before an initial rise, which shows strong amalgamation. Along with stratigraphic-base-level rise, preservation increases and during the highest rate of base-level rise even levees can enter the depositional record, a sign of high accommodation space.

At macroscale, in general, cycles of base-level fall are documented, which indicate a high preservation potential with plenty of accommodation space. Such a "high accommodation" setting with preservation of fall hemicycles is most common in distal parts of an alluvial depositional environment due to sediment volume partitioning (Cross and Lessenger, 1998).

To support this interpretation of depositional cycles, total and spectral natural gamma-ray logs were measured. The logs can be interpreted only when the composition of the rocks is known. Usually, ^{40}K is the main source of radioactivity, which corresponds to clay (or K-feldspar) content (Davies andElliott, 1996). Here, spectral data suggest that uranium and thorium make a significant contribution to the total counts reading, commonly 10 to 30%, in some cases up to 50–60%. Uranium and thorium are randomly distributed and not related to any specific lithology. Further, the data confirm that ^{40}K is responsible for the observed gamma-ray depositional cyclicity. Peaks are related to the silt fraction as well as to metamorphic and nonmetamorphic claystone clasts in gravelly channel deposits. An influence of limestone clasts to the spectral signature can be excluded.

However, depositional cycles in the distal-alluvial-plain sedimentary system are not adequately represented in the gamma-ray log. In particular, lowest gamma-ray emission is not related to the gravelly channels, as one would expect for pure siliciclastic systems. Here, lowest gamma-ray emissions are shifted to fine, but clast-free, quartz-rich sandstones on top of the channels, i.e., first overbank unit after the channel body. Due to the detrital carrier of ^{40}K, silt units may produce the highest peaks. As a result, sand, gravel, and clay layers cannot be distinguished properly with gamma-ray data in this succession.

Nevertheless, the log patterns for the distal-alluvial-plain system show clear trends that are characterized by highly serrated funnel and bell shapes. In many cases, these trends are not suitable to carry out depositional cyclicity, because they do not reliably reflect rock types or depositional energy, but rock composition. Our example demonstrates that gamma-ray patterns should be interpreted with great caution for lithological or sequence stratigraphical interpretations and that an outcrop control is essential.

Delta-Top Sedimentary System

In the lower Haojiagou Formation, clear trends at three hierarchical levels were found (Fig. 10): An ideal microscale hemicycle consists of coarsening-upward or fining-upward successions between 2 and 6 m thick. The fining-upward hemicycle starts with small-channel (CH) or crevasse-splay (CS) architectural elements and terminates with a floodplain (FF) or swamp (SW) architectural element Hereby, the crevasse succession shows a prominent fining-upward trend until interbedding with floodplain units takes place. The floodplain further intercalates with coal beds of the swamp architectural element. A coarsening-upward hemicycle is organized vice versa. However, in almost all cases the coal swamps show an alluvial top, due to crevasse-splay activity. These cycles correspond to fluvial cycles, and like those, they are laterally not traceable and therefore considered to represent autocycles.

A typical mesocycle is 15 to 25 m thick and composed of proportions of these microcycles. Which proportion is developed or preserved is strongly related to the position within the mesocycle. Accordingly, during high A/S conditions (e.g., rise to fall turnaround) fine-grained, low-energy architectural elements are preserved in the microcycles and units are not or slightly amalgamated. Low A/S conditions (e.g., fall to rise turnaround) are represented by coarse-grained, high-energy proportions of microscale cycles.

At least the whole delta-top depositional environment is part of an overall fining- and coarsening-upward cycle (macroscale). Its rise hemicycle comprises about 30–40 m of stratigraphical thickness, and its corresponding fall hemicycle about 60 m. Thickest, most frequent, and purest coal seams are developed around the rise hemicycle and rise-to-fall turnaround, which means formation and preservation of coal during transgressive and high lake level (Hamilton and Tadros, 1994).

All depositional units are well preserved, and no significant erosional features are observed. Even channels are only slightly incised and display a sheetlike geometry. Commonly, the fall hemicycle is preserved at all scales. Therefore, the overall preservation potential is very high, and due to the dominance of coal swamps (SW) the delta-top depositional environment seems to be undersupplied and endowed with much accommodation space to preserve any kind of clastic input.

Hence, rise and fall hemicycles are commonly well preserved at all scales, but with changing proportions. During a higher-order rise to fall turnaround, cycles of lower-order are mostly symmetrical. In contrast, during a higher-order fall to rise turnaround, lower-order rise cycles are more prominent. In addition, during a rise to fall turnaround lower-order cycles get thicker and vice versa.

Again, gamma-ray logs showed some surprising results with respect to standard lithology. As expected, clean coal units show the lowest values (< 40 cps). The highest values (> 45 up to > 60cps) are produced from clay-rich floodplain or crevasse-splay units. Crevasse channels show intermediate values from 35 to 50 cps. There is neither a clear sand line nor a clear clay line, so to distinguish between sandy and clayey units is almost impossible. Clayey coal beds are also not distinguishable from sand units by gamma-ray logs. In general, an outcrop resistivity log is highly recommended for this section. Nevertheless, the best fit is to establish first a coal-bed line of about 40 cps. Then the stratigraphical column in between these low-emission peaks should be subdivided into weakly serrated funnel and bell-shaped units. For each of these smallscale units, individual sand and clay lines should be established and interpreted in the standard way.

For sequence stratigraphical interpretation, mesoscale and macroscale depositional cycles might be useful. Both are regionally traceable, which makes it likely to assume a regional control, e.g., by lake-level fluctuations. Hence, thick, grouped and clean coal beds can be used as a proxy for the maximum retrogradation of the fluvial systems or the maximum progradation of the lake environment, which means a base-level rise to fall turnaround point (Hamilton and Tadros, 1994). This may result in a ground-water level remaining close to the surface of the delta plain. The interbedded coarsest clastic units indicate a fall to rise turnaround point.

Delta-Front Sedimentary System

In Fig. 11 (section of upper Huangshanjie Formation), lithological trends at microscale show a fining-upward from thick gravelly channel beds to thin coarse sand layers. Some cycles have a coarsening trend toward the top. This is due to an architectural-element succession from channels (CH) to mouth bars (MB) and to beach (BH), coastal mudflats (MF), and rippled sand sheet elements (RS).

Mesoscale cyclicity starts again with partly amalgamated channels that are thinning upwards. This coincides with an increase of coastal mudflats (MF), mouth bars (MB), and rippled sand sheets (RS) towards the top, which displays an overall fining-upward trend. At macroscale, thick medium-grained rippled-sand-sheet units (RS) dominate towards the top, while amalgamated channel complexes dominate the base, which reveals a clear overall fining-up trend.

The delta-front depositional system differs significantly from all other systems because a large portion of the bed load is dropped here due to the sudden decline in stream power when the fluvial system enters the lake. This leads to an oversupply preserving only the coarsest parts of the architectural elements and the depositional cyclicity. Most of the fine sand, silt, and mud bypass into the lake. Nevertheless, there is sufficient accommodation space created to preserve fall hemicycles on the microscale. This may be due to autocyclic depositional-lobe switching, which is supported by poor preservation of fall hemicycles in the mesoscale and the overall fining-up at the macroscale. Cycle symmetry goes along with preservation patterns. At mesoscale, strongly asymmetric cycles with prominent rise hemicycles are observed, but at microscale, cycles are developed more symmetrically.

Concerning gamma-ray logs, beach deposits have the lowest values around 40 to 45 cps followed by rippled sand sheets around 45 to 50 cps. Gravelly channels have intermediate values of 50 to 60 cps, and as expected, coastal mudflats or floodplains show peaks with 60 to 65 cps. Again, the natural gamma-ray emissions do not reflect microscale and macroscale grain-size trends, but they confirm the overall trend towards less clay content at macroscale, so it is possible in the delta-front system to detect higher clay contents and clean sands. As in other depositional environments, it is not possible to use these serrated bell, egg, and funnel log shapes for environmental interpretation (e.g., Cant, 1992; Hamilton and Galloway, 1989; Serra and Sulpice, 1975).

All scales of cycles show that the fluvial system retrogrades more and more towards the top and deltaic conditions increasingly dominate the sedimentary record, as indicated by increased intercalations of typical delta-front architectural elements. This causes an overall fining-up trend at macroscale. In contrast, in marine delta systems coarsening-up trends are reported at the macroscale, which is due to a prograding clastic source that suddenly switches away to another delta lobe (Reading, 1996). Then, rapid flooding follows due to subsidence, which cannot be compensated by clastic input. In contrast, this lacustrine delta system seems to be accommodation-controlled as sediment input can fill up the foreshore area with only one coarse-grained bed of about 0.5–3 m thickness. This little accommodation space suppressed the formation of new delta lobes and thus typical coarsening-upward trends. Instead, an approaching channel directly settled on the shallow lacustrine architectural elements. If the sediment source for the recently filled area switched away, slow subsidence resulted in reflooding, forming the observed general pattern of deltaic fining-upward cycles, e.g., the more symmetric microcycles.

Nevertheless, base-level-fall turnaround points can be set at maximum grain size of channels or alternatively at mouth-bar deposits. A base-level-rise turnaround is represented by (the top of) rippled sand sheets as pure shallow lacustrine deposits or alternatively by a coastal mudflat.

In the marine literature many examples exist for deltaic fining-upward successions, e.g., in the Holocene Mekong delta (Ta et al. 2002) and in subtidal to intertidal estuarine environments (Reineck and Singh, 1980). Meckel (1975) described fining-upward successions in an estuarine environment from the Colorado Delta. For a lacustrine delta setting, Tye and Coleman (1989) described a similar process of rapid filling of flat interchannel freshwater lakes from the Mississippi delta plain. An area of 29 km^2 was filled in only 12 years.

Delta-Slope Sedimentary System

Microcycles (2–6 m thick) show fining- and coarsening-upward and compose mesoscale cycles of 8–25 m thickness. Although this environment has the largest proportion of coarsening-upward trends, fining-upward slightly dominates. The composition of these cycles displays a wide range of architectural elements; their type and kind is related to their position in the sequence with respect to higher-order cycles (Fig. 12).

At mesoscale, subaqueous channels and levees accumulate at distinct levels on top of a package of delta foreset and subaqueous fan sediments 5–10 m thick, showing mostly an overall coarsening-upward trend and representing the fall mesoscale hemicycles. Some architectural elements around the fall-to-rise turnaround point are part of the delta-front and fluvial environment. Rise mesoscale hemicycles are organized in the opposite way, so that the complete cycle appears quite symmetrical (Fig. 12). This symmetry shifts towards a dominance of rise hemicycles as distance to the delta front decreases, i.e., in an overall progradational trend of the section. Nevertheless, in the whole environment both hemicycles can be preserved, but rise and fall conditions are superimposed by higher-order cycles.

As in the previous depositional environments, total counts of natural gamma-ray behave almost independently from grain-size distribution and thus depositional energy (Fig. 12). Nevertheless, it shows prominent cyclicity. Highest total values (> 160 cps) are measured in siltstone and fine-grained sandstone units in the upper part of a

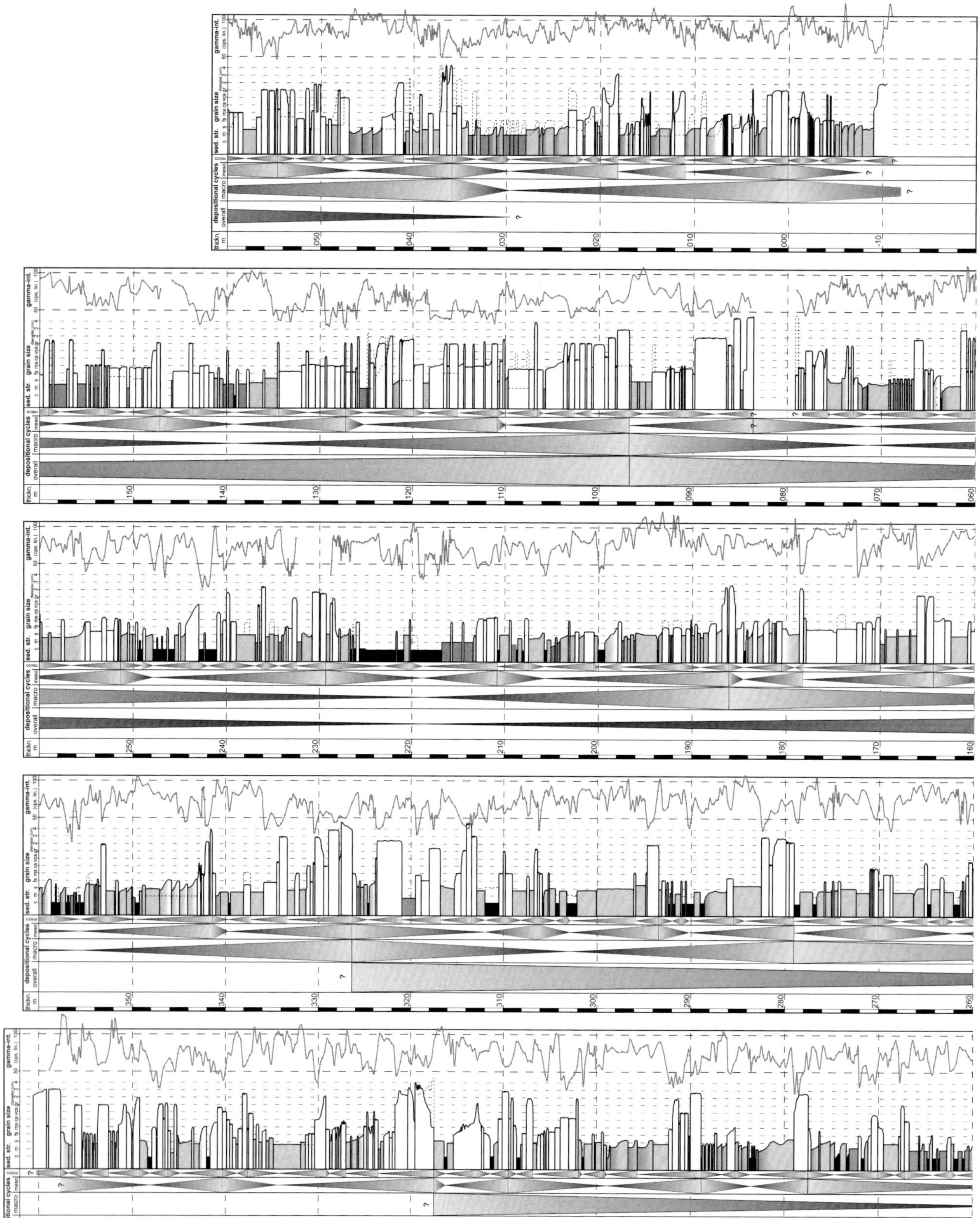

FIGURE 13.—Simplified main litholog and natural gamma log of the Haojiagou Valley section. Note the fourfold depositional cyclicity. Gamma-ray log and litholog do not coincide in the classical way in all parts of the section. For details, see text. Relogged and extended with gamma-ray data after Ashraf et al. (1999, 2001).

mesoscale rise cycle and not in claystones at the rise turnaround point (average 110–140 cps), which is rather what one would expect. Gravelly units show low and intermediate values. Spectral measurements show that the high total counts in siltstones and fine-grained sandstones are caused by ^{40}K and not by U or Th. Compositional data of gravel and siltstone lithofacies types suggest that the carriers of the ^{40}K are metamorphic and nonmetamorphic claystone clasts. This means that natural gamma-ray intensity is no clear predictor of lithologic properties, inasmuch as it is not possible to detect any sand or clay lines here (Figs. 10–13). U/Th ratios are independent of total counts gamma-ray log, which is governed by ^{40}K, but also fail to predict grain-size trends (Fig. 12).

The middle part of the lower Huangshanjie Formation, starting from 20 m upwards, shows a higher U/Th ratio, decreasing again towards 10 m depth (strongly serrated funnel shape) and increasing again towards 2 m depth (serrated bell shape) (Fig. 12). Units with a higher U/Th ratio (DF, CH, MB, and SF architectural elements) show at least one or more criteria for enhanced sediment input such as a wide grain-size spectrum, graded beds, rapid deposition, and increased thickness. Especially mouth-bar (MB) and coarse-grained (= proximal) parts of subaqueous-fan (SF) elements seem to have higher U/Th ratios. At the same time, there exist similar units (e.g., CH, LV, CS), which do not show higher U/Th ratios, e.g., the sand/gravel beds from 20 to 22 m or around 2 m depth.

We assume that besides a detrital component, which would follow grain-size spectra, U was redistributed by dissolution and precipitation. Most probably, this already occurred when river water entered the lake and water chemistry changed suddenly, e.g., a drop in oxygen content up to reducing conditions and/or an increase in pH and organic content. In this case, a higher U/Th ratio reflects increased terrestrial influence coinciding with increased freshwater influx.

Despite its subaqueous setting, the principle of changing A/S conditions also apply to this depositional environment, because stream power from underflow processes or other currents limit accommodation space and refocuses sediments in the lake (Einsele, 2000; Reading, 1996; Schäfer, 2004). An increase of hyperpycnal flow or subaqueous channel activity may indicate an increase of sediment flux to the lake system that coincides with a reduction of accommodation space due to increased stream power at the delta slope. Hence, a fall in stratigraphic-base-level indicates increased sediment supply as well as limitation of accommodation space.

In conclusion, depositional cyclicity may display retrogradation or progradation of the delta and/or fluvial system. Channels indicate an advance (increase of sediment supply) and the most distal (pro-) delta sediments a retreat (decreased sediment supply). At scales of lower order, this can be also interpreted as a lateral facies shift. Hence, progradation or lake transgression in a sequence stratigraphical sense can be traced only by correlations throughout the whole delta, at an overall stratigraphical scale or best by regional correlations of different delta systems in order to quantify lake-level fluctuations independently from local changes in sediment supply. Here, the observed cyclicity is superimposed at a macroscale and overall scale with fall cycles. This is why we consider the delta to prograde into the lake.

STACKING PATTERNS

1-D Overall Log Section

Regarding the whole section (Fig. 13) a fourth, overall hierarchy of depositional cycles can be added to the three scales discussed so far (Figs. 9–12). This overall scale shows a fall-to-rise turnaround at the base of the delta-front depositional system (Fig. 11), and a rise-to-fall turnaround at the delta-top depositional environment (Fig. 10). Another fall turnaround is interpreted within the distal-alluvial-plain depositional environment (Fig. 9). As valid for the macroscale, mesoscale, and microscale cycles this overall cycle superimposes the lower-order cycles in terms of lithofacies and architectural-element distribution and maximum grain sizes.

Mostly 2–4 microcycles are stacked to one mesocycle rise and 2 or 3 mesocycles are stacked in a macrocycle rise. Regarding the fall hemicycles, 1–3 (exceptionally 5) microcycles build up a mesocycle fall and 1–4 mesocycles a macrocycle fall. Looking at a whole cycle on average 3–5 microcycles build a mesocycle, truncation results in the development of only two microcycles. The same pattern is observed for the stacking of mesoscale into macroscale cycles. For the overall scale there is not enough statistical evidence, but here four macroscale cycles build a complete overall scale cycle (two for rise and two for fall hemicycles).

2-D Wall Panels

Terrestrial sedimentary environments show the greatest lateral heterogeneity of all depositional environments (Einsele, 2000, and references therein). Hence, the lateral traceability and spatial distribution of architectural elements play an important role in sequence stratigraphical analyses. To complement our 1-D logging data, we present some characteristic wall panels for each depositional environment.

Fig. 14A shows loosely spaced channels in thick overbank units (FF, LV, CS) in the upper Haojiagou Formation. Most of the channels are weakly incised and show a flat shape. They concentrate in distinct levels that are laterally traceable, but they never form continuous layers due to amalgamation. Interbedded crevasse splays connect them. Because the lithofacies makeup appears quite similar during this stratigraphical interval, we exclude a change of the sedimentary system and interpret the concentration of channels as mesoscale cyclicity due to changes in accommodation space, e.g., lake level.

The mesoscale architecture of a delta-top environment of the lower Haojiagou Formation is displayed in Fig. 14B. In general, large sheetlike coal seams and floodplain units that are traceable over large distances dominate. Feeder channels and crevasse splays are rare. Usually they are not directly connected, but distal parts amalgamate. It is likely that in periods of higher crevasse activity or when less accommodation space is available the coal swamp system collapsed due to poisoning with clastic input. At the same time, this preserved the underlying coal swamp or mire. Hence, we consider this process externally controlled and increased crevasse and floodplain deposition as progradation of the fluvial system (base-level fall).

Fig. 14C illustrates the central portion of the delta-front depositional environment represented by the upper Huangshanjie Formation. Clearly visible is the sheetlike appearance and hence the high lateral interconnectedness between coarse-grained thick channel units (CH) and thinner fine-grained beach sands (BH) and rippled sand sheets (RS). Vertically, coastal mudflats (CM) and levees (LV) separate them. In some horizons, channels concentrate and amalgamate while also some incision takes place. Incised channels are filled up with levee and coastal-mudflat sediments.

In these levels, accommodation space is low or for a limited time even negative, while supply is high. Channel bases directly overlying foreshore sediments, channel incision and fill patterns, and the presence of levees and coastal mudflats requires a quick readjustment of A/S changes. Channels, rippled sand sheets, and coastal mudflats form a distinct mesoscale cyclicity that is traceable laterally over more than one kilometer. Because of this traceability it potentially can have a sequence stratigraphical relevance. Of course, we cannot decide whether this environment represents a complete reorganization of the paleo-landscape or simply the entry point of a new and large fluvial system. From stacking patterns, however, we can state that this unit represents the overall base-level fall to rise turnaround because the highest sediment supply together with the most limited accommoda-

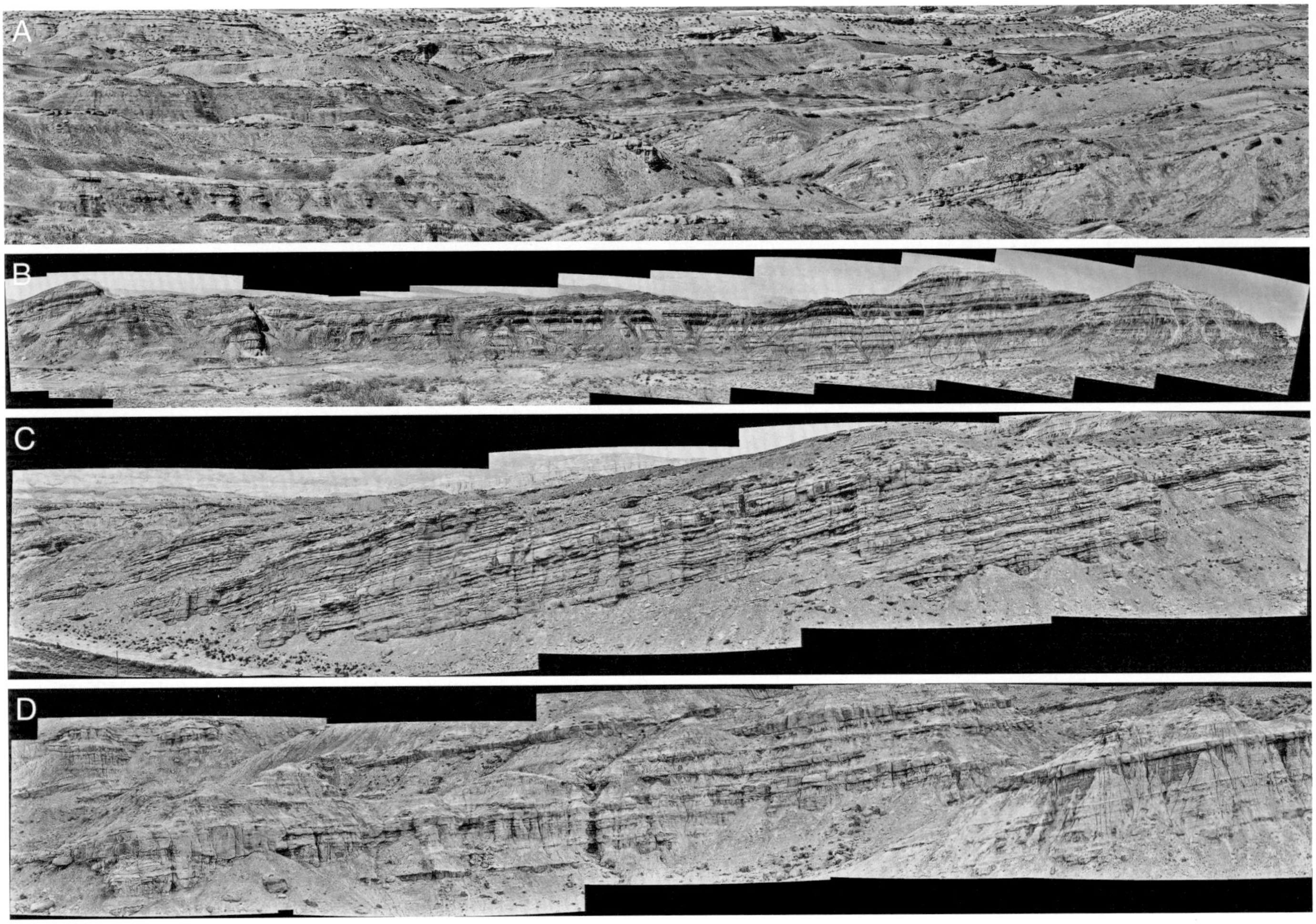

FIGURE 14.—14A. Wall panel from the distal-alluvial-plain geosystem, represented by the upper Haojiagou Formation. Stacking analyses reveal concentration of isolated gravelly channels in distinct levels forming a mesoscale depositional cyclicity.

Figure 14B. Wall panel from the delta-top geosystem displayed by the lower Haojiagou Formation. Coal seams and their alluvial tops are systematically interbedded and classified as mesoscale depositional cycles.

Figure 14C. Wall panel of the upper Huangshanjie Formation showing a delta-front depositional setting with interaction of shoreline deposits and shallow lacustrine architectural elements. Stacking analyses reveal amalgamation of gravelly channel complexes in distinct levels forming a mesoscale depositional cyclicity.

Figure 14D. Wall panel of the lower Huangshanjie Formation representing a delta-slope and prodelta setting. Loosely spaced subaqueous channels with subaqueous levees intercalate with sheetlike event layers showing turbidite-like features resulting mostly from gravity-driven underflow processes during flood events. A higher frequency and more prominent feeder channels reflect a mesoscale depositional cyclicity.

tion space is observed here. Due to the rapid drop and rise in base-level, we conclude that this was controlled by a severe drop in lake level with subsequent flooding.

Finally, the lower Huangshanjie Formation exhibits the transition of a delta-front (higher portion) to delta-slope (lower portion) depositional system (Fig. 14D). In the 2-D view, isolated ribbon-like subaqueous channels (w/t 1:5–1:45) are associated with crevasse splays and levee units. Channels showing w/t ratios typically < 15 commonly reflects bank resistance and rapid filling, although some are associated with base-level rise (Gibling, 2006). They intercalate with sheetlike subaqueous channels, delta-foreset, and subaqueous-fan architectural elements (w/t 1:150–1:400). These broad sheets do not fit to any data reported by Gibling (2006) for fluvial systems, which support interpretation of subaqueous fan deposits. Note, geometry relationships and lithofacies makeup differ significantly from the distal alluvial-plain environment (Fig. 14A), but in both cases channels are laterally not connected and show aggradational patterns.

These lateral and vertical stacking patterns and geometries support our interpretation of lithofacies and architectural elements that gravity-driven underflow events are the dominant depositional process (for details see Hornung et al., 2011, in press). During flood events, when the river is heavily charged with sediments, fluvial channels extent downslope as hyperpycnal flow, also producing a distal run-out facies. During waning flood conditions, settling of fractionated suspension may produce massive beds with normal grading. Architectural elements concentrate at mesoscale level, which is determined by variations in A/S ratio. We interpret this cyclicity as allogenetic, but regarding the large dimensions of the system, we cannot rule out delta-lobe switching. At least the macroscale can be used for regional correlations.

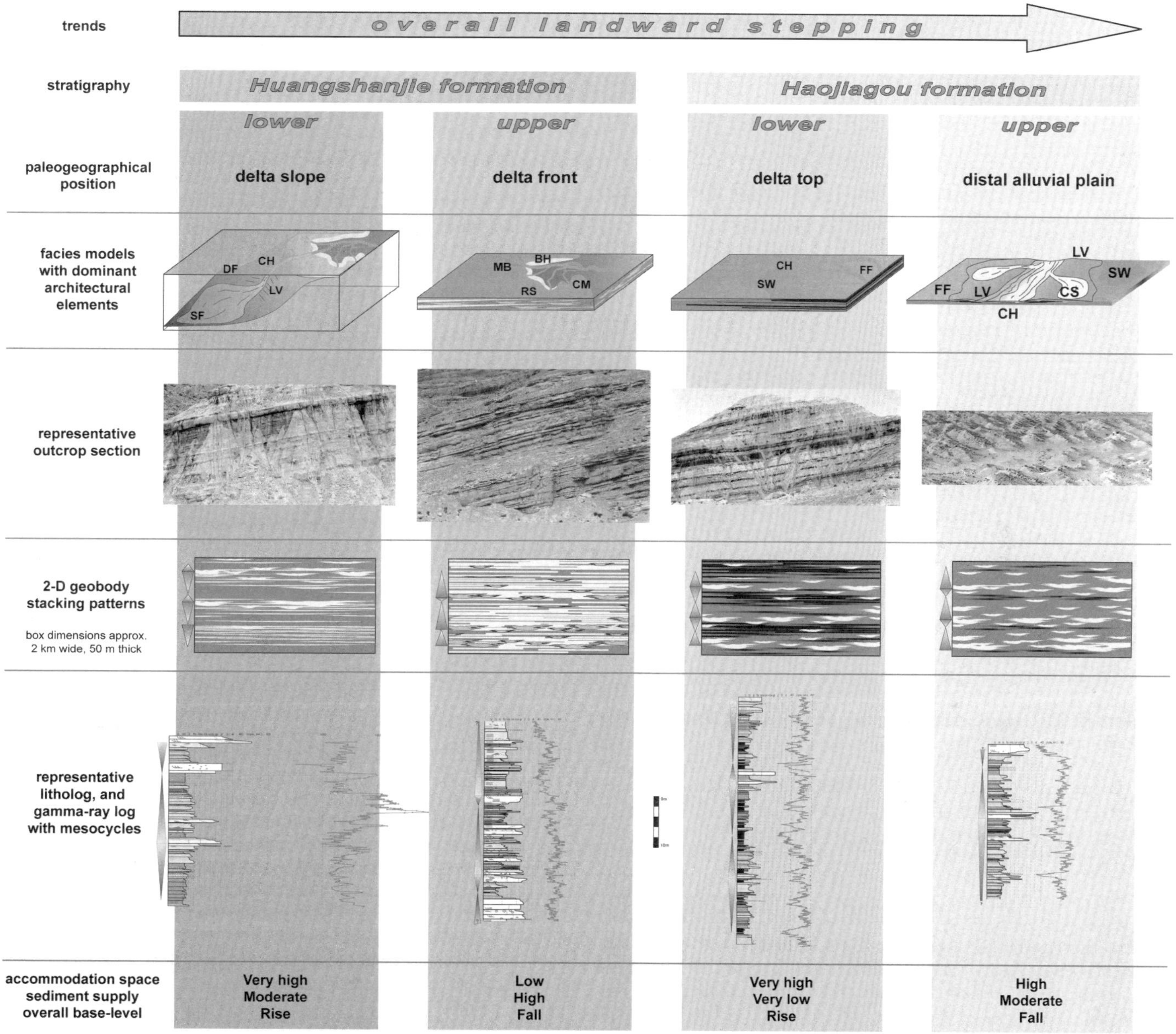

FIGURE 15.—The various geosystems at the Haojiagou section and their four depositional environments. Sketch displays typical portions of the succession, but note that most of the section is represented by transitional stages between these end members highlighted here. For detail see text, and for more extended facies model descriptions (top part of figure) see Hornung et al. (2011).

DISCUSSION

We could demonstrate that our portion of the upper Triassic succession represents an overall landward-stepping trend due to lacustrine delta progradation (Fig. 15). Thus, from a sedimentological point of view, the established subdivision of the two formations is based upon two principally different depositional environments: the Huangshanjie Formation is lacustrine deltaic (Norian age) whereas the Haojiagou Formation is dominantly continental and marginal lacustrine (Rhaetian age). Both formations can be further subdivided into an upper and lower unit (member) according to their depositional environments:

1. Lower Huangshanjie Formation (lowermost unit): lacustrine prodelta and delta-slope environment
2. Upper Huangshanjie Formation (lower middle unit): lacustrine coastal-dominated delta-front environment
3. Lower Haojiagou Formation (upper middle unit): lacustrine delta top with fluvial and backswamp environments
4. Upper Haojiagou Formation (uppermost unit): distal alluvial-plain environment.

Depositional Dynamics

Two principal types of coarse-grained deltas are distinguished in the literature; fan deltas and braid deltas. Fan deltas are defined as an alluvial fan prograding into a standing body of water, and braid deltas as a braid plain doing the same (e.g., McPherson et al., 1987). However, distributary and converging channels and a braided pattern of fluvial topsets characterize braid deltas. Diagnostic features of fan deltas are angular and very unsorted highly amalgamated sediments of mass-flow or alluvial origin, which spread over the whole subaerial division (for details see e.g., Bowman et al., 2007; Blair and McPherson, 2008). All of these features are not adequately developed in the Upper Triassic Haojiagou section. The main difference is the presence of an extensive subaerial floodplain with levees and other overbank elements that were never reported for these delta types, which are almost entirely composed of bedload sediments. In addition, no typical braid patterns are observable. So far, no high-gradient, high-supply deltas are described in the literature which are of mixed-load character and fit our observations. Thus, as outlined in the companion paper by Hornung et al. (2011, in press), we introduce a new type of coarse-grained, mixed-load delta not described so far. Both delta top and delta slope show a stacking of coarse-grained bedload sediments with extensive fine-grained suspended-load sediments and/or coal seams at the delta top. Underflow events dominate the subaqueous depositional processes, i.e., bottom currents last far out into the lake, also controlling distal parts. Even bedload up to gravels can be transferred at the delta slope as hyperconcentrated flows through channels.

Relationships of A to S through Stratigraphy

Although this high-resolution study gives detailed implications about temporal changes in sediment supply and accommodation space, we cannot be unambiguously separate regional (allogenetic) from local (autogenetic) controls. To do this is a matter of scale and data spacing, but parallel sections several kilometers away or well logs are not available yet in the same quality.

Symmetrical cycles indicate plenty of accommodation space and a balanced A/S ratio (and vice versa). Depending on total subsidence cycles can get very thick under these conditions. When the A/S ratio is unbalanced, cycles become incomplete, which in turn imply a weak preservation potential. Much accommodation space and weak sediment supply may result in condensed sections or sediment starvation. Negative accommodation space causes sediment bypass and incision (Cross and Lessenger, 1998; Hornung, 1999). Sediment supply is directly related to depositional energy when initial averaged grain-size distributions are assumed, i.e., in gravel deposits most of the finer-grained sediments are bypassed (Einsele and Hinderer, 1997; Einsele and Hinderer, 2001). According to these simplified depositional rules, we assessed the relationship of A to S through stratigraphy.

Distinct symmetrical cycles suggest plenty of accommodation space with moderate sediment supply in the lowermost Huangshanjie Formation (early Norian), but sediment influx was somehow unsteady because of abruptly intercalated coarse-grained units (Fig. 12). Later, cycles tend to become asymmetrical and less preserved as fall hemicycles are discriminated or missing due to the decrease in the creation of accommodation space. This tendency culminates in the upper Huangshanjie Formation (late Norian) and is reflected by slight base incisions, amalgamated architectural elements, and truncated cycles (Fig. 11). At the same time sediment supply increases as coarse-grained architectural elements are favored, which indicate bypass of most of the accompanying finer-grained sediment. During deposition of the upper Huangshanjie Formation, trends were opposing when sediment supply reached its maximum and accommodation space its minimum at the base (Fig. 11). During deposition of the lower Haojiagou Formation (early Rhaetian), sediment supply dramatically dropped further and simultaneously accommodation space reached its maximum. This is reflected by fine-grained, complete and symmetrical cycles, as well as regionally extensive coal deposits (Fig. 10). Towards the Triassic–Jurassic boundary (upper Haojiagou Formation, late Rhaetian), both sediment supply and accommodation space were moderate and balanced, which allowed completely preserved cycles (Fig. 9).

Implications for Marine vs. Lacustrine Sequence Stratigraphical Concepts

The strong link of A and S in a lacustrine depositional environment represents a major difference from the marine realm (Bohacs et al., 2000; Pla-Pueyo et al., 2009). Factors controlling S are river discharge, which increases sediment transport capacity, and mechanical erosion, which is largely a matter of relief and thus tectonic uplift (Einsele and Hinderer, 1997). Factors controlling A on a basin-wide scale are relative depth of the lake (subsidence + sediment infill + lake level), and sill height, which is mainly a function of tectonic control (Bohacs et al., 2003). In the case of a closed hydrology, increased river discharge lifts the lake level, possibly with some phase shift due to groundwater adjustment (Gierlowski-Kordesch, written communication). This process is reversible, and stronger seasonality can increase temporal sediment load and may produce high-frequency lake-level fluctuations. For open lake hydrology, these processes are limited to the sill level and A and S are more independent. Nevertheless, in both cases climate control is much more pronounced and more closely linked in lacustrine sequence stratigraphy because of the strong coupling with catchment processes (e.g., river discharge, erosion; e.g., Scholz et al., 1990; Juhász et al., 1997; Leeder et al., 2002, Johnson and Graham, 2004; Bohacs et al., 2000). This also applies to the coupling of climate and tectonics, which act on long-term and short-term periods, respectively (Pla-Pueyo et al., 2009). This results in a number of sedimentary patterns making comparisons between marine and lacustrine stratigraphic sequences delicate, e.g., absence of typical coarsening-upward sequences, which are well known from the marine realm (Bohacs et al., 2000).

In consequence, terms like "maximum flooding surface", "highstand", or "lowstand" would not describe the condition of the system adequately, because these terms refer only to the role of the lake level such as in marine sequence stratigraphy (see also Bohacs et al., 2000). The concept of stratigraphic-base-level seem to reveal a better genetic understanding of such intracontinental systems because it much more emphasizes the role of limited accommodation space linked to unsteady sediment supply as shown in this study (see also Hornung and Aigner 1999; Hornung and Aigner, 2002b). Nevertheless, there are some analogues to make both concepts comparable (Fig. 6): "Sequence boundaries" of the marine concept correspond to a final stage of fall hemicycles in the stratigraphic-base-level concept. The "maximum flooding surface (mfs)" corresponds to the rise turnaround point and the "lowstand systems tract (LST)" to the fall turnaround. Consequently, a rise hemicycle correlates fairly to the "late lowstand (LLST)" and mainly to the "transgressive systems tract (TST)" and a fall hemicycle to the "highstand (HST)" and "regressive systems tract" (Hornung, 1999; Catuneanu et al., 2009). In marginal positions "highstand" and "lowstand systems tracts" do not exist, because these conditions result in immediate sediment bypass, and do not enter the depositional record, due to weak preservation potential. Hence, like most other authors (e.g., Hamilton and Tadros, 1994; Flint et al., 1995; Diessel et al., 2000; Farrell, 2001), we use the coal seams as an indicator of base-level-rise deposits (in marine terminology "transgressive" systems tract) and put a rise-to-fall turnaround (marine "mfs") there or in pure lacustrine clay layers above them. As indicators of base-level fall we use stacked channel units (see also Ramon and Cross, 1997).

Controls on Deposition

Carroll and Bohacs (1999) and Bohacs et al. (2000) developed a very useful classification scheme to characterize ancient lake-basin types and to balance their tectonic and climatic controls. Three end members are distinguished: the underfilled, balanced filled, and overfilled lake basin. According to their scheme, the large lake in the Junggar basin during the Late Triassic can be classified as balanced filled, i.e., hydrologically open and relatively shallow, due to the strong and smallscale vertical heterogenities observed. Based on prominent examples like the Green River Formation, this lake type is characterized by strong autochthonous deposition of carbonates. This feature is absent in the Junggar lake, most probably due to strong siliciclastic input. Our classification is in line with Bohacs et al. (2000), who already characterized the Permian Junggar lake as shallow balanced filled.

Characterization of lake type and trends in the depositional record support climate reconstructions by Ashraf et al. (2001) and Ashraf and Wang (2004) from the pollen record showing an increasing humidity and seasonality. This in turn might be accelerated sediment supply largely balancing increasing accommodation space and spreading of coal mires in the lower Haojiagou Formation. The long persistence of a large lake in the Junggar Basin from the Permian to the Jurassic, thus more than 50 million years and covering up to more than 300,000 km^2, needs a stable tectonic boundary. Tectonic movements of the hinterland are a significant control that affects sediment production (Platt, 1989; Leeder, 1997; Einsele, 2000; Einsele and Hinderer, 1997). Based on seismic and literature data, Lawrence (1990) compiled the tectonic evolution of the ancestral Tian Shan and sketched sediment distribution towards the basin center (Fig. 2). For the Triassic, he shows that the equilibrium point between erosion and deposition continuously moves towards the hinterland and then suddenly turns back, creating an erosional discontinuity reaching far out into the basin. Zhang et al. (1984), Watson et al. (1987), Coleman (1989), Lawrence (1990), Sengör et al. (1993), Otto (1997), Ritts (1998), and Greene et al. (2004) interpret this as an uplift pulse around the Triassic–Jurassic boundary, thus beyond the top of our section.

On the other hand, large sediment volumes and coarse-grained sediment supply need an accentuated relief in the hinterland. This is in line with the geodynamic reconstructions assuming a repeated compression of the ancestral Tian Shan due to far-field stresses by the collision of the Qintang Block at the southern margin of the Tarim Basin forming the Kun Lun orogenic belt. From the Late Permian to the Late Triassic the extensional Late Carboniferous to Permian back-arc faulting at the southern margin of the Junggar Basin and the Turpan Basin was reversed to compressional thrust tectonics (Mattern et al., 1996; Wartes et al., 2002, Allen et al., 1991). This geodynamic model can well explain the long-lasting tectonic forcing of the system, providing both relatively high sediment supply and high subsidence at low but persistent rates. Most probably, an isostatically balanced system became established over this long period by flexural loading of the thrusted sheets, converting the southern margin of the Junggar Basin into a foreland basin.

Gamma-Ray Patterns

Surprisingly, we found that throughout our section, gamma-ray cycles are often incoherent to lithologic cycles. This is of particular importance for correlation of outcrop to subsurface sequences and for deducing lithology from well logs. Spectral gamma-ray shows that this anomaly is mainly caused by ^{40}K, which is detritally carried by metamorphic and nonmetamorphic claystones and mixes up as intraclasts with gravel deposits. In addition, clay pollution of organic coal seams also may mask primary lithologies.

The U/Th ratio behaves independently of total counts gamma data (Fig. 12) and thus represents an essential tool to assess depositional environments. ^{232}Thorium is almost insoluble in higher salinities (0.00064 ppb), in freshwater (0.01–1 ppb), and even during weathering processes (Sakaguchi et al., 2006; Langmuir and Herman, 1980). The mobility of thorium rises many orders of magnitude when the pH is below 7 due to inorganic complexing and when the pH is over 8.5 due to strong complexes with organic matter, which may occur in swamps, waterlogged sediments, soils, and organic-rich surface waters (Langmuir and Herman, 1980). It can be detritally transported as the heavy mineral monazite and ferromanganese oxyhydroxides (Blake et al., 1996) in the fine sand fraction > 64 μm (Nyakairu and Koeberl, 2002). ^{238}Uranium is fairly soluble in freshwater (U^{6+}) and acidic conditions, but is also transported as detrital heavy-mineral grains (U^{4+}) and adsorbed to sediment particles (Sakaguchi et al., 2006). U^{6+} is easily precipitated and redistributed with organic matter. This is favored at the freshwater–lake water interface and at a fluctuating groundwater level (Yamamoto and Sakaguchi, 2006; Blake et al., 1996). An example is given in Edgington et al. (1996), where higher U/Th rates in sediments of Lake Baikal are related to freshwater inputs during interglacial periods. Phedorin et al. (2000) and Goldberg et al. (2000) relate these findings to orbital parameters controlling interglacial vs. glacial periods. Blake et al. (1996) documented seasonal changes in uranium activity ratios in lake sediments and use them as a long-term proxy for changes in catchment hydrology. Davies and Elliott (1996) detected sequence stratigraphic surfaces with the U/Th ratio. Most probably, the documented increase of the U/Th ratio in the lower Haojiagou Formation is caused by the interaction of fresh-water influx with oxygen-deficient and/or organic-rich lake water. Thus, increased U counts might be used as a proxy for fresh-water input in such a transitional environment.

CONCLUSIONS

We studied a complete Upper Triassic lacustrine fluvio-deltaic sequence, which shows a long-term progradational trend from delta-slope, delta-front, delta-top, up to distal alluvial-plain environments. Due to excellent outcrops along the Haojiagou Valley at the southern margin of the Junggar Basin (northwestern China), we were able to log continuously a 505-m-long section and 220 m of parallel sections and map architectural elements by photomosaic techniques. A quantitative, high-resolution analogue study contributes to a better understanding of depositional dynamics and preservation potential and thus improves reservoir prediction. In addition, we were able to document distinct differences in the architecture, and sequence stratigraphy of lacustrine versus marine deltas, and propose a new coarse-grained mixed-load delta type.

Sequence stratigraphical analysis was carried out using the concept of stratigraphic-base-level (ratio of accommodation space to sediment supply), which worked out to be most suitable for landlocked successions (e.g., Martinsen et al., 1999; Hornung and Aigner, 1999; Hornung and Aigner, 2002b). The four hierarchies of fall and rise cycles found in this sequence are closely linked, and the upper hierarchy determines the symmetry or asymmetry of the lower-order cycles, and thus the lithological record and preservation potential. In addition, due to the lacustrine setting, there is a close linkage of sediment supply with accommodation space (Pla-Pueyo et al., 2009). Periods of higher sediment supply were linked to decreased accommodation space and vice versa. This means that the preservation potential is also strongly coupled to depositional sequences, as well as to the environmental setting. Marine genetic terms like "highstand" and "lowstand" are misleading in such cases (e.g., lakes with open hydrology) and should be abandoned in landlocked systems.

The application of sequence stratigraphical concepts has vast consequences for basin-wide correlations and depositional predictions (Diessel et al., 2000). Detailed studies may contribute to the ongoing process of developing appropriate sequence stratigraphic models for different depositional environments also, those not part of the classical shelf setting (e.g., Lambiase, 1990; Olsen, 1996, Olsen et al., 1996, Carroll and Bohacs, 1999; Catuneanu et al., 2009; Pla-Pueyo et al., 2009; Martinsen et al., 1999; Alonso-Zarza, 2003; Hornung and Aigner, 2002b, Cross et al., 1993; Cross and Lessenger, 1998; Shanley and McCabe, 1994; Homewood et al., 2000; Schlager, 1993; Jervey, 1988). Our study demonstrates that sedimentological and petrophysical outcrop logging at sub-decimeter scale is the key to reveal preservation potential, cycle symmetry, and stacking patterns and thus provide the basis for sequence stratigraphy. Such data challenge applications of established marine sequence stratigraphical concepts on continental settings (Keighley et al., 2003) for which they were never intended. For lacustrine settings Carroll and Bohacs (1999), Bohacs et al. (2000), Bohacs et al. (2003), and Bohacs et al. (2007) already pointed out some fundamental differences, providing a helpful classification scheme for lake-basin types based on fundamental depositional controls. However, their sequence stratigraphic concept is very close to the marine approach. In general, it is difficult and complex to relate depositional cyclicity to any specific or a couple of primary controls, inasmuch as they are mostly working in tandem anyway (Pla-Pueyo et al., 2009). Therefore, we find it appropriate to relate the sedimentary record solely to changes in the ratio of accommodation space to sediment supply (stratigraphic base-level), in that this sequence stratigraphical model does not require any further analysis of the controls.

A surprising but very stimulating outcome of this study was the result from spectral gamma-ray measurements. A first observation is that total gamma-ray counts do not follow grain sizes as expected, but can show irregular grain-size-independent patterns, e.g., coarse-grained deposits yield enhanced counts. This is due to ^{40}K-carrying mudstone components. Sedimentation in channels extended far down the delta slope. It seems that during flood events gravelly sediments from the mountain range mixed with fine-grained sediments of the floodplain, creating hyperconcentrated flows into the lake, which transported coarse-grained sediments by a close interaction of current and mass-flow processes beyond the delta front. Secondly, we could demonstrate that the U/Th ratio serves as an indicator to separate landlocked subaerial from lacustrine subaqueous sediments independently of grain sizes and architectural elements. Both findings should stimulate studies in other lacustrine deltas and should underline that interpretation of gamma logs must done with caution.

ACKNOWLEDGMENTS

For financial, logistical, and political support, we are much obliged to the Government of Xingjian Province, No.1 Regional Geological Surveying Party Bureau of Geology and Mineral Resources of Xingjian, Sino-German Centre for Science Promotion, Jilin University, Academia Sinica, Nanjing Institute of Geology and Paleontology (Chinese Academy of Science), Max-Planck-Gesellschaft, Deutsche Forschungsgemeinschaft, and the Sino-German Cooperation Group. In particular, thanks are given to Rahman Ashraf (University of Tübingen and Kabul) and Volker Mosbrugger (Senckenberg Museum and Research Institute) for extremely valuable support and the invitation to join their group. Jens Hornung thanks his teammates Andreas Matzke, Franziska Grossmann, Fritz Lörcher, Henrik Stöhr, Michael Maisch, and our Chinese colleagues from which we received great cooperation. Special thanks are due to Prof. Sun Ge (head of the Paleontological Research Department of Jilin University) for making this study possible. Especially, we thank Prof. Zuo and Prof. Wang (No.1 Regional Geological Surveying Party, Bureau of Geology and Mineral Resources of Xingjian) for hosting, catering, providing cars, and technical equipment. We thank also the provided survey staff (drivers, workers, etc.) for their effort. Lie Jie and Miao Yunan are thanked very much for their friendship, excellent help in practical fieldwork, and the organization of the daily arrangements during the whole stay (two months). With Prof. Yongdong Wang, Jens Hornung had many fruitful discussions in the field, and Mr. Cheng (PetroChina) guided us during a field trip through the Haojiagou section. Prof. Wu supported our work with a pile of data from his own and other studies.

In the review process, we received helpful suggestions from Elizabeth Gierlowski-Kordesch, Stephanie K. Davidson, Colin P. North, and an anonymous reviewer. This paper significantly benefits from their input. Elizabeth Gierlowski-Kordesch took a lot of effort to polish the English language, and John Southard did the final editing of the paper, which we very much appreciate.

REFERENCES

Aigner, T., Asprion, U., Hornung, J., Junghans, W.D., and Kostrewa, R., 1996, Integrated outcrop analogue studies for triassic alluvial reservoirs: Examples from Southern Germany: Journal of Petroleum Geology, v. 19, p. 393–406.

Aigner, T., Asprion, U., Heinz, J., Hornung, J., and Schauer, M., 1998, Dynamische Stratigraphie: Anwendung in Reservoir-, Aquifer- und Rohstoff-Geologie: Erlanger Geologische Abhandlungen, Sonderband v. 2, p.1.

Aigner, T., Heinz, J., Hornung, J., and Asprion, U., 1999, A hierarchical process-approach to reservoir heterogeneity: examples from outcrop analogs: Bulletin du Centre de Recherches Elf Exploration Production, v. 22, p. 1–11.

Allen, M.B., Windley, B.F., Zhang, C., Zhao, Z.Y., and Wang, G.R., 1991, Basin evolution within and adjacent to the Tien-Shan Range, NW China: Geological Society of London, Journal, v. 148, p. 369–378.

Allen, M.B., Windley, B.F., and Zhang, C., 1993, Palaeozoic collisional tectonics and magmatism of the Chinese Tien Shan, central Asia: Tectonophysics, v. 220, p. 89–115.

Allen, M.B., Sengör, A.M.C., and Natal'in, B.A., 1995, Junggar, Turfan and Alakol Basins as Late Permian to Early Triassic extensional structures in a sinistral shear zone in the Altaid orogenic collage, central-Asia: Geological Society of London, Journal, v. 152, p. 327–338.

Alonso-Zarza, A.M., 2003, Palaeoenvironmental significance of palustrine carbonates and calcretes in the geological record: Earth-Science Reviews, v. 60, p. 261–298.

Anderson, D.S., and Cross, T.A., 2001, Large-scale cycle architecture in continental strata, Hornelen basin (Devonian), Norway: Journal of Sedimentary Research, v. 71, p. 255–271.

Ashraf, A.R., Ge, S., Xinfu, W., Uhl, D., Che, L., and Mosbrugger, V., 1999, The Triassic–Jurassic boundary in the Junggar Basin (NW China)—Preliminary Palynostratigraphic Results: Acta Palaeobotanica Supplementum, v. 2, p. 85–91.

Ashraf, A.R., and Wang, X.F., 2001, Palynostratigraphic analysis of the Huangshanjie, Haojiagou and Badaowan formations in the Junggar Basin (NW-China), *in* Sun, G., Mosbrugger, V., Ashraf, A.R., and Wang, Y.D., eds., The Advanced Study of Prehistoric Life and Geology of Junggar Basin, Xinjiang, China: Proceedings of Sino-German Cooperation Symposium on the Prehistory Life and Geology of Junggar Basin, Xingjiang, China: Urumqi 2001, p. 40–64.

Ashraf, A.R., Sun, Y., Li, J., Sun, G., and Mosbrugger, V., 2004, Palynostratigraphical analysis of the Huangshanjie, Haojiagou, Badaowan-, Sangonghe- and Xishhanyao Formation (Upper Triassic–Middle Jurassic) in the Southern Junggar Basin (NW-China), *in* Sun, G., Mosbrugger, V., Ashraf, A.R., and Sun, Y., eds., Proceedings of Sino-German Cooperation Symposium on Paleontology, Geological Evolution and Environmental Changes of Xinjiang, Jilin University, China: Urumqi 2004, p. 41–44.

Ashraf, A., Sun, Y., Sun, G., Uhl, D., Mosbrugger, V., Li, J., and Herrmann, M., 2010, Triassic and Jurassic palaeoclimate development in the Junggar Basin, Xinjiang, Northwest China—a review and additional lithological data, *in* Martin, T., Sun, G., Mosbrugger, V., eds., Triassic-Jurassic biodiversity, ecosystems, and climate in the Junggar Basin, Xinjiang, Northwest China: Palaeobiodiversity and Palaeoenvironments, Special volume 90, Berlin, Springer, p. 187–201.

Ayers, W.B., 1986, Lacustrine and fluvial–deltaic depositional systems, Fort

Union Formation (Paleocene), Powder River basin, Wyoming and Montana: American Association of Petroleum Geologists, Bulletin, v. 70, p. 1651–1673.

Back, S., De Batist, M., Strecker, M.R., and Vanhauwaert, P., 1999, Quaternary depositional systems in northern Lake Baikal, Siberia: Journal of Geology, v. 107, p. 1–12.

Bally, A.W., Chou, I.M., Clayton, R., Eugster, H.P., Kidwell, S., Meckel, L.D., Ryder, R.T., Watts, A.B., and Wilson, A.A., 1986, Notes on sedimentary basins in China—report of the American sedimentary basins delegation to the People's Republic of China: U.S. Geological Survey, Open-File Report 86-327, 108 p.

Bartov, Y., Stein, M., Enzel, Y., Agnon, A., and Reches, Z., 2002, Lake levels and sequence stratigraphy of Lake Lisan, the late Pleistocene precursor of the Dead Sea: Quaternary Research, v. 57, p. 9–21.

Bian, W., Hornung, J., Liu, Z., Wang, P., and Hinderer, M., 2010, Sedimentary and palaeoenvironmental evolution of the Junggar Basin, Xinjiang, Northwest China *in* Martin, T., Sun, G., Mosbrugger, V., eds., Triassic-Jurassic biodiversity, ecosystems, and climate in the Junggar Basin, Xinjiang, Northwest China: Palaeobiodiversity and Palaeoenvironments, Special volume 90, Berlin, Springer, p. 175–186.

Blair, T.C., and McPherson, J.G., 2008, Quaternary sedimentology of the Rose Creek fan delta, Walker Lake, Nevada, USA, and implications to fan-delta facies models: Sedimentology, v. 55, p. 579–615.

Blake, W.H., Plater, A.J., and Boyle, J.F., 1998, Seasonal trends in the uranium-series isotopic signatures of lake waters and sediments: Hawes Water, northwest England: Journal of Paleolimnology, v. 20, p. 1–14.

Blum, M.D., and Törnqvist, T.E., 2000, Fluvial responses to climate and sea-level change: a review and look forward: Sedimentology, v. 47, p. 2–48.

Bohacs, K., Carroll, A.R., Neal, J.E., and Mankiewicz, P.J., 2000, Lake-basin type, source potential, and hydrocarbon character: an integrated-sequence-stratigraphic-geochemical framework, *in* Gierlowski-Kordesch, E.H., and Kelts, K.R., eds., Lake Basins Through Space and Time: American Association of Petroleum Geologists, Studies in Geology, p. 3–34.

Bohacs, K.M., Carroll, A.R., and Neal, J.E., 2003, Lessons from large lake systems—Thresholds, nonlinearity, and strange attractors, *in* Chan, M.A., and Archer, A.W., eds., Extreme Depositional Environments: Mega End Members in Geologic Time: Geological Society of America, Special Paper, p. 75–90.

Bohacs, K.M., Grabowski, G., and Carroll, A.R., 2007, Lithofacies architecture and variations in expression of sequence stratigraphy within representative intervals of the Green River Formation, greater Green River Basin, Wyoming and Colorado: The Mountain Geologist, v. 44, p. 39–60.

Bowman, D., Shachnovich-Firtel, Y., and Devora, S., 2007, Stream channel convexity induced by continuous base level lowering, the Dead Sea, Israel: Geomorphology, v. 92, p. 60–75.

Cant, D.J., 1992, Subsurface facies analysis, *in* Walker, R.G., and James, N.P., eds., Facies Models: Response to Sea Level Change: St. John's, Newfoundland, Geological Association of Canada, p. 27–45.

Carroll, A.R., Liang, Y.H., Graham, S.A., Xiao, X.C., Hendrix, M.S., Chu, J.C., and McKnight, C.L., 1990, Junggar Basin, Northwest China—Trapped Late Paleozoic Ocean: Tectonophysics, v. 181, p. 1–14.

Carroll, A.R., Graham, S.A., Hendrix, M.S., Ying, D., and Zhou, D., 1995, Late Paleozoic tectonic amalgamation of northwestern China—sedimentary record of the northern Tarim, northwestern Turpan, and southern Junggar Basins: Geological Society of America, Bulletin, v. 107, p. 571–594.

Carroll, A.R., and Bohacs, K.M., 1999, Stratigraphic classification of ancient lakes: Balancing tectonic and climatic controls: Geology, v. 27, p. 99–102.

Carroll, A.R., and Wartes, M.A., 2003, Organic carbon burial by large Permian lakes, northwest China, *in* Chan, M.A., and Archer, A.W., eds., Extreme Depositional environments: Mega End Members in Geologic Time: Geological Society of America, Special Paper, p. 91–104.

Catuneanu, O., Abreub, V., Bhattacharya, J.P., Blum, M.D., Dalrymple, R.W., Eriksson, P.G., Fielding, C.R., Fisher, W.L., Galloway, W.E., Gibling, M.R., Giles, K.A., Holbrook, J.M., Jordan, R., Kendall, C.G.St.C., Macurda, B., Martinsen, O.J., Miall, A.D., Neal, J.E., Nummedal, D., Pomar, L., Posamentier, H.W., Pratt, B.R., Sarg, J.F., Shanley, K.W., Steel, R.J., Strasser, A., Tucker, M.E., and Winker, C., 2009, Towards the standardization of sequence stratigraphy: Earth-Science Reviews, v. 92, p. 1–33.

Cogne, J.P., Chen, Y., Courtillot, V., Rocher, F., Wang, G.Q., Bai, M.X., and You, H.Z., 1995, A paleomagnetic study of mesozoic sediments from the Junggar and Turfan Basins, northwestern China: Earth and Planetary Science Letters, v. 133, p. 353–366.

Coleman, R.G., 1989, Continental growth of Northwest China: Tectonics, v. 8, p. 621–635.

Cross, T.A., Baker, M.R., Chapin, M.A., Clark, M.S., Gardner, M.H., Hanson, M.S., Lessenger, M.A., Little, L.D., McDonough, K.J., Sonnenfeld, M.D., Valasek, D.W., Williams, M.R., and Witter, D.N., 1993, Applications of high-resolution sequence stratigraphy to reservoir analysis. Subsurface reservoir characterization from outcrop observations: Paris, Editions Technip, p. 11–33.

Cross, T.A., and Lessenger, M.A., 1997, Correlation strategies for clastic wedges, *in* Coalson, E.B., Osmond, J.C., and Williams, E.T., eds., Innovative Applications of Petroleum Technology in the Rocky Mountain Area: Denver, Rocky Mountain Association of Geologists, p. 183–203.

Cross, T.A., and Lessenger, M.A., 1998, Sediment volume partitioning: rationale for stratigraphical model evaluation and high-resolution stratigraphical correlation, *in* Gradstein, F.M., Sandvik, K.O., and Milton, N.J., eds., Sequence Stratigraphy—Concepts and Applications: Norwegian Petroleum Society, Special Publication, p. 171–195.

Csato, I., Kendall, C.G.St.C., Nairn, A.E.M., and Baum, G.R., 1997, Sequence stratigraphic interpretations in the southern Dead Sea basin, Israel: Geological Society of America, Bulletin, v. 109, p. 1485–1501.

Dam, G., and Surlyk, F., 1992, Forced regressions in a large wave-dominated and storm-dominated anoxic lake, Rhaetian–Sinemurian Kap Stewart Formation, East Greenland: Geology, v. 20, p. 749–752.

Davies, S.J., and Elliott, T., 1996, Spectral gamma ray characterization of high resolution sequence stratigraphy: examples from Upper Carboniferous fluvio-deltaic systems, County Clare, Ireland: Geological Society of London, Special Publication 104, p. 25–35.

Diessel, C., Boyd, R., Wadsworth, J., Leckie, D., and Chalmers, G., 2000, On balanced and unbalanced accommodation/peat accumulation ratios in the Cretaceous coals from Gates Formation, Western Canada, and their sequence-stratigraphic significance: International Journal of Coal Geology, v. 43, p. 143–186.

Dreyer, T., 1993, Quantified fluvial architecture in ephemeral stream deposits of the Esplugafreda Formation (Paleocene), Tremp–Graus Basin, northern Spain, *in* Dreyer, T., Marzo, M., and Puigdefábregas, C., eds., Alluvial Sedimentation, International Association of Sedimentologists, Special Publication, p. 337–362.

Edgington, D.N., Robbins, J.A., Colman, S.M., Orlandini, K.A., and Gustin, M.P., 1996, Uranium-series disequilibrium, sedimentation, diatom frustules, and paleoclimate change in Lake Baikal: Earth and Planetary Science Letters, v. 142, p. 29–42.

Einsele, G., and Hinderer, M., 1997, Terrestrial sediment yield and the lifetimes of reservoirs, lakes, and larger basins: Geologische Rundschau, v. 86, p. 288–310.

Einsele, G., 2000, Sedimentary Basins, Evolution, Facies and Sediment Budget: Berlin, Springer-Verlag, 792 p.

Einsele, G., and Hinderer M., 2001, The world's large lake basins as denudation-accumulation systems and implications for their lifetimes: Journal of Paleolimnology 26, 355–372.

Escalona, A., and Mann, P., 2006, Sequence-stratigraphic analysis of Eocene clastic foreland basin deposits in central Lake Maracaibo using high-resolution well correlation and 3-D seismic data: American Association of Petroleum Geologists, Bulletin, v. 90, p. 581–623.

Farrell, K.M., 2001, Geomorphology, facies architecture, and high-resolution, non-marine sequence stratigraphy in avulsion deposits, Cumberland Marshes, Saskatchewan: Sedimentary Geology, v. 139, p. 93–150.

Flint, S., Aitken, J., and Hampson, G., 1995, Application of sequence stratigraphy to coal-bearing coastal plain successions: implications for the UK Coal Measures, *in* Whateley M.K.G., and Spears D.A., eds., European Coal Geology: Geological Society of London, Special Publication 82, p. 1–16.

Gibling, M.R., 2006, Width and thickness of fluvial channel bodies and valley fills in the geological record: A literature compilation and classification: Journal of Sedimentary Research, v. 76, p. 731–770.

GILBERT, G.K., 1885, The topographic features of lake shores, US Geological Survey, Annual Report 5, p. 75–123.

GOLDBERG, E.L., PHEDORIN, M.A., GRACHEV, M.A., BOBROV, V.A., DOLBNYA, I.P., KHLYSTOV, O.M., LEVINA, O.V., AND ZIBOROVA, G.A., 2000, Geochemical signals of orbital forcing in the records of paleoclimates found in the sediments of Lake Baikal: Nuclear Instruments & Methods in Physics Research Section A, Accelerators Spectrometers Detectors and Associated Equipment, v. 448, p. 384–393.

GOLONKA, J., ROSS, M.I., AND SCOTESE, C.R., 1994, Phanerozoic paleogeographical and paleoclimatic modelling maps, *in* Embry, A.F., Beauchamp, B., and Glass, D.J., eds., Pangea: Global Environments and Resources: Canadian Society of Petroleum Geologists, p. 1–47.

GRAHAM, S.A., BRASSELL, S., CARROLL, A.R., XIAO, X., DEMAISON, G., MCKNIGHT, C.L., LIANG, Y., CHU, J., AND HENDRIX, M.S., 1990, Characteristics of selected petroleum source rocks, Xianjiang–Uygur Autonomous Region, northwest China: American Association of Petroleum Geologists, Bulletin, v. 74, p. 493–512.

GRASTY, R.I., HOLMAN, P.B., AND BLANCHARD, Y.B., 1991, Transportable calibration pads for ground and airborne gamma-ray spectrometers: Geological Survey of Canada, Paper 90–23, p. 1–26.

GREENE, T.J., ZINNIKER, D., MOLDOWAN, J.M., KEMING, C., AND AIGUO, S., 2004, Controls of oil family distribution and composition in nonmarine petroleum systems: A case study from the Turpan–Hami basin, northwestern China: American Association of Petroleum Geologists, Bulletin, v. 88, p. 447–481.

HAMILTON, D.S., AND GALLOWAY, W.E., 1989, New exploration techniques in the analysis of diagenetically complex reservoir sandstones, Sydney Basin, NSW: The APEA Journal, v. 29, p. 235–257.

HAMILTON, D.S., AND TADROS, N.Z., 1994, Utility of coal seams as genetic stratigraphic sequence boundaries in nonmarine basins—an example from the Gunnedah Basin, Australia: American Association of Petroleum Geologists, Bulletin, v. 78, p. 267–286.

HENDRIX, M.S., GRAHAM, S.A., CARROLL, A.R., SOBEL, E.R., MCKNIGHT, C.L., SCHULEIN, B.J., AND WANG, Z.X., 1992, Sedimentary record and climatic implications of recurrent deformation in the Tian–Shan—Evidence from Mesozoic strata of the north Tarim, south Junggar, and Turpan Basins, northwest China: Geological Society of America, Bulletin, v. 104, p. 53–79.

HENDRIX, M.S., 2000, Evolution of Mesozoic sandstone compositions, southern Junggar, northern Tarim, and western Turpan basins, northwest China: A detrital record of the ancestral Tian Shan: Journal of Sedimentary Research, v. 70, p. 520–532.

HINDERER, M., 1999, Klimagesteuerte Denudations-Sedimentakkumulations-Systeme: Massenbilanzen, Modellierung und Anwendung [unpublished Habilitationsschrift thesis]: Universität Tübingen, Tübingen, 175 p.

HINDERER, M., AND EINSELE, G., 2001, The world's large lake basins as denudation–accumulation systems and implications for their lifetimes: Journal of Paleolimnology, v. 26, p. 355–372.

HOMEWOOD, P.W., GUILLOCHEAU, F., ESCHARD, R., AND CROSS, T.A., 1992, Corrélations haute résolution et stratigraphie génétique: une démarche intégrée: Bulletin des Centres de Recherches Exploration-Production, Elf-Aquitaine, v. 16, p. 357–381.

HOMEWOOD, P.W., MAURIAUD, P., AND LAFONT, F., 2000, Best practices in sequence stratigraphy for explorationists and reservoir engineers: Elf EP editions, Pau, 81 p.

HORNUNG, J., 1999, Dynamische Stratigraphie, Reservoir- und Aquifer-Sedimentologie einer alluvialen Ebene: Der Stubensandstein in Baden-Württemberg (Obere Trias, Mittlerer Keuper): Tübinger Geowissenschaftliche Arbeiten (TGA), Reihe A, v. 56, 156 p.

HORNUNG, J., AND AIGNER, T., 1999, Reservoir and aquifer characterization of fluvial architectural elements: Stubensandstein, Upper Triassic, southwest Germany: Sedimentary Geology, v. 129, p. 215–280.

HORNUNG, J., AND AIGNER, T., 2002a, Reservoir architecture in a terminal alluvial plain: An outcrop analogue study (Upper Triassic, Southern Germany)—Part 1: Sedimentology and petrophysics: Journal of Petroleum Geology, v. 25, p. 3–29.

HORNUNG, J., AND AIGNER, T., 2002b, Reservoir architecture in a terminal alluvial plain: An outcrop analogue study (upper Triassic, Southern Germany) part II: Cyclicity, controls and models: Journal of Petroleum Geology, v. 25, p. 151–178.

HORNUNG, J., ASHRAF, A., HINDERER, M., BIAN, W., WANG, P., GE, S., AND MOSBRUGGER, V., 2011, Depositional dynamics of a coal bearing lacustrine delta: A case study from the Junggar-Basin (NW-China, Upper Triassic): Palaeobiodiversity and Palaeoenvironments.

HORTON, B.K., AND SCHMITT, J.G., 1996, Sedimentology of a lacustrine fan-delta system, Miocene Horse Camp Formation, Nevada, USA: Sedimentology, v. 43, p. 133–155.

HUANG, T.K., 1978, An outline of the tectonic characteristics of China: Eclogae Geologicae Helvetiae, v. 71, p. 611–635.

JERVEY, M.T., 1988, Quantitative geological modeling of siliciclastic rock sequences and their seismic expression, *in* Wilgus, C.K., Hastings, B.S., Kendall, C.G.St.C., Posamentier, H.W., Ross, C.A., and Van Wagoner, J.C., eds., Sea Level Change: An Integrated Approach: SEPM, Special Publication 42, p. 47–69.

JOHNSON, C.L., AND GRAHAM, S.A., 2004, Sedimentology and reservoir architecture of a synrift lacustrine delta, Southeastern Mongolia: Journal of Sedimentary Research, v. 74, p. 770–785.

JUHÁSZ, E., KOVÁCS, L.O., MÜLLER, P., TÓTH-MAKK, Á., PHILLIPS, L., AND LANTOS, M., 1997, Climatically driven sedimentary cycles in the Late Miocene sediments of the Pannonian Basin, Hungary: Tectonophysics, v. 282, p. 257–276.

KASSE, C., VANDENBERGHE, J., VAN HUISSTEDEN, J., BOHNCKE, S.J.P., AND BOS, J.A.A., 2003, Sensitivity of Weichselian fluvial systems to climate change (Nochten mine, eastern Germany): Quaternary Science Reviews, v. 22, p. 2141–2156.

KEIGHLEY, D., FLINT, S., HOWELL, J., AND MOSCARIELLO, A., 2003, Sequence stratigraphy in lacustrine basins: A model for part of the Green River Formation (Eocene), southwest Uinta Basin, Utah, USA: Journal of Sedimentary Research, v. 73, p. 987–1006.

KIDDER, T.R., 2006, Climate change and the Archaic to Woodland transition (3000–2500 Cal B.P.) in the Mississippi River Basin: American Antiquity, v. 71, p. 195–231.

LAMBIASE, J.J., 1990, A model for tectonic control of lacustrine stratigraphic sequences in continental rift basins, *in* Katz, B., ed., Lacustrine Basin Exploration: Case Studies and Modern Analogs: American Association of Petroleum Geologists, Memoir 50, p. 265–276.

LANGMUIR, D., AND HERMAN, J.S., 1980, The mobility of thorium in natural-waters at low-temperatures: Geochimica et Cosmochimica Acta, v. 44, p. 1753–1766.

LAWRENCE, S.R., 1990, Aspects of the petroleum geology of the Junggar Basin, Northwest China: Geological Society of London, Special Publications 50, p. 545–557.

LEE, K.Y., 1985, Geology of the petroleum and coal deposits in the Junggar (Zhungaer) basin, Xinjiang Uygur Zizhiqu, northwest China: U.S. Geological Survey, Open-File report, p. 85–230.

LEEDER, M.R., 1997, Sedimentary basins: Tectonic recorders of sediment discharge from drainage catchments: Earth Surface Processes and Landforms, v. 22, p. 229–237.

LEEDER, M.R., HARRIS, T., AND KIRKBY, M.J., 1998, Sediment supply and climate change: implications for basin stratigraphy: Basin Research, v. 10, p. 7–18.

LIGOUIS, B., 2001, Organic petrography, geochemistry, stratigraphy, facies geology and basin analysis in the Triassic and Jurassic of the Junggar Basin (NW China), *in* Sun, G., Mosbrugger, V., Ashraf, A.R., and Wang, Y.D., eds., The Advanced Study of Prehistoric Life and Geology of Junggar Basin, Xinjiang, China: Proceedings of Sino-German Cooperation Symposium on the Prehistory Life and Geology of Junggar Basin, Xingjiang, China: Urumqi 2001, p. 104–113.

MAISCH, M.W., MATZKE, A.T., PFRETZSCHNER, H.U., SUN, G., STOHR, H., AND GROSSMANN, F., 2001, The fossil vertebrate faunas of the Toutunhe and Qigu formations of the southern Junggar Basin and their biostratigraphical and palecological implications, *in* Sun, G., Mosbrugger, V., Ashraf, A.R., and Wang, Y.D., eds., The Advanced Study of Prehistoric Life and Geology of Junggar Basin, Xinjiang, China: Proceedings of Sino-German Cooperation Symposium on the Prehistory Life and Geology of Junggar Basin, Xingjiang, China: Urumqi 2001, p. 83–94.

MAISCH, M.W., MATZKE, A.T., PFRETZSCHNER, H.U., SUN, G., STOHR, H., AND GROSSMANN, F., 2004, Fossil vertebrates from the Middle and Upper Jurassic of the Southern Junggar Basin (NW China)—results of the Sino–German

Expeditions 1999–2000: Neues Jahrbuch für Geologie und Paläontologie, Monatshefte, p. 297–313.

Martinsen, O.J., Ryseth, A., Helland-Hansen, W., Flesche, H., Torkildsen, G., and Idil, S., 1999, Stratigraphic base level and fluvial architecture: Ericson Sandstone (Campanian), Rock Springs Uplift, SW Wyoming, USA: Sedimentology, v. 46, p. 235–259.

Mattern, F., Schneider, W., Li, Y., and Li, X., 1996, A traverse through the western Kunlun (Xinjiang, China): Tentative geodynamic implications for the Paleozoic and Mesozoic: Geologische Rundschau, v. 85, p. 705–722.

McPherson, J.G., Shanmugam, G., and Moiola, R.J., 1987, Fan-deltas and braid deltas: Varieties of coarse-grained deltas: Geological Society of America, Bulletin, v. 99, p. 331–340.

Meckel, L.D., 1975, Holocene sand bodies in the Colorado delta area, northern Gulf of California, *in* Broussard, M.C., ed., Deltas, Models for Exploration, Houston Geological Society, Houston, Texas, p. 239–265.

Miall, A.D., 1996, The Geology of Fluvial Deposits, Sedimentary Facies, Basin Analysis and Petroleum Geology: Berlin, Springer-Verlag, 582 p.

Miall, A.D., 2006, Reconstructing the architecture and sequence stratigraphy of the preserved fluvial record as a tool for reservoir development: A reality check: American Association of Petroleum Geologists, Bulletin, v. 90, p. 989–1002.

Moore, T.C., Klitgord, K.D., Golmshtok, A.J., and Weber, E., 1997, Sedimentation and subsidence patterns in the central and north basins of Lake Baikal from seismic stratigraphy: Geological Society of America, Bulletin, v. 109, p. 746–766.

Nyakairu, G.W.A., and Koeberl, C., 2002, Variation of mineral, chemical, and rare earth element composition in size fractions of clay-rich sediments from the Kajjansi and Ntawo clay deposits, central Uganda: Chemie Der Erde—Geochemistry, v. 62, p. 73–86.

Olsen, P.E., Kent, D.V., Cornet, B., Witte, W.K., and Schlische, R.W., 1996, High-resolution stratigraphy of the Newark rift basin (early Mesozoic, eastern North America): Geological Society of America, Bulletin, v. 108, p. 40–77.

Olsen, P.E., 1996, Tectonic, climatic, and biotic modulation of lacustrine ecosystems—Examples from Newark Supergroup of eastern North America, *in* Katz, B., ed., Lacustrine Basin Exploration: Case Studies and Modem Analogs, American Association Petroleum Geologists, Memoir 50, p. 209–224.

Otto, S.C., 1997, Mesozoic–Cenozoic history of deformation and petroleum systems in sedimentary basins of Central Asia: implications of collisions on the Eurasian margin: Petroleum Geoscience, v. 3, p. 327–341.

Overeem, I., Kroonenberg, S.B., Veldkamp, A., Groenesteijn, K., Rusakov, G.V., and Svitoch, A.A., 2003, Small-scale stratigraphy in a large ramp delta: recent and Holocene sedimentation in the Volga delta, Caspian Sea: Sedimentary Geology, v. 159, p. 133–157.

Pfretzschner, H.U., Martin, T., and Maisch, M., 2004, A new docodont from the Totunhe formation (Middle Jurassic) of the Junggar Basin, *in* Sun, G., Mosbrugger, V., Ashraf, A.R., and Sun, Y., eds., Proceedings of the Sino-German Cooperation Symposium on Paleontology, Geological Evolution and Environmental Changes of Xinjiang, Jilin University, China: Urumqi 2004, p. 9–12.

Phedorin, M.A., Goldberg, E.L., Bobrov, V.A., Khlystov, O.M., and Grachev, M.A., 2000, Multi-wavelength synchrotron radiation XRF determination of U and Th in sedimentary cores from Lake Baikal: Geostandards Newsletter—The Journal of Geostandards and Geoanalysis, v. 24, p. 217–226.

Pla-Pueyo, S., Gierlowski-Kordesch, E.H., Viseras, C., and Soria, J.M., 2009, Major controls on sedimentation during the evolution of a continental basin: Pliocene–Pleistocene of the Guadix Basin (Betic Cordillera, southern Spain): Sedimentary Geology, v. 219, p. 97–114.

Platt, N.H., 1989, Climatic and tectonic controls on sedimentation of a Mesozoic lacustrine sequence—the Purbeck of the Western Cameros Basin, northern Spain: Palaeogeography, Palaeoclimatology, Palaeoecology, v. 70, p. 187–197.

Qi, S., Zhang, Y., Cheng, X., Lu, H., and Zhan, J., 1994, Jurassic biostratigraphy and ecological environment of Junggar Basin: Exploration and Research Institute of Xinjiang Petroleum Administration, Nanjing Geological and Paleontological Institute, Academia Sinica, 89 p.

Ramon, J.C., and Cross, T., 1997, Characterization and prediction of reservoir architecture and petrophysical properties in fluvial channel sandstones, Middle Magdalena Basin, Colombia: Ciencia, Technología, Futuro, v. 1, p. 19–46.

Reading, H.G., 1996, Sedimentary Environments; Processes, Facies and Stratigraphy: Oxford, U.K. Blackwell Science, 688 p.

Reineck, M.E., and Singh, J.B., 1980, Depositional Sedimentary Environments: Berlin, Springer, 551 p.

Ritts, B.D., 1998, Mesozoic tectonics and sedimentation, and petroleum systems of the Qaidam and Tarim basins, NW China [unpublished PhD. dissertation]: Stanford University, Stanford, California, 691 p.

Sakaguchi, A., Yamamoto, M., Sasaki, K., and Kashiwaya, K., 2006, Uranium and thorium isotope distribution in an offshore bottom sediment core of the Selenga Delta, Lake Baikal, Siberia: Journal of Paleolimnology, v. 35, p. 807–818.

Schäfer, A., 2005, Klastische Sedimente, Fazies und Sequenzstratigraphie: Elsevier, Spektrum Akademischer Verlag, 414 p.

Schellhorn, R., Maisch, M.W., Wings, O., Schwarz, D., and Sun, G., 2007, Crocodyliform remains from the Qigu formation in the southern Junggar Basin (NW China), *in* Sun, G., Mosbrugger, V., Dun, Y.W., and Bruch, A., eds., Proceedings of the International Symposium for Sino-German Cooperation on Geology and Environmental Changes in Northern China, Urumqi, China, p. 11–15.

Schlager, W., 1993, Accommodation and supply—a dual control on stratigraphic sequences: Sedimentary Geology, v. 86, p. 111–136.

Schneider, W., 1996, The coal-bearing Jurassic at the southern margin of the Junggar Basin, Xinjiang. Sedimentary facies and environments: Geo-wissenschaften, v. 14, p. 285–287.

Scholz, C.A., Rosendahl, B.R., and Scott, D.L., 1990, Development of coarse-grained facies in lacustrine rift basins—Examples from east Africa: Geology, v. 18, p. 140–144.

Scholz, C.A., and Finney, B.P., 1994, Late Quaternary sequence stratigraphy of Lake Malawi (Nyasa), Africa: Sedimentology, v. 41, p. 163–179.

Scholz, C.A., and Hutchinson, D.R., 2000, Stratigraphic and structural evolution of the Selenga Delta Accommodation Zone, Lake Baikal Rift, Siberia: International Journal of Earth Sciences, v. 89, p. 212–228.

Sengör, A.M.C., 1985, East Asian Tectonic Collage: Nature, v. 318, p. 16–17.

Serra, O., and Sulpice, L., 1975, Sedimentological analysis of shale–sand series from well logs: Society of Professional Well Log Analysts, 16th Annual Symposium Transactions, 23 p.

Shanley, K.W., and Mccabe, P.J., 1994, Perspectives on the sequence stratigraphy of continental strata: American Association of Petroleum Geologists, Bulletin, v. 78, p. 544–568.

Sharps, R.W., 1990, Paleomagnetism and tectonics of northwest China [unpublished PhD. dissertation]: Stanford University, Stanford, California, 250 p.

Smith, A.G., and Briden, J.C., 1977, Mesozoic and Cenozoic Palaeocontinental Maps: Cambridge, U.K., Cambridge University Press, 63 p.

Strecker, U., Steidtmann, J.R., and Smithson, S.B., 1999, A conceptual tectonostratigraphic model for seismic facies migrations in a fluvio-lacustrine extensional basin: American Association of Petroleum Geologists, Bulletin, v. 83, p. 43–61.

Sun, G., Mosbrugger, V., Li, J., and Cheng, X.S., 2001, Late Triassic flora of the Junggar Basin, Xingjiang, China, *in* Sun, G., Mosbrugger, V., Ashraf, A.R., and Wang, Y.D., eds., The Advanced Study of Prehistoric Life and Geology of Junggar Basin, Xinjiang, China. Proceedings of Sino-German Cooperation Symposium on the Prehistory Life and Geology of Junggar Basin, Xingjiang, China: Urumqi 2001.

Ta, T.K.O., Nguyen, V.L., Tateishi, M., Kobayashi, I., Tanabe, S., and Saito, Y., 2002, Holocene delta evolution and sediment discharge of the Mekong River southern Vietnam: Quaternary Science Reviews, v. 21, p. 1807–1819.

Tang, Z.H., Parnell, J., and Longstaffe, F.J., 1997, Diagenesis and reservoir potential of Permian–Triassic fluvial/lacustrine sandstones in the southern Junggar basin, northwestern China: American Association of Petroleum Geologists, Bulletin, v. 81, p. 1843–1865.

Tye, R.S., and Coleman, J.M., 1989, Depositional processes and stratigraphy of fluvially dominated lacustrine deltas—Mississippi delta plain: Journal of Sedimentary Petrology, v. 59, p. 973–996.

Wartes, M.A., Carroll, A.R., and Greene, T.J., 2002, Permian sedimentary record of the Turpan–Hami basin and adjacent regions, northwest China:

Constraints on postamalgamation tectonic evolution: Geological Society of America, Bulletin, v. 114, p. 131–152.

WATSON, M.P., HAYWARD, A.B., PARKINSON, D.N., AND ZHANG, Z.M., 1987, Plate tectonic history, basin development and petroleum source rock deposition onshore China: Marine and Petroleum Geology, v. 4, p. 205–225.

WHEELER, H.E., 1964, Baselevel, lithosphere surface, and time–stratigraphy: Geological Society of America, Bulletin, v. 75, p. 599–609.

WINDLEY, B.F., ALLEN, M.B., ZHANG, C., ZHAO, Z.Y., AND WANG, G.R., 1990, Paleozoic accretion and Cenozoic redeformation of the Chinese Tien-Shan Range, central Asia: Geology, v. 18, p. 128–131.

WINGS, O., FOWLE, D.W., MAISCH, M.W., MARTIN, T., PFRETZSCHNER, H.U., AND DUN, G., 2007, Dinosaur teeth from the Jurassic Qigu and Shishugou Formations of the Junggar Basin (Xinjiang, China) and their palaeogeographic implications, *in* Sun, G., Mosbrugger, V., Dun, Y.W., and Bruch, A., eds., Proceedings of the International Symposium for Sino-German Cooperation on Geology and Environmental Changes in Northern China: Urumqi, p. 5–10.

WOOD, S.H., 1994, Seismic expression and geological significance of a lacustrine delta in Neogene deposits of the western Snake River plain, Idaho: American Association of Petroleum Geologists, Bulletin, v. 78, p. 102–121.

YAMAMOTO, M., AND SAKAGUCHI, A., 2006, Uranium (^{238}U and ^{234}U) and thorium (^{232}Th and ^{230}Th) records in the bottom: Kanazawa University research paper, published online, http://hdl.handle.net/2297/2385.

ZHANG, Z.M., LIOU, J.G., AND COLEMAN, R.G., 1984, An outline of the plate-tectonics of China: Geological Society of America, Bulletin, v. 95, p. 295–312.

ZHOU, D., 1997, Studies in the tectonics of China: Extensional tectonics of the northern margin of the South China Sea; amalgamation and uplift of the Tian Shan; and wedge extrusion model for the Altyn Tagh fault [Unpublished PhD. dissertation]: Stanford University, Stanford, California, 354 p.

PEDOGENIC AND NON-PEDOGENIC CALCRETES IN THE DEVONIAN RIDGEWAY CONGLOMERATE FORMATION OF SW WALES, UK: A CAUTIONARY TALE

ROBERT D. HILLIER
Department of Geology, National Museum of Wales, Cardiff, CF10 3NP, UK
Email:robhillier@aol.com

SUSAN B. MARRIOTT
School of Earth Sciences, University of Bristol, Queen's Road, Bristol, BS8 1RJ, UK

AND

BRIAN P.J. WILLIAMS
Department of Geology and Petroleum Geology, University of Aberdeen, Kings College, Aberdeen, AB24 3UE, UK

ABSTRACT: Calcic pedocomplexes in the Siluro-Devonian Old Red Sandstone (ORS) of the Anglo-Welsh Basin (UK) have traditionally been interpreted as paleosols developed in dryland depositional environments. Their recognition has been used to indicate a range of controls, including climate, landscape stability, sedimentation rate, soil residence time, and proximity to alluvial channels (the pedofacies concept). A study of the Devonian Ridgeway Conglomerate Formation (RCF) in Pembrokeshire, southwest Wales, has, however challenged some of these notions, recognizing that many calcretes were not developed in soil horizons. The RCF was deposited as part of a dryland alluvial fan and axial fluvial valley complex. Regionally, structural blocks and basins were defined by a series of extensional faults, with the RCF being deposited in a half-graben as a hanging-wall alluvial fan. The RCF is heterolithic, comprising conglomerates, sandstones, and mudstones that reflect differences in processes, suggesting sheetfloods, low-relief lateral accretion, and cohesive debris flows across the alluvial fan.

Pedogenic calcretes are common in all areas of the fan and axial fluvial zone. In mudstone and sandstone-grade lithofacies they comprise common horizonated nodules and subhorizontal crystallaria sheets in association with pedogenic indicators such as drab haloes, desiccation cracks, and ped textures. Also observed are both horizontal and vertical root traces, some of which have been the focus of micrite nodule growth (rhizogenic calcretes). Wedge-shaped peds are absent. Pedogenic profiles display upward-increasing percentages of nodules, and may be capped by blocky, massive calcrete and laminated micrite that developed in small ponded areas. In gravel-grade lithofacies, the pedogenic expression is different, and comprises carbonate-coated clasts with pendant and pore-occluding calcrete fabrics.

Pedogenic calcretes are best developed in proximal areas of the fan, possibly on terraces adjacent to fan-channel entrenchment zones (the pedofacies concept). Proximal fan areas may also have had increased soil residence times due to reduced sedimentation rates compared to distal fan and axial fluvial valley zones.

In distal fan and axial alluvial zones, thin layerbound micritic groundwater calcretes are common, typically being sharp based with upper surfaces comprising vertical and cylindrical nodules that possibly developed in the capillary-fringe zone. Inclined heterolithic bedsets, the deposits of laterally accreted ephemeral channels also commonly contain layer-bound micritic calcretes, again interpreted as having a groundwater origin. Lake-margin calcretes comprising centimeter-thick, laminated micrite, represent possible calcretized matgrounds in fan-toe, ephemeral ponds.

The identification of common non-pedogenic calcretes in the RCF begs the question: how much of the ORS calcretes are similarly non-pedogenic in nature? Our analysis may act as a cautionary check for subsurface work where carbonate horizons in alluvial suites are being modelled solely in accordance with the pedofacies concept.

KEY WORDS: calcrete, Old Red Sandstone, pedogenesis, paleosol, alluvial fan, Devonian

INTRODUCTION

Near surface, terrestrial accumulations of calcium carbonate (calcrete; Lamplugh 1902; Goudie 1973) are common in modern and ancient soil profiles as well as shallow-water carbonate systems (Wright and Tucker 1991; Alonso-Zarza 2003). Recent calcic soils or pedogenic calcrete are prominent in climate zones where a seasonal moisture deficit occurs, facilitating $CaCO_3$ precipitation. Low solubility of $CaCO_3$ coupled with aridity inhibits subsequent leaching (Machette 1985; Retallack et al. 2004). Calcic horizons in soil profiles are characterized by various morphologies and degrees of development reflecting texture and composition of parent material, climate, and precipitation regime (Gile 1975, 1977; Machette 1985). Calcretes also occur below the zone of soil formation in the vadose or phreatic groundwater zones (Carlisle 1983; Wright and Tucker 1991; Williams and Krause 1998; Tandon and Kumar 1999). Here, displacive and replacive $CaCO_3$ precipitates from carbonate-rich, mobile groundwater in porous, permeable intervals.

Calcretes are well documented in the stratigraphic record back to Precambrian times (Retallack 1986). Their recognition has conventionally been linked to paleosols that are interpreted to have originally formed in a warm to arid climate typified by a low, seasonal rainfall (Allen 1973, 1986; Retallack 2005). The level of calcrete development has aided interpretation of sediment accumulation rates (soil residence times; Marriott and Wright 1993, 2006; Wright and Marriott 1996), landscape stability, and proximity to adjacent fluvial channel sandstones (the pedofacies concept; Bown and Kraus 1987, Kraus and Bown 1993; Kraus 1999). In rock-record descriptions, few examples of non-pedogenic calcretes are recognized (e.g., Wright and Tucker 1991; Rasbury et al. 2006; Khalaf 2007; Jutras et al. 2007). The

From River To Rock Record: The Preservation Of Fluvial Sediments And Their Subsequent Interpretation
SEPM Special Publication No. 97, Copyright © 2011
SEPM (Society for Sedimentary Geology), ISBN 978-1-56576-305-0, p. 311–325.

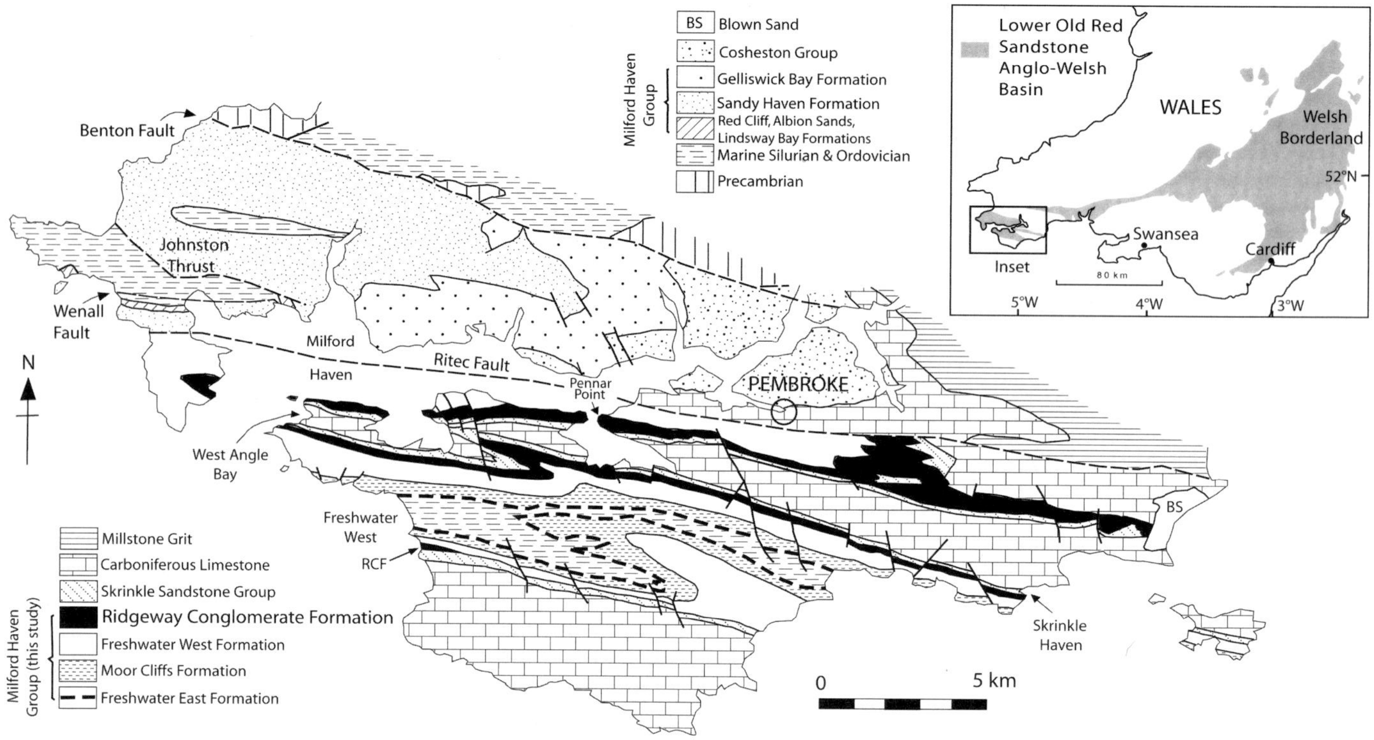

FIGURE 1.—Geological map of southern Pembrokeshire illustrating Ridgeway Conglomerate Formation exposure (modified after Hillier and Williams 2006). Localities described in text are at Skrinkle Haven, Freshwater West, Pennar Point, and West Angle Bay. Inset shows present extent of Lower Old Red Sandstone Anglo-Welsh Basin exposure across Wales and the Welsh Borderland.

distinction between pedogenic and non-pedogenic calcretes in the rock record is difficult, because products from different processes may be superficially similar, and pedogenic features may also overprint earlier non-pedogenic calcretes (Wright and Tucker 1991; Pimentel et al. 1996; Rasbury et al. 2006). In addition, a universally accepted model to distinguish between the two types has yet to be formulated (Mack et al. 2000).

In this paper we describe the field occurrence and petrographic signature of calcretes in the Devonian Old Red Sandstone (ORS) Ridgeway Conglomerate Formation of SW Wales, UK. We challenge the previously held notion that all calcretes in the ORS are pedogenic in origin. We discuss the criteria for the recognition of pedogenic and non-pedogenic calcretes in the Formation. This work should provide a cautionary check to future paleoenvironmental interpretations that use calcretes as a proxy indicator of paleoclimate and paleopedogenesis.

GEOLOGICAL CONTEXT

The Anglo-Welsh Basin Old Red Sandstone

The Siluro-Devonian Anglo-Welsh Basin (UK) contains the ORS continental basin-fill succession, which attains a thickness of up to 4.4 km (Fig. 1; Friend et al. 2000; Barclay et al. 2005; Hillier and Williams 2006). The Basin is an erosional remnant of a wider cover that extended across southern Britain, developing externally to the Caledonian Mountain belt (Simon and Bluck 1982; Allen and Crowley 1983; Soper and Woodcock 2003). It comprises fluvial and alluvial fan deposits of a significant drainage network, with minor marine incursions (Allen 1974; Allen and Williams 1979, 1982; Allen et al. 1982; Barclay et al. 1994; Hillier and Williams 2004; Hillier et al. 2007; Marriott and Wright 1993, 2004; Love and Williams 2000; Williams and Hillier 2004). To the south and east of the Basin lay the contemporaneous Devonian Rheic Ocean coastline. In southern Pembrokeshire, extensional faulting modified ORS sedimentation as early as Ludlow times (Hillier 2000; Hillier and Williams 2004), a consequence of dextral transpression associated with the Rheic convergence of the Avalonian and Armorican terranes (Friend et al. 2000). Tectonism continued throughout the Devonian with significant deformation occurring during the Emsian-Eifelian Acadian Orogeny (Friend et al. 2000; Dewey and Strachan 2003; Soper and Woodcock 2003).

The Lower ORS (Figure. 2) succession is mudrock dominated, comprising deposits predominantly of a low-gradient floodplain in a semi arid, dryland environment (Allen and Williams 1979, 1982; Marriott and Wright 1993, 2004; Love and Williams 2000). Short-lived fluctuations to a more humid climate may have occurred (Thomas et al. 2006; Hillier and Williams 2007; Hillier et al. 2007). Fluvial sandstone bodies of varying thickness and geometries occur throughout the Lower ORS (see review in Williams and Hillier 2004). The basin lay in tropical geodetic latitudes of approximately 17 ± 5° S during the Devonian (Channel et al. 1992). The Lower ORS contains evidence of early terrestrialization, preserving vascular plants, palynomorphs, vertebrate communities, and terrestrial trace fossils (Dineley 1999a, 1999b; Edwards and Richardson 2004; Morrissey and Braddy 2004; Barclay et al. 2005).

The Devonian Ridgeway Conglomerate Formation

The Ridgeway Conglomerate Formation (RCF) was deposited as part of a dryland alluvial fan and axial fluvial valley complex of

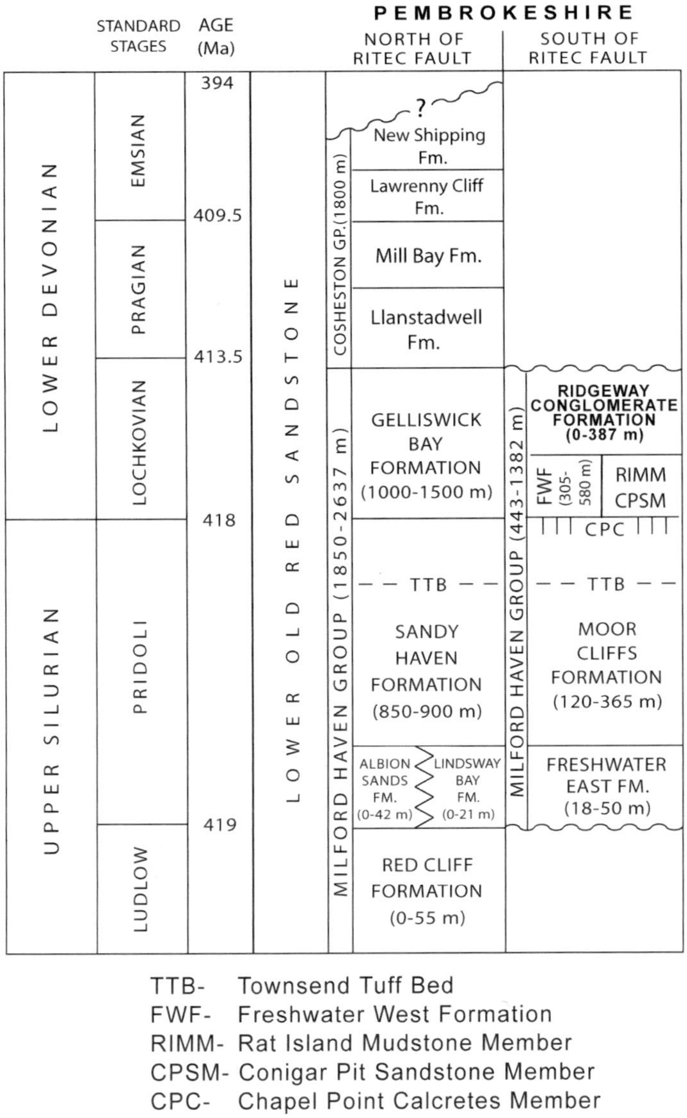

FIGURE 2.—Lower Old Red Sandstone stratigraphy of southern Pembrokeshire (modified after Hillier and Williams 2007).

probable latest Lochkovian to earliest Pragian age (Fig. 3; Dixon 1921; Dixon 1933a; Dixon 1933b: Williams 1964; Williams et al. 1982; Hillier and Williams 2007). The alluvial fan association was deposited on the hangingwall of a tilt faultblock, its deposits thickening into the Ritec Fault, indicating the control this fault had on its development. The RCF is heterolithic, comprising exotic and intraformational clast conglomerates, sandstones, and poorly sorted mudstones with common floating, outsized clasts up to pebble-grade grain size. The fan drained a southerly hinterland underlain by lower Paleozoic and Precambrian rocks (Cope and Bassett 1987). Preserved deposits reflect various ephemeral processes operating over the fan surface, including sheetfloods, confined flow within low-relief laterally accreted channels, and cohesive debris flows (Hillier and Williams 2007).

The alluvial fans appear to have prograded northward, interfingering with deposits of an axial fluvial facies association. The latter is dominated by upward-thinning and -fining inclined heterolithic stratification. Bedload transport of pedogenic aggregates was an important process in depositing some of the mudstones (Ékes 1993).The axial fluvial association also contains thinly laminated interbedded mudstones and sandstones that were deposited in ephemeral floodout zones and axial-valley lakes at the toe of the fan system. These ephemeral lakes possibly developed under subhumid climate episodes, or were sustained by baseflow through permeable fan deposits (Hillier and Williams 2007). Sedimentary structures in this mudstone-dominated lacustrine association include pinstripe lamination, matgrounds, possible algal "roll-up" structures, calcretized peetees, and root traces (Hillier and Williams 2007). Periods of semi arid climate and/or active tectonic conditions are represented by fan gravels prograding into the axial system and the presence of calcretes of various stages of development.

OLD RED SANDSTONE PEDOGENIC CALCRETES

Pedogenic calcretes have been recognized in many studies of the Anglo-Welsh ORS, being described in detail by Allen (1973, 1974, 1986), Marriott and Wright (1993, 2006), and Love and Williams (2000). The paleosols that contain them have been interpreted as calcic Vertisols at varying stages of development. Most conspicuous is a lower B_k or C_k horizon containing pale calcrete nodules at varying stages of development (Fig. 4). Carbonate morphology varies greatly from fine filaments or crystallaria (both parallel or perpendicular to bedding) to subspherical or columnar nodules up to 0.3 m in length. Structural B horizons comprising wedge-shaped peds of intersecting slickensided fractures reflect seasonal shrinking and swelling of the clay-rich soil profile (Fig. 4). The structural horizon may distort pre-existing nodules, commonly preserving fanned prismatic fabrics. The A horizon, where present, contains desiccation cracks, drab haloes, and root traces (Hillier et al. 2008).

Complete profiles are rarely preserved, and most commonly attain thicknesses of 1 to 3 m. Thicker profiles of up to 15 m have been recorded (Allen 1986; Love and Williams 2000). Preserved profiles are more commonly truncated by erosion surfaces often strewn with reworked-calcrete-nodule conglomerates. Reactivation of fractures in the structural horizon often led to deformed erosion surfaces and conglomerate lags (Marriott and Wright 2006). Composite or compound profiles reflect horizon overprinting related to slow aggradation or rapid burial of the soil profile (Wright and Marriott 1996). Allen (1986) proposed that the average duration for paleosol development was 3×10^4 years; Wright and Marriott (1996) considered that calcrete nodules of stage II development (per Machette 1985) formed during 10^4 to 10^5 years.

Love and Williams (2000) demonstrated that the degree of paleosol development can vary laterally over 10 to 30 m of section, with the stage of development of calcrete nodules increasing into topographic lows preserved in the paleosol profiles. They also showed that the calcretes with the highest stage of development in the section (mostly Stage III) were observed in locations farthest away from channel sandbodies in both vertical and lateral directions. This spatial distribution is consistent with the pedofacies concept (Bown and Kraus 1987), where soil maturity (stages of development as a function of time) increases in areas of low net sediment accumulation rate. In this example, low net sediment accumulation rate is inferred for areas distant from contemporaneous fluvial channels.

A thick, well-developed calcic pedocomplex, the Chapel Point Calcretes Member, crops out across the Anglo-Welsh Basin at a position close to the Silurian-Devonian boundary (Fig. 2: Allen and Williams 1981, 1982; Barclay et al 2005). Its development may indicate a period of geomorphic stability across the basin, or a period of extended aridity. Variations in its development presumably reflect differential subsidence and soil residence times linked to fault-bounded depocenters and terraces (Williams and Hillier 2004).

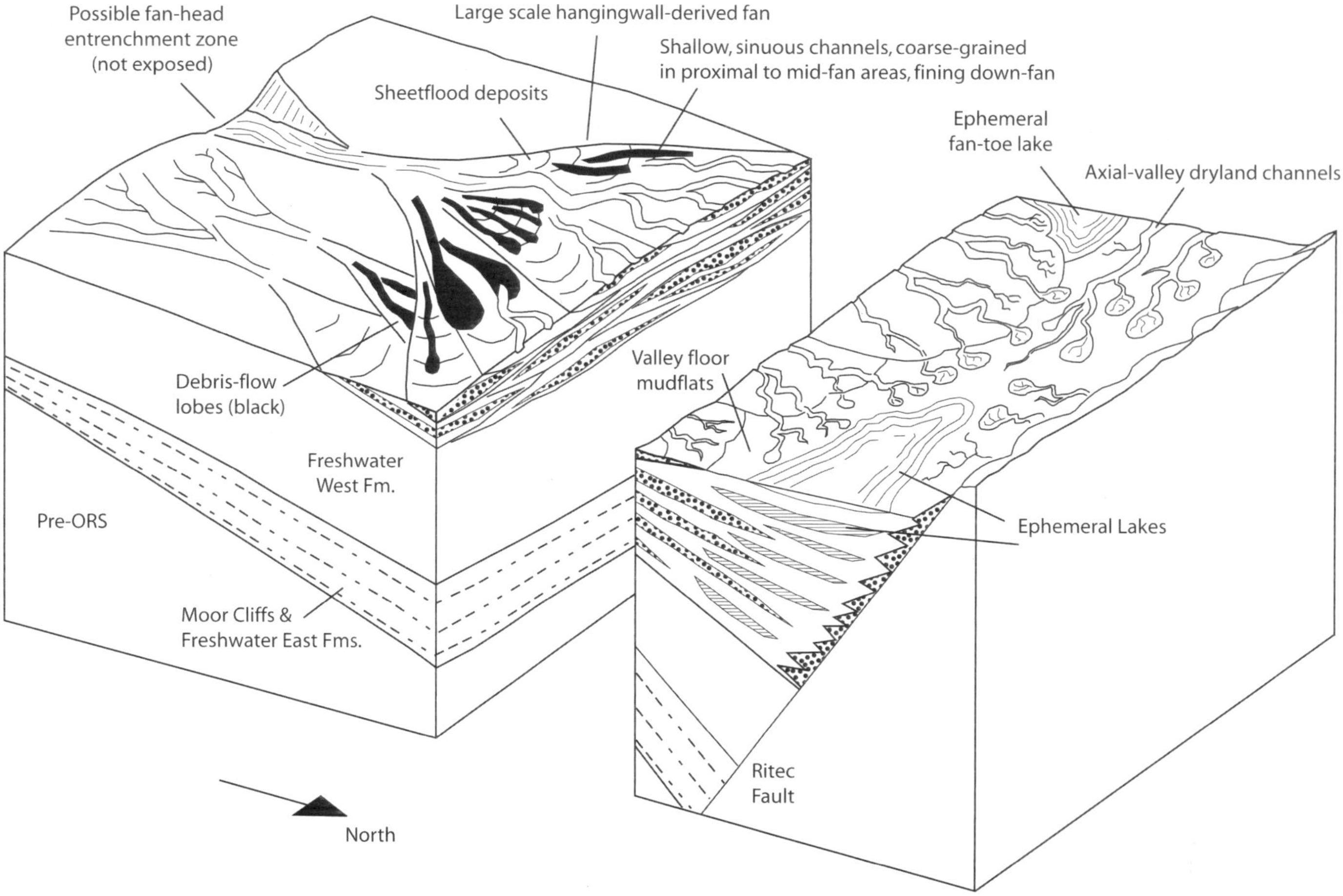

FIGURE 3.—Paleogeomorphology of the Ridgeway Conglomerate Formation alluvial fan and axial-valley complex (modified after Hillier and Williams 2007).

METHODOLOGY

This study predominantly comprised section logging of RCF calcretes in coastal exposures at Skrinkle Haven, Freshwater West, West Angle Bay and Pennar Point (Fig. 1). Full graphic logs of measured sections are presented in Hillier and Williams (2007).

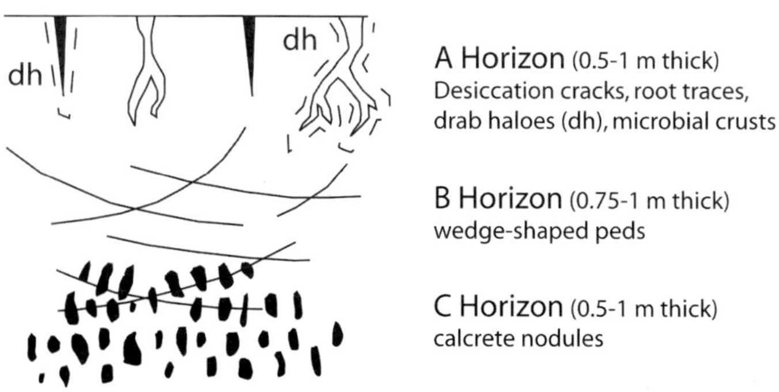

FIGURE 4.—Simplified cumulate profile of typical Old Red Sandstone Vertisol, modified after Marriott and Wright (2006).

Carbonate percentage was visually estimated whilst logging. At all four locations, the RCF is continuously exposed, though outcrop quality is variable, with structural deformation (jointing, cleavage, and folding) increasing northward with proximity to the Ritec Fault. Inland, few exposures exist (see Hillier and Williams 2007).

The study was supplemented by petrographic thin-section analysis of calcretes from Skrinkle Haven and Freshwater West. In addition, bulk and clay mineralogy of two mudstone samples from Freshwater West (45 and 92 m above the base of the RCF) were analyzed by X-ray powder diffraction by Dr. S. Hillier (Macaulay Institute, Aberdeen, UK). Three grams of each sample were wet ground in a McCrone mill and spray dried to produce random powders (Hillier 2002). X-ray powder diffraction (XRPD) patterns were recorded on a Siemens D5000 diffractometer from 2 to 75°2θ using Cobalt Kα radiation. Quantitative phase analysis (QPA) was made by a full-pattern RIR method, as described in detail in Omotoso et al., (2006).

Clay fractions of < 2 μm were obtained by timed sedimentation, prepared as oriented mounts using the filter peel transfer technique and scanned from 2 to 45° 2θ in the air-dried state after glycolation, and after heating to 300°C for one hour. Clay minerals identified were quantified using a mineral intensity factor approach based on calculated XRPD patterns. Further details can be found in Hillier (2003).

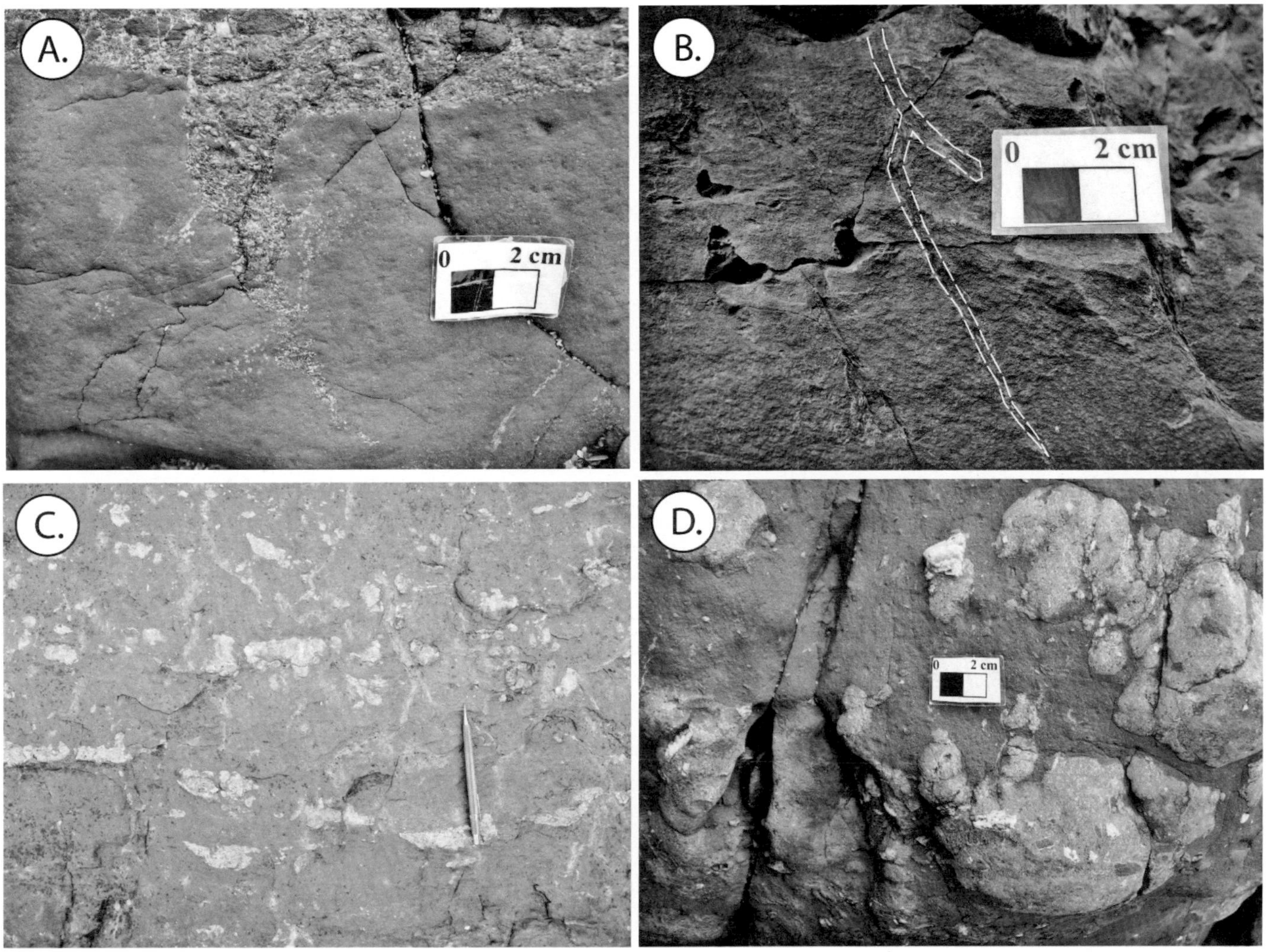

FIGURE 5.—Ridgeway Conglomerate Formation Vertisols. A) A/B horizon—sandstone-filled desiccation cracks, Freshwater West. B) A/B horizon—siltstone-filled root trace, Pennar Point. Note downward bifurcation and tapering. C) B/C—horizon with isolated calcrete nodules and drab haloes surrounding downward-bifurcating siltstone-filled root traces (West Angle Bay). D) Large micritic nodules preserved in gritty mudstone debrite bed, Freshwater West.

RIDGEWAY CONGLOMERATE FORMATION PEDOGENIC CALCRETES

Mudstone and Sandstone-Grade Lithofacies

Description: Common in mudstone-prone intervals of the RCF in both the alluvial fan and axial valley lithofacies are sandstone or siltstone-filled desiccation cracks up to 0.9 m long (Fig. 5A). Both sandstone and mudstone beds are penetrated by horizontal and vertical root traces that taper and branch along their length (Fig. 5B). Root-trace fills are heterolithic in nature, and often have a red claystone lining. Both desiccation cracks and root traces may be surrounded by low-chroma gray-green drab haloes (Fig. 5C). Mudstones and sandstones are commonly bioturbated by backfilled *Beaconites barretti* burrows.

Ubiquitous in the formation are subspherical pink, buff, or pale gray micritic carbonate nodules typically 10 to 20 mm in diameter, though reaching up to 0.3 m in diameter in proximal areas of the alluvial fan facies association (Fig. 5D). Nodules are commonly horizonated, with nodule-rich zones up to 5 m thick being observed. Nodule-bearing horizons crosscut depositional architecture elements such as inclined heterolithic bedsets, and are generally not restricted to individual beds. Nodules are observed at various stages of development (Stages II to V of Machette 1985, though most commonly stages II to III; Hillier and Williams 2007; Figs. 6A, B).

A number of nodules have vertical and horizontal geometries similar in size and morphology to the root traces previously described (Fig. 6C). We interpret these as root-cast rhizoliths (Klappa 1980). Simple columnar vertical tubules up to 10 mm wide and 150 mm long are possible calcretized burrow fills. Also commonly observed are millimeter-scale sparry calcite crystallaria comprising a boxwork fabric in mudstone units (Figs. 6A, D). In a small number of profiles at Freshwater West, micrite nodules are crosscut by later dense arrays of subhorizontal sparry fracture fills. These are typically 1 to 3 mm in diameter but may locally exceed 10 mm, and comprise en echelon, anastamosing sets.

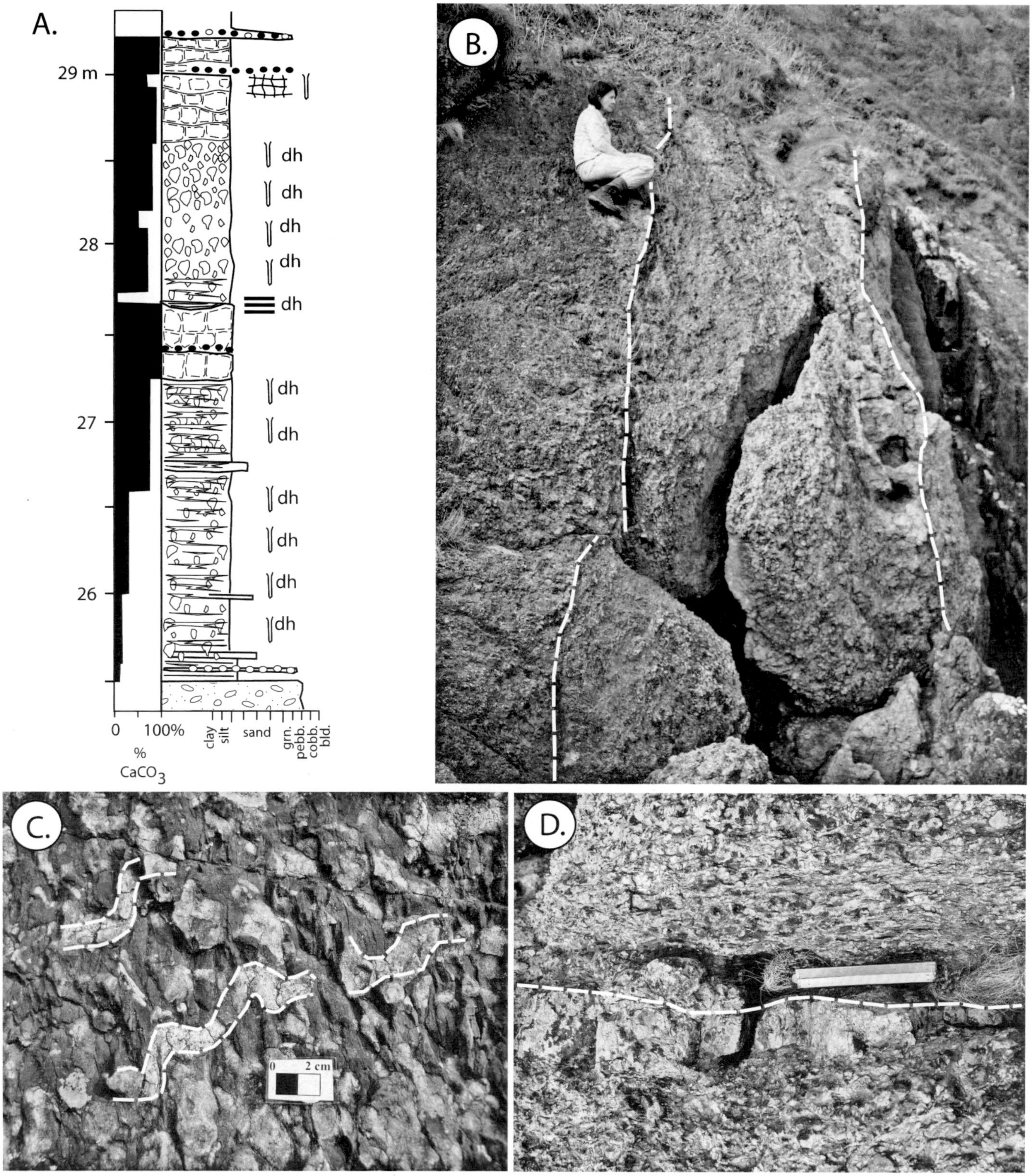

FIGURE 6.—A) Pedogenic calcrete profiles at Skrinkle Haven displaying upward-increase in carbonate development. Note absence of wedge-shaped peds (for key see Figure 10). Log height in meters above base of formation. B) Image of upper Vertisol profile, (stratigraphic top to right). Note well developed carbonate horizon with vuggy porosity that caps the profile. C) Isolated micrite nodules with calcretized, ptygmatic root trace (way-up to right). D) Upward increase in calcrete content; earlier micrite nodules overprinted by later subhorizontal micrite fracture fills. Profile topped by massive micrite zone, scale bar 0.25 m.

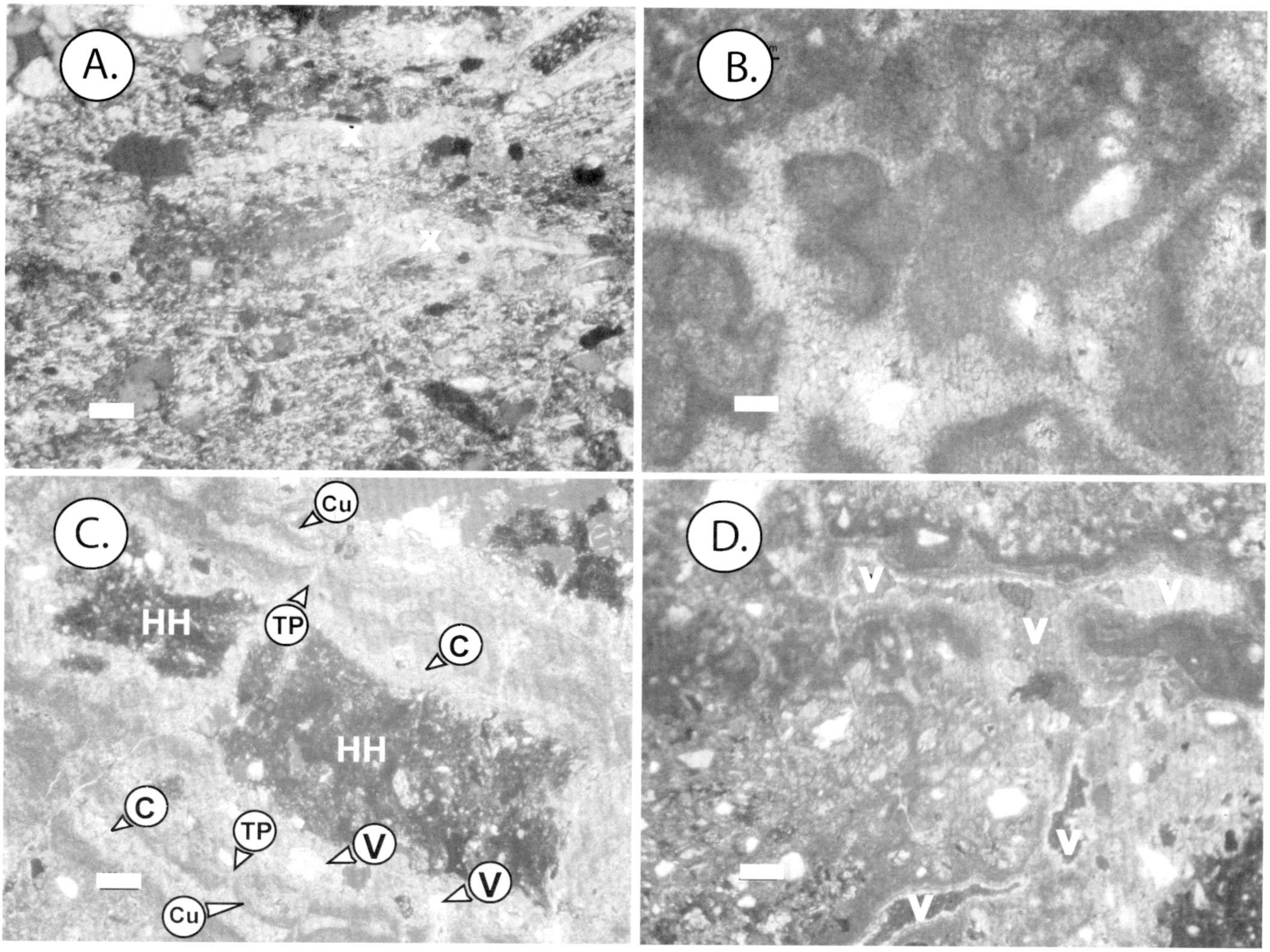

FIGURE 7.—Petrographic images of Ridgeway Conglomerate Formation calcretes, scale bar 100 microns in all images. A) Blocky soil structure with sparry calcite crystallaria (X). B) Pedogenic calcrete nodular alpha fabric with fractures or voids filled with sparry calcite. C) Laminar calcrete, layers of micrite and sparry calcite crystallaria (C), voids filled with sparry calcite (V), tepee structures (TP), and curls (Cu). Remaining patches of host sediment are indicated by HH. D) Pedogenic calcrete nodules overprinted by vadose-zone calcite cement in voids (V).

Interpretation: We interpret these deposits as calcic paleosols at various stages of development. Interpreted A horizons comprise common desiccation cracks and root traces, the latter commonly being folded, presumably by the process of differential compaction. Although the root traces may have had a vascular plant source, Hillier et al. (2008) used circumstantial evidence to suggest that their producer was the enigmatic giant fungus *Prototaxites*. Gray or green drab haloes were presumably created by pseudogleying in association with seasonal wetting and drying cycles. The common *Beaconites barretti* burrows were probably created by myriapods (Morrissey and Braddy 2004). Horizonated carbonate nodules are interpreted as the B_k or C_k horizon in a calcic paleosol. The boxwork crystallaria in the mudstones are thought to relate to the pedogenic processes that lead to development of an angular to subangular blocky ped structure in the B horizons of clay-rich soils. These structures have been shown to be crack patterns related to seasonal wetting and drying cycles that also give rise to surface desiccation cracks (Sleeman 1963; Brewer 1964). The crystallaria (Fig. 7A) formed in situ as crystal sheets by crystallization from solution in planes between the blocky peds, with the force of crystallization forcing the peds apart, accentuating the boxwork texture (Brewer 1964).

Conspicuous by their absence in the RCF are wedge-shaped peds that are common in structural B horizons, below upper horizons with desiccation cracks in other ORS paleosols. Wedge-shaped peds are curved, intersecting slickensided surfaces that are a characteristic feature of some modern soils that are rich in smectite and other swelling clays. They are considered to be caused by shear stress set up by repeated shrinking and swelling episodes due to marked seasonal differences in soil moisture that occur in climates marked by distinct wet and dry seasons, or in areas that suffer prolonged periods of waterlogging followed by prolonged drought (see review in Wilding and Tessier 1988).

Although slickensided fracture surfaces were reported by Ékes (1993) from the RCF at Skrinkle Haven and West Angle Bay, we interpret these as being tectonically generated because they are observed in close proximity to small-scale faults at these localities. A possibility is that the southerly-sourced, exotic clast provenance to the RCF brought a clay-mineral assemblage that was not conducive to shrink swell processes in the soil profile. To address this, bulk and clay mineralogy of two mudstone samples from Freshwater West (45 and 92 m above the base of the RCF) were analyzed by X-ray powder diffraction. Clay fractions from both samples are identical, being dominated by a mixture of illite and highly illitic mixed-layer illite smectite. Chlorite is also present as a dioctahedral type. Both samples are similar to mudstones from other ORS formations in the Anglo-Welsh Basin that contain common wedge-shaped peds (Hillier et al. 2006), so a provenance control on the absence of these structures in the RCF can be discounted.

Development of a structural B horizon also depends on movement (translocation) and accumulation of swelling clay minerals (Birkeland 1999) at a suitable depth. Studies by Yaalon and Kalmar (1978), on slickenside distribution in modern soils in seasonally wet and dry climates, allowed them to suggest that the optimum depth for distinctive wedge-shaped peds to occur was 1.5 to 2 m below the depth of surface cracking. Below this depth their development was prevented by overburden pressure and less variation in soil moisture. In shallow soils desiccation cracking is likely to overprint and obliterate any slickenside development. In the RCF, mudstone units are relatively thin or are associated with debris flows (Hillier and Williams 2007), so although the clay mineralogy is similar to other ORS formations, desiccation or repeated movement are likely to have obliterated any slickenside development.

However, shrink swell cycles could be represented in the RCF by other morphologies, such as the boxwork structures described above. Also, there are dense arrays of subhorizontal sparry fracture fills that superficially resemble rhizogenic calcretes developed around root mats (see Wright et al. 1995; Alonso-Zarza 1999; Alonso-Zarza et al.1998a; Rasbury et al. 2006). Petrographic analysis (Fig. 7B) reveals the micritic calcrete alpha fabric in (almost) discrete nodules surrounded by further micrite and cut by fractures filled with sparry calcite. The fractures could be related to reworking or merely represent stages in development of the calcrete nodule as voids were filled at different times as the host sediment was gradually displaced. It is possible that the subhorizontal fractures also represent shrink swell stresses set up within a structural B horizon in the paleosol profiles. This interpretation differs from the commonly observed wedge-shaped peds in other ORS formations (see above).

Pedogenic Calcrete Profiles in Mudstone and Sandstone-Grade Lithofacies

Description: Profiles developed in the RCF typically show upwardly increasing percentages of $CaCO_3$ reflecting increasing occurrence and fusion of nodules (Fig. 6A). Two profiles from Skrinkle Haven are capped by layers of dense, tabular to wavy-bedded $CaCO_3$. The beds contain vertical fractures and are rubbly in parts. Internally, blocks are massive in nature, and contain little primary host sediment (Fig. 6B). This is shown in Figure 7C, where patches of host sediment are incorporated within a laminar calcrete with fractures filled by sparry calcite as crystallaria. The tops of these densely cemented zones are marked by erosional surfaces strewn with intraformational and extraformational conglomerate lags. At the top of one profile 27.7 m above the base of the RCF (Figure 6A), a thin 15 mm thick laminar carbonate is preserved. This is micritic, and is developed within a low-relief topographic hollow eroded into the underlying massive carbonate horizon.

Interpretation: Laminar calcrete can develop by accretion on already indurated surfaces such as accumulations of coalescent calcrete nodules or bedrock or precipitation in ponds on an impermeable surface (Wright 1989; Wright and Wilson 1987). Such layers represent hard-pan formation, and are the most highly developed pedogenic calcretes in the RCF. Brecciation may reflect the displacive growth of carbonate, thermal expansion, or wetting and drying cycles (Wright and Tucker 1991). Although root traces have been observed penetrating dense calcrete horizons elsewhere in the ORS Anglo-Welsh Basin (Hillier et al. 2008), and may be a cause of brecciation (Wright and Tucker 1991), no supportive evidence was observed here. Figure 7C shows the micritic laminae from this calcrete that could be biogenic in origin (see below), with possible teepees and curls, separated by spar-filled voids and crystallaria. Although a variety of mechanisms can lead to the development of laminar calcrete including root mats (Wright et al. 1988; Wright et al. 1996; Alonzo-Zarza 1999), or lichens (Klappa 1979), their occurrence in topographic lows leads us to favor an origin of carbonate precipitation from ponded water above the hardpan surface, possibly linked to cyanobacterial activity (Wright and Wilson 1987; Verrecchia et al. 1995).

Pedogenic Calcrete in Gravel-Grade Lithofacies

Description: Pore-occluding $CaCO_3$ cements are observed in conglomerate bedsets up to 1 m thick at Skrinkle Haven and Freshwater West. Cements preferentially coat the undersides of clasts (pendant, meniscus fabrics) or completely occlude porosity (Fig. 8A, B). Petrographically the cements comprise sparry calcite.

Interpretation: Similar cements are observed on modern alluvial fans (Gile et al. 1966), and represent the calcrete pedogenic expression in coarse-grained deposits (Machette 1985). They were termed gully-bed cements by Lattman (1973) from studies of fans in southern Nevada. In the latter, $CaCO_3$ saturated runoff infiltrates surface gravels, precipitating cement around clast margins, with thickest development occurring on their undersides. Coatings tend to be thickest around largest clasts, and on the underside of limestone pebbles that may also display dissolution features on their upper surfaces (Treadwell-Steitz and McFadden 2000). No evidence to support the former statement has been found in this study, and limestone clasts, although present in the RCF, represent a minority component of clast composition (Williams 1964). The process of pore occlusion occurs rapidly on modern fans, cementing recently developed gravels in as little as one to two years (Lattman 1973).

Pendant fabrics have been inferred to resemble stalactite formation; however, Brock and Buck (2005) consider their genesis to be more complex. They believe that voids develop at the clast pendant contact and remain open despite contemporaneous carbonate precipitation. In this model, most recent cements precipitate closest to the clast margin.

Tops to cemented zones in the RCF are commonly abrupt, coinciding with the top of the conglomerate bed, whereas bases are more diffuse, an observation made in modern fan gravels from SE Spain (Alonzo-Zarza et al. 1998b). This observation may be related to the downward infiltration of $CaCO_3$ -saturated waters precipitating cement in surficial layers first. Subsequent pore occlusion and reduction in permeability would therefore limit further downward expansion of $CaCO_3$ authigenesis.

The carbonate-cemented gravels described by other workers commonly have thick horizontal laminae capping profiles; these, however, are not observed in the RCF, though they may have been deposited and subsequently eroded. Although exceptions exist (see discussion in Hillgärtner et al. 2001), pendant fabrics are generally indicative of vadose-zone pedogenesis (Dunham 1971), that is above the water table (the phreatic or groundwater-saturated pore space).

RIDGEWAY CONGLOMERATE FORMATION NON-PEDOGENIC CALCRETES

Layer-Bound, Non-Inclined Groundwater Calcrete

Description: Common in all areas of the RCF are layer-bound micritic calcretes hosted in siltstone and sandstone beds (Fig. 9A).

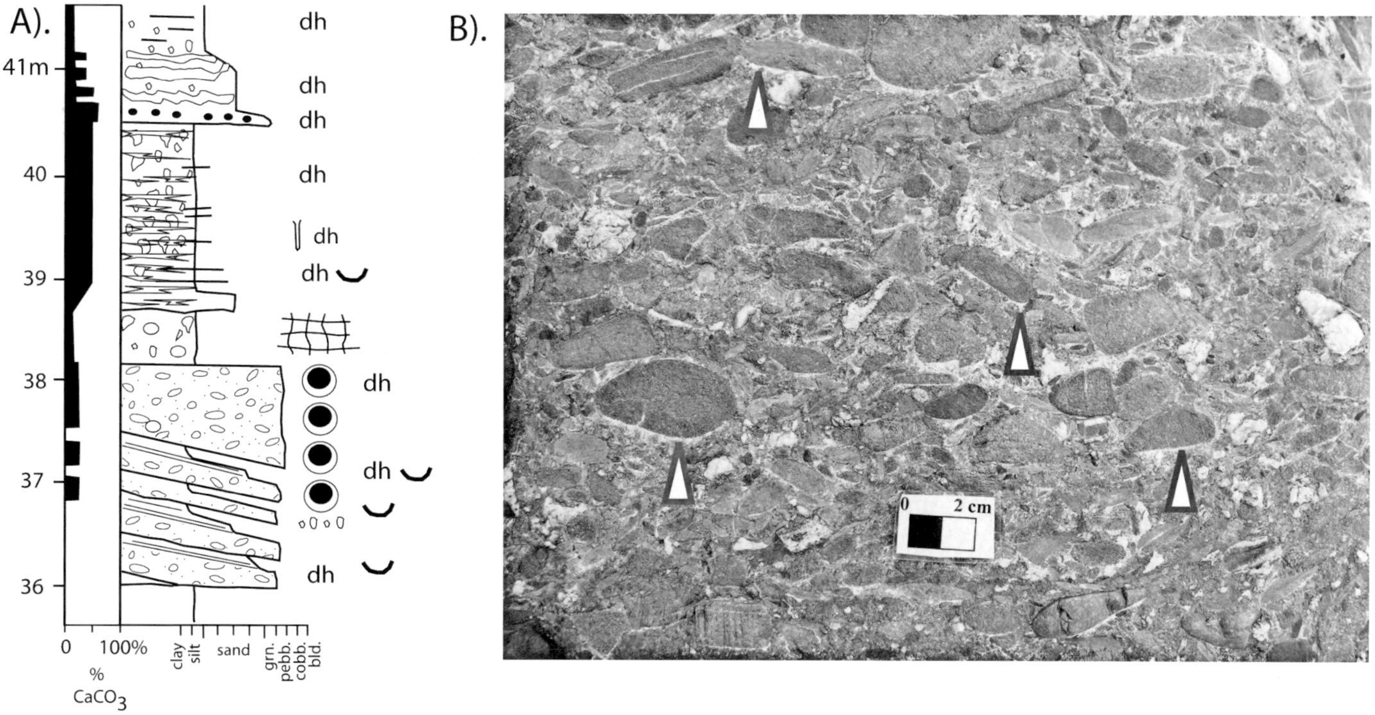

FIGURE 8.—A) Alluvial fan conglomerate bedsets with pendant pedogenic calcrete, Freshwater West (for key see Figure 10). Log height in meters above base of formation. B) Meniscate geometries to sparry pedogenic cements (arrowed), image taken at 37.75 m on Figure 8A log.

Calcrete margins are commonly diffusive across bed boundaries, but occasionally are sharp. Upper boundaries often exhibit upward-directed flares, tubules, and mammiform projections of calcrete (Figure 9B). Host rocks do not contain pedogenic features described above, although interbedded mudstones may, and some of these have been overprinted by vadose sparry calcite precipitation (Fig. 7D).

Interpretation: Similar beds were observed from the Plio-Pleistocene Palomas Basin by Mack et al. (2000), who interpret the calcrete as developing from groundwater precipitation below the water table. Upwardly directed tubular zones were possibly formed in the zone of capillary rise above the water table (Birkeland 1999). Better developed calcretes up to 1 m thick are locally preserved in proximal alluvial fan facies associations. Calcrete bed bases are well defined, whereas upper

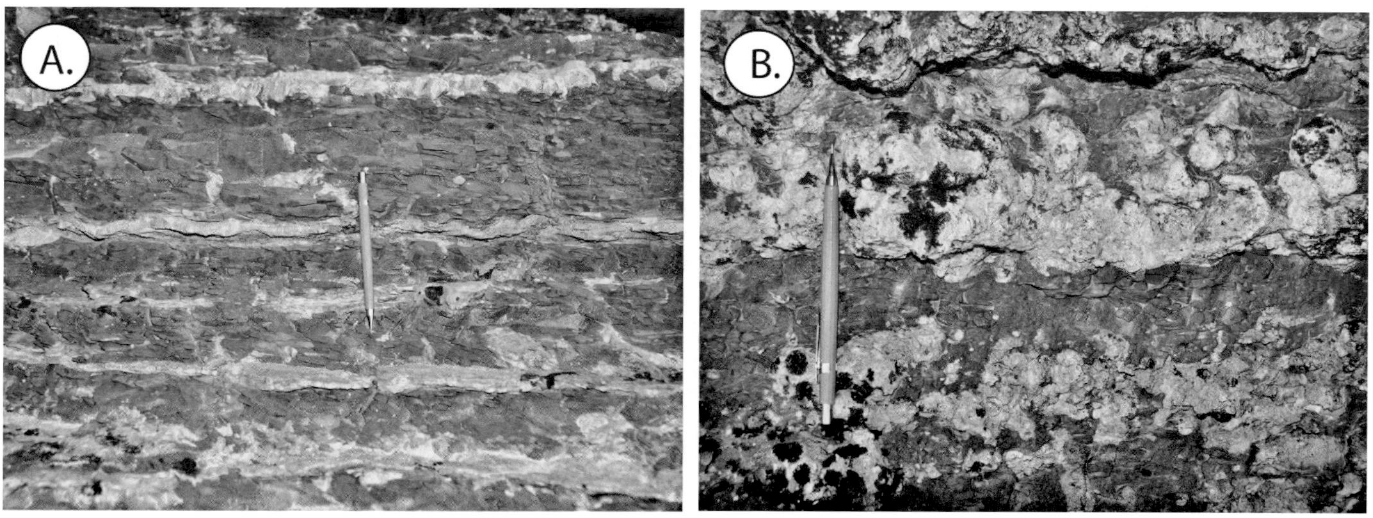

FIGURE 9.—Layer-bound, non-inclined groundwater calcrete, Skrinkle Haven. A) Thin-bedded micrite-cemented sandstone beds; note sharp base and tops to cemented zones. B) Sharp-based micritic calcrete with upward-directed projections.

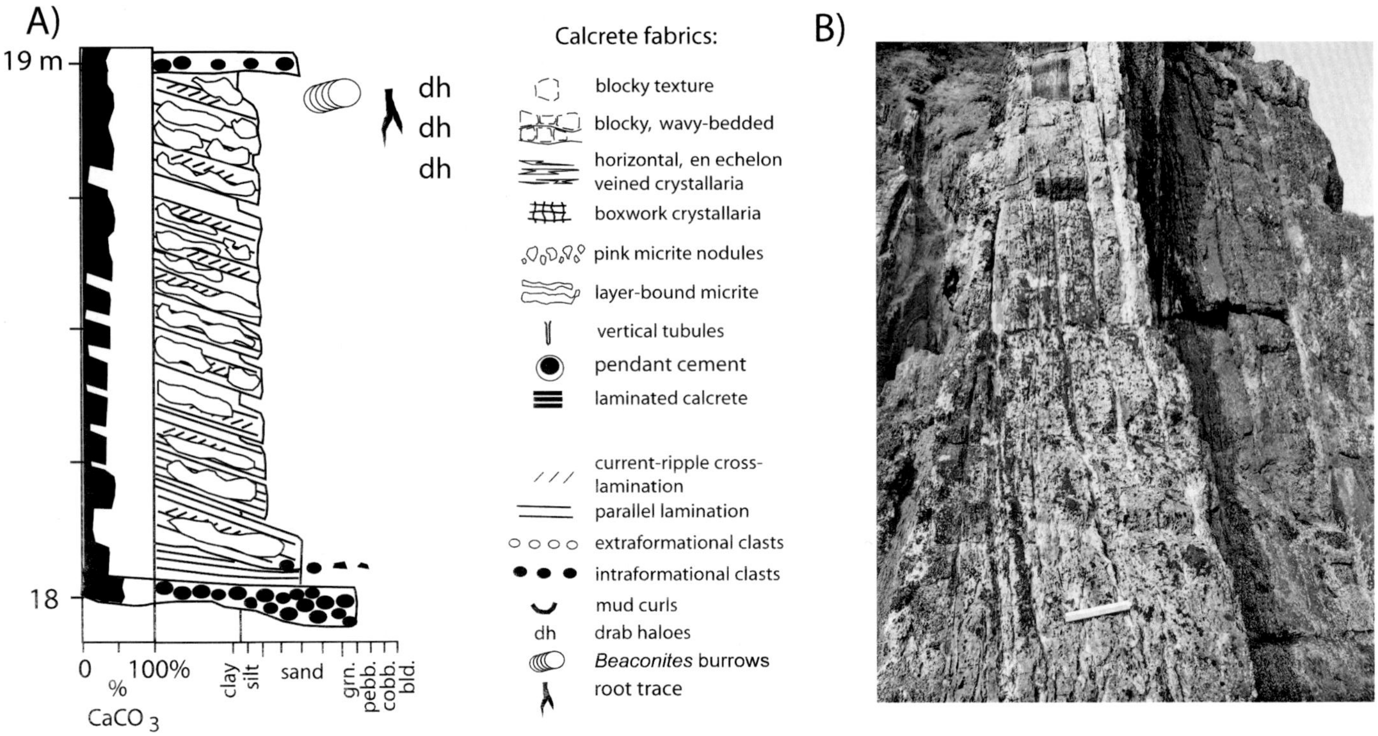

FIGURE 10.—Layer-bound inclined, groundwater calcrete, Skrinkle Haven. A) Graphic log through inclined heterolithic stratification illustrating calcrete development restricted to very fine to fine-grained sandstone bedsets; log height in meters above base of formation. Carbonate percentage visually estimated. B) Image of Inclined heterolithic stratification detailed in Part A; way-up to right, scale bar 0.25 m.

bed surfaces are more diffuse in nature, grading upward into mudstones that may contain pedogenic features including carbonate nodules and drab haloes. Internally, the thick massive beds display textures that indicate nodule amalgamation within the host sediment. Pseudo-horizontal, wavy surfaces are defined by thin, laterally discontinuous remanent host sediment layers. The carbonate fills only the original pore spaces, not forcing the host sediment grains apart as displacive pedogenic carbonate tends to do (Birkeland 1999). Beds are similar to the "thick massive beds" of Mack et al. (2000) who interpreted them as representing precipitation from laterally flowing, shallow groundwater common in modern alluvial fans (Stokes et al. 2007).

Layer-Bound, Inclined Groundwater Calcrete

Description: Bedsets of upward-fining and -thinning sandstones and mudstones are common in the RCF, with carbonate nodules in such deposits often layer-bound, and hosted predominantly in sandstone beds (Figs. 10A, B). Carbonate fabrics define elongate nodules that mirror bedset geometries. Nodules may be discrete or commonly amalgamate, with carbonate constituting up to 40% in sandstone bedsets.

Interpretation: Inclined bedsets are interpreted as deposits of lateral-accretion surfaces of ephemeral, dryland channels in both the fan and adjacent axial alluvial facies associations (inclined heterolithic stratification, IHS; Hillier and Williams 2007). Although pedogenic features are observed in the IHS bedsets (desiccation cracks, root traces, drab haloes), we interpret the layer-bound nature of the calcretes as reflecting precipitation from groundwater in more permeable bedsets. These calcretes differ from those generated by pedogenic processes in that the latter are horizonated, and cross cut depositional architectural elements such as IHS bedsets.

Lake-Margin Calcrete

Description: Common at West Angle Bay are thin-bedded heterolithic bedsets comprising finely laminated and interbedded sandstone, siltstone and claystone (Hillier and Williams 2007). Individual siltstone and sandstone beds exhibit planar laminated and current-ripple cross-lamination. Small-scale flame structures and desiccation cracks are common, as are wrinkle-marked bedding surfaces. Calcretes here are less well developed than in other parts of the RCF, but comprise layer-bound micrite up to 15 mm in thickness, with well developed laminated fabrics. The layers develop polygonal-like patterns with positive overfolded margins (Fig. 11A, B).

Interpretation: Deposition is believed to have taken place in small-scale, ephemeral lakes or "billabongs" in the axial alluvial facies association (Hillier and Williams 2007). The laminated calcretes are interpreted as representing calcretized matgrounds that developed at the margins or bottoms of the ephemeral lakes. Their morphology is similar to alpha peetee structures created by flotation of loosely attached microbial mats that become folded when subjected to current-induced shear stress, gas upbuilding, or desiccation (Gerdes et al, 1993; Behr 2002; Noffke et al. 2003). An association with wrinkle marks supports a microbial interpretation. Isolated carbonate nodules, rhizoliths, and drab haloes also occur in this association, implying that pedogenic processes overprinted lake-margin calcrete formation following desiccation of the lake system.

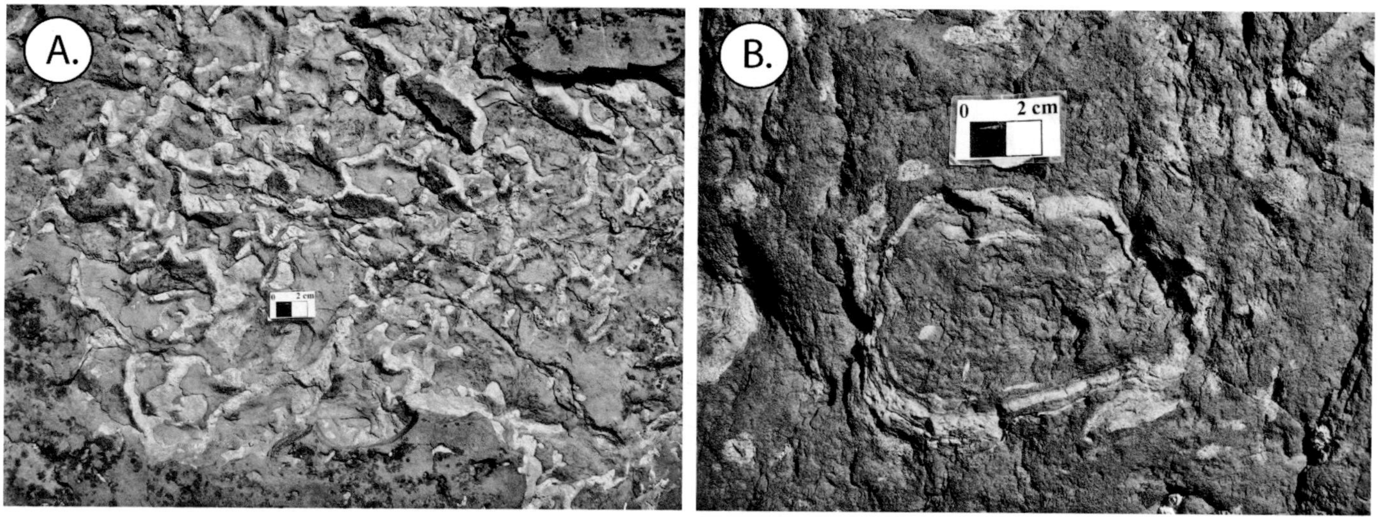

FIGURE 11.—Non-pedogenic lake-margin calcrete association, loose blocks at West Angle Bay. A) Micritic, layer-bound calcrete with upturned, polygonal margins; possible micritized matground. B) Finely laminated micrite representing possible calcretized matground.

DISCUSSION

Patterns of Calcrete Development in the Ridgeway Conglomerate Formation

Proximal areas: The most proximal areas preserved in the RCF alluvial fan facies association crop out at Freshwater West. Here conglomerate bedsets constitute the highest percentage of lithofacies, and contain litharenite clasts with the largest maximum pebble diameter (Hillier and Williams 2007). Also present are numerous mudstones with outsized floating clasts interpreted as debrites. Interbeds of conglomerates and sandstones interpreted as streamflow deposits led to the interpretation that the Freshwater West section was deposited in a mid-fan setting of mixed-process alluvial fan (Hillier and Williams 2007). Calcretes at Freshwater West exhibit some of the greatest degree of development in the RCF and indeed the Lower ORS of the Anglo-Welsh Basin (Fig. 12). Pedogenic horizons in mudstone and sandstone-grade lithofacies contain well preserved A to C horizons, the latter up to Stage V of Machette (1985). Also common are well-developed horizons of subhorizontal micritic fracture fills. Conglomerate bedsets in proximal areas also have well developed pendant pedogenic fabrics.

The well-developed nature of the pedogenic horizons here may have a number of causes. Firstly, the section thickness (115 m thick) is a quarter of that preserved in distal, axial fluvial reaches (Hillier and Williams 2007). It is tempting to infer that the hanging-wall fan zone was accommodation limited, with greatest development of accommodation space occurring in the axial fluvial valley zone proximal to the Ritec Fault. If this hypothesis is correct, proximal hanging-wall areas would have had increased soil residence times that would have facilitated the formation of better developed pedogenic calcrete (Fig. 12). Development of pedogenic calcrete would also be greatest if the fan was entrenched in its most proximal areas. Channel entrenchment would have greatly controlled the areas of the fan surface where active sedimentation would take place (McCraw 1968; Wright and Alonso-Zarza 1990). Areas of high sedimentation rates would have low soil residence times and therefore poorly developed pedogenic calcrete; those areas with low sedimentation rates such as terraces marginal to active channels would facilitate higher levels of pedogenic calcrete development in as much as they would have increased soil residence times (the pedofacies concept of Bown and Kraus 1987; Wright and Alonso-Zarza 1990).

There is no direct evidence at Freshwater West that channel entrenchment took place, though this may be due to its interpreted "mid" fan location (Hillier and Williams 2007). Strong calcrete development could also occur over the entire fan surface if sediment supply was reduced through tectonic and or climatic forcing (Stokes et al. 2007). The absence at present of any correlative horizons in the RCF to confirm temporal and spatial dimensions to pedogenic calcretes makes this impossible to ascertain.

Layer-bound groundwater calcrete in non-inclined heterolithic bedsets are also common (Fig. 12). Proximal areas of the fan would have created a focus for groundwater funnelling through permeable gravels, as documented in Quaternary deposits of Almeria, SE Spain (Stokes et al. 2007). In the Palomas Basin, the hanging-wall catchment area is an order of magnitude greater than that of the footwall, with greater relief. This possibly explains the restriction of thick, massive groundwater calcretes to hanging-wall areas (Mack et al. 2000).

Of note at Freshwater West is the lack of reworked calcrete nodule clasts in conglomerate beds despite the common occurrence of calcretes in the section. The conglomerates are dominated by exotic extraformational lithologies, and mudstone rip-up clasts. This must provide information about the provenance region that the system drained, which is presumed to have had high relief, and we can infer

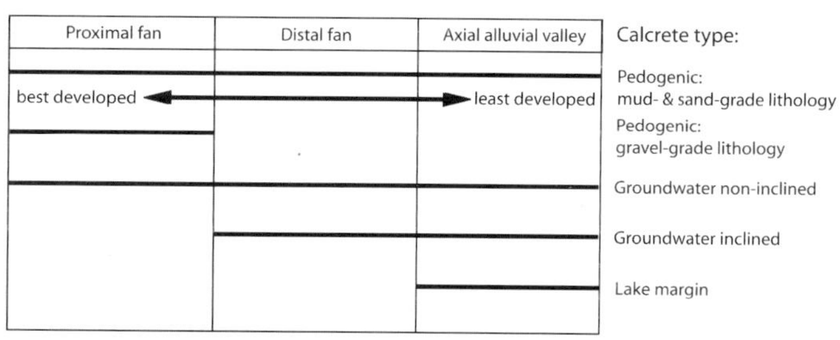

FIGURE 12.—Schematic summary of development of pedogenic and non-pedogenic calcrete in the Ridgeway Conglomerate Formation fan and axial-valley zones.

TABLE 1.—Summary of criteria to distinguish between various calcrete origins in the Ridgeway Conglomerate Formation.

Pedogenic—mud to sand-grade lithology	Pedogenic—gravel-grade lithology	Groundwater	Lake margin
Presence of soil profile and usually A horizons above the calcrete horizon. Diffuse contacts.	Layer-bound occurrence with no evidence of upper soil horizons.	Layer-bound with no evidence of upper soil horizons.	Layer-bound with no evidence of upper soil horizons.
Micritic. needle-fiber, sparry crystallaria, large rhombic crystals.	Microspar cement.	Micritic, sparry and cement Microspar cement	Micrite
Displacive growth of nodules	Pore-filling; non-displacive.	Pore-filling; non-displacive	Pore-filling non-displacive
Dense microfabrics, often associated with biogenic features (rhizoconcretions needle-fiber calcite, burrows)	No biogenic association	No biogenic association	Associated with root traces, matgrounds, burrows and rhizoconcretions.
Host sediment can have a range of grain sizes	Only in conglomerate	Unlikely in fine-grained sediment	Present in wide range of grain sizes (claystone through to coarse-grained sandstone).

either that soils here were thin or absent, or did not contain calcrete horizons.

Distal Axial-Valley Area: Pedogenic calcretes in the axial alluvial valley complex are less developed than those in proximal fan areas, typically of Stage I to II of Machette (1985). Presumably this is linked to the increased accommodation-space development, sedimentation rate, and corresponding reduction in soil residence time in the axial valley as discussed above. It could also relate to the finer grain size of the parent sediment in the axial valley compared with the fan, since calcrete develops more rapidly in gravel-rich (> 50% gravel) substrates (Gile et al. 1966; Machette 1985; Wright and Tucker 1991).

Non-pedogenic, layer-bound inclined and non-inclined groundwater calcretes are common (Fig. 12). This zone may have had an elevated groundwater level to facilitate development of groundwater calcrete, although other pedogenic features one might expect from this (e.g., gleyed or mottled horizons) are not observed. It is possible that the groundwater levels fluctuated greatly in this zone as a result of fan-toe discharge variations, and that formation of groundwater calcrete was rapid, developing following raised water-table levels. An alternative is that the groundwater calcretes were formed at depth in aquifers beneath the zone of groundwater fluctuation and below the solum (zone of soil formation). This is because permanent saturation prevents transfer processes between horizons due to lack of water movement (Ellis and Mellor 1995). This scenario may explain the absence of pedogenic features indicative of prolonged periods of high water tables (e.g., groundwater gleys, gleysols). The occurrence of lake-margin calcretes is seen only in the axial-valley zone (Fig. 12). Hillier and Williams (2007) hypothesized that the lakes were sourced by fan-toe springs and/or fluvial discharge events.

The Pedofacies Concept: In calcic paleosols, the pedofacies concept (Bown and Kraus 1987; Kraus 1987; Kraus and Bown 1993) can be used to predict proximity to channel sandbodies in the rock record. The predictive model relies on the recognition of increasing or decreasing paleosol maturity patterns (or stages of paleosol development) both horizontally and vertically. Areas with low net sedimentation rates and hence increased soil maturity may be expected in areas distal to alluvial channel fairways. In the rock record, therefore, morphological changes in paleosol development may aid the prediction of proximity to the alluvial-channel zone. Extreme care must be undertaken to ensure that modelling is firmly based on the presence and degree of development of pedogenic calcrete and excludes non-pedogenic calcrete. This then clearly links soil development to residence time associated with likely sedimentation rates related to distance from the channel bank and supports the conclusions of Pimentel et al. (1996) based on their analysis of Paleogene calcretes from Portugal.

As discussed above, no unified model to distinguish pedogenic from non-pedogenic calcretes currently exists in field-based studies. Guidelines for discrimination are, however, provided by Mack et al. (2000). A summary of features detailing our observations from the RCF is given in Table 1; we recommend that the features are not used in isolation to interpret calcrete origin and suggest that the inferred sedimentological context, and any patterns to calcrete development and morphology, be taken into account in the study area. The presence or absence of inferred soil profiles or horizonation is crucial in the interpretation of pedogenic versus non-pedogenic origins. This may be difficult to ascertain in the rock record because surficial soil layers may be removed by erosion, and later pedogenic processes may overprint pre-existing calcretes of any origin. Layer-bound, or lithology-bound calcretes with no upper soil profiles may point to non-pedogenic origins.

The problem of correct diagnosis is profound in subsurface studies that may rely on limited datasets, using percentage $CaCO_3$ identified from petrophysical analysis (e.g., Williams et al. 1993; Hanneman et al. 1994). Where core material exists it may be a useful to adopt a mindset of doubt in interpreting formative development processes. Clearly not all calcretes are pedogenic in origin, and as such it would be always appropriate to challenge the notion that preserved calcretes are pedogenic in nature, and to establish objective criteria (e.g., context, grain size, horizonation, morphology, micromorphology, petrography) in each case to try to distinguish processes of calcrete development. In addition, paleosols preserved in the rock record commonly overprint earlier profiles, or are truncated by erosion surfaces. As such the gradualistic model of Bown and Kraus (1987) may not always be appropriate to apply in these circumstances.

CONCLUSIONS

1. Calcretes of both pedogenic and non-pedogenic origin are present in the RCF.
2. Pedogenic calcretes are common in the formation. In mudstone and sandstone-grade lithofacies, A and B horizons where preserved contain desiccation cracks, root traces (some of which are calcretized), microbial crusts, *Beaconites* burrows, and drab haloes. C horizon calcrete nodules at varying stages of development are observed. Early formed, pink micrite nodules are often overprinted by subhorizontal sparry fracture fills. Profiles exhibit an upward increase in carbonate content, and may be capped by dense, blocky carbonate layers and laminated calcrete, the latter interpreted as forming in small ponds in association with cyanobacteria.
3. In proximal alluvial fan conglomerates, sparry, pendant calcrete fabrics occlude porosity and are interpreted as the pedogenic expression in coarse-grained host lithologies.
4. Paleosols in the RCF do not contain wedge-shaped peds commonly observed in other formations of the Anglo-Welsh Basin ORS, though the structural B horizon could be reflected in smaller structures such as boxwork patterns and microfractures.
5. Layer-bound groundwater micritic calcretes are common in both inclined and non-inclined bedsets. Cementation is restricted to sandstones and coarse siltstones, and margins to cemented intervals are commonly abrupt.
6. Lake-margin calcretes comprise thin-bedded laminated micrites with polygonal upturned margins interpreted as calcretized matgrounds.
7. The recognition of non-pedogenic calcretes in the RCF opens the possibility that other calcretes previously interpreted as being pedogenic in origin in the ORS may also have originated by non-pedogenic processes.
8. We advocate caution in the use of calcretes as indicators of paleoclimate, landscape evolution and in models to predict the location of alluvial sandbodies. In each case it would be beneficial to generate objective criteria (e.g., context, grainsize, horizonation, morphology, petrography) to distinguish between different processes of calcrete development.

ACKNOWLEDGMENTS

This paper has benefited from discussion with Prof V.P. Wright. We sincerely thank Dr. S. Hillier for geochemical analysis undertaken at the Macaulay Institute, and also acknowledge V. Davies and S. Howells for fieldwork assistance. We thank W. J. Autin, J.B. Southard, and an anonymous referee for thorough, constructive comments.

REFERENCES

ALLEN, J.R.L., 1973, Compressional structures (patterned ground) in Devonian pedogenic limestones: Nature, v. 243, p. 84–86.

ALLEN, J.R.L., 1974, The Devonian rocks of Wales and the Welsh Borderland, *in* Owen, T.R., ed., The Upper Palaeozoic and Post Palaeozoic rocks of Wales: Cardiff, University of Wales Press, p. 47–84.

ALLEN, J.R.L., 1986, Pedogenic calcretes in the Old Red Sandstone facies (late Silurian—early Carboniferous) of the Anglo-Welsh area, southern Britain, *in* Wright, V.P., ed., Palaeosols: Their Recognition and Interpretation: Oxford, U.K., Blackwell, p. 56–86.

ALLEN, J.R.L., AND CROWLEY, S.J., 1983, Lower Old Red Sandstone fluvial dispersal systems in the British Isles: Royal Society of Edinburgh, Transactions, Earth Sciences, v. 74, p. 61–68.

ALLEN, J.R.L., AND WILLIAMS, B.P.J., 1979, Interfluvial drainage on Siluro-Devonian alluvial plains in Wales and the Welsh Borders: Geological Society of London, Journal, v. 136, p. 361–366.

ALLEN, J.R.L., AND WILLIAMS, B.P.J., 1981, Sedimentology and stratigraphy of the Townsend Tuff Bed (Lower Old Red Sandstone) in South Wales and the Welsh Borders: Geological Society of London, Journal, v. 138, p. 15–29.

ALLEN, J.R.L., AND WILLIAMS, B.P.J., 1982, The architecture of an alluvial suite: rocks between the Townsend Tuff and Pickard Bay Tuff Beds (early Devonian), southwest Wales: Royal Society, London, Philosophical Transactions, v. B 287, p. 51–89.

ALLEN, J.R.L., THOMAS, R.G., AND WILLIAMS, B.P.J., 1982, The Old Red Sandstone north of Milford Haven, *in* Bassett, M.G., ed., Geological Excursions in Dyfed, Southwest Wales: National Museum of Wales, Cardiff, p. 123–149.

ALONSO-ZARZA, A.M., 1999, Initial stages of laminar calcrete formation by roots: examples from the Neogene of central Spain: Sedimentary Geology, v. 126, p. 177–191.

ALONSO-ZARZA, A.M., 2003, Palaeoenvironmental significance of palustrine carbonates and calcretes in the geological record: Earth-Science Reviews, v. 60, p. 261–298.

ALONSO-ZARZA, A.M., SANZ, M.E., CALVO, J.P, AND ESTEVEZ, P., 1998a, Calcified root cells in Miocene pedogenic carbonates of the Madrid Basin: evidence for the origin of Microcodium b: Sedimentary Geology, v. 116, p. 81–97.

ALONSO-ZARZA, A.M., SILVA, P.G., GOY, J.L., AND ZAZO, C., 1998b, Fan-surface dynamics and biogenic calcrete development: interactions during ultimate phases of fan evolution in the semiarid SE Spain (Murcia): Geomorphology, v. 24, p. 147–167.

BARCLAY, W.J., RATHBONE, P.A., WHITE D.E., AND RICHARDSON, J.B., 1994, Brackish water faunas from the St. Maughans Formation: the Old Red Sandstone section at Ammons Hill, Hereford and Worcester, UK re-examined: Geological Journal, v. 29, p. 369–379.

BARCLAY, W.J., BROWNE, M.A.E., MCMILLAN, A.A., PICKETT, E.A., STONE, P., AND WILBY, P.R., 2005, The Old Red Sandstone of Great Britain: Peterborough, Geological Conservation Review Series, No. 31, Joint Nature Conservation Committee.

BEHR, H-J., 2002, Magadiite and Magadi Chert: a critical analysis of the silica sediments in the Lake Magadi Basin, Kenya, *in* Renault, R.W., and Ashley, G.M., eds., Sedimentation in Continental Rifts: SEPM Special Publication 73, p. 257–273.

BIRKELAND, P.W., 1999, Soils and Geomorphology: New York, Oxford University Press, 448 p.

BOWN, T.M., AND KRAUS, M.J., 1987, Integration of channel and floodplain suites in aggrading alluvial systems. I. Developmental sequences and lateral relations of Lower Eocene Alluvial paleosols, Willwood Formation, Bighorn Basin, Wyoming: Journal of Sedimentary Petrology, v. 57, p. 587–601.

BREWER, R., 1964, Fabric and Mineral Analysis of Soils: Sydney, John Wiley and Sons, 470 p.

BROCK, A.L., AND BUCK, B.J., 2005, A new formation process for calcic pendants from Pahranagat Valley, Nevada, USA, and implication for dating Quaternary landforms: Quaternary Research, v. 63, p. 359–367.

CARLISLE, D., 1983, Concentration of uranium and vanadium in calcretes and gypcretes, *in* R.C.L. Wilson, ed., Residual Deposits: Geological Society of London, Special Publication 11, p. 185–195.

CHANNEL, J.E.T., MCCABE, C., AND WOODCOCK, N.H., 1992, An Early Devonian (pre-Acadian) magnetization component recorded in the Lower Old Red Sandstone of South Wales (UK): Geophysical Journal International, v. 108, p. 883–894.

COPE, J.C.W., AND BASSETT, M.G., 1987, Sediment sources and Palaeozoic history of the Bristol Channel area: Proceedings of the Geologists' Association, v. 98, p. 315–330.

DEWEY, J.F., AND STRACHAN, R.A., 2003, Changing Silurian-Devonian relative plate motion in the Caledonides: sinistral transpression to sinistral transtension: Geological Society of London, Journal, v. 160, p. 219–229.

DINELEY, D.L., 1999a, Late Silurian fossil fishes sites of the Welsh Borders, *in* Dineley, D.L., and Metcalfe, S.J., eds., Fossil Fishes of Great Britain: Nature Conservancy Council, Geological Conservation Review 16, p. 65–106.

DINELEY, D.L. 1999b, Early Devonian fossil fishes sites of the Welsh Borders, *in*

Dineley, D.L., and Metcalfe, S.J., eds., Fossil Fishes of Great Britain: Nature Conservancy Council, Geological Conservation Review 16, p.109–144.

Dixon, E.E.L., 1921, Geology of the South Wales Coalfield, Part XIII. The country around Pembroke and Tenby: Memoir of the Geological Survey, UK.

Dixon, E.E.L., 1933a, Some recent stratigraphic work and its bearing on south Pembrokeshire problems: Proceedings of the Geologists' Association, v. 44, p. 217–225.

Dixon, E.E.L., 1933b, Notes on the geological succession in south Pembrokeshire: Proceedings of the Geologists' Association, v. 44, p. 402–411.

Dunham, R.J., 1971, Meniscus cement, *in* Bricker, O.P., ed., Carbonate Cements: Baltimore, John Hopkins University Press, Studies in Geology 19, p. 297–300.

Edwards, D., and Richardson, J.B., 2004, Silurian and Lower Devonian plant assemblages from the Anglo-Welsh Basin: a palaeobotanical and palynological synthesis: Geological Journal, v. 39, p. 375–402.

Ékes, C., 1993, Bedload transported pedogenic mud aggregates in the Lower Old Red Sandstone in southwest Wales: Geological Society of London, Journal, v. 150, p. 469–472.

Ellis, S., and Mellor, A., 1995, Soils and Environment: London, Routledge. 364 p.

Friend, P.F., Williams, B.P.J., Ford, M., and Williams, E.A., 2000, Kinematics and dynamics of Old Red Sandstone basins, *in* Friend, P.F. and Williams, B.P.J., eds., New Perspectives on the Old Red Sandstone: Geological Society of London, Special Publication 180, p. 29–60.

Gerdes, G., Claes, M., Dunajtschik-Piewak, K., Riege, H., Krumbein, W.E., and Reineck, H.-E., 1993, Contribution of microbial mats to sedimentary surface structures: Facies, v. 29, p. 61–74.

Gile, L.H., 1975, Holocene soils and soil-geomorphic relations in an arid region of southern New Mexico: Quaternary Research, v. 5, p. 321–360.

Gile, L.H., 1977, Holocene soils and soil-geomorphic relations in a semi-arid region of southern New Mexico: Quaternary Research, v. 7, p. 112–132.

Gile, L.H., Peterson, F.F., and Grossman, R.B., 1966, Morphological and genetic sequences of carbonate accumulations in desert soils: Soil Science, v. 100, 347–360.

Goudie, A.S., 1973, Duricrusts in Tropical and Subtropical Landscapes, Oxford, U.K., Clarendon Press, 192 p.

Hanneman, D.L., Wideman, C.J., and Halvorson, J.W., 1994, Calcic paleosols, their use in subsurface stratigraphy: American Association of Petroleum Geologists, Bulletin, v. 78, p. 1360–1371.

Hillgärtner, H., Dupraz, C., and Hug, W. 2001, Microbially induced cementation of carbonate sands: are micrite meniscus cements good indicators of vadose diagenesis?: Sedimentology, v. 48, p. 117–131.

Hillier, R.D. 2000, Silurian marginal marine sedimentation and the anatomy of the marine-Old Red Sandstone transition in Pembrokeshire, SW Wales, *in* Friend, P.F., and Williams, B.P.J., eds., New Perspectives on the Old Red Sandstone: Geological Society of London, Special Publication 180, p 343–354.

Hillier, R.D., and Williams, B.P.J., 2004, Sedimentation and tectonics: the marine Silurian-basal Lower Old Red Sandstone transition in southwest Wales: Geological Journal, v. 39, p. 237–256.

Hillier, R.D., and Williams, B.P.J., 2006, The alluvial Old Red Sandstone basins, *in* Brenchley, P.J., and Rawson, P.F., eds., The Geology of England and Wales, Second Edition: Bath, Geological Society of London, p. 155–172.

Hillier, R.D., and Williams, B.P.J., 2007, The Ridgeway Conglomerate Formation of SW Wales, and its implications. The end of the Lower Old Red Sandstone?: Geological Journal, v. 42, p. 55–83.

Hillier, R.D., Marriott, S.B., Wright, V.P., and Williams, B.P.J., 2007, Possible climate variability the Lower Old Red Sandstone Conigar Pit Sandstone Member (early Devonian), south Wales. UK: Sedimentary Geology, v. 202, p.35–57.

Hillier, R.D., Edwards, D., and Morrissey, L.B., 2008, Sedimentological evidence of rooting structures in the Lower Old Red Sandstone, Anglo-Welsh Basin: Palaeogeography, Palaeoclimatology, Palaeoecology, v. 270, p. 366–380.

Hillier, S., 2002, Spray drying for X-ray powder diffraction specimen preparation: International Union of Crystallography, Commission on Powder Diffraction Newsletter No 27, p. 7–9.

Hillier, S., 2003, Quantitative analysis of clay and other minerals in sandstones by X-ray powder diffraction (XRPD), *in* Worden, R., ed., Clay Mineral Cements in Sandstones: International Association of Sedimentologists, Special Publication 34, p. 213–251.

Hillier, S., Wilson, M.J., and Merriman, R.J., 2006, Clay mineralogy of the Old Red Sandstone and Devonian sedimentary rocks of Wales, Scotland and England: Clay Minerals, v. 41, p. 433–471.

Jutras, P., Macrae, R.A., and Utting, J., 2007, Visean tectonostratigraphy and basin architecture beneath the Pennsylvanian New Brunswick platform of eastern Canada: Bulletin of Canadian Petroleum Geology, v. 55, p. 217–236.

Khalaf, F.I., 2007, Occurrences and genesis of calcretes in the Mio-Pleistocene fluviatile sequence in Kuwait, northeastern Arabian Peninsula: Sedimentary Geology, v. 199, p. 129–139.

Klappa, C.F., 1979, Lichen stromatolites: criterion for subaerial exposure and a mechanism for the formation of laminar calcrete (caliche): Journal of Sedimentary Petrology, v. 49, p. 387–400.

Klappa, C.F., 1980, Rhizoliths in terrestrial carbonates: classification, recognition, genesis and significance: Sedimentology, v. 27, p. 613–629.

Kraus, M.J., 1987, Integration of channel and floodplain suites, II: vertical relations of alluvial paleosols: Journal of Sedimentary Petrology, v. 57, p.602–612.

Kraus, M.J., 1999, Paleosols in clastic sedimentary rocks: their geological applications: Earth-Science Reviews, v. 47, p. 41–70.

Kraus, M.J., and Bown, T.M., 1993, Palaeosols and sandbody prediction in alluvial sequences, *in* North, C.P., and Prosser, D.J., eds., Characterisation of Fluvial and Aeolian Reservoirs: Geological Society of London, Special Publication 73, p. 23–31.

Lamplugh, G.W., 1902, Calcrete: Geological Magazine, v. 9, p. 75.

Lattman, L.H., 1973, Calcium carbonate cementation of alluvial fans in southern Nevada: Geological Society of America, Bulletin, v. 84, p. 3013–3028.

Love, S.E., and Williams, B.P.J., 2000, Sedimentology, cyclicity and floodplain architecture in the Lower Old Red Sandstone of SW Wales, *in* Friend, P.F., and Williams, B.P.J., eds., New Perspectives on the Old Red Sandstone: Geological Society of London, Special Publication 180, p. 371–388.

McCraw, J.D., 1968, The soil pattern of some New Zealand alluvial fans: 9th International Congress on Soil Science, Transactions, v. 4, p. 631–640.

Machette, M.N., 1985, Calcic soils of the southwestern United States, *in* Weide, D.L., ed., Soils and Quaternary Geology of the South West United States: Geological Society of America, Special Paper 203, p. 1–21.

Mack, G.H., Cole, D.R., and Trevino, L., 2000, The distribution and descrimination of shallow, authigenic carbonate in the Pliocene-Pleistocene Palomas Basin, southern Rio Grande Rift: Geological Society of America, Bulletin, v. 112, p. 643–656.

Marriott, S.B., and Wright, V.P., 1993, Palaeosols as indicators of geomorphic stability in two Old Red Sandstone alluvial suites, south Wales: of Geological Society of London, Journal, v. 150, p. 1109–1120.

Marriott, S.B., and Wright, V.P., 1996, Sediment recycling on Siluro-Devonian floodplains: Geological Society of London, Journal, v. 153, p. 661–664.

Marriott, S.B., and Wright, V.P., 2004, Mudrock deposition in an ancient dryland system: Moor Cliffs Formation, Lower Old Red Sandstone, southwest Wales, UK: Geological Journal, v. 39, p. 277–298.

Marriott, S.B., and Wright, V.P., 2006, Investigating paleosol completeness and preservation in mid-Paleozoic alluvial paleosols: a case study in paleosol taphonomy from the Lower Old Red Sandstone, *in* Alonso-Zarza, A.M., and Tanner, L.H., eds., Paleoevnironmental Record and Applications of Calcretes and Palustrine Carbonates: Geological Society of America, Special Paper 416, p. 43–52.

Morrissey, L.B., and Braddy, S.J., 2004, Terrestrial trace fossils from the Lower Old Red Sandstone of southwest Wales: Geological Journal, v. 39, p. 315–336.

Noffke, N., Gerdes, G., and Klenke, T., 2003, Benthic cyanobacteria and their influence on the sedimentary dynamics of peritidal depositional systems (siliciclastic, evaporitic salty, and evaporitic carbonatic): Earth-Science Reviews, v. 62, p. 163–176.

Omotoso, O., McCarty, D.K., Hillier, S., and Kleeberg, R., 2006, Some successful approaches to quantitative mineral analysis as revealed by the 3rd Reynolds Cup contest: Clay and Clay Minerals, v. 54, p. 748–760.

Pimentel, N.L., Wright, V.P., and Azevedo, T.M., 1996, Distinguishing early groundwater alteration effects from pedogenesis in ancient alluvial basins: examples from the Palaeogene of southern Portugal: Sedimentary Geology, v. 105, p.1–10.

Retallack, G.J., 1986, The fossil record of soils, *in* Wright, V.P., ed., Palaeosols: their Recognition and Interpretation: Oxford, U.K., Blackwell, p. 1–57.

Retallack, G., 2005, Pedogenic carbonate proxies for amount and seasonality of precipitation in paleosols: Geology, v. 33, p. 333–336.

Retallack, G., Wynn, J.G., and Fremd, T.J., 2004, Glacial-interglacial-scale paleoclimatic change without large ice sheets in the Oligocene of central Oregon: Geology, v. 32, p. 297–300.

Rasbury, E.T., Gierlowski-Kordesch, E.H., Cole, J.M., Sookdeo, C., Spartaro, G., and Nienstedt, J., 2006, Calcite cement stratigraphy of a nonpedogenic calcrete in the Triassic New Haven Arkose (Newark Supergroup) *in* Alonso-Zarza, A.M., and Tanner, L.H., eds., Paleoenvironmental Record and Applications of Calcretes and Palustrine Carbonates: Geological Society of America, Special Paper 416, p. 203–222.

Simon, J.B., and Bluck, B.J., 1982, Palaeodrainage of the southern margin of the Caledonide mountain chain in the northern British Isles: Royal Society of Edinburgh, Transactions, Earth Sciences, v. 73, p. 11–15.

Sleeman, J.R., 1963, Cracks, peds and their surfaces in some soils of the Riverine Plain, New South Wales: Australian Journal of Soil Research, v. 1, p. 91–102.

Soper, N.J., and Woodcock, N.H., 2003, The lost Lower Old Red Sandstone of England and Wales: a record of post-Iapetan flexure or Early Devonian transtension?: Geological Magazine, v. 140, p. 627–647.

Stokes, M., Nash, D.J., and Harvey, A.M., 2007, Calcrete "fossilisation" of alluvial fans in SE Spain: the roles of groundwater, pedogenic processes and fan dynamics in calcrete development: Geomorphology, v. 85, p. 63–84.

Tandon, S.K., and Kumar, S., 1999, Semi-arid/arid zone calcretes: a review, *in* Singhvi, A.K., and Derbyshire, E., eds., Palaeoevnironmental Reconstruction in Arid Lands: Rotterdam, A.A. Balkema, p. 109–152.

Thomas, R.G., Williams, B.P.J., Morrissey, L.B., Barclay, W.J., and Allen, K.C., 2006, Enigma variations: the stratigraphy, provenance, palaeoseismicity and depositional history of the Lower Old Red Sandstone Cosheston Group, south Pembrokeshire, Wales: Geological Journal, v. 41, p. 481–586.

Treadwell-Steitz, C., and McFadden, L.D., 2000, Influence of parent material and grain size on carbonate coatings in gravelly soils, Palo Duro Wash, New Mexico: Geoderma, v. 94, p. 1–22.

Verrecchia, E.P., Freytet, P., Verrecchia, K.E., and Dumont, J.L., 1995, Spherulites in calcrete laminar crusts: biogenic $CaCO_3$ precipitation as a major contributor to crust formation: Journal of Sedimentary Research, v. A65, p. 690–700.

Wilding, L.P. and Tessier, D., 1988, Genesis of Vertisols: shrink swell phenomena, *in* Vertisols: their Distribution, Properties, Classification and Management: Texas A&M University, Technical Monograph 18 SMSS, p. 55–81.

Williams, B.P.J., 1964, The stratigraphy, petrology and sedimentology of the Ridgeway Conglomerate and associated formations in south Pembrokeshire. Unpublished PhD. Thesis, University of Wales, Swansea, 245 p.

Williams, B.P.J., and Hillier, R.D., 2004, Variable alluvial sandstone architecture within the Lower Old Red Sandstone, southwest Wales: Geological Journal, v. 39, p. 257–275.

Williams, B.P.J., Allen, J.R.L., and Marshall, J.D., 1982, Old Red Sandstone facies of the Pembroke Peninsula, south of the Ritec Fault, *in* Bassett, M.G., ed., Geological Excursions in Dyfed, Southwest Wales: National Museum of Wales, Cardiff, p. 151–174.

Williams, B.P.J., Love, S.E., and Davies, D.K., 1993, Fluvial channel reservoirs: prediction and location using floodplain paleosols, American Association of Petroleum Geologists abstracts, New Orleans, p. 201.

Williams, C.A., and Krause, F.F., 1998, Pedogenic-phreatic carbonates on a Middle Devonian (Givetian) terrigenous alluvial—deltaic plain, Gilwood Member (Watt Mountain Formation), north-central Alberta, Canada: Sedimentology, v. 45, p. 1105–1124.

Wright, V.P., 1989, Terrestrial stromatolites and laminar calcretes: a review: Sedimentary Geology, v. 65, p. 1–13.

Wright, V.P., and Alonzo-Zarza, A.M., 1990, Pedostratigraphic models for alluvial fan deposits: a tool for interpreting ancient sequences: Geological Society of London, Journal, v. 147, p. 8–10.

Wright, V.P., and Tucker, M.E., 1991, Calcretes: an introduction, *in* Wright, V.P., and Tucker, M.E., eds., Calcretes: International Association of Sedimentologists, Reprint Series 2, p. 1–22.

Wright, V.P., and Marriott, S.B., 1996, A quantitative approach to soil occurrence in alluvial deposits and its application to the Old Red Sandstone of Britain: Geological Society of London, Journal, v. 153, p. 907–913.

Wright, V.P., and Wilson, R.C.L., 1987, A terra-rosa like complex from the Upper Jurassic of Portugal: Sedimentology, v. 34, p. 259–273.

Wright, V.P., Platt, N.H., and Wimbledon, W.A. 1988, Biogenic laminar calcretes: evidence of calcified root-mat horizons in palaeosols: Sedimentology, v. 35, p. 603–620.

Wright, V.P., Platt, N.H., Marriott, S.B., and Beck, V.H., 1995, A classification of rhizogenic (root-formed) calcretes, with examples from the Upper Jurassic—Lower Cretaceous of Spain and Upper Cretaceous of southern France: Sedimentary Geology, v. 100, p. 143–158.

Wright, V.P., Beck, V.H., and Sanz-Montero, M.E., 1996, Spherulites in calcrete laminar crusts: biogenic $CaCO_3$ precipitation as a major contributor to crust formation- Discussion: Journal of Sedimentary Research, v. 66, p. 1040–1041.

Yaalon, D.H., and Kalmar, D., 1978, Dynamics of cracking and swelling clay soils: displacement of skeletal grains, optimal depth of slickensides and rate of intra-pedonic turbation: Earth Surface Processes, v. 3, p. 31–42.

ALLUVIAL FACIES DISTRIBUTIONS IN CONTINENTAL SEDIMENTARY BASINS—DISTRIBUTIVE FLUVIAL SYSTEMS

GARY S. WEISSMANN

University of New Mexico, Department of Earth and Planetary Sciences, MSC03 2040, 1 University of New Mexico, Albuquerque, NM 87131-0001, U.S.A.

ADRIAN J. HARTLEY

Department of Geology & Petroleum Geology, School of Geosciences, University of Aberdeen, Aberdeen, AB24 3UE, U.K.

GARY J. NICHOLS

Department of Earth Sciences, Royal Holloway, University of London, Egham, Surrey, TW20 0EX, U.K.

LOUIS A. SCUDERI, MICHELLE E. OLSON, HOLLY A. BUEHLER, AND LAUREN C. MASSENGILL

University of New Mexico, Department of Earth and Planetary Sciences, MSC03 2040, 1 University of New Mexico, Albuquerque, NM 87131-0001, U.S.A.

ABSTRACT: Fluvial depositional patterns in modern continental sedimentary basins from different tectonic settings are dominated by distributive fluvial systems (DFSs). A review of satellite imagery from over 700 modern continental sedimentary basins from different tectonic and climatic settings shows that rivers on DFSs are generally not confined to valleys and have a clear apex from which active and abandoned channelbelts radiate outward to form a positive topographic feature centered on this apex. Channels have no tributaries on the DFSs and commonly decrease in size downstream on the DFSs due to bifurcation, infiltration, and evaporation. In contrast, fluvial systems in sedimentary basins that are confined within a valley, such as those held between adjacent or opposing DFSs, incised into a DFSs, or that lie in an axial position in a basin, typically have less area for floodplain development and display coarse-grained-dominated facies. These confined rivers appear to be more similar to tributary rivers that are commonly described for present fluvial facies models. Because sediments deposited in continental sedimentary basins are those that constitute the rock record, this work suggests that many of the fluvial rocks observed may have formed on DFSs. This strongly suggests the need to develop new facies models that account for processes on DFSs.

KEY WORDS: distributive fluvial systems, fans, fluvial planform, continental sedimentary basins.

INTRODUCTION

When we think of fluvial systems and their preservation in the rock record, we typically draw upon our experiences with rivers in degradational settings most common in the world (e.g., river systems in valleys with many tributaries) and apply concepts developed from these rivers to interpret the rock record. Indeed, most facies models that we use to interpret ancient fluvial systems for applications in petroleum development, aquifer characterization, and paleogeographic and paleoclimatic reconstruction (e.g., Miall, 1996; Bridge, 2006) have been developed using tributary rivers that exist outside active sedimentary basins or, if the river system lies within a sedimentary basin, the facies models developed typically do not place the studied reach into the context of the sedimentary basin (e.g., position in the basin, basin type and size, etc.). We ask: Are these the appropriate rivers to study when attempting to evaluate facies distributions in the rock record?

To address this question, Weissmann et al. (2010) and Hartley et al. (2010) described fluvial systems and form in over 700 modern continental sedimentary basins from a wide range of tectonic settings. They showed that *distributive fluvial systems* (DFSs), commonly called megafans, fluvial fans, and alluvial fans in the literature, dominate the fluvial depositional patterns in active continental sedimentary basins, with a much smaller proportion of the fluvial area in these basins covered by rivers that are generally confined, for example: (1) held in valleys incised into the DFS deposits, (2) held between larger DFSs in an interfan area, (3) held between opposing DFSs in an axial position in the basin, or (4) held between a DFS and the basin edge in an axial position in the basin. DFSs are defined by Hartley et al. (2010) as having: (1) a clear apex from which channel systems (both active and abandoned) diverge downstream, (2) the presence of a distinguishable positive topographic feature centered on the apex with slopes decreasing away from the apex both laterally and downstream, (3) a river that commonly displays distributive characteristics, often

TABLE 1.—Landsat 7 Bands (after http://landsat.gsfc.nasa.gov/about/etm_.html)

Channel & Wavelength	Band	Resolution (m)
1 = 0.45 - 0.515 μm	BLUE	30
2 = 0.525 - 0.605 μm	GREEN	30
3 = 0.63 - 0.69 μm	RED	30
4 = 0.75 - 0.90 μm	NEAR INFRARED	30
5 = 1.55 - 1.75 μm	MIDDLE INFRARED	30
7 = 2.09 - 2.35 μm	MIDDLE INFRARED	30
6 = 10.4 - 12.5 μm	THERMAL INFRARED	60
8 = 0.52 - 0.90 μm	PANCHROMATIC	15

From River To Rock Record: The Preservation Of Fluvial Sediments And Their Subsequent Interpretation
SEPM Special Publication No. 97, Copyright © 2011
SEPM (Society for Sedimentary Geology), ISBN 978-1-56576-305-0, p. 327–355.

TABLE 2.—A partial list of rivers and studies that describe deposits of these rivers that are commonly used to develop facies models.

River	Citation	Braided Meandering or Anastamosed	In a basin?
Amite River, Louisiana	McGowen and Garner, 1970.	Meandering	Yes, Mississippi Embayment
Babbage River, Yukon	Forbes, 1983.	Gravel meandering	No
Beaton River, BC	Nanson, 1980.	Meandering	No
Bella Coola River, BC	Church, 1983; Desoleges and Church, 1987.	Gravel meandering	No
Bijou Creek, Colorado	McKee et al., 1967.	Braided	No
Brahmaputra River, Assam	Goswami, 1985; Sarma, 2005.	Braided	Yes, foreland
Brahmaputra River, Bangladesh	Coleman 1969; Bristow 1987, 1993; Thorne, et al., 1993; Richardson et al., 1996; Richardson and Thorne, 1998; McLelland et al., 1999; Ashworth et al., 2000; Best et al., 2003.	Braided	Yes, remnant ocean basin periphery
Brazos River, Texas	Bernard et al., 1962; McGowen and Garner, 1970.	Meandering	Yes, passive margin
Calamus River, NB	Bridge et al., 1986; Bridge et al., 1998; Bridge and Gabel, 1992; Gabel, 1993; Leclair and Bridge, 2001.	Braided	No
Canterbury Plains, NZ	Ashworth et al., 1999.	Braided	Yes, passive margin
Carson River, NV	Peakall 1998	Meandering, DFS	Yes
Colorado River, Texas	McGowen and Garner, 1970; Blum, 1993	Meandering	Yes, Passive Margin
Congaree River, SC	Levey, 1978	Meandering	No
Devils River	Kochel and Baker, 1988	Canyon	No
Donjek River, Alaska	Williams and Rust, 1969; Rust, 1972; Miall, 1978; Reinfelds and Nanson, 1993.	Braided	No
Endrick River, Scotland	Bluck, 1971	Braided	No
Kicking Horse River, BC	Smith, 1974	Braided	No
Lake Eyre Basin, Australia	Nanson and Croke, 1992; Nadon, 1994	Anastamosed	Yes (cratonic)
Madison River, MT	Leeder and Alexander, 1987; Alexander and Leeder, 1990; Gawthorpe et al. 1993; Alexander, et al., 1994;Peakall et al. 2000;	Meandering	Yes
Magdalena River, Colombia	Smith, 1986	Anastamosed, DFS	Yes
Mississippi River	Fisk, 1944, 1947; Farrell, 1987; Jordan and Pryor, 1992; Harbor 1998; LeClair and Bridge, 2001	Meandering	Yes, Mississippi Embayment
Mountain streams, Banff	Smith and Smith 1980	Braided	No
Muddy Creek, WY	Dietrich and Smith, 1984	Meandering	No
Niobrara River, NB	Ethridge et al, 1999; Bristow et al. 1999; Skelly et al., 2003;	Braided	No
North Fork Toutle River, Washington	Dinehart, 1989; Dinehart, 1992	Braided	No
Pecos River, TX	Kochel and Baker, 1988	Canyon	No
Platte River, Nebraska and Colorado	Smith 1971, 1972; Miall, 1978; Blodgett and Stanley, 1980; Crowley, 1983	Braided	No
Rhine River	Carling et al., 2000a, Carling et al., 2000b	Braided	No
Rhine-Meuse Delta, Netherlands	Törnqvist, 1993, 1994; Julien and Klaasen,1995; Ten Brinke et al, 1999; Törnqvist et al., 2000; Stouthamer and Berendsen, 2000, 2001; Berendsen and Stouthamer, 2001; Stouthamer, 2001; Julien et al., 2002; Wilbers and Ten Brinke, 2003;	Meandering / anastomosed	Yes (passive margin)
Rio Grande, TX	Kochel and Baker, 1988	Canyon	No
Rio Puerco, NM	Shepherd, 1987	Meandering	Yes, rift, but degradational
River Allier	Kleinhans et al. 2002	Meandering	No
River Culm, Devon, UK	Nicholas and McLelland 1999	Meandering	No
River Feshie, Scotland	Ferguson and Werrity, 1983	Braided	No
River Waal	Kleinhans et al., 2002	Meandering	Yes (passive margin)
Sagavanirktok River	Lunt et al., 2004a; Lunt et al., 2004b; Lunt and Bridge, 2004.	Braided	Yes (Passive Margin)

TABLE 2.—*Continued.*

River	Citation	Braided Meandering or Anastamosed	In a basin?
Saskatchewan River	Cant, 1978; Cant and Walker, 1978; Campbell and Hendry, 1987; Smith and Pérez-Arlucea, 1994; Smith et al., 1998; Pérez-Arlucea and Smith, 1999; Morozova and Smith, 1999, 2000; Farrell, 2001; Slingerland and Smith, 2004	Braided Meandering	No, but aggrading in Cumberland Marshes
Scott River, Alaska	Boothroyd and Ashley, 1975; Boothroyd and Nummedal, 1978; Miall, 1978.	Braided	No
South Esk River, Scotland	Bridge and Jarvis, 1982; Bridge et al. 1995;	Meandering	No
Squamish River, BC	Brierlery and Hickin, 1991	Braided	No
Sukunka, BC	Rice and Church, 2001	Braided	No
Tana River, Norway	Collinson, 1970	Braided	No
Toutle River, North Fork, WA	Dinehart, 1989, 1992	Braided	No
Trinity, Colorado, Nueces Rivers, TX	Aslan and Blum 1999	Meandering	No
Trollheim fan	Collinson, 1986; Miall, 1978; Blair and McPherson, 1992;	Alluvial Fan	Yes
Wabash River, US	Jackson, 1976a, 1976b, 1976c, 1978	Meandering	No
William River Delta, Saskatchewan, Canada	Smith and Smith, 1984; Bridge, et al., 1995	Braided	No

bifurcating into smaller channels, (4) no tributaries join the DFS below the apex, and (5) evidence for abandoned channels with a divergent or arcuate pattern away from the apex. In using the term distributive, we do not intend to imply that all channel belts are simultaneously active on the DFS, though several channel belts may be active. Rather, the channel belts, both active and abandoned, form a broad pattern that effectively distributes sediment across the DFS over time.

Weissmann et al. (2010) also suggest that significant differences between DFS river systems and tributary systems exist, namely that: (1) channels are not confined in a valley on the open DFS, with channel systems radiating outward from the DFS apex, and (2) channel size commonly decreases downstream on the DFS, though this decrease in channel size is not always observed. The axial streams in a basin, if confined laterally, and rivers that are incised into the DFS or held between DFSs (e.g., interfan locations) appear to have character similar to tributary systems.

In this paper, we expand on concepts presented by Weissmann et al. (2010), showing additional examples of rivers in modern continental sedimentary basins and describing in greater detail some of our initial findings on differences between rivers on DFSs and those of tributary systems. In this work, we make observations of river systems that are depositing sediment in continental sedimentary basins that are generally not affected by the Quaternary sea-level fluctuations; thus, we do not evaluate fluvial deposition along aggradational passive margins. We also do not evaluate aggradation that occurs within relatively narrow incised valleys (e.g., Archer and Greb, 1995; Blum, 1993; Blum and Price, 1998; Blum and Aslan, 2006). We recognize that some fluvial rocks were deposited in these settings; however, in this work we provide an alternative interpretation for much of the fluvial rock record.

We begin by reviewing fluvial morphology in the tributary systems that are commonly used in developing the current generation of fluvial facies models. We then present several examples of fluvial form and distribution in modern continental sedimentary basins found in different tectonic and climatic settings, contrasting the morphology of these river systems with the river systems in degradational settings (we call these "tributary" systems in this paper to distinguish them from the distributive systems). In this discussion, we develop hypotheses on large-scale facies distributions in the basins based on our initial observations from satellite imagery. However, we recognize that all of these hypotheses must be tested through a more thorough image analysis closely coupled with on-the-ground studies of the fluvial basin fill. We conclude with observations that may help us build a sedimentary basin context within which we can understand fluvial facies on a scale larger than channel-belt facies distributions.

DATA USED FOR STUDY

Google Earth® was initially used to identify modern continental sedimentary basins and DFSs on every continent except Antarctica. From our initial review we identified and cataloged 724 modern continental sedimentary basins (Weissmann et al., 2010) and 415 large DFSs (e.g., DFSs longer than 30 km from apex to toe) (Hartley et al., 2010) in our database. This initial database of basin and fan features, with a data structure that includes locality name, and latitude, longitude, and elevation for the feature, served as our template for adding additional information and as a guide for documenting further analysis.

In order to analyze individual features (hereafter termed "database objects" or "objects"), we populate the database with a combination of Landsat imagery and Shuttle Radar Topography Mission (SRTM) elevation data for each of the sedimentary basins and large DFSs. We selected Landsat 7 multispectral imagery as a base for our analysis of each database object because of the combination of spectral (seven channels plus a broad visible to near infrared panchromatic band; See Table 1), spatial (as high as 15 meters) and temporal (every 16 days) resolution. The Landsat 7 satellite operates in a sun-synchronous, near polar orbit providing repeat coverage approximately every 16 days with a swath width of 185 km, which is comparable to many of the objects we are including in the database. This allows for monitoring of active fan surfaces on a regular basis. Earlier Landsat data (Landsat 1–3, 1972–1982; Landsat 4–5, 1982 to present) can be incorporated into our data analysis stream and our database structure on an as-needed basis. Ancillary data (reports, journal articles, map sheets, etc.) can be linked directly to each database object.

Topography for each of the features is derived from Shuttle Radar Topography Mission (SRTM) data sets. The SRTM, consisting of a specially modified radar system that was flown on the Space Shuttle in February 2000, obtained elevation data for over 80% of the world. We utilize SRTM 90 m^2 grid cell data (3 arc seconds) for all of our sites where data were available. This was augmented with 30 m^2 grid cell data (1 arc second) in the United States. Absolute SRTM vertical accuracy for the entire globe is 10 meters. Relative accuracy for any local area is 6 meters. For both vertical measures 90% of the observations fall within 10 meters of the true elevation. Absolute SRTM horizontal accuracy is 20 meters for the entire globe. Relative horizontal accuracy for any local area is 15 meters. For both horizontal measures the values represent a circle in which 90% of the observations fall (Smith and Sandwell, 2003; Farr et al., 2007).

The SRTM digital elevation data were processed to fill data voids due to radar shadowing and missing values. After processing, the SRTM data were added to the database referenced to each database object. For a sedimentary basin object larger than a single Landsat scene or SRTM Digital Elevation Model (DEM) we used ARCINFO (ESRI, 2009) to mosaic multiple images and DEMs into a single file. As with individual images, the mosaicked files were referenced to the appropriate database object (or objects if multiple basins and fans were covered). All imagery, digital elevation data, and derived layers entered into the database are georeferenced to a common projection (Albers equal area).

Characteristics of fluvial system morphology within the sedimentary basins and on the DFSs are analyzed using both Landsat imagery and SRTM elevation data. Slope, aspect, and a range of hydrologically significant measures of surface curvature, stream profile, and stream basin characteristics (ESRI, 2009) will ultimately be added to the database structure for each database object. Vector-based streamlines of flow direction, channel form, and channel sinuosity as well as additional data layers for geology, vegetation, and other surface (and subsurface) mapping can be added.

CHARACTERISTICS OF TRIBUTARY FLUVIAL SYSTEMS COMMONLY USED IN FACIES MODELS

A review of modern river systems commonly used to develop current facies models (e.g., Miall, 1996; Bridge, 2006) shows that most of our understanding of fluvial successions is derived from tributary rivers that lie outside continental sedimentary basins (Table 2). These rivers typically lie within degradational valleys and form a tributary network. Additionally, most of the fluvial geomorphology literature that is used by fluvial sedimentologists (e.g., Leopold et al., 1964; Allen, 1965; Schumm, 1977, 1986, 2005; Knighton, 1998; Schumm et al., 2000) focuses on examples from degradational fluvial systems that lie outside active sedimentary basins. Notable exceptions exist, where studies have been conducted in sedimentary basins (Table 2). However, these studies typically focus on facies distributions found within the channel belt and do not place these deposits into the broader context of basin position and juxtaposition to other facies assemblage types.

Several river systems listed in Table 2 lie along a passive margin or in the Mississippi Embayment; however, these rivers presently exist in relatively narrow valleys and are tributary in form. Though we focus on river systems in continental sedimentary basins, we include these rivers in Table 2 since they have been used to interpret the rock record in settings that are not restricted to an incised-valley fill. We recognize that these valley-fill systems are represented in the rock record, but we hypothesize that these have form that is different from rivers that fill broad sedimentary basins. Thus, depositional patterns observed in these settings may not be analogs for much of the fluvial rock record.

In this section, we present examples of modern fluvial systems with both meandering and braided morphology. In presenting this work, our intention is not to criticize the excellent research conducted on these rivers, but rather to attempt to provide a larger-scale context for fluvial systems within a sedimentary basin. We believe that these studies of fluvial systems in degradational settings have validity in terms of in-channel processes and products at the scale of bar forms, macroforms, and channel belts. However, they do not inform us about the way the macroform-scale deposits stack into the overall 3D basin-fill architecture. Unfortunately, at this early stage in these studies, we do not have the data to directly compare deposits from rivers held within continental sedimentary basins versus the degradational rivers commonly used in facies models. We expect to conduct such work in the near future, and hope others will also conduct such studies on modern DFSs. Collecting such data represents the next stage in this research.

Meandering (Single-Channel) Rivers

Commonly cited meandering rivers in the United States include the Wabash River (e.g., Jackson, 1975, 1976a, 1976b, 1978) and the lower Mississippi River (e.g., Fisk, 1944, 1947; Farrell, 1987; Autin et al., 1991; Jordan and Pryor, 1992; Saucier, 1994; Aslan and Autin 1998; Blum, et al., 2000; Rittenour et al., 2005, 2007) (Fig. 1). In both systems, the channel belts are held inside broad valleys (Figs. 1B, 1D) and the river cannot migrate laterally beyond the valley sides. Though some constructional environments exist in these valleys (Blum et al., 2000; Rittenour et al., 2005, 2007), the overall position of the river in a large valley indicates net degradation over recent time. Several tributary streams enter these rivers along their course, so that the rivers become larger in a downstream direction.

The Wabash River does not lie in an active sedimentary basin; thus, these deposits will not be preserved in the long-term rock record, a feature noted by Jackson (1978) with respect to preservation potential. This river is held within a valley that is completely filled by an amalgamated meander belt (Jackson, 1976a) (Figs. 1A, 1B). The overbank deposits form a thin veneer on top of the point bars. Because this river is held within a valley, most of the deposits are coarse-grained-dominated since the river continually reworks its deposits as it sweeps from side to side in the valley (Jackson, 1978).

In the case of the lower Mississippi River valley, Holocene meandering-river deposits lay between terraces composed of late Pleistocene or early Holocene braided-river deposits or as a relatively thin veneer on top of these deposits (Figs. 1C, 1D) (Fisk, 1944; Jordan and Pryor, 1992; Blum et al., 2000; Rittenour et al., 2005, 2007). The meandering-river deposits consist of several broad meander belts that are held within fine-grained floodbasin deposits (Fisk, 1944, 1947; Farrell, 1987; Jordon and Pryor, 1992; Saucier, 1994; Blum, 2000; Rittenour et al., 2007). A complex valley cut and fill occurred along this river due to Quaternary climate variability and sea-level change (e.g, Blum and Aslan, 2006). Deposits that are preserved will therefore represent valley-fill systems but are not necessarily representative of the fluvial fill in sedimentary basins without incised-valley-fill systems.

The lateral migration of the river system within its meander belt produced an amalgamated pattern of point-bar and channel deposits (Fig. 1C). Chute and neck cutoff avulsion styles are most common, where a meander loop is abandoned when the channel shifts into an older scroll bar or cuts across a thin neck at the meander-loop base. Beneath the scroll bars are sand deposits left behind as the point bar developed; thus these meander belt deposits are primarily coarse grained capped by a relatively thin veneer of fine floodplain deposits (Fisk, 1944; Mann and Thomas, 1968; Farrell, 1987; Aslan and Autin, 1998).

In the southern part of the lower Mississippi valley, floodbasins fill the areas between meander belt alluvial ridges (Farrell, 1987; Aslan and Autin, 1998, 1999; Aslan et al., 2005). These floodbasins consist of backswamp and crevasse-splay deposits (Farrell, 1987), with records of nodal avulsions captured by these sediments (Aslan and Autin, 1998; Aslan et al., 2005). In this area, the river appears to be slowly aggrading, as shown by the presence of Holocene deposits over Pleistocene

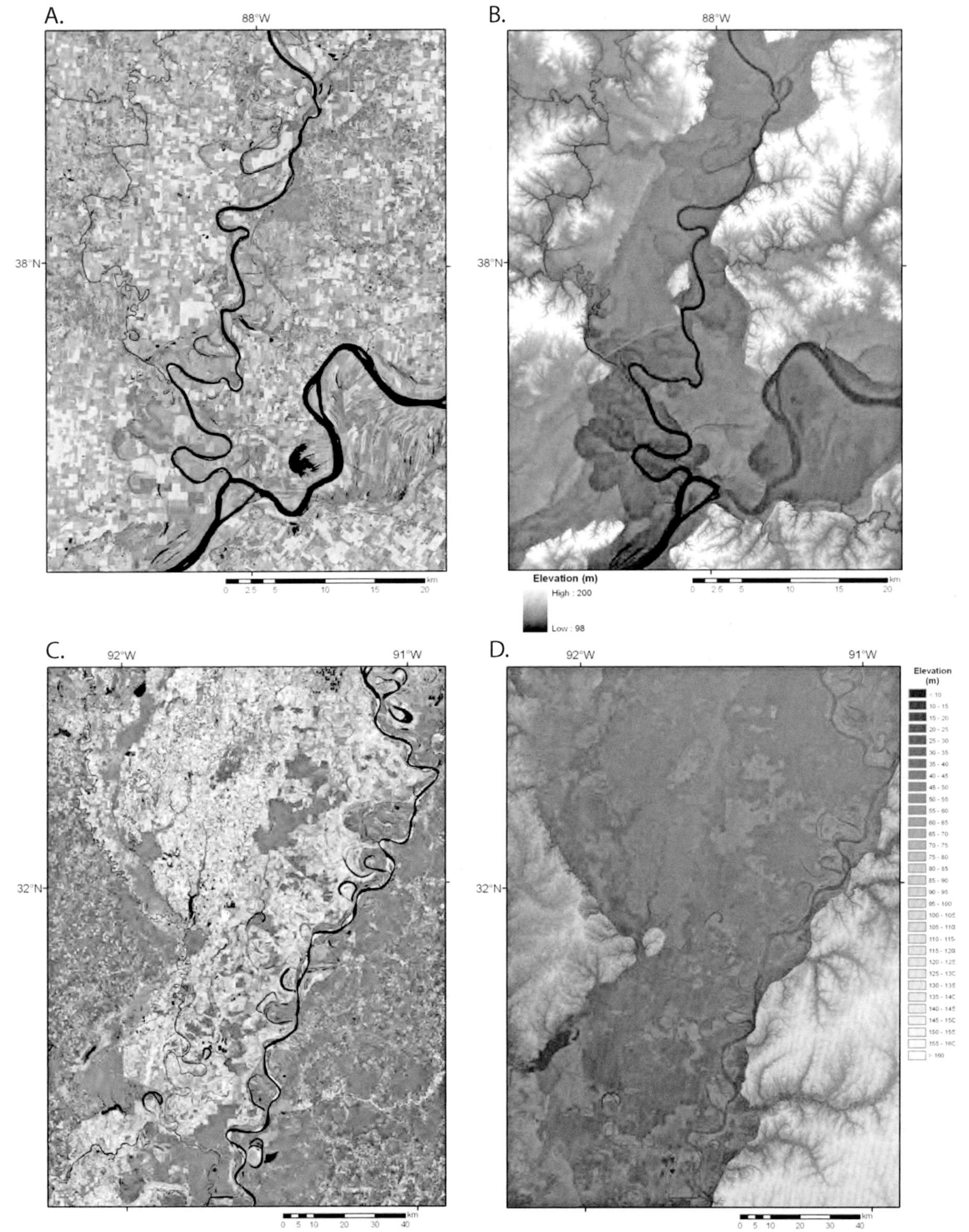

FIGURE 1.—Examples of meandering river systems in degradational (tributary) settings. A) A false color Landsat mosaic of the Wabash River Valley, Illinois, Indiana, USA. False color from red—band 4, green—band 5, and blue—band 7. B) A digital elevation model (DEM) of the Wabash River Valley derived from SRTM 1-arcsec data. C) A false color Landsat mosaic of the lower Mississippi River Valley, Louisiana, Mississippi, USA. False color from red—band 7, green—band 5, and blue—band 4. D) A digital elevation model (DEM) of the Lower Mississippi River Valley derived from SRTM 1-arcsec data.

braided-river deposits. Blum and Aslan (2006) describe the expected distribution of sedimentary facies in such an incised valley fill.

Generally, meandering rivers in degradational settings appear to have an amalgamated form and increase in size downstream (e.g., Leopold et al., 1964; Schumm, 1977; Knighton, 1998). Though floodplain deposits are observed surrounding these meander belts, the overall preservation potential of these fine-grained deposits is limited because the river sweeps back and forth across its valley, reworking these deposits.

Braided (Multichannel) Rivers

Commonly cited studies on modern braided rivers include those conducted on the Platte River in Colorado and Nebraska, USA (Smith, 1970, 1971; Blodgett and Stanley, 1980; Crowley, 1983), the Calamus River in Nebraska, USA (Bridge et al., 1986), the Donjek River in the Yukon Territory, Canada (Williams and Rust, 1969), the Kicking Horse River in British Columbia, Canada (Smith, 1974), and the Brahmaputra River in Bangladesh (Best and Ashworth, 1997; Sarma, 2005) and Assam, India (Goswami, 1985; Sarma, 2005) (Fig. 2). Except for the Brahmaputra River, all of these rivers lie in large river valleys and are generally in degradational settings (Fig. 2). None lie in active sedimentary basins except for the upper Brahmaputra River in Assam, India, and the lower Brahmaputra River downstream from the Himalayan forebulge in Bangladesh. As with the meandering-river examples, flanking terraces that surround these rivers indicate past aggradational events. However, the overall position of the river inside a larger valley indicates degradation.

The Platte River has been described as a "model" system for braided-river sedimentology (Miall, 1977). However, as with other High Plains rivers, the Platte River is degradational (Blodgett and Stanley, 1980). It is mainly a sandy bedload-dominated stream (Smith, 1971) due, in part, to the lack of fines in much of the sediment source area (Figs. 2A, B). Fine-grained deposits that exist are limited and most commonly occur along the relatively narrow floodplains along the channel, along the downstream margins of transverse-bar and longitudinal-bar deposits, and draped along the tops of traverse bars in upward-fining successions (Smith, 1970, 1971; Blodgett and Stanley, 1980; Crowley, 1983).

The Calamus River sits in a gently sinuous terrace-bounded valley (Figs. 2D, E). The modern channel cuts into both Holocene ancestral-river deposits that make up the terraces and a Quaternary eolian dune field at various points along the valley. It has been estimated that degradation has been taking place for the last ~ 3000 years and is most likely still occurring today (Brice, 1964). The Calamus floodplain is 100–150 m wide and consists of abandoned channel bars and channels at different stages of fill while the Calamus channel, sinuous and undivided upstream, becomes wider, shallower, and braided as it nears its confluence with the North Loup River (Bridge et al., 1986). The suspended load is composed mostly of sand with very little silt and clay due to source limitations (Bridge et al., 1986). Fine-grained sediment in this system is rare and is found mainly on the downstream part of bars and as channel fills (Bridge et al., 1986). The Calamus River is confined to its valley, and therefore a lack of extensive overbank deposits is attributed to "the dominance of lateral channel migration

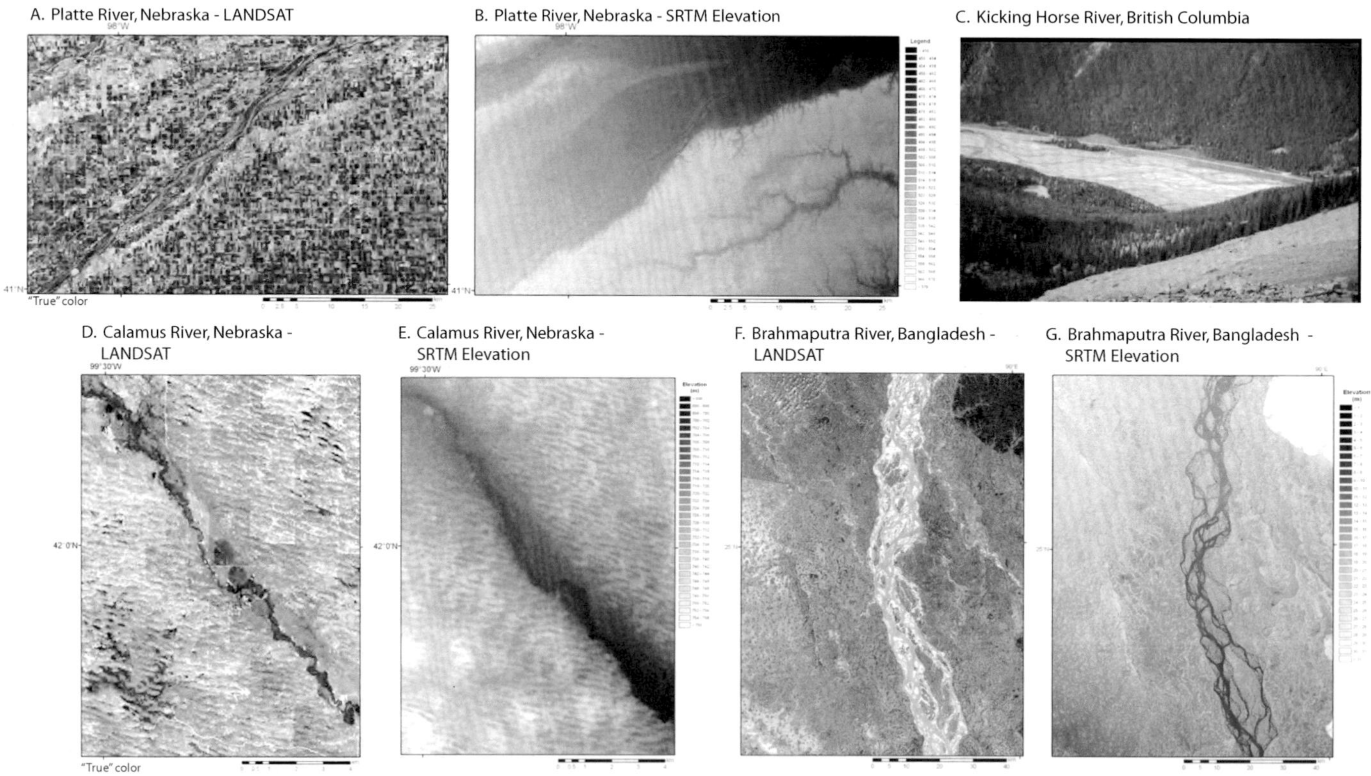

FIGURE 2.—Examples of braided river systems in degradational (tributary) settings. A) A true color Landsat mosaic of the Platte River, Nebraska, USA. B) A digital elevation model (DEM) of the Platte River Valley, Nebraska, derived from SRTM 1-arcsec data. C) A photograph of the Kicking Horse River near Field, British Columbia. D) A true-color Landsat mosaic of the Calamus River, Nebraska, USA. E) A digital elevation model (DEM) of the Calamus River area derived from SRTM 1-arcsec data. F) A true-color Landsat mosaic of the lower Brahmaputra River in Bangladesh. G) A digital elevation model (DEM) of the Lower Brahmaputra River area derived from SRTM 3-arcsec data.

over vertical accretion, and the paucity of fine-grained suspended load" (Bridge et al., 1986, p. 867).

The upper Kicking Horse River, a tributary of the Columbia River, flows through a steep glacial valley that is bordered by debris fans (Smith, 1974) (Fig. 2C). The sedimentology of the Kicking Horse, described by Smith (1974), is limited to channel deposits and reflects a coarse-grained-dominated system where fine-grained sediment is found on the downstream portions of migrating bars and overlying coarse deposits. Fine-grained beds, deposited from suspension, are thin (generally less than 2.5 cm thick) and discontinuous.

The glacially fed Donjek River is held within a straight, steep valley made up of neoglacial till, outwash sands and gravels, and loess (Denton and Stuiver, 1967; Williams and Rust, 1969). Within the valley, the river has four topographic levels each of which is active during various levels of discharge (Williams and Rust, 1969). Fine-grained facies are most common in abandoned channels and on upper surfaces of lateral and point bars, indicating a gradational relationship between facies (Williams and Rust, 1969).

In all these examples, the rivers consist of multiple channels that are separated by compound mid-channel bars that are active during high-stage flows. At high stage, the rivers in each system can completely fill the active valley (wall to wall). The channel-belt deposits are dominated by coarse-grained sediments with floodplain fines held temporarily on bar tops and abandoned channels until the next flood event moves them farther downstream. A striking lack of fines is observed in these modern systems (Williams and Rust, 1969; Smith, 1970, 1971, 1974; Blodgett and Stanley, 1980; Bridge et al., 1986). Yet the record of ancient rivers interpreted to be braided have significant fines associated with the channel-belt deposits, e.g., the New Red Sandstone in the Hebridean Province, Scotland (Steel, 1974) and the Old Red Sandstone of Scotland (e.g., Friend, 1978; Bluck, 1980, 1986).

Studies on braided rivers that cross modern sedimentary basins have been used in the development of some facies models. Notably, recent work by Lunt and Bridge (2004) and Lunt et al. (2004a, 2004b) on the Sagavinirktok River in Alaska describe channel-belt deposits (both active and abandoned) on the landward side of a passive-margin basin near Prudhoe Bay. Several studies on the Brahmaputra River in Assam, India, also evaluated depositional form in this portion of the Himalayan Foreland Basin (Goswami, 1985; Sarma, 2005; Singh, S.K., 2007). Finally, the Brahmaputra River in Bangladesh is aggradational as it enters the marginal marine basin south of the Himalayan forebulge (Figs. 2F, G). Many of the studies performed on the Brahmaputra River (e.g., Coleman, 1969; Bristow, 1987; Best and Ashworth, 1997; Best et al., 2003) have been carried out on large sections of the river in Bangladesh just above its confluence with the Ganges River. In this section, the Brahmaputra sits atop fine Quaternary fluvial sediment deposits with Pleistocene alluvial terraces east and west of the River (Coleman, 1969). Most of these studies, however, focus on the facies distributions inside the channel belts and on channel migration and do not place the deposits in the broader context of position in the sedimentary basin or controls related to the form of the sedimentary basin.

LARGE-SCALE FLUVIAL FORM IN CONTINENTAL SEDIMENTARY BASINS

In this section, we present a sample of river-deposit form selected from the global database of 724 modern continental sedimentary basins (Weissmann, et al., 2010) (Figs. 3, 4). The basins lie within all continental tectonic settings—compressional (e.g., foreland, back-bulge, piggyback, forearc), extensional (e.g., rift), and strike-slip (e.g., transtensional, transpressional) settings. Additionally, these basins

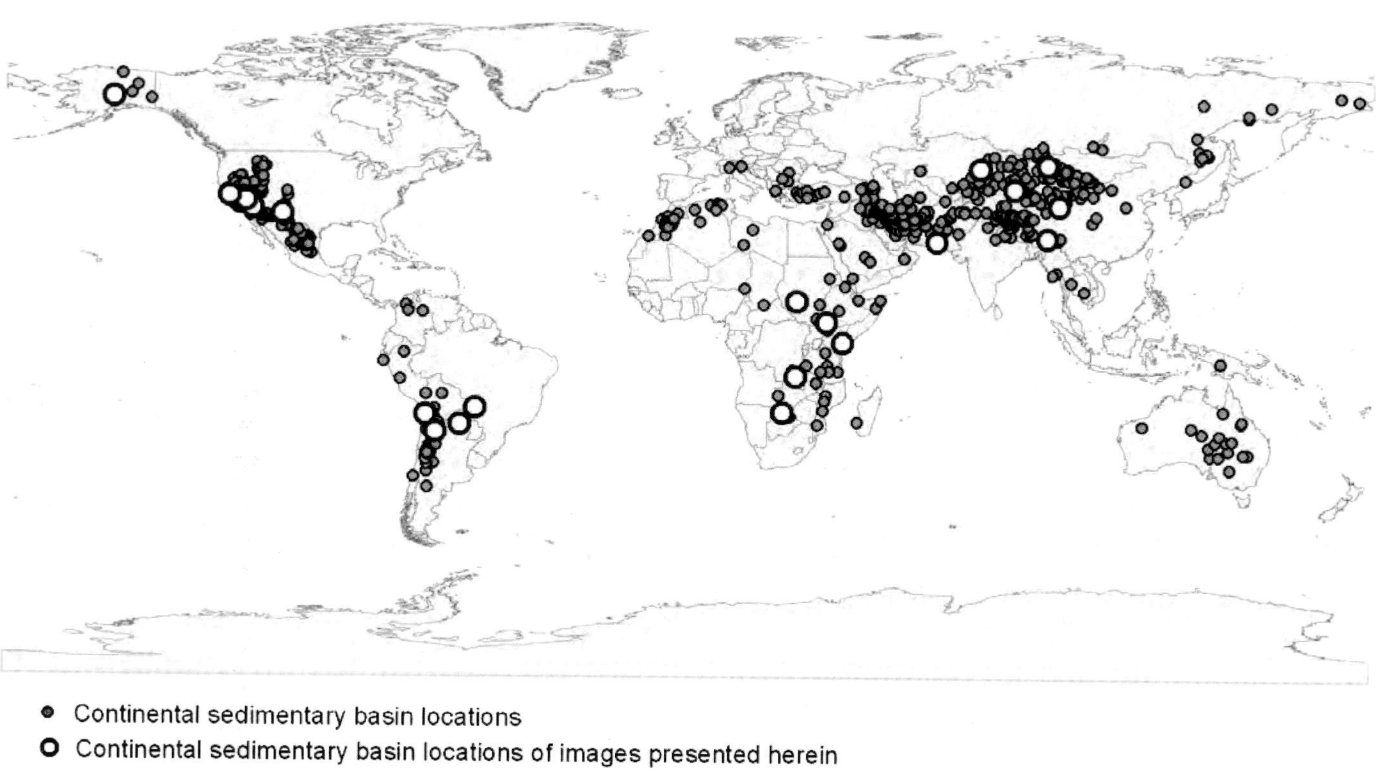

FIGURE 3.—Locations of continental sedimentary basins around the world, with the large dots showing locations of basins presented in Figure 4 (modified from Weissmann, et al., 2010).

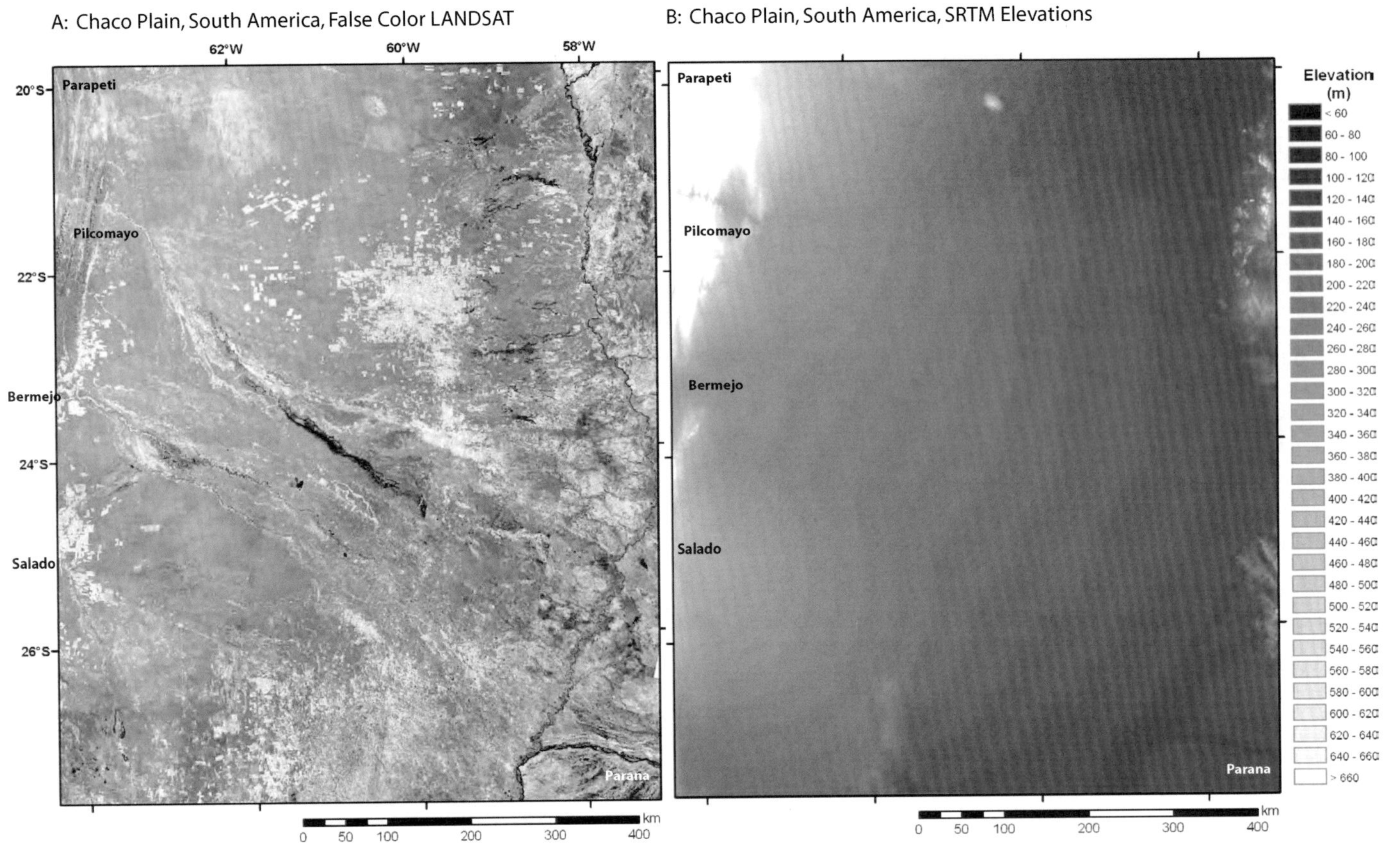

FIGURE 4.—Images and DEMs of continental sedimentary-basin examples from around the world (see locations in Figure 3). A) A false-color Landsat mosaic of the Chaco Plain, Andean Foreland Basin, central South America. False color from red—band 7, green—band 5, and blue—band 4. B) A digital elevation model (DEM) of the Chaco Plain, Andean Foreland Basin, central South America, derived from SRTM 3-arcsec data. C) A false-color Landsat mosaic of the northern Chaco Plain, Andean Foreland Basin, Bolivia and Paraguay. False color from red—band 7, green—band 5, and blue—band 4. D) A false-color Landsat mosaic of the axial Paraguay River, Andean Foreland Basin, Paraguay and Brazil. This axial river is confined between opposing DFSs in the basin and forms an amalgamated meander belt that is similar in form to tributary meander belts. False color from red—band 4, green—band 5, and blue—band 7. E) A false-color Landsat mosaic of the upper Brahamputra River system, Himalayan Foreland Basin, Assam, India. False color from red—band 7, green—band 5, and blue—band 4. F) A false-color Landsat mosaic of the McKinley River DFS, Alaska Range Foreland Basin, Alaska. False color from red—band 4, green—band 5, and blue—band 7. G) A true-color Landsat mosaic of the Ili River DFS, foreland basin to the Tien Shan, Kazakhstan. H) A true-color Landsat mosaic of DFSs entering the southern side of the Tarim Basin, China. I) A false-color Landsat mosaic of the Pantanal Basin, Brazil and Paraguay. False color from red—band 4, green—band 5, and blue—band 7. J) A digital elevation model (DEM) of the Pantanal Basin derived from SRTM 3-arcsec data. K) A true-color Landsat mosaic of several piggy-back basins in the Andes Mountains, Argentina. L) A false-color Landsat mosaic of DFSs entering a piggyback basin in the Sulaiman Montains of Pakistan. False color from red—band 7, green—band 5, and blue—band 4. M) A true-color Landsat mosaic of DFSs crossing the Andean Forearc Basin, Chile. N) A true-color Landsat mosaic of the central San Joaquin Basin, California. Lateral DFSs enter the basin from both the east and west sides with the axial San Joaquin River held between the DFSs. Where the DFSs do not fill the horizontal accommodation in the basin, the San Joaquin River forms an axial DFS (marked by yellow arrows). O) A true-color Landsat mosaic of the Rio Grande Rift, New Mexico, USA. P) A false-color Landsat mosaic of the rift basin holding the Okavango, Kwando, and Zambezi DFS, Botswana and Namibia. False color from red—band 4, green—band 5, and blue—band 7. Q) A false-color Landsat mosaic of DFSs formed in the East Africa Rift in the Democratic Republic of Congo. False color from red—band 4, green—band 5, and blue—band 7. R) A false-color Landsat mosaic of the East Africa Rift, Kenya. False color from red—band 7, green—band 5, and blue—band 4. S) A true-color Landsat mosaic of basins in the Basin and Range Province, California, USA. Notice how the axial river forms a DFS where horizontal accommodation is available. T) A false-color Landsat mosaic of a rift flank basin to the East Africa Rift, Sudan. False color from red—band 4, green—band 5, and blue—band 7. U) A true-color Landsat mosaic of a rift flank basin to the East Africa Rift, Somalia. V) A false-color Landsat mosaic of a strike-slip basin along the Kunlun Fault, China. False color from red—band 4, green—band 5, and blue—band 7. W) A true-color Landsat mosaic of Death Valley, California, USA. Notice how the axial Amargosa River forms a DFS where accommodation is available. X) A true-color Landsat mosaic of several basins in the Gobi–Altai region, Mongolia.

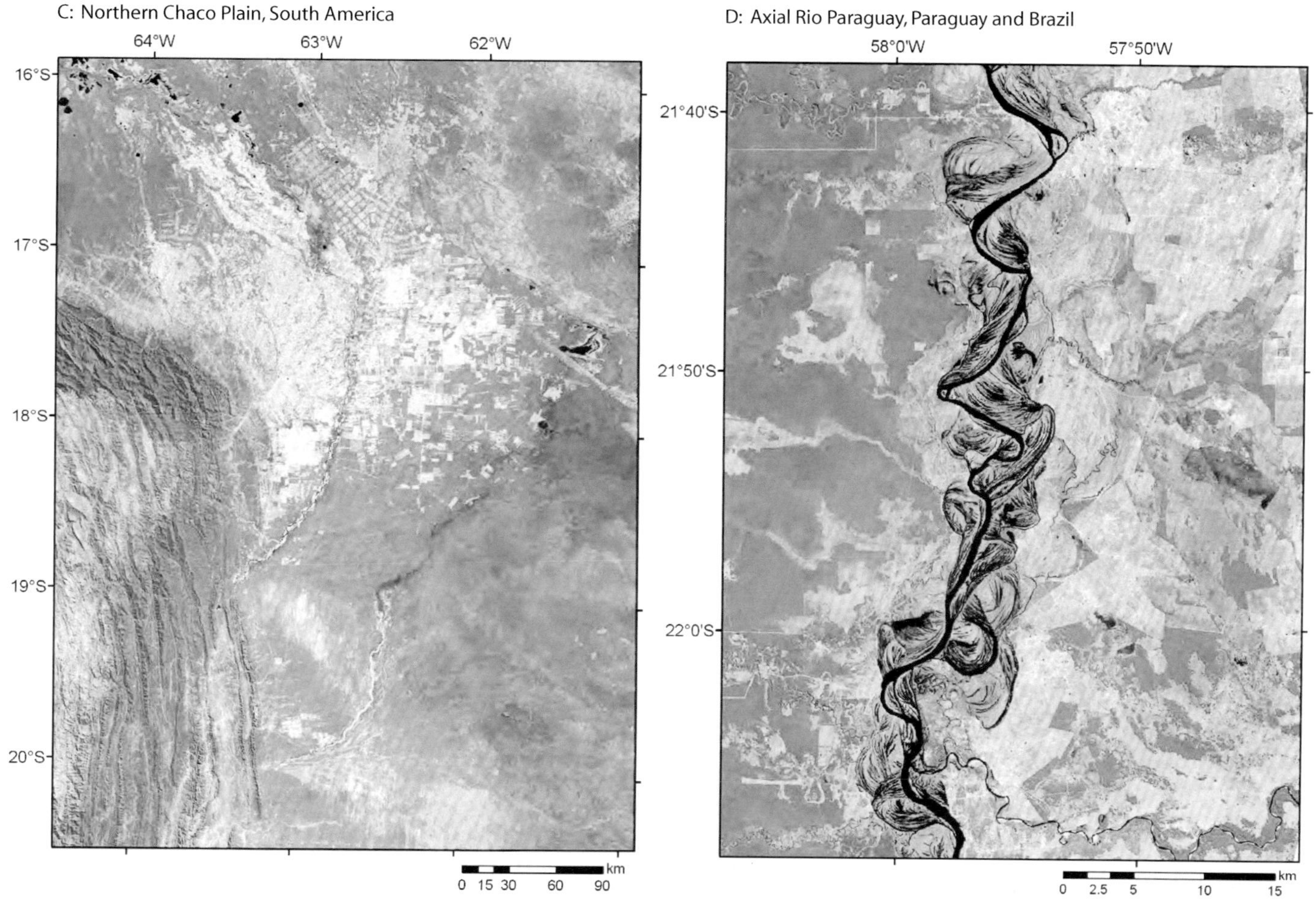

FIGURE 4.—*Continued.*

occur in all climatic regimes and are sampled from both endorheic and exorheic basins (Hartley et al., 2010).

In *ALL* of the sedimentary basins studied, without exception, DFSs dominate the fluvial basin fill (Fig. 4; Weissmann et al., 2010) with a small portion of the basin covered by the axial river system. These DFSs terminate in axial fluvial systems, eolian dune fields, freshwater lakes, or playas, depending on climate, position in a basin, and whether the basin is open or closed (Hartley et al., 2010). As noted earlier, we purposely exclude detailed description of fluvial form in marine marginal settings, thus generally avoiding the influences of sea-level change and cycles of incision and aggradation associated with relatively narrow incised valleys in coastal settings, such as those described by Blum (1993).

Basins in Compressional Settings

Foreland Basins: Many workers have shown that DFSs commonly occur in most large foreland basins that are present today (e.g., Geddes, 1960; Gole and Chitale, 1966; Wells and Dorr, 1987; Mohindra et al., 1992; Iriondo, 1993; Sinha and Friend, 1994; Sinha, 1996; Horton and DeCelles, 1997, 2001; Gupta, 1997; DeCelles and Cavazza, 1999; Kale, 2002; Gibling et al., 2005; Sinha et al., 2005; Leier, et al., 2005; Singh, I.B., 2007; Chakraborty and Ghosh, 2010; Chakraborty et al., in press). In these basins, large megafans cover most of the basin area, with smaller "fluvial fans" from foothills-fed rivers filling in areas between the megafans (DeCelles and Cavazza, 1999; Horton and DeCelles, 2001; Sinha and Friend, 1994). On the megafans, two types of rivers are observed—mountain-fed and plains-fed rivers (Sinha and Friend, 1994). Mountain-fed rivers are the main trunk streams that are sourced in the highlands outside the basin, while plains-fed rivers are sourced within the basin and have headwaters on the DFS surface. The plains-fed rivers are fed by precipitation runoff and groundwater discharge. Typically, these exist as underfit streams in abandoned mountain-fed channels. Because these DFSs are being deposited in tectonically active basins, fluvial form and processes may be locally influenced by tectonic activity (e.g., Lahiri, 1996).

The largest DFSs observed in our study are present in the Chaco Plain of the Andean Foreland Basin in Bolivia, Paraguay, and Argentina (Hartley et al., 2010; Weissmann et al., 2010), with the Pilcomayo DFS having a length of over 700 km and covering an area greater than 210,000 km^2 (Iriondo, 1993) (Figs. 4A, 4B, 5). Smaller DFSs exist to the north of this area, primarily because the foreland basin appears to be narrower in this region (Fig. 4C). On each of the DFSs in this basin, the modern channel and a large number of abandoned channels form a radial pattern from the location where the river enters the sedimentary basin (e.g., the DFS apex). The digital elevation model (DEM), derived from SRTM 3-arcsec data, shows that each DFS forms a sediment cone that has a convex-upward profile parallel to the mountain front and a low-gradient concave-upward profile downfan (Fig. 4B), in a morphology similar to that described for alluvial fans (e.g., Blair and McPherson, 1994).

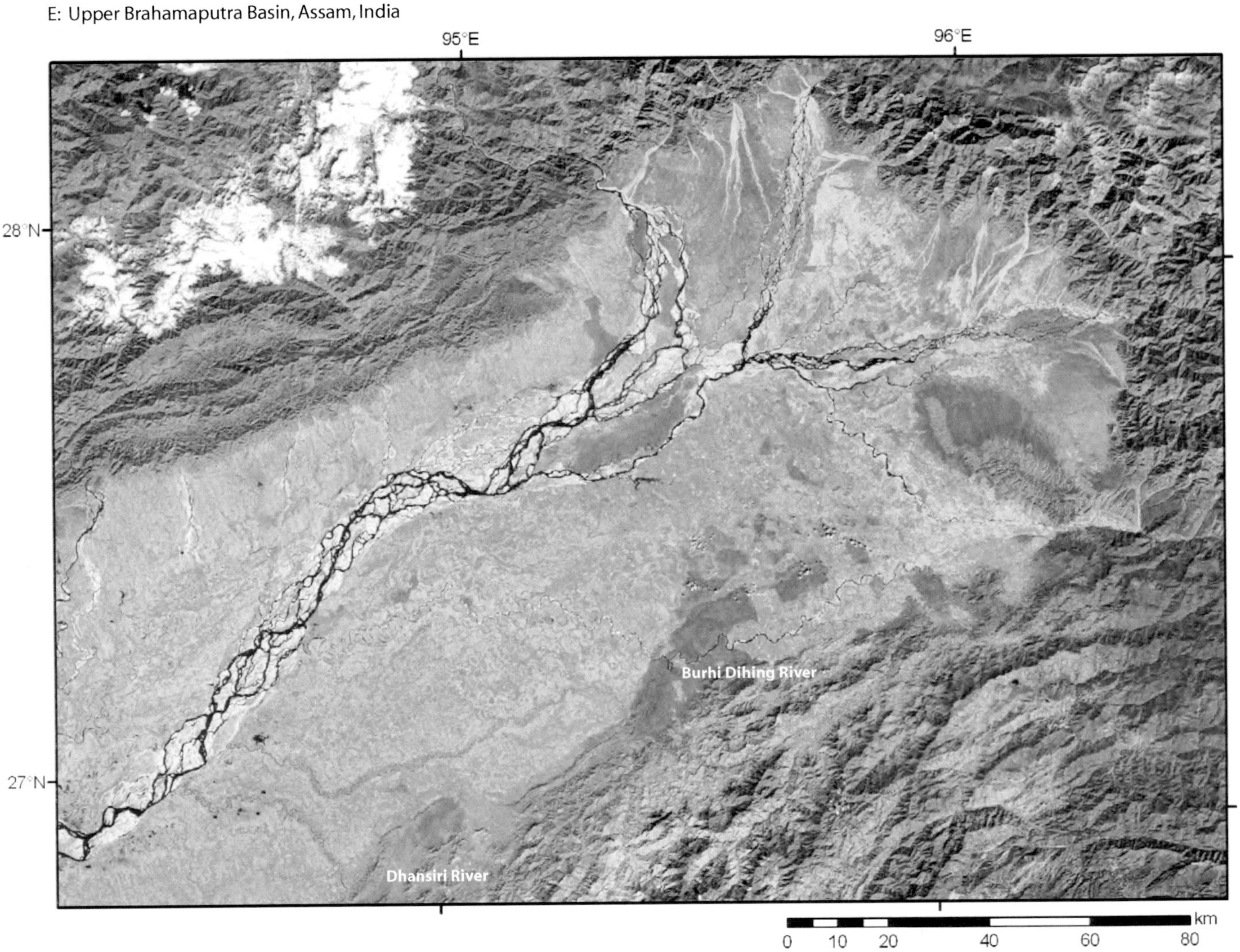

FIGURE 4.—*Continued.*

Both mountain-fed and plains-fed rivers are seen on the Chaco Plain DFS surfaces, with foothills-fed rivers forming smaller DFS between the large DFSs (Horton and DeCelles, 2001; Figs. 4A, 5). Few of the mountain-fed rivers directly reach the axial river in the basin, with many of the rivers disappearing into swamps on the distal DFSs. Mountain-fed rivers range in form from braided, multichannel systems near the apex to single-channel meandering in medial positions. Most of the mountain-fed channel systems in this area decrease in size downfan due to bifurcation, evaporation, and infiltration into the groundwater (Fig. 5). For example, the channel belt near the apex of the Pilcomayo DFS is about 1500 m wide. On the medial DFS (approximately 150 km downfan from the apex), the channel belt diminishes to approximately 220 m wide. Farther downfan, channels bifurcate and individual channels are approximately 30–60 m wide (just at the resolution of the satellite imagery). The "plains-fed" rivers appear to lie within abandoned mountain-fed channels, with water entering these channels from local rainfall or from groundwater discharge. The plains-fed rivers and the greatly diminished mountain-fed rivers are tributary to the axial Paraguay River.

Based on apparent facies distributions observed in the satellite imagery, the bulk of the depositional area on the Pilcomayo DFS is dominated by fine-grained floodplain deposits with discrete channel deposits held between them. The proportion of channel-belt deposits appears to increase near the apex, with a higher proportion of fine-grained sediment present distally. Additionally, many of the distal channels have few exposed sand bars and the river appears to be muddy. Meander belts on the medial portion appear to have a simple form (as opposed to the amalgamated form), and chute-cutoff and neck-cutoff avulsions are therefore rare. Avulsion deposits may also be very common (e.g., Smith et al., 1989). Ancient examples of channel fill and avulsion deposits from this type of DFS may be present in the Chinle Formation in New Mexico (Fig. 6) and other similar fine-grained fluvial successions (e.g., Dubiel et al., 1991; Kraus and Aslan, 1993; Kraus, 1996; Kraus and Davies-Vollum, 2004; Uba et al., 2005), in the Carboniferous coal Measures of NW Europe (e.g., Fielding, 1984; Hartley, 1993), and in the upper Cretaceous Blackhawk Formation, Utah (Hampson et al., 2005).

Between each of the large DFSs of the Chaco Plain are smaller DFSs from foothills-fed rivers exiting the Andes (e.g., Horton and DeCelles, 2001; Figs. 4A, 5). These DFSs fill the interfan area between the larger DFSs and show similar downfan channel size decrease, radial pattern of abandoned channels, and distributions of floodplain material, but at

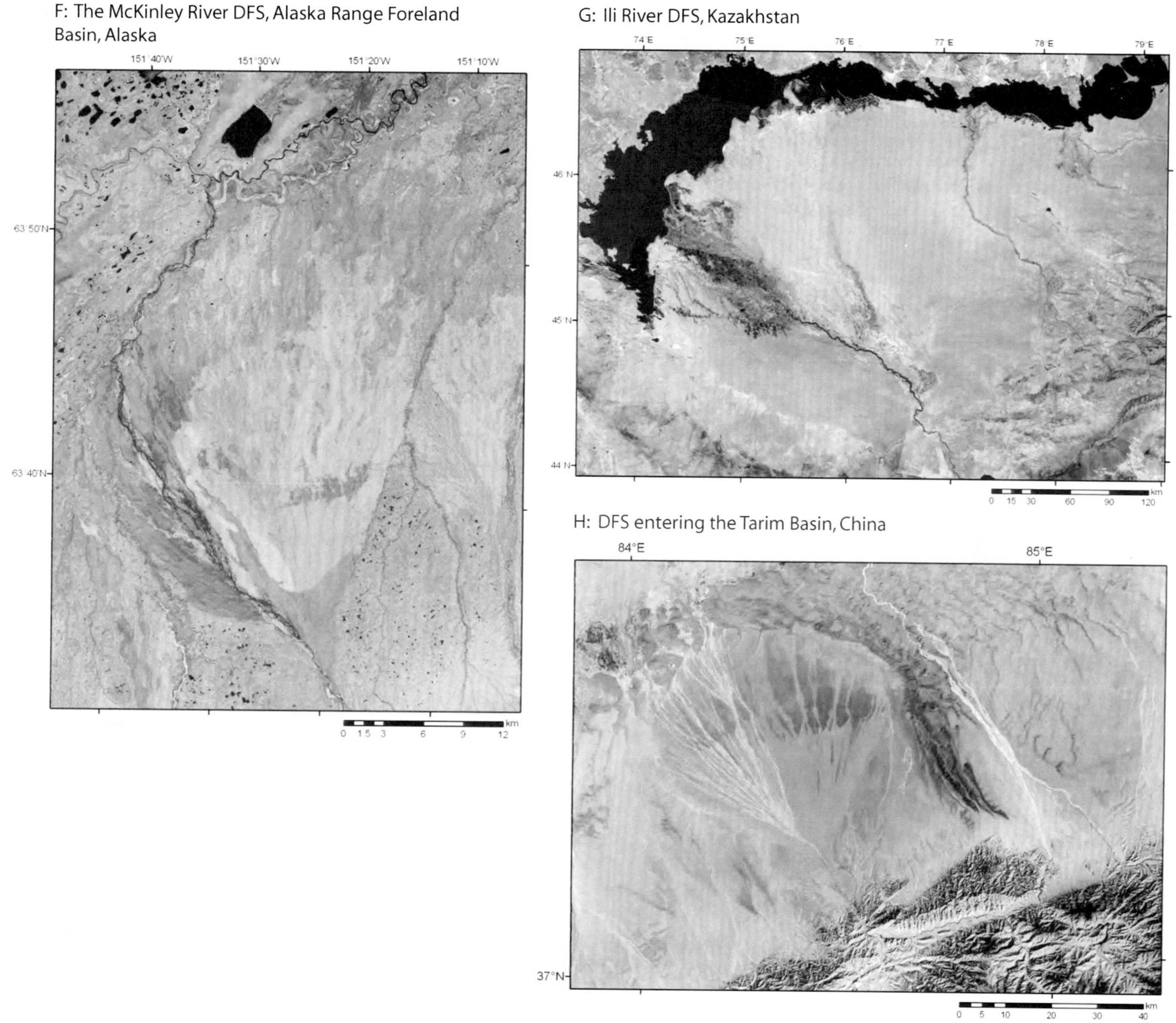

FIGURE 4.—*Continued.*

a smaller scale. These rivers are typically confined between the two adjacent large DFSs downstream from the fluvial fan.

The axial Paraguay River lies on the eastern side of the Chaco Plain (Figs. 4A, 4D, 5) and is confined between either the basin edge to the east and the large DFSs to the west or between opposing DFSs in places where rivers enter the basin from the east (e.g., between the Pilcomayo and the Paraná DFSs). Over time, the river has swept back and forth due to the confinement within this valley, so that deposits display the "classic" amalgamated meander belt form observed in the meandering tributary systems of degradational settings (Fig. 1). This river increases in size downstream (Pasquini and Depetris, 2007). Much like the confined tributary meandering rivers (Fig. 1), we expect deposits beneath the axial system to consist primarily of an amalgamation of point-bar remnants.

Rivers entering the eastern Himalayan Foreland Basin in Assam, India, also form large DFSs (Fig. 4E), coalescing to form a large fluvial plain (Karmakar and Rao, 1999). Rivers entering the basin from the Himalayas generally display a braided form across their entire DFS, though some channel belts become more meandering down-fan (Fig. 4E). Several active and abandoned channel belts are present on each of the DFSs, forming a radial pattern of channelbelts from the DFS apex. Many of the abandoned channels hold active plains-fed streams that are not directly connected to the mountain front. As in the Andean Foreland Basin, these DFSs have a convex-upward form similar to much smaller alluvial fans. Contrasting with the Andean Foreland DFSs, most of the mountain-fed rivers cross their DFS and are directly tributary to the axial Brahmaputra River.

Notably, as the channel belts on the DFSs spread out and bifurcate distally, space is made available for preservation of floodplain fines. Thus, distally we observe the presence of discrete (but potentially wide) braided-channel-belt deposits that are surrounded by floodplain

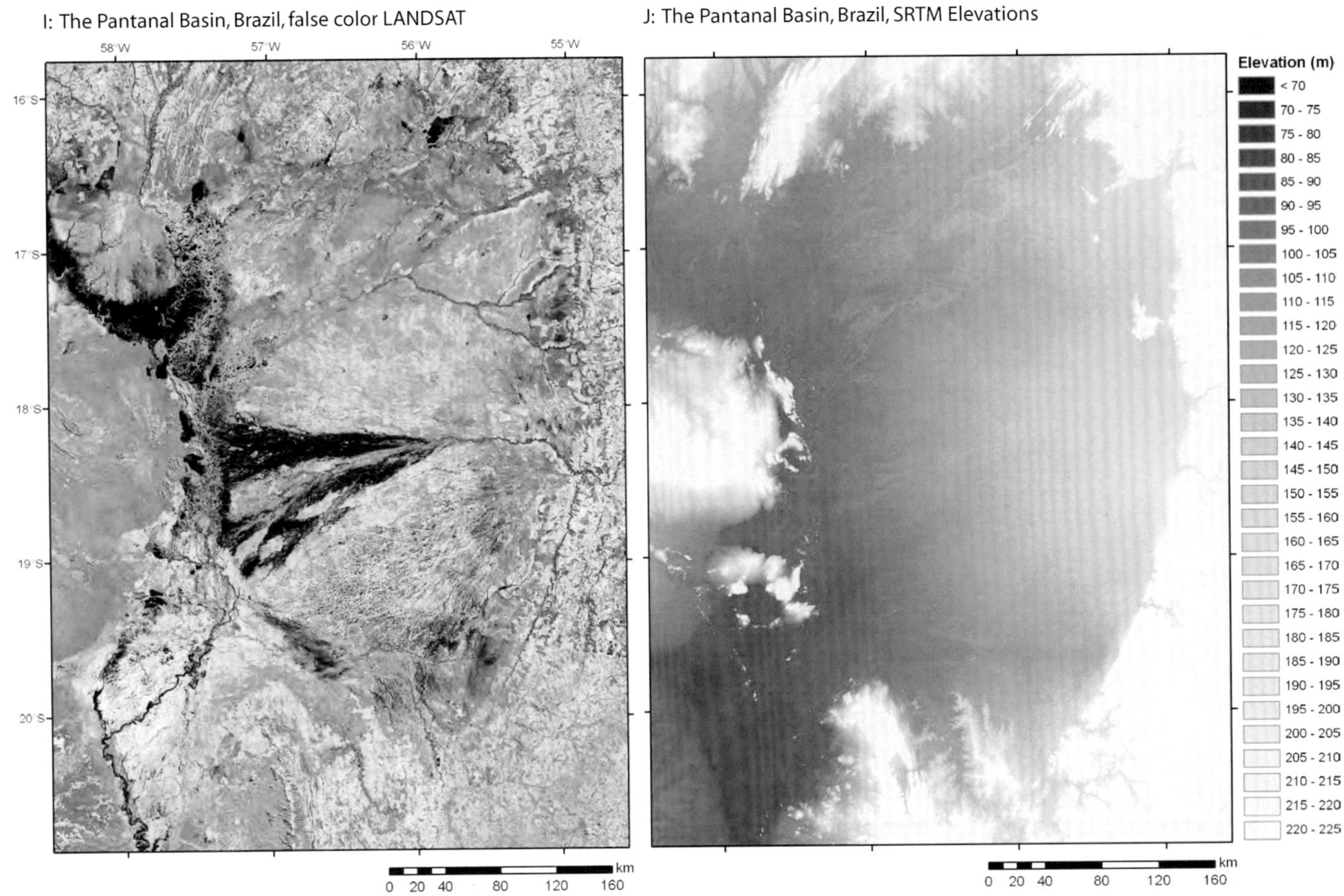

FIGURE 4.—*Continued.*

fines. As we move toward the apex, we observe a greater proportion of channel-belt deposits.

Entering the eastern Himalayan Foreland Basin from the forebulge side to the south are the Burhi Dihing and Dhansiri Rivers (Fig. 4E) (Sarma, 2005). These rivers have meandering form from the DFS apex to the axial Brahmaputra, probably due to lower sediment supply and stream discharge from the Indian highlands and low gradients in this part of the basin. Multiple abandoned meandering channels are present across the Burhi Dihing DFS in a pattern that radiates outward from the apex. Most of the meander belts display a "simple" form, where the meander belt consists of a single channel with some chute-cutoff or neck-cutoff avulsions (Sarma, 2005). These meander belts do not display the amalgamated form observed in confined settings. Avulsions on this DFS appear to be dominated by nodal avulsion style, causing the channel to switch back and forth across its DFS. We expect deposits on this DFS to consist of discrete sandbodies that display relatively wide lateral-accretion units (not amalgamated and truncated sections) held in floodplain fines. However, it appears that many of the simple meander-belt deposits may be stacked and connected.

The Brahmaputra River forms a classic braided channel belt along the axis of the Assam Basin (e.g., Gilfellon et al., 2003; Sarma, 2005; Singh, S.K., 2007). This river is held between opposing DFSs (some of which are presently inactive) in the basin and covers a relatively small area compared to that of the DFSs (about 12%). During seasonal floods, this river completely fills its valley (Goswami, 1985; Singh, S.K., 2007), and fine-grained bar-top deposits are stored only temporarily until the next flood. Thus, these axial deposits are coarse-grained dominated with minimal preservation of fine-grained sediment (Goswami, 1985; Sarma 2005).

The fluvial form in other foreland basins is consistently dominated by large DFSs. For example, the foreland basin north of the Alaska Range holds several large DFSs while rivers outside the basin are held in tributary channels (Fig. 4F; Weissmann et al., 2010). Likewise, the Ili River, in a foreland basin to the Tien Shan, Kazakhstan, forms a large DFS that terminates in Lake Balkhash (Fig. 4G). Several other smaller rivers also form DFS as they enter this basin. Several DFS at different scales are also present around the Tarim Basin, a large foreland basin developed between the Tien Shan and the Kunlun Shan in northwest China (Fig. 4H). Most of these DFSs terminate in a large eolian dune field (the second largest dune field in the world; Sun et al., 2009), and thus may offer an excellent modern analog for fluvial–eolian interaction in sedimentary basins.

Finally, rivers in the west-central Himalayan Foreland basin of the Ganges Plain form several large, coalescing DFSs (Geddes, 1960; Gole and Chitale, 1966; Wells and Dorr, 1987; Mohindra et al., 1992; Sinha and Friend, 1994; Sinha, 1996; Gupta, 1997; DeCelles and Cavazza, 1999; Kale, 2002; Gibling et al., 2005; Sinha et al., 2005; Leier, et al., 2005; Singh, I.B., 2007; Weissmann et al., 2010; Hartley et al., 2010; Chakraborty and Ghosh, 2010; Chakraborty et al., in press). Gibling et al. (2005) described how monsoon strength cyclicity results in a

K: Piggy-back basins in the Andes Mountains, Argentina.

L: Piggy-back basin in Pakistan

FIGURE 4.—*Continued.*

stratigraphic record of incision and aggradation, thus forming regional paleosol-bounded sequences in the DFSs deposits similar to those described by Weissmann et al. (2002), Weissmann et al. (2005), and Fontana et al. (2008).

Other Basins Associated with Forelands: The Pantanal Basin of Brazil has been described as either a backbulge basin (Horton and DeCelles, 1997) or a reactivated forebulge depression (Ussami et al., 1999) associated with the Andean Foreland development. Several

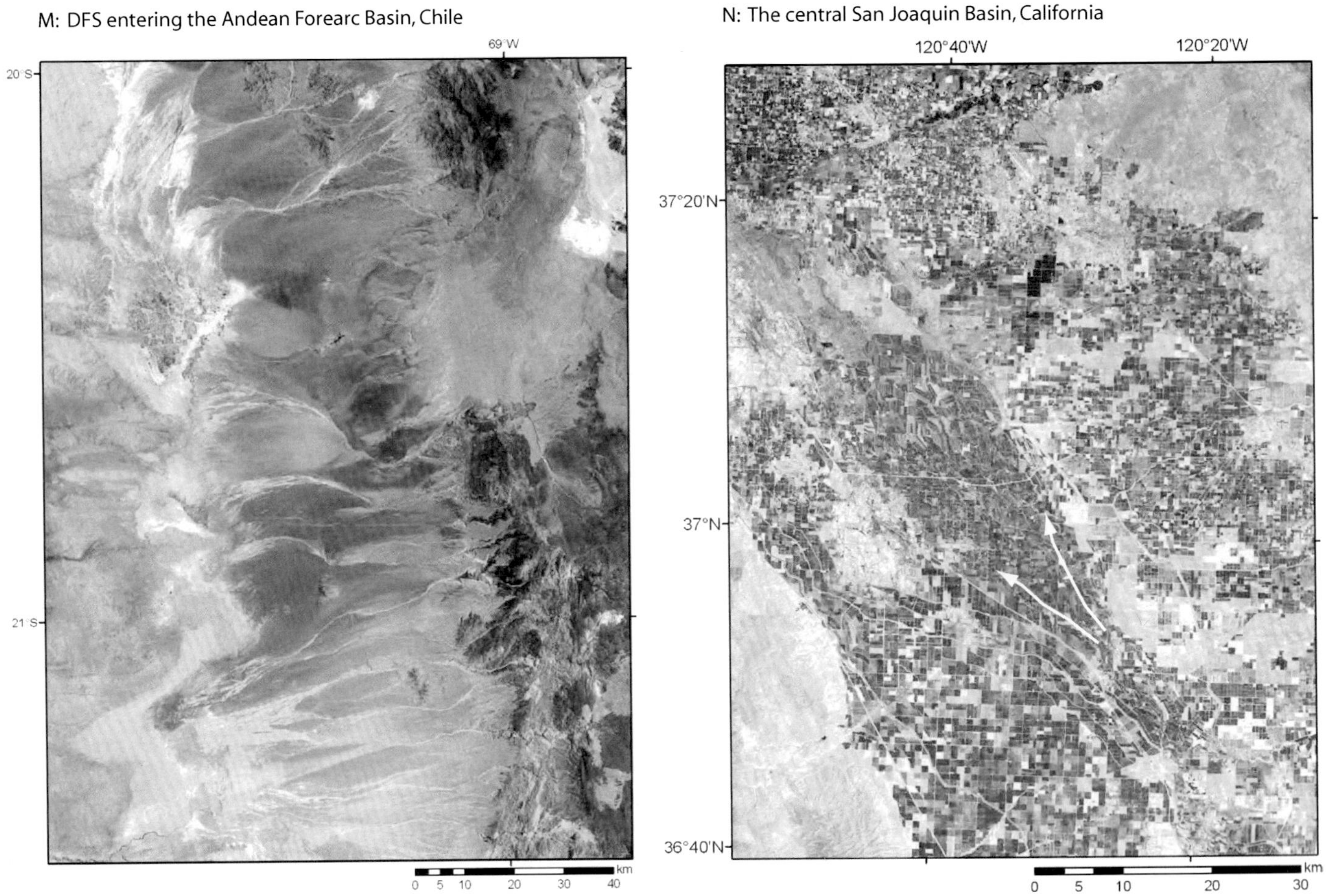

FIGURE 4.—*Continued.*

DFSs develop as rivers enter this basin, with the largest being the Taquari River (Fig. 4I; deSouza et al, 2002; Assine and Soares, 2004; Assine, 2005; Assine and Silva, 2009). The Taquari River is forming a large, round DFS that is presently incised in the proximal to medial reaches (the upper 100 km of the DFS). A radiating pattern of paleochannels is clearly apparent on the abandoned upper surface of the DFS, originating from an apex located where the river enters the basin from the Brazilian highlands (Assine and Soares, 2004). The SRTM DEM clearly shows the convex-upward profile that is characteristic of fan deposits (Fig. 4J). The geomorphic features of this DFS were described by deSouza et al. (2002) and Assine (2005).

Where the active Taquari River leaves its incised valley at mid-fan (e.g., the intersection point), it is developing a broad depositional lobe (deSouza et al., 2002). The river below this intersection point avulses through nodal avulsion processes and appears to currently be undergoing an avulsion to a new position located north of the present channel (Assine, 2005). The mature meander belt below the intersection point displays a single channel form that has very few chute or neck cutoffs. Thus this channel belt probably will not develop an amalgamated form similar to confined tributary rivers (Fig. 7A). Well developed natural levees are shown by lighter colors in the satellite image, with darker wetlands surrounding this alluvial ridge (Fig. 7A). This channel system diminishes in size down-fan, ultimately terminating in splays in the distal portion of the DFS (Fig. 7B; deSouza et al., 2002; Assine, 2005). Due to backwater effects from the Paraguay River, most of the sediment that enters the Taquari River DFS at the apex is deposited somewhere on the DFS (deSouza et al., 2002).

The Taquari River deposits above the intersection point contrast sharply with those in the unconfined area below the intersection point (Fig. 7C; deSouza et al., 2002; Assine, 2005). The incision above the intersection point is sufficiently deep so that the river cannot overtop this valley under the present flow regime (Assine, 2005). In this confined reach, the meander belt consists of the amalgamated form, being dominated by chute-cutoff and neck-cutoff avulsions as the river sweeps back and forth inside its narrow valley. No space exists in this valley for nodal avulsions. This form is more reminiscent of the confined tributary rivers of Fig. 1.

The paleo–channel belts that are present on the abandoned DFS surface appear to have form similar to the channel system below the intersection point (e.g., simple meander belt form without point-bar amalgamation), except that these paleochannels appear to be larger than the modern channel system. Tracing paleo–channel belts down-fan, however, we observe similar bifurcation into splays (Fig. 7D).

The Taquari River DFS appears to be experiencing incision and aggradational cyclicity that is similar to climate-driven cycles described by Weissmann et al. (2002), Weissmann et al. (2005), Gibling et al. (2005), and Fontana et al. (2008). Such cyclic depositional patterns may be common on DFSs and may lead to development of continental sequence stratigraphic models, though the

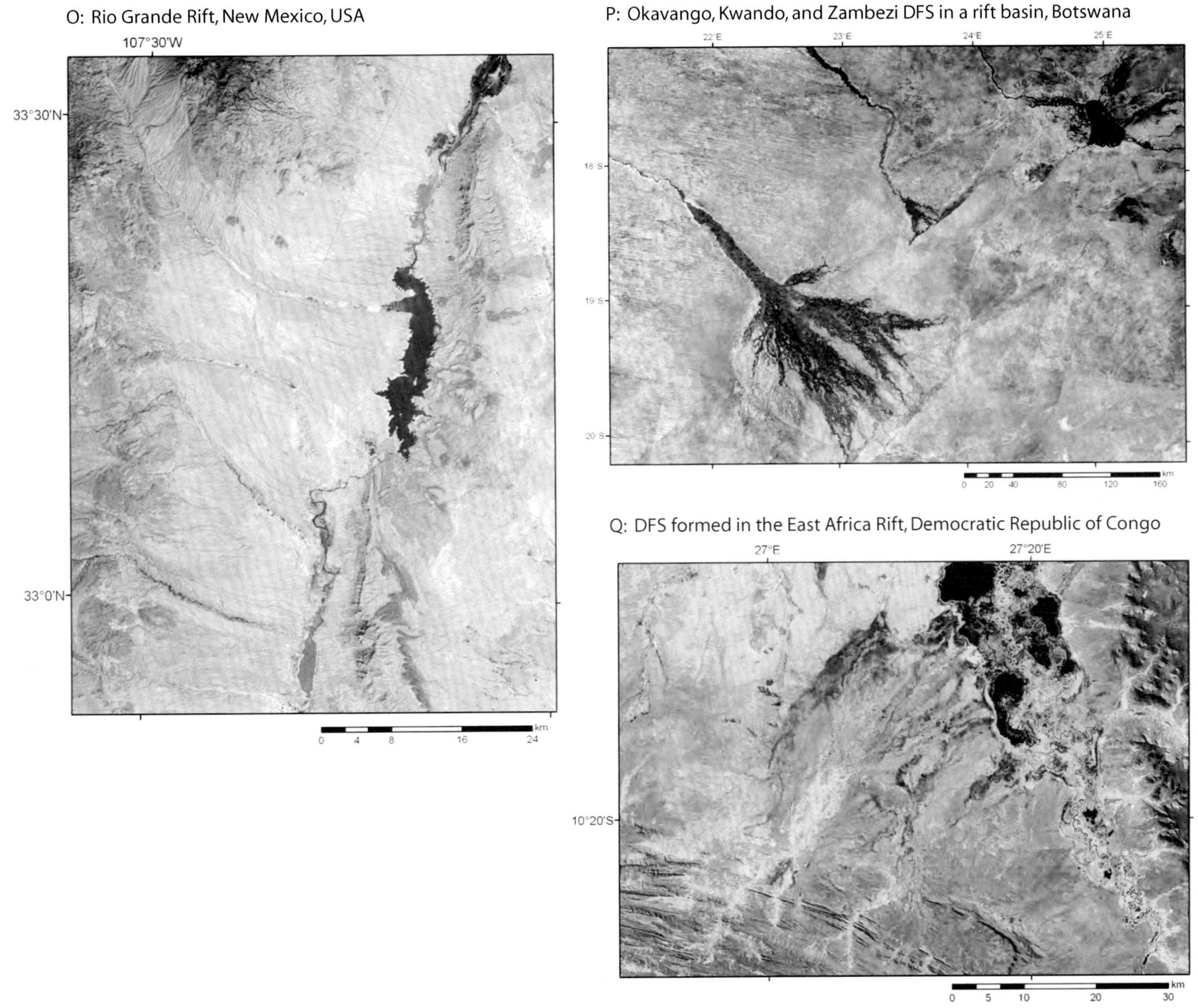

FIGURE 4.—*Continued.*

extreme Quaternary climate shifts may create exaggerated depositional cycles compared to those of other time periods.

Piggyback Basins: Rivers entering piggyback basins (Ori and Friend, 1984) around the world also develop DFSs (e.g., Howard et al., 2006) (Figs. 4K, L). Piggyback basins are restricted to tectonically active mountain ranges and tend to be relatively narrow with the long axis of the basin trending parallel to the basin-bounding thrusts. Most identified modern piggyback basins are in relatively arid settings, and thus DFSs in these basins lack significant vegetation and are dominated by braided, laterally mobile channel belts or sheetflood and debris-flow processes. Two types of DFSs occur in piggyback basins: lateral systems derived from the uplifted basin flanks, and axial systems fed by the lateral systems. Axial DFSs are developed where the lateral systems or other topographic features allow expansion of the fluvial system across the basin floor (Hartley et al., 2010). In many instances, these axial DFSs terminate in either a playa or perennial lake. In most examples, lateral systems dominate the basin, with the axial system restricted to a single channel developed between opposing lateral systems. Sheetflood and debris-flow deposition appear to be important on the lateral systems, whereas sheetflood and streamflow processes appear to dominate in the axial area (Ramos et al., 2002). Because these basins are located in actively deforming mountain belts, they have limited preservation potential due to either uplift and deformation and/or stream capture from drainage networks developed at lower altitudes. Where preserved (e.g., Ori and Friend, 1984), the deposits of the lateral systems are typically from small alluvial fans with those of the axial system similar to braided streams with a flashy discharge.

Forearc Basins: DFSs form in the continental forearc basins that developed on the west coast of South America (Fig. 4M). The apices of these examples occur where the rivers cross the thrust-bounded margin of the Precordillera of northern Chile, lose confinement, and create large DFSs. The hyperarid climate present in the current forearc has

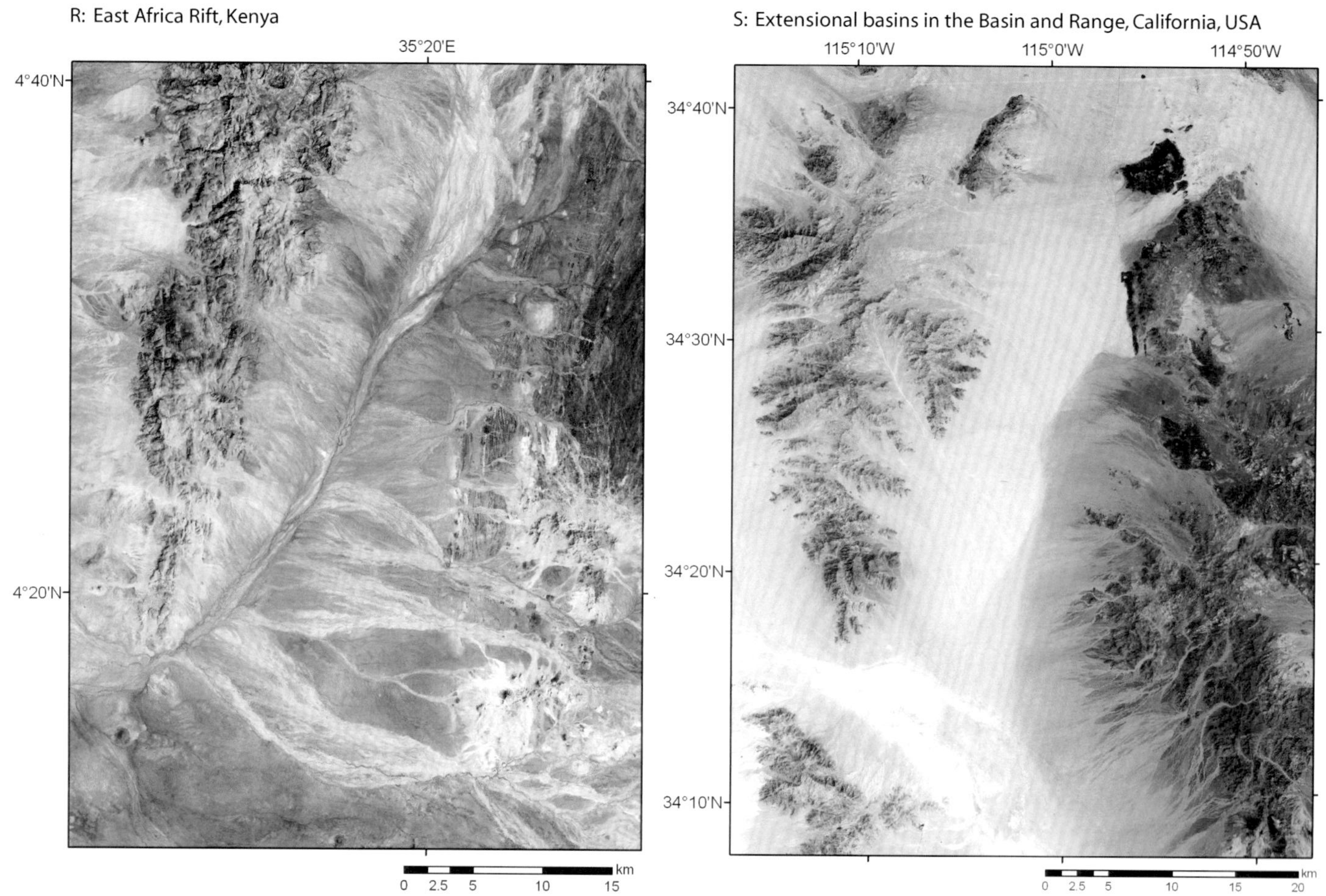

FIGURE 4.—*Continued.*

preserved these fans, which were largely built in the Miocene (Evenstar et al., 2009). Increased aridity in the Atacama throughout the Late Cenozoic (Hartley and Chong, 2002) has resulted in a progressive decrease in fan activity, with older fan deposits now largely incised by modern channel systems. The incised modern fan channels have a limited lateral extent compared to the main fan bodies and terminate in evaporative playa systems. Sediments preserved at the toes of these large-scale fan systems comprise pebbly sandstones restricted to individual channel belts interbedded with evaporites, diatomites, and mudstones (e.g., Saéz et al., 1999).

Other Compressional Sedimentary Basins: The San Joaquin Basin in California initially formed as a forearc basin but now acts more like a foreland basin due to the transpression of the San Andreas strike-slip system and uplift of the Coast Ranges to the west of the basin (Bartow, 1991; Lettis and Unruh, 1991; Unruh, 1991). Large DFSs fill the basin from both sides, with the San Joaquin River axial system held between DFSs in the northern portion of the basin (Fig. 4N) (Arkley, 1962; Janda, 1966; Marchand, 1977; Lettis, 1988; Weissmann et al., 2002; Weissmann et al., 2005). Overall, the DFSs dominate the depositional area in the San Joaquin Basin, with DFSs filling approximately 85% of the basin, the axial San Joaquin River filling approximately 10% of the basin, and lacustrine and delta deposits filling the remainder of the valley.

Several DFSs on the east side of the basin have rivers that are directly connected to formerly glaciated portions of the Sierra Nevada, and thus these rivers experienced dynamic changes in sediment supply and discharge throughout the Quaternary (Janda, 1966; Marchand, 1977; Marchand and Allwardt, 1981; Lettis, 1988). Weissmann et al. (2002) and Weissmann et al. (2005) described sequence development on these DFSs due to the glacial–interglacial cyclicity. During periods of glacial outwash, the intersection point moved toward the DFS apex and widespread aggradation occurred. Incision and exposure of the upper DFS surface occurred during interglacial periods, thus forming paleosol-bounded stratigraphic sequences. Similar sequences developed on DFSs that are not linked to glaciated regions in the Sierra Nevada; however, the sequences appear to be more subtle and the DFSs tend to be much smaller and do not fill the accommodation available in the basin (Weissmann et al., 2005). Similar glacially driven sequences have been described by Fontana et al. (2008) for DFS on the Venetian–Friulian Plain of northeast Italy.

The axial San Joaquin River is held between opposing DFSs along most of the basin. However, where the nonglaciated DFSs are present and the horizontal accommodation is not filled in the basin, the axial San Joaquin River spreads out across the basin floor in a distributive network (Fig. 4N). Though anthropogenic modification is extreme in this basin, canals and controlled stream courses follow the old San Joaquin River channels. In these unconfined reaches, several active channels are held between relatively large areas of floodplain deposits. Prior to human alteration, these areas were very swampy. In contrast, along the confined reaches of the San Joaquin River, deposits tend to be coarse-grained dominated (Lettis, 1988).

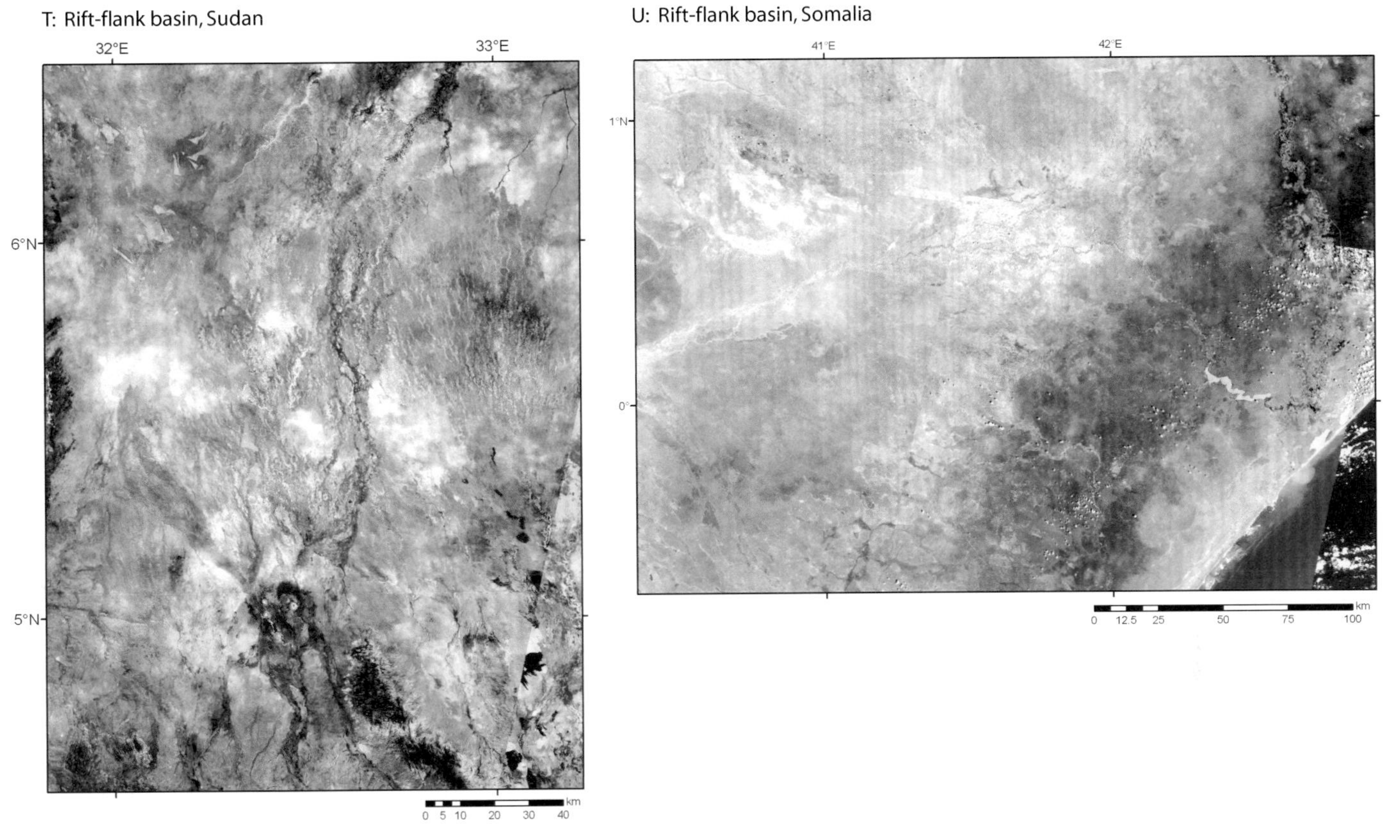

FIGURE 4.—*Continued.*

Basins in Extensional Settings

Rift and Rift-Associated Basins: Distributive systems have long been recognized in extensional settings (e.g., Leeder and Gawthorpe, 1987; Gawthorpe and Leeder, 2000; Mack et al., 2003). Two types of DFSs can be recognized: lateral and axial. Lateral systems are derived from both the hanging-wall and footwall of active half graben. Hanging wall systems are larger and in many cases may dominate the basin fill in half graben (e.g., Leeder and Gawthorpe, 1987; Frostick and Reid, 1987; Mack et al., 2003). Axial DFSs occur at the toes of lateral systems, from which they are often fed and run parallel to the basin-bounding faults. Depending on the climate, axial systems may either terminate in a playa or a permanent lake in an endorheic basin. Examples of these types of systems are known from the Plio-Pleistocene of the Rio Grande Rift and Mimbres area in New Mexico (Mack et al., 1997; Mack et al., 2006).

In the Rio Grande rift between 5 Ma and 0.8 Ma, much of the fill of half graben comprised DFSs derived laterally from the hanging-wall dip slope (Fig. 4O; Mack and Leeder, 1999; Mack et al., 2006). In contrast, footwall-derived alluvial fans cover only a small part of the basin. Within the half graben the axial ancestral Rio Grande River covered a relatively small part of the basin-fill as lateral development of the river was constrained by both the hanging-wall-derived and footwall-derived DFSs (e.g., Leeder and Mack, 2001). Facies models have been developed for these DFS deposits by Mack et al. (2002), who documented a series of discrete single and multistory channel bodies interbedded with overbank mudstones. The channel bodies consist of pebbly sandstones and sandstones displaying lateral accretion.

The southern Rio Grande Rift system also provides some well documented examples of axial fluvial systems that "spill over" into adjacent half graben to form DFSs. For example, the Rio Mimbres forms a Quaternary to modern DFS where the axial fluvial system flowed into an internally drained half graben to form a terminal DFS (Mack et al., 1997). The orientation of the DFS varies depending on the entry point of the trunk river into the basin. Where the trunk river runs parallel to the basin-bounding fault, a series of basin-axis-parallel channel sand bodies are developed (Mack et al., 1997). In contrast, if the trunk channel enters the basin on the hanging wall dip slope, a large-scale laterally derived DFS is developed with channel sandbodies nearly orthogonal to the basin axis.

The well-documented Okavango DFS (Stanistreet and McCarthy, 1993; Stanistreet et al., 1993; McCarthy et al., 2002; Gumbricht et al., 2004) and associated DFSs of Namibia developed to the northeast of the Okavango are similar to the lateral "spillover" examples from the Rio Grande Rift (Fig. 4P). In contrast to the Rio Grande Rift, these examples occur in basins that are in the initial stages of rift development through propagation of the East African Rift System into the Kalahari Basin (Fig. 4P; Scholz, 1976; Shaw and Thomas, 1993). The apex of the Okavango DFS is located at the western fault-bounded margin of the basin where the main channel crosses the fault. The main trunk channel has a meandering planform and bifurcates into a series of smaller meandering and anastamosing channels that diminish in size downstream due to evaporation and infiltration (Stanistreet et al., 1993). Only two percent of the original discharge that enters the Okavango DFS at the apex leaves the basin (Wilson and Dincer, 1976; McCarthy et al., 2002) in the form of a small channel that crosses the eastern basin-margin faults and terminates in a lake. The Okavango represents an example of a lateral DFS developed in a full graben.

Northeast of the Okavango, two smaller DFSs, the Kwando and Zambezi, are developed (Fig. 4P). Again the apices of both of these

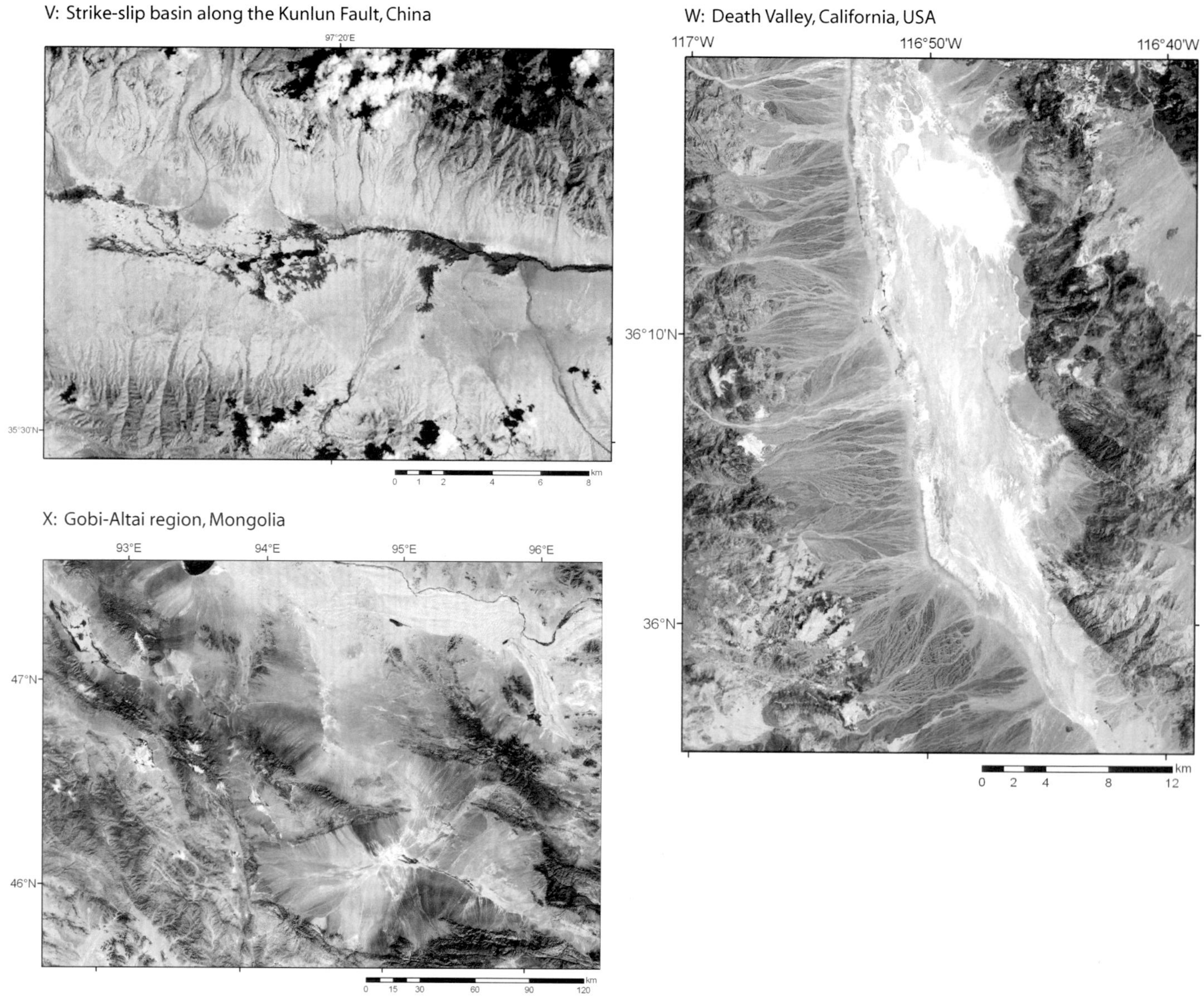

FIGURE 4.—*Continued.*

examples are located where the trunk system crosses a fault. In the Kwando example, the river channels decrease in size down-fan through bifurcation, evaporation, and infiltration and terminate in an axial river system that eventually joins the Zambezi along a basin-bounding fault. Facies developed on the Kwando DFS are likely to be similar to those of the Okavango DFS. The apex of the Zambezi DFS occurs where the main channel crosses a fault. Subsequent channel switching from this point has resulted in the development of the DFS which is dominated by relatively large meandering channels up to 500 m wide. It is interesting to note that the channel planform changes from braided to meandering across the fault.

Additional examples from the East African Rift system serve to highlight the importance of climate in controlling DFS characteristics. In the example illustrated in Fig. 4Q from the southern part of the Democratic Republic of the Congo, DFSs form where rivers enter the rift basin on the northern edge of the Mweru–Tshangalele Rift, a southwestern extension of the main western rift. In the relatively wet climate of this basin, the DFSs terminate in either perennial lakes, wetlands, or the axial river systems. Fluvial form on these DFSs are typically meandering near the DFS apex, bifurcating into smaller splay channels (below the resolution of the Landsat imagery) within 4–5 km below the DFS apex. Down-fan, the splays and/or wetlands coalesce into vague channel belts. However, clear channels are not present until the distal portions of the DFSs. This may be due to discharge of groundwater in these areas. Further evaluation is needed to determine the reason behind this morphology.

In the drier climatic zones, for example those in the rift in northern Kenya and southern Ethiopia, DFSs that have processes commonly associated with alluvial fans (e.g., sheetflood, debris flow, and shallow braided channels) are developed (e.g., Jungerius et al., 2002; Quade et al., 2004). An example illustrated in Fig. 4R shows a series of lateral DFSs (alluvial fans) developed on the flanks of a small half graben. Footwall-derived fans from the west are limited in their development by an axial fluvial system which is largely sourced from the north. Fans derived from the hanging wall dip slope to the east increase in size to the south as the basin widens. The lateral fans display a braided planform with channels bifurcating downstream while the axial system becomes increasingly confined to the south as the lateral fan sizes increase and constrain the system.

Further examples of DFS development in arid climates can be seen in the Basin and Range province of the western United States (e.g., Beaty, 1963, 1989; Wells et al., 1987; Harden et al., 1991; Jansson et al., 1993; Harvey and Wells, 2003; Harvey, 2005) (Fig. 4S). DFSs are developed as both lateral and axial systems. Lateral systems are represented by alluvial fans that fill the available accommodation space in many basins with the axial system restricted to a relatively minor braided stream. In some endorheic basins, sufficient space may be available for axial systems to expand and form DFSs that terminate in a playa (Fig. 4S). Alluvial fans comprise a combination of sheetflood and debris-flow deposits (e.g., Beaty, 1963, 1989; Harvey and Wells, 2003; Harvey, 2005) and the axial systems comprise a mix of braided-stream deposits. Most of the fill of the half graben in the Basin and Range Province are alluvial-fan deposits, with the axial system comprising a smaller proportion of the fill (e.g., Harvey, 2005).

The general characteristics of DFS development in rift basins, as seen from the examples illustrated above, is relatively simple, with laterally derived DFSs dominating much of the basin fill and axial systems forming a relatively minor component. Hanging wall-derived DFSs tend to be larger than footwall-derived DFSs due to differences in the size of drainage basin area (Frostick and Reid, 1987), and the accommodation space available for fan expansion into the basin. Differences do exist between DFSs developed in arid climates and those in more humid ones. DFSs in arid climates tend to be dominated by braided, sheetflood and debris-flow deposition whereas those in more humid climates tend to display a meandering planform.

Rift-Flank Basins: DFSs are developed on rift-basin flanks where regional domal uplift associated with mantle-plume emplacement results in the generation of a radial drainage away from the center of the dome. This is illustrated in Fig. 4T, where a number of northward draining DFSs are developed on the northern flank of the East African Dome in southern Sudan. DFSs developed in this tectonic setting

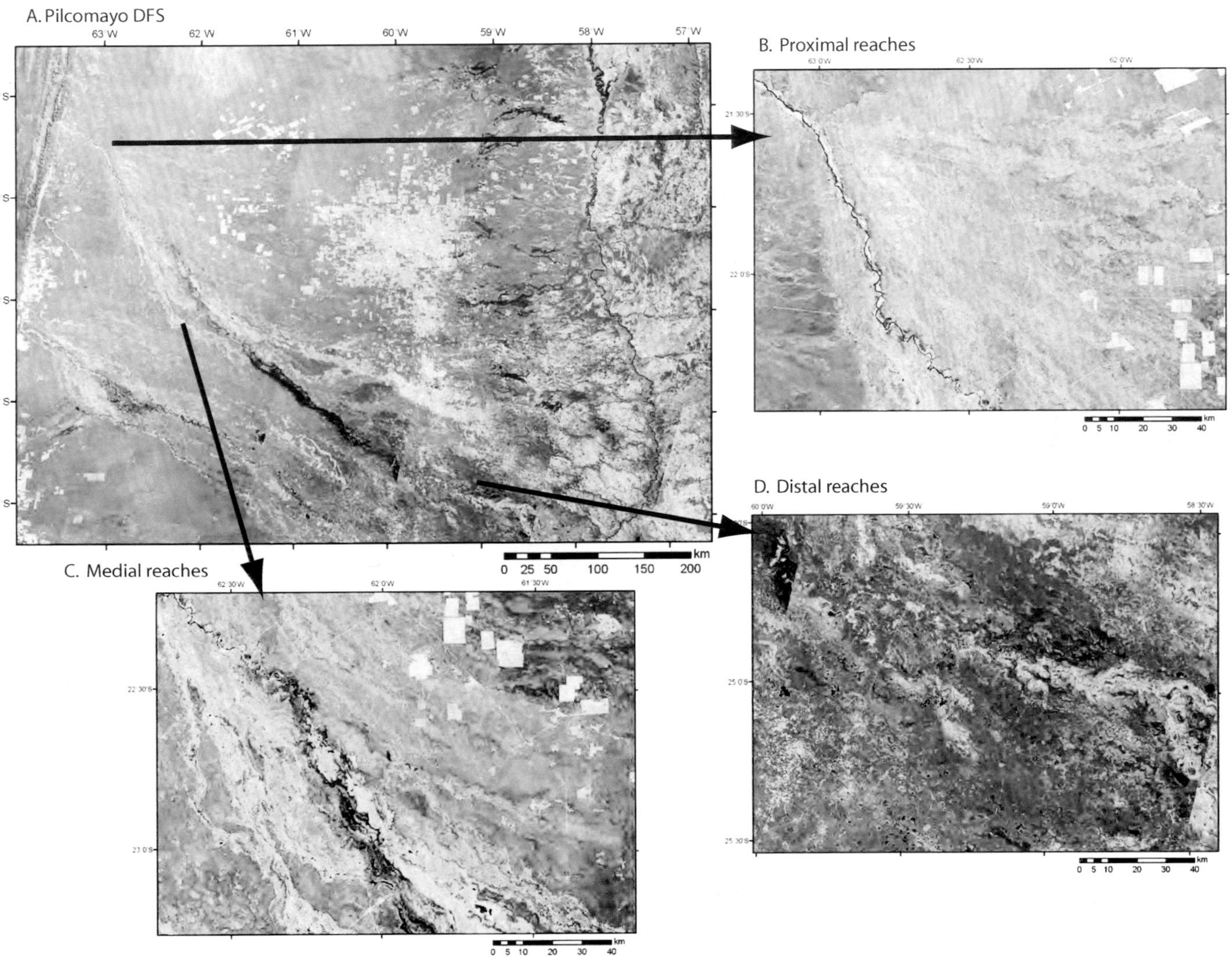

FIGURE 5.—False-color Landsat mosaics showing details along the Pilcomayo DFS, Bolivia, Paraguay, and Argentina. A) The full Pilcomayo DFS. B) The braided Pilcomayo River crossing the proximal Pilcomayo DFS. C) The smaller channels of the medial Pilcomayo DFS, with several bifurcations apparent. D) Splays and swamps of the distal Pilcomayo DFS; in this area, channels are just at or below the resolution of the Landsat imagery.

commonly have a radial length > 30 km and sometimes more than 100 km. The apices of DFSs occur at different points down the rift-dome flanks but at similar elevations (the examples illustrated in Fig. 4T range between 474 and 542 m) but are not associated with abrupt changes in elevation. Modern channel planforms on the southern Sudan examples are meandering and widths are generally less than 75 m. Active channels occupy only a small percentage of the modern land surface (~ 5%); however, as a result of switching, channels are likely to form a larger proportion of the overall DFS deposit. The location of these DFSs in a tropical climate means that the overbank area is heavily vegetated. We expect the deposits of these systems to be representative of "simple" meander belts where nodal avulsions result in channel switching to produce discrete sandbodies that display relatively wide lateral-accretion units (not amalgamated and truncated sections) held in extensive floodplain fines with well developed soils.

Another example of rift-flank DFS development can be seen on the eastern margin of the East African Dome. In this example, a number of DFSs are developed between the edge of the dome and the Indian Ocean (Fig. 4U). All of these DFSs emerge from an incised channel system with apices located at different elevations, reflecting variations in stream power, discharge, sediment load, etc., for individual fluvial systems. Indeed different phases of DFS development can be observed in Fig. 4U. A number of incised, older DFSs are located upstream of present-day apices, indicating that DFS apices have migrated down-dip. Channel planform is variable, with some displaying straight channels that bifurcate downstream and become meandering while others comprise meandering channels that bifurcate and splay downstream. Radial lengths are greater than 40 km, with some approaching 100 km. Most channels either terminate in splays within wetlands or join an axial channel that eventually leaves the basin.

An important difference between DFSs developed on rift flanks and those in rift basins is that rift-flank DFSs tend to be larger (> 30 km in radius). The reason for this difference is that rift-flank DFSs are not restricted by basin width while the DFSs that develop inside the rift basins are restricted by a basin that is typically less than 30 km across.

Basins in Strike-slip Settings

Examples of DFSs developed in strike-slip basins are shown in Figs. 4V, 4W, and 4X. One of the basins associated with the sinistral Kunlun Fault in the Qinghai region of central China is shown in Fig. 4W. The Kunlun Fault formed in response to continental collision between India and China in the late Cenozoic (Jolivet et al., 2003). A series of pull-apart basins are associated with the fault. In each of the basins along the Kunlun Fault, lateral DFSs dominate the fluvial depositional pattern, with the axial fluvial system held between opposing DFSs. These DFSs show deposits that are similar in form to arid-climate alluvial fans.

Death Valley in California is a pull-apart basin that has been described extensively (e.g., see reviews of Blair and McPherson, 1994 and Knott et al., 2005) (Fig. 4W). DFSs developed in Death Valley take

FIGURE 6.—A Chinle Formation (Triassic) outcrop at Ghost Ranch, New Mexico. The person at the base of the meander-belt strata is approximately 1.75 m tall (marked by arrow).

two main forms, lateral alluvial fans derived from both the hanging wall and footwall and an axial DFS associated with the termination of the Amargosa River (Fig. 4W). DFSs formed on the hanging wall dip slope, which is modified locally by faults, flank the western margin of the basin and have radii of 6 to 10 km. Sedimentary processes on these fans are dominated by sheetflood and debris flow (Blair and McPherson, 1994; Staley, et al., 2006; Volker et al., 2007). Footwall-derived fans have radial lengths that range from 0.3 to 3 km and are dominated by debris-flow processes with minor sheetflood deposits.

The axial Amargosa River enters the basin from the south and comprises a low-relief braided fluvial system 70 m wide with up to 2 m of relief at the apex. Downstream the channel system bifurcates into a series of low-relief straight to meandering mud-filled channels that decrease in width and relief and eventually form splays or terminate in a playa.

While the facies distributions do display some asymmetry, with larger hanging wall fans being two to three times bigger than footwall-derived fans, most of the basin fill comprises playa and the fine-grained portion of the axial Amargosa DFS. In this particular example, it is likely that a combination of arid climate and rapid subsidence associated with the Black Mountain Fault Zone makes a situation where deposition cannot keep pace with subsidence such that fine-grained facies dominate the basin-fill succession at present.

The Gobi Altai region of Mongolia contains a number of transpressional basins associated with a broad corridor of eastward sinistral strike-slip displacements and NE-SW shortening (Figure 4X; Cunningham, 1996; Howard et al., 2006). The basins contain alluvial-fan, fluvial, playa, and eolian deposits of Miocene and younger age (Owen et al., 1997) and the majority are dominated by DFSs (Fig. 4X). Basins are bounded by active thrusts which often have a strong strike-slip component, and the fault-bounded blocks are elevated 1 to 2 km above adjacent basin floors (Cunningham, 1996). DFSs are developed in all of the basins in Fig. 4X. In many basins they take the form of large-scale alluvial fans with radii ranging from 20 to 60 km. As with other strike-slip basins, two styles of DFSs are developed: lateral and axial. The lateral systems dominate most of the basins and in some endorheic basins are the only style of DFS developed (e.g., the basin in the southern part of Fig. 4X). These DFSs typically have a single active channel with a braided planform which bifurcates downstream. A number of the DFSs in the southern basin in Fig. 4X are affected by active faults that break the DFS surface, resulting in the formation of new apices and incision up dip from these points. Axial systems are generally restricted to a small proportion of the basin area except for the well developed axial system in the basin in the NE of Fig. 4X. In this example a single braided channel passes downstream into a single meandering channel. This channel then bifurcates into a number of meandering channels, one of which terminates in a series of splays, while the other channels rejoin to form a single channel where an eolian dune field is developed basinwards.

Strike-slip-basin and rift-basin depositional systems are consistently similar in terms of being dominated by lateral systems represented by alluvial fans in dry climates or meandering-dominated DFSs in "wetter" rifts, with occasional axial systems if there is enough space for them to develop. Axial systems in these strike-slip and rift basins may be braided or meandering, with some meandering systems present in arid basins (e.g., Figs. 4P and 4X), such that planform may be independent of climate (Hartley et al., 2010).

DISCUSSION

Through the examples presented in several basins from various tectonic settings, we have shown that DFSs dominate the fluvial depositional area in continental sedimentary basins. This is true for all 724 modern sedimentary basins thus far identified (Weissmann et al., 2010). Though the characteristics of the channel deposits of DFS rivers may be similar to those described in current facies models, the larger-scale architecture are considered to be different. Rivers on DFSs commonly decrease in size downfan (though a few examples exist where this is not the case, e.g., the Kosi DFS) and migrate through nodal avulsions across the active DFS. Because of the radial pattern of channelbelt deposits, coarse-grained channelbelt deposits overlap or are closely spaced near the DFS apex and have a greater separation downfan. Thus, we expect higher proportions of overlapping coarse-grained deposits in the proximal reaches of the DFS. This has been shown for several ancient DFS examples (e.g., Galloway 1980; Tyler and Ethridge, 1983; Bluck, 1986; Nichols and Fisher, 2007; Cain and Mountney, 2009).

Where rivers are confined in a sedimentary basin (e.g., in incised valleys, axial positions, or between larger DFSs), rivers appear to switch back and forth across the confined valley. Thus, the rivers develop an amalgamated form if meandering, or consist primarily of coarse-grained bar forms if braided. Rivers that are not confined to one position on the DFS (e.g., on an open fan) appear to commonly develop "simple" meander belts or a more open braided form. Thus, we expect these different depositional styles to produce different facies architectures. If this proves to be correct, we may be able to develop a more predictive large-scale architectural model for fluvial systems based on sandbody geometry and form.

An example of different controls on sandbody style is shown on the Taquari River DFS (Fig. 7), where the DFS is incised in the proximal to medial reaches but active below an intersection point. The character of fluvial style appears to be significantly different above and below the intersection point. In this system, we expect that preserved meander-belt deposits in the confined incised valley reach would be coarse-grained dominated, with preservation of fines possible only where shifting of the channel belt results in isolated patches of floodplain deposits (e.g., *valley fills* of Gibling, 2006). This sand body would be relatively narrow (e.g., the width of the incised valley, 2–4 km wide in this case) and display an amalgamated form similar to that of tributary rivers. The sand body may be multistory as the aggrading system fills the valley during the next aggradational period. Similar forms were described by Weissmann et al. (2002) and Weissmann et al. (2005) for DFSs in the San Joaquin Basin of California and by Fontana et al. (2008) for DFSs along the southern Alpine foreland basin of Italy.

Below the intersection point, we expect preserved sand bodies to consist of discrete ribbon channels that are surrounded by floodplain deposits (similar to Chinle Formation deposits shown in Fig. 6). Additionally, since the channel shifts regularly through nodal avulsions (Assine 2005), avulsion deposits similar to those described by Smith et al. (1989) in the Cumberland Marshes of Saskatchewan, Aslan and Autin (1998, 1999) for the lower Mississippi River, Makaske et al. (2007) for the Rhine–Meuse system, and Kraus (1996) and Kraus and Davies-Vollum (2004) for the Eocene Willwood Formation in Wyoming may dominate the fine-grained deposits around these meander-belt deposits. These different channel belt forms may help explain the types of sand-body geometries that have been observed in the rock record (e.g., Gibling, 2006).

In addition to differences in the channel-belt deposits above and below the intersection point, it appears that the soils developing in the depositional lobe will contrast with those on the exposed, inactive DFS. Because the channel belt can shift across the active depositional lobe relatively rapidly (Assine 2005), soils will not have much time to develop. Thus, we expect to observe relatively immature soils in this area, with many of these soils displaying a hydromorphic form. In contrast, the exposed upper DFS surface is high and well drained, and it will remain exposed until the next aggradational phase on the DFS (Weissmann et al., 2002; Weissmann et al., 2005). Therefore, these soils will gain more maturity and may form a "continental sequence boundary" (e.g., Weissmann et al., 2002; Weissmann et al., 2005; Gibling et al., 2005; Fontana et al., 2008). This type of exposure

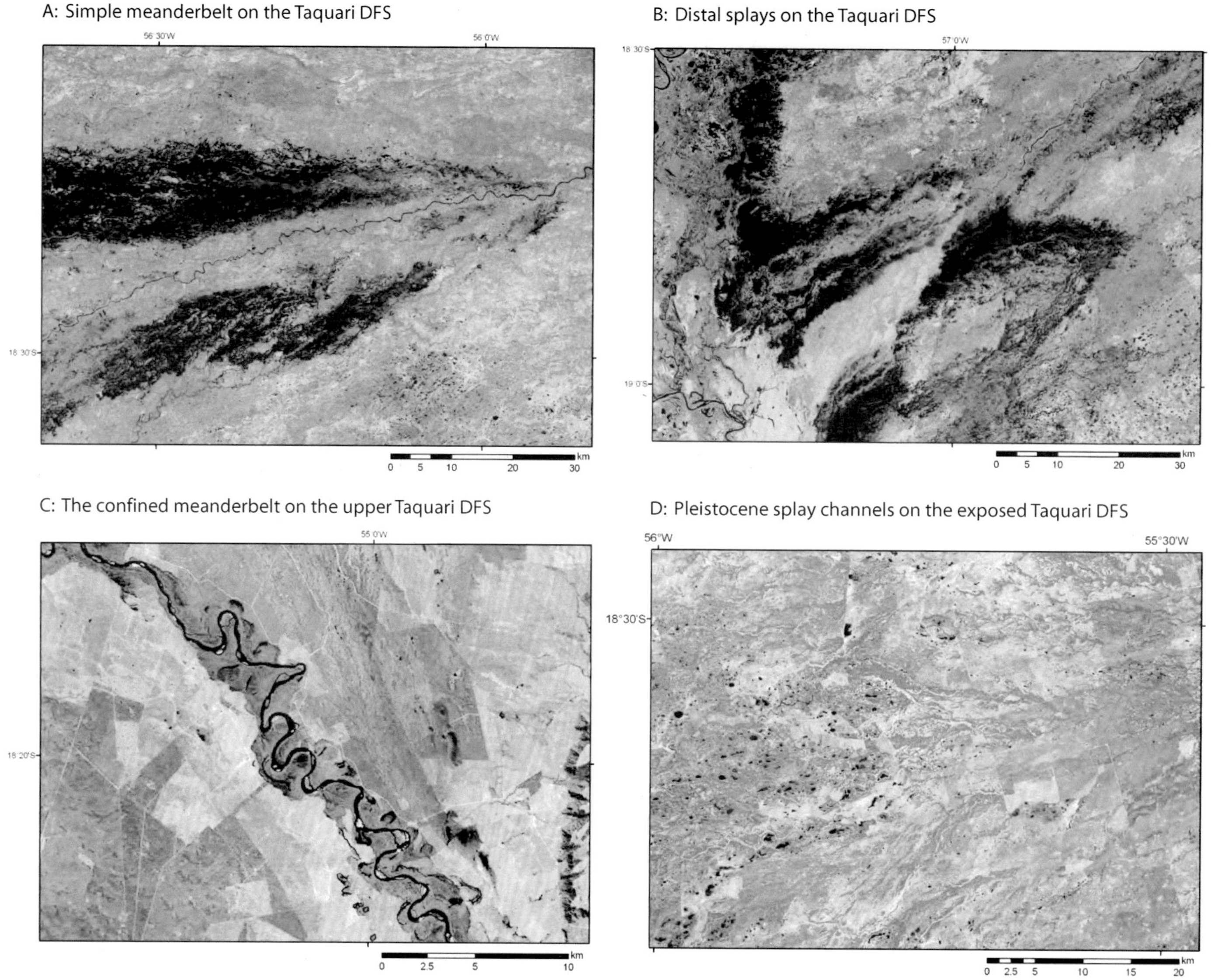

FIGURE 7.—Examples of meander belt style from the Taquari River DFS, Pantanal Basin, Brazil. A) The simple meander belt located below the intersection point. The lighter color surrounding the meander belt is levee deposits, while the dark regions are swamps. B) Distal splays on the Taquari River DFS. C) The confined, amalgamated meander belt on the upper DFS where the river is held in an incised valley. D) Splay and bifurcated channel deposits on the exposed Pleistocene surface of the Taquari River DFS.

surface may explain laterally extensive, mature paleosols that are commonly observed in fluvial successions in the rock record (e.g., Bown and Kraus, 1987; Kraus, 1997, 1999; Dubiel et al., 1991).

On several DFSs we observe spring lines along the distal DFS (Fig. 8). Channel patterns and soils developing above this spring line will probably differ significantly from those below the spring line. Moisture and drainage differences on the DFSs would produce differences in the soil morphology, as well.

Focused fieldwork on several DFSs is needed to test these observations from satellite imagery. If our interpretations can be supported, we may be able to use channel-belt and paleosol character to define (1) position in a sedimentary basin and (2) shifts of depositional patterns on DFSs through time from the rock record.

An issue with any study that uses modern geomorphic facies distributions to interpret rock record facies distributions is the fact that we live in the Quaternary—a relatively unusual period in Earth's history. Many of the fluvial patterns that we observe are adjusted to, or are in the process of adjusting to, the present interglacial conditions. Additionally, many of the landforms and exposed DFSs surfaces are remnant from aggradational periods related to past glacial events (e.g., Zehfuss, et al., 2001; Weissmann, et al., 2002; Weissmann et al., 2005; Fontana et al., 2008) or relicts related to past climatic conditions that favored aggradation (e.g., Gibling et al., 2005). Thus, we are faced with constructing facies models from river systems that may not be analogs for ancient climates.

Even with these limitations, the distributive nature of fluvial systems appears to be related to lack of confinement as the river enters the sedimentary basin and climatic cyclicity appears to affect incision and aggradational events on DFSs. DFSs occur in all climatic settings and all continental sedimentary basins, regardless of tectonic setting. Therefore, we hypothesize that DFSs have dominated the fill of continental sedimentary basins in the past.

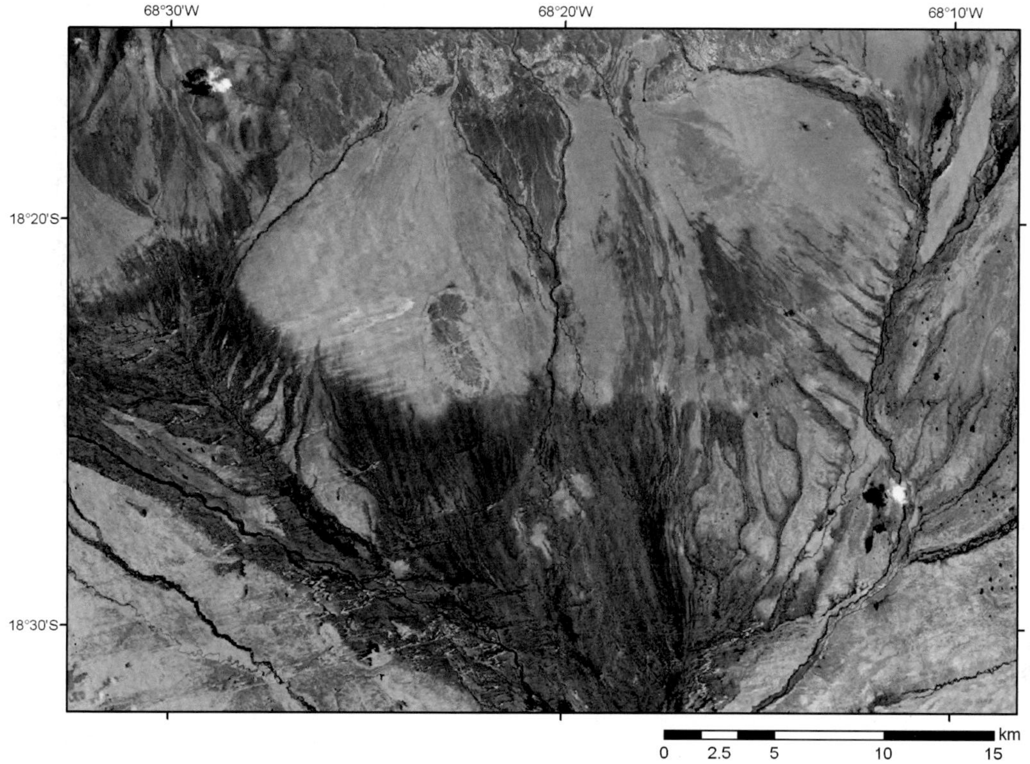

FIGURE 8.—Spring lines (shown by green and blue areas) located on the distal portion of DFSs in the Oruro Department, Bolivia. False color image from red—band 7, green—band 5, blue—band 4.

The internal stratigraphic structure of the DFS deposits, however, may differ between ancient and modern examples. For example, deep incision into DFSs followed by incised-valley fill, caused by extreme changes in sediment supply and stream discharge due to the river's direct connection to a glaciated upland drainage basin (e.g., Weissmann, et al., 2002; Weissmann et al., 2005; Fontana et al., 2008), may not have occurred as commonly in the geologic past. Rather, more subtle changes in sediment supply and stream discharge may have controlled DFS depositional cyclicity (e.g, more similar to nonglaciated DFSs described by Weissmann et al., 2005, or to the Ganges systems described by Gibling et al., 2005).

Significant additional work is needed to more completely evaluate the role that Quaternary climate plays on facies distributions in fluvial systems. Our point in this work, however, is that we should be evaluating fluvial systems that occur in sedimentary basins, as opposed to those in degradational settings, since these have a higher probability of being preserved in the rock record. Additionally, outcrop and subsurface studies should be conducted that work toward understanding fluvial basin fill in the context of position in the sedimentary basin in which they were deposited. Through a combination of geomorphic analysis of fluvial systems in sedimentary basins and rock-record studies, we can build a better understanding of the evolution of fluvial basin fills.

SUMMARY

In this paper, we present observations and hypotheses developed from an initial review and extensive compilation of imagery from continental sedimentary basins around the world. Extensive field work on many of these DFSs is needed in order to evaluate these hypotheses and build an understanding of the implications of the DFS observation on rock-record character and form. The interpretations produced from satellite imagery and study of modern systems are limited in that (1) they only represent a snapshot in time and (2) this snapshot is from the Quaternary, an arguably unusual time relative to the rest of Earth's history due to the rapid climate shifts of glacial–interglacial cycles. Thus, evaluations of large-scale stratigraphic architecture in ancient systems are required in order to understand how these systems evolve through time. A significant challenge facing us in stratigraphic studies, however, is the limited outcrop sizes relative to the size of large DFSs we observe. While most outcrop exposures are small compared to the size of DFSs that formed the deposits, there are a few exceptions (e.g., the Cenozoic of the Ebro Basin, Spain, and the Permian of Utah) where proximal–distal variations in the deposits of DFSs can be traced over tens of kilometers (Hirst and Nichols, 1986; Nichols, 1987; Nichols and Fisher, 2007; Cain and Mountney, 2009).

Through this compilation of basins and a review of imagery from these basins, we observe that modern basin fill is dominated by DFSs, with a relatively minor portion of the fluvial basin fill covered by axial fluvial systems. Thus, since basin subsidence is required for preservation of sedimentary rocks, we expect that much of the fluvial rock record from continental sedimentary basins will be derived from DFSs. Though the DFS interpretation may not be appropriate for all fluvial sedimentary successions (e.g., incised-valley fill, such as those describe by Blum and Aslan, 2006, or other valley-fill successions such as those described by Potter, 1978), the evidence presented herein that shows the prevalence of DFSs in continental sedimentary basin fill indicates that we need to recognize that much of the ancient fluvial record may have developed on DFSs. Based on this work, a new breed of facies models must be developed in a basin context, accounting for the architecture developed on DFSs and the more areally limited axial systems.

ACKNOWLEDGMENTS

This work has been supported by National Aeronautics and Space Administration (NASA) Earth System Enterprise grant #NNS04AB25G and National Science Foundation (NSF; OCI-0753336), both grants to Louis Scuderi. Any opinions, findings, and conclusions or recommendations are those of the authors and do not necessarily reflect the views of the National Aeronautics and Space Administration or the National Science Foundation. The latter stages of this work were supported by the Fluvial Systems Research Group DFS consortium sponsored by ConocoPhillips, Total, BG, and Chevron. Discussions with Justin Wilkinson are greatly appreciated as they helped us improve this manuscript, and the DFSs that Ruth Schrichte showed Gary Weissmann long ago set some of the foundations for this work. Review comments from Brian Bluck, Peter McCabe, and the editors also helped us focus and improve the manuscript.

REFERENCES:

Alexander, J., Bridge, J.S., Leeder, M.R., Collier, R.E.L., and Gawthorpe, R.L., 1994, Holocene meander belt evolution in an active extensional basin, southwestern Montana: Journal of Sedimentary Research, v. B64, p. 542–559.

Alexander, J., and Leeder, M.R., 1990, Geomorphology and surface tilting in an active extensional basin, SW Montana, USA: Geological Society of London, Journal, v. 147, p. 461–467.

Allen, J.R.L., 1965, A review of the origin and characteristics of recent alluvial sediments: Sedimentology, v. 5, p. 89–191.

Archer, A.W., and Greb, S.F., 1995, An Amazon-scale drainage system in the early Pennsylvanian of central North America: Journal of Geology, v. 103, p. 611–628.

Arkley, R.J., 1962, The geology, geomorphology, and soils of the San Joaquin Valley in the vicinity of the Merced River, California: California Division of Mines and Geology, Bulletin 182, p. 25–31.

Ashworth, P.J., Best, J.L., Peakall, J., and Lorsong, J.A., 1999, The influence of aggradation rate on braided alluvial architecture: field study and physical-scale modeling of the Ashburton River gravels, Canterbury Plains, New Zealand, *in* Smith, N.D., and Rogers, J., eds., Fluvial Sedimentology VI: International Association of Sedimentologists, Special Publication 28, p. 333–346.

Ashworth, P.J., Best, J.L., Roden, J.E., Bristow, C.S., and Klassen, G.J., 2000, Morphological evolution and dynamics of a large, sand braid-bar, Jamuna River, Bangladesh: Sedimentology, v. 47, p. 533–555.

Aslan, A., and Autin, W.J., 1998, Holocene flood-plain soil formation in the southern lower Mississippi Valley: implications for interpreting alluvial paleosols: Geological Society of America, Bulletin, v. 110, p. 433–449.

Aslan, A., and Autin, W.J., 1999, Evolution of the Holocene Mississippi River floodplain, Ferriday, Louisiana: insights on the origin of fine-grained floodplains: Journal of Sedimentary Research, v. 69, p. 800–815.

Aslan, A., Autin, W.J., and Blum, M.D., 2005, Causes of river avulsion: insights from the late Holocene avulsion history of the Mississippi River, U.S.A.: Journal of Sedimentary Research, v. 75, p. 650–664, doi: 10.2110/jsr.2005.053.

Aslan, A., and Blum, M.D., 1999, Contrasting styles of Holocene avulsion, Texas Gulf Coastal Plain, USA, *in* Smith, N.D., and Rogers, J., eds., Fluvial Sedimentology VI: International Association of Sedimentologists, Special Publication 28, p. 193–209.

Assine, M.L., 2005, River avulsions on the Taquari megafan, Pantanal wetland, Brazil: Geomorphology, v. 70, p. 357–371.

Assine, M.L., and Silva, A., 2009, Contrasting fluvial styles of the Paraguay River in the northwestern border of the Pantanal wetland, Brazil: Geomorphology, v. 113, p. 189–199, doi: 10.1016/j.geomorph.2009.03.012.

Assine, M.L., and Soares, P.C., 2004, Quaternary of the Pantanal, west-central Brazil: Quaternary International, v. 114, p. 23–34.

Autin, W.J., Burns, S.F., Miller, B.J., Saucier, R.T., and Sneed, J.I., 1991, Quaternary geology of the lower Mississippi Valley, *in* Morrison, R.B., ed., Quaternary Nonglacial Geology; Conterminous U.S.: Boulder, Colorado, Geological Society of America, The Geology of North America, v. K-2, p. 547–581.

Bartow, J.A., 1991, The Cenozoic evolution of the San Joaquin Valley, California: U.S. Geological Survey, Professional Paper 1501, 40 p.

Beaty, C.B., 1963, Origin of alluvial fans, White Mountains, California and Nevada: Association of American Geographers, Annals, v. 53, p. 516–535.

Beaty, C.B., 1989, Great big boulders I have known: Geology, v. 17, p. 349–352.

Berendsen, H.J.A., and Stouthamer, E., 2001, Paleogeographic development of the Rhine–Meuse delta, The Netherlands: Assen, Van Gorcum, 268 p.

Bernard, H.A., LeBlanc, R.J., and Major, C.J., 1962, Recent and Pleistocene geology of southeast Texas, *in* Rainwater, E.H., and Zingula, R.P., eds, Geology of the Gulf Coast and Central Texas. Geological Society of America guidebook for 1962 annual meeting, p. 175–224.

Best, J.L., and Ashworth, P.J., 1997, Scour in large braided rivers and the recognition of sequence stratigraphic boundaries: Nature, v. 387, p. 275–277.

Best, J.L., Ashworth, P., Bristow, C., and Roden, J., 2003, Three-dimensional sedimentary architecture of a large, mid-channel sand braid bar, Jamuna River, Bangladesh: Journal of Sedimentary Research, v. 73, p. 516–530.

Blair, T.C., and McPherson, J.G., 1992. The Trollheim alluvial fan and facies model revisited: Geological Society of America, Bulletin, v. 104, p. 762–769.

Blair, T.C., and McPherson, J.G., 1994, Alluvial fans and their natural distinction from rivers based on morphology, hydraulic processes, sedimentary processes, and facies assemblages: Journal of Sedimentary Research, v. A64, p. 450–489.

Blodgett, R.H., and Stanley, K.O., 1980, Stratification, bed forms, and discharge relations of the Platte River system, Nebraska: Journal of Sedimentary Petrology, v. 50, p. 139–148.

Bluck, B.J., 1971, Sedimentation in the meandering River Endrick: Scottish Journal of Geology, v. 7, p. 93–138.

Bluck, B.J., 1980, Structure, generation, and preservation of upward fining, braided stream cycles in the Old Red Sandstone of Scotland: Royal Society of Edinburgh, Transactions, Earth Sciences, v. 71, p. 29–46.

Bluck, B.J., 1986, Upward coarsening sedimentation units and facies lineages, Old Red Sandstone, Scotland: Royal Society of Edinburgh, Transactions, Earth Sciences, v. 77, p. 251–264.

Blum, M.D., 1993, Genesis and architecture of incised valley fill sequences: A late Quaternary example from the Colorado River, Gulf Coastal Plain of Texas, *in* Weimer, P., and Posamentier, H.W., eds., Siliciclastic Sequence Stratigraphy: Recent Developments and Applications: American Association of Petroleum Geologists, Memoir 58, p. 259–283.

Blum, M.D., and Aslan, A., 2006, Signatures of climate vs. sea-level change within incised valley-fill successions: Quaternary examples from the Texas Gulf Coast: Sedimentary Geology, v. 190, p. 177–211.

Blum, M.D., and Price, D.M., 1998, Quaternary alluvial plain construction in response to glacio-eustatic and climatic controls, Texas Gulf Coastal Plain, *in* Shanley, K.J., and McCabe, P.J, eds., Relative Role of Eustasy, Climate, and Tectonism on Continental Rocks: SEPM, Special Publication 59, p. 31–48.

Blum, M.D., Guccione, M.J., Wysocki, D.A., Robnett, P.C., and Rutledge, E.M., 2000, Late Pleistocene evolution of the lower Mississippi River valley, southern Missouri to Arkansas, Geological Society of America, Bulletin, v. 112, p. 221–235.

Boothroyd, J.C., and Ashley, G.M., 1975, Process, bar morphology, and sedimentary structures on braided outwash fans, northeastern Gulf of Alaska, *in* Jopling, A.V., and McDonald, B.D., eds., Glacio-Fluvial and Glaciolacustrine Sedimentation. SEPM, Special Publication 23, p. 193–222.

Boothroyd, J.C., and Nummedal, D., 1978, Proglacial braided outwash: a model for humid alluvial-fan deposits, *in* Miall, A.D., ed., Fluvial Sedimentology: Canadian Society of Petroleum Geologists, Memoir 5, p. 641–668.

Bown, T.M., and Kraus, M.J., 1987, Integration of channel and floodplain suites: I. Developmental sequence and lateral relations of alluvial paleosols: Journal of Sedimentary Petrology, v. 57, p. 587–601.

Brice, J.C., 1964, Channel patterns and terraces of the Loup Rivers in Nebraska: U. S. Geological Survey, Professional Paper: 422-D, p. D1-D41.

Bridge, J.S., 2006, Fluvial facies models: recent developments, *in* Posamentier, H.W., and Walker, R.G., eds., Facies Models Revisited: SEPM, Special Publication 84, p. 85–170.

Bridge, J.S., and Gabel, S.L., 1992, Flow and sediment dynamics in a low sinuosity, braided river: Calamus River, Nebraska Sandhills: Sedimentology, v. 39, p. 125–142.

Bridge, J.S., and Jarvis, J., 1982, The dynamics of a river bend: a study in flow and sedimentary processes: Sedimentology, v. 29, p. 499–541.

Bridge, J.S., Smith, N.D., Trent, F., Gabel, S.L., and Bernstein, P., 1986, Sedimentology and morphology of a low-sinuosity river; Calamus River, Nebraska Sand Hills: Sedimentology, v. 33, p. 851–870.

Bridge, J.S., Alexander, J., Collier, R.E.L., Gawthorpe, R.L., and Jarvis, J., 1995, Ground penetrating radar and coring used to document the large-scale structure of point-bar deposits in 3-D: Sedimentology, v. 42, p. 839–852.

Bridge, J.S., Collier, R.E.L., and Alexander, J., 1998, Large-scale structure of Calamus River deposits (Nebraska, USA) revealed using ground penetrating radar: Sedimentology, v. 45, p. 977–986.

Brierlery, G.J., and Hickin, E.J., 1991, Channel platform as a noncontrolling factor in fluvial sedimentology; the case of the Squamish River floodplain, British Columbia: Sedimentary Geology, v. 75, p. 67–83.

Bristow, C.S., 1987, Brahmaputra river: channel migration and deposition, *in* Ethridge, F.G., Flores, R.M., and Harvey, M.D., eds., Recent Developments in Fluvial Sedimentology: SEPM, Special Publication 39, p. 63–74.

Bristow, C.S., 1993, Sedimentary structures exposed in bar tops in the Brahmaputra River, Bangladesh, *in* Best, J.L., and Bristow, C.S. eds.., Braided Rivers: Geological Society of London, Special Publication 75, p. 277–289.

Bristow, C.S., Skelly, R.L., and Ethridge, F.G., 1999, Crevasse splays from the rapidly aggrading, sand-bed, braided Niobrara River, Nebraska: effect of base-level rise: Sedimentology, v. 46, p. 1029–1047.

Cain, S.A., and Mountney, N.P., 2009, Spatial and temporal evolution of a terminal fluvial fan system: the Permian Organ Rock Formation, South-east Utah, USA: Sedimentology, v. 56, p. 1774–1800, doi:10.1111/j.1365-3091.2009.01057.x.

Campbell, J.E., and Hendry, H.E., 1987, Anatomy of a gravelly meander lobe in the Saskatchewan River, near Nipawin, Canada, *in* Ethridge, F.G., Flores, R.M., and Harvey, M.D., eds., Recent Developments in Fluvial Sedimentology: SEPM, Special Publication 39, p. 179–189.

Cant, D.J., 1978, Bedforms and bar types in the South Saskatchewan River: Journal of Sedimentary Petrology, v. 48, p. 1321–1330.

Cant, D.J., and Walker, R.G., 1978, Fluvial processes and facies sequences in the sandy braided South Saskatchewan River, Canada: Sedimentology, v. 25, p. 625–648.

Carling, P.A., Golz, E., Orr, H.G., and Radecki-Pawlik, A., 2000a, The morphodynamics of fluvial sand dunes in the River Rhine, near Mainz, Germany. I. Sedimentology and morphology: Sedimentology, v. 47, p. 227–252.

Carling, P.A., Williams, J.J., Golz, E., and Kelsey, A.D., 2000b, The morphodynamics of fluvial sand dunes in the River Rhine, near Mainz, Germany. II. Hydrodynamics and sediment transport: Sedimentology, v. 47, p. 253–278.

Chakraborty, T., and Ghosh, P., 2010, The geomorphology and sedimentology of the Tista megafan, Darjeeling Himalaya: Implications for megafan building processes: Geomorphology, v. 115, p. 252–266, doi: 10.1016/j.geomorph.2009.06.035.

Chakraborty, T., Kar, R., Ghosh, P., and Basu, S., in press, Kosi Megafan: Historical records, geomorphology and the recent avulsion of the Kosi River: Quaternary International, doi: 10.1016/j.quaint.2009.12.002.

Church, M., 1983, Pattern of instability in a wandering gravel bed channel, *in* Collinson, J.D., and Lewin, J., eds., Modern and Ancient Fluvial Systems: International Association of Sedimentologists, Special Publication 6, p. 169–180.

Coleman, J.M, 1969, Brahmaputra River: Channel processes and sedimentation: Sedimentary Geology, v. 3, p. 129–239.

Collinson, J.D., 1970, Bedforms of the Tana River, Norway: Geografiska Annaler, v. 52A, p. 31–56.

Collinson, J.D., 1986, Alluvial sediments, *in* Reading, H.G., ed., Sedimentary Environments and Facies: Oxford, U.K., Blackwell Scientific, p. 20–62.

Crowley, K.D., 1983, Large-scale bed configurations (macroforms), Platte, River Basin, Colorado and Nebraska: Primary structures and formative processes: Geological Society of America, Bulletin, v. 94, p. 117–133.

Cunningham, W.D., 1996, Late Cenozoic transpression in southwestern Mongolia and the Gobi Altai–Tien Shan connection: Earth and Planetary Science Letters, v. 140, p. 67–82.

DeCelles, P.G., and Cavazza, W., 1999, A comparison of fluvial megafans in the Cordilleran (Upper Cretaceous) and modern Himalayan foreland basin systems: Geological Society of America, Bulletin, v. 111, p. 1315–1334.

Denton, G.H., and Stuiver, M., 1967, Late Pleistocene glacial stratigraphy and chronology, northwestern Saint Elias Mountains, Yukon Territory, Canada: Geological Society of America, Bulletin, v. 78, p. 485–510.

Desoleges, J.R., and Church, M., 1987, Channel and floodplain facies in a wandering gravel-bed river, *in* Ethridge, F.G., Flores, R.M., and Harvey, M.D., eds., Recent Developments in Fluvial Sedimentology: SEPM, Special Publication 39, p. 99–109.

deSouza, O.C., Araujo, M.R., Mertes, L.A.K., and Melack, J.M., 2002, Form and process along the Taquari River alluvial fan, Pantanal, Brazil: Zeitschrift für Geomorphologie, N.F., Supplementband, v. 129, p. 73–107.

Dietrich, W.E., and Smith, J.D., 1984, Bedload transport in a river meander: Water Resources Research, v. 20, p. 1355–1380.

Dinehart, R.L., 1989, Dune migration in a steep coarse-bedded stream: Water Resources Research, v. 25, p. 911–923.

Dinehart, R.L., 1992, Evolution of coarse gravel bed forms: field measurements at flood stage: Water Resources Research, v. 28, p. 2667–2689.

Dubiel, R.F., Parrish, J.T., Parrish, J.M., and Good, S.C., 1991, The Pangaean Megamonsoon—evidence from the upper Triassic Chinle Formation, Colorado Plateau: Palaios, v. 6, p. 347–370.

ESRI, 2009, ArcINFO. Environmental Systems Research Institute (ESRI), Redlands, California.

Ethridge, F.G., Skelly, R.L., and Bristow, C.S., 1999, Avulsion and crevassing in the sandy, braided Niobrara River: complex response to base-level rise and aggradation, *in* Smith, N.D., and Rogers, J., eds., Fluvial Sedimentology VI: International Association of Sedimentologists, Special Publication 28, p. 179–181.

Evenstar, L.A., Hartley, A.J., Stuart, F.M., Mather, A.E., Rice, C.M., and Chong, G., 2009, Multiphase development of the Atacama planation surface recorded by cosmogenic ^{3}He exposure ages: implications for uplift and Cenozoic climate change in western South America: Geology, v. 37, p. 27–30.

Farr, T.G., Rosen, P.A., Caro, E., Crippen, R., Duren, R., Hensley, S., Kobrick, M., Paller, M., Rodriguez, E., Roth, L., Seal, L, Shaffer, S., Shimada, J., Umland, J., Werner, M., Oskin, M., Burbank, D., and Alsdorf, D., 2007, The Shuttle Radar Topography Mission: Reviews of Geophysics, v. 45, RG2004, doi: 10.1029/2005RG000183.

Farrell, K.M., 1987, Sedimentology and facies architecture of overbank deposits of the Mississippi River, False River Region, Louisiana, *in* Ethridge, F.G., Flores, R.M., and Harvey, M.D., eds., Recent Developments in Fluvial Sedimentology; Contributions from the Third International Fluvial Sedimentology Conference: SEPM, Special Publication 39, p. 112–120.

Farrell, K.M., 2001, Geomorphology, facies architecture, and high resolution, non-marine sequence stratigraphy in avulsion deposits, Cumberland marshes, Saskatchewan: Sedimentary Geology, v. 139, p. 93–150.

Ferguson, R.I., and Werrity, A., 1983, Bar development and channel changes in the gravelly River Feshie, Scotland, *in* Collinson, J.D., and Lewin, J., eds., Modern and Ancient Fluvial Systems: International Association of Sedimentologists, Special Publication 6, p. 181–193.

Fielding, C.R., 1984, A coal depositional model for the Durham Coal Measures of NE England: Geological Society of London, Journal, v. 141, p. 919–931.

Fisk, H.N., 1944, Geological investigation of the alluvial valley of the lower Mississippi River: Vicksburg, Mississippi, U.S., Army Corps of Engineers, Mississippi River Commission, 78 p.

Fisk, H.N., 1947, Fine-grained alluvial deposits and their effect on Mississippi River activity: U.S. Army Corps of Engineers, Mississippi River Commission, 82 p.

Fontana, A., Mozzi, P., and Bondesan, A., 2008, Alluvial megafans in the Venetian–Friulian Plain (north-eastern Italy): Evidence of sedimentary and erosive phases during Late Pleistocene and Holocene: Quaternary International, v. 189, p. 71–90, doi: 10.1016/j.quaint.2007.08.044.

Forbes, D.L., 1983, Morphology and sedimentology of a sinuous gravel-bed channel system, lower Babbage River, Yukon coastal plain, Canada, *in* Collinson, J.D., and Lewin, J., eds., Modern and Ancient Fluvial Systems:

International Association of Sedimentologists, Special Publication 6, p. 195–206.

FRIEND, P.F., 1978, Distinctive features of some ancient river systems, *in* Miall, A.D., ed., Fluvial Sedimentology: Canadian Society of Petroleum Geologists, Memoir 5, p. 531–542.

FROSTICK, L.E., AND REID, I., 1987, A new look at rifts: Geology Today, v. 3, p. 122–126.

GABEL, S.L., 1993, Geometry and kinematics of dunes during steady and unsteady flows in the Calamus River, Nebraska, USA: Sedimentology, v. 40, p. 237–269.

GALLOWAY, W.E., 1980, Deposition and early hydrologic evolution of Westwater Canyon wet alluvial-fan system, *in* Rautman, C.A., ed., Geology and Mineral Technology of the Grants Uranium Region 1979: New Mexico Bureau of Mines and Mineral Resources, Memoir 38, p. 59–69.

GAWTHORPE, R.L., AND LEEDER, M.R., 2000, Tectono-sedimentary evolution of active extensional basins: Basin Research, v. 12, p. 195–218.

GAWTHORPE, R.L., COLLIER, R.E.L., ALEXANDER, J., BRIDGE, J.S., AND LEEDER, M.R., 1993, Ground penetrating radar: application to sandbody geometry and heterogeneity studies, *in* North, C.J., and Prosser, D.J., eds., Characterization of Fluvial and Aeolian Reservoirs: Geological Society of London, Special Publication 73, p. 421–432.

GEDDES, A., 1960, The alluvial morphology of the Indo-Gangetic Plain: its mapping and geographical significance: Institute of British Geographers, Transactions and Papers, v. 28, 253–276.

GIBLING, M.R., 2006, Width and thickness of fluvial channel bodies and valley fills in the geological record: a literature compilation and classification: Journal of Sedimentary Research, v. 76, p. 731–770, doi: 10.2110/jsr.2006.060.

GIBLING, M.R., TANDON, S.K., SINHA, R., AND JAIN, M., 2005, Discontinuity-bounded alluvial sequences of the southern Gangetic Plains, India: aggradation and degradation in response to monsoonal strength: Journal of Sedimentary Research, v. 75, p. 369–385.

GILFELLON, G.B., SARMA, J.N., AND GOHAIN, K., 2003, Channel and bed morphology of a part of the Brahmaputra River in Assam: Geological Society of India, Journal, v. 62, p. 227–235.

GOLE, C.V., AND CHITALE, S.V., 1966, Inland delta building activity of Kosi River: American Society of Civil Engineers, Proceedings, Journal of the Hydraulics Division, Paper 4722, v. 92, p. 111–126.

GOSWAMI, D.C., 1985, Brahmaputra River, Assam, India: physiography, basin denudation, and channel aggradation: Water Resources Research, v. 21, p. 959–978.

GUMBRICHT, T., MCCARTHY, J., AND MCCARTHY, T.S., 2004, Channels, wetlands and islands in the Okavango delta, Botswana, and their relation to hydrological and sedimentological processes: Earth Surface Processes and Landforms, v. 29, p. 15–29.

GUPTA, S., 1997, Himalayan drainage patterns and the origin of fluvial megafans in the Ganges foreland basin: Geology, v. 23, p. 11–14.

HAMPSON, G.J., DAVIES, W., DAVIES, S.J., HOWELL, J.A., AND ADAMSON, K.R., 2005, Use of spectral gamma-ray data to refine subsurface fluvial stratigraphy, Late Cretaceous strata in the book Cliffs, Utah, U.S.A.: Geological Society of London, Journal, v. 162, p. 603–621.

HARBOR, D.J., 1998, Dynamics of bedforms in the lower Mississippi River: Journal of Sedimentary Research, v. 68, p. 750–762.

HARDEN, J.W., TAYLOR, E.M., HILL, C., MARK, R.K., MCFADDEN, L.D., REHEIS, M.C., SOWERS, J.M., AND WELLS, S.G., 1991, Rates of soil development from four soil chronosequences in the southern Great Basin: Quaternary Research, v. 35, p. 383–399.

HARTLEY, A.J., 1993, A depositional model for the mid-Westphalian A to late Westphalian B coal measures of South Wales: Geological Society of London, Journal, v. 150, p. 1121–1136.

HARTLEY, A.J., AND CHONG, G., 2002, Late Pliocene age for the Atacama Desert: Implications for the desertification of western South America: Geology, v. 30, p. 43–46, doi: 10.1130/0091–7613(2002)030<043:LPAFTA>2.0CO;2.

HARTLEY, A.J., WEISSMANN, G.S., NICHOLS, G.J., AND WARWICK, G.L., 2010, Large distributive fluvial systems: characteristics, distribution and controls on development: Journal of Sedimentary Research, v. 80, p. 167–183, doi: 10.2110/jsr.2010.016.

HARVEY, A.M., 2005, Differential effects of base-level, tectonic setting and climatic change on Quaternary alluvial fans in the northern Great Basin, Nevada, USA, *in* Harvey, A.M., Mather, A.E., and Stokes, M., eds., Alluvial Fans: Geomorphology, Sedimentology, Dynamics: Geological Society of London, Special Publication 251, p. 117–131.

HARVEY, A.M., AND WELLS, S.G., 2003, Late Quaternary variations in alluvial fan sedimentologic and geomorphic processes, Soda Lake basin, eastern Mojave Desert, California, *in* Enzel, Y., Wells, S.G., and Lancaster, N., eds., Paleoenvironments and Paleohydrology of the Mojave and Southern Great Basin Deserts: Geological Society of America, Special Paper 368, p. 207–230.

HIRST, J.P.P., AND NICHOLS, G.J., 1986, Thrust tectonic controls on alluvial sedimentation patterns, southern Pyrenees, *in* Allen, P.A., and Homewood, P., eds., Foreland Basins: International Association of Sedimentologists, Special Publication 8, p. 153–164.

HORTON, B.K., AND DECELLES, P.G., 1997, The modern foreland basins system adjacent to the Central Andes: Geology, v. 25, p. 895–898.

HORTON, B.K., AND DECELLES, P.G., 2001, Modern and ancient fluvial megafans in the foreland basin system of the central Andes, southern Bolivia: implications for drainage network evolution in fold–thrust belts: Basin Research, v. 13, p. 43–63.

HOWARD, J.P., CUNNINGHAM, W.D., AND DAVIES, S.J., 2006, Competing processes of clastic deposition and compartmentalized inversion in an actively evolving transpressional basin, western Mongolia: Geological Society of London, Journal, v. 163, p. 657–670.

IRIONDO, M., 1993, Geomorphology and late Quaternary of the Chaco (South America): Geomorphology, v. 7, p. 289–303.

JACKSON, R.G., II, 1975, Hierarchical attributes and a unifying model of bed forms composed of cohesionless material and produced by shearing flow: Geological Society of America, Bulletin, v. 86, p. 1523–1533.

JACKSON, R.G., II, 1976a, Depositional model of point bars in the lower Wabash River: Journal of Sedimentary Petrology, v. 46, p. 579–594.

JACKSON, R.G., II, 1976b, Velocity–bed-form–texture patterns of meander bends in the lower Wabash River of Illinois and Indiana: Geological Society of America, Bulletin, v. 86, p. 1511–1522.

JACKSON, R.G., 1976c, Large-scale ripples of the Lower Wabash River: Sedimentology, v. 23, p. 593–623.

JACKSON, R.G., 1978, Preliminary evaluation of lithofacies models for meandering alluvial streams, *in* Miall, A.D., ed., Fluvial Sedimentology: Canadian Society of Petroleum Geologists, Memoir 5, p. 543–576.

JANDA, R.J., 1966, Pleistocene history and hydrology of the upper San Joaquin River, California: Ph.D. Thesis, University of California, Berkeley, 293 p.

JANSSON, P., JACOBSON, D., AND HOOKE, R.L., 1993, Fan and playa areas in southern California and adjacent parts of Nevada: Earth Surface Processes and Landforms, v. 18, p. 109–119.

JOLIVET, M., BRUNEL, M., SEWARD, D., XU, Z., YANG, J., MALAVIEILLE, J., ROGER, F., LEYRELOUP, A., ARNAUD, N., AND WU, C., 2003, Neogene extension and volcanism in the Kunlun Fault Zone, northern Tibet: new constraints on the age of the Kunlun Fault: Tectonics, v. 22, 1052, doi:10.1029/2002TC001428.

JORDAN, D.W., AND PRYOR, W.A., 1992, Hierarchical levels of heterogeneity in a Mississippi River meander belt and application to reservoir systems: American Association of Petroleum Geologists, Bulletin, v. 76, p. 1601–1624.

JULIEN, P.Y., AND KLAASEN, G.J., 1995, Sand-dune geometry of large rivers during floods: American Society of Civil Engineers, Journal of Hydraulic Engineering, v. 128, p. 1042–1050.

JULIEN, P.Y., KLAASSEN, G.J., TEN BRINKE, W.B.M., AND WILBERS, A.W.E., 2002, Case study: bed resistance of Rhine River during 1998 flood: American Society of Civil engineers, Journal of Hydraulic Engineering, v. 128, p. 1042–1050.

JUNGERIUS, P.D., MATUNDURA, J., AND VAN DE ANCKER, J.A.M., 2002, Road construction and gully erosion in west Pokot, Kenya: Earth Surface Processes and Landforms, v. 27, p. 1237–1247, doi: 10.1002/esp.423.

KALE, V.S., 2002, Fluvial geomorphology of Indian rivers: an overview: Progress in Physical Geography, v. 26, p. 400–433.

KARMAKAR, R.M., AND RAO, A.E.V., 1999, Soils on different physiographic units in lower Brahmaputra Valley zone of Assam: I. Characterization and classification: Indian Society of Soil Science, Journal, v. 47, p. 761–767.

KLEINHANS, M.G., WILBERS, A.W.E., DESWAAF, A. AND VAN DEN BERG, J.H., 2002,

Sediment supply–limited bedforms in sand–gravel rivers: Journal of Sedimentary Research, v. 72, p. 629–640.

Knighton, D., 1998, Fluvial Forms and Processes, A New Perspective: London, Arnold, 383 p.

Knott, J.R., Sarna-Wojciecki, A.M., Machette, M.N., and Klinger, R.E., 2005, Upper Neogene stratigraphy and tectonics of Death Valley: a review: Earth-Science Reviews, v. 73, p. 245–270.

Kochel, R.C., and Baker, V.R., 1988, Paleoflood analysis using slack water deposits, *in* Baker, V.R., Kochel, R.C., and Patton, P.C., eds., Flood Geomorphology: New York, Wiley, p. 357–376.

Kraus, M.J, 1996, Avulsion deposits in lower Eocene alluvial rocks, Bighorn Basin, Wyoming: Journal of Sedimentary Research, v. 66, p. 354–366.

Kraus, M.J., 1997, Lower Eocene alluvial paleosols: pedogenic development, stratigraphic relationships, and paleosol/landscape associations: Palaeogeography, Palaeoclimatology, Palaeoecology, v. 129, p. 387–406.

Kraus, M.J., 1999, Paleosols in clastic sedimentary rocks: their geologic applications: Earth-Science Reviews, v. 47, p. 41–70.

Kraus, M.J., and Aslan, A., 1993, Eocene hydromorphic paleosols: significance for interpreting ancient floodplain processes: Journal of Sedimentary Petrology, v. 63, p. 453–463.

Kraus, M.J., and Davies-Vollum, K.S., 2004, Mudrock-dominated fills formed in avulsion splay channels: examples from the Willwood Formation, Wyoming: Sedimentology, v. 51, p. 1127–1144.

Lahiri, S, 1996, Channel pattern as signature of neotectonic movements—A case study from Brahmaputra Valley in Assam: Photonirvachak, Journal of the Indian Society of Remote Sensing, v. 24, p. 265–272.

Leclair, S.F., and Bridge, J.S., 2001, Quantitative interpretation of sedimentary structures formed by river dunes: Journal of Sedimentary Research, v. 71, p. 713–716.

Leeder, M.R., and Alexander, J.A., 1987, The origin and tectonic significance of asymmetrical meander belts: Sedimentology, v. 34, p. 217–226.

Leeder, M.R., and Gawthorpe, R.L., 1987, Sedimentary models for extensional tilt-block/half-graben basins, *in* Coward, M.P., Dewey, J.F., and Hancock, P.L., eds., Continental Extensional Tectonics: Geological Society of London Special Publication 28, p. 139–152.

Leeder, M.R., and Mack, G.H., 2001, Lateral erosion ('toe-cutting') of alluvial fans by axial rivers: implications for basin analysis and architecture: Geological Society of London, Journal, v. 158, p. 885–893.

Leier, A.L., DeCelles, P.G., and Pelletier, J.D., 2005, Mountains, monsoons, and megafans: Geology, v. 33, p. 289–292.

Leopold, L.B., Wolman, M.G., and Miller, J.P., 1964, Fluvial Processes in Geomorphology. San Francisco, W.H. Freeman and Company, 522 p.

Lettis, W.R., 1988, Quaternary geology of the northern San Joaquin Valley, *in* Graham, S.A., ed., Studies of the Geology of the San Joaquin Basin: SEPM, Pacific Section, v. 60, p. 333–351.

Lettis, W.R., and Unruh, J.R., 1991, Quaternary geology of the Great Valley, California, *in* Morrison, R.B., ed., Geological Society of America, The Geology of North America, v. K-2, p. 164–176.

Levey, R.A., 1978, Bedform distribution and internal stratification of coarse-grained point bars, upper Congaree River, S.C., *in* Miall, A.D., ed., Fluvial Sedimentology: Canadian Society of Petroleum Geologists, Memoir 5, p. 105–127.

Lunt, I.A., and Bridge, J.S., 2004, Evolution and deposits of a gravelly braid bar and a channel fill, Sagavanirktok River, Alaska: Sedimentology, v. 51, p. 415–432.

Lunt, I.A., Bridge, J.S., and Tye, R.S., 2004a, Development of a 3-D depositional model of braided river gravels and sands to improve aquifer characterization, *in* Bridge, J.S., and Hyndman, D., eds., Aquifer Characterization: SEPM, Special Publication 80, p. 139–169.

Lunt, I.A., Bridge, J.S., and Tye, R.S., 2004b, A quantitative, three-dimensional depositional model of gravelly braided rivers: Sedimentology, v. 51, p. 377–414.

Mack, G.H., and Leeder, M.R., 1999, Climatic and tectonic controls on alluvial-fan and axial-fluvial sedimentation in the Plio-Pleistocene Palomas half graben, southern Rio Grande Rift: Journal of Sedimentary Research, v. 69, p. 635–652.

Mack, G.H., Love, D.W., and Seager, W.R., 1997, Spillover models for axial rivers in regions of continental extension: the Rio Mimbres and Rio Grande in the southern Rio Grande Rift, U.S.A.: Sedimentology, v. 44, p. 637–652.

Mack, G.H., Leeder, M.R., and Salyards, S.L., 2002, Temporal and spatial variability of alluvial-fan and axial-fluvial sedimentation in the Plio-Pleistocene Palomas half graben, southern Rio Grande rift, New Mexico, U.S.A., *in* Renaut, G.W., and Ashley, G.M., eds., Sedimentation in Continental Rifts: SEPM, Special Publication 73, p. 165–177.

Mack, G.H., Seager, W.R., and Leeder, M.R., 2003, Synclinal-horst basins: examples from the southern Rio Grande Rift and southern transition zone of southwestern New Mexico, U.S.A.: Basin Research, v. 15, p. 365–377.

Mack, G.H., Seager, W.R., Leeder, M.R., Pérez-Arlucea, M., and Salyards, S.L., 2006, Pliocene and Quaternary history of the Rio Grande, the axial river of the southern Rio Grande rift, New Mexico, USA: Earth-Science Reviews, v. 79, p. 141–162.

Makaske, B., Berendsen, H.J.A., and Van Ree, M.H.M., 2007, Middle Holocene avulsion-belt deposits in the central Rhine–Meuse Delta, the Netherlands: Journal of Sedimentary Research, v. 77, p. 110–123.

Mann, C.J., and Thomas, W.A., 1968, The ancient Mississippi River: Association of Gulf Coast Geological Societies, Transactions, v. 18, p. 187–204.

Marchand, D.E., 1977, The Cenozoic history of the San Joaquin Valley and the adjacent Sierra Nevada as inferred from the geology and soils of the eastern San Joaquin Valley, *in* Singer, M.J., ed., Soil Development, Geomorphology, and Cenozoic History of the Northeastern San Joaquin Valley and Adjacent Areas, California: University of California Press, Guidebook for Joint Field Session, Soil Science Society of America and Geological Society of America, p. 39–50.

Marchand, D.E., and Allwardt, A., 1981, Late Cenozoic stratigraphic units, northeastern San Joaquin Valley, California: U.S. Geological Survey, Bulletin 1470, 70 p.

McCarthy, T.S., Smith, N.D., Ellery, W.N., and Gumbricht, T., 2002, The Okavango Delta—semiarid alluvial-fan sedimentation related to incipient rifting, *in* Renaut, G.W., and Ashley, G.M., eds., Sedimentation in Continental Rifts: SEPM, Special Publication 73, p. 179–193.

McGowen, J.H., and Garner, L.E., 1970, Physiographic features and stratification types of coarse grained point bars: Modern and ancient examples: Sedimentology, v. 14, p. 77–111.

McKee, E.D., Crosby, E.J., and Berryhill, H.L., Jr., 1967, Flood deposits, Bijou Creek, Colorado, June 1965: Journal of Sedimentary Petrology, v. 37, p. 829–851.

McLelland, S.J., Ashworth, P.J., Best, J.L., Roden, J., and Klaasen, G.J., 1999, Flow structure and transport of sand-grade suspended sediment around an evolving braid bar, Jamuna River, Bangladesh, *in* Smith, N.D., and Rogers, J., eds., Fluvial Sedimentology VI: International Association of Sedimentologists, Special Publication 28, p. 43–57.

Miall, A.D., 1977, A review of the braided-river depositional environment: Earth-Science Review, v. 13, p. 1–62.

Miall, A.D., 1978, Lithofacies type and vertical profile models in braided river deposits: a summary, *in* Miall, A.D., ed., Fluvial sedimentology: Canadian Society of Petroleum Geologists, Memoir 5, p. 597–604.

Miall, A.D., 1996, The Geology of Fluvial Deposits: Sedimentary Facies, Basin Analysis and Petroleum Geology: Berlin, Springer, 582 p.

Mohindra, R., Parkash, B., and Prasad, J., 1992, Historical geomorphology and pedology of the Gandak megafan, Middle Gangetic plains, India: Earth Surface Processes and Landforms, v. 17, p. 643–662.

Morozova, G.S., and Smith, N.D., 1999, Holocene avulsion history of the lower Saskatchewan fluvial system, Cumberland Marshes, Saskatchewan-Manitoba, Canada, *in* Smith, N.D., and Rogers, J., eds., Fluvial Sedimentology VI: International Association of Sedimentologists, Special Publication 28, p. 231–249.

Morozova, G.S., and Smith, N.D., 2000, Holocene avulsion styles and sedimentation patterns of the Saskatchewan River, Cumberland Marshes, Canada: Sedimentary Geology, v. 130, p. 81–105.

Nadon, G., 1994, The genesis and recognition and anastomosed fluvial deposits: data from the St Mary River Formation, southwestern Alberta: Journal of Sedimentary Research, v. B64, p. 451–463.

Nanson, G.C., 1980, Point bar and floodplain formation of the meandering

Beatton River, northeastern British Columbia, Canada: Sedimentology, v. 27, p. 3–29.

Nanson, G.C., and Croke, J.C., 1992, A genetic classification of floodplains: Geomorphology, v. 4, p. 459–486.

Nicholas, A.P., and McLelland, S.J., 1999, Hydrodynamics of a floodplain recirculation zone investigated by field monitoring and numerical simulation, *in* Marriott, S.B., and Alexander, J., eds., Floodplains: Interdisciplinary Approaches: Geological Society of London, Special Publication 163, p. 15–26.

Nichols, G.J., 1987, Structural controls on fluvial distributary systems—the Luna System, Northern Spain, *in* Ethridge, F.G., Flores, R.M., and Harvey, M.D., eds., Recent Developments in Fluvial Sedimentology: SEPM, Special Publication 39, p. 269–277.

Nichols, G.J., and Fisher, J.A., 2007, Processes, facies, and architecture of fluvial distributary system deposits: Sedimentary Geology, v. 195, p. 75–90.

Ori, G.G., and Friend, P.F., 1984, Sedimentary basins formed and carried piggyback on active thrust sheets: Geology, v. 12, p. 475–478.

Owen, L.A., Windley, B.F., Cunningham, W.D., Badamgarov, J. and Dorjnamjaa, D., 1997, Quaternary alluvial fans in the Gobi Desert, southern Mongolia: Evidence for neotectonics and climate change: Journal of Quaternary Science, v. 12, p. 239–252.

Pasquini, A.I., and Depetris, P.J., 2007, Discharge trends and flow dynamics of South American rivers draining the southern Atlantic seaboard: an overview: Journal of Hydrology, v. 333, p. 385–399.

Peakall, J., 1998, Axial river evolution in response to half-graben faulting: Carson River, Nevada: Journal of Sedimentary Research, v. 68, p. 788–799.

Peakall, J., Leeder, M., Best, J., and Ashworth, P., 2000, River response to lateral ground tilting: a synthesis and some implications for the modeling of alluvial architecture in extensional basins: Basin Research, v. 12, p. 413–424.

Pérez-Arlucea, M., and Smith, N.D., 1999, Deposional patterns following the 1870s avulsion of the Saskatchewan River (Cumberland Marshes, Saskatchewan): Journal of Sedimentary Research, v. 69, p. 62–73.

Potter, P.E., 1978, Significance and origin of big rivers: Journal of Geology, v. 86, p. 13–33.

Quade, J., Levin, N., Semaw, S., Stout, D., Renne, P., Rogers, M., and Simpson, S., 2004, Paleoenvironments of the earliest stone toolmakers, Gona, Ethiopia: Geological Society of America, Bulletin, v. 116, p. 1529–1544, doi: 10.1130/B25358.1.

Ramos, E., Busquets, P., and Vergès, J., 2002, Interplay between longitudinal fluvial and transverse alluvial fan systems and growing thrusts in a piggyback basin (SE Pyrenees): Sedimentary Geology, v. 146, p. 105–131.

Reinfelds, I., and Nanson, G., 1993, Formation of braided river floodplains, Waimakariri River, New Zealand: Sedimentology, v. 40, p. 1113–1127.

Rice, S.P., and Church, M., 2001, Longitudinal profiles in simple alluvial systems: Water Resources Research, v. 37, p. 417–426.

Richardson, W.R.R., and Thorne, C.R., 1998, Secondary currents and channel changes around a braid bar in the Brahmaputra River, Bangladesh: American Society of Civil Engineers, Journal of Hydraulic Engineering, v. 124, p. 325–328.

Richardson, W.R.R., Thorne, C.R., and Mahmood, S., 1996, Secondary flow and channel changes around a bar in the Brahmaputrea River, Bangladesh, *in* Ashworth, P.J., Bennett, S.J., Best, J.L., and McLelland, S.J., eds., Coherent Flow Structures in Open Channels: Chichester, U.K., Wiley, p. 519–543.

Rittenour, T.M., Goble, R.J., and Blum, M.D., 2005, Development of an OSL chronology for Late Pleistocene channel belts in the lower Mississippi valley, USA: Quaternary Science Reviews, v. 24, p. 2539–2554.

Rittenour, T.R., Blum, M.D., and Goble, R.J., 2007, Fluvial evolution of the lower Mississippi River valley during the last 100 k.y. glacial cycle: Response to glaciation and sea-level change: Geological Society of America, Bulletin, v. 119, p. 586–608, doi: 10.1130/B25934.1.

Rust, B.R., 1972, Structure and process in a braided river: Sedimentology, v. 18, p. 221–245.

Saéz, A., Cabrera, L., Jensen, A., and Chong, G., 1999, Late Neogene lacustrine record and palaeogeography in the Quillagua–Llamara basin, Central Andean fore-arc (northern Chile): Palaeogeography, Palaeoclimatology, Palaeoecology, v. 151, p. 5–37, doi: 10.1016/S0031-0182(99)00013-9.

Sarma, J.N., 2005, Fluvial process and morphology of the Brahmaputra River in Assam, India: Geomorphology, v. 70, p. 226–256.

Saucier, R.T., 1994, Geomorphology and Quaternary geologic history of the lower Mississippi Valley: U.S. Army Corps of Engineers, Waterways Experiment Station, 2 volumes, 363 p.

Scholz, C.H., 1976, Rifting in the Okavango Delta: Natural History, v. 85, p. 34–43.

Schumm, S.A., 1977, The Fluvial System: New York, Wiley-Interscience, 338 p.

Schumm, S.A., 1986, Alluvial river response to active tectonics, *in* Active Tectonics: National Academy Press, Washington, D.C., p. 80–94.

Schumm, S.A., 2005, River variability and complexity: Cambridge, U.K., Cambridge University Press, 220 p.

Schumm, S.A., Dumont, J.F., and Holbrook, J.M., 2000, Active Tectonics and Alluvial Rivers: Cambridge, U.K., Cambridge University Press, 276p.

Shaw, P.A., and Thomas, D.S.G., 1993, Geomorphological processes, environmental change and landscape sensitivity in the Kalahari region of Southern Africa, *in* Thomas, D.S.G., and Allison, R.J., eds., Landscape Sensitivity: Chichester, John Wiley & Sons, p. 83–95.

Shepherd, R.G., 1987, Lateral accretion surfaces in ephemeral-stream point bars, Rio Puerco, New Mexico, *in* Ethridge, F.G., Flores, R.M., and Harvey, M.D., eds., Recent Developments in Fluvial Sedimentology: SEPM, Special Publication 39, p. 93–98.

Singh, I.B., 2007, The Ganga River, *in* Gupta, A., ed., Large Rivers: Geomorphology and Management: Chichester, John Wiley & Sons, Ltd., p. 347–371.

Singh, S.K., 2007, Erosion and weathering in the Brahmaputra River system, *in* Gupta, A., ed., Large Rivers: Geomorphology and Management: Chichester, John Wiley & Sons, Ltd., p. 373–393.

Sinha, R., 1996, Channel avulsion and floodplain structure in the Gandak–Kosi interfan, north Bihar plains, India: Zeitschrift für Geomorphologie, N.F., Supplementband, v. 103, p. 249–268.

Sinha, R., and Friend, P.F., 1994, River systems and their sediment flux, Indo-Gangetic plains, Northern Bihar, India: Sedimentology, v. 41, p. 825–845.

Sinha, R., Jain, V., Babu, G.P., and Ghosh, S., 2005, Geomorphic characterization and diversity of the fluvial systems of the Gangetic Plains: Geomorphology, v. 70, p. 207–225.

Skelly, R.L., Bristow, C.S., and Ethridge, F.G., 2003, Architecture of channel-belt deposits in an aggrading shallow sandbed braided river: the lower Niobrara River, northeast Nebraska: Sedimentary Geology, v. 158, p. 249–270.

Slingerland, R.L., and Smith, N.D., 2004, River avulsions and their deposits: Annual Review of Earth and Planetary Sciences, v. 32, p. 255–283.

Smith, B., and Sandwell, D., 2003, Accuracy and resolution of shuttle radar topography mission data: Geophysical Research Letters, v. 30, p. 1467–1470, doi:10.1029/2002GL016643.

Smith, D.G., 1986, Anastomosing river deposits, sedimentation rates and basin subsidence, Magdalena River, northwestern Colombia, South America: Sedimentary Geology, v. 46, p. 177–196.

Smith, N.D., 1970, The braided stream depositional environment: comparison of the Platte River with some Silurian clastic rocks, north-central Appalachians: Geological Society of America, Bulletin, v. 81, p. 2993–3013.

Smith, N.D., 1971, Transverse bars and braiding in the Lower Platte River, Nebraska: Geological Society of America, Bulletin, v. 82, p. 3407–3420.

Smith, N.D., 1972, Some sedimentological aspects of planar cross-stratification in a sandy braided river: Journal of Sedimentary Petrology, v. 42, p. 624–634.

Smith, N.D., 1974, Sedimentology and bar formation in the upper Kicking Horse River, a braided outwash stream: Journal of Geology, v. 81, p. 205–223.

Smith, N.D., and Pérez-Arlucea, M., 1994, Fine-grained splay deposition in the avulsion belt of the lower Saskatchewan River, Canada: Journal of Sedimentary Research, v. B64, p. 159–168.

Smith, N.D., and Smith, D.G., 1984, William River, an outstanding example of channel widening and braiding caused by bed-load addition: Geology, v. 12, p. 78–82.

Smith, N.D., Cross, T.A., Dufficy, J.P., and Clough, S.R., 1989, Anatomy of an avulsion: Sedimentology, v. 36, p. 1–23.

Smith, N.D., Slingerland, R.L., Pèrez-Arlucea, M., and Morozova, G.S., 1998, The 1870s avulsion of the Saskatchewan River: Canadian Journal of Earth Sciences, v. 35, p. 453–466.

Staley, D.M., Wasklewicz, T.A., and Blaszczynski, J.S., 2006, Surficial patterns of debris flow deposition on alluvial fans in Death Valley, CA, using airborne laser SWATH mapping data: Geomorphology, v. 74, p. 152–163.

Stanistreet, I.G., and McCarthy, T.S., 1993, The Okavango fan and the classification of subaerial fan systems: Sedimentary Geology, v. 85, p. 115–133.

Stanistreet, I.G., Cairncross, B., and McCarthy, T.S., 1993, Low sinuosity and meandering bedload rivers of the Okavango Fan: channel confinement by vegetated levees without fine sediment: Sedimentary Geology, v. 85, p. 135–156.

Steel, R.J., 1974, New Red Sandstone floodplain and piedmont sedimentation in the Hebridean province, Scotland: Journal of Sedimentary Petrology, v. 44, p. 336–357.

Stouthamer, E., 2001, Sedimentary products of avulsion in the Holocene Rhine–Meuse Delta, The Netherlands: Sedimentary Geology, v. 145, p. 73–92.

Stouthamer, E., and Berendsen, H.J.A., 2000, Factors controlling the Holocene avulsion history of the Rhine–Meuse delta (The Netherlands): Journal of Sedimentary Research, v. 70, p. 1051–1064.

Stouthamer, E., and Bernedsen, H.J.A., 2001, Avulsion frequency, avulsion duration, and interavulsion period of Holocene channel belts in the Rhine–Meuse delta, The Netherlands: Journal of Sedimentary Research, v. 71, p. 589–598.

Sun, J., Zhang, Z., and Zhang, L., 2009, New evidence on the age of the Taklimakan Desert: Geology, v. 37, p. 159–162.

Ten Brinke, W.B.M., Wilbers, A.W.E., and Wesseling, C., 1999, Dune growth, decay and migration rates during a large magnitude flood at a sand and mixed sand–gravel bed in the Dutch Rhine river system, *in* Smith, N.D., and Rogers, J., eds., Fluvial Sedimentology VI: International Association of Sedimentologists, Special Publication 28, p. 15–32.

Thorne, C.R., Russell, A.P.G., and Alam, M.K., 1993, Planform pattern and channel evolution of Brahmaputra River, Bangladesh, *in* Best, J.L., and Bristow, C.S., eds., Braided Rivers: Geological Society of London, Special Publication 75, p. 257–276.

Törnqvist, T.E., 1993, Holocene alternation of meandering and anastomosing fluvial systems in the Rhine–Meuse Delta (central Netherlands) controlled by sea-level rise and subsoil erodibility: Journal of Sedimentary Petrology, v. 63, p. 683–693.

Törnqvist, T.E., 1994, Middle and late Holocene avulsion history of the River Rhine (Rhine–Meuse Delta, Netherlands): Geology, v. 22, p. 711–714.

Törnqvist, T.E., Wallinga, J., Murray, A.S., DeWolf, H., Cleveringa, P., and DeGans, W., 2000, Response of the Rhine–Meuse system (west-central Netherlands) to the last Quaternary glacio-eustatic cycles: Global Planetary Change, v. 27, p. 89–111.

Tyler, N., and Ethridge, F.G., 1983, Depositional setting of the Salt Wash Member of the Morrison Formation, southwest Colorado: Journal of Sedimentary Petrology, v. 53, p. 67–82.

Uba, C.E., Heubeck, C., and Hulka, C., 2005, Facies analysis and basin architecture of the Neogene Subandean synorogenic wedge, southern Bolivia: Sedimentary Geology, v. 180, p. 91–123.

Unruh, J.R., 1991, The uplift of the Sierra Nevada and implications for the Cenozoic epeirogeny in the western Cordillera: Geological Society of America, Bulletin, v. 103, p. 1395–1404.

Ussami, N., Shiraiwa, S., Dominguez, J.M.L., 1999, Basement reactivation in a sub-Andean foreland flexural bulge: the Pantanal wetland, SW Brazil: Tectonics, v. 18, p. 25–39.

Volker, H.X., Wasklewicz, T.A., and Ellis, M.A., 2007, A topographic fingerprint to distinguish alluvial fan formative processes: Geomorphology, v. 88, p. 34–45.

Weissmann, G.S., Mount, J.F., and Fogg, G.E., 2002, Glacially driven cycles in accommodation space and sequence stratigraphy of a stream-dominated alluvial fan, Central Valley, California: Journal of Sedimentary Research, v. 72, p. 240–251.

Weissmann, G.S., Bennett, G., and Lansdale, A.L., 2005, Factors controlling sequence development on Quaternary fluvial fans, San Joaquin Basin, California, USA, *in* Harvey, A., Mather, A., and Stokes, M., eds., Alluvial Fans: Geomorphology, Sedimentology, Dynamics: Geological Society of London, Special Publication 251, p. 169–186.

Weissmann, G.S., Hartley, A.J., Nichols, G.J., Scuderi, L.A., Olson, M., Buehler, H., and Banteah, R., 2010, Fluvial form in modern continental sedimentary basins: Distributive Fluvial Systems (DFS): Geology, v. 38, p. 39–42, doi: 10.1130/G30242.1.

Wells, N.A., and Dorr, J.A., 1987, Shifting of the Kosi River, northern India: Geology, v. 15, p. 204–207.

Wells, S.G., McFadden, L.D., and Dohrenwend, J.C., 1987, Influence of late Quaternary climatic changes on geomorphic and pedogenic processes on a desert piedmont, eastern Mojave Desert, California: Quaternary Research, v. 27, p. 130–146.

Wilbers, A.W.E., and Ten Brinke, W.B.M., 2003, The response of subaqueous dunes to floods in sand and gravel bed reaches of the Dutche Rhine: Sedimentology, v. 50, p. 1013–1034.

Williams, P.F., and Rust, B.R., 1969, The sedimentology of a braided river: Journal of Sedimentary Petrology, v. 39, p. 649–670.

Wilson, B.H., and Dincer, T., 1976, An introduction to the hydrology and hydrography of the Okavango Delta, *in*, Symposium on the Okavango Delta, Botswana: Botswana Society, Gaborone, p. 33–48.

Zehfuss, P.H., Bierman, P.R., Gillespie, A.R., Burke, R.M., and Caffee, M.W., 2001, Slip rates on the Fish Springs fault, Owens Valley, California, deduced from cosmogenic ^{10}Be and ^{26}Al and soil development on fan surfaces: Geological Society of America, Bulletin, v. 113, p. 241–255.

Long Term: Allogenically Dominated?

PRESERVATION OF A LONG-LIVED FLUVIAL SYSTEM IN A MOUNTAIN CHAIN: THE TAGLIAMENTO VALLEY (SOUTHEASTERN ITALIAN ALPS)

GIOVANNI MONEGATO AND CRISTINA STEFANI
Dipartimento di Geoscienze, Università di Padova, Via Gradenigo 6, 35131 Padova, Italy
e-mail: giovanni.monegato@unipd.it

ABSTRACT: The continental succession filling the valley of the Tagliamento River (eastern Southern Alps) is one of the best examples of long-lived fluvial system in the Alpine Chain. The sedimentary record, divided into a number allostratigraphic units, covers the Messinian–Quaternary time interval and allows the unravelling of the evolution of the drainage basin in the most tectonically active sector of the Alps. The timing of the tectonic activity was compared with the ages of the sedimentary units and its influence on the sedimentation along the valley, and the preservation of the sedimentary bodies was evaluated. After the deep incision determined by the Messinian Salinity Crisis the sedimentation along the valley was firstly driven by the late Zanclean marine transgression; subsequently the strike-slip deformation along the valley, caused by the western segment of the Idrija Fault, created reaches of local subsidence, in which there were unit superposition, or uplift, where the units are nested. Strike-slip movement contributed to the northeastern migration of the trunk valley at its present location. Finally, the onset of the Quaternary glaciations determined the increase of sediment discharge of the catchment and enhanced the erosion during the glacier advances.

KEY WORDS: facies analysis, Messinian to Quaternary, Tagliamento River, Tectonics, Southeastern Alps

INTRODUCTION

Mountain valleys in uplifted orogens usually act as transfer zones (*sensu* Schumm 1977) and have very low storage potential due to dominant erosion (Schumm 1993). It is for this reason that the preservation and distribution of valley-fill sediments are commonly scattered and discontinuous. Moreover, valley fills inside mountain chains are usually characterized by coarse-grained texture, and the remnants are difficult to correlate each other along the valley and also with the fine-grained continental and marine units in the relative foreland basins.

Nevertheless, instances of long-lived fluvial successions in a uplifted mountain belts often occur in intermountain basins (e.g., Nemes et al. 1997; Kuhlemann 2000; Dunkl et al. 2005), abandoned valley reaches (e.g., De Franco et al. 2004; Sölva et al. 2004), and valleys whose spatial stability has been tectonically influenced (e.g., Vincent and Elliott 1997; Schumm et al. 2000; Vincent 2001; Aldinucci et al. 2007; Hou et al. 2008).

Base-level variations as well as climate change, tectonic uplift, and fluvial hydraulics are the main factors controlling deposition along a valley (e.g., Schumm 1977; Schumm and Ethridge 1994; Jones et al. 1999). Climate change and tectonic uplift tend to become dominant in the proximal part of a drainage system, where, as suggested by Schumm (1993), the sea-level changes are weakly felt.

Taking into account a modern mountain belt affected by Quaternary glaciations such as the Alps, the valley-fill record is mainly related to the Last Glacial Maximum and the subsequent deglaciation, while older sedimentary units of Plio-Quaternary age are scarce and scattered (e.g., Preusser and Schlüchter 2004; Ostermann et al. 2006; Keil and Neubauer 2009).

This study regards the example of a long-lived fluvial record in the Alpine chain. The Tagliamento conglomerate (TC) and the valley fill related to the last glacial cycle are linked to a fluvial system in the eastern Southern Alps, whose evolution has been active since the Late Miocene onwards (Monegato et al. 2010). The aim of the paper is to discuss the preservation of the sedimentary record of the Tagliamento river system during the last 6 Myr and its importance in unravelling the drainage evolution of the chain and the tectonic deformations affecting the valley.

GEOMORPHOLOGICAL AND GEOLOGICAL SETTING

The Tagliamento drainage basin is located in the eastern Southern Alps (Fig. 1) and covers an area of about 2300 km^2 (Fig. 2). It can be split into two main sub-basins related to the Tagliamento *s.s.* and to the Fella River, its main tributary. The drainage basin is characterized by roughly E-W oriented trunk valleys linked by narrow N-S reaches (Fig. 2). The catchment is also characterized by wind gaps with fluvial remnants, which witness ancient tracks of the main streams. These geomorphologic remains are represented by the saddles of Chiampon, Chianzutan, Cima Corso, and Pura Pass, located in the Tagliamento catchment (Fig. 3). The main abandoned wind gap is represented by the valley of Lake Cavazzo (Fig. 3), N-S trending and parallel to the present-day trunk upstream of the valley mouth.

The modern drainage basin of the Tagliamento River cuts the eastern Southalpine Chain (ESC), whose exhumation started during the Serravallian (Castellarin et al. 1992; Stefani et al. 2007), and whose tectonic structures (Fig. 4) divide the catchment into three adjoining sectors with different source rocks. From north to south they are:

- Sector A: confined between the northern watershed and the E-W structures of the Fella–Sava fault where Paleozoic units of the so-called *Palaeocarnic Chain* crop out (Selli 1963; Spalletta et al. 1982; Venturini 1990; Venturini et al. 2002);
- Sector B: confined between the Fella–Sava and the Tagliamento–Fella alignments where Permo-Triassic sedimentary rocks are exposed (Carulli et al. 1987; Carulli et al. 2000; Venturini et al. 2002);
- Sector C: includes the Carnic and Julian Prealps, where upper Triassic to Miocene formations crop out (Massari et al. 1986; Carulli et al. 2000).

The Tagliamento River system was affected by a dramatic deepening during the sea-level drop of the Messinian Salinity Crisis, as were most of the streams flowing into the Mediterranean (e.g., Bini et al. 1978; Clauzon 1982; Felber et al. 1992; Gargani 2004), and was subsequently filled during the late Zanclean transgression. The Messinian canyon of

From River To Rock Record: The Preservation Of Fluvial Sediments And Their Subsequent Interpretation
SEPM Special Publication No. 97, Copyright © 2011
SEPM (Society for Sedimentary Geology), ISBN 978-1-56576-305-0, p. 359–374.

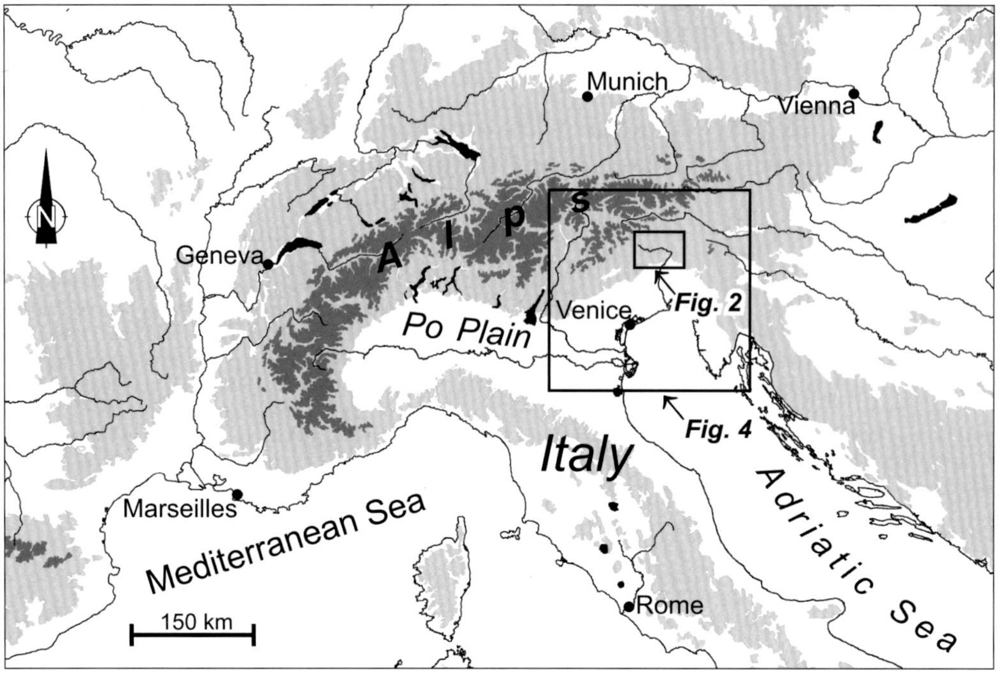

FIGURE 1.—Map of the European Alps and surroundings with location of the Tagliamento catchment (see details in Fig. 2)

the Tagliamento River has been recognized in a series of commercial seismic lines across the Friulian plain (Zanferrari et al. 2008; Ghielmi et al. 2009); it is hundreds of meters deep and several kilometers wide, and has a NE-SW trend.

Quaternary glaciations are documented in the Tagliamento catchment, even though most of the evidence is related to the last glacial advance during the late Pleistocene (Monegato et al. 2007). The extension of the older glaciations is poorly tested because of the scarcity of preserved deposits (Castiglioni 2004; Zanferrari et al. 2008).

However, for the increase of sediment discharge during the Gelasian in the Eastern Alps the importance of Quaternary glaciations are debated (Kuhlemann et al. 2001; Kuhlemann et al. 2002; Mancin et al. 2009). The discharge rate from the eastern Southern Alps calculated by Mancin et al. (2009) shows a change from less than 0.4 mm/a to 0.6 mm/a in the early Gelasian (Fig. 5); subsequently the discharge rate had grown till 1.2 mm/a at 1.5 Ma; for the later Pleistocene no data are available. According to these authors the cause of the enhancement of discharge could have been an early expansion of glaciation in the Alps.

TECTONIC SETTING

The eastern Southalpine Chain (ESC, Fig. 4) is a SE-verging thrust–fold belt developed during the Neoalpine tectonic phase (Serravallian to Present) of the Alpine Orogenesis (e.g., Doglioni and Bosellini 1987; Castellarin et al. 1992; Castellarin and Cantelli 2000). This chain was formed by the indentation of the Adriatic foreland beneath the Alpine Chain (Fig. 4) and the contemporary dextral strike-slip movement of the Insubric Lineament (Massari 1990; Ratschbacher et al. 1991; Fodor et al. 1998). The Neoalpine Event is usually subdivided into (1) the Serravallian–Messinian event (Castellarin et al. 1992), which recorded the most significant uplift rate, when the main stress direction varied from NNW-SSE to NW-SE and (2) the Pliocene–Quaternary events, which had a principal N-S stress direction; in the eastern part of the ESC, the local stress was distributed from NNW-SSE to ENE-WSW (Fig. 4), due to the sinking of the wedge-shaped foreland beneath the eastern South Alpine front, caused by the pushing northwards of the Adriatic foreland (Fig. 4, Zanferrari et al. in press).

In the eastern Southern Alps the Pliocene–Quaternary tectonic phase has been subdivided into three events, recognized in the frontal part of the chain (Caputo et al. 2003; 2010), during the early Zanclean, the Plio-Pleistocene boundary and second half or the Pleistocene (Fig. 5). The deformation inside the chain is marked by the presence of strike-slip faults that could be related to the Idrija Fault (Fig. 4) (M.E. Poli, personal communication). This is the main strike-slip structure located between the eastern Southern Alps and the Dinarides, and the recent activity of the Idrija Fault is witnessed by the general seismicity of the ESC, as indicated by recent earthquakes (Galadini et al. 2005; Burrato et al. 2008).

From the Zanclean onwards the Southern Alpine foreland was also affected by the tectonic load of northeast-migrating Apennine chain (Fig. 4; Carminati et al. 2003; Ghielmi et al. 2009), where, from the Messinian onwards, five main compressional events have been recognized (Fig. 5; Boccaletti and Sani 1998) and could have influenced the subsidence in the Venetian plain since the Pliocene (Carminati et al. 2003; Mancin et al. 2009).

METHODS

The valley-fill deposits of the Tagliamento River have been the subject of geomorphological, sedimentological, compositional, and structural analyses (Zanferrari et al. in press). Detailed field mapping at the scale 1:5.000 and collection of data from the present Tagliamento River were undertaken. Facies analyses were carried out on outcrops at the decimeter scale. Facies associations were made, and are summarized in Table 1, where the occurrence also for each unit is marked. For each of the sedimentary units and for the present-day river some longitudinal profiles were constructed along the lower to middle reaches (Fig. 6). Arenite composition were studied following the Gazzi-Dickinson procedures (Gazzi, 1966; Dickinson, 1970) in order to delineate changes in supply. For each section 500 points (arenites) or 300 points (sands) were determined, using a 0.5 mm grid spacing. The

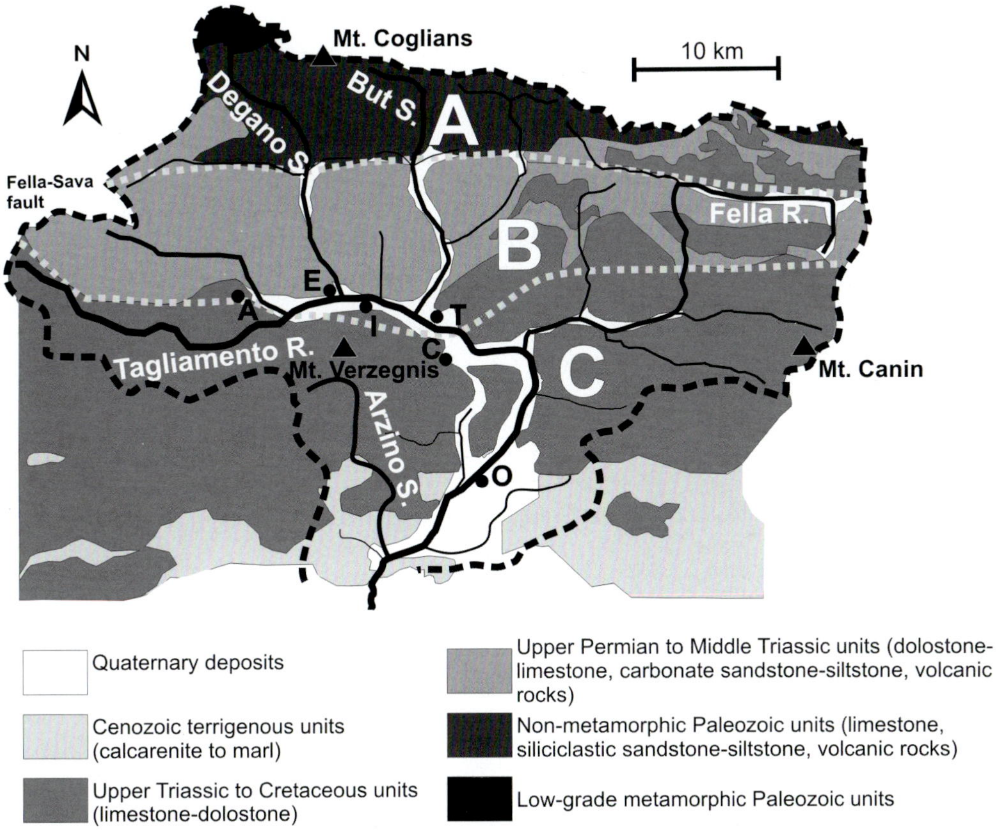

FIGURE 2.—Geological sketch map of the Tagliamento drainage basin subdivided into three main sectors whose borders are drawn in dotted gray lines; Fella–Sava Fault is the boundary between A and B sectors. White capital letters refer to the three sectors with different lithologic units in which the catchment can be subdivided (see text for explanation). Location in black capital letters, A, Ampezzo; C, Cavazzo Carnico; E, Enemonzo; I, Invillino; O, Osoppo; T, Tolmezzo.

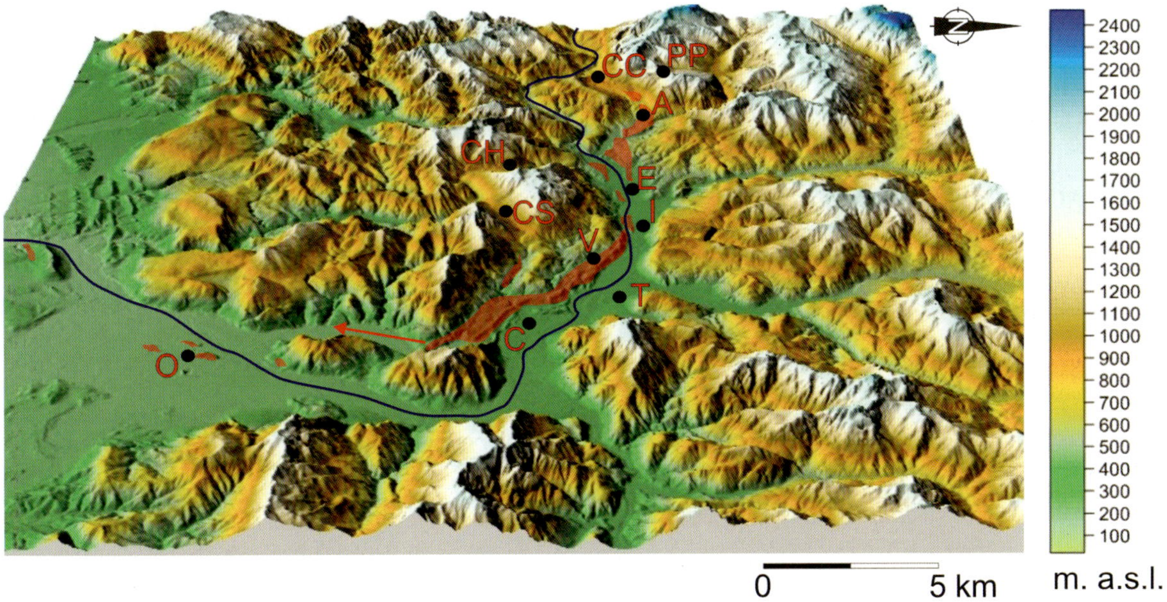

FIGURE 3.—DTM of the Tagliamento valley showing the present-day channel (blue line) and the distribution of the TC outcrops (pale red); red arrow indicates the flow direction in Cavazzo Lake valley. A, Ampezzo; C, Cavazzo Carnico; CC, Cima Corso Saddle; CH, Chiampon Saddle; CS, Chianzutan Saddle; E, Enemonzo; I, Invillino; O, Osoppo; PP, Pura Pass; T, Tolmezzo; V, Verzegnis.

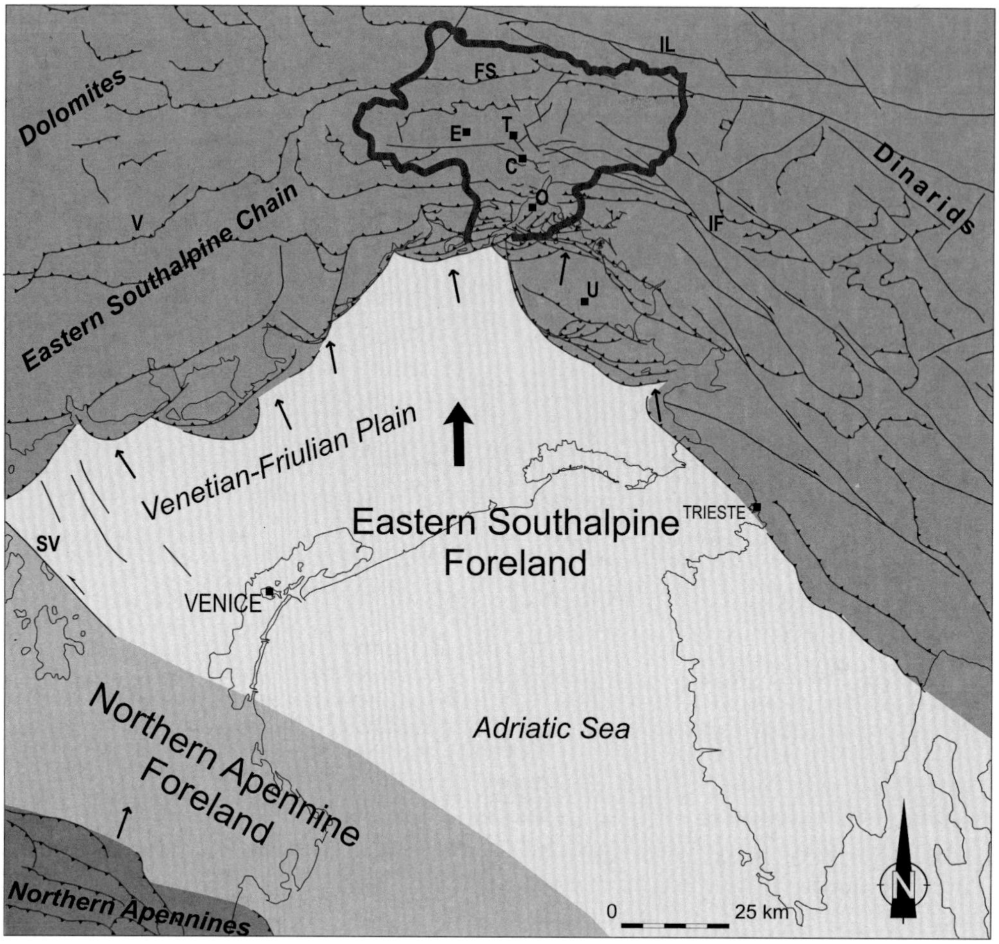

FIGURE 4.—Regional tectonic sketch of northeastern Italy and western Slovenia, modified from Zanferrari et al. (in press), with the location of the watershed of the Tagliamento drainage basin (thick black line). Thick arrow indicates the main stress direction; local stress is marked by thin arrows. FS, Fella–Sava Fault; IF, Idrija Fault; IP, Insubric Lineament; SV, Schio–Vicenza Fault; V, Valsugana Line. Main locations: C, Cavazzo Carnico; E, Enemonzo; O, Osoppo; T, Tolmezzo; U, Udine.

results are shown in the Q+F, L, CE plot (Fig. 7) suggested by Lugli et al. (2007) for sediments derived from carbonate-rich source areas.

THE TAGLIAMENTO VALLEY FILL: STRATIGRAPHY AND DEFORMATION

The Tagliamento valley fill consists of various stratigraphic units, which can be grouped into:

a) the Tagliamento conglomerate (TC), b) old glacial remnants scattered along the valley slopes, which are not considered in the present work, c) the Spilimbergo unit, related to the Last Glacial Maximum (LGM), and d) the postglacial valley fill.

The TC is a succession dominated by cobble- to pebble-size conglomerates, with subordinate sandstones, mudstones, and breccias that unconformably overlie bedrock strata, which are disposed in imbricated thrust systems. The TC succession has been internally subdivided into five allostratigraphic units (*sensu* NACSN 1983), which are, in chronological order starting from the oldest: Faeit, Ambiesta, Cesclans, Ampezzo, and Invillino (Fig. 5, Table 2). These units are 30 to 100 m thick, bounded by angular unconformities or nested due to deepening of the trunk valley. Changes in facies association are slightly across the bounding surfaces, dominance of fluvial associations is understandable, even though the facies-association assemblage differs from one unit to the other. In addition, provenance variations are pronounced, especially among the oldest units (Fig. 7), indicating the evolution of the drainage basin through time (Monegato 2006).

The main outcrops appear in a series of ridges (Fig. 3) that, taken together, represent a belt 37 km long, 0.6–1.2 km wide, and 200 m thick, running along the modern Tagliamento valley (Fig. 8), while the respective deltaic deposits are located at the mouth of the trunk valley in the isolated Osoppo hills (Fig. 3). The ridges are generally continuous along the valley (Fig. 3), interrupted only by short, narrow gorges, whereas in the central and in the lower reaches of the trunk valley, deposits are more discontinuous.

The TC is dominated by a carbonate fraction exceeding 60%. Calcite cement is always very abundant as columnar isopachous calcite and as equant calcite in the remaining spaces. The abundance of carbonate rocks determined the availability of diluted carbonate in the groundwaters, and early cementation of the sedimentary bodies themselves preceded compaction as a result of burial (e.g, James 1985), as evidenced by calcite cement in clast-supported framework. The early cementation is also highlighted by the fact that deformation structures (folds and joints) that affected single units before the erosion

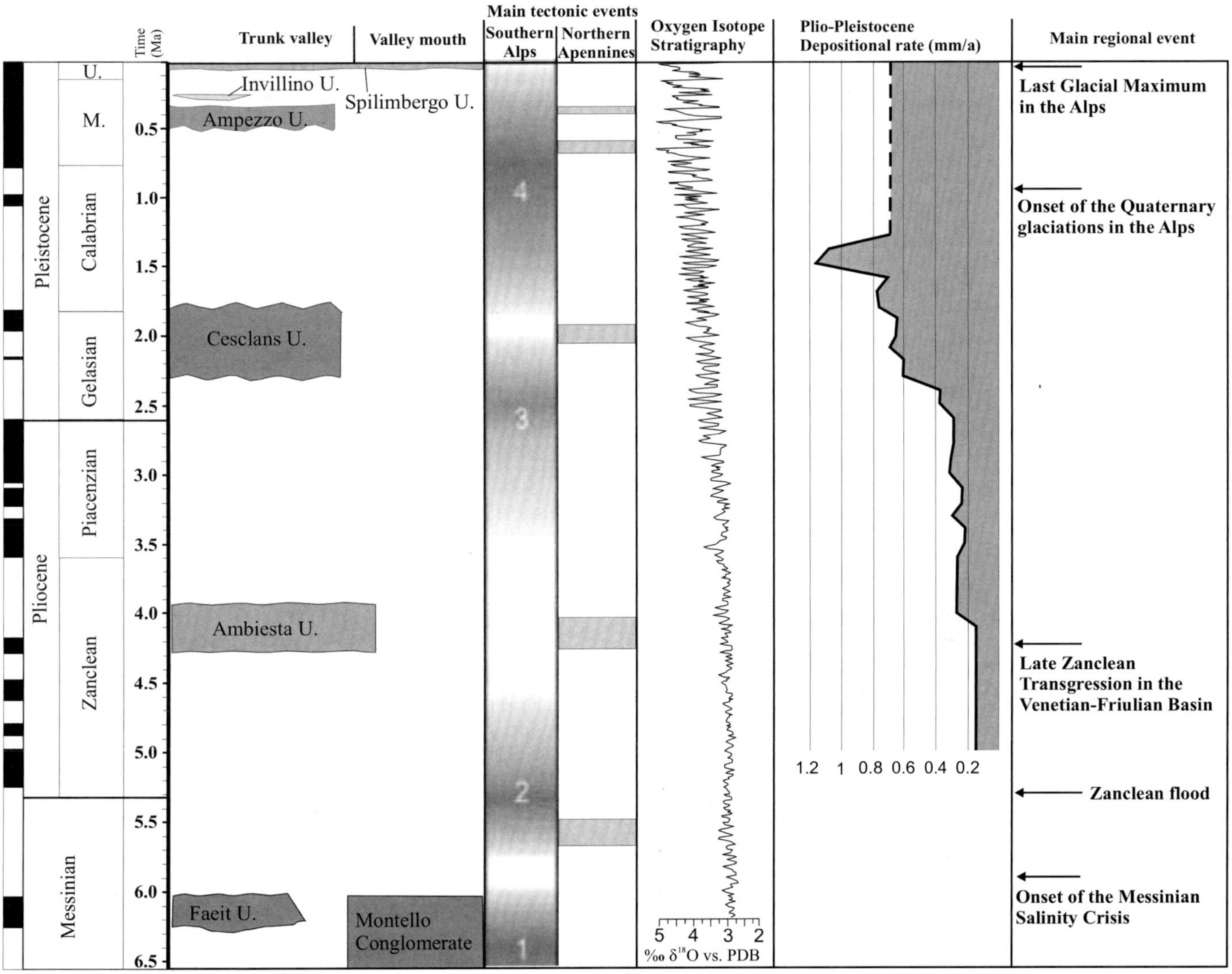

FIGURE 5.—Chronostratigraphic scheme of the preserved units linked to the Tagliamento River. Chronostratigraphy is after Gibbard et al. (2009). Eastern Southern Alps tectonic events are from Caputo et al. (2010). Northern Apennines tectonic events are from Boccaletti and Sani (1998). Oxigen Isotopic Stratigraphy (benthic records) is after Lisiecki and Raymo (2005). Sediment discharge rate is from Mancin et al. (2009). Onset of the MSC and Zanclean flood is from Roveri et al. (2008). Late Zanclean transgression is from Ghielmi et al. (2009). Onset of the Quaternary glaciations in the Alps is from Muttoni et al. (2003).

and the deposition of the younger ones (Table. 2) would be unlikely in case of loose deposits.

In detail, the Faeit Unit is characterized by fluvial conglomerate B1 laterally intertongued with A1–A2 breccias; the petrographic signature, lacking in metamorphic rocks as well as quartz-rich Paleozoic sandstones, indicates that the catchment was still confined south of the Fella–Sava Line (Fig. 2) and matches with that of the topmost deposits of the Montello Conglomerate (Messinian) in the piedmont plain. The conglomerate bodies are heavily deformed by two fold systems (Table 2) and preserved at higher elevation than the other units of the TC (Fig. 6).

The Ambiesta Unit is also heavily deformed by two fold systems (Table 2, Fig. 9A); B1–B3 fluvial associations prevail, with lateral intertonguing with A1 breccias. This is the unit with generally better development of fine-grained deposits (B3), common in two sectors of the trunk valley, near Enemonzo and close to Cavazzo Carnico. The Gilbert-type delta body of Osoppo marks both the mouth of the valley and the base level for the fluvial sedimentation. In addition, provenance indicates an important change in the drainage network, with a widening northwards of the Fella–Sava Line (Monegato et al. 2010).

The Cesclans Unit is folded only by one fold system (Table 2, Fig. 9B); this fact is visible in the angular unconformity with the underlying unit. The B1 association (Fig. 10A) is intertongued with A2 breccia bodies (Fig. 10B). Fine-grained facies of the B3 association are still common above those of the Ambiesta Unit. Provenance does not differ from the previous unit (Fig. 7), indicating a stasis in the drainage evolution.

The Ampezzo Unit lies on an angular unconformity above the previous unit, partially nested in the upstream reach of the trunk valley. The unit is made up of B1–B2 fluvial conglomerates intertongued with A1 breccias (Fig. 10C) and A3 conglomerates. Deltaic associations (D1–D4) concern the Gilbert-type deltas near Enemonzo and Cavazzo

TABLE 1.—Summary of the recognized facies associations; facies code according to Miall (1996), Eyles et al. (1983), and Vincent (2001).

Code	Description	Interpretation	Occurrence
A1	Disorganized clast-supported breccia, massive gravel facies (Gb and Gms). Clasts are normally angular to subangular, oligomictic and rich in up to 1 m outsized boulders. The sedimentary bodies are characterized by thick bedding and an overall thickness of up to 100 m; laterally banked up on the bedrock and intertongued with other facies associations.	Deposits connected to the dynamics of the valley slopes: grain flow, rock avalanches and landslides. A1 is interbedded with alluvial facies associations (Fig. 10C).	Valley sides, especially at the bottom of the northern side of Carnic Prealps.
A2	Disorganized oligomictic conglomerates (Gms) with minor disorganized clast-supported breccia (Gb). Clasts are mostly subangular (Fig. 11B), with rare up to 1 m outsize boulders; beds are tabular, up to 1 m thick, with a few hundreds of meters of lateral continuity. Their dip ranges from 15 to 30°. Moreover, this association is commonly preserved on the valley sides, where it is intertongued with fluvial facies associations.	The depositional products of restricted catchment area tributaries of the trunk valley, discharged during flood events in an alluvial-fan setting. They could be compared with the Type 1 alluvial fan of Blair and McPherson (1994).	The thickest sedimentary bodies, 30 m, crop out close to Cavazzo Carnico (Fig. 10B).
A3	Moderately well sorted, clast-supported, subrounded conglomerates (Gms and Gh) with planar stratification; laminated sandstone lenses are interbedded (Sh). Beds are decimeteRs to meters thick, characterized by tabular shape and basal erosional surfaces; small 3–5 m wide and 1–2 m thick scour-and-fill structures are common. Conglomerates are polymictic with an imbricated fabric.	This association is interpreted as alluvial-fan deposits accumulated by flow expansion at the junction of tributary streams with the main valley. The association occurs at the junction of the Tagliamento River with its main tributaries, also testified by outcrop location.	It occurs at the junction between the Tagliamento River and Bût Stream, near Tolmezzo.
B1	Clast-supported, poorly to well-sorted polymictic conglomerates, with subrounded to rounded clasts showing imbricated fabric (Fig. 10A). B1 is dominated by crudely planar to low-angle bedding (Gh), with minor massive bodies (Gms) and lags (Glg). Beds have a basal erosional surface. The generally tabular shape of the beds is punctuated by shallow channels, several meters wide and up to 1 m thick. Local massive to laminated sandstone units (Sh, Sr, Sm) are interbedded with the conglomerates.	B1 is interpreted as a shallow-channel braided system dominated by longitudinal bars (e.g., "Scott type" of Miall and Gibling 1978; Miall 1996). The bodies interfingered with the C facies association could be ascribed to outwash-plain braided streams.	B1 facies association is the most common and occurs along the Tagliamento valley from Ampezzo to Osoppo.
B2	Well-sorted polymictic conglomerates, with mostly subrounded clasts. Bodies are characterized by a basal erosional concave-up surface, few tens of meters wide and with an average thickness of 5–8 m. Low-angle bedding conglomerates (Gh-Gp) with minor sandy layers (Sh) are common; minor trough cross-bedded (Gt) and channel-lag (Glg) intervals occur, filling small channels. Conglomerates are normally organized in an imbricated fabric.	B2 is interpreted as a deeply-strongly confined braided system (e.g.: "Donjek type" of Miall and Gibling 1978; Miall 1996) with occurrence of side bars.	It occurs along the Tagliamento valley, especially in the Ampezzo Unit, as well as in the topsets of the Gilbert-type delta of Osoppo, related to the Ambiesta Unit.
B3	The B3 facies association (Fig. 9A) is dominated by a planar-bedded sandstone facies (Sh, Sr, St), with meters-thick intervals; sandstone packages are tabular, normally without a basal erosional surface, with average thickness of a few decimeters. Laminated to massive siltstone drapes (Fl) occur; local scours are filled by matrix- to clast-supported massive conglomerate (Gsc). Occurrence of weakly developed paleosols (P) with calcrete.	B3 represents overbank deposits formed in a floodplain in periods of moderate discharge as well as in scanty areas (Reinfelds and Nanson 1993).	B3 is located in two areas along the Tagliamento valley close to Enemonzo and Cavazzo Carnico (Fig. 6).

TABLE 1.—Continued.

Code	Description	Interpretation	Occurrence
C	Matrix-supported polymictic diamicton (Dmm), with stretched and faceted subangular to subrounded clasts. Minor clast-supported massive to crudely bedded conglomerate (Gms, Gh), lenses and beds a few meters thick, are present. The diamicton commonly drapes the bedrock.	This association represents glacial deposits, mostly lodgement till, locally interfingering with B1 association.	C association occurs in the upstream part of the valley fill as diamictite. Loose diamicton is preserved all along the valley.
D1	Gravel bars (Gp) normally showing gentle primary dip, with minor trough cross-bedded gravel facies (Gt) and sandstone layers (Sh, Sr). D1 bed packages are up to 10 m wide and about 5 m thick, with erosional concave-up base.	Topset beds of a Gilbert-type delta (Postma 1990)	D1 is present at the top of the delta bodies of Invillino, Enemonzo, and Osoppo.
D2	Graded and sorted clino-stratified conglomerates, associated with massive sandstone layers (Sm), with dip angles ranging from 15° to 35°, depending on the clast size of the deposit. Long axes (a) of the clasts are oriented parallel to the flow direction. Beds are normally up to 1 m thick.	Foreset beds of a Gilbert-type delta (Postma 1990)	D2 in visible at Invillino, Enemonzo, Cavazzo Carnico, and Osoppo.
D3	Well sorted to poorly sorted inclined conglomerates (Gms) and coarse massive to laminated sandstones (Sm-Sh). Massive coarse conglomerates fill scour troughs (Gsc).	Toeset beds of a Gilbert-type delta	D3 is present in the delta bodies of Enemonzo and Osoppo.
D4	Laminated sandstones (Sh) and mudstones (Fl, Fm), with planar or gently inclined beds a few centimeters thick.	Bottomset beds of a Gilbert-type delta (Postma 1990)	D4 is preserved at Cavazzo Carnico (Fig. 10D), and Osoppo.

Carnico (Fig. 10D). Provenance analysis indicates an increase in the noncarbonate lithic component (Fig. 7), suggesting erosion enhancement along the northern watershed, where Paleozoic successions crop out (Fig. 2). The unit is not folded but only faulted by WNW-ESE strike-slip fault systems (Fig. 9C, D).

Finally, the Invillino Unit is nested inside the older units (Fig. 6) in the central part of the trunk valley. The B1 facies association is prevalently associated with A3 conglomerates. In the upstream reach the C association is common, indicating a phase of glacial advance. D1–D2 deltaic associations near Invillino suggest the presence of a lake along the valley during the deposition of the unit. Deformation is weak, and vertical fault planes show a WNW-ESE strike.

Loose and strongly weathered glacial deposits are located along the valley slopes in scattered remnants at higher elevation than the younger LGM deposits (Zanferrari et al. in press). Because of their discontinuity and uncertain age, they are not discussed in the present paper.

The Splimbergo Unit (Fig. 5), related to the LGM (Monegato et al. 2007), is widely present in the drainage basin and generally consists of loose till, ice-contact, glaciolacustrine, and periglacial deposits with thickness ranging from some meters to tens of meters. These deposits occur along the valley slopes and on the plateaus.

The postglacial deposits are made up of loose fluvial gravels along the valley and scattered lacustrine successions related to landslide damming of the trunk valleys.

AGE

The lack of chronological data relating to the TC has caused the irresolution of its age attributions by previous authors, who generally have considered it as a stack of two depositional units, if not a single sedimentary body. The age ascription of the TC differs among various authors from Pliocene (Taramelli 1875, 1881; Stefanini 1915) to middle Pleistocene (Penck and Brückner 1909; Gortani and Desio 1926; Feruglio 1929; Gortani 1935; Carulli et al. 2000). However, none of these age interpretations have been supported by constraining data and derive from speculation on pebble composition, degree of cementation, or stratigraphic position.

The allostratigraphic subdivision of the TC succession, strengthened by petrographic and structural analyses, is here supported by some chronological constraints, which are summarized in Table 2.

The clast composition of the older Faeit Unit (Fig. 7) could well be linked with that of the topmost part of the Montello Conglomerate in the piedmont plain (Zanferrari et al. 2008); the structural deformation affecting the unit, as well as the highest topographic level of the basal surface along the valley (Fig. 6), strongly suggest a Messinian age (Fig. 5).

As regards the Ambiesta Unit, which comprises the deltaic deposits of Osoppo previously ascribed to the late Zanclean (Monegato et al. 2006; Dalla Vecchia 2008), it represents the infilling of the valley during a transgression phase. The latter is well documented westwards by pelitic marine sedimentation in correspondence of main fluvial valleys (e.g., Favero and Grandesso 1983). This transgression was determined by active flexural subsidence caused by a major tectonic event in the Northern Apennines (Fig. 5; Ghielmi et al. 2009). Moreover, the remarkable change in composition between the Faeit and Ambiesta units points to important changes in the drainage basin which occurred during the Messinian Salinity Crisis (Monegato et al. 2010).

The two angular unconformities bounding the Cesclans Unit indicate a long time span and the occurrence of tectonic phases before and after its deposition. The first, which deformed the older units, is in agreement with the principal stress direction of the tectonic phase in the eastern Southern Alps at the Plio-Pleistocene boundary (Fig. 5; Caputo et al. 2003; Caputo et al. 2010). The second may be related to local stress during the following Pleistocene phase (Fig. 5). Moreover, compositional data indicate that the lower Pleistocene conglomerate

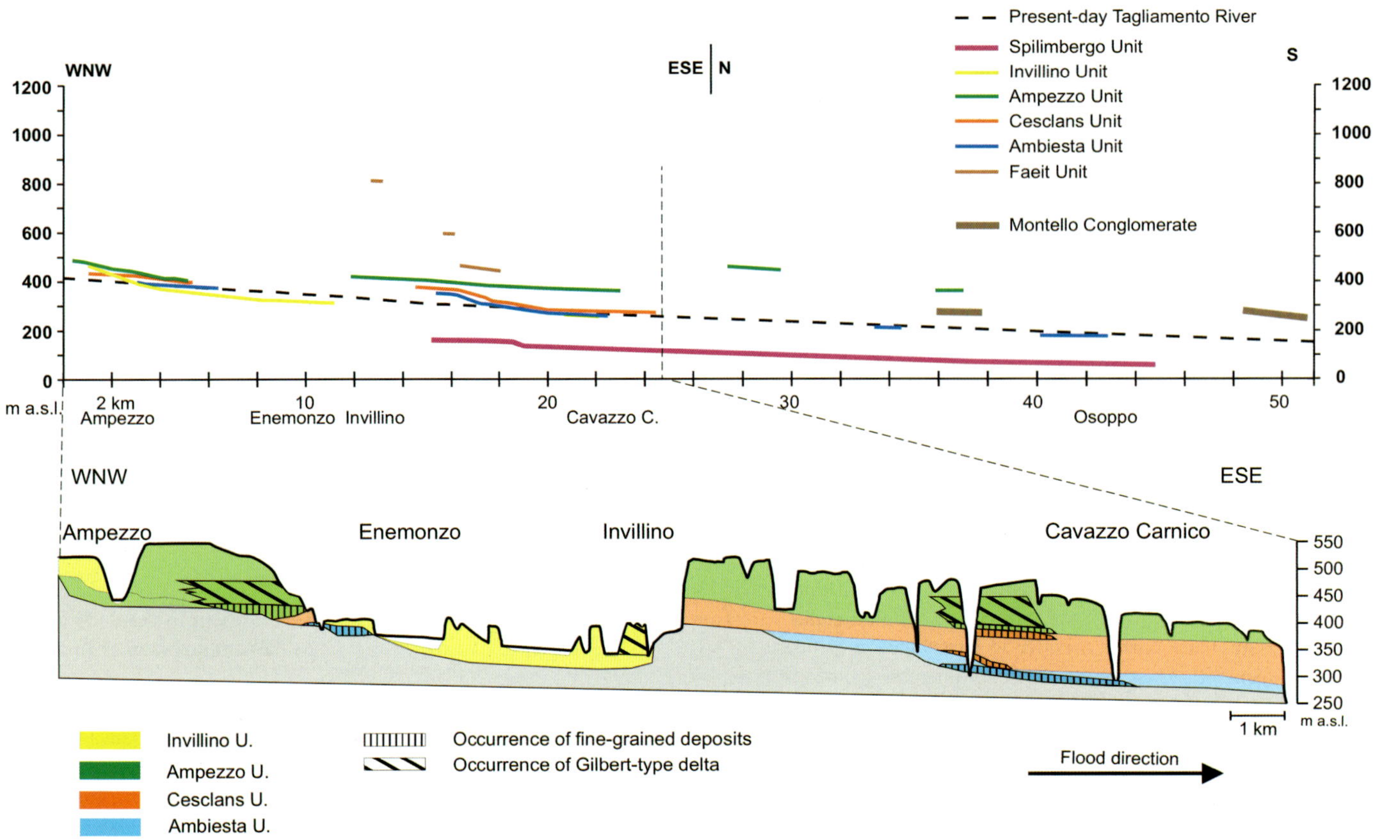

FIGURE 6.—Longitudinal profile of the Tagliamento valley showing the elevation of the basal boundaries of the described units. Longitudinal section of the TC along the Tagliamento valley, where are marked the occurrence of fine-grained facies associations and Gilbert-type deltas. The tectonic deformation along the valley produced a reach dominated by incision and nesting of the unit (from Ampezzo to Invillino) and another with superposition of the same (Cavazzo Carnico area).

located at the valley mouth (Zanferrari et al. 2008) could be related to this unit, and whose age is in agreement with its stratigraphic position and deformation.

The pollen assemblage in the lacustrine deposits of the basal part of the Ampezzo Unit indicates a warm phase of Middle Pleistocene (Monegato 2006); this age is in agreement with its stratigraphical position and the weaker structural deformations (Table 2).

The Invillino Unit, nested in the previous ones, is the youngest unit of the TC succession in which the glacial features appear to confirm a Middle Pleistocene age too.

The Spilimbergo unit related to the Tagliamento glacier has been dated at 22–18 ^{14}C kyr BP using several wood remnants found in the end-moraine system (Monegato et al. 2007).

At present, the riverbed of the Tagliamento valley is filled by a 120-m-thick gravel unit of Late Pleistocene–Holocene age, related to the glacial withdrawal after the Last Glacial Maximum. These younger deposits constitute the modern analogue for the rock record of the valley fill represented by the TC.

EVOLUTION OF THE TAGLIAMENTO VALLEY AND FACTORS CONTROLLING THE PRESERVATION OF THE VALLEY FILL

The preservation of the long sedimentary record of the Tagliamento valley during the last 6 My (Fig. 5) is a uncommon event in the Messinian–Quaternary alpine drainage systems, where most of the records are related to the Last Glacial Maximum, while older deposits are very discontinuous and mostly of undetermined age (e.g., Preusser and Schlüchter 2004; Ostermann et al. 2006; Keil and Neubauer 2009).

During the Messinian the drainage developed in a low-rate-uplift mountain belt, according to apatite fission-track data (Monegato et al. 2010); subsequently the expansion of the drainage network and its deep carving during the Messinian Salinity Crisis created a deep canyon, in which the river was confined during the later periods, and also led to the abandonment of the Chianzutan wind gap during the Messinian–Zanclean transition (Fig. 11A, B). However, the Messinian imprint has been recognized in many southern alpine valleys (Bini et al. 1978; Felber et al. 1992), in which though the subsequent sedimentary record is not well preserved or is lacking.

The second main migration of the trunk valley (Fig. 11 B, C) occurred before the deposition of the Invillino Unit, which is characterized by clearly glacial features (Table 2). The carving caused by the glacier happened along the direction of the E-W faults in the Tolmezzo area (Fig. 4); the combination of long-term tectonic activity and glacial erosion is the most likely reason for this important change in the main drainage. This may also explain the widening of the trunk valley (Fig. 11D), which was enlarged during this period from an average of 500 m to the present 1500 m of width (Fig. 11E). The marginal position of the main remnants of the older valley fill (Fig. 3), as well as the deactivation of the whole N-S reach now occupied by the Lake Cavazzo, allowed the preservation from

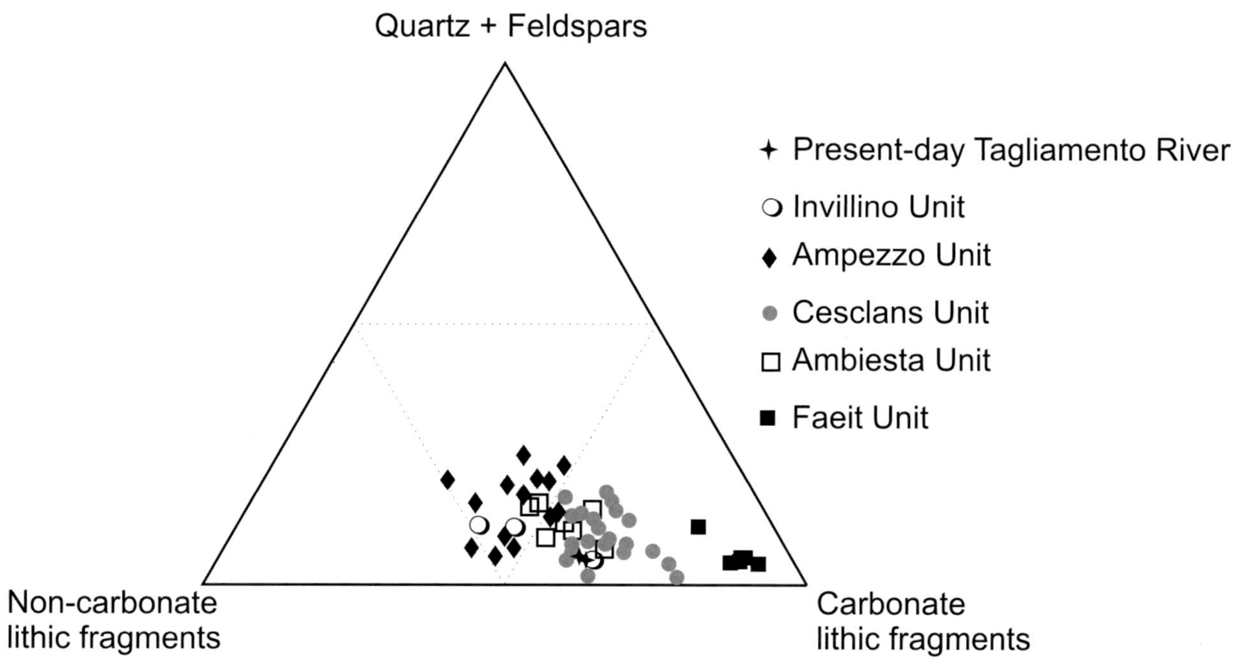

FIGURE 7.—Composition of sand and sandstones of the Tagliamento valley fill; Q+F, L and CE diagram was considered as suggested by Lugli et al. (2007) for source areas characterized by great amounts of carbonate rock fragments. Data are from Monegato (2006).

erosion of a 24-km-long reach away from the valley axis, which is usually affected by channel incision, subglacial quarrying, or glacial widening of the valley section (e.g., Brocklehurst and Whipple 2006). In addition, early cementation probably enhanced the resistance to the erosion soon after their deposition. Strike-slip faults on the conglomerate ridges suggest that the tectonic activity also continued during the late Quaternary; this activity may have contributed to create a new reach characterised by local tectonic subsidence south of Tolmezzo and filled by coarse fluvial deposits after the last glaciation (Fig. 11E).

During the considered time bracket, in the eastern Southalpine Chain the tectonic activity has been more intense than in the other sectors of the Southern Alps, and it is still active (Galadini et al. 2005; Burrato et al. 2008); hence tectonics could have been the main driving factor for the preservation of both the sedimentary fill and the ancient wind gaps of the Tagliamento valley.

Looking at the distribution of the units of the TC and their basal surfaces along the valley, in the central part of the E-W reach a displacement is visible (Fig. 6); this is evidenced by the different elevation of the basal surfaces, which represented the ancient profiles of the valley, and is remarkable in the older units, becoming weaker towards the younger (Fig. 6). The longitudinal section shows also a nesting of the units upstream of Invillino, while downstream all the units of the TC are piled up. Furthermore, facies analysis indicates the occurrence of fine-grained deposits upstream and downstream of Invillino (Fig. 6), as well as the location of the Gilbert-type deltas along the valley (Fig. 6). These evidences match with the general prediction models by Ouchi (1985) and Holbrook and Schumm (1999) in case of tectonic uplift along a valley, where the reaches upstream and downstream are prone to fine-grained aggradation, while the uplifted reach is affected by erosion.

Recently, geophysical research has shown a step 100 m high transverse to the flow direction in the bedrock below the Late Pleistocene valley fill (Barnaba et al. 2010), which has no evidence effect on the coarse-grained riverbed sediments above. This step could be related to an E-W tectonic structure along the valley.

The cause of the deformation along the E-W reaches of the valley are faults showing evidence of local strike-slip (Fig. 4); they could be related to a western segment of the Idrija Fault, which might have migrated towards the west during the Plio-Quaternary time (M.E. Poli, personal communication). The probable curved geometry of the strike-slip-fault-created valley reaches affected by tectonic uplift, where the units are nested, and others by smooth tectonic subsidence, where the units are superimposed. This fact also allowed the preservation of wind gaps (Chianzutan and Cima Corso saddles, Fig. 3) that are abandoned reaches of the ancient trunk valley.

From the Messinian to the Middle Pleistocene the progressive stream captures and northeasternward migrations of the trunk valley (Fig.11A, C) were mainly caused by the tectonic activity, as testified by angular unconformities between different units. This deformation almost matches with the phases of tectonic activity in the eastern Southern Alps (Fig. 5; Caputo et al. 2003; Caputo et al. 2010). Meanwhile the late Zanclean tectonic event in the Northern Apennines (Boccaletti and Sani 1998) controlled the subsidence and the base level of the Venetian–Friulian Plain (Ghielmi et al. 2009). The high base level of late Zanclean, which reached the mouth of the valley near Osoppo (Fig. 3), determined the sedimentation of the Ambiesta Unit along the valley.

During the Gelasian, the second Plio-Quaternary tectonic event determined the presence of local subsidence in the reach close to Cavazzo Carnico; this event could have driven the increase of production of detritus, even if the early extent of glaciations has been also questioned (Fig. 5; Mancin et al. 2009). The Cesclans Unit was deposited along the valley and is related to this phase (Fig. 5), as is indicated by the abundance of A1–A2 facies-association breccias intertongued with the fluvial conglomerates. During the same period a coarse alluvial fan, related to the Tagliamento River, developed in the piedmont plane (Zanferrari et al. 2008). The deformation affecting the Cesclans Unit points to a tectonic cause for the increase of sediment

TABLE 2.—Main characters of the stratigraphic units.

Stratigraphic unit		Age	Distribution	Facies associations	Bounding unconformities	Tectonic features
TAGLIAMENTO CONGLOMERATE	Faeit	Early Messinian	Chianzutàn saddle, Faeit valley (Fig. 3)	B1 with associated A1-A2	Lower angular discordances and onlap on the bedrock; upper diachronous erosion surface	Folds with N-S and 105/10° axes and pressure-solution cleavage
	Ambiesta	Early Pliocene	From Cavazzo C. to Verzegnis and near Enemonzo Osoppo Hills (Fig. 3)	B1-B3 with associated A1	Lower angular discordances and onlap on the bedrock; upper erosion surface marked by angular unconformities	Folds with 200/10° axes (Fig. 9A), NNE-SSW axis NNE dipping; SE-verging reverse faults
	Cesclans	Late Pliocene – Early Pleistocene	From Cavazzo C. to Verzegnis and from Enemonzo to Ampezzo (Fig. 3)	B1-B3 with associated A2 with lesser A1	Lower angular unconformity and lateral onlap on the bedrock; upper erosion surface marked by angular unconformities	Syncline with N110° axis and reverse faults (Fig. 9B); monocline dipping at 30/30°; WNW-ESE strike-slip faults
	Ampezzo	Middle Pleistocene	From Cavazzo C. to Verzegnis and from Enemonzo to Ampezzo (Fig. 3)	B2-B1 with associated A3, A1 near Cavazzo C. and D1–4 in the Ambiesta creek and Socchieve	Lower angular unconformity and lateral onlap on the bedrock; upper diachronous erosion surface	WNW-ESE strike-slip faults (Fig. 9C), sub-vertical joint systems (Fig. 9D)
	Invillino	Middle Pleistocene	From Invillino to Ampezzo (Fig. 3)	B1 with associated A3, C at Ampezzo and Invillino, D1–4 at Invillino	Lower angular discordances with the bedrock; upper diachronous erosion surface	WNW-ESE strike-slip faults, N-S vertical joint systems
Spilimbergo		Late Pleistocene	Present-day Tagliamento valley	B1, C	Lower erosion surface on the bedrock and older units; upper diachronous erosion surface	None
Post-glacial		Holocene	Present-day Tagliamento valley	B1 with associated A3 and A1 D4	Lower erosion surface on older units; upper topographic surface	None

discharge during the Gelasian in the catchment; this fact could be extended to explain the higher sediment discharge rate of the whole eastern Southalpine Chain (Fig. 5), also because the early extent of glaciations has not been documented yet for the Alps (Muttoni et al. 2003; Muttoni et al 2007).

Later, at the end of the Early Pleistocene, the onset of the major Quaternary glaciations in the Alps (Muttoni et al. 2003) led to an increase of sediment discharge rate from the Eastern Alps (Kuhlemann et al. 2001; Kuhlemann et al. 2002). The climate change also affected the Tagliamento drainage basin; nonetheless the time when the glaciers developed in this sector of the Alpine Chain is still undetermined due the lack of proxies (Castiglioni 2004). Meanwhile another tectonic phase affected the Southern Alps during the middle Pleistocene (Fig. 5; Caputo et al. 2010). The Ampezzo Unit could be the remnant of these tectonic and climatic events (Fig. 5), even if sedimentary evidence of glacial advances has not yet been found; in fact, the lacustrine deposits of Cavazzo Carnico are related to a warm phase of the middle Pleistocene (Monegato 2006).

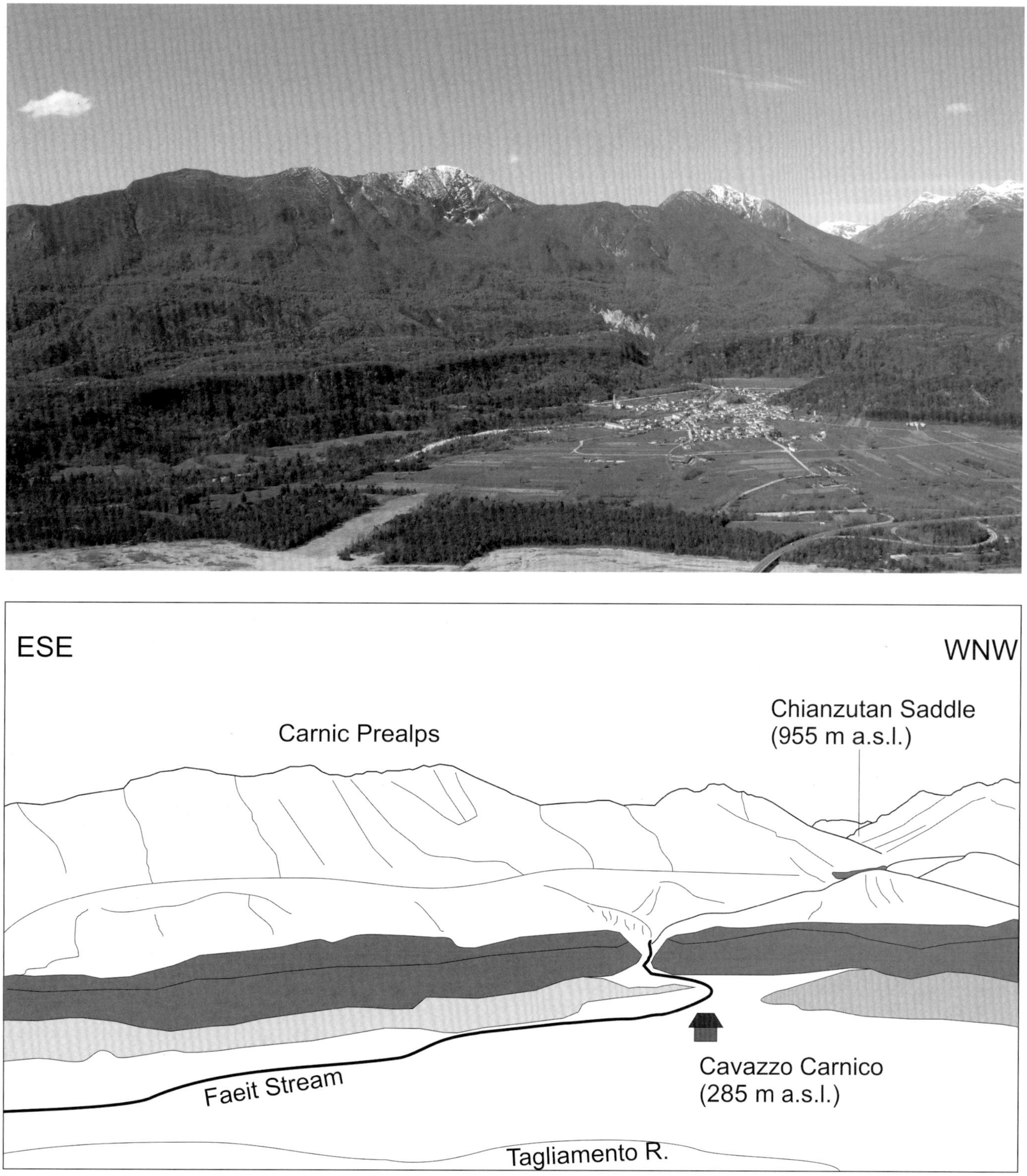

FIGURE 8.—Panoramic view and line drawing of the TC ridge (dark gray) at the northern feet of the Carnic Prealps close to Cavazzo Carnico village, located on Triassic bedrock in the left side of the paleovalley (pale gray).

The development of glaciers in the Tagliamento catchment is witnessed by glacial facies association of the Invillino Unit and then by the widespread glacial deposits related to the last glaciation. The increase of sediment discharge and the change in valley width indicate that the real climatic control on sediment production and accumulation happened in the middle Pleistocene. However, the thickness of sediments accumulated along the Tagliamento valley south of Tolmezzo (Barnaba et al. 2010) after the withdrawal of the LGM

FIGURE 9.—A) deformed sandstone and conglomerates of facies association B3 in the Ambiesta Unit near Enemonzo; B) tilted conglomerate of the southern flank of the fold affecting the Cesclans Unit near Cavazzo Carnico; C) strike-slip fault affecting the deltaic conglomerate (facies association D2) of the Ampezzo Unit near Enemonzo; D) WNW-ESE joints affecting subhorizontal conglomerates of the Ampezzo Unit near Enemonzo.

glaciers suggests that local subsidence has been affecting this reach until recent times. This subsidence can be related to transtension along the Idrija Fault towards the west, activity which has been persistent through the considered time span.

CONCLUSIONS

The Tagliamento valley fill is the most widespread long-lived fluvial record preserved in the Southern Alps. Stratigraphic reconstructions, along with facies and geomorphologic analyses, highlighted the evolution of the trunk valley during the last 6 Myr. The characterization of drainage anomalies, along with deformation structures affecting the sedimentary bodies and the presence of several wind gaps lead to estimate the causes of the preservation of the valley fill during the Messinian–Quaternary time interval. These are primarily related to the tectonic activity in the eastern Southern Alps with the interaction of main eustatic and climatic events, affecting the whole Alpine Chain.

In the first instance the Messinian Salinity Crisis triggered the upwards deep incision of the trunk valley, in which the river flowed in the subsequent period, and caused the expansion of the drainage basin, as evidenced by change in provenance.

Moreover, the tectonic activity of the eastern Southern Alps has been characterized by four main events during the Messinian–Quaternary time span. These events determined the creation of accommodation space in reaches of the trunk valley, as well as the deformation of the valley fill, and contributed to the displacement of wind gaps, in which by early-cemented coarse-grained deposits remained. The preservation of most of the valley fills along the E-W reaches could be imputed to the westwards propagation of the Idrija Fault.

The tectonic activity also contributed to enhance detrital production during the Gelasian, before the onset of the Quaternary glaciations. In addition, during the late Zanclean a marine transgression driven by flexuration of the Southalpine foreland due to the tectonic load of the Northern Apennines determined the infill of the valley upstream from the Osoppo delta.

Finally, climate influence seems to become important only with the onset of the Quaternary glaciations in the Alps, which started at the end

FIGURE 10.—A) Gravel bars of facies association B1 in the Cesclans Unit near Cavazzo C.; B) breccia of facies association A2 in the Cesclans Unit; C) landslide breccia of facies association A1 in the Ampezzo Unit near Cavazzo C.; D) fine-grained laminated bottomset of the Gilbert-type delta of Cavazzo Carnico (Ampezzo Unit).

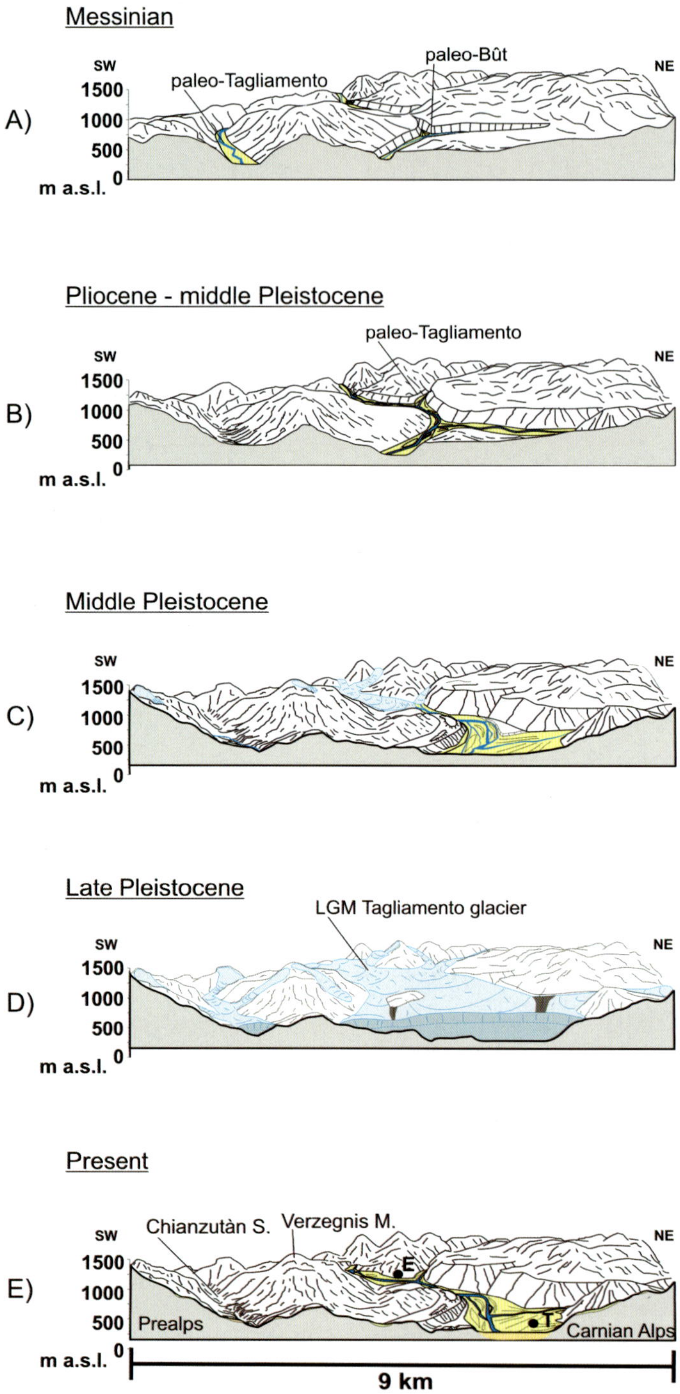

FIGURE 11.—Cartoons of the evolution of the Tagliamento valley from the Messinian onwards. At the Messinian (A)–Early Pliocene (B) boundary the trunk valley migrated towards the northeast and filled during the late Zanclean transgression, when the river formed a delta body at Osoppo. The valley remained confined in this setting (B) until the end of the early Pleistocene and had been affected by temporary episodes of damming that created lakes and deltaic bodies. The valley shape changed at the onset of the Quaternary glaciations; these determined the widening of the channel and its further migration northeastwards (C). The subglacial quarrying during the LGM (D) determined the present-day setting of the river (E). E, Enemonzo; T, Tolmezzo.

of the Early Pleistocene and influenced the increase of sediment production in the Tagliamento basin from the middle Pleistocene.

ACKNOWLEDGMENTS

The authors are very grateful to M. Ghinassi for constructive sedimentological discussions, to M.E. Poli and A. Zanferrari for structural considerations and to G. Fedetto for English review. We thank M. Pérez-Arlucea and S. Jones for their constructive and detailed revisions and editors S.K. Davidson, C.P. North, and S. Leleu for their thorough comments. Financial support by CARG-FVG (Geological Mapping Project in the Friuli Region: 049 "Gemona" sheet) and by CPD A058234 of Padova University (CS) is appreciated.

REFERENCES

ALDINUCCI, M., GHINASSI, M., AND SANDRELLI, F., 2007, Climatic and tectonic signature in the fluvial infill of a late Pliocene Valley (Siena Basin, Northern Apennines, Italy): Journal of Sedimentary Research, v. 77, p. 398–414.

BARNABA, C., MARELLO, L., VUAN, A., PALMIERI, F., ROMANELLI, M., PRIOLO, E., AND BREITENBERG, C., 2010, The buried shape of an alpine valley from gravity surveys, seismic and ambient noise analysis: Geophysical Journal International, v. 180, p. 715–733.

BINI, A., CITA, M.B., AND GAETANI, M., 1978, Southern Alpine lakes. Hypothesis of an erosional origin related to the Messinian entrenchment: Marine Geology, v. 27, p. 271–288.

BLAIR, T.C., AND MCPHERSON, J.G., 1994, Alluvial fans and their natural distinction from rivers based on morphology, hydraulic processes, sediment processes, and facies assemblages: Journal of Sedimentary Research, v. A64, p. 450–489.

BOCCALETTI, M., AND SANI, F., 1998. Cover thrust reactivations related to internal basement involvement during Neogene–Quaternary evolution of the northern Apennines: Tectonics, v. 17, p. 112–130.

BROCKLEHURST, S.H., AND WHIPPLE, K.X., 2006, Assessing the relative efficiency of fluvial and glacial erosion through simulation of fluvial landscapes: Geomorphology, v. 75, p. 283–299.

BURRATO, P., POLI, M.E., VANNOLI, P., ZANFERRARI, A., BASILI, R., AND GALADINI, F., 2008, Sources of Mw 5+ earthquakes in northeastern Italy and western Slovenia: An updated view based on geological and seismological evidence: Tectonophysics, v. 453, p.157–176.

CAPUTO, R., POLI, M.E., AND ZANFERRARI, A., 2003, Neogene–Quaternary twist tectonics in the eastern Southern Alps, Italy: Memorie di Scienze Geologiche, v. 54, p. 155–158.

CAPUTO, R., POLI, M.E., AND ZANFERRARI, A., 2010, Neogene–Quaternary tectonic stratigraphy of the eastern Southern Alps, NE Italy: Journal of Structural Geology, v. 32, p. 1009–1027.

CARMINATI, E., DOGLIONI, C., AND SCROCCA, D., 2003, Apennines subduction-related subsidence of Venice (Italy): Geophysical Research Letters, v. 30, p. 1717, doi:10.1029/2003GL017001.

CARULLI, G.B., FRIZZO, P., LONGO SALVADOR, G., SEMENZA, E., BIANCHIN, G., MANTOVANI, F., AND MEZZACASA, G., 1987, La geologia della zona fra il T. Chiarzò e il F. Fella (Alpi Carniche): Giornale di Geologia, v. 49, p.1–32.

CARULLI, G.B., COZZI, A., LONGO SALVADOR, G., PERNARCIC, E., PODDA, F., AND PONTON, M., 2000, Carta Geologica delle Prealpi Carniche: Udine, Museo Friulano Storia Naturale.

CASTELLARIN, A., AND CANTELLI, L., 2000, Neo-Alpine evolution of the eastern Southern Alps: Journal of Geodynamics, v. 30, p. 251–274.

CASTELLARIN, A., CANTELLI, L., FESCE, A.M., MERCIER, J., PICOTTI, V., PINI, G.A., PROSSER, G., AND SELLI, L., 1992, Alpine compressional tectonics in the southern Alps. Relations with the N-Apennines: Annales Tectonicae, v. 6, p. 62–94.

CASTIGLIONI, G.B., 2004, Quaternary glaciations in the eastern sector of the Italian Alps, *in* Elhers, J., and Gibbard, P.L., eds., Quaternary Glaciations—Extent and Chronology—Part I: Europe: Amsterdam, Elsevier, p. 209–215.

CLAUZON, G., 1982, Le canyon messinien du Rhone: une preuve décisive du dessicated deep-basin model [Hsü, Cita et Ryan, 1973]: Societé Géologique de France Bulletin, v. 24, p. 597–610.

DALLA VECCHIA, F.M., 2008, Vertebrati fossili del Friuli: Pubblicazione n. 50, Udine, Edizioni Museo Friulano di Storia Naturale, 304 p.

DE FRANCO, R., BERRA, F., BIELLA, G., BONIOLO, G., CAIELLI, G., CORSI, A., FORCELLA, F., LAZZATI, F., LOZEJ, A., MORRONE, A., AND TONDI, R., 2004, Late Neogene–Quaternary evolution of the intermontane Clusone Basin (Southern Alps, Italy): integration of seismic and geological data: Journal of Quaternary Science, v. 19, p. 409–421.

DICKINSON, W.R., 1970, Interpreting detrital modes of greywacke and arkose: Journal of Sedimentary Petrology, v. 40, p. 695–707.

DOGLIONI, C., AND BOSELLINI, A., 1987, Eoalpine and mesoalpine tectonics in the Southern Alps: Geologische Rundschau, v. 76, p. 735–754.

DUNKL, I., KUHLEMANN, J., REINECKER, J., AND FRISCH, W., 2005, Cenozoic relief evolution of the Eastern Alps—constraints from apatite fission track age-provenance of Neogene intramontane sediments: Austrian Journal of Earth Sciences, v. 98, p. 92–105.

EYLES, N., EYLES, C.H., AND MIALL, A.D., 1983, Lithofacies types and vertical profile models; an alternative approach to the description and environmental interpretation of glacial diamict and diamictite sequences: Sedimentology, v. 30, p. 393–410.

FAVERO, V., AND GRANDESSO, P., 1983, Nuovi affioramenti di Pliocene marino nei dintorni di Bassano del Grappa (Vicenza): Società Geologica Italiana, Memorie, v. 24, p. 71–77.

FELBER, M., FREI, W., AND HEITZMANN, P., 1992, Seismic evidence of pre-Pliocene valley formation and filling in the region of Novazzano (Southern Ticino, Switzerland): Eclogae Geologiae Helvetiae, v. 84, p.753–761.

FERUGLIO, E., 1929, Nuove ricerche sul Quaternario del Friuli: Giornale di Geologia, v. 4, p. 1–36.

FODOR, L., JELEN, B., MARTOR, E., SKABERNE, D., CAR, J., AND VRABEC, M., 1998, Miocene–Pliocene tectonic evolution of the Slovenian Periadriatic fault: implication for Alpine–Carpathian extrusion model: Tectonics, v. 17, p. 690–709.

GALADINI, F., POLI, M.E., AND ZANFERRARI, A., 2005, Seismogenic sources potentially responsible for earthquakes with $M \geq 6$ in the eastern Southern Alps (Thiene–Udine sector, NE Italy): Geophysical Journal International, v. 161, p. 739–762.

GARGANI, J., 2004, Modelling of the erosion in the Rhone valley during the Messinian crisis (France): Comptes rendus Geoscience, v. 121, p. 13–22.

GAZZI, P., 1966, Le arenarie del flysch sopracretaceo dell'Appennino modenese: correlazioni con il Flysch di Monghidoro: Mineralogica et Petrographica Acta, v. 12, p. 69–97.

GHIELMI, M., MINERVINI, M., NINI, C., ROGLEDI, S., ROSSI, M., AND VIGNOLO, A., 2009, Sedimentary and tectonic evolution in the eastern Po Plain and Northern Adriatic Sea Area from Messinian to Middle Pleistocene (Italy): Società Geologica Italiana, Rendiconti online, v. 9, p. 32–35.

GIBBARD, P.L., HEAD, M.J., AND WALKER, M.J.C. AND THE SUBCOMMISSION ON QUATERNARY STRATIGRAPHY, 2009, Formal ratification of the Quaternary System / Period and the Pleistocene Series / Epoch with a base at 2.58 Ma: Journal of Quaternary Science, v. 25, p. 96–102.

GORTANI, M., 1935, I terrazzi del bacino montano del Tagliamento e nelle valli contigue: Giornale di Geologia, v. 9, p. 1–41.

GORTANI, M., AND DESIO, A., 1926, Note illustrative della Carta geologica delle Tre Venezie. Foglio "Pontebba": Venezia, Ufficio Idrografico Regio Magistrato Acque di Venezia.

HOLBROOK, J., AND SCHUMM, S.A., 1999, Geomorphic and sedimentary response of rivers to tectonic deformation: a brief review and critique of a tool for recognizing subtle epeirogenic deformation in modern and ancient settings: Tectonophysics, v. 305, p. 287–306.

HOU, B., FRAKES, L.A., SANDIFORD, M., WORRALL, L., KEELING, J., AND ALLEY, N.F., 2008, Cenozoic Eucla Basin and associated palaeovalleys, southern Australia—Climatic and tectonic influences on landscape evolution, sedimentation and heavy mineral accumulation: Sedimentary Geology, v. 203, p. 112–130.

JAMES, W.C., 1985, Early diagenesis, Atherton Formation (Quaternary): A guide for understanding early cement distribution and grain modifications in nonmarine deposits: Journal of Sedimentary Petrology, v. 55, p. 135–146.

JONES, S., FROSTICK, L.E., AND ASTIN, T.H., 1999, Climatic and tectonic controls on fluvial incision and aggradation in the Spanish Pyrenees: Geological Society of London, Journal, v. 156, p. 761–769.

KEIL, M., AND NEUBAUER, F., 2009, Initiation and development of a fault-controlled, orogen-parallel overdeepened valley: the Upper Enns Valley, Austria: Austrian Journal of Earth Sciences, v. 102, p. 80–90.

KUHLEMANN, J., 2000, Post-collisional sediment budget of circum-Alpine basins (Central Europe): Memorie di Scienze Geologiche, v. 52, p. 1–91.

KUHLEMANN, J., FRISCH, W., DUNKL, I., AND SZÉKELY, B., 2001, Quantifying tectonic versus erosive denudation by the sediment budget: the Miocene core complexes of the Alps: Tectonophysics, v. 330, p. 1–23.

KUHLEMANN, J., FRISCH, W., SZÉKELY, B., DUNKL, I., AND KAZMER, M., 2002, Post-collisional sediment budget history of the Alps: tectonic versus climatic control: International Journal of Earth Science, v. 91, p. 818–837.

LISIECKI, L.E., AND RAYMO, M.E., 2005, A Pliocene–Pleistocene stack of 57 globally distributed benthic $\delta^{18}O$ records: Paleoceanography, v. 20, p. 1–17.

LUGLI, S., MARCHETTI DORI, S., AND FONTANA, D., 2007, Alluvial sand composition as a tool to unravel late Quaternary sedimentation of the Modena Plain, northern Italy, *in* Arribas, J., Critelli, S., and Johnsson, M.J., eds., Sedimentary Provenance and Petrogenesis: Perspectives from Petrography and Geochemestry: Geological Society of America, Special Publication 420, p. 57–72.

MANCIN, N., DI GIULIO, A., AND COBIANCHI, M., 2009, Tectonic vs. climate forcing in the Cenozoic sedimentary evolution of a foreland basin (Eastern Southalpine system, Italy): Basin Research, v. 21, p. 799–823.

MASSARI, F., 1990, The foredeeps of the Northern Adriatic margin: evidence of diachroneity in deformation of the Southern Alps: Rivista Italiana di Paleontologia e Stratigrafia, v. 96, p. 351–380.

MASSARI, F., GRANDESSO, P., STEFANI, C., AND ZANFERRARI, A., 1986, The Oligo-Miocene Molasse of the Veneto- Friuli region, Southern Alps; Giornale di Geologia, v. 48, p. 235–255.

MIALL, A.D., 1996, The Geology of Fluvial Deposits: Berlin, Springer-Verlag, 582 p.

MIALL, A.D., AND GIBLING, M.R., 1978, The Siluro-Devonian clastic wedge of Somerset Island, Arctic Canada, and some regional paleogeographic implications: Sedimentary Geology, v. 21, p. 85–127.

MONEGATO, G., 2006, Le successioni continentali conglomeratiche messiniano-pleistoceniche nel bacino del Fiume Tagliamento: Ph.D Thesis, University of Padova, 160 p.

MONEGATO, G., ZANFERRARI, A., AND GLIOZZI, E., 2006, Stratigraphy, sedimentology and age of the Osoppo Hill deposits, *in* Melis, R., Romano, R., and Fonda, G., eds., Giornate di Paleontologia 2006, guida alle escursioni: Trieste, EUT, p. 81–83.

MONEGATO, G., RAVAZZI, C., DONEGANA, M., PINI, R., CALDERONI, G., AND WICK, L., 2007, Evidence of a two-fold glacial advance during the last glacial maximum in the Tagliamento end moraine system (eastern Alps): Quaternary Research, v. 68, p. 284–302.

MONEGATO, G., STEFANI, C., AND ZATTIN, M., 2010, From present rivers to old terrigenous sediments: the evolution of the drainage system in the eastern Southern Alps: Terra Nova, v. 22, p. 218–226.

MUTTONI, G., CARCANO, C., GARZANTI, E., GHIELMI, M., PICCIN, A., PINI, R., ROGLEDI, S., AND SCIUNNACH, D., 2003, Onset of major Pleistocene glaciations in the Alps: Geology, v. 31, p. 989–992.

MUTTONI, G., RAVAZZI, C., BREDA, M., PINI, R., LAJ, C., KISSEL, C., MAZAUD, A., AND GARZANTI, E., 2007, Magnetostratigraphic dating of an intensifcation of glacial activity in the southern Italian Alps during Marine Isotope Stage 22: Quaternary Research, v. 67, p. 161–173.

NACSN (NORTH AMERICAN COMMISSION ON STRATIGRAPHIC NOMENCLATURE), 1983, North American Stratigraphic Code: American Association of Petroleum Geologists, Bulletin, v. 67, p. 841–875.

NEMES, F., NEUBAUER, F., CLOETHING, S., AND GENSER, J., 1997, The Klagenfurt Basin in the Eastern Alps: an intra-orogenic decoupled flexural basin?: Tectonophysics, v. 282, p. 189–203.

OSTERMANN, M., SANDERS, D., AND KRAMERS, J., 2006, $^{230}Th/^{234}U$ ages of calcite cements of the proglacial valley fills of Gamperdona and Bürs (Riss ice age, Vorarlberg, Austria): geological implications: Austrian Journal of Earth Science, v. 99, p. 31–41.

OUCHI, S., 1985, Response of alluvial rivers to slow active tectonic movement: Geological Society of America, Bulletin, v. 96, p. 504–515.

PENCK, A., AND BRÜCKNER, E., 1909, Die Alpen im Eiszeitalter; Leipzig, Tauchnitz, 1199 p.

POSTMA, G., 1990, Depositional architecture and facies of river and fan deltas: a synthesis, *in* Colella, A., and Prior, D.B., eds., Coarse-Grained Deltas: International Association of Sedimentologists, Special Publication 10, p. 13–27.

PREUSSER, F., AND SCHLÜCHTER, C., 2004, Dates from an important early Late Pleistocene ice advance in the Aare Valley, Switzerland: Eclogae Geologiae Helvetiae, v. 97, p. 245–253.

RATSCHBACHER, L., FRISCH, W., AND LINZER, H.G., 1991, Lateral extrusion in the eastern Alps, part 2: structural analysis: Tectonics, v. 10, p. 257–271.

REINFELDS, I., AND NANSON, G.C., 1993, Formation of braided river floodplains, Waimakariri River, New Zealand: Sedimentology , v. 40, p. 1113–1127.

ROVERI, M., BERTINI, A., COSENTINO, D., DI STEFANO, A., GENNARI, R., GLIOZZI, E., GROSSI, F., IACCARINO, S.M., LUGLI, S., MANZI, V., AND TAVIANI, M., 2008, A high-resolution stratigraphic framework for the latest Messinian events in the Mediterranean area: Stratigraphy, v. 5, p. 323–342.

SCHUMM, S.A., 1977, The Fluvial System: New York, Wiley, 338 p.

SCHUMM, S.A., 1993, River response to baselevel change: implications for sequence stratigraphy: Journal of Geology, v. 101, p. 279–294.

SCHUMM, S.A., AND ETHRIDGE, F.G., 1994, Origin, evolution, and morphology of fluvial valleys, *in* Darlymple, R.W., Boyd, R., and Zaitlin, B.A., eds., Incised-Valley Systems: Origin and Sedimentary Sequences: SEPM, Special Publication 51, p. 11–27.

SCHUMM, S.A., DUMONT, J.F., AND HOLBROOK, J.M., 2000, Active Tectonics and Alluvial Rivers, Cambridge, U.K., Cambridge University Press, 276 p.

SELLI, R., 1963, Schema geologico delle Alpi Carniche e Giulie occidentali: Giornale di Geologia, v. 30, p.1–136.

SÖLVA, H., STÜWE, K., AND STRAUSS, P., 2004, The Drava River and Pohorje Mountain Range (Slovenia): Geomorphological Interactions: Naturwissenschaftlichen Verein für Steiermark, Mitteilungen v. 134, p. 57–66.

SPALLETTA, C., VAI, G.B., AND VENTURINI, C., 1982, La Catena Paleocarnica, *in* Castellarin, A., and Vai, G.B., eds., Guida alla geologia del Sudalpino centro-orientale: Guide geologiche regionali: Società Geologica Italiana, p 281–292.

STEFANI, C., FELLIN, M.G., ZATTIN, M., ZUFFA, G.G., DALMONTE, C., MANCIN, N., AND ZANFERRARI, A., 2007, Provenance and paleogeographic evolution in a multi-source foreland: the Cenozoic Venetian–Friulian Basin (NE Italy): Journal of Sedimentary Research, v. 77, p. 867–887.

STEFANINI, G., 1915, Il Neogene Veneto: Istituti di Geologia e Mineralogia della Regia Università di Padova, Memorie, v. 3, p. 340–624.

TARAMELLI, T., 1875, Dei terreni morenici e alluvionali del Friuli: Regio Istituto Tecnico di Udine, Annali, v. 8, p. 1–91.

TARAMELLI, T., 1881, Carta geologica del Friuli: Pavia, Fusi.

VENTURINI, C., 1990, Geologia delle Alpi Carniche centro-orientali: Museo Friulano di Storia Naturale, Udine, v. 36, 220 p.

VENTURINI, C., PONDRELLI, M., DELZOTTO, S., FONTANA, C., AND DISCIENZA, K., 2002, Carta Geologica delle Alpi Carniche: Firenze, S.EL.CA.

VINCENT, S.J., 2001, The Sis palaeovalley: a record of proximal fluvial sedimentation and drainage basin development in response to Pyrenean mountain building: Sedimentology, v. 48, p. 1235–1276.

VINCENT, S.J., AND ELLIOTT, T., 1997, Long-lived transfer zone paleovalleys in mountain belts: an example from the Tertiary of the Spanish Pyrenees: Journal of Sedimentary Research, v. 67, p. 303–310.

ZANFERRARI, A., AVIGLIANO, R., GRANDESSO, P., MONEGATO, G., PAIERO, G., POLI, M.E., AND STEFANI, C., 2008, Note illustrative della Carta Geologica d'Italia alla scala 1:50.000: Foglio 065 "Maniago": APAT-Servizio Geologico d'Italia—Regione Autonoma Friuli Venezia Giulia.

ZANFERRARI, A., POLI, M.E., MONEGATO, G., MASETTI, D., ROMANO, R., PODDA, F., RUSSO, S., PONTON, M., AND STEFANI, C., in press, Note illustrative della Carta Geologica d'Italia alla scala 1:50.000: Foglio 049 "Gemona del Friuli": APAT-Servizio Geologico d'Italia—Regione Autonoma Friuli Venezia Giulia.

LATERAL AND VERTICAL VARIABILITY OF CHANNEL BELT STACKING DENSITY AS A FUNCTION OF SUBSIDENCE AND SEDIMENT SUPPLY: FIELD EVIDENCE FROM THE INTRAMOUNTAINE KÖRÖS BASIN, HUNGARY

ANNAMÁRIA NÁDOR
Geological Institute of Hungary, H-1143 Budapest, Stefánia 14.,
e-mail:nador@mafi.hu
AND
ORSOLYA SZTANÓ
Eötvös Loránd University Budapest, Dept. of Geology, H-1117 Budapest, Pázmány Péter 1/c,
e-mail: sztano@ludens.elte.hu

ABSTRACT: Late Neogene–Quaternary inversion of the Pannonian basin created subsidence of several sub-basins coupled with uplift of the surrounding mountainous regions, which facilitated accumulation of Plio-Pleistocene fluvial sequences up to a thousand meters thick. The location of the main channel belt zones in the Pleistocene fluvial succession in the Körös basin at the eastern part of the Pannonian basin was determined by analysing well-log patterns from more than fifty water-prospecting wells, and their proportion to floodplain alluvium has been quantified also. The main aim of the paper is to compare channel belt stacking density patterns from Körös basin data sets with various model predictions on alluvial architecture and thus contribute to the better understanding on disparities between them. The vertical changes of the channel/floodplain ratios along seven sections show that the proportion of channels is generally lower for the deeper parts of the sections, indicating that subsidence rate was relatively high, creating enough accommodation to store fluvial deposits, including their thick vertical aggradational floodplain sediments, while the proportion of channel belt deposits increased at the expense of floodplain sediments in the upper 300–180 m part of the sedimentary succession representing the last ~ 1.3 million years, indicating less accommodation available and/or more sediment supply to the basin. This pattern is principally in agreement with the LAB models on alluvial architecture (Leeder 1978; Alexander and Leeder 1987; Bridge and Leeder 1979) concerning the inverse correlation of stacking density and subsidence.

The analysis of lateral variability on changes in the ratio of accommodation to sediment supply as a control on stacking density showed that the proportion of channel belt deposits is higher close to the northwestern and southwestern margins of the basin, especially in the upper part of the sedimentary succession, and values become significantly lower coming closer to the depocenter to the south, showing that rivers do not necessarily flow towards the subsidence maximum as suggested by the LAB model. The thickest point-bar units (11–20 m) at the northern and western margin of the basin were deposited by the paleo–Tisza River, which was forced to flow along a transtensional strike-slip fault zone. Another somewhat smaller river, the paleo–Kettős-Körös at the southeastern margin of the basin was fed by high sediment supply from the uplifting Apuseni Mountains, which thus could pass without deviating towards the laterally subsiding basin center, proving model predictions of Hickson et al. (2005).

Our results clearly indicate that the primary control on channel density was driven by variations in subsidence and uplift of the hinterland, thus sediment supply. The increased channel stacking density since about 1.3 Ma in the Körös basin is related to a waning compression in the Pannonian–Carpathian realm, manifested in decreased subsidence of the Körös basin coupled with the erosion-driven isostatic uplift of the Apuseni Mountains source area. The Körös basin example shows that basin evolution is more complex than all variables controlling fluvial architecture could be described by a single model.

KEY WORDS: channel belt stacking density, subsidence, neotectonics, sediment supply, Pannonian basin

INTRODUCTION

The Great Hungarian Plain (GHP), in the central part of the Pannonian basin within the Alp-Carpathian orogen, occupies an area of about 100 000 km^2 and forms one of the largest alluvial plains in Europe (Fig. 1). Late Neogene subsidence of several sub-basins coupled with uplift of the surrounding mountainous regions facilitated accumulation of Plio-Pleistocene fluvial sequences up to a thousand meters thick in the GHP. The area is a key locality in understanding fluvial processes in a continental intramountaine setting and responses to climatic and tectonic forcing. The Late Pleistocene drainage-pattern development of the GHP and its allogenic controls has been reviewed and discussed recently by several authors (Gábris 1998; Gábris and Nagy 2005; Timár et al. 2005; Nagy et al. 2007; Nádor et al. 2007; Gábris and Nádor 2007). Although these papers discuss the fluvial archive only of the topmost 10–20 meters, they provide valuable information on the deposits, style, and dimensions of the Late Pleistocene rivers, which are thought be large-scale analogues for Pleistocene sedimentation on the GHP, as the depositional environment did not change significantly over the last 2.5 million years. Only a few papers have discussed the alluvial stratigraphy of the overall Pleistocene fluvial succession using modern sedimentological concepts and basin evolution models (e.g. Borsy 1990; Thamó-Bozsó et al. 2002; Nádor et al. 2003; Juhász et al. 2004).

The main geological context of fluvial sedimentation during the Quaternary and its allogenic controls have been outlined by earlier workers, as summarized in Rónai (1985): vertical tectonic movements have resulted in the development of several sub-basins accommodating thick alluvial successions, while climatic changes related to glacial–interglacial periods during the Pleistocene exerted a control on fluvial sedimentation through variations in discharge and sediment yield, manifested in the cyclic pattern of fluvial deposits nearly 500 m thick.

In addition to the borehole data of low spatial coverage, analysis of well-logs of thousands of water-prospecting wells on the Great Hungarian Plain provide valuable information on the thickness and facies distribution of Quaternary fluvial sediments (Urbancsek 1960, 1965). Based on subsurface facies analysis from well-logs in the Körös basin, located at the eastern part of the GHP, Juhász et al. (2004) identified depositional elements, such as channel fills, pointbars, sandy sheet-flood deposits, floodplain and floodbasin deposits and performed

From River To Rock Record: The Preservation Of Fluvial Sediments And Their Subsequent Interpretation
SEPM Special Publication No. 97, Copyright © 2011
SEPM (Society for Sedimentary Geology), ISBN 978-1-56576-305-0, p. 375–392.

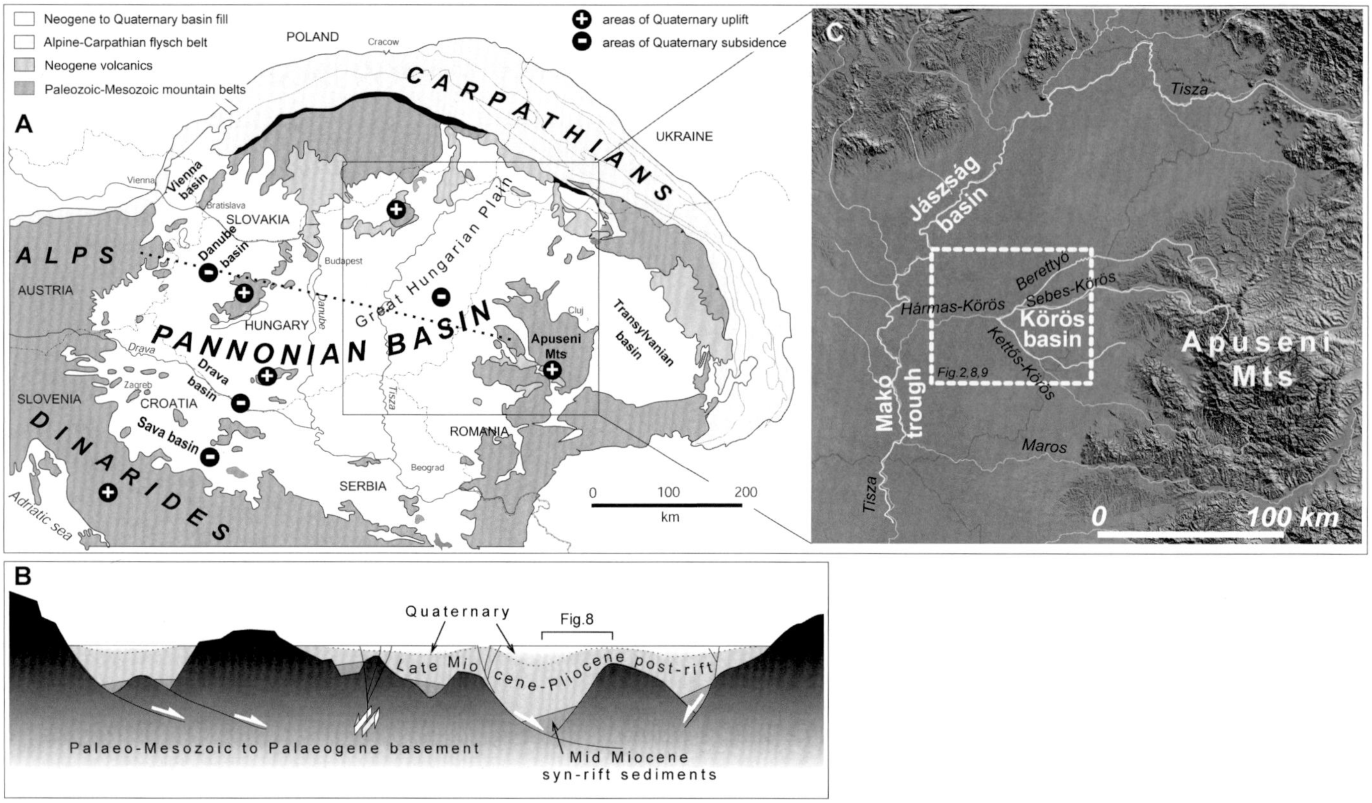

FIGURE 1.—The Pannonian Basin (A) is situated within the Carpathian arc. Dotted line shows the location of the cross section (B) with the main tectonostratigraphic units. The eastern part is the Great Hungarian Plain, an alluvial plain existing since the Pliocene. The Körös basin and other Neogene to Quaternary depressions (C) have been subsiding within the Pannonian Basin due to basin inversion, providing long fluvial archives.

a large-scale paleo-environmental reconstruction by showing channel belts, wetlands, and the rim of an alluvial fan in the area. However, Juhász et al. (2004) did not consider the lateral and vertical variability of facies associations and their lithofacies map was drawn on the basis of "average" composition of logs. If a borehole was dominated by floodplain sediments, it was considered as "floodplain", not taking into account occasional depth intervals with a higher proportion of channels. Depicted "channel belts" were wide zones where actual channel belt deposits were relatively common, but these do not reflect the migration of channel belts through the basin.

In this study we provide a new interpretation and model for the Pleistocene fluvial basin-fill complex in the Körös basin (Fig. 1), partly using and reinterpreting the previous dataset of Juhász et al. (2004), and partly incorporating new wells into the analysis (Fig. 2). Channel belt sediments have been identified by analyzing well-log patterns from more than fifty water-prospecting wells, and their proportion to floodplain alluvium has been quantified also. The main aim was to locate the main channel belt zones at different time slices and relate the lateral and vertical changes of channel/floodplain proportions to variations in accommodation throughout the studied basin. To discuss allogenic controls on variations in ratio of sediment supply to accommodation, seismic sections have provided new information on the thickness of Quaternary sediments and neotectonic movements of the area. On the basis of reviewing previous ideas on cyclicity of fluvial sediments (Rónai 1985; Nádor et al. 2003; Juhász et al. 2004), we demonstrated that channel belt stacking density is negligible to climatic signals at the scale of observation, whose influence operated on a lower-scale magnitude.

GEOLOGICAL SETTING

Back-arc style extension of the Pannonian basin system (Fig. 1) started in the Early to Middle Miocene times, coeval with the thrusting of the Carpathian belt (Horváth and Royden 1981), and resulted in a complex basin-floor structure (Csontos et al. 1992; Csontos and Vörös 2004) characterized by elongated basement highs and narrow depressions. These developed during mid-Miocene rifting and were rearranged by several compressional and extensional tectonic phases during Neogene basin development (Horváth 1995, Horváth and Cloetingh 1996; Tari et al. 1999). These depressions gradually lost connectedness to the Paratethys sea, and a single large depression developed by the Late Miocene, which was occupied by the brackish to freshwater Lake Pannon (Magyar et al. 1999). The uplift and erosion of the surrounding Alpine–Carpathian mountain belt supplied a significant amount of sediments via large fluvio-deltaic systems into the Lake Pannon (Bérczi and Phillips 1985; Bérczi et al. 1988; Juhász 1991, 1994), resulting in the accumulation of a post-rift sedimentary succession 5000–7000 m thick (Horváth and Royden 1981; Horváth 1993). River deltas that entered the basin from the northwest and northeast prograded to the south, so finally the lake basin filled. The youngest, southern part of the lake in front of the northern foothills of the Dinarids persisted as late as ca. 4.5 Ma (Magyar et al. 1999). During the Pliocene most of the intra-Carpathian area was an alluvial plain with shallow ponds and marshes, and only small, relatively low hills and some volcanoes in Transdanubia and north of the GHP were present.

The syn-rift and post-rift subsidence of the Pannonian basin was interrupted by several compressional events. A late-stage compres-

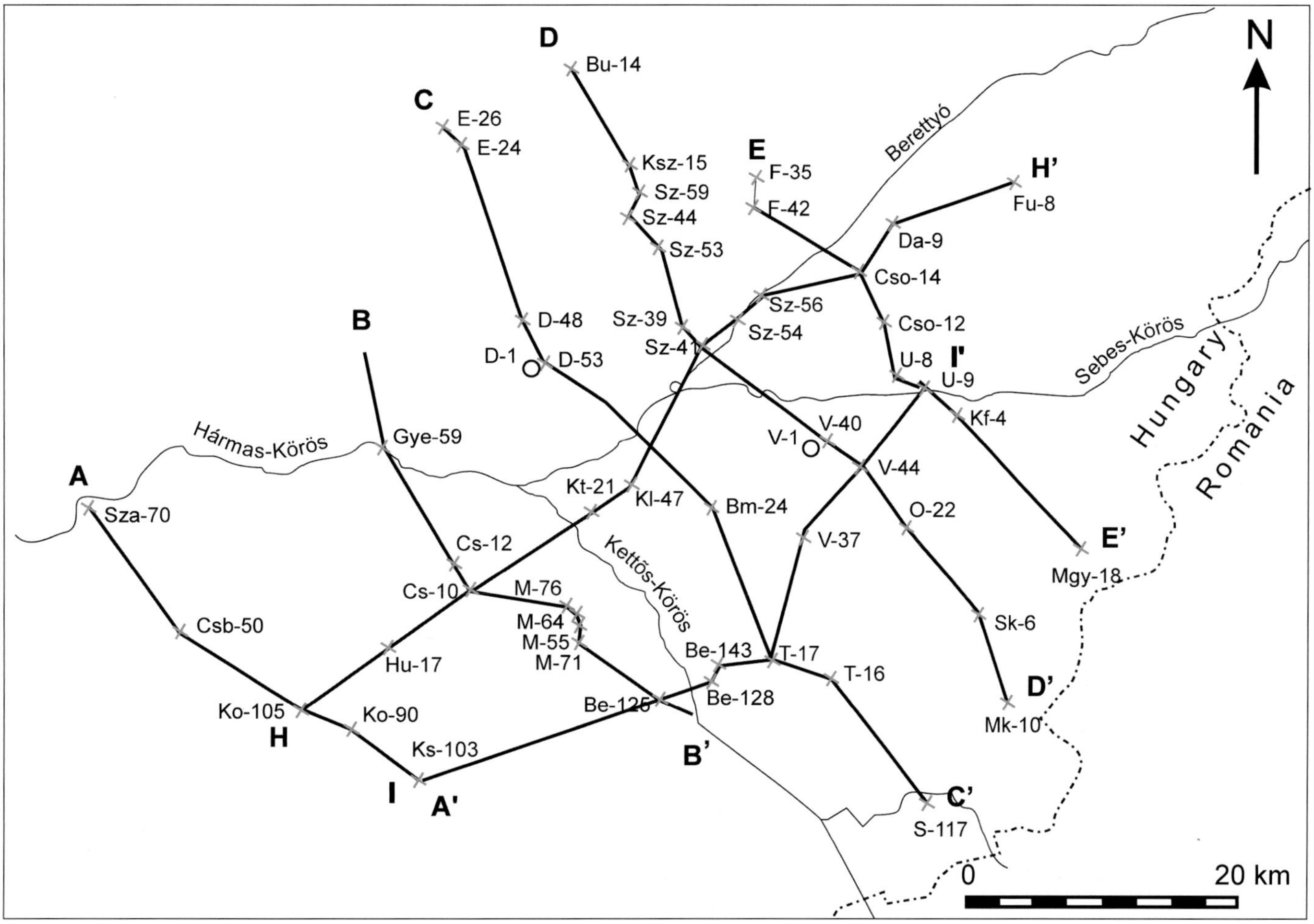

FIGURE 2.—Location of studied water-prospecting wells and geological profiles in the Körös basin.

sional basin inversion started ca. 9–5 Ma ago (Bada et al. 2006a; Sztanó et al. 2007; Uhrin et al. 2009) and resulted in the gradual uplift of the hilly areas in the western part of the Pannonian basin (Transdanubia) and at the basin margins, while former basins (Great Hungarian Plain, Danube basin, Drava and Sava basins, Vienna basin) continued to subside at various rates (Tari et al. 1999) (Fig. 1). The subsiding basins were continuously fed by sediments, which were carried by the rivers draining the uplifted areas. In this way the basin-fill successions record the uplift history as well. This tectonic inversion, being intensified with the onset of the Quaternary, is assumed to be the manifestation of large-wavelength lithospheric folding, as intraplate compressional stress gradually built up within the Carpathian arc (Horváth and Cloetingh 1996; Cloetingh et al. 1999; Bada and Horváth 2001).

The differential subsidence of the Great Hungarian Plain had a major control on drainage-pattern development, and consequently, on basin-scale facies distribution. Although modern tectonic models were not available in the early–mid twentieth century, it was realized that alluvial fans prograded from the slightly elevated margins towards the subsiding depressions in the basin interior and thus played an important role in filling up the basin (Sümeghy 1944). There were three main subsiding areas during the Pleistocene on the GHP (Fig. 1). The largest and deepest was the Makó trough, along the southern reach of the present Tisza River, which served as the main depocenter of the entire GHP during the Early and Middle Pleistocene. This depression was filled mainly by coarse-grained sediments transported by the paleo-Danube from the northwest, and subordinately by fine-grained sediments transported by the paleo-Tisza from the northeast, depositing a Quaternary sedimentary succession 700–800 m thick (Franyó 1992). The Jászság basin was a relatively small depression on the northern part of the GHP which was filled with fine-grained sandy–silty sediments brought by the smaller rivers and creeks draining from the uplifting mountains on the north.

The target area of this study is the Körös basin, at the eastern part of the Great Hungarian Plain, with an area of about 400 km^2 (Fig. 1), and one of the lowest-lying parts of the GHP. The Körös basin was the third major depression that has been subsiding continuously during the Plio-Pleistocene, resulting in the accumulation of fluvial sequences over 500 m thick during the Quaternary. The main modern rivers of the area are the Berettyó, the Sebes, and the Kettős–Körös Rivers, which display a transverse pattern, oriented approximately east–west. After their confluence at the western part of the Körös basin, they continue as the Hármas–Körös River to the low-gradient, south-directed, axial Tisza River in the central part of the GHP (Figs. 1, 2). Quaternary fluvial sediments of the Körös basin have been deposited by the paleo–Körös Rivers, draining the Apuseni Mountains, and by the paleo–Tisza River, coming from the NE Carpathians (Thamó-Bozsó et al. 2002;

Thamó-Bozsó et al. 2007). Both river systems are supposed to have been relatively low-energy, meandering systems (Gábris and Nádor 2007 and references therein).

Large-scale facies mapping (Juhász et al. 2004) showed that several parts of the Körös basin are dominated by stacked, multistory sandy channel fills, while the central and deepest part of the basin is characterized by muddy floodbasin sediments. In the eastern part of the basin the distal rim of alluvial fans were outlined where sandy sheetflood deposits alternate with floodplain sediments.

METHODS

Water-prospecting wells have been drilled in vast number all over the GHP. Their depth depends on the location of good aquifers (sandy units), but generally they are several hundred meters deep. In the Körös basin area fifty-one wells were selected from many hundreds on the basis of the following criteria: (1) preferably they should penetrate the entire Pleistocene fluvial succession, i.e., should be 400–500 m deep, (2) good quality logs—SP, resistivity, and in some cases gamma ray, (3), equal spatial coverage. We made five NW–SE-trending and two NE–SW-trending profiles along the selected wells (Fig. 2). The average spacing between the wells was 3–8 km.

Based on the analysis of well-log shapes and stacking patterns a lithological interpretation was done in each well, which also provided information on the depositional environment. The described units, such as channels, floodplain sediments with a thickness from a few meters to a few tens of meters correspond to depositional units according to the hierarchical classification of Miall (1996), formed on a timescale of 10^3–10^4 years. Logs obviously do not reveal the architectural elements of the individual depositional units, such as different channel bars and macroforms. In order to complete the lithological characterization of the fluvial units identified from well-logs, we used the lithological description of similar, near-surface fluvial sediments from sand-pit and clay-pit sections in the area and lithological information from two cored deep boreholes from Dévaványa (D-1) and Vésztő (V-1) (Nádor et al. 2003). Although these descriptions are not directly related to the studied units, this is the only available information on the lithological control.

In order to determine the proportion of channel belt deposits to floodplain alluvium, we measured and added the total thickness of all sandy channel fills regardless of their amalgamated, blocky, or upward-fining character in each well, as well as the thickness of overbank fines. Thin crevasse sands of the floodplain and channel-fills less than 3 m thick were neglected in this calculation, because the resolution of the well-logs does not allow the quantification of their thickness. To characterize the ratio of channel belt and floodplain deposits with a number, we plotted the cumulative thickness of channel belt deposit to the total thickness of the respected intervals; thus numbers between 0 and 1 characterize the proportion of channel belt units in the overall sequence (0 no channels, 1 only channel deposits). In some boreholes, where the majority of the upper or lower part of the borehole dataset was missing, ratios were not calculated.

Two regional seismic sections and several others have been analyzed in order to get a wider picture of the geometry of the entire Neogene basin-fill complex. On the basis of these sections the previous maps on the thickness of Quaternary sediments of the area (Franyó 1992; Magyari et al. 2009) have been revised; moreover, the sections provided essential information on the major structural elements of the basin, which exerted a strong control on alluvial architecture.

THICKNESS OF QUATERNARY SEDIMENTS IN THE KÖRÖS BASIN

In the past decades several maps showing the thickness of Quaternary sediments have been edited for the Great Hungarian Plain (Fig. 3A; Urbancsek 1965; Rónai 1985; Franyó 1992) based on borehole data (mostly water-prospecting wells). All of these maps were based on oversimplified lithostratigraphic considerations, i.e., the base of the Quaternary was marked at the boundary between fluvial ("Quaternary") and marshy ("Pliocene") sediments, i.e., at the lowermost appearance of significant channel sands. However, the paleogeographical evolution of the area was more complex. The GHP was a very low-relief alluvial plain with shallow ponds and marshes during the Pliocene. Because only relatively low hills surrounded the GHP at that time, only small rivers were assumed to flow into/across the basin, except for one or two trunk rivers (paleo-Danube, paleo-Tisza). The accumulation of fluvial sequences encompassing a higher proportion of sandy channel sediments was related to the intensification of uplift of the marginal areas, providing an increased sediment supply to the basin via fluvial systems, and subsidence of local depressions, all resulting from late Neogene basin inversion. However, these processes were not necessarily contemporaneous with the Pliocene–Pleistocene boundary in chronostratigraphic terms, and especially did not occur in a uniform spatial distribution over the entire GHP. This is illustrated by two magnetostratigraphically dated cores in the Körös basin (Figs. 3, 4). In the D-1 borehole the chronostratigraphic Pliocene–Pleistocene boundary (2.58 Ma) is at a depth of 416 m bsl, while the first major channel sand occurs at a depth of 500 m bsl. In the V-1 borehole the depth of the Matuyama–Gauss magnetic reversal is at 482 m bsl, while significant sandy channel fills occur as deep as 650 m bsl. This indicates that some major rivers crossed the Körös basin during the Pliocene also. A similar pattern characterizes the deepest depression of the GHP, the Makó trough (Fig. 1), where coarse-grained fluvial sediments deposited by the paleo–Danube are 700–800 m thick. The thickness of the underlying Pliocene fine-grained alluvial sediments also attains about thousand meters, indicating a high rate of subsidence since the Pliocene.

Recently the Quaternary thickness map of the GHP was revised by Magyari et al. (2009) on the basis of a dense borehole pattern, which has been cross-correlated along a systematic network of regional sections; furthermore, some geostatistical methods were also applied. The geometry of the Quaternary basin fill showed a pattern similar to the previous maps (Urbancsek 1965; Rónai 1985; Franyó 1992) i.e., the maximum depth of Pleistocene was around 450 m in the Körös basin, located south of the present Sebes–Körös River near to its confluence with the Kettős–Körös River (Fig. 3A). It is important to emphasize that modern seismic profiles were not available during this work and some old sections did not reveal reflections from the upper 500–600 ms.

During this work we incorporated two regional seismic sections and several short sections whose interpretation, together with the analyzed borehole logs, made it possible to develop a more precise picture of the thickness of the Plio-Pleistocene fluvial sediments (Fig. 3B). Contrary to the previous maps (Fig. 3A; e.g., Franyó 1992; Magyari et al. 2009) the seismic profiles (see later, Fig. 8) clearly show that reflections are gently dipping towards the south, outlining the center of the Körös basin along the Kettős–Körös River, where the maximum thickness of Quaternary sediments can be as much as 1000 m. Thus the investigated boreholes cover mainly the northern part of the basin, and its deepest central depression—where the base Quaternary lies—was not attained by boreholes. It is important to note that the confluence of the present rivers (Berettyó, Sebes and Kettős–Körös Rivers), which was supposed to be dragged by a topographic depression above the subsidence maximum (Juhász et al. 2004), is not located above the basin center with highest subsidence rate, but farther to the north–northwest. This means that the current local topographic minimum does not necessarily coincide with the depocenter and, or subsidence maximum. Our results on the distribution on channel belt stacking density also support this, as discussed later.

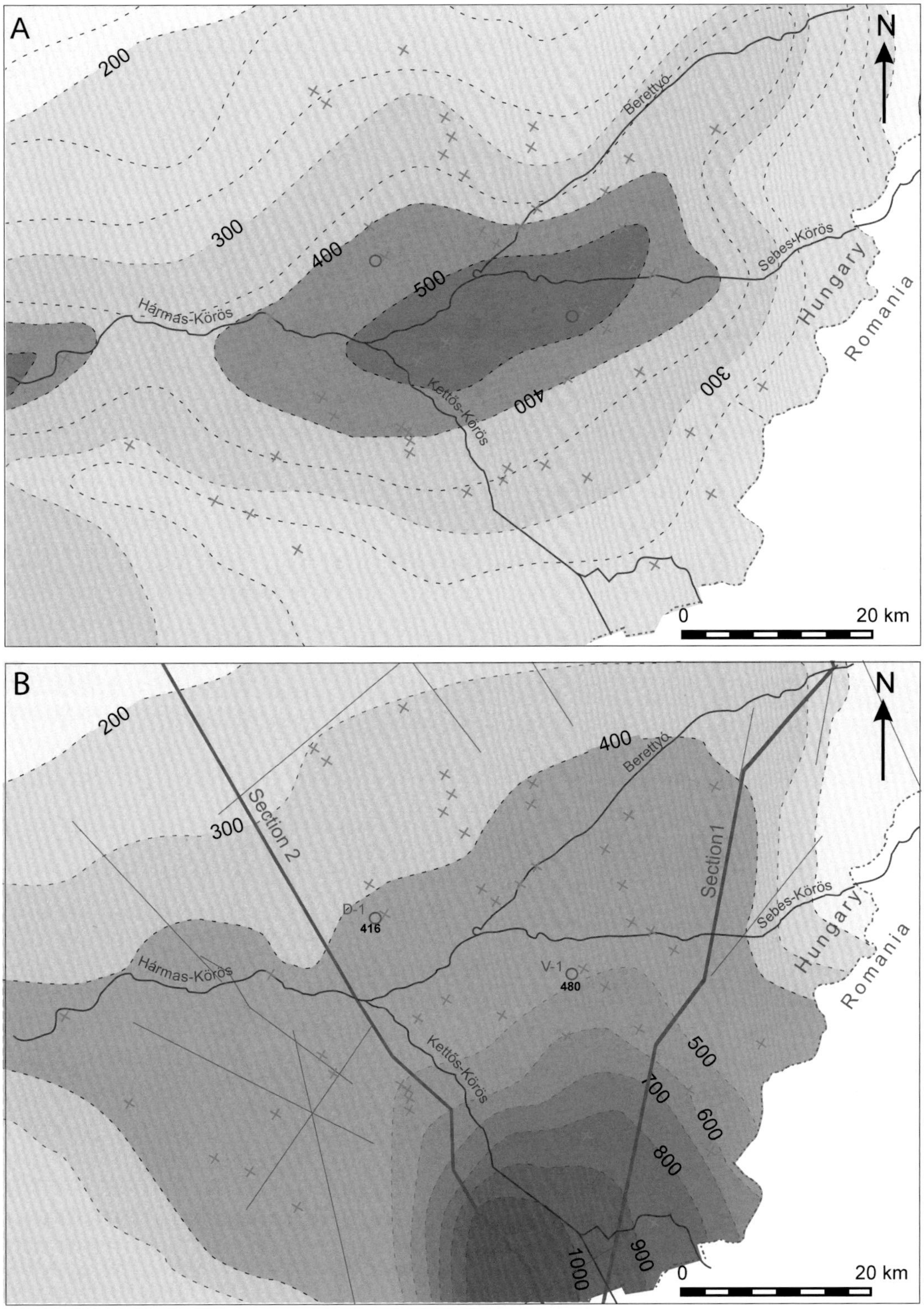

FIGURE 3.—A. Thickness of Quaternary deduced from borehole data, where traditionally the base of the Pleistocene was put at the first thick and coarse-grained bed (Franyó 1992).B. In this study the thickness of Quaternary is drawn from combination of borehole data and seismic lines showing that the basin depocenter is farther to the south and accommodates a Pleistocene fluvial succession nearly 1000 m thick.

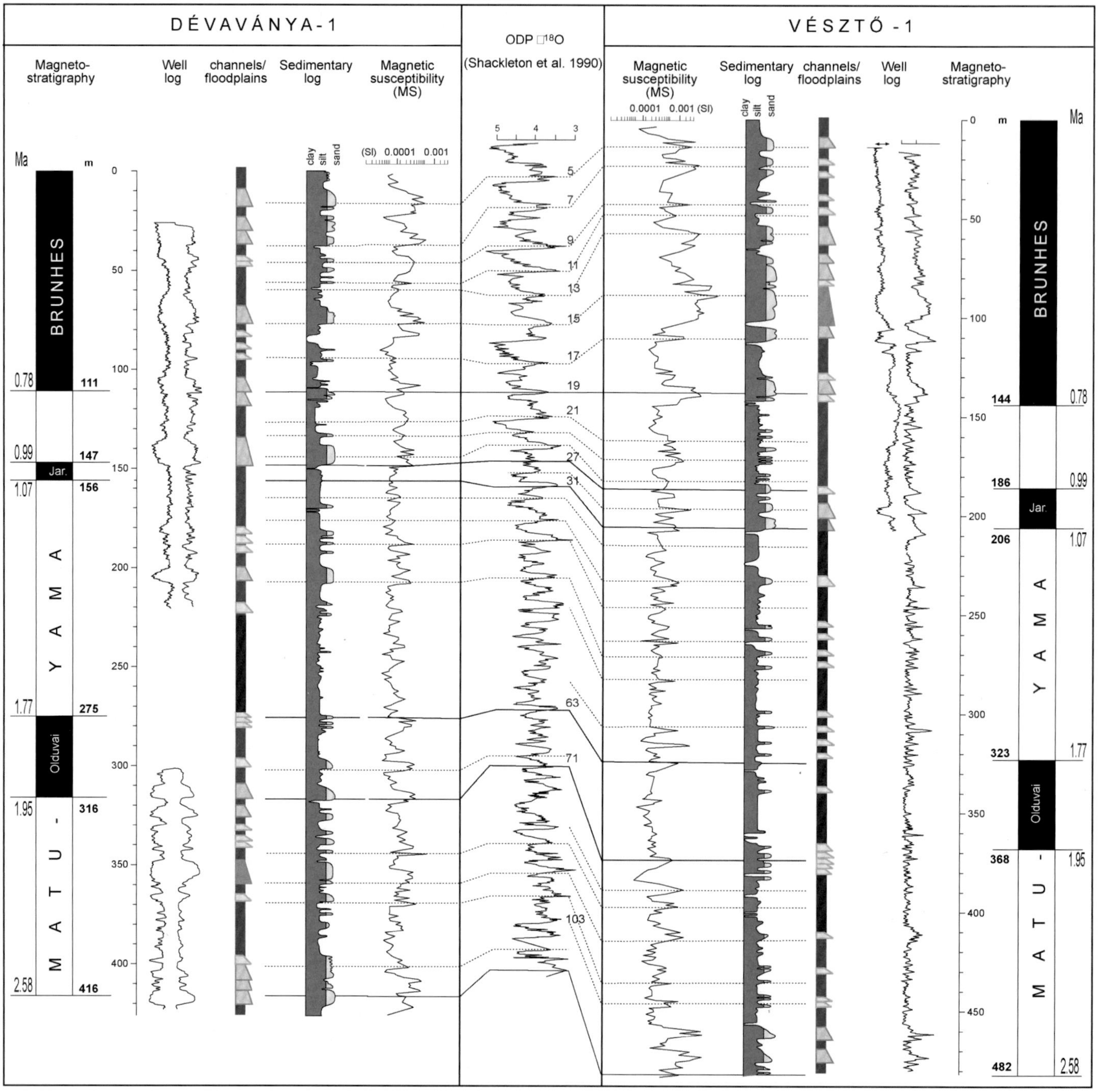

FIGURE 4.—Two magnetostratigraphically dated borehole sections of D-1 and V-1 from the Körös basin with well-logs, sedimentary logs, depositional interpretations, and magnetic-susceptibility data. The latter correlates well with marine oxygene isotope proxy of ODP 677, indicating Milankovitch-scale climatic control on sedimentation (Nádor et al. 2003). Note that depositional elements (channels, floodplains) do not show any correspondence to climatic cycles. For discussion see text.

DEPOSITIONAL UNITS

On the basis of the well-log characters two main depositional units were identified: (1) sand-prone channels and (2) mud-prone floodplains. Juhász et al. (2004) distinguished more detailed facies associations and possible channel types, but without any lithological or geometrical control these are rather hypothetical. Here we summarize only those main types which are relevant in determining the lateral and vertical variations of channel belt distribution in the encompassing thick floodplain deposits.

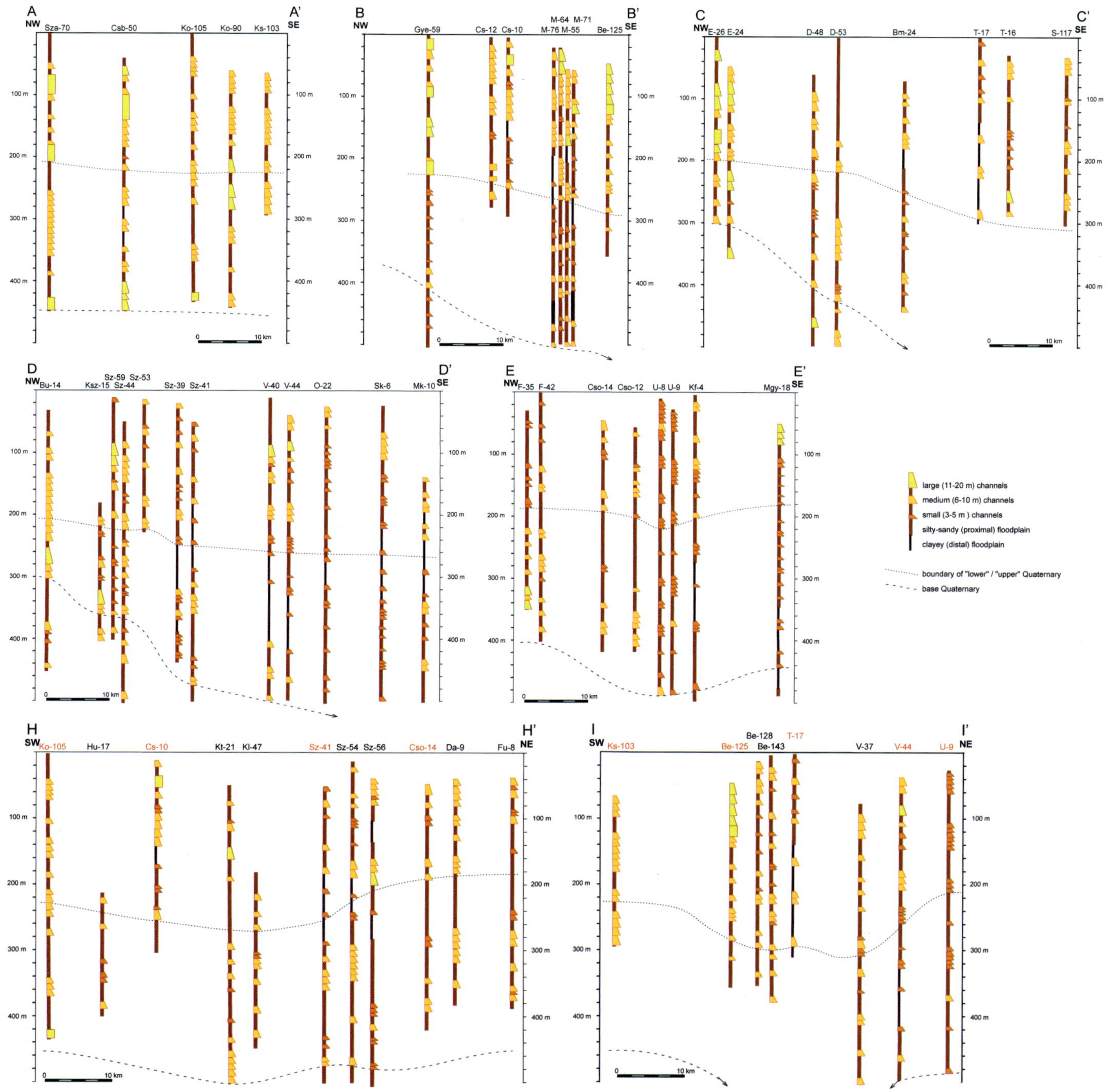

FIGURE 5.—Simplified log interpretations and sections showing channel and floodplain deposits. Small, medium, and large channels were distinguished. The "upper" and "lower" parts of the Quaternary succession are indicated with a dashed line (for discussion see text). This horizon, correlated to D-1 and V-1 (Fig. 4) boreholes in stratigraphic terms, is somewhat older than 1 Ma, marking an increase in proportion of channels in the majority of the wells. For location of sections see Fig. 2. Sections A–E are in dip direction, H–I are in basin strike direction.

Channels

The channel environment is represented by two distinct log shapes. Massive sand bodies occur rarely in the wells and have cylindrical, or bell-shaped, slightly serrated log patterns. They have a sharp base indicating downcutting into the underlying clayey–silty sediments. It is difficult to estimate the thickness of an individual channel sand body, inasmuch the thickness of these units varies from a few meters to 15–20 meters and they often represent multistory, amalgamated channel fills partly eroding each other.

The second type is more common and is composed of upward-fining sequences starting with sandy deposits capped by silty–clayey successions (Fig. 4). These are interpreted as point-bar successions showing the lateral migration of meandering channels thus indicate the channel belt zones. The thickness of point-bar sequences fall into three basic categories: small (3–5 m), medium (6–10 m) and large (11–20 m). These channel deposits either occur as solitary units encompassed by floodplain sediments 20–50 meter thick, or are superimposed on each other and form amalgamated channel complexes 20–100 m thick (e.g., Figs. 5A, B, C, D). Solitary channel deposits are built up of small (and to a lesser extent medium-size) point-bar sequences, while amalgamated channel deposits several tens of meters thick are composed mostly of medium-scale and large-scale point-bars.

Based on field observations from nearby sand pits, the near-surface fluvial channel-center deposits are light brown fine- to medium-grained sands, sometimes cross-laminated, with abundant muscovite grains and rare mottling. The channel margin is composed of light- to dark-brown sandy silt to silty sand in the outcrops with frequent occurrences of yellow to brown coloured mottles, carbonate nodules, and abundant mica grains. Shell fragments are common in the sandy layers and thin lamination in silty parts.

It is important to note that both massive, amalgamated sand bodies and bar sequences refer to a channel environment. Although amalgamated or blocky units may have been formed by braided systems (as suggested by Juhász et al. 2004), due to lack of direct evidence (sedimentary facies) we do not intend to draw unfounded conclusions on the style of the rivers. Instead, these deposits are simply regarded as channel systems, where the difference is mostly related to variations in accommodation and/or sediment input rates, as is discussed later. If accommodation is low, or sediment input is high, than channels are forced to move horizontally and thus cannibalize previous channel deposits, and only the partly eroded blocky sand units are preserved in the rock record. If accommodation is high, or sediment input is low, than solitary channel units develop, and full upward-fining sequences may be preserved between thick floodplain suits, because vertical accretion is the dominant process.

Floodplains

Two types of floodplain deposits can be distinguished. The first produces a serrated log pattern, and is composed dominantly of alternating silt and clay. They may contain thin (max. 1–2 m thick) sand intercalations which appear as thin "spikes" (Fig. 4). These sandy intercalations are interpreted either as crevasse sands, or as minor tributaries of the main channels and indicate the proximal part of the floodplain. Sections characterized by tens of meters of monotonous overbank fines, mostly clays, may have deposited in the distal parts of the floodplain, probably in low-lying marshes or ponds. It is important to emphasize that abandoned clayey channel fills a few meters thick overlying relatively thin active sandy channel fills are not distinguished from overbank fines.

Clay-pit sections in the area expose similar floodplain and backswamp sediments which are light- to dark-gray, brown silty clay to clay, with light to dark brown faint mottles, patches of dark-colored organic accumulations, and occasional thin lamination. They may contain carbonate-coated roots ($<$ 1 mm thick and $\sim$ 0.5 cm long) and isolated carbonate nodules, broken shell fragments (bivalves and planispiral mollusks), which may accumulate in large numbers at some locations, forming highly fossiliferous layers. Clay translocation features are common.

DISTRIBUTION OF DEPOSITIONAL UNITS AT BASIN SCALE

After determining depositional units in each studied well, five NW–SE-trending and two NE–SW-trending basin strike profiles were edited (Fig. 2). The simplified profiles show that the entire fluvial basin fill is characterized by floodplain sediments encompassing a large number of widely distributed channel belt deposits (Fig. 5). Massive muds, several tens of meters thick without any major sand occur in the area between the present Sebes- and Kettős Körös Rivers (Fig. 2), especially in the lower half of the sequences (Fig. 5, B, C, D, E, H, I).

The areal distribution of channel belt deposits is not uniform in the basin. The thickest point-bar units, often amalgamated, and thicker blocky sand bodies are restricted to the northwestern and western part of the study area (e.g., wells Sza-70, Csb-50, Ko-90—Section A; Gye-59, Be-125—Section B; wells E-26 and E-24—Section C). The medium-size point-bar successions, solitary-, or amalgamated, are the most frequent types and occur almost everywhere in every time slice in the basin (e.g., Section H). The smallest point-bar units (3–5 m thick) occur as solitary channel deposits within thick overbank fines and are found close to the central part of the basin, especially in the lower half of the Quaternary sequence (e.g., wells Gye-59, M-76, -64, -55, 71—Section B; well Bm-24—Section C; wells V-40, -44, Sk-6—Section D; Mgy-18—Section E).

From the thickness of channel and floodplain deposits, the proportion of channels was calculated for the boreholes, clearly indicating differences between the upper and lower parts of the successions. The plots (Fig. 6) show that the proportion of channel belt deposits is generally higher in the upper parts of almost all sections; values range mostly between 0.4 and 0.8, while the deeper parts of the basin-fill sediments are characterized by lower values of 0.2–0.4. In those cases where the channel proportions of the upper part are relatively low, the proportion in the lower parts is even lower (Fig. 6, Section D). Exceptionally low or high values in both intervals call the attention to marked differences within the basin. The boundary between the so-called "upper" and "lower" parts of the Quaternary succession depicted is at about 230 m depth on average, but in the marginal areas (Fig. 5 Sections A, E, and northern part of Sections B, C, and D), where the thickness of the whole Quaternary is less than 400 m, the boundary is between 180 and 200 m, while in the deepest parts of the basin (Fig. 5 Sections B, C, D, I) it may attain 300 m. This boundary runs ca. 50 m below the 1 Ma horizon in the magnetostratigraphically dated V-1 and D-1 boreholes (Fig. 4).

In Section A, running at the southwestern margin of the basin, the proportion of channel belt deposits is high both in the upper part (ranges between 0.55 and 0.75) and in the lower parts (0.45–0.65), so the splitting of this section to "upper" and "lower" parts is rather based on depth considerations of this horizons compared to other parts of the basin. An exception is well Ko-105, where both values drop (Fig. 6A). The channel belt deposits are composed mostly of superimposed medium- to large-scale point-bar sequences up to a cumulative thickness of 100 m and blocky sand units 30–40 m thick with only a subordinate appearance of floodplain fines (Fig. 5A).

In Section B, the channel proportion is also relatively high in the upper part (0.4–0.7) but remains as low as 0.2–0.4 in the lower part (Fig. 6B). Not only the number of the channels but also the thickness of individual channel deposits is obviously decreased (Fig. 5B) in the lower part, especially strikingly in wells Gy-59 and M-76, -64, -55, -71, where solitary channel deposits represented by small-scale point-bars are divided by floodplain sediments 10–30 m thick, while the upper part is characterized by medium- to large-scale amalgamated point-bar sequences occasionally over a thickness of 100 m.

In the upper part of Section C, the proportion of channel belt deposits ranges between 0.6 and 0.8 at the northwestern margin of the basin, where they are composed mostly of superimposed medium- to large-scale point-bar sequences and blocky sand units. As the section continues towards the basin center, the ratio of channel belt deposits drops to a value of 0.2–0.4 and those deposits are represented mostly by small- to medium-scale point-bar sequences. Note the long, virtually channel-free interval in the upper part of well D-53. The lower parts of the sequences are characterized by relatively low

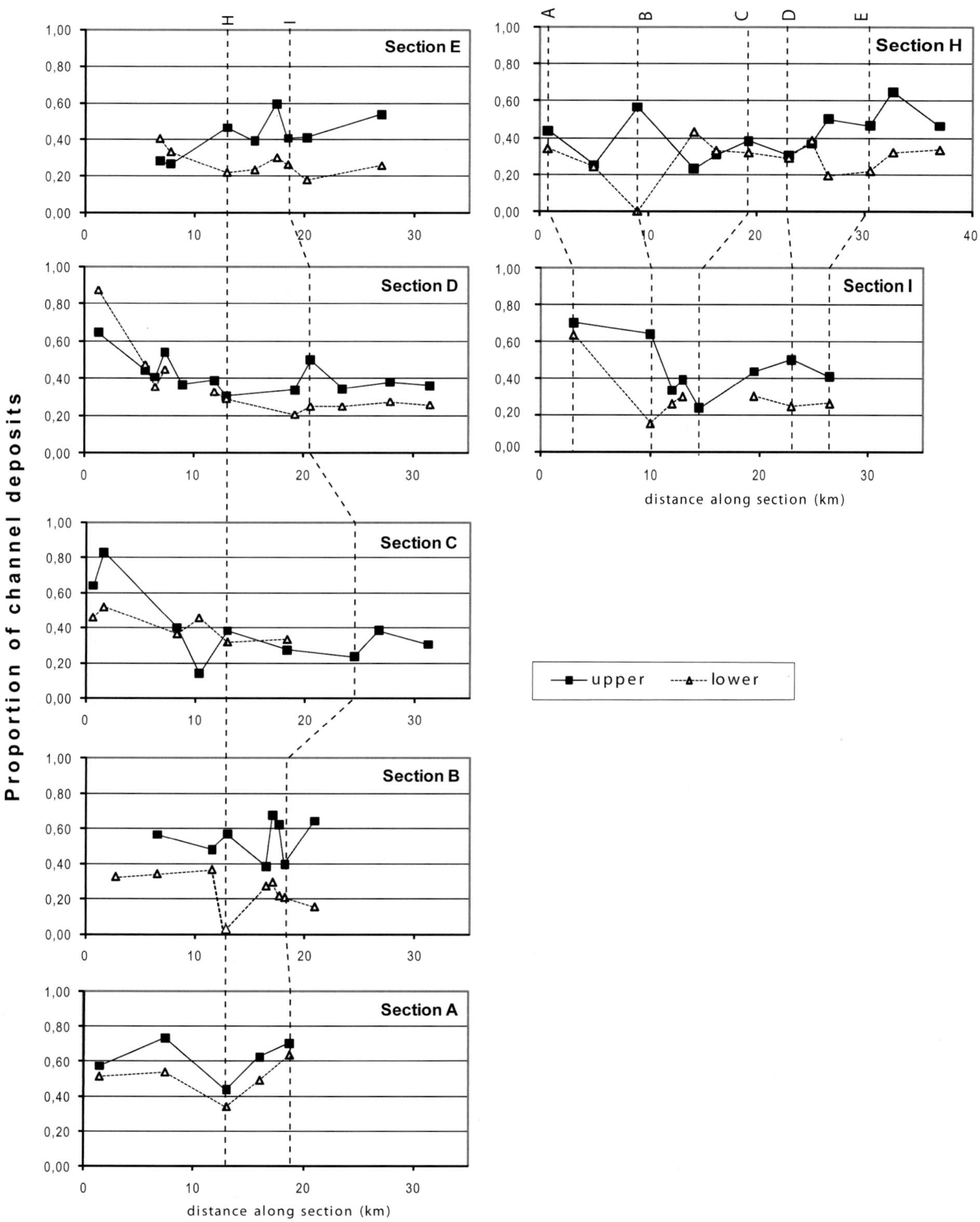

FIGURE 6.—Proportion of channel deposits in the "lower" and "upper" parts of the sections. The boundary between the "lower" and "upper" parts between ca. 200 and 300 m was chosen on the basis of detailed calculations on sand/mud proportions in successive segments of the wells. In the majority of the wells this ratio is higher in the upper parts. Plots are oriented as corresponding sections; for identification of wells see Fig. 5.

proportions of channel belt deposits, with a value around 0.4 along the entire section. Similarly to Section B, the deeper part is characterized by solitary small (sometimes medium-) size point-bar sequences divided by thick floodplain deposits (Figs. 5C, 6C).

In Section D a similar decrease in proportion of channel deposits from northwest to southeast is present from 0.65 to rather low values (below 0.4) even in the upper part (Fig. 6D). This is because in Section D large channels are rare, and are missing particularly in the lower part except for the marginal Bu-14 borehole (Fig. 5D). Both D and E sections are characterized by mostly small-scale point-bar sequences, amalgamated, or as solitary ones in the floodplain alluvium (Figs. 5D, E). In Section E, while channel proportions (0.4–0.2) decrease to southeast in the lower part, as in Sections C and D, the values are slightly increasing in the upper part.

Sections H and I run perpendicular to Sections A–E in a southwest–northeast direction (Fig. 2). Section H runs northwest of the central depression of the basin. Its upper part is characterized by a proportion of channel belt deposits between 0.3 and 0.6, while this values drops to 0.2–0.4 for the lower part (Fig. 6H). Section I crosses near the deepest part of the basin. Its southwestern marginal part holds a high channel belt proportion around 0.7 in the upper part, but approaching the basin center this value drops below 0.4 (Fig. 6I). In both sections the curves showing the proportion of channel belt to floodplain deposits are nearly identical for the upper and lower parts, with a value around 0.3 close to the deepest part of the basin, while towards the margins the upper parts are characterized again by a higher proportion of channel belt deposits.

The pattern described above shows that proportion of channels is generally lower for the deeper parts of the sections, representing deposition before ca. 1 Ma, and channel proportion increased afterwards. The geographic position with respect to the basin center further influences this general picture. The proportion of channel belt deposits is higher close to basin margins, where the thickness of the Quaternary sedimentary succession is around or below 400 m, and these values are significantly lower approaching the depocenter to the south.

DISCUSSION

The proportion and spatial distribution of various types of alluvial deposits (e.g., channel belt and overbank) in sedimentary basins are determined by the interaction of sediment supply and available accommodation. The amount and type of sediments produced are controlled by the lithology of the source areas, topographic relief, climate, and vegetation. Space to put sediments depends on base-level changes, driven by eustasy and/or tectonics (due to the intramountaine setting of the Körös basin, eustatic changes are not relevant in this case). The manner in which sediment fills the available space depends on the flow regime and type of the channels (e.g., braided, meandering), which are in turn controlled by water, sediment supply, intrabasinal tectonism, and local base level. Controls on channel movement along the floodplain determine the spatial distribution of channel belt deposits. These can be either allocyclic, governed by tectonic, climatic, and/or eustatic changes independent of processes in the sedimentary basin, or autocyclic, related to factors such as seasonal floods, progradation of levees, and avulsions (Allen 1978; Miall 1996; Bridge 2003). From this cascade of processes we are mainly discussing how the subsidence history of the basin influences the spatial distribution of alluvial deposits.

Overview of Experimentally Derived Models of Alluvial Architecture

Focusing on the geometry and distribution of channel belt sand deposits and floodplain sediments in avulsion-dominated alluvial basins, the classical models of alluvial architecture (Leeder 1978; Alexander and Leeder 1987; Bridge and Leeder 1979), henceforth referred to as the LAB model, established a basic connection between subsidence, channel avulsion, and depositional stacking patterns of channel belt deposits. These models related thick aggradational overbank fines with separated channels for areas of high temporal (vertical) subsidence creating an increased accommodation. They also supposed that channels are drawn to areas of subsidence maximum, i.e. although the channel stacking density is low, the major (largest) channels occur here. If subsidence rate reduces in time and accommodation space becomes limited, stacking density of channels increases and channels cannibalize the old deposit, amalgamated channel complexes are common, and the sedimentary succession is characterized by a high proportion of channel and crevasse-splay sediments at the expense of floodplain (Fig. 7). These models also assumed a constant avulsion rate and that channel belt stacking density and hence connectedness is increasing with high sedimentation rate. Thus the stratigraphic arrangement, stacking density, and interconnectedness of amalgamated channel sand bodies record relative changes in subsidence rates where other controls are held constant, as also concluded earlier by Allen (1978).

Even the early models (Alexander and Leeder 1987) noted that sediment supply is also a key factor in determining alluvial architecture and that rivers can be deflected from the subsidence maximum if sediment supply is high and sedimentation rate compensates subsidence. The position of channel belts would be determined by local variations in topography; however, the total thickness of the deposit would be the greatest over subsidence maximum. Later three-dimensional models (Mackey and Bridge 1995) incorporated differential sedimentation and slope change between the channel-belt and the floodplain as a control on avulsion and showed that high sediment supply may cause different architectural patterns as predicted by the LAB model, i.e., high aggradation rates tend to shift channel belt deposition away from subsidence maxima. They concluded that the competition of slopes downvalley and across valley is the main driver of avulsion. However, they did not examine the influence of avulsion type, or the relative importance of local sedimentation versus flow depth as a control on sand-body thickness.

Studying downstream changes in alluvial architecture, Heller and Paola (1996) considered the effects of four basic variables: (1) the geometry of subsidence (fore-tilted and back-tilted basins), (2) whether avulsions takes place locally (the river reoccupies its previous channel downstream), or regionally (cf. "nodal" avulsion of Leeder 1978), (3) thickness of channel belt sand bodies, and (4) relationship between avulsion frequency and local sedimentation rate within the active channel belt. For the last variable they differentiated three major scenarios and found that if avulsion frequency increases at rates slower than the increase in sedimentation rate, then stacking density increases with decreasing sedimentation rate, similar to LAB model predictions. If avulsion frequency increases linearly with sedimentation rate, there is no change in stacking pattern with changes in sedimentation rate. Finally if avulsion frequency increases faster than sedimentation rate, then stacking patterns become more dense with increasing sedimentation rates, the opposite of that predicted by the LAB models. Their final conclusion was that although subsidence strongly influences the rate at which alluvial architecture changes in the downstream direction, the relationship between avulsion frequency and sedimentation rate determines whether stacking patterns become more or less dense downstream.

In their three-dimensional experiments, Hickson et al. (2005) varied subsidence rate and pattern along with rates of water supply and sediment supply. They found that the principal architectural signature of lateral variability in subsidence rate is stratal tilting and channel belts are not attracted to the subsidence maximum, as predicted by the LAB model. They hypothesized that the effect of variable subsidence geometry is felt by the fluvial system only if subsidence proceeds faster

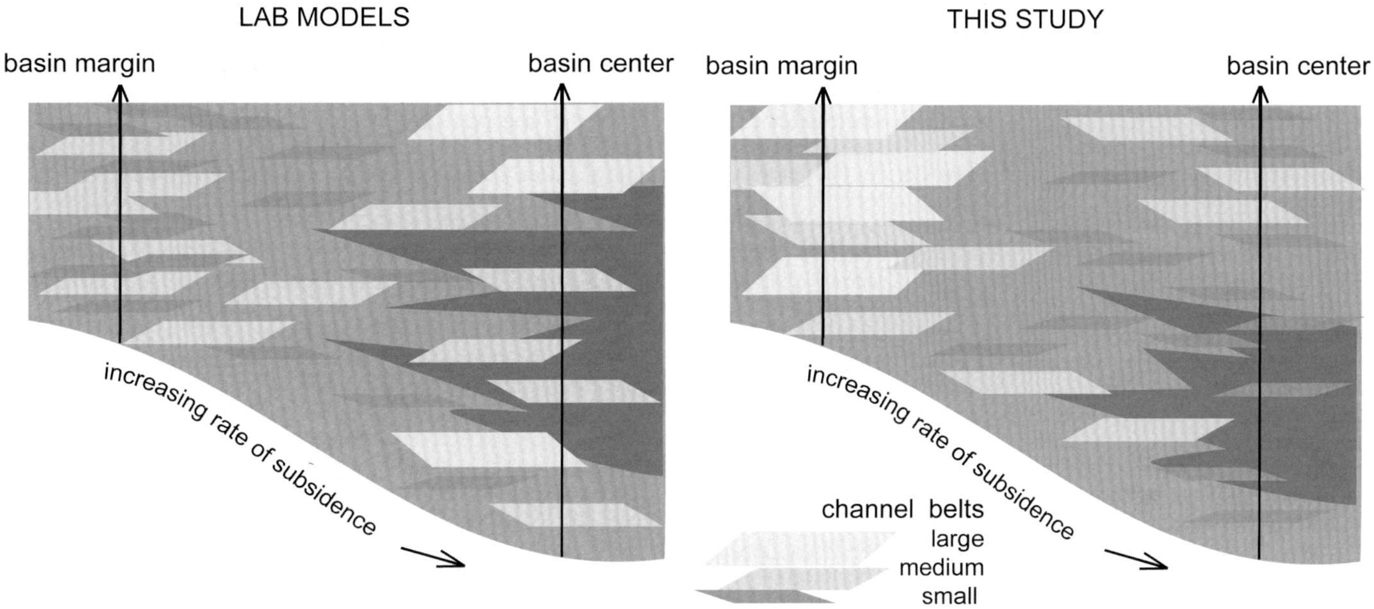

FIGURE 7.—Comparison of channel belt stacking density predicted by the early models (Leeder 1978; Alexander and Leeder 1987; Bridge and Leeder 1979) and based on field data from the Körös basin in this study. Vertical stacking density corresponds to rate of subsidence, in agreement with the LAB models. Lateral changes in stacking density, however, show that the amalgamated large channels in the Körös basin are found at the basin margin and not above the subsidence maximum as predicted by the LAB model. This is explained by local factors: tectonic trapping along the northwestern margin and by high sediment supply at the southwestern margin.

than the river can adjust to the formation of a topographic low via the deposition of overbank material in the form of splays and sand sheets. In other words, fluvial channels are attracted to lateral subsidence maximum only if cross-stream slope associated with tectonics exceeds downstream slope.

Field Tests on Channel-Belt Stacking density and Their Disparity with Experimentally Driven Models

The controls on avulsion-related stacking are difficult to constrain because of the difficulty in determining sedimentation rates in the rock record with sufficient accuracy. Due to the little observational data, large-scale avulsion is still a poorly understood process despite the large number of hypothetical works (e.g., Alexander and Leeder 1987; Heller and Paola 1996, Bridge 2003 and references therein). However, many workers have used the LAB model to interpret relative subsidence of sedimentary basins (e.g., Behrensmeyer and Tauxe 1982; Blakey and Gubitosa 1984; Shuster and Steidtmann 1987; Shanley and McCabe 1993), and the petroleum industry also widely applied it, because prediction of sand-body connectedness plays a major role in exploration strategies. Some study results, such as on the Tertiary deposits of the Bighorn Basin, Wyoming (Kraus 1987), were consistent with the LAB model. In contrast, others, such as Read and Dean (1982) on the Carboniferous deposits in the Namurian Basin, Scotland, produced results opposite to LAB model predictions, where regions with high sedimentation rate coincided with increases in sand/shale ratio.

Because results of field studies are ambiguous, more quantitative data are needed to test models of alluvial architecture. The most promising sections for a quantitative approach are datable young deposits, which thus can offer the best correlation between alluvial architecture and changes in sedimentation rate. Several field tests of the relationship between changes in sedimentation rate and channel belt stacking density have been done on Holocene and Late Pleistocene fluvial systems (e.g., Rhine–Meuse delta, Törnqvist 1995, Berendsen and Stouthamer 2002, Cohen et al. 2002; Rio Grande Rift, Leeder et al. 1996; Carson River, Nevada, Peakall 1998). Rivers in the Rhine–Meuse delta system were strongly influenced by sea-level changes, and the studied successions cover mostly the Holocene. The Rio Grande and the Carson River are in semiarid settings, therefore the sediment supply was low.

However, the disparity between experimental models and field data is a broader and more fundamental problem. As case studies provide examples from different geological settings with a great variety of controls on fluvial sedimentation (e.g., basin tectonics, slope profiles, discharges, sediment supply rates, sea-level changes, etc.), it is obvious that none of them proves, or denies fully model predictions. Nature is more complex than all variables controlling fluvial architecture could be described by a single, even highly sophisticated model, and a generalized reason could be given for the disparity with model predictions. As hopefully increasing number of well constrained field tests from various environments, as well as improvement of models will eliminate the differences in the future.

Körös Basin Field Test Results

The investigation of the spatial distribution of channel belt deposits in the Körös basin provides an excellent field data set from a long fluvial record spanning most of the Quaternary to test model predictions of the effects of temporal (vertical) and lateral changes in subsidence on channel belt stacking patterns. This is a high sediment-supply system under a temperate climate.

River Systems during the Quaternary: Point-bar sequences with different sizes provide valuable information on the sizes of the former rivers filling up different parts of the basin. The thickness of a single point-bar unit is roughly equivalent to the bankfull depth of the former channel (Allen 1970). The three distinct groups of point-bar sequences mark three different sizes of rivers that filled the basin. The thickest point-bar units (11–20 m) at the northern and western margin of the basin refer to the presence of large rivers. During most of the Pleistocene, the paleo–Tisza River was flowing along a northeast–southwest-striking depression along the western margin of the Körös basin (Gábris 2002). Therefore the thickest point-bar sequences are regarded as the remnants of the paleo–Tisza River system.

The modern Tisza River, draining the entire Eastern Carpathians, has a bankfull depth of about 8 m. Reconstructions on the drainage-pattern development on the GHP (Gábris and Nádor 2007; Nagy et al. 2007) showed that during the Late Pleistocene a large river with channel depth of about 8 m and a width of 150–200 m (Sztanó et al. 2002) followed a course in the northern margin of the GHP, and shared the East Carpathian drainage area with the paleo-Tisza. The dimensions of this river suggest that average channel-forming discharge might have attained that of the modern Tisza River, though the drainage area was only about half (Cserkész-Nagy et al. 2008). This indicates that rivers of the GHP (including the paleo-Tisza, flowing along the western margin of the Körös basin) were significantly larger during the Pleistocene than at present. This is further supported by the medium-size point-bar sequences of the Körös basin, which refer to ancient rivers with a bankfull depth of 6–10 m, equalling the size of the present Tisza River. These deposits are common all over the basin; therefore it should be suspected that discharge, as well as sediment delivered from the uplifting mountain chain, was significantly larger than at present.

The smallest point-bar successions with 3–5 m thickness, indicating former rivers with a bankfull depth of a few meters, are comparable with sizes of the present tributaries of the Tisza River (e.g., Körös, Berettyó). These channels are concentrated mostly in the basin center, outlining a drainage pattern similar to the present.

Climatic Control on Sediment Supply: Before discussing the spatial and temporal relation of channel proportions to subsidence and basin morphology, climatic control on discharge and sediment supply is addressed, which has initiated a long-lasting debate on the hierarchy of fluvial cycles and their autocyclic and/or allocyclic origin. In the Körös basin 11 large sedimentary cycles, 30 to 60 m thick on average, were recognized in several fully cored boreholes at a depth between 100 and 1200 m (Rónai 1985 and references therein), and were correlated with the classical Alpine glacial and interglacial periods described by Penck and Brückner (1909). Based on well-log analysis in the Körös basin, Juhász et al. (2004) outlined upward-fining fluvial cycles of 40–100 m thickness, starting with multistory channel fills, followed by an alternating sandy–muddy succession, capped by silty–clayey floodplain sediments. These cycles were observed only in a restricted number of boreholes close to the basin margins and could not be correlated regionally. The controls on formation of these cycles remained open, but Juhász et al. (2004) suggested a climatic origin, in that six of them were differentiated, and therefore were thought to correspond to 400 ky eccentricity cycles.

Based on detailed cyclostratigraphic study of two continuous and magnetostratigraphically dated cores 500 m deep, Nádor et al. (2003) revealed that systematic variations in magnetic susceptibility correspond to Milankovitch-scale cyclicity of 40 and 100 ky, which also showed correspondence to the $\delta^{18}O$ variations of the ODP 677 site; thus a much larger number of fluvial cycles characterize the deposits (Fig. 4). The depositional model of Nádor et al. (2003) suggested a periodic change in sediment flux to the basin: an increased sediment flux happened during interglacials due to the increased discharge and transport capacity of the rivers, while a decreased sediment supply to the distal parts of the basin occurred during glacials due to storage of sediments near to the source area. However, a systematic correspondence between the cycles and the large-scale depositional elements (i.e., channels, floodplains) identified from well-logs (Fig. 4) is not found. On floodplain areas magnetic susceptibility and grain-size variations correlate due to higher fluxes of coarse-grained magnetic minerals during floods, also imaged by well-logs. At the same time the varying amount of magnetic minerals in channel belts are masked within the overall sandy lithology; thus climatic signals remain hidden in these parts of the well-logs. Therefore these cycles may occur either in floodplains or in channel belts, which means that the actual location of depositional elements is not controlled by climate.

Vertical Changes of Accommodation/Sediment Supply as a Control on Stacking Density: Comparing our observations on channel stacking density in space and time with the various model predictions (Fig. 7), first we address the temporal relations of channel proportions to basin morphology, subsidence, and sediment supply.

The overall trend that the proportion of channel belt deposits is higher close to basin margins and decreases towards the center justifies the LAB models (Leeder 1978; Alexander and Leeder 1987; Bridge and Leeder 1979) concerning the inverse correlation of stacking density and subsidence (Fig. 7).

Regarding the temporal (vertical) variation in channel belt stacking density, it was observed that the proportion of channels is generally lower for the deeper, and higher in the upper parts of the sections (Figs. 5, 6). The overall trend that the lower parts of the sections in the Körös basin are dominated mainly by a low proportion of channel belt sediments indicates that subsidence rate was relatively high, creating enough accommodation to store fluvial deposits, including their thick vertical aggradational floodplain sediments, in agreement with the LAB models. There was a marked change in the accommodation/sediment supply ratio (A/S) at several parts of the basin, when the upper 300–180 m of sedimentary succession started to be deposited. Most of the sections show that the proportion of channel belt deposits increased at the expense of floodplain, which indicate that there was perhaps less accommodation available and/or more sediment supply to the basin.

To quantify differences between subsidence and sediment accumulation rates for the "upper" and "lower" parts of the studied sections with higher and lower sand proportions respectively throughout the basin, precise dating is necessary. This is available in only two magnetostratigraphically dated boreholes (D-1 and V-1, Figs. 2, 4), where the boundary between the upper and lower units is running ca. 50 m below the 1 Ma horizon and cannot be extended basinwide. In addition the thickness map of the Quaternary constructed on the basis of seismic profiles (Fig. 3B) combined with the two dated boreholes (Fig. 4) revealed that many of the wells (see sections in Fig. 5) did not reach the base of the Quaternary. Furthermore both the D-1 and V-1 boreholes are situated on the northern flank of the basin (Fig. 3), therefore accumulation rates calculated from these two locations do not represent differences between the basin center and margins.

The upward decrease of the A/S ratio is particularly expressed in the northwestern and southwestern marginal areas, where the lower parts are characterized by low channel belt proportions (0.2–0.4), while in the upper 300–180 m the channel proportions increase to ~ 0.6 or higher, represented by amalgamated large- and medium-size point-bar sequences (Section B, northern parts of Sections C and D, southwestern parts of Sections I and H, Figs. 5, 6). In the basin center, where the thickness of Quaternary sediments is the highest, both the lower and upper parts are characterized by massive muds, several tens of meters thick, with scattered and mostly isolated channel belt sediments; thus the proportion of channel belt deposits ranges between 0.2 and 0.4 in the lower parts and also remains fairly low (0.3–0.5) in the upper parts (Figs. 5, 6, central and southeastern parts of Sections C, D, and the central parts of Section I). Although there is a slight increase in channel stacking density in the basin center, this signal is suppressed by the effect of high subsidence.

Lateral Changes of Accommodation/Sediment Supply as a Control on Stacking Density: The next question to be discussed is the anomalous distribution of major channel belts throughout the basin, especially in the upper part of the sedimentary sequence.

The highest proportion of channel belts, as well as large channels with the thickest, amalgamated point-bar units and thicker blocky sand bodies are found along the northwestern, western, and southwestern low-subsidence margins of the basin, instead of outlining drainage lines towards the deepest center of the basin with subsidence maximum as predicted by the early LAB models (Fig. 7). This pattern can be explained theoretically either by local effects that trapped rivers along certain strikes along the margins (i.e. tectonic control), or by an increased sediment supply, which compensates subsidence, as, for example, in the case of axial river systems in foreland basins, which bypass the subsidence maximum near the thrust front.

The tectonic control on drainage-pattern development has been discussed by a great number of authors, e.g., Burbank and Pinter (1999), Schumm et al. (2000), and references therein. In the Körös basin region several tectonic studies (Pöstyényi 1993; Szentgyörgyiné 1988) showed a sinistral stike-slip zone at the northwestern margin, which joins a larger Pleistocene left lateral strike-slip zone towards the northeast, identified by Pogácsás et al. (1989). Some elements of these sinistral strike-slip fault zones could be also identified from seismic sections (Fig. 8). Above the basement nonconformity Miocene synrift and postrift sediments are seen; their distribution was controlled largely by synsedimentary extensional faults. Due to a regional compressional event at about the Miocene–Pliocene boundary (Sztanó et al. 2007), uplift of the northeastern part of the basin took place, which resulted in shift of the depocenter to the south (Fig. 8, Section 1). The increased subsidence at the southern part and the coeval uplift at the northern part was pronounced during the Pliocene, but continued to a smaller extent as differential subsidence during the Quaternary. As compression made former fault zones rejuvenated, mainly as transpressional strike-slip systems (Windhoffer et al. 2004), some of them were still active during the Quaternary. The uppermost few hundred milliseconds are nulled on the seismic sections, but from the geometry of the uppermost reflections both the base of the Quaternary and the geometry of the Lower Quaternary strata can be deduced. In case of transpressional strike-slip zones numbered as 1, 2, and 3 in Figure 8, their activity can be supposed even close to the surface. Local anomalies (minor increases) in channel proportion occur near to these fault zones (Section B: M-64, -55, Section C: T-16, Section D: V-44, Section E: U-8) (Fig. 9B). Minor topographic elevation above these transpressional zones may have locally influenced accommodation, thus channel stacking density, without changing the courses of rivers themselves.

The distribution of the northern belt of large sand-prone channels (paleo-Tisza) may be related to these recently active fault zones. This was proved for the youngest sediments of the area. Based on OSL dating and analysis of thickness anomalies of uppermost Pleistocene sediments in the Körös basin, Nádor et al. (2007) demonstrated that a sinistral strike-slip fault zone was active till 14 ky ago and channelled the paleo-Tisza. It is very likely that the Middle and Late Pleistocene paleo-Tisza (Fig. 9A) was also trapped in this tectonically active zone, and therefore did not flow towards the basin center. Another reason for not avulsing towards the subsidence maximum might be the little topographic difference between the low-subsidence marginal parts and the central parts with an increased subsidence, analogous to the present situation. Lateral channel shifts and incision of new channels might be constrained in the zone of sandy substratum with a channel floor topographically at a significantly lower position than the surface of the neighboring clayey, non-erodible floodplain (Fig. 10). Moreover, because the paleo-Tisza was one of the largest drainage systems of the GHP during the Pleistocene, with a significant discharge and sediment supply (as shown by the size of its point-bar deposits), the high sediment supply could have also contributed to counterbalance the lateral subsidence of the basin center, which thus failed to attract the river.

The distribution of channel belt deposits (Fig. 9) shows that a somewhat smaller river system (paleo-Kettős-Körös) could have coexisted along the southwestern margin of the basin during the last 1–1.3 Ma, as shown by the amalgamated medium-size point-bar sequences and their relatively high proportion to floodplain sediments. The course of this river system was clearly not influenced by the tectonic structures described above, inasmuch as it crosses several sinistral strike-slip zones reconstructed from the seismic sections (Fig. 9). Another important feature is that this channel belt is found along the side of the southwestern flank of the depression, but not in the deepest parts. Referring to the experimental results of Hickson et al. (2005) this pattern can occur if cross-stream slope associated with lateral subsidence does not exceed downstream slope. To keep the downstream fluvial slope relatively steep, a relatively fast uplift of the upstream part with a high sediment supply has to be assumed. The drainage area is in the Apuseni Mountains, which was characterized by an erosion-driven, isostatic uplift (Sanders 1998), which is also reflected by the heavy-mineral spectra in the Körös basin sedimentary succession (Thamó-Bozsó et al. 2002). This direct sediment flux from the southeast might have fed the paleo-Kettős-Körös and provided a sediment supply high enough to compensate the lateral subsidence of the basin center, which failed to attract this river as well.

Once a relatively coarse-grained, sand-prone sediment delivery system was set up on the southwestern flank of the basin, it provided a substratum for the next generation of streams to incise more easily than into the fine-grained floodplain clays in the central part of the basin, even though it was topographically lower (Fig. 10). Amalgamated channel belt deposits show that supplied sediment exceeded the available accommodation on these areas and channels were forced to cannibalize each other.

So what has happened during the last 1–1.3 million years that altered alluvial sedimentation, decreased rate of subsidence, and/or increased sediment supply in the Körös basin? According to regional neotectonic models (Bada et al. 2006b and references therein), decreased subsidence of the basin areas implies a waning compression. The uplift of the Apuseni Mountains and the related increase of sediment flux to the basin might also have been connected to the relief of compressional stress field (Sanders 1998). Both of these processes together or separately might have resulted in the increased channel stacking density since about 1.3 Ma in the Körös basin.

CONCLUSIONS

Investigation of the spatial distribution of channel belt deposits in the Körös basin provides an excellent field data set to test previous model predictions of alluvial architecture. Our detailed analysis of the spatial distribution of the various depositional elements has two major conclusions. The lower part of the fluvial sedimentary succession was deposited under conditions of high accommodation and/or low sediment supply, as shown by the relatively low proportion of channel belt sediments. Fluvial sediments representing the last ~ 1.3 million years were deposited under decreasing accommodation and/or increased sediment supply, as shown by the increasing proportion of channel belt deposits throughout most parts of the basin. This clearly indicates that the primary control on the channel density was subsidence and/or uplift of the hinterland, driven by tectonics (basin inversion) and not climate. These are principally in agreement with LAB model predictions (Leeder 1978; Alexander and Leeder 1987;

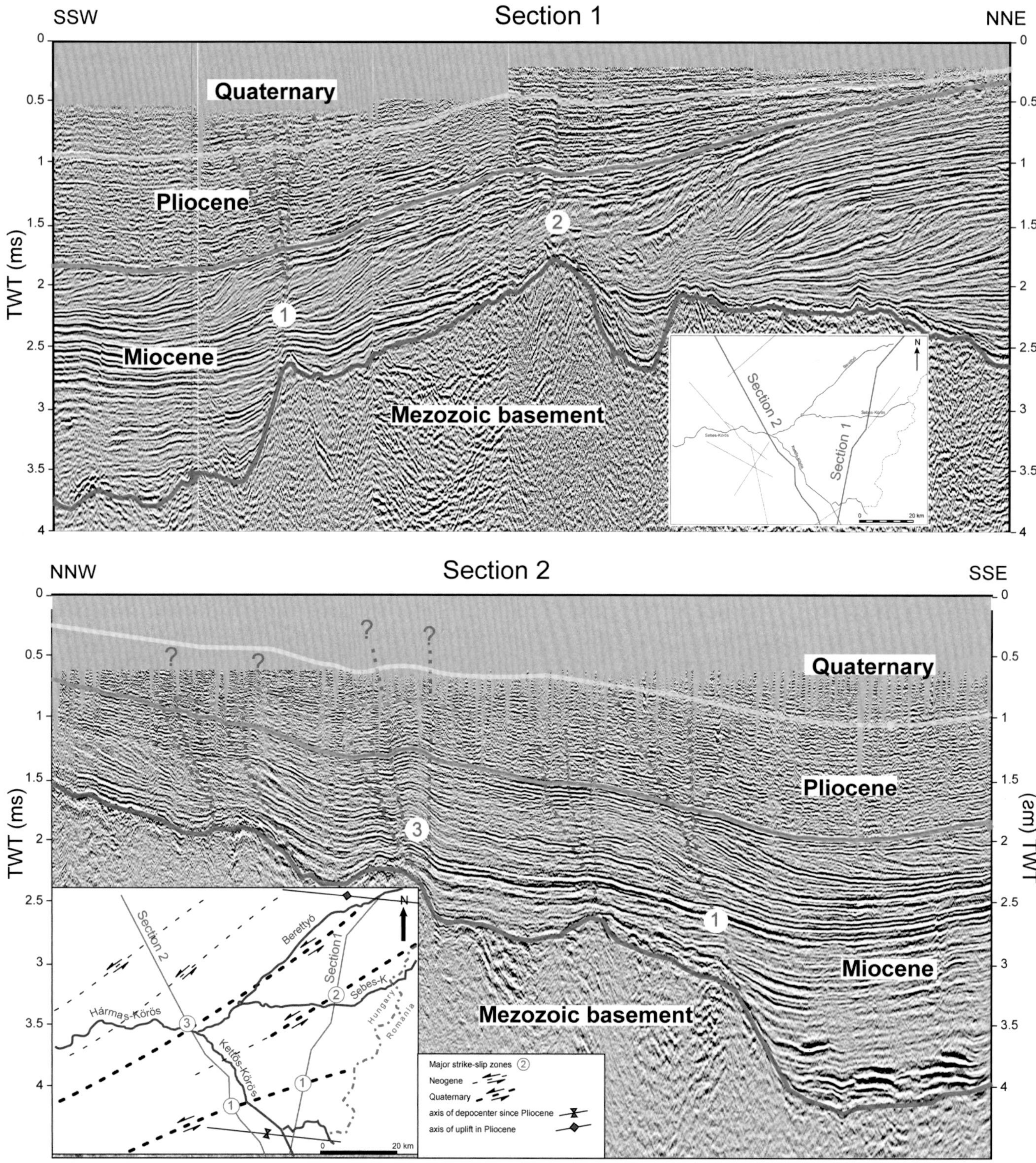

FIGURE 8.—Regional seismic sections crossing the Körös Basin. Two major and several short seismic lines helped to identify the base Quaternary above the thick Pliocene succession. The major strike-slip zones appear on the sections as positive flower structures indicating a strongly transpressional character during the Pliocene to Quaternary. These were basically formed above mid-Miocene extension-related fault systems in the basement, when compression began (cf. Windhoffer et al. 2004).

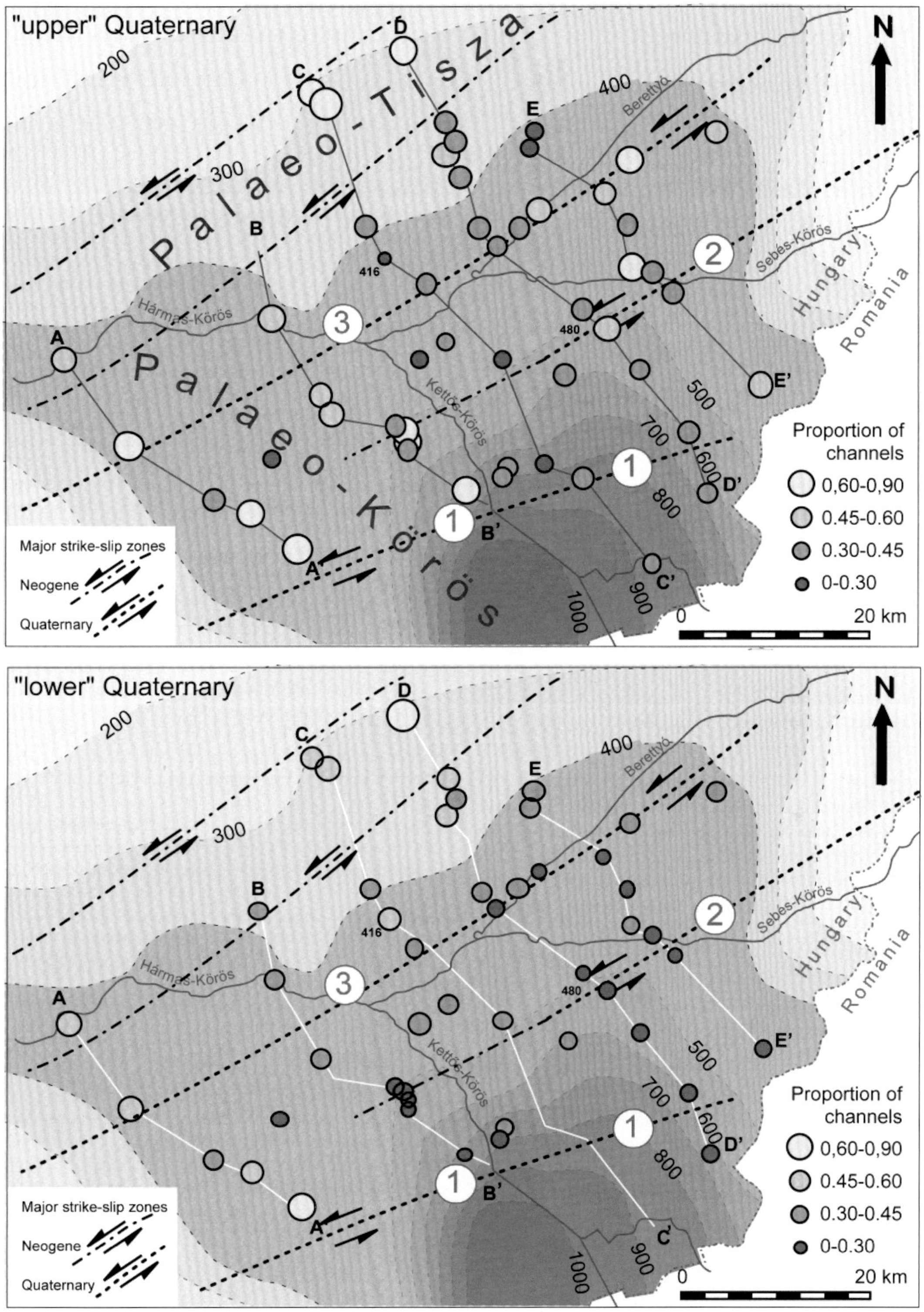

FIGURE 9.—Comparison of neotectonic structural elements, thickness of the Quaternary basin fill, and channel proportions for the "lower" (A) and for the "upper" Quaternary (B). For discussion see text.

Bridge and Leeder 1979) which relate thick aggradational overbank fines with separated channels for areas of high temporal (vertical) subsidence.

Our second conclusion is that rivers do not necessarily flow towards a subsidence maximum as suggested by the LAB model. The anomalous appearance of major river systems at the basin margins was related to a marked tectonic control exerted by several transtensional strike-slip fault zones that directed the paleo-Tisza. The isostatic

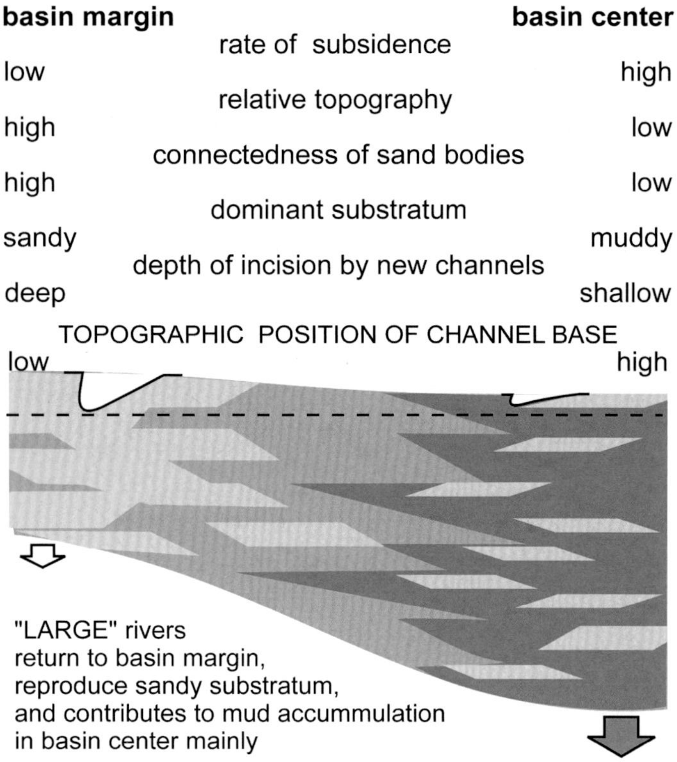

FIGURE 10.—Substratum lithology controls explaining deflection of large rivers away from the subsidence maximum.

uplift of the Apuseni Mountains catchment area provided increased sediment supply to the ancestral Kettős-Körös River along the southwestern margin of the basin, which could pass without deviating towards the laterally subsiding basin center. Thus this supports model predictions of Hickson et al. (2005) emphasizing the role of sediment supply in alluvial architecture.

The increased channel stacking density since about 1.3 Ma in the Körös basin is related to waning compression in the region, manifested in decreased subsidence of the Körös basin coupled with erosion-driven uplift of the Apuseni Mountains and related increase of sediment flux to the basin. The conclusion we drew from a sedimentological approach can contribute to a better understanding of the neotectonic evolution of the Pannonian–Carpathian realm and also highlights that disparity between experimental models and field data on alluvial architecture is a broader and more fundamental problem. Multistage basin evolution is more complex than if all variables controlling fluvial architecture could be described by a single, even sophisticated model. Phases of basin evolution with good understanding of controlling factors, such as rate and geometry of subsidence and the sediment supply can be approached by experimental models, but the complex evaluation of successive phases needs a combination of constraints as well as considerations of local effects.

ACKNOWLEDGMENTS

Seismic sections were provided by Hungarian National Oil Company (MOL Zrt). We thank Emese M. Bordy (Rhodes University, South Africa) and Martin Gibling (Dalhouse University, Canada) for thorough discussions on various aspects of fluvial sedimentation and its controls. This research is a contribution to the FLAG (Fluvial Archives Group) activities. Our special thanks are forwarded to John Holbrook and Darrel Long, whose comments greatly improved the manuscript.

REFERENCES

ALEXANDER, J., AND LEEDER, M.R., 1987, Active tectonic controlon alluvial architecture, *in* Ethridge, F.G., Flores, R.M., and Harvey, M.D., eds., Recent Development in Fluvial Sedimentology: SEPM Special Publication 39, p. 243–252.

ALLEN, J.R.L., 1970, Studies in fluviatile sedimentation: a comparision of fining upwards cyclothems with special reference to coarse-member composition and interpretation: Journal of Sedimentary Petrology, v. 40, p. 298–323.

ALLEN, J.R.L., 1978, Studies in fluviatile sedimentation: an exploratory quantative model for the architecture of avulsion-controlled alluvial suites: Sedimentary Geology, v. 21, p. 129–147.

BADA, G., AND HORVÁTH, F., 2001, On the structure and tectonic evolution of the Pannonian basin and surrounding orogens: Acta Geologica Hungarica, v. 44, p. 301–327.

BADA, G., HORVÁTH, F., TÓTH, L., FODOR, L., TIMÁR, G., AND CLOETINGH, S., 2006 a, Societal aspects of ongoing deformation in the Pannonian region, *in* Pinter, N., Grenerczy, Gy., Weber, J., Stein, S., and Medak, D., eds., The Adria Microplate: GPS Geodesy, Tectonics, and Hazards: NATO Science Series: IV: Earth and Environmental Sciences, Springer-Verlag, v. 61, p. 385–402.

BADA, G., GRENERCZY, GY., TÓTH, L., HORVÁTH, F., STEIN, S., CLOETINGH, S., WINDHOFFER, G., FODOR, L., PINTER, N., AND FEJES, I., 2006 b, Motion of Adria and ongoing inversion of the Pannonian basin: Seismicity, GPS velocities and stress transfer, *in* Stein, S., and Mazzotti, S., eds., Continental Intraplate Earthquakes: Science, Hazard, and Policy Issues: Geoloigcal Society of America, Special Paper 425, p. 243–262.

BÉRCZI, I., AND PHILLIPS, R.L., 1985, Processes and depositional environments within deltaic–lacustrine sediments, Pannonian Basin, Southeast Hungary: Geophysical Transactions, v. 31, p. 55–74.

BÉRCZI, I., HÁMOR, G., JÁMBOR, Á., AND SZENTGYÖRGYI, K., 1988, Neogene sedimentation in Hungary, *in* Royden, L.H., and Horváth, F., eds., The Pannonian Basin: American Association of Petroleum Geologists Memoir 45, p. 57–67.

BEHRENSMEYER, A.K., AND TAUXE, L., 1982, Isochronous fluvial systems in Miocene deposits of northern Pakistan: Sedimentology, v. 29, p. 331–352.

BERENDSEN, H.J.A., AND STOUTHAMER, E., 2002, Palaeogeographic evolution and avulsion history of the Holocene Rhine–Meuse delta, The Netherlands: Netherlands Journal of Geosciences, v. 81, p. 97–112.

BLAKEY, R.C., AND GUBITOSA, R., 1984, Controls of sandstone body geometry and architecture in the Chinle Formation (Upper Triassic), Colorado Plateau: Sedimentary Geology, v. 38, p. 51–86.

BORSY, Z., 1990, Evolution of the alluvial fans of the Alföld, *in* Rachocki, A.H., and Church, M., eds., Alluvial fans, A Field Approach, New York, John Wiley and Sons, p. 229–248.

BRIDGE, J.S., 2003, Rivers and Floodplains: Forms, Processes and Sedimentary Record: Oxford, UK, Blackwell Publishing, 486 p.

BRIDGE, J.S., AND LEEDER, M.R., 1979, A simulation model of alluvial stratigraphy: Sedimentary Geology, v. 26, p. 617–644.

BURBANK, D.W., AND PINTER, N., 1999, Landscape evolution: the interactions of tectonics and surface processes: Basin Research, v. 11, p. 1–6.

CLOETINGH, S., BUROV, E., AND POLIAKOV, A., 1999, Lithosphere folding: Primary response to compression? (from central Asia to Paris basin): Tectonics, v. 18, p. 1064–1083.

COHEN, K.M., STOUTHAMER, E., AND BERENDSEN, H.J.A., 2002, Fluvial deposits as a record for Late Quaternary neotectonic activityin the Rhine–Meuse delta, The Netherlands: Netherland Journal of Geosciences v. 81, p. 389–405.

CSERKÉSZ-NAGY Á.T., TÓTH T., AND SZTANÓ O., 2008, Reconstruction of a Pleistocene meandering river and its climatical background in the NE part of the Great Hungarian Plain by using UHR seismic images: FLAG Biennial Meeting Budapest, Abstract Book, p. 20.

CSONTOS, L. AND VÖRÖS, A., 2004, Mesozoic plate tectonic reconstruction of the Carpathian region: Palaeogeography, Palaeoclimatology, Palaeoecology, v. 210, p. 1–56.

CSONTOS, L., NAGYMAROSY, A., HORVÁTH, F., AND KOVAC, M., 1992, Tertiary evolution of the Intra-Carpathian area: a model: Tectonophysics, v. 208, p. 221–241.

Franyó, F., 1992, Thickness of Quaternary sediments 1:200 000: Manuscript, Geological Institute of Hungary.

Gábris, Gy., 1998, Late Glacial and Post Glacial development of drainage network and the paleohydrology in the Great Hungarian Plain, *in* Bassa, L., and Kertész, Á., eds., Windows on Hungarian Geography: Akadémiai Kiadó, Budapest, p. 23–36.

Gábris, Gy., 2002, A Tisza helyváltozásai (The shifting Tisza River), *in*: Mészáros, R., Schweitzer, F., and Tóth, J., eds., Jakucs László, a tudós, az ismeretterjesztő és a művész: Magyar Tudományos Akadémia Földrajzi Kutatóintézet, Pécsi Tudományegyetem kiadása, Pécs, p. 91–105.

Gábris, Gy., and Nagy, B., 2005, Climate and tectonically controlled river style changes on the Sajó–Hernád alluvial fan (Hungary), *in*: Harvey, A.M., Mather, A.E., and Stokes, M., eds., Alluvial Fans: Geomorphology, Sedimentology, Dynamics: Geological Society of London, Special Publications 251, p. 61–67.

Gábris, Gy., and Nádor, A., 2007, Long-term fluvial archives in Hungary: response of the Danube and Tisza rivers to tectonic movements and climatic changes during the Quaternary: a review and new synthesis: Quaternary Science Reviews, v. 26, p. 2785–2782.

Heller, P.L., and Paola, C., 1996, Downstream changes in alluvial architecture: an exploration of controls on channel-stacking pattrens: Journal of Sedimentary Research, v. 66, p. 297–306.

Hickson, T.A., Sheets, B.A., Paola, C., and Kelberer, M., 2005, Experimental test of tectonic controls on three-dimensional alluvial facies architecture: Journal of Sedimentary Research, v. 75, p. 710–722.

Horváth, F., 1993, Towards a mechanical model for the formation of the Pannonian basin: Tectonophysics, v. 226, p. 333–357.

Horváth, F., 1995, Phases of compression during the evolution of the Pannonian basin and its bearing on hydrocarbon exploration: Marine and Petroleum Geology, v. 12, p. 837–844.

Horváth, F., and Royden, L.H., 1981, Mechanism for formation of the intra-Carpathian basins: A review: Earth Evolutionary Sciences, v. 1, p. 307–316.

Horváth, F., and Cloetingh, S., 1996, Stress induced late-stage subsidence anomalies in the Pannonian Basin, *in* Cloetingh, S., Ben Avraham, Z., Sassi, W., and Horváth, F., eds., Dynamics of Extensional Basins and Inversion Tectonics: Tectonophysics, v. 266, p. 287–300.

Juhász, Gy., 1991, Sedimentological and lithostratigraphical framework of the Pannonian (s.l.) sequence in the Hungarian Plain, Eastern Hungary: Acta Geologica Hungarica, v. 34, p. 53–72.

Juhász, Gy., 1994, Magyarországi neogén medencerészek pannóniai s. l. üledéksorának összehasonlító elemzése (Comparison of the Pannonian s.l. sedimentary successions of the Neogene sub-basins in Hungary): Földtani Közlöny, v. 124, p. 341–365.

Juhász, Gy., Müller, P., and Tóth-Makk, Á., 2004. Alluvial architecture and fluvial cycles in Quaternary deposits in a continental interior basin, E Hungary: Geologica Croatica, v. 57, p. 171–190.

Kraus, M.J., 1987, Integration of channel and floodplain suites, II. Vertical relations of alluvial paleosols: Journal of Sedimentary Petrology, v. 57, p. 602–612.

Leeder, M.R., 1978, A quantitative stratigraphic model for alluvium, with special reference to channel deposit density and interconnectedness, *in* Miall, A.D., ed., Fluvial Sedimentology: Canadian Society of Petroleum Geologists, Memoir 5, p. 587–596.

Leeder, M.R., Mack, G.H., Peakall, J., and Salyards, S.L., 1996, First quantitative test of alluvial stratigraphic models: Southern Rio Grande rift, New Mexico: Geology, v. 24, p. 87–90.

Mackay, S.D., and Bridge, J.S., 1995, Three-dimensional model of alluvial stratigraphy: theory and application: Journal of Sedimentary Research, v. 65, p. 7–31.

Magyar, I., Geary, D.H., and Müller, P., 1999, Paleogeographic evolution of the Late Miocene Lake Pannon in Central Europe: Palaeogeography, Palaeoclimatology, Palaeoecology, v. 147, p. 151–167.

Magyari, Á., Unger, Z., Tóth-Makk, Á., Babinszki, E., Kercsmár, Zs., Koloszár, L., Marsi, I., Nádor, A., and Tullner, T., 2009, The Quaternary basement map of the Great Hungarian Plain: (abstract): 11th Mining, Metallurgy and Geology Conference, Sighetu Marmatiei, April 2–5, 2009. Abstracts, p. 162.

Miall, A.D., 1996, The Geology of Fluvial Deposits: Berlin, Springer-Verlag, 582 p.

Nádor, A., Lantos, M., Tóth-Makk Á., and Thamó-Bozsó, E., 2003, Milankovitch-scale multi-proxy records for the fluvial sediments of the last 2.6 Ma from the Pannonian Basin, Hungary: Quaternary Science Reviews, v. 22, p. 2157–2175.

Nádor, A., Thamó-Bozsó, E., Magyari, Á., and Babinszki, E., 2007, Fluvial responses to tectonics and climate change during the Late Weichselian in the eastern part of the Pannonian Basin (Hungary): Sedimentary Geology, v. 202, p. 174–192.

Nagy, Á.T., Tóth, T., and Sztanó, O., 2007, A "harmadik folyó"—pleisztocén folyóvizi üledékek ultranagy felbontású szeizmikus szelvényeken a Tisza Tiszadob-Martfű közötti szakaszán (The "third river", Pleistocene fluvial sediments on ultra-high resolution seismic images between Tiszadob and Martfű): Földtani Közlöny, v. 137, p. 239–260.

Peakall, J., 1998, Axial river evolution in response top half-graben faulting, Carson River, Nevada, U.S.A.: Journal of Sedimentary Research, v. 68, p. 788–799.

Penck, A., and Brückner, E., 1909, Die Alpen im Eiszeitalter: Leipzig, Tauchnitz.

Pogácsás, Gy., Lakatos, L., Barvitz, A., Vakarcs, G. and Farkas, Cs., 1989, Pliocén–Quarter oldaleltolódások a Nagyalföldön (Pliocene–Quaternary transcurrent fault in the Great Hungarian Plain): Általános Földtani Szemle, v. 24, p. 149–169.

Pöstyényi, F., 1993, Szeizmikus feldolgozási jelentés, Dévaványa-Kelet. Kézirat. (Report on seismic interpretation, Dévaványa-East): Manuscript. Országos Földtani és Geofizikai Adattár T. 19875.

Read, W.A., and Dean, J.M., 1982, Quantitative relationships between numbers of fluvial cycles, bulk lithological composition and net subsidence in a Scottish Namurian basin: Sedimentology, v. 29, p. 181–200.

Rónai, A., 1985, Az Alföld negyedidőszaki földtana (The Quaternary of the Great Hungarian Plain): Geologica Hungarica, Series Geologica, v. 21, 446 p.

Sanders, C., 1998, Tectonics and erosion. Competitive forces in a compressive orogene—A fission track study of the Romanian Carpathians: Ph.D. Thesis, Vrije Universitiet, Amsterdam, 204 p.

Schumm, S.A., Dumont, J.F., and Holbrook, J.M., 2000, Active Tectonics and Alluvial Rivers: Cambridge, U.K., Cambridge University Press, 276 p.

Shackleton, N.J., Berger, A., and Peltier, W.R., 1990, An alternative astronomical calibration of the Lower Pleistocene timescale based on ODP Site 677: Royal Society of Edinburgh Transactions: Earth Sciences, v. 81, p. 251–261.

Shanley, K.W., and McCabe, P.J., 1993, Alluvial architecture in a sequence stratigraphic framework: a case history form the Upper Cretaceous of southern Utah, USA: *in* Flint, S., and Byrant, I. D., eds., The Geological Modeling of Hydrocarbon Reservoirs: International Association of Sedimentologists, Special Publication 15, p. 21–56.

Shuster, M.W., and Steidtmann, J.R., 1987, Fluvial-sandstone architecture and thrust-induced subsidence, northern Green River Basin, Wyoming, *in* Ethridge, F.G., Flores, R.M., and Harvey, M.D. eds., Recent Developments in Fluvial Sedimentology: SEPM Special Paper 39, p. 279–285.

Sümeghy, J., 1944, A Tiszántúl. Magyar tájak földtani leírása VI. (Transtiscia. Geologische Beschreibung Ungarischer Landschaften. VI.): Magyar Királyi Földtani Intézet kiadása, 208 p.

Szentgyörgyi, K-né, 1988, Szarvas (Endrőd-II) lehatároló kutatási fázisú földtani zárójelentése. (Final report on the Szarvas (Endrőd-II) area): Manuscript, Országos Földtani és Geofizikai Adattár T. 14798.

Sztanó O., Tóth T., Magyari O., Magyari Á., and Horváth F., 2002, Alluvial architecture from ultra high-resoltution single-channel seismic survey of the meandering Tisza river, Pannonian Basin, Hungary (abstract): 16th International Sedimentological Congress, Abstract Volume, p. 357–359.

Sztanó O., Magyar I., and Horváth F., 2007, Changes of water depth in the Late Miocene Lake Pannon revisited: the end of an old legend: Geophysical Research Abstracts, v. 9, 38897836.

Tari, G., Dövényi, P., Dunkl, I., Horváth, F., Lenkey, L., Stefanescu, M., Szafián, P., and Tóth, T., 1999, Lithospheric structure of the Pannonian basin derived from seismic, gravity and geothermal data, *in* Durand, B., Jolivet, L.,

Horváth, F., and Séranne, M., eds., The Mediterranean Basins: Tertiary extension within the Alpine Orogen: Geological Society of London, Special Publication 156, p. 215–250.

Thamó-Bozsó, E., Kercsmár, Zs., and Nádor, A., 2002, Tectonic control on changes in sediment supply on Quaternary alluvial systems, Körös sub-basin, SE Hungary, *in* Jones, S.J., and Frostick, L.E., eds., Sediment Flux to Basins: Causes, Controls and Consequences: Geological Society of London, Special Publication 191, p. 37–53.

Thamó-Bozsó, E., Murray, A. S., Nádor, A., and Magyari, Á., 2007, Investigation of river network evolution using luminescence dating and heavy mineral analysis of Late-Quaternary fluvial sands from the Great Hungarian Plain: Quaternary Geochronology, v. 2, p. 168–173.

Timár, G., Sümegi, P., and Horváth, F., 2005, Late Quaternary dynamics of the Tisza River: Evidence of climatic and tectonic controls: Tectonophysics, v. 410, p. 97–110.

Törnqvist, T.E., 1995, Alluvial architecture of the Quaternary Rhine–Meuse system in The Netherlands—discussion: Netherlands Journal of Geosciences, v. 74, p. 183–186.

Uhrin, A., Magyar, I., Sztanó, O., and Fodor L., 2009, Shelf-margin progradation in Lake Pannon as a tool for reconstructing Late Miocene tectonic evolution of the Pannonian Basin: 7th Meeting of the Central European Tectonic Studies Group (CETeG), Pécs. Abstract Volume, p. 33.

Urbancsek, J., 1960, Az alföldi ártézi kutak fajlagos vízhozama és abból levonható vízföldtani és ősföldrajzi következtetések (Water productivity of the wells on the Great Hungarian Plain and its implications to hydrogeology and palaeogeography): Hidrológiai Közlöny, v. 1960/5, p. 398–403.

Urbancsek, J., 1965. Az Alföld negyedkori földtani képződményeinek mélyszerkezete (Geology and tectonics of the Quaternary sediments on the Great Hungarian Plain): Hidrológiai Közlöny, v. 1965/3, p. 111–124.

Windhoffer, G., Bada, G., Nieuwland, D., Wórum, G., Horváth, F., and Cloething, S., 2004, On the mechanics of basin formation in the Pannonian basin: Inferences from analogue and numerical modelling: Tectonophysics, v. 410, p. 389–415.

DEPOSYSTEMS, PALEOSOLS, AND CLIMATIC VARIABILITY IN A CONTINENTAL SYSTEM: THE UPPER TRIASSIC CHINLE FORMATION, COLORADO PLATEAU, U.S.A.

RUSSELL F. DUBIEL
U.S. Geological Survey, MS 939 Box 25046 DFC, Denver, Colorado, 80225 USA
AND
STEPHEN T. HASIOTIS
Department of Geology, University of Kansas, 1475 Jayhawk Blvd., 120 Lindley Hall, Lawrence, Kansas, 66045 USA

ABSTRACT: The Upper Triassic Chinle Formation of the Colorado Plateau in the western United States was deposited as a continental fluvial-floodplain-lacustrine-eolian depositional system near the west coast of Pangea just north of the paleoequator. The deposystem evolved in response to regional tectonics and long-term climate change. Clastic and volcanic sediment was supplied by uplifted highlands and arc magmatism on the margins of the basin. Climate changed from a Pangean megamonsoon regime in the early Late Triassic to an arid climate setting at the close of the Triassic. Chinle fluvial discharge and sediment load varied through time in response to climate, producing degradational and aggradational cycles. Degradation eroded paleovalleys, and aggradation filled them with both meandering and braided fluvial systems, as well as varying amounts of floodplain mudstones. Paleosols developed on paleovalley fills, interfluves, and floodplain siltstones and mudstones, all of which combine to define several scales of degradational and aggradational cycles. An upward stratigraphic trend in Chinle paleosols from Oxisols and Gleysols, to Alfisols (Argillisols) and Vertisols, culminating in Inceptisols (Prototsols) and Aridisols (Calcisols), supports previous interpretations that climate evolved from monsoonal or humid conditions at the beginning of Chinle deposition to progressively more arid climate conditions at the end of deposition.

KEY WORDS: sequence stratigraphy, climate, ichnofossils, fluvial-lacustrine

INTRODUCTION

By applying sequence stratigraphic principles, sequence boundaries and flooding surfaces have been used to define genetic units and their cyclic stacking patterns in continental rocks. Studies of fluvial-floodplain successions have recognized stratigraphic cycles, aggradational-degradational cycles, and fluvial aggradational cycles (e.g., Wright and Marriott 1993; Shanley and McCabe 1994; Bourquin et al. 1998; Bourquin et al. 2009; Hasiotis and Honey 2000; Atchley et al. 2004; Pace et al. 2009). These types of studies generally attributed cyclic sedimentation to allogenic and autogenic controls, including tectonics, sea level or local base level, sediment supply and discharge, or climate. Distinguishing between these controls often has been problematic.

Many studies have examined continental cyclicity using fluvial sandstone and sandbody architecture, whereas others have used floodplain mudstones and paleosols. These studies of paleosol-bearing alluvial successions also have identified cyclic, often hierarchical, deposition (e.g., McCarthy and Plint 1998; Kraus and Aslan 1999; Kraus 2002; Atchley et al. 2004), attributing larger-scale alluvial cycles to either tectonic activity, changes in eustatic sea-level, local base-level, or climate (Blakey and Gubitosa 1984; Kraus 1987, 2002; Wright and Marriott 1993; Atchley et al. 2004) and smaller-scale cycles to such processes as channel avulsion and migration (Kraus and Aslan 1999; Atchley et al. 2004; Cleveland et al., 2007). Kraus (2002) suggested that floodplain paleosols vary at several spatial and temporal scales and cited studies that presented paleosol variability at basin scales, observing that such studies require basin-scale exposures and a means to correlate units across the basin. Noting these caveats, Kraus (2002) and Kraus and Aslan (1999) indicated that floodplain- to basin-scale paleosol variability could be used to interpret controls that determine facies and architecture in continental basins.

The Upper Triassic Chinle Formation of the Colorado Plateau in the western United States is a succession of fluvial, floodplain, lacustrine, and eolian strata deposited in a large continental cratonic basin (Stewart et al. 1972; Blakey and Gubitosa 1983; Dubiel 1987a, 1994). A few previous efforts have applied sequence stratigraphic interpretations to the Chinle and identified sequence boundaries (Lucas 1997; Lucas et al. 1997; Demko et al., 1998; Demko 2003; Beer 2005), whereas others have identified cyclic fluvial-floodplain sedimentation and paleosols (Kraus and Middleton 1987; Dubiel 1987a, 1994; Therrien and Fastovsky 2000; Boucher 2004; Beer 2005; Prochnow et al., 2006; Cleveland et al., 2007, 2008). The sequence boundaries and paleosols led to interpretations that inferred controls on sedimentation by both allogenic and autogenic processes, including tectonics, base level, and climate.

This concept paper explores basin-scale stratigraphic trends in paleosol variability in the Chinle Formation. Observations of Chinle sections throughout the depositional basin indicate that paleosols in fluvial, floodplain, lacustrine, and eolian units exhibit a general trend in stratigraphic succession from predominantly Oxisols and Gleysols at the base of the Chinle, through Alfisols (Argillisols) and Vertisols, and finally to Inceptisols (Protosols) and Aridisols (Calcisols) at the top of the section. Integrating identifications of sequence boundaries and paleosols defines a hierarchy of degradational-aggradational cycles within the Chinle. The pattern of both the paleosol and degradational-aggradational cycles indicates that climate change operating at several orders of magnitude had a major influence on deposition in the basin.

STRATIGRAPHY

Triassic rocks of the Colorado Plateau (Fig. 1) generally unconformably overlie Permian sedimentary strata. Locally, the rocks unconformably overlie Pennsylvanian strata, and in a few places they lie directly on Precambrian igneous rocks of the uplifts. Triassic stratigraphy on the Colorado Plateau (Fig. 2A) is relatively straightforward due to the lack of significant structural disruption. The nomenclature applied to rocks of the same age, however, varies

From River To Rock Record: The Preservation Of Fluvial Sediments And Their Subsequent Interpretation
SEPM Special Publication No. 97, Copyright © 2011
SEPM (Society for Sedimentary Geology), ISBN 978-1-56576-305-0, p. 393–421.

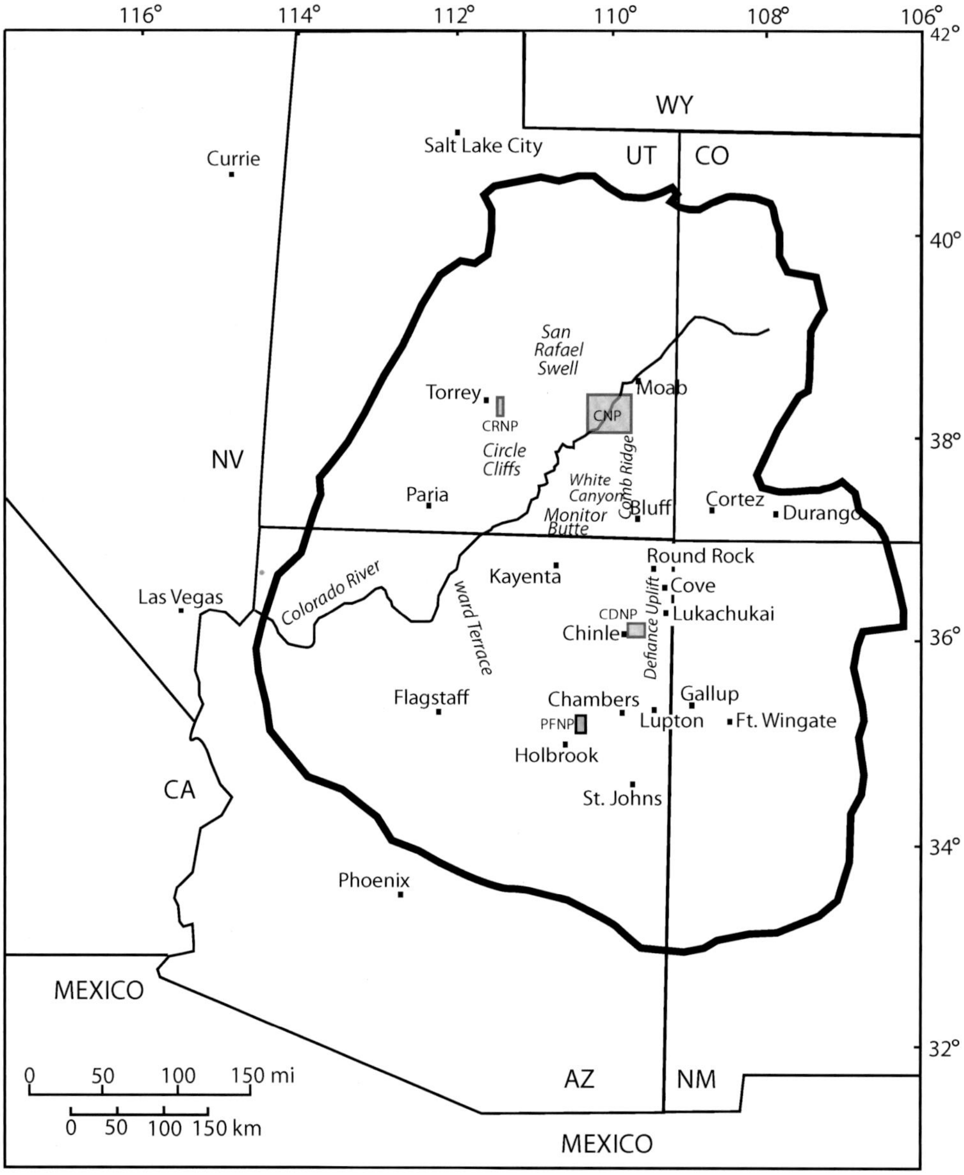

FIGURE 1.—Map of the Colorado Plateau region showing locations of towns, national parks, and major geographic features referred to in the text and in subsequent figures (CRNP, Capitol Reef National Park, Utah; CNP, Canyonlands National Park, Utah; PFNP, Petrified Forest National Park, Arizona; CDNM, Canyon De Chelly National Monument, Arizona).

between regions due to its historical development over a large geographic area that includes several distinct geologic provenances. Upper Triassic stratigraphy of the Western Interior is complex, and the reader is referred to the literature for a complete review of the history of its nomenclature on the Colorado Plateau (e.g., Stewart et al. 1972; Dubiel 1994).

Upper Triassic continental rocks are assigned to the Chinle Formation throughout the Colorado Plateau (Stewart et al. 1972). The Chinle is equivalent in part to the Dolores Formation in southwestern Colorado (Blodgett 1984, 1988). The Chinle is more than 1600 ft (500 m) thick (Fig. 2B) near the depocenter of the Chinle basin (Stewart et al. 1972; Dubiel 1989a), around the Four Corners area where Colorado, New Mexico, Utah, and Arizona meet. Basal strata were deposited in paleovalleys eroded in the underlying Moenkopi Formation and locally into the Permian De Chelly Sandstone in northern Arizona (Stewart et al. 1972; Blakey 1974). The Chinle varies in thickness in part due to the relief of the paleovalleys, to variable subsidence within the basin, and to uneven erosional truncation beneath the Triassic-Jurassic unconformity (Pipiringos and O'Sullivan 1978). In ascending order, the Chinle consists of the Shinarump, Monitor Butte, Moss Back, Petrified Forest, Owl Rock, and Rock Point or Church Rock Members (Fig. 2). Locally, a mottled unit is present at the base of the Chinle, and it is formally named the Temple Mountain Member only in the San Rafael Swell in central Utah (Fig. 1). In northern Arizona at Petrified Forest National Park (Fig. 1), numerous informal sandstone units are named within the Petrified Forest Member (Stewart et al. 1972; Therrien and Fastovsky, 2000; Martz and Parker 2010).

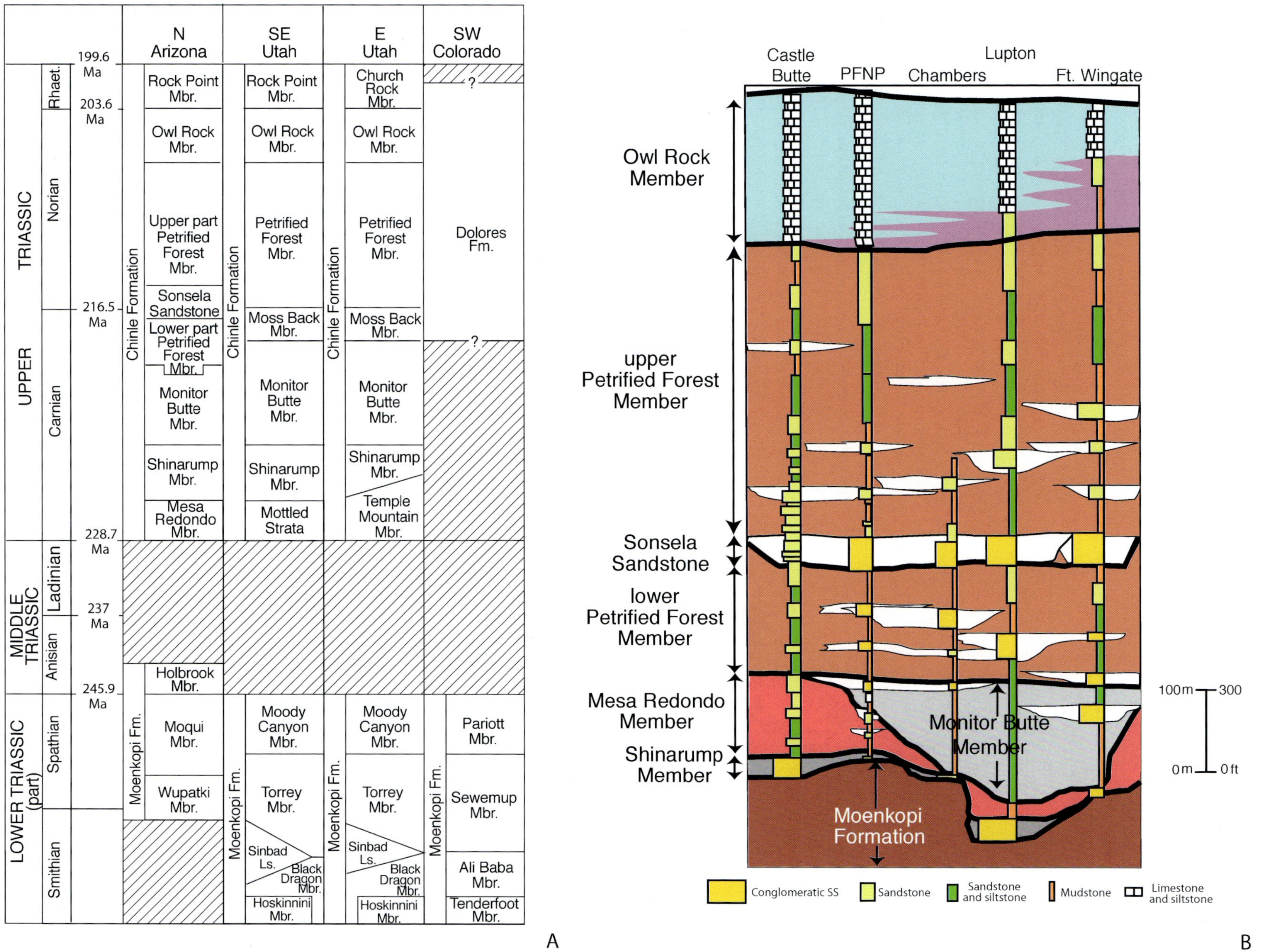

FIGURE 2.—A) Generalized stratigraphic correlation chart of Triassic strata in northern Arizona and southeastern and eastern Utah (modified from Dubiel 1994). B) Measured sections of the Chinle Formation in and adjacent to Petrified Forest National Park, northern Arizona, showing lithology and complex cut-and-fill nature of the members (modified from Dubiel et al. 1999).

Mottled strata or the Shinarump Member and equivalents typically lie at or near the base of the Chinle. The Shinarump fills paleovalleys eroded into the underlying rocks, and thus it has an irregular distribution and is not present everywhere (Blakey and Gubitosa 1983; Dubiel 1987a, 1991). The distinctive mottling in the Temple Mountain Member at the base of the Chinle in the San Rafael Swell is now considered to be pedogenically modified Moenkopi and Shinarump strata, similar to the mottled strata of other areas (Stewart et al. 1972; Dubiel 1987a; Dubiel et al. 1991; Beer 2005).

The Monitor Butte Member, the laterally equivalent Mesa Redondo Member, or the lower part of the Petrified Forest Member overlie the Shinarump. These units were deposited over the Shinarump as the basal paleovalleys filled, and they extend over a much wider geographic area, as do the succeeding members of the Chinle. The Moss Back Member is restricted to the paleovalley cut in underlying Monitor Butte strata in southeastern Utah (Stewart et al. 1972; Blakey and Gubitosa 1983; Dubiel et al. 1991; Dubiel 1994; Beer 2005). The Petrified Forest Member overlies the Moss Back and is widespread on the Colorado Plateau. The Sonsela Sandstone occurs stratigraphically about in the middle of the member in northern Arizona and northwestern New Mexico. Recent work suggests that the Sonsela in Petrified Forest National Park can be subdivided into several units with local names (Woody 2003, 2006; Martz and Parker 2010).

The Owl Rock Member overlies the Petrified Forest and also extends over a wide region on and beyond the Colorado Plateau. The Owl Rock is recognized as a sequence of distinctive pastel-colored siltstones and limestones extending from the Colorado Plateau into western Utah, northern Arizona, and southern Nevada. In the Four Corners area, the Church Rock or Rock Point Members of the Chinle Formation overlie the Owl Rock (Stewart et al. 1972; Blakey and Gubitosa 1983; Dubiel 1989a; Hasiotis 1991). O'Sullivan (1970) and Dubiel (1989a) discussed the complexities and uncertainties regarding Church Rock and Rock Point stratigraphy.

PALEOGEOGRAPHIC AND PALEOCLIMATIC SETTING

During the Triassic, the supercontinent Pangea was at its greatest size—a single landmass distributed equally between the northern and southern hemispheres (Fig. 3A) (Ziegler et al. 1983; Parrish 1985). This configuration maximized monsoonal circulation (Parrish and Peterson 1988; Peterson 1988; Kutzbach and Gallimore 1989), a global climatic pattern that had a strong influence on the distribution of continental deposits both in coastal and interior basins (Dubiel et al. 1991). During the Triassic, what is now the Colorado Plateau of the western United States was near the western equatorial coast of Pangea (Fig. 3A), where the effects of the tropical monsoonal climate were pronounced. Triassic rocks throughout the western United States (Fig. 3B) record the effects of tectonics, sea-level fluctuations, and climate on sedimentation. Concomitant with the end of Early Triassic marine

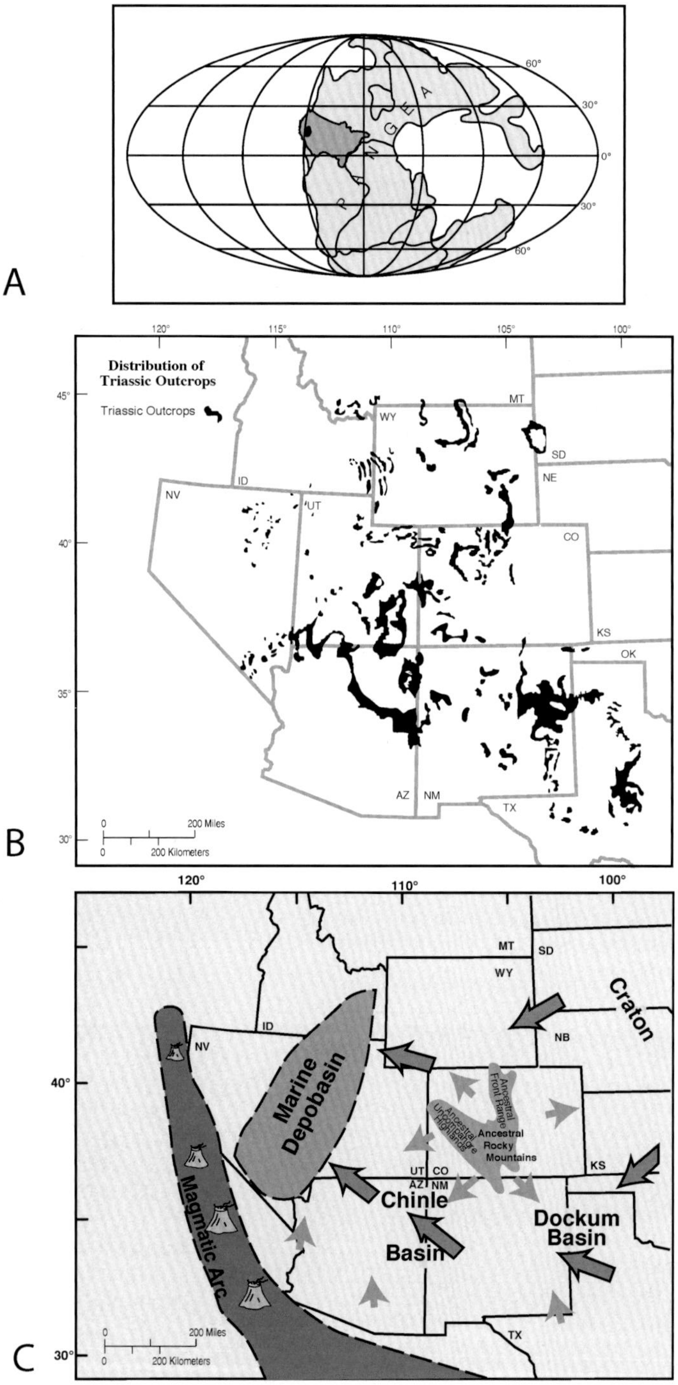

FIGURE 3.—A) Schematic reconstruction of Pangea showing the approximate Triassic location of the Colorado Plateau (black) and the United States (light shading) (modified from Dubiel et al. 1991). Derived from reconstruction of Induan (earliest Triassic) paleogeography by Ziegler et al. (1983). B) Map of the Western Interior of the United States showing generalized Triassic rock outcrops (modified from Dubiel 1994). Note three primary Triassic outcrop areas: the Colorado Plateau, the Dockum Basin in New Mexico and west Texas, and the states of Wyoming, northeast Utah, and northwest Colorado. By comparing outcrop distribution with the paleogeographic reconstructions in Figure 11, one can ascertain

←

the degree of control in reconstructing individual depositional systems. Outcrops compiled from 1:500,000-scale and 1:1,000,000-scale geologic maps of each state and other sources including: McKee et al. 1959; MacLachlan 1972; Stewart et al. 1972; McGowen et al. 1983; and Dubiel 1989b, 1992. C) Map showing general tectonic setting of the Triassic Western Interior (compiled and modified from sources including: Dickinson 1981; Blakey and Gubitosa 1983; Dubiel 1987a, 1989b; and Dubiel et al. 1991) as well as general pattern of major fluvial systems in the Triassic.

sedimentation, an evolving magmatic arc along the west margin of Pangea produced a Late Triassic continental basin (Lawton 1994), contributed volcanic detritus (Riggs et al. 1994; Riggs et al. 1996; Riggs et al. 2003), and recycled uplifted, upper Paleozoic clastic detritus to the region from the south and from the ancestral Rocky Mountains (Fig. 3C).

The Lower and Middle Triassic Moenkopi Formation was deposited in marine to marginal-marine settings on the west coast of Pangea that graded eastward into sabkha, deltaic, coastal-plain, and continental beds sourced from the craton on the east and from uplifted highlands (Blakey 1974, 1989; Blakey et al. 1993). Middle Triassic rocks on the Colorado Plateau are only locally preserved due to nondeposition or extensive erosion (Stewart et al. 1972). Parrish (1993a) stated that the Triassic would have experienced the greatest influx of seasonal moisture due to the development of the megamonsoon. It would have required abundant precipitation and fluvial flow to the coast to transport and deposit the volume of sediment represented by the Moenkopi coastal-plain and deltaic strata (Blakey 1974, 1989). The dramatic increase in Moenkopi deltaic and nearshore-marine clastic sediments shed from the continent may herald the predicted increase of rainfall and discharge that delivered sediment to the coast in comparison to the prior Permian arid climate depositional systems on the west coast of Pangea (Blakey 1996; Dubiel et al. 1996).

Upper Triassic rocks of the western United States were deposited in continental rift-rim or back-arc basins located about 5–15° north of the paleoequator near the west coast of Pangea (e.g., Van der Voo et al. 1976; Dickinson 1981; Pindell and Dewey 1982; Bazard and Butler, 1991; Lawton 1994). An evolving magmatic arc on the western margin of the Triassic continent (Fig. 3C) provided volcanic ash and clastic sediment to the Chinle depositional basin (Stewart et al. 1972; Blakey and Gubitosa 1983; Busby-Spera 1988). Clastic sediment was supplied northward and northeastward from southern highlands that lay in southern Arizona and northern Mexico near the volcanic arc, and southward, westward, and northwestward from the ancestral Rocky Mountains (Fig. 3C) (Stewart et al. 1972; Blakey and Gubitosa 1983; Dubiel 1987a, 1989a, 1989b).

Upper Triassic continental deposition covered the Colorado Plateau and extended to the east and north onto the craton. Upper Triassic continental clastics of the Dockum basin in eastern New Mexico and western Texas (Fig. 3C) were sourced in the Ouachita uplift on the south and east (Dickinson 1981; McGowan et al. 1983; Lehman and Schnabel 1992). In addition, the volume of sediment supplied to the basin as well as evidence from paleocurrent indicators suggest that sediment was also derived from the craton westward through the Dockum basin (Blakey and Gubitosa 1983, 1984), as well as southwestward from the Ouachita Mountains. Good and Dubiel (1992) proposed, on the basis of sandbody size and architecture, that Chinle fluvial systems must have had a drainage area larger than just the Chinle basin and that it probably included the Dockum basin. Supporting this interpretation, Lucas (1993a, 1993b), on the basis of stratigraphic continuity from the Chinle basin to the Dockum basin, indicated that the two were connected. Good (1993) provided evidence of molluscan faunal continuity between Chinle and Dockum bivalves. Detrital Cambrian zircons in basal Chinle Formation fluvial conglomerates in eastern Nevada (Gehrels et al. 1993) indicate a fluvial provenance east of both the Chinle and Dockum basins, a strong possibility in that the craton to the north and east of the Chinle and Dockum basins exposed Permian strata and granitic basement that could have been sources of the clastic detritus to the Dockum and Chinle basins. Riggs et al. (1996) and Dickinson and Gehrels (2008) provided further evidence, also based on detrital zircons, that Chinle depositional systems, and in particular the Chinle trunk river system, extended significantly to the east of the two depositional basins and onto the craton.

Late Triassic deposystems on the Colorado Plateau are dominated by continental strata that include stream, floodplain, lake, and marsh deposits, and at the close of the Triassic by playa-lake, loess, eolian sand-sheet, and eolian dune deposits (Stewart et al. 1972; Blakey and Gubitosa 1983; Dubiel et al. 1991; Dubiel 1994). The prominence of Late Triassic arid climate continental systems in the northern Western Interior is a result of the slightly higher paleolatitudes of this region on Pangea, a location in the middle latitudes just north of the greatest influx of monsoonal precipitation. Triassic continental systems in the Chinle basin that are dominated by streams, lakes, and marshes indicate abundant precipitation, in contrast to preceding Permian Cutler Group and subsequent Early Jurassic Wingate Sandstone deposystems, which were dominated by eolian ergs, arid-region lakes, and arid-climate coastal-marine systems (Blakey 1994, 1996; Dubiel et al. 1996). This larger-scale climatic change indicates two major controls: (1) the northward migration of the Colorado Plateau from paleolatitudes at 0–5° north of the paleoequator in the Permian to about 30° north of the paleoequator in the Jurassic; and (2) the demise of the Pangean megamonsoonal climate, associated with the breakup of Pangea in the latest Triassic and earliest Jurassic.

DEPOSITIONAL ENVIRONMENTS

Sedimentological studies of the Chinle Formation provide evidence that bears on the interpretation of depositional environments and paleoclimate during the Late Triassic. In contrast to the predominance of eolian deposition on the Colorado Plateau in the Permian and Jurassic Periods (e.g., Peterson, 1988; Blakey, 1996), Late Triassic deposition was characterized by a fluvial-deltaic-lacustrine system of extensive marshes, bogs, rivers, floodplains, and lakes (e.g., Stewart et al. 1972; Blakey and Gubitosa 1983; Dubiel 1987a, 1989b; Dubiel et al. 1991; Hasiotis 1991; Dubiel 1994; Demko 1995a, 1995b; Beer 2005). Previous reports have described Chinle lithofacies and interpreted depositional environments (Table 1) (Stewart et al. 1972; Blakey and Gubitosa 1983; Dubiel 1987a, 1991, 1994; Demko et al. 1998; Tanner 2003; Beer 2005; Prochnow et al. 2005; Prochnow et al. 2006; Cleveland et al. 2007); the depositional setting of stratigraphic units, significant sequence stratigraphic surfaces, and paleosols are summarized below.

Sedimentological studies throughout the Colorado Plateau have documented a complex fluvial-lacustrine system in a climate with significant precipitation but punctuated by seasonally dry periods, with an increase in arid climate conditions and environments in the upper part of the Chinle (Blakey and Gubitosa, 1983; Dubiel 1987a, 1987b, 1989a, 1994; Dubiel et al. 1991; Hasiotis 1991; Dubiel and Hasiotis 1994a, 1994b). Similar interpretations of wet conditions punctuated by dry periods were proposed for the Chinle in Petrified Forest National Park, Arizona (Bown et al. 1983; Therrien and Fastovsky 2000), in the area around Moab, Utah (Cleveland et al. 2007), and for the correlative Upper Triassic Dolores Formation in southwestern Colorado (Blodgett 1984, 1988). Other studies have highlighted the arid nature of Chinle climates, but still noted the seasonality of precipitation (Tanner 2003; Tanner and Lucas 2006). Although all of these studies describe the overall trend toward more arid conditions upward in the Chinle, continued study suggests that there were not only seasonal variations in moisture and precipitation and a long-term drying trend, but also intermediate-term variations in wet and dry conditions. This pattern is expressed in the variation of fluvial and floodplain deposition and pedogenic modification and bioturbation in paleosols and crevasse splays (Kraus and Middleton 1987; Hasiotis 1991; Hasiotis and Mitchell 1993; Hasiotis et al. 1993; Hasiotis and Dubiel 1994, 1995a, 1995b; Therrien and Fastovsky 2000; Beer 2005).

Sub-Chinle Unconformity

The Chinle Formation fills paleovalleys eroded into the underlying Moenkopi Formation, recognized by the scale of the incisions that

TABLE 1.—Summary of Chinle Formation Lithofacies.

Lithofacies	Sedimentary structures	Interpretation
Yellow, gray, to white, coarse-grained to conglomeratic quartzose sandstone	Thick-bedded, horizontally laminated and massive to planar-tabular crossbeds	Bedload deposition on fluvial in-channel bars; typical in paleovalleys; clasts may be extrabasinal quartz, quartzite, and chert, or intrabasinal carbonate nodules
Yellow, gray, to white, medium-grained, quartzose sandstone	Thin- to thick-bedded, medium- to large-scale trough cross-stratification, locally associated with epsilon cross stratification	Mixed-load and suspended-load deposits in sinuous fluvial channels, lateral-accretion stratification
Yellow, gray, to white medium-grained, quartzose sandstone	Massive, locally associated with epsilon cross stratification	Mixed-load and suspended-load deposition in sinuous fluvial channels, lateral-accretion stratification
Yellow-orange, medium-grained quartzose sandstone	Trough-crossbedded, associated with epsilon cross stratification	Mixed-load and suspended-load deposits in sinuous fluvial channels, lateral-accretion stratification
Brown to gray, medium-grained sandstone	Planar-tabular and medium- to large-scale trough cross-stratification, with less abundant horizontal lamination and locally abundant, overturned trough crossbeds	Fluvial channel deposition in braided (low-sinuosity) river to meandering (high-sinuousity) river system
Green, limestone nodule-bearing conglomerate	Thin- to thick-bedded, horizontally laminated, and massive to planar-tabular crossbedded	Channel-lag deposits created by migration of the cutbank cannibalizing carbonate-nodule-rich paleosols; typical in crevasse splays and small streams
Green, fine- to medium-grained sandstone	Steeply dipping, thick bedded, large-scale cross stratification	Delta-front foreset beds deposited in lacustrine Gilbert-type deltas
Green, fine- grained sandstone	Thin-bedded, horizontally laminated and climbing-ripple cross-stratified, capped with symmetrical ripples	Crevasse-splay deposition associated with high flow regime
Green, fine-grained sandstone	Very thin-bedded, interbedded with red mudstone	Crevasse-splay deposition associated with low flow regime
Green, interbedded sandstone and siltstone	Thin to very thin bedded, horizontally to ripple laminated	Levee deposition, produced by overbank flooding
Tan, calcareous fine-grained sandstones	Thin-bedded, horizontally laminated or massive	Lacustrine deposits
Reddish-orange, fine-grained sandstone	Large-scale, planar crossbeds; surfaces locally with adhesion ripples and mudcracks	Eolian sand sheets and small eolian dunes
Red to brown siltstone	Massive	Deposition on levee and floodplain; pedogenically modified
Reddish-orange siltstone	Massive to weakly horizontally stratified	Wind-blown loess; locally pedogenically modified
Red, lavender, gray, and purple mudstone	Thin- to thick-bedded, horizontally laminated to massive	Levee and floodplain deposition; locally pedogenically modified
Orange-brown pelleted mudstone	Thin- to thick-bedded	Reworked, pedogenically modified mudstone deposited in lake-margin and sublittoral lake setting
Gray, carbonaceous mudstone	Thin- to thick-bedded, horizontally laminated to massive; with fragmented and whole carbonized plant fossils	Deposition in dysoxic abandoned channels and floodplain scour
Gray to purple mudstone	Thin- to thick-bedded, horizontally laminated to massive	Oxic to dysoxic back-swamp, marsh, floodplain lake, or perennial open lake
Black mudstone	Thin- to thick-bedded, laminated to massive	Dysoxic to anoxic, organic-rich lake and marsh
Coal	Laminated to massive with or without plant fragments	Bogs formed in abandoned channel and perennial marshe and swamp
Tan to gray limestone	Structureless, micritic, thin to very thick bedded, locally with hematite-stained clay.	Well-oxygenated floodplain lake or perennial open lake
Tan to gray and green limestone with coalesced carbonate nodules	Structureless, micritic, thin to very thick bedded with hematite-stained clay	Subaerially exposed, pedogenically modified carbonate deposition in intermittent open lake

contain Shinarump sandstones and the onlapping relation of overlying stratigraphic units (Stewart et al. 1972; Blakey and Gubitosa 1983, 1984; Dubiel 1991). Blakey (1974) reported that a scour in the Circle Cliffs area of central Utah is 300 ft (100 m) deep and 5 mi (3 km) across. Locally, in the southern Circle Cliffs and at Monitor Butte in northern Arizona, the basal beds of Shinarump contain clasts of Moenkopi siltstone as much as 2 ft (65 cm) across. The Moenkopi clasts in the Shinarump were derived from the adjacent walls or floor of the paleovalley.

Mottled Strata

Mottled strata typically occur at the top of the Moenkopi Formation and locally in older rocks that underlie the Chinle Formation. Commonly they occur as the basal unit of the Chinle Formation (Fig. 4), and locally similar mottling occurs stratigraphically higher in the Chinle in the Monitor Butte Member. Collectively these units are referred to as the mottled strata, mottled member, purple mottled unit, or Temple Mountain Member of the Chinle Formation. The mottled strata are generally silicified and are characterized by large, irregularly shaped mottles of dark purple, lavender, yellow, and white (Fig. 4A-F). The mottled coloration extends into the Moenkopi as much as 25 ft (7.5 m), decreasing downward in intensity. Characteristic features in these mottled units include indistinct horizonation, rare rhizoliths, and large cylindrical traces. As observed in thin section and identified by x-ray diffraction, the mottled coloration is due to variation in the relative concentrations and composition of iron-bearing minerals (Dubiel 1987a). The dark-purple areas contain coarse-grained hematite cement that has completely replaced the rock matrix, leaving only quartz sand and silt grains, and locally abundant kaolinite. In the lavender mottles, hematite has only partially replaced the clay and cement matrix and the hematite grain size is finer. Limonite and goethite (iron oxyhydroxides) color the matrix in yellow mottles, and hematite is completely absent in the white areas (Dubiel 1987a).

Ubiquitous in the mottled strata are large cylindrical trace fossils that are as much as 4 in. (14 cm) in diameter and as much as 5 to 10 ft (2–3 m) long (Fig. 4B, F). Although generally restricted to the mottled strata, the trace fossils locally extend downward into underlying sandstones of the Chinle (Fig. 3F) and in some places as much as 4 ft (1.3 m) into the underlying Moenkopi (Fig. 4B). Where present in the Shinarump Member sandstones at the base of the Chinle, bedding is obscured, and the mottles are locally oriented vertically, where they are aligned along the large trace fossils (Fig. 4F).

The repeated episodes of bioturbation and multiple episodes of alternating oxidation-reduction have obscured much of the original sedimentary fabric. The mottled purple and white coloration at the top of the Moenkopi Formation is restricted to the flanks of the paleovalleys that are filled with Chinle deposits (Dubiel 1987a; Beer 2005), whereas the mottled Shinarump sandstones occupy the lowest parts of the paleovalley fill.

Shinarump Member

The Shinarump Member is characterized by white to yellow and gray, medium- to coarse-grained and conglomeratic sandstone that displays complex cut-and-fill structures, lenticular internal scour surfaces, and locally abundant large-scale lateral-accretion bedding (Fig. 5 A–F). The Shinarump unconformably overlies the Moenkopi or older rocks (Stewart et al. 1972). Generally, the unconformity is a broad paleovalley, but locally it is a narrow scour and rarely a narrow slot-like incision (Fig. 5A). The Shinarump is a laterally extensive tabular unit or forms isolated lenticular sandbodies that are isolated or that can be stacked into a composite sandbody (Fig. 5B–E) (Blakey and Gubitosa 1984). Large-scale trough cross-stratification and less abundant tabular-planar cross-stratification, as well as lateral-accretion stratification with crossbed paleocurrent directions perpendicular to inclined bedding are common features (Dubiel 1987a, 1994). Although characterized by these sedimentary structures, many outcrops are massive in their entirety. Sandstones grade laterally into siltstone and mudstone that contain carbonized, whole and fragmented plant fossils.

The Shinarump was deposited in fluvial systems under conditions of rising base level in the lowest parts of paleovalleys cut into the Moenkopi Formation or older rocks (Stewart et al. 1972; Blakey and Gubitosa 1983, 1984; Dubiel 1987a, 1994). The transition from massive, conglomeratic, and tabular-planar stratified sandstone into medium-grained, trough cross-stratified sandstone represents a change from primarily bedload deposition in low-sinuosity streams with transverse bars to mixed-load deposition in low-sinuosity fluvial systems dominated by sand waves and local point bars (Blakey and Gubitosa 1984; Dubiel 1987a, 1991, 1994). The basal Shinarump was deposited as a confined-valley fluvial system whose lateral migration was impeded (Blakey and Gubitosa 1984). As the valleys were filled, Shinarump streams were less confined and migrated laterally between the valley walls (Blakey and Gubitosa 1984; Dubiel 1991).

The mapped pattern of Shinarump outcrops on the Colorado Plateau, as well as paleocurrent measurements from channel systems that can be traced across three-dimensional canyon cuts, provides evidence for the existence of both trunk river systems and tributary fluvial systems. The major fluvial axis of Shinarump deposition in the Chinle basin was to the northwest through the Defiance uplift and the Four Corners area (Stewart et al. 1972; Blakey and Gubitosa 1983; Dubiel 1987a, 1989b, 1994). Tributary fluvial systems flowed off the ancestral Rocky Mountain highlands as well as the magmatic arc to the south and converged with the trunk fluvial system (Fig. 3C). The Dockum basin and the craton to the east were also tributary drainage areas of Chinle fluvial systems. A second trunk Shinarump system trended northward and westward from the ancestral Rockies in northwestern Colorado (Stewart et al. 1972; Shropshire 1974; Dubiel 1992; Dubiel et al. 1992). Recent reconstructions based on analysis of detrital zircons support the interpretation that these trunk river systems drained a large portion of the continental interior (Riggs et al. 1996; Dickinson and Gehrels 2008).

Mottled strata formed on Shinarump fluvial sands within the paleovalleys were locally cut out by overlying Shinarump fluvial sandbodies, indicating that those paleosols with crayfish burrows developed during periods of landscape stability before the site was reoccupied by active fluvial deposition. Where the sandbodies contain abundant crayfish burrows, the sediments were located adjacent to an active channel that experienced alternating, and possibly seasonal, fluctuations in water table. Presumably, the lateral migration of Shinarump fluvial channels within the valley walls reworked much of the fine-grained sediment that may have been deposited, so that pedogenic modification affected only those coarse-grained fluvial deposits.

The upper contact of the Shinarump at Capitol Reef National Park (CRNP) and at Monitor Butte in southeast Utah along the San Juan River (Fig. 1) was illustrated by Stewart et al. (1972) as being characterized by erosional scours as much 20 to 30 ft (6–10 m) deep. At other locales in southeastern Utah, the upper part of the Shinarump displays the deeply weathered mottled paleosols. Dubiel (1991), Demko (2003), and Beer (2005) interpreted the upper contact of the Shinarump as a regionally extensive sequence boundary. This interpretation implies that the aggrading fluvial conditions that deposited the Shinarump changed to a degradational regime in which there was nondeposition or removal of Shinarump strata.

Monitor Butte Member

The Monitor Butte Member overlies the Shinarump throughout much of the Colorado Plateau. The Monitor Butte contains yellow,

FIGURE 4.—Photographs showing purple and white mottled Oxisols, Gleysols, and lateritic paleosols, including plinthite, in the uppermost Moenkopi Fm. and lowermost Chinle Fm. A) Purple- and white-mottled Oxisol developed on the Moenkopi and overlain by slope-forming Monitor Butte Member at Millard Canyon, western CNP. There is no basal Shinarump fluvial sandstone at this locality, which is interpreted as an interfluve area between trunk rivers in the Shinarump drainage system. B) Shinarump fluvial sandstone unconformably overlying the Moenkopi at Long's Canyon on the west side of CRNP. The purple and white Oxisol is developed primarily on Shinarump strata. The Shinarump is also extensively bioturbated by 3–6 ft (1–3 m) long crayfish burrows that extend down into the underlying Moenkopi, indicating that the Moenkopi was not lithified at the time of Shinarump deposition or pedogenic modification. C) Lateritic pedogenic modification of successive sandstone and siltstone units at the base of the Chinle at White Canyon, Utah. Note dark-red nodular iron concretions. Hammer and glasses for scale. D) Oxisol and plinthitic texture developed on the Mesa Redondo Member basal strata of the Chinle in PFNP. Note reticulate pattern.

brown, and red sandstone and siltstone; gray to black, organic-rich mudstone; green, bentonitic, silty sandstone and mudstone; red, calcareous mudstone; and tan to pink and green limestone. These diverse lithofacies represent complexly interfingered continental depositional environments. The Monitor Butte overlies the scour surfaces or the mottled paleosol at the top of the Shinarump fluvial sandstones.

The most prominent facies of the Monitor Butte Member contains green, bentonitic sandstones, siltstones, and mudstones (Fig. 6A–D). These strata commonly exhibit large-scale foreset bedding as much as 50 ft (15 m) thick, lateral accretion bedding as much as 25 ft (8 m) thick, and sets of shallow dipping, coarsening-upward beds with abundant climbing-ripple lamination (Dubiel 1987a, 1994). The green beds contain finely comminuted black plant fragments and intact plant specimens, including true ferns (Ash et al. 1982) and casts of giant horsetails (*Equisetites* sp.) as much as 8 in (20 cm) in diameter (Holt 1947; Ash 1967, 1972).

The black mudstones typically contain conchostracans, fish scales, and fish bones, whereas the gray mudstones have a lower organic carbon content and contain abundant calcareous ostracodes and plant fragments (Dubiel 1987a). Black mudstones are as thick as 50 ft (15 m) and in places contain coal beds as thick as 10 in (13 cm) (Fig. 6E, F). Locally, the mudstones are interbedded with tan, thin-bedded limestones that contain unionid bivalves and small vertebrate bones (Dubiel 1987a, 1987b; Parrish and Good 1987). Overlying the organic-rich strata are laterally extensive gray to red mudstones (Fig. 6G) that exhibit pedogenic horizonation, reddened and clay-rich profiles, crosscutting curvilinear fractures, slickensided angular peds, and locally contain carbonate nodules.

The black mudstones and thin coals in the basal part of the Monitor Butte Member represent freshwater bog and peat deposits, deposited in paludal and lacustrine marsh settings. The red and gray mudstones are paleosols developed on fluvial floodplain and lacustrine deposits. They exhibit reddened, clay-rich B-horizons with prismatic and subangular to angular blocky ped structures typical of Alfisols (Argillisols; Mack et al. 1993). Elsewhere these red mudstones exhibit abundant slickensides and crosscutting pseudoanticlinal features composed of subangular to prismatic peds and wedge-shaped ped structures. Such typical paleosol features resulted from paleosol movement due to alternate wetting and drying, and a relatively high percentage of smectite clays. The top of the Monitor Butte is marked by a regionally extensive red mudstone as much as 15 ft (5 m) thick that exhibits blocky to angular prismatic fabric, clay slickensides on ped faces, crosscutting curvilinear fractures, trace fossils assigned to *Scoyenia gracilis*, locally abundant crayfish burrows assigned to *Camborygma eumekenomos*, and isolated carbonate nodules (Dubiel 1987a, 1991, 1994). This series of paleosols was developed on Monitor Butte sediments in interfluve areas under degradational conditions associated with the cutting of the paleovalley that contains the overlying Moss Back fluvial sandstones (Dubiel 1994; Beer 2005). (See following section on Paleosols and Fig. 10 for description and discussion of paleosol features).

The Monitor Butte Member was deposited as an extensive system of fluvial and deltaic distributary channels and splays, and lacustrine delta deposits (Dubiel 1983, 1987a, 1994). Fluvial and deltaic systems prograded into these lakes and filled the remnant topography of the Moenkopi paleovalleys. Bentonitic clays and the presence of altered lithic clasts (Schultz 1963) and relict glass shards (Waters and Granger 1953) indicate that volcanic ash was a major component of the sediment, probably supplied from the magmatic arc that lay off to the west of the Chinle basin (Fig. 3C). The green strata of the Monitor Butte are restricted to the central parts of the paleovalley fills and overlie the scours in the top of the Shinarump fluvial sandstones, indicating that lake development was likely related to rise in base level in the fluvial system. One possible mechanism for such a rise in base level (or lake level) may have been lateral migration or aggradation of the major trunk system and damming and flooding of the tributary drainages. High groundwater tables and lacustrine-marsh environments indicate rapid channel aggradation and subsequent damming of tributary streams to form dendritic lakes. These types of lakes were suggested by Pierson (1984) to be similar to those formed in the last century on the Samala River in Guatemala (Kuenzi et al. 1979). This same process can be observed in many modern tropical river systems, such as the Amazon in South America (Fig. 6 H) (Archer 2005), which contains examples of tributary rivers that are flooded by lakes due to aggradation or migration of adjacent trunk rivers. At the close of Monitor Butte deposition, degradation cut a paleovalley into the top of the lacustrine and fluvial strata, and laterally extensive Vertisols, calcic Vertisols, and calcic Argillisols developed on the interfluves (Dubiel 1987a, 1991; Beer 2005).

Moss Back Member

The Moss Back Member crops out in a wide belt between White Canyon, Utah, and Canyonlands National Park (CNP), Utah (Figs. 1, 7A–D) (Stewart et al. 1972; Blakey and Gubitosa 1983; Dubiel 1987a, 1991, 1994; Beer 2005). The Moss Back overlies the Monitor Butte Member (Fig. 2) in the area of White Canyon, and it overlies the Moenkopi Formation to the north in Canyonlands National Park and near Moab, Utah. The Moss Back consists of brown to gray medium-grained sandstone and carbonate-nodule conglomerate as much as 100 ft (30 m) thick. It fills the broad Cottonwood paleovalley cut into the underlying Monitor Butte Member (Fig. 7A) (Blakey and Gubitosa 1983). Sedimentary structures include abundant tabular-planar and large-scale trough crossbedding and less abundant horizontal lamination. The sandbodies are lenticular but are stacked into a broad sheet that exhibits multiple internal scours. Large-scale lateral accretion stratification as thick as 60 ft (20 m), with large and small cross beds that trend at a high angle to the inclined bedding, is present in White Canyon, Utah (Fig. 7B) and the southern part of Canyonlands National Park.

Lithology and sedimentary structures in the Moss Back indicate that deposition was by fluvial processes in a northwest-trending trunk river that aggraded and filled the paleovalley in the top of the Monitor Butte Member (Stewart et al. 1972; Blakey and Gubitosa 1983). Overall aggradation, however, was punctuated by variations in style from bed-load-dominated low-sinuosity rivers to bed-load and suspended-load high-sinuosity rivers. The lenticular, coarse-grained sandstones and conglomerate were deposited primarily as bedload in low-sinuosity channels, but the thick lateral accretion sets indicate that at times these

←

Circled scale is 4 in (10 cm). E) Purple and white lateritic paleosol with plinthite nodules developed on basal Chinle sandstone and extending down about 6 ft (2 m) into Moenkopi at St. Johns, Arizona. Pocket knife (4 in; 10 cm) for scale. F) Strongly developed vertical mottles in Shinarump fluvial sandstone at Buckhorn Wash in the northern San Rafael Swell, UT. The vertical iron concentrations are adjacent to and surround white iron-depleted zones that are aligned along 6–8 ft (2–3.5 m) long crayfish burrows.

FIGURE 5.—Photographs of the Shinarump Member of the Chinle Formation. A) At Canyon De Chelly National Monument in northeastern Arizona, the Shinarump unconformably overlies the Permian De Chelly Sandstone along the White House Ruins trail and fills a steep-sided paleo-gully cut into the eolian dune sets of the De Chelly. This fill indicates that the Shinarump trunk fluvial system at this locality was deposited in a preexisting paleovalley that had small gullies eroded into the flanks of the valley walls. Subsequent rise of baselevel and Shinarump fluvial deposition filled the paleovalley to and above the level of the paleogulley. Geologist circled for scale. B) Shinarump fluvial sandstones unconformably overlying marine siltstones of the Moenkopi Formation on the west side of Capital Reef National Park, Utah. The Shinarump is more than 60 ft (20 m) thick and exhibits both large-scale downstream-accreting bar forms and lateral-accretion bedding, indicating a succession of both braided and meandering streams. C) Shinarump fluvial sandstones at Colt Mesa in the southern Circle Cliffs, Utah. The 75 ft (25 m)-thick sandstone comprises three stories: the lower third exhibits distinct lateral accretion strata about 25 ft (7.5 m) thick

formed in a high-sinuosity river (Dubiel 1994; Hazel 2000). Carbonate-nodule-bearing paleosols in the underlying Monitor Butte Member were the source of carbonate clasts in Moss Back conglomerates.

Petrified Forest Member

Lavender and brown sandstones and variegated mudstones of the Petrified Forest Member overlie the Moss Back Member. The Petrified Forest Member is present throughout the Colorado Plateau, with the thickest exposures near the town of Chinle and at Petrified Forest National Park (PFNP) in northern Arizona (Figs. 1, 8A–D) (Stewart et al. 1972). Bentonitic sandstones and thin carbonate-nodule conglomerates typically exhibit large-scale trough cross-stratification, large-scale internal scour surfaces, and lateral-accretion bedding. Sandstones grade laterally into red, gray, and purple mudstones that contain a variety of pedogenic structures (platy, blocky, subangular blocky) and pedogenic carbonate nodules. Interfingered with the sandstones are finer grained bentonitic strata that contain abundant vertebrate remains, lungfish toothplates, gastropods, and unionid bivalves (Stewart et al. 1972; Dubiel 1987a; Parrish and Good 1987). Isolated outcrops of black, organic-carbon-rich and conchostracan-bearing mudstone are present in the member.

Originally the Petrified Forest Member consisted of the lower and upper members separated by the Sonsela Sandstone Bed (Stewart et al., 1972; Deacon, 1990). Recent work in PFNP has highlighted the stratigraphic complexity of fluvial sandstones and associated floodplain mudstones in the Petrified Forest Member and in the Sonsela (Therrien and Fastovsky 2000; Woody 2003, 2006; Parker 2006; Martz and Parker 2010). Sedimentologic and paleontologic studies have now recognized numerous informally named fluvial sandstones within these two units (Parker 2006; Woody 2006; Martz and Parker 2010). Each of the locally named sandstones exhibits similar sedimentary features, the most prominent of which are large-scale lateral-accretion bedding and a conglomeratic base within scours cut into underlying units. The abundance of lateral-accretion bedding, numerous crevasse-splay deposits, and the presence of thick, stacked floodplain mudstones that completely encase the sandstones indicate deposition by high-sinuosity suspended-load streams that were subject to frequent avulsion events.

The Petrified Forest Member was deposited after the Moss Back filled the paleovalleys, so that the rivers were no longer laterally restricted. Consequently, streams meandered across the floodplains, and extensive crevasse-splay deposits and floodplain mudstones are more common than in previous units in the Chinle. The presence of altered glass shards (Waters and Granger 1953) and bentonitic mudstones (Schultz 1963) indicate that volcanic ash from the magmatic arc (Fig. 3C) continued to form a significant component of the sediment. Scour surfaces within the mudstones at PFNP (Kraus and Middleton 1987) attest to fluctuations in base level in floodplain sections similar to those identified at the base of fluvial sandstones. These fluctuations are thought to reflect alternations in floodplain accumulation and degradation similar to those expressed in the fluvial sandstones.

Many of the floodplain mudstones in the Petrified Forest Member exhibit physical evidence of pedogenic modification (Fig. 8E–J). In addition, abundant trace fossils that indicate subaerial exposure of fluvial sandstones and overbank deposits support the interpretation of extensive pedogenesis (e.g., Hasiotis and Dubiel 1993, 1995a, 1995b; Hasiotis et al. 1998; Martin and Hasiotis 1998; Wahl et al. 1998; Hasiotis and Martin 1999). The paleosol maturity ranges from minor reddening, clay accumulation, and ped formation in B horizons, to a series of cumulic clay-rich B horizons as thick as 25 ft (7.5 m) (Fig. 8E).

Owl Rock Member

The Owl Rock Member, which overlies the Petrified Forest Member (Figs. 2, 9A), contains limestones 4 in (10 cm) to 6.5 ft (2 m) thick that can be traced laterally for several tens of miles (kilometers). They typically display a mottled coloration and knobby-weathered texture, representing extensive bioturbation, an interpretation supported by the presence of abundant cylindrical trace fossils interpreted to be crayfish burrows (Hasiotis and Mitchell 1989, 1993; Hasiotis et al. 1993). Locally the limestones contain ostracodes and charophytes, and abundant bones and teeth from a variety of freshwater fish and aquatic reptiles (Kirby 1981, 1993). Interbedded with the limestones are thick-bedded, massive siltstones that exhibit few physical sedimentary structures but that locally contain abundant rhizoliths and trace fossils. Fine-grained units are distinctly nonbentonitic as compared to the lower Chinle units, indicating that the contribution of volcanic ash from the magmatic arc had waned.

Locally at the base of the Owl Rock Member, and higher in the unit at scattered localities within the Chinle basin, are lenses of intrabasinal conglomerate (Fig. 9B), formed primarily from reworked pedogenic carbonate nodules from the underlying Petrified Forest Member or adjacent Owl Rock strata (Dubiel, 1987a; Dubiel and Good, 1991). The conglomerates are crossbedded and exhibit large-scale, low-angle lateral-accretion stratification (Fig. 9C). Abundant in the Owl Rock are laterally extensive limestones that exhibit a variety of depositional and pedogenic fabrics (Dubiel 1987a, 1994; Tanner 2000). Interbedded with the limestones are pink and gray siltstones (Fig. 9B, C, D) as well as rare red siltstones and very fine grained sandstones (Fig. 9B). The siltstones are massive (unbedded), laterally extensive tabular units that exhibit conchoidal fracture. They are similar in color, bedding, and grain size to strata interpreted as loessite beds in Triassic and Permian rocks (Chan, 1999; Soreghan et al., 2002). Owl Rock depositional settings are interpreted as low- to high-sinuosity streams with extensive floodplains that alternated with lacustrine and marsh settings, similar to

←

dipping to the right, characteristic of a meandering fluvial system, whereas the upper two stories are composed of less distinct, coarser-grained lateral-accretion sets dipping into the outcrop. D) Outcrop of fluvial sandstones of the Shinarump at Colt Mesa in the southern Circle Cliffs, Utah. The lower part of the outcrop displays a scour cut into the underlying Moenkopi Formation filled with thick-bedded lateral-accretion strata dipping to the left, indicating meandering-stream deposition. The upper stories display lateral-accretion bedding that dips into the outcrop. E) Shinarump fluvial sandstones at White Canyon in southeastern Utah unconformably overlying the Moenkopi. The adit is a former uranium mine. The Shinarump displays large-scale lateral-accretion bedding indicative of a meandering stream. F) Agua Zarca Member (Shinarump Member equivalent in New Mexico) fluvial sandstones at the base of the Chinle at Ft. Wingate, northwestern New Mexico. The fluvial sandstones are extensively bioturbated by crayfish burrows and exhibit pervasive purple and white mottling due to both the burrowing and pedogenic modification. The crayfish burrowing and the mottling indicate successive fluctuating water tables, which produced a lateritic soil profile.

FIGURE 6.—Photographs of the Monitor Butte Member and modern analogues. A) Exposure in Capitol Reef National Park. The Shinarump fills paleovalleys cut into marine siltstones of the underlying Moenkopi Fm. The Shinarump is in turn cut by a paleovalley that is filled by green, micaceous sandstone and siltstone of the Monitor Butte. B) The Monitor Butte exhibits foreset and topset bedding interpreted as lacustrine deltaic deposition. C) Foreset and topset beds of lacustrine delta in the Monitor Butte at Hite, Utah. The height of the foresets indicates a lake depth of approximately 45 ft (15 m). D) Inclined sandstones at CRNP interpreted as lateral accretion bedding on the basis of crossbed paleocurrents perpendicular to the inclined beds. They fill a paleovalley cut into the Shinarump, which unconformably fills a paleovalley cut into the Moenkopi. E) Black, organic-rich laminated marsh mudstones of the Monitor Butte Member in White Canyon, Utah. Hammer circled for scale. F) Bedded black mudstones of the Monitor Butte in Red Canyon, Utah, deposited in a lacustrine and fresh-water marsh system within a drowned fluvial valley. Geologist for scale. G) Chinle strata in CRNP, Utah. Both the Monitor Butte and Petrified Forest Members of the Chinle display variegated floodplain mudstones, which have been pedogenically modified, forming Alfisols with distinct reddish B horizons. H) NASA photo of the Amazon Basin showing the damming of tributary streams by trunk rivers, forming dendritic lakes on the tributaries. To the south, the Solimões River has meandered in its valley. The levee and splay deposits have constricted the flow, damming several tributary streams and forming dendritic lakes in the tributary valleys.

the modern Florida Everglades (Dubiel 1987a, 1994; Tanner 2000). Floodplain sedimentation deposited thick mudstones and siltstones that exhibit paleosol profiles as thick as 25 ft (7.5 m) and characterized by horizonation, bioturbation by burrowing crayfish, and development of Bk horizons with nodular to rhizolithic carbonates nodules. The interbedding of fluvial conglomerate with thick paleosols indicates that alluvial sedimentation alternated with periods of degradation and pedogenesis (Kirby 1981, 1993; Dubiel, 1987a; Dubiel and Hasiotis, 1994a,b; Tanner, 2000).

Owl Rock strata were deposited in fluvial and marsh to lacustrine environments that developed in response to continued subsidence in the basin with a reduction in volcanic ash input (Stewart et al. 1972; Dubiel 1987a, 1994). The occurrence of fluvial-channel conglomerates and lateral accretion deposition on point bars, bioturbation by crayfish, and the presence of fish and other aquatic vertebrates in both the limestones and the siltstones all indicate that seasonally abundant precipitation and perennial streams continued to characterize the depositional environments.

FIGURE 6.—Continued.

Rock Point and Church Rock Members

Orange to red and brown sandstones and siltstones of the Church Rock and Rock Point Members overlie the Owl Rock Member (Fig. 9E, F). Because the Church Rock has a limited distribution in southeastern Utah (Stewart et al. 1972), this discussion concentrates on the Rock Point Member, which is prominent in northern Arizona (Dubiel 1989a). Sandstones are fine to medium grained and are either structureless or contain faint, trough cross-stratification. Locally, basal beds contain limestone-nodule and siliciclastic-pebble conglomerate lenses. Many of the units contain small but abundant meniscate backfilled trace fossils assigned to *Naktodemasis bowni* (e.g., Smith et al. 2008) and locally abundant lavender, downward-bifurcating trace fossils, which are interpreted as rhizoliths (e.g., Dubiel 1992; Kraus and Hasiotis 2006). Locally at the top of the Rock Point, mudstones contain desiccation cracks as much as 3 in (1.5 cm) wide and 3 ft (1 m) deep that are filled with sandstone of the overlying Lower Jurassic Wingate Sandstone.

Rock Point sandstones weather as thin- to thick-bedded ledges, whereas siltstones and mudstones form steep slopes. Siltstones are tabular, laterally extensive, and contain faint horizontal laminations or are structureless. Locally, sandstones in the upper part of the Rock Point exhibit large-scale crossbeds as thick as 10 ft (3 m) with upward-coarsening laminae, one of the few criteria diagnostic of eolian deposition. These beds become more abundant and thicker upward in the section to the contact between the Chinle and the overlying Lower Jurassic Wingate Sandstone (Fig. 8E) (Stewart et al. 1972; Dubiel 1989a).

The Rock Point Member was deposited as eolian dunes and sand sheets with minor loess, and in minor fluvial channels (Dubiel 1987a, 1994). The large-scale, eolian cross-stratification and the large mudcracks indicate that drier periods and desiccation were prominent during deposition of the uppermost part of the Chinle. Examination of petrographic thin sections of the Rock Point indicates that the red rocks, in contrast to the variegated colors of the lower Chinle, contain greater amounts of coarse-grained diagenetic hematite (Dubiel 1987a). The red color by itself is not *de facto* evidence of arid climates, but merely reflects the increased hematite content (Dubiel and Smoot 1994). This increase in hematite content may be related to the lack of organic carbon in the rocks available to inhibit oxidation. Both hematite development and the decreased organic carbon are thought to reflect deposition in the more oxygenated playa and eolian environments. The upper part of Rock Point Member has an increase in abundance and thickness of eolian dune deposits (Fig. 8E). The

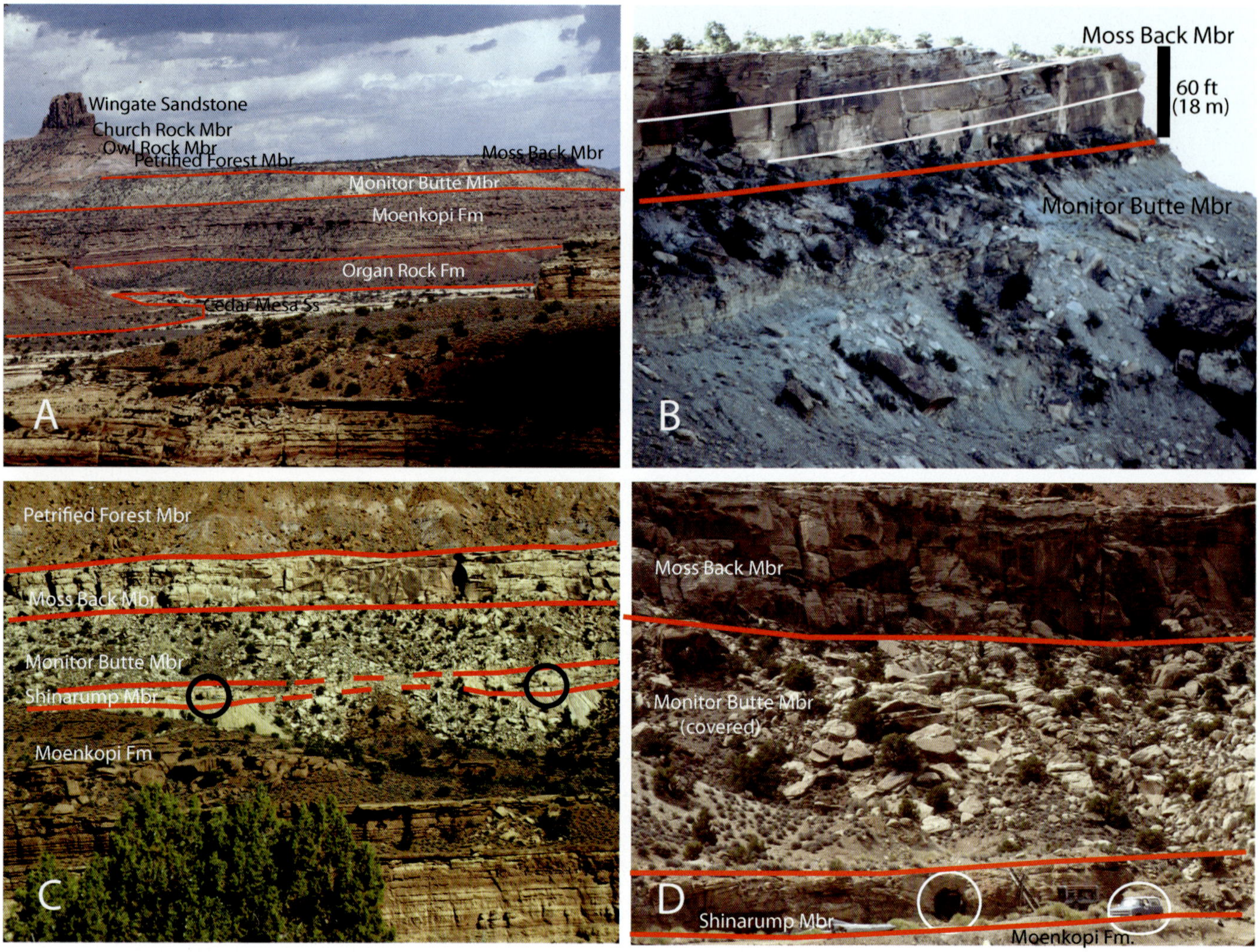

FIGURE 7.—Moss Back Member fluvial sandstones. A) Paleovalley cut into the Monitor Butte Member and filled with fluvial sandstone of the Moss Back Member at Jacob's Chair, White Canyon, Utah. B) Large-scale, inclined bedding interpreted as lateral accretion on the basis of crossbed paleocurrent direction perpendicular to inclined strata in Moss Back fluvial sandstones at White Canyon, Utah. The height of the Moss Back lateral-accretion sets at this locality indicates a meandering stream with depth of about 60 ft (18 m). C) Moss Back Member at Happy Jack Mine, White Canyon, Utah, showing the Moss Back and Shinarump fluvial deposits both lying within successive paleovalleys. Mine adits (circled for scale) are 10 ft (3 M) high. D) Detail at Happy Jack Mine showing the disparity in thickness between the fluvial deposits of the Moss Back and Shinarump Members in White Canyon, Utah. Mine adit (10 ft or 3 m high) and truck circled for scale.

loessites and eolianites indicate a distinct shift to overall semiarid and arid climates, but the repeated alternations reflect a continued cyclicity of more arid and less arid periods on time scales longer than seasonal.

Sedimentologic evidence suggests that conditions of lower but fluctuating water table developed during the final stages of Chinle deposition. Minor ephemeral fluvial-channel sandstones, playa sandstones and siltstones, and eolian sandsheet, loess, and dune deposits of the Rock Point Member were deposited under more extended dry periods, albeit punctuated by slightly less dry intervals. The predominance of eolian sandsheet, loess, and dune deposits in the uppermost part of the Rock Point points to significant climatic drying at the close of the Late Triassic, a scenario suggested by several studies (e.g., McKee et al. 1959; Blakey and Gubitosa 1983, 1984; Dubiel 1987a, 1987b, 1989a) and supported by more recent work (Dubiel et al. 1991; Dubiel 1994; Tanner 2000, 2003; Tanner and Lucas 2006; Cleveland et al.,2008). The overlying eolian erg deposits of the Wingate Sandstone attest to the transition to predominantly eolian sedimentation and arid climatic conditions in the Early Jurassic (Dubiel 1987a, 1994; Blakey 1994).

PALEOSOLS

Alluvial, marsh (paludal), lacustrine, playa, and eolian deposits of the Chinle Formation exhibit a variety of paleosols (Table 2). The description of Chinle paleosols used in this paper draws on several classification schemes and recent studies as well as comparison to modern tropical and subtropical soils (Buringh 1968, 1979; Mohr et al. 1972; Soil Survey Staff, 1975; MacFarlane 1976; Young 1976; Mack et al. 1993; Tardy 1997; Kraus 1999; Retallack 2001; Cecil and Dulong 2003). Several studies have described general features of Chinle paleosols (Kraus and Middleton 1987; Dubiel 1987a, 1994; Dubiel and Hasiotis 1994a, 1994b; Hasiotis and Dubiel 1994, 1995a; Demko et al.

1998; Hasiotis et al. 1998; Atchley et al. 2004; Cleveland et al. 2007, 2008), but most have not presented a comprehensive evaluation of paleosol features, their classification, or the stratigraphic distribution across the entire Chinle Basin. A few recent studies have documented specific paleosols in the Chinle, focusing on single members or on localized areas (Tanner 2000; Therrien and Fastovsky 2000; Beer 2005; Prochnow et al. 2006). Tanner and Lucas (2006) provided a summary of calcareous Chinle paleosols in which they focused on calcretes and their paleoclimatic implications.

The present study documents the major paleosols types within members of the Chinle Formation and highlights an upward stratigraphic trend in the range of paleosols in the members. In addition, the characteristics of the paleosols are described that, coupled with other sedimentologic and paleontologic evidence, provide for interpretations of the climates that produced them. Although several studies have alluded to the general climatic influence on Chinle paleosols, they have mostly described the climate in terms of annual precipitation. The focus herein is on the features and types of paleosols that indicate the seasonal input of precipitation to the tropical depositional system expressed as the number of wet months in a year (e.g., Cecil and Dulong 2003). Wet months are defined as months in which rainfall exceeds evapotranspiration (e.g., Cecil and Dulong 2003). The overall trend in climate is interpreted on the basis of the development of individual paleosols and on longer time scales as inferred from the succession of paleosols and their stratigraphic position. Other orders of paleosols are present locally in each member, but generally one paleosol order predominates. Oxisols and Gleysols (*sensu* Mack et al., 1993) are most common at the base of the section, whereas Histosols and Alfisols or Argillisols (*sensu* Mack et al. 1993) are common in the Monitor Butte Member. Alfisols and Vertisols characterize many floodplain mudstones in the Petrified Forest Member, with a change to calcic Vertisols at the top of the unit. Pedogenic carbonate in paleosol profiles is even more pronounced in the Owl Rock Member, and those paleosols are characterized as Calcisols. In the Rock Point Member, paleosols generally are immature and represent Entisols and Inceptisols or Protosols (*sensu* Mack et al. 1993), which locally contain pedogenic carbonate and fabrics typical of Aridisols.

Within individual stratigraphic units, the typical paleosol order and its characteristic pedogenic features can be represented by a general paleosol profile (Fig. 10A–D). In the upper beds of the Lower Triassic Moenkopi Formation are deeply weathered Oxisols and Gleysols that display variable redoximorphic color mottling associated with the redistribution of iron sesquioxides due to intense weathering and fluctuating water tables (Fig. 10A). The same is true for basal fluvial strata of the Shinarump, Temple Mountain, and Monitor Butte Members of the Chinle Formation. The paleosols developed as deep-weathering soil profiles on subaerially exposed interfluves under an ever-wet to monsoonal climate regime in a tropical latitudinal setting (Buringh 1968, 1979; Mohrt al. 1972; Young 1976; Cecil and Dulong 2003). Locally the Oxisols display classic translocated iron as nodules and mottles lower in the profile or in reticulate fracture networks that are characteristic of plinthitic fabric and are referred to as plinthites (MacFarlane 1976; Tardy 1997). Despite the intense pedogenesis, locally these paleosols also exhibit faint vertic features typical of soils that experienced a seasonal input of moisture (Fig. 8H), which is indicative of paleosols that developed under different climates and are thus polygenetic in formation.

Laterites and plinthites exhibit redoximorphic features in highly weathered tropical soils (Buringh 1979; Tardy 1997), commonly occurring as dark-red redox concretions that form platy, polygonal, or reticulate patterns. Generally, plinthite forms in a soil horizon that is saturated with water for some period of time during the year (Tardy 1997). Initially, iron is segregated in the form of red or dark-red redox concretions (Fig. 4C, D, E, F). These concretions are not considered plinthite unless enough segregation of iron by repeated wetting and drying has taken place to permit their irreversible hardening upon exposure by erosion or landscape degradation. Plinthite concretions are coherent enough to be separated readily from the surrounding soil (Tardy 1997). Plinthite commonly occurs within and above reticulate and mottled horizons (Fig. 4D). Plinthite concretions are observed in basal mottled units or Oxisols and Gleysols of the Chinle, and locally they occur as clasts in fluvial conglomerates that have scoured into underlying Oxisols. Locally, the Oxisols exhibit vertically oriented color mottles associated with multiple generations of deep crayfish burrows (Fig. 4F), indicative of fluctuating water tables and bioturbation within the sediment at the time of deposition.

The size, morphology, composition, and dispersion of the mottles closely resemble redoximorphic features formed by fluctuating water tables in modern soils, particularly in tropical and subtropical regions (Mohr et al. 1972; Young 1976). Locally, where clay content is higher within these units, extensive periods of standing water produce surface gleying, where iron- and manganese-bearing minerals become reduced. In particular, the paleosols resemble the purple- and white-mottled, iron-rich concretionary Oxisols and laterites in groundwater-modified modern tropical soils (Buringh 1968, 1979; McFarlane 1976; Retallack 1997, 2001; Tardy 1997), as well as paleosols and laterites described from the Triassic Budra Formation, southwestern Sinai (Goldbery and Beyth 1984). These pedogenic characteristics categorize these types of paleosols as Gleysols in the classification scheme of Mack et al. (1993). The large trace fossils are interpreted as the casts of crayfish burrows and are assigned to the ichnotaxon *Camborygma eumekenomos* (Hasiotis and Mitchell 1989, 1993; Hasiotis et al. 1993; Hasiotis and Dubiel 1993, 1994).

Both the crayfish burrows and the mottled coloration of the mottled strata are thought to indicate fluctuating water tables and pedogenic modification within the sediments in response to seasonal rainfall and flooding produced by the Triassic monsoonal climate (Dubiel 1987a; Dubiel et al. 1991; Hasiotis and Mitchell 1993; Hasiotis et al. 1993; Hasiotis and Dubiel 1994). The association of crayfish burrows within paleosols believed to have formed due to fluctuating water tables supports the interpretation that the burrows were created during seasonally dry periods as crayfish burrowed down to the lowered water table (Hasiotis and Mitchell 1993; Hasiotis et al. 1993; Hasiotis and Dubiel 1994). Iron within the sediments was redistributed by pedogenesis in response to the fluctuating water tables that produced seasonally alternating oxidizing and reducing conditions.

Where developed on the Moenkopi, the mottled strata represent relative landscape stability after paleovalley incision and long-term paleosol development on interfluve areas between paleovalleys. Similarly, they represent times of intermediate-term stability and nondeposition within the Shinarump fluvial system, where they are developed on fluvial sandstones (Dubiel 1987a, 1991; McCarthy and Plint 1998; Beer 2005). A discussion of sequence-boundary development and paleosol superposition in the Chinle is presented in a subsequent section, and is based in part on previous studies (Dubiel 1987a, 1991; Dubiel et al. 1999; Beer 2005).

The Monitor Butte Member contains several types of distinct paleosols. The lacustrine-deltaic depositional facies are locally associated with thin coals (Dubiel 1987a; 1994; Beer 2005) that can be classified as Histosols. These organic-rich paleosols indicate high fresh water tables within marshes that preserved peat deposits. The Monitor Butte floodplain mudstones contain paleosols with reddened, clay-rich B horizons that are classified as Alfisols or Argillisols (Fig. 10B). At the top of the Monitor Butte, there is a regionally extensive red paleosol with isolated carbonate nodules that forms the interfluves of the paleovalley in which the Moss Back was deposited. In places, the paleosol contains angular peds with slickensided surfaces. This regional paleosol is classified as an Argillisol, Vertisol, and calcic Vertisol that required a relatively long time period for development on

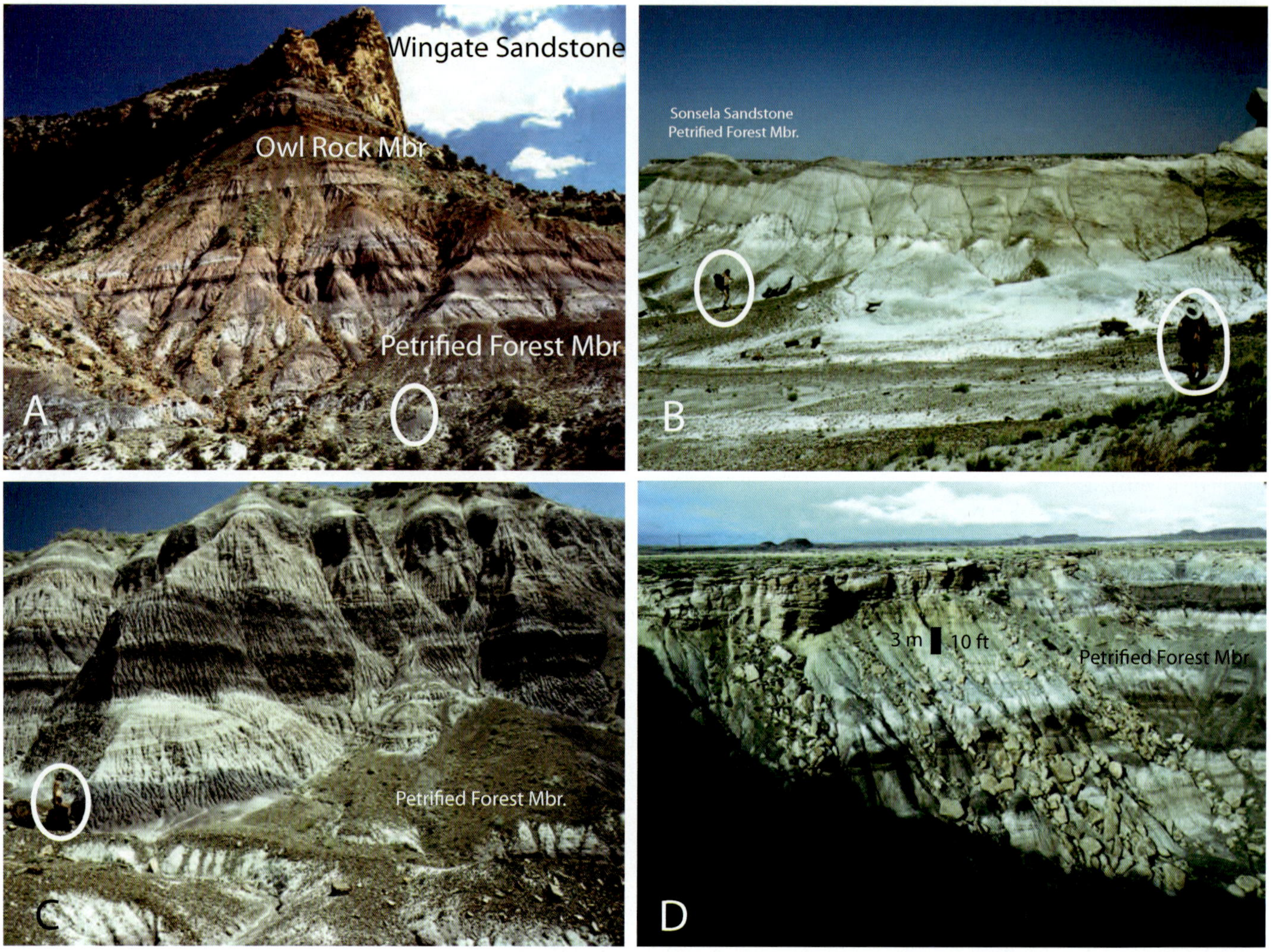

FIGURE 8.—Petrified Forest Member and paleosols. A) Interbedded variegated sandstones, siltstones, and mudstones of the Petrified Forest Member at the Hillside Mine in Red Canyon, Utah. Fluvial sandstones are interspersed with pedogenically modified floodplain mudstones with dark red clay-rich B horizons typical of Alfisols. Geologist circled for scale. B) Sonsela Sandstone of the Petrified Forest Member at Blue Mesa in PFNP, Arizona. The Sonsela is composed of conglomeratic sandstone with large-scale compound bar forms, interpreted as a braided-stream deposit. Geologists, circled for scale. C) White, crossbedded conglomeratic sandstones interbedded with purple to lavender and gray mudstones, interpreted as crevasse splay deposits and paleosols, respectively. Geologist circled for scale. D) Fluvial channel sandstones and floodplain mudstones with paleosols at Rainbow Forest in PFNP. The channel deposits display large-scale heterolithic strata interpreted as lateral-accretion bedding on the basis of crossbedding paleocurrents perpendicular to the large-scale inclined bedding, indicative of meandering-stream deposition. E) Red mudstone, 30 ft (9 m) thick, exhibiting large-scale crosscutting curvilinear slickensided fractures indicative of a Vertisol, below large-scale lateral-accretion bedding in a carbonate-nodule conglomerate within a crevasse-splay channel. Geologist in center for scale. F) Large-scale curvilinear slickensided fractures within a 10 ft (3 m)-thick vertic paleosol in PFNP. The upper 1 ft (30 cm) of the paleosol is bleached green from reducing solutions that passed through the overlying fluvial sandstone. G) Vertic features in a 6 ft (2 m)-thick paleosol in Black Forest area of PFNP. The A horizon is lavender; the B horizon is dark red and contains crosscutting bowl-shaped fractures highlighted by carbonate cement, as well as small carbonate nodules. Scale circled is 4 in (10 cm). H) Bowl-shaped Vertic fractures developed in a coarse-grained fluvial sandstone at Ft. Wingate, New Mexico. Circled scale is 4 in (10 cm). I) Mudstones and calcareous paleosol horizons in the upper part of the Petrified Forest Member at Round Rock, northern Arizona. J) Calcareous nodules aligned along vertic fractures below a fluvial sandstone at Ward Terrace, Arizona. Geologist for scale.

the interfluves associated with the degradation and incision of the Cottonwood paleovalley.

The Moss Back Member does not contain significant paleosols, inasmuch as it is a fluvial sandstone unit deposited within the Cottonwood paleovalley, but the overlying Petrified Forest Member contains abundant paleosols. Once the Moss Back filled the paleovalleys, subsequent fluvial systems were able to migrate across the landscape, and a variety of paleosols developed within Petrified Forest fluvial deposits and floodplain strata. Most abundant in the floodplain strata are paleosols that exhibit distinct horizonation with

FIGURE 8.—Continued.

absent or pale gray to lavender, organic-rich A horizons and thin to thick red to purple argillic B horizons (Fig. 10B). These paleosols are characterized as Alfisols (Soil Survey Staff 1975; Retallack 1997, 2001) or Argillisols (Mack et al. 1993).

Other Petrified Forest paleosols are thick red paleosols characterized by curvilinear bowl-shaped fractures and pseudoanticlinal structures, slickensided, angular-blocky to prismatic peds, abundant insect trace fossils, and isolated to coalesced carbonate nodules (Fig. 8E–H). These

FIGURE 9.—Owl Rock and Rock Point Members. A) Owl Rock Member limestones and siltstones overlain by red sandstones and siltstones of the Rock Point Member and the Jurassic Wingate Sandstone at Round Rock, northern Arizona. The Owl rock units were deposited in a lacustrine marsh and paludal setting, whereas the red sandstones and siltstones of the Rock Point and Wingate were deposited as eolian dunes and loess deposits. Geologist circled for scale. B) Owl Rock Member near Round Rock, northern Arizona, with basal fluvial carbonate-nodule conglomerate overlying Petrified Forest Member. Geologist circled for scale. C) Interbedded limestones and siltstones of the Owl Rock Member at Cove, Arizona. Limestones were deposited as lacustrine and marsh deposits in a shallow lake and as limestone-nodule conglomerates in lateral-accretion sets as meandering-stream deposits. D) Fluvial limestone-nodule conglomerates and pedogenically

characteristics are evidence of repeated wetting and drying. The thickness of the clay-rich horizons and the prominence of vertic features such as slickensides and arcuate fractures indicate that these tropical-latitude paleosols are Vertisols (Fig. 10C) (Buringh 1968, 1979; Mohr et al. 1972; Young 1976) and calcic Vertisols (Mack et al. 1993). They formed in mud-rich floodplain settings under conditions of seasonal rainfall with a wet season that lasted several months (Buringh 1968, 1979; Mohr et al. 1972). In fact, Vertisols with these prominent features can be associated with a specific paleoclimate regime – that is, one marked by seasonality in precipitation (Blodgett 1988; Retallack 2001).

There is an upward gradation in the Petrified Forest Member to paleosols that have a greater concentration of carbonate nodules, coalesced nodules, and less prominent vertic features (Fig. 8I, J). These paleosols contain reddened B horizons and prominent Bk horizons formed by isolated nodules that are prominent lower in the section, grading dominantly coalesced nodules at the top, thus varying from Argillisols, calcic Argillisols, calcic Vertisols, to Calcisols. The increased abundance in carbonate nodules indicates that climate become progressively drier (i.e., increasingly less rainfall or increase solar insolation) toward the close of Petrified Forest Member deposition, but the local persistence of vertic features in these calcic paleosols indicates that precipitation was still seasonal, but probably lesser in amount with a longer dry season. Blodgett (1984, 1988) pointed out, with respect to the equivalent Upper Triassic Dolores Formation in southwestern Colorado, that isolated carbonate nodules form in clay-rich tropical soils as a feature of alternating wet and dry seasonality and do not indicate aridity. The depth of the carbonates nodules in the profile and the thickness of the profiles affected by vertic fractures indicate that water tables fluctuated as much as 15 to 20 ft (4–6 m) within the floodplain.

In contrast to these vertic paleosols in the lower and middle part of the Petrified Forest Member, much of the upper part of the Petrified Forest Member as well as the overlying Owl Rock Member contain paleosols that are characterized by clay-rich red B horizons with abundant isolated to coalesced carbonate nodules but with few vertic features (Fig. 9A-D). The paleosols exhibit only rare vertic fractures, but locally display abundant bioturbation attributed to crayfish (Fig. 9D) (Hasiotis and Mitchell 1993; Hasiotis et al. 1993; Dubiel and Hasiotis 1994a, 1994b) as well as distinct rhizoliths (Tanner 2000). The prominence of significant Bc or Bk horizons in these units indicates that they represent Aridisols or Calcisols (Figs. 9 D, 10D). These strata have been interpreted as palustrine and lacustrine carbonates with superimposed paleosols, an environment similar to the modern Florida Everglades (Dubiel 1987a, 1987b, 1989b, 1994; Hasiotis 1991; Hasiotis and Dubiel 1993; Tanner 2000).

These uppermost Petrified Forest and Owl Rock Member paleosols are interpreted as calcic Alfisols (i.e., Argillisols) and Calcisols. Tanner (2000) suggested that Owl Rock palustrine carbonates may have formed under climates that range from subhumid to semiarid, with drier conditions indicated by the presence of brecciated fabrics, coated grains, and the occurrence of well-developed calcrete horizons and Alfisols. The development of Bk horizons in the floodplain paleosols attests to a significant dry season, possibly a dry subhumid climate but not necessarily a semiarid one. The recognition of possible loess deposits in the Owl Rock Member suggests that at times, climate changed to a distinctly drier phase during Owl Rock deposition to deposit windblown sand and dust. The larger-scale interbedding of floodplain and lacustrine siltstone and limestone with eolian and loess siltstones in the Owl Rock indicates longer-term climatic fluctuations, on the scale of the degradational-aggradational alternations of climate within the fluvial sandstone and floodplain mudstone successions in the lower Chinle.

At the top of the Chinle, dominantly red Rock Point Member strata exhibit only immature paleosols classified as Entisols and Inceptisols or Protosols (*sensu* Mack et al. 1993). Sandstones and mudstones were deposited by fluvial systems, on lacustrine or playa mudflats traversed by small fluvial systems, and as eolian sandsheets, loess, and dunes (Dubiel 1987a, 1989a, 1989b; 1994). Rare downward-bifurcating rhizoliths with lavender haloes are present and are locally associated with meniscate backfilled burrows (*Naktodemasis* isp.). The beds exhibit blocky prismatic fabric, extensive carbonate cementation, and desiccation cracks. These carbonate-bearing paleosols are mainly Aridisols (Fig. 10D), immature paleosols that developed on a variety of arid-setting eolian strata.

DEGRADATIONAL-AGGRADATIONAL CYCLES

Sequence boundaries at the bases of fluvial sequences are recognized in several Chinle studies. Major surfaces at the base of fluvial paleovalley fills have been documented in the lower part of the Chinle (Dubiel 1987a, 1991; Demko et al. 1998, Demko et al. 2005; Demko 2003; Beer 2005) and in the upper part of the Chinle in northeastern New Mexico (Cleveland et al. 2007). Interfluve paleosols have also been interpreted to indicate sequence boundaries in continental sequences (McCarthy and Plint 1998), and in fact, may represent the major sequence events in successions of fluvial strata (Hasiotis and Honey 2000). Interfluve paleosols have also been identified as significant surfaces and sequence boundaries in the Chinle (Dubiel 1987a, 1991; Demko et al. 1998; Beer 2005). Deeply weathered paleosols that developed on fluvial deposits within paleovalleys and lapping onto paleovalley walls in the Chinle are an additional indicator of landscape stability, periods of nondeposition, or degradation and fluvial downcutting within the paleovalley-fill strata. These within-valley, deeply weathered paleosols are also documented as sequence boundaries in the Chinle fluvial depositional system (Demko 1995a, 2003; Demko et al. 1998; Demko et al. 2005; Beer 2005). The sequence boundaries that are evident at the bases of fluvial valley fills in the Chinle, such as the bases of the Shinarump and Moss Back Members (Figs. 5A-D, 7A, B), Sonsela Sandstone of the Petrified Forest Member (Fig. 8B), Owl Rock (Fig. 9B), and Rock Point Members (Fig. 9E) all indicate periods of degradation associated with decreased sediment load and stream discharge.

←

modified floodplain mudstones of the Owl Rock Member at Ward Terrace, northern Arizona. Geologists circled bottom left and right for scale. The 20 ft (6 m)-thick floodplain mudstone in the lower half of the exposure has a 10 ft-thick carbonate-rich B horizon, representing about 20,000 to 40,000 years for pedogenic development (Birkeland, 1984). E) Red eolian dune sandstones and loessite siltstones of the Rock Point Member and Wingate Sandstone at Round Rock, northern Arizona. Geologist circled for scale. F) Upper part of the Rock Point Member overlain by eolian dune sets of the Wingate Sandstone at Lukachukai, northern Arizona. Note how the eolian dune beds in the upper Chinle become more numerous and thicker upward in the section toward the contact with the Wingate Sandstones. This progression indicates a repetitive alternation of dry eolian dune deposition with slightly more humid but still arid loess deposition. This alternation probably indicates alternating seasonality, along with a gradual and continual drying of the Triassic climate to a more arid regime that deposited the eolian dune deposits of the Wingate.

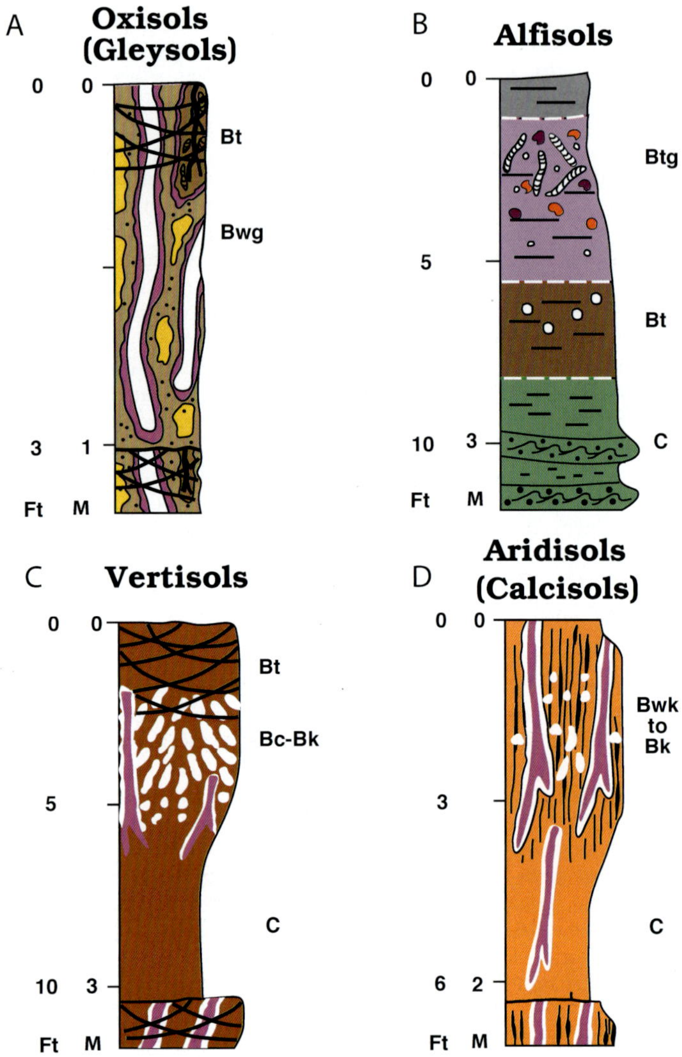

FIGURE 10.—Paleosol models for the Chinle Formation. A) Oxisols, Gleysols, and lateritic paleosols all exhibit profiles 3–5-ft (1–1.5 m) thick characterized by white and purple mottles of alternating iron-poor and iron-rich zones, respectively. Red areas are dominated by hematite cement, and yellow mottles contain mainly goethite. White and purple mottles are commonly vertically elongated along crayfish burrows, which are generally 3 to 6 ft (1–2 m) long and 8 in (19 cm) in diameter. B) Alfisols exhibit profiles 3–25 ft (1–7.5 m) thick and are characterized by gray to lavender A horizons and dark red clay-rich B horizons, locally with small carbonate nodules and abundant trace fossils of *Scoyenia gracilis*. C) Vertisols range in thickness from 5 to 25 ft (1.5–7.5 m), and contain cross-cutting arcuate fractures with slickensides, with locally abundant carbonate nodules. D) Aridisols are from 3 to 6 ft (1–1.75 m) thick and are characterized by blocky to prismatic peds and rare to locally abundant carbonate nodules. They also contain abundant rhizoliths, which are locally altered lavender and are distinguished from the crayfish burrows in other paleosols by a downward-bifurcating pattern and an intricate network of dark secondary and tertiary, finer-scale rhizoliths.

Fluvial channel, floodplain, and crevasse-splay strata in the Chinle were deposited as aggradational units within paleovalleys that had been cut into underlying units under degradational conditions. Sequence boundaries are marked by both the bases of the paleovalleys and by deep-weathering paleosols developed on interfluve areas of the paleovalley walls (e.g., McCarthy and Plint 1998). This stratigraphic pattern has previously been reported in floodplain mudstone sequences of the Chinle Formation in the Petrified Forest Member in PFNP and attributed to possible thermotectonic pulses in the basin (Kraus and Middleton 1987). Fluvial aggradational cycles have also been described in the Chinle in northern New Mexico and attributed to tectonic pulses (Cleveland et al. 2007). Similar degradational-aggradational cycles in fluvial sandstones have been noted in studies of ancient fluvial successions, and they have been recognized in modern settings influenced by the Asian summer monsoon (e.g., Tandon and Gibling 1997).

The vertical succession of interpreted depositional environments provides a depositional model characterized by fluvial-floodplain, lacustrine-deltaic, palustrine, and eolian systems (Fig. 11A-F), representing a mega-aggradational cycle in the Chinle (Fig. 12) that lies on the TR-3 unconformity and is overlain by the J-0 unconformity (Pipiringos and O'Sullivan 1978). Each of the members of the Chinle represents a major degradational-aggradational cycle hundreds of feet (tens to hundreds of meters) thick that was deposited on an erosional scour or paleovalley cut into the underlying units (Fig. 12). The progression from valley-fill fluvial systems in the Shinarump Member to a marsh, lake, and delta complex in the Monitor Butte Member and back to paleovalley-fill fluvial systems in the Moss Back Member are examples of the major alternating degradational-aggradational cycles in the lower part of the Chinle. The major degradational-aggradational cycles are bounded below and above by sequence boundaries, and they are equivalent to the continental sequences recognized previously in the lower Chinle (Demko et al. 1998; Demko et al. 2005; Demko 2003: Beer 2005) and herein for the entire Chinle.

The Shinarump, Monitor Butte/Temple Mountain, and Moss Back Members represent three cycles of degradation and subsequent aggradation, each being deposited within paleovalleys (Fig. 12). The initial period of degradation is marked by deeply weathered interfluve Oxisols and Gleysols that were developed either on the Moenkopi Formation or on Shinarump and/or Temple Mountain Member fluvial sandstones (Fig. 4). Fluvial Shinarump sandstones display several smaller-scale, or higher-frequency cut-and-fill episodes (Fig. 5C). A second period of incision and degradation is marked by truncation of the Shinarump and correlative paleosols and pedogenically modified strata (Fig. 6A, B) (Stewart et al. 1972; Dubiel 1991; Beer 2005). The Monitor Butte Member and correlative Temple Mountain Member overlie this unconformity and consist of aggradational high-sinuosity fluvial deposition and fluvial-lacustrine-deltaic deposition as interbedded sandstones, mudstones, and paleosols with histic and argillic diagnostic characteristics. These lacustrine and deltaic strata also exhibit higher frequency cut-and-fill episodes as a result of fluctuating lake levels and clastic fill (Fig. 6D) (Beer 2005; Demko et al. 2005). The fluvial sandstones of the Moss Back Member represent a third major degradational-aggradational cycle (Fig. 7A,C). The subsequent succession from fluvial and floodplain systems in the Petrified Forest Member to basal fluvial paleovalley-fill and then lacustrine-palustrine strata in the Owl Rock Member represent two additional major degradational-aggradational cycles (Fig. 9A). The succession of fluvial, sandsheet, loess, and eolian dune strata in the Rock Point Member represents the final major degradational-aggradational cycle at the top of the Chinle (Fig. 9E).

PALEOCLIMATE CYCLICITY

This study interprets the degradational and aggradational cycles in the Chinle to reflect several orders of climatic variability, whereas other

TABLE 2.—Summary of Chinle Formation Pedofacies. Paleosol Designation: (*) Soil Survey Staff (1975) and (+) Mack et al (1993).

Pedofacies	Paleopedologic features	Interpretation	Paleosol Designation(s)
Purple mottled strata	Quasivertical, irregularly shaped to elongate color mottles of dark purple, lavender, yellow, and white	Plinthite texture of highly leached soil profile by highly fluctuating water table and soil moisture with wetting and drying; coloring after crayfish burrows and rhizoliths, also enhancing meniscate backfilled burrows assigned to *Naktodemasis bowni*	Oxisol* Gleysol+
Red to red- brown blocky mudstone or claystone	Subangular blocky, blocky, and prismatic structure, locally with green, purple, and yellow mottles; locally with small, isolated carbonate nodules	Pedogenic structures typical of illuviated clays in a B horizon; mottles represent redoximorphic coloration around roots, burrows, and fractures; carbonate nodules developed by pedogenic accumulation	Alfisol* or Ultisol* Argillisol+
Red to red- brown fractured mudstone or claystone	Subangular blocky, blocky, and prismatic structures bounded by sets of bowl-shaped crosscutting pseudoanticlinal fractures, with green, purple, and yellow mottles; locally with small, isolated carbonate nodules	Pedogenic structures typical of shrink and swell clays that produce mukkara, typical in B horizons; mottles represent redoximorphic coloration around roots, burrows, and fractures; carbonate nodules developed by pedogenic accumulation	Vertisol*+ Calcic Vertisol+
Carbonaceous black mudstone to coaly black mudstone	Thin to thick beds of carbonaceous black shale with evidence of *in situ* plant material such as rhizoliths and steinkerns	Accumulation of plant debris in areas of high water table that produce waterlogged soils; rare to very rare and absent bioturbation	Histosol*+
Pale quartz sandstone with bands of iron-stained sandstone or silty to clayey quartz sandstone	Hematite to goethite stained layers of massive quartz sandstone, locally associated with or overlying silt to clay coated sandstone or thin, discontinuous bedded silty clay	Spodic horizon developed by eluviation of clays and iron oxide minerals to zones of accumulation where this material is stable	Spodosol*+
Burrow- and rhizolith-disrupted fabric	Primary sedimentary structures, bedding, and bed sets disrupted by rhizoliths and burrows, including shallow penetrating footprints	Incipient pedogenesis of fluvial, lacustrine, eolian, or volcanic ash-fall deposits produced by animal and plant bioturbation; remnant bedding indicates immature paleosols; massive, variegated colored units with visible bioturbation are more mature paleosols	Entisol* Inceptisol* Protosol+
Carbonate-nodule-bearing orange-red or reddish-orange siltstone, mudstone, and sandstone	Small nodular, laminar, to platy carbonate in layers, rare to abundant; locally grains coated in carbonate, with carbonate forming cylindrical concretions and hollow tubes	Pedogenically illuviated and reprecipitated carbonate; carbonate engulfing contained roots to form tubes or producing solid concretions; roots and burrows present with or without pedogenic carbonate	Aridisol*
Carbonate-nodule-bearing limy siltstone and limestone	Small to large nodular, laminar, platy, to massive; with or without autoclastic brecciation; with or without tubular or cylindrical structures; generally associated with hematite stained silt and clay infilling fractures, tubes, cylinders, and horizontal layers	Pedogenically illuviated and reprecipitated carbonate with pedogenically produced silt and clay fractions resulting from the dissolution of impure limestone; with roots and burrows of organisms containing reprecipitated carbonate	Calcisol+

studies have interpreted some of these cycles in the Chinle as being due to thermal changes or tectonic pulses in the basin (Kraus and Middleton 1987; Cleveland et al. 2007). Although interpretations of paleoclimate based solely on paleosols can be problematic (Retallack 2001), integrating interpretations from paleosol data with sedimentologic, paleontologic, and ichnologic evidence (e.g., Hasiotis et al. 2007) provides a more robust model for characterizing the climate and the distribution of rainfall at the time of deposition of the Chinle

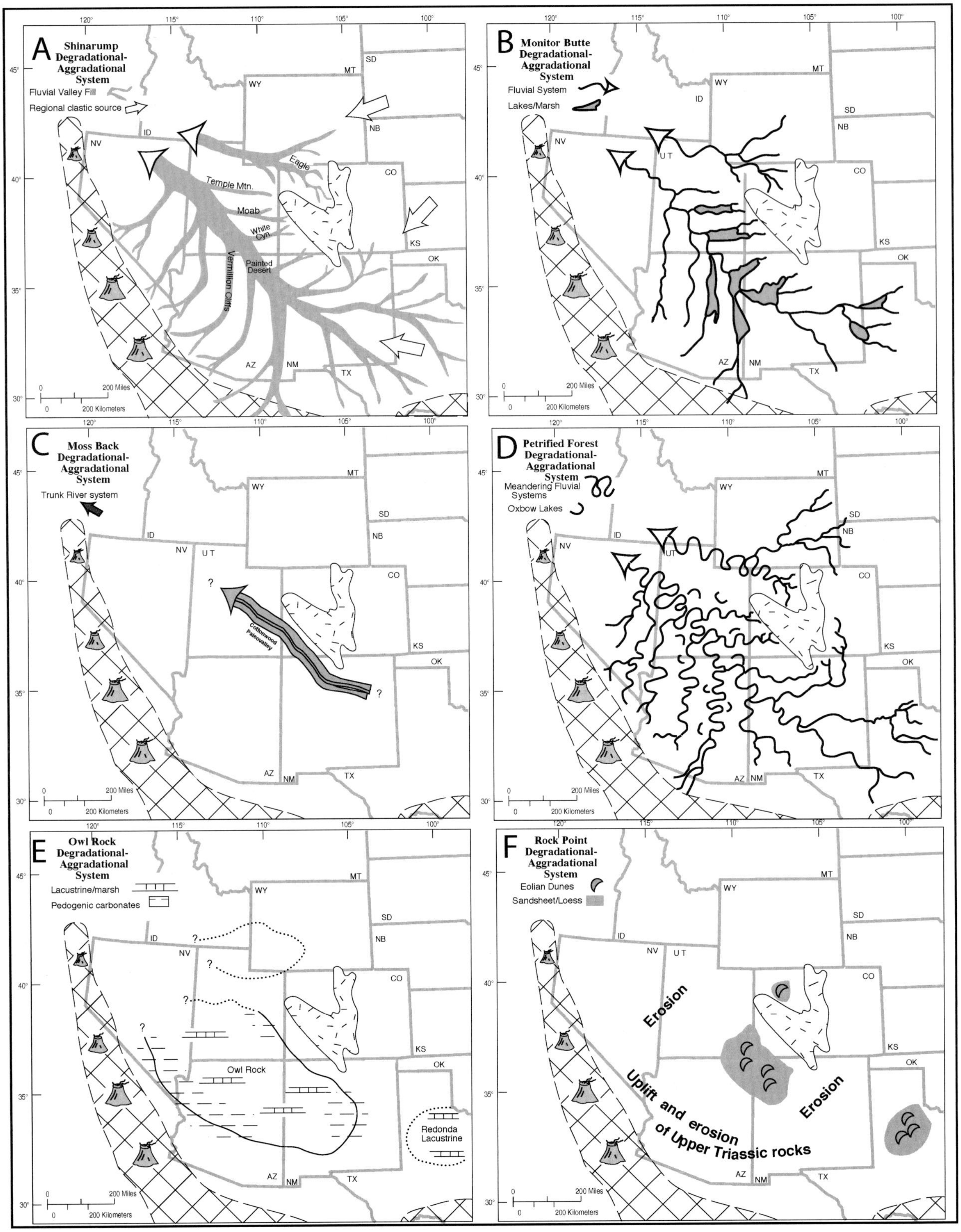

FIGURE 11.—Paleogeographic maps of the Triassic Western Interior basin (modified from Dubiel, 1994). A) Map showing Shinarump Member paleovalley system; paleovalley names are referenced in the text. B) Map showing Monitor Butte Member fluvial-lacustrine-marsh system with lakes that were dammed by lateral migration of trunk or tributary streams. C) Map showing Moss Back Member trunk fluvial system. D) Map showing Petrified Forest Member meandering fluvial system. E) Map showing Owl Rock Member lacustrine and palustrine system. F) Map showing Church Rock and Rock Point Members playa and eolian system.

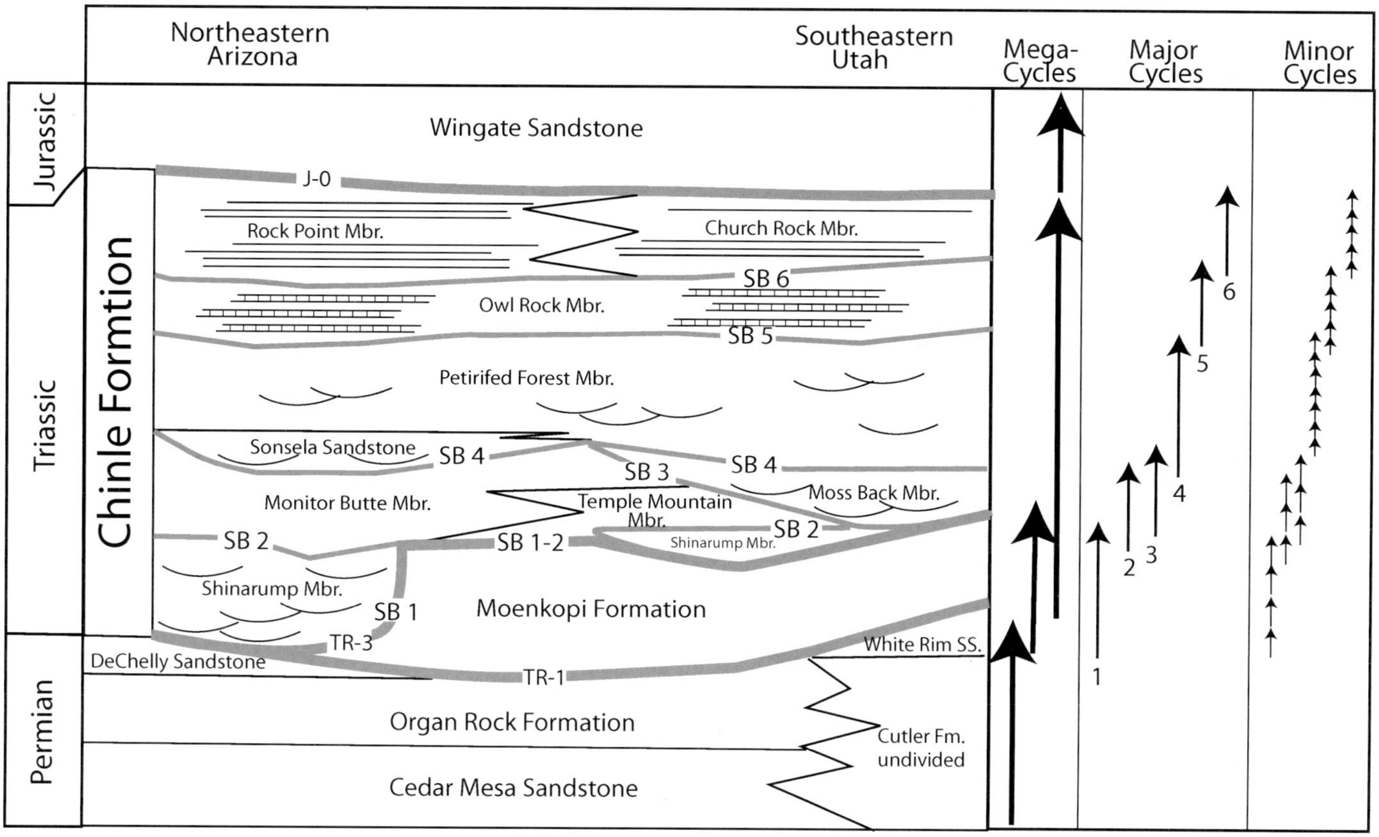

FIGURE 12.—Conceptual sequence stratigraphic diagram showing unconformities (heavy lines), stratigraphic units, sequence boundaries (thin lines), and degradational-aggradational cycles (arrows). Mega-cycles are defined between regional unconformities. Unconformities are from Pipiringos and O'Sullivan (1978; TR-1, TR-3, and J-0). Major cycles are defined between sequence boundaries for the Chinle. Sequence boundaries (thin lines) depicting erosion and paleovalleys are adapted from Stewart et al. (1972), Dubiel (1987a, 1991), Beer (2005), and the present study. Major cycles are numbered to correspond to strata between numbered sequence boundaries (SB). Sequence boundary (SB) numbering is extended from scheme of Beer (2005). Minor cycles are depicted schematically, and correspond to fluvial sandstone packages and corresponding floodplain mudstones and paleosols.

Formation. As pointed out by Cecil (1990), the major input of clastic sediment in tropical fluvial systems occurs under alternating wet and dry seasons. We propose that likewise on longer time scales, changes from drier intervals to wetter intervals also produce higher sediment load. This type of variability within the Chinle basin was produced by changes in the annual intensity of the megamonsoon, as well as on decadal to millennial and longer time scales. Changes in the strength of the monsoon affected stream power and sediment load, as well as affecting aggradation or degradation of the fluvial systems, similar to that described for the modern monsoon in India (Sinha 2005; Gibling et al. 2005).

The succession of depositional systems (Fig. 13A), sequence boundaries (Fig. 12), and paleosols developed both on interfluve strata as well as in valley fills and on floodplain deposits provides the evidence to schematically reconstruct paleoclimate variation on several scales in the Chinle tropical depositional system. Seasonality of rainfall, especially in the tropics, is more important than the total amount of rainfall in determining weathering, pedogenesis, bioturbation, vegetation cover, and sediment yield (Fig. 13B) (Cecil and Dulong 2003; Hasiotis et al. 2007). By incorporating paleosol and ichnological evidence in this study, the emphasis is placed on rainfall in the wet season and the number of wet months per year as an indicator of a tropical monsoonal climate for the early part of Chinle deposition, followed by a change to semihumid or semiarid climates in the latter part of Chinle deposition. Other studies have interpreted a semihumid to semiarid climate for the Chinle (Tanner and Lucas 2006; Cleveland et al. 2007); those interpretations may reflect study of local areas where the entire Chinle stratigraphic section is not exposed.

Chinle paleosols exhibit a distinct stratigraphic trend (Fig. 13C) from Oxisols, Gleysols, and Histosols reflecting fluctuating, high water tables, wetter climates, and long periods of landscape stability at the base of the Chinle, to Alfisols (Argillisols) and Vertisols indicating alternating wet-dry seasonality and greater fluctuating soil moisture and water tables in the middle of the Chinle, and to calcic Vertisols, Calcisols, and Inceptisols (Aridisols) characteristic of drier climatic conditions and highly fluctuating, deep water tables at the close of Chinle deposition (Fig. 13C). This progression in the major paleosol types implies a systematic change in the number of wet months – that is, the seasonal distribution of rainfall – in the climate through time (Fig. 13D). Because the number of wet months is in part related to latitude and monsoonal climates, the change in paleosols orders is interpreted to signal the northward migration of the Colorado Plateau and the Chinle basin within Pangea from monsoonal-wet equatorial latitudes in the Early Triassic to semiarid and arid mid-latitudes in the Late Triassic (Fig. 13C). Similar development of a stratigraphic Oxisol-Histosol-Calcisol progression in foreland-basin deposits has

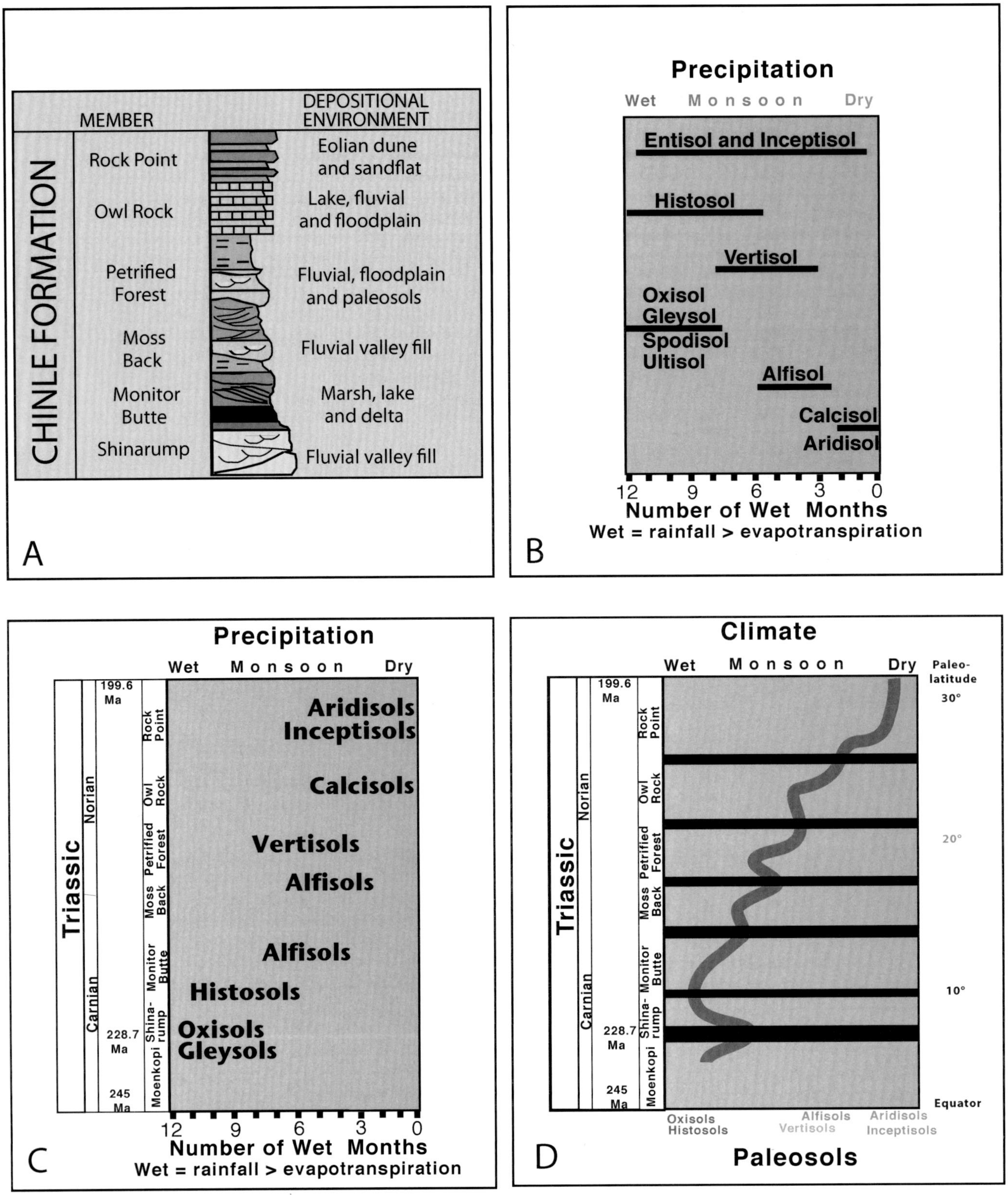

FIGURE 13.—A) Lithostratigraphic column and interpreted depositional environments for members of the Chinle Formation. B) Diagram illustrating the relation between the number of wet months per year versus type of paleosol formed at tropical latitudes (modified from Cecil and Dulong, 2003). C) Diagram illustrating the general stratigraphic distribution of major paleosol orders in the Chinle Formation, plotted against the seasonal distribution of precipitation as the number of wet months per year (adapted from Cecil and Dulong, 2003; ages from

been described for the northward migration of India under monsoonal conditions from the equatorial region to drier mid-latitudes in the Paleogene (Singh et al. 2009).

Each of the major aggradational-degradational cycles is interpreted to have been produced by long-term fluctuations in climate, between drier periods that resulted in diminished fluvial discharge and sediment load, and subsequent erosion of paleovalleys, and wetter cycles that produced increased discharge and sediment load and the filling of the paleovalleys, similar to the modern monsoonal river systems in India (Gibling et al. 2005). Factors influencing these large-scale fluctuations include the rate and locus of tectonic subsidence in the basin, rate of sediment supply from highland and volcanic sources, and amount and seasonal distribution of precipitation as a result of long-term climate.

Superimposed on these major degradational-aggradational cycles are shorter-term, or higher-frequency, minor cycles (Figs. 12, 13C). These minor cycles are on the scale of fluvial sandstones and related paleosols that represent variations in fluvial style from low-sinuosity to high-sinuosity in the Chinle members. Smaller allogenic and autogenic cycles are represented by individual fluvial sandstones and adjacent paleosols, and continued work in this study or by others will undoubtedly document additional cycles. The successions of paleosols developed on Chinle floodplain mudstones indicate that these smaller cycles of paleosol development are on the order of 10^2 to 10^6 years.

One other observation for cyclic climate change affecting sedimentation at several scales of magnitude in the Chinle, rather than tectonic or autocyclic controls, is the alternation of sandsheet and eolian sandstones in the upper part of the Rock Point Member (Fig. 9F). Despite the climate becoming dry enough to deposit eolian sandsheet, loess, and dune strata, the regular alternations of eolian lithofacies (Fig. 9F) (Dubiel 1989a) demonstrate more arid conditions for the eolian dunes alternating with less arid conditions for the eolian sandsheet deposits. These facies alternations indicate that climate cyclicity continued to affect Chinle deposition even under semiarid to arid climatic settings.

SUMMARY

The continental strata of the Chinle Formation contain several orders of degradational-aggradational cycles as expressed in the sequence stratigraphy and sequence boundaries recognized from fluvial sandstones and valley fills and from the succession of paleosols. These degradational-aggradational cycles are interpreted to be the result of changes in climate on several orders of magnitude. Variations in fluvial discharge and clastic sediment load in topical settings has been attributed principally to the influence of climate and particularly to the amount of seasonal rainfall in terms of the number of wet months per year (Cecil 1990; Cecil and Dulong 2003).

The Chinle Formation on the Colorado Plateau displays sequences of valley-fill, lacustrine-marsh, alluvial-floodplain, lacustrine-palustrine, and playa-eolian environments that reflect long-term ($\sim 10^6$ to 10^7 yr) climatic change from an ever-wet and wet-dry monsoonal climate at the start of Chinle deposition to one of increasingly arid climate at the close of Chinle deposition. This trend corresponds to the migration of the Colorado Plateau area from tropical latitudes just north of the paleoequator on Pangea and within the monsoonal belt in the Triassic northward into the drier mid-latitudes in the Jurassic. It may also reflect the disruption of the global monsoonal circulation system by the breakup of Pangea in the Late Triassic and Early Jurassic (Parrish 1993a, 1993b). Pedogenic modification of sedimentary deposits under this long-term climate change produced an overall succession from Oxisols and Gleysols in the underlying Moenkopi Formation and basal members of the Chinle, to Alfisols (Argillisols) and Vertisols in the middle members of the Chinle, and finally to Inceptisols (Protosols) and Aridisols in the upper part of the Chinle. Within these major paleosol orders a spectrum of structural and textural pedogenic fabrics and bioturbation patterns record spatial and temporal variations in parent material and also indicate changes in the seasonality of total precipitation expressed as the number of wet months per year. These paleosol variations span time scales of 10^2 to 10^6 yr, arising from perturbations in the strength and duration of the monsoon, which affected the number of wet months per year and the rainfall per month. Invertebrate ichnofossils, such as crayfish burrows, in paleosols provide one criterion to infer paleohydrologic conditions and seasonal water-table fluctuations, and to interpret climate on the scale of 10^0 to 10^2 yr.

ACKNOWLEDGMENTS

In addition to our own work, this paper draws on the research, publications, and expertise of many colleagues. We thank others who, through discussions and their own research, have influenced our studies of the Chinle: Ron Blakey, Larry Middleton, and Nancy Riggs, Northern Arizona University; our dissertation advisor Mary Kraus, University of Colorado; and Tim Demko, ExxonMobil Upstream Research Company. We appreciate constructive reviews and comments by Mark Kirschbaum and Dick Keefer (U.S. Geological Survey) of an earlier draft of the paper. Critiques by Sylvie Bourquin (Université de Rennes, France), Greg Nadon (Ohio University), and editors Stephanie Davidson, Colin North, and Sophie Leleu (University of Aberdeen) provided numerous suggestions to focus the concepts expressed in this paper.

REFERENCES

Archer, A.W., 2005, A review of Amazonian depositional systems, *in* Blum, M.D., Marriott, S.B., and Leclair, S.F., eds., Fluvial Sedimentology VII: International Association of Sedimentologists, Special Publication 35, p. 17–39.

Ash, S.R., 1967, The Chinle (Upper Triassic) megaflora, Zuni Mountains, New Mexico: New Mexico Geological Society, 18th Annual Field Conference Guidebook, p. 125–131.

Ash, S.R., 1972, Plant megafossils in the Chinle Formation, *in* Breed, C. S., and Breed, W.J., eds., Investigations in the Triassic Chinle Formation: Museum of Northern Arizona, Bulletin 47, p. 23–44.

Ash, S.R., Litwin, R.J., and Traverse, A., 1982, The Upper Triassic fern *Phlebopteris smithii* (Daugherty) Arnold and its spores: Palynology, v. 6, p. 202–219.

Atchley, S.C., Nordt, L.C., and Dworkin, S.I., 2004, Eustatic control on alluvial sequence stratigraphy: a possible example from the Cretaceous–Tertiary transition of the Tornillo Basin, Big Bend National Park, West Texas: Journal of Sedimentary Research, v. 74, p. 391–404.

←

International Commission on Stratigraphy 2009). D) Diagram illustrating the overall trend of drying climate through time in the Chinle based on sedimentologic evidence and the type of paleosols dominant in individual members. The curve represents the Chinle Mega-Degradational-Aggradational cycle, and the perturbations of the curve correspond to Major Degradational-Aggradational cycles (see Figure 12 and text discussion). Heavy black horizontal lines schematically depict missing section inferred from unconformities and sequence boundaries; see Figures 2 and 12, and discussion in text.

Bazard, D.R., and Butler, R.F., 1991, Paleomagnetism of the Chinle and Kayenta Formations, New Mexico and Arizona: Journal of Geophysical Research, v. 96, p. 9847–9871.

Beer, J.J., 2005, Sequence stratigraphy of fluvial and lacustrine deposits in the lower part of the Chinle Formation, south central Utah, United States: Paleoclimatic and tectonic implications: MS Thesis, University of Minnesota, 169 p.

Birkeland, P.W., 1984, Soils and Geomorphology: New York, Oxford University Press, 372 p.

Blakey, R.C., 1974, Stratigraphic and depositional analysis of the Moenkopi Formation, southeastern Utah: Utah Geological and Mineral Survey, Bulletin 104, 81 p.

Blakey, R.C., 1989, Triassic and Jurassic geology of the southern Colorado Plateau, *in* Jenny, J.P., and Reynolds, S.J., eds., Geologic Evolution of Arizona: Arizona Geological Society, Digest v. 17, p. 369–396.

Blakey, R.C., 1994, Paleogeographic and tectonic controls on some Lower and Middle Jurassic erg deposits, Colorado Plateau, *in* Caputo, M.V., Peterson, J.A., and Franczyk, K.J., eds., Mesozoic Systems of the Rocky Mountain Region, USA: Denver, SEPM, Rocky Mountain Section, p. 273–298.

Blakey, R.C., 1996, Permian eolian deposits, sequences, and sequence boundaries, Colorado Plateau, *in* Longman, M.W., and Sonnenfeld, M.D., eds., Paleozoic Systems of the Rocky Mountain Region: Rocky Mountain Section, SEPM (Society for Sedimentary Geology), p. 405–426.

Blakey, R.C., and Gubitosa, R., 1983, Late Triassic paleogeography and depositional history of the Chinle Formation, southern Utah and northern Arizona, *in* Reynolds, M.W., and Dolly, E.D., Mesozoic Paleogeography of West Central United States: Rocky Mountain Section, Society of Economic Paleontologists and Mineralogists, Rocky Mountain Paleogeography Symposium 2, p. 57–76.

Blakey, R.C., and Gubitosa, R., 1984, Controls of sandstone body geometry and architecture in the Chinle Formation (Upper Triassic), Colorado Plateau: Sedimentary Geology, v. 38, p. 51–86.

Blakey, R.C., Basham, E.L., and Cook, M.J., 1993, Early and Middle Triassic paleogeography, Colorado Plateau and vicinity, *in* Morales, M., ed., Aspects of Mesozoic Geology and Paleontology of the Colorado Plateau: Museum of Northern Arizona, Bulletin 59, p. 13–26.

Blodgett, R.H., 1984, Nonmarine depositional environments and paleosol development in the Upper Triassic Dolores Formation, southwestern Colorado, *in* Brew, D.C., ed., Field Trip Guidebook: Geological Society of America, Rocky Mountain Section, 37th Annual Meeting, Durango, Colorado, p. 46–92.

Blodgett, R.H., 1988, Calcareous paleosols in the Triassic Dolores Formation, southwestern Colorado, *in* Reinhardt, J., and Sigleo, W., eds., Paleosols and Weathering through Geologic Time: Geological Society of America, Special Paper 216, p. 103–121.

Boucher, T.E., 2004. Influence of halokinesis on fluvial architecture and paleosol development: the late Triassic Chinle formation at Castle Valley (Paradox Basin), southeastern Utah: M.S. Thesis, Baylor University, Waco, TX, 65 p.

Bourquin, S., Rigollet, C., and Bourges, P., 1998, High-resolution sequence stratigraphy of alluvial fan–delta environment: Stratigraphic and geodynamic implications – an example from Keuper Chaunoy Sandstones, Paris Basin: Sedimentary Geology, v. 121, p. 207–237.

Bourquin, S., Guillocheau, F., and Péron, S., 2009, Braided rivers within an arid alluvial plain (example from the Lower Triassic, western Germanic Basin): Recognition criteria and expression of stratigraphic cycles: Sedimentology, v. 56, p. 2235–2264.

Bown, T.M., Kraus, M.J., and Middleton, L.T., 1983, Triassic fluvial systems, Petrified Forest National Park, Arizona (abstract): Symposium on Southwestern Geology and Paleontology, Museum of Northern Arizona, 1 p.

Buringh, P., 1968, Introduction to the study of tropical and subtropical soils: Wageningen, The Netherlands, Centre for Agricultural Publishing and Documentation, Agricultural University of Wageningen, 118 p.

Buringh, P., 1979, Introduction to the Study of Soils in Tropical and Subtropical regions: Wageningen, The Netherlands, Centre for Agricultural Publishing and Documentation, Agricultural University of Wageningen, Third Edition, 124 p.

Busby-Spera, C., 1988, Speculative tectonic model for the early Mesozoic arc of the southwest Cordilleran United States: Geology, v. 16, p. 1121–1125.

Cecil, C.B., 1990, Paleoclimate controls on stratigraphic repetition of chemical and siliciclastic rocks: Geology, v. 18, p. 533–536.

Cecil, C.B., and Dulong, F.T., 2003, Precipitation models for sediment supply in warm climates, *in* Cecil, C.B., and Edgar, N.T., Climate Controls on Stratigraphy: SEPM, Special Publication 77, p. 21–27.

Chan, M.J., 1999, Triassic loessite of north-central Utah: Stratigraphy, petrophysical character, and paleoclimate implications: Journal of Sedimentary Research, v. 69, p. 477–485.

Cleveland, D.M., Atchley, S., and Nordt, L.C., 2007, Continental sequence stratigraphy of the Upper Triassic Chinle strata, northern New Mexico, USA: Allocyclic and autocyclic origins of paleosols-bearing alluvial successions: Journal of Sedimentary Research, v. 77, p. 909–924.

Cleveland, D.M., Nordt, L.C., and Atchley, S., 2008, Paleosols, trace fossils, and precipitation estimates of the uppermost Triassic strata in northern New Mexico: Palaeogeography, Palaeoclimatology, Palaeoecology, v. 257, p. 421–444.

Deacon, M.W., 1990, Depositional analysis of the Sonsela sandstone bed, Chinle Formation, northeast Arizona and northwest New Mexico: Unpublished Master's thesis, Northern Arizona University, Flagstaff, 128 p.

Demko, T.M., 1995a, Taphonomy of fossil plants in the Upper Triassic Chinle Formation: Ph.D. dissertation, University of Arizona, Tucson, 274 p.

Demko, T.M., 1995b, Taphonomy of fossil plants in Petrified Forest National Park, Arizona, *in* Boaz, D., ed., Fossils of Arizona: Mesa Southwest Paleontological Society and Mesa Southwest Museum, Mesa AZ, Proceedings, p. 37–52.

Demko, T.M., 2003, Sequence stratigraphy of a fluvial-lacustrine succession in Triassic Lower Chinle Formation, Central Utah, USA: Geological Society of America, Abstracts with Programs, v. 35, p. 426.

Demko, T.M., Dubiel, R.F., and Parrish, J.T., 1998, Plant taphonomy in incised valleys: Implications for interpreting paleoclimate from fossil plants: Geology, v. 26, p. 1119–1122.

Demko, T.M., Nicoll, K., Beer, J.J., Hasiotis, S.T., and Park, L.E., 2005, Mesozoic Lakes of the Colorado Plateau: Geological Society of America, Field Trip Guidebook, Field Guide 6, p. 329–356.

Dickinson, W.R., 1981, Plate tectonic evolution of the southern Cordillera, *in* Dickinson, W.R., and Payne, W.D., eds., Relations of Tectonics to Ore Deposits in the Southern Cordillera: Arizona Geological Society, Digest v. 14, p. 113–135.

Dickinson, W.R., and Gehrels, G.E., 2008, U-PB Ages of detrital zircons in relation to paleogeography: Triassic paleodrainage networks and sediment dispersal across southwest Laurentia: Journal of Sedimentary Research. v. 78, p. 745–764.

Dubiel, R.F., 1983, Sedimentology of the lower part of the Upper Triassic Chinle Formation, southeastern Utah: U.S. Geological Survey, Open-File Report 83–459, 48 p.

Dubiel, R.F., 1987a, Sedimentology of the Upper Triassic Chinle Formation, southeastern Utah: Ph.D. Dissertation, Boulder, University of Colorado, 132 p.

Dubiel, R.F., 1987b, Sedimentology of the Upper Triassic Chinle Formation, southeastern Utah: Paleoclimatic implications, *in* Morales, M., and Elliott, D.K., eds., Triassic Continental Deposits of the American Southwest: Journal of the Arizona-Nevada Academy of Science, v. 22, p. 35–45.

Dubiel, R.F., 1989a, Sedimentology and revised nomenclature of the Upper Triassic Chinle Formation and the Lower Jurassic Wingate Sandstone, northwestern New Mexico and northeastern Arizona, *in* Anderson, O.J., Lucas, S.G., Love, D.W., and Cather, S.M., eds., Southeastern Colorado Plateau: New Mexico Geological Society, 40th Field Conference Guidebook, p. 213–223.

Dubiel, R.F., 1989b, Depositional and climatic setting of the Upper Triassic Chinle Formation, Colorado Plateau, *in* Lucas, S.G., and Hunt, A.P., eds., Dawn of the Age of dinosaurs in the American Southwest: New Mexico Museum of Natural History, 1989 Spring Field Conference Guidebook, p. 171–187.

Dubiel, R.F., 1991, Architectural facies analysis of non-marine depositional systems in the Upper Triassic Chinle Formation, southeastern Utah, *in* Miall, A.D., and Tyler, N., eds., The Three-Dimensional Facies Architecture of

Terrigenous Clastic Sediments and Its Implications for Hydrocarbon Discovery and Recovery: SEPM, Concepts in Sedimentology and Paleontology, v. , p. 104–110.

Dubiel, R.F., 1992, Sedimentology and depositional history of the Upper Triassic Chinle Formation in the Uinta-Piceance and Eagle basins, northwestern Colorado and northeastern Utah: U.S. Geological Survey Bulletin 1787-W, 25 p.

Dubiel, R.F., 1994, Triassic deposystems, paleogeography, and paleoclimate of the Western Interior, *in* Caputo, M.V., Peterson, J.A., and Franczyk, K.J., eds., Mesozoic Systems of the Rocky Mountain Region, USA: Denver, SEPM (Society for Sedimentary Geology), Rocky Mountain Section, p. 133–168.

Dubiel, R.F., and Good, S.C., 1991, Paleovalley fills and lacustrine deposition in the Owl Rock Member of the Chinle Formation (Upper Triassic), Defiance uplift, Arizona (abstract): Geological Society of America, Abstracts with Programs, v. 23, p. 18.

Dubiel, R.F., and Hasiotis, S.T., 1994a, Integration of sedimentology, paleosols, and trace fossils for paleohydrologic and paleoclimatic interpretations in a tropical alluvial system (abstract): Geological Society of America, Annual Meeting, Seattle, WA, Abstracts with Programs, v. 26, no. 7, p. A-502.

Dubiel, R.F., and Hasiotis, S.T., 1994b, Paleosols and rhizofacies as indicators of climate change and groundwater fluctuations: The Upper Triassic Chinle Formation (abstract): Geological Society of America, Abstracts with Programs, v. 26, no. 6, p. 11–12.

Dubiel, R.F., and Smoot, J.P., 1994, Criteria for interpreting paleoclimate from red beds - A tool for Pangean reconstructions, *in* Beauchamp, B., Embry, A.F., and Glass, D., eds., Pangea: Global Environments and Resources: Canadian Society of Petroleum Geologists, Memoir 17, p. 295–310.

Dubiel, R.F., Parrish, J.T., Parrish, J.M., and Good, S.C., 1991, The Pangaean megamonsoon - Evidence from the Upper Triassic Chinle Formation Colorado Plateau: Palaios, v. 6, p. 347–370.

Dubiel, R.F., Skipp, G., and Hasiotis, S.T., 1992, Continental depositional environments and tropical paleosols of the Upper Triassic Chinle Formation, Eagle basin, western Colorado, *in* Flores, R.M., ed., SEPM Mid-Year Meeting on the Mesozoic of the Western Interior, Fieldtrip Guidebook, p. 21–37.

Dubiel, R.F., Huntoon, J.E., Condon, S.M., and Stanesco, J.D., 1996, Permian deposystems, paleogeography, and paleoclimate of the Paradox Basin and vicinity, *in* Longman, M.W., and Sonnenfeld, M.D., eds., Paleozoic Systems of the Rocky Mountain Region: Rocky Mountain Section, SEPM (Society for Sedimentary Geology), p. 427–443.

Dubiel, R.F., Hasiotis, S.T., and Demko, T.M., 1999, Incised valley fills in the lower part of the Chinle Formation, Petrified Forest National Park, Arizona: Complete measured sections and regional stratigraphic implications for Upper Triassic rocks, *in* Santucci, V., and McClelland, L., eds., National Park Service Paleontologic Research Volume, Geologic Resource Division Technical Report NPS/NRGRD/GRDTR-99/03, p. 78–84.

Gehrels, G.E., Dickinson, W.R., and Riggs, N., 1993, Detrital zircon provenance link of the Chinle Formation and the Auld Lang Syne Group in Nevada (abstract): Geological Society of America, Abstracts with Programs, v. 25, p. 41.

Gibling, M.R., Tandon, S.K., Sinha, R., and Jain, M., 2005, Discontinuity-bounded alluvial sequences of the southern Gangetic plains, India: Aggradation and degradation in response to monsoonal strength: Journal of Sedimentary Research, v. 75, p. 369–385.

Goldbery, R., and Beyth, M., 1984, Lateritization and groundwater alteration phenomena in the Triassic Budra Formation, southwestern Sinai: Sedimentology, v. 31, p. 575–594.

Good, S.C., 1993, Molluscan paleobiology of the Upper Triassic Chinle Formation, Arizona and Utah: Ph.D. Dissertation, Boulder, University of Colorado, 276 p.

Good, S.C., and Dubiel, R.F., 1992, Architecture of fluvial sandstones of the Upper Triassic Chinle Formation, southeaster Utah: SEPM 1992 Theme Meeting, Mesozoic of the Western Interior, Abstracts with Program, p. 27.

Hasiotis, S.T., 1991, Paleontology, sedimentology, and paleoecology of the Upper Triassic Chinle Formation in the Canyonlands Area, southeastern Utah: Unpublished Master's Thesis, Department of Geology, University of Buffalo, 354 p.

Hasiotis, S.T., and Dubiel, R.F., 1993, Crayfish burrows and their paleohydrologic significance - Upper Triassic Chinle Formation, Ft. Wingate, New Mexico, *in* Lucas, S.G. and Morales, M., eds., The Nonmarine Triassic Symposium: New Mexico Museum of Natural History, Bulletin 3, Field Guidebook, p. G24–G26.

Hasiotis, S.T., and Dubiel, R.F., 1994, Ichnofossil tiering in Triassic alluvial paleosols: Implications for Pangean continental rocks and paleoclimate, *in* Embry, A.F., Beauchamp, B., and Glass, D., eds., Carboniferous to Jurassic Pangea: Global Environments and Resources: Canadian Society of Petroleum Geologists, Memoir 17, p. 311–317.

Hasiotis, S.T., and Dubiel, R.F., 1995a, Continental trace fossils, Petrified Forest National Park, Arizona: Tools for paleohydrologic and paleoecosystem reconstructions, *in* Santucci. V., and McClelland, L., eds., National Park Service Paleontologic Research, Technical Report NPS/NRPO/NRTR-95/16, p. 82–88.

Hasiotis, S.T., and Dubiel, R.F., 1995b, Termite (Insecta: Isoptera) nest ichnofossils from the Triassic Chinle Formation, Petrified Forest National Park, Arizona. Ichnos, v. 4, p. 119–130.

Hasiotis, S.T., Dubiel, R.F., and Demko, T.M., 1998, A holistic approach to reconstructing Triassic paleoecosystems: Using ichnofossils and paleosols as a basic framework, *in* Santucci, V., and McClelland, L., eds., National Park Service Paleontologic Research, Geologic Resource Division Technical Report NPS/NRGRD/GRDTR-98/01, p. 122–124.

Hasiotis, S.T., and Honey, J.G., 2000, Paleohydrologic and stratigraphic significance of crayfish burrows in continental deposits: Examples from several Paleocene Laramide basins in the Rocky Mountains: Journal of Sedimentary Research, v. 70, p. 127–139.

Hasiotis, S.T., Kraus, M.J., and Demko, T.M., 2007, Climate controls on continental trace fossils, *in* Miller, W. III, ed., Trace Fossils—Concepts, Problems, Prospects: Amsterdam, Elsevier Press, p. 172–195.

Hasiotis, S.T., and Martin, A., 1999, Probable reptile nests from the Upper Triassic Chinle Formation, Petrified Forest National Park, Arizona, *in* Santucci, V. and McClelland, L., National Park Service Paleontological Research, v. 4, p. 85–90.

Hasiotis, S.T., and Mitchell, C.E., 1989, Lungfish burrows in the Upper Triassic Chinle and Dolores formations, Colorado Plateau-Discussion: New evidence suggests origin by a burrowing decapod crustacean: Journal of Sedimentary Petrology, v. 59, p. 871–875.

Hasiotis, S.T., and Mitchell, C.E., 1993, A comparison of crayfish burrow morphologies: Triassic and Holocene fossil, paleo- and neo-ichnological evidence, and the identification of their burrowing signatures: Ichnos, v. 2, p. 291–314.

Hasiotis, S.T., Mitchell, C.E., and Dubiel, R.F., 1993, Application of burrow morphologic interpretations: Permian and Holocene lungfish burrow morphologies and their comparison to Upper Triassic Chinle Formation burrow localities in western Colorado: Ichnos, v. 2, p. 315–333.

Hazel, J.E., 2000, Sedimentary response to intrabasinal salt tectonism in the Upper Triassic Chinle Formation, Paradox Basin, Canyonlands, Utah: U.S. Geological Survey, Bulletin 2000-F, 34 p.

Holt, E.L., 1947, Upright trunks of *Neocalamites* from the Upper Triassic of western Colorado: Journal of Geology, v. 55, p. 511–513.

International Commission on Stratigraphy, 2009, International Stratigraphic Chart: http://www.stratigraphy.org/column.php?id=Chart/Time%20Scale, August 2009.

Kirby, R.G., 1981, vertebrate fauna from the Upper Triassic Owl Rock Member of the Chinle Formation in Northern Arizona: M.S. Thesis, Northern Arizona University, Flagstaff, 476 p.

Kirby, R.E., 1993, Relationships of Late Triassic basin evolution and faunal replacement events in the southwestern United States: Perspectives from the upper part of the Chinle Formation in northern Arizona, *in* Lucas, S.G., and Morales, M., eds., The Nonmarine Triassic: New Mexico Museum of Natural History and Science, Bulletin 3, p. 233–242.

Kraus, M.J., 1987, Integration of channel and floodplain suites, II. Vertical relations of alluvial paleosols: Journal of Sedimentary Petrology, v. 57, p. 602–612.

Kraus, M.J., 1999, Palaeosols in clastic sedimentary rocks: their geologic applications: Earth-Science Reviews, v. 47, p. 41–70.

Kraus, M.J., 2002, Basin-scale changes in floodplain paleosols: Implications for interpreting alluvial architecture: Journal of Sedimentary Research, v. 72, p. 500–509.

KRAUS, M.J., AND ASLAN, A., 1999, Palaeosol sequences in floodplain environments: hierarchical approach, *in* Thiry, M., and Simon-Coincon, R., eds., Palaeoweathering, Palaeosurfaces and Related Continental Deposits: International Association of Sedimentologists, Special Publication 27, p. 303–321.

KRAUS, M.J., AND HASIOTIS, S.T., 2006, Significance of different modes of rhizolith preservation to interpreting paleoenvironmental and paleohydrologic settings: examples from Paleogene paleosols, Bighorn basin, Wyoming: Journal of Sedimentary Research, v. 76, p. 633–646.

KRAUS, M.J., AND MIDDLETON, L.T., 1987, Dissected paleotopography and base-level changes in a Triassic fluvial system: Geology, v. 15, p. 18–21.

KUENZI, W.D., HORST, O.H., AND MCGEHEE, R.V., 1979, Effect of volcanic activity on fluvial-deltaic sedimentation in a modern arc-trench gap: Geological Society of America, Bulletin, v. 90, p. 827–838.

KUTZBACH, J.E., AND GALLIMORE, R.G., 1989, Pangaean climates: Megamonsoons of the megacontinent: Journal of Geophysical Research, v. 94, p. 3341–3357.

LAWTON, T.F., 1994, Tectonic and sedimentologic framework of the Western Interior, *in* Peterson, J.A., and Caputo, M., eds., Mesozoic Systems of the Western Interior: Rocky Mountain Section, SEPM (Society for Sedimentary Geology), Mesozoic Paleogeography Volume, p. 1–25.

LEHMAN, T.M., AND SCHNABLE, J.P., 1992, Triassic sedimentation in west Texas and eastern New Mexico: SEPM, 1992 Theme Meeting, Mesozoic of the Western Interior, Abstracts Volume, p. 39.

LUCAS, S.G., 1993a, The Upper Triassic Chinle Group, western United States, *in* Lucas, S.G., and Morales, M., eds., The Nonmarine Triassic: New Mexico Museum of Natural History and Science, Bulletin 3, p. G2-G3.

LUCAS, S.G., 1993b, The Chinle Group - Revised stratigraphy and biochronology of Upper Triassic nonmarine strata in the western United States, *in* Morales, M., ed., Aspects of Mesozoic Geology and Paleontology of the Colorado Plateau: Museum of Northern Arizona, Bulletin 59, p. 27–50.

LUCAS, S.G., 1997, The Upper Triassic Chinle Group, western United States: A nonmarine standard for Late Triassic time, *in* Dickins, J.M., ed., Late Palaeozoic and Early Mesozoic Circum-Pacific Events and Their Global Correlation: Cambridge, U.K., Cambridge University Press, p. 209–228.

LUCAS, S.G., HECKERT, A.B., ESTEP, J.W., AND ANDERSON, O., 1997, Stratigraphy of the Upper Triassic Chinle Group, Four Corners region, *in* Anderson, O.J., Kues, B., and Lucas, S.G., eds., Mesozoic Geology and Paleontology of the Four Corners region: New Mexico Geological Society, Guidebook, p. 81–108.

MACFARLANE, M.J., 1976, Laterite and Landscape: London, Academic Press, 151 p.

MACK, G.H., JAMES, W.C., AND MONGER, H.C., 1993, Classification of paleosols: Geological Society of America, Bulletin, v. 105, p. 129–136.

MACLACHLAN, M.E., 1972, Triassic system, *in* Mallory, W.W., ed., Geologic Atlas of the Rocky Mountain region: Rocky Mountain Association of Geologists, p. 166–176.

MARTZ, J.W., AND PARKER, W.G., 2010, Revised lithostratigraphy of the Sonsela Member (Chinle Formation, Upper Triassic) in the southern part of Petrified Forest National Park, northeastern Arizona: PLoS ONE (online journal) doi:10.1371/journal.pone.0009329, 26 p.

MARTIN, A., AND HASIOTIS, S.T., 1998, Vertebrate tracks and their significance in the Chinle Formation (Late Triassic), Petrified Forest National Park, Arizona, *in* Santucci, V. L. and McClelland, L., eds., National Park Service Paleontological Research, Technical Report NPS/NRGRD/GRDTR-98/01, p. 138–143.

MCCARTHY, P.J., AND PLINT, A.G., 1998, Recognition of interfluve sequence boundaries: Integrating paleopedology and sequence stratigraphy: Geology, v. 26, p. 387–390.

MCGOWEN, J.H., GRANATA, G.E., AND SENI, S.J., 1983, Depositional setting of the Triassic Dockum Group, Texas panhandle and eastern New Mexico, *in* Reynolds, M.W., and Dolly, E.D., Mesozoic Paleogeography of West Central United States: Rocky Mountain Section, Society of Economic Paleontologists and Mineralogists, Rocky Mountain Paleogeography Symposium 2, p. 13–38.

MCKEE, E.D., ORIEL, S.S., KETNER, K.B., MACLACHLAN, M.E., GOLDSMITH, J.W., MACLACHLAN, J.C., AND MUDGE, M.R., 1959, Paleotectonic maps of the Triassic System: U.S. Geological Survey, Miscellaneous Geologic Investigations Map I-300, p. 13–14.

MOHR, E.C.J., VAN BAREN, F.A., AND VAN SCHUYLENBORGH, J., 1972, Tropical Soils: The Hague, Netherlands, Mouton-Ichtiar Baru-Van Hoeve, 480 p.

O'SULLIVAN, R.B., 1970, The upper part of the Upper Triassic Chinle Formation and related rocks, southeastern Utah and adjacent areas: U.S. Geological Survey Professional Paper 644-E, 33 p.

PACE, D.W., GASTALDO, R.A., AND NEVELING, J., 2009, Early Triassic aggradational and degradational landscapes of the Karoo Basin and evidence for climate oscillation following the Permian-TR event: Journal of Sedimentary Research, v. 79, p. 316–331.

PARKER, W.G., 2006, The stratigraphic distribution of major fossil localities in Petrified Forest National Park, Arizona, *in* Parker, W.G., Ash, S.R., and Irmis, R.B., eds., A Century of Research at Petrified Forest National Park: Museum of Northern Arizona, Bulletin 62, p. 46–61.

PARRISH, J.T., 1985, Latitudinal distribution of land and shelf and absorbed radiation during the Phanerozoic: U.S. Geological Survey, Open-File Report 85–31, 21 p.

PARRISH, J.T., 1993a, Climate of the supercontinent Pangea: Journal of Geology, v. 101, p. 215–233.

PARRISH, J.T., 1993b, Mesozoic climates of the Colorado Plateau, *in* Morales, M., ed., Aspects of Mesozoic Geology and Paleontology of the Colorado Plateau: Museum of Northern Arizona, Bulletin 59, p. 1–12.

PARRISH, J.T., AND PETERSON, F., 1988, Wind directions predicted from global circulation models and wind directions determined from eolian sandstones of the western United States–A comparison: Sedimentary Geology, v. 56, p. 261–282.

PARRISH, J.M., AND GOOD, S.C., 1987, Preliminary report on vertebrate and invertebrate fossil occurrences, Chinle Formation (Upper Triassic), southeastern Utah, *in* Campbell, J.A., ed., Geology of Cataract Canyon and Vicinity: Four Corners Geological Society, Tenth Field Conference Guidebook, p. 109–115.

PETERSON, F., 1988, Pennsylvanian to Jurassic eolian transportation systems in the western United States: Sedimentary Geology, v. 56, p. 207–260.

PIERSON, L.C., 1984, Depositional environments of the Shinarump and Monitor Butte Members of the Chinle Formation, in the White Canyon area of southeastern Utah: M.S. Thesis, Flagstaff, Northern Arizona University, 106 p.

PINDELL, J., AND DEWEY, J.F., 1982, Permo-Triassic reconstruction of western Pangaea and the evolution of the Gulf of Mexico/Caribbean region: Tectonics, v. 1, p. 179–211.

PIPIRINGOS, G.N., AND O'SULLIVAN, R.B., 1978, Principal unconformities in Triassic and Jurassic rocks, western interior United States–A preliminary survey: U.S. Geological Survey, Professional Paper 1035-A, 29 p.

PROCHNOW, S.J., NORDT, L.C., ATCHLEY, S.C., HUDEC, M., AND BOUCHER, T.E., 2005, Triassic paleosol catenas associated with a salt-withdrawal minibasin in southeastern Utah, U.S.A.: Rocky Mountain Geologist, v. 40, p. 25–49.

PROCHNOW, S.J., NORDT, L.C., ATCHLEY, S.C., AND HUDEC, M.R., 2006, Multi-proxy paleosol evidence for middle and late Triassic climate trends in eastern Utah: Palaeogeography, Palaeoclimatology, Palaeoecology, v. 232, p. 53–72.

RETALLACK, G.J., 1997, A Colour Guide to Paleosols: New York, John Wiley & Sons, 175 p.

RETALLACK, G.J., 2001, Soils of the Past; An Introduction to Paleopedology, Second Edition: Oxford, U.K., Blackwell Science, 404 p.

RIGGS, N.R., ASH, S.R., AND MATTINSON, J.M., 1994, Isotopic dating of a nonvolcanic continental sequence, Chinle Formation, Arizona (abstract): Rocky Mountain Section, Geological Society of America, Abstracts with Programs, v. 26, p. 61.

RIGGS, N.R., LEHMAN, T.M., GEHRELS, G.E., AND DICKINSON, W.R., 1996, Detrital zircon link between headwaters and terminus of the Upper Chinle-Dockum paleoriver system: Science, v. 273, 97–100.

RIGGS, N.R., ASH, S.R., BARTH, A.P., GEHRELS, G.E., AND WOODEN, J.L., 2003, Isotopic age of the Black Forest Bed, Petrified Forest Member, Chinle Formation, Arizona: Geological Society of America Bulletin, v. 115, p. 1315–1323.

SCHULTZ, L.G., 1963, Clay minerals in Triassic rocks of the Colorado Plateau: U.S. Geological Survey Bulletin 1147-C, C1-C47.

Shanley, K.W., and McCabe, P.J., 1994, Perspectives on the sequence stratigraphy of continental strata: American Association of Petroleum Geologists, Bulletin, v. 78, p. 544–568.

Shropshire, K.L., 1974, The Chinle and Jelm Formations (Triassic) of north-central Colorado: Ph.D. Dissertation, Boulder, University of Colorado, 229 p.

Singh, B.P., Tandon, S.K., Singh, G.P., and Pawar, J.S., 2009, Palaeosols in early Himalayan foreland basin sequences demonstrate latitudinal shift-related long-term climate change: Sedimentology, v. 56, p. 1464–1487.

Sinha, R., 2005, Why do Gangetic rivers aggrade or degrade?: Current Science, v. 89, p. 836–840.

Smith, J.J., Hasiotis, S.T., Kraus, M.J., and Woody, D., 2008, Morphology and paleoenvironmental implications of adhesive meniscate burrows (AMB), Paleogene Willwood Formation, Bighorn Basin, Wyoming: Journal of Paleontology, v. 82, p. 267–278.

Soil Survey Staff, 1975, Soil taxonomy, A basic system of soil classification for making and interpreting soil surveys: United States Department of Agriculture, Natural Resources Conservation Service, Agriculture Handbook No. 436, 754 p.

Soreghan, G.S., Elmore, R.D., and Lewchuk, M.T., 2002, Sedimentologic-magnetic record of western Pangean climate in upper Paleozoic loessite (lower Cutler beds), Utah: Geological Society of America, Bulletin, v. 114, p. 1019–1035.

Stewart, J.H., Poole, F.G., and Wilson, R.F., 1972, Stratigraphy and origin of the Upper Triassic Chinle Formation and related Upper Triassic strata in the Colorado Plateau region: U.S. Geological Survey Professional Paper 690, 336 p.

Tandon, S.K., and Gibling, M.R., 1997, Calcretes at sequence boundaries in Upper Carboniferous cyclothems of the Sydney basin, Atlantic Canada: Sedimentary Geology, v. 112, p. 43–67

Tanner, L.H., 2000, Palustrine-lacustrine and alluvial facies of the (Norian) Owl Rock Formation (Chinle Group), Four Corners region, southwestern U.S.A.: Implications for Late Triassic paleoclimate: Journal of Sedimentary Research, v. 70, p. 1280–1289.

Tanner, L.H., 2003, Pedogenic features of the Chinle Group, Four Corners region: Evidence of Late Triassic aridification, *in* Lucas, S.G., Semken, S.C., Berglof, W.R., and Ulmer-Scholle, D.S., eds., Geology of the Zuni Plateau: New Mexico Geological Society, 54th Annual Field Conference Guidebook, p. 269–280.

Tanner, L.H., and Lucas, S.G., 2006, Calcareous paleosols of the Upper Triassic Chinle Group, Four Corners region, southwestern United States: Climatic implications, *in* Alonso-Zarza, A.M., and Tanner, L.H., eds., Paleoenvironmental Record and Applications of Calcretes and Palustrine Carbonates: Geological Society of America, Special Paper 416, p. 53–74.

Tardy, Y., 1997, Petrology of Laterites and Tropical soils (translated by V.A.K. Sarma): Rotterdam, Netherlands, A. A. Balkema, 408 p.

Therrien, F., and Fastovsky, D.E., 2000, Paleoenvironments of early theropods, Chinle Formation (Late Triassic), Petrified Forest National Park, Arizona: Palaios, v. 15, p. 194–211.

Van der Voo, R., Mauk, F.J., and French, R.B., 1976, Permian-Triassic continental configurations and the origin of the Gulf of Mexico: Geology, v. 4, p. 177–180.

Wahl, A., Martin, A., and Hasiotis, S.T., 1998, Vertebrate coprolites and coprophagy traces, Chinle Formation (Upper Triassic), Petrified Forest National Park, Arizona, *in* Santucci, V.L. and McClelland, L., eds., National Park Service Paleontological Research, Technical Report NPS/NRGRD/GRDTR-98/01, p. 144–148.

Waters, A.C., and Granger, H.C., 1953, Volcanic debris in uraniferous sandstones and its possible bearing on the origin and precipitation of uranium: U.S. Geological Survey, Circular 224, 26 p.

Woody, D.T., 2003, Geologic reassessment of the Sonsela Member of the Chinle Formation, Petrified Forest National Park, Arizona: Unpublished Master's thesis, Museum of Northern Arizona, Flagstaff, 205 p.

Woody, D.T., 2006, Revised stratigraphy of the lower Chinle Formation (Upper Triassic) of Petrified Forest National park, Arizona, *in* Parker, W.G., Ash, S.R., and Irmis, R.B., eds., A Century of Research at Petrified Forest National Park: Museum of Northern Arizona, Bulletin 62, p. 17–45.

Wright, V.P., and Marriott, S.B., 1993, The sequence stratigraphy of fluvial depositional systems: the role of floodplain sediment storage: Sedimentary Geology, v. 86, p. 203–210.

Young, A., 1976, Tropical Soils and Soil Survey: Cambridge, U.K., Cambridge University Press, 468 p.

Ziegler, A.M., Scotese, C.R., and Barrett, S.F., 1983, Mesozoic and Cenozoic paleogeographic maps, *in* Brosche, P., and Sundermann, J., eds., Tidal Friction and the Earth's Rotation, II.: Berlin, Springer-Verlag, p. 240–252.

ALLUVIAL VALLEYS AND ALLUVIAL SEQUENCES: TOWARDS A GEOMORPHIC ASSESSMENT

MARTIN R. GIBLING

Department of Earth Sciences, Dalhousie University, Halifax, NS B3H 3J5, Canada

CHRISTOPHER R. FIELDING

Department of Geosciences, 214 Bessey Hall, University of Nebraska-Lincoln, NE 68588-0340, U.S.A.

AND

RAJIV SINHA

Department of Civil Engineering, Indian Institute of Technology, Kanpur 208016, Uttar Pradesh, India

ABSTRACT: Criteria for identifying alluvial-valley fills in the geological record have commonly been derived from coastal areas, where high-magnitude glacioeustatic fluctuations generate predictable systems tracts. In contrast, inland alluvial valleys formed in response to climatic, tectonic, and intrinsic events of varied magnitude and duration. Sequence boundaries and architecture reflect a variety of causes, typically mediated through changes in the river's equilibrium profile and equilibrium floodplain height. We evaluate valley and sequence models using Quaternary examples from southern Asia and Australia. For these areas, proxy records are well established for climate and sea level, but the scarcity of records for tectonic activity and intrinsic channel changes hamper an interpretation of stratigraphic events.

On the large alluvial plains of northern India and central Australia, climatic changes have radically affected sediment and water discharge on timescales as short as 10^3 to 10^4 years. Periods of incision and aggradation generate discontinuity-bounded sequences (aggradational–degradational rhythms), and reaches may vary between valley and channel morphology over short periods and distances. Across Asia, changes in monsoonal precipitation have caused near-synchronous incision and aggradation in varied tectonic settings. However, mature interfluve paleosols correlative with valley bases may be local or absent on these large plains, where small interfluve channels, eolian dunes, and lakes aggrade while mainstem valleys incise. Gullied floodplains, adjacent to valleys and tributaries, and terrace deposits are useful indicators of incised systems. Climatic effects appear to be underrepresented in interpretations of ancient alluvial sequences in general.

Angular unconformities, channel diversion, and incised and tilted fans near major faults reflect tectonic controls on architecture, and soft-sediment deformation and rare surface ruptures testify to earthquake events. In smaller basins such as the Rio Grande Rift of southwestern USA, alluvial architecture reflects both tectonic and climatic controls. Channel-body amalgamation may not be a good proxy for tectonically induced subsidence, especially for megafans at upland exits where a low degree of amalgamation may reflect rapid accumulation rather than rapid subsidence and accommodation creation.

Multistory valleys and extensive multistory and single-story sheets are present in inland and coastal alluvial areas. However, below the Australian continental shelf, seismic profiles identify entrenched channels and channel belts rather than the valley fills that are commonly assumed to form under such conditions. This example shows that sea-level fall and river extensions need not generate deeply incised fluvial valleys in coastal settings.

KEY WORDS: alluvial valleys, sequences, systems tracts, interfluves, aggradation, incision, climate, monsoon, tectonics

INTRODUCTION

Since the late 1970s, ancient valley fills have received considerable attention because sand-prone valley fills constitute hydrocarbon reservoirs and aquifers. A large body of research has shown that downstream coastal valleys have reasonably predictable fills because cutting and filling respond strongly to base-level transit cycles driven by rapid, high-magnitude glacioeustatic changes, at least for Quaternary systems. Numerous sequence-stratigraphic models have linked valley dynamics with sea-level fall and rise (e.g., Shanley and McCabe 1994; Catuneanu 2006), and many coastal-valley fills have been described (Boyd et al. 2006; Dalrymple 2006).

We know much less about upstream alluvial valleys beyond the range of sea-level influence, where valley cutting and filling reflects the complex interplay of tectonics, climate, sediment supply, and river drainage changes. Despite their economic potential, upstream valley fills have proved difficult to identify in the rock record because valley and extra-valley materials are similar. Additionally, even for some Quaternary systems, key forcing factors lack proxy records that would allow reliable identification of controlling factors.

In this paper, we review some ancient alluvial-valley fills (Table 1), and discuss sequence-stratigraphic models for alluvial systems (Fig. 1) and some problems of distinguishing channels and valleys in Quaternary systems. We then discuss criteria for identifying alluvial-valley fills, elaborating on criteria set out by Fielding and Gibling (2005) and Gibling (2006), and drawing in particular on our own experience of Quaternary river systems and their stratigraphic record in Asia and Australia, which are illustrated from photographs in the accompanying DVD. Finally, we compare upstream (alluvial) and downstream (coastal) valley systems, and suggest guidelines for recognizing alluvial-valley fills and applying appropriate sequence models to alluvial successions. Although many earlier researchers have reached conclusions similar to those set out here, we hope that the present paper will serve a useful purpose in applying knowledge from several well-documented geomorphic settings to the rock record.

The Asian Quaternary study areas share a monsoonal setting and lie as much as 1200 km upstream of the present tidal limit (Fig. 2A). They include tectonically active areas of the Himalayan Foreland Basin drained by the Ganga and Brahmaputra rivers, especially the Ganga Plains of northern India. Information is also included from the Sabarmati River in the Khambat Graben of northwestern India, an active extensional basin. The Australian rivers (Fig. 3) include the Burdekin, Fitzroy, and other rivers that drain eastward from the Great Dividing Range and extended during the late Quaternary across the

From River To Rock Record: The Preservation Of Fluvial Sediments And Their Subsequent Interpretation
SEPM Special Publication No. 97, Copyright © 2011
SEPM (Society for Sedimentary Geology), ISBN 978-1-56576-305-0, p. 423–447.

TABLE 1.—Selected literature examples of alluvial valleys from the geological record (Neoproterozoic to Pleistocene). The entries for coastal alluvial tracts represent predominantly fluvial fills and are a small sample of a large literature set.

Formation & Age	Thickness	Width	Fill	Comments	Authors
Inland Alluvial Tracts					
a) Inland valleys incised into bedrock					
Sis Cgl., Eocene – Oligocene, Spain	1.4 km	6–7.5 km	Coarsening up: conglomerates, channel sandstones, overbank fines, lacustrine limestone, and coal. Fining-up large-scale cycles. Many channel bodies within complex valley fill.	Valley along a transfer zone in thrust belt, long-lived (20 m.y.). Fining-up cycles due to sediment pulses from tectonism. Marginal unconformities pass into central disconformities. Syntectonic valley development. Topography probably <120 m during active life.	Vincent & Elliott 1997; Vincent 1999, 2001
Ocejo Fm., Pennsylvanian, Spain	25–450 m	75–1200 m+	Complex fills, conglomerate abundant, sandstone, mudstone, coal. Filled sinkholes on karsted base.	Large valleys filled by screes, debris and stream flows, floodplains, lakes, and mires. Trends follow basement structures, and incision probably linked to tectonics. Valley forms tightened by deformation.	Iwaniw 1984; Alonso 1989
Chinle Fm., Triassic, SW USA	15–90 m	1–8 km	Multistory conglomerate and sandstone bodies, minor mudstone at higher levels. Stories 1–10 m thick.	Valleys on major unconformity, with steep walls. Braided and locally meandering rivers, pass up into widespread sheets as valleys filled. Storey width similar to valley width, so streams were not underfit.	Blakey & Gubitosa,1984; Dubiel 2009
Kyrock Sandstone, Pennsylvanian, E USA	66 m	5 km	Conglomerates, pebbly sandstone, mud clasts. Fining-up cycles to 6 m thick. Shales with coal above.	Valley on Miss./Penn. Unconformity, in foreland basin. Stepped, terraced margins, and braided and meandering channels. Length >83 km, dendritic tributary pattern. Top sandstones extend beyond valley.	Sedimentation Seminar 1978
b) Inland valleys incised into strata of similar age					
Miocene, Italy	60 m	2 km	Coarse massive conglomerate, with planar cross-stratified conglomerate downvalley. Sandstone at top, some fines.	Braided-river channels a few metres deep with highly concentrated flows. Stacked sequences. Cut into lake clays in extensional basin. Incision and high sediment supply linked to igneous updoming.	Pascucci et al. 2006

TABLE 1.—*Continued.*

Formation & Age	Thickness	Width	Fill	Comments	Authors
Blairmore Gp., Cretaceous, Alberta, Canada	8.5 - 100 m	220 m - 22 km	Coarse conglomerate, with basal grooves and flutes, mudstone blocks, sandstone more common above.	Braided-river channels with gravel sheets and diagonal bars; high gradient and supply, variable discharge. Cut in fines, in foreland basin. Source 400 km distant, sources shifted to yield short-lived sediment pulses.	Leckie & Krystinik 1995
Brigham Gp, Neoproterozoic, W USA	>60 m	100s m to >2 km	Conglomerates (some matrix-supported), pebbly sandstones, nested channel sets, minor siltstone.	Stream and debris flows, possible estuarine levels, overlain by fluvial/ marine beds. Linked to Varanger glacial event, but tectonic events also possible.	Levy et al. 1994
Whiteclay Gravel Beds, Miocene, USA	20 m	300 m	Stratified sandstone and coarse conglomerate. Multistory fill, with aggradational stacking and incision.	Valley margins steep to vertical, locally overhanging. Traceable for 35 km, with sharp bends. Fault-controlled fill of a slot-canyon, with fault reactivation and probable stacking. Channels were up to 4 m deep and tens of metres wide.	Fielding et al. 2007
Prasident Sandstone, Pennsylvanian, Germany	15–50 m	100s m to >35 km	Multistory sandstones, with mudstone clasts. Massive siltstone from bank collapse.	Low-sinuosity channels in foreland basin. Thickest zones (30–45 m) at margins and where bodies amalgamated.	Hampson et al. 1999
Joggins Fm., Pennsylvanian, ECanada	To 6.5 m	>350 m	Multistory sandstone / mudstone bodies, with slumps, pedogenic clasts. 3 or more stories 1.6–4.4 m thick.	Small valley fills on alluvial plain, with stepped margins. No correlative paleosols or erosional surfaces within adjacent floodplain deposits, and not apparently part of regional stratigraphic surfaces.	Rygel & Gibling 2006
Coastal Alluvial Tracts					
Dunvegan Fm., Cretaceous, W. Canada	15–40 m	1–10 km	Multistory sandstones with mudstone clasts, up to 50% mudstone in seaward locations. Stories up to 8 m thick. Topmost stories heterolithic.	Meandering rivers within valleys, traceable for 330 km downstream across delta area of 50,000 km^2 within foreland basin. Steep margins cut into marine beds downdip and non-marine beds updip. Incision due to eustasy, but rivers deflected by faults and forebulge.	Plint 2002; Plint & Wadsworth 2003, 2006

TABLE 1.—*Continued.*

Formation & Age	Thickness	Width	Fill	Comments	Authors
Pleistocene, Indonesia	To 41 m	To 5 km	Sands to clays with shell fragments above, and some floodplain deposits within valley. Data from 3D seismic.	High- and low-sinuosity and braided systems, on alluvial/coastal plain, adjacent to volcanic arc. Traceable for >200 km, channels within valleys and meander scroll-bars and terraces prominent. Small incised tributary valleys, with dendritic patterns.	Posamentier 2001
Pleistocene, Malay Basin	40 m	>11.5 km	Composition not known. Data from 3D seismic.	Valley fill has meander belt with scroll bars at base, and incised tributaries with gullied margins. Associated fluvial systems include braided rivers, small meander belts, and ribbon bodies in 57 m interval of coastal plain.	Miall 2002

continental shelf. We also discuss the Channel Country rivers that drain internally to Lake Eyre, as well as rivers of the Riverine Plain that drain westward from the Great Dividing Range.

In discussing alluvial models, there are both advantages and disadvantages in choosing Quaternary examples from India and Australia. As an advantage, we have been able to identify key areas where detailed alluvial information can be linked to the geomorphic setting. The chronology of the strata is known with sufficient certainty to test incision and alluviation events against some forcing factors. In particular, climatic and alluvial events can be compared with unusual certainty—something that is rarely possible for the older sedimentary record.

As a disadvantage, the examples represent a limited set of global tectonic and climatic settings. The Asian plains experience the formidable Southwest Indian Monsoon, and the Australian sites experience major tropical cyclonic storms. Climatic factors may be less effective in areas where extreme events are less frequent. Although the Himalayan Foreland Basin is tectonically active, some of our field sites lie near the distal margin of the foreland basin and currently experience only modest tectonic activity. The Australian inland plains are largely inactive tectonically, although neotectonic activity has been recorded locally. In order to include a more balanced set of examples, we refer to other Neogene alluvial units, several in active extensional basins, for which exceptionally good datasets exist. A broader issue, discussed briefly, is whether the Quaternary icehouse period characterized by strong glacial and interglacial rhythms is relevant to much of the rock record.

ANCIENT ALLUVIAL-VALLEY FILLS

Table 1 lists literature examples of the relatively few alluvial-valley fills in upstream settings in the literature for which detailed information is available. In addition to our Indian and Australian case studies, much literature is available for Neogene alluvial strata in inland settings beyond the range of oceanic effects. Table 2 presents a small selection of this literature, chosen to represent datasets of sufficiently high resolution to allow an assessment of controlling factors. Although the assessments generally refer to stratigraphic intervals within the basin fills, some individual alluvial events can be compared with proxy records.

Inland Alluvial Tracts

Rock units in Table 1 that represent inland alluvial tracts range in age from Neoproterozoic to Pleistocene. Several examples are valley fills on bedrock unconformities, which are relatively easy to identify (Gibling 2006), and their fills are dominated by conglomerates laid down by flood flows, hyperconcentrated flows, and debris flows. In contrast, few valley fills have been identified in alluvial-plain successions where valley fills are less distinctive and where their bases may not correspond to prominent paleosols or other floodplain surfaces (Rygel and Gibling 2006). In all examples in Table 1, valley dimensions are considerably greater than those of the component channels—commonly an order of magnitude larger and probably much more in large valley fills such as the 450-m-thick fills of the Ocejo Formation (Iwaniw 1984). However, Joggins Formation valley dimensions are little larger than those of their component channels (Rygel and Gibling 2006). Valley fills are also important components of alluvial fan and volcanoclastic apron deposits (e.g., Smith 1988).

Where the original authors discussed controls on valley incision and filling, a link to tectonic activity is commonly the preferred explanation. Many of the examples in Table 1 follow fault lines or were influenced by deformation (Vincent 2001; Fielding et al. 2007), and their fills may reflect high rates of sediment supply caused, for example, by hinterland drainage changes and igneous doming (Leckie and Krystinik 1995; Pascucci et al. 2006). For none of these examples has a clear link been demonstrated between cutting and filling and climatic changes.

Both tectonic and climatic effects are prominent in the examples of Table 2. In the Rio Grande Rift in New Mexico (Mack et al. 2006), basin-bounding and intrabasinal faults influenced alluvial architecture during a period of aggradation, whereas glacial–interglacial cycles

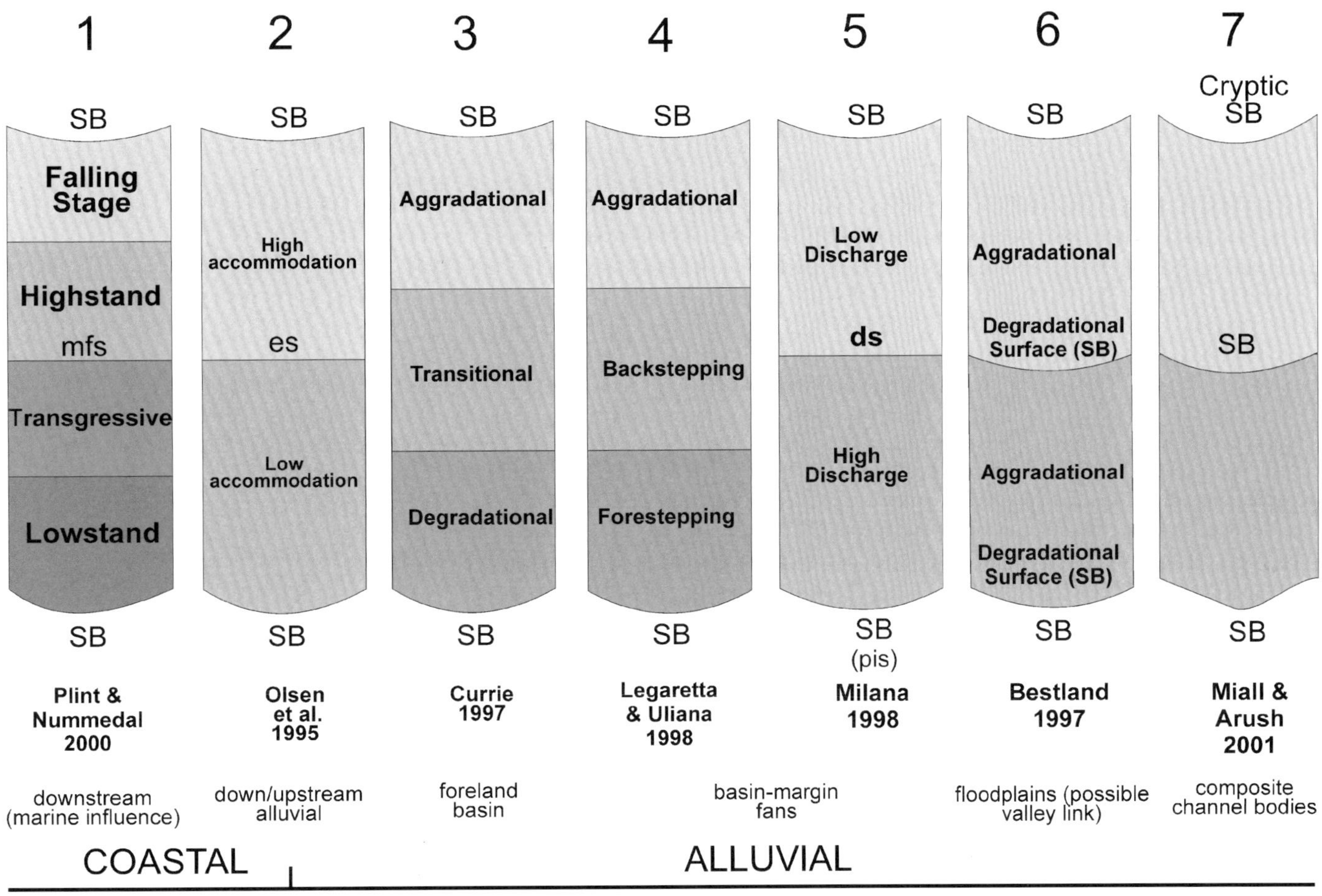

FIGURE 1.—Literature examples of sequence-stratigraphic models and systems tracts for alluvial strata. Most examples include reference to valley fills and floodplain deposits. Systems tracts in adjacent columns are not intended to be equivalent. SB = sequence boundary; mfs = maximum flooding surface; es = expansion surface; ds = downlap surface; pis = proximal incision surface. For Column 1, Wright and Marriott (1993) referred the main phase of alluvial buildup to the transgressive systems tract. For column 2, see also Martinsen et al. (1999), McCarthy et al. (1999), and Catuneanu (2006). For Column 6, see also Törnqvist et al. (2003). For Column 5, see also Weissmann et al. (2002). For Column 6, see also Bull (1991) and Gibling et al. (2005). All terms were identified as systems tracts, apart from those in Column 6.

were important during a later degradational phase when tectonic activity was reduced. Climatic effects were prominent in the St. David Formation of Arizona (Smith 1994) based on correlation between facies and paleosol stable isotope values. In the Pannonian Basin of Hungary (Nádor et al. 2003), cyclic alluvial patterns recorded by grain size and magnetic susceptibility provide a good correlation with Milankovitch effects. For all three areas, magnetostratigraphy provided age control.

Coastal Alluvial Tracts

For ancient coastal plains, valley fills that consist of fluvial strata or contain both fluvial and marine strata have been described by many authors. Three examples are shown in Table 1. Cretaceous valley fills of the Dunvegan Formation of western Canada are especially noteworthy because drainage systems were mapped across a large delta and for more than 300 km downstream (Plint 2002). The studies by Posamentier (2001) and Miall (2002) are of particular interest because they represent valley fills documented using 3D seismic profiles from (currently) offshore areas. Both studies illustrate the heterogeneity of alluvial systems in coastal areas, as well as demonstrating high-resolution geomorphic features such as incised tributaries and underfit valleys. Other detailed bedrock examples were recorded by Willis (1997) and Garrison and van den Bergh (2006).

These fills lie in Segment 3 of incised-valley systems (Boyd et al. 2006), landward of the transgressive estuarine–marine limit and extending to the inland limit of sea-level control on fluvial systems. Measured from the average highstand shoreline, the landward limit of coastal onlap due to sea-level rise is typically less than 400 km inland, varying with hinterland sediment supply and the gradient of the alluvial surface (Blum and Törnqvist 2000).

Table 2 presents information for three Quaternary coastal tracts where controlling factors have been especially thoroughly evaluated. For the Gulf of Mexico, both glacioeustasy and climate exerted a major control on alluvial architecture (Blum and Aslan 2006). In coastal areas of northern Italy, subsidence rates varied greatly and, in concert with glacioeustasy, influenced the creation of accommodation and architectural patterns (Amorosi and Milli 2001). However, Amorosi et al. (2008) used palynological datasets to demonstrate that finer-scale alluvial architecture corresponds to changes in climatic parameters. For the Rhine–Meuse Delta and the Rhine

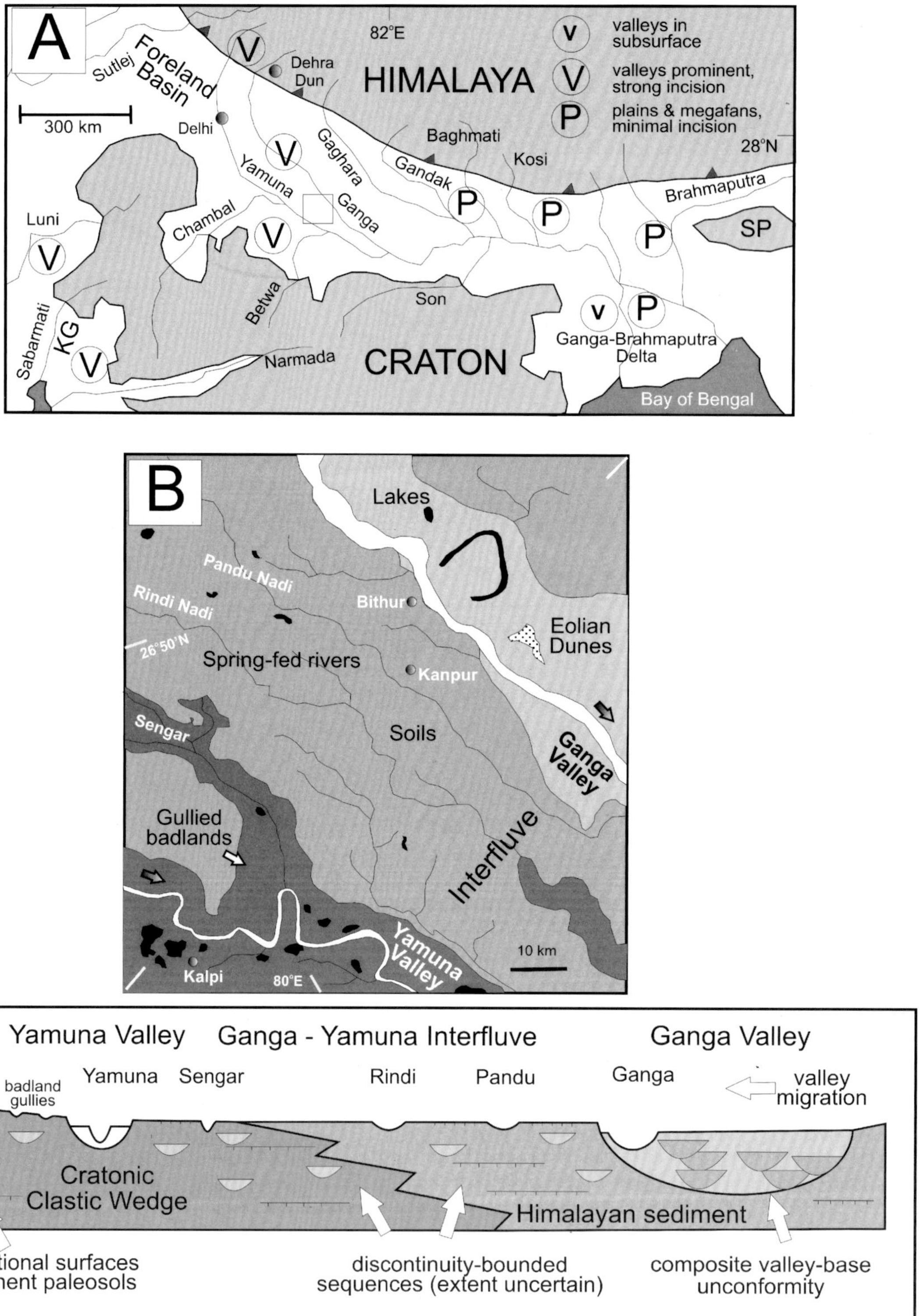

FIGURE 2.—Alluvial plains of southern Asia, for which geomorphic and Quaternary information have been drawn on in this study. A) Major rivers of the Himalayan Foreland Basin (referred to here as the Ganga Plains) and adjacent areas, to show general distribution of incised and non-incised reaches. The Yamuna, Ganga, and Gaghara rivers are separated by extensive interfluves traversed by small channels sourced on the plains by springs from monsoon rains. Some data are derived from structurally controlled depressions (duns) within the Himalayan foothills. KG = Khambat Graben, SP = Shillong Plateau. B) Landform map for Ganga–Yamuna interfluve (box in Part A), to illustrate the range of deposystems present; modified from Gibling et al. (2005). C) Schematic cross section of southern Ganga Plains (line of section shown by white lines in Part B) to show relationships of valleys, channels, and associated surfaces; modified from Sinha et al. (2009).

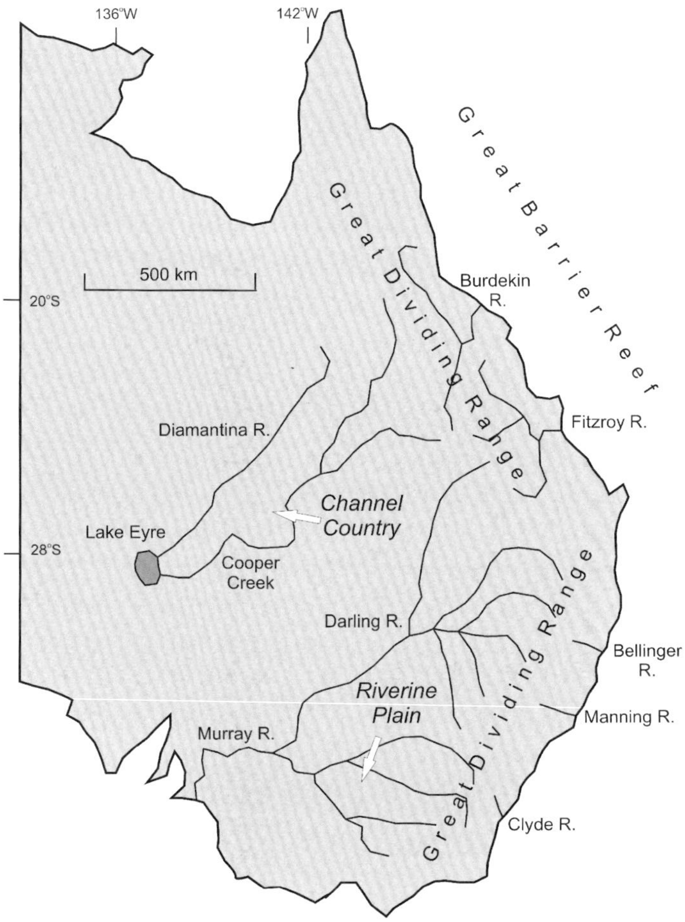

FIGURE 3.—Rivers of eastern Australia discussed in the text.

Valley, local faults and climate influenced deposition, although the overwhelming effect of glacioeustasy in coastal areas is evident (Erkens et al. 2009; Hijma et al. 2009).

ALLUVIAL VALLEYS AND SEQUENCE-STRATIGRAPHIC MODELS

Figure 1 illustrates some models that have applied sequence-stratigraphic concepts to alluvial successions, emphasizing studies that have given particular attention to inland areas. Using systems tracts identified downstream (Column 1), several studies have traced bounding surfaces into upstream alluvium based on tidal indicators, stacking patterns, and alluvial architecture (Shanley and McCabe 1991). Studies have commonly linked valley fills and multistory channel bodies to low-accommodation periods and channel bodies isolated within floodplain deposits to high-accommodation periods. In a widely used model, Wright and Marriott (1993) attributed most downstream alluvial accumulation to the transgressive systems tract. McCarthy et al. (1999) and Plint and Wadsworth (2003) identified low- and high-accommodation periods driven largely by sea level (Column 2). In their examples, valley fills and prominent paleosols (low-accommodation indicators) alternate with lacustrine, wetland, and cumulative floodplain deposits (high-accommodation indicators).

The concept of low- and high-accommodation systems tracts was extended to areas far inland by Martinsen et al. (1999). They identified expansion surfaces (Column 2) where sediments overlap valleys to cover the adjoining plain, and also recognized the balance between accommodation (A) and sediment supply (S) through an A/S ratio designed to represent stratigraphic base level. Catuneanu (2006) adopted low- and high-accommodation systems tracts as a robust framework for many alluvial systems. For a foreland basin, Currie (1997) identified degradational, transitional, and aggradational systems tracts (Column 3), with valley fills developed on degradational surfaces. Currie noted that systematic changes in systems-tract thickness imply that subsidence rates controlled accommodation. For fault-bounded basins, Gawthorpe et al. (1994) noted the importance of sediment supply and accommodation, with subsidence rates at hanging walls sufficiently high to cancel out all but the most rapid sea-level oscillations. In such settings, sequence boundaries may be poorly developed, and aggradational successions are prominent (Davies and Gibling 2003).

For basin-margin fans, two South American studies identified systems tracts that reflect alternate sediment accumulation and incision (Columns 4 and 5) and incorporate progradational and downlap stratal geometries. Legaretta and Uliana (1998) identified forestepping, backstepping and aggradational systems tracts, attributed to climate or tectonism, with valleys filled during backstepping periods. Flume experiments and field studies by Milana (1998) identified sequences generated as a result of alternate low and high discharge, without change in subsidence rate. Although they did not specify systems tracts, Weissmann et al. (2002) and Bennett et al. (2006) studied Californian fans influenced by alpine glaciation, and mapped widespread sequences ~ 15–30 m thick, bounded by valley fills and prominent paleosols. Incision took place under conditions of declining sediment:water ratio during late glacial and interglacial periods, causing loss of accumulation space, a basinward shift of deposition, and prolonged soil formation on the fan surfaces. During the main glacial periods, aggradation took place under conditions of increased sediment supply that filled the valleys and expanded over the fan surfaces with an increase in accumulation space. Long-term subsidence allowed preservation of stacked sequences. For fans in Montana, Ritter et al. (1995) identified synchronous aggradation and entrenchment of fans in varied tectonic settings, linked to climatically induced variation in discharge and sediment load.

For floodplain-dominated deposits in a variety of tectonic settings, numerous authors have recognized discontinuity-bounded sequences linked in many cases to valley fills (Column 6). The sequences represent alternate aggradation and degradation and have commonly been attributed to changes in climate and vegetation (Bull 1991; Bestland 1997; Gibling et al. 2005). Cryptic sequence boundaries (Column 7) may be recognized in stacked fluvial channel bodies or valley fills on the basis of changes in petrography and paleoflow direction (Miall and Arush 2001).

For alluvial settings, some authors have linked accommodation to "stratigraphic base level", which defines the upper limit to deposition (Shanley and McCabe 1994; Blum and Törnqvist 2000). However, as Blum and Törnqvist point out, upper limits in a fluvial regime are defined by equilibrium floodplain heights, which depend on discharge regimes and sediment supply. The strata enter preservation space (net accommodation) when they subside below the river's equilibrium profile. This profile represents the base level of erosion or maximum depth of incision, and has also been termed "geomorphic base level". Sediment accumulates in inland valleys when the equilibrium profile rises above the valley floor. Blum and Törnqvist (2000) suggested that the term "stratigraphic base level" should be abandoned because it is the same as the upper level of accommodation and does not control anything. In the Quaternary examples discussed below, lakes are mainly small and uncommon, and their local base levels do not appear to have influenced river courses.

TABLE 2.—Selected Neogene alluvial case studies for which the authors evaluated controls on alluvial accumulation and architecture. Avulsion of alluvial reaches is a prominent feature of all these case studies.

Formation, Age and Location	Tectonic Controls	Climatic Controls	Glacioeustatic Controls	References
Inland Alluvial Tracts				
Palomas and Camp Rice Fms., New Mexico, USA, Pliocene to Pleistocene (Rio Grande Rift)	Fundamental during aggradational period (5 to 0.78 Ma), linked to changes in basin symmetry and growth, and activity on basin-bounding and intrabasinal faults. Multistoreyed channel bodies border major faults, due to tectonic drawdown of axial sediment. Axial river and alluvial fan deposits intercalated, due to channel relocation. Condensed paleosols and floodplain deposits prominent away from faults. Minimal incision.	Fundamental during degradational period with reduced tectonic activity (0.78 Ma to present). Cyclic alluvial-fan deposits with calcic paleosols linked to glacial / interglacial cycles; incision prominent.	Out of range of oceanic effects.	Leeder et al. 1996; Mack and Leeder 1999; Mack and Madoff 2005; Mack et al. 2006.
St. David Fm., Arizona, USA, Pliocene to Pleistocene (extensional basin)	Period of minimal tectonic activity.	Broad correlation of sedimentation rate, channel geometry and facies abundance with climate change as deduced from pedogenic-carbonate isotope data.	Out of range of oceanic effects.	Smith 1994
Pannonian Basin, Hungary, Quaternary (two cores each nearly 500 m long)	Active subsidence regime, but sediment changes linked to tectonics do not correlate with sedimentary cycles.	General correlation of channel and floodplain (grain-size) cycles with 40 and 100 kyr Milankovitch patterns, suggesting link to glacial / interglacial cycles.	Out of range of oceanic effects.	Nádor et al. 2003
Coastal Alluvial Tracts				
Coastal plains and deltas, Italy, Quaternary (largely core data)	Rapid subsidence locally allows accumulation of thick successions; subsidence rates influence whether accommodation is created at different stages in the glacioeustatic cycles.	Glacial and interglacial changes delineated by pollen assemblages in cores, and correlate with changes in facies architecture.	Fundamental influence on architecture, controlling valley fills, lowstand paleosols, systems tracts, and stacking patterns.	Amorosi et al. 1999a, 1999b; Amorosi and Milli 2001; Amorosi et al. 2008
Coastal and alluvial plains, Gulf of Mexico, USA (Quaternary)	Steady subsidence.	Controls allostratigraphic units, terrace strata and architecture through changes in discharge regimes and sediment concentration. Interaction between climatically controlled factors and glacioeustasy results in composite, cross-cutting and superimposed valleys with adjacent mature paleosols, with fills representing 100 kyr cycles.	Controls architecture and stacking patterns within incised valley downstream from apex of alluvial plain.	Blum 1994; Blum et al. 1994; Blum and Price 1998; Blum and Aslan 2006
Rhine-Meuse Delta and Rhine Valley, Netherlands and Germany (Quaternary)	Local faults and tectonic influence on subsidence regimes, avulsion locations, deformed longitudinal profiles, and channel orientation and diversion.	Controlled terrace formation and cut-and-fill cycles in some areas and during some periods.	Fundamental control on incision during sea-level fall and on alluvial style during sea-level rise.	Cohen et al. 2002; Törnqvist et al. 2003; Stouthamer and Berendsen 2007; Erkens et al. 2009; Hijma et al. 2009

From this overview, we identify some themes that are explored below:

- Sequence boundaries have commonly been identified in upstream alluvial areas, where they are marked by valley bases, erosional floodplain surfaces, and paleosols.
- Numerous authors have linked systems tracts to tectonically created accommodation, although the interplay of climate, source-area tectonism, and basin subsidence is frequently noted (Catuneanu 2006, p. 247). The systems tracts are typically tens to hundreds of meters thick (thick enough for regional correlation), and the identification of regional thickness variations has probably encouraged researchers to link sequence architecture to tectonically generated accommodation. In other cases, however, sequences have been referred to climatic factors, and a rise in the fluvial equilibrium profile has been noted to increase accommodation space.

MODERN VALLEYS: THE ISSUE OF FLOW CONFINEMENT

Valleys are commonly understood to be landforms that are larger (typically both deeper and wider) than the channels that they contain. Following this understanding, large floods should be confined within valleys but not necessarily within the component channels. In a widely accepted definition that draws on a comparison of valley and channel sizes, Zaitlin et al. (1994) and Boyd et al. (2006) defined an incised-valley system as a "fluvially-eroded, elongate topographic low that is typically larger than a single channel form, and is characterized by an abrupt seaward shift of depositional facies across a regionally mappable sequence boundary at its base." Flood-frequency analysis should assist in distinguishing valleys, where flow is confined to the conduit, from channels, which frequently inundate interchannel areas.

Valleys might be expected to inundate the adjacent plains less frequently than about every thirty years. For a set of gauged alluvial channels, Williams (1978) showed that the adjoining floodplains were inundated every 1.5 years on average, with recurrence intervals from 1.01 to 32 years (see also Andrews 1980). Thus, drainage systems identified as "channels" should inundate their floodplains on timescales of a year to a few decades.

However, flood frequency is difficult to apply, for several reasons. Firstly, surveyed cross sections with gauging stations are rare in large valleys and for multichannel systems, and few gauging records extend for more than a century. In an analysis based on gauging records back to 1921 for the catastrophic 1981 Laingsburg valley flood in South Africa, Zawada (1994) noted that the flow represents a 10,000 year event if the 1981 flood is excluded, but represents a 400 year event if the flood is included. Secondly, many rivers experience periods of enhanced and reduced flood magnitude and frequency over decades to millennia (Graf et al. 1991), and flood-frequency estimates may apply only to short periods. Finally, anthropogenic structures such as canals and dams have modified water and sediment discharge in many rivers.

In monsoonal areas that experience strongly seasonal flow, assessments of valley prominence may be misleading. Rivers may appear to be small entities within large valleys at low-flow stage. However, they may fill or overtop the conduit margins during high-stage flow, rising to levels that a casual visitor can scarcely envisage.

Satellite imagery offers a means of testing whether fluvial conduits are acting (over short timescales) as valleys that routinely confine floods or as channels that frequently inundate their surroundings. For the Ganga Plains, flood maps created by the Dartmouth Flood Observatory, New Hampshire, USA, provide an exceptional view of areas covered by floodwaters from 1985 onwards. These maps can be viewed on the observatory website at http://www.dartmouth.edu/~floods/ where they can be accessed from the Master Index of DFO Rapid Response Inundation Maps (as file DFO 2007–115, FL-2007–000096-IND; summary sheets E70N30, E80N30, and E90N30). Coupled with field observations, satellite images for the Ganga Plains allow researchers to delineate areas with strong or minimal incision (Fig. 2A). On the western Ganga Plains near to and downstream from Delhi (Fig. 2A), the Ganga and Yamuna rivers are considered to occupy valleys in most reaches. They are almost entirely restricted to narrow courses, while spring-fed monsoonal rivers cause minor floods across the extensive plain (interfluve) between them. These observations accord well with anecdotal evidence from villagers that floods do not cover interfluves along parts of the Yamuna River (Tandon et al. 2006). The Gaghara and Rapti rivers and spring-fed rivers on the Ganga–Gaghara interfluve are also largely confined but flood more extensively in places.Because these Himalayan-sourced rivers are largely restricted to their valleys, the spring-fed interfluve channels may behave as independent systems, largely detached from the big rivers.

In contrast, on the eastern Ganga Plains (Fig. 2A), the Gandak, Baghmati, and Kosi rivers show extreme flooding, inundating enormous areas for prolonged periods (Jain and Sinha 2003, 2005). The braided Brahmaputra River has flooded much of the ground between bedrock areas of the Himalayan Foothills and the Shillong Plateau (Fig. 2A).

Beyond a purely morphological assessment, trends of greater or lesser flow confinement and valley prominence on the Ganga Plains appear to correlate with trends in precipitation and unit stream power of the rivers (Sinha et al. 2005; Tandon et al. 2006). Precipitation decreases systematically westward on the plains as a result of monsoonal airflows moving inland and westward up the foreland basin from the Bay of Bengal (Bookhagen and Burbank 2006). Closer to the Bay of Bengal, rivers such as the Gandak and Kosi have low slopes, lie within a zone of high rainfall, and show large water and sediment discharges, yielding low estimates of stream power (6.4–20 W/m^2). They occupy an aggradational zone where channels show minimal incision and inundate broad floodplains every year. Farther from the Bay of Bengal, rivers such as the Ganga and Yamuna have relatively high slopes, lie within a zone of modest rainfall, and show modest discharges of both sediment and water, yielding high estimates of stream power (up to 43 W/m^2). They occupy deeply incised valleys over long reaches. The Brahmaputra responds to very high precipitation and sediment yield from the Shillong Plateau and the eastern Himalayan syntaxis, and constitutes a braidbelt within a broad, tectonically controlled bedrock valley.

These observations suggest that large basins are likely to show more varied expressions of fluvial style than small basins. Although these northern Indian rivers occupy the same alluvial plain, the area is so vast that regional differences in precipitation, slope, and tectonics may cause substantial differences in fluvial behavior and, particularly, in the degree of incision and prominence of valleys. Smaller basins would be expected to show less regional variation in this respect.

CRITERIA FOR IDENTIFYING ALLUVIAL VALLEYS

In a major review of estuarine and incised-valley systems, Boyd et al. (2006) set out 14 criteria applicable to estuarine and, potentially, alluvial valleys. Table 3 lists these criteria and provides an assessment of their application to alluvial valleys in inland areas. We discuss some of these criteria in more detail below, in the light of Indian and Australian Quaternary systems, and illustrate sequence architecture for southern Asian alluvium in Figure 4.

EROSIONAL PALEOTOPOGRAPHY (CRITERION 1A OF TABLE 3)

Some, but not all, inland alluvial valleys show prominent paleotopography.

TABLE 3.—Applicability of criteria for estuarine and incised valley systems (Boyd et al. 2006) to inland alluvial valleys. Comments are based on the modern alluvial areas discussed in the text, principally in southern Asia and Australia, and on documented case studies in the ancient record (Table 1). See text for further discussion.

Criteria for Estuarine and Incised Valley Systems (Boyd et al. 2006)	Application to Inland Alluvial Valleys
1a. Erosional paleotopographic feature, base truncates underlying strata.	Erosional, base truncates strata.
1b. Erosional relief commonly >10 m.	Size is variable, and valleys with erosional relief of only a few metres are common. Some coastal and inland valleys show minimal erosional relief, whereas confluence scours may show deep incision.
2. Base and walls constitute sequence boundary, with correlative interfluve soils.	Base and walls constitute sequence boundary. Interfluve soils may border valleys, but may not be widespread. Key surfaces may not be traceable across broad interfluves, where small rivers, lakes and dunes may aggrade as largely independent systems.
3. Incised tributary valleys	Common. Large areas of gullied floodplains (badlands) with alluvial and colluvial gully fills may border valleys.
4. Base marks basinward facies shift, with proximal facies over distal. Filled mainly during base-level rise or (in fluvial systems) due to change from low to high accommodation style.	No basinward facies shift, as valley filled largely with alluvial sediment similar to associated strata. Filled mainly when sediment supply exceeds transport capacity (balance of discharge of water and sediment changes), as a result of tectonic or climatic causes. Results in local or regional rise of equilibrium profile (temporary base level of erosion, but not base level related to standing-water bodies). Valley filling not usually related to tectonically induced changes in accommodation; however, thickness changes of large-scale sediment packages may indicate accommodation effects. Temporary sediment buildup related to equilibrium floodplain heights and megafan topography. Landslides may promote filling in bedrock areas.
5. Fill onlaps base and walls due to filling on rising base level.	Sediment buildup may onlap margins, but usually due to change in sediment/water discharge (see #4). Fill may be promoted by base-level rise where valleys impeded by tributary inputs.
6. Associated with major sequence stratigraphic surfaces, including sequence boundary, maximum flooding surface, and ravinement surfaces.	Basal erosion surface, adjacent paleosols, and floodplain erosion surfaces fit criteria for sequence boundaries; stacked fills may contain cryptic sequence boundaries. Maximum flooding and ravinement surfaces not present. Systems tracts may be applicable in some circumstances (Columns 2–5 of Figure 1), but systems tracts of Column 1 inappropriate.
7a. Component channels much smaller than valley.	As for left-hand column.
7b. Floodplain and terrace surfaces may be present.	Terraces may be difficult to identify in ancient record.
8.-13. Criteria concerning estuarine facies.	Not applicable.
14. Valleys tend to follow downwarped areas, avoid upwarped zones.	Some valleys profoundly affected by tectonic activity: incision along fault lines, erosion through upwarped and tilted strata, diversion and avulsion due to tectonic elevation changes. Floodplains uplifted by earthquakes may be abandoned.

Below many modern coastal and shelf areas worldwide, valley fills with strong erosional paleotopography are prominent, and have been cut and filled largely as a consequence of glacioeustasy linked to the abundance of supplied sediment, which may vary considerably by catchment (Blum and Aslan 2006). Although there is much local variation, the timing and duration of cutting and filling correlates in general terms with glacial and interglacial chronology. For example, below the Ganga–Brahmaputra Delta, where modern channels are not incised (Fig. 2A), buried valleys extend for at least 350 km inland from the modern shoreline, filled with some 90 m of sandy sediment (Goodbred and Kuehl 2000a). The valleys filled during the lowstand of Marine Isotope Stage (MIS) 2 and the succeeding transgression, as sea-level rise promoted accommodation increase and backfilling under conditions of high sediment supply, eventually generating an even delta plain (Umitsu 1993).

In contrast, incision in the Indian alluvial plains far inland reflects multiple causes, and the history and duration of valley cutting and filling is much less predictable. Over timescales of thousands to a few tens of thousands of years, individual drainages have varied in their degree of incision, and may in some cases have alternated between "channels" and "valleys" over relatively short timescales. To illustrate the contrast with downstream glacioeustatic controls, we discuss below evidence for incision induced by climatic and tectonic factors in different parts of the Indian plains, as well as an example from inland Australia.

Climatic Controls

There is abundant evidence that climate is a major driver of valley dynamics in southern Asia. Comparison of dated sections with modeled and proxy precipitation records (Gibling et al. 2005; Bookhagen et al. 2006) suggests that fluvial activity has responded strongly to variations in the Southwest Indian Monsoon, the intensity of which reflects insolation and, to a lesser degree, glacial boundary

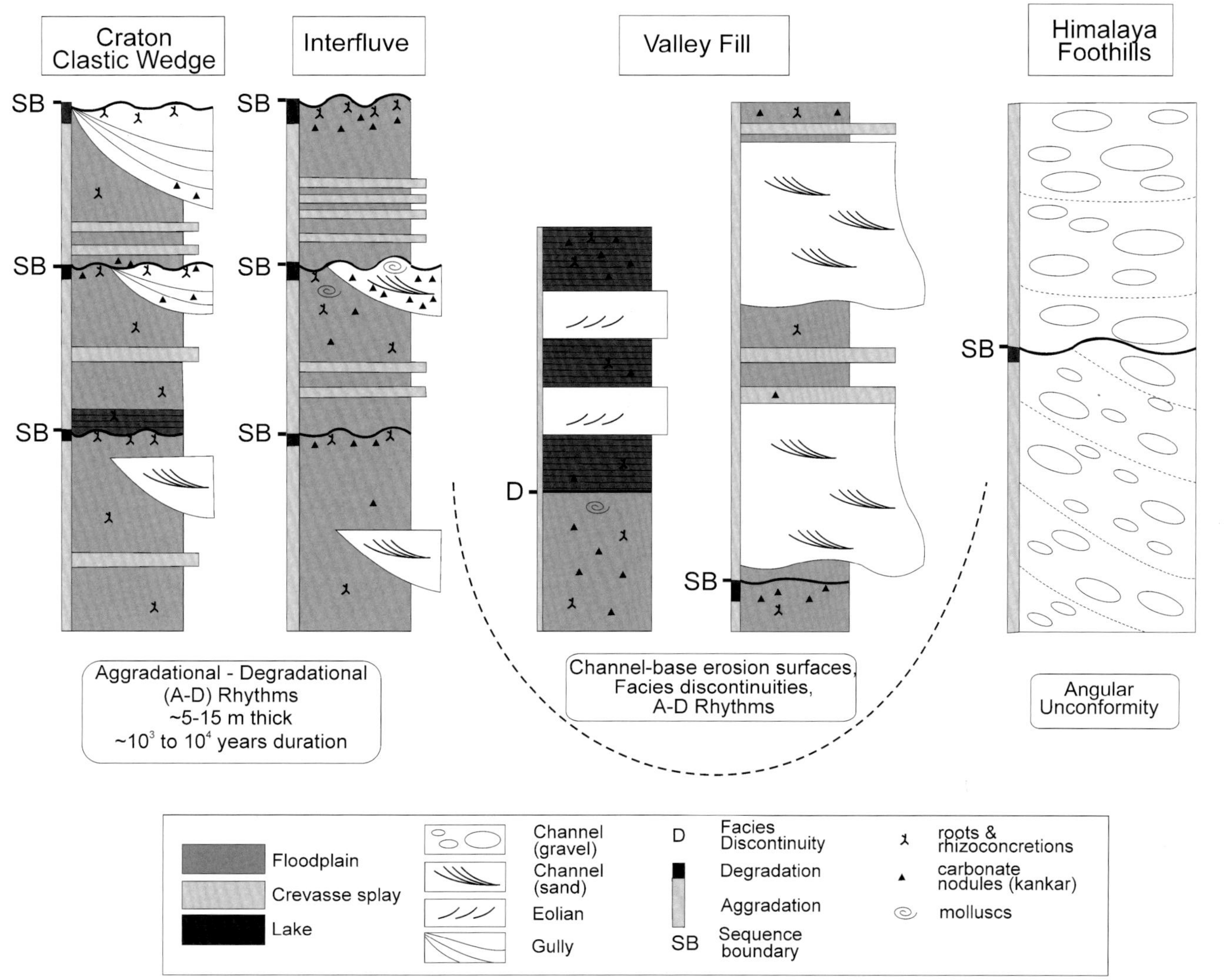

FIGURE 4.—Alluvial sequence patterns suggested by research to date in the western Ganga Plains and Himalayan Foothills. Based on information in Gibling et al. (2005), Sinha et al. (2007), Roy et al. (2007), Sinha et al. (2009), and unpublished data. No columns are shown for large megafans and river tracts in the eastern Ganga Plains (Gandak, Baghmati, and Kosi rivers) because few exposures, subsurface information, and age dates are available: see text for some details.

conditions in the North Atlantic (Prell and Kutzbach 1987; Tandon et al., 2006, Fig. 2). Discharge variations have probably crossed thresholds for fluvial geomorphic response and have caused rapid, widespread incision and aggradation.

Rivers across southern Asia incised when monsoon precipitation and river discharge intensified following the Last Glacial Maximum (LGM) of MIS 2 (see Tandon et al. 2006, Table 1, for evidence that supports this). Because incision is observed from mountain valleys, from rivers in the foreland basin, and from rivers in the Khambat Graben (Srivastava et al. 2001; Pratt-Sitaula et al. 2004; Tandon et al. 2006), this incision cannot be attributed to local tectonic activity. During peak monsoon intensity between 11 and 7 ka, large sediment volumes swept through drainage systems to accumulate in the Ganga–Brahmaputra Delta (Goodbred and Kuehl 2000a, 2000b; Goodbred 2003). Many inland valleys aggraded after ~7 ka when reduced discharge decreased erosive capacity and promoted deposition.

Rapid incision and aggradation is well illustrated by the Sabarmati River of NW India (Fig. 5), where 40 m cliffs cut through late Quaternary strata (Srivastava et al. 2001). Dating of the topmost cliff strata and the oldest alluvium trenched in the adjacent valley brackets incision between about 14 and 3.5 ka, spanning the period of post-LGM monsoon intensification. Sediment has been accumulating within the valley since 3.5 ka. Although the Sabarmati follows a tectonic lineament, the evidence suggests that incision primarily reflects enhanced monsoonal discharge.

It remains unclear whether the Sabarmati and other valleys will persist or will fill completely under the current regime of reduced discharge. However, many valleys may fill over unusually short periods. The huge sediment loads of many Asian rivers (Hovius 1998) reflect rapid weathering and denudation in humid tropical uplands, and earthquake activity releases enormous sediment volumes virtually instantaneously: the 1950 Assam earthquake released 47 billion m^3 of sediment and caused rapid aggradation of thick channel fills (Goswami 1985).

FIGURE 5.—Incised valley of the Sabarmati River near Mahudi in northwest India (Fig. 2A). The cliffs are about 40 m high, and topmost strata have been dated at 12 ± 2 ka, with a date of 4.5 ± 1 ka from a trench in the sand bar at right, suggesting incision between about 14 and 3.5 ka (Srivastava et al. 2001).

As noted above, the Gandak, Baghmati, and Kosi rivers of the eastern Ganga Plains are not presently incised (DVD 1–3). Due to a paucity of cliff exposures and drill cores, their history is poorly known, and it is not clear whether valley fills form part of these successions. However, aspects of the successions may reflect climatic events in the hinterland, especially the availability of glacial materials (Singh et al. 1993). Below the Kosi Megafan, undated borehole cuttings show the presence of a lower gravel and sand unit > 60 m thick, which fines distally and probably represents glacier-fed detritus. Above, an 8–10 m upper unit of sand and mud with fining-up channel units represents the recent westward sweep of the Kosi River across the megafan, along with the deposits of spring-fed channels (Singh et al. 1993). Similarly, for the Baghmati River (Sinha et al. 2005), undated borehole cuttings indicate the presence of a lower sand sheet, with an upper sheet of sand and mud laid down by anastomosing channels similar to those of the modern river.

Tectonic Controls

Tectonic effects on alluvium are also prominent locally. In the Himalayan Foothills, piedmont alluvial fans border strands of the Main Boundary Thrust in an intermontane depocentre or "dun" (Mukerji 1990). At Pinjaur Dun in the Sutlej River catchment (Fig. 2A, DVD 4), the fans are dissected by narrow valleys up to 40 m deep, and have been steepened and tilted below the thrust, suggesting that upwarping led to incision (Tandon et al. 2006; Singh and Tandon 2007). The topmost fan strata are dated at 24 ± 4.5 ka, and later monsoon intensification may have played a part in the incision. Tectonic activity also promoted incision of alluvial terraces in Nepal (Lavé and Avouac 2000). At Dehra Dun (Fig. 2A), piedmont pebble and cobble gravels rest with angular unconformity on underlying gravels (Fig. 4), indicating that tectonic tilting promoted degradation and generated sequence boundaries.

On the main alluvial plains south of the Himalaya, there is little evidence that frequent tectonic events have caused incision. In north Bihar, earthquakes (such as the 1934 event of 8.4 magnitude) take place along faults that do not break the surface, and digital elevation models show that the Baghmati River has responded to tectonic tilting without incision (Jain and Sinha 2005). The 2001 Bhuj earthquake of northwest India (magnitude 7.7), located on a contractional structure west of the Sabarmati River (Fig. 2A), did not produce ground-surface rupture but produced liquefaction in the form of sand blows, sand-blow craters, and lateral spreading over a 400 km radius, as well as causing local subsidence and uplift (Hengesh and Lettis 2002). An 1819 earthquake (magnitude ~ 8) on a fault north of Bhuj , however, did produce surface rupture, and diverted the Indus River, producing a lake on the downthrown side of the fault (Hengesh and Lettis 2002).

Although commonly linked to faults by some researchers, cliffed reaches on the Ganga Plains do not necessarily correlate with fault lines. The Ganga River forms a prominent cliff line along its southern margin (Fig. 2B, DVD 5 and 6), and lies within a strongly asymmetric valley with a less distinctive northern margin. There is no indication that the cliff line is fault-controlled, and the asymmetry may best be explained by the progressive southward migration of the channel in response to thrusting and deformation at the Himalayan Front to the north (Tandon et al. 2006).

Soft-sediment deformation is prominent locally. On the Ganga Plains near Kalpi (Fig. 2B), sediment dikes terminate at an erosional discontinuity in floodplain strata with dewatered sands, and calcite veins are prominent elsewhere (DVD 7 and 8). These features are indicative of earthquake activity, which may have caused uplift or river diversion and created the discontinuity (Gibling et al. 2005), in a manner perhaps analogous to the effects of earthquakes in northwest India.

Example from Inland Australia

Climatic fluctuations are implicated in the behavior of many inland Australian systems. In Quaternary deposits of the Channel Country (Fig. 3), where rivers drain internally to Lake Eyre, muddy topstratum deposits are associated with the modern anastomosing rivers. At shallow depth, widespread sand and gravel sheets extend to > 40 m depth (DVD 9), and their upper parts show paleochannels scaled to much larger discharge than at present, with meander-belt scroll bars (DVD 10) and lateral-accretion sets. These coarse-grained deposits represent aggressive sediment transport during periods of elevated precipitation and discharge, especially during MIS 7, 5, and 3, which were pluvially enhanced in Australia (Nanson et al. 1992; Gibling et al. 1998; Nanson et al. 2008). Although Lake Eyre was a perennial lake up to 25 m deep during early MIS 5, subsequent perennial periods were less effective and alternated with prolonged ephemeral periods (Magee et al. 1995). There is no evidence that Lake Eyre highstands affected events upstream in the Channel Country some hundreds of kilometers upstream. A similar link to precipitation and discharge phases is also evident for the Riverine Plain (Fig. 3), part of the Murray River catchment in southeastern Australia (Page et al. 2009).

Very low subsidence rates in this cratonic setting have generally resulted in amalgamated coarse-grained sheets that may represent much of the later Pleistocene and probably contain cryptic sequence boundaries of climatic origin. Neotectonic activity has influenced fluvial courses in several inland areas, for example in association with the Innamincka Dome of the Lake Eyre basin and the Cadell Block of the Riverine Plain (Nanson et al. 2008; Page et al. 2009).

In summary, the southern Asian and Australian examples indicate that:

- Alluvial systems show rapid alternation between incision and aggradation, especially under monsoonal forcing, and drainage systems may alternate between valley and channel states, or at least between deeply and modestly incised states.
- Many alluvial events correlate well with monsoonal proxy records. Some climatic events of exceptionally high magnitude, such as post-

LGM monsoon intensification and later decline, generated continent-wide incision and aggradation.

- Tectonic activity has generated valleys and sequences close to thrust fronts, and locally causes sediment disruption and upwarping on the open plains. However, few proxy records exist for fault movement and tectonically induced subsidence, and it is difficult in most cases to evaluate the degree to which tectonic events caused stratigraphic events.

SCALE OF BASAL RELIEF (CRITERION 1B)

Valley models typically draw on evidence for deep incision that created prominent sequence boundaries. Our observations indicate that some, but not all, inland and coastal valleys show deep scour and prominent erosional relief. Because incision characterizes virtually all fluvial conduits (Salter 1993), erosional margins alone are not diagnostic of valleys, especially because their prominence may reflect the most recent major flood (Wolman and Miller 1960; Schumm and Lichty 1963; Nash 1994). Furthermore, some important fluvial surfaces may show only modest incision. As discussed later, flume-tank models of Strong and Paola (2008) suggest that the final valley form need not correspond to any particular surface generated during incision, and that valleys may not be characterized by prominent incised boundaries. We highlight here case studies from Australian and Asian coastal systems and inland Australian examples that illustrate these two caveats.

Some Key Fluvial Surfaces Have Minimal Incision

Channels with minimal incision have been imaged on seismic profiles below Australian continental shelves, and may routinely have flooded the adjacent plains during lowstand periods. Several large rivers crossed the low-gradient shelf of the Great Barrier Reef of northeastern Australia during sea-level lowstands. The subsea extension of the Burdekin River has conduits up to 2500 m wide and 20 m deep, with anabranches where the gradient was especially low and with more notable incision in distal areas (Fielding et al. 2003, 2005). In seismic profiles, the conduits show a simple cross-sectional geometry, with lateral-accretion deposits formed in meandering channels. The fluvial bodies show only modest degrees of incision, being essentially a single story thick although wider than a typical story (Fig. 6). Similar features are also observed in the subsea extension of the Fitzroy River (Ryan et al. 2007), where conduits are up to 1600 m wide and 20 m deep, with high-sinuosity reaches and lateral-accretion deposits. Incision is most pronounced in inshore and distal areas, with a modestly incised mid-shelf zone.

In southern Australia, the subsea extension of the Murray River has conduits up to 1000 m wide and 20 m deep (Hill et al. 2009). Point-bar deposits indicate that some reaches were meandering, and composite channel bodies represent amalgamation of deposits generated during multiple sea-level cycles. Onshore, the lower Murray River was influenced by fault lines and neotectonic activity, and channels are incised up to 130 m below the adjacent plain. The downstream reduction in width and depth reflects the low shelf gradient, less neotectonic activity, and relatively short periods of subaerial exposure and fluvial activity through the glacial cycle.

For all three of these river systems, flow was not always confined to the conduits, potentially allowing the river to avulse and relocate within a channel belt. Thus, interfluves are likely to have experienced flooding, militating against the formation of well developed paleosols. Fielding et al. (2003) and Ryan et al. (2007) considered these conduits to be "entrenched" rather than "incised", denoting a degree of basal erosion consistent with "channels" rather than "valleys".

These observations contrast with Gulf of Mexico systems. In this setting, incision below highstand shorelines tended to fix the river

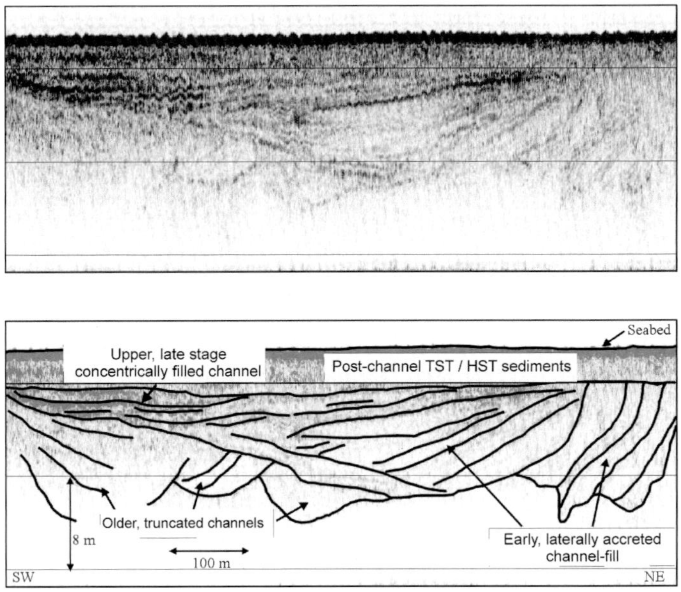

FIGURE 6.—Seismic profile (uninterpreted above, and interpreted below) for subsea channel fills of the Burdekin River, northeast Australia. Note that the bulk of the fill comprises a single set of lateral-accretion deposits.

positions, resulting in "stratigraphic partitioning" between the conduits and interfluves (Blum and Aslan 2006). Flow appears to have been confined largely to the conduits, allowing the development of prominent paleosols on interfluves.

The modest degree of incision of some rivers in their mid-shelf reaches may reflect long periods of falling sea level as glacial ice built up. Consequently, coastal rivers tend to be graded to a mid-shelf position, rather than to the lowest sea levels at times of maximum ice buildup where proximity to the shelf break would have promoted incision (Blum and Aslan 2006; Hill et al. 2009). Gulf of Mexico rivers were probably graded for much of their later existence to a level 40 to 85 m below present sea level (Blum and Aslan 2006, their Fig. 4B). Additionally, situations in which the gradient of the alluvial system exceeds that of the shelf may result in alluvial aggradation without prominent incision, as well as detachment of the alluvial system from the shoreline (Petter and Muto 2008).

Incision was also modest for lowstand alluvial systems below the broad, low-gradient shelf of the East China Sea. Here, an extensive fluvial sheet averages 30 m thick and is > 330 km wide, and rests on an erosional surface (Wellner and Bartek 2003). Wellner and Bartek (2003) emphasized the importance of two factors in creating this sheet-like fluvial body: a) shelf physiography, where a low-gradient shelf damped down incision and promoted lateral migration of channels, and b) high-frequency variation in precipitation and discharge, especially reduced discharge later in MIS 2, which reduced the energy of the system and limited incision.

Similar wide fluvial sheets with only modest incision have been noted in the ancient record of coastal fluvial systems. Holbrook (2001) recorded a Cretaceous composite valley fill 28 m thick and more than 87 km wide in the southwestern USA, with less than 15 m of basal relief. In Queensland, Australia, Allen and Fielding (2007) documented Permian sequence boundaries covered by braided-fluvial sheets with minimal basal incision. The strata were laid down in a slowly subsiding cratonic basin subject to glacioeustasy, where the low gradients of the alluvial plain and shelf damped down incision and led to alluvial sheets in which any valleys would have been very shallow. Low rates of relative sea-level fall may also generate sheet-like bodies (Strong and Paola 2008).

Some Channel Confluences Show Exceptionally Deep Incision

Channel confluences may be sites of deep scour. In the Brahmaputra River, depth soundings show that scours at the confluences of anabranches and of tributary and main channels may be as much as five times the mean channel depth (Best and Ashworth 1997). This contrasts strongly with surveyed cross sections of modern rivers and experimental studies, for which maximum channel depth is commonly less than twice the mean depth (Gibling 2006, his Table 9). In the Brahmaputra example, deep incision is an intrinsic effect and does not denote a sequence boundary. Similarly large confluence scours have been identified in the Triassic Hawkesbury Sandstone of Australia (Miall and Jones 2003) and the Morrison Formation of the southwest USA (Cowan 1991).

CORRELATIVE SURFACES ON ALLUVIAL PLAINS (CRITERION 2)

Interfluve paleosols that correlate with episodes of incision are common in coastal plains and in some smaller alluvial basins. However, they may be absent or present only locally in large alluvial basins such as the Ganga Plains. Interfluves between valleys on large alluvial plains are complex geomorphic areas, and the time available for soil development is locally variable.

For coastal plains, deep valley incision driven by large, prolonged sea-level falls commonly results in reduced overbank flooding and prolonged periods of minimal sedimentation on adjacent interfluves. Depending on climatic factors, prominent soils may form on these lowstand surfaces, correlative with sequence-bounding valley bases. Quaternary lowstand paleosols are prominent on the coastal plains of Italy (Amorosi et al. 1999a; Amorosi et al. 2008) and the Gulf of Mexico (Blum and Price 1998; Blum and Aslan 2006). Tandon and Gibling (1997), Feldman et al. (2005), and Fischbein et al. (2009) documented Pennsylvanian bedrock examples, from an icehouse period similar to the late Quaternary.

In contrast, on the Ganga Plains, the Yamuna, Ganga, and Gaghara rivers pass through Himalayan exits with a quasi-regular spacing (Hovius 1996), so that the rivers are separated by low-relief interfluves tens to more than 100 km wide and hundreds of kilometers in extent downslope (Fig. 2A, B). Despite their apparent simplicity, these broad alluvial surfaces are complex entities formed by amalgamation of alluvium from different sources. For the Ganga–Yamuna Interfluve (Fig. 2C), petrographic studies show that the interfluve represents an amalgamation of: a) a clastic wedge sourced from the cratonic Chambal and Betwa rivers, either a transversely directed megafan or the product of cratonic axial rivers at the basin margin (Sinha et al. 2009), and b) Himalayan alluvium that onlaps the cratonic wedge.

Interfluves between the major rivers exhibit a complex mosaic of landforms. It is unlikely that a single correlative surface could be traced across an interfluve or correlated between adjacent interfluves and clastic wedges. The interfluves are traversed by small channels (Sengar, Rindi Nadi and Pandu Nadi in Fig. 2B) that arise on the interfluves from springs replenished by monsoon rains. Source-bordering eolian dunes are also present locally on the plains (DVD 11 to 15), with extensive sandhills east of Delhi (Tandon et al. 2006). Lakes are minor components of this interfluve, but large lakes or tals (DVD 16) and many small oxbow lakes are common elsewhere, recording more than 20,000 years of lake history (Sharma et al. 2004; Sinha et al. 2005).

Aggradational–degradational rhythms are commonly prominent below the interfluves. Although these rhythms may correspond to incisional events in the mainstem Ganga and Yamuna valleys, they more probably record migration or avulsion of small plains-fed channels, climatic and vegetational effects, or possibly local tectonic events. Such rhythms are prominent architectural features below the Ganga–Yamuna Interfluve, as indicated from cliffs more than 30 m high and drill cores up to 50 m long. In large cliffs at Kalpi dated at ~ 119 to 34 ka, aggradational–degradational rhythms are ~ 15 m thick, and consist largely of floodplain deposits with immature paleosols and small channel bodies deposited by spring-fed rivers. These deposits are interrupted by erosional discontinuities marked by groundwater and pedogenic carbonates, carbonate gravel, and gully fills (Figs. 4, 7, and DVD 17 and 18; Gibling et al. 2005; Sinha et al. 2006). Some floodplain deposits are capped by prominent, red paleosols rich in carbonate nodules that record periods of minimal sedimentation or erosion (Fig. 4; Sinha et al. 2007). It is not known how far individual discontinuities and paleosols extend at any of these locations, but we suspect that most are of local extent. Adjacent to the Ganga and Sengar rivers, carbonate-rich paleosols cap some cliff successions (Gibling et al. 2005), and may correlate with the adjacent valley margins. Thick calcretes are present on alluvial surfaces in the Thar Desert of northwestern India (Dhir et al. 2004; DVD 19 and 20).

Too little is known about the subsurface stratigraphy of megafans and interfluves in the eastern Ganga Plains to determine whether similar rhythmic successions are present. On the Gandak megafan (Mohindra et al. 1992), calcareous paleosols are common in the topmost strata, and could be present at depth, and zones rich in carbonate nodules are present at depth below the Baghmati plains (Sinha et al. 2005).

In contrast to the situation on large alluvial plains, small alluvial basins may show important paleosol units that correlate with incision along major drainage axes. In the Plio-Pleistocene successions of the Rio Grande Rift, stacked petrocalcic paleosols form condensed sections that record prolonged periods of minimal sedimentation, allowing the formation of mature paleosols (Mack and Leeder 1999; Mack et al. 2006). These authors linked the paleosol units to several causes in different parts of the basin and at different times: a) basin tilting along bounding faults which draws drainages to the more rapidly subsiding areas; b) movement of rivers into other basin strands, leaving floodplains inactive; c) incision on alluvial fans, and d) a period of prolonged degradation and floodplain stasis as tectonic activity waned.

FIGURE 7.—Cliff 33 m high at Kalpi on the Yamuna River, India (Fig. 2B). Arrows denote prominent, carbonate-cemented erosional surfaces that divide floodplain and channel deposits into aggradational–degradational rhythms (sequences) from water's edge to cliff top. Person is circled.

In summary,

- Sequences with thicknesses of 5–15 m and short durations (~ 10^3 to 10^4 years) appear to characterize floodplain deposits in the Ganga–Yamuna regions of the Ganga Plains; they may have multiple causes and need not be related to trunk-valley incision.
- Valley-bordering paleosols are present on floodplain surfaces locally but, because valley incision and aggradation may be short-lived, prolonged soil formation may be unusual.
- Interfluve channels, lakes, and eolian dunes may aggrade during periods of trunk-system incision. Thus, models that invoke prolonged, widespread stasis of interfluve surfaces appear unrealistic for large alluvial plains.
- Wide interfluves may constitute an alluvial system that operates independently of the major rivers. Thus, events in the two systems need not correspond, although both systems could be affected simultaneously by climatic or tectonic events.
- In view of the complex geomorphic patterns of large alluvial plains, it is unlikely that key surfaces will be basinwide. In contrast, mature paleosols may be widespread in smaller basins, especially where tectonic effects are important.

INCISED TRIBUTARIES AND GULLIES (CRITERION 3)

Our observations agree with the view of Posamentier (2001) and Miall (2002) that incised tributaries constitute good evidence for the presence of incised valleys, implying that the valleys are not flooding across their margins. Using seismic time slices, these authors observed good examples in Miocene fluvial strata offshore Indonesia and in the Malay Basin (Table 1).

The Ganga Plains show excellent examples of incised tributaries and zones of floodplain degradation bordering valleys. The Yamuna Valley is flanked for some 250 km by gullied badlands (DVD 21) that also extend northwards along the tributary Sengar River, which is incised near the valley (Figs. 2B, 7). Small gullied areas flank the southern margin of the Ganga Valley but are too small to be shown in Figure 2B. Along the Chambal and Betwa rivers to the south, badlands cover large areas, including a 5,000 km^2 tract along the Chambal River, where some gullies are > 10 m deep (Haigh 1984; DVD 22). Badland gully fills of alluvial and colluvial material, up to 10 m thick, are present in outcrops (Fig. 4, DVD 18; Gibling et al. 2005), and gullied zones may be correlative with some incised valleys. However, gullying has been variously attributed to climate change, local uplift, and human activity, although gullied tracts predate the late medieval Mughal period and are not caused solely by recent agricultural practices (Haigh 1984).

Triassic and Cenozoic formations in the western USA contain excellent ancient analogues for these gullied and badland landscapes (Retallack 1986; Kraus and Middleton 1987; Bestland 1997; Bestland et al. 1997), linked in some cases to incised rivers. In these examples, extensive floodplain erosional surfaces (some of them terraced) are marked by prominent paleosols and covered by alluvial and colluvial material with weakly developed paleosols. Retallack (1986) linked some paleosols and erosional surfaces to deeply incised channels, and some erosional surfaces border channel bodies that supplied sediment to cover and "heal" the landscape surfaces (Kraus and Middleton 1987). Bestland (1997) noted that the Cenozoic successions represent aggradational pulses separated by degradational events, and inferred alternate landscape stabilization and destabilization in response to climatic instability and vegetational change. Kraus and Middleton (1987) inferred periods of upwarping to explain the erosional surfaces.

Incised tributary valleys are well represented in Pennsylvanian strata in midcontinental USA (Feldman et al. 1995; Kvale and Archer 2007; Fischbein et al. 2009). They form part of mappable drainage networks and feed into much larger trunk fluvial systems, typically with thick, coarse-grained alluvial fills.

DIMENSIONS OF VALLEYS AND COMPONENT CHANNELS (CRITERION 7A)

Our observations concur with those of Schumm and Ethridge (1994), who noted that many paleovalley fills are an order of magnitude larger than average modern coastal-plain channels. Ancient valley fills commonly record this contrast (Table 1), as do coastal valleys (Plink-Bjørklund and Steel 2006). Blakey and Gubitosa (1984) noted that, within relatively narrow Triassic valleys, channels were much shallower than the host valleys but channel and valley widths matched and the channels were not underfit.

On bedrock unconformities, it is clear that valleys were incised prior to deposition of the existing fill, and that their original and present dimensions were similar. In contrast, Blum and Price (1998) and Blum and Aslan (2006) described "composite valley-fill unconformities" (Fig. 8) below the Gulf of Mexico coastal plain, formed by the amalgamation of numerous smaller valleys as a result of avulsion. Numerous ancient examples are known (Willis 1997; Holbrook 2001). The dimensions of a composite valley fill are much greater than the instantaneous dimensions of the valley, and a failure to appreciate this may lead to erroneous assumptions about incision processes and (paleo)flow volumes.

Analogue experiments have demonstrated the creation of composite valley fills. Strong and Paola (2008) conducted tank experiments on valley cutting and filling, modeling a passive-margin setting where subsidence increases basinward and using a constant sediment and water supply to isolate the effects of coastal base-level change. They observed that erosional surfaces are diachronous and amalgamated, and that valley-base unconformities form over a period so prolonged that some deposits above the unconformity are older than some deposits below. Although the stratigraphic form of the deposit resembles a valley with a prominent basal surface, this surface did not exist as a topographic valley form at any given time. Strong and Paola (2008) distinguished stratigraphic valleys (the valley-form erosional surface preserved in the stratigraphic record) from topographic valleys (the topography at an instant in time).

The Ganga Valley upstream from Kanpur (Fig. 2B) is a stratigraphic valley that contains smaller channel bodies and has a probable composite unconformity at its base. Numerous smaller channel bodies penetrated during drilling lie within floodplain and eolian deposits and

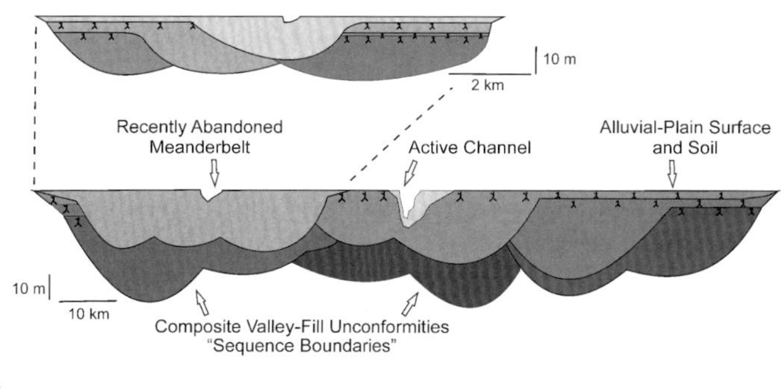

FIGURE 8.—Composite valley fills, based on Texas coastal plain bordering the Gulf of Mexico; modified from Blum and Price (1998).

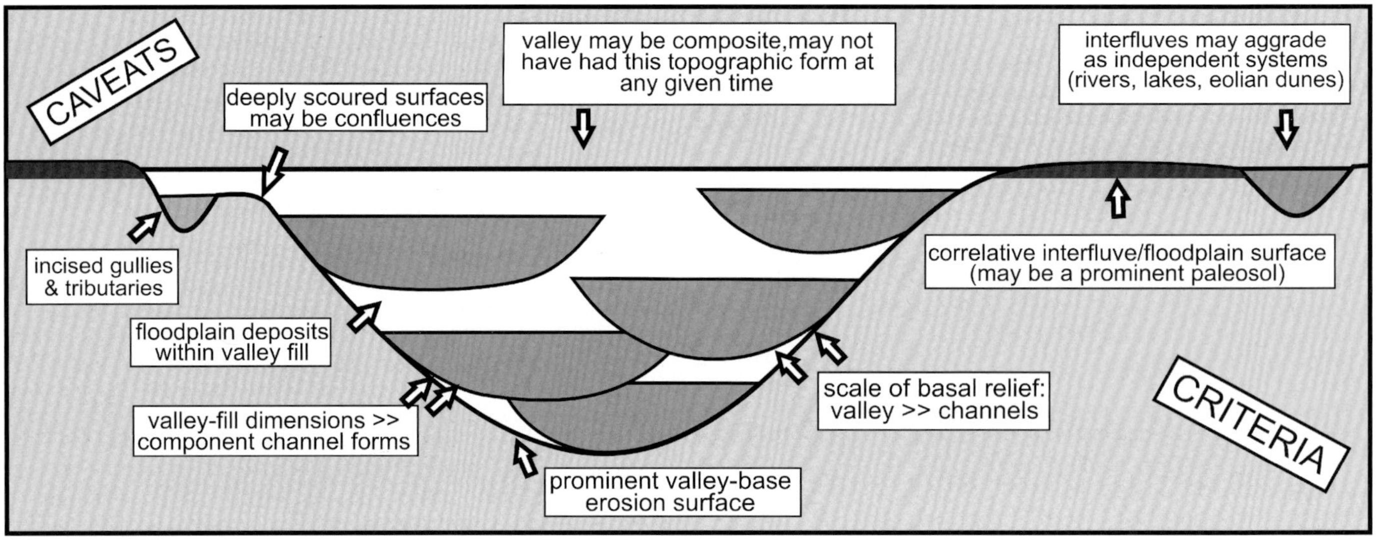

FIGURE 9.—Guidelines for identifying alluvial valley fills in the geological record. Text in the lower part of the diagram shows criteria that have been shown to be valid in certain areas, whereas text in the upper part lists caveats that may influence the decision to identify a valley fill. See Table 3 and text for details.

date back to ~60 ka (Fig. 4; Gibling et al. 2005; Roy et al. 2007; Sinha et al. 2007). Prominent paleosols with carbonate nodules in the cores mark discontinuities. Cliff exposures near Bithur consist of floodplain deposits capped by lacustrine and eolian strata dated at 27.5 to 25.2 ka (Fig. 4). These facies suggest that Ganga Valley channels became underfit and floodplains were replaced by shallow lakes and eolian dunes during the Last Glacial Maximum, when proxy records suggest reduced monsoonal precipitation. The top of the floodplain unit is not a degradational surface but marks a prominent facies discontinuity. Just below the valley near Bithur, partially amalgamated channel bodies dated at < 10 ka record renewed Ganga discharge and Holocene aggradation.

Similar evidence for reduced discharge is also seen along some reaches of the Yamuna Valley, where floodplain deposits are capped by a prominent lacustrine unit (DVD 23 and 24). The unit has been dated at 15.5 ka radiocarbon years (Williams and Clarke 1995; Tandon et al. 2006), indicating reduced discharge in MIS 2.

Age dates suggest that channel bodies below the northern part of the Ganga Valley are older, supporting a progressive southward migration of the Ganga River, suggested also by the river's location at the southern valley margin. Thus, the Ganga valley has migrated southward, creating a composite, diachronous valley fill with dimensions that greatly exceed the instantaneous topography. The valley fill contains internal sequence boundaries (carbonate-rich paleosols) and facies discontinuities (Fig. 4).

Many alluvial valleys in the rock record follow tectonic lineaments that may have been sufficiently long-lived to generate large stratigraphic valleys that contain smaller channels. The Cenozoic Sis Conglomerate of the Spanish Pyrenees is a remarkable sediment body 1.4 km thick and 7.5 km wide, formed along a transverse fault during thrusting (Table 1; Vincent 2001). The sediment body includes tilted stratal sets and angular unconformities. The valley fill evolved for some 25 million years, generating a thick stratal package that includes floodplain fines, lakes, and coal, but the topography at any given time did not exceed 120 m (S. Vincent, written communication, 2005). The base of the Sis Conglomerate thus represents a stratigraphic valley but not a topographic valley.

TERRACES (CRITERION 7B)

We concur with the view that terraces provide important evidence for valley recognition. In modern valleys such as those across much of northwest Europe, flights of terraces commonly result from repeated incision and aggradation superimposed on long-term downcutting (Bridgland, 2000; Bridgland et al., 2006). The Somme Valley of France was generated by incision over about a million years as a result of prolonged uplift, during which aggradational periods generated widespread terraces that probably represent climatic variations (Antoine et al. 2003). Terraces are infrequently identified in the older record (Holbrook 2009). Quaternary subsea examples have been imaged in seismic profiles (Thomas and Anderson 1994; Simms et al. 2006), and strath terraces cut into the substrate were identified off Brittany by Menier et al. (2006). Although difficult to identify in the ancient record, older examples include the studies of Sedimentation Seminar (1978), Bestland (1997), and Posamentier (2001).

In southern Asia, terraces constitute an important facies component in upland and lowland valleys (Lavé and Avouac 2000; Monecke et al. 2001; Tandon et al. 2006; Gibling et al. 2008), and are especially prominent in deep valleys draining the Himalaya (DVD 25 to 28). In some cases, episodes of accumulation and incision have been linked to variations in monsoonal strength (Bookhagen et al. 2006). Terraces along an alluvial reach are commonly assumed to represent a system-wide alluviation episode. However, terraces along the Belan River in northern India (DVD 29 to 31) and the Bellinger River in Australia may have similar elevations but different ages, and thus represent local intra-system variation (Cohen and Nanson 2008; Gibling et al. 2008). For the Clyde and Manning Rivers of coastal Australia, strong precipitation and discharge events can cause catastrophic stripping of floodplain deposits (Nanson 1986), which contributes to local variation in terrace preservation (DVD 32 and 33).

EVALUATION OF GUIDELINES FOR RECOGNIZING ALLUVIAL-VALLEY FILLS

It is apparent from the above discussion that alluvial and coastal valleys show points of similarity and difference (Table 3; Fig. 9). The differences primarily reflect the huge influence of sea level on downstream systems, which forces incision, mediates valley filling, and generates a predictable set of systems tracts related to base-level transit cycles that operate in the Milankovitch band. (However, the architecture of the fills largely reflects sediment delivery from upstream: Blum and Aslan 2006).

In contrast, upstream alluvial systems are subject to multiple influences over varied and commonly short timescales. These influences include a) precipitation and, in some settings, meltwater availability, b) tectonics, including regional subsidence and fault motions, and c) intrinsic drainage changes such as avulsion, river capture, and channel and valley migration. Away from coastal areas, the recognition of sequence boundaries may be problematic because allocyclic and autocyclic events do not necessarily generate recognizable signals (Neal and Abreu 2009). Monsoonal precipitation changes are somewhat predictable because they respond to insolation and glacial effects, but tectonic and intrinsic events have a low level of predictability. As with coastal valleys, fills may be composite and may contain sequence boundaries and major discontinuities.

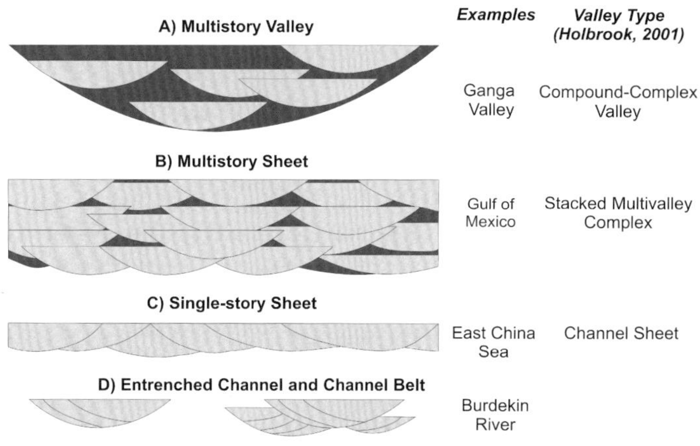

FIGURE 10.—Schematic diagram to illustrate Quaternary valley fills and incised channel bodies discussed in the text. See Table 4 for references to Quaternary examples. Note that nested valley forms may be present in some examples but are not shown in the diagram.

ARCHITECTURE AND GEOMETRY OF ALLUVIAL-VALLEY FILLS

In an important definition paper, Holbrook (2001) divided valley fills into those with a valley-form (concave-up) basal unconformity and those with a smooth to nearly smooth basal unconformity (Table 4). In the latter group, relief on the basal surface was similar to or less than the thickness of larger overlying channel-fill elements. In essence, Holbrook's system recognizes lensoid and sheet-like valley fills, although scales and aspect ratios of the two groups were not set out. Holbrook divided each group into four types (Table 4), based on the presence or absence of nested valley fills with the alluvial body (whether aggradation was interrupted by episodes of local or regional incision), as well as by stacking patterns (lateral or vertical amalgamation of valley fills).

Our evaluation of Quaternary valley fills includes information about their architecture that, in some cases, allows comparison with Holbrook's classification. Good information is available from Quaternary fluvial systems such as the Burdekin River, where fluvial conduits are filled and for which seismic profiles are available. However, insufficient information is available for many inland areas of the Indian and Australian plains, where valleys are still evolving and for which cores and cliff outcrop allow only a limited evaluation of valley architecture.

For well-documented Asian and Australian examples, we identify multistory valleys, multistory sheets, and single-story sheets, which correspond with three of Holbrook's types (Fig. 10; Table 4). We use more general terms than those of Holbrook because it is not always possible to be certain if incision surfaces were local or regional—a key aspect of his classification. However, high-resolution age dating indicates that monsoon intensification following the Last Glacial Maximum led to fluvial incision in many Asian rivers. The Ganga Valley record contains sequence boundaries that probably represent other regional climatic events, suggesting that it is a compound–complex valley in Holbrook's system. Fluvial deposits below and bordering the Gulf of Mexico constitute a stacked multivalley complex, and those below the East China Sea constitute a channel sheet formed during and since the LGM. The subsurface sand sheets of the Australian Channel Country include components that date back to nearly 750 ka (Nanson et al. 2008) and are probably stacked multivalley complexes.

Additionally, we identify entrenched channels and channel belts typified by former river reaches below continental shelves in Australia (Fig. 10). These are not technically valley fills because they scale with individual channels in the system (Fig. 6). Fluvial deposits of this type appear to be common below some offshore areas, related to numerous factors that include shelf slope and the depth of the shelf edge. They show that valleys with prominent incision were not universally generated during periods of lowered sea level.

For rapidly subsiding basins with a high sediment supply, Holbrook (2001) suggested that incisional periods would have been few and allomembers may not be present. Although subsurface data are sparse, current information suggests that megafan deposits of the Himalayan Foreland Basin consist largely of aggradational channel and interchannel successions, in accord with examples from the ancient record (Davies and Gibling 2003).

ALLUVIAL SEQUENCE MODELS: UNDERSTANDING ACCOMMODATION

For large alluvial plains, especially in foreland basins, we follow Blum and Törnqvist (2000) in suggesting that accommodation terminology should be applied with caution. Sequences with low- and high-accommodation systems tracts have been widely regarded as the prime form of stratal packaging in inland alluvial systems (Column 2 of Fig. 1; Catuneanu 2006), and have commonly been linked to tectonism acting through subsidence (e.g., Currie 1997). Although this may be true in some cases, we suggest that many fine-scale aspects of stratigraphic architecture in these deposits may reflect climatic effects.

Researchers have tended to favor tectonically driven accommodation scenarios to explain alluvial successions for several reasons: 1) accommodation models are familiar to many of us because of their evident application to coastal valleys influenced by sea-level change, 2) some alluvial basins experience active tectonism, and 3) accommodation can readily be explored by mapping stratal thickness variation. Regional subsidence, uplift and subsidence associated with individual faults, and forebulge dynamics may greatly influence the development of foreland and other basins (Catuneanu et al. 1997; Plint and Wadsworth 2006). However, decrease and increase in space for sediment accumulation has also been interpreted to reflect changing balance of water and sediment discharge (Columns 4 and 5 of Fig. 1; Weissmann et al. 2002). If the terms "accommodation" and "base level" are used in alluvial studies, it needs to be clear whether subsidence or change in the "fluvial equilibrium profile" is being invoked to explain accommodation.

We do not mean to minimize tectonic controls, the importance of which is evident in many parts of Asia and Australia and in other areas (Table 2). However, it is by no means clear that the many prominent

TABLE 4.—Valley types identified by Holbrook (2001), with Quaternary examples identified in the present study (see Fig. 10). Entrenched Channels and Channel Belts are not technically valley fills, and were not part of Holbrook's (2001) system.

Valley Types	Definition	Quaternary Examples Identified in Present Study
Valley-form Basal Unconformity		
Simple	Single valley scour, fill represents consistent aggradation	
Complex	Valley fill interrupted by local incision	
Compound	Valley fill interrupted by regionally significant incision	
Compound-Complex	Valley fill interrupted by local and regionally significant incision	**Ganga Valley, India** (Tandon et al. 2006; Sinha et al. 2007): Valley fill contains numerous channel fills; incision during monsoon intensification (~20 ka B.P. to present) was a regional event in Asia; sequence boundaries within valley fill may reflect earlier regional incision events; valley form migrates laterally due to advancing Himalayan thrust front.
Smooth to Near-smooth Basal Unconformity		
Channel Sheet	Rests on regional erosion surface, fill lacks valley-scale surfaces; aggradation may cause stacking of channel belts	**East China Sea** (Wellner & Bartek 2003): Multi-storey fluvial sheet >330 km wide and 30 m average thickness (maximum 53 m). Formed since last glacial cycle (~20 ka B.P. to present). Fed by Changjiang (Yangtze) and Huanghe (Yellow) rivers. Interaction of glacioeustasy and climatic factors on a low-gradient shelf. Contains local autocyclic incisions and possible regional incision / aggradation phases due to high-frequency climatic events. Valley underfilled, with capping marine beds.
Stacked Channel Sheet	Two or more superimposed channel sheets with no intervening open-marine strata	
Multivalley Complex	Laterally amalgamated valley fills	
Stacked Multivalley Complex	Vertically amalgamated valley fills with no intervening open-marine strata	**Gulf of Mexico, Texas Coastal Plain** (Blum & Aslan 2006): Vertically and laterally amalgamated valley complexes, representing succession of 100 kyr valley fills. Interaction of glacioeustasy and climatic factors.
Additional Responses to Glacioeustatic Events (not included by Holbrook 2001)		
Entrenched Channels and Channel Belts	Channel body or belt with evidence for some incision, relatively narrow and not amalgamated into sheet	**Burdekin River, Australia** (Fielding et al. 2003): Individual channel body or a few amalgamated channel bodies. Not technically a valley fill because accretion surfaces accord with channel dimensions. Represents lowstand extension across shelf.

surfaces visible in alluvial successions in the rock record represent tectonic effects. The Asian Quaternary case studies cited above, as well as studies elsewhere (Bull 1991; Ritter et al. 1995; Bestland 1997; Weissmann et al. 2002), indicate that climatic effects have been generally undervalued in interpreting the stratigraphic record, as suggested by Blum (1994).

On alluvial plains under the strong driving force of the monsoon, many key alluvial surfaces and packages show a first-order correlation with patterns of precipitation and discharge, and some (such as post-LGM incision caused by monsoon intensification) may be of continental extent. Effects intrinsic to channel systems are also very important. Wells and Dorr (1987) noted that the Kosi River can rise 9 m in a day, and annual floods are commonly large enough to shift channels (Jain and Sinha 2003). We believe that, together, climatic and intrinsic effects are capable of generating a fine-scale architecture that may dominate interfluves and clastic wedges (Fig. 4), perhaps including some megafans.

Given the many uncertainties in assigning sequences to particular causes, it may be best to recognize aggradational–degradational rhythms as a "ruling motif" (Column 7, Fig. 1), considering their correlation to valley fills where such can be identified. This nongenetic terminology allows for varied explanations without prejudging the causes. Despite the short duration and low thickness of many such rhythms, they can legitimately be described as "sequences" because they have erosional boundaries. As discussed by Schlager (2004) and Neal and Abreu (2009), sequences are scale-independent and fractal in nature, and their duration and thickness should not be used to erect sequence hierarchies.

Although not necessarily inappropriate, the use of systems-tract terminology to alluvial successions may be problematic. The concept of systems tracts originated in marine basins where stratal bodies can be linked to global sea-level change (although local modifications have been widely noted, e.g., Martinsen and Helland-Hansen 1995). Where seismic records are available, observable features such as progradation, aggradation, and retrogradation should be applied in describing "accommodation successions", rather than "highstand" and "lowstand", which imply a linkage to base-level transit cycles (Neal and Abreu 2009).

The identification of systems tracts in an alluvial basin tends to imply a basinwide effect, whether the authors intended this or not. However, alluvial regions such as the Ganga Plains are huge areas with many geomorphic elements (orogenic and cratonic sources; megafans, interfluves and axial drainage systems) and large precipitation gradients. In basins of this scale, basinwide bounding surfaces will be a rarity, and most prominent surfaces may be of local extent. Additionally, systems tracts were initially designed to record parallel responses to forcing factors through a range of depositional systems (alluvial, coastal, deep sea), rather than for one system (alluvial). In these respects, "nonmarine" sequence stratigraphy may need to throw off some of the shackles of a marine-based framework.

There are also dangers in inferring accommodation from the degree of amalgamation of channel bodies in the bedrock record. Amalgamated bodies are commonly used to infer a low-accommodation setting or period, whereas bodies isolated in floodplain deposits are used to infer a high-accommodation setting or period. The link is relevant to many coastal valleys (Blum and Törnqvist 2000), and is also valid in some alluvial contexts such as cratonic settings with slow subsidence rates (Leckie and Cheel 1997; White and Leckie 1999; Allen and Fielding 2007). However, several studies have noted that alluvial aggradation rates may vary without change in base-level or upstream controls, with potential implication for channel-stacking patterns (Holbrook et al. 2006; Petter and Muto 2008).

Linkage between amalgamation and accommodation may be misleading or incorrect in other cases. Such a linkage may be particularly misleading in the case of megafans (distributive fluvial systems) which are dominant in many inland basins (Weissmann et al. 2010). Megafans originate where sediment-charged rivers at mountain exits deposit mud and sand as flow spreads out and decelerates, and their deposits typically exhibit channel bodies isolated in mudstone and rarely amalgamated (Nichols 1987; Hirst 1991). Not all megafans are located in rapidly subsiding settings with high accommodation. For example, the Kosi and Gandak megafans lie in a zone of rapid subsidence near the Himalayan Front (Fig. 2A), but a cratonic clastic wedge (probable megafan) at the southern fringe of the foreland basin (Fig. 2C) was laid down in a slowly subsiding region. Thus, the low degree of amalgamation in megafan deposits may typically reflect rapid sand and mud deposition, rather than tectonically driven accommodation. Additionally, tough floodplain muds and paleosols in seasonal climates may prevent channel enlargement, generating fixed channel bodies through local geomorphic controls (Gibling 2006).

To avoid inappropriate inferences of accommodation, Patterson (2009) recommended the use of the nongenetic terms amalgamated, semi-amalgamated, and non-amalgamated channel complexes to characterize channel-body architecture.

APPLICATION OF QUATERNARY OBSERVATIONS TO DEEP TIME

We have drawn on Quaternary systems in evaluating models for alluvial-valley fills. However, insights from these systems may be less applicable or even inappropriate when applied to other parts of the geological record.

The current icehouse period with its high-magnitude and high-frequency changes in sea level and climate differs considerably from earlier greenhouse periods, with considerable effects on marine and coastal systems (Dalrymple 2006; Sømme et al. 2009). A few studies have discussed how these changes would affect fluvial systems. Willis (1997) suggested that, under greenhouse conditions with modest sea-level changes, valley fills would tend to be dominated by coarser fluvial input. The prominence of valleys in coastal areas may reflect in part the magnitude of relative sea-level change (Rygel et al. 2008). Gibling (2006) noted that Jurassic and Cretaceous valley fills, formed during a greenhouse period with little or no glacial ice, were in general smaller than those of the Carboniferous and Permian, when the buildup and melting of glacial ice caused high-magnitude changes in sea level.

Precambrian and Cambrian to Silurian fluvial systems, which predated the advent of a widespread cover of vascular plants, show substantial differences from modern systems (Davies and Gibling, 2010a, 2010b; Long, this volume). Among many differences in sediment transport and storage, a prime contrast lies in the enhanced strength of river banks colonized by vegetation, leading to contrasts in alluvial architecture. The deposits of pre-vegetational river channels typically comprise broad sheets with little evidence of channel margins and limited floodplain material. Meandering rivers with systematic point-bar migration and stable floodplains appear to have been rare prior to the advent of riparian vegetation, but increased greatly in abundance once terrestrial surfaces had been colonized (Cotter, 1978; Davies and Gibling, 2010a). Valleys are likely to be difficult to identify in pre-vegetational deposits, and pre-vegetational river systems may have responded very differently than Quaternary systems to forcing factors.

CONCLUSIONS

The discipline of sequence stratigraphy was initially designed to analyze coastal and shallow marine strata, and many concepts appropriate to these settings have been applied to updip alluvial systems, including criteria for identifying alluvial-valley fills. However, there are considerable differences in the factors that govern architecture in alluvial and coastal-marine settings. In the coastal realm, high-magnitude glacioeustatic fluctuations, on a 10^4 to 10^5 year timescale in the late Quaternary, result in sequences with coastal valleys, correlative interfluve paleosols, and a predictable succession of systems tracts. In contrast, a combination of climatic, tectonic, and intrinsic factors of varied magnitude and duration generate inland alluvial valleys. Where precipitation shows strong secular variation, as in the monsoonal setting of southern Asia, climate is a strong forcing factor that, over short timescales of 10^3 to 10^4 years, may be the dominant factor in creating the alluvial architecture.

As a result of these differing controls, criteria for the recognition of valley fills in coastal settings are not always applicable in updip alluvial settings. Although both coastal and alluvial valleys are commonly deeply incised, the degree of incision varies in space and time across inland tracts due to short-term episodes of incision (sequence-boundary formation) and aggradation that are linked to changes in the river's equilibrium profile and equilibrium floodplain height. These events generate aggradational–degradational rhythms—a nongenetic term that may be useful in inland settings. Mature paleosols commonly cover the areas between coastal valleys but, across wide alluvial plains, eolian, lacustrine, and channel facies may build up while mainstem valleys incise. Thus, traceable sequence boundaries may be restricted to local areas in large basins, although they may be areally important in small basins. Useful criteria for confirming incision include the presence of gullied surfaces, which are prominent adjacent to many valleys and their tributaries, and terraces, although these may be difficult to recognize in the ancient record.

ACKNOWLEDGMENTS

We are grateful to Mike Blum and Andrew Miall for thoughtful reviews that were of great assistance to us, and to Stephanie Davidson, Sophie Leleu, and Colin North for their helpful editorial assistance. We thank many colleagues for discussion on the concepts set out here. MRG is grateful for financial support from a Discovery Grant of the Natural Sciences and Engineering Research Council of Canada, and for a travel grant from the organizers of the "From River to Rock Record" conference at Aberdeen.

REFERENCES

Allen, J.P., and Fielding, C.R., 2007, Sequence architecture within a low-accommodation setting: An example from the Permian of the Galilee and Bowen basins, Queensland, Australia: American Assocation of Petroleum Geologists, Bulletin, v. 91, p. 1503–1539.

Alonso, J.L., 1989, Fold reactivation involving angular unconformable sequences: theoretical analysis and natural examples from the Cantabrian Zone (Northwest Spain): Tectonophysics, v. 170, p. 57–77.

Amorosi, A., and Milli, S., 2001, Late Quaternary depositional architecture of Po and Tevere river deltas (Italy) and worldwide comparison with coeval deltaic successions: Sedimentary Geology, v. 144, p. 357–375.

Amorosi, A., Colalongo, M.L., Pasini, G., and Preti, D., 1999a, Sedimentary response to Late Quaterary sea-level changes in the Romagna coastal plain (northern Italy): Sedimentology, v. 46, p. 99–121.

Amorosi, A., Colalongo, M.L., Fusco, F., Pasini, G., and Fiorini, F., 1999b, Glacio-eustatic control of continental-shallow marine cyclicity from Late Quaterary deposits of the southeastern Po Plain, Northern Italy: Quaternary Research, v. 52, p. 1–13.

Amorosi, A., Pavesi, M., Ricci Lucchi, M., Sarti, G., and Piccin, A., 2008, Climatic signature of cyclic fluvial architecture from the Quaternary of the central Po Plain, Italy: Sedimentary Geology, v. 209, p. 58–68.

Andrews, E.D., 1980, Effective and bankfull discharges of streams in the Yampa River basin, Colorado and Wyoming: Journal of Hydrology, v. 46, p. 311–330.

Antoine, P., Munaut, A.-V., Limondin-Lozouet, N., Ponel, P., Duperon, J., and Duperon, M., 2003, Response of the Selle River to climatic modifications during the Lateglacial and Early Holocene (Somme Basin—Northern France): Quaternary Science Reviews, v. 22, p. 2061–2076.

Bennett, V.G.L., Weissmann, G.S., Baker, G.S., and Hyndman, D.W., 2006, Regional-scale assessment of a sequence-bounding paleosol on fluvial fans using ground-penetrating radar, eastern San Joaquin Valley, California: Geological Society of America Bulletin, v. 118, p. 724–732.

Best, J.L., and Ashworth, P.J., 1997, Scour in large braided rivers and the recognition of sequence stratigraphic boundaries: Nature, v. 387, p. 275–277.

Bestland, E.A., 1997, Alluvial terraces and paleosols as indicators of Early Oligocene climate change (John Day Formation, Oregon): Journal of Sedimentary Research, v. 67, p. 840–855.

Bestland, E.A., Retallack, G.J., and Swisher C.C., III, 1997, Stepwise climate change recorded in Eocene–Oligocene paleosol sequences from central Oregon: Journal of Geology, v. 105, p. 153–172.

Blakey, R.C., and Gubitosa, R., 1984, Controls of sandstone body geometry and architecture in the Chinle Formation (Upper Triassic), Colorado Plateau: Sedimentary Geology, v. 38, p. 51–86.

Blum, M.D., 1994, Genesis and architecture of incised valley fill sequences: A late Quaternary example from the Colorado River, Gulf Coastal Plain of Texas, in Weimer, P., and Posamentier, H.W., eds., Siliciclastic Sequence Stratigraphy: Recent Developments and Applications: American Association of Petroleum Geologists, Memoir 58, p. 259–283.

Blum, M.D. and Aslan, A., 2006, Signatures of climate vs. sea-level change within incised valley-fill successions: Quaternary examples from the Texas Gulf Coast: Sedimentary Geology, v. 190, p. 177–211.

Blum, M.D., and Price, D.M., 1998, Quaternary alluvial plain construction in response to glacio-eustatic and climatic controls, Texas Gulf coastal plain, in Shanley, K.W., and McCabe, P.J., eds., Relative Role of Eustasy, Climate, and Tectonism in Continental Rocks: SEPM, Special Publication 59, p. 31–48.

Blum, M.D., and Törnqvist, T.E., 2000, Fluvial responses to climate and sea-level change: a review and look forward: Sedimentology, v. 47, p. 2–48.

Blum, M.D., Toomey, R.S., III, and Valastro, S., Jr., 1994, Fluvial response to Late Quaternary climatic and environmental change, Edwards Plateau, Texas: Palaeogeography, Palaeoclimatology, Palaeoecology, v. 108, p. 1–21.

Bookhagen, B., and Burbank, D.W., 2006, Topography, relief, and TRMM-derived rainfall variations along the Himalaya: Journal of Geophysical Research, v. 33, L08405, doi:10.1029/2006GL026037, p. 1–5.

Bookhagen, B., Fleitmann, D., Nishiizumi, K., Strecker, M.R., and Thiede, R.C., 2006, Holocene monsoonal dynamics and fluvial terrace formation in the northwest Himalaya, India: Geology, v. 34, p. 601–604.

Boyd, R., Dalrymple, R.W., and Zaitlin, B.A., 2006, Estuarine and incised-valley facies models, in Posamentier, H.W., and Walker, R.G., eds., Facies Models Revisited: SEPM, Special Publication 84, p. 171–236.

Bridgland, D.R., 2000, River terrace systems in north-west Europe: an archive of environmental change, uplift and early human occupation: Quaternary Science Reviews, v. 19, p. 1293–1303.

Bridgland, D.R., Antoine, P., Limondin-Lozouet, N., Santisteban, J.K., Westaway, R., and White, M.J., 2006, The Palaeolithic occupation of Europe as revealed by evidence from the rivers: data from IGCP 449: Journal of Quaternary Science, v. 21, p. 437–455.

Bull, W.B., 1991, Geomorphic Responses to Climate Change: Oxford, U.K., Oxford University Press, 326 p.

Catuneanu, O., 2006, Principles of Sequence Stratigraphy: Amsterdam, Elsevier, 375 p.

Catuneanu, O., Sweet, A.R., and Miall, A.D., 1997, Reciprocal architecture of Bearpaw T-R sequences, uppermost Cretaceous, Western Canada Sedimentary Basin: Bulletin of Canadian Petroleum Geology, v. 45, p. 75–94.

Cohen, K.M., Stouthamer, E., and Berendsen, H.J.A., 2002, Fluvial deposits as a record for Late Quaternary neotectonic activity in the Rhine–Meuse delta, The Netherlands: Netherlands Journal of Geosciences, v. 81, p. 389–405.

Cohen, T.J., and Nanson, G.C., 2008, Topographically associated but chronologically disjunct late Quaternary floodplains and terraces in a partly confined valley, south-eastern Australia: Earth Surface Processes and Landforms, v. 33, p. 424–443.

Cotter, E., 1978, The evolution of fluvial style, with special reference to the Central Appalachian Paleozoic, in Miall, A.D., ed., Fluvial Sedimentology: Canadian Society of Petroleum Geologists Memoir 5, p. 361–383.

Cowan, E.J., 1991, The large-scale architecture of the fluvial Westwater Canyon Member, Morrison Formation (Upper Jurassic), San Juan Basin, New Mexico, in Miall, A.D., and Tyler, N., eds., The Three-Dimensional Facies Architecture of Terrigenous Clastic Sediments and its Implications for Hydrocarbon Discovery and Recovery: SEPM, Concepts in Sedimentology and Paleontology, p. 80–93.

Currie, B.S., 1997, Sequence stratigraphy of nonmarine Jurassic–Cretaceous rocks, central Cordilleran foreland-basin system: Geological Society of America, Bulletin, v. 109, p. 1206–1222.

Dalrymple, R.W., 2006, Incised valleys in time and space: an introduction to the volume and an examination of the controls on valley formation and filling, in Dalrymple, R.W., Leckie, D.A., and Tillman, R.W., eds., Incised Valleys in Time and Space: SEPM, Special Publication 85, p. 5–12.

Davies, S.J., and Gibling, M.R., 2003, Architecture of coastal and alluvial deposits in an extensional basin: the Carboniferous Joggins Formation of eastern Canada: Sedimentology, v. 50, p. 415–439.

Davies, N.S., and Gibling, M.R., 2010a, Cambrian to Devonian evolution of alluvial systems: The sedimentological impact of the earliest land plants: Earth-Science Reviews, v. 98, p. 171–200.

Davies, N.S., and Gibling, M.R., 2010b, Paleozoic vegetation and the Siluro-Devonian rise of fluvial lateral accretion sets: Geology, v. 38, p. 51–54.

Dhir, R.P., Tandon, S.K., Sareen, B.K., Ramesh, R., Rao, T.K.G., Kailath, A.J., and Sharma, N., 2004, Calcretes in the Thar Desert: genesis, chronology and paleoenvironment: Indian Academy of Science, Proceedings (Earth and Planetary Science), v. 113, p. 473–515.

Erkens, G., Dambeck, R., Volleberg, K.P., Bouman, M.T.I.J., Bos, J.A.A., Cohen, K.M., Wallinga, J., and Hoek, W.Z., 2009, Fluvial terrace formation in the northern Upper Rhine Graben during the last 20 000 years as a result of allogenic controls and autogenic evolution: Geomorphology, v. 103, p. 476–495.

Feldman, H.R., Gibling, M.R., Archer, A.W., Wightman, W.G., and Lanier, W.P., 1995, Stratigraphic architecture of the Tonganoxie Paleovalley Fill (Lower Virgilian) in northeastern Kansas: American Association of Petroleum Geologists, Bulletin, v. 79, p. 1019–1043.

Feldman, H.R., Franseen, E.K., Joeckel, R.M., and Heckel, P.H., 2005, Impact of longer-term modest climate shifts on architecture of high-frequency sequences (cyclothems), Pennsylvanian of mid-continent U.S.A.: Journal of Sedimentary Research, v. 75, p. 350–368.

Fielding, C.R., and Gibling, M.R., 2005, Distinguishing between channel and valley fills: definitions, diagnostic criteria and dimensional data (abstract): 8th International Conference on Fluvial Sedimentology, Delft, Netherlands, Abstracts, p. 101.

Fielding, C.R., Trueman, J.D., Dickens, G.R., and Page, M., 2003, Anatomy of the buried Burdekin River channel across the Great Barrier Reef shelf: how does a major river operate on a tropical mixed siliciclastic/carbonate margin during sea level lowstand?: Sedimentary Geology, v. 157, p. 291–301.

Fielding, C.R., Trueman, J.D., Dickens, G.R., and Page, M., 2005, Geomorphology and internal architecture of the ancestral Burdekin River across the Great Barrier Reef shelf, northeast Australia, in Blum, M.D., Marriott, S.B., and Leclair, S.F., eds., Fluvial Sedimentology VII, International Association of Sedimentologists, Special Publication 35, p. 321–347.

Fielding, C.R., LaGarry, H.E., LaGarry, L.A., Bailey, B.E., and Swinehart, J.B., 2007, Sedimentology of the Whiteclay Gravel Beds (Ogallala Group) in northwestern Nebraska, USA: structurally controlled drainage promoted by Early Miocene uplift of the Black Hills Dome: Sedimentary Geology, v. 202, p. 58–71.

Fischbein, S.A., Joeckel, R.M., and Fielding, C.R., 2009, Fluvial–estuarine reinterpretation of large, isolated sandstone bodies in epicontinental cyclothems, Upper Pennsylvanian, northern Midcontinent, USA, and their significance for understanding late Paleozoic sea-level fluctuations: Sedimentary Geology, v. 216, p. 15–28.

Garrison, J.R., Jr., and van den Bergh, T.C.V., 2006, Effects of sedimentation rate, rate of relative rise in sea level, and duration of sea-level cycle on the filling of incised valleys: examples of filled and "overfilled" incised valleys from the upper Ferron Sandstone, Last Chance Delta, east-central Utah, U.S.A., in Dalrymple, R.W., Leckie, D.A., and Tillman, R.W., eds., Incised Valleys in Time and Space: SEPM, Special Publication 85, p. 239–279.

Gawthorpe, R.L., Fraser, A.J., and Collier, R.E.L., 1994, Sequence stratigraphy in active extensional basins: implications for the interpretation of ancient basin-fills: Marine and Petroleum Geology, v. 11, p. 642–658.

Gibling, M.R., 2006, Width and thickness of fluvial channel bodies and valley fills in the geological record: A literature compilation and classification: Journal of Sedimentary Research, v. 76, p. 731–770.

Gibling, M.R., Nanson, G.G., and Maroulis, J.C., 1998, Anastomosing river sedimentation in the Channel Country of central Australia: Sedimentology, v. 45, p. 595–619.

Gibling, M.R., Tandon, S.K., Sinha, R., and Jain, M., 2005, Discontinuity-bounded alluvial sequences of the southern Gangetic Plains, India: aggradation and degradation in response to monsoonal strength: Journal of Sedimentary Research, v. 75, p. 369–385.

Gibling, M.R., Sinha, R., Roy, N.G., Tandon, S.K., and Jain, M., 2008, Quaternary fluvial and eolian deposits on the Belan River, India: paleoclimatic setting of Paleolithic to Neolithic archeological sites over the past 85,000 years: Quaternary Science Reviews, v. 27, p. 391–410.

Goodbred, S.L., Jr., 2003, Response of the Ganges dispersal system to climate change: a source-to-sink view since the last interstade: Sedimentary Geology, v. 162, p. 83–104.

Goodbred, S.L., Jr., and Kuehl, S.A., 2000a, The significance of large sediment supply, active tectonism, and eustasy on margin sequence development: Late Quaternary stratigraphy and evolution of the Ganges–Brahmaputra delta: Sedimentary Geology, v. 133, p. 227–248.

Goodbred, S.L., Jr., and Kuehl, S.A., 2000b, Enormous Ganges–Brahmaputra sediment discharge during strengthened early Holocene monsoon: Geology, v. 28, p. 1083–1086.

Goswami, D.C., 1985, Brahmaputra River, Assam, India: Physiography, basin denudation, and channel aggradation: Water Resources Research, v. 21, p. 959–978.

Graf, J.B., Webb, R.H., and Hereford, R., 1991, Relation of sediment load and flood-plain formation to climatic variability, Paria River drainage basin, Utah and Arizona: Geological Society of America, Bulletin, v. 103, p. 1405–1415.

Haigh, M.J., 1984, Ravine erosion and reclamation in India: Geoforum, v. 15, p. 543–561.

Hampson, G., Stollhofen, H., and Flint, S., 1999, A sequence stratigraphic model for the Lower Coal Measures (Upper Carboniferous) of the Ruhr district, north-west Germany: Sedimentology, v. 46, p. 1199–1231.

Hengesh, J.V., and Lettis, W.R., 2002, Geologic and tectonic setting, in Jain, S.K., Lettis, W.R., Murty, C.V.R., and Bardet, J.-P., eds., Bhuj, India Earthquake of January 26, 2001 Reconnaissance Report: Earthquake Spectra, Supplement A to v. 18, p. 7–22.

Hijma, M.P., Cohen, K.M., Hoffmann, G., van der Spek, A.J.F., and Stouthamer, E., 2009, From river valley to estuary: the evolution of the Rhine mouth in the early to middle Holocene (western Netherlands, Rhine–Meuse delta): Netherlands Journal of Geosciences, v. 88, p. 13–53.

Hill, P.J., De Deckker, P., Von Der Borch, C., and Murray-Wallace, C.V., 2009, Ancestral Murray River on the Lacepede Shelf, southern Australia: Late Quaternary migrations of a major river outlet and strandline development: Australian Journal of Earth Sciences, v. 56, p. 135–157.

Hirst, J.P.P., 1991, Variations in alluvial architecture across the Oligo-Miocene Huesca fluvial system, Ebro Basin, Spain, in Miall, A.D., and Tyler, N., eds., The Three-Dimensional Facies Architecture of Terrigenous Clastic Sediments and its Implications for Hydrocarbon Discovery and Recovery: SEPM, Concepts in Sedimentology and Paleontology, p. 111–121.

Holbrook, J., 2001, Origin, genetic interrelationships, and stratigraphy over the continuum of fluvial channel-form bounding surfaces: an illustration from middle Cretaceous strata, southeastern Colorado: Sedimentary Geology, v. 144, p. 179–222.

Holbrook, J., 2009, A persisting preference for progressive piling: Rock vs. river biases in the medium-term processes and architectural complexity of channel-belt stacking: From River to Rock Record, Programme and Abstracts, Aberdeen, U.K.

Holbrook, J., Scott, R.W., and Oboh-Ikuenobe, F.E., 2006, Base-level buffers and buttresses: a model for upstream versus downstream control of fluvial geometry and architecture within sequences: Journal of Sedimentary Research, v. 76, p. 162–174.

Hovius, N., 1996, Regular spacing of drainage outlets from linear mountain belts: Basin Research, v. 8, p. 29–44.

Hovius, N., 1998, Controls on sediment supply by large rivers, in Shanley, K.W., and McCabe, P.J., eds., Relative Role of Eustasy, Climate, and Tectonism in Continental Rocks: SEPM, Special Publication 59, p. 3–16.

Iwaniw, E., 1984, Lower Cantabrian basin margin deposits in NE Leon, Spain—a model for valley-fill sedimentation in a tectonically active, humid climatic setting: Sedimentology, v. 31, p. 91–110.

Jain, V., and Sinha, R., 2003, Hyperavulsive-anabranching Baghmati river system, north Bihar plains, eastern India: Zeitschrift für Geomorphologie, v. 47, p. 101–116.

Jain, V., and Sinha, R., 2005, Response of active tectonics on the alluvial Baghmati River, Himalayan foreland basin, eastern India: Geomorphology, v. 70, p. 339–356.

Kraus, M.J., and Middleton, L.T., 1987, Dissected paleotopography and base-level changes in a Triassic fluvial sequence: Geology, v. 15, p. 18–21.

Kvale, E.P., and Archer, A.W., 2007, Paleovalley fills: Trunk vs. tributary: American Assocation of Petroleum Geologists, Bulletin, v. 91, p. 809–821.

Lavé, J., and Avouac, P., 2000, Active folding of fluvial terraces across the Siwaliks Hills, Himalayas of central Nepal: Journal of Geophysical Research, v. 105, p. 5735–5770.

Leckie, D.A. and Cheel, R.J., 1997, Sedimentology and depositional history of Lower Cretaceous coarse-grained clastics, southwest Alberta and southeast British Columbia: Bulletin of Canadian Petroleum Geology, v. 45, p. 1–24.

Leckie, D.A., and Krystinik, L.F., 1995, Cretaceous igneous-clast conglomerate in the Blairmore Group, Rocky Mountain Foothills and adjacent subsurface (Bow Island Formation), Alberta, Canada: Bulletin of Canadian Petroleum Geology, v. 43, p. 320–342.

Leeder, M.R., Mack, G.H., Peakall, J., and Salyards, S.L., 1996, First quantitative test of alluvial stratigraphic models: Southern Rio Grande rift, New Mexico: Geology, v. 24, p. 87–90.

LEGARRETA, L., AND ULIANA, M.A., 1998, Anatomy of hinterland depositional sequences: Upper Cretaceous fluvial strata, Neuquen Basin, west-central Argentina, in Shanley, K.W., and McCabe, P.J., eds., Relative Role of Eustasy, Climate, and Tectonism in Continental Rocks: SEPM, Special Publication 59, p. 83–92.

LEVY, M., CHRISTIE-BLICK, N., AND LINK, P.K., 1994, Neoproterozoic incised valleys of the eastern Great Basin, Utah and Idaho: fluvial response to changes in depositional base level, in Dalrymple, R.W., Boyd, R., and Zaitlin, B.A., eds., Incised-Valley Systems: Origin and Sedimentary Sequences: SEPM, Special Publication 51, p. 369–382.

MACK, G.H., AND LEEDER, M.R., 1999, Climatic and tectonic controls on alluvial-fan and axial-fluvial sedimentation in the Plio-Pleistocene Palomas Half Graben, southern Rio Grande Rift: Journal of Sedimentary Research, v. 69, p. 635–652.

MACK, G.H., AND MADOFF, R.D., 2005, A test of models of fluvial architecture and palaeosol development: Camp Rice Formation (Upper Pliocene–Lower Pleistocene), southern Rio Grande rift, New Mexico, USA: Sedimentology, v. 52, p. 191–211.

MACK, G.H., SEAGER, W.R., LEEDER, M.R., PÉREZ-ARLUCEA, M., AND SALYARDS, S.L., 2006, Pliocene and Quaternary history of the Rio Grande, the axial river of the southern Rio Grande rift, New Mexico, USA: Earth-Science Reviews, v. 79, p. 141–162.

MAGEE, J.W., BOWLER, J.M., MILLER, G.H., AND WILLIAMS, D.L.G., 1995, Stratigraphy, sedimentology, chronology and palaeohydrology of Quaternary lacustrine deposits at Madigan Gulf, Lake Eyre, South Australia: Palaeogeography, Palaeoclimatology, Palaeoecology, v. 113, p. 3–42.

MARTINSEN, O.J., AND HELLAND-HANSEN, W., 1995, Strike variability of clastic depositional systems: Does it matter for sequence-stratigraphic analysis?: Geology, v. 23, p. 439–442.

MARTINSEN, O.J., RYSETH, A., HELLAND-HANSEN, W., FLESCHE, H., TORKILDSEN, G., AND IDIL, S., 1999, Stratigraphic base level and fluvial architecture: Ericson Sandstone (Campanian), Rock Springs Uplift, SW Wyoming, USA: Sedimentology, v. 46, p. 235–259.

MCCARTHY, P.J., FACCINI, U.F., AND PLINT, A.G., 1999, Evolution of an ancient coastal plain: palaeosols, interfluves and alluvial architecture in a sequence stratigraphic framework, Cenomanian Dunvegan Formation, NE British Columbia, Canada: Sedimentology, v. 46, p. 861–891.

MENIER, D., REYNAUD, J.-Y., PROUST, J.-N., GUILLOCHEAU, F., GUENNOC, P., BONNET, S., TESSIER, B., AND GOUBERT, E., 2006, Basement control on shaping and infilling of valleys incised at the southern coast of Brittany, France, in Dalrymple, R.W., Leckie, D.A., and Tillman, R.W., eds., Incised Valleys in Time and Space: SEPM, Special Publication 85, p. 37–55.

MIALL, A.D., 2002, Architecture and sequence stratigraphy of Pleistocene fluvial systems in the Malay Basin, based on seismic time-slice analysis: American Association of Petroleum Geologists, Bulletin, v. 86, p. 1201–1216.

MIALL, A.D., AND ARUSH, M., 2001, Cryptic sequence boundaries in braided fluvial successions: Sedimentology, v. 48, p. 971–985.

MIALL, A.D., AND JONES, B.G., 2003, Fluvial architecture of the Hawkesbury Sandstone (Triassic), near Sydney, Australia: Journal of Sedimentary Research, v. 73, p. 531–545.

MILANA, J.P., 1998, Sequence stratigraphy in alluvial settings: A flume-based model with applications to outcrop and seismic data: American Association of Petroleum Geologists, Bulletin, v. 82, p. 1736–1753.

MOHINDRA, R., PARKASH, B., AND PRASAD, J., 1992, Historical geomorphology and pedology of the Gandak Megafan, Middle Gangetic Plains, India: Earth Surface Processes and Landforms, v. 17, p. 643–662.

MONECKE, K., WINSEMANN, J., AND HANISCH, J., 2001, Climatic response of Quaternary alluvial deposits in the upper Kali Gandaki valley (West Nepal): Global and Planetary Change, v. 28, p. 293–302.

MUKERJI, A.B., 1990, The Chandigarh Dun alluvial fans: An analysis of the process–form relationship, in Rachocki, A.H., and Church, M., eds., Alluvial Fans: A Field Approach: New York, Wiley, p. 131–149.

NÁDOR, A., LANTOS, M., TOTH-MAKK, A., AND THAMO-BOZSO, E., 2003, Milankovitch-scale multi-proxy records from fluvial sediments of the last 2.6 Ma, Pannonian Basin, Hungary: Quaternary Science Reviews, v. 22, p. 2157–2175.

NANSON, G.C., 1986, Episodes of vertical accretion and catastrophic stripping: A model of disequilibrium flood-plain development: Geological Society of America, Bulletin, v. 97, p. 1467–1475.

NANSON, G.C., PRICE, D.M., AND SHORT, S.A., 1992, Wetting and drying of Australia over the past 300 ka: Geology, v. 20, p. 791–794.

NANSON, G.C., PRICE, D.M., JONES, B.G., MAROULIS, J.C., COLEMAN, M., BOWMAN, H., COHEN, T.J., PIETSCH, T.J., AND LARSEN, J.R., 2008, Alluvial evidence for major climate and flow regime changes during the middle and late Quaternary in eastern central Australia: Geomorphology, v. 101, p. 109–129.

NASH, D.B., 1994, Effective sediment-transporting discharge from magnitude-frequency analysis: Journal of Geology, v. 102, p. 79–95.

NEAL, J., AND ABREU, V., 2009, Sequence stratigraphy hierarchy and the accommodation succession method: Geology, v. 37, p. 779–782.

NICHOLS, G.J., 1987, Structural controls on fluvial distributary systems—the Luna System, northern Spain, in Ethridge, F.G., Flores, R.M., and Harvey, M.D., eds., Recent Developments in Fluvial Sedimentology: SEPM, Special Publication 39, p. 269–277.

OLSEN, T., STEEL, R., HOGSETH, K., SKAR, T., AND RØE, S.-L., 1995, Sequential architecture in a fluvial succession: sequence stratigraphy in the Upper Cretaceous Mesaverde Group, Price Canyon, Utah: Journal of Sedimentary Research, v. B65, p. 265–280.

OVERPECK, J., ANDERSON, D., TRUMBORE, S., AND PRELL, W., 1996, The southwest Indian monsoon over the last 18000 years: Climate Dynamics, v. 12, p. 213–225.

PAGE, K.J., KEMP, J., AND NANSON, G.C., 2009, Late Quaternary evolution of Riverine Plain paleochannels, southeastern Australia: Australian Journal of Earth Sciences, v. 56, p. S19-S33.

PASCUCCI, V., GIBLING, M.R., AND SANDRELLI, F., 2006, Valley formation and filling in response to Neogene magmatic doming of Elba Island, Tuscany, Italy, in Dalrymple, R.W., Leckie, D.A., and Tillman, R.W., eds., Incised Valleys in Time and Space: SEPM, Special Publication 85, p. 327–343.

PATTERSON, P.E., 2009, Alluvial hierarchy: description of alluvial strata in the rock record: From River to Rock Record, Aberdeen, Scotland, U.K., Abstract.

PETTER, A.L., AND MUTO, T., 2008, Sustained alluvial aggradation and autogenic detachment of the alluvial river from the shoreline in response to steady fall of relative sea level: Journal of Sedimentary Research, v. 78, p. 98–111.

PLINK-BJØRKLUND, P., AND STEEL, R.J., 2006, Incised valleys on an Eocene coastal plain and shelf, Spitsbergen—part of a linked shelf-slope system, in Dalrymple, R.W., Leckie, D.A., and Tillman, R.W., eds., Incised Valleys in Time and Space: SEPM, Special Publication 85, p. 281–307.

PLINT, A.G., 2002, Paleo-valley systems in the Upper Cretaceous Dunvegan Formation, Alberta and British Columbia: Bulletin of Canadian Petroleum Geology, v. 50, p. 277–296.

PLINT, A.G., AND NUMMEDAL, D., 2000, The falling stage systems tract: recognition and importance in sequence stratigraphic analysis, in Hunt, D.R., and Gawthorpe, R.L., eds., Sedimentary Responses to Forced Regressions: Geological Society of London, Special Publication 172, p. 1–17.

PLINT, A.G., AND WADSWORTH, J.A., 2003, Sedimentology and palaeogeomorphology of four large valley systems incising delta plains, western Canada Foreland Basin: implications for mid-Cretaceous sea-level changes: Sedimentology, v. 50, p. 1147–1186.

PLINT, A.G., AND WADSWORTH, J.A., 2006, Delta-plain paleodrainage patterns reflect small-scale fault movement and subtle forebulge uplift: Upper Cretaceous Dunvegan Formation, Western Canada Foreland Basin, in Dalrymple, R.W., Leckie, D.A., and Tillman, R.W., eds., Incised Valleys in Time and Space: SEPM, Special Publication 85, p. 219–237.

POSAMENTIER, H.W., 2001, Lowstand alluvial bypass systems: Incised vs. unincised: American Association of Petroleum Geologists, Bulletin, v. 85, p. 1771–1793.

PRATT-SITAULA, B., BURBANK, D.W., HEIMSATH, A., AND OJHA, T., 2004, Landscape disequilibrium on 1000–10,000 year scales, Marsyandi River, Nepal, central Himalaya: Geomorphology, v. 58, p. 223–241.

PRELL, W.L., AND KUTZBACH, J.E., 1987, Monsoon variability over the past 150,000 years: Journal of Geophysical Research, v. 92, p. 8411–8425.

RETALLACK, G.J., 1986, Fossil soils as grounds for interpreting long-term controls on ancient rivers: Journal of Sedimentary Petrology, v. 56, p. 1–18.

RITTER, J.B., MILLER, J.R., ENZEL, Y., AND WELLS, S.G., 1995, Reconciling the

roles of tectonism and climate in Quaternary alluvial fan evolution: Geology, v. 23, p. 245–248.
Roy, N.G., Sinha, R., Gibling, M.R., Jain, M., and Murray, A.S., 2007, Valley-fill stratigraphy in the southern Ganga basin: southward migration and monsoonal forcing: Quaternary International, v. 167–168 Supplement, p. 388–389.
Rust, B.R., and Nanson, G.C., 1986, Contemporary and palaeo channel patterns and the late Quaternary stratigraphy of Cooper Creek, southwest Queensland, Australia: Earth Surface Processes and Landforms, v. 11, p. 581–590.
Ryan, D.A., Bostock, H.C., Brooke, B.P., and Marshall, J.F., 2007, Bathymetric expression of the Fitzroy River palaeochannel, northeast Australia: Response of a major river to sea-level change on a semi-rimmed, mixed siliciclastic–carbonate shelf: Sedimentary Geology, v. 201, p. 196–211.
Rygel, M.C., and Gibling, M.R., 2006, Natural geomorphic variability recorded in a high-accommodation setting: fluvial architecture of the Pennsylvanian Joggins Formation of Atlantic Canada: Journal of Sedimentary Petrology, v. 76, p. 1230–1251.
Rygel, M.C., Fielding, C.R., Frank, T.D., and Birgenheier, L.P., 2008, The magnitude of Late Paleozoic glacioeustatic fluctuations: A synthesis: Journal of Sedimentary Research, v. 78, p. 500–511.
Salter, T., 1993, Fluvial scour and incision: models for the development of realistic reservoir geometries, in North, C.P., and Prosser, D.J., eds., Characterization of Fluvial and Aeolian Reservoirs: Geological Society of London, Special Publication 73, p. 33–51.
Schlager, W., 2004, Fractal nature of stratigraphic sequences: Geology v. 32, p. 85–188.
Schumm, S.A., and Ethridge, F.G., 1994, Origin, evolution and morphology of fluvial valleys, in Dalrymple, R.W., Boyd, R., and Zaitlin, B.A., eds., Incised-Valley Systems: Origin and Sedimentary Sequences: SEPM, Special Publication 51, p. 11–27.
Schumm, S.A., and Lichty, R.W., 1963, Channel widening and flood-plain construction along Cimarron River in Southwestern Kansas: U.S. Geological Survey, Professional Paper 352-D, p. 71–88.
Sedimentation Seminar, 1978, Sedimentology of the Kyrock Sandstones (Pennsylvanian) in the Brownsville Paleovalley, Edmonson and Hart counties, Kentucky: Lexington, Kentucky, Kentucky Geological Survey, Series X, p. 24.
Shanley, K.W., and McCabe, P.J., 1991, Predicting facies architecture through sequence stratigraphy – an example from the Kaiparowits Plateau, Utah: Geology, v. 19, p. 742–745.
Shanley, K.W., and McCabe, P.J., 1994, Perspectives on the sequence stratigraphy of continent strata: American Association of Petroleum Geologists, Bulletin, v. 78, p. 544–568.
Sharma, S., Joachimski, M., Sharma, M., Tobschall, H.J., Singh, I.B., Sharma, C., Chauhan, M.S., and Morgenroth, G., 2004, Lateglacial and Holocene environmental changes in Ganga plain, Northern India: Quaternary Science Reviews, v. 23, p. 145–159.
Simms, A., Anderson, J.B., Taha, Z.P., and Rodriguez, A.B., 2006, Overfilled versus underfilled incised valleys: examples from the Quaternary Gulf of Mexico, in Dalrymple, R.W., Leckie, D.A., and Tillman, R.W., eds., Incised Valleys in Time and Space: SEPM, Special Publication 85, p. 117–139.
Singh, H., Parkash, B., and Gohain, K., 1993, Facies analysis of the Kosi megafan deposits: Sedimentary Geology, v. 85, p. 87–113.
Singh, V., and Tandon, S.K., 2007, Evidence and consequences of tilting of two alluvial fans in the Pinjaur dun, Northwestern Himalayan Foothills: Quaternary International, v. 159, p. 21–31.
Sinha, R., Gibling, M.R., Jain, V., and Tandon, S.K., 2005, Sedimentology and avulsion patterns of the anabranching Baghmati River in the Himalayan foreland basin, India, in Blum, M.D., and Marriott, S.B., eds., Fluvial Sedimentology VII: International Association of Sedimentologists, Special Publication 35, p. 181–196.
Sinha, R., Tandon, S.K., Sanyal, P., Gibling, M.R., Stuben, D., Berner, Z., and Ghazanfari, P., 2006, Calcretes from a Late Quaternary interfluve in the Ganga Plains, India: Carbonate types and isotopic systems in a monsoonal setting: Palaeogeography, Palaeoclimatology, Palaeoecology, v. 242, p. 214–239.
Sinha, R., Bhattacharjee, P.S., Sangode, S.J., Gibling, M.R., Tandon, S.K., Jain, M., and Godfrey-Smith, D., 2007, Valley and interfluve sediments in the Southern Ganga plains, India: Exploring facies and magnetic signatures: Sedimentary Geology, v. 201, p. 386–411.
Sinha, R., Kettanah, Y., Gibling, M.R., Tandon, S.K., Jain, M., Bhattacharjee, P.S., Dasgupta, A.S., and Ghazanfari, P., 2009, Craton-derived alluvium as a major sediment source in the Himalayan Foreland Basin of India: Geological Society of America, Bulletin, v. 121, p. 1596–1610.
Smith, G.A., 1988, Sedimentology of proximal to distal volcaniclastics dispersed across an actic foldbelt: Ellensburg Formation (late Miocene), central Washington: Sedimentology, v. 35, p. 953–977.
Smith, G.A., 1994, Climatic influences on continental deposition during late-stage filling of an extensional basin, southeastern Arizona: Geological Society of America, Bulletin, v. 106, p. 1212–1228.
Sømme, T.O., Helland-Hansen, W., and Granjeon, D., 2009, Impact of eustatic amplitude variations on shelf morphology, sediment dispersal, and sequence stratigraphic interpretation: Icehouse versus greenhouse systems: Geology, v. 37, p. 587–590.
Srivastava, P., Shukla, U.K., Mishra, P., Sharma, M., Sharma, S., Singh, I.B., and Singhvi, A.K., 2000, Luminescence chronology and facies development of Bhur sands in the interfluve region of Central Ganga Plain, India: Current Science, v. 78, p. 498–503.
Srivastava, P., Juyal, N., Singhvi, A.K., Wasson, R.J., and Bateman, M.D., 2001, Luminescence chronology of river adjustment and incision of Quaternary sediments in the alluvial plain of the Sabarmati River, north Gujarat, India: Geomorphology, v. 36, p. 217–229.
Stouthamer, E., and Berendsen, H.J.A., 2007, Avulsion: The relative roles of autogenic and allogenic processes: Sedimentary Geology, v. 198, p. 309–325.
Strong, N., and Paola, C., 2008, Valleys that never were: time surfaces versus stratigraphic surfaces: Journal of Sedimentary Research, v. 78, p. 579–593.
Tandon, S.K., and Gibling, M.R., 1997, Calcretes at sequence boundaries in Upper Carboniferous cyclothems of the Sydney Basin, Atlantic Canada: Sedimentary Geology, v. 112, p. 43–67.
Tandon, S.K., Gibling, M.R., Sinha, R., Singh, V., Ghazanfari, P., Dasgupta, A.S., Jain, M., and Jain, V., 2006, Alluvial valleys of the Gangetic Plains, India: Timing and causes of incision, in Dalrymple, R.W., Leckie, D.A., and Tillman, R.W., eds., Incised Valleys in Time and Space: SEPM, Special Publication 85, p. 15–35.
Thomas, M.A., and Anderson, J.B., 1994, Sea-level controls on the facies architecture of the Trinity–Sabine incised-valley system, Texas continental shelf, in Dalrymple, R.W., Boyd, R., and Zaitlin, B.A., eds., Incised-Valley Systems: Origin and Sedimentary Sequences: SEPM, Special Publication 51, p. 63–82.
Törnqvist, T.E., Wallinga, J., and Busschers, F.S., 2003, Timing of the last sequence boundary in a fluvial setting near the highstand shoreline—Insights from optical dating: Geology, v. 31, p. 279–282.
Umitsu, M., 1993, Late Quaternary sedimentary environments and landforms in the Ganges delta: Sedimentary Geology, v. 83, p. 177–186.
Vincent, S.J., 2001, The Sis palaeovalley: a record of proximal fluvial sedimentation and drainage basin development in response to Pyrenean mountain building: Sedimentology, v. 48, p. 1235–1276.
Waelbroeck, C., Labeyrie, L., Michel, E., Duplessy, J.C., McManus, J.F., Lambeck, K., Balbon, E., and Labracherie, M., 2002, Sea-level and deep water temperature changes derived from benthic foraminifera isotopic records: Quaternary Science Reviews, v. 21, p. 295–305.
Weissmann, G.S., Mount, J.F., and Fogg, G.E., 2002, Glacially-driven cycles in accumulation space and sequence stratigraphy of a stream-dominated alluvial fan, San Joaquin Valley, California, U.S.A.: Journal of Sedimentary Research, v. 72, p. 240–251.
Weissmann, G.S., Hartley, A.J., Nichols, G.J., Scuderi, L.A., Olson, M., Buehler, H., and Banteah, R., 2010, Fluvial form in modern continental sedimentary basins: Distributive fluvial systems: Geology, v. 38, p. 39–42.
Wellner, R.W., and Bartek, L.R., 2003, The effect of sea level, climate, and shelf physiography on the development of incised-valley complexes: a modern example from the East China Sea: Journal of Sedimentary Research, v. 73, p. 926–940.
Wells, N.A., and Dorr, J.A., Jr., 1987, Shifting of the Kosi River, northern India: Geology, v. 15, p. 204–207.
White, J.M., and Leckie, D.A., 1999, Palynological age constraints on the

Cadomin and Dalhousie formations in SW Alberta: Bulletin of Canadian Petroleum Geology, v. 47, p. 199–222.

WILLIAMS, G.P., 1978, Bank-full discharge of rivers: Water Resources Research, v. 14, p. 1141–1154.

WILLIAMS, M.A.J., AND CLARKE, M.F., 1995, Quaternary geology and prehistoric environments in the Son and Belan valleys, north central India: Geological Society of India, Memoir 32, p. 282–308.

WILLIAMS, M.A.J., PAL, J.N., JAISWAL, M., AND SINGHVI, A.K., 2006, River response to Quaternary climatic fluctuations: evidence from the Son and Belan valleys, north-central India: Quaternary Science Reviews, v. 25, p. 2619–2631.

WILLIS, B.J., 1997, Architecture of fluvial-dominated valley-fill deposits in the Cretaceous Fall River Formation: Sedimentology, v. 44, p. 735–757.

WOLMAN, M.G., AND MILLER, J.P., 1960, Magnitude and frequency of forces in geomorphic processes: Journal of Geology, v. 68, p. 54–74.

WRIGHT, V.P., AND MARRIOTT, S.B., 1993, The sequence stratigraphy of fluvial depositional systems: the role of floodplain sediment storage: Sedimentary Geology, v. 86, p. 203–210.

ZAITLIN, B.A., DALRYMPLE, R.W., AND BOYD, R., 1994, The stratigraphic organization of incised-valley systems associated with relative sea-level change, in Dalrymple, R.W., Boyd, R., and Zaitlin, B.A., eds., Incised-Valley Systems: Origin and Sedimentary Sequences: SEPM, Special Publication 51, p. 45–60.

ZAWADA, P.K., 1994, Palaeoflood hydrology of the Buffels River, Laingsburg, South Africa: was the 1981 flood the largest?: South African Journal of Geology, v. 97, p. 21–32.

DIGITAL APPENDIX

The following supplementary figures are available in electronic format either on the DVD accompanying this volume or through the digital archive on the SEPM website at http://www.sepm.org/

See Figure 2A for Asian locations and Figure 3 for Australian locations.

S1. Upstream view of the Gandak River and sand flats near Patna, India. Note the flat nature of the landscape and the minimal degree of incision of the channel into adjacent alluvium, in accord with the prolonged and widespread flooding experienced in this area. The sand bars are in use for agriculture during the dry season. Photo by M.R.Gibling.

S2. Upstream view of Baghmati River anabranch at Benibad near Muzzafarpur, India. Note the flat nature of the landscape and the minimal degree of incision of the channel into adjacent alluvium, in accord with the prolonged and widespread flooding experienced in this area. Scientists are standing on a low levee a few meters high and tens of meters wide that borders the channel. See details in Sinha et al. (2005). Photo by M.R.Gibling.

S3. Upstream view of Baghmati River anabranch near Nawada, India. Scientists are standing below a crevasse channel (at lower right), a meter-thick splay from which buried an orchard during the previous season's flooding. This site provides field evidence for the frequent overbank flooding indicated in the flood map. See details in Sinha et al. (2005). Photo by M.R.Gibling.

S4. Alluvial fan lobe in foreground, where a temple is built on the lobe, and in the middle distance. The lobe is cut by an incised channel about 40 m deep, shallowing distally. The site lies within the Himalayan Foothills of India, in the intermontane valley of Pinjaur Dun, bounded to the north (left) by the Main Boundary Thrust and to the south (right) by the Siwalik Hills. The topmost strata of the lobe date to ~24 ka B.P., and incision post-dates this period of deposition. Fans in Pinjaur Dun are tilted, and incision was a response to uplift on the Main Boundary Fault (Tandon et al. 2006; Singh and Tandon 2007). Photo by M.R.Gibling.

S5. Cliffs about 30 m high along the southern margin of the Ganga Valley near Kannauj, northwest of Kanpur, India (Text-figure 2B). Downstream view. The cliffs are cut into late Quaternary alluvium which dates back to about 27 ka B.P. downstream at Bithur. See details in Gibling et al. (2005) and Sinha et al. (2007). Photo by M.R.Gibling.

S6. Cliffs about 20 m high along the southern margin of the Ganga Valley near Kannauj, northwest of Kanpur, India (Text-Figure 2B). Downstream view. Note the shallow gullies cutting through the cliffs, part of a narrow belt of badlands bordering the valley in this area (Sinha et al. 2007). Photo by M.R.Gibling.

S7. Clastic dikes of red clay, with some white calcite precipitates, cutting through silty alluvium in cliffs along the Yamuna Valley near Kalpi, India. The dikes terminate below a discontinuity within the Quaternary section, just above the level of the photo. Note the distorted stratification in the alluvium which, along with the dikes, suggests deformation due to earthquake activity. Measuring stick is 50 cm long. See details in Gibling et al. (2005). Photo by M.R.Gibling.

S8. Near-vertical calcite vein that transects ~5 m of alluvium in terrace cuts, at Chillahia on the Belan River, India. Hammer is 30 cm long. See details in Gibling et al. (2008). Photo by M.R.Gibling.

S9. Fluvial sand sheet exposed in bank of anabranch, Cooper Creek, Australia. The topstratum of the bank is dark mud attributed to anastomosing rivers similar to those of the present day. The pale substratum is an unusual exposure of a widespread sand sheet proven from augering, drilling and trenches across much of the Channel Country. The sand sheet dates from MIS 3 back to about 750,000 years before present, and represents periods of enhanced fluvial discharge and sediment transport linked to secular variation in precipitation (Gibling et al. 1998; Nanson et al. 2008). Undercutting of the soft sand has caused bank migration, toppled trees, and resulted in unusual amounts of sand on the adjacent point bar. Photo by M.R.Gibling.

S10. Relict meanders and point bars with scroll-bar topography on alluvial surface, Cooper Creek, Australia. White bar is 2 km long. The relict topography represents the top of a near-surface sand sheet attributed to an earlier period of enhanced fluvial activity. See Rust and Nanson (1986) for details. Aerial photo CAB 7059 Run 4 Number 141, General Manager, Australian Surveying and Land Information Group, Canberra.

S11. Source-bordering eolian mound in Ganga Valley (Gahira Bypass Section), near Unnao north of Kanpur, India. The silts were dated to early-mid Holocene by Srivastava et al. (2000). Photo by M.R.Gibling.

S12. Eolian mound in Ganga Valley, showing thick units of poorly stratified silt, with a vegetated cap and paleosol. Photo by M.R.Gibling.

S13. Close-up of yellow silt in eolian mound, Ganga Valley (DVD 12), showing poor stratification and weakly defined red layers, suggesting incipient paleosols. Trowel is 15 cm long. Photo by M.R.Gibling.

S14. Source-bordering eolian mound of weakly stratified sand and silt in the Belan Valley, India. Eolian deposits in the valley yielded dates from about 14 to 7 ka B.P., representing a period of climatic instability as the monsoon intensified following the Last Glacial Maximum (Gibling et al. 2008). Photo by M.R.Gibling.

S15. Poorly stratified sand with disseminated shell fragments in eolian mound, Belan Valley (DVD 14). Lens cap is 5 cm in diameter. Photo by M.R.Gibling.

S16. Baraila Tal (large lake), Bihar, India. Abundant pale mounds in the foreground are the "chimneys" of crustacean burrows, extending up to 15 cm above the mud surface. Mounds in the distance are piles of lake reeds collected by local farmers. See Sinha et al. (2005). Photo by M.R.Gibling.

S17. Carbonate-cemented erosional surface (discontinuity) within alluvial cliffs along Yamuna Valley near Kalpi, India. Several similar surfaces, which mark both pedogenic and groundwater carbonate accumulation, divide the cliff succession into aggradational-degradational rhythms (see text). Hammer is 30 cm long. See Gibling et al.

(2005) and Sinha et al. (2007) for a description of these rhythms. Photo by M.R.Gibling.

S18. Colluvial gully fills up to 10 m thick in cliffs along the Yamuna Valley near Kalpi, India. Strata filling a major gully dip systematically to the right across most of the illustrated cliff face, and are cut by two successive gully fills at right of the photo. Cliff is 10 m high, and the topmost cliff strata represent the cultural level of human occupation. The gully forms and fills are analogous to modern badland features that border the modern Yamuna River. See Gibling et al. (2005) for a detailed account of these gully fills, which yielded a date of 36 ka B.P. Photo by M.R.Gibling.

S19. Thick calcrete on terrace surface at Benar near Jodhpur, in the Thar Desert of northwest India (north of the Luni River in Text-figure 2A. See Dhir et al. (2004) for details. Photo by M.R.Gibling.

S20. Close-up of calcrete in DVD 19 to show closely packed nodules. Scale is 9 cm long. Photo by M.R.Gibling.

S21. Gullied badlands behind incised valley of the Yamuna River near Kalpi, India. Gullies are well vegetated and up to 15 m deep, and they dissect late Quaternary alluvium along a broad belt bordering the river. Photo by M.R.Gibling.

S22. Gullied cliff face along the Betwa River near Kotra, India (Text-figure 2A). Gullies are up to 15 m deep and form part of a wide belt bordering the river. They cut into an undated alluvial succession that contains erosional surfaces and comprises stacked aggradational and degradational rhythms. The succession contains fluvial carbonate gravels derived from erosion of floodplain deposits. Tough material in foreground is an older, well cemented alluvial unit. See description in Gibling et al. (2005). Photo by M.R.Gibling.

S23. Gullied cliff face along the Yamuna Valley at Tilauli near Allahabad, India. The main part of the cliff consists of red-brown floodplain silt and clay. At the cliff top is an indurated, well stratified yellow silt 3 m thick that is rich in gastropods and interpreted as a lacustrine unit. Shells yielded a date of 15.5 radiocarbon years B.P. (Williams and Clarke 1995), calibrated to ~18.8 ka B.P. The change from floodplain to lake deposition at the cliff top represents a period of reduced discharge on the alluvial plain. Subsequent incision by the Yamuna River post-dates the lacustrine unit, and is attributed to enhanced discharge during intensification of the Southwest Indian Monsoon following the Last Glacial Maximum (Tandon et al. 2006). Photo by M.R.Gibling.

S24. Close-up of lacustrine unit at Tilauli (DVD 23). The unit is 3 m thick, and has partly slumped down the cliff in the foreground. Note gullied landscape and Yamuna Valley in distance. Photo by M.R.Gibling.

S25. Alluvial terrace (centre-right) where the Gandak River has cut through alluvium from a tributary valley near Kalopani, Nepal Himalaya. The higher cliff at centre-left is largely cut through bedrock. Hut in fields at centre-right of photo is 5 m high. Photo by M.R.Gibling.

S26. Alluvial terrace from the tributary Miristi Khola, incised by the Gandak River north of Tatopani, Nepal Himalaya. House on terrace top is 8 m high. See Monecke et al. (2001) for a description of the alluvium. Photo by M.R.Gibling.

S27. Terraces at base of Gandak Valley south of the Main Central Thrust near Tatopani, Nepal Himalaya. The valley cuts through the Annapurna and Dhaulagiri Massifs, 8 km above sea-level, parts of which are visible in the distance. The valley is about 6.5 km deep. People on bridge in foreground for scale. Photo by M.R.Gibling.

S28. Close-up of terraces in DVD 27 to show coarse-grained alluvium and occurrence of terraces in small rock-cut embayments. People on bridge in foreground for scale. Photo by M.R.Gibling.

S29. Alluvial terraces of the Belan River, India, with Kaimur Hills of the Indian Craton visible to the south. The river at left is cut into Proterozoic quartzites, and the terrace succession at right commences with channel-base groundwater calcrete, which projects as ledges into the river. People are standing on the top of the calcrete. The alluvium in the cliff yielded dates of about 90 to 20 ka B.P. (Williams et al. 2006). Photo by M.R.Gibling.

S30. Alluvial terrace and cultivated inset terrace, Belan River, India. The alluvium at this location yielded dates back to about 13 ka B.P. (Gibling et al. 2008), a quite different spectrum of ages from the terrace succession shown in DVD 29, about 5 km upstream. Photo by M.R.Gibling.

S31. Strath (bedrock) terrace on Proterozoic quartzite, with thin alluvial cover, Chopani-Mando, Belan River, India. The terrace level on the far bank of the river was the site of Palaeolithic and Mesolithic settlements (Gibling et al. 2008). Photo by M.R.Gibling.

S32. Charity Creek on the Manning River, New South Wales, Australia, looking upstream. At the bend at the extreme left, the alluvial terrace has been partially stripped away by flood events in the 1968 to 1978 period, leaving an expanse of gravel. Site of "catastrophic stripping" described by Nanson (1986). Photo by M.R.Gibling.

S33. Close-up of Charity Creek site at river bend at extreme left in DVD 32, looking downstream. The alluvial terrace has been stripped away over a width of about 50 m, leaving an incised terrace margin and a gravel expanse. Photo by M.R.Gibling.